Special Triangles

Name	*Characteristic*	*Examples*
Right Triangle	Triangle has a right angle.	
Isosceles Triangle	Triangle has two equal sides.	$AB = BC$
Equilateral Triangle	Triangle has three equal sides.	$AB = BC = CA$
Similar Triangles	Corresponding angles are equal; corresponding sides are proportional.	$A = D,\ B = E,\ C = F$ $\dfrac{AB}{DE} = \dfrac{AC}{DF} = \dfrac{BC}{EF}$

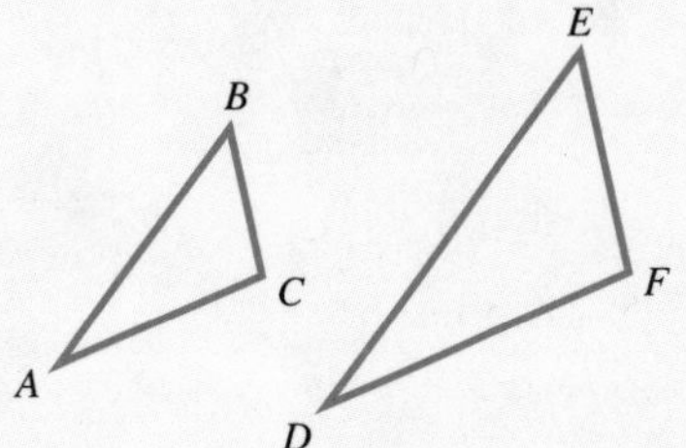

INTERMEDIATE ALGEBRA

EIGHTH EDITION

INTERMEDIATE ALGEBRA

EIGHTH EDITION

Margaret L. Lial
American River College

John Hornsby
University of New Orleans

An Imprint of Addison Wesley Longman, Inc.

Reading, Massachusetts • Menlo Park, California • New York • Harlow, England
Don Mills, Ontario • Sydney • Mexico City • Madrid • Amsterdam

Publisher: Jason Jordan

Acquisitions Editor: Jennifer Crum

Project Manager: Kari Heen

Developmental Editor: Terry McGinnis

Managing Editor: Ron Hampton

Production Supervisor: Kathleen A. Manley

Production Services: Elm Street Publishing Services, Inc.

Compositor: Typo-Graphics

Art Editor: Jennifer Bagdigian

Art Development: Meredith Nightingale

Artists: Precision Graphics, Darwin Hennings, Gary Torrisi, and Bob Giuliani

Marketing Manager: Craig Bleyer

Prepress Buyer: Caroline Fell

Manufacturing Coordinator: Evelyn Beaton

Text and Cover Designer: Susan Carsten

Cover Illustration: © Peter Siu/SIS

Library of Congress Cataloging-in-Publication Data

Lial, Margaret L.
Intermediate algebra.—8th ed. / Margaret L. Lial, John Hornsby.
p. cm.
Includes index.
ISBN 0-321-03646-8 (student ed.)
ISBN 0-321-04132-1 (annot. inst. ed.)
1. Algebra. I. Hornsby, John. II. Title.

QA152.2.L52 2000
512.9—dc21 99-24945
CIP

Printed in the U.S.A.

123456789-VH-02 01 00 99

Contents

CHAPTER 6 Rational Expressions 361

CHAPTER 7 Roots and Radicals 432

CHAPTER 8 Quadratic Equations and Inequalities 499

Preface

The eighth edition of *Intermediate Algebra* is designed for college students who have completed a course in introductory algebra or who require review of the basic concepts of algebra before taking additional courses in mathematics, science, business, nursing, or computer science. The primary objectives of this text are to familiarize students with mathematical symbols and operations and to consistently reinforce function and graphing concepts so important in later mathematics courses. Polynomial, rational, radical, exponential, and logarithmic equations and both linear and nonlinear systems are covered to allow students to solve a wide variety of related applications featuring real data.

This revision of *Intermediate Algebra* reflects our ongoing commitment to creating the best possible text and supplements package using the most up-to-date strategies for helping students succeed. One of these strategies, evident in our new Table of Contents and consistent with current teaching practices, involves the early introduction of functions and graphs of linear equations. The decision to place this material earlier in the text resulted from overwhelming feedback gathered from *Intermediate Algebra* users across the country. Chapter 3, Graphs, Linear Equations, and Functions, provides the groundwork for the function/graph emphasis that continues throughout the book and allows the use of functions to model real data in interesting applications. Chapter 4, Systems of Linear Equations, follows naturally after Chapter 3 and extends previous work with both linear equations and problem solving. Consistent with our approach to functions and graphs, in Chapters 5–9 we now introduce graphs of polynomial, rational, radical, and quadratic functions, as well as logarithmic and exponential functions when the corresponding material on expressions and equations is presented.

Other up-to-date pedagogical strategies to foster student success include a strong emphasis on vocabulary and problem solving, an increased number of real world applications in both examples and exercises, and a focus on relevant industry themes throughout the text.

Another strategy for student success, an exciting new CD-ROM called "Pass the Test," debuts with this edition of *Intermediate Algebra.* Directly correlated to the text's content, "Pass the Test" helps students master concepts by providing interactive pretests, chapter tests, section reviews, and InterAct Math tutorial exercises. To support an increased emphasis on graphical manipulation, the CD-ROM also includes a graphing tool that can be used for open-ended, student-directed exploration of number lines and coordinate graphs, as well as for exercises relevant to the graphing content throughout the book.

The *Student's Study Guide and Journal,* redesigned and enhanced with an optional journal feature for those who would like to incorporate more writing in their mathematics curriculum, provides an additional strategy for student success.

Although *Intermediate Algebra,* Eighth Edition, integrates many new elements, it also retains the time-tested features of previous editions: learning objectives for each section, careful exposition, fully developed examples, Cautions and Notes, and design features that highlight important definitions, rules, and procedures. Since the hallmark of any mathematics text is the quality of its exercise sets, we have carefully developed exercise sets that provide ample opportunity for drill and, at the same time, test conceptual understanding. In preparing this edition we have also addressed the standards of the National Council of Teachers of Mathematics and the American Mathematical Association of Two-Year Colleges, incorporating many new exercises focusing on concepts, writing, graph interpretation, technology use, collaborative work, and analysis of data from a wide variety of sources in the world around us.

Content Changes

We have fine-tuned and polished presentations of topics throughout the text based on user and reviewer feedback. Some of the content changes you may notice include the following:

- We consistently emphasize problem solving using a six-step problem-solving method, first introduced in Chapter 2 and continually reinforced in examples and exercise sets throughout the text.
- Chapter 3 of this edition is former Chapter 7, retitled Graphs, Linear Equations, and Functions. The presentation of functions in Section 3.5 now includes new examples with more visual representations of function concepts.
- Chapter 4, Systems of Linear Equations, is former Chapter 8. New real-world applications have been added to Section 4.3. Material on determinants and Cramer's rule is consolidated in Section 4.5.
- We now introduce graphs of simple polynomial functions in Chapter 5, rational functions in Chapter 6, and radical functions in Chapter 7.
- The material on integer exponents is consolidated in Section 5.1. Rational exponents and radicals are introduced in just one section, 7.1.
- Quadratic functions and graphs are presented earlier in the text when quadratic equations are solved (Chapter 8) rather than with the material on conic sections, as in previous editions.
- The material on inverse, exponential, and logarithmic functions (Chapter 9) is now placed before conic sections (Chapter 10) instead of after it.

New Features

We believe students and instructors will welcome the following new features:

Industry Themes To help motivate the material, each chapter features a particular industry that is presented in the chapter opener and revisited in examples and exercises in the chapter. Identified by special icons, these examples and exercises incorporate sourced data, often in the form of graphs and tables. Featured industries include business, health care, entertainment, sports, transportation, and others. (See pages 1, 125, and 207.)

New Examples and Exercises We have added 25% more real application problems with data sources. These examples and exercises often relate to the industry themes. They are designed to show students how algebra is used to describe and interpret data in everyday life. (See pages 7, 61, and 234.)

Television networks have been losing viewers to cable programming since 1982, as the two graphs below show.

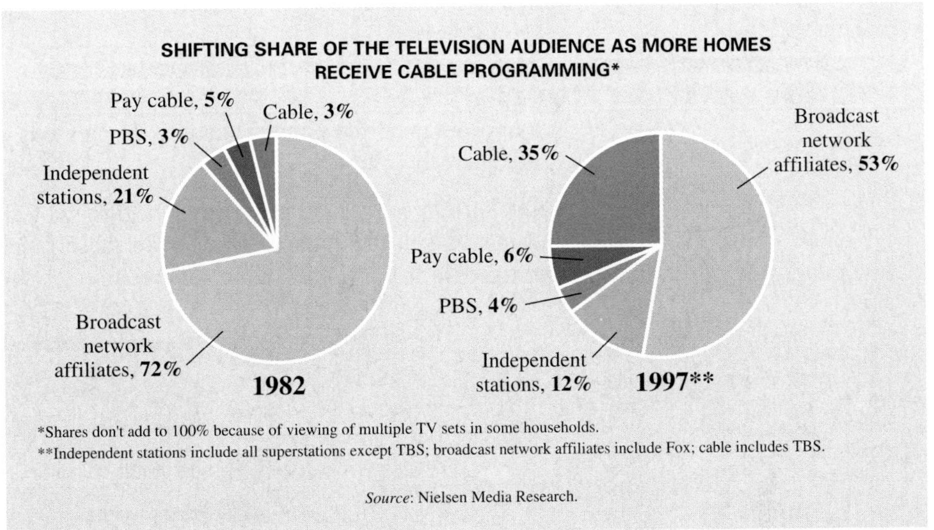

Use these graphs for Exercises 49–52.

49. In a typical group of 50,000 television viewers, how many would have watched cable in 1982?

50. In 1982, how many of a typical group of 110,000 viewers watched independent stations?

51. How many of a typical group of 35,000 viewers in 1997 watched cable?

52. In a typical group of 65,000 viewers, how many watched independent stations in 1997?

Technology Insights Exercises Technology is part of our lives, and we assume that all students of this text have access to scientific calculators. *While graphing calculators are not required for this text,* it is likely that students will go on to courses that use them. For this reason, we have included Technology Insights exercises in selected exercise sets. These exercises illustrate the power of graphing calculators and provide an opportunity for students to interpret typical results seen on graphing calculator screens. (See pages 134, 171, and 292.)

Mathematical Journal Exercises While we continue to include conceptual and writing exercises that require short written answers, new journal exercises have been added that ask students to fully explain terminology, procedures, and methods, document their understanding using examples, or make connections between concepts. Instructors who wish to incorporate a journal component in their classes will find these exercises especially useful. For the greatest possible flexibility, both writing exercises and journal exercises are indicated with icons in the Annotated Instructor's Edition, but not in the Student Edition. (See pages 15, 181, and 288.)

Group Activities Appearing at the end of each chapter, these activities allow students to apply the industry theme of the chapter to its mathematical content in a collaborative setting. (See pages 35, 111, and 264.)

Test Your Word Power To help students understand and master mathematical vocabulary, this new feature has been incorporated at the end of each chapter. Key terms from the chapter are presented with four possible definitions in multiple-choice format. Answers and examples illustrating each term are provided at the bottom of the appropriate page. (See pages 37, 196, and 350.)

HALLMARK FEATURES

We have retained the popular features of previous editions of the text. Some of these features are as follows:

Learning Objectives Each section begins with clearly stated numbered objectives, and material in the section is keyed to these objectives. In this way students know exactly what is being covered in each section.

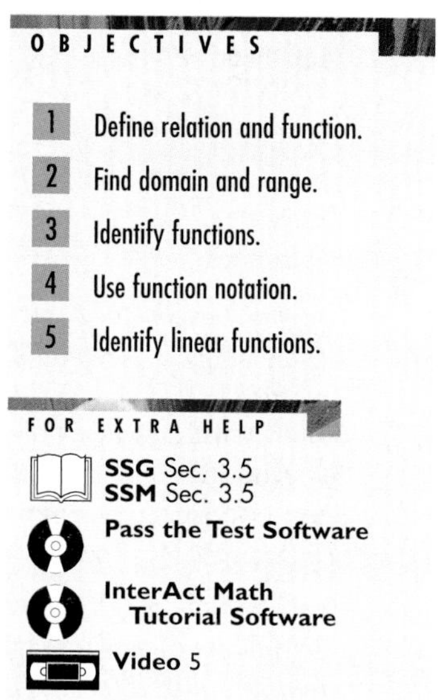

Cautions and Notes We often give students warnings of common errors and emphasize important ideas in Cautions and Notes that appear throughout the exposition.

Connections Retained from the previous edition, Connections boxes have been streamlined and now often appear at the beginning or the end of the exposition in selected sections. They continue to provide connections to the real world or to other mathematical concepts, historical background, and thought-provoking questions for writing or class discussion. (See pages 51, 69, and 168.)

Problem Solving Increased emphasis has been given to our six-step problem-solving method to aid students in solving application problems. This method is continually reinforced in examples and exercises throughout the text. (See pages 64, 71, and 227.)

Ample and Varied Exercise Sets Students in intermediate algebra require a large number and variety of practice exercises to master the material. This text contains approximately 6000 exercises, including about 1700 review exercises, plus numerous conceptual and writing exercises, journal exercises, and challenging exercises that go beyond the examples. More illustrations, diagrams, graphs, and tables now accompany the exercises. Multiple-choice, matching, true/false, and completion exercises help to provide variety. Exercises suitable for calculator use are marked with a calculator icon, in both the Student Edition and the Annotated Instructor's Edition. (See pages 11, 145, and 215.)

Relating Concepts Previously titled Mathematical Connections, these sets of exercises often appear near the end of selected section exercise sets. They tie together topics and highlight the relationships among various concepts and skills. For example, they may show how algebra and geometry are related, or how a graph of a linear equation in two variables is related to the solution of the corresponding linear equation in one variable. Instructors have told us that these sets of exercises make great collaborative activities for small groups of students. (See pages 27, 58, and 101.)

Ample Opportunity for Review Each chapter concludes with a Chapter Summary that features Key Terms and Symbols, Test Your Word Power, and a Quick Review of each section's content. Chapter Review Exercises keyed to individual sections are included as well as mixed review exercises and a Chapter Test. Following every chapter after Chapter 1, there is a set of Cumulative Review Exercises that covers material going back to the first chapter. Students always have an opportunity to review material that appears earlier in the text, and this provides an excellent way to prepare for the final examination in the course. (See pages 196–206 and 350–360.)

SUPPLEMENTS

Our extensive supplements package includes the Annotated Instructor's Edition, testing materials, study guides, solutions manuals, CD-ROM software, videotapes, and a Web site. For more information on these and other helpful supplements, contact your Addison Wesley Longman sales representative.

FOR THE INSTRUCTOR

Annotated Instructor's Edition (ISBN 0-321-04132-1)
For immediate access, the Annotated Instructor's Edition provides answers to all text exercises and Group Activities in color in the margin or next to the corresponding exercise, as well as Chalkboard Examples and Teaching Tips. To assist instructors in assigning homework, additional icons not shown in the Student Edition indicate journal exercises, writing exercises, and challenging exercises ▲.

CHALKBOARD EXAMPLE

Simplify $\dfrac{\frac{1}{2} \cdot 10 - 6 + \sqrt{9}}{\frac{5}{6} \cdot 12 - 3(2)^2}$.

Answer: -1

TEACHING TIP A student once told one of the authors that she remembered the rule involving division by 0 as follows: "It is *unde*fined when 0 is *under*neath."

Exercises designed for calculator use, are indicated in both the Student Edition and the Annotated Instructor's Edition.

Instructor's Solutions Manual (ISBN 0-321-06198-5)
The *Instructor's Solutions Manual* provides solutions to all even-numbered exercises, including answer art, and lists of all writing, journal, challenging, Relating Concepts, and calculator exercises.

Answer Book (ISBN 0-321-06199-3)
The *Answer Book* contains answers to all exercises and lists of all writing, journal, challenging, Relating Concepts, and calculator exercises. Instructors may ask the bookstore to order multiple copies of the *Answer Book* for students to purchase.

Printed Test Bank (ISBN 0-321-06197-7)
The *Printed Test Bank* contains short answer and multiple-choice versions of a placement test and final exam; six forms of chapter tests for each chapter, including four open-response (short answer) and two multiple-choice forms; 10 to 20 additional exercises per objective for instructors to use for extra practice, quizzes, or tests; answer keys to all of the above listed tests and exercises; and lists of all writing, journal, challenging, Relating Concepts, and calculator exercises.

TestGen-EQ with QuizMaster EQ (ISBN 0-321-06203-5)
This fully networkable software presents a friendly graphical interface which enables professors to build, edit, view, print and administer tests. Tests can be printed or easily exported to HTML so they can be posted to the Web for student practice.

For the Student

Student's Study Guide and Journal (ISBN 0-321-06201-9)
The *Student's Study Guide and Journal* contains a "Chart Your Progress" feature for students to track their scores on homework assignments, quizzes, and tests, additional practice for each learning objective, section summary outlines that give students additional writing opportunities and help with test preparation, and self-tests with answers at the end of each chapter. A manual icon at the beginning of each section in the Student Edition identifies section coverage.

Student's Solutions Manual (ISBN 0-321-06200-0)
The *Student's Solutions Manual* provides solutions to all odd-numbered exercises (journal and writing exercises excepted). A manual icon at the beginning of each section in the Student Edition identifies section coverage.

InterAct Math Tutorial Software (ISBN 0-321-06202-7) (Student Version)
This tutorial software correlates with every odd-numbered exercise in the text. The program is highly interactive with sample problems and interactive guided solutions accompanying every exercise. The program recognizes common student errors and provides customized feedback with sophisticated answer recognition capabilities. The management system (InterAct Math Plus) allows instructors to create, administer, and track tests, and to monitor student performance during practice sessions.

"Real to Reel" Videotapes (0-321-05660-4)
This videotape series provides separate lessons for each section in the book. A videotape icon at the beginning of each section identifies section coverage. All objectives, topics, and problem-solving techniques are covered and content is specific to *Intermediate Algebra,* Eighth Edition.

"Pass the Test" Interactive CD-ROM (ISBN 0-321-06205-1)
This dual-platform MAC/WIN CD helps students to master the course content by providing interactive pre-tests, chapter tests, section reviews, and InterAct tutorial exercises. After studying a chapter in class, students take a pre-test to determine what areas in that chapter need additional work. They are then directed to section reviews and tutorial exercises for continued practice. Students continue to take chapter tests and practice their skills until they have mastered the chapter. A unique graphing tool is provided for exploring the relationship between graphs and their algebraic representation.

World Wide Web Supplement (www.LialAlgebra.com)
Students can visit the Web site to explore additional real world applications related to the chapter themes, look up words in a complete glossary, and work through graphing calculator tutorials. The tutorials consist of step-by-step procedures as well as practice exercises for mastering basic graphing calculator skills.

MathXL (http://www.mathxl.com)
Available on-line with a pre-assigned ID and password by ordering a new copy of *Intermediate Algebra,* Eighth Edition, with ISBN 0-201-68161-7, MathXL helps students prepare for tests by allowing them to take practice tests that are similar to the chapter tests in their text. Students also get a personalized study plan that identifies strengths and pinpoints topics where more review is needed. For more information on subscriptions, contact your Addison Wesley Longman sales representative.

Math Tutor Center
Available free to any student who purchases a new Lial/Hornsby, *Intermediate Algebra,* Eighth Edition text ordered with ISBN 0-201-66331-7, the Addison Wesley Longman Math Tutor Center is staffed by qualified mathematics instructors who provide students with tutoring on text examples, exercises, and problems. Tutoring assistance is provided by telephone, fax, and e-mail and is available five days a week, seven hours a day. If a student purchases a used book, a registration number for the tutoring service may be obtained for a fee by calling our toll-free customer service number, 1-800-447-2226 and requesting ISBN 0-201-44461-5. Registration for the service is active for one or more of the following time periods depending on the course duration: Fall (8/31–1/31), Spring (1/1–6/30), or Summer (5/1–8/31). The Math Tutor Center service is also available for other Addison Wesley Longman textbooks in developmental math, precalculus math, liberal arts math, applied math, applied calculus, calculus, and introductory statistics. For more information, please contact your Addison Wesley Longman sales representative.

Spanish Glossary (ISBN 0-321-01647-5)
This book includes math terms that would be encountered in Basic Math through College Algebra.

ACKNOWLEDGMENTS

For a textbook to last through eight editions, it is necessary for the authors to rely on comments, criticisms, and suggestions of users, nonusers, instructors, and students. We are grateful for the many responses that we have received over the years. We wish to thank the following individuals who reviewed this edition of the text:

Jose Alvarado, University of Texas–Pan American; Judy Barclay, Cuesta College; Marc D. Campbell, Daytona Beach Community College; Terry Cheng, Irvine Valley College; Charles N. Curtis, Missouri Southern State College; Glenn DiStefano, Louisiana State University at Alexandria; Warrene C. Ferry, Jones County Junior College; Steve Grosteffon, Santa Fe Community College; Adam Hall, Belleville Area College; Judy Kasabian, El Camino College; Inessa Levi, University of Louisville; Mitchel Levy, Broward Community College–Central; Ellen Lund, San Jacinto College; Eric Matsuoka, Leeward Community College; Philip Meyer, Skyline College; Michael Montano, Riverside Community College; Lionel Mordecai, Southwestern College; Martin Peres, Broward Community College–North; Robert J. Rapalje, Seminole Community College; Dale Rice, Belleville Area College; Dale Rohm, University of Wisconsin–Stevens Point; Joyce Saxon, Morehead State University; Richard Semmler, Northern Virginia Community College; Steve Shattuck, Central Missouri State University; Kathleen A. Smith, University of Central Arkansas; Elizbeth Suco, Miami Dade Community College–Wolfson Campus; and Bettie A. Truit, Black Hawk Community College.

No author can complete a project of this magnitude without the help of many other individuals. Our sincere thanks go to Jenny Crum of Addison Wesley Longman who coordinated the package of texts of which this book is a part. Other dedicated staff at Addison Wesley Longman who worked long and hard to make this revision a success include Jason Jordan, Kari Heen, Susan Carsten, Meredith Nightingale, and Kathy Manley.

While Terry McGinnis has assisted us for many years "behind the scenes" in producing our texts, she has contributed far more to these revisions than ever. There is no question that these books are improved because of her attention to detail and consistency, and we are most grateful for her work above and beyond the call of duty. Kitty Pellissier continues to do an outstanding job in checking the answers to exercises. Many thanks to Jenny Bagdigian who coordinated the art programs for the books.

Cathy Wacaser and Ann Sargent of Elm Street Publishing Services provided their usual excellent production work. They are the best in the business. As usual, Paul Van Erden created an accurate, useful index. Becky Troutman prepared the Index of Applications. We are also grateful to Tommy Thompson who made suggestions for the feature "For the Student: 10 Ways to Succeed with Algebra," to Vickie Aldrich and Lucy Gurrola who wrote the Group Activity features, and Janis Cimperman of St. Cloud University and Steve C. Ouellette of the Walpole, Massachusetts State Public Schools.

To these individuals and all the others who have worked on these books for 30 years, remember that we could not have done it without you. We hope that you share with us our pride in these books.

Margaret L. Lial
John Hornsby

An Introduction to Calculators

There is little doubt that the appearance of handheld calculators nearly three decades ago and the later development of scientific and graphing calculators have changed the methods of learning and studying mathematics forever. Where the study of computations with tables of logarithms and slide rules made up an important part of mathematics courses prior to 1970, today the widespread availability of calculators make their study a topic only of historical significance.

Most consumer models of calculators are inexpensive. At first, however, they were costly. One of the first consumer models available was the Texas Instruments SR-10, which sold for about $150 in 1973. It could perform the four operations of arithmetic and take square roots, but could do very little more.

Today calculators come in a large array of different types, sizes, and prices. *For the course for which this textbook is intended, the most appropriate type is the scientific calculator,* which costs $10–$20.

In this introduction, we explain some of the features of scientific and graphing calculators. However, remember that calculators vary among manufacturers and models, and that while the methods explained here apply to many of them, they may not apply to your specific calculator. For this reason, it is important to remember that *this introduction is only a guide, and is not intended to take the place of your owner's manual.* Always refer to the manual in the event you need an explanation of how to perform a particular operation.

Scientific Calculators

Scientific calculators are capable of much more than the typical four-function calculator that you might use for balancing your checkbook. Most scientific calculators use *algebraic logic.* (Models sold by Texas Instruments, Sharp, Casio, and Radio Shack, for example, use algebraic logic.) A notable exception is Hewlett Packard, a company whose calculators use *Reverse Polish Notation* (RPN). In this introduction, we explain the use of calculators with algebraic logic.

Arithmetic Operations

To perform an operation of arithmetic, simply enter the first number, press the operation key ([+], [−], [×], or [÷]), enter the second number, and then press the [=] key. For example, to add 4 and 3, use the following keystrokes.

[4] [+] [3] [=] [7]

Change Sign Key

The key marked [±] allows you to change the sign of a display. This is particularly useful when you wish to enter a negative number. For example, to enter −3, use the following keystrokes.

[3] [±] [−3]

Memory Key

Scientific calculators can hold a number in memory for later use. The label of the memory key varies among models; two of these are [M] and [STO]. [M+] and [M−] allow you to add to or subtract from the value currently in memory. The memory recall key, labeled MR, RM, or RCL, allows you to retrieve the value stored in memory.

Suppose that you wish to store the number 5 in memory. Enter 5, then press the key for memory. You can then perform other calculations. When you need to retrieve the 5, press the key for memory recall.

If a calculator has a constant memory feature, the value in memory will be retained even after the power is turned off. Some advanced calculators have more than one memory. It is best to read the owner's manual for your model to see exactly how memory is activated.

Clearing/Clear Entry Keys

These keys allow you to clear the display or clear the last entry entered into the display. They are usually marked C and CE. In some models, pressing the C key once will clear the last entry, while pressing it twice will clear the entire operation in progress.

Second Function Key

This key is used in conjunction with another key to activate a function that is printed *above* an operation key (and not on the key itself). It is usually marked 2nd. For example, suppose you wish to find the square of a number, and the squaring function (explained in more detail later) is printed above another key. You would need to press 2nd before the desired squaring function can be activated.

Square Root Key

Pressing the square root key, $\sqrt{x}$, will give the square root (or an approximation of the square root) of the number in the display. For example, to find the square root of 36, use the following keystrokes.

3 6 $\sqrt{x}$ 6

The square root of 2 is an example of an irrational number (Chapter 7). The calculator will give an approximation of its value, since the decimal for $\sqrt{2}$ never terminates and never repeats. The number of digits shown will vary among models. To find an approximation of $\sqrt{2}$, use the following keystrokes.

2 $\sqrt{x}$ 1.4142136   An approximation

Squaring Key

This key, x^2, allows you to square the entry in the display. For example, to square 35.7, use the following keystrokes.

3 5 . 7 x^2 1274.49

The squaring key and the square root key are often found on the same key, with one of them being a second function (that is, activated by the second function key, described above).

Reciprocal Key

The key marked $1/x$ is the reciprocal key. (When two numbers have a product of 1, they are called *reciprocals.* See Chapter 1.) Suppose that you wish to find the reciprocal of 5. Use the following keystrokes.

5 $1/x$ 0.2

Inverse Key

Some calculators have an inverse key, marked INV. Inverse operations are operations that "undo" each other. For example, the operations of squaring and taking the square root are inverse operations. The use of the INV key varies among different models of calculators, so read your owner's manual carefully.

Exponential Key

The key marked x^y or y^x allows you to raise a number to a power. For example, if you wish to raise 4 to the fifth power (that is, find 4^5, as explained in Chapter 1), use the following keystrokes.

Root Key

Some calculators have this key specifically marked $\sqrt[x]{y}$ or $\sqrt[y]{x}$; with others, the operation of taking roots is accomplished by using the inverse key in conjunction with the exponential key. Suppose, for example, your calculator is of the latter type and you wish to find the fifth root of 1024. Use the following keystrokes.

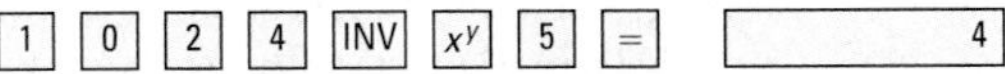

Notice how this "undoes" the operation explained in the exponential key discussion above.

Pi Key

The number π is an important number in mathematics. It occurs, for example, in the area and circumference formulas for a circle. By pressing the π key, you can display the first few digits of π. (Because π is irrational, the display shows only an approximation.) One popular model gives the following display when the π key is pressed: 3.1415927.

Methods of Display

When decimal approximations are shown on scientific calculators, they are either *truncated* or *rounded.* To see how a particular model is programmed, evaluate 1/18 as an example. If the display shows .0555555 (last digit 5), it truncates the display. If it shows .0555556 (last digit 6), it rounds off the display.

When very large or very small numbers are obtained as answers, scientific calculators often express these numbers in scientific notation (Chapter 5). For example, if you multiply 6,265,804 by 8,980,591, the display might look like this:

5.6270623 13.

The "13" at the far right means that the number on the left is multiplied by 10^{13}. This means that the decimal point must be moved 13 places to the right if the answer is to be expressed in its usual form. Even then, the value obtained will only be an approximation: 56,270,623,000,000.

Graphing Calculators

Graphing calculators are becoming increasingly popular in mathematics classrooms. While you are not expected to have a graphing calculator to study from this book, we do include a feature in many exercise sets called *Technology Insights* that asks you to interpret typical graphing calculator screens. These exercises can help to prepare you for future courses where graphing calculators may be recommended or even required.

Basic Features

Graphing calculators provide many features beyond those found on scientific calculators. In addition to the typical keys found on scientific calculators, they have keys that can be used to create graphs, make tables, analyze data, and change settings. One of the major differences between graphing and scientific calculators is that a graphing calculator has a larger viewing screen with graphing capabilities. The screens below illustrate the graphs of $y = x$ and $y = x^2$.

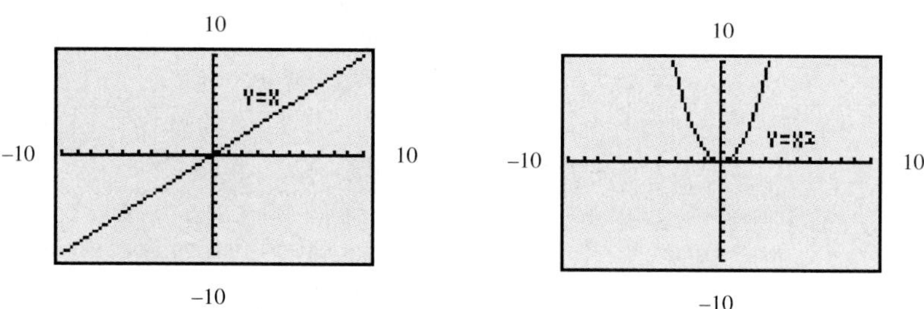

If you look closely at the screens, you will see that the graphs appear to be "jagged" rather than smooth, as they should be. The reason for this is that graphing calculators have much lower resolution than a computer screen. Because of this, graphs generated by graphing calculators must be interpreted carefully.

Editing Input

The screen of a graphing calculator can display several lines of text at a time. This feature allows you to view both previous and current expressions. If an incorrect expression is entered, an error message is displayed. The erroneous expression can be viewed and corrected by using various editing keys, much like a word-processing program. You do not need to enter the entire expression again. Many graphing calculators can also recall past expressions for editing or updating. The screen on the left below shows how two expressions are evaluated. The final line is entered incorrectly, and the resulting error message is shown in the screen on the right.

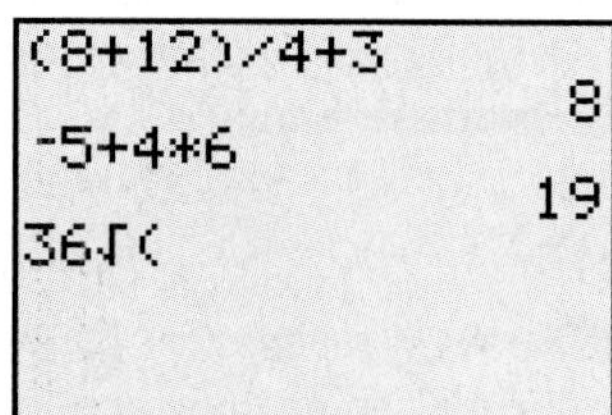

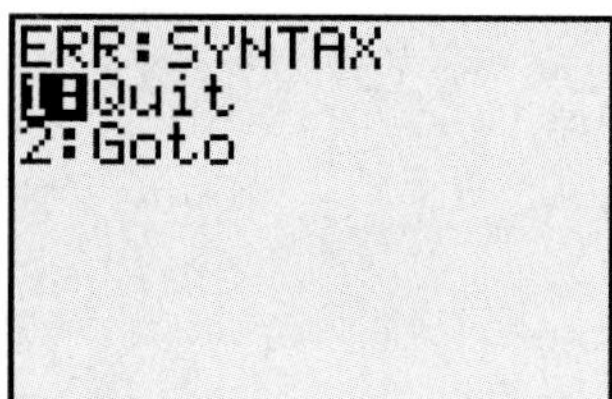

Order of Operations

Arithmetic operations on graphing calculators are usually entered as they are written in mathematical equations. For example, to evaluate $\sqrt{36}$ on a typical scientific calculator, you would first enter 36 and then press the square root key. As seen above, this is not the correct syntax for a graphing calculator. To find this root, you would first press the square root key, and then enter 36. See the screen on the left at the top of the next page. The order of operations on a graphing calculator is also important, and current models assist the user by inserting parentheses when typical errors might occur. The open parenthesis that follows the square root symbol is automatically entered by the calculator, so that an expression such as $\sqrt{2 \times 8}$ will not be calculated incorrectly as $\sqrt{2} \times 8$. Compare the two entries and their results in the screen on the right.

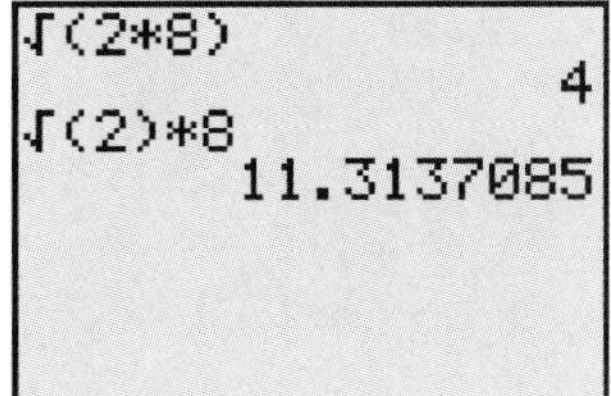

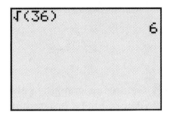

Viewing Windows

The viewing window for a graphing calculator is similar to the viewfinder in a camera. A camera usually cannot take a photograph of an entire view of a scene. The camera must be centered on some object and can only capture a portion of the available scenery. A camera with a zoom lens can photograph different views of the same scene by zooming in and out. Graphing calculators have similar capabilities. The xy-coordinate plane is infinite. The calculator screen can only show a finite, rectangular region in the plane, and it must be specified before the graph can be drawn. This is done by setting both minimum and maximum values for the x- and y-axes. The scale (distance between tick marks) is usually specified as well. Determining an appropriate viewing window for a graph is often a challenge, and many times it will take a few attempts before a satisfactory window is found.

The screen on the left shows a "standard" viewing window, and the graph of $y = 2x + 1$ is shown on the right. Using a different window would give a different view of the line.

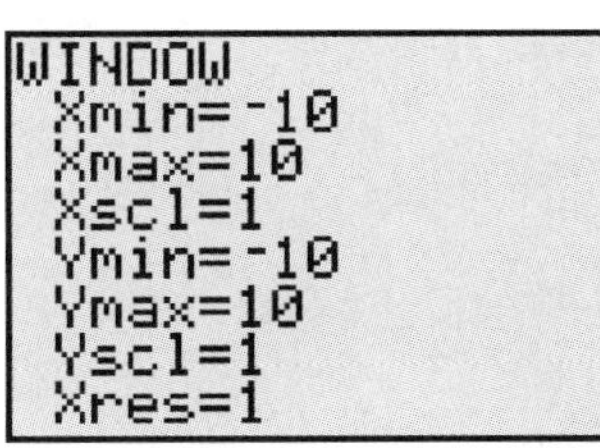

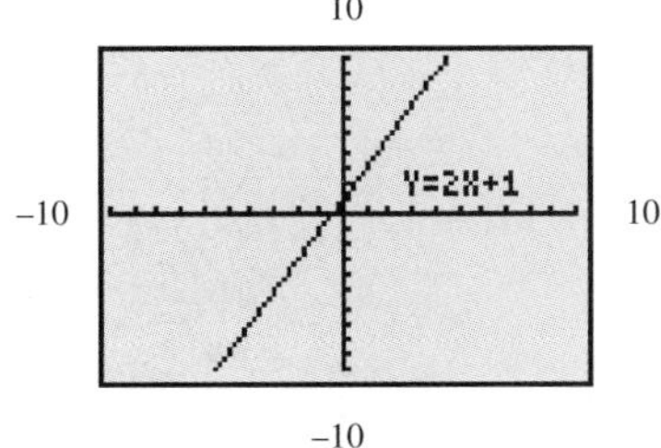

Locating Points on a Graph: Tracing and Tables

Graphing calculators allow you to trace along the graph of an equation, and, while doing this, display the coordinates of points on the graph. See the screen on the left at the top of the next page, which indicates that the point (2, 5) lies on the graph of $y = 2x + 1$. Tables for equations can also be displayed. The screen on the right shows a partial table for this same equation. Note the middle of the screen, which indicates that when $x = 2$, $y = 5$.

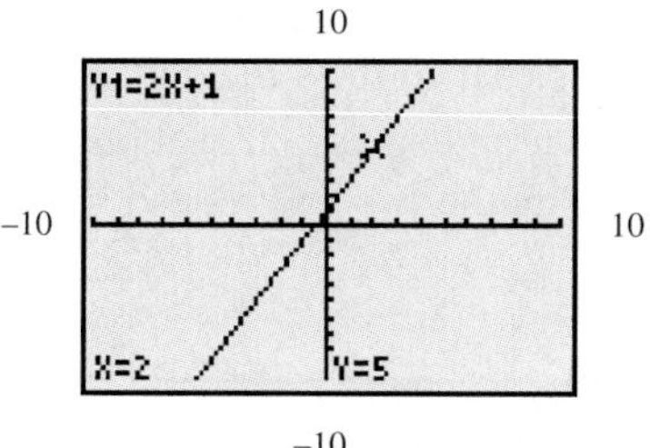

X	Y1	
-1	-1	
0	1	
1	3	
2	5	
3	7	
4	9	
5	11	

Y1=2X+1

Additional Features

There are many features of graphing calculators that go far beyond the scope of this book. These calculators can be programmed, much like computers. Many of them can solve equations at the stroke of a key, analyze statistical data, and perform symbolic algebraic manipulations. Mathematicians from the past would have been amazed by today's calculators. Many important equations in mathematics cannot be solved by hand. However, their solutions can often be approximated using a calculator. Calculators also provide the opportunity to ask "What if . . . ?" more easily. Values in algebraic expressions can be altered and conjectures tested quickly.

Final Comments

Despite the power of today's calculators, they cannot replace human thought. **In the entire problem-solving process, your brain is the most important component.** Calculators are only tools, and like any tool, they must be used appropriately in order to enhance our ability to understand mathematics. Mathematical insight may often be the quickest and easiest way to solve a problem; a calculator may neither be needed nor appropriate. By applying mathematical concepts, you can make the decision whether or not to use a calculator.

Review of the Real Number System 1

Economics

- **1.1** Basic Terms
- **1.2** Operations on Real Numbers
- **1.3** Properties of Real Numbers

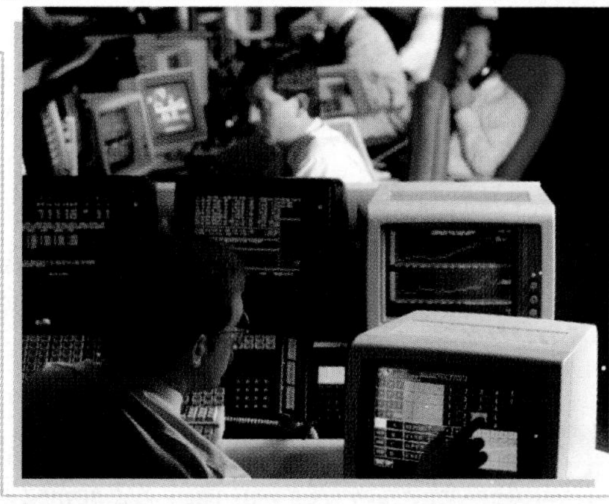

We are living in an age of instant communication. Much of the information we encounter is given in the form of graphs and tables. For example, the graph shown below indicates the inflation-adjusted annual percentage change in U.S. exports during the period 1987–1998. From the graph we can tell that first there was an increase in this rate, followed by a sharp decrease. In 1993, a wave of Asian trade sparked another general increasing trend, but in 1997 the collapse of Asian economies led to yet another decrease. From the graph, how much did the annual rate decline from 1988–1993?

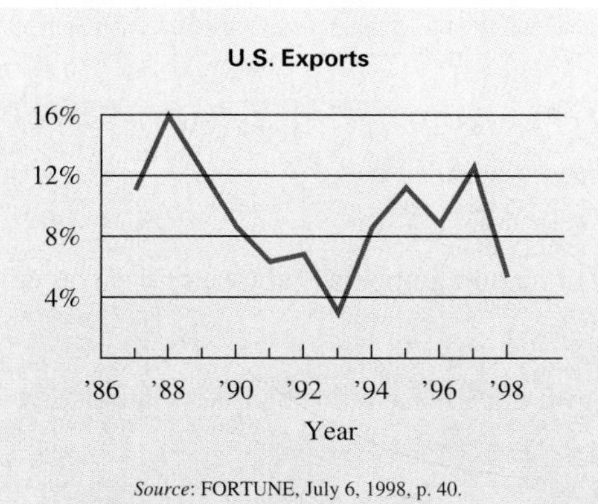

Source: FORTUNE, July 6, 1998, p. 40.

Merchandise Trade Balance of the United States

Year	*Millions of Dollars*
1992	−84,501
1993	−115,744
1994	−151,098
1995	−158,703
1996	−166,597

Source: U.S. Bureau of the Census, *U.S. Merchandise Trade,* series FT, monthly.

Tables such as the one above also provide information in a concise manner. The United States is dependent on goods and services supplied by other countries (in particular, Canada and Japan). The table indicates the total U.S. merchandise trade balance (exports minus imports) in millions of dollars over the five-year period from 1992–1996.

Throughout this book we use graphs and tables to illustrate a variety of mathematical concepts, as well as many types of applications. In this chapter we use economic data to reinforce some of the basic concepts of intermediate algebra.

INTERNET Visit our Web site at www.LialAlgebra.com

1.1 Basic Terms

OBJECTIVES

1. Write sets.
2. Know the common sets of numbers.
3. Use number lines.
4. Find additive inverses.
5. Use absolute value.
6. Use inequality symbols.
7. Graph sets of real numbers.

FOR EXTRA HELP

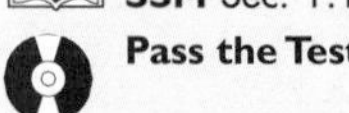
SSG Sec. 1.1
SSM Sec. 1.1

Pass the Test Software

InterAct Math Tutorial Software

Video 1

This chapter reviews some of the basic symbols and rules of algebra that are studied in elementary algebra.

OBJECTIVE 1 Write sets. A **set** is a collection of objects called the **elements** or **members** of the set. In algebra, the elements in a set are usually numbers. Set braces, $\{\quad\}$, are used to enclose the elements. For example, 2 is an element of the set $\{1, 2, 3\}$.

A set can be defined either by listing or by describing its elements. For example,

$$S = \{\text{Oregon, Ohio, Oklahoma}\}$$

defines the set S by *listing* its elements. The same set might be *described* by saying that set S is the set of all states in the United States whose names begin with the letter "O."

Set S above has a finite number of elements. Some sets contain an infinite number of elements, such as

$$N = \{1, 2, 3, 4, 5, 6, \ldots\},$$

where the three dots show that the list continues in the same pattern. Set N is called the set of **natural numbers,** or **counting numbers.** A set containing no elements, such as the set of natural numbers less than 1, is called the **empty set,** or **null set,** usually written $\emptyset$. The empty set may also be written as $\{\quad\}$.

CAUTION: Do not write $\{\emptyset\}$ for the empty set; $\{\emptyset\}$ is a set with one element (the empty set) and thus is not an empty set. Also, the number 0 is not the same as the empty set; do not use the symbol $\emptyset$ for zero. Use only the notations $\emptyset$ or $\{\quad\}$ for the empty set.

To write the fact that 2 is an element of the set $\{0, 1, 2, 3\}$, we use the symbol $\in$ (read "is an element of"):

$$2 \in \{0, 1, 2, 3\}.$$

The number 2 is also an element of the set of natural numbers N above, so we may write

$$2 \in N.$$

To show that 0 is *not* an element of set N, we draw a slash through the symbol $\in$:

$$0 \notin N.$$

Two sets are equal if they contain exactly the same elements. For example,

$$\{1, 2\} = \{2, 1\},$$

because the sets contain the same elements. (Order doesn't matter.) On the other hand, $\{1, 2\} \neq \{0, 1, 2\}$ ($\neq$ means "is not equal to") since one set contains the element 0 while the other does not.

In algebra, letters called **variables** are often used to represent numbers or to define sets of numbers. For example,

$$\{x \mid x \text{ is a natural number between 3 and 15}\}$$

(read "the set of all elements x such that x is a natural number between 3 and 15") defines the set

$$\{4, 5, 6, 7, \ldots, 14\}.$$

The notation $\{x \mid x$ is a natural number between 3 and 15$\}$ is an example of **set-builder notation.**

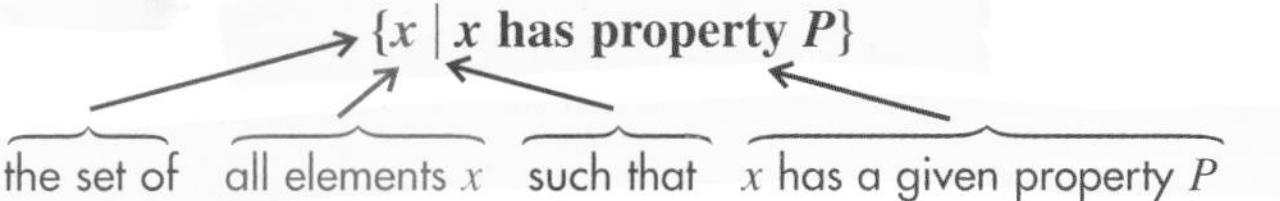

EXAMPLE 1 Listing the Elements in a Set

List the elements in the set.

(a) $\{x \mid x$ is a natural number less than 4$\}$
The natural numbers less than 4 are 1, 2, and 3. The given set is

$$\{1, 2, 3\}.$$

(b) $\{y \mid y$ is one of the first five even natural numbers$\} = \{2, 4, 6, 8, 10\}$

(c) $\{z \mid z$ is a natural number at least 7$\}$
The set of natural numbers at least 7 is an infinite set; write it with three dots as

$$\{7, 8, 9, 10, \ldots\}.$$

EXAMPLE 2 Using Set-Builder Notation to Describe a Set

Use set-builder notation to describe the set.

(a) $\{1, 3, 5, 7, 9\}$
There are often several ways to describe a set with set-builder notation. One way to describe this set is

$$\{y \mid y \text{ is one of the first five odd natural numbers}\}.$$

(b) $\{5, 10, 15, \ldots\}$
This set can be described as $\{d \mid d$ is a multiple of 5 greater than 0$\}$.

OBJECTIVE 2 **Know the common sets of numbers.** The following sets of numbers will be used throughout this book.

Sets of Numbers

Natural Numbers or Counting Numbers	$\{1, 2, 3, 4, 5, 6, 7, 8, \ldots\}$
Whole Numbers	$\{0, 1, 2, 3, 4, 5, 6, \ldots\}$
Integers	$\{\ldots, -3, -2, -1, 0, 1, 2, 3, \ldots\}$
Rational Numbers	$\left\{\frac{p}{q} \,\middle|\, p \text{ and } q \text{ are integers, with } q \neq 0\right\}$ or $\{x \mid x$ has a terminating or repeating decimal representation$\}$
Irrational Numbers	$\{x \mid x$ is a real number that is not rational$\}$ or $\{x \mid x$ has a nonterminating, nonrepeating decimal representation$\}$
Real Numbers	$\{x \mid x$ is represented by a point on a number line$\}$*

*An example of a number that is not represented by a point on a number line is $\sqrt{-1}$. This number, which is not a real number, is discussed in Chapter 7.

Examples of irrational numbers include most square roots, such as $\sqrt{7}$, $\sqrt{11}$, $\sqrt{2}$, and $-\sqrt{5}$. (Some square roots *are* rational: $\sqrt{16} = 4$, $\sqrt{100} = 10$, and so on.) Another irrational number is π, the ratio of the circumference of a circle to its diameter. All irrational numbers are real numbers.

Real numbers can also be defined in terms of decimals. By repeated subdivisions, any decimal can be located (at least in theory) as a point on a number line. Because of this, the set of real numbers can be defined as the set of all decimals. Also, the set of rational numbers can be shown to be the set of all repeating or terminating decimals. A bar over the series of numerals that repeat indicates a repeating decimal. For example, some decimal forms of rational numbers are $.\overline{6} = \frac{2}{3}$ (the 6 repeats), $.25 = \frac{1}{4}$, $.2 = \frac{1}{5}$, $.\overline{142857} = \frac{1}{7}$ (the block of digits 142857 repeats), $.4\overline{3} = \frac{13}{30}$ (the 3 repeats), and so on. The set of irrational numbers is the set of decimals that do not repeat and do not terminate.

The relationships among these various sets of numbers are shown in Figure 1; in particular, the figure shows that the set of real numbers includes both the rational and the irrational numbers.

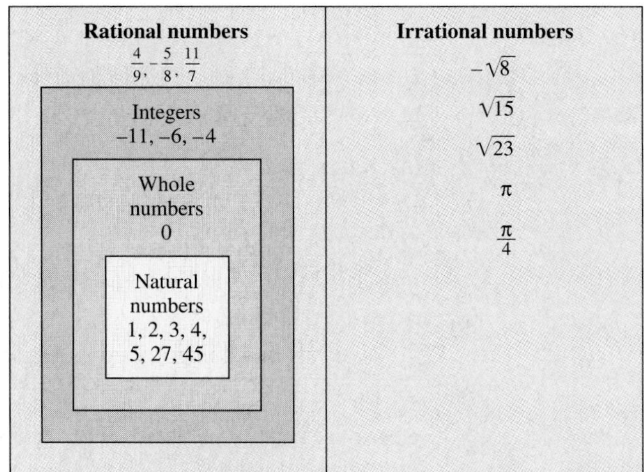

All numbers shown here are real numbers.

(a)

Irrational numbers
Real numbers
Rational numbers
Integers
Whole numbers
Positive numbers
Zero
Negative integers
Non-integer rational numbers

(b)

Figure 1

EXAMPLE 3 Identifying Examples of Number Sets

(a) $0, \quad \frac{2}{3}, \quad -\frac{9}{64}, \quad \frac{28}{7}$ (or 4), $\quad 2.45,$ and $\quad 1.\overline{37}$ are rational numbers.

(b) $\sqrt{3}, \quad \pi, \quad -\sqrt{2},$ and $\quad \sqrt{7} + \sqrt{3}$ are irrational numbers.

(c) $-8, \quad \frac{12}{2}, \quad -\frac{3}{1},$ and $\quad \frac{75}{5}$ are integers.

(d) All the numbers in parts (a), (b), and (c) above are real numbers.

(e) $\frac{4}{0}$ is undefined, since the definition of rational number requires the denominator to be nonzero. (However $\frac{0}{4}$ equals 0, which is a real number.)

OBJECTIVE 3 **Use number lines.** Number lines provide a way to picture a set of numbers. To construct a **number line,** choose any point on a horizontal line and label it 0. Then choose a point to the right of 0 and label it 1. The distance from 0 to 1 establishes a scale that can be used to locate more points, with positive numbers to the right of 0 and negative numbers to the left of 0. A number line is shown in Figure 2.

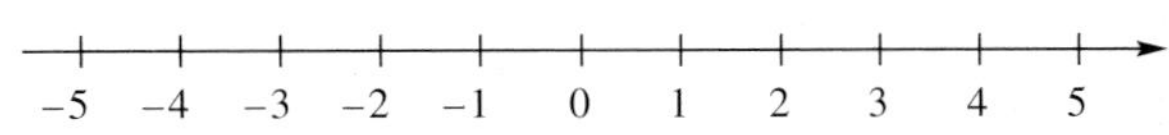

Figure 2

Each number is called the **coordinate** of the point that it labels, while the point is the **graph of the number.** A number line with several selected points graphed on it is shown in Figure 3.

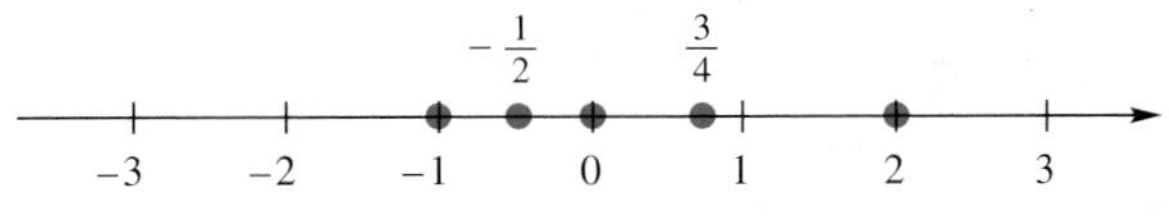

Figure 3

OBJECTIVE 4 **Find additive inverses.** Two numbers that are the same distance from 0 on the number line but on opposite sides of 0 are called **additive inverses, negatives,** or **opposites** of each other. For example, 5 is the additive inverse of -5, and -5 is the additive inverse of 5.

Additive Inverse

For any number a, the number $-a$ is the **additive inverse** of a.

The number 0 is its own additive inverse. The sum of a number and its additive inverse is always 0.

The symbol "−" can be used to indicate any of the following three things:

1. a negative number, such as -9 or -15;
2. the additive inverse of a number, as in " -4 is the additive inverse of 4";
3. subtraction, as in $12 - 3$.

In writing the number $-(-5)$, the symbol "−" is being used in two ways: the first − indicates the additive inverse of -5, and the second indicates a negative number, -5. Since the additive inverse of -5 is 5, then $-(-5) = 5$. This example suggests the following property.

Double Negative

For any number a,

$$-(-a) = a.$$

Numbers written with a positive or negative sign, such as $+4$, $+8$, -9, and -5, are called **signed numbers.** Positive numbers can be called signed numbers even if the positive sign is left off. The following table shows several signed numbers and the additive inverse of each.

Number	*Additive Inverse*
6	-6
-4	$-(-4)$, or 4
$\frac{2}{3}$	$-\frac{2}{3}$
-8.7	8.7
0	0

OBJECTIVE 5 Use absolute value. The **absolute value** of a number a, written $|a|$, is the distance on the number line from 0 to a. For example, the absolute value of 5 is the same as the absolute value of -5, since each number lies five units from 0. See Figure 4. That is,

$$|5| = 5 \quad \text{and} \quad |-5| = 5.$$

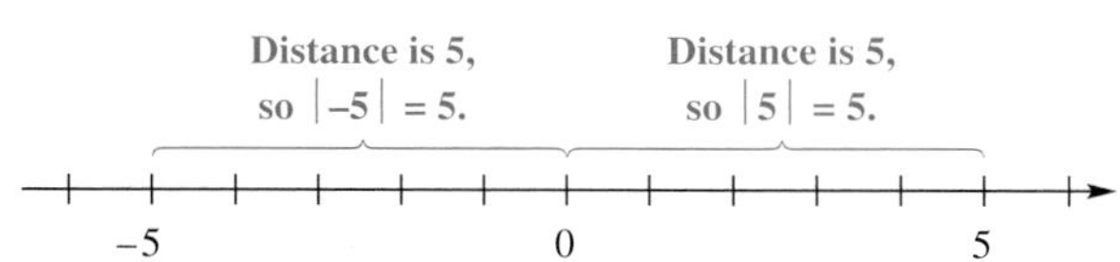

Figure 4

Since absolute value represents distance, and since distance is never negative, **the absolute value of a number is never negative.**

The formal definition of absolute value is as follows.

Absolute Value

$$|a| = \begin{cases} a & \text{if } a \text{ is positive or 0} \\ -a & \text{if } a \text{ is negative} \end{cases}$$

The second part of this definition, $|a| = -a$ if a is negative, is tricky. If a is a *negative* number, then $-a$, the additive inverse or opposite of a, is a positive number, and $|a|$ is positive. See Example 4(b) that follows.

EXAMPLE 4 Evaluating Absolute Value Expressions

Find the indicated value.

(a) $|2| = 2$

(b) $|-2| = -(-2) = 2$

(c) $|0| = 0$

(d) $-|8|$

Evaluate the absolute value first. Then find the additive inverse.

$$-|8| = -(8) = -8$$

(e) $-|-8|$

Work as in part (d): $|-8| = 8$, so

$$-|-8| = -(8) = -8.$$

(f) $|-2| + |8|$

Evaluate each absolute value first, and then add.

$$|-2| + |8| = 2 + 8 = 10$$

EXAMPLE 5 Comparing Rates of Change in Industries

Absolute value is useful in applications comparing size without regard to sign. The projected annual rate of employment change (in percent) in some of the fastest growing and most rapidly declining industries from 1994–2005 is shown below.

Industry (1994–2005)	*Percent Rate of Change*
Health services	5.7
Computer and data processing services	4.9
Child day care services	4.3
Footware, except rubber and plastic	−6.7
Household audio and video equipment	−4.2
Luggage, handbags, and leather products	−3.3

Source: U.S. Bureau of Labor Statistics.

What industry in the list is expected to see the greatest change? The least change?

We want the greatest change, without regard to whether the change is an increase or a decrease. Look for the number in the list with the largest absolute value. That number is found in footware, since $|-6.7| = 6.7$. Similarly, the least change is in the luggage, handbags, and leather products industry: $|-3.3| = 3.3$.

OBJECTIVE 6 Use inequality symbols. A statement that two numbers are *not* equal is called an **inequality.** For example, the numbers $\frac{5}{7}$ and 3 are not equal. Write this inequality as

$$\frac{5}{7} \neq 3,$$

where the slash through the equals sign is read "is not equal to."

When two numbers are not equal, one must be less than the other. The symbol $<$ means "is less than." For example, $8 < 9$, $-6 < 15$, and $0 < \frac{4}{3}$. Similarly, "is greater than" is written with the symbol $>$. For example, $12 > 5$, $9 > -2$, and $\frac{6}{5} > 0$. The number line in Figure 5 shows the graphs of the numbers 4 and 9. The graphs show that $4 < 9$. Starting at 4, add the positive number 5 to get 9. As this example suggests, $a < b$ means that there exists a positive number c such that $a + c = b$.

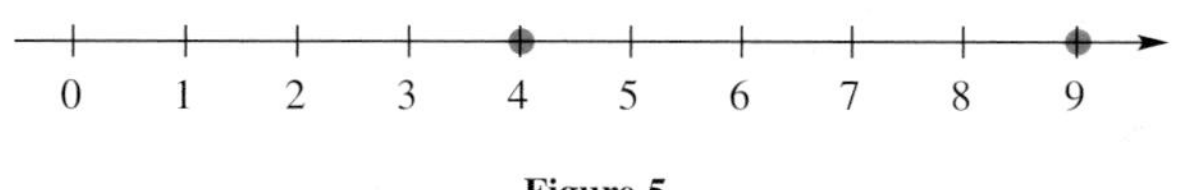

Figure 5

On the number line, the lesser of two given numbers is always located to the left of the other. Also, if a is less than b, then b is greater than a. The geometric definitions of $<$ and $>$ are as follows.

Definitions of $<$ and $>$

On the number line,

$a < b$ if a is to the left of b; $b > a$ if b is to the right of a.

We can use a number line to determine order. As shown on the number line in Figure 6, -6 is located to the left of 1. For this reason, $-6 < 1$. Also, $1 > -6$. From the same number line, $-5 < -2$, or $-2 > -5$.

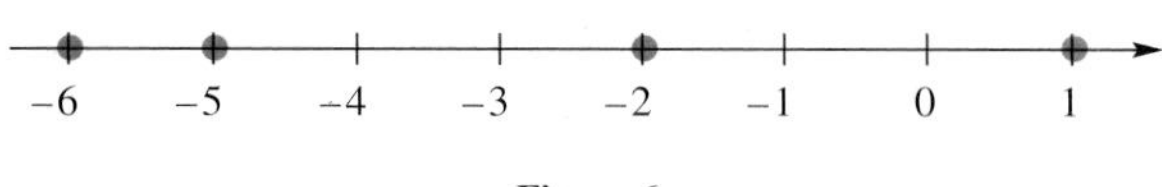

Figure 6

The following box summarizes results about positive and negative numbers in both words and symbols.

Words	*Symbols*
Every negative number is less than 0.	If a is negative, then $a < 0$.
Every positive number is greater than 0.	If a is positive, then $a > 0$.
0 is neither positive nor negative.	

In addition to the symbols $<$ and $>$, the symbols $\leq$ and $\geq$ often are used.

Symbols of Inequality

Meaning	*Example*
$<$ is less than	$-4 < -1$
$>$ is greater than	$3 > -2$
$\leq$ is less than or equal to	$6 \leq 6$
$\geq$ is greater than or equal to	$-8 \geq -10$

EXAMPLE 6 **Interpreting Inequality Symbols**

The following table shows several inequalities and why each is true.

Statement	*Reason*
$6 \leq 8$	$6 < 8$
$-2 \leq -2$	$-2 = -2$
$-9 \geq -12$	$-9 > -12$
$-3 \geq -3$	$-3 = -3$
$6 \cdot 4 \leq 5(5)$	$24 < 25$

In the last line, recall that the dot in $6 \cdot 4$ indicates the product 6×4, or 24. Also, 5(5) means 5×5, or 25. The statement is thus $24 \leq 25$, which is true.

OBJECTIVE 7 **Graph sets of real numbers.** Inequality symbols and variables are used to write sets of real numbers. For example, the set $\{x \mid x > -2\}$ consists of all the real numbers greater than -2. On a number line, we show the elements of this set (the set of all real numbers to the right of -2) by drawing a line from -2 to the right. We use a parenthesis at -2 to indicate that -2 is not an element of the given set. The result, shown in Figure 7, is called the **graph of the set** $\{x \mid x > -2\}$.

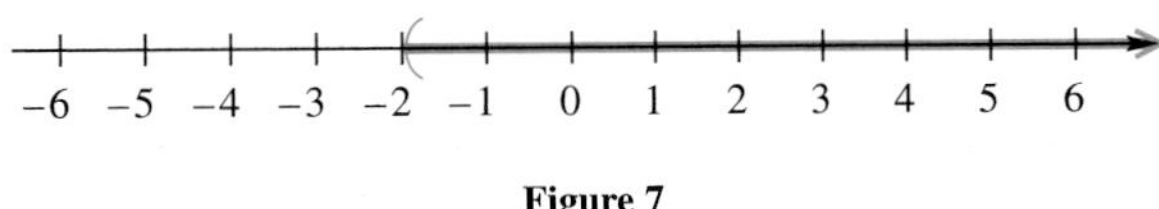

Figure 7

The set of numbers greater than -2 is an example of an **interval** on the number line. A simplified notation, called **interval notation,** is used for writing intervals. Using this notation, the interval of all numbers greater than -2 is written $(-2, \infty)$. The infinity symbol ∞ does not indicate a number; it shows that the interval includes all real numbers greater than -2. The left parenthesis indicates that -2 is not included. A parenthesis is always used next to the infinity symbol in interval notation. The set of all real numbers is written in interval notation as $(-\infty, \infty)$.

EXAMPLE 7 **Graphing an Inequality Written in Interval Notation**

Write $\{x \mid x < 4\}$ in interval notation and graph the interval.

The interval is written $(-\infty, 4)$. The graph is shown in Figure 8 on the next page. Since the elements of the set are all real numbers *less* than 4, the graph extends to the left.

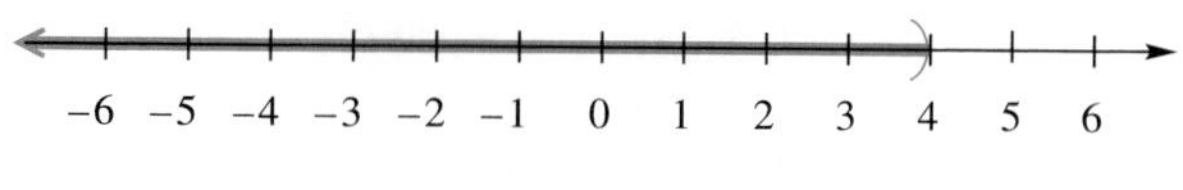

Figure 8

The set $\{x \mid x \leq -6\}$ contains all real numbers less than or equal to -6. To show that -6 itself is part of the set, a *square bracket* is used at -6, as shown in Figure 9. In interval notation, this set is written $(-\infty, -6]$.

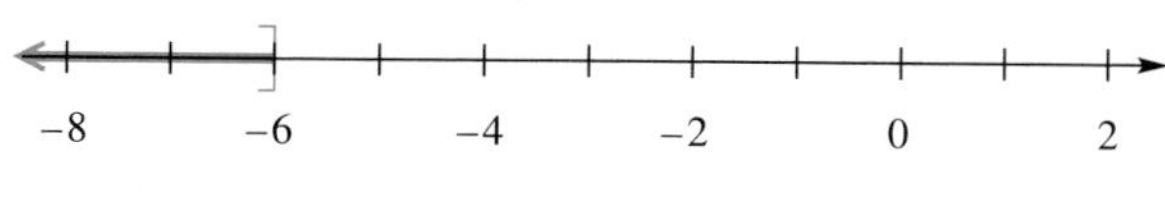

Figure 9

EXAMPLE 8 Graphing an Inequality Written in Interval Notation

Write $\{x \mid x \geq -4\}$ in interval notation and graph the interval.

This set is written in interval notation as $[-4, \infty)$. The graph is shown in Figure 10. A square bracket is used at -4 since -4 is part of the set.

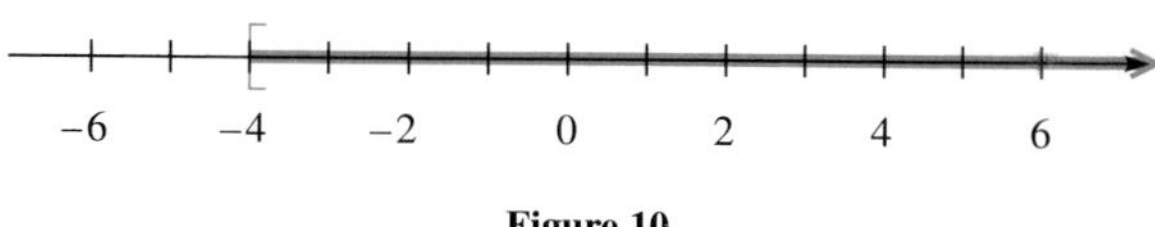

Figure 10

In a previous course you may have graphed $\{x \mid x > -2\}$ using an open circle instead of a parenthesis at -2. Also, you may have graphed $\{x \mid x \geq -4\}$ using a solid dot instead of a bracket at -4.

It is common to graph sets of numbers that are *between* two given numbers. For example, the set $\{x \mid -2 < x < 4\}$ is made up of all those real numbers between -2 and 4, but not the numbers -2 and 4 themselves. This set is written in interval notation as $(-2, 4)$. The graph has a heavy line between -2 and 4 with parentheses at -2 and 4. See Figure 11. The inequality $-2 < x < 4$ is read "-2 is less than x and x is less than 4," or "x is between -2 and 4."

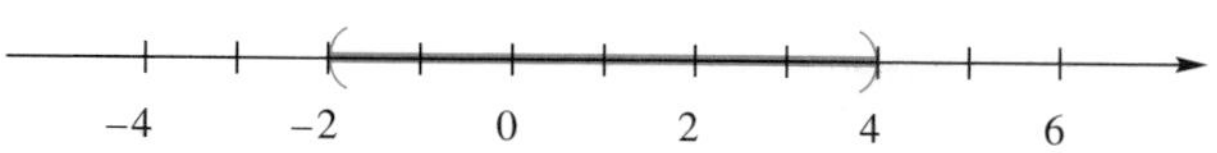

Figure 11

EXAMPLE 9 Graphing a Three-Part Inequality

Write $\{x \mid 3 < x \leq 10\}$ in interval notation and graph the interval.

Use a parenthesis at 3 and a square bracket at 10 to get $(3, 10]$ in interval notation. The graph is shown in Figure 12. Read the inequality $3 < x \leq 10$ as "3 is less than x and x is less than or equal to 10," or "x is between 3 and 10, excluding 3 and including 10."

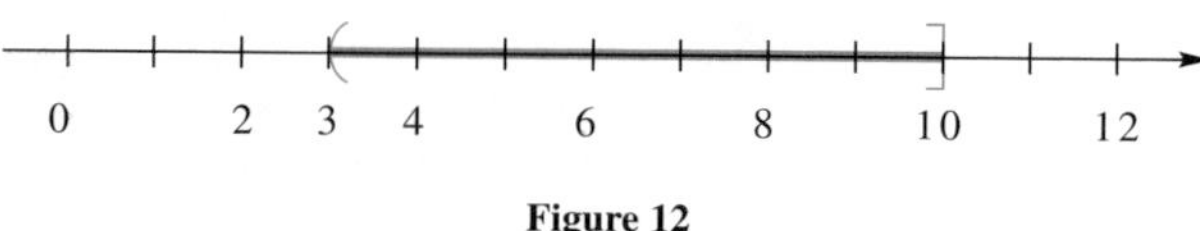

Figure 12

1.1 EXERCISES

Decide whether each statement about real numbers is true or false. If it is false, tell why.

1. Division of a nonzero number by zero is undefined.

2. If zero is divided by a nonzero number, the result is zero.

3. Every number has a positive additive inverse.

4. Every number has a positive absolute value.

5. The absolute value of a negative number is its additive inverse.

6. $-12, \frac{2}{7}$, and $.8\overline{3}$ are all rational numbers.

Write each set by listing its elements. See Example 1.

7. $\{y \mid y$ is a natural number greater than $5\}$

8. $\{x \mid x$ is a natural number greater than or equal to $11\}$

9. $\{z \mid z$ is an integer less than or equal to $4\}$

10. $\{p \mid p$ is an integer less than $3\}$

11. $\{a \mid a$ is an even integer greater than $8\}$

12. $\{k \mid k$ is an odd integer less than $1\}$

13. $\{x \mid x$ is an irrational number that is also rational$\}$

14. $\{r \mid r$ is a number that is both positive and negative$\}$

15. $\{p \mid p$ is a number whose absolute value is $4\}$

16. $\{w \mid w$ is a number whose absolute value is $7\}$

17. A student claimed that $\{x \mid x$ is a natural number greater than $3\}$ and $\{y \mid y$ is a natural number greater than $3\}$ actually name the same set, even though different variables are used. Was this student correct?

18. A student claimed that $\{\emptyset\}$ and $\emptyset$ name the same set. Was this student correct?

Write each set using set-builder notation. (More than one description is possible.) See Example 2.

19. $\{4, 8, 12, 16, \ldots\}$

20. $\{\ldots, -6, -3, 0, 3, 6, \ldots\}$

21. $\{2, 4, 6, 8\}$

22. $\{11, 12, 13, 14\}$

*Which elements of the given set are **(a)** natural numbers, **(b)** whole numbers, **(c)** integers, **(d)** rational numbers, **(e)** irrational numbers, **(f)** real numbers, **(g)** undefined? See Example 3.*

23. $\left\{-8, -\sqrt{5}, -.6, 0, \frac{1}{0}, \frac{3}{4}, \sqrt{3}, 4, 5, \frac{13}{2}, 17, \frac{40}{2}\right\}$

24. $\left\{-9, -\sqrt{6}, -.7, 0, \frac{2}{0}, \frac{6}{7}, \sqrt{7}, 3, 8, \frac{21}{2}, 13, \frac{75}{5}\right\}$

Graph the elements of the set on a number line.

25. $\{-3, -1, 0, 4, 6\}$

26. $\{-4, -2, 0, 3, 5\}$

27. $\left\{-\frac{2}{3}, 0, \frac{4}{5}, \frac{12}{5}, \frac{9}{2}, 4.8\right\}$

28. $\left\{-\frac{6}{5}, -\frac{1}{4}, 0, \frac{5}{6}, \frac{13}{4}, 5.2, \frac{11}{2}\right\}$

29. Explain the difference between the graph of a number and the coordinate of a point.

30. Explain why the real numbers .36 and $.\overline{36}$ have different points as graphs on a number line.

Find the value of each expression. See Example 4.

31. $|-8|$

32. $|-11|$

33. $-|5|$

34. $-|17|$

35. $-|-2|$

36. $-|-18|$

37. $-|4.5|$

38. $-|12.6|$

39. $|-2| + |3|$
40. $|-16| + |12|$
41. $|-9| - |-3|$
42. $|-10| - |-5|$
43. $|-9| + |-13|$
44. $|-13| + |-21|$
45. $|-1| + |-2| - |-3|$
46. $|-6| + |-4| - |-10|$

Refer to Example 5 for Exercises 47 and 48.

The table shows the percent change for annual new factory orders for durable goods in the United States during the period from 1988–1992.

Year	Percent Change
1988	10.2
1989	3.3
1990	−.4
1991	−4.6
1992	5.4

Source: U.S. Census Bureau.

47. In which year was the percent change the greatest?
48. In which year was the percent change the least?

Refer to a number line to answer true or false to each statement.

49. $-6 < -2$
50. $-4 < -3$
51. $-.32 < -\frac{4}{3}$
52. $-\frac{5}{3} > -1.2$

TECHNOLOGY INSIGHTS (EXERCISES 53–56)

Graphing calculators have the capability of determining whether a statement of equality or inequality is true or false. The calculator returns a 1 if the statement is true. It returns a 0 if the statement is false. See the screen.

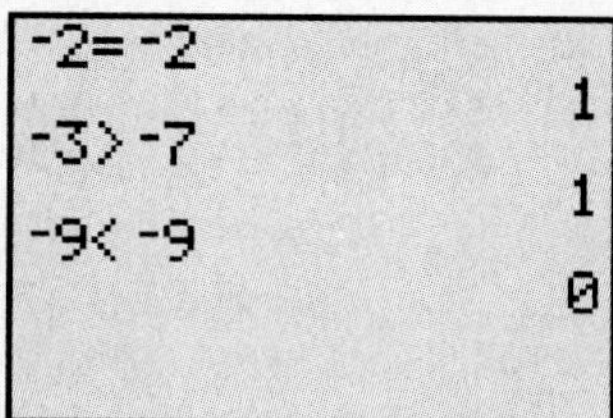

Decide whether the calculator would return a 1 or a 0 in each of the following screens.

53.

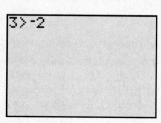

54.
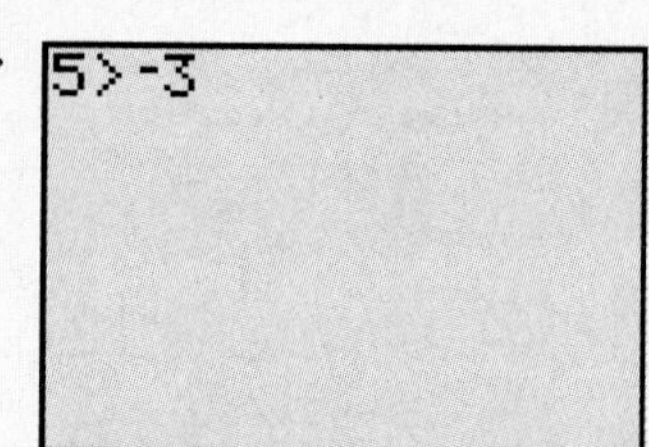

TECHNOLOGY INSIGHTS (EXERCISES 53–56) (CONTINUED)

55. -3≥-3

56. 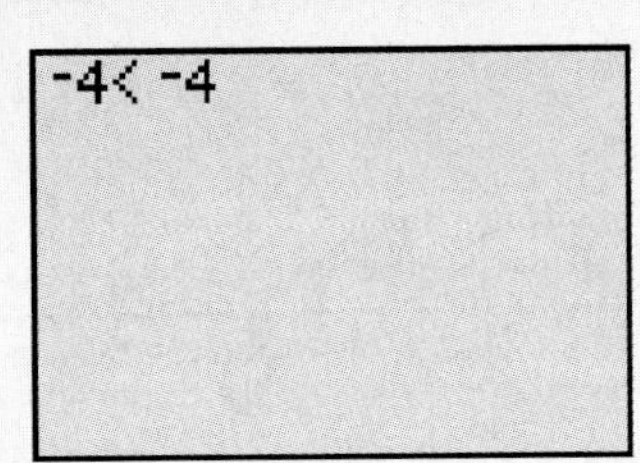

57. An inequality of the form "$a < b$" may also be written "$b > a$." Write $-3 < 2$ using this alternative form, and explain why both inequalities are true.

58. If $x > 0$ is a false statement for a given value of x, then is $x < 0$ necessarily a true statement? If not, explain why.

Use an inequality symbol to write each statement. See Example 6.

59. 6 is less than 11.

60. -4 is less than 12.

61. 4 is greater than x.

62. 7 is greater than y.

63. $3t - 4$ is less than or equal to 10.

64. $5x + 4$ is greater than or equal to 19.

65. 5 is greater than or equal to 5.

66. -3 is less than or equal to -3.

67. t is between -3 and 5.

68. r is between -4 and 12.

69. $3x$ is between -3 and 4, including -3 and excluding 4.

70. $5y$ is between -2 and 6, excluding -2 and including 6.

71. $5x + 3$ is not equal to 0.

72. $6x + 7$ is not equal to -3.

The slash symbol, /, is used to obtain the *negation* of the meaning of a symbol. For example, if $a = b$ is true, then $a \neq b$ is false. The slash symbol is also used to negate inequality: "$a \not< b$" is read "a is not less than b" and "$a \not> b$" is read "a is not greater than b." The chart shows how these symbols are equivalent to $\geq$ and $\leq$, respectively.

Symbolism with Slash	*Equivalent Statement*
$a \not< b$	$a \geq b$
$a \not> b$	$a \leq b$

Write an equivalent statement based on the explanation above.

73. $3 \not< 2$

74. $4 \not> 5$

75. $-3 \not> -3$

76. $-6 \not< -6$

77. $5 \geq 3$

78. $6 \leq 7$

Sea level refers to the surface of the ocean. The depth of a body of water such as an ocean or sea can be expressed as a negative number, representing average depth in feet below sea level. On the other hand, the altitude of a mountain can be expressed as a positive number, indicating its height in feet above sea level. The charts on the next page give selected depths and heights.

Bodies of Water	Average Depth in Feet (as a negative number)
Pacific Ocean	−12,925
South China Sea	−4,802
Gulf of California	−2,375
Caribbean Sea	−8,448
Indian Ocean	−12,598

Mountain	Altitude in Feet (as a positive number)
McKinley	20,320
Point Success	14,150
Matlalcueyetl	14,636
Ranier	14,410
Steele	16,644

Source: The World Almanac and Book of Facts.

79. List the bodies of water in order, starting with the deepest and ending with the shallowest.

80. List the mountains in order, starting with the shortest and ending with the tallest.

81. Decide whether the statement is true or false: The absolute value of the depth of the Pacific Ocean is greater than the absolute value of the depth of the Indian Ocean.

82. Decide whether the statement is true or false: The absolute value of the depth of the Gulf of California is greater than the absolute value of the depth of the Caribbean Sea.

Write each set using interval notation and graph the set on a number line. See Examples 7–9.

83. $\{x \mid x > -2\}$ **84.** $\{x \mid x < 5\}$ **85.** $\{x \mid x \leq 6\}$

86. $\{x \mid x \geq -3\}$ **87.** $\{x \mid 0 < x < 3.5\}$ **88.** $\{x \mid -4 < x < 6.1\}$

89. $\{x \mid 2 \leq x \leq 7\}$ **90.** $\{x \mid -3 \leq x \leq -2\}$ **91.** $\{x \mid -4 < x \leq 3\}$

92. $\{x \mid 3 \leq x < 6\}$ **93.** $\{x \mid 0 < x \leq 3\}$ **94.** $\{x \mid -1 \leq x < 6\}$

The graph shows the number of workplace age discrimination charges filed in the United States in the years 1990–1997. Use the graph to answer the following questions.

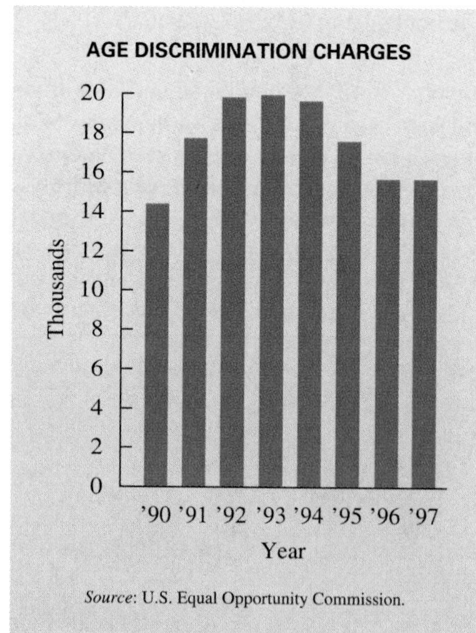

95. During which years was the number of charges filed greater than 18,000?

96. During which years was the number of charges filed less than or equal to 16,000?

97. If x represents the number of charges in the year 1991 and y represents the number of charges in the year 1997, which of the following is true: $x < y$ or $x > y$?

98. If x represents the number of charges in 1990 and y represents the number of charges in 1994, which is true: $x < y$ or $x > y$?

99. List the sets of numbers introduced in this section. Give a short explanation, including three examples, for each set.

100. List at least five symbols introduced in this section, and give a true statement involving each one.

1.2 Operations on Real Numbers

OBJECTIVES

1. Add and subtract real numbers.
2. Find the distance between two points.
3. Multiply and divide real numbers.
4. Use exponents and square roots.
5. Learn the order of operations.
6. Evaluate expressions for given values of variables.

FOR EXTRA HELP

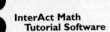

SSG Sec. 1.2
SSM Sec. 1.2

Pass the Test Software

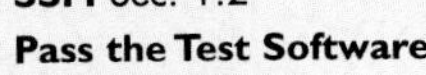
InterAct Math Tutorial Software

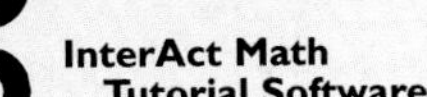
Video 1

In this section we review the rules for adding, subtracting, multiplying, and dividing real numbers.

OBJECTIVE 1 **Add and subtract real numbers.** The answer to an addition problem is called the **sum.** The rules for adding real numbers follow.

Adding Real Numbers

Like Signs
To add two numbers with the *same* sign, add their absolute values. The sign of the answer (either + or −) is the same as the sign of the two numbers.

Unlike Signs
To add two numbers with *different* signs, subtract the absolute values of the numbers. The answer is positive if the positive number has the larger absolute value. The answer is negative if the negative number has the larger absolute value.

For example, to add -12 and -8, first find their absolute values:

$$|-12| = 12 \qquad \text{and} \qquad |-8| = 8.$$

Since these numbers have the *same* sign, add their absolute values: $12 + 8 = 20$. Give the sum the sign of the two numbers. Since both numbers are negative, the sign is negative and

$$-12 + (-8) = -20.$$

Find $-17 + 11$ by subtracting the absolute values, since these numbers have different signs.

$$|-17| = 17 \qquad \text{and} \qquad |11| = 11$$

$$17 - 11 = 6$$

Give the result the sign of the number with the larger absolute value.

$$-17 + 11 = -6$$

Negative since $|-17| > |11|$

EXAMPLE 1 Adding Real Numbers

Add.

(a) $(-6) + (-3) = -(6 + 3) = -9$

(b) $(-12) + (-4) = -(12 + 4) = -16$

(c) $4 + (-1) = 3$

(d) $-9 + 16 = 7$

(e) $-\frac{1}{4} + \frac{2}{3} = -\frac{3}{12} + \frac{8}{12} = \frac{5}{12}$

(f) $-16 + 12 = -4$

(g) $-\frac{7}{8} + \frac{1}{3} = -\frac{21}{24} + \frac{8}{24} = -\frac{13}{24}$

(h) $-2.3 + 5.6 = 3.3$

The answer to a subtraction problem is called the **difference.** Thus, the difference between 7 and 5 is 2. To see how subtraction should be defined, compare the two statements below.

$$7 - 5 = 2$$
$$7 + (-5) = 2$$

In a similar way,

$$9 - 3 = 9 + (-3).$$

That is, to subtract 3 from 9, add the additive inverse of 3 to 9. These examples suggest the following rule for subtraction.

Definition of Subtraction

For all real numbers *a* and *b*,

$$a - b = a + (-b).$$

(Change the sign of the second number and add.)

EXAMPLE 2 Subtracting Real Numbers

Subtract.

Change to addition.
Change sign of second number.

(a) $6 - 8 = 6 + (-8) = -2$

Changed to addition.
Sign changed.

(b) $-12 - 4 = -12 + (-4) = -16$

(c) $-10 - (-7) = -10 + [-(-7)]$ This step is often omitted.
$= -10 + 7$
$= -3$

(d) $8.43 - (-5.27) = 8.43 + 5.27 = 13.70$

When a problem with both addition and subtraction is being worked, add and subtract in order from the left, as in the following example. Remember to work inside parentheses or brackets first.

EXAMPLE 3 Adding and Subtracting Real Numbers

Perform the operations.

(a) $15 - (-3) - 5 - 12 = (15 + 3) - 5 - 12$
$= 18 - 5 - 12$
$= 13 - 12$
$= 1$

(b) $-9 - [-8 - (-4)] + 6 = -9 - [-8 + 4] + 6$
$= -9 - [-4] + 6$
$= -9 + 4 + 6$
$= -5 + 6$
$= 1$

OBJECTIVE 2 Find the distance between two points. The number line in Figure 13 shows several points. To find the distance between the points 4 and 7, we subtract the numbers: $7 - 4 = 3$. Since distance is never negative, we must be careful to subtract in such a way that the answer is not negative. Or, to avoid this problem altogether, we can take the absolute value of the difference. Then the distance between 4 and 7 is either

$$|7 - 4| = 3 \quad \text{or} \quad |4 - 7| = 3.$$

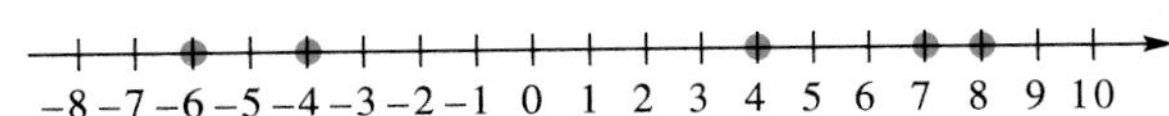

Figure 13

Distance

The **distance** between two points on a number line is the absolute value of the difference between the numbers.

EXAMPLE 4 Finding Distance between Points on the Number Line

Find the distance between the following pairs of points from Figure 13.

(a) 8 and -4

Find the absolute value of the difference of the numbers, taken in either order.

$$|8 - (-4)| = 12 \quad \text{or} \quad |-4 - 8| = 12$$

(b) -4 and -6

$$|-4 - (-6)| = 2 \quad \text{or} \quad |-6 - (-4)| = 2$$

OBJECTIVE 3 Multiply and divide real numbers. A **product** is the answer to a multiplication problem. For example, 24 is the product of 8 and 3. The rules for products of real numbers are given below.

Multiplying Real Numbers

Like Signs
The product of two numbers with the *same* sign is positive.

Unlike Signs
The product of two numbers with *different* signs is negative.

EXAMPLE 5 Multiplying Real Numbers

Multiply.

(a) $-3(-9) = 27$

(b) $-\frac{3}{4}\left(-\frac{5}{3}\right) = \frac{5}{4}$

(c) $7 \cdot 9 = 63$

(d) $-.05(.3) = -.015$

(e) $\frac{2}{3}(-3) = -2$

(f) $-\frac{5}{8}\left(\frac{12}{13}\right) = -\frac{15}{26}$

Addition and multiplication are the basic operations on real numbers. Subtraction was defined in terms of addition, and similarly, division is defined in terms of multiplication. The result of dividing two numbers is called the **quotient.** The quotient of the real numbers a and b $(b \neq 0)$ is the real number q such that $a = bq$. That is,

$$\frac{a}{b} = q \quad \text{only if} \quad a = bq.$$

For example,

$$\frac{36}{9} = 4 \quad \text{since} \quad 36 = 9 \cdot 4.$$

Also,

$$\frac{-12}{-2} = 6 \quad \text{since} \quad -12 = -2(6).$$

In a division such as $\frac{35}{-5} = -7$, -7 is the correct quotient because it answers the question, "What number multiplied by -5 gives the product 35?" Now consider $\frac{5}{0}$. There is *no* number whose product with 0 gives 5. On the other hand, $\frac{0}{0}$ would be satisfied by *every* real number, because any number multiplied by 0 gives 0. When dividing, we always want a unique quotient, and therefore *division by 0 is undefined.* However, dividing 0 by a nonzero number gives the quotient 0.

Here are some examples.

$$\frac{0}{6} = 0 \qquad \frac{15}{0} \text{ is undefined.} \qquad \frac{0}{0} \text{ is undefined.}$$

CAUTION Remember that division by 0 is always undefined.

Division of two numbers can be restated as multiplication.

Dividing Real Numbers

If a and b are real numbers and $b \neq 0$, then

$$\frac{a}{b} = a \cdot \frac{1}{b}.$$

If $b \neq 0$, $\frac{1}{b}$ is the **reciprocal** of b. This rule for division is the reason we "multiply by the reciprocal of the denominator" when a division problem involves fractions.

EXAMPLE 6 Dividing Real Numbers

Find the quotient.

(a) $\dfrac{24}{-6} = 24\left(-\dfrac{1}{6}\right) = -4$

(b) $\dfrac{-\frac{2}{3}}{-\frac{1}{2}} = -\dfrac{2}{3}\left(-\dfrac{2}{1}\right) = \dfrac{4}{3}$

Since division is equivalent to multiplication by the reciprocal, the rules for the sign of the quotient are the same as for the sign of the product.

The rules for multiplication and division suggest the results given below.

Equivalent Forms of a Fraction

The fractions $\dfrac{-x}{y}$, $-\dfrac{x}{y}$, and $\dfrac{x}{-y}$ are equivalent.

Also, the fractions $\dfrac{x}{y}$ and $\dfrac{-x}{-y}$ are equivalent. (Assume $y \neq 0$.)

Every fraction has three signs: the sign of the numerator, the sign of the denominator, and the sign of the fraction itself. As shown above, changing any two of these three signs does not change the value of the fraction. Changing only one sign, or changing all three, *does* change the value.

OBJECTIVE 4 Use exponents and square roots. A **factor** of a given number is any number that divides evenly (without remainder) into the given number. For example, 2 and 6 are factors of 12 since $2 \cdot 6 = 12$. Other factors of 12 include 4 and 3, 12 and 1, -4 and -3, -12 and -1, and -6 and -2. A number is in **factored form** if it is expressed as a product of two or more numbers.

In algebra, exponents are used as a way of writing the products of repeated factors. For example, the product $2 \cdot 2 \cdot 2 \cdot 2 \cdot 2$ is written

$$2 \cdot 2 \cdot 2 \cdot 2 \cdot 2 = 2^5.$$

The number 5 shows that 2 appears as a factor five times. The number 5 is the **exponent,** 2 is the **base,** and 2^5 is an **exponential expression.** Multiplying out the five 2s gives

$$2^5 = 2 \cdot 2 \cdot 2 \cdot 2 \cdot 2 = 32.$$

Definition of Exponent

If a is a real number and n is a natural number,

$$a^n = \underbrace{a \cdot a \cdot a \ldots a}_{n \text{ factors of } a}.$$

EXAMPLE 7 Evaluating an Exponential Expression

Evaluate the exponential expression.

(a) $5^2 = 5 \cdot 5 = 25$

Read 5^2 as "5 squared."

(b) $\left(\frac{2}{3}\right)^3 = \left(\frac{2}{3}\right)\left(\frac{2}{3}\right)\left(\frac{2}{3}\right) = \frac{8}{27}$

Read $\left(\frac{2}{3}\right)^3$ as "$\frac{2}{3}$ cubed."

(c) $2^6 = 2 \cdot 2 \cdot 2 \cdot 2 \cdot 2 \cdot 2 = 64$

Read 2^6 as "2 to the sixth power."

Be careful when evaluating an exponential expression with a negative sign. Compare the results in the next example.

EXAMPLE 8 Evaluating Exponential Expressions with Negative Signs

Evaluate the exponential expression.

(a) $(-3)^5 = (-3)(-3)(-3)(-3)(-3) = -243$

(b) $(-2)^6 = (-2)(-2)(-2)(-2)(-2)(-2) = 64$ Base is -2.

(c) $-2^6 = -(2 \cdot 2 \cdot 2 \cdot 2 \cdot 2 \cdot 2) = -64$ Base is -2.

Example 8 suggests the following generalizations.

The product of an odd number of negative factors is negative.

The product of an even number of negative factors is positive.

As shown by Examples 8(b) and (c), it is important to distinguish between $-a^n$ and $(-a)^n$.

$$-a^n = -1\underbrace{(a \cdot a \cdot a \cdots a)}_{n \text{ factors of } a} \qquad \text{Base is } a.$$

$$(-a)^n = \underbrace{(-a)(-a) \cdots (-a)}_{n \text{ factors of } -a} \qquad \text{Base is } -a.$$

CONNECTIONS

In Example 7, we used the terms "squared" and "cubed" to refer to powers of 2 and 3, respectively. The term "squared" comes from the figure of a square, which has the same measure for both length and width, as shown in the figure. Similarly, the term "cubed" comes from the figure of a cube. As shown in the figure, the length, width, and height of a cube have the same measure. This use of terminology illustrates a connection between algebra and geometry.

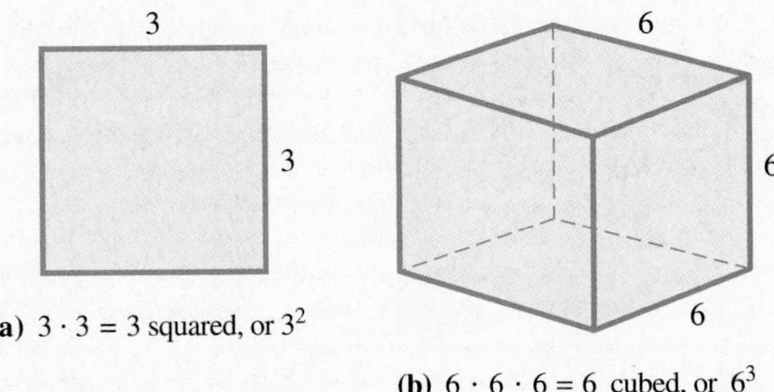

(a) $3 \cdot 3 = 3$ squared, or 3^2

(b) $6 \cdot 6 \cdot 6 = 6$ cubed, or 6^3

FOR DISCUSSION OR WRITING

Why do you suppose there is no special terminology similar to the words squared and cubed for powers that are higher than three?

As we saw in Example 7, $5^2 = 5 \cdot 5 = 25$, so that 5 squared is 25. The opposite of squaring a number is called taking its **square root.** For example, a square root of 25 is 5. Another square root of 25 is -5, since $(-5)^2 = 25$. Thus, 25 has two square roots, 5 and -5. The positive square root of a number is written with the symbol $\sqrt{}$. For example, the positive square root of 25 is written $\sqrt{25} = 5$. The negative square root of 25 is written $-\sqrt{25}$. Since the square of any nonzero real number is positive, a number like $\sqrt{-25}$ is not a real number.

EXAMPLE 9 Finding Square Roots

Find each root.

(a) $\sqrt{36} = 6$ since 6 is positive and $6^2 = 36$.

(b) $\sqrt{0} = 0$ since $0^2 = 0$.

(c) $\sqrt{\dfrac{9}{16}} = \dfrac{3}{4}$ since $\frac{3}{4}$ is positive and $\left(\frac{3}{4}\right)^2 = \frac{9}{16}$.

(d) $\sqrt{100} = 10$ since 10 is positive and $10^2 = 100$.

(e) $-\sqrt{100} = -10$

(f) $\sqrt{-100}$ is not a real number.

Notice the difference among the expressions in parts (d), (e), and (f). Part (d) shows the positive square root of 100, part (e) shows the negative square root of 100, and part (f) shows the square root of -100, which is not a real number.

The symbol $\sqrt{}$ is used only for the *positive* square root, except that $\sqrt{0} = 0$.

OBJECTIVE 5 Learn the order of operations. Given a problem such as $5 + 2 \cdot 3$, should 5 and 2 be added first or should 2 and 3 be multiplied first? When a problem involves more than one operation, we use the following order of operations. (This is the order used by computers and many calculators.)

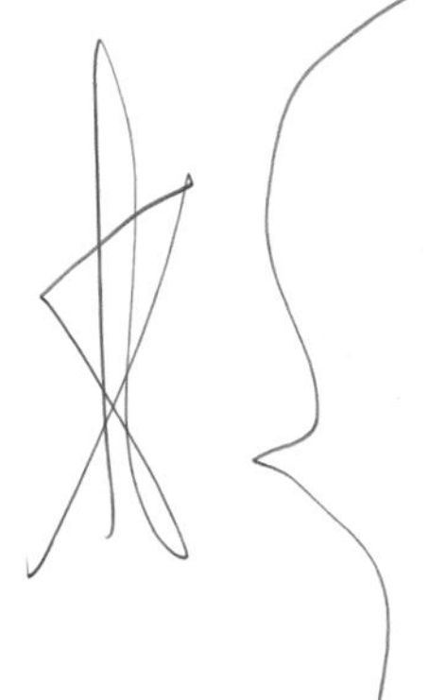

Order of Operations

If parentheses, square brackets, or fraction bars are present:

Step 1 Work separately above and below any division bar.

Step 2 Use the rules that follow within each set of parentheses or square brackets. Start with the innermost set and work outward.

If no parentheses or brackets are present:

Step 1 Evaluate all powers and roots.

Step 2 Do any multiplications or divisions in the order in which they occur, working from left to right.

Step 3 Do any additions or subtractions in the order in which they occur, working from left to right.

EXAMPLE 10 Using Order of Operations

Simplify.

(a) $5 + 2 \cdot 3$

To simplify this expression, first multiply and then add.

$$\begin{aligned} 5 + 2 \cdot 3 &= 5 + 6 && \text{Multiply.} \\ &= 11 && \text{Add.} \end{aligned}$$

(b) $24 \div 3 \cdot 2 + 6$

Follow the rules for order of operations.

$$\begin{aligned} 24 \div 3 \cdot 2 + 6 &= 8 \cdot 2 + 6 && \text{Divide.} \\ &= 16 + 6 && \text{Multiply.} \\ &= 22 && \text{Add.} \end{aligned}$$

In Example 10(b), notice that multiplications and divisions are done *in the order they appear from left to right.* The division was done before the multiplication, because it was encountered first.

EXAMPLE 11 Using Order of Operations

Simplify $4 \cdot 3^2 + 7 - (2 + 8)$.

Work inside the parentheses first.

$$4 \cdot 3^2 + 7 - (2 + 8) = 4 \cdot 3^2 + 7 - 10$$

Simplify powers and roots. Since $3^2 = 3 \cdot 3 = 9$,

$$4 \cdot 3^2 + 7 - 10 = 4 \cdot 9 + 7 - 10.$$

Do all multiplications or divisions, working from left to right.

$$4 \cdot 9 + 7 - 10 = 36 + 7 - 10$$

Finally, do all additions or subtractions, working from left to right.

$$36 + 7 - 10 = 43 - 10$$
$$= 33$$

EXAMPLE 12 Using Order of Operations

Simplify $\frac{1}{2} \cdot 4 + (6 \div 3 \cdot 7)$.

Work inside the parentheses first, doing the division before the multiplication.

$$\frac{1}{2} \cdot 4 + (6 \div 3 \cdot 7) = \frac{1}{2} \cdot 4 + (2 \cdot 7) \quad \text{Divide.}$$
$$= 2 + (14) \quad \text{Multiply.}$$
$$= 16 \quad \text{Add.}$$

EXAMPLE 13 Using Order of Operations

Simplify $\frac{5 + (-2^3)(2)}{6 \cdot \sqrt{9} - 9 \cdot 2}$.

The division bar is also a grouping symbol. The numerator and denominator must always be calculated separately before dividing.

$$\frac{5 + (-2^3)(2)}{6 \cdot \sqrt{9} - 9 \cdot 2} = \frac{5 + (-8)(2)}{6 \cdot 3 - 9 \cdot 2} \quad \text{Evaluate powers and roots.}$$
$$= \frac{5 - 16}{18 - 18} \quad \text{Multiply.}$$
$$= \frac{-11}{0} \quad \text{Subtract.}$$

Since division by 0 is not possible, the given expression is undefined.

OBJECTIVE 6 **Evaluate expressions for given values of variables.** An expression is *evaluated* by substituting the given numerical values for the variables.

EXAMPLE 14 Evaluating Expressions

Evaluate the expression when $m = -4$, $n = 5$, and $p = -6$.

(a) $5m - 9n$

Replace m with -4 and n with 5.

$$5m - 9n = 5(-4) - 9(5) = -20 - 45 = -65$$

(b) $\frac{m + 2n}{4p} = \frac{-4 + 2(5)}{4(-6)} = \frac{-4 + 10}{-24} = \frac{6}{-24} = -\frac{1}{4}$

(c) $-3m^2 + n^3$

Replace m with -4 and n with 5.

$$\begin{aligned} -3m^2 + n^3 &= -3(-4)^2 + 5^3 \\ &= -3(16) + 125 \\ &= -48 + 125 = 77 \end{aligned}$$

When evaluating expressions, it is a good idea to use parentheses around any negative numbers and fractions that are substituted for variables. Notice the placement of the parentheses in Example 14(c) assures that -4 is squared, giving a positive result. Writing (-4^2) would lead to -16, which would be incorrect.

1.2 EXERCISES

Complete each statement and give an example.

1. The sum of a positive number and a negative number is 0 if ______.
2. The sum of two positive numbers is a ______ number.
3. The sum of two negative numbers is a ______ number.
4. The sum of a negative number and a positive number is negative if ______.
5. The sum of a negative number and a positive number is positive if ______.
6. The difference between two positive numbers is negative if ______.
7. The difference between two negative numbers is negative if ______.
8. The product of two numbers with like signs is ______.
9. The product of two numbers with unlike signs is ______.
10. The quotient formed by any nonzero number divided by 0 is ______, and the quotient formed by 0 divided by any nonzero number is ______.

Add or subtract as indicated. See Examples 1–3.

11. $13 + (-4)$
12. $19 + (-13)$
13. $-6 + (-13)$
14. $-8 + (-15)$
15. $-\frac{7}{3} + \frac{3}{4}$
16. $-\frac{5}{6} + \frac{3}{8}$
17. $-.125 + .312$
18. $-.235 + .455$
19. $-8 - (-12) - (2 - 6)$
20. $-3 + (-14) + (-5 + 3)$
21. $\left(-\frac{5}{4} - \frac{2}{3}\right) + \frac{1}{6}$
22. $\left(-\frac{5}{8} + \frac{1}{4}\right) - \left(-\frac{1}{4}\right)$
23. $(-.382) + (4 - .6)$
24. $(3 - 2.94) - (-.63)$

The sketch shows a number line with several points labeled. Find the distance between each pair of points. See Example 4.

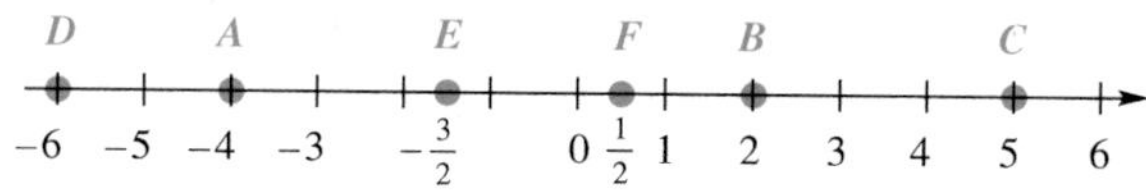

25. A and B
26. A and C
27. D and F
28. E and C

29. Give an example of a difference between two negative numbers that is equal to 5. State the rule for determining the sign of the answer after subtraction has been changed to addition.

30. Give an example of a sum of a positive number and a negative number that is equal to 4. State the rule for determining the sign of the answer when adding two numbers with different signs.

31. A statement that is often heard is "Two negatives give a positive." When is this true? When is it not true? Give a more precise statement that conveys this message.

32. Explain why the reciprocal of a nonzero number must have the same sign as the number.

Multiply or divide. See Examples 5 and 6.

33. $(-15)(-3)$ 34. $(-12)(-4)$ 35. $\frac{3}{4}(-20)(-12)$

36. $-\frac{2}{5}(-15)(-3)$ 37. $-3.45(-2.14)$ 38. $-2.4(-2.45)$

39. $\frac{-100}{-25}$ 40. $\frac{-300}{-60}$ 41. $\frac{\frac{12}{13}}{-\frac{4}{3}}$ 42. $\frac{\frac{5}{6}}{-\frac{1}{30}}$ 43. $\frac{5}{0}$ 44. $\frac{-1}{0}$

Decide whether each statement is true or false. If the statement is false, explain why.

45. $(-2)^7$ is a negative number.

46. $(-2)^8$ is a positive number.

47. The product of 8 positive factors and 8 negative factors is positive.

48. The product of 3 positive factors and 3 negative factors is positive.

49. $-4^6 = (-4)^6$

50. $-4^7 = (-4)^7$

51. $\sqrt{16}$ is a positive number.

52. $3 + 5 \cdot 6 = 3 + (5 \cdot 6)$

53. In the exponential -3^5, -3 is the base.

54. $\sqrt{a}$ is positive for all positive numbers a.

Evaluate. See Examples 7–9.

55. $\sqrt{121}$ 56. $\sqrt{361}$ 57. $.28^3$ 58. $.91^3$

59. $\left(-\frac{7}{10}\right)^2$ 60. $-\left(\frac{7}{10}\right)^2$ 61. $-\sqrt{900}$ 62. $-\sqrt{400}$

63. Why is it incorrect to say that $\sqrt{16}$ is equal to 4 or -4?

64. Explain why $\sqrt{-1000}$ is not a real number.

Find the following roots on a calculator. Show as many digits as your calculator displays.

65. $\sqrt{18,499}$ 66. $\sqrt{432.8}$ 67. $\sqrt{93.26}$ 68. $\sqrt{8.93}$

69. **(a)** If a is a positive number, is $-\sqrt{-a}$ positive, negative, or not a real number?
(b) If a is a positive number, is $-\sqrt{a}$ positive, negative, or not a real number?

70. Explain the rules for order of operations in your own words.

Perform each operation where possible, using the order of operations. See Examples 10–13.

71. $-7(-3) - (-2^3)$ 72. $-4 - 3(-2) + 5^2$

73. $|-6 - 5|(-8) + 3^2$ 74. $(-6 - 3)|-2 - 3|$

75. $(-8-5)(-2-1)$

76. $\dfrac{(-10+4)\cdot(-3)}{-7-2}$

77. $\dfrac{(-6+3)\cdot(-4)}{-5-1}$

78. $\dfrac{2(-5+3)}{-2^2}-\dfrac{(-3^2+2)3}{3-(-4)}$

79. $\dfrac{2(-5)+(-3)(-2^2)}{-6+5+1}$

80. $\dfrac{3(-4)+(-5)(-2)}{2^3-2+(-6)}$

81. $-\dfrac{1}{4}[3(-5)+7(-5)+1(-2)]$

82. $\dfrac{5-3\left(\dfrac{-5-9}{-7}\right)-6}{-9-11+3\cdot 7}$

83. $\dfrac{-4\left(\dfrac{12-(-8)}{3\cdot 2+4}\right)-5(-1-7)}{-9-(-7)-[-5-(-8)]}$

84. Write a paragraph explaining how you would evaluate the expression $(a+2b)(-3b^2+\sqrt{c})$, if $a=4$, $b=5$, and $c=16$.

Evaluate each expression if $a=-3$, $b=64$, and $c=6$. See Example 14.

85. $3a+\sqrt{b}$

86. $-2a-\sqrt{b}$

87. $\sqrt{b}+c-a$

88. $\sqrt{b}-c+a$

89. $\dfrac{2c+a^3}{4b+6a}$

90. $\dfrac{3c+a^2}{2b-c^2}$

TECHNOLOGY INSIGHTS (EXERCISES 91–92)

91. The screen indicates that -3 is stored in A and 6 is stored in C. What answer will the calculator give for the last line in the display?

```
-3→A:6→C
                6
4A³+2C
```

92. The screen indicates that $\frac{1}{2}$ is stored in X and $-\frac{1}{4}$ is stored in Y. What answer will the calculator give for the last line in the display?

```
1/2→X: -1/4→Y
             -.25
(X*Y)²*128
```

Solve the problems in Exercises 93–96.

93. The highest temperature ever recorded in Juneau, Alaska, was 90°F. The lowest temperature ever recorded there was -22°F. What is the difference between these two temperatures? (*Source: The World Almanac and Book of Facts.*)

94. On August 10, 1936, a temperature of 120°F was recorded in Arkansas. On February 13, 1905, Arkansas recorded a temperature of -29°F. What is the difference between these two temperatures? (*Source: The World Almanac and Book of Facts.*)

95. Telescope Peak, altitude 11,049 feet, is next to Death Valley, 282 feet below sea level. Find the difference between these altitudes. (*Source: The World Almanac and Book of Facts.*)

96. The surface of the Dead Sea has altitude 1299 feet below sea level. A stunt pilot is flying 80 feet above that surface. How much altitude must she gain to clear a 3852-foot pass by 225 feet? (*Source: The World Almanac and Book of Facts.*)

Use the graph of California exports to answer the questions in Exercises 97 and 98.

97. What is the difference between the January and February changes?

98. What is the difference between the changes in April and May?

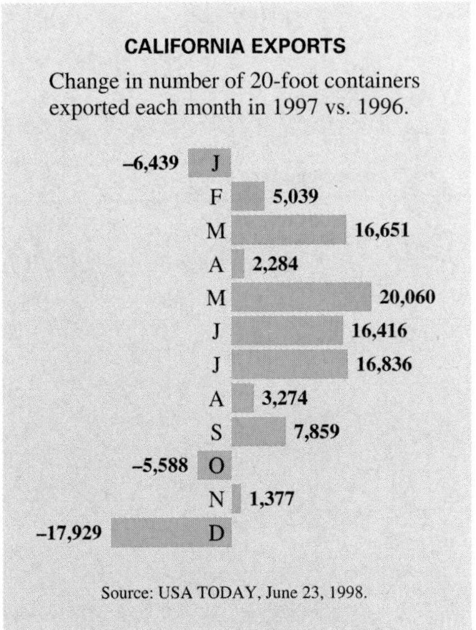

Source: USA TODAY, June 23, 1998.

RELATING CONCEPTS (EXERCISES 99–102)

In Section 1.1 we discussed the meanings of $a < b$, $a = b$, and $a > b$. Choose two numbers a and b such that $a < b$.

Work Exercises 99–102 in numerical order.

99. Find the difference $a - b$.

100. How does the answer in Exercise 99 compare to 0? (Is it greater than, less than, or equal to 0?)

101. Repeat Exercise 99 with different values for a and b.

102. How does the answer in Exercise 101 compare to 0? Based on your observations in these exercises, complete the following statement: If $a < b$, then $a - b$ _____ 0.

Did you make the connection that $a < b$ implies the difference $a - b$ is negative?

1.3 Properties of Real Numbers

OBJECTIVES

1. Use the distributive property.
2. Use the inverse properties.
3. Use the identity properties.
4. Use the commutative and associative properties.
5. Use the multiplication property of 0.

In this section we discuss basic properties of real numbers. These properties give results that occur consistently in work with numbers, so they have been generalized to apply to expressions with variables as well.

OBJECTIVE 1 **Use the distributive property.** Notice that

$$2(3 + 5) = 2 \cdot 8 = 16$$

and

$$2 \cdot 3 + 2 \cdot 5 = 6 + 10 = 16$$

so

$$2(3 + 5) = 2 \cdot 3 + 2 \cdot 5.$$

This idea is illustrated by the divided rectangle in Figure 14.

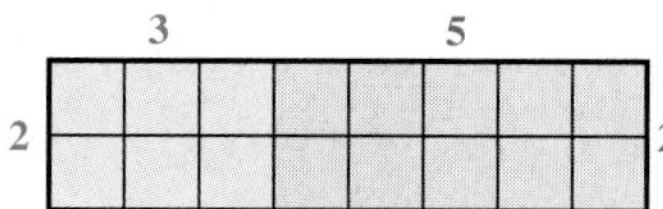

Area of left part is $2 \cdot 3 = 6$.
Area of right part is $2 \cdot 5 = 10$.
Area of total is $2(3 + 5) = 16$.

Figure 14

Similarly,

$$-4[5 + (-3)] = -4(2) = -8$$

and

$$-4(5) + (-4)(-3) = -20 + 12 = -8$$

so

$$-4[5 + (-3)] = -4(5) + (-4)(-3).$$

These examples suggest the **distributive property of multiplication with respect to addition,** or simply the **distributive property.**

Distributive Property

For any real numbers a, b, and c,

$$a(b + c) = ab + ac \quad \text{and} \quad (b + c)a = ba + ca.$$

The distributive property can also be written

$$ab + ac = a(b + c).$$

This property is important because it provides a way to change a *product* $a(b + c)$ to a *sum* $ab + ac$ or a sum to a product. When the form $a(b + c) = ab + ac$ is used, we sometimes refer to it as "removing parentheses" or "expanding."

EXAMPLE 1 Using the Distributive Property

Use the distributive property to rewrite the expression.

(a) $3(x + y)$

In the statement of the property, let $a = 3$, $b = x$ and $c = y$. Then

$$3(x + y) = 3x + 3y.$$

(b) $-2(5 + k) = -2(5) + (-2)(k)$

$= -10 - 2k$

(c) $4x + 8x$
Use the second form of the property.

$$4x + 8x = (4 + 8)x = 12x$$

(d) $3r - 7r = 3r + (-7)r$ Definition of subtraction
$= [3 + (-7)]r$ Distributive property
$= -4r$

(e) $5p + 7q$
Since there is no common factor here, we cannot use the distributive property to simplify the expression.

As illustrated in Example 1(d), the distributive property can also be used for subtraction, so that

$$a(b - c) = ab - ac.$$

OBJECTIVE 2 **Use the inverse properties.** In Section 1.1 we saw that the additive inverse of a number a is $-a$. For example, 3 and -3 are additive inverses, as are -8 and 8. The number 0 is its own additive inverse. In Section 1.2, we saw that two numbers with a product of 1 are reciprocals. Another name for a reciprocal of a number is its **multiplicative inverse.** This is similar to the idea of an additive inverse. Thus, 4 and $\frac{1}{4}$ are multiplicative inverses, and so are $-\frac{2}{3}$ and $-\frac{3}{2}$. (Note that a pair of reciprocals has the same sign.) These properties are called the **inverse properties** of addition and multiplication.

Inverse Properties

For any real number a, there is a single real number $-a$, such that

$$a + (-a) = 0 \quad \text{and} \quad -a + a = 0.$$

For any nonzero real number a, there is a single real number $\frac{1}{a}$ such that

$$a \cdot \frac{1}{a} = 1 \quad \text{and} \quad \frac{1}{a} \cdot a = 1.$$

OBJECTIVE 3 **Use the identity properties.** The numbers 0 and 1 each have a special property. Zero is the only number that can be added to any number to get that number. That is, adding 0 leaves the identity of a number unchanged. For this reason, 0 is called the **identity element for addition.** In a similar way, multiplying by 1 leaves the identity of any number unchanged, so 1 is the **identity element for multiplication.** The following **identity properties** summarize this discussion.

Identity Properties

For any real number a,

$$a + 0 = 0 + a = a$$

and

$$a \cdot 1 = 1 \cdot a = a.$$

The identity property for 1 is especially useful in simplifying algebraic expressions.

EXAMPLE 2 Using the Identity Property $1 \cdot a = a$

Use the identity property for 1 to rewrite the expression so the distributive property can be used.

(a) $12m + m$

$$\begin{aligned} 12m + m &= 12m + 1m && \text{Identity property} \\ &= (12 + 1)m && \text{Distributive property} \\ &= 13m \end{aligned}$$

(b)
$$\begin{aligned} y + y &= 1y + 1y && \text{Identity property} \\ &= (1 + 1)y && \text{Distributive property} \\ &= 2y \end{aligned}$$

(c)
$$\begin{aligned} -(m - 5n) &= -1(m - 5n) && \text{Identity property} \\ &= -1 \cdot m + (-1)(-5n) && \text{Distributive property} \\ &= -m + 5n \end{aligned}$$

Expressions such as $12m$ and $5n$ from Example 2 are examples of *terms.* A **term** is a number or the product (or quotient) of a number and one or more variables raised to powers. Terms with exactly the same variables raised to exactly the same powers are called **like terms.** The number in the product is called the **numerical coefficient** or just the **coefficient.** For example, in the term $5p$, the coefficient is 5.

OBJECTIVE 4 Use the commutative and associative properties. In the expression

$$-2m + 5m + 3 - 6m + 8,$$

there are five terms. Notice that $(-2m)(5m)(3)(-6m)(8)$ is just one term, since it is a *product* of numbers and variables.

Simplifying expressions as in Examples 2(a) and (b) is called **combining like terms.** Only like terms may be combined. To combine like terms in the expression

$$-2m + 5m + 3 - 6m + 8$$

we need two more properties. We are familiar with the fact that

$$3 + 9 = 12 \quad \text{and} \quad 9 + 3 = 12.$$

Also,

$$3 \cdot 9 = 27 \quad \text{and} \quad 9 \cdot 3 = 27.$$

Furthermore, notice that

$$(5 + 7) + (-2) = 12 + (-2) = 10$$

and

$$5 + [7 + (-2)] = 5 + 5 = 10.$$

Also,

$$(5 \cdot 7)(-2) = 35(-2) = -70$$

and

$$(5)[7(-2)] = 5(-14) = -70.$$

These observations suggest the following properties.

Commutative Properties

For any real numbers a and b,

$$a + b = b + a$$
$$ab = ba.$$

Associative Properties

For any real numbers a, b, and c,

$$a + (b + c) = (a + b) + c$$
$$a(bc) = (ab)c.$$

The associative properties are used to *regroup* the terms of an expression. The commutative properties are used to change the *order* of the terms in an expression.

EXAMPLE 3 Using the Commutative and Associative Properties

Use the properties to combine like terms.

$-2m + 5m + 3 - 6m + 8$

$= (-2m + 5m) + 3 - 6m + 8$ — Order of operations

$= 3m + 3 - 6m + 8$ — Distributive property

By the order of operations, the next step would be to add $3m$ and 3, but they are unlike terms. To get $3m$ and $-6m$ together, use the associative and commutative properties. Begin by inserting parentheses and brackets according to the order of operations.

$[(3m + 3) - 6m] + 8$

$= [3m + (3 - 6m)] + 8$ — Associative property

$= [3m + (-6m + 3)] + 8$ — Commutative property

$= [(3m + [-6m]) + 3] + 8$ — Associative property

$= (-3m + 3) + 8$ — Distributive property

$= -3m + (3 + 8)$ — Associative property

$= -3m + 11$ — Add.

In practice, many of the steps are not written down, but you should realize that the commutative and associative properties are used whenever the terms in an expression are rearranged in order to combine like terms.

EXAMPLE 4 Using the Properties of Real Numbers

Use the properties to simplify the expression.

(a) $5y^2 - 8y^2 - 6y^2 + 11y^2$

$$\begin{aligned} 5y^2 - 8y^2 - 6y^2 + 11y^2 &= (5 - 8 - 6 + 11)y^2 \\ &= 2y^2 \end{aligned}$$

(b) $-2(m - 3)$

$$\begin{aligned} -2(m - 3) &= -2(m) - (-2)(3) \\ &= -2m + 6 \end{aligned}$$

(c) $3x^3 + 4 - 5(x^3 + 1) - 8$

First use the distributive property to eliminate the parentheses.

$$3x^3 + 4 - 5(x^3 + 1) - 8 = 3x^3 + 4 - 5x^3 - 5 - 8$$

Next use the commutative and associative properties to rearrange the terms; then combine like terms.

$$\begin{aligned} &= 3x^3 - 5x^3 + 4 - 5 - 8 \\ &= -2x^3 - 9 \end{aligned}$$

(d) $8 - (3m + 2)$

Think of $8 - (3m + 2)$ as $8 - 1(3m + 2)$.

$$\begin{aligned} 8 - 1(3m + 2) &= 8 - 3m - 2 \\ &= 6 - 3m \end{aligned}$$

(e)

$$\begin{aligned} (3x)(5)(y) &= [(3x)(5)]y && \text{Order of operations} \\ &= [3(x \cdot 5)]y && \text{Associative property} \\ &= [3(5x)]y && \text{Commutative property} \\ &= [(3 \cdot 5)x]y && \text{Associative property} \\ &= (15x)y \\ &= 15(xy) && \text{Associative property} \\ &= 15xy \end{aligned}$$

As mentioned earlier, many of these steps usually are not written out.

OBJECTIVE 5 **Use the multiplication property of 0.** The additive identity property gives a special property of 0, namely that $a + 0 = a$ for any real number a. The **multiplication property of 0** gives a special property of 0 that involves multiplication: The product of any real number and 0 is 0.

Multiplication Property of 0

For all real numbers a,

$$a \cdot 0 = 0 \quad \text{and} \quad 0 \cdot a = 0.$$

1.3 EXERCISES

Choose the correct response in Exercises 1–3.

1. The identity element for addition is

(a) $-a$ **(b)** 0 **(c)** 1 **(d)** $\frac{1}{a}$.

2. The identity element for multiplication is

(a) $-a$ **(b)** 0 **(c)** 1 **(d)** $\frac{1}{a}$.

3. The coefficient in the term $-8yz^2$ is

(a) -8 **(b)** y **(c)** z^2 **(d)** $-8y$.

Complete each statement.

4. Like terms are defined to be ______.

5. The distinction between the commutative and associative properties is that ______.

6. The multiplication property of 0 states that ______.

Use the properties of real numbers to simplify each expression. See Examples 1 and 2.

7. $5k + 3k$

8. $6a + 5a$

9. $-9r + 7r$

10. $-4n + 6n$

11. $-8z + 4w$

12. $-12k + 3r$

13. $-a + 7a$

14. $-s + 9s$

15. $2(m + p)$

16. $3(a + b)$

17. $-12(x - y)$

18. $-10(p - q)$

19. $-5(2d + f)$

20. $-2(3m + n)$

Simplify each expression by removing parentheses and combining like terms. See Examples 1–4.

21. $4x + 3x + 7 + 19$

22. $5m + 9m + 8 + 14$

23. $-12y + 4y + 3 + 2y$

24. $-5r - 9r + 8r - 5$

25. $3(k + 2) - 5k + 6 + 3$

26. $5(r - 3) + 6r - 2r + 4$

27. $.25(8 + 4p) - .5(6 + 2p)$

28. $.4(10 - 5x) - .8(5 + 10x)$

29. $-(2p + 5) + 3(2p + 4) - 2p$

30. $-(7m - 12) - 2(4m + 7) - 8m$

31. $2 + 3(2z - 5) - 3(4z + 6) - 8$

32. $-4 + 4(4k - 3) - 6(2k + 8) + 7$

TECHNOLOGY INSIGHTS (EXERCISES 33–36)

Give the results of the calculations shown on the screens.

33. For P = 2

34. For X = 3

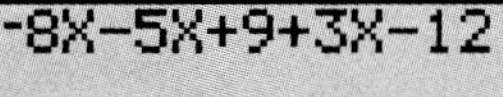

(continued)

TECHNOLOGY INSIGHTS (EXERCISES 33-36) (CONTINUED)

35. For M = −2

```
-2(M+1)+3(M-4)
```

36. For A = −4

```
6(A-5)-4(A+6)
```

Complete each statement so that the indicated property is illustrated. Simplify the answer, if possible.

37. $5x + 8x =$ ____________ (distributive property)

38. $9y - 6y =$ ____________ (distributive property)

39. $5(9r) =$ ____________ (associative property)

40. $-4 + (12 + 8) =$ ____________ (associative property)

41. $5x + 9y =$ ____________ (commutative property)

42. $-5 \cdot 7 =$ ____________ (commutative property)

43. $1 \cdot 7 =$ ____________ (identity property)

44. $-12x + 0 =$ ____________ (identity property)

45. $-\frac{1}{4}ty + \frac{1}{4}ty =$ ____________ (inverse property)

46. $-\frac{9}{8}\left(-\frac{8}{9}\right) =$ ____________ (inverse property)

47. $8(-4 + x) =$ ____________ (distributive property)

48. $3(x - y + z) =$ ____________ (distributive property)

49. $0(.875x + 9y - 88z) =$ ____________ (multiplication property of 0)

50. $0(35t^2 - 8t + 12) =$ ____________ (multiplication property of 0)

51. Give an "everyday" example of a commutative operation and of one which is not commutative.

52. Give an "everyday" example of inverse operations.

53. Replace x with 5 to show that $2 + 6x \neq 8x$.

54. Replace x with 5 to show that $4x - x \neq 4$.

Use the distributive property to calculate the following values mentally.

55. $96 \cdot 19 + 4 \cdot 19$

56. $27 \cdot 60 + 27 \cdot 40$

57. $58 \cdot \frac{3}{2} - 8 \cdot \frac{3}{2}$

58. $8.75(15) - 8.75(5)$

59. $4.31(69) + 4.31(31)$

60. $\frac{8}{5}(17) + \frac{8}{5}(13)$

RELATING CONCEPTS (EXERCISES 61–66)

While it may seem that simplifying the expression $3x + 4 + 2x + 7$ to $5x + 11$ is fairly easy, there are several important steps that require mathematical justification. These steps are usually done mentally. For now, provide the property that justifies the statement in the simplification. (These steps could be done in other orders.)

61. $3x + 4 + 2x + 7 = (3x + 4) + (2x + 7)$

62. $= 3x + (4 + 2x) + 7$

63. $= 3x + (2x + 4) + 7$

64. $= (3x + 2x) + (4 + 7)$

65. $= (3 + 2)x + (4 + 7)$

66. $= 5x + 11$

Did you make the connection that it is the properties of real numbers that allow us to combine terms in an expression?

67. Write a paragraph explaining the various properties introduced in this section. Give examples.

68. Are there *any* different numbers that satisfy the statement $a - b = b - a$? Give an example if your answer is yes.

69. By the distributive property, $a(b + c) = ab + ac$. This property is more completely named the distributive property of multiplication with respect to addition. Is there a distributive property of addition with respect to multiplication? That is, does

$$a + (b \cdot c) = (a + b)(a + c)$$

for all real numbers, a, b, and c? To find out, try various sample values of a, b, and c.

70. Explain how the distributive property is used in combining like terms. Give an example.

CHAPTER 1 GROUP ACTIVITY

How Americans Spend Their Money

Objective: Construct and read bar graphs and circle graphs.

Graphs and tables, first discussed in the introduction to this chapter, are a great way of presenting information. They allow you to make comparisons and approximations as well as draw conclusions more quickly than reading a page of text with the same information.

Listed in the table on the next page are common personal consumption expenditures of Americans during the years 1991–1997.

(continued)

Personal Consumption Expenditures (in billions of dollars)

Category	*1991*	*1992*	*1993*	*1994*	*1995*	*1996*	*1997*
Food and Tobacco	$693.8	$709.5	$733.4	$761.7	$783.8	$805.2	$832.3
Clothing, Accessories, Jewelry	265.7	283.5	298.1	312.7	323.4	338.0	353.3
Personal Care	59.1	63.1	65.1	68.4	71.9	75.0	79.4
Housing	616.5	646.8	672.8	712.7	750.3	787.4	829.8
Household Operation	448.4	470.6	504.1	535.0	562.8	592.8	620.7
Medical Care	668.7	733.2	785.5	826.1	871.6	912.4	957.3
Personal Business	318.9	341.7	357.4	370.4	389.1	416.2	459.1
Transportation	436.8	471.5	504.0	542.2	572.3	611.6	636.4
Recreation	292.0	310.8	340.2	370.2	402.5	432.3	462.9
Education and Research	86.1	93.1	98.5	104.7	112.2	119.7	129.4
Religious and Welfare Activities	104.1	115.6	121.3	131.2	139.8	151.1	157.6
TOTAL	$3990.1	$4239.4	$4480.4	$4735.3	$4979.7	$5241.7	$5518.2

Source: Bureau of Economic Analysis, United States Department of Commerce.

A. Have each person in the group construct a bar graph for one category of personal consumption expenditures over the seven-year period shown in the table.

B. Have each person in the group construct a circle graph for one year shown in the table. Include all eleven categories of personal consumption expenditures. Use approximate values, as needed.

C. As a group, examine the graphs you have constructed.

1. Discuss any conclusions you can draw from them.
2. What do the graphs tell you about how personal consumption of Americans changed over the seven years?
3. Write a paragraph that summarizes the group's conclusions.

CHAPTER 1 SUMMARY

KEY TERMS

1.1 **set**
elements (members)
empty set (null set)
variable
set-builder notation
number line
coordinate
graph
additive inverse (negative, opposite)
signed numbers
absolute value
inequality
interval
interval notation

1.2 **sum**
difference
product
quotient
reciprocal
factor
factored form
exponent
base
exponential expression
square root

1.3 **multiplicative inverse**
identity element for addition
identity element for multiplication
term
like terms
coefficient (numerical coefficient)
combining like terms

NEW SYMBOLS

Symbol	Meaning
$\{a, b\}$	set containing the elements a and b
$\emptyset$	the empty set
$\in$	is an element of (a set)
$\notin$	is not an element of
$\neq$	is not equal to
$\{x \mid x$ **has property** $P\}$	the set of all x, such that x has property P
$\lvert x \rvert$	the absolute value of x
$<$	is less than
$\leq$	is less than or equal to
$>$	is greater than
$\geq$	is greater than or equal to
(a, ∞)	the interval $\{x \mid x > a\}$
$(-\infty, a)$	the interval $\{x \mid x < a\}$
$(a, b]$	the interval $\{x \mid a < x \leq b\}$
a^m	m factors of a
$\sqrt{a}$	the square root of a

TEST YOUR WORD POWER

See how well you have learned the vocabulary in this chapter. Answers, with examples, are given at the bottom of the next page.

1. The **empty set** is a set
(a) with 0 as its only element
(b) with an infinite number of elements
(c) with no elements
(d) of ideas.

2. A **variable** is
(a) a symbol used to represent an unknown number
(b) a value that makes an equation true
(c) a solution of an equation
(d) the answer in a division problem.

3. A **coordinate** is
(a) the number that corresponds to a point on a number line
(b) the graph of a number
(c) any point on a number line
(d) the distance from 0 on a number line.

4. The **absolute value** of a number is
(a) the graph of the number
(b) the reciprocal of the number
(c) the opposite of the number
(d) the distance between 0 and the number on a number line.

5. **Interval notation** is
(a) a portion of a number line
(b) a special notation for describing a point on a number line
(c) a way to use symbols to describe an interval on a number line
(d) a notation to describe unequal quantities.

6. The **reciprocal** of a number a is
(a) a
(b) $\frac{1}{a}$
(c) $-a$
(d) 1.

7. A **factor** is
(a) the answer in an addition problem
(b) the answer in a multiplication problem
(c) one of two or more numbers that are added to get another number
(d) any number that divides evenly into a given number.

8. An **exponential expression** is
(a) a number that is a repeated factor in a product
(b) a number or a variable written with an exponent
(c) a number that shows how many times a factor is repeated in a product
(d) an expression that involves addition.

(continued)

TEST YOUR WORD POWER

9. A **term** is
(a) a numerical factor
(b) a number or a product or quotient of numbers and variables raised to powers
(c) one of several variables with the same exponents
(d) a sum of numbers and variables raised to powers.

10. A **numerical coefficient** is
(a) the numerical factor in a term
(b) the number of terms in an expression
(c) a variable raised to a power
(d) the variable factor in a term.

QUICK REVIEW

CONCEPTS	EXAMPLES
1.1 BASIC TERMS	
Sets of Numbers	
Natural Numbers $\{1, 2, 3, 4, \ldots\}$	10, 25, 143
Whole Numbers $\{0, 1, 2, 3, 4, \ldots\}$	0, 8, 47
Integers $\{\ldots, -2, -1, 0, 1, 2, \ldots\}$	$-22, -7, 0, 4, 9$
Rational Numbers $\left\{\frac{p}{q} \,\middle\vert\, p \text{ and } q \text{ are integers, } q \neq 0\right\}$, or $\{x \mid x$ has a terminating or repeating decimal representation$\}$	$-\frac{2}{3}, -.14, 0, 6, \frac{5}{8}, .\overline{3}$
Irrational Numbers $\{x \mid x$ is a real number that is not rational$\}$ or $\{x \mid x$ has a nonterminating, nonrepeating decimal representation$\}$	$\pi, .121121112\ldots, \sqrt{3}, -\sqrt{22}$
Real Numbers $\{x \mid x$ is represented by a point on a number line$\}$	$-3, .7, \pi, -\frac{2}{3}$
1.2 OPERATIONS ON REAL NUMBERS	
Addition *Same sign:* Add the absolute values. The sum has the same sign as the numbers.	Add: $-2 + (-7) = -(2 + 7) = -9$.
Different signs: Subtract the absolute values. The answer has the sign of the number with the larger absolute value.	Add: $-5 + 8 = 8 - 5 = 3$. Add: $-12 + 4 = -(12 - 4) = -8$.
Subtraction Change the sign of the second number and add.	Subtract: $-5 - (-3) = -5 + 3 = -2$.

Answers to Test Your Word Power

1. (c) *Example:* The set of whole numbers less than 0 is the empty set, written $\emptyset$. **2.** (a) *Examples: a, b, c* **3.** (a) *Example:* The point graphed three units to the right of 0 on a number line has coordinate 3. **4.** (d) *Examples:* $|2| = 2$ and $|-2| = 2$ **5.** (c) *Examples:* $(-\infty, 5]$, $(1, \infty)$, $[-3, 3)$ **6.** (b) *Examples:* 3 is the reciprocal of $\frac{1}{3}$; $-\frac{5}{2}$ is the reciprocal of $-\frac{2}{5}$. **7.** (d) *Example:* 2 and 5 are factors of 10 since both divide evenly (without remainder) into 10; other factors of 10 are $-10, -5, -2, -1, 1$, and 10. **8.** (b) *Examples:* 3^4 and x^{10} **9.** (b) *Examples:* $6, \frac{x}{2}, -4ab^2$ **10.** (a) *Examples:* The term 3 has numerical coefficient 3, $8z$ has numerical coefficient 8, and $-10x^3y$ has numerical coefficient -10.

CONCEPTS	EXAMPLES
Multiplication	
Same sign: The product is positive.	Multiply: $(-3)(-8) = 24$.
Different signs: The product is negative.	Multiply: $(-7)(5) = -35$.
Division	
Same sign: The quotient is positive.	Divide: $\frac{-15}{-5} = 3$.
Different signs: The quotient is negative.	Divide: $\frac{-24}{12} = -2$.
The product of an even number of negative factors is positive.	$(-5)^6$ is positive.
The product of an odd number of negative factors is negative.	$(-5)^7$ is negative.
Order of Operations	
If parentheses, square brackets, or fraction bars are present:	Perform the indicated operations.
Step 1 Work separately above and below any fraction bar.	$\frac{12+3}{5\cdot 2} = \frac{15}{10} = \frac{3}{2}$
Step 2 Use the rules that follow within each set of parentheses or square brackets. Start with the innermost set and work outward.	
If no parentheses or brackets are present:	$(-6)[2^2 - (3+4)] + 3 = (-6)[2^2 - 7] + 3$
Step 1 Evaluate all powers and roots.	$= (-6)[4 - 7] + 3$
Step 2 Do any multiplications or divisions in the order in which they occur, working from left to right.	$= (-6)[-3] + 3$ $= 18 + 3$
Step 3 Do any additions or subtractions in the order in which they occur, working from left to right.	$= 21$

1.3 PROPERTIES OF REAL NUMBERS

CONCEPTS	EXAMPLES
For any real numbers, *a*, *b*, and *c*:	
Distributive Property	
$a(b + c) = ab + ac$	$12(4 + 2) = 12 \cdot 4 + 12 \cdot 2$
Inverse Properties	
$a + (-a) = 0$ and $-a + a = 0$	$5 + (-5) = 0$
$a \cdot \frac{1}{a} = 1$ and $\frac{1}{a} \cdot a = 1 \quad (a \neq 0)$	$-\frac{1}{3}(-3) = 1$
Identity Properties	
$a + 0 = a$ and $0 + a = a$	$-32 + 0 = -32$
$a \cdot 1 = a$ and $1 \cdot a = a$	$17.5(1) = 17.5$
Commutative Properties	
$a + b = b + a$	$9 + (-3) = -3 + 9$
$ab = ba$	$6(-4) = (-4)6$
Associative Properties	
$a + (b + c) = (a + b) + c$	$7 + (5 + 3) = (7 + 5) + 3$
$a(bc) = (ab)c$	$-4(6 \cdot 3) = (-4 \cdot 6)3$
Multiplication Property of 0	
$a \cdot 0 = 0$ and $0 \cdot a = 0$	$47 \cdot 0 = 0 \qquad 0(-18) = 0$

CHAPTER 1 REVIEW EXERCISES

For help with any of these exercises, look in the section given in brackets.

[1.1] *Graph each set on a number line.*

1. $\left\{-4, -1, 2, \frac{9}{4}, 4\right\}$

2. $\left\{-5, -\frac{11}{4}, -.5, 0, 3, \frac{13}{3}\right\}$

Find the value of each expression.

3. $|-16|$

4. $|23|$

5. $-|-4|$

Let set $S = \left\{-9, -\frac{4}{3}, -\sqrt{10}, 0, \frac{5}{3}, \sqrt{7}, \frac{12}{3}\right\}$. *Simplify the elements of S as necessary and then list the elements that belong to the specified set.*

6. Whole numbers

7. Integers

8. Rational numbers

9. Real numbers

Write each set by listing its elements.

10. $\{x \mid x$ is a natural number between 3 and 9$\}$

11. $\{y \mid y$ is a whole number less than 4$\}$

Write true *or* false *for each inequality.*

12. $4 \cdot 2 \leq |12 - 4|$

13. $2 + |-2| > 4$

14. The table gives the 1995 net trade balance, in millions of dollars, for selected U.S. merchandise trade partners.

Country	*Net Balance (millions of dollars)*
France	−2,971
Germany	−14,909
Australia	7,100
Japan	−33,865
South Korea	72

Source: U.S. Dept. of Commerce, *Survey of Current Business,* June 1996.

A negative balance means that imports exceeded exports, while a positive balance means that exports exceeded imports.

(a) Which country had the greatest discrepancy between exports and imports?

(b) Which country had the smallest discrepancy between exports and imports?

Write in interval notation and graph.

15. $\{x \mid x < -5\}$

16. $\{x \mid -2 < x \leq 3\}$

[1.2] *Add or subtract, as indicated.*

17. $-\frac{5}{8} - \left(-\frac{7}{3}\right)$

18. $-\frac{4}{5} - \left(-\frac{3}{10}\right)$

19. $-5 + (-11) + 20 - 7$

20. $-9.42 + 1.83 - 7.6 - 1.9$

21. $-15 + (-13) + (-11)$

22. $-1 - 3 - (-10) + (-7)$

23. $\frac{3}{4} - \left(\frac{1}{2} - \frac{9}{10}\right)$

24. State in your own words how to determine the sign of the sum of two numbers.

25. How is subtraction related to addition?

Find each product or quotient.

26. $2(-5)(-3)(-3)$ 27. $-\frac{3}{7}\left(-\frac{14}{9}\right)$ 28. $\frac{-38}{-19}$ 29. $\frac{-2.3754}{-.74}$

[1.1] *Use the table to answer* true *or* false *for each statement. Use absolute value for comparisons.*

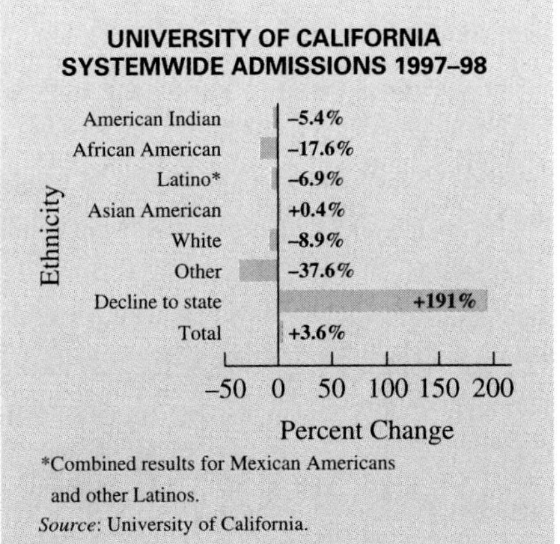

*Combined results for Mexican Americans and other Latinos.
Source: University of California.

30. The absolute value of the percent change for Latinos was greater than that for African Americans.

31. The algebraic sum of the percent change for American Indians and that for Asian Americans is a negative number.

32. The largest percent change shown is 37.6%.

33. Which one of the following is undefined? $\frac{5}{7-7}$ or $\frac{7-7}{5}$

[1.2] *Evaluate each expression. If it is not a real number, say so.*

34. $\left(\frac{3}{7}\right)^3$ 35. $(1.7)^2$ 36. $\sqrt{400}$ 37. $\sqrt{-64}$

38. $-14\left(\frac{3}{7}\right) + 6 \div 3$ 39. $\frac{-5(3)^2 + 9(\sqrt{4}) - 5}{6 - 5(-\sqrt{4})}$

Let $k = -4$, $m = 2$, *and* $n = 16$, *and evaluate each expression.*

40. $4k - 7m$ 41. $-3\sqrt{n} + m + 5k$

42. In order to evaluate $(3 + 2)^2$, should you work within the parentheses first, or should you square 3 and square 2 and then add?

43. By replacing a with 4 and b with 6, show that $(a + b)^2 \neq a^2 + b^2$.

[1.3] *Use the properties of real numbers to simplify each expression.*

44. $2q + 19q$ 45. $13z - 17z$ 46. $-m + 6m$

47. $5p - p$ 48. $-2(k + 3)$ 49. $6(r + 3)$

50. $-3y + 6 - 5 + 4y$ 51. $2a + 3 - a - 1 - a - 2$

52. $-2(k - 1) + 3k - k$ 53. $-3(4m - 2) + 2(3m - 1) - 4(3m + 1)$

Complete each statement so that the indicated property is illustrated. Simplify the answer, if possible.

54. $2x + 3x =$ ____________________
(distributive property)

55. $-4 \cdot 1 =$ ____________________
(identity property)

56. $2(4x) =$ ______________________
(associative property)

57. $-3 + 13 =$ ______________________
(commutative property)

58. $-3 + 3 =$ ______________________
(inverse property)

59. $5(x + z) =$ ______________________
(distributive property)

60. $0 + 7 =$ ______________________
(identity property)

61. $8 \cdot \frac{1}{8} =$ ______________________
(inverse property)

62. $\frac{9}{28} \cdot 0 =$ ______________________
(multiplication property of 0)

MIXED REVIEW EXERCISES*

Perform the indicated operations.

63. $\left(-\frac{4}{5}\right)^4$

64. $-\frac{5}{8}(-40)$

65. $\frac{75}{-5}$

66. $9(2m + 3n)$

67. $-25\left(-\frac{4}{5}\right) + 3^3 - 32 \div \sqrt{4}$

68. $(-5)^3$

69. $-(3k - 4h)$

70. $-8 + |-14| + |-3|$

71. $-\sqrt{25}$

72. $\frac{6 \cdot \sqrt{4} - 3 \cdot \sqrt{16}}{-2 \cdot 5 + 7(-3) - 10}$

73. $-4.6(2.48)$

74. $-\frac{10}{21} \div -\frac{5}{14}$

75. $-\frac{2}{3}[5(-2) + 8 - 4^3]$

76. $-(-p + 6q) - (2p - 3q)$

77. $-2(3k^2 + 5m)$ if $k = -4$ and $m = 2$

78. $.8 - 4.9 - 3.2 + 1.14$

79. $-|-8| + |-3|$

80. -3^2

81. $-\frac{4.64}{.16}$

82. $-\frac{2}{3} - \left(\frac{1}{6} - \frac{5}{9}\right)$

83. $-2x + 5 - 4x + 1$

84. $\frac{4m^3 - 3n}{7k^2 - 10}$ if $m = 2$, $n = 16$, and $k = -4$

RELATING CONCEPTS (EXERCISES 85–94)

Evaluate the expression $\frac{2}{3}x - y^2 - 3z$ *for* $x = 5$, $y = -4$, *and* $z = 1$. *Then respond to the questions or statements in Exercises 85–94,* ***working them in order.***

85. What is the value of the expression for these particular values of x, y, and z?

86. Is the value of the expression greater than -16 or less than -16?

87. Is the value of the expression **(a)** an integer? **(b)** a rational number?

88. What is the absolute value of the expression?

---

*The order of exercises in this final group does not correspond to the order in which topics occur in the chapter. This random ordering should help you prepare for the chapter test in yet another way.

RELATING CONCEPTS (EXERCISES 85–94) (CONTINUED)

89. Is the square root of the value of the expression a real number?

90. What is the square of the expression?

91. Give the additive inverse of the value of the expression.

92. Give the multiplicative inverse of the value of the expression.

93. If parentheses are placed around the first two terms of the expression, will you obtain the same answer? If not, what is the new answer?

94. If parentheses are placed around the last two terms of the expression, will you obtain the same answer? If not, what is the new answer?

Did you make the connection that the topics of this chapter are interrelated?

CHAPTER 1 TEST

1. Graph $\left\{-3, .75, \frac{5}{3}, 5, 6.3\right\}$ on a number line.

Let $A = \left\{-\sqrt{6}, -1, -.5, 0, 3, 7.5, \frac{24}{2}\right\}$. *First simplify each element as needed and then list the elements from A that belong to the set.*

2. Whole numbers

3. Integers

4. Rational numbers

5. Real numbers

Write each set in interval notation, and graph it.

6. $\{x \mid x < -3\}$

7. $\{y \mid -4 < y \leq 2\}$

Perform the indicated operations.

8. $-6 + 14 + (-11) - (-3)$

9. $10 - 4 \cdot 3 + 6(-4)$

10. $7 - 4^2 + 2(6) + (-4)^2$

11. $\dfrac{10 - 24 + (-6)}{\sqrt{16}(-5)}$

12. $\dfrac{-2[3 - (-1 - 2) + 2]}{\sqrt{9}(-3) - (-2)}$

13. $\dfrac{8 \cdot 4 - 3^2 \cdot 5 - 2(-1)}{-3 \cdot 2^3 + 1}$

Results for selected industries from FORTUNE *magazine's 1998 Customer Satisfaction Index are shown on the next page. Use these data to answer the questions in Exercises 14–16.*

Industry	*1997 Score*	*Change from 1996*
Beverages (beer)	81	2.5%
Gasoline	78	1.3%
U.S. Postal Service	69	−6.8%
Fast food, pizza, carryout	66	−5.7%
Broadcasting (national news)	62	−11.4%
Internal Revenue Service	54	8.0%

Source: FORTUNE, February 16, 1998, p. 166.

14. What are the largest and smallest changes from 1996 given in the list?

15. Which changes have the largest and smallest absolute values?

16. Is the difference between the change for the Postal Service and the change for the Internal Revenue Service positive or negative? Show the work that led to your answer.

Find each indicated root. If the number is not real, say so.

17. $\sqrt{196}$ **18.** $-\sqrt{225}$ **19.** $\sqrt{-16}$

20. For the expression $\sqrt{a}$, under what conditions will its value be **(a)** positive **(b)** not real **(c)** 0?

21. Evaluate $\dfrac{8k + 2m^2}{r - 2}$ if $k = -3$, $m = -3$, and $r = 25$.

22. Use the properties of real numbers to simplify $-3(2k - 4) + 4(3k - 5) - 2 + 4k$.

23. How does the subtraction sign affect the terms $-4r$ and 6 when simplifying $(3r + 8) - (-4r + 6)$? What is the simplified form?

Match each statement with the appropriate property. Answers may be used more than once.

24. $6 + (-6) = 0$

25. $-2 + (3 + 6) = (-2 + 3) + 6$

26. $5x + 15x = (5 + 15)x$

27. $13 \cdot 0 = 0$

28. $-9 + 0 = -9$

29. $4 \cdot 1 = 4$

30. $(a + b) + c = (b + a) + c$

A. distributive property

B. inverse property

C. identity property

D. associative property

E. commutative property

F. multiplication property of 0

Linear Equations and Inequalities

2

Linear equations describe quantities that increase or decrease at the same rate. Many quantities in applications have this characteristic. In this chapter, the theme is entertainment and leisure, so many of the applications will come from these fields.

Entertainment

Americans are spending more time and money on entertainment and leisure than ever. One beneficiary of this entertainment boom is Broadway. A recent study by the League of American Theaters and Producers gave the increase, shown in the graph, in the number of theater tickets sold (in millions) from the 1990–1991 season through the 1996–1997 season.

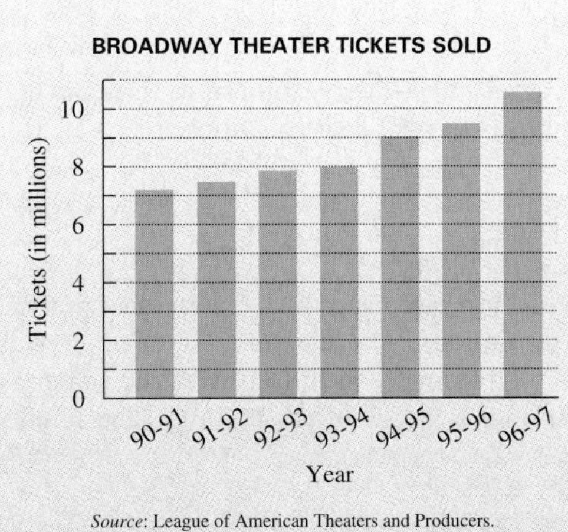

Source: League of American Theaters and Producers.

As the graph indicates, theater attendance has increased from approximately 7.3 million in 1990–1991 to approximately 10.6 million in 1996–1997. Although the annual increases were not exactly equal, they are fairly close. Based on the annual increases shown in the graph, how many theater tickets do you think were sold for the 1997–1998 season? In the exercises for Section 2.1, we give a linear equation that approximates the number of tickets sold in a given year.

Visit our Web site at www.LialAlgebra.com

2.1 Linear Equations in One Variable

OBJECTIVES

1. Define linear equations.
2. Solve linear equations using the addition and multiplication properties of equality.
3. Solve linear equations using the distributive property.
4. Solve linear equations with fractions or decimals.
5. Identify conditional equations, contradictions, and identities.

FOR EXTRA HELP

SSG Sec. 2.1
SSM Sec. 2.1
Pass the Test Software
InterAct Math Tutorial Software
Video 2

To solve a real-world problem, we must first set up a **mathematical model,** in other words, a mathematical description of the situation. Constructing such a model requires a good understanding of the situation to be modeled and a familiarity with relevant mathematical techniques. In this chapter we look at the mathematics of *linear models,* which are used for data whose graphs can be approximated by a straight line.

An **algebraic expression** is the result of adding, subtracting, multiplying, and dividing (except by 0), or taking roots on any collection of variables and numbers. Some examples of algebraic expressions include

$$8x + 9, \qquad \sqrt{y} + 4, \qquad \text{and} \qquad \frac{x^3y^8}{z}.$$

OBJECTIVE 1 Define linear equations. Applications of mathematics often lead to **equations,** statements that two algebraic expressions are equal.

Linear Equation

A **linear equation in one variable** can be written in the form

$$ax = b,$$

where a and b are real numbers, with $a \neq 0$.

Examples of linear equations include

$$3x = -2, \qquad y - 3 = 5, \qquad \text{and} \qquad 2k + 5 = 10.$$

A linear equation is also called a **first-degree equation,** since the highest power on the variable is one. Some examples of nonlinear equations are

$$|x - 3| = 5, \qquad \frac{8}{x} = -22, \qquad \text{and} \qquad \sqrt{x} = 6.$$

If the variable in an equation is replaced by a real number that makes the statement true, then that number is a **solution** of the equation. For example, 8 is a solution of the equation $y - 3 = 5$, since replacing y with 8 gives a true statement. An equation is *solved* by finding its **solution set,** the set of all solutions. The solution set of the equation $y - 3 = 5$ is $\{8\}$.

OBJECTIVE 2 Solve linear equations using the addition and multiplication properties of equality. **Equivalent equations** are equations with the same solution set. Equations are generally solved by starting with a given equation and producing a series of simpler equivalent equations. For example,

$$8x + 1 = 17, \qquad 8x = 16, \qquad \text{and} \qquad x = 2$$

are all equivalent equations since each has the same solution set, $\{2\}$. We use the addition and multiplication properties of equality to produce equivalent equations.

Addition and Multiplication Properties of Equality

Addition Property of Equality
For all real numbers a, b, and c, the equations

$$a = b \quad \text{and} \quad a + c = b + c$$

are equivalent. (The same number may be added to both sides of an equation without changing the solution set.)

Multiplication Property of Equality
For all real numbers, a, b, and c, where $c \neq 0$, the equations

$$a = b \quad \text{and} \quad ac = bc$$

are equivalent. (Both sides of an equation may be multiplied by the same nonzero number without changing the solution set.)

Because subtraction and division are defined in terms of addition and multiplication, respectively, these properties can be extended: The same number may be subtracted from both sides of an equation, and both sides may be divided by the same nonzero number.

EXAMPLE 1 Solving a Linear Equation

Solve $4y - 2y - 5 = 4 + 6y + 3$.

$$
\begin{aligned}
2y - 5 &= 7 + 6y && \text{Combine terms.}\\
2y - 5 + 5 &= 7 + 6y + 5 && \text{Add 5.}\\
2y &= 12 + 6y\\
2y - 6y &= 12 + 6y - 6y && \text{Subtract } 6y.\\
-4y &= 12\\
\frac{-4y}{-4} &= \frac{12}{-4} && \text{Divide by } -4.\\
y &= -3
\end{aligned}
$$

To be sure that -3 is the solution, check by substituting back into the *original* equation (*not* an intermediate one).

$$
\begin{aligned}
4y - 2y - 5 &= 4 + 6y + 3 && && \text{Original equation}\\
4(-3) - 2(-3) - 5 &= 4 + 6(-3) + 3 && ? && \text{Let } y = -3.\\
-12 + 6 - 5 &= 4 - 18 + 3 && ? && \text{Multiply.}\\
-11 &= -11 && && \text{True}
\end{aligned}
$$

Since a true statement is obtained, -3 is the solution. The solution set is $\{-3\}$.

We use the following steps to solve a linear equation in one variable. (Not all equations require all these steps.)

Solving a Linear Equation in One Variable

Step 1 **Clear fractions.** Eliminate any fractions by multiplying both sides by a common denominator.

Step 2 **Simplify each side separately.** Simplify each side of the equation as much as possible by using the distributive property to clear parentheses and by combining like terms as needed.

Step 3 **Isolate the variable terms on one side.** Use the addition property to get all terms with variables on one side of the equation and all numbers on the other.

Step 4 **Isolate the variable.** Use the multiplication property to get an equation with just the variable (with coefficient 1) on one side.

Step 5 **Check.** Check by substituting back into the original equation.

OBJECTIVE 3 Solve linear equations using the distributive property. In Example 1 we did not use Step 1 and the distributive property in Step 2 as given above. Many other equations, however, will require one or both of these steps, as shown in the next examples.

EXAMPLE 2 Using the Distributive Property to Solve a Linear Equation

Solve $2(k - 5) + 3k = k + 6$.

Step 1 Since there are no fractions in this equation, Step 1 does not apply.

Step 2 Use the distributive property to simplify and combine terms on the left side of the equation.

$$2(k - 5) + 3k = k + 6$$

$$2k - 10 + 3k = k + 6 \qquad \text{Distributive property}$$

$$5k - 10 = k + 6 \qquad \text{Combine like terms.}$$

Step 3 Next, use the addition property of equality.

$$5k - 10 + 10 = k + 6 + 10 \qquad \text{Add 10.}$$

$$5k = k + 16$$

$$5k - k = k + 16 - k \qquad \text{Subtract } k.$$

$$4k = 16 \qquad \text{Combine like terms.}$$

Step 4 Use the multiplication property of equality to get just k on the left.

$$\frac{4k}{4} = \frac{16}{4} \qquad \text{Divide by 4.}$$

$$k = 4$$

Step 5 Check that the solution set is $\{4\}$ by substituting 4 for k in the original equation.

OBJECTIVE 4 Solve linear equations with fractions or decimals. When fractions or decimals appear as coefficients in equations, our work can be made easier if we multiply both sides of the equation by the least common denominator (LCD) of all the fractions. This is an application of the multiplication property of equality, and it produces an equivalent equation with integer coefficients. The next examples illustrate this idea.

EXAMPLE 3 Solving a Linear Equation with Fractions

Solve $\frac{x+7}{6} + \frac{2x-8}{2} = -4$.

Start by eliminating the fractions. Multiply both sides by the LCD, 6.

Step 1 $$6\left[\frac{x+7}{6} + \frac{2x-8}{2}\right] = 6(-4)$$

Step 2 $$6\left(\frac{x+7}{6}\right) + 6\left(\frac{2x-8}{2}\right) = 6(-4) \quad \text{Distributive property}$$

$$x + 7 + 3(2x - 8) = -24$$

$$x + 7 + 6x - 24 = -24 \quad \text{Distributive property}$$

$$7x - 17 = -24 \quad \text{Combine like terms.}$$

Step 3 $$7x - 17 + 17 = -24 + 17 \quad \text{Add 17.}$$

$$7x = -7$$

Step 4 $$\frac{7x}{7} = \frac{-7}{7} \quad \text{Divide by 7.}$$

$$x = -1$$

Step 5 Check by substituting -1 for x in the original equation.

$$\frac{x+7}{6} + \frac{2x-8}{2} = -4$$

$$\frac{-1+7}{6} + \frac{2(-1)-8}{2} = -4 \quad ? \quad \text{Let } x = -1.$$

$$\frac{6}{6} + \frac{-10}{2} = -4 \quad ?$$

$$1 - 5 = -4 \quad \text{True}$$

The solution checks, so the solution set is $\{-1\}$.

In later sections we solve problems involving interest rates and concentrations of solutions. These problems involve percents that are converted to decimal numbers. The equations that are used to solve such problems involve decimal coefficients. We can clear these decimals by multiplying by the largest power of 10 necessary to obtain integer coefficients. The next example shows how this is done.

EXAMPLE 4 Solving a Linear Equation with Decimals

Solve $.06x + .09(15 - x) = .07(15)$.

Since each decimal number is given in hundredths, multiply both sides of the equation by 100. (This is done by moving the decimal points two places to the right.)

$$.06x + .09(15 - x) = .07(15)$$

$$6x + 9(15 - x) = 7(15) \quad \text{Multiply by 100.}$$

$$6x + 9(15) - 9x = 105 \quad \text{Distributive property}$$

$$-3x + 135 = 105 \quad \text{Combine like terms.}$$

$$-3x + 135 - 135 = 105 - 135 \quad \text{Subtract 135.}$$

$$-3x = -30$$

$$\frac{-3x}{-3} = \frac{-30}{-3} \quad \text{Divide by } -3.$$

$$x = 10$$

Check to verify that the solution set is $\{10\}$.

CAUTION When multiplying the term $.09(15 - x)$ by 100 in Example 4, do not multiply both .09 and $15 - x$ by 100. This step is not an application of the distributive property, but of the associative property. The correct procedure is

$$100[.09(15 - x)] = [100(.09)](15 - x) \quad \text{Associative property}$$
$$= 9(15 - x). \quad \text{Multiply.}$$

OBJECTIVE 5 Identify conditional equations, contradictions, and identities. All the equations above had a solution set containing one element; for example, $2(k - 5) + 3k = k + 6$ has solution set $\{4\}$. Some equations that appear to be linear have no solutions, while others have an infinite number of solutions. The chart below gives the names of these types of equations.

Type of Linear Equation	*Number of Solutions*
Conditional	One
Contradiction	None, solution set $\emptyset$
Identity	Infinite, solution set {all real numbers}

The next example shows how to recognize these types of equations.

EXAMPLE 5 Identifying Conditional Equations, Identities, and Contradictions

Solve each equation. Decide whether it is a conditional equation, an identity, or a contradiction.

(a) $5x - 9 = 4(x - 3)$

Work as in the previous examples.

$$5x - 9 = 4(x - 3)$$
$$5x - 9 = 4x - 12 \quad \text{Distributive property}$$
$$5x - 9 - 4x = 4x - 12 - 4x \quad \text{Subtract } 4x.$$
$$x - 9 = -12 \quad \text{Combine like terms.}$$
$$x - 9 + 9 = -12 + 9 \quad \text{Add } 9.$$
$$x = -3$$

The solution set, $\{-3\}$, has one element, so $5x - 9 = 4(x - 3)$ is a *conditional equation.*

(b) $5x - 15 = 5(x - 3)$

Use the distributive property on the right side.

$$5x - 15 = 5x - 15$$

Both sides of the equation are *exactly the same,* so any real number would make the equation true. For this reason, the solution set is the set of all real numbers, and the equation $5x - 15 = 5(x - 3)$ is an *identity.*

(c) $5x - 15 = 5(x - 4)$
Use the distributive property.

$$5x - 15 = 5x - 20 \qquad \text{Distributive property}$$

$$5x - 15 - 5x = 5x - 20 - 5x \qquad \text{Subtract } 5x.$$

$$-15 = -20 \qquad \text{False}$$

Since the result, $-15 = -20$, is *false,* the equation has no solution. The solution set is $\emptyset$. The equation $5x - 15 = 5(x - 4)$ is a *contradiction.*

CONNECTIONS

A mathematical model that can be used to predict the value of one variable given the value of another can be found by using pairs of data. This process is called *curve fitting.* In the next section, we will see how linear equations like those discussed below are developed from data.

By studying the winning times in the 500-meter speed-skating event at Olympic games back to 1900, it was found that the winning times for men were closely approximated by the equation

$$y_m = 46.338 - .097x,$$

where y_m is the time in seconds needed to win for men, with x the Olympic year and $x = 0$ corresponding to 1900. (For example, 1994 winning times would be estimated by replacing x with $1994 - 1900 = 94$.) The corresponding equation for women is

$$y_w = 57.484 - .196x.$$

FOR DISCUSSION OR WRITING

1. Find the year in which **(a)** $y_m = 42.458$; **(b)** $y_w = 44.940$.
2. Use the equations to predict the winning times for men and women in 1994.

2.1 EXERCISES

1. Which of the following equations are linear equations in x?

   (a) $3x + x - 1 = 0$ **(b)** $8 = x^2$ **(c)** $6x + 2 = 9$ **(d)** $\frac{1}{2}x - \frac{1}{x} = 0$

2. Which of the equations in Exercise 1 are not linear equations in x? Explain why.
3. Decide whether 6 is a solution of $3(x + 4) = 5x$ by substituting 6 for x. If it is not a solution, explain why.
4. Use substitution to decide whether -2 is a solution of $5(x + 4) - 3(x + 6) = 9(x + 1)$. If it is not a solution, explain why.
5. If two equations are equivalent, they have the same ______.

6. The equation $4[x + (2 - 3x)] = 2(4 - 4x)$ is an identity. Let x represent the number of letters in your last name. Is this number a solution of this equation? Check your answer.

7. The expression $.06(10 - x)(100)$ is equivalent to
(a) $.06 - .06x$ **(b)** $60 - 6x$ **(c)** $6 - 6x$ **(d)** $6 - .06x$

8. Describe in your own words the steps used to solve a linear equation.

Solve each equation. See Examples 1 and 2.

9. $7k + 8 = 1$ **10.** $5m - 4 = 21$ **11.** $8 - 8x = -16$ **12.** $9 - 2r = 15$

13. $7y - 5y + 15 = y + 8$ **14.** $2x + 4 - x = 4x - 5$

15. $12w + 15w - 9 + 5 = -3w + 5 - 9$ **16.** $-4t + 5t - 8 + 4 = 6t - 4$

17. $2(x + 3) = -4(x + 1)$ **18.** $4(y - 9) = 8(y + 3)$

19. $3(2w + 1) - 2(w - 2) = 5$ **20.** $4(x - 2) + 2(x + 3) = 6$

21. $2x + 3(x - 4) = 2(x - 3)$ **22.** $6y - 3(5y + 2) = 4(1 - y)$

23. $6p - 4(3 - 2p) = 5(p - 4) - 10$ **24.** $-2k - 3(4 - 2k) = 2(k - 3) + 2$

25. $-[2z - (5z + 2)] = 2 + (2z + 7)$ **26.** $-[6x - (4x + 8)] = 9 + (6x + 3)$

27. $-3m + 6 - 5(m - 1) = 4m - (2m - 4) - 9m + 5$

28. $4(k + 2) - 8k - 5 = -3k + 9 - 2(k + 6)$

29. $-[3y - (2y + 5)] = -4 - [3(2y - 4) - 3y]$

30. $2[-(y - 1) + 4] = 5 + [-(6y - 7) + 9y]$

31. $-(9 - 3a) - (4 + 2a) - 3 = -(2 - 5a) + (-a) + 1$

32. $-(-2 + 4x) - (3 - 4x) + 5 = -(-3 + 6x) + x + 1$

33. $2(-3 + m) - (3m - 4) = -(-4 + m) - 4m + 6$

34. To solve the linear equation $.05y + .12(y + 5000) = 940$, we can multiply both sides by a power of 10 so that all coefficients are integers. What is the smallest power of 10 that will accomplish this goal?

35. Suppose that in solving the equation

$$\frac{1}{3}y + \frac{1}{2}y = \frac{1}{6}y,$$

you begin by multiplying both sides by 12, rather than the *least* common denominator, 6. Should you get the correct solution anyway? Explain.

Solve each equation. See Examples 3 and 4.

36. $\frac{3x}{4} + \frac{5x}{2} = 13$ **37.** $\frac{8y}{3} - \frac{2y}{4} = -13$

38. $\frac{x - 8}{5} + \frac{8}{5} = -\frac{x}{3}$ **39.** $\frac{2r - 3}{7} + \frac{3}{7} = -\frac{r}{3}$

40. $\frac{4t + 1}{3} = \frac{t + 5}{6} + \frac{t - 3}{6}$ **41.** $\frac{2x + 5}{5} = \frac{3x + 1}{2} + \frac{-x + 7}{2}$

42. $.05y + .12(y + 5000) = 940$ **43.** $.09k + .13(k + 300) = 61$

44. $.02(50) + .08r = .04(50 + r)$ **45.** $.20(14{,}000) + .14t = .18(14{,}000 + t)$

46. $.05x + .10(200 - x) = .45x$ **47.** $.08x + .12(260 - x) = .48x$

48. The equation $x + 2 = x + 2$ is called a(n) ________________, because its solution set is {all real numbers}. The equation $x + 1 = x + 2$ is called a(n) ________________, because it has no solutions.

49. Which one of the following is a conditional equation?
(a) $2x + 1 = 3$ **(b)** $x = 3x - 2x$ **(c)** $2(x + 2) = 2x + 2$
(d) $5x - 3 = 4x + x - 5 + 2$

50. Explain the distinction between a conditional equation, an identity, and a contradiction.

Decide whether each equation is conditional, an identity, or a contradiction. Give the solution set. See Example 5.

51. $-2p + 5p - 9 = 3(p - 4) - 5$

52. $-6k + 2k - 11 = -2(2k - 3) + 4$

53. $6x + 2(x - 2) = 9x + 4$

54. $-4(x + 2) = -3(x + 5) - x$

55. $-11m + 4(m - 3) + 6m = 4m - 12$

56. $3p - 5(p + 4) + 9 = -11 + 15p$

57. $7[2 - (3 + 4r)] - 2r = -9 + 2(1 - 15r)$

58. $4[6 - (1 + 2m)] + 10m = 2(10 - 3m) + 8m$

Decide whether or not each pair of equations is equivalent. If not equivalent, explain why.

59. $5x = 10$ and $\dfrac{5x}{x + 2} = \dfrac{10}{x + 2}$

60. $x + 1 = 9$ and $\dfrac{x + 1}{8} = \dfrac{9}{8}$

61. $y = -3$ and $\dfrac{y}{y + 3} = \dfrac{-3}{y + 3}$

62. $m = 1$ and $\dfrac{m + 1}{m - 1} = \dfrac{2}{m - 1}$

63. $k = 4$ and $k^2 = 16$

64. $p^2 = 36$ and $p = 6$

Work each of the following problems related to the chapter theme.

65. The mathematical model

$$y = .55x - 42.5$$

approximates the Broadway tickets sold from 1990–1997. In the model, $x = 90$ corresponds to the 1990–1991 season, $x = 91$ corresponds to the 1991–1992 season, and so on, and y is the number of tickets sold in millions.
(a) Based on this model, how many tickets were sold in 1995–1996? (*Hint:* 1995–1996 corresponds to $x = 95$.)
(b) In what season did the number of tickets sold reach 7.9 million? (*Hint:* The number of tickets sold in millions corresponds to y.)

66. The graph showing the number of tickets sold for Broadway shows in the 1990s was introduced at the beginning of this chapter. For the period shown in the graph, between which two seasons did ticket sales increase the most? In which seasons were ticket sales closest to the same?

67. The chart shows the number of tickets (in millions) sold to theatergoers under age 18.

Season	*Millions of Tickets*
1990–1991	.5
1996–1997	1.1

Source: League of American Theaters and Producers.

If $x = 90$ corresponds to the 1990–1991 season, and so on, the model

$$y = .1x - 8.5$$

approximates the number under age 18 attending the theater. Use the model to estimate the number of these youths attending in 1993–1994. In what season did these ticket sales amount to .75 million?

68. Use the model in Exercise 65 to find the number of tickets sold in 1994–1995. Is your answer a good approximation of the number shown in the graph at the beginning of this chapter?

2.2 Formulas

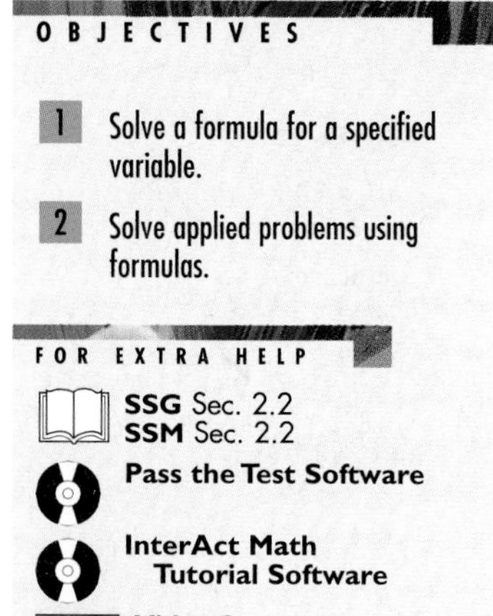

OBJECTIVES
1 Solve a formula for a specified variable.
2 Solve applied problems using formulas.
FOR EXTRA HELP
SSG Sec. 2.2
SSM Sec. 2.2
Pass the Test Software
InterAct Math Tutorial Software
Video 2

Models for many applied problems already exist; they are called *formulas.* A formula is a mathematical expression in which letters are used to describe a relationship. Some formulas that we will be using are

$$d = rt, \qquad I = prt, \qquad \text{and} \qquad P = 2L + 2W.$$

A list of the formulas used in algebra is given inside the covers of the book.

OBJECTIVE 1 Solve a formula for a specified variable. In some applications, the necessary formula is solved for one of its variables, which may not be the unknown number to be found. For example, the formula $I = prt$ says that interest on a loan or investment equals principal (amount borrowed or invested) times rate (percent) times time at interest (in years). If you want to know how long it will take for an investment at a stated interest rate to earn a predetermined amount of interest, it would help to first solve the formula for t. This process is called **solving for a specified variable.** Notice how the steps used in the following examples are very similar to those used in solving a linear equation. When you are solving for a specified variable, the key is to treat that variable as if it were the only one; treat all other variables like numbers.

EXAMPLE 1 Solving for a Specified Variable

Solve the formula $I = prt$ for t.

Solve this formula for t by assuming that I, p, and r are constants (having a fixed value) and that t is the variable. Then use the properties of the previous section as follows.

$$I = prt$$

$$I = (pr)t \qquad \text{Associative property}$$

$$\frac{I}{pr} = \frac{(pr)t}{pr} \qquad \text{Divide by } pr.$$

$$\frac{I}{pr} = t$$

The result is a formula for t, time in years.

While the process of solving for a specified variable uses the same steps as solving a linear equation from Section 2.1, the following additional suggestions may be helpful.

Solving for a Specified Variable

Step 1 Get all terms containing the specified variable on one side of the equation and all terms without that variable on the other side.

Step 2 If necessary, use the distributive property to combine the terms with the specified variable. The result should be the product of a sum or difference and the variable.

Step 3 Divide both sides by the factor that is the coefficient of the specified variable.

EXAMPLE 2 Solving for a Specified Variable

Solve the formula $P = 2L + 2W$ for W.

This formula gives the relationship between the perimeter (distance around) a rectangle, P, the length of the rectangle, L, and the width of the rectangle, W. See Figure 1.

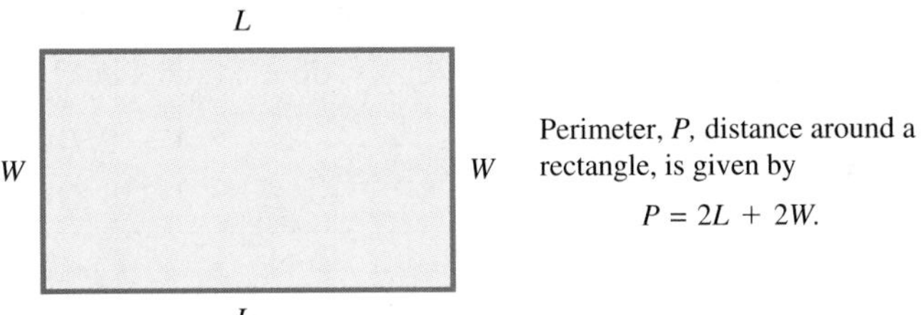

Figure 1

Solve the formula for W by getting W alone on one side of the equals sign. To begin, subtract $2L$ from both sides.

$$P = 2L + 2W$$

Step 1 $$P - 2L = 2L + 2W - 2L \quad \text{Subtract } 2L.$$

$$P - 2L = 2W$$

Step 2 is not needed here.

Step 3 $$\frac{P - 2L}{2} = \frac{2W}{2} \quad \text{Divide both sides by 2.}$$

$$\frac{P - 2L}{2} = W$$

In the result for Example 2, do not try to eliminate the 2 in the numerator and denominator. The 2 in the numerator is not a *factor* of the entire numerator.

A rectangular solid has the shape of a box, but is solid. See Figure 2. The labels H, W, and L represent the height, width, and length of the figure, respectively.

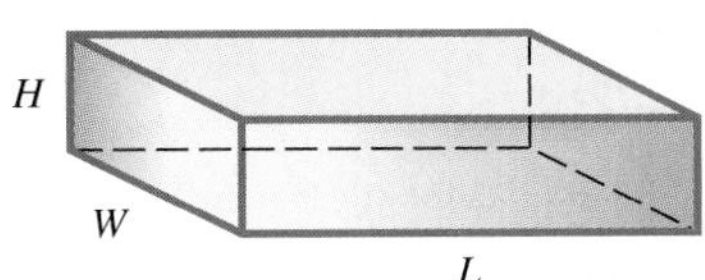

Figure 2

The surface area of any solid three-dimensional figure is the total area of its surface. For a rectangular solid, the surface area A is

$$A = 2HW + 2LW + 2LH.$$

EXAMPLE 3 Using the Distributive Property to Solve for a Specified Variable

Given the surface area, height, and width of a rectangular solid, write a formula for the length.

Step 1 To solve for the length L, write the equation so that only the terms involving L appear on the right side.

$$A = 2HW + 2LW + 2LH$$

$$A - 2HW = 2LW + 2LH \qquad \text{Subtract } 2HW.$$

Step 2 Next, use the distributive property on the right to write $2LW + 2LH$ so that L, the specified variable, is a factor.

$$A - 2HW = L(2W + 2H) \qquad \text{Distributive property}$$

Step 3 Finally, divide both sides by the coefficient of L, which is $2W + 2H$.

$$\frac{A - 2HW}{2W + 2H} = L \quad \text{or} \quad L = \frac{A - 2HW}{2W + 2H}$$

The most common error in working a problem like Example 3 is not using the distributive property correctly. We must write the expression so that the specified variable is a *factor;* then we can divide by its coefficient in the final step.

OBJECTIVE 2 Solve applied problems using formulas. The next example uses the distance formula, $d = rt$, which relates d, the distance traveled, r, the rate or speed, and t, the travel time.

EXAMPLE 4 Finding Average Speed

Janet Branson found that on the average it took her $\frac{3}{4}$ hour each day to drive a distance of 15 miles to work. What was her average speed?

Find the speed r, by solving $d = rt$ for r.

$$d = rt$$

$$\frac{d}{t} = \frac{rt}{t} \qquad \text{Divide by } t.$$

$$\frac{d}{t} = r$$

Notice that only Step 3 was needed to solve for r in this example. Now find the speed by substituting the given values of d and t into this formula.

$$r = \frac{d}{t}$$

$$r = \frac{15}{\frac{3}{4}} \qquad \text{Let } d = 15,\ t = \tfrac{3}{4}.$$

$$r = 15 \cdot \frac{4}{3}$$

$$r = 20$$

Her average speed was 20 miles per hour.

PROBLEM SOLVING

As seen in Example 4, it may be convenient to first solve for a specific unknown variable before substituting the given values. This is particularly useful when we wish to substitute several different values for the same variable.

When a consumer loan is paid off ahead of schedule, the finance charge is smaller than if the loan were paid off over its scheduled life. By one method, called *the rule of 78,* the amount of *unearned interest* (finance charge that need not be paid) is given by

$$u = f \cdot \frac{k(k + 1)}{n(n + 1)},$$

where u is the amount of unearned interest (money saved) when a loan scheduled to run n payments is paid off k payments ahead of schedule. The total scheduled finance charge is f.

Actually, if uniform monthly payments are made, the rule results in more interest in the early months of the loan than is strictly proper. Thus any refund for early repayment would be less by this rule than it really should be. However, the method is still often used, and its accuracy is acceptable for fairly short-term consumer loans. (With tables and calculators readily available, very accurate interest and principal allocations of loan payments are easily determined.)

The next example illustrates the use of the rule of 78.

EXAMPLE 5 Using the Rule of 78

Juan Ortega is scheduled to pay off a loan in 36 monthly payments. If the total finance charge is \$360, and the loan is paid off after 24 payments (that is, 12 payments ahead of schedule), find the unearned interest, u.

A calculator is helpful in this problem. Here $f = 360$, $n = 36$, and $k = 12$. Use the formula above.

$$u = f \cdot \frac{k(k + 1)}{n(n + 1)} = 360 \cdot \frac{12(13)}{36(37)} = 42.16 \text{ (to the nearest cent)}$$

A total of \$42.16 of the \$360 finance charge need not be paid. The total finance charge for this loan is \$360 − \$42.16 = \$317.84.

CONNECTIONS

Formulas are an important part of applied mathematics, since they provide relationships among several quantities in a particular setting. Formulas are found in geometry, the mathematics of finance, branches of science, and many other fields.

The formula

$$A = \frac{24f}{b(p + 1)}$$

(continued)

CONNECTIONS (CONTINUED)

gives the approximate annual interest rate for a consumer loan paid off with monthly payments. Here f is the finance charge on the loan, p is the number of payments, and b is the original amount of the loan.

FOR DISCUSSION OR WRITING

Find the approximate annual interest rate for an installment loan to be repaid in 24 monthly installments. The finance charge on the loan is \$200 and the original loan balance is \$1920. Compare your answer with the interest rate if the same loan is spread over 36 payments with the same finance charge. What is the interest rate if the finance charge is \$225 for the same loan paid in 24 installments?

2.2 EXERCISES

RELATING CONCEPTS (EXERCISES 1-6)

Consider the following equations:

First Equation	*Second Equation*
$x = \dfrac{5x + 8}{3}$	$t = \dfrac{bt + k}{c} \quad (c \neq 0).$

Solving the second equation for t follows the same logic as solving the first equation for x. When solving for t, we treat all other variables as though they were constants.

The following group of exercises illustrates the "parallel logic" of solving for x and solving for t. ***Work the exercises in order*** *with parts (a) and (b) side by side.*

1. **(a)** Clear the first equation of fractions by multiplying through by 3.
   (b) Clear the second equation of fractions by multiplying through by c.
2. **(a)** Get the terms involving x on the left side of the first equation by subtracting $5x$ from both sides.
   (b) Get the terms involving t on the left side of the second equation by subtracting bt from both sides.
3. **(a)** Combine like terms on the left side of the first equation. What property allows us to write $3x - 5x$ as $(3 - 5)x = -2x$?
   (b) Write the expression on the left side of the second equation so that t is a factor. What property allows us to do this?
4. **(a)** Divide both sides of the first equation by the coefficient of x.
   (b) Divide both sides of the second equation by the coefficient of t.
5. Look at your answer for the second equation. What restriction must be placed on the variables? Why is this necessary?
6. Why were no restrictions needed in the answer for the first equation?

Did you make the connection that solving an equation with several variables for a specified variable uses exactly the same steps as solving an equation with just one variable?

7. When a formula is solved for a particular variable, several different equivalent forms may be possible. If we solve $A = \frac{1}{2}bh$ for h, one possible correct answer is

$$h = \frac{2A}{b}.$$

Which one of the following is *not* equivalent to this?

(a) $h = 2\left(\frac{A}{b}\right)$ **(b)** $h = 2A\left(\frac{1}{b}\right)$ **(c)** $h = \frac{A}{\frac{1}{2}b}$ **(d)** $h = \frac{\frac{1}{2}A}{b}$

8. One source for geometric formulas gives the formula for the perimeter of a rectangle as $P = 2L + 2W$, while another gives it as $P = 2(L + W)$. Are these equivalent? If so, what property justifies their equivalence?

Solve each formula for the specified variable. See Examples 1–3.

9. $d = rt$; for t (distance)

10. $I = prt$; for r (simple interest)

11. $A = bh$; for b (area of a parallelogram)

12. $P = 2L + 2W$; for L (perimeter of a rectangle)

13. $P = a + b + c$; for a (perimeter of a triangle)

14. $V = LWH$; for W (volume of a rectangular solid)

15. $A = \frac{1}{2}bh$; for h (area of a triangle)

16. $C = 2\pi r$; for r (circumference of a circle)

17. $S = 2\pi rh + 2\pi r^2$; for h (surface area of a right circular cylinder)

18. $A = \frac{1}{2}(B + b)h$; for B (area of a trapezoid)

19. $\mathrm{C} = \frac{5}{9}(\mathrm{F} - 32)$; for F (Fahrenheit to Celsius)

20. $\mathrm{F} = \frac{9}{5}\mathrm{C} + 32$; for C (Celsius to Fahrenheit)

21. $A = 2HW + 2LW + 2LH$; for H (surface area of a rectangular solid)

22. $V = \frac{1}{3}Bh$; for h (volume of a right pyramid)

Solve each of the following for the specified variable. In each case, use the distributive property to factor as necessary. See Example 3.

23. $2k + ar = r - 3y$; for r

24. $4s + 7p = tp - 7$; for p

25. $w = \frac{3y - x}{y}$; for y

26. $c = \frac{-2t + 4}{t}$; for t

27. Use the formula $2x - mx = z - m$ to explain the steps to follow in solving for m.

28. Suppose the formula $A = 2HW + 2LW + 2LH$ is "solved for L" as follows.

$$A = 2HW + 2LW + 2LH$$
$$A - 2LW - 2HW = 2LH$$
$$\frac{A - 2LW - 2HW}{2H} = L$$

While there are no algebraic errors here, what is wrong with the final line, if we are interested in solving for L?

Solve each problem. Refer to the inside covers of this book if you need to look up a formula. See Example 4.

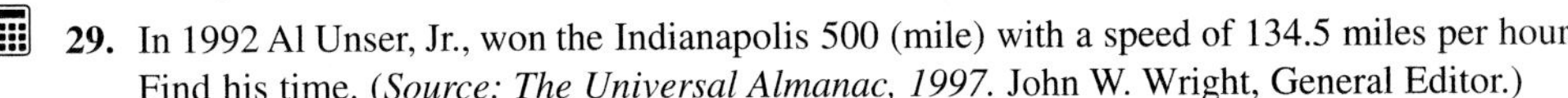

29. In 1992 Al Unser, Jr., won the Indianapolis 500 (mile) with a speed of 134.5 miles per hour. Find his time. (*Source: The Universal Almanac, 1997.* John W. Wright, General Editor.)

30. In 1975, rain shortened the Indianapolis 500 to 435 miles. It was won by Bobby Unser, who averaged 149.2 miles per hour. What was his time? (*Source: The Universal Almanac, 1997.* John W. Wright, General Editor.)

31. The lowest temperature ever recorded in Arizona was $-40°C$ on January 7, 1971. Find the corresponding Fahrenheit temperature.

32. The melting point of brass is 900°C. Find the corresponding Fahrenheit temperature.

33. The base of the Great Pyramid of Cheops is a square whose perimeter is 920 meters. What is the length of each side of this square? (*Source: Atlas of Ancient Archaeology,* edited by Jacquetta Hawkes, McGraw-Hill Book Co., 1974.)

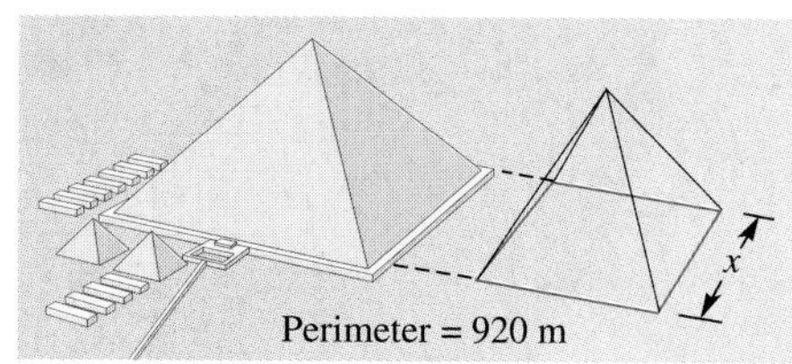

34. The Peachtree Plaza Hotel in Atlanta is in the shape of a cylinder with radius 46 meters and height 220 meters. Find its volume to the nearest hundred thousand cubic meters.

35. Faye Korn traveled from Kansas City to Louisville, a distance of 520 miles, in 10 hours. Find her rate in miles per hour.

36. The distance from Melbourne to London is 10,500 miles. If a jet averages 500 miles per hour between the two cities, what is its travel time in hours?

37. The surface area of a soda can is 86.125π square centimeters. The radius of the can is 3.25 centimeters. Find the height of the can.

38. A cylindrical-shaped oatmeal box has a radius of 4 inches and a surface area of 88π square inches. What is its height?

39. How does the volume of a soda can change if the radius is doubled?

40. How does the volume of a soda can change if the height is halved?

41. A cord of wood contains 128 cubic feet of wood. If a stack of wood is 4 feet wide and 4 feet high, how long must it be if it contains exactly 1 cord?

42. Give two sets of possible dimensions for a stack of wood that contains 1.5 cords. (See Exercise 41.)

43. In order to purchase fencing to go around a yard, would you need to use perimeter or area to decide how much to buy?

44. In order to purchase sod for a lawn, would you need to use perimeter or area to decide how much to buy?

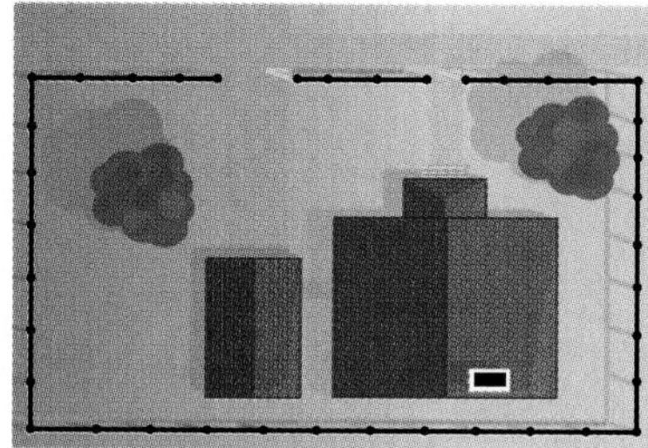

Solve each problem.

45. A mixture of alcohol and water contains a total of 36 ounces of liquid. There are 9 ounces of pure alcohol in the mixture. What percent of the mixture is water? What percent is alcohol?

46. A mixture of acid and water is 35% acid. If the mixture contains a total of 40 liters, how many liters of pure acid are in the mixture? How many liters of pure water are in the mixture?

47. A real estate agent earned \$6300 commission on a property sale of \$210,000. What is her rate of commission?

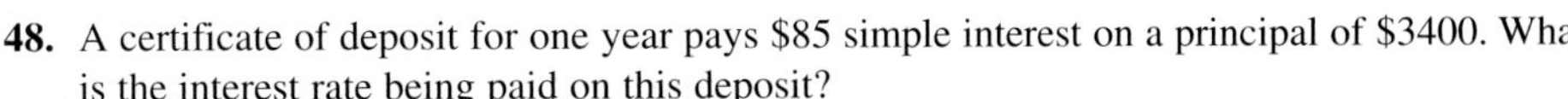

48. A certificate of deposit for one year pays \$85 simple interest on a principal of \$3400. What is the interest rate being paid on this deposit?

Television networks have been losing viewers to cable programming since 1982, as the two graphs below show.

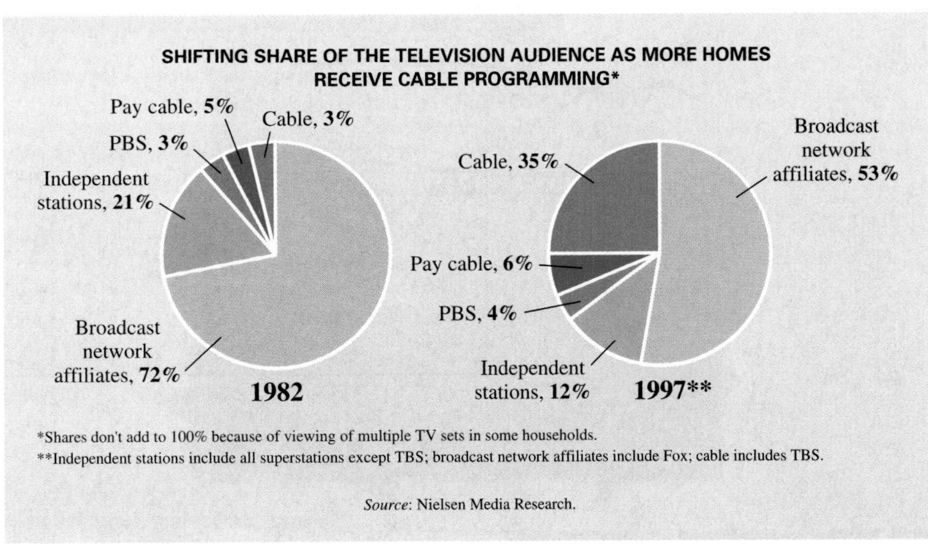

Use these graphs for Exercises 49–52.

49. In a typical group of 50,000 television viewers, how many would have watched cable in 1982?

50. In 1982, how many of a typical group of 110,000 viewers watched independent stations?

51. How many of a typical group of 35,000 viewers in 1997 watched cable?

52. In a typical group of 65,000 viewers, how many watched independent stations in 1997?

Refer to the formula for the rule of 78 to solve each of the following. See Example 5.

53. Rhonda Alessi bought a new Ford and agreed to pay it off in 36 monthly payments. The total finance charge is \$700. Find the unearned interest if she pays the loan off 4 payments ahead of schedule.

54. Paul Lorio bought a car and agreed to pay in 36 monthly payments. The total finance charge on the loan was \$600. With 12 payments remaining, Paul decided to pay the loan in full. Find the amount of unearned interest.

55. The finance charge on a loan taken out by Vic Denicola is \$380.50. If there were 24 equal monthly installments needed to repay the loan, and the loan is paid in full with 8 months remaining, find the amount of unearned interest.

56. Adrian Ortega is scheduled to repay a loan in 24 equal monthly installments. The total finance charge on the loan is \$450. With 9 payments remaining, he decides to repay the loan in full. Find the amount of unearned interest.

The formula

$$A = \frac{24f}{b(p + 1)}$$

gives the approximate annual interest rate for a consumer loan paid off with monthly payments. Here f is the finance charge on the loan, p is the number of payments, and b is the original amount of the loan. Solve the following problems using the formula above.

57. Find the approximate annual interest rate for an installment loan to be repaid in 24 monthly installments. The finance charge on the loan is \$200, and the original loan balance is \$1920.

58. Find the approximate annual interest rate for an automobile loan to be repaid in 36 monthly installments. The finance charge on the loan is \$740 and the amount financed is \$3600.

TECHNOLOGY INSIGHTS (EXERCISES 59–60)

In Exercises 59 and 60 the screen shows a formula with values for each variable except one and a numerical answer for the formula.

Use that answer to find a value for the remaining variable.

59.

```
4→W
              4
2L+2W
             28
```

60.

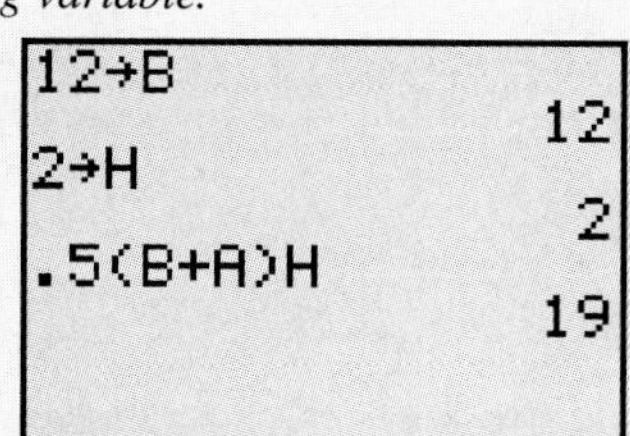

2.3 Applications of Linear Equations

OBJECTIVES

1. Translate from word expressions to mathematical expressions.
2. Write equations from given information.
3. Solve problems about unknown numerical quantities.
4. Solve problems about percents, simple interest, and mixture.
5. Solve problems about angles.

FOR EXTRA HELP

SSG Sec. 2.3
SSM Sec. 2.3
Pass the Test Software
InterAct Math Tutorial Software
Video 2

OBJECTIVE 1 Translate from word expressions to mathematical expressions. Producing a mathematical model of a real situation often involves translating verbal statements into mathematical statements. Although the problems we will be working with are simple ones, the methods we use will also apply to more difficult problems encountered later.

PROBLEM SOLVING

Usually there are key words and phrases in the verbal problem that translate into mathematical expressions involving addition, subtraction, multiplication, and division. Translations of some commonly used expressions follow.

Translating from Words to Mathematical Expressions

Verbal Expression	*Mathematical Expression*
Addition	
The **sum** of a number and 7	$x + 7$
6 **more than** a number	$x + 6$
3 **plus** 8	$3 + 8$
24 **added to** a number	$x + 24$
A number **increased by** 5	$x + 5$
The **sum** of two numbers	$x + y$

PROBLEM SOLVING (CONTINUED)

Verbal Expression	*Mathematical Expression*
Subtraction	
2 **less than** a number	$x - 2$
12 **minus** a number	$12 - x$
A number **decreased by** 12	$x - 12$
The **difference between** two numbers	$x - y$
A number **subtracted from** 10	$10 - x$
Multiplication	
16 **times** a number	$16x$
Some number **multiplied by** 6	$6x$
$\frac{2}{3}$ **of** some number (used only with fractions and percent)	$\frac{2}{3}x$
Twice (2 times) a number	$2x$
The **product** of two numbers	xy
Division	
The **quotient** of 8 and some number	$\frac{8}{x}$ $(x \neq 0)$
A number **divided by** 13	$\frac{x}{13}$
The **ratio** of two numbers or the **quotient** of two numbers	$\frac{x}{y}$ $(y \neq 0)$

CAUTION Because subtraction and division are not commutative operations, it is important to correctly translate expressions involving them. For example, "2 less than a number" is translated as $x - 2$, *not* $2 - x$. "A number subtracted from 10" is expressed as $10 - x$, not $x - 10$. For division, it is understood that the number doing the dividing is the denominator, and the number that is divided is the numerator. For example, "a number divided by 13" and "13 divided into x" both translate as $\frac{x}{13}$. Similarly, "the quotient of x and y" is translated as $\frac{x}{y}$.

OBJECTIVE 2 Write equations from given information. The symbol for equality, $=$, is often indicated by the word "is." In fact, since equal mathematical expressions represent different names for the same number, any words that indicate the idea of "sameness" indicate translation to $=$.

The table in the last Problem Solving box listed mathematical *expressions* that correspond to phrases. *Equations* correspond to sentences. Be sure you understand the difference.

EXAMPLE 1 Translating Sentences into Equations

Translate the following verbal sentences into equations.

Verbal Sentence	*Equation*
Twice a number, decreased by 3, is 42.	$2x - 3 = 42$
If the product of a number and 12 is decreased by 7, the result is 105.	$12x - 7 = 105$
A number divided by the sum of 4 and the number is 28.	$\frac{x}{4+x} = 28$
The quotient of a number and 4, plus the number, is 10.	$\frac{x}{4} + x = 10$

PROBLEM SOLVING

Throughout this book we will be examining different types of applications. While there is no one method that will allow us to solve all types of applied problems, the six steps listed below are helpful.

Solving an Applied Problem

Step 1 **Determine what you are asked to find.** Read the problem carefully. Decide what is given and what must be found. Choose a variable and write down exactly what it represents.

Step 2 **Write down any other pertinent information.** If there are other unknown quantities, express them using the variable. Draw figures or diagrams and use charts, if they apply.

Step 3 **Write an equation.** Write an equation expressing the relationships among the quantities given in the problem.

Step 4 **Solve the equation.** Use the methods of earlier sections to solve the equation.

Step 5 **Answer the question(s) of the problem.** Reread the problem and make sure that you answer the question(s) posed. In some cases, you will need to give more than just the solution of the equation.

Step 6 **Check.** Check your solution using the original words of the problem. Be sure your answer makes sense.

OBJECTIVE 3 **Solve problems about unknown numerical quantities.** The next examples illustrate the six steps for problem solving.

EXAMPLE 2 Finding Unknown Numerical Quantities

The Perry brothers, Jim and Gaylord, were two outstanding pitchers in the major leagues during the past few decades. Together, they won 529 games. Gaylord won 99 games more than Jim. How many games did each brother win?

Step 1 We are asked to find the number of games each brother won. We must choose a variable to represent the number of wins of one of the men.

Let j = the number of wins for Jim.

Step 2 We must also find the number of wins for Gaylord. Since he won 99 games more than Jim,

$$j + 99 = \text{Gaylord's number of wins.}$$

Step 3 The sum of the number of wins is 529, so we can now write an equation.

$$\underbrace{j}_{\text{Jim's wins}} + \underbrace{(j + 99)}_{\text{Gaylord's wins}} = \underbrace{529}_{529}$$

Step 4 Solve the equation.

$$j + (j + 99) = 529$$
$$2j + 99 = 529 \quad \text{Combine like terms.}$$
$$2j = 430 \quad \text{Subtract 99.}$$
$$j = 215 \quad \text{Divide by 2.}$$

Step 5 Since j represents the number of Jim's wins, Jim won 215 games. Gaylord won $j + 99 = 215 + 99 = 314$ games.

Step 6 314 is 99 more than 215, and the sum of 314 and 215 is 529.

The words of the problem are satisfied, and our solution checks.

Remember to answer all questions asked in an applied problem. In Example 2, we were asked for the number of wins for *each* brother, so there was an extra step at the end to find Gaylord's number.

OBJECTIVE 4 Solve problems about percents, simple interest, and mixture. Percent means per one hundred. For example, 17% means $\frac{17}{100}$, or .17, and 109% means 1.09.

EXAMPLE 3 Solving a Percent Problem

In 1998 there were 212 long distance area codes in the United States. This was an increase of 147% over the number when the area code plan originated in 1947. How many area codes were there in 1947? (*Source:* Pacific Bell Telephone Directory.)

Step 1 Find: the number of area codes in 1947

Given: the number in 1947, increased by 147%, was 212.

Let x = the number of area codes in 1947;

Step 2 $1.47x$ = the increase.

Step 3 From the given information, we get the equation:

$$\text{the number in 1947} + \text{the increase} = 212.$$
$$x + 1.47x = 212$$

Step 4 Solve the equation.

$$1x + 1.47x = 212 \quad \text{Identity property}$$
$$2.47x = 212 \quad \text{Combine terms.}$$
$$x = 86 \quad \text{Divide by 2.47. (Use a calculator if you wish.)}$$

Steps 5 and 6 There were 86 area codes in 1947. Check that the increase, $212 - 86 = 126$, is 147% of 86.

CAUTION Watch out for two common errors that occur in solving problems like the one in Example 3. First, do not try to find 147% of the number of codes in 1998 and then subtract from that amount. Second, write the equation correctly. The decimal 1.47 *must be multiplied by x,* as shown in the example.

The next example shows how to solve a mixture problem involving different concentrations. Notice how a sketch and a chart are used.

EXAMPLE 4 Solving a Mixture Problem

A chemist must mix 8 liters of a 40% solution of potassium chloride with some 70% solution to get a mixture that is a 50% solution. How much of the 70% solution should be used?

The information in the problem is illustrated in Figure 3.

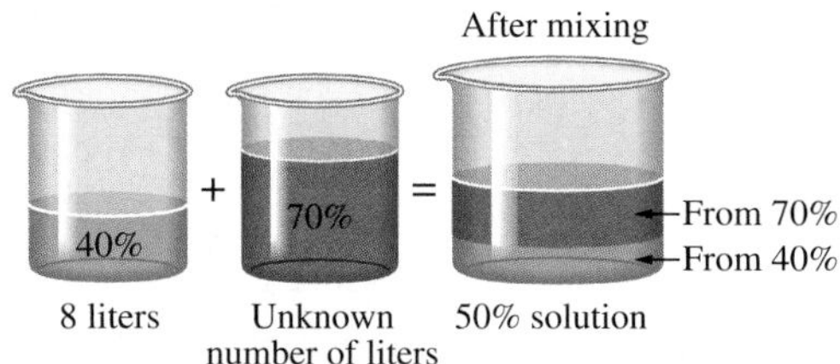

Figure 3

Let $x =$ the number of liters of the 70% solution that should be used. The information in the problem is used to get the following chart.

Strength	*Liters of Solution*	*Liters of Pure Potassium Chloride*
40%	8	$.40(8) = 3.2$
70%	x	$.70x$
50%	$8 + x$	$.50(8 + x)$

Sum must equal

The numbers in the right-hand column were found by multiplying the strengths and the numbers of liters. The number of liters of pure potassium chloride in the 40% solution plus the number of liters in the 70% solution must equal the number of liters in the 50% solution.

$$3.2 + .70x = .50(8 + x)$$
$$3.2 + .70x = 4 + .50x \quad \text{Distributive property}$$
$$.20x = .8 \quad \text{Subtract 3.2 and } .50x.$$
$$x = 4 \quad \text{Divide by } .20.$$

The chemist must use 4 liters of the 70% solution. ■

EXAMPLE 5 Solving an Investment Problem

After winning the state lottery, Michael Chin has $40,000 to invest. He will put part of the money in an account paying 4% simple interest, and the remainder into stocks paying 6% simple interest. His accountant tells him the total annual income from these investments should be $2040. How much should he invest at each rate?

Let $x =$ the amount invested at 4%;

then $40{,}000 - x =$ the amount invested at 6%.

The formula for interest is $I = prt$. Here the time, t, is 1 year. Make a chart to organize the given information.

Percent as a Decimal	*Amount Invested*	*Interest in One Year*
.04	x	.04x
.06	40,000 − x	.06(40,000 − x)
		2040

The last column gives the equation.

Interest at 4% + interest at 6% = total interest

$$.04x + .06(40{,}000 - x) = 2040.$$

Solve the equation.

$$.04x + 2400 - .06x = 2040 \quad \text{Distributive property}$$
$$2400 - .02x = 2040 \quad \text{Combine terms.}$$
$$-.02x = -360 \quad \text{Subtract 2400.}$$
$$x = 18{,}000 \quad \text{Divide by } -.02.$$

Chin should invest $18,000 at 4% and $40,000 − $18,000 = $22,000 at 6%. Check by finding the annual interest at each rate; they should add up to $2040. ■

In Example 5, we chose to let the variable represent the amount invested at 4%. Students often ask, "Can I let the variable represent the other unknown?" The answer is yes. The equation will be different, but in the end the two answers will be the same. In Exercise 7 you are asked to rework Example 5, letting the variable equal the amount invested at 6%.

The investment problems in this chapter deal with *simple interest.* In most "real-world" applications, *compound interest* is used. However, more advanced methods (covered in a later chapter) are needed for compound interest problems, so we will deal only with simple interest until then.

OBJECTIVE 5 Solve problems about angles. Recall that a basic unit of measure of angles is the **degree**; an angle that measures one degree (1°) is $\frac{1}{360}$ of a complete revolution. See Figure 4.

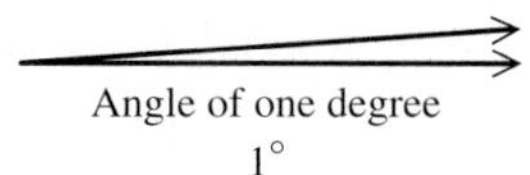

Figure 4

Figures 5 and 6 show some special angles, with their names. A **right angle** measures 90°; a **straight angle** measures 180°. If the sum of the measures of two angles is 90°, the angles are **complementary angles,** and they are called **complements** of each other. If the sum of the measures of two angles is 180°, the angles are **supplementary angles,** and they are called **supplements** of each other.

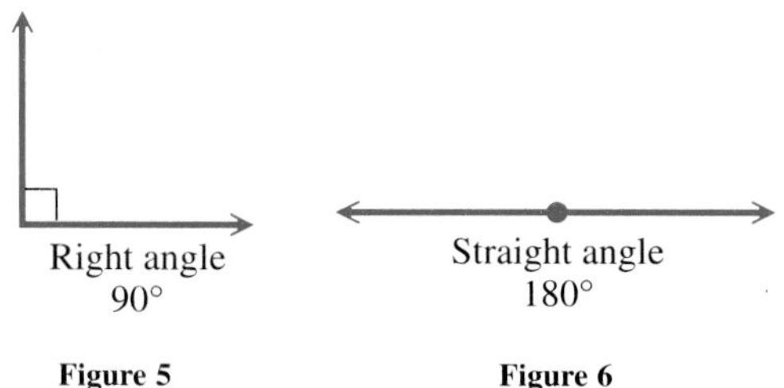

Figure 5 Figure 6

One of the important results of Euclidean geometry (the geometry of the Greek mathematician Euclid) is that the sum of the angle measures of any triangle is 180°. This property, along with the others mentioned above, is used in the next example.

EXAMPLE 6 **Solving Problems about Angle Measures**

(a) Find the value of x, and determine the measure of each angle in Figure 7.

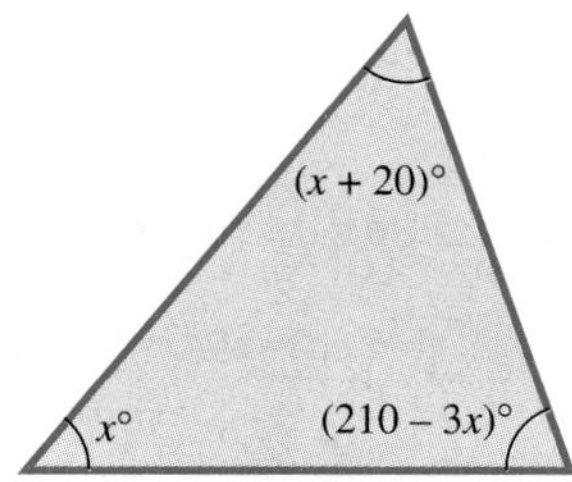

Figure 7

Since the three marked angles are angles of a triangle, their sum must be 180°. Write the equation indicating this, and then solve.

$$x + (x + 20) + (210 - 3x) = 180$$

$$-x + 230 = 180 \quad \text{Combine like terms.}$$

$$-x = -50 \quad \text{Subtract 230.}$$

$$x = 50 \quad \text{Divide by } -1.$$

One angle measures 50°, another measures $x + 20 = 50 + 20 = 70°$, and the third measures $210 - 3x = 210 - 3(50) = 60°$. Since $50° + 70° + 60° = 180°$, the answers are correct.

(b) The supplement of an angle measures 10° more than three times its complement. Find the measure of the angle.

Let $x =$ the degree measure of the angle.

Then, $180 - x =$ the degree measure of its supplement,

and $90 - x =$ the degree measure of its complement.

Now use the words of the problem to write the equation.

Supplement	measures	10 more than	three times its complement
$180 - x$	$=$	$10 +$	$3(90 - x)$

Solve the equation.

$$180 - x = 10 + 270 - 3x \quad \text{Distributive property}$$
$$180 - x = 280 - 3x$$
$$2x = 100 \quad \text{Add } 3x\text{; subtract } 180.$$
$$x = 50 \quad \text{Divide by } 2.$$

The angle measures 50°. Since its supplement (130°) is 10° more than three times its complement (40°), that is

$$130 = 10 + 3(40)$$

is true, the answer checks.

CONNECTIONS

Probably the most famous study of problem-solving techniques was developed by George Polya (1888–1985). Among his many publications was the modern classic *How to Solve It.* In this book, Polya proposed a four-step process for problem solving.

POLYA'S FOUR-STEP PROCESS FOR PROBLEM SOLVING

1. **Understand the problem.** You must first decide what you are to find.
2. **Devise a plan.** Here are some strategies that may prove useful.

   Problem-Solving Strategies
   If a formula applies, use it.
   Write an equation and solve it.
   Draw a sketch.
   Make a table or a chart.
   Look for a pattern.
   Use trial and error.
   Work backward.

   We used the first of these strategies in the previous section. In this section we used the next three strategies.

(continued)

CONNECTIONS (CONTINUED)

3. **Carry out the plan.** This is where the algebraic techniques you are learning in this book can be helpful.
4. **Look back and check.** Is your answer reasonable? Does it answer the question that was asked?

FOR DISCUSSION OR WRITING

Compare Polya's four steps with the six steps for problem solving given earlier. Which of our steps correspond with each of Polya's steps?

2.3 EXERCISES

Decide whether each of the following translates into an expression or an equation.

1. the sum of a number and 6
2. 75% of a number
3. $\frac{2}{3}$ of a number is 12.
4. 7 is 3 more than a number.
5. the ratio of a number and 5
6. 15 divided by a number is 3.
7. Rework Example 5, letting the variable equal the amount invested at 6%. Compare your answers with those found in Example 5. Are they the same?
8. Explain why $13 - x$ is *not* a correct translation of "13 less than a number."

Translate each verbal phrase into a mathematical expression. Use x to represent the unknown number. See the Problem Solving box at the beginning of the section.

9. a number decreased by 18
10. 12 more than a number
11. the product of 9 less than a number and 6 more than the number
12. the quotient of a number and 6
13. the ratio of 12 and a nonzero number
14. $\frac{6}{7}$ of a number
15. Write a few sentences describing the six steps for problem solving.
16. Which one of the following is *not* a valid translation of "30% of a number"?

    (a) $.30x$ **(b)** $.3x$ **(c)** $\frac{3x}{10}$ **(d)** $.30$

Let x represent the unknown, and write an equation for each sentence. See Example 1.

17. If the quotient of a number and 6 is added to twice the number, the result is 8 less than the number.
18. If the product of a number and -4 is subtracted from the number, the result is 9 more than the number.
19. When $\frac{2}{3}$ of a number is subtracted from 12, the result is 10.
20. When 75% of a number is added to 6, the result is 3 more than the number.

Use the six-step problem-solving method to solve each problem. See Example 2.

21. In 1996 the two most popular places where book buyers shopped were large chain bookstores and small chain/independent bookstores. In a sample of book buyers, 70 more shopped at large chain bookstores than at small chain/independent bookstores. A total of 442 book buyers shopped at these two types of stores. Complete the following problem-solving steps to find how many buyers shopped at each type of bookstore. (*Source*: Book Industry Study Group.)

Step 1 Let $x =$ _____.

Step 2 $x - 70 =$ _____.

Step 3 _____ + _____ $= 442$

Step 4 $x =$ _____

Step 5 There were _____ large chain bookstore shoppers and _____ small chain/independent shoppers.

Step 6 The number of _____ was _____ more than the number of _____, and the total number of these two bookstore types was _____.

22. According to figures provided by the Air Transport Association of America, the Boeing B747–400 and the McDonnell Douglas L1011–100/200 are among the air carriers with maximum passenger seating. The Boeing seats 110 more passengers than the McDonnell Douglas, and together the two models seat 696 passengers. (*Source: The World Almanac and Book of Facts.*) Write out problem-solving Steps 1–6 as in Exercise 21 to find the seating capacity of each model.

23. The two top women-owned businesses in a recent year were Ingram Industries, Nashville, owned by Martha Ingram, and TLC Beatrice International, New York, owned by Loida Nicolas Lewis. Together their sales totaled \$13.7 billion. Ingram Industries' sales were \$9.3 billion more than TLC Beatrice International's sales. What were the sales for each company? (*Source: The Universal Almanac, 1997,* John W. Wright, General Editor.)

24. In a recent year, the two U.S. industrial corporations with the highest profits were General Motors and General Electric. Their profits together totaled \$13.5 billion. General Electric profits were \$.3 billion less than General Motors profits. What were the profits for each corporation? (*Source: The Universal Almanac, 1997,* John W. Wright, General Editor.)

25. Babe Ruth and Rogers Hornsby were two great hitters. Together they got 5803 base hits in their careers. Hornsby got 57 more hits than Ruth. How many base hits did each get? (*Source:* David S. Neft and Richard M. Cohen, *The Sports Encyclopedia: Baseball, 1997.* St. Martins Griffin; New York.)

26. In the 1996 presidential election, Bill Clinton and Bob Dole together received 538 electoral votes. Clinton received 220 more votes than Dole. How many votes did each candidate receive? (*Source:* Congressional Quarterly, Inc.)

Refer to the accompanying graph to help you answer the questions in Exercises 27–30.

27. In 1990, video rental revenue was \$.27 billion more than twice video sales revenue. Together, these sales amounted to \$9.81 billion. What was the revenue from each of these sources?

28. Video rentals accounted for \$.15 billion more than video sales in a later year. At that time, combined sales were \$14.83 billion. In what year was this true, and what was the revenue from each source?

29. Compare your answers to Exercises 27 and 28 with the graph. Do they correspond to the information shown there? Remember that a graph such as this provides a visual estimate, not exact values.

30. From the graph, what seems to be the trend for video rentals versus video sales?

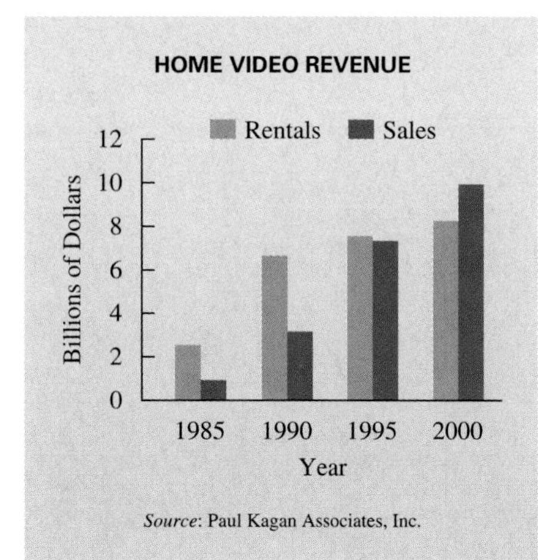

Solve each problem. See Example 3.

31. Composite scores on the ACT exam rose from 20.6 in 1990 to 20.9 in 1996. What percent increase is this? (*Source:* The American College Testing Program.)

32. In 1996, the number of participants in the ACT exam was 925,000. Earlier, in 1990, a total of 817,000 took the exam. What percent increase was this? (*Source:* The American College Testing Program.)

33. In 1985, the average tuition for public 4-year universities in the United States was \$1386 for full-time students. By 1996, it had risen approximately 227%. What was the approximate cost in 1996? (*Source:* National Center for Education Statistics, U.S. Dept. of Education.)

34. The consumer price index (CPI) in 1996 was 156.9. This represented a 3.0% increase from 1995. What was the CPI in 1995? (*Source:* U.S. Bureau of Labor Statistics.)

35. At the end of a day, Jeff Hornsby found that the total cash register receipts at the motel where he works amounted to \$1650.78. This included the 8% sales tax charged. Find the amount of the tax.

36. Dongming Wei sold his house for \$159,000. He got this amount knowing that he would have to pay a 6% commission to his agent. What amount did he have after the agent was paid?

Solve each problem. See Example 4.

37. In a chemistry class, 12 liters of a 12% alcohol solution must be mixed with a 20% solution to get a 14% solution. How many liters of the 20% solution are needed?

Strength	*Liters of Solution*	*Liters of Alcohol*
12%	12	
20%		
14%		

38. How many liters of a 10% alcohol solution must be mixed with 40 liters of a 50% solution to get a 40% solution?

Strength	*Liters of Solution*	*Liters of Alcohol*
	x	
	40	
40%		

39. A medicated first aid spray on the market is 78% alcohol by volume. If the manufacturer has 50 liters of the spray containing 70% alcohol, how much pure alcohol should be added so that the final mixture is the required 78% alcohol? (*Hint:* Pure alcohol is 100% alcohol.)

40. How much water must be added to 3 gallons of a 4% insecticide solution to reduce the concentration to 3%? (*Hint:* Water is 0% insecticide.)

41. It is necessary to have a 40% antifreeze solution in the radiator of a certain car. The radiator now holds 20 liters of 20% solution. How many liters of this should be drained and replaced with 100% antifreeze to get the desired strength? (*Hint:* The number of liters drained is equal to the number of liters replaced.)

42. A tank holds 80 liters of a chemical solution. Currently, the solution has a strength of 30%. How much of this should be drained and replaced with a 70% solution to get a final strength of 40%?

Solve the problem. See Example 5.

43. Jason Jordan earned $12,000 last year by giving tennis lessons. He invested part at 3% simple interest and the rest at 4%. He earned a total of $440 in interest. How much did he invest at each rate?

Percent as a Decimal	*Amount Invested*	*Interest in One Year*
.03		
.04		
	12,000	440

44. Lakeisha Holiday won $60,000 in a slot machine in Las Vegas. She invested part at 2% simple interest and the rest at 3%. She earned a total of $1600 in interest. How much was invested at each rate?

Percent as a Decimal	*Amount Invested*	*Interest in One Year*
.02	x	$.02x$
	60,000	

45. Amado Carillo invested some money at 4.5% simple interest and $1000 less than twice this amount at 3%. His total annual income from the interest was $1020. How much was invested at each rate?

46. Kari Heen invested some money at 3.5% simple interest, and $5000 more than 3 times this amount at 4%. She earned $1440 in interest. How much did she invest at each rate?

47. Ed Moura has $29,000 invested in stocks paying 5%. How much additional money should he invest in certificates of deposit paying 2% so that the average return on the two investments is 3%?

48. Terry McGinnis placed $15,000 in an account paying 6%. How much additional money should she deposit at 4% so that the average return on the two investments is 5.5%?

Fill in the blank or blanks with the correct response.

49. The sum of the measures of the angles of any triangle is ______ degrees.

50. If two angles are complementary, the sum of their measures is ______ degrees.

51. If two angles are supplementary, the sum of their measures is ______ degrees.

52. The measure of a straight angle is ______ degrees, while the measure of a right angle is ______ degrees.

Solve each problem. See Example 6.

53. Find the measure of each angle in the triangle.

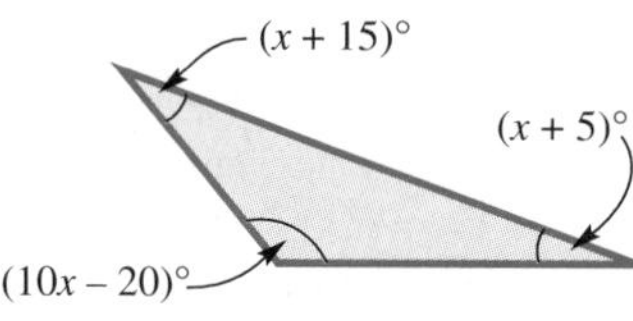

54. Find the measure of each angle in the triangle.

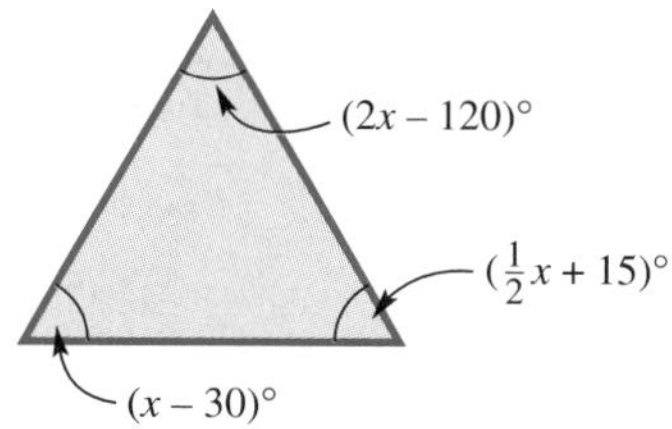

55. Find the measure of each marked angle.

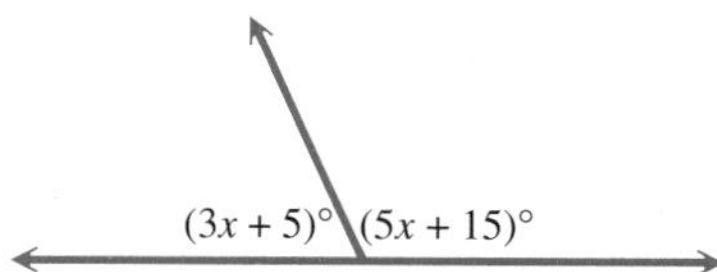

56. Together, the two marked angles form a right angle. Find the measure of each angle.

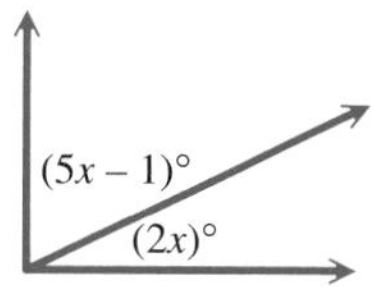

RELATING CONCEPTS (EXERCISES 57–61)

Consider the following two problems.

Problem A

Jack has \$800 invested in two accounts. One pays 5% interest per year and the other pays 10% interest per year. The amount of yearly interest is the same as he would get if the entire \$800 was invested at 8.75%. How much does he have invested at each rate?

Problem B

Jill has 800 liters of acid solution. She obtained it by mixing some 5% acid with some 10% acid. Her final mixture of 800 liters is 8.75% acid. How much of each of the 5% and 10% mixtures did she use to get her final mixture?

In Problem A, let x represent the amount invested at 5% interest, and in Problem B, let y represent the amount of 5% acid used. ***Work the following problems in sequence.***

57. **(a)** Write an expression in x that represents Jack's amount of money invested at 10% in Problem A.

(b) Write an expression in y that represents Jill's amount of 10% acid mixture used in Problem B.

58. **(a)** Write expressions that represent the amount of interest Jack earns per year at 5% and at 10%.

(b) Write expressions that represent the amount of pure acid in Jill's 5% and 10% acid mixtures.

59. **(a)** The sum of the two expressions in part (a) of Exercise 58 must equal the total amount of interest earned in one year. Write an equation representing this fact.

(b) The sum of the two expressions in part (b) of Exercise 58 must equal the amount of pure acid in the final mixture. Write an equation representing this fact.

60. **(a)** Solve Problem A.

(b) Solve Problem B.

RELATING CONCEPTS (EXERCISES 57–61) (CONTINUED)

61. Write a paragraph explaining the similarities between the solution processes used in solving Problems A and B.

Did you make the connection that these two problems are both examples of "mixture problems" and are solved in the same way?

2.4 More Applications of Linear Equations

OBJECTIVES

1. Solve problems about different denominations of money.
2. Solve problems about uniform motion.
3. Solve problems about geometric figures.

FOR EXTRA HELP

SSG Sec. 2.4
SSM Sec. 2.4
Pass the Test Software
InterAct Math Tutorial Software
Video 3

In this section, we discuss three additional types of problems, listed in the Objectives. The same six problem-solving steps are used.

OBJECTIVE 1 Solve problems about different denominations of money. These problems are very similar to the simple interest problems in Section 2.3.

PROBLEM SOLVING

In problems involving money, use the basic fact that

$$\begin{bmatrix}\text{Number of monetary}\\ \text{units of the same kind}\end{bmatrix} \times [\text{Denomination}] = \begin{bmatrix}\text{Total monetary}\\ \text{value}\end{bmatrix}.$$

For example, 30 dimes have a monetary value of $30(.10) = 3.00$ dollars. Fifteen five-dollar bills have a value of $15(5) = 75$ dollars.

EXAMPLE 1 Solving a Problem about Denominations of Money

For a bill totaling \$5.65, a cashier received 25 coins consisting of nickels and quarters. How many of each type of coin did the cashier receive?

Let $x =$ the number of nickels received;

then $25 - x =$ the number of quarters received.

The information for this problem may be arranged in a chart.

Number of Coins	*Denomination*	*Value*
x	.05	$.05x$
$25 - x$	.25	$.25(25 - x)$

Multiply the numbers of coins by the denominations, and add the results to get 5.65.

$$\begin{aligned} .05x + .25(25 - x) &= 5.65 \\ 5x + 25(25 - x) &= 565 && \text{Multiply by 100.} \\ 5x + 625 - 25x &= 565 && \text{Distributive property} \\ -20x &= -60 && \text{Combine terms; subtract 625.} \\ x &= 3 && \text{Divide by } -20. \end{aligned}$$

The cashier has 3 nickels and $25 - 3 = 22$ quarters. Check to see that the total value of these coins is \$5.65.

Be sure that your answer is reasonable when working problems like Example 1. Since you are dealing with a number of coins, an answer can be neither negative nor a fraction.

OBJECTIVE 2 Solve problems about uniform motion.

PROBLEM SOLVING

Uniform motion problems use the distance formula, $d = rt$. In this formula, when rate (or speed) is given in miles per hour, time must be given in hours. To solve such problems, draw a sketch to illustrate what is happening in the problem, and make a chart to summarize the given information. (*Note:* Difficulties often arise in uniform motion problems because a sketch is not drawn. The sketch will help to set up the equation.)

EXAMPLE 2 Solving a Motion Problem

Two cars leave the same place at the same time, one going east and the other west. The eastbound car averages 40 miles per hour, while the westbound car averages 50 miles per hour. In how many hours will they be 300 miles apart?

A sketch shows what is happening in the problem: The cars are going in *opposite* directions. See Figure 8.

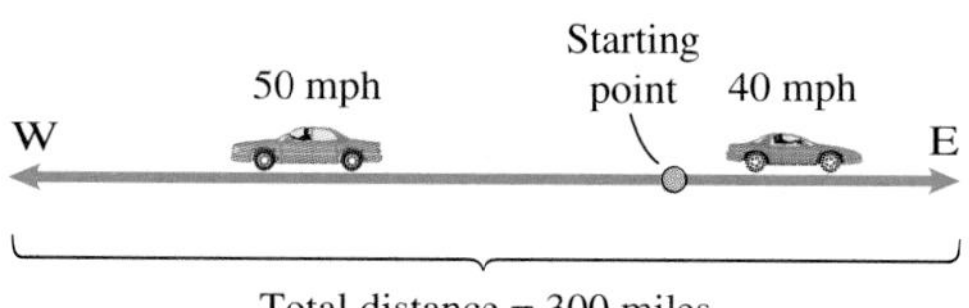

Figure 8

Let x represent the time traveled by each car. Summarize the information of the problem in a chart.

	Rate	Time	Distance
Eastbound car	40	x	$40x$
Westbound car	50	x	$50x$

We find the distances traveled by the cars, $40x$ and $50x$, from the formula $d = rt$. When the expressions for rate and time are entered in the chart, *fill in the distance expression by multiplying rate by time.*

From the sketch in Figure 8, the sum of the two distances is 300.

$$
\begin{aligned}
40x + 50x &= 300 \\
90x &= 300 && \text{Combine terms.} \\
x &= \frac{300}{90} && \text{Divide by 90.} \\
x &= \frac{10}{3} && \text{Lowest terms}
\end{aligned}
$$

The cars travel $\frac{10}{3} = 3\frac{1}{3}$ hours, or 3 hours and 20 minutes. ■

Do not write 300 as the distance for each car in Example 2. Three hundred miles is the *total* distance traveled.

Example 2 involved motion in opposite directions. The next example deals with motion in the same direction.

EXAMPLE 3 Solving a Motion Problem

Jeff Bezzone can bike to work in $\frac{3}{4}$ hour. When he takes the bus, the trip takes $\frac{1}{4}$ hour. If the bus travels 20 miles per hour faster than Jeff rides his bike, how far is it to his workplace?

Although the problem asks for a distance, it is easier here to let x be Jeff's speed when he rides his bike to work. Then the speed of the bus is $x + 20$. By the distance formula, for the trip by bike,

$$d = rt = x \cdot \frac{3}{4} = \frac{3}{4}x,$$

and by bus,

$$d = rt = (x + 20) \cdot \frac{1}{4} = \frac{1}{4}(x + 20).$$

Summarize the information of the problem in a chart.

	Rate	*Time*	*Distance*
Bike	x	$\frac{3}{4}$	$\frac{3}{4}x$
Bus	$x + 20$	$\frac{1}{4}$	$\frac{1}{4}(x + 20)$

The key to setting up the correct equation is to understand that the distance in each case is the same. See Figure 9.

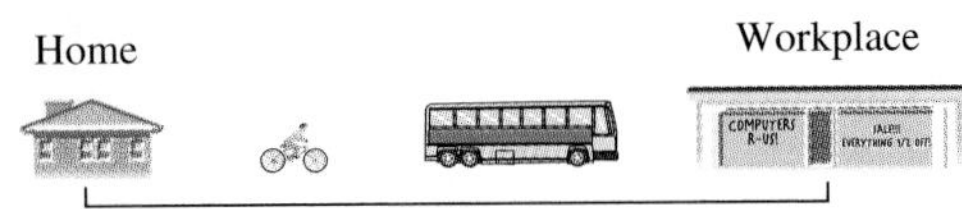

Figure 9

Since the distance is the same in both cases,

$$\frac{3}{4}x = \frac{1}{4}(x + 20).$$

Solve the equation. First multiply on both sides by 4.

$$4\left(\frac{3}{4}x\right) = 4\left(\frac{1}{4}\right)(x + 20)$$

$$3x = x + 20 \quad \text{Multiply; distributive property}$$

$$2x = 20 \quad \text{Subtract } x.$$

$$x = 10 \quad \text{Divide by 2.}$$

Now answer the question in the problem. The required distance is given by

$$d = \frac{3}{4}x = \frac{3}{4}(10) = \frac{30}{4} = 7.5.$$

Check by finding the distance using

$$d = \frac{1}{4}(x + 20) = \frac{1}{4}(10 + 20) = \frac{30}{4} = 7.5,$$

the same result. The required distance is 7.5 miles.

As mentioned in Example 3, it was easier to let the variable represent a quantity other than the one that we were asked to find. This is the case in some problems. It takes practice to learn when this approach is the best, and practice means working lots of problems!

OBJECTIVE 3 Solve problems about geometric figures. Problems about geometric figures usually require one of the formulas covered in Section 2.2.

PROBLEM SOLVING

When applied problems deal with geometric figures, the appropriate formula may provide the equation for the problem as shown in the next example.

EXAMPLE 4 Solving a Problem about a Geometric Figure

A label is in the shape of a rectangle. The length of the rectangle is 1 centimeter more than twice the width. The perimeter is 110 centimeters. Find the length and the width. See Figure 10.

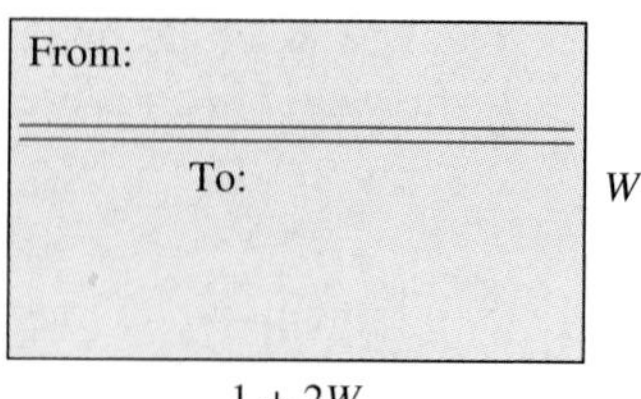

Figure 10

Let $W =$ the width

$2W + 1 =$ the length (one more than twice the width).

The perimeter of a rectangle is given by the formula $P = 2L + 2W$. Replace P in this formula with 110 and L with $1 + 2W$, giving

$$P = 2L + 2W$$
$$110 = 2(1 + 2W) + 2W$$
$$110 = 2 + 4W + 2W \quad \text{Distributive property}$$
$$110 = 2 + 6W \quad \text{Combine terms.}$$
$$108 = 6W \quad \text{Subtract 2.}$$
$$18 = W. \quad \text{Divide by 6.}$$

The width of the label is 18 centimeters, and the length is $1 + 2W = 1 + 2(18) = 1 + 36 = 37$ centimeters.

2.4 EXERCISES

1. In Section 2.3, Example 3, suppose the answers were fractions. Are these answers reasonable? Explain.
2. Suppose one of the answers in Section 2.3, Example 3, was a negative number. Is this reasonable? Explain.
3. Consider this problem. Two cars that are 200 miles apart travel towards one another on a straight highway. One is traveling at 40 miles per hour, and the other is traveling at 50 miles per hour. In how many hours will they be 65 miles apart? What would you choose to have the variable represent? Explain.
4. Refer to the problem stated in Exercise 3. What would you put in a chart for the distance traveled by each car?

	Rate	*Time*	*Distance*
Car 1	40		
Car 2	50		

Solve each problem.

5. What amount of money is found in a coin purse containing 21 nickels and 14 dimes?
6. Capetown, South Africa, and Miami are 7700 miles apart. How long would it take a jet averaging 480 miles per hour to travel between the two cities?
7. Anh Nguyen traveled from Louisville to Kansas City, a distance of 520 miles, in 10 hours. What was his rate in miles per hour?
8. A square has perimeter 20 inches. What would be the perimeter of an equilateral triangle whose sides each measure the same length as the side of the square?

Use the graph to answer the questions in Exercises 9 and 10.

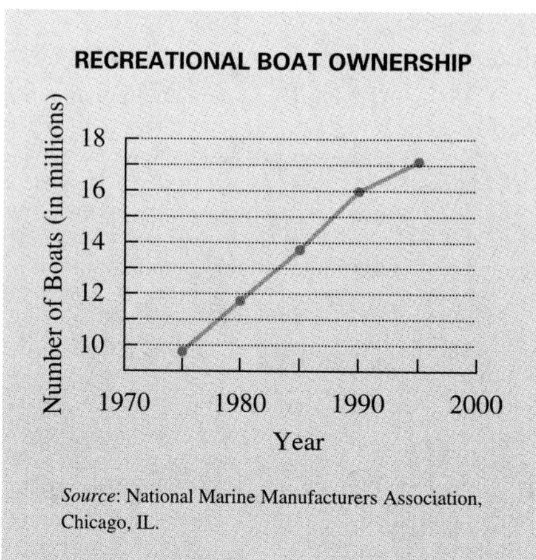

9. What is the best estimate of the number of recreational boats owned in 1975?
(a) 8 million **(b)** 9 million **(c)** 9.7 million **(d)** 10 million

10. The increase in boat ownership slowed during which 5-year period?
(a) 1975–1980 **(b)** 1980–1985 **(c)** 1985–1990 **(d)** 1990–1995

Solve each problem. See Example 1.

11. Sam Abo-zahrah has a box of coins that he uses when playing poker with his friends. The box currently contains 44 coins, consisting of pennies, dimes, and quarters. The number of pennies is equal to the number of dimes, and the total value is \$4.37. How many of each denomination of coin does he have in the box?

Number of Coins	Denomination	Value
x	.01	$.01x$
x		
	.25	

12. Roma Sherry found some coins while looking under her sofa pillows. There were equal numbers of nickels and quarters, and twice as many half-dollars as quarters. If she found \$2.60 in all, how many of each denomination of coin did she find?

Number of Coins	Denomination	Value
x	.05	$.05x$
x		
$2x$	.50	

13. The school production of *Our Town* was a big success. For opening night, 410 tickets were sold. Students paid \$3 each, while nonstudents paid \$7 each. If a total of \$1650 was collected, how many students and how many nonstudents attended?

14. A total of 550 people attended a Boston Pops concert. Floor tickets cost \$40 each, while balcony tickets cost \$28 each. If a total of \$20,800 was collected, how many of each type of ticket were sold?

15. At the Sacramento Monarchs home games, Row 1 seats cost \$35 each and Row 2 seats cost \$30 each. The 105 seats in these rows were sold out for the season. The total receipts for them were \$3420. How many of each type of seat were sold? (*Source:* Sacramento Monarchs.)

16. In the nineteenth century, the United States minted half-cent coins. What decimal number would represent this denomination?

Solve each problem. (Hint: Change minutes to hours.) See Examples 2 and 3.

17. A train leaves Little Rock, Arkansas, and travels north at 85 kilometers per hour. Another train leaves at the same time and travels south at 95 kilometers per hour. How long will it take before they are 315 kilometers apart?

	Rate	*Time*	*Distance*
First train	85	t	
Second train			

18. Two steamers leave a port on a river at the same time, traveling in opposite directions. Each is traveling 22 miles per hour. How long will it take for them to be 110 miles apart?

	Rate	*Time*	*Distance*
First steamer		t	
Second steamer	22		

19. Nancy and Mark commute to work, traveling in opposite directions. Nancy leaves the house at 8:00 A.M. and averages 35 miles per hour. Mark leaves at 8:15 A.M. and averages 40 miles per hour. At what time will they be 140 miles apart?

20. Jeff leaves his house on his bicycle at 8:30 A.M. and averages 5 miles per hour. His wife, Joan, leaves at 9:00 A.M., following the same path and averaging 8 miles per hour. At what time will Joan catch up with Jeff?

21. When Tri drives his car to work, the trip takes 30 minutes. When he rides the bus, it takes 45 minutes. The average speed of the bus is 12 miles per hour less than his speed when driving. Find the distance he travels to work.

22. Latoya can get to school in 15 minutes if she rides her bike. It takes her 45 minutes if she walks. Her speed when walking is 10 miles per hour slower than her speed when riding. How far does she travel to school?

23. A pleasure boat on the Mississippi River traveled from Baton Rouge to New Orleans with a stop at White Castle. On the first part of the trip, the boat traveled at an average speed of 10 miles per hour. From White Castle to New Orleans the average speed was 15 miles per hour. The entire trip covered 100 miles. How long did the entire trip take if the two parts each took the same number of hours?

24. Steve leaves Nashville to visit his cousin David in Napa, 80 miles away. He travels at an average speed of 50 miles per hour. One-half hour later David leaves to visit Steve, traveling at an average speed of 60 miles per hour. How long after David leaves will they meet?

25. In a run for charity Janet runs at a speed of 5 miles per hour. Paula leaves 10 minutes after Janet and runs at 6 miles per hour. How long will it take for Paula to catch up with Janet?

26. Read over Example 3 in this section. The solution of the equation is 10. Why is 10 miles per hour not the answer to the problem?

Solve each problem. See Example 4.

27. The John Hancock Center in Chicago has a rectangular base. The length of the base measures 65 feet less than twice the width. The perimeter of this base is 860 feet. What are the dimensions of the base?

28. The John Hancock Center (Exercise 27) tapers as it rises. The top floor is rectangular and has perimeter 520 feet. The width of the top floor measures 20 feet more than one-half its length. What are the dimensions of the top floor?

29. The Bermuda Triangle supposedly causes trouble for aircraft pilots. It has a perimeter of 3075 miles. The shortest side measures 75 miles less than the middle side, and the longest side measures 375 miles more than the middle side. Find the lengths of the three sides.

30. The Vietnam War Memorial in Washington, D.C., is in the shape of two sides of an isosceles triangle. If the two walls of equal length were joined by a straight line of 438 feet, the perimeter of the resulting triangle would be 931.5 feet. Find the lengths of the two walls.

31. A farmer wishes to enclose a rectangular region with 210 meters of fencing in such a way that the length is twice the width and the region is divided into two equal parts, as shown in the figure. What length and width should be used?

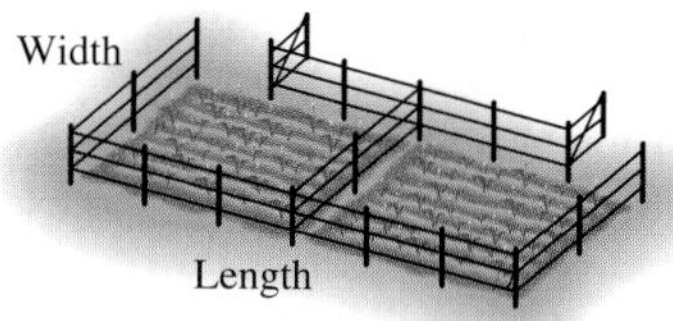

32. Joshua Rogers has a sheet of tin 12 centimeters by 16 centimeters. He plans to make a box by cutting equal squares out of each of the four corners and folding up the remaining edges. How large a square should he cut so that the finished box will have a length that is 5 centimeters less than twice the width?

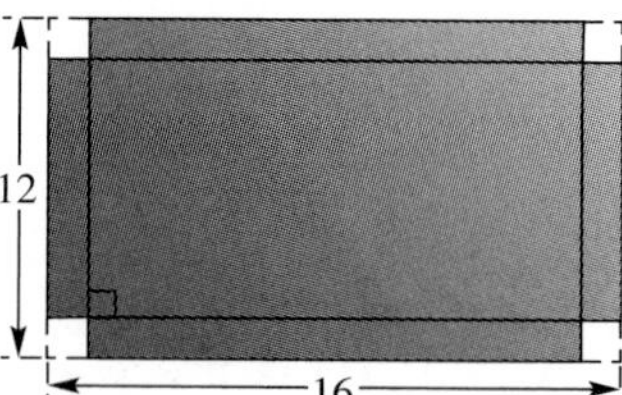

Consecutive integers are integers that follow each other in counting order, such as 8, 9, and 10, or 33, 34, and 35. Suppose we wish to solve the following problem.

Find three consecutive integers such that the sum of the first and third, increased by 3, is 50 more than the second.

Let x represent the first of the unknown integers. Then $x + 1$ will be the second, and $x + 2$ will be the third. The equation we need can be found by going back to the words of the original problem.

Sum of the first and third	increased by 3	is	50 more than the second.
↓	↓	↓	↓
$x + (x + 2)$	$+ 3$	$=$	$(x + 1) + 50$

The solution of this equation is 46 so the first integer is $x = 46$, the second is $x + 1 = 47$, and the third is $x + 2 = 48$. The three integers are 46, 47, and 48. Check by substituting these numbers back into the words of the original problem.

Solve each problem.

33. Two pages facing each other in this book have 153 as the sum of their page numbers. What are the two page numbers?

34. If I add my current age to the age I will be next year on this date, the sum will be 87 years. How old will I be ten years from today?

35. Find three consecutive integers such that the sum of the first and twice the second is 17 more than twice the third.

36. Find four consecutive integers such that the sum of the first three is 60 more than the fourth.

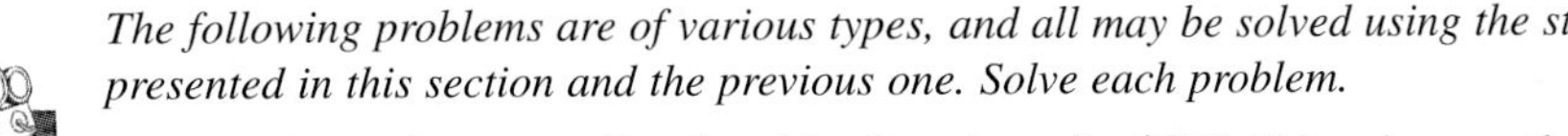

The following problems are of various types, and all may be solved using the strategies presented in this section and the previous one. Solve each problem.

37. An electronics store offered a videodisc player for \$255. This price was the sale price, after the regular price had been discounted 40%. What was the regular price?

38. After a discount of 30%, the sale price of *The Parents' Guide to Kids' Sports* was \$6.27. What was the regular price of the book? (Give your answer to the nearest 5¢.)

39. The length of a rectangle is 3 inches more than its width. If the length were decreased by 2 inches and the width were increased by 1 inch, the perimeter of the resulting rectangle would be 24 inches. Find the dimensions of the original rectangle.

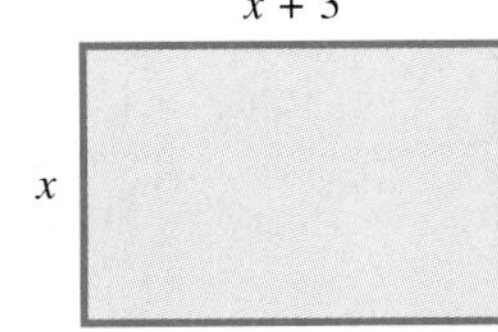

40. A monthly phone bill includes a monthly charge of \$10 for local calls plus an additional charge of 50¢ for each toll call in a certain area. A federal tax of 5% is added to the total bill. If all calls were local or within the 50¢ area and the total bill was \$17.85, find the number of toll calls made.

41. A grocer buys lettuce for \$5.20 a crate. Of the lettuce he buys, 10% cannot be sold. If he charges 40¢ for each head he sells and makes a profit of 10¢ on each head he buys, how many heads of lettuce are in the crate?

42. In 1995 admissions to spectator sports (in millions of dollars) were \$.5 million more than in 1992. The 1992 admissions were \$2.5 million more than in 1985. Total admissions for 1985 and 1995 were \$7.6 million. What were 1995 admissions? (*Source: Survey of Current Businesses,* May, 1997.)

2.5 Linear Inequalities in One Variable

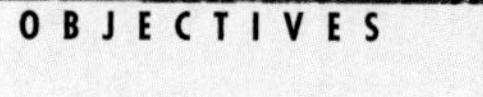

OBJECTIVES

1. Solve linear inequalities using the addition property.
2. Solve linear inequalities using the multiplication property.
3. Solve linear inequalities having three parts.
4. Solve applied problems using linear inequalities.

FOR EXTRA HELP

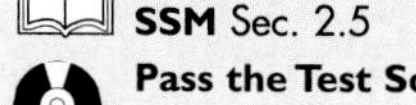

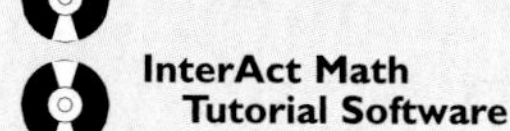

Video 3

An **inequality** says that two expressions are *not* equal. Solving inequalities is similar to solving equations.

Linear Inequality

A **linear inequality in one variable** can be written in the form

$$ax < b,$$

where a and b are real numbers, with $a \neq 0$.

Examples of linear inequalities include

$$x + 5 < 2, \qquad y - 3 \geq 5, \qquad \text{or} \qquad 2k + 5 \leq 10.$$

(Throughout this section we give the definitions and rules only for $<$, but they are also valid for $>$, $\leq$, and $\geq$.)

OBJECTIVE 1 Solve linear inequalities using the addition property. An inequality is solved by finding all numbers that make the inequality true. Usually, an inequality has an infinite number of solutions. These solutions, like the solutions of equations, are found by producing a series of simpler equivalent inequalities. **Equivalent inequalities** are inequalities with the same solution set. We use the addition and multiplication properties of inequality to produce equivalent inequalities.

Addition Property of Inequality

For all real numbers a, b, and c, the inequalities

$$a < b \qquad \text{and} \qquad a + c < b + c$$

are equivalent. (The same number may be added to both sides of an inequality without changing the solution set.)

As with equations, the addition property can be used to *subtract* the same number from both sides of an inequality.

EXAMPLE 1 Using the Addition Property of Inequality

Solve $x - 7 < -12$.

Add 7 to both sides.

$$x - 7 + 7 < -12 + 7$$
$$x < -5$$

The solution set, $(-\infty, -5)$, is graphed in Figure 11.

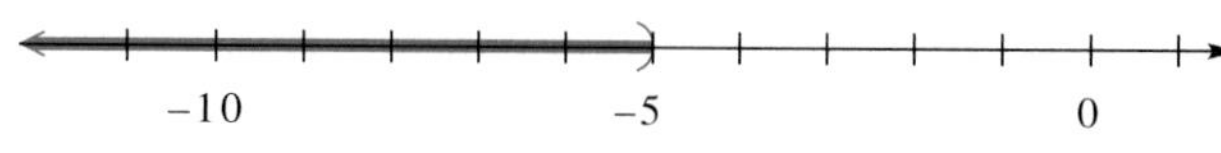

Figure 11

Interval notation, which is used to write solution sets of inequalities, was discussed briefly in Section 1.1. We now summarize the various types of intervals.

Type of Interval	Set	Interval Notation	Graph
	$\{x \mid a < x\}$	(a, ∞)	a
Open interval	$\{x \mid a < x < b\}$	(a, b)	a b
	$\{x \mid x < b\}$	$(-\infty, b)$	b
	$\{x \mid x$ is a real number$\}$	$(-\infty, \infty)$	
	$\{x \mid a \le x\}$	$[a, \infty)$	a
Half-open interval	$\{x \mid a < x \le b\}$	$(a, b]$	a b
	$\{x \mid a \le x < b\}$	$[a, b)$	a b
	$\{x \mid x \le b\}$	$(-\infty, b]$	b
Closed interval	$\{x \mid a \le x \le b\}$	$[a, b]$	a b

EXAMPLE 2 Using the Addition Property of Inequality

Solve the inequality $14 + 2m \le 3m$, and graph the solution set.

First, subtract $2m$ from both sides.

$$14 + 2m \le 3m$$

$$14 + 2m - 2m \le 3m - 2m \qquad \text{Subtract } 2m.$$

$$14 \le m \qquad \text{Combine like terms.}$$

The inequality $14 \le m$ (14 is less than or equal to m) can also be written $m \ge 14$ (m is greater than or equal to 14). Notice that in each case, the inequality symbol points to the smaller expression, 14. The solution set in interval notation is $[14, \infty)$. The graph is shown in Figure 12.

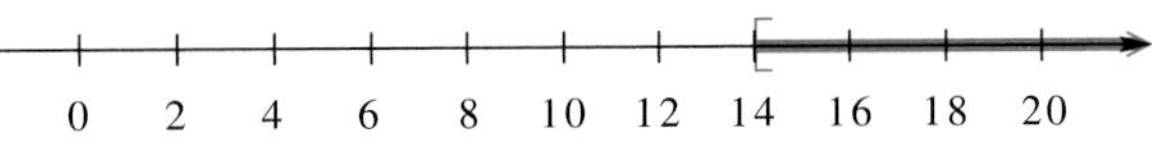

Figure 12

CAUTION Errors often occur in graphing inequalities when the variable term is on the right side. (This is probably due to the fact that we read from left to right.) To guard against such errors, it is a good idea to rewrite these inequalities so that the variable is on the left, as discussed in Example 2.

OBJECTIVE 2 Solve linear inequalities using the multiplication property. An inequality such as $3x \le 15$ can be solved by dividing both sides by 3. This is done with the multiplication property of inequality, which is a little more involved than the corresponding property for equations. To see how this property works, start with the true statement

$$-2 < 5.$$

Multiply both sides by, say, 8.

$$-2(8) < 5(8)$$

$$-16 < 40 \qquad \text{True}$$

This gives a true statement. Start again with $-2 < 5$, and this time multiply both sides by -8.

$$-2(-8) < 5(-8)$$

$$16 < -40 \qquad \text{False}$$

The result, $16 < -40$, is false. To make it true, change the direction of the inequality symbol to get

$$16 > -40.$$

As these examples suggest, multiplying both sides of an inequality by a *negative* number reverses the direction of the inequality symbol. The same is true for dividing by a negative number, since division is defined in terms of multiplication.

Multiplication Property of Inequality

For all real numbers a, b, and c, with $c \neq 0$,

(a) the inequalities

$$a < b \quad \text{and} \quad ac < bc$$

are equivalent if $c > 0$;

(b) the inequalities

$$a < b \quad \text{and} \quad ac > bc$$

are equivalent if $c < 0$. (Both sides of an inequality may be multiplied by a *positive* number without changing the direction of the inequality symbol. Multiplying or dividing by a *negative* number reverses the inequality symbol.)

Remember to reverse the direction of the inequality symbol when multiplying or dividing by a negative number.

EXAMPLE 3 Using the Multiplication Property of Inequality

Solve each inequality, then graph its solution set.

(a) $5m \leq -30$

Use the multiplication property to divide both sides by 5. Since $5 > 0$, do *not* reverse the inequality symbol.

$$5m \leq -30$$

$$\frac{5m}{5} \leq \frac{-30}{5} \quad \text{Divide by 5.}$$

$$m \leq -6$$

The solution set, graphed in Figure 13, is the interval $(-\infty, -6]$.

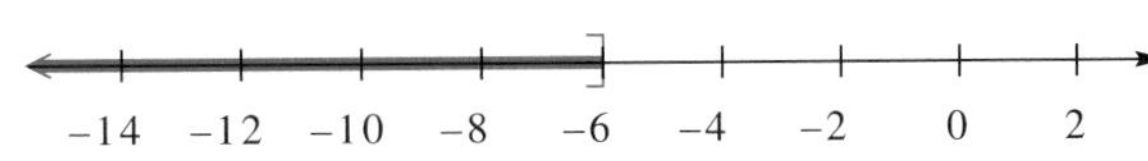

Figure 13

(b) $-4k \leq 32$

Divide both sides by -4. Since $-4 < 0$, the inequality symbol must be reversed.

$$-4k \leq 32$$

$$\frac{-4k}{-4} \geq \frac{32}{-4} \quad \text{Divide by } -4 \text{ and reverse symbol.}$$

$$k \geq -8$$

Figure 14 shows the graph of the solution set, $[-8, \infty)$.

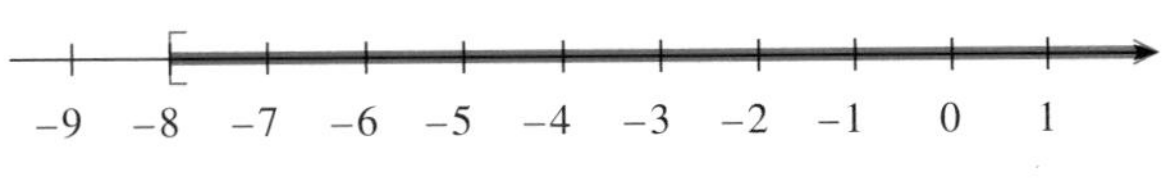

Figure 14

The steps used in solving a linear inequality are given below.

Solving a Linear Inequality

Step 1 **Simplify each side separately.** Simplify each side of the inequality as much as possible by using the distributive property to clear parentheses and by combining like terms as needed.

Step 2 **Isolate the variable terms on one side.** Use the addition property of inequality to get all terms with variables on one side of the inequality and all numbers on the other side.

Step 3 **Isolate the variable.** Use the multiplication property to change the inequality to the form $x < k$ or $x > k$.

Remember: Reverse the direction of the inequality symbol **only** when **multiplying or dividing both sides of an inequality by a negative number.**

EXAMPLE 4 Solving a Linear Inequality Using the Distributive Property

Solve $-3(x + 4) + 2 \geq 8 - x$. Graph the solution set.

Step 1	$-3x - 12 + 2 \geq 8 - x$	Clear parentheses.
	$-3x - 10 \geq 8 - x$	Combine like terms.
Step 2	$-3x - 10 + x \geq 8 - x + x$	Add x on both sides.
	$-2x - 10 \geq 8$	
	$-2x - 10 + 10 \geq 8 + 10$	Add 10 on both sides.
	$-2x \geq 18$	
Step 3	$\dfrac{-2x}{-2} \leq \dfrac{18}{-2}$	Divide by -2, change $\geq$ to $\leq$.
	$x \leq -9$	

Figure 15 shows the graph of the solution set $(-\infty, -9]$.

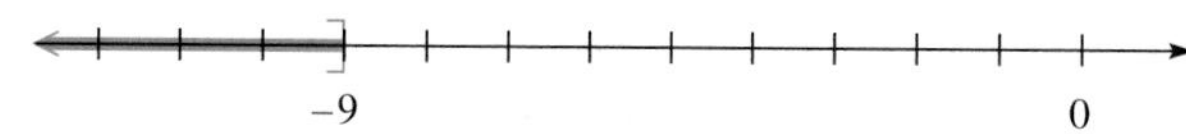

Figure 15

EXAMPLE 5 Solving a Linear Inequality with Fractions

Solve $-\frac{2}{3}(r - 3) - \frac{1}{2} < \frac{1}{2}(5 - r)$, and graph the solution set.

To clear fractions, multiply both sides by the common denominator, 6. Then use the distributive property, and combine terms.

	$-\frac{2}{3}(r - 3) - \frac{1}{2} < \frac{1}{2}(5 - r)$	
	$-4(r - 3) - 3 < 3(5 - r)$	Multiply by 6, the LCD.
Step 1	$-4r + 12 - 3 < 15 - 3r$	Distributive property
	$-4r + 9 < 15 - 3r$	Combine like terms.
Step 2	$3r - 4r + 9 < 3r + 15 - 3r$	Add $3r$ to both sides.
	$-r + 9 < 15$	
	$-r + 9 - 9 < 15 - 9$	Subtract 9.
	$-r < 6$	

Step 3 To solve for *r*, multiply both sides of the inequality by -1. Since -1 is negative, change the direction of the inequality symbol.

$(-1)(-r) > (-1)(6)$	Multiply by -1, change $<$ to $>$.
$r > -6$	

The solution set, $(-6, \infty)$, is graphed in Figure 16.

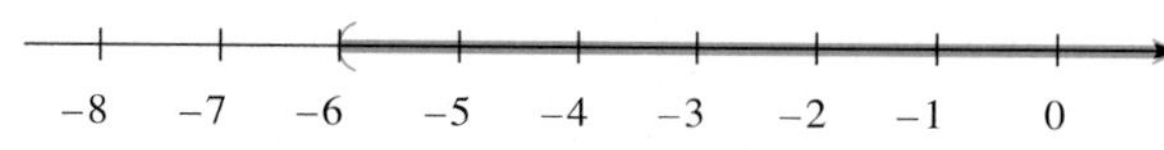

Figure 16

OBJECTIVE 3 Solve linear inequalities having three parts. For some applications, it is necessary to work with an inequality such as

$$3 < x + 2 < 8,$$

where $x + 2$ is *between* 3 and 8. To solve this inequality, we subtract 2 from each of the three parts of the inequality, giving

$$3 - 2 < x + 2 - 2 < 8 - 2$$
$$1 < x < 6.$$

The solution set, (1, 6), is graphed in Figure 17.

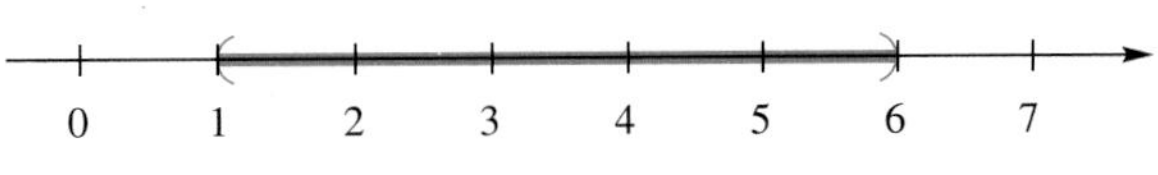

Figure 17

When using inequalities with three parts like the one above, it is important to have the numbers in the correct positions. It would be *wrong* to write an inequality as $8 < x + 2 < 3$, since this would imply that $8 < 3$, a false statement. In general, three-part inequalities are written so that the symbols point in the same direction, and they both point toward the smaller number.

EXAMPLE 6 Solving a Three-Part Inequality

Solve the inequality $-2 \leq 3k - 1 \leq 5$, and graph the solution set.

To begin, we add 1 to each of the three parts.

$$-2 + 1 \leq 3k - 1 + 1 \leq 5 + 1 \qquad \text{Add 1.}$$
$$-1 \leq 3k \leq 6$$
$$\frac{-1}{3} \leq \frac{3k}{3} \leq \frac{6}{3} \qquad \text{Divide by 3.}$$
$$-\frac{1}{3} \leq k \leq 2$$

A graph of the solution set, $\left[-\frac{1}{3}, 2\right]$, is shown in Figure 18.

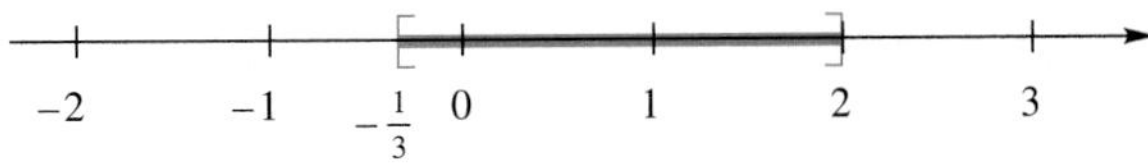

Figure 18

The types of solutions to linear equations or linear inequalities are shown in the box that follows.

Solution Sets of Linear Equations and Inequalities

Equation or Inequality	*Typical Solution Set*	*Graph of Solution Set*
Linear equation $ax = b$	$\{p\}$	p
Linear inequality $ax < b$	$(-\infty, p)$ or (p, ∞)	p p
Three-part inequality $b < ax < c$	(p, q)	p q

OBJECTIVE 4 Solve applied problems using linear inequalities. In addition to the familiar "is less than" and "is greater than," the expressions "is no more than," and "is at least" also denote inequalities.

PROBLEM SOLVING

Expressions for inequalities sometimes appear in applied problems. The chart below shows how to interpret these expressions.

Word Expression	*Interpretation*	*Word Expression*	*Interpretation*
a is at least b	$a \geq b$	a is at most b	$a \leq b$
a is no less than b	$a \geq b$	a is no more than b	$a \leq b$

The final example shows how an applied problem is solved using a linear inequality.

EXAMPLE 7 Solving an Application Using a Linear Inequality

A rental company charges \$15 to rent a chain saw, plus \$2 per hour. Rusty Brauner can spend no more than \$35 to clear some logs from his yard. What is the maximum amount of time he can keep the rented saw?

Let h = the number of hours he can rent the saw. He must pay \$15, plus \$2h, to rent the saw for h hours, and this amount must be *no more than* \$35.

$$\underbrace{15 + 2h}_{\text{Cost of renting}} \quad \underbrace{\leq}_{\text{is no more than}} \quad \underbrace{35}_{\text{35 dollars.}}$$

$$15 + 2h - 15 \leq 35 - 15 \qquad \text{Subtract 15.}$$

$$2h \leq 20$$

$$h \leq 10 \qquad \text{Divide by 2.}$$

He can keep the saw for a maximum of 10 hours. (Of course, he may keep it for less time, as indicated by the inequality $h \leq 10$.)

2.5 EXERCISES

Match each inequality with the correct graph or interval notation.

1. $x \leq 3$
2. $x > 3$
3. $x < 3$
4. $x \geq 3$
5. $-3 \leq x \leq 3$
6. $-3 < x < 3$

A. 0 3

B. 0 3

C. $(3, \infty)$

D. $(-\infty, 3]$

E. $(-3, 3)$

F. $[-3, 3]$

7. Explain how to determine whether to use parentheses or brackets when graphing the solution set of an inequality.
8. Describe the steps used to solve a linear inequality. Explain when it is necessary to reverse the inequality symbol.

Solve each inequality. Give the solution set in both interval and graph forms. See Examples 1–5.

9. $4x + 1 \geq 21$
10. $5t + 2 \geq 52$
11. $\dfrac{3k - 1}{4} > 5$
12. $\dfrac{5z - 6}{8} < 8$
13. $-4x < 16$
14. $-2m > 10$
15. $-\dfrac{3}{4}r \geq 30$
16. $-1.5y \leq -\dfrac{9}{2}$
17. $-1.3m \geq -5.2$
18. $-2.5y \leq -1.25$
19. $\dfrac{2k - 5}{-4} > 5$
20. $\dfrac{3z - 2}{-5} < 6$
21. $y + 4(2y - 1) \geq y$
22. $m - 2(m - 4) \leq 3m$
23. $-(4 + r) + 2 - 3r < -14$
24. $-(9 + k) - 5 + 4k \geq 4$
25. $-3(z - 6) > 2z - 2$
26. $-2(y + 4) \leq 6y + 16$
27. $\dfrac{2}{3}(3k - 1) \geq \dfrac{3}{2}(2k - 3)$
28. $\dfrac{7}{5}(10m - 1) < \dfrac{2}{3}(6m + 5)$
29. $-\dfrac{1}{4}(p + 6) + \dfrac{3}{2}(2p - 5) < 10$
30. $\dfrac{3}{5}(k - 2) - \dfrac{1}{4}(2k - 7) \leq 3$
31. $3(2x - 4) - 4x < 2x + 3$
32. $7(4 - x) + 5x < 2(16 - x)$
33. $8\left(\dfrac{1}{2}x + 3\right) < 8\left(\dfrac{1}{2}x - 1\right)$
34. $10x + 2(x - 4) < 12x - 10$
35. A student solved the inequality $5x < -20$ by dividing both sides by 5 and reversing the direction of the inequality symbol. His reasoning was that since -20 is a negative number, reversing the direction of the symbol was required. Is this correct? Explain why or why not.
36. Which is the graph of $-2 < x$?

(a) −2

(b)

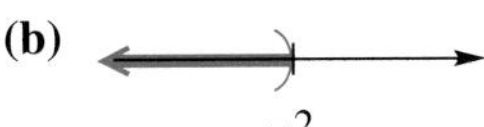

−2

(c)

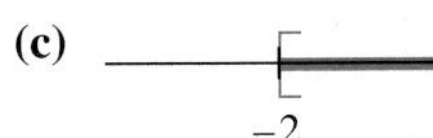

−2

(d)

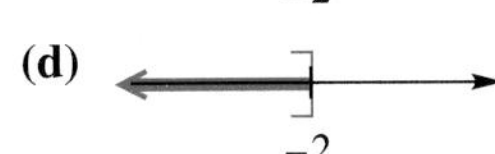

−2

Solve the inequality. Give the solution set in both interval and graph forms. See Example 6.

37. $-4 < x - 5 < 6$
38. $-1 < x + 1 < 8$
39. $-9 \leq k + 5 \leq 15$
40. $-4 \leq m + 3 \leq 10$
41. $-6 \leq 2z + 4 \leq 16$
42. $-15 < 3p + 6 < -12$

43. $-19 \le 3x - 5 \le 1$

44. $-16 < 3t + 2 < -10$

45. $-1 \le \frac{2x - 5}{6} \le 5$

46. $-3 \le \frac{3m + 1}{4} \le 3$

47. $4 \le 5 - 9x < 8$

48. $4 \le 3 - 2x < 8$

Answer the questions in Exercises 49–52 based on the given graph.

49. In which months did the percent of tornadoes exceed 7.7%?

50. In which months was the percent of tornadoes at least 12.9%?

51. The data used to determine the graph were based on the number of tornadoes sighted in the United States during the last twenty years. A total of 17,252 tornadoes were reported. In which months were fewer than 1500 reported?

52. How many more tornadoes occurred during March than October? (Use the total given in Exercise 51.)

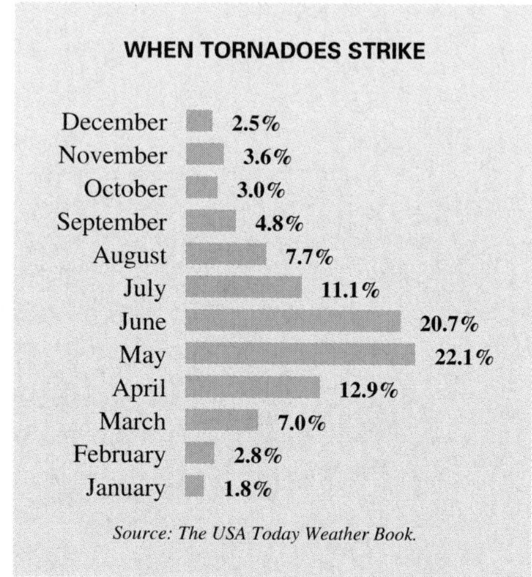

Solve each problem. See Example 7.

53. In a midwestern city, taxicabs charge \$1.50 for the first $\frac{1}{5}$ mile and \$.25 for each additional $\frac{1}{5}$ mile. Dantrell Davis has only \$3.75 in his pocket. What is the maximum distance he can travel (not including a tip for the cabbie)?

54. Eleven years ago taxicab fares in the city in Exercise 53 were \$.90 for the first $\frac{1}{7}$ mile and \$.10 for each additional $\frac{1}{7}$ mile. Based on the information given there and the answer you found, how much farther could Dantrell Davis have traveled at that time?

55. Margaret Westmoreland earned scores of 90 and 82 on her first two tests in English Literature. What score must she make on her third test to keep an average of 84 or greater?

56. Susan Carsten scored 92 and 96 on her first two tests in Methods in Teaching Art. What score must she make on her third test to keep an average of 90 or greater?

57. A couple wishes to rent a car for one day while on vacation. Ford Automobile Rental wants \$35.00 per day and 14¢ per mile, while Chevrolet-For-A-Day wants \$34.00 per day and 16¢ per mile. After how many miles would the price to rent the Chevrolet exceed the price to rent a Ford?

58. Jane and Terry Brandsma went to Long Island for a week. They needed to rent a car, so they checked out two rental firms. Avis wanted \$28 per day, with no mileage fee. Downtown Toyota wanted \$108 per week and 14¢ per mile. How many miles would they have to drive before the Avis price is less than the Toyota price?

A product will produce a profit only when the revenue R from selling the product exceeds the cost C of producing it. In Exercises 59 and 60 find the smallest whole number of units x that must be sold for the business to show a profit for the item described.

59. Peripheral Visions, Inc. finds that the cost to produce x studio quality videotapes is $C = 20x + 100$, while the revenue produced from them is $R = 24x$ (C and R in dollars).

60. Speedy Delivery finds that the cost to make x deliveries is $C = 3x + 2300$, while the revenue produced from them is $R = 5.50x$ (C and R in dollars).

RELATING CONCEPTS (EXERCISES 61-65)

Work Exercises 61–65 in order.

61. Solve the linear equation $5(x + 3) - 2(x - 4) = 2(x + 7)$, and graph the solution on a number line.

62. Solve the linear inequality $5(x + 3) - 2(x - 4) > 2(x + 7)$, and graph the solutions on a number line.

63. Solve the linear inequality $5(x + 3) - 2(x - 4) < 2(x + 7)$, and graph the solutions on a number line.

64. Graph all the solution sets of the equation and inequalities in Exercises 61–63 on the same number line. What set do you obtain?

65. Based on the results of Exercises 61–63, complete the following using a conjecture (educated guess): The solution set of $-3(x + 2) = 3x + 12$ is $\{-3\}$, and the solution set of $-3(x + 2) < 3x + 12$ is $(-3, \infty)$. Therefore the solution set of $-3(x + 2) > 3x + 12$ is ______.

Did you make the connection that the solution sets of the $<$ and $>$ inequalities that correspond to a linear equation are separated by the solution of the equation?

66. Suppose that $-1 < x < 5$. Complete this inequality: ______ $< -x <$ ______.

67. Suppose that $4 < y < 1$. What can you say about y?

68. What is wrong with writing a statement of inequality as $10 > 5 < 8$? What three-part inequality would avoid this problem?

TECHNOLOGY INSIGHTS (EXERCISES 69-72)

The calculator screen shows a graph of the x-values that make the inequality $-3 < 3x + 6 < 18$ true. Thus, from the graph, the solution set is the interval $(-3, 4)$. (Verify this.)

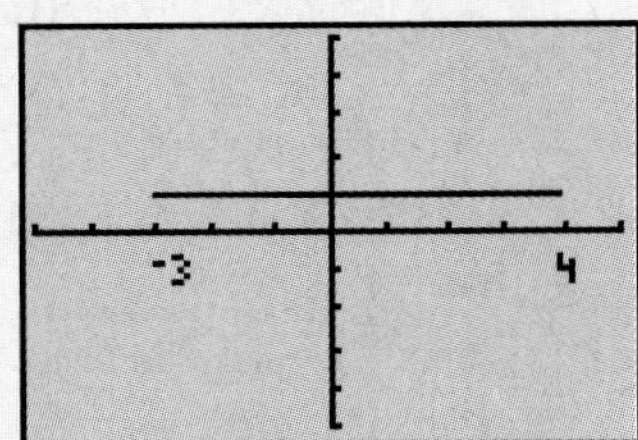

Use each screen to give the solution set of the inequality in interval notation for Exercises 69–72. Note the calculator does not show endpoints, so you must decide from the inequality

(continued)

TECHNOLOGY INSIGHTS (EXERCISES 69–72) (CONTINUED)

whether endpoints are included or excluded. If the graph goes to one side of the screen (as in Exercise 69 or 70), you may assume there is no endpoint in that direction.

69. $\dfrac{2x-5}{-8} \geq 1 - x$

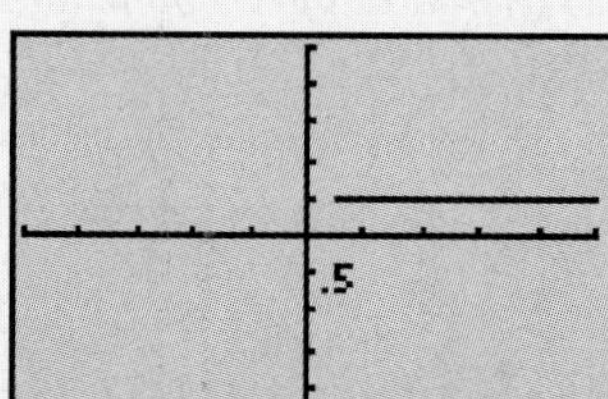

70. $2(x + 5) - 3x + 1 \geq 5$

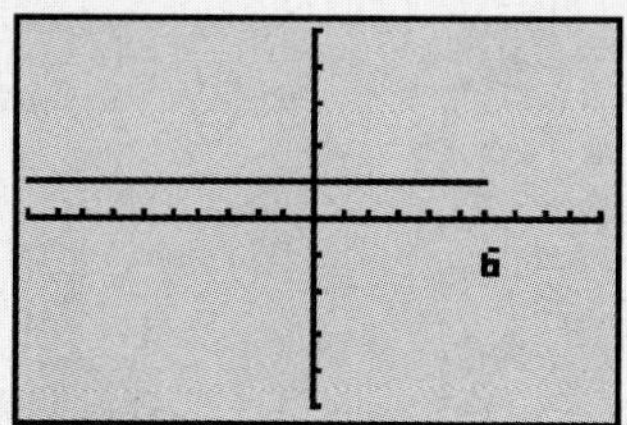

71. $-10 > 3x + 2 > -16$

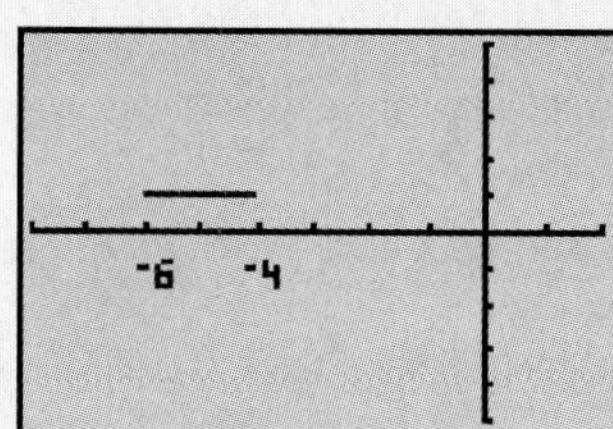

72. $6 < 3x + 3 < 15$

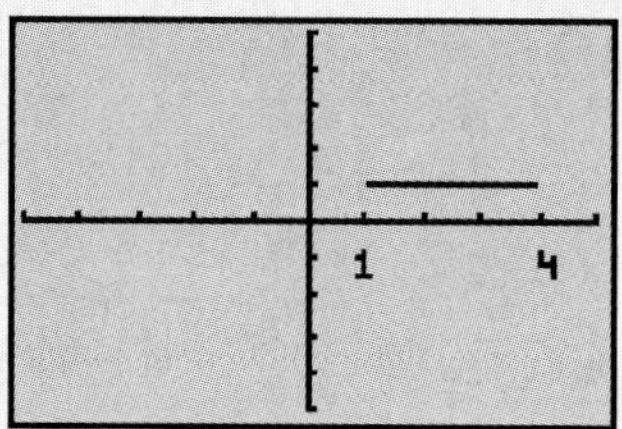

2.6 Set Operations and Compound Inequalities

OBJECTIVES

1. Find the intersection of two sets.
2. Solve compound inequalities with the word *and.*
3. Find the union of two sets.
4. Solve compound inequalities with the word *or.*

FOR EXTRA HELP

SSG Sec. 2.6
SSM Sec. 2.6
Pass the Test Software
InterAct Math Tutorial Software
Video 3

In this section we discuss the use of the words *and* and *or* as they relate to sets and inequalities.

OBJECTIVE 1 Find the intersection of two sets. The intersection of two sets is defined using the word *and.*

Intersection of Sets

For any two sets A and B, the **intersection** of A and B, symbolized $A \cap B$, is defined as follows:

$$A \cap B = \{x \mid x \text{ is an element of } A \textbf{ and } x \text{ is an element of } B\}.$$

EXAMPLE 1 Finding the Intersection of Two Sets

Let $A = \{1, 2, 3, 4\}$ and $B = \{2, 4, 6\}$. Find $A \cap B$.

The set $A \cap B$ contains those elements that belong to both A *and* B at the same time: the numbers 2 and 4. Therefore,

$$A \cap B = \{1, 2, 3, 4\} \cap \{2, 4, 6\}$$
$$= \{2, 4\}.$$

A **compound inequality** consists of two inequalities linked by a connective word such as *and* or *or*. Examples of compound inequalities are

$$x + 1 \le 9 \quad \text{and} \quad x - 2 > 3$$

and

$$2x \ge 4 \quad \text{or} \quad 3x - 6 < 5.$$

OBJECTIVE 2 **Solve compound inequalities with the word *and*.** Use the following steps.

Solving Inequalities with *and*

Step 1 Solve each inequality in the compound inequality individually.

Step 2 Since the inequalities are joined with *and*, the solution will include all numbers that satisfy both solutions in Step 1 at the same time (the intersection of the solution sets).

The next examples show how to solve a compound inequality with *and*.

EXAMPLE 2 Solving a Compound Inequality with *and*

Solve the compound inequality

$$x + 1 \le 9 \quad \text{and} \quad x - 2 \ge 3.$$

Step 1 Solve each inequality in the compound inequality individually.

$$x + 1 \le 9 \quad \text{and} \quad x - 2 \ge 3$$
$$x \le 8 \quad \text{and} \quad x \ge 5$$

Step 2 Since the inequalities are joined with the word *and*, the solution set will include all numbers that satisfy both solutions in Step 1 at the same time. Thus, the compound inequality is true whenever $x \le 8$ and $x \ge 5$ are both true. The top graph in Figure 19 shows $x \le 8$ and the bottom graph shows $x \ge 5$.

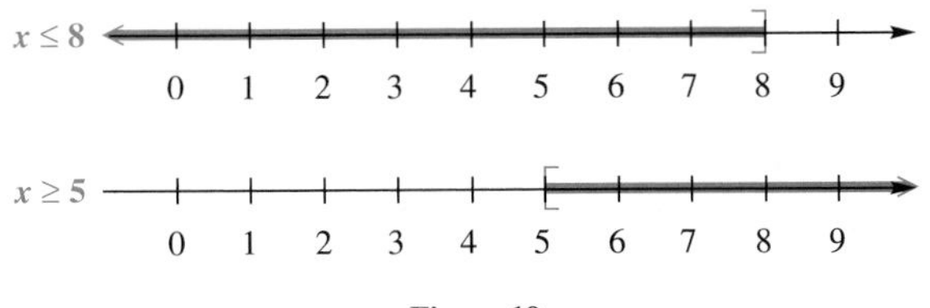

Figure 19

Next, we find the intersection of the two graphs in Figure 19 to get the solution set of the compound inequality. The solution set consists of all numbers between 5 and 8, including both 5 and 8. This is the intersection of the two graphs, written in interval notation as [5, 8]. See Figure 20.

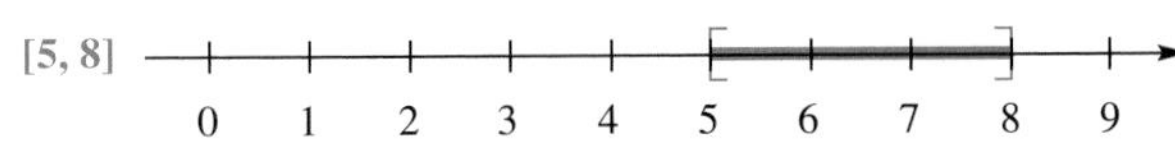

Figure 20

EXAMPLE 3 Solving a Compound Inequality with *and*

Solve the compound inequality

$$-3x - 2 > 4 \quad \text{and} \quad 5x - 1 \le -21.$$

Step 1 Solve each inequality separately.

$$\begin{aligned} -3x - 2 &> 4 & \text{and} \quad 5x - 1 &\le -21 \\ -3x &> 6 & \text{and} \quad 5x &\le -20 \\ x &< -2 & \text{and} \quad x &\le -4 \end{aligned}$$

The graphs of $x < -2$ and $x \le -4$ are shown in Figure 21.

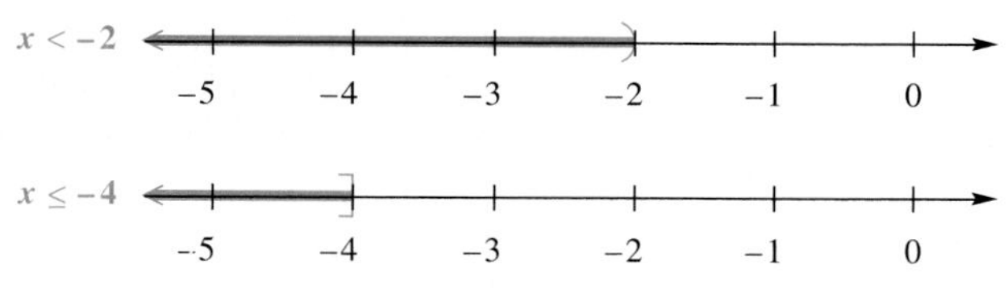

Figure 21

Step 2 Now find all values of x that satisfy both conditions; that is, the real numbers that are less than -2 and also less than or equal to -4. As shown by the graph in Figure 22, the solution set is $(-\infty, -4]$.

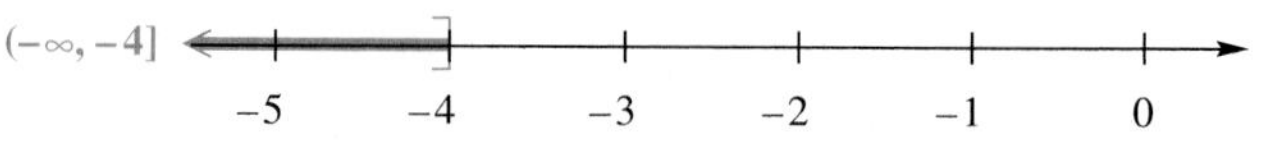

Figure 22

EXAMPLE 4 Solving a Compound Inequality with *and*

Solve $x + 2 < 5$ and $x - 10 > 2$.

First solve each inequality separately.

$$\begin{aligned} x + 2 &< 5 & \text{and} \quad x - 10 &> 2 \\ x &< 3 & \text{and} \quad x &> 12 \end{aligned}$$

The graphs of $x < 3$ and $x > 12$ are shown in Figure 23.

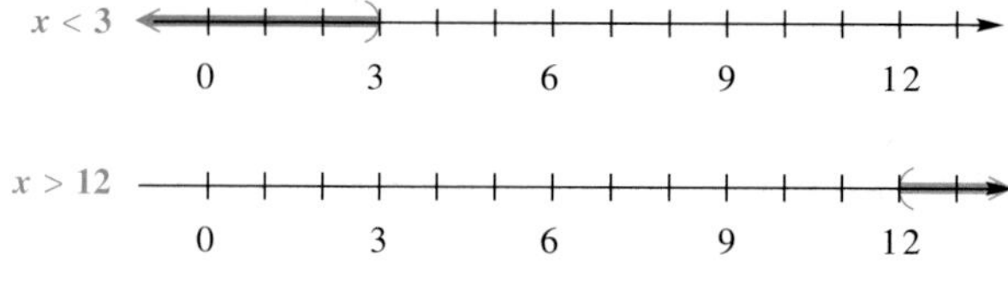

Figure 23

There is no number that is both less than 3 *and* greater than 12, so the given compound inequality has no solution. The solution set is $\emptyset$. See Figure 24.

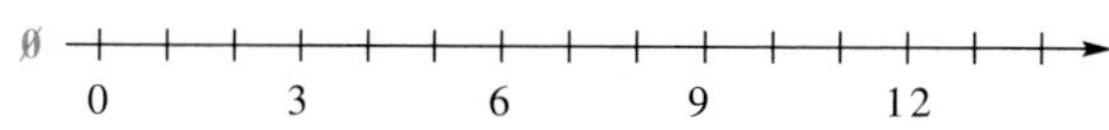

Figure 24

OBJECTIVE 3 **Find the union of two sets.** The union of two sets is defined using the word *or*.

Union of Sets

For any two sets A and B, the **union** of A and B, symbolized $A \cup B$, is defined as follows:

$$A \cup B = \{x \mid x \text{ is an element of } A \textbf{ or } x \text{ is an element of } B\}.$$

EXAMPLE 5 **Finding the Union of Two Sets**

Find the union of the sets $A = \{1, 2, 3, 4\}$ and $B = \{2, 4, 6\}$.

We begin by listing all the elements of set A: 1, 2, 3, 4. Then we list any additional elements from set B. In this case the elements 2 and 4 are already listed, so the only additional element is 6. Therefore,

$$A \cup B = \{1, 2, 3, 4\} \cup \{2, 4, 6\} = \{1, 2, 3, 4, 6\}.$$

The union consists of all elements in either A *or* B (or both).

Notice in Example 5, that even though the elements 2 and 4 appeared in both sets A and B, they are only written once in $A \cup B$.

OBJECTIVE 4 **Solve compound inequalities with the word *or*.** Use the following steps.

Solving Inequalities with *or*

Step 1 Solve each inequality in the compound inequality individually.

Step 2 Since the inequalities are joined with *or*, the solution will include all numbers that satisfy either one or both of the solutions in Step 1 (the union of the solution sets).

The next examples show how to solve a compound inequality with *or*.

EXAMPLE 6 **Solving a Compound Inequality with *or***

Solve $6x - 4 < 2x$ or $-3x \leq -9$.

Solve each inequality separately (Step 1).

$$\begin{aligned} 6x - 4 &< 2x \quad &\text{or} \quad -3x &\leq -9 \\ 4x &< 4 \\ x &< 1 \quad &\text{or} \quad x &\geq 3 \end{aligned}$$

The graphs of these two inequalities are shown in Figure 25.

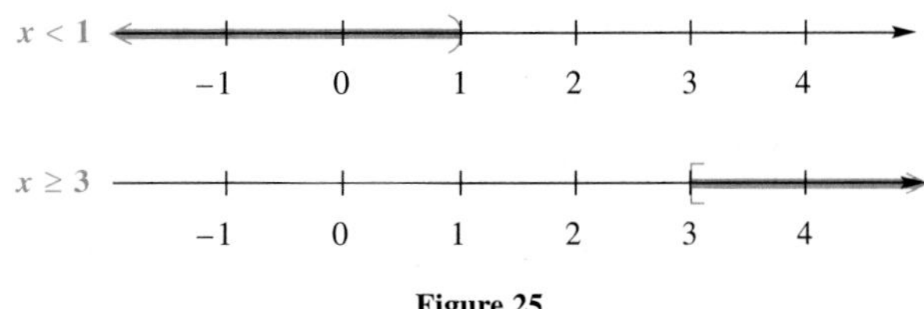

Figure 25

Since the inequalities are joined with *or,* find the union of the two sets (Step 2). The union is shown in Figure 26, and is written $(-\infty, 1) \cup [3, \infty)$.

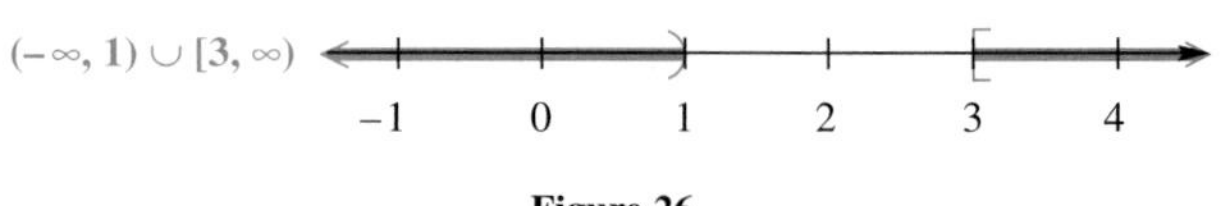

Figure 26

When inequalities are used to write the solution set in Example 6, it *should* be written as

$$x < 1 \text{ or } x \geq 3,$$

which keeps the numbers 1 and 3 in their order on the number line. Writing $3 \leq x < 1$ would imply that $3 \leq 1$, which is **FALSE.** Remember, there is no short-cut way to write the solution of such a union.

EXAMPLE 7 Solving a Compound Inequality with *or*

Solve $-4x + 1 \geq 9$ or $5x + 3 \geq -12$.

Solve each inequality separately.

$$\begin{aligned} -4x + 1 &\geq 9 & \text{or} \quad 5x + 3 &\geq -12 \\ -4x &\geq 8 & \text{or} \quad 5x &\geq -15 \\ x &\leq -2 & \text{or} \quad x &\geq -3 \end{aligned}$$

The graphs of these two inequalities are shown in Figure 27.

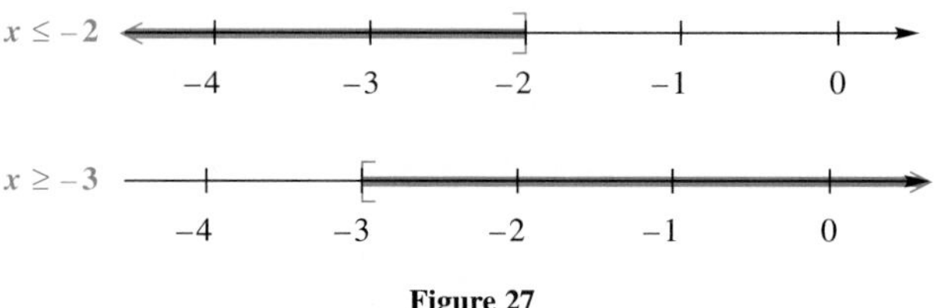

Figure 27

By taking the union, we obtain every real number as a solution, since every real number satisfies at least one of the two inequalities. The set of all real numbers is written in interval notation as $(-\infty, \infty)$ and graphed in Figure 28.

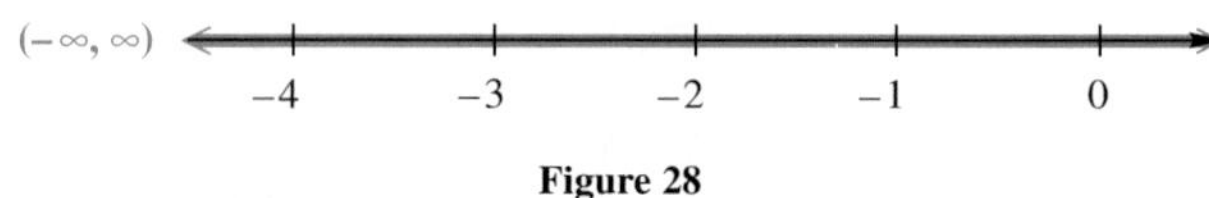

Figure 28

EXAMPLE 8 Applying Intersection and Union

The five highest-grossing films of all time (adjusted for inflation) are listed on the next page. Gross income is given through May 4, 1997.

Five All-Time Highest-Grossing Films

Film	*Admissions*	*Gross Income*
Gone with the Wind	197,548,731	$869,214,425
Star Wars	176,063,374	$774,992,216
E.T.	135,987,938	$598,346,929
The Ten Commandments	131,000,000	$576,400,000
The Sound of Music	130,571,938	$574,514,287

Source: Exhibitor Relations Co.

List the elements of the following sets.

(a) the set of top five films with admissions greater than 170,000,000 *and* gross greater than $800,000,000

The only film that satisfies both conditions is *Gone with the Wind,* so the set is

{*Gone with the Wind*}.

(b) the set of top five films with admissions less than 170,000,000 *or* gross greater than $700,000,000

Here, a film that satisfies at least one of the conditions is in the set. This set includes all five films:

{*Gone with the Wind, Star Wars, E.T., The Ten Commandments, The Sound of Music*}.

2.6 EXERCISES

Decide whether the statement is true or false. If it is false, explain why.

1. The union of the solution sets of $x + 1 = 5$, $x + 1 < 5$, and $x + 1 > 5$ is $(-\infty, \infty)$.
2. The intersection of the sets $\{x \mid x \geq 7\}$ and $\{x \mid x \leq 7\}$ is $\emptyset$.
3. The union of the sets $(-\infty, 8)$ and $(8, \infty)$ is $\{8\}$.
4. The intersection of the sets $(-\infty, 8]$ and $[8, \infty)$ is $\{8\}$.
5. The intersection of the set of rational numbers and the set of irrational numbers is $\{0\}$.
6. The union of the set of rational numbers and the set of irrational numbers is the set of real numbers.

Let $A = \{1, 2, 3, 4, 5, 6\}$, $B = \{1, 3, 5\}$, $C = \{1, 6\}$, and $D = \{4\}$. Specify each of the following sets. See Examples 1 and 5.

7. $B \cap A$	**8.** $A \cap B$	**9.** $A \cap D$	**10.** $B \cap C$
11. $B \cap \emptyset$	**12.** $A \cap \emptyset$	**13.** $A \cup B$	**14.** $B \cup D$
15. $B \cup C$	**16.** $C \cup B$	**17.** $C \cup D$	**18.** $D \cup C$

19. Use the sets *A*, *B*, and *C* for Exercises 7–18 to show that $A \cap (B \cap C)$ is equal to $(A \cap B) \cap C$. This is true for any choices of sets. What property does this illustrate? (*Hint:* See Section 1.3.)
20. Repeat Exercise 19, showing that $A \cup (B \cup C)$ is equal to $(A \cup B) \cup C$.
21. Write a few sentences showing how the concept of intersection can be applied to a real-life situation.

22. A compound inequality uses one of the words *and* or *or.* Explain how you will determine whether to use *intersection* or *union* when graphing the solution set.

Two sets are specified by graphs. Graph the intersection of the two sets.

23.

24. 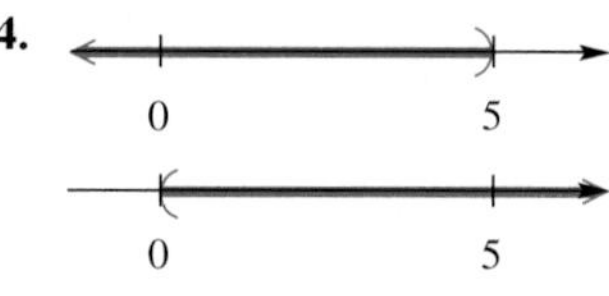

25.

26. 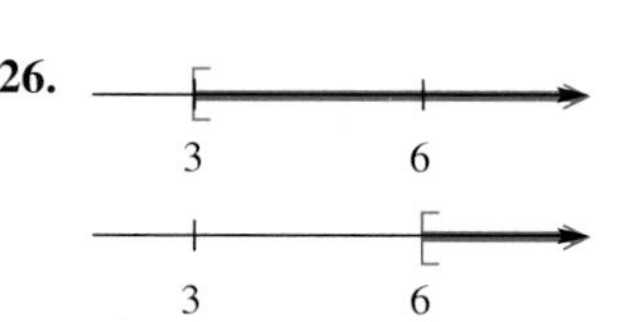

For each compound inequality, give the solution set in both interval and graph forms. See Examples 2–4.

27. $x - 3 \le 6$ and $x + 2 \ge 7$

28. $x + 5 \le 11$ and $x - 3 \ge -1$

29. $-3x > 3$ and $x + 3 > 0$

30. $-3x < 3$ and $x + 2 < 6$

31. $3x - 4 \le 8$ and $-4x + 1 \ge -15$

32. $7x + 6 \le 48$ and $-4x \ge -24$

Two sets are specified by graphs. Graph the union of the two sets.

33.

34. 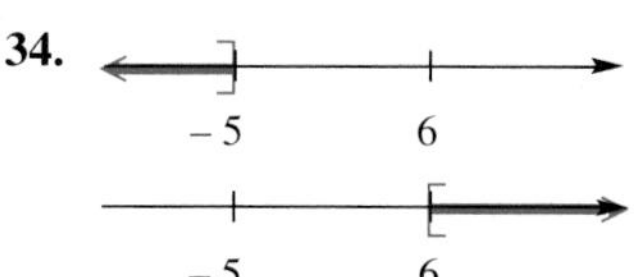

35.

36. 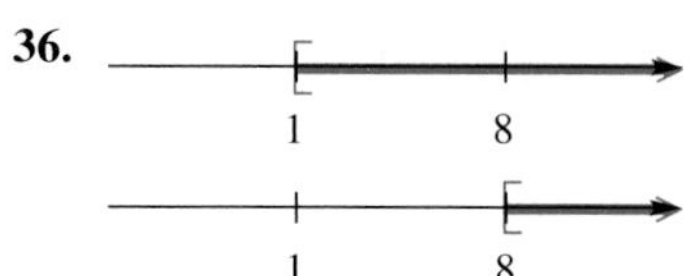

For each compound inequality, give the solution set in both interval and graph forms. See Examples 6 and 7.

37. $x + 2 > 7$ or $1 - x > 6$

38. $x + 1 > 3$ or $x + 4 < 2$

39. $x + 1 > 3$ or $-4x + 1 > 5$

40. $3x < x + 12$ or $x + 1 > 10$

Express each set in the simplest interval form.

41. $(-\infty, -1] \cap [-4, \infty)$

42. $[-1, \infty) \cap (-\infty, 9]$

43. $(-\infty, -6] \cap [-9, \infty)$

44. $(5, 11] \cap [6, \infty)$

45. $(-\infty, 3) \cup (-\infty, -2)$

46. $[-9, 1] \cup (-\infty, -3)$

47. $[3, 6] \cup (4, 9)$

48. $[-1, 2] \cup (0, 5)$

For each compound inequality, decide whether intersection or union should be used. Then give the solution set in both interval and graph forms. See Examples 2, 3, 4, 6, and 7.

49. $x < -1$ and $x > -5$

50. $x > -1$ and $x < 7$

51. $x < 4$ or $x < -2$

52. $x < 5$ or $x < -3$

53. $-3x \le -6$ or $-3x \ge 0$

54. $2x - 6 \le -18$ and $2x \ge -18$

55. $x + 1 \ge 5$ and $x - 2 \le 10$

56. $-8x \le -24$ or $-5x \ge 15$

Use the graphs to answer the questions in Exercises 57 and 58.

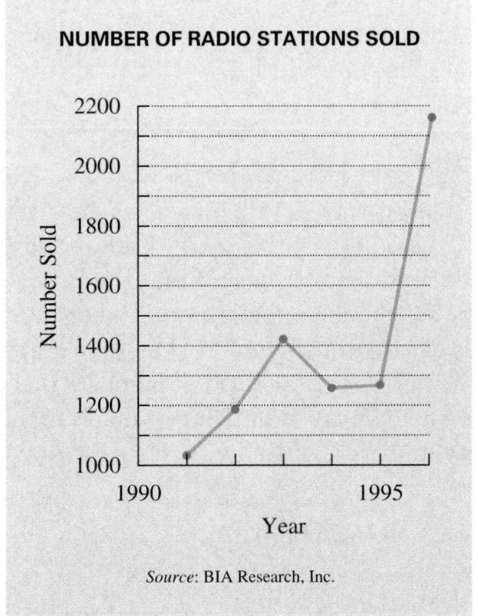

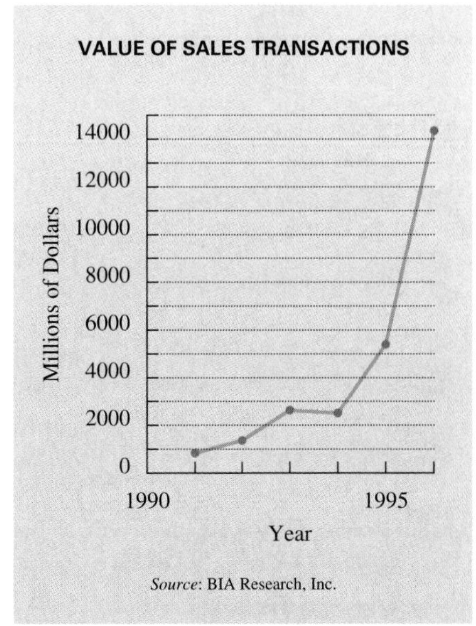

57. In which years did the number of stations sold exceed 1200 *and* the value of transactions exceed 3000 (million dollars)?

58. In which years was the number of stations sold greater than 1300 *or* the value of transactions less than 3000 (million dollars)?

RELATING CONCEPTS (EXERCISES 59–64)

The figures represent the backyards of neighbors Luigi, Maria, Than, and Joe. Find the area and the perimeter of each yard. Suppose that each resident has 150 feet of fencing and enough sod to cover 1400 square feet of lawn. Give the name or names of the residents whose yards satisfy the following descriptions.

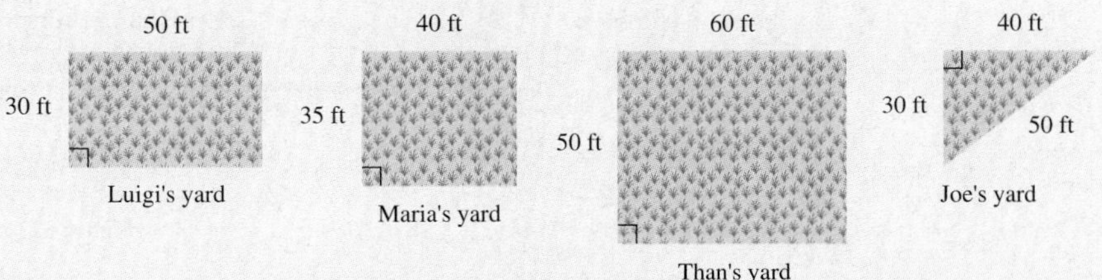

59. The yard can be fenced *and* the yard can be sodded.

60. The yard can be fenced *and* the yard cannot be sodded.

61. The yard cannot be fenced *and* the yard can be sodded.

62. The yard cannot be fenced *and* the yard cannot be sodded.

63. The yard can be fenced *or* the yard can be sodded.

64. The yard cannot be fenced *or* the yard can be sodded.

Did you make the connection that intersections and unions are useful in applied as well as theoretical situations?

2.7 Absolute Value Equations and Inequalities

OBJECTIVES

1. Use the distance definition of absolute value.
2. Solve equations of the form $|ax + b| = k$, for $k > 0$.
3. Solve inequalities of the form $|ax + b| < k$ and of the form $|ax + b| > k$, for $k > 0$.
4. Solve absolute value equations that involve rewriting.
5. Solve equations of the form $|ax + b| = |cx + d|$.
6. Solve special cases of absolute value equations and inequalities.

FOR EXTRA HELP

SSG Sec. 2.7
SSM Sec. 2.7
Pass the Test Software
InterAct Math Tutorial Software
Video 3

Sometimes a model requires that a linear expression must be positive. We use absolute value to ensure that this will happen. Such situations are expressed as absolute value equations or inequalities.

OBJECTIVE 1 Use the distance definition of absolute value. In Chapter 1 we saw that the absolute value of a number x, written $|x|$, represents the distance from x to 0 on the number line. For example, the solutions of $|x| = 4$ are 4 and -4, as shown in Figure 29. Since absolute value represents distance from 0, it is reasonable to interpret the solutions of $|x| > 4$ to be all numbers that are *more* than 4 units from 0. The set $(-\infty, -4) \cup (4, \infty)$ fits this description. Figure 30 shows the graph of the solution set of $|x| > 4$. The solution set of $|x| < 4$ consists of all numbers that are *less* than 4 units from 0 on the number line. Another way of thinking of this is to think of all numbers between -4 and 4. This set of numbers is given by $(-4, 4)$, as shown in Figure 31.

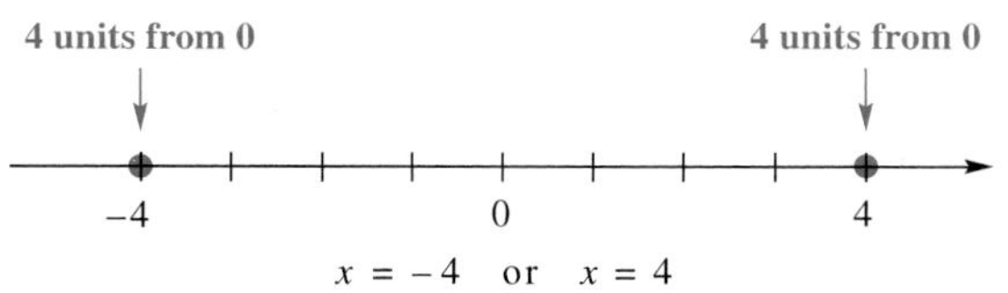

Figure 29

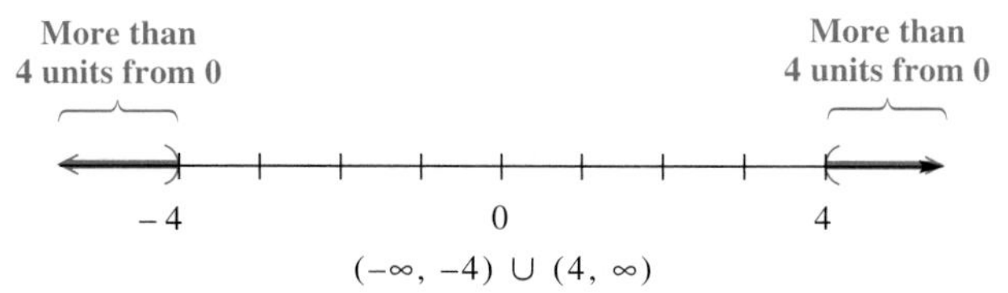

Figure 30

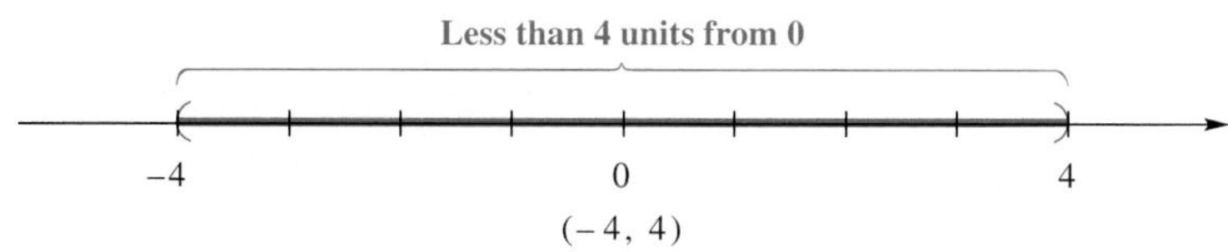

Figure 31

The equation and inequalities just described are examples of **absolute value equations and inequalities.** They involve the absolute value of a variable expression, and generally take the form

$$|ax + b| = k, \qquad |ax + b| > k, \qquad \text{or} \qquad |ax + b| < k,$$

where k is a positive number. We solve them by rewriting them as compound equations or inequalities.

OBJECTIVE 2 Solve equations of the form $|ax + b| = k$, for $k > 0$. The first example shows how to solve a typical absolute value equation. Remember that since absolute value refers to distance from the origin, most absolute value equations will have two cases.

EXAMPLE 1 Solving an Absolute Value Equation

Solve $|2x + 1| = 7$.

For $|2x + 1|$ to equal 7, $2x + 1$ must be 7 units from 0 on the number line. This can happen only when $2x + 1 = 7$ or $2x + 1 = -7$. We solve this compound equation as follows.

$$\begin{aligned} 2x + 1 &= 7 \quad &\text{or}\quad 2x + 1 &= -7 \\ 2x &= 6 &\text{or}\quad 2x &= -8 \\ x &= 3 &\text{or}\quad x &= -4 \end{aligned}$$

The solution set is $\{-4, 3\}$. Its graph is shown in Figure 32.

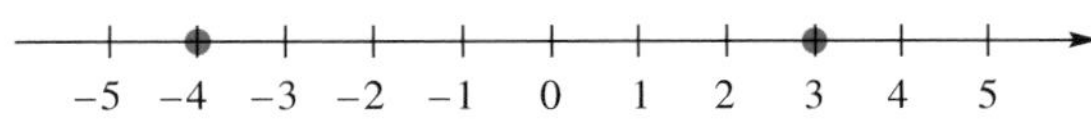

Figure 32

OBJECTIVE 3 Solve inequalities of the form $|ax + b| < k$ and of the form $|ax + b| > k$, for $k > 0$.

EXAMPLE 2 Solving an Absolute Value Inequality with >

Solve $|2x + 1| > 7$.

This absolute value inequality must be rewritten as

$$2x + 1 > 7 \quad \text{or} \quad 2x + 1 < -7,$$

because $2x + 1$ must represent a number that is *more* than 7 units from 0 on either side of the number line. Now we solve the compound inequality.

$$\begin{aligned} 2x + 1 &> 7 \quad &\text{or}\quad 2x + 1 &< -7 \\ 2x &> 6 &\text{or}\quad 2x &< -8 \\ x &> 3 &\text{or}\quad x &< -4 \end{aligned}$$

The solution set, $(-\infty, -4) \cup (3, \infty)$, is graphed in Figure 33. Notice that the graph consists of two intervals.

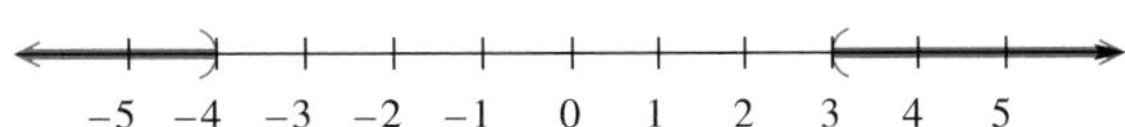

Figure 33

EXAMPLE 3 Solving an Absolute Value Inequality with <

Solve $|2x + 1| < 7$.

Here the expression $2x + 1$ must represent a number that is less than 7 units from 0 on the number line. Another way of thinking of this is to realize that $2x + 1$ must be between -7 and 7. This is written as the three-part inequality

$$-7 < 2x + 1 < 7.$$

We solved such inequalities in Section 2.5 by working with all three parts at the same time.

$$\begin{aligned} -7 < 2x + 1 &< 7 \\ -8 < 2x &< 6 \qquad \text{Subtract 1 from each part.} \\ -4 < x &< 3 \qquad \text{Divide each part by 2.} \end{aligned}$$

The solution set is $(-4, 3)$, and the graph consists of a single interval, as shown in Figure 34.

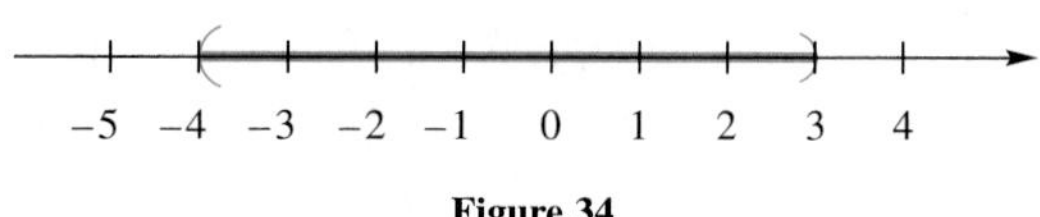

Figure 34

Look back at Figures 32, 33, and 34. These are the graphs of $|2x + 1| = 7$, $|2x + 1| > 7$, and $|2x + 1| < 7$. If we find the union of the three sets, we get the set of all real numbers. This is because for any value of x, $|2x + 1|$ will satisfy one and only one of the following: It is equal to 7, greater than 7, or less than 7.

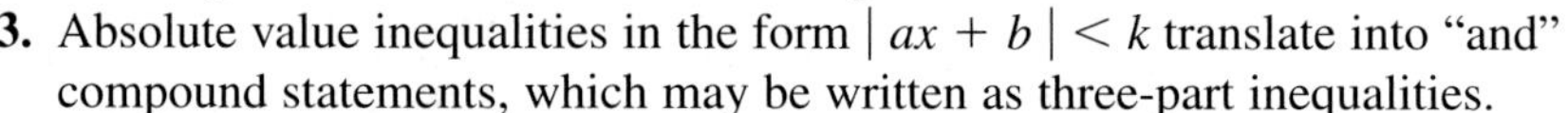

When solving absolute value equations and inequalities of the types in Examples 1, 2, and 3, remember the following.

1. The methods described apply when the constant is alone on one side of the equation or inequality and is *positive.*

2. Absolute value inequalities written in the form $|ax + b| > k$ and absolute value equations translate into "or" compound statements.

CAUTION **3.** Absolute value inequalities in the form $|ax + b| < k$ translate into "and" compound statements, which may be written as three-part inequalities.

4. An "or" statement *cannot* be written in three parts. It would be *incorrect* to use

$$-7 > 2x + 1 > 7$$

in Example 2, since this would imply that $-7 > 7$, which is false.

OBJECTIVE 4 Solve absolute value equations that involve rewriting. Sometimes an absolute value equation or inequality requires some rewriting before it can be set up as a compound statement, as shown in the next example.

EXAMPLE 4 Solving an Absolute Value Equation Requiring Rewriting

Solve the equation $|x + 3| + 5 = 12$.

First get the absolute value expression alone on one side of the equals sign. Do this by subtracting 5 on each side.

$$|x + 3| + 5 - 5 = 12 - 5 \quad \text{Subtract 5.}$$

$$|x + 3| = 7$$

Then use the method shown in Example 1.

$$x + 3 = 7 \quad \text{or} \quad x + 3 = -7$$

$$x = 4 \quad \text{or} \quad x = -10$$

Check that the solution set is $\{4, -10\}$ by substituting in the original equation.

Use a similar method to solve an absolute value *inequality* that requires rewriting.

The methods used in Examples 1, 2, and 3 are summarized here.

Solving Absolute Value Equations and Inequalities

Let k be a positive number, and p and q be two numbers.

1. To solve $|ax + b| = k$, solve the compound equation

$$ax + b = k \quad \text{or} \quad ax + b = -k.$$

The solution set is of the form $\{p, q\}$, with two numbers.

2. To solve $|ax + b| > k$, solve the compound inequality

$$ax + b > k \quad \text{or} \quad ax + b < -k.$$

The solution set is of the form $(-\infty, p) \cup (q, \infty)$, which consists of two separate intervals.

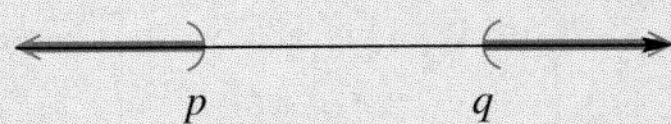

3. To solve $|ax + b| < k$, solve the compound inequality

$$-k < ax + b < k.$$

The solution set is of the form (p, q), a single interval.

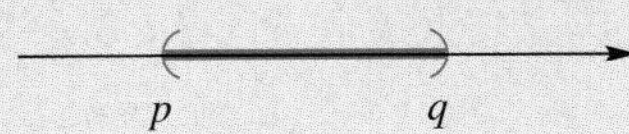

Some people prefer to write the compound statements in parts 1 and 2 of the summary as

$$ax + b = k \quad \text{or} \quad -(ax + b) = k$$

and

$$ax + b > k \quad \text{or} \quad -(ax + b) > k.$$

These forms are equivalent to those we give in the summary and produce the same results.

OBJECTIVE 5 Solve equations of the form $|ax + b| = |cx + d|$. For two expressions to have the same absolute value, they must either be equal or be negatives of each other.

Solving $|ax + b| = |cx + d|$

To solve an absolute value equation of the form

$$|ax + b| = |cx + d|,$$

solve the compound equation

$$ax + b = cx + d \quad \text{or} \quad ax + b = -(cx + d).$$

EXAMPLE 5 Solving an Equation with Two Absolute Values

Solve the equation $|z + 6| = |2z - 3|$.

This equation is satisfied either if $z + 6$ and $2z - 3$ are equal to each other, or if $z + 6$ and $2z - 3$ are negatives of each other. Thus, we have

$$z + 6 = 2z - 3 \quad \text{or} \quad z + 6 = -(2z - 3).$$

Solve each equation.

$$\begin{aligned} 6 + 3 &= 2z - z & \text{or} \quad z + 6 &= -2z + 3 \\ & & 3z &= -3 \\ 9 &= z & \text{or} \quad z &= -1 \end{aligned}$$

The solution set is $\{9, -1\}$.

OBJECTIVE 6 Solve special cases of absolute value equations and inequalities. When a typical absolute value equation or inequality involves a *negative* constant or *zero* alone on one side, use the properties of absolute value to solve. Keep in mind the following.

1. The absolute value of an expression can never be negative: $|a| \geq 0$ for all real numbers a.
2. The absolute value of an expression equals 0 only when the expression is equal to 0.

The next two examples illustrate these special cases.

EXAMPLE 6 Solving Special Cases of Absolute Value Equations

Solve each equation.

(a) $|5r - 3| = -4$

Since the absolute value of an expression can never be negative, there are no solutions for this equation. The solution set is $\emptyset$.

(b) $|7x - 3| = 0$

The expression $7x - 3$ will equal 0 *only* for the solution of the equation

$$7x - 3 = 0.$$

The solution of this equation is $\frac{3}{7}$. The solution set is $\{\frac{3}{7}\}$. It consists of only one element.

EXAMPLE 7 Solving Special Cases of Absolute Value Inequalities

Solve each of the following inequalities.

(a) $|x| \geq -4$

The absolute value of a number is never negative. For this reason, $|x| \geq -4$ is true for *all* real numbers. The solution set is $(-\infty, \infty)$.

(b) $|k + 6| - 3 < -5$

Add 3 to both sides to get the absolute value expression alone on one side.

$$|k + 6| < -2$$

There is no number whose absolute value is less than -2, so this inequality has no solution. The solution set is $\emptyset$.

(c) $|m - 7| + 4 \leq 4$

Adding -4 to both sides gives

$$|m - 7| \leq 0.$$

The value of $|m - 7|$ will never be less than 0. However, $|m - 7|$ will equal 0 when $m = 7$. Therefore, the solution set is $\{7\}$.

CONNECTIONS

Absolute value is used to find the relative error of a measurement in science, engineering, manufacturing, and other fields. If x_t represents the true value of a measurement and x represents the measured value, then the *relative error in x* equals the absolute value of the difference between x_t and x divided by x_t. That is,

$$\text{relative error in } x = \left| \frac{x_t - x}{x_t} \right|.$$

In many situations in the work world, the relative error must be less than some predetermined amount. For example, suppose a machine filling *quart* milk cartons is set for a relative error no greater than .05. Here $x_t = 32$ ounces, the relative error $= .05$ ounce, and we must find x, given

$$\left| \frac{32 - x}{32} \right| = \left| 1 - \frac{x}{32} \right| \leq .05.$$

FOR DISCUSSION OR WRITING

With this tolerance level, how many ounces may a carton contain?

2.7 EXERCISES

Keeping in mind that the absolute value of a number can be interpreted as the distance between the graph of the number and 0 on the number line, match each absolute value equation or inequality with the graph of its solution set.

Choices

1. (a) $|x| = 5$
(b) $|x| < 5$
(c) $|x| > 5$
(d) $|x| \leq 5$
(e) $|x| \geq 5$

A.
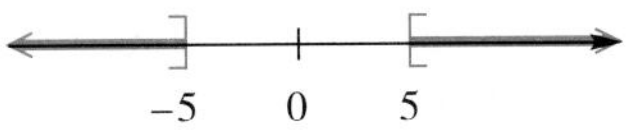

B.
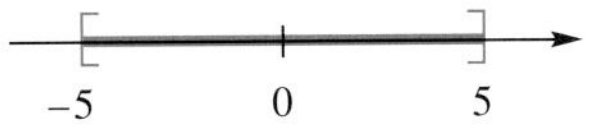

C.
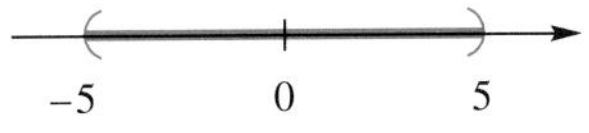

D.
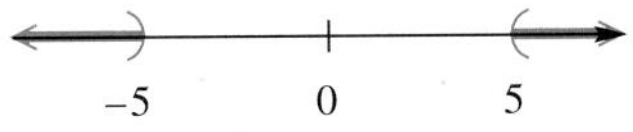

E.
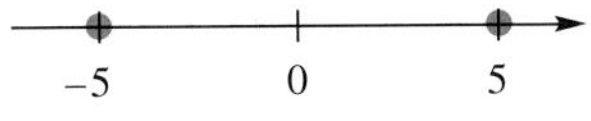

2. **(a)** $|x| = 9$
(b) $|x| > 9$
(c) $|x| \geq 9$
(d) $|x| < 9$
(e) $|x| \leq 9$

A. −9 0 9

B. −9 0 9

C. −9 0 9

D. −9 0 9

E. −9 0 9

3. How many solutions will $|ax + b| = k$, have if **(a)** $k = 0$; **(b)** $k > 0$; **(c)** $k < 0$?

4. The graph of the solution set of $|2x + 1| = 9$ is given here.

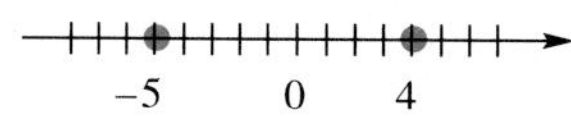

Without actually doing the algebraic work, graph the solution set of each inequality, referring to the graph above.
(a) $|2x + 1| < 9$ **(b)** $|2x + 1| > 9$

Solve each equation. See Example 1.

5. $|x| = 12$ **6.** $|k| = 14$ **7.** $|4x| = 20$ **8.** $|5x| = 30$

9. $|y - 3| = 9$ **10.** $|p - 5| = 13$ **11.** $|2x + 1| = 7$ **12.** $|2y + 3| = 19$

13. $|4r - 5| = 17$ **14.** $|5t - 1| = 21$ **15.** $|2y + 5| = 14$ **16.** $|2x - 9| = 18$

17. $\left|\frac{1}{2}x + 3\right| = 2$ **18.** $\left|\frac{2}{3}q - 1\right| = 5$ **19.** $\left|1 - \frac{3}{4}k\right| = 7$ **20.** $\left|2 - \frac{5}{2}m\right| = 14$

21. Decide when to use *and* and when to use *or* to solve each of the following. Assume k is a positive number.
(a) $|ax + b| = k$ **(b)** $|ax + b| < k$ **(c)** $|ax + b| > k$

22. The graph of the solution set of $|3y - 4| < 5$ is given here.

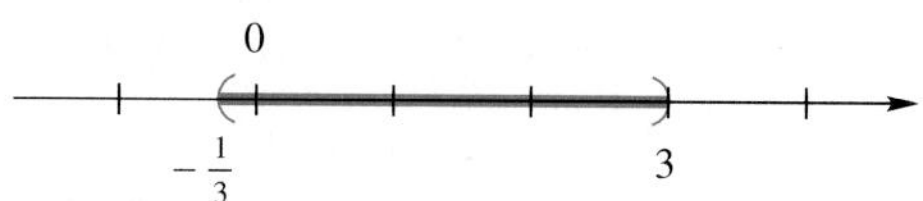

Without actually doing the algebraic work, graph the solution set of each of the following, referring to the graph above.
(a) $|3y - 4| = 5$ **(b)** $|3y - 4| > 5$

Solve each inequality and graph the solution set. See Example 2.

23. $|x| > 3$ **24.** $|y| > 5$ **25.** $|k| \geq 4$ **26.** $|r| \geq 6$

27. $|t + 2| > 10$ **28.** $|4x + 1| \geq 21$ **29.** $|3 - x| > 5$ **30.** $|5 - x| > 3$

Solve each inequality and graph the solution set. See Example 3. (Hint: Compare your answers to those in Exercises 23–30.)

31. $|x| \leq 3$ **32.** $|y| \leq 5$ **33.** $|k| < 4$ **34.** $|r| < 6$

35. $|t + 2| \leq 10$ **36.** $|4x + 1| < 21$ **37.** $|3 - x| \leq 5$ **38.** $|5 - x| \leq 3$

Decide which method you should use to solve each absolute value equation or inequality. Find the solution set and graph. See Examples 1–3.

39. $|-4 + k| > 9$ **40.** $|-3 + t| > 8$ **41.** $|r + 5| > 20$ **42.** $|3x - 1| < 8$

43. $|7 + 2z| = 5$ **44.** $|9 - 3p| = 3$ **45.** $|3r - 1| \le 11$ **46.** $|2s - 6| \le 6$

47. $|-6x - 6| \le 1$ **48.** $|-2x - 6| \le 5$ **49.** $|3x - 1| \ge 8$ **50.** $|r + 5| \le 20$

51. Write an absolute value equation in the variable x that states that the distance between x and 4 is equal to 9.

52. Write an absolute value inequality in the variable x that states that the distance between $2x$ and -3 is **(a)** greater than 4 **(b)** less than 4.

Solve each equation or inequality. Give the solution set in set notation for equations and in interval notation for inequalities. See Example 4.

53. $|x + 4| + 1 = 2$ **54.** $|y + 5| - 2 = 12$ **55.** $|2x + 1| + 3 > 8$

56. $|6x - 1| - 2 > 6$ **57.** $|x + 5| - 6 \le -1$ **58.** $|r - 2| - 3 \le 4$

Solve each equation. See Example 5.

59. $|3x + 1| = |2x + 4|$ **60.** $|7x + 12| = |x - 8|$

61. $\left|m - \frac{1}{2}\right| = \left|\frac{1}{2}m - 2\right|$ **62.** $\left|\frac{2}{3}r - 2\right| = \left|\frac{1}{3}r + 3\right|$

63. $|6x| = |9x + 1|$ **64.** $|13y| = |2y + 1|$

65. $|2p - 6| = |2p + 11|$ **66.** $|3x - 1| = |3x + 9|$

Solve each equation or inequality. See Examples 6 and 7.

67. $|12t - 3| = -8$ **68.** $|13w + 1| = -3$ **69.** $|4x + 1| = 0$

70. $|6r - 2| = 0$ **71.** $|2q - 1| < -6$ **72.** $|8n + 4| < -4$

73. $|x + 5| > -9$ **74.** $|x + 9| > -3$ **75.** $|7x + 3| \le 0$

76. $|4x - 1| \le 0$ **77.** $|5x - 2| \ge 0$ **78.** $|4 + 7x| \ge 0$

79. $|10z + 7| > 0$ **80.** $|4x + 1| > 0$

RELATING CONCEPTS (EXERCISES 81–84)

The ten tallest buildings in Kansas City, Missouri, are listed along with their heights.

Building	*Height (in feet)*
One Kansas City Place	626
AT&T Town Pavilion	590
Hyatt Regency	504
Kansas City Power and Light	476
City Hall	443
Federal Office Building	413
Commerce Tower	402
City Center Square	402
Southwest Bell Telephone	394
Pershing Road Associates	352

Source: The World Almanac and Book of Facts.

(continued)

RELATING CONCEPTS (EXERCISES 81–84) (CONTINUED)

Use the information in the chart to ***work Exercises 81–84 in order.***

81. To find the average of a group of numbers, we add the numbers and then divide by the number of items added. Use a calculator to find the average of the heights.

82. Let k represent your answer for Exercise 81. Then for a height x, the expression $|x - k| < 50$ says that the height is within 50 feet of the average height k. Using your results from Exercise 81, list the buildings that are within 50 feet of the average.

83. Repeat Exercise 82, but find the buildings that are within 75 feet of the average.

84. **(a)** Write an absolute value inequality that describes the height of a building that is *not* within 75 feet of the average.

(b) Solve the inequality you wrote in part (a).

(c) Use the result of part (b) to find the buildings that are not within 75 feet of the average.

(d) Confirm that your answer to part (c) makes sense by comparing it with your answer to Exercise 83.

Did you make the connection that absolute value statements are useful when comparing quantities with an average?

To determine when a manufacturing process is out of control, sample measurements x are compared to the required measurement x_1. Whenever $|x - x_1| \geq k$, for a predetermined value k, the process is out of control. See the Connections box at the end of this section.

85. A box of oatmeal must contain 16 ounces. The machine that fills the boxes is set to fill them within .5 ounce of 16 ounces.

(a) Write an absolute value equation to describe this situation.

(b) Solve the equation for x.

(c) For what values of x is the machine out of control?

86. A machine produces .25-inch bolts with an error of not more than .002 inch. Answer parts (a)–(c) of Exercise 85 for this machine.

SUMMARY Exercises on Solving Linear and Absolute Value Equations and Inequalities

Students often have difficulty distinguishing between the various types of equations and inequalities introduced in this chapter. This section of miscellaneous equations and inequalities provides practice in solving all such types. You might wish to refer to the boxes in this chapter that summarize the various methods of solution.

Solve each equation or inequality.

1. $4z + 1 = 49$

2. $|m - 1| = 6$

3. $6q - 9 = 12 + 3q$

4. $3p + 7 = 9 + 8p$

5. $|a + 3| = -4$

6. $2m + 1 \leq m$

7. $8r + 2 \geq 5r$

8. $4(a - 11) + 3a = 20a - 31$

9. $2q - 1 = -7$

10. $|3q - 7| - 4 = 0$

11. $6z - 5 \leq 3z + 10$

12. $|5z - 8| + 9 \geq 7$

13. $9y - 3(y + 1) = 8y - 7$

14. $|y| \geq 8$

15. $9y - 5 \geq 9y + 3$

16. $13p - 5 > 13p - 8$

17. $|q| < 5.5$

18. $4z - 1 = 12 + z$

19. $\frac{2}{3}y + 8 = \frac{1}{4}y$

20. $-\frac{5}{8}y \geq -20$

21. $\frac{1}{4}p < -6$

22. $7z - 3 + 2z = 9z - 8z$

23. $\frac{3}{5}q - \frac{1}{10} = 2$

24. $|r - 1| < 7$

25. $r + 9 + 7r = 4(3 + 2r) - 3$

26. $6 - 3(2 - p) < 2(1 + p) + 3$

27. $|2p - 3| > 11$

28. $\frac{x}{4} - \frac{2x}{3} = -10$

29. $|5a + 1| \leq 0$

30. $5z - (3 + z) \geq 2(3z + 1)$

31. $-2 \leq 3x - 1 \leq 8$

32. $-1 \leq 6 - x \leq 5$

33. $|7z - 1| = |5z + 3|$

34. $|p + 2| = |p + 4|$

35. $|1 - 3x| \geq 4$

36. $\frac{1}{2} \leq \frac{2}{3}r \leq \frac{5}{4}$

37. $-(m + 4) + 2 = 3m + 8$

38. $\frac{p}{6} - \frac{3p}{5} = p - 86$

39. $-6 \leq \frac{3}{2} - x \leq 6$

40. $|5 - y| < 4$

41. $|y - 1| \geq -6$

42. $|2r - 5| = |r + 4|$

43. $8q - (1 - q) = 3(1 + 3q) - 4$

44. $8y - (y + 3) = -(2y + 1) - 12$

45. $|r - 5| = |r + 9|$

46. $|r + 2| < -3$

47. $2x + 1 > 5$ or $3x + 4 < 1$

48. $1 - 2x \geq 5$ and $7 + 3x \geq -2$

CHAPTER 2 GROUP ACTIVITY

A Trip to the Theater

Objective: Write and solve application problems.

As theater attendance increased on Broadway, people across the country also attended Broadway productions on tour. In 1999 *Les Misérables* went on tour. The following ticket prices are from the March 1999 performance in Albuquerque, New Mexico.

Seating	*Ticket Price*
Balcony	\$31
Floor	\$46
Best floor seats	\$51
Handicapped on floor (may be accompanied by one friend at this price)	\$31
College student	Half price

Source: University of New Mexico Ticket Office.

(continued)

A. Create your application problem.

1. Decide on the number of people in your theater party.
2. Pick ticket prices for two different seatings. Decide how many tickets of each price to use in your problem.
3. Find the total cost for all the tickets used by your theater party.
4. Devise a story to go with your problem.
5. Write the problem. The number of people for each type of seating will be your unknown, so do not include these numbers. Be sure to include the total number of people, the types of seating with costs, and the total cost.

B. Exchange problems with another group. Solve each other's problems, using the six-step problem solving method from this chapter. Show each step.

C. Return papers. Check to see whether or not the other group solved your problem correctly.

CHAPTER 2 SUMMARY

KEY TERMS

2.1 **mathematical model**
algebraic expression
equation
linear equation in one variable (first-degree equation)
solution
solution set
equivalent equations
conditional equation
contradiction
identity

2.2 **formula**

2.3 **right angle**
degree
straight angle
complementary angles (complements)
supplementary angles (supplements)

2.5 **inequality**
linear inequality in one variable
equivalent inequalities

2.6 **intersection**
compound inequality
union

2.7 **absolute value equation**
absolute value inequality

NEW SYMBOLS

$1°$	one degree
$\cap$	set intersection
$\cup$	set union

TEST YOUR WORD POWER

See how well you have learned the vocabulary in this chapter. Answers, with examples, are given at the bottom of the next page.

1. An **algebraic expression** is
(a) an expression that uses any of the four basic operations or the operation of taking roots on any collection of variables and numbers
(b) an expression that contains fractions
(c) an equation that uses any of the four basic operations or the operation of taking roots on any collection of variables and numbers
(d) an equation in algebra.

2. An **equation** is
(a) an algebraic expression
(b) an expression that contains fractions
(c) an expression that uses any of the four basic operations or the operation of taking roots on any collection of variables and numbers
(d) a statement that two algebraic expressions are equal.

3. A **solution set** is the set of numbers that
(a) make an expression undefined
(b) make an equation false
(c) make an equation true
(d) make an expression equal to 0.

4. **Complementary angles** are angles
(a) formed by two intersecting lines
(b) whose sum is 90°
(c) whose sum is 180°
(d) formed by parallel lines.

5. **Supplementary angles** are angles
(a) formed by two intersecting lines
(b) whose sum is 90°
(c) whose sum is 180°
(d) formed by perpendicular lines.

TEST YOUR WORD POWER

6. Interval notation is
- **(a)** a portion of a number line
- **(b)** a special notation for describing a point on a number line
- **(c)** a way to use symbols to describe an interval on a number line
- **(d)** a notation to describe unequal quantities.

7. The **intersection** of two sets A and B is the set of elements that belong
- **(a)** to both A and B
- **(b)** to either A or B, or both
- **(c)** to either A or B, but not both
- **(d)** to just A.

8. The **union** of two sets A and B is the set of elements that belong
- **(a)** to both A and B
- **(b)** to either A or B, or both
- **(c)** to either A or B, but not both
- **(d)** to just B.

QUICK REVIEW

CONCEPTS	EXAMPLES
2.1 LINEAR EQUATIONS IN ONE VARIABLE	
Solving a Linear Equation in One Variable	Solve the equation.
Step 1 Clear fractions.	$4(8 - 3t) = 32 - 8(t + 2)$
Step 2 Simplify each side separately.	$32 - 12t = 32 - 8t - 16$ $32 - 12t = 16 - 8t$
Step 3 Isolate the variable terms on one side.	$32 - 12t + 12t = 16 - 8t + 12t$ $32 = 16 + 4t$ $32 - 16 = 16 + 4t - 16$ $16 = 4t$
Step 4 Isolate the variable.	$\frac{16}{4} = \frac{4t}{4}$ $4 = t$
Step 5 Check.	The solution set is $\{4\}$. This can be checked by substituting 4 for t in the original equation.
2.2 FORMULAS	
Solving a Formula for a Specified Variable	Solve for h: $A = \frac{1}{2}bh$.
Step 1 Get all terms with the specified variable on one side and all terms without that variable on the other side.	$A = \frac{1}{2}bh$
Step 2 If necessary, use the distributive property to combine terms with the specified variable.	$2A = 2\left(\frac{1}{2}bh\right)$ $2A = bh$
Step 3 Divide both sides by the factor that is the coefficient of the specified variable.	$\frac{2A}{b} = h$

Answers to Test Your Word Power

1. (a) *Examples:* $\frac{3y - 1}{2}$, $6 + \sqrt{2x}$, $4a^3b - c$ **2.** (d) *Examples:* $2a + 3 = 7$, $3y = -8$, $x^2 = 4$ **3.** (c) *Example:* $\{8\}$ is the solution set of $2x + 5 = 21$. **4.** (b) *Example:* Angles with measures 35° and 55° are complementary angles. **5.** (c) *Example:* Angles with measures 112° and 68° are supplementary angles. **6.** (c) *Examples:* $(-\infty, 5]$, $(1, \infty)$, $[-3, 3)$ **7.** (a) *Example:* If $A = \{2, 4, 6, 8\}$ and $B = \{1, 2, 3\}$, $A \cap B = \{2\}$. **8.** (b) *Example:* Using the above sets A and B, $A \cup B = \{1, 2, 3, 4, 6, 8\}$.

CONCEPTS	EXAMPLES

2.3 APPLICATIONS OF LINEAR EQUATIONS

Solving an Applied Problem

Step 1 Determine what you are asked to find.

How many liters of 30% alcohol solution and 80% alcohol solution must be mixed to obtain 100 liters of 50% alcohol solution?

Let x = number of liters of 30% solution needed;

$100 - x$ = number of liters of 80% solution needed.

Step 2 Write down any other pertinent information.

Summarize the information of the problem in a chart.

Liters	*Concentration*	*Liters of Pure Alcohol*
x	.30	$.30x$
$100 - x$	.80	$.80(100 - x)$
100	.50	$.50(100)$

Step 3 Write an equation.

The equation is

$$.30x + .80(100 - x) = .50(100).$$

Step 4 Solve the equation.
Step 5 Answer the question(s) of the problem.
Step 6 Check.

Solving this equation gives $x = 60$; 60 liters of 30% alcohol and $100 - 60 = 40$ liters of 80% alcohol should be used. Check this result.

2.4 MORE APPLICATIONS OF LINEAR EQUATIONS

To solve a uniform motion problem, draw a sketch and make a chart. Use the formula $d = rt$.

Two cars start from towns 400 miles apart and travel toward each other. They meet after 4 hours. Find the speed of each car if one travels 20 miles per hour faster than the other.

Let x = speed of the slower car in miles per hour;

$x + 20$ = speed of the faster car.

Use the information in the problem, and $d = rt$ to complete the chart.

	r	t	d
Slower car	x	4	$4x$
Faster car	$x + 20$	4	$4(x + 20)$

A sketch shows that the sum of the distances, $4x$ and $4(x + 20)$, must be 400.

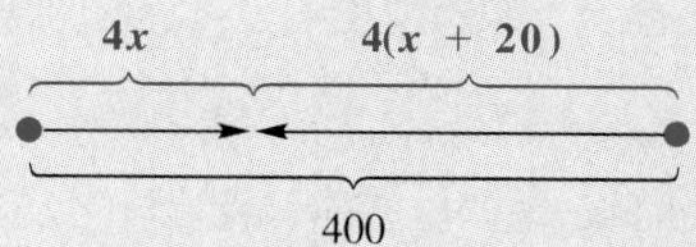

The equation is

$$4x + 4(x + 20) = 400.$$

Solving this equation gives $x = 40$. The slower car travels 40 miles per hour and the faster car travels $40 + 20 = 60$ miles per hour.

CONCEPTS	EXAMPLES

2.5 LINEAR INEQUALITIES IN ONE VARIABLE

Solving Linear Inequalities in One Variable

Step 1 Simplify each side of the inequality by clearing parentheses and combining like terms.

Step 2 Get all terms with variables on one side and all terms without variables on the other side.

Step 3 Write the inequality in the form $x < k$ or $x > k$.

If an inequality is multiplied or divided by a *negative* number, the inequality symbol *must be reversed.*

Solve $3(x + 2) - 5x \leq 12$.

$$3x + 6 - 5x \leq 12$$
$$-2x + 6 \leq 12$$
$$-2x \leq 6$$
$$\frac{-2x}{-2} \geq \frac{6}{-2}$$
$$x \geq -3$$

The solution set is $[-3, \infty)$ and is graphed below.

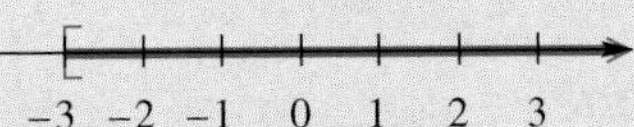

2.6 SET OPERATIONS AND COMPOUND INEQUALITIES

Solving a Compound Inequality

Step 1 Solve each inequality in the compound inequality individually.

Step 2 If the inequalities are joined with *and,* the solution set is the intersection of the two individual solution sets. If the inequalities are joined with *or,* the solution set is the union of the two individual solution sets.

Solve $x + 1 > 2$ and $2x < 6$.

$$x + 1 > 2 \quad \text{and} \quad 2x < 6$$
$$x > 1 \quad \text{and} \quad x < 3$$

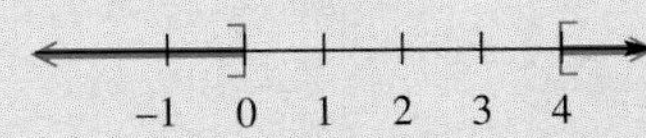

The solution set is $(1, 3)$.

Solve $x \geq 4$ or $x \leq 0$.

The solution set is $(-\infty, 0] \cup [4, \infty)$.

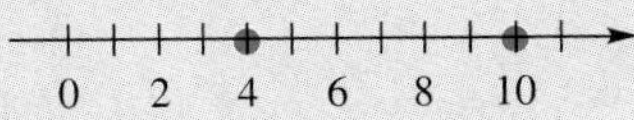

2.7 ABSOLUTE VALUE EQUATIONS AND INEQUALITIES

Solving Absolute Value Equations and Inequalities

Let k be a positive number.

To solve $|ax + b| = k$, solve the compound equation

$$ax + b = k \quad \text{or} \quad ax + b = -k.$$

Solve $|x - 7| = 3$.

$$x - 7 = 3 \quad \text{or} \quad x - 7 = -3$$
$$x = 10 \quad \text{or} \quad x = 4$$

The solution set is $\{4, 10\}$.

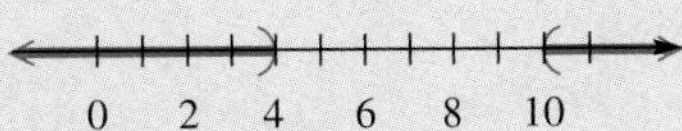

To solve $|ax + b| > k$, solve the compound inequality

$$ax + b > k \quad \text{or} \quad ax + b < -k.$$

Solve $|x - 7| > 3$.

$$x - 7 > 3 \quad \text{or} \quad x - 7 < -3$$
$$x > 10 \quad \text{or} \quad x < 4$$

The solution set is $(-\infty, 4) \cup (10, \infty)$.

To solve $|ax + b| < k$, solve the compound inequality

$$-k < ax + b < k.$$

Solve $|x - 7| < 3$.

$$-3 < x - 7 < 3$$
$$4 < x < 10 \qquad \text{Add 7.}$$

(continued)

CONCEPTS	EXAMPLES
	The solution set is $(4, 10)$. 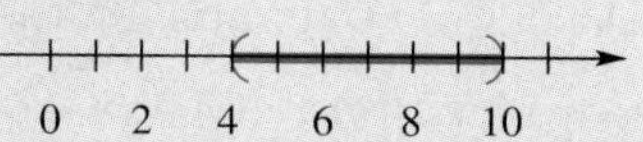
To solve an absolute value equation of the form $\lvert ax + b \rvert = \lvert cx + d \rvert$ solve the compound equation $ax + b = cx + d$ or $ax + b = -(cx + d)$.	Solve $\lvert x + 2 \rvert = \lvert 2x - 6 \rvert$. $x + 2 = 2x - 6$ or $x + 2 = -(2x - 6)$; $x = 8$ or $x = \frac{4}{3}$. The solution set is $\left\{\frac{4}{3}, 8\right\}$.

CHAPTER 2 REVIEW EXERCISES

[2.1] *Solve each equation.*

1. $-(8 + 3y) + 5 = 2y + 6$
2. $-\frac{3}{4}x = -12$
3. $\frac{2q + 1}{3} - \frac{q - 1}{4} = 0$
4. $5(2x - 3) = 6(x - 1) + 4x$

Decide whether the given equation is conditional, an identity, or a contradiction. Give the solution set.

5. $7r - 3(2r - 5) + 5 + 3r = 4r + 20$
6. $8p - 4p - (p - 7) + 9p + 6 = 12p - 7$
7. $-2r + 6(r - 1) + 3r - (4 - r) = -(r + 5) - 5$

[2.2] *Solve each formula for the specified variable.*

8. $V = LWH$; for H
9. $A = \frac{1}{2}(B + b)h$; for b

Solve each equation for x.

10. $M = -\frac{1}{4}(x + 3y)$
11. $P = \frac{3}{4}x - 12$

Find the unknown value.

12. The area of a mural that is in the shape of a rectangle is 132 square feet. Its length is 16.5 feet. What is its width?

16.5 feet

13. If Kyung Ho deposits \$1200 at 3% simple annual interest, how long will it take for the deposit to earn \$126?
14. Find the simple interest rate that Francesco Castellucio is getting, if a principal of \$30,000 earns \$7800 interest in 4 years.
15. If a child has a fever of 40 degrees Celsius, what is the child's temperature in Fahrenheit?

16. The formula for the area of a trapezoid is $A = \frac{1}{2}h(B + b)$. Suppose that you know the values of B, b, and h. Write *in words* the procedure you would use to find A. It may begin as follows: "To find the area of this trapezoid, I would first add . . . ."

[2.3] *Write each word phrase as a mathematical expression, using x as the variable.*

17. the difference between 9 and twice a number

18. the product of 4 and a number, subtracted from 8

Solve each problem.

19. A number is decreased by 35%, giving 260. Find the number.

20. According to the Wilderness Society, in the early 1990s there were 1631 breeding pairs of northern spotted owls in California and Washington. California had 289 more pairs than Washington. How many pairs were there in each state?

21. Segments of the U.S. weight loss and diet control market expanded dramatically between 1996 and 1999. The health club market rose from \$16.18 billion to \$17.76 billion. What percent increase does this represent?

22. Kevin Connors invested some money at 6% and \$4000 less than this amount at 4%. Find the amount invested at each rate if his total annual interest income is \$840.

23. How many liters of a 20% solution of a chemical should be mixed with 15 liters of a 50% solution to get a mixture that is 30% chemical?

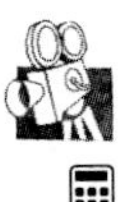

24. The number (in millions) of each of several satellite TV systems in use as of July 1997 is shown in the pie chart.
(a) To the nearest tenth of a percent, what percent of the total systems in use were Primestar?
(b) Which system has the smallest market share?

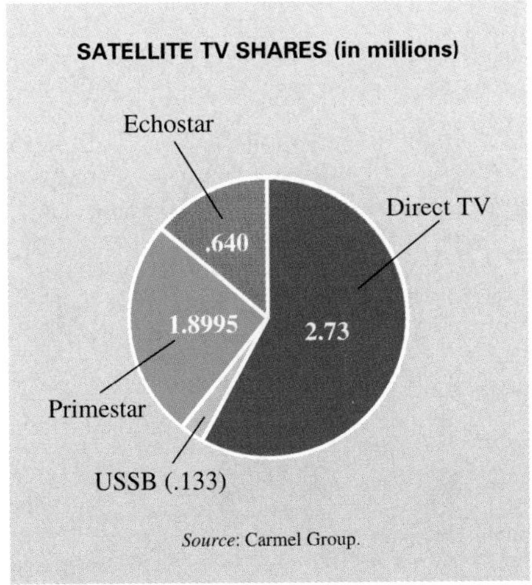

25. The complement of an angle measures 10° less than one-fifth of its supplement. Find the measure of the angle.

26. Find the measure of each angle in the triangle.

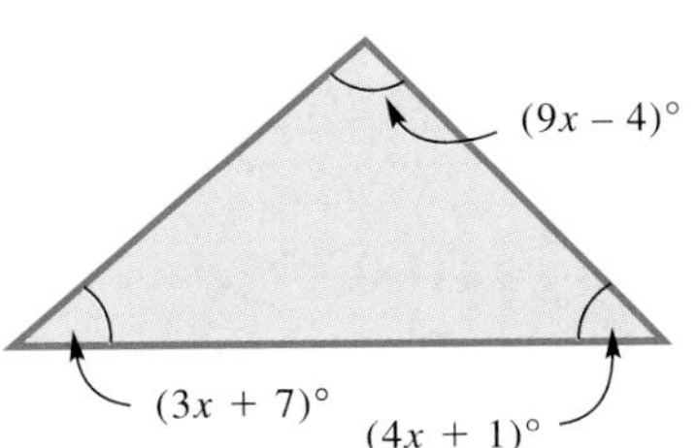

27. Find the measure of each marked angle.

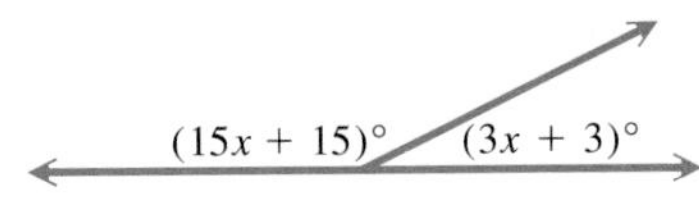

[2.4] *Solve each problem.*

28. The hit movie *Titanic* earned more in Europe than in the United States. As of September 1998, the movie had earned about $686 million in Europe. The average movie ticket in the United States is $4.59. In London, it costs $10.59 and in Frankfurt, $8.42. If 74,000,000 tickets were sold in London and Frankfurt, how many would need to be sold in each country to earn $686,000,000? Round answers to the nearest million. (*Source: Parade Magazine,* September 13, 1998, p. 25.)

29. Suzanne Gainey, who runs a candy shop, wishes to mix 30 pounds of candy worth $6 per pound with candy worth $3 per pound to get a mixture worth $5 per pound. How much of the $3 candy should be used?

30. Which of the following choices is the best *estimate* for the average speed on a trip of 405 miles that lasted 8.2 hours?

(a) 50 miles per hour **(b)** 30 miles per hour
(c) 60 miles per hour **(d)** 40 miles per hour

31. An 85-mile trip to the beach took the Rodriguez family 2 hours. During the second hour, a rainstorm caused them to average 7 miles per hour less than their speed during the first hour. Find their average speed for the first hour.

32. Two cars leave towns 230 kilometers apart at the same time, traveling directly toward one another. One car travels 15 kilometers per hour slower than the other. They pass one another 2 hours later. What are their speeds?

33. The perimeter of a triangle is 34 inches. The middle side is twice as long as the shortest side. The longest side is 2 inches less than three times the shortest side. Find the lengths of the three sides.

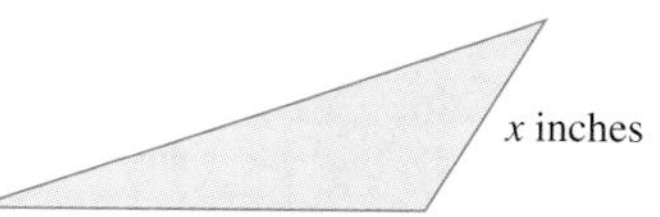

34. The sum of two consecutive integers is 105. What are the integers?

[2.5] *Solve each inequality. Express the solution set in interval form.*

35. $-\frac{2}{3}k < 6$ **36.** $-5x - 4 \geq 11$ **37.** $\frac{6a + 3}{-4} < -3$

38. $5 - (6 - 4k) \geq 2k - 7$ **39.** $8 \leq 3y - 1 < 14$ **40.** $\frac{5}{3}(m - 2) + \frac{2}{5}(m + 1) > 1$

41. To pass algebra, a student must have an average of at least 70% on five tests. On the first four tests, a student has grades of 75%, 79%, 64%, and 71%. What possible grades on the fifth test would guarantee a passing grade in the class?

42. While solving the inequality $10x + 2(x - 4) < 12x - 13$, a student did all the work correctly and obtained the statement $-8 < -13$. The student did not know what to do at this point, because the variable "disappeared." How would you help the student interpret this result?

[2.6] *Let* $A = \{a, b, c, d\}$, $B = \{a, c, e, f\}$, *and* $C = \{a, e, f, g\}$. *Find the set.*

43. $A \cap B$ **44.** $A \cap C$ **45.** $B \cup C$ **46.** $A \cup C$

Solve each compound inequality. Graph the solution set.

47. $x \leq 4$ and $x < 3$ **48.** $x + 4 > 12$ and $x - 2 < 1$

49. $x > 5$ or $x \leq -1$ **50.** $x - 4 > 6$ or $x + 3 \leq 18$

Express each union or intersection in simplest interval form.

51. $(-3, \infty) \cap (-\infty, 4)$ **52.** $(-\infty, 6) \cap (-\infty, 2)$

53. $(4, \infty) \cup (9, \infty)$ **54.** $(1, 2) \cup (1, \infty)$

55. According to the Bureau of Labor Statistics, the following are the median weekly earnings of full-time workers by occupation for men and women.

Occupation	*Men*	*Women*
Managerial and professional specialty	\$ 852	\$616
Mathematical and computer scientists	\$1005	\$754
Waiters and waitresses	\$ 300	\$264
Bus drivers	\$ 482	\$354

Give the occupation that satisfies each description.

(a) The median earnings for men are less than \$900 *and* for women are greater than \$500.

(b) The median earnings for men are greater than \$900 *or* for women are greater than \$600.

[2.7] *Solve each absolute value equation.*

56. $|x| = 7$

57. $|3k - 7| = 8$

58. $|z - 4| = -12$

59. $|4a + 2| - 7 = -3$

60. $|3p + 1| = |p + 2|$

61. $|2m - 1| = |2m + 3|$

Solve each absolute value inequality. Give the solution set in interval form.

62. $|-y + 6| \leq 7$

63. $|2p + 5| \leq 1$

64. $|x + 1| \geq -3$

65. $|5r - 1| > 9$

66. $|11x - 3| \leq -2$

67. $|11x - 3| \leq 0$

68. Write an inequality that states that the distance between x and 14 is greater than 12.

MIXED REVIEW EXERCISES*

Solve.

69. $5 - (6 - 4k) > 2k - 5$

70. $S = 2HW + 2LW + 2LH$; for L

71. $x < 3$ and $x \geq -2$

72. $-4(3 + 2m) - m = -3m$

73. A newspaper recycling collection bin is in the shape of a box, 1.5 feet wide and 5 feet long. If the volume of the bin is 75 cubic feet, find the height.

74. $|3k + 6| \geq 0$

75. The sum of the smallest and largest of three consecutive integers is 47 more than the middle integer. What are the integers?

76. $|2k - 7| + 4 = 11$

77. $.05x + .03(1200 - x) = 42$

78. $|p| < 14$

79. $\frac{3}{4}(a - 2) - \frac{1}{3}(5 - 2a) < -2$

80. $-4 < 3 - 2k < 9$

81. $-.3x + 2.1(x - 4) \leq -6.6$

82. The supplement of an angle measures 25° more than twice its complement. Find the measure of the angle.

83. A loan has a finance charge of \$450. The loan was scheduled to run for 24 months. Find the unearned interest if the loan is paid off with 5 payments left.

*The order of exercises in this final group does not correspond to the order in which topics occur in the chapter. This random ordering should help you prepare for the chapter test.

84. An automobile averaged 45 miles per hour for the first part of a trip and 50 miles per hour for the second part. If the entire trip took 4 hours and covered 195 miles, for how long was the average speed 45 miles per hour?

85. $|5r - 1| > 14$

86. $x \geq -2$ or $x < 4$

87. How many liters of a 20% solution of a chemical should be mixed with 10 liters of a 50% solution to get a 40% mixture?

88. $|m - 1| = |2m + 3|$

89. $\frac{3y}{5} - \frac{y}{2} = 3$

90. $|m + 3| \leq 1$

91. $|3k - 7| = 4$

92. In the 1940 presidential election, Franklin Roosevelt and Wendell Willkie together received 531 electoral votes. Roosevelt received 367 more votes than Willkie in the landslide. How many votes did each man receive? (*Source:* Congressional Quarterly, Inc., Washington, D.C.)

93. $5(2x - 7) = 2(5x + 3)$

In Exercises 94 and 95, sketch the graph of each solution set.

94. $x > 6$ and $x < 8$

95. $-5x + 1 \geq 11$ or $3x + 5 \geq 26$

96. The solution set of $|3x + 4| = 7$ is shown on the number line.

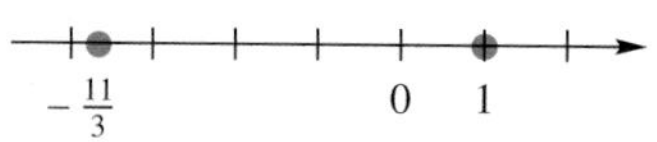

(a) What is the solution set of $|3x + 4| \geq 7$?
(b) What is the solution set of $|3x + 4| \leq 7$?

CHAPTER 2 TEST

Solve each equation.

1. $3(2y - 2) - 4(y + 6) = 3y + 8 + y$

2. $.08x + .06(x + 9) = 1.24$

3. $\frac{x + 6}{10} + \frac{x - 4}{15} = \frac{x + 2}{6}$

4. Decide whether the equation

$$3x - (2 - x) + 4x + 2 = 8x + 3$$

is *conditional*, an *identity*, or a *contradiction*. Give its solution set.

5. Solve for v: $-16t^2 + vt - S = 0$.

Solve each problem.

6. The Daytona 500 (mile) race was shortened to 450 miles in 1974 due to the energy crisis. In that year it was won by Richard Petty, who averaged 140.9 miles per hour. What was Petty's time? (*Source:* NASCAR.)

7. In a certain South Dakota county, 6118 residents live in poverty. This represents 63.1% of the population of the county. What is the population of the county?

8. Moses Tanui from Kenya won the 1996 men's Boston marathon with a rate of 12.19 miles per hour. The women's race was won by Uta Pippig from Germany, who ran at 10.69 miles per hour. Pippig's time was .3 hour longer than Tanui's. Find their winning times. (*Source: The Universal Almanac, 1997,* John W. Wright, General Editor.)

9. Craig Bleyer invested some money at 3% simple interest and some at 5% simple interest. The total amount of his investments was $28,000, and the interest he earned during the first year was $1240. How much did he invest at each rate?

10. Find the measure of each angle.

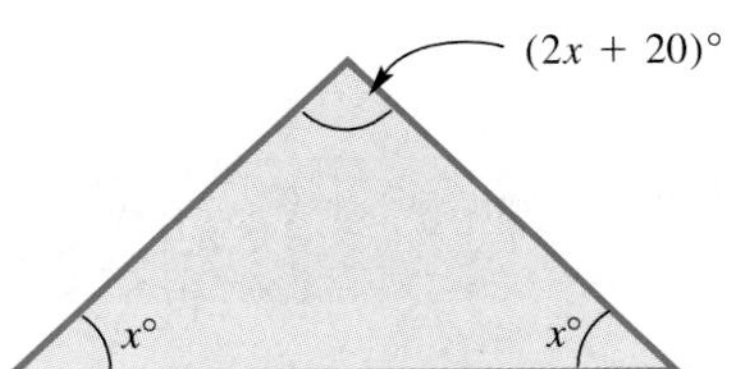

11. The graph shows CD-ROM sales for 1991–1994 in millions.
   (a) How much greater were 1994 sales than 1991 sales?
   (b) Reference books are the most popular use for the CD-ROM format. In 1993, they represented 43% of the 16.5 million CD-ROMs sold that year. About how many CD-ROMs sold in 1993 were reference books?

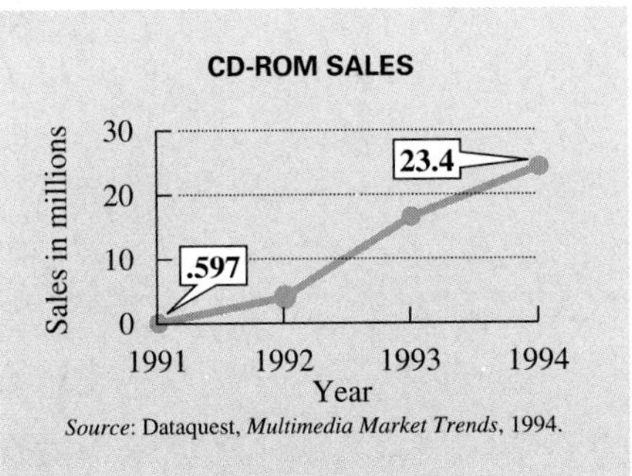

12. What is the special rule that must be remembered when multiplying or dividing both sides of an inequality by a negative number?

Solve each inequality. Give the solution set in both interval and graph forms.

13. $4 - 6(x + 3) \leq -2 - 3(x + 6) + 3x$

14. $-\frac{4}{7}x > -16$

15. $-6 \leq \frac{4}{3}x - 2 \leq 2$

16. Which one of the following inequalities is equivalent to $x < -3$?
   (a) $-3x < 9$ (b) $-3x > -9$ (c) $-3x > 9$ (d) $-3x < -9$

17. The graph shows the number (in millions) of U.S. citizen departures to Europe. For the period shown in the graph, between which two years did the number of departures increase the most? For which two years did they stay the same?

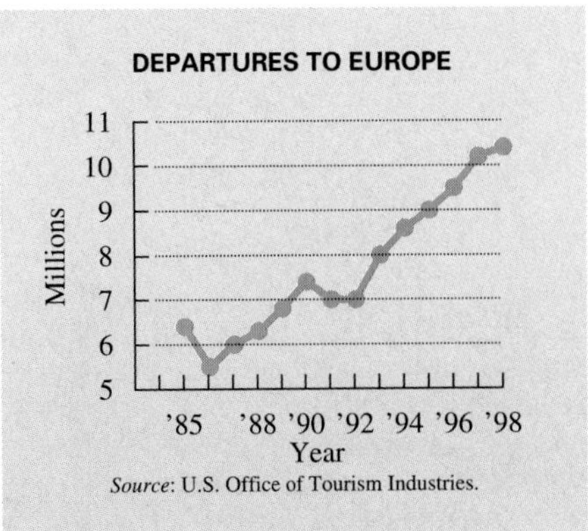

18. Let $A = \{1, 2, 5, 7\}$ and $B = \{1, 5, 9, 12\}$. Find
(a) $A \cap B$ **(b)** $A \cup B$.

19. Solve each compound inequality.
(a) $3k \geq 6$ and $k - 4 < 5$ **(b)** $-4x \leq -24$ or $4x - 2 < 10$

Solve each absolute value equation or inequality.

20. $|4x - 3| = 7$ **21.** $|4x - 3| > 7$ **22.** $|4x - 3| < 7$

23. $|3 - 5x| = |2x + 8|$ **24.** $|-3x + 4| - 4 < -1$

25. If $k < 0$, what is the solution set of
(a) $|5x + 3| < k$ **(b)** $|5x + 3| > k$ **(c)** $|5x + 3| = k$?

CUMULATIVE REVIEW EXERCISES CHAPTERS 1–2

From now on, each chapter will conclude with a set of cumulative review exercises designed to cover the major topics from the beginning of the course. This feature allows the student to constantly review topics that have been introduced up to that point.

Let $A = \left\{-8, -\frac{2}{3}, -\sqrt{6}, 0, \frac{4}{5}, 9, \sqrt{36}\right\}$. *Simplify the elements of A as necessary and then list the elements that belong to the set.*

1. Natural numbers **2.** Whole numbers **3.** Integers

4. Rational numbers **5.** Irrational numbers **6.** Real numbers

Add or subtract, as indicated.

7. $-\frac{4}{3} - \left(-\frac{2}{7}\right)$ **8.** $|-4| - |2| + |-6|$

9. $(-2)^4 + (-2)^3$ **10.** $\sqrt{25} - 5(-1)^0$

Evaluate each expression.

11. $(-3)^5$ **12.** $\left(\frac{6}{7}\right)^3$ **13.** $\left(-\frac{2}{3}\right)^3$ **14.** -4^6

15. Which one of the following is not a real number: $-\sqrt{36}$ or $\sqrt{-36}$?

16. Which one of the following is undefined: $\frac{4 - 4}{4 + 4}$ or $\frac{4 + 4}{4 - 4}$?

Evaluate if $a = 2$, $b = -3$, *and* $c = 4$.

17. $-3a + 2b - c$ **18.** $-2b^2 - 4c$ **19.** $-8(a^2 + b^3)$ **20.** $\frac{3a^3 - b}{4 + 3c}$

Use the properties of real numbers to simplify each expression.

21. $-7r + 5 - 13r + 12$ **22.** $-(3k + 8) - 2(4k - 7) + 3(8k + 12)$

Identify the property of real numbers illustrated by each equation.

23. $(a + b) + 4 = 4 + (a + b)$ **24.** $4x + 12x = (4 + 12)x$

25. $-9 + 9 = 0$

26. What is the reciprocal, or multiplicative inverse, of $-\frac{2}{3}$?

Solve each equation.

27. $-4x + 7(2x + 3) = 7x + 36$

28. $-\frac{3}{5}x + \frac{2}{3}x = 2$

29. $.06x + .03(100 + x) = 4.35$

30. $P = a + b + c$; for b

Solve each inequality. Give the solution set in both interval and graph forms.

31. $3 - 2(x + 7) \leq -x + 3$

32. $-4 < 5 - 3x \leq 0$

33. $2x + 1 > 5$ or $2 - x > 2$

34. $|-7k + 3| \geq 4$

Solve each problem.

35. Kathy Manley invested some money at 7% interest and the same amount at 10%. Her total interest for the year was \$150 less than one-tenth of the total amount she invested. How much did she invest at each rate?

36. A dietician must use three foods, A, B, and C, in a diet. He must include twice as many grams of food A as food C, and 5 grams of food B. The three foods must total at most 24 grams. What is the largest amount of food C that the dietician can use?

37. Laurie Reilly got scores of 88 and 78 on her first two tests. What score must she make on her third test to keep an average of 80 or greater?

38. Jack and Jill are running in the Fresh Water Fun Run. Jack runs at 7 miles per hour and Jill runs at 5 miles per hour. If they start at the same time, how long will it be before Jack is $\frac{1}{4}$ mile ahead of Jill?

39. How much pure alcohol should be added to 7 liters of 10% alcohol to increase the concentration to 30% alcohol?

40. A coin collection contains 29 coins. It consists of pennies, nickels, and quarters. The number of quarters is 4 less than the number of nickels, and the face value of the collection is \$2.69. How many of each denomination are there in the collection?

Clark's rule is a formula used in reducing drug dosage according to weight from the recommended adult dosage to a child dosage. It is as follows.

$$\frac{\text{Weight of child in pounds}}{150} \times \text{adult dose} = \text{child's dose}$$

41. Find a child's dosage if the child weighs 55 pounds and the recommended adult dosage is 120 milligrams.

42. Find a child's dosage if the child weighs 75 pounds and the recommended adult dosage is 40 drops.

43. Since 1975, the number of daily newspapers has steadily declined. According to the graph,

(a) by how much did the number of daily newspapers decrease between 1990 and 1995?

(b) by what *percent* did the number of daily newspapers decrease from 1990 to 1995?

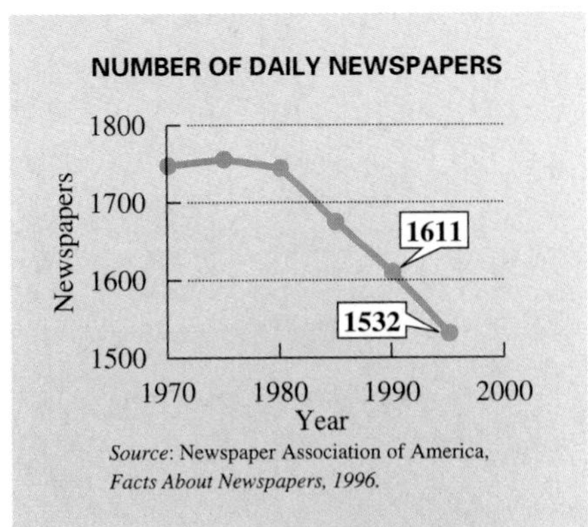

Source: Newspaper Association of America, *Facts About Newspapers, 1996.*

44. The body-mass index, or BMI, of a person is given by the formula

$$\text{BMI} = \frac{704 \times (\text{weight in pounds})}{(\text{height in inches})^2}.$$

Ken Griffey, Jr., is listed as being 6 feet, 3 inches tall and weighing 205 pounds. What is his BMI? (*Source: Readers Digest,* October 1993.)

Graphs, Linear Equations, and Functions

Graphs are widely used in the media because they present a lot of information in an easy-to-understand form. As the saying goes, "A picture is worth a thousand words." The pie graph shows 1996 U.S. energy production by source. Which source produced the most energy in 1996? The line graph shows U.S. energy consumption in the years 1991–1996. By how many quadrillion BTUs did U.S. energy consumption increase from 1991–1996?* We return to these graphs later in the chapter.

Energy

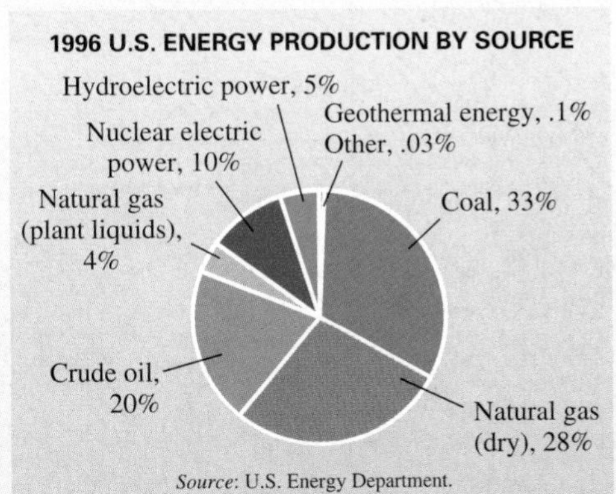

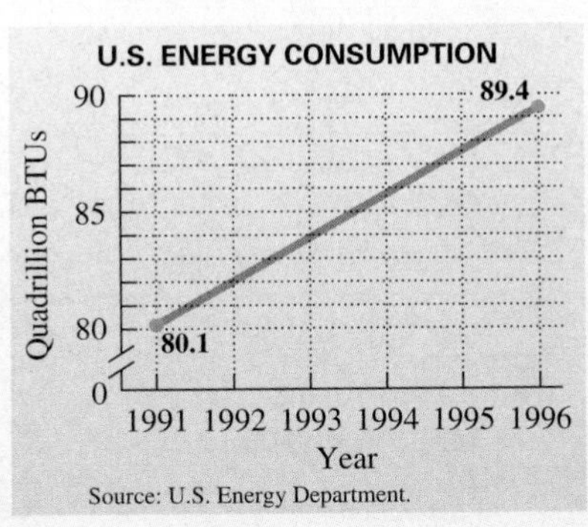

These graphs each depict some aspect of energy in the United States. Many of the applications in this chapter are based on data about energy, the theme of this chapter. We will use this theme as we work with linear equations in two variables, whose graphs are straight lines.

*BTU stands for British Thermal Unit and is a measure of energy consumption.

3.1 The Rectangular Coordinate System

OBJECTIVES

1. Plot ordered pairs.
2. Find ordered pairs that satisfy a given equation.
3. Graph lines.
4. Find x- and y-intercepts.
5. Recognize equations of vertical or horizontal lines.
6. Use a graphing calculator to graph an equation.

FOR EXTRA HELP

SSG Sec. 3.1
SSM Sec. 3.1
Pass the Test Software
InterAct Math Tutorial Software
Video 4

René Descartes, a seventeenth-century French mathematician, is credited with giving us an indispensable method of locating a point on a plane. It seems Descartes, who was lying in bed ill, was watching a fly crawl about on the ceiling near a corner of the room. It occurred to him that the location of the fly on the ceiling could be described by determining its distances from the two adjacent walls. In this chapter, we use this insight to plot points and graph equations in two variables.

OBJECTIVE 1 **Plot ordered pairs.** Each of the pairs of numbers (1, 2), $(-1, 5)$, and (3, 7) is an example of an **ordered pair;** that is, a pair of numbers written within parentheses in which the order of the numbers is important. We graph an ordered pair using two perpendicular number lines that intersect at their zero points, as shown in Figure 1. The common zero point is called the **origin.** The position of any point in this plane is determined by referring to the horizontal number line, the ***x*-axis,** and the vertical number line, the ***y*-axis.** The first number in the ordered pair indicates the position relative to the x-axis, and the second number indicates the position relative to the y-axis. The x-axis and the y-axis make up a **rectangular** (or **Cartesian**, for Descartes) **coordinate system.**

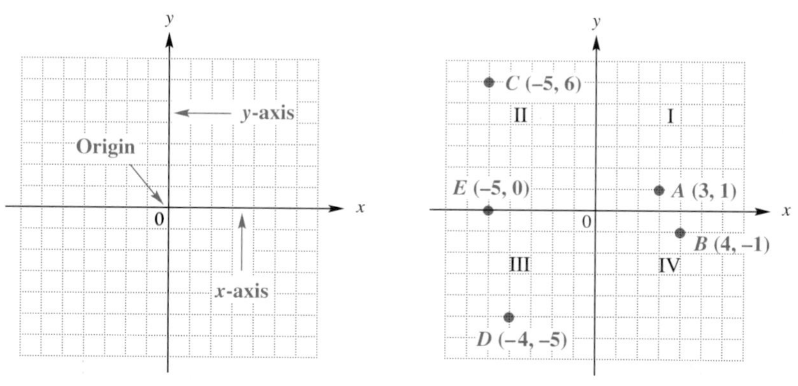

Figure 1 Figure 2

To locate, or **plot,** the point on the graph that corresponds to the ordered pair (3, 1), we move three units from zero to the right along the x-axis, and then one unit up parallel to the y-axis. The point corresponding to the ordered pair (3, 1) is labeled A in Figure 2. The point $(4, -1)$ is labeled B, $(-5, 6)$ is labeled C, and $(-4, -5)$ is labeled D. Point E corresponds to $(-5, 0)$. The phrase "the point corresponding to the ordered pair (3, 1)" is often abbreviated "the point (3, 1)." The numbers in an ordered pair are called the **coordinates** of the corresponding point.

The parentheses used to represent an ordered pair are also used to represent an open interval (introduced in Chapter 1). The context of the discussion tells whether ordered pairs or open intervals are being discussed.

The four regions of the graph shown in Figure 2 are called **quadrants I, II, III,** and **IV,** reading counterclockwise from the upper right quadrant. The points of the x-axis and y-axis themselves do not belong to any quadrant. For example, point E in Figure 2 belongs to no quadrant.

OBJECTIVE 2 Find ordered pairs that satisfy a given equation. Each solution to an equation with two variables includes two numbers, one for each variable. To keep track of which number goes with which variable we write the solutions as ordered pairs, with the x-value given first. For example, we can show that $(6, -2)$ is a solution of $2x + 3y = 6$ by substitution.

$$
\begin{aligned}
2x + 3y &= 6 \\
2(6) + 3(-2) &= 6 \quad ? \qquad \text{Let } x = 6,\ y = -2. \\
12 - 6 &= 6 \quad ? \\
6 &= 6 \qquad \text{True}
\end{aligned}
$$

Since the pair of numbers $(6, -2)$ makes the equation true, it is a solution. On the other hand, since

$$2(5) + 3(1) = 10 + 3 = 13 \neq 6,$$

$(5, 1)$ is not a solution of the equation $2x + 3y = 6$.

To find ordered pairs that satisfy an equation, select any number for one of the variables, substitute it into the equation for that variable, and then solve for the other variable.

Some other ordered pairs satisfying $2x + 3y = 6$ are $(0, 2)$ and $(3, 0)$. Since any real number could be selected for one variable and would lead to a real number for the other variable, equations with two variables usually have an infinite number of solutions.

EXAMPLE 1 Completing Ordered Pairs

Complete the table of ordered pairs for $2x + 3y = 6$.

x	y
-3	
	-4

First let $x = -3$ and substitute into the equation to find y. Then let $y = -4$ and substitute to find x.

$$
\begin{aligned}
2x + 3y &= 6 \\
2(-3) + 3y &= 6 \qquad \text{Let } x = -3. \\
-6 + 3y &= 6 \\
3y &= 12 \\
y &= 4
\end{aligned}
$$

The ordered pair is $(-3, 4)$.

$$
\begin{aligned}
2x + 3y &= 6 \\
2x + 3(-4) &= 6 \qquad \text{Let } y = -4. \\
2x - 12 &= 6 \\
2x &= 18 \\
x &= 9
\end{aligned}
$$

The ordered pair is $(9, -4)$.

These pairs lead to the following completed table.

x	y
-3	4
9	-4

OBJECTIVE 3 **Graph lines.** The **graph of an equation** is the set of points corresponding to all ordered pairs that satisfy the equation. It gives a "picture" of the equation. Since most equations with two variables are satisfied by an infinite number of ordered pairs, their graphs include an infinite number of points. To graph an equation, we plot a number of ordered pairs that satisfy the equation until we have enough points to suggest the shape of the graph. For example, to graph $2x + 3y = 6$, plot all the ordered pairs mentioned above. These are shown in Figure 3(a). The resulting points appear to lie on a straight line. If all the ordered pairs that satisfy the equation $2x + 3y = 6$ were graphed, they would form a straight line. The equation $2x + 3y = 6$ is called a **first-degree** equation, because it has no term with one variable to a power greater than one. The graph of any first-degree equation in two variables is a straight line. The graph of $2x + 3y = 6$ is the line shown in Figure 3(b).

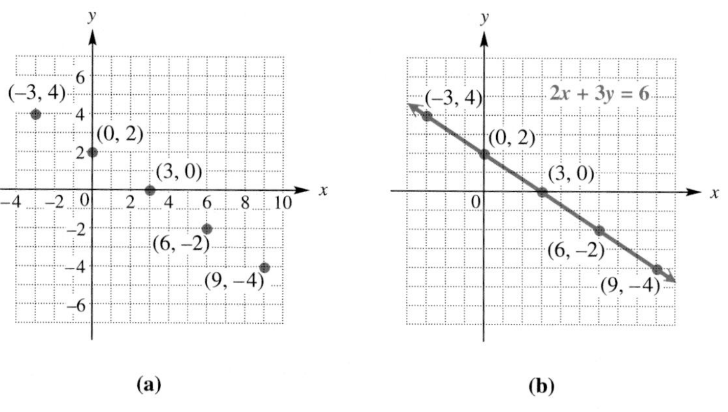

Figure 3

Since first-degree equations with two variables have straight-line graphs, they are called *linear equations in two variables.* (We discussed linear equations in one variable in Chapter 2.)

Standard Form of a Linear Equation in Two Variables

A **linear equation in two variables** can be written in the form

$$Ax + By = C \quad (A \text{ and } B \text{ not both } 0).$$

This form is called **standard form.**

OBJECTIVE 4 **Find x- and y-intercepts.** A straight line is determined if any two different points on the line are known, so finding two different points is enough to graph the line. Two useful points for graphing are the x- and y-intercepts. The ***x*-intercept** is the point (if any) where the line intersects the x-axis; likewise, the ***y*-intercept** is the point (if any) where the line intersects the y-axis. In Figure 3(b), the y-value of the point where the line intersects the x-axis is 0. Similarly, the x-value of the point where the line intersects the y-axis is 0. This suggests a method for finding the x- and y-intercepts.

Intercepts

Let $y = 0$ to find the x-intercept; let $x = 0$ to find the y-intercept.

EXAMPLE 2 Finding Intercepts

Find the x- and y-intercepts of $4x - y = -3$, and graph the equation.

Find the x-intercept by letting $y = 0$.

$$4x - 0 = -3 \quad \text{Let } y = 0.$$
$$4x = -3$$
$$x = -\frac{3}{4} \quad x\text{-intercept is } \left(-\tfrac{3}{4}, 0\right).$$

For the y-intercept, let $x = 0$.

$$4(0) - y = -3 \quad \text{Let } x = 0.$$
$$-y = -3$$
$$y = 3 \quad y\text{-intercept is } (0, 3).$$

The intercepts are the two points $\left(-\frac{3}{4}, 0\right)$ and $(0, 3)$. These ordered pairs are shown in the table with Figure 4. Use these points to draw the graph, as shown in Figure 4.

x	y
$-\frac{3}{4}$	0
0	3

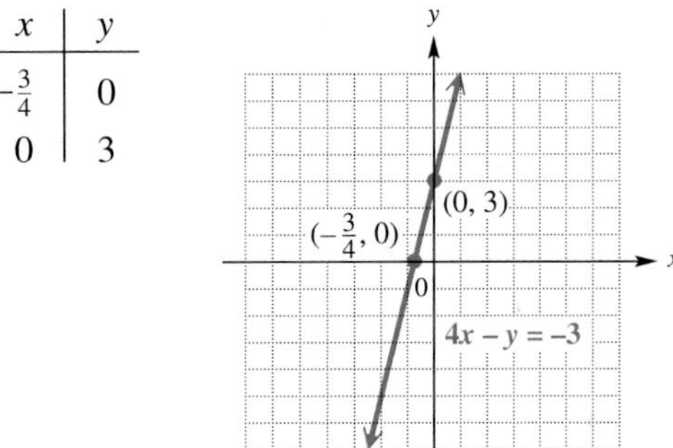

Figure 4

While two points, such as the two intercepts in Figure 4, are sufficient to graph a straight line, it is a good idea to use a third point to guard against errors. Verify that $(-1, -1)$ also lies on the graph of $4x - y = -3$.

OBJECTIVE 5 Recognize equations of vertical or horizontal lines. A graph can fail to have an x-intercept or a y-intercept, which is why the phrase "if any" was added when discussing intercepts.

EXAMPLE 3 Graphing a Horizontal Line

Graph $y = 2$.

Writing $y = 2$ as $0x + 1y = 2$ shows that any value of x, including $x = 0$, gives $y = 2$, making the y-intercept $(0, 2)$. Since y is always 2, there is no value of x corresponding to $y = 0$, and so the graph has no x-intercept. The graph, shown with a table of ordered pairs in Figure 5, is a horizontal line.

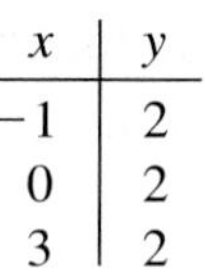

x	y
-1	2
0	2
3	2

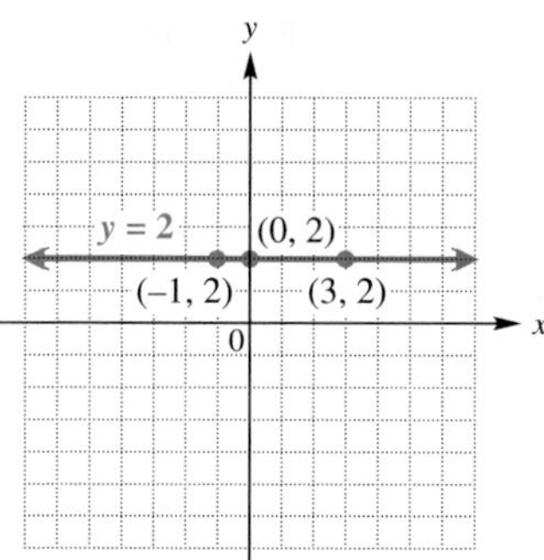

Figure 5

EXAMPLE 4 Graphing a Vertical Line

Graph $x + 1 = 0$.

The form $1x + 0y = -1$ shows that every value of y leads to $x = -1$, and so no value of y makes $x = 0$. The graph, therefore, has no y-intercept. The only way a straight line can have no y-intercept is to be vertical, as shown in Figure 6.

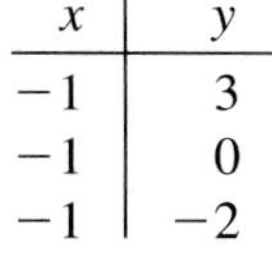

x	y
-1	3
-1	0
-1	-2

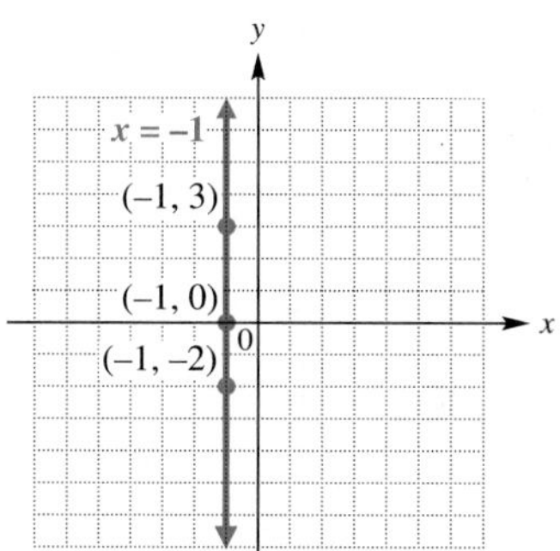

Figure 6

Some lines have both the x-intercept and the y-intercept at the origin.

EXAMPLE 5 Graphing a Line That Passes through the Origin

Graph $x + 2y = 0$.

Find the x-intercept by letting $y = 0$.

$$x + 2y = 0$$

$$x + 2(0) = 0 \qquad \text{Let } y = 0.$$

$$x + 0 = 0$$

$$x = 0 \qquad x\text{-intercept is } (0, 0).$$

To find the y-intercept, let $x = 0$.

$$x + 2y = 0$$

$$0 + 2y = 0 \qquad \text{Let } x = 0.$$

$$y = 0 \qquad y\text{-intercept is } (0, 0).$$

Both intercepts are the same ordered pair, (0, 0). (This means that the graph goes through the origin.) To find another point to graph the line, choose any number for x (or for y), say $x = 4$, and solve for y.

$$
\begin{aligned}
x + 2y &= 0 \\
4 + 2y &= 0 \qquad \text{Let } x = 4. \\
2y &= -4 \\
y &= -2
\end{aligned}
$$

This gives the ordered pair $(4, -2)$. These two points lead to the graph shown in Figure 7.

x	y
0	0
4	-2

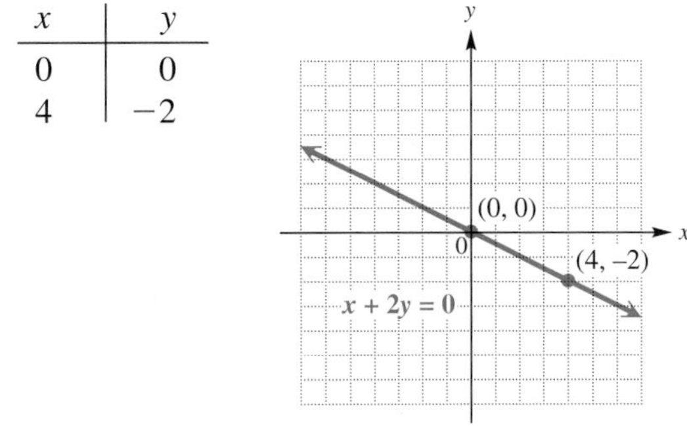

Figure 7

OBJECTIVE 6 **Use a graphing calculator to graph an equation.** When graphing by hand, we first set up a rectangular coordinate system, then plot points and draw the graph. Similarly, when graphing with a graphing calculator, we first tell the calculator how to set up a rectangular coordinate system. This involves choosing the minimum and maximum x- and y-values that will determine the viewing screen. In the screen shown in Figure 8, we have chosen minimum x- and y-values of -10 and maximum x- and y-values of 10. The *scale* on each axis determines the distance between the tick marks; in the view shown, the scale is 1 for both axes. We will refer to this as the *standard window.*

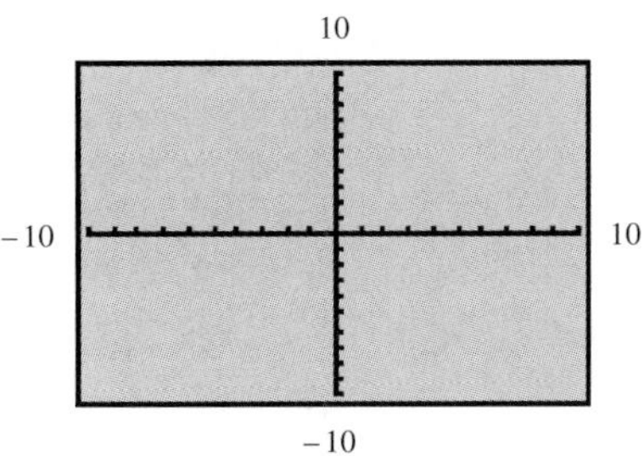

Figure 8

To graph an equation, it is usually necessary to solve the equation for y in order to enter it into the calculator. Once the equation is graphed, graphing calculators allow us to find the intercepts of the graph or any other point on the graph easily.

EXAMPLE 6 Graphing a Linear Equation with a Graphing Calculator and Finding the Intercepts

Use a graphing calculator to graph $4x - y = 3$.

Because we want to be able to see the intercepts on the screen, we can use them to determine an appropriate window. Here, the x-intercept is $(.75, 0)$ and the y-intercept is $(0, -3)$. Although many choices are possible, we choose the *standard window.* We must solve the equation for y to enter it into the calculator.

$$
\begin{aligned}
4x - y &= 3 \\
-y &= -4x + 3 && \text{Subtract } 4x \text{ on each side.} \\
y &= 4x - 3 && \text{Multiply both sides by } -1.
\end{aligned}
$$

The graph is shown in Figures 9 and 10, which also give the intercepts at the bottoms of the screens.

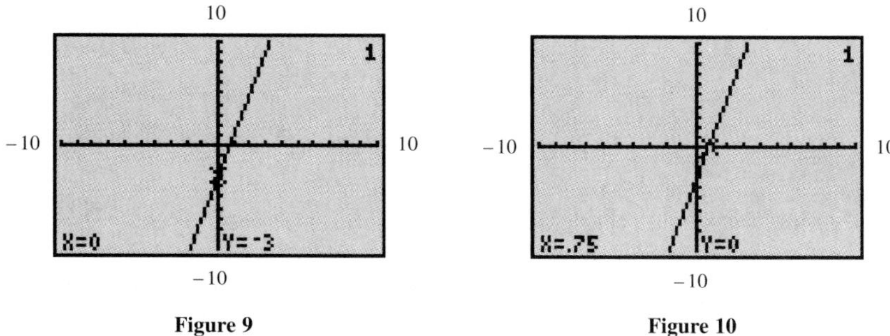

Figure 9 Figure 10

Some calculators have the capability of locating the x-intercept (called "Root" or "Zero") with a special function. Consult your owner's manual.

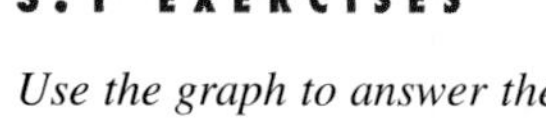

3.1 EXERCISES

Use the graph to answer the questions.

1. The graph indicates the total area of solar collectors used for heating pools from 1991–1995.

   (a) In which consecutive years did the numbers increase?

   (b) Between which two years was the decrease the greatest?

   (c) In what year was the number about 6000?

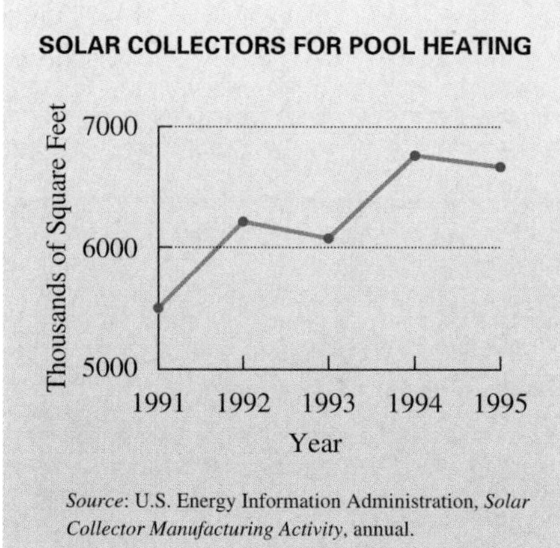

2. The graph shows the number of solar collectors used to heat water from 1991–1995.

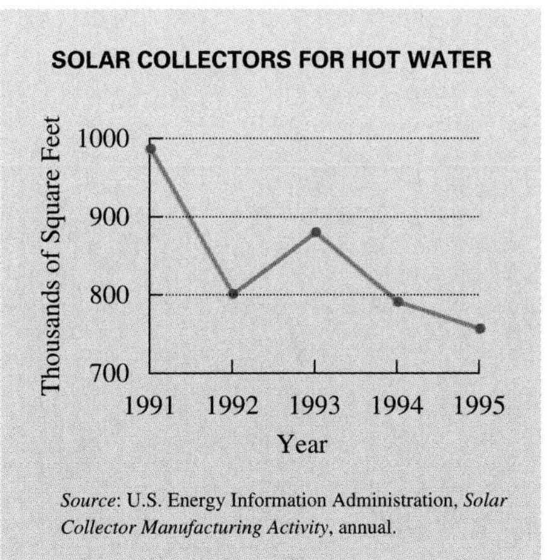

(a) In what year was the number of solar collectors used to heat water the greatest?
(b) In what one-year period was the decrease the greatest?
(c) In which year was the number greater than in the previous year?

3. If you were to use a bar graph or pie graph to show the information given in Exercises 1 and 2, which one would you choose?

4. What is another name for the rectangular coordinate system? For whom is it named?

Fill in each blank with the correct response.

5. The point with coordinates (0, 0) is called the _____ of a rectangular coordinate system.

6. For any value of x, the point $(x, 0)$ lies on the _____-axis.

7. To find the x-intercept of a line, we let _____ equal 0 and solve for _____.

8. The equation _____ = 4 has a horizontal line as its graph.
(x or y)

9. To graph a straight line we must find a minimum of _____ points.

10. The point (_____, 4) is on the graph of $2x - 3y = 0$.

Name the quadrant, if any, in which each point is located.

11. (a) (1, 6)
(b) (−4, −2)
(c) (−3, 6)
(d) (7, −5)
(e) (−3, 0)

12. (a) (−2, −10)
(b) (4, 8)
(c) (−9, 12)
(d) (3, −9)
(e) (0, −8)

13. Use the given information to determine the possible quadrants in which the point (x, y) must lie.

(a) $xy > 0$ (b) $xy < 0$ (c) $\frac{x}{y} < 0$ (d) $\frac{x}{y} > 0$

14. What must be true about the coordinates of any point that lies along an axis?

Locate the following points on a rectangular coordinate system.

15. (2, 3)
16. (−1, 2)
17. (−3, −2)
18. (1, −4)
19. (0, 5)
20. (−2, −4)
21. (−2, 4)
22. (3, 0)
23. (−2, 0)
24. (3, −3)

Complete the given table for each equation, and then graph the equation. See Example 1.

25. $x - y = 3$

x	y
0	
	0
5	
2	

26. $x - y = 5$

x	y
0	
	0
1	
3	

27. $x + 2y = 5$

x	y
0	
	0
2	
	2

28. $x + 3y = -5$

x	y
0	
	0
1	
	−1

29. $4x - 5y = 20$

x	y
0	
	0
2	
	−3

30. $6x - 5y = 30$

x	y
0	
	0
3	
	−2

31. Explain why the graph of $x + y = k$ cannot pass through quadrant III if $k > 0$.

32. Explain how to determine the intercepts of and graph the linear equation $4x - 3y = 12$.

Find the x-intercept and the y-intercept. Then graph each equation. See Examples 2–5.

33. $2x + 3y = 12$

34. $5x + 2y = 10$

35. $x - 3y = 6$

36. $x - 2y = -4$

37. $\frac{2}{3}x - 3y = 7$

38. $\frac{5}{7}x + \frac{6}{7}y = -2$

39. $y = 5$

40. $y = -3$

41. $x = 2$

42. $x = -3$

43. $x + 4 = 0$

44. $y + 2 = 0$

45. $x + 5y = 0$

46. $x - 3y = 0$

47. $2x = 3y$

48. $-\frac{2}{3}y = x$

49. $3y = -\frac{4}{3}x$

50. $4y = 3x$

51. Track qualifying records at North Carolina Motor Speedway from 1965–1998 are approximated by the linear equation $y = 1.22x + 118$, where y is the speed (in miles per hour) in year x. In the model $x = 0$ corresponds to 1965, $x = 10$ corresponds to 1975, and so on. Use the model to approximate the speed of the 1995 winner, Hut Stricklin. (*Source:* NASCAR.)

52. According to information provided by Families USA Foundation, the national average family health care cost in dollars between 1980 and 2000 (projected) can be approximated by the linear model $y = 382.75x + 1742$, where $x = 0$ corresponds to 1980 and $x = 20$ corresponds to 2000. Based on this model, what would be the expected national average health care cost in 2000?

TECHNOLOGY INSIGHTS (EXERCISES 53–58)

53. The screen shows the graph of one of the equations below, along with the coordinates of a point on the graph. Which one of the equations is it?

(a) $x + 2y = 4$

(b) $-3x + 5y = 15$

(c) $y = 4x - 2$

(d) $y = -2$

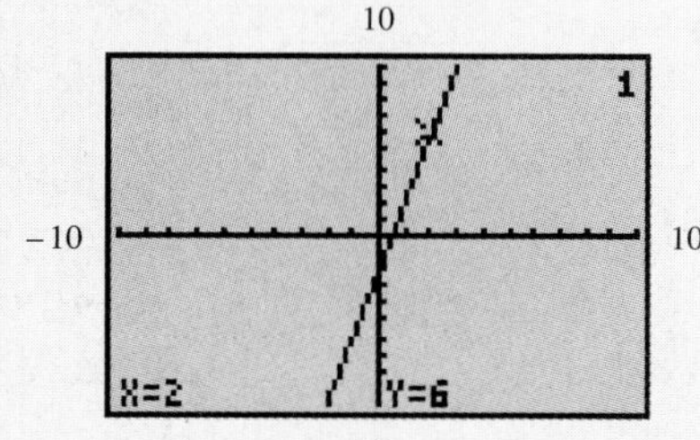

TECHNOLOGY INSIGHTS (EXERCISES 53–58) (CONTINUED)

54. The screens show the graph of one of the equations below. Two views of the graph are given, along with the intercepts. Which one of the equations is it?

(a) $3x + 2y = 6$ **(b)** $-3x + 2y = 6$

(c) $-3x - 2y = 6$ **(d)** $3x - 2y = 6$

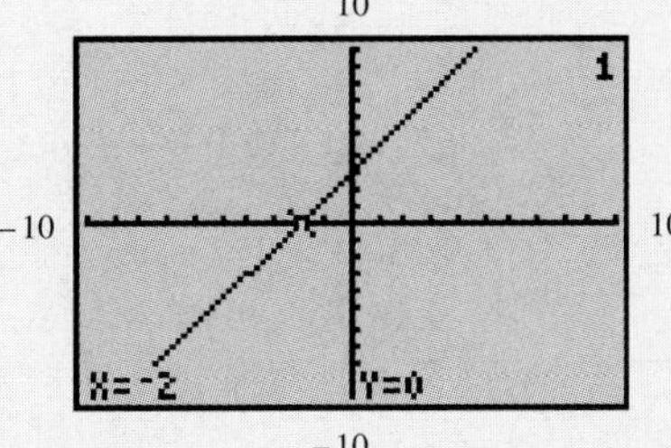

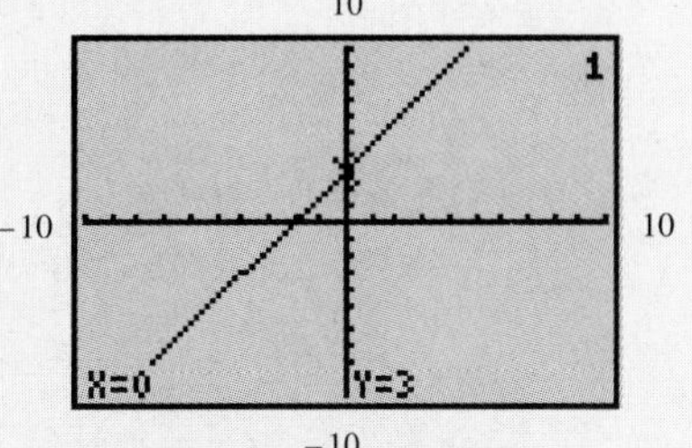

55. The table of ordered pairs shown was generated by a graphing calculator with a *table* feature.

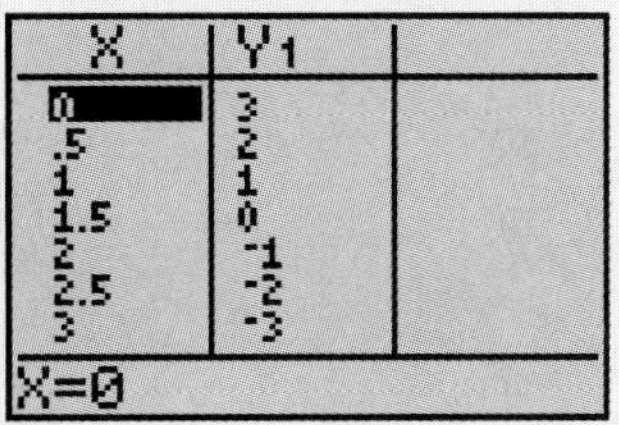

X	Y1
0	3
.5	2
1	1
1.5	0
2	-1
2.5	-2
3	-3

X=0

(a) What is the x-intercept?

(b) What is the y-intercept?

(c) Which one of the equations below corresponds to this table of values?

A. $Y_1 = 2X - 3$

B. $Y_1 = -2X - 3$

C. $Y_1 = 2X + 3$

D. $Y_1 = -2X + 3$

56. Refer to the model equation in Exercise 52. A portion of its graph is shown on the accompanying screen, along with the coordinates of a point on the line displayed at the bottom. How is this point interpreted in the context of the model?

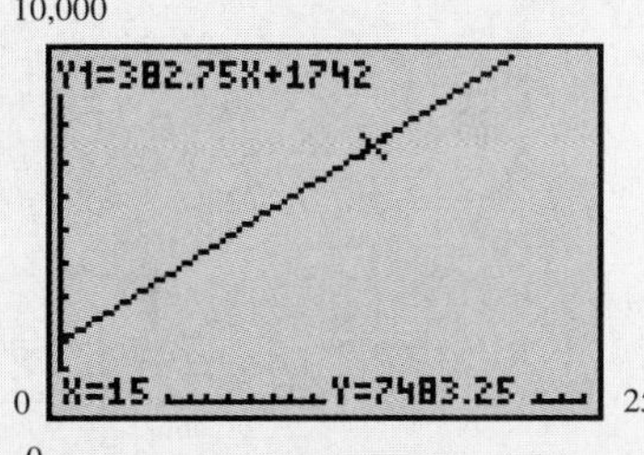

57. The screens each show the graph of $x + y = 15$ (which was entered as $y = -x + 15$). However, different viewing windows are used. Which one of the two windows do you think would be more useful for this graph? Why?

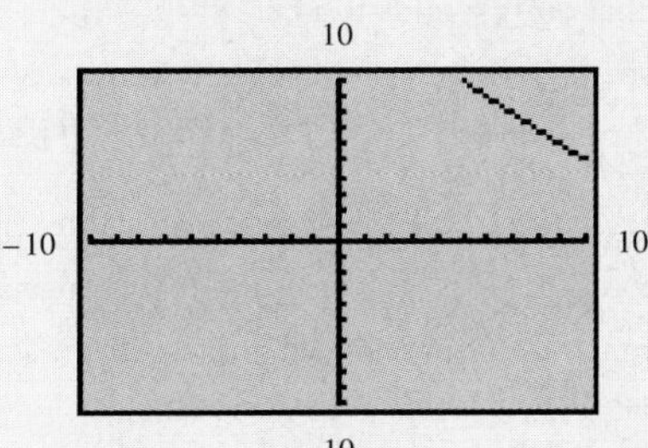

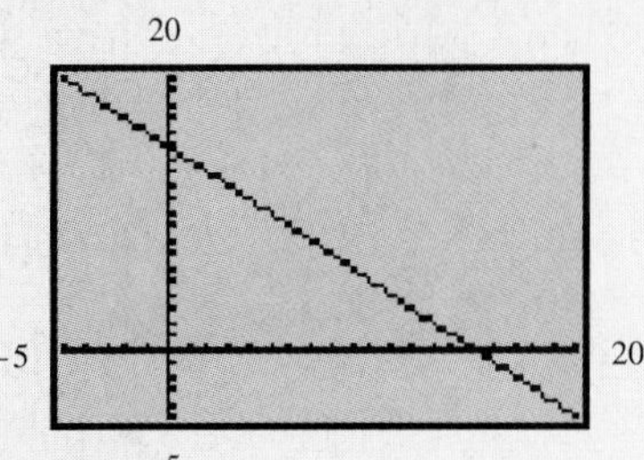

(continued)

TECHNOLOGY INSIGHTS (EXERCISES 53–58) (CONTINUED)

58. The screen shows the graph of $x + 2y = 0$. It was sketched in a traditional manner in Example 5. In what form should you enter the equation into the calculator?

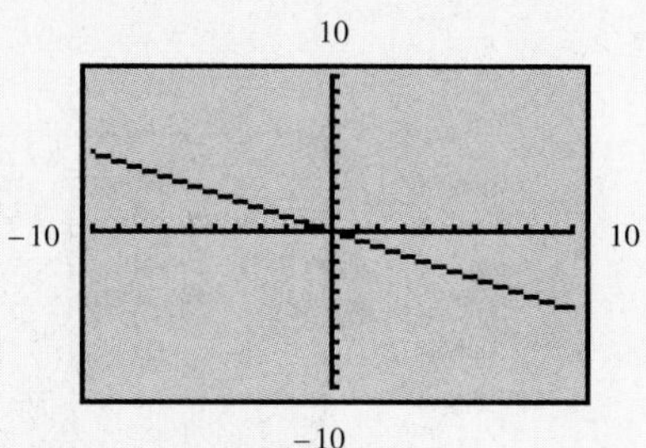

RELATING CONCEPTS (EXERCISES 59–64)

If the endpoints of a line segment are known, the coordinates of the midpoint of the segment can be found. The figure shows the coordinates of the points P and Q. Let $\overline{PQ}$ represent the line segment with endpoints at P and Q.

To derive a formula for the midpoint of $\overline{PQ}$, ***work Exercises 59–64 in order.***

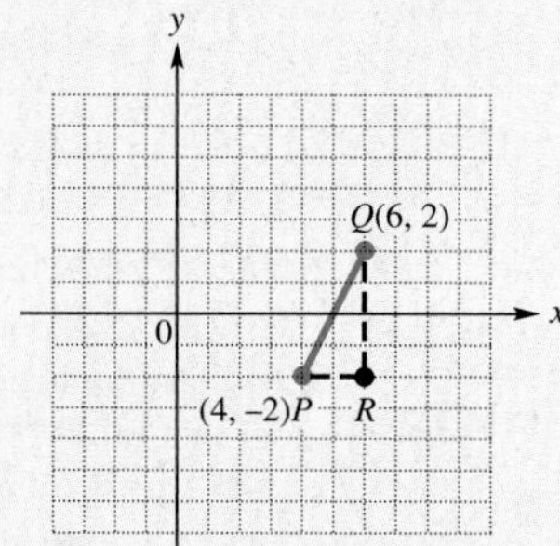

59. In the figure, R is the point with the same x-coordinate as Q and the same y-coordinate as P. Write the ordered pair that corresponds to R.

60. From the graph, determine the coordinates of the midpoint of $\overline{PR}$.

61. From the graph, determine the coordinates of the midpoint of $\overline{QR}$.

62. The x-coordinate of the midpoint M of $\overline{PQ}$ is the x-coordinate of the midpoint of $\overline{PR}$ and the y-coordinate is the y-coordinate of the midpoint of $\overline{QR}$. Write the ordered pair that corresponds to M.

63. The average of two numbers is found by dividing their sum by 2. Find the average of the x-coordinates of points P and Q. Find the average of the y-coordinates of points P and Q.

64. Comparing your answers to Exercises 62 and 63, what connection is there between the coordinates of P and Q and the coordinates of M?

Did you make the connection that the coordinates of M are the averages of the x- and y-coordinates of P and Q?

3.2 The Slope of a Line

OBJECTIVES

1. Find the slope of a line given two points on the line.
2. Find the slope of a line given the equation of the line.
3. Graph a line given its slope and a point on the line.
4. Use slope to decide whether two lines are parallel or perpendicular.
5. Solve problems involving average rate of change.

FOR EXTRA HELP

SSG Sec. 3.2
SSM Sec. 3.2
Pass the Test Software
InterAct Math Tutorial Software
Video 4

Slope (steepness) is used in many practical ways. The slope of a hill (sometimes called the *grade*) is often given as a percent. For example, a 10% (or $\frac{10}{100} = \frac{1}{10}$) slope means the hill rises 1 unit for every 10 horizontal units. Stairs and roofs have slopes too, as shown in Figure 11.

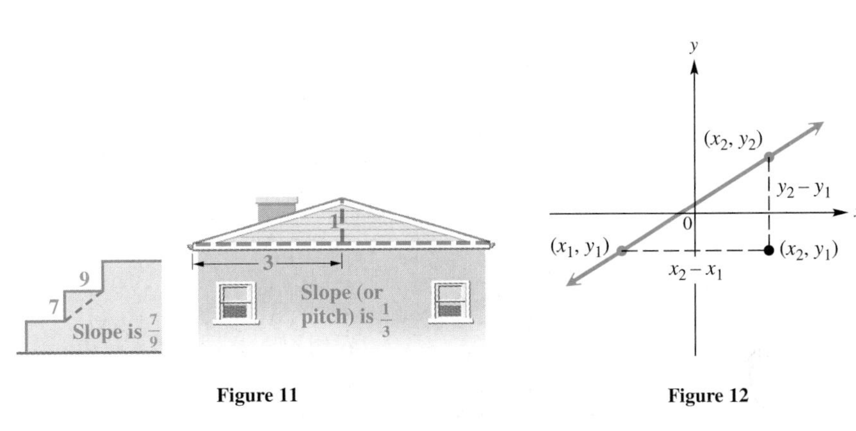

Figure 11 **Figure 12**

In each example that we mentioned, slope is the ratio of vertical change, or **rise,** to horizontal change, or **run.** A simple way to remember this is to think "slope is rise over run."

OBJECTIVE 1 Find the slope of a line given two points on the line. To get a formal definition of the slope of a line, it is convenient to use *subscripted* variables to designate two different points on the line. To differentiate between the points, we write them as (x_1, y_1) and (x_2, y_2). See Figure 12. (The small numbers 1 and 2 in these ordered pairs are called subscripts. Read (x_1, y_1) as "x-sub-one, y-sub-one.")

As we move along the line in Figure 12 from (x_1, y_1) to (x_2, y_2), the y-value changes from y_1 to y_2, an amount equal to $y_2 - y_1$. As y changes from y_1 to y_2, the value of x changes from x_1 to x_2 by the amount $x_2 - x_1$. The ratio of the change in y to the change in x (the rise over the run) is called the **slope** of the line, with the letter m traditionally used for the slope.

Slope

The slope of the line through the distinct points (x_1, y_1) and (x_2, y_2) is

$$m = \frac{\text{change in } y}{\text{change in } x} = \frac{y_2 - y_1}{x_2 - x_1} \quad (x_1 \neq x_2).$$

EXAMPLE 1 Using the Definition of Slope

Find the slope of the line through the points $(2, -1)$ and $(-5, 3)$.

If $(2, -1) = (x_1, y_1)$ and $(-5, 3) = (x_2, y_2)$, then

$$m = \frac{y_2 - y_1}{x_2 - x_1}$$
$$= \frac{3 - (-1)}{-5 - 2} = \frac{4}{-7} = -\frac{4}{7}.$$

See Figure 13. On the other hand, if $(2, -1) = (x_2, y_2)$ and $(-5, 3) = (x_1, y_1)$, the slope would be

$$m = \frac{-1 - 3}{2 - (-5)} = \frac{-4}{7} = -\frac{4}{7},$$

the same answer.

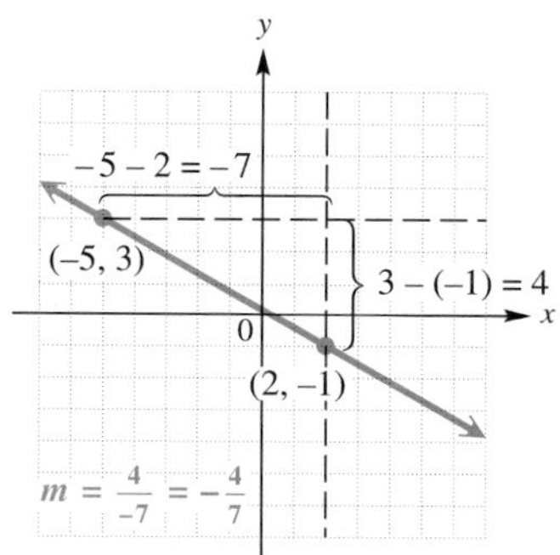

Figure 13

Example 1 suggests that the slope is the same no matter which point is considered first. Also, using similar triangles from geometry, it can be shown that the slope is the same no matter which two different points on the line are chosen.

In calculating the slope, be careful to subtract the y-values and the x-values in the *same* order.

Correct		Incorrect	
$\frac{y_2 - y_1}{x_2 - x_1}$ or	$\frac{y_1 - y_2}{x_1 - x_2}$	$\frac{y_2 - y_1}{x_1 - x_2}$ or	$\frac{y_1 - y_2}{x_2 - x_1}$

Also, remember that the change in y is the *numerator* and the change in x is the *denominator.*

OBJECTIVE 2 Find the slope of a line given the equation of the line. When the equation of a line is given, one way to find the slope is to use the definition of slope by first finding two different points on the line.

EXAMPLE 2 Finding the Slope of a Line

Find the slope of the line $4x - y = 8$.

The intercepts can be used as the two different points needed to find the slope. Let $y = 0$ to find that the x-intercept is $(2, 0)$. Then let $x = 0$ to find that the y-intercept is $(0, -8)$. Use these two points in the slope formula. The slope is

$$m = \frac{-8 - 0}{0 - 2} = \frac{-8}{-2} = 4.$$

EXAMPLE 3 Finding the Slope of a Line

Find the slope of each of the following lines.

(a) $x = -3$

By inspection, $(-3, 5)$ and $(-3, -4)$ are two points that satisfy the equation $x = -3$. Use these two points to find the slope.

$$m = \frac{-4 - 5}{-3 - (-3)} = \frac{-9}{0}$$

Since division by 0 is undefined, the slope is undefined. This is why the definition of slope includes the restriction $x_1 \neq x_2$.

(b) $y = 5$

Find the slope by selecting two different points on the line, such as $(3, 5)$ and $(-1, 5)$, and by using the definition of slope.

$$m = \frac{5 - 5}{3 - (-1)} = \frac{0}{4} = 0$$

As shown in Section 3.1, $x = -3$ has a graph that is a vertical line, and $y = 5$ has a graph that is a horizontal line. Generalizing from those results and the results of Example 3 above, we can make the following statements about vertical and horizontal lines.

Slopes of Vertical and Horizontal Lines

The slope of a vertical line is undefined; the slope of a horizontal line is 0.

The slope of a line also can be found directly from its equation. Look again at the equation $4x - y = 8$ in Example 2. Solve this equation for y.

$$\begin{aligned} 4x - y &= 8 && \text{Equation from Example 2} \\ -y &= -4x + 8 && \text{Subtract } 4x \text{ from each side.} \\ y &= 4x - 8 && \text{Multiply both sides by } -1. \end{aligned}$$

Notice that the slope, 4, we found using the slope formula in Example 2 is the same number as the coefficient of x in the equation $y = 4x - 8$. We will see in the next section that this always happens, *as long as the equation is solved for y.*

EXAMPLE 4 Finding Slope from an Equation

Find the slope of the graph of $3x - 5y = 8$.

Solve the equation for y.

$$\begin{aligned} 3x - 5y &= 8 \\ -5y &= -3x + 8 && \text{Subtract } 3x \text{ from both sides.} \\ y &= \frac{3}{5}x - \frac{8}{5} && \text{Divide each side by } -5. \end{aligned}$$

The slope is given by the coefficient of x, so the slope is $\frac{3}{5}$.

OBJECTIVE 3 **Graph a line given its slope and a point on the line.** Examples 5 and 6 show how to graph a straight line by using the slope and one point on the line.

EXAMPLE 5 Using the Slope and a Point to Graph a Line

Graph the line that has slope $\frac{2}{3}$ and goes through the point $(-1, 4)$.

First locate the point $(-1, 4)$ on a graph as shown in Figure 14. Then, from the definition of slope,

$$m = \frac{\text{change in } y}{\text{change in } x} = \frac{2}{3}.$$

Move *up* 2 units in the y-direction and then 3 units to the *right* in the x-direction to locate another point on the graph (labeled P). The line through $(-1, 4)$ and P is the required graph.

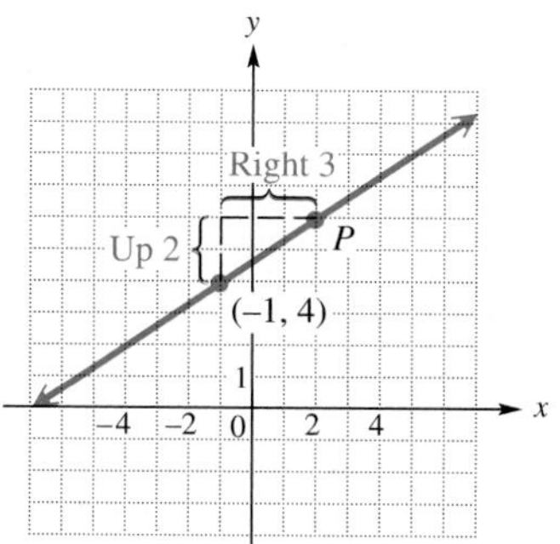

Figure 14

EXAMPLE 6 Using the Slope and a Point to Graph a Line

Graph the line through $(2, 1)$ that has slope $-\frac{4}{3}$.

Start by locating the point $(2, 1)$ on the graph. Find a second point on the line by using the definition of slope.

$$\text{slope} = \frac{\text{change in } y}{\text{change in } x} = \frac{-4}{3}$$

Move *down* 4 units from $(2, 1)$ and then 3 units to the *right.* Draw a line through this second point and $(2, 1)$, as shown in Figure 15.

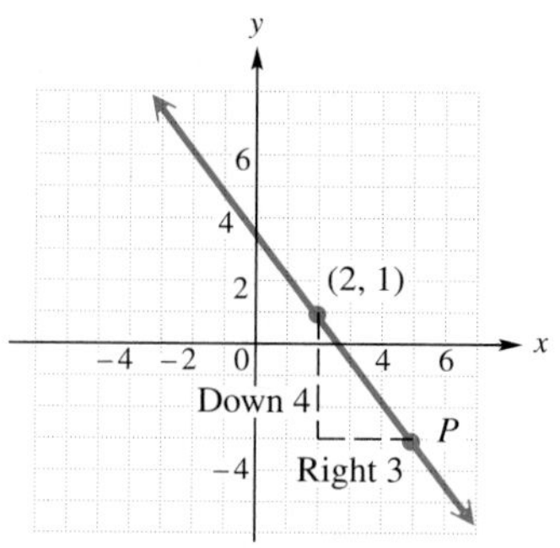

Figure 15

The slope also could be written as

$$\frac{\text{change in } y}{\text{change in } x} = \frac{4}{-3}.$$

In this case the second point is located *up* 4 units and 3 units to the *left*. Verify that this approach produces the same line.

In Example 5, the slope of the line is the *positive* number $\frac{2}{3}$. The graph of the line in Figure 14 goes up from left to right. The line in Example 6 has a *negative* slope, $-\frac{4}{3}$. As Figure 15 shows, its graph goes down from left to right. These facts suggest the following generalization.

A positive slope indicates that the line goes up from left to right; a negative slope indicates that the line goes down from left to right.

Figure 16 shows lines of positive, zero, negative, and undefined slopes.

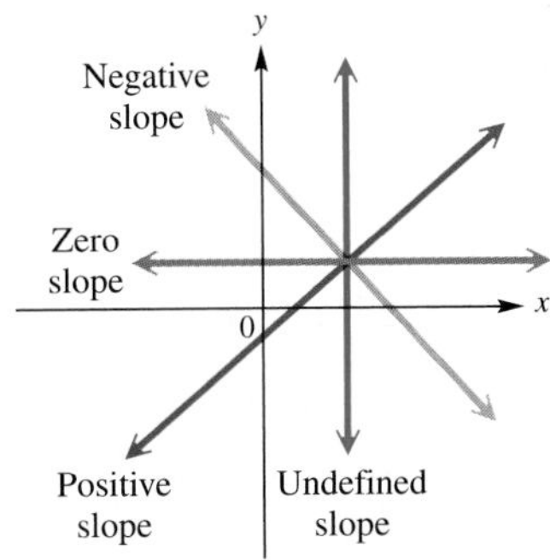

Figure 16

OBJECTIVE 4 Use slope to decide whether two lines are parallel or perpendicular. The slopes of a pair of parallel or perpendicular lines are related in a special way. The slope of a line measures the steepness of the line. Since parallel lines have equal steepness, their slopes must be equal. Also, lines with the same slope are parallel.

Slopes of Parallel Lines

Two nonvertical lines with the same slope are parallel; two nonvertical parallel lines have the same slope.

EXAMPLE 7 Determining Whether Two Lines Are Parallel

Are the lines L_1, through $(-2, 1)$ and $(4, 5)$, and L_2, through $(3, 0)$ and $(0, -2)$, parallel?

The slope of L_1 is

$$m_1 = \frac{5 - 1}{4 - (-2)} = \frac{4}{6} = \frac{2}{3}.$$

The slope of L_2 is

$$m_2 = \frac{-2 - 0}{0 - 3} = \frac{-2}{-3} = \frac{2}{3}.$$

Since the slopes are equal, the lines are parallel.

To see how the slopes of perpendicular lines are related, consider a nonvertical line with slope $\frac{a}{b}$. If this line is rotated 90°, the vertical change and the horizontal change are exchanged and the slope is $-\frac{b}{a}$, since the horizontal change is now negative. See Figure 17. Thus, the slopes of perpendicular lines have product -1 and are negative reciprocals of each other. For example, if the slopes of two lines are $\frac{3}{4}$ and $-\frac{4}{3}$, then the lines are perpendicular because $\frac{3}{4}(-\frac{4}{3}) = -1$.

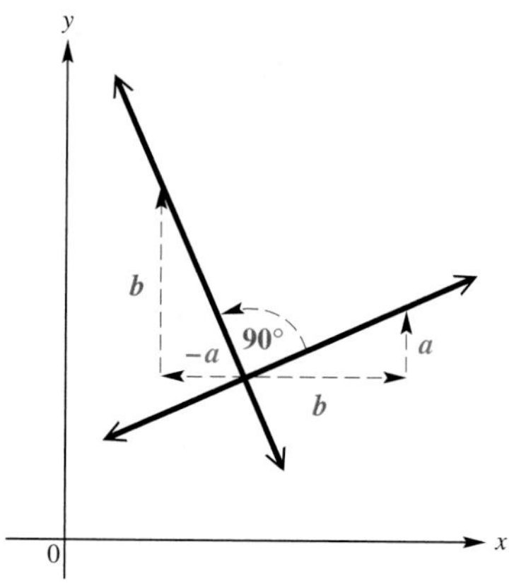

Figure 17

Slopes of Perpendicular Lines

If neither is vertical, perpendicular lines have slopes that are negative reciprocals; that is, their product is -1. Also, lines with slopes that are negative reciprocals are perpendicular.

EXAMPLE 8 Determining Whether Two Lines Are Perpendicular

Are the lines with equations $2y = 3x - 6$ and $2x + 3y = -6$ perpendicular?

Find the slope of each line by first solving each equation for y.

$$2y = 3x - 6 \qquad\qquad 2x + 3y = -6$$

$$y = \frac{3}{2}x - 3 \qquad\qquad 3y = -2x - 6$$

$$y = -\frac{2}{3}x - 2$$

$$\text{Slope: } \frac{3}{2} \qquad\qquad \text{Slope: } -\frac{2}{3}$$

Since the product of the slopes of the two lines is $\frac{3}{2}(-\frac{2}{3}) = -1$, the lines are perpendicular.

In Example 8, alternatively, we could have found the slope of each line by using intercepts and the slope formula.

EXAMPLE 9 Interpreting Graphing Calculator Graphs of Parallel or Perpendicular Lines

Decide from their graphs whether the two lines are parallel or perpendicular.

(a) The graphs of the equations $2x - 3y = -3$ and $7x - 10y = 30$

The graphs are shown in Figure 18. Based strictly on the graphs, we might conclude that the lines are parallel. However, checking their slopes algebraically, we get

$$\begin{aligned} 2x - 3y &= -3 \\ -3y &= -2x - 3 \\ y &= \frac{2}{3}x + 1 \end{aligned} \qquad \text{and} \qquad \begin{aligned} 7x - 10y &= 30 \\ -10y &= -7x + 30 \\ y &= \frac{7}{10}x - 3. \end{aligned}$$

The slopes are $\frac{2}{3}$ and $\frac{7}{10}$, which are not equal. Thus, the lines are *not* parallel.

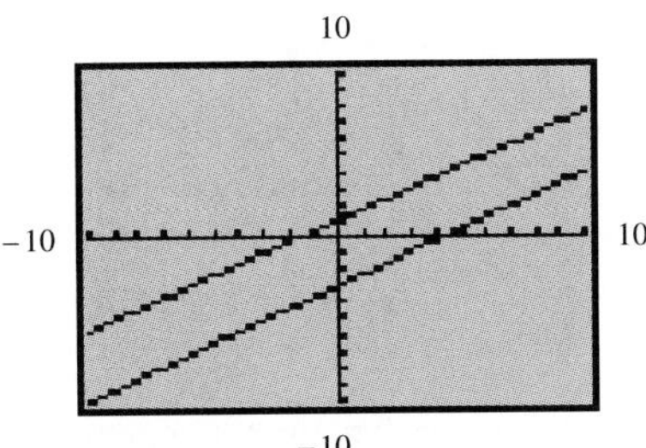

The graphs are *not* parallel, though they may appear to be.

Figure 18

(b) The graphs of the equations $2y = 3x - 6$ and $2x + 3y = -6$

In Example 8, we saw algebraically that the lines with these equations are perpendicular. If we graph them in the standard window, however, as in Figure 19(a), they do not appear to be perpendicular. To get a more realistic view, we use a *square window.* Graphing calculators have a function that will set a square window for the user. See Figure 19(b). Again, we cannot rely completely on what we see on the screen—we must understand the mathematical theory to draw the correct conclusion.

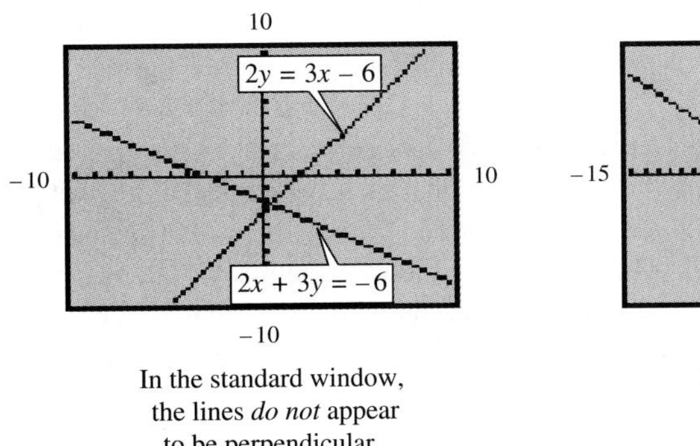

In the standard window, the lines *do not* appear to be perpendicular.

(a)

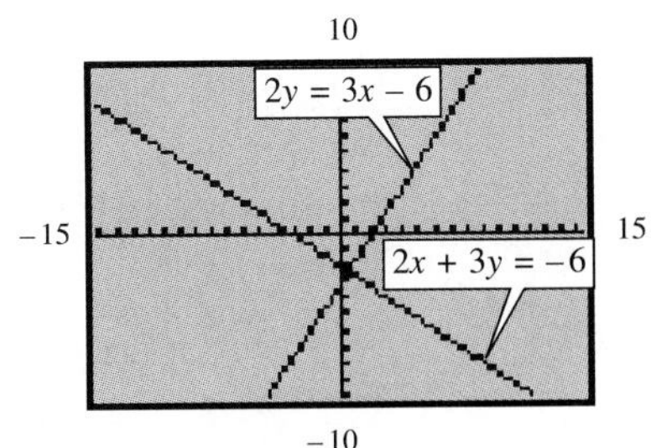

In the square window, the lines *do* appear to be perpendicular.

(b)

Figure 19

CONNECTIONS

On a trip from San Francisco, a passenger kept track every half hour of the distance traveled, with the following results for the first three hours.

Time in Hours	0	.5	1	1.5	2	2.5	3
Distance in Miles	0	20	48	80	104	126	150

From the distance formula, $d = rt$, solving for rate (or speed) gives

$$\text{Average speed} = \frac{\text{Distance}}{\text{Time}}.$$

For example, the average speed over the time interval from $t = 0$ to $t = 3$ is

$$\text{Average speed} = \frac{150 - 0}{3 - 0} = 50,$$

or 50 miles per hour. Similarly, the average speed from $t = 1$ to $t = 2$ is

$$\text{Average speed} = \frac{104 - 48}{2 - 1} = 56,$$

or 56 miles per hour. If the time in hours is represented by x and the distance is represented by y, then we can write the following formula for average speed.

$$\text{Average speed} = \frac{y_2 - y_1}{x_2 - x_1}.$$

Of course, x and y may represent other quantities, so this formula can be used to find the rate of change of any quantity with respect to another quantity. As shown in this section, this formula gives the slope of a line, where it represents the average change in y as x changes by 1 unit.

FOR DISCUSSION OR WRITING

Use the formula developed above to find the average speed from $t = 2$ to $t = 3$. How might we find the speed at some particular instant? Why do you think the average speed keeps changing?

OBJECTIVE 5 Solve problems involving average rate of change. We know that the slope of a line is the ratio of the change in y (vertical change) to the change in x (horizontal change). This idea can be applied to real-life situations as discussed in the previous Connections box. The slope gives the average rate of change in y per unit of change in x, where the value of y depends on the value of x. The next example further illustrates this idea of average rate of change. We assume a linear relationship between x and y.

EXAMPLE 10 Interpreting Slope as Average Rate of Change

The graph in Figure 20 approximates the U.S. production of crude oil in the years 1991–1995. Find the average rate of change in production in quadrillion BTUs per year.

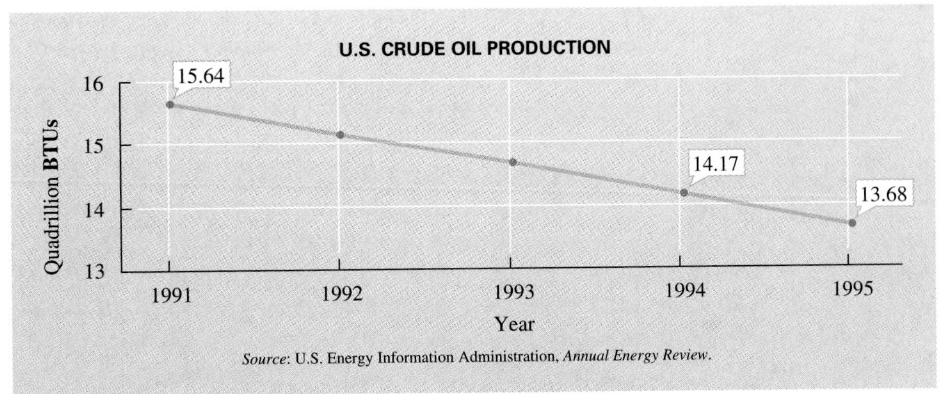

Figure 20

To use the slope formula, we need two pairs of data. From the graph, if $x = 1991$, then $y = 15.64$ and for $x = 1995$, $y = 13.68$, so we have the ordered pairs (1991, 15.64) and (1995, 13.68). By the slope formula,

$$\text{Average rate of change} = \frac{y_2 - y_1}{x_2 - x_1} = \frac{13.68 - 15.64}{1995 - 1991} = -.49.$$

This means that production of crude oil *decreased* by .49 quadrillion BTUs each year in the period from 1991–1995.

3.2 EXERCISES

1. A ski slope drops 30 feet for every horizontal 100 feet.

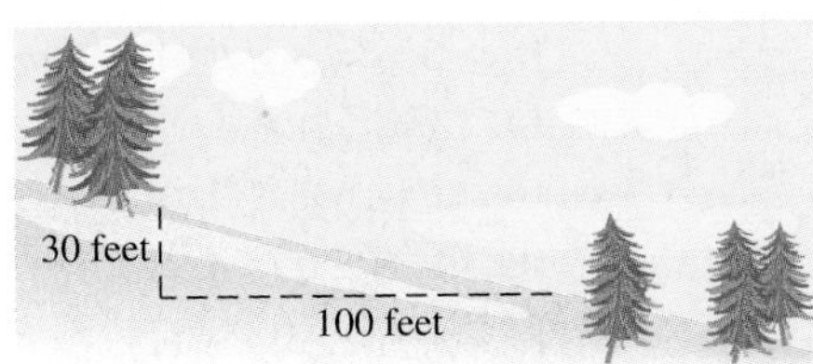

Which of the following express its slope? (There are several correct choices.)

(a) $-.3$ **(b)** $-\frac{3}{10}$ **(c)** $-3\frac{1}{3}$ **(d)** $-\frac{30}{100}$ **(e)** $-\frac{10}{3}$

2. A hill has a slope of $-.05$. How many feet in the vertical direction correspond to a run of 50 feet?

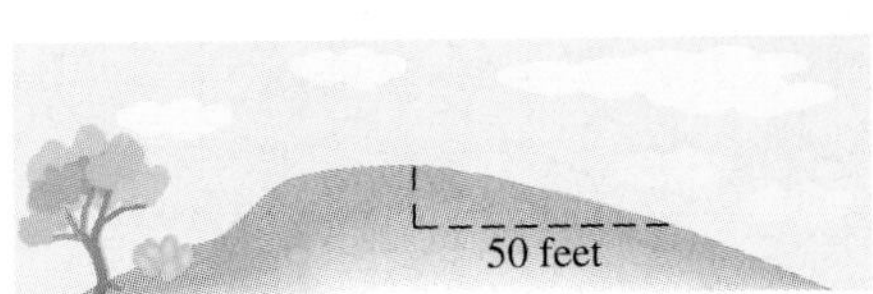

Use the given figure to determine the slope of the line segment described, by counting the number of units of "rise," the number of units of "run," and then finding the quotient.

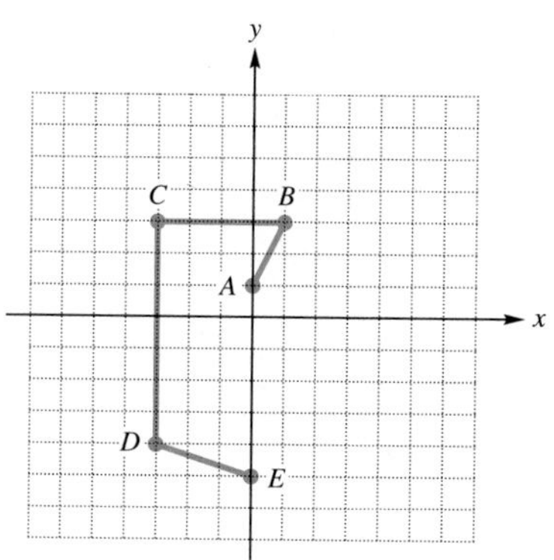

3. AB **4.** BC **5.** CD **6.** DE

Find each slope. See Example 1.

7. $m = \dfrac{6 - 2}{5 - 3}$

8. $m = \dfrac{5 - 7}{-4 - 2}$

9. $m = \dfrac{4 - (-1)}{-3 - (-5)}$

10. $m = \dfrac{-6 - 0}{0 - (-3)}$

11. $m = \dfrac{-5 - (-5)}{3 - 2}$

12. $m = \dfrac{7 - (-2)}{-3 - (-3)}$

13. Which of the following forms of the slope formula are correct? Explain why the incorrect forms are wrong.

(a) $\dfrac{y_1 - y_2}{x_2 - x_1}$ **(b)** $\dfrac{y_1 - y_2}{x_1 - x_2}$ **(c)** $\dfrac{x_2 - x_1}{y_2 - y_1}$ **(d)** $\dfrac{y_2 - y_1}{x_2 - x_1}$

Find the slope of the line through the pair of points by using the slope formula. See Example 1.

14. $(-2, -3)$ and $(-1, 5)$

15. $(-4, 3)$ and $(-3, 4)$

16. $(-4, 1)$ and $(2, 6)$

17. $(-3, -3)$ and $(5, 6)$

18. $(2, 4)$ and $(-4, 4)$

19. $(-6, 3)$ and $(2, 3)$

Find the slope of each line.

20.

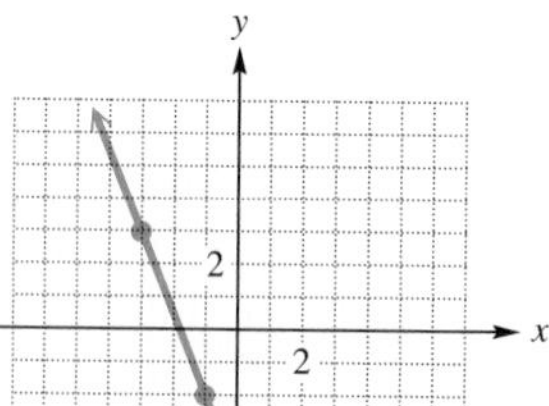

21.

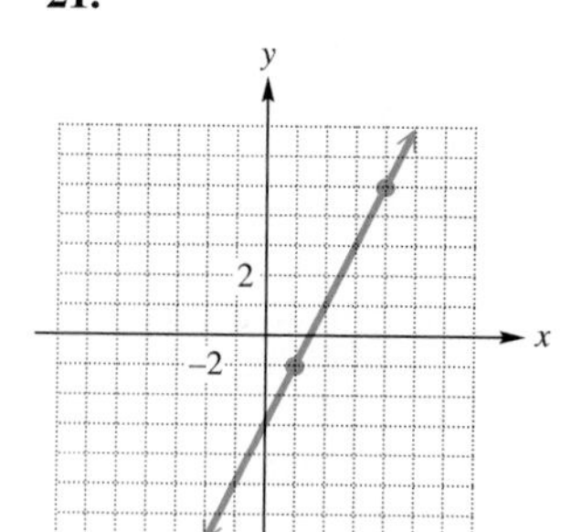

22.

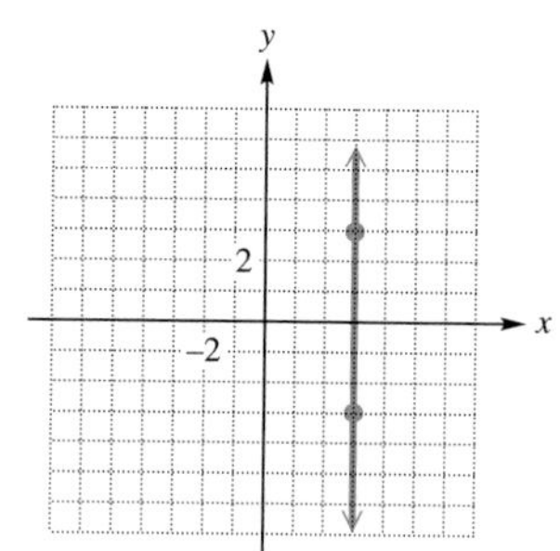

Based on the figure shown here, determine which line satisfies the given description.

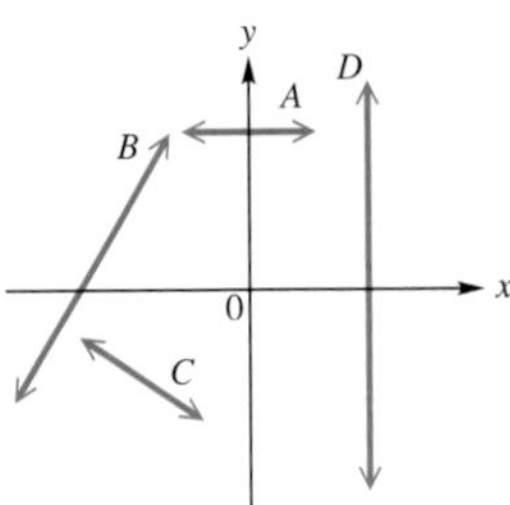

23. The line has positive slope.

24. The line has negative slope.

25. The line has slope 0.

26. The line has undefined slope.

Find the slope of the line and sketch the graph. See Examples 1–4.

27. $x + 2y = 4$
28. $x + 3y = -6$
29. $-x + y = 4$
30. $-x + y = 6$
31. $6x + 5y = 30$
32. $3x + 4y = 12$
33. $5x - 2y = 10$
34. $4x - y = 4$
35. $y = 4x$
36. $y = -3x$
37. $x - 3 = 0$
38. $y + 5 = 0$

39. A vertical line has equation ______ $= c$ for some constant c; a horizontal line has equation ______ $= d$ for some constant d.

40. Explain the meaning of *slope.* Give examples.

41. Explain the procedure for graphing a straight line using its slope and a point on the line.

Use the methods shown in Examples 5 and 6 to graph the line described.

42. Through $(-4, 2)$; $m = \frac{1}{2}$
43. Through $(-2, -3)$; $m = \frac{5}{4}$
44. Through $(0, -2)$; $m = -\frac{2}{3}$
45. Through $(0, -4)$; $m = -\frac{3}{2}$
46. Through $(-1, -2)$; $m = 3$
47. Through $(-2, -4)$; $m = 4$
48. $m = 0$; through $(2, -5)$
49. Undefined slope; through $(-3, 1)$
50. Undefined slope; through $(-4, 1)$
51. $m = 0$; through $(5, 3)$

52. If a line has slope $-\frac{4}{9}$, then any line parallel to it has slope ______, and any line perpendicular to it has slope ______.

Decide whether the two lines are parallel, perpendicular, or neither. See Examples 7 and 8.

53. $3x = y$ and $2y - 6x = 5$
54. $2x + 5y = -8$ and $6 + 2x = 5y$
55. $4x + y = 0$ and $5x - 8 = 2y$
56. $x = 6$ and $6 - x = 8$
57. $4x - 3y = 8$ and $4y + 3x = 12$
58. $2x = y + 3$ and $2y + x = 3$
59. $4x - 3y = 5$ and $3x - 4y = 2$
60. $5x - y = 7$ and $5x = 3 + y$

61. The line through $(4, 6)$ and $(-8, 7)$ and the line through $(7, 4)$ and $(-5, 5)$

62. The line through $(9, 15)$ and $(-7, 12)$ and the line through $(-4, 8)$ and $(-20, 5)$

Use the concept of slope to solve each problem.

63. The upper deck at the new Comiskey Park in Chicago has produced, among other complaints, displeasure with its steepness. It's been compared to a ski jump. It is 160 feet from home plate to the front of the upper deck and 250 feet from home plate to the back. The top of the upper deck is 63 feet above the bottom. What is its slope?

64. When designing the new arena in Boston to replace the old Boston Garden, architects were careful to design the entrance ramps so that circus elephants would be able to march up the ramps. The maximum grade (or slope) that an elephant will walk on is 13%. Suppose that such a ramp was constructed with a horizontal run of 150 feet. What would be the maximum vertical rise the architects could use?

In Exercises 65–70, refer to Example 10.

65. The market for international phone calls during the ten-year period from 1986 to 1995 is depicted in the accompanying bar graph. The tops of the bars approximate a straight line. Assuming that the traffic volume at the beginning of this period was 18 billion minutes and at the end was 60 billion minutes, what was the average rate of change for the ten-year period?

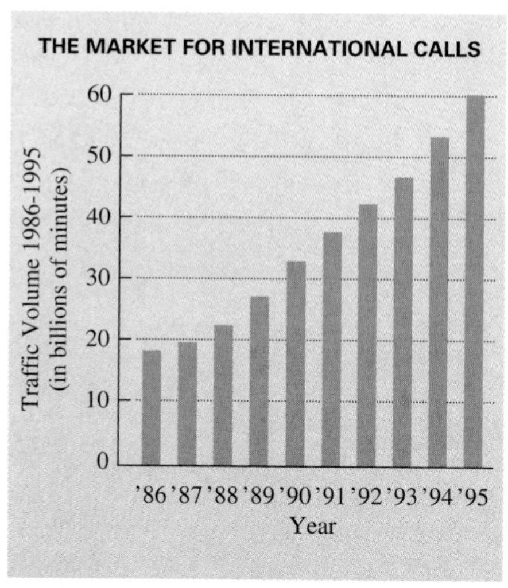

66. On the third day of a rotation diet, Lynn Elliott weighed 92.5 kilograms. By the eleventh day, she weighed 90.9 kilograms. What was her average rate of weight loss per day?

67. The table gives data for the graph in Example 10, Figure 20. Use these data to find the average rate of change in crude oil production from 1991 to 1992; 1992 to 1994; 1993 to 1995. What do your answers suggest about the average rate of change?

U.S. Crude Oil Production

Year	*Quadrillion BTUs*
1991	15.64
1992	15.15
1993	14.66
1994	14.17
1995	13.68

Source: U.S. Energy Information Administration, *Annual Energy Review.*

68. The table gives book publishers' net dollar sales from 1995–2000. (Sales are estimated for the years 1998–2000.)

Book Publishers' Sales

Year	*Sales (in millions)*
1995	19,000
1996	20,000
1997	21,000
1998	22,000
1999	23,000
2000	24,000

Source: Book Industry Study Group.

(a) Find the average rate of change from 1995–1996; 1995–1999; 1998–2000. What do you notice about your answers? What does this tell you?

(b) Calculate the rates of change in part (a) as percents. What do you notice?

69. The graph shows how estimated undeveloped capacity of hydroelectric power (in millions of kilowatts) in the U.S. Pacific Division decreased from 1991–1996.

Year	Kilowatts (in millions)
1991	26.2
1992	26.2
1993	26.2
1994	26.1
1995	24.0
1996	22.9

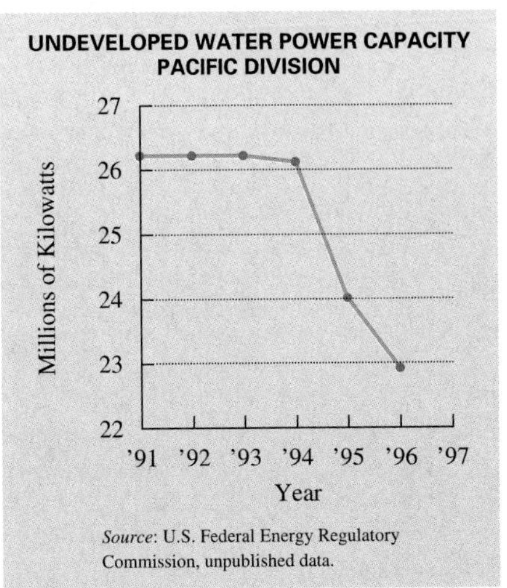

(a) Use the information given in the table of values for 1991 and 1996 to determine the average rate of change in capacity per year from 1991 through 1996.

(b) Use the information given for 1993 and 1996 to determine the average rate of change in capacity per year over that period.

(c) What do you notice about the answers for parts (a) and (b)? Explain.

70. The long-distance telephone market has changed rapidly and continuously since it was opened up to competition. The fourth-quarter market shares of AT&T for even-numbered years since 1984 are shown in the table.

AT&T Fourth-Quarter Market Shares

Year	Market Share (%)
1984	87.7
1986	81.5
1988	73.9
1990	66.5
1992	60.6
1994	57.6
1996	53.8

Source: Federal Communications Commission.

(a) Find the average annual rate of change of market share for each pair of successive years in the table, 1984 and 1986, 1986 and 1988, and so on.

(b) Plot these points on a grid using the years as x-values and the market share as y-values. Do the points appear to lie approximately on a line?

(c) How do the average rates you found in part (a) compare with your answer to part (b)?

TECHNOLOGY INSIGHTS (EXERCISES 71–74)

71. The graphing calculator–generated table shows several ordered pairs that lie on the graph of a line. Find the slope of the line.

X	Y1
-3	-13
-2	-10
-1	-7
0	-4
1	-1
2	2
3	5

X= -3

72. The table shown was generated by a graphing calculator. It gives several points that lie on the graph of a line. What is the slope of the line?

X	Y1
2	-.5
3	1.25
4	3
5	4.75
6	6.5
7	8.25
8	10

X=8

73. The graphing calculator screen shows two lines. One is the graph of $Y_1 = -2x + 3$ and the other is the graph of $Y_2 = 3x - 4$. Which is which?

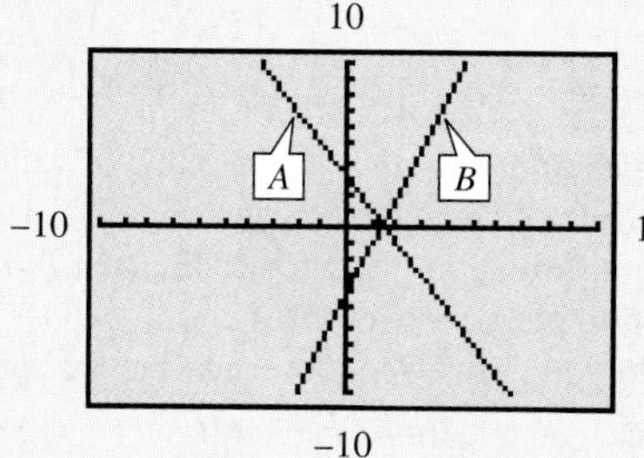

74. The graphing calculator screen shows two lines. One is the graph of $Y_1 = 2x - 5$ and the other is the graph of $Y_2 = 4x - 5$. Which is which?

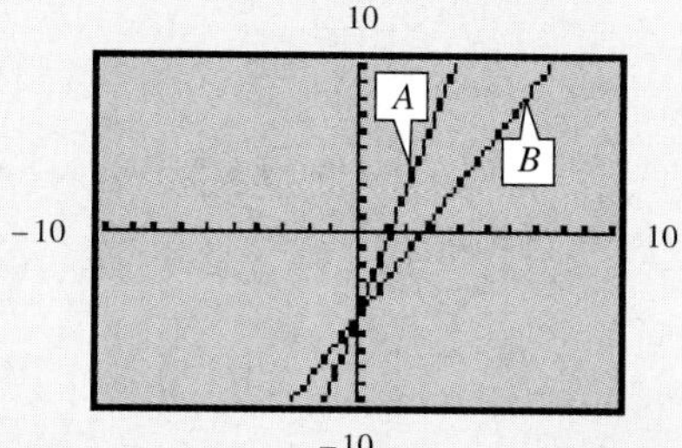

Solve each problem using your knowledge of the slopes of parallel and perpendicular lines.

75. Show that $(-13, -9)$, $(-11, -1)$, $(2, -2)$, and $(4, 6)$ are the vertices of a parallelogram. (*Hint:* A parallelogram is a four-sided figure with opposite sides parallel.)

76. Is the figure with vertices at $(-11, -5)$, $(-2, -19)$, $(12, -10)$, and $(3, 4)$ a parallelogram? Is it a rectangle? (A rectangle is a parallelogram with a right angle.)

RELATING CONCEPTS (EXERCISES 77–82)

In these exercises we investigate a method of determining whether three points lie on the same straight line. (Such points are said to be **collinear.**) The points we consider are $A(3, 1)$, $B(6, 2)$, and $C(9, 3)$.

Work Exercises 77–82 in order.

77. Find the slope of segment AB.

78. Find the slope of segment BC.

79. Find the slope of segment AC.

80. If slope of segment AB = slope of segment BC = slope of segment AC, then A, B, and C are collinear. Use the results of Exercises 77–79 to show that this statement is satisfied.

RELATING CONCEPTS (EXERCISES 77–82) (CONTINUED)

81. Use the slope formula to determine whether the points $(1, -2)$, $(3, -1)$, and $(5, 0)$ are collinear.

82. Repeat Exercise 81 for the points $(0, 6)$, $(4, -5)$, and $(-2, 12)$.

Did you make the connection that if the slopes of line segments connecting pairs of points are the same, the points are collinear?

3.3 Linear Equations in Two Variables

OBJECTIVES

1. Write the equation of a line given its slope and a point on the line.
2. Write the equation of a line given two points on the line.
3. Write the equation of a line given its slope and y-intercept.
4. Find the slope and y-intercept of a line given its equation.
5. Write the equation of a line parallel or perpendicular to a given line.
6. Apply concepts of linear equations in two variables to real data.
7. Use a graphing calculator to solve linear equations in one variable.

FOR EXTRA HELP

SSG Sec. 3.3
SSM Sec. 3.3

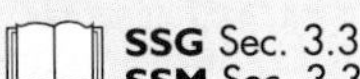

Pass the Test Software

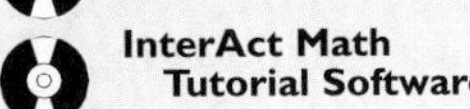

InterAct Math Tutorial Software

Video 4

Many real-world situations can be described by straight-line graphs. This section shows how to write a linear equation that satisfies given conditions.

OBJECTIVE 1 **Write the equation of a line given its slope and a point on the line.** A straight line is a set of points in the plane such that the slope between any two points is the same. In Figure 21, point P is on the line through P_1 and P_2 if the slope of the line through points P_1 and P equals the slope of the line through points P and P_2.

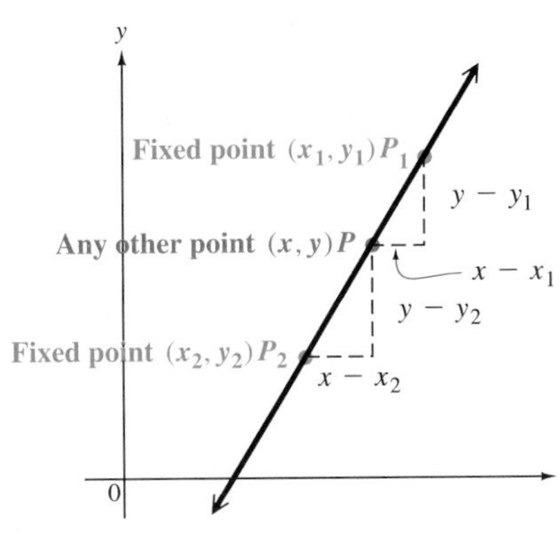

Figure 21

Setting these slopes equal to m gives

$$\frac{y - y_1}{x - x_1} = \frac{y - y_2}{x - x_2} = m,$$

$$\frac{y - y_1}{x - x_1} = m$$

$$y - y_1 = m(x - x_1). \qquad \text{Multiply both sides by } x - x_1.$$

This last equation is the *point-slope form* of the equation of the line. To use this form to write the equation of a line, we need to know the coordinates of a point (x_1, y_1) and the slope of the line, m.

Point-Slope Form

The **point-slope form** of the equation of a line is

Slope ↓

$$y - y_1 = m(x - x_1).$$

↑ Given point ↑

EXAMPLE 1 Using the Point-Slope Form

Find the equation of the line with slope $\frac{1}{3}$, going through the point $(-2, 5)$.

Use the point-slope form of the equation of a line, with $(x_1, y_1) = (-2, 5)$ and $m = \frac{1}{3}$.

$$y - y_1 = m(x - x_1)$$

$$y - 5 = \frac{1}{3}[x - (-2)] \qquad \text{Let } y_1 = 5, m = \tfrac{1}{3}, x_1 = -2.$$

$$y - 5 = \frac{1}{3}(x + 2)$$

$$3y - 15 = x + 2 \qquad \text{Multiply by 3.}$$

or

$$x - 3y = -17 \qquad \text{Combine terms.}$$

In Section 3.1, we defined *standard form* for a linear equation as

$$Ax + By = C.$$

In addition, from now on, let us agree that A, B, and C will be integers with no common factor (except 1) and $A \geq 0$. For example, the final equation found in Example 1, $x - 3y = -17$, is written in standard form.

The definition of "standard form" is not standard from one text to another. Any linear equation can be written in many different (all equally correct) forms. For example, the equation $2x + 3y = 8$ can be written as $2x = 8 - 3y$, $3y = 8 - 2x$, $x + \frac{3}{2}y = 4$, $4x + 6y = 16$, and so on. In addition to writing it in the form $Ax + By = C$ with $A \geq 0$, let us agree that the form $2x + 3y = 8$ is preferred over any multiples of both sides, such as $4x + 6y = 16$. (To write $4x + 6y = 16$ in standard form, divide both sides by 2.)

OBJECTIVE 2 Write the equation of a line given two points on the line. To find an equation of a line when two points on a line are known, first use the slope formula to find the slope of the line. Then use the slope with either of the given points in the point-slope form.

EXAMPLE 2 Finding an Equation of a Line When Two Points Are Known

Find an equation of the line through the points $(-4, 3)$ and $(5, -7)$.

First find the slope, using the definition.

$$m = \frac{-7 - 3}{5 - (-4)} = -\frac{10}{9}$$

Use either $(-4, 3)$ or $(5, -7)$ as (x_1, y_1) in the point-slope form of the equation of a line. If $(-4, 3)$ is used, then $-4 = x_1$ and $3 = y_1$.

$$y - y_1 = m(x - x_1) \qquad \text{Point-slope form}$$

$$y - 3 = -\frac{10}{9}[x - (-4)] \qquad \text{Let } y_1 = 3,\ m = -\tfrac{10}{9},\ x_1 = -4.$$

$$y - 3 = -\frac{10}{9}(x + 4)$$

$$9(y - 3) = -10(x + 4) \qquad \text{Multiply by 9.}$$

$$9y - 27 = -10x - 40 \qquad \text{Distributive property}$$

$$10x + 9y = -13 \qquad \text{Standard form}$$

Verify that if $(5, -7)$ were used, the same equation would be found.

CONNECTIONS

Earlier examples and exercises gave equations that described real data. Now we are able to show how such equations can be found. The process of writing an equation whose graph approximates a set of data is called *data-fitting.* For example, the bar graph shows the number of multimedia personal computers (PCs), in millions, in U.S. homes.

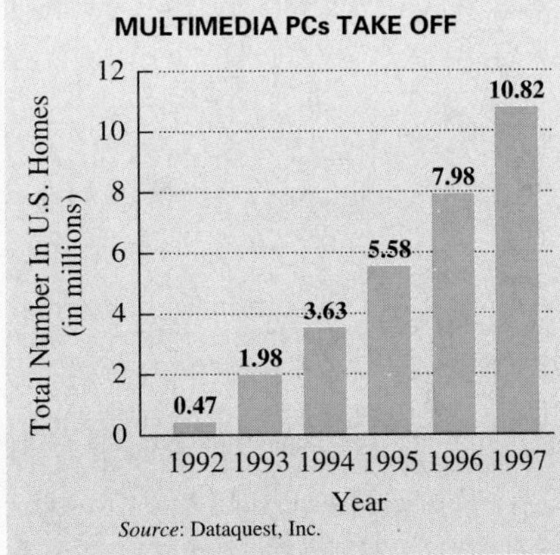

The data shown in the bar graph increase linearly, that is, a straight line could be drawn through the tops of any two bars that is very close to the top of each bar. This indicates that we can write a linear equation to approximate the data, allowing us to estimate the number of PCs for years other than those shown in the graph. If we let $x = 0$ represent 1992, then $x = 1$ represents 1993, $x = 2$ represents 1994, and so on. The ordered pair for 1993 is $(1, 1.98)$ and for 1996 is $(4, 7.98)$. The slope of the line through these two points is

$$\frac{y_2 - y_1}{x_2 - x_1} = \frac{7.98 - 1.98}{4 - 1} = 2.$$

(continued)

CONNECTIONS (CONTINUED)

Using (1, 1.98) and $m = 2$, we get the following equation of the line.

$$\begin{aligned} y - y_1 &= m(x - x_1) & \\ y - 1.98 &= 2(x - 1) & \text{Let } m = 2, x_1 = 1, y_1 = 1.98. \\ y - 1.98 &= 2x - 2 & \text{Distributive property} \\ y &= 2x - .02 & \text{Add } 1.98. \end{aligned}$$

The equation tells us that the number of multimedia PCs y (in millions) in year x is given by $y = 2x - .02$.

FOR DISCUSSION OR WRITING

Use the equation found above to predict the number of multimedia PCs in 1998. (Recall that 1992 is represented by $x = 0$.) According to the equation, in what year were there no multimedia PCs? What does the $-.02$ in the equation indicate about the number of multimedia PCs? What might cause this equation to become very inaccurate in future years?

Notice that the point-slope form does not apply to a vertical line, since the slope of a vertical line is undefined. A vertical line through the point (c, d) has equation $x = c$.

A horizontal line has slope 0. From the point-slope form, the equation of a horizontal line through the point (c, d) is

$$\begin{aligned} y - y_1 &= m(x - x_1) & \\ y - d &= 0(x - c) & y_1 = d, x_1 = c \\ y - d &= 0 & \\ y &= d. & \end{aligned}$$

In summary, horizontal and vertical lines have the following special equations.

Equations of Vertical and Horizontal Lines

The vertical line through (c, d) has equation $\boldsymbol{x = c}$, and the horizontal line through (c, d) has equation $\boldsymbol{y = d}$.

OBJECTIVE 3 Write the equation of a line given its slope and y-intercept. Suppose a line has slope m and y-intercept $(0, b)$. Using the point-slope form, the equation of the line is

$$\begin{aligned} y - y_1 &= m(x - x_1) & \\ y - b &= m(x - 0) & x_1 = 0, y_1 = b \\ y &= mx + b. & \text{Add } b. \end{aligned}$$

When the equation is solved for y, the coefficient of x is the slope, m, and the constant b is the y-value of the y-intercept. Because this form of the equation shows the slope and the y-intercept, it is called the *slope-intercept form.*

Slope-Intercept Form

The equation of a line with slope m and y-intercept $(0, b)$ is written in **slope-intercept form** as

$$y = mx + b.$$

$$\underset{\text{Slope}}{\uparrow} \qquad \underset{y\text{-intercept is } (0,\, b).}{\uparrow}$$

EXAMPLE 3 Using the Slope-Intercept Form

Find an equation of the line with slope $-\frac{4}{5}$ and y-intercept $(0, -2)$.

Here $m = -\frac{4}{5}$ and $b = -2$. Substitute these values into the slope-intercept form.

$$y = mx + b$$

$$y = -\frac{4}{5}x - 2$$

OBJECTIVE 4 Find the slope and y-intercept of a line given its equation. If the equation of a line is written in slope-intercept form, the coefficient of x is the slope and the constant leads to the y-intercept.

EXAMPLE 4 Finding the Slope and y-intercept from the Equation

Find the slope and y-intercept of the graph of $3y + 2x = 9$.

Write the equation in slope-intercept form by solving for y.

$$3y + 2x = 9$$

$$3y = -2x + 9$$

$$y = -\frac{2}{3}x + 3 \qquad \text{Slope-intercept form}$$

Slope ↑ ↑ y-intercept is $(0, 3)$.

From the slope-intercept form, the slope is $-\frac{2}{3}$ and the y-intercept is $(0, 3)$.

The slope-intercept form of a linear equation is the most useful for several reasons. Every linear equation (of a nonvertical line) has a *unique* (one and only one) slope-intercept form. In Section 3.5 we will study *linear functions,* which are defined by the slope-intercept form. Also, this is the form we must use when graphing a line with a graphing calculator.

OBJECTIVE 5 Write the equation of a line parallel or perpendicular to a given line. As mentioned in the previous section, parallel lines have the same slope and perpendicular lines have slopes with product -1.

EXAMPLE 5 Finding Equations of Parallel or Perpendicular Lines

Find the equation in slope-intercept form of the line passing through the point $(-4, 5)$, and **(a)** parallel to the line $2x + 3y = 6$; **(b)** perpendicular to the line $2x + 3y = 6$.

(a) The slope of the graph of $2x + 3y = 6$ can be found by solving for y.

$$2x + 3y = 6$$

$$3y = -2x + 6 \quad \text{Subtract } 2x \text{ on both sides.}$$

$$y = -\frac{2}{3}x + 2 \quad \text{Divide both sides by 3.}$$

↑ Slope (points to $-\frac{2}{3}$)

The slope is given by the coefficient of x, so $m = -\frac{2}{3}$. This means that the required equation of the line through $(-4, 5)$ and parallel to $2x + 3y = 6$ also has slope $-\frac{2}{3}$. Now use the point-slope form, with $(x_1, y_1) = (-4, 5)$ and $m = -\frac{2}{3}$.

$$y - 5 = -\frac{2}{3}[x - (-4)] \quad y_1 = 5,\ m = -\tfrac{2}{3},\ x_1 = -4$$

$$y - 5 = -\frac{2}{3}(x + 4)$$

$$y - 5 = -\frac{2}{3}x - \frac{8}{3} \quad \text{Distributive property}$$

$$y = -\frac{2}{3}x - \frac{8}{3} + \frac{15}{3} \quad \text{Add } 5 = \tfrac{15}{3} \text{ to both sides.}$$

$$y = -\frac{2}{3}x + \frac{7}{3} \quad \text{Combine like terms.}$$

We did not clear fractions after the substitution step here because we want the equation in slope-intercept form—that is, solved for y.

(b) To be perpendicular to the line $2x + 3y = 6$, a line must have a slope that is the negative reciprocal of $-\frac{2}{3}$, which is $\frac{3}{2}$. Use the point $(-4, 5)$ and slope $\frac{3}{2}$ in the point-slope form.

$$y - 5 = \frac{3}{2}[x - (-4)] \quad y_1 = 5,\ m = \tfrac{3}{2},\ x_1 = -4$$

$$y - 5 = \frac{3}{2}(x + 4)$$

$$y - 5 = \frac{3}{2}x + 6 \quad \text{Distributive property}$$

$$y = \frac{3}{2}x + 11 \quad \text{Add 5 to both sides.}$$

A summary of the various forms of linear equations follows.

Summary of Forms of Linear Equations

$y - y_1 = m(x - x_1)$	**Point-slope form** Slope is m. Line passes through (x_1, y_1).
$y = mx + b$	**Slope-intercept form** Slope is m. y-intercept is $(0, b)$.
$Ax + By = C$	**Standard form** $(A \geq 0)$ Slope is $-\frac{A}{B}$. $(B \neq 0)$ x-intercept is $\left(\frac{C}{A}, 0\right)$. $(A \neq 0)$ y-intercept is $\left(0, \frac{C}{B}\right)$. $(B \neq 0)$
$x = c$	**Vertical line** Undefined slope x-intercept is $(c, 0)$.
$y = d$	**Horizontal line** Slope is 0. y-intercept is $(0, d)$.

OBJECTIVE 6 Apply concepts of linear equations in two variables to real data.

EXAMPLE 6 Determining a Linear Equation to Describe Real Data

Suppose that it is time to fill up your car with gasoline. You drive into your local station and notice that 89-octane gas is selling for \$1.20 per gallon. Experience has taught you that the final price you pay can be determined by the number of gallons you buy multiplied by the price per gallon (in this case, \$1.20). As you pump the gas you observe two sets of numbers spinning by: one is the number of gallons you have pumped, and the other is the price you pay for that number of gallons.

The table below uses ordered pairs to illustrate this situation.

Number of Gallons Pumped	*Price for This Number of Gallons*
0	\$0.00 = 0(\$1.20)
1	\$1.20 = 1(\$1.20)
2	\$2.40 = 2(\$1.20)
3	\$3.60 = 3(\$1.20)
4	\$4.80 = 4(\$1.20)

If we let x denote the number of gallons pumped, then the price y that you pay can be found by the linear equation $y = 1.20x$, where y is in dollars. This is a simple, realistic application of linear equations. Theoretically, there are infinitely many ordered pairs (x, y) that satisfy this equation, but in this application we are limited to nonnegative values for x, since we cannot have a negative number of gallons. There is also a practical maximum value for x in this situation, which varies from one car to another. What do you think determines this maximum value?

In Example 6, the ordered pair (0, 0) satisfied the equation, so the equation has the form $y = mx$, where $b = 0$. If a realistic situation involves an initial charge plus a charge per unit, the equation will have the form $y = mx + b$, where $b \neq 0$.

EXAMPLE 7 Determining an Equation and Interpreting Ordered Pairs That Satisfy It

Suppose that you can get a car wash at the gas station in Example 6 if you pay an additional \$3.00.

(a) Write an equation that defines the price you will pay.

Since an additional \$3.00 will be charged, you will pay $1.20x + 3.00$ dollars for x gallons of gas and a car wash. Thus, if y represents the price, the equation is $y = 1.2x + 3$. (We deleted the unnecessary zeros.)

(b) Interpret the ordered pairs (5, 9) and (10, 15) in relation to the equation from part (a).

The ordered pair (5, 9) indicates that the price of 5 gallons of gas and a car wash is \$9.00. Similarly, (10, 15) indicates that the price of 10 gallons of gas and a car wash is \$15.00.

OBJECTIVE 7 Use a graphing calculator to solve linear equations in one variable. In Section 3.1, we saw how a graphing calculator is used to graph a linear equation in two variables. Figure 22 shows the graph of $y = -4x + 7$. From the values at the bottom of the screen, we see that when $x = 1.75$, $y = 0$. This means that $x = 1.75$ satisfies the equation $-4x + 7 = 0$, a linear equation in one variable. Therefore, the solution set of $-4x + 7 = 0$ is $\{1.75\}$. This can be verified using the algebraic method, shown in Section 2.1. (The word "Zero" indicates that the x-intercept has been located.)

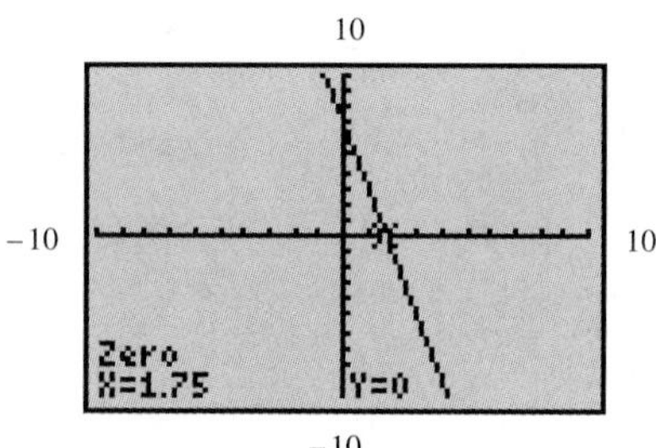

Figure 22

EXAMPLE 8 Solving an Equation with a Graphing Calculator

Use a graphing calculator to solve $-2x - 4(2 - x) = 3x + 4$.

Begin by writing the equation as an equivalent equation with 0 on one side. Do this by subtracting $3x$ and 4 from both sides.

$$-2x - 4(2 - x) - 3x - 4 = 0$$

Then graph $y = -2x - 4(2 - x) - 3x - 4$ and find the x-intercept. Notice that the viewing window must be altered from the one shown in Figure 22 because the x-intercept does not lie in the interval $[-10, 10]$. As seen in Figure 23, the x-intercept of the line is $(-12, 0)$, and thus the solution or zero of the equation is -12. The solution set is $\{-12\}$.

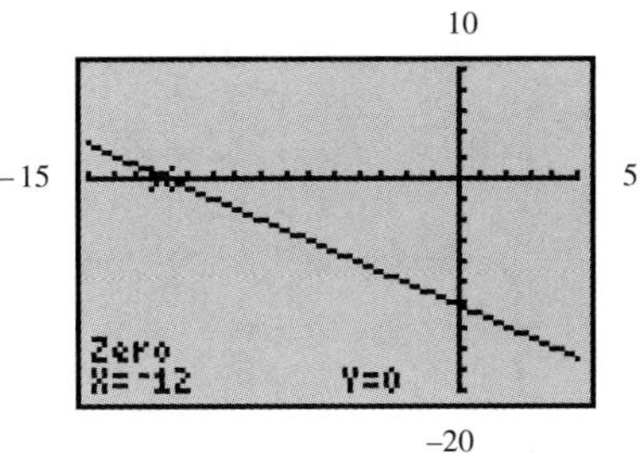

Figure 23

3.3 EXERCISES

1. The following equations all represent the same line. Which one is in standard form as defined in the text?

   (a) $3x - 2y = 5$

   (b) $2y = 3x - 5$

   (c) $\frac{3}{5}x - \frac{2}{5}y = 1$

   (d) $3x = 2y + 5$

2. Which one of the following equations is in point-slope form?

   (a) $y = 6x + 2$

   (b) $4x + y = 9$

   (c) $y - 3 = 2(x - 1)$

   (d) $2y = 3x - 7$

3. Which of the equations in Exercise 2 is in slope-intercept form?

4. Write the equation $y + 2 = -3(x - 4)$ in slope-intercept form.

5. Write the equation from Exercise 4 in standard form.

6. Write the equation $10x - 7y = 70$ in slope-intercept form.

Match each equation with the graph that it most closely resembles. (Hint: Determining the signs of m and b will help you make your decision.)

7. $y = 2x + 3$

8. $y = -2x + 3$

9. $y = -2x - 3$

10. $y = 2x - 3$

11. $y = 2x$

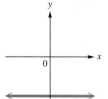

12. $y = -2x$

13. $y = 3$

14. $y = -3$

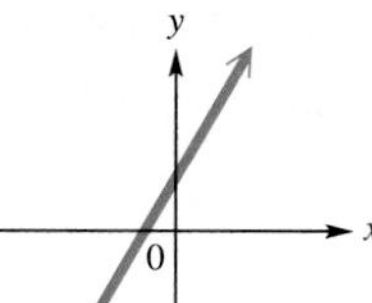

B.

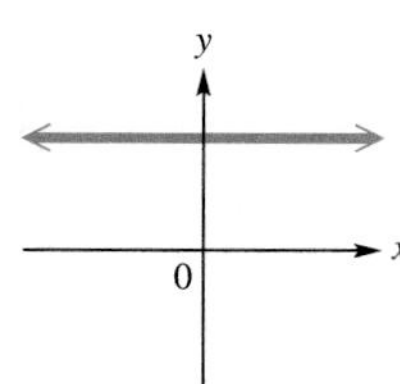

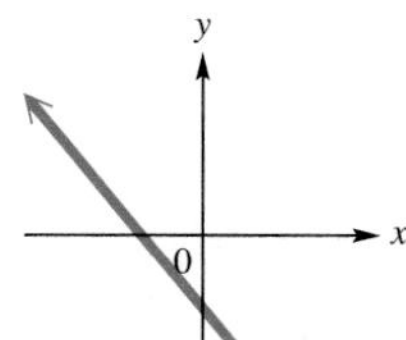

D.

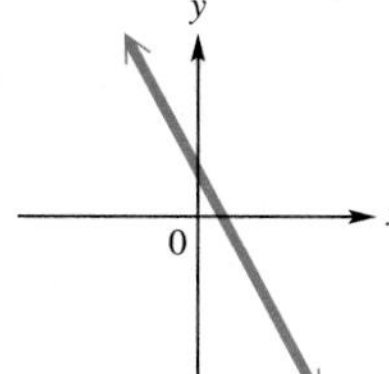

E.

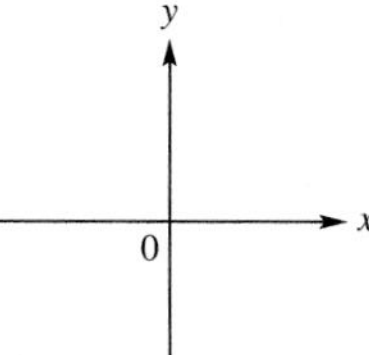

F.

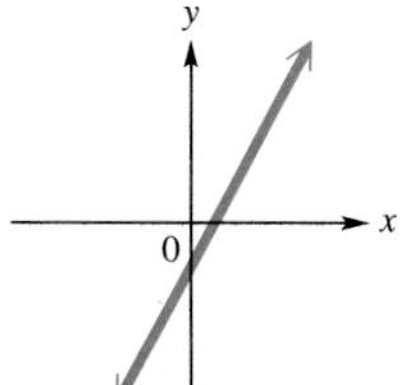

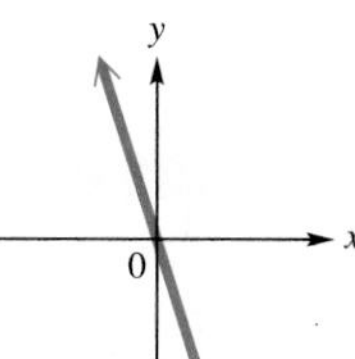

H.

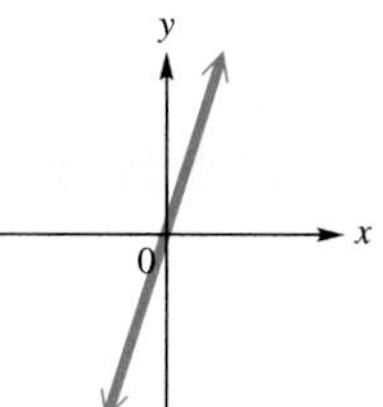

Write the equation in slope-intercept form of the line satisfying the given conditions. See Example 1.

15. Through $(-2, 4)$; slope $-\frac{3}{4}$

16. Through $(-1, 6)$; slope $-\frac{5}{6}$

17. Through $(5, 8)$; slope -2

18. Through $(12, 10)$; slope 1

19. Through $(-5, 4)$; slope $\frac{1}{2}$

20. Through $(7, -2)$; slope $\frac{1}{4}$

21. x-intercept $(3, 0)$; slope 4

22. x-intercept $(-2, 0)$; slope -5

23. In your own words, list all the forms of linear equations in two variables, and describe when each form should be used.

24. Explain why the point-slope form of an equation cannot be used to find the equation of a vertical line.

Write an equation that satisfies the given conditions.

25. Through $(9, 5)$; slope 0

26. Through $(-4, -2)$; slope 0

27. Through $(9, 10)$; undefined slope

28. Through $(-2, 8)$; undefined slope

29. Through $(.5, .2)$; vertical

30. Through $\left(\frac{5}{8}, \frac{2}{9}\right)$; vertical

31. Through $(-7, 8)$; horizontal

32. Through $(2, 7)$; horizontal

Write the slope-intercept form (if possible) of the equation of the line passing through the two points. See Example 2.

33. $(3, 4)$ and $(5, 8)$

34. $(5, -2)$ and $(-3, 14)$

35. $(6, 1)$ and $(-2, 5)$

36. $(-2, 5)$ and $(-8, 1)$

37. $\left(-\frac{2}{5}, \frac{2}{5}\right)$ and $\left(\frac{4}{3}, \frac{2}{3}\right)$

38. $\left(\frac{3}{4}, \frac{8}{3}\right)$ and $\left(\frac{2}{5}, \frac{2}{3}\right)$

39. $(2, 5)$ and $(1, 5)$

40. $(-2, 2)$ and $(4, 2)$

41. $(7, 6)$ and $(7, -8)$

42. $(13, 5)$ and $(13, -1)$

43. $(1, -3)$ and $(-1, -3)$

44. $(-4, 6)$ and $(5, 6)$

Find the equation in slope-intercept form of the line satisfying the given conditions. See Example 3.

45. $m = 5; \quad b = 15$

46. $m = -2; \quad b = 12$

47. $m = -\frac{2}{3}; \quad b = \frac{4}{5}$

48. $m = -\frac{5}{8}; \quad b = -\frac{1}{3}$

49. Slope $\frac{2}{5}$; y-intercept $(0, 5)$

50. Slope $-\frac{3}{4}$; y-intercept $(0, 7)$

Write an equation in slope-intercept form of the line shown in the graph. (Hint: Use the indicated points to find the slope.)

51.

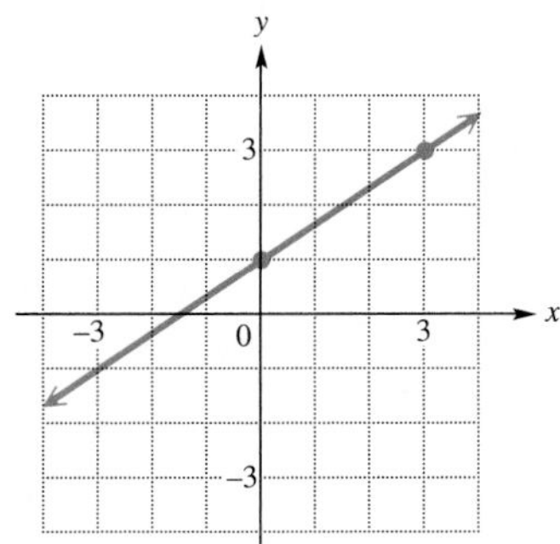

52.

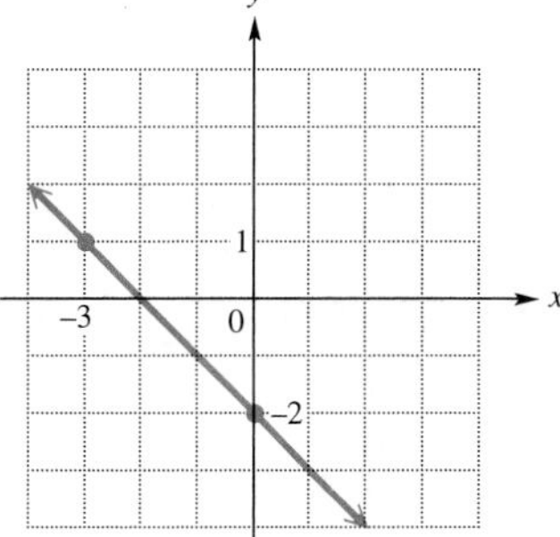

For each equation ***(a)*** *write in slope-intercept form,* ***(b)*** *give the slope of the line, and* ***(c)*** *give the y-intercept. See Example 4.*

53. $x + y = 12$

54. $x - y = 14$

55. $5x + 2y = 20$

56. $6x + 5y = 40$

57. $2x - 3y = 10$

58. $4x - 3y = 10$

Write an equation in slope-intercept form of the line satisfying the given conditions. See Example 5.

59. Through $(7, 2)$; parallel to $3x - y = 8$

60. Through $(4, 1)$; parallel to $2x + 5y = 10$

61. Through $(-2, -2)$; parallel to $-x + 2y = 10$

62. Through $(-1, 3)$; parallel to $-x + 3y = 12$

63. Through $(8, 5)$; perpendicular to $2x - y = 7$

64. Through $(2, -7)$; perpendicular to $5x + 2y = 18$

65. Through $(-2, 7)$; perpendicular to $x = 9$

66. Through $(8, 4)$; perpendicular to $x = -3$

Write an equation in the form $y = mx + b$ for each of the following situations. Then give three ordered pairs that satisfy the equation with x-values of 1, 5, and 10. See Examples 6 and 7.

67. It costs a $15 flat fee to rent a chain saw, plus $3 per day starting with the first day. Let x represent the number of days rented, so y represents the charge to the user (in dollars).

68. It costs a borrower $.05 per day for an overdue book, plus a flat $.50 charge for all books borrowed. Let x represent the number of days the book is overdue, so y represents the total fine to the tardy user.

Many real-world situations can be modeled by straight-line graphs. One way to find the equation of such a line is to use two typical data points from the information provided, with the point-slope form of the equation of a line. Because of the usefulness of the slope-intercept form, such equations are often given in the form $y = mx + b$. In Exercises 69–72, assume that the situation described can be modeled by a straight-line graph, and use the information to find the slope-intercept form of the equation of the line.

69. The line graph at the beginning of this chapter shows a straight line that approximates U.S. energy consumption in quadrillion BTUs for the years 1991–1996. Here $x = 1$ represents 1991, $x = 2$ represents 1992, and so on. Use the ordered pairs (1, 80.1) and (6, 89.4) from the graph to write an equation of the line. (*Source:* U.S. Energy Department.)

70. The table gives nuclear power production in quadrillion BTUs per year for three years.

Nuclear Power

Year	*Quadrillion BTUs*
1993	6.52
1994	6.84
1995	7.19

Source: U.S. Energy Information Administration, *Annual Energy Outlook 1996.*

(a) Use the values for 1993 and 1995, with $x = 3$ representing 1993 and $x = 5$ representing 1995, to write an equation of a line through these points.

(b) Use the equation to estimate production in 1994. How does your result compare to the value for 1994 in the table?

71. Median household income of African Americans increased in recent years, as shown in the bar graph.

(a) Use the information given for the years 1993 and 1997, letting $x = 3$ represent 1993, $x = 7$ represent 1997, and y represent the median income, to write an equation that models median household income.

(b) Use the equation to approximate the median income for 1995. How does your result compare to the actual value, $23,583?

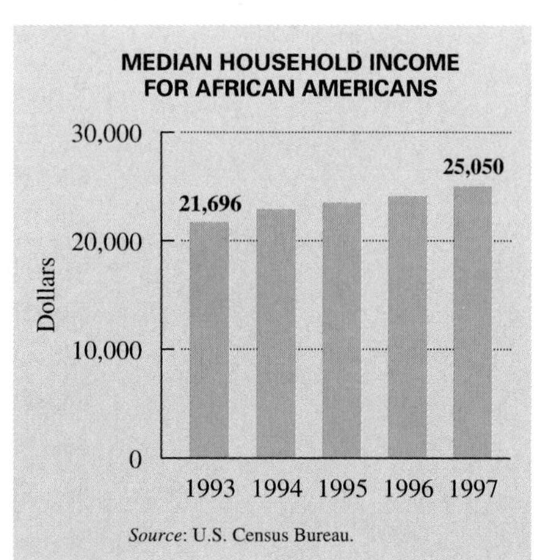

72. The bar graph shows median household income for Hispanics.

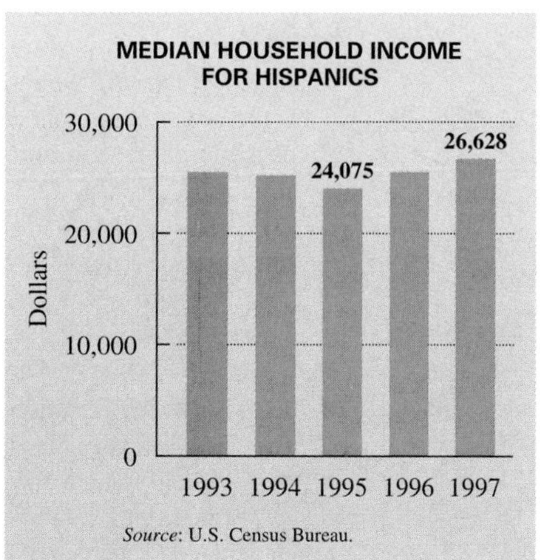

(a) Use the information for the years 1995 and 1997 to write an equation. Let $x = 5$ represent 1995, $x = 7$ represent 1997, and y represent the median income.

(b) Looking at the graph, would you expect the equation from part (a) to give good approximations for 1993 and 1994 income? Would the equation give a reasonable approximation for 1998? Explain.

TECHNOLOGY INSIGHTS (EXERCISES 73-78)

In Exercises 73–76, do the following.

(a) *Simplify and rewrite the equation so that the right side is 0. Then replace 0 with y.*

(b) *The graph of the equation for y is shown with each exercise. Use the graph to determine the solution of the given equation. See Example 8.*

(c) *Solve the equation using the methods of Chapter 2.*

73. $2x + 7 - x = 4x - 2$

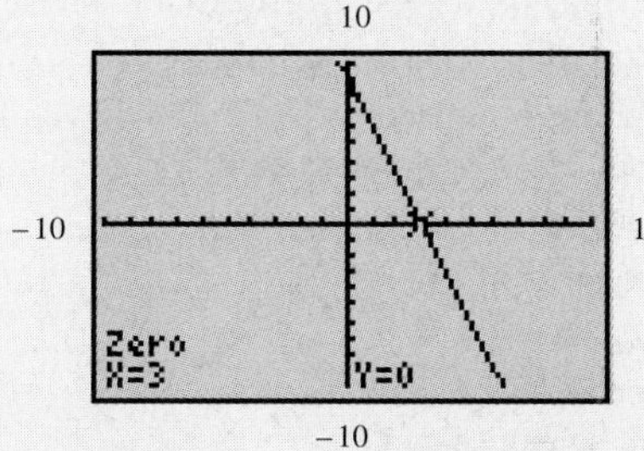

74. $7x - 2x + 4 - 5 = 3x + 1$

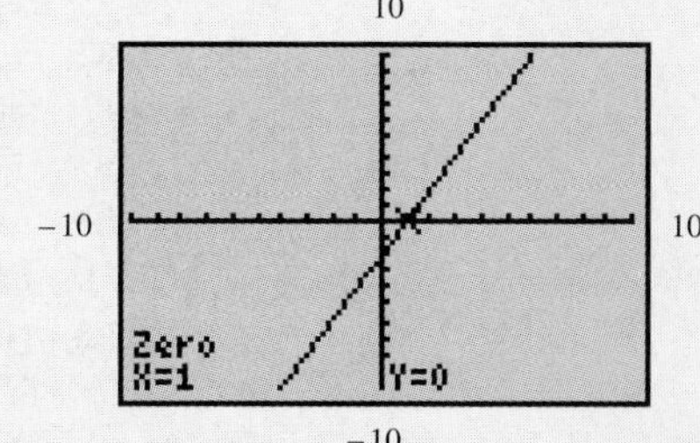

75. $3(2x + 1) - 2(x - 2) = 5$

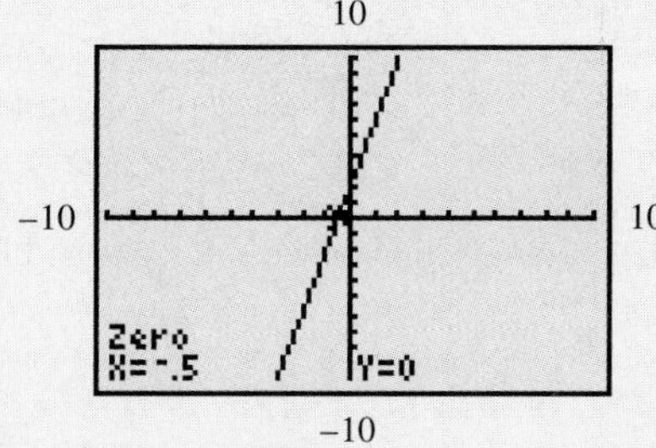

76. $4x - 3(4 - 2x) = 2(x - 3) + 6x + 2$

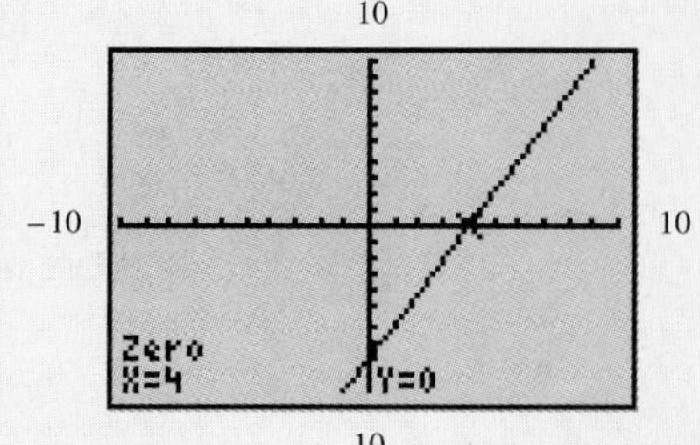

(continued)

TECHNOLOGY INSIGHTS (EXERCISES 73–78) (CONTINUED)

77. The graph of Y_1 is shown in the *standard viewing window.* Which is the only choice that could possibly be the solution of the equation $Y_1 = 0$?
(a) -15 **(b)** 0
(c) 5 **(d)** 15

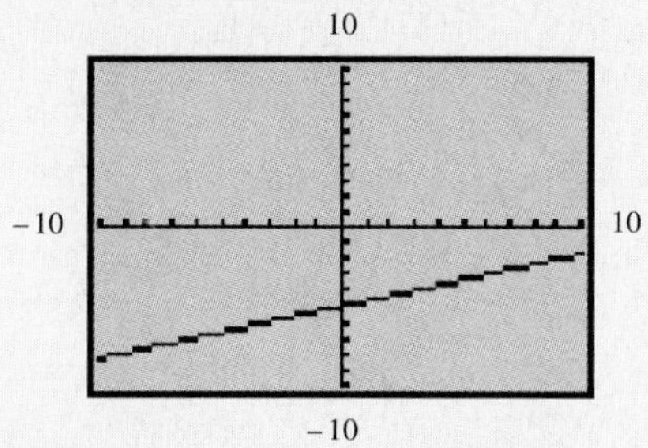

78. (a) Solve the equation $-2(x - 5) = -x - 2$ using the methods of Chapter 2.
(b) Explain why the standard viewing window of a graphing calculator cannot provide graphical support of the solution found in part (a). What minimum and maximum x-values would make it possible for the solution to be seen?

RELATING CONCEPTS (EXERCISES 79–84)

In Section 2.2 we learned how formulas can be applied to problem solving. In Exercises 79–84, we will see how the formula that relates the Celsius and Fahrenheit temperatures is derived.

Work Exercises 79–84 in order.

79. There is a linear relationship between Celsius and Fahrenheit temperatures. When C = 0°, F = _____°, and when C = 100°, F = _____°.

80. Think of ordered pairs of temperatures (C, F), where C and F represent corresponding Celsius and Fahrenheit temperatures. The equation that relates the two scales has a straight-line graph that contains the two points determined in Exercise 79. What are these two points?

81. Find the slope of the line described in Exercise 80.

82. Now think of the point-slope form of the equation in terms of C and F, where C replaces x and F replaces y. Use the slope you found in Exercise 81 and one of the two points determined earlier, and find the equation that gives F in terms of C.

83. To obtain another form of the formula, use the equation you found in Exercise 82 and solve for C in terms of F.

84. The equation found in Exercise 82 is graphed on the graphing calculator screen shown here. Observe the display at the bottom, and interpret it in the context of this group of exercises.

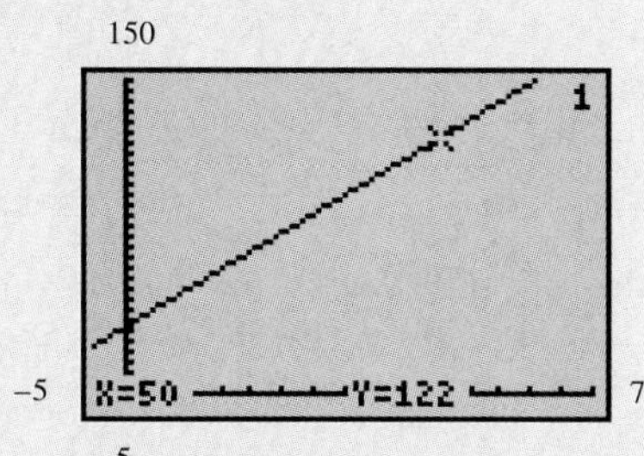

Did you make the connection between Celsius and Fahrenheit temperatures?

3.4 Linear Inequalities in Two Variables

OBJECTIVES

1 Graph linear inequalities in two variables.

2 Graph the intersection of two linear inequalities.

3 Use a graphing calculator to solve linear inequalities.

FOR EXTRA HELP

SSG Sec. 3.4
SSM Sec. 3.4

Pass the Test Software

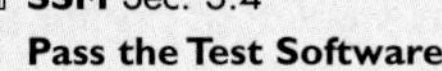

InterAct Math Tutorial Software

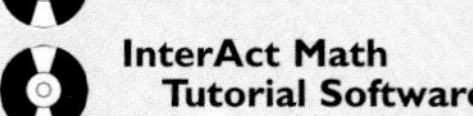

Video 5

OBJECTIVE 1 Graph linear inequalities in two variables. Linear inequalities in one variable were graphed on the number line in Chapter 2. In this section linear inequalities in two variables are graphed on a rectangular coordinate system.

Linear Inequality

An inequality that can be written as

$$Ax + By < C \quad \text{or} \quad Ax + By > C,$$

where A, B, and C are real numbers and A and B are not both 0, is called a **linear inequality in two variables.**

Also, $\leq$ and $\geq$ may replace $<$ and $>$ in the definition.

A line divides the plane into three regions: the line itself and the two half-planes on either side of the line. Recall that the graphs of linear inequalities in one variable are intervals on the number line that sometimes include an endpoint. The graphs of linear inequalities in two variables are *regions* in the real number plane and may include a *boundary line.* The **boundary line** for the inequality $Ax + By < C$ or $Ax + By > C$ is the graph of the *equation* $Ax + By = C$. To graph a linear inequality, follow these steps.

Graphing a Linear Inequality

Step 1 **Draw the boundary.** Draw the graph of the straight line that is the boundary. Make the line solid if the inequality involves $\leq$ or $\geq$; make the line dashed if the inequality involves $<$ or $>$.

Step 2 **Choose a test point.** Choose any point not on the line as a test point.

Step 3 **Shade the appropriate region.** Shade the region that includes the test point if it satisfies the original inequality; otherwise, shade the region on the other side of the boundary line.

EXAMPLE 1 Graphing a Linear Inequality

Graph $3x + 2y \geq 6$.

First graph the straight line $3x + 2y = 6$. The graph of this line, the boundary of the graph of the inequality, is shown in Figure 24. The graph of the inequality $3x + 2y \geq 6$ includes the points of the line $3x + 2y = 6$, and either the points *above* the line $3x + 2y = 6$ or the points *below* that line. To decide which, select any point not on the line $3x + 2y = 6$ as a test point. The origin, (0, 0), is often a good choice. Substitute the values from the test point (0, 0) for x and y in the inequality $3x + 2y > 6$.

$$3(\mathbf{0}) + 2(\mathbf{0}) > 6 \quad ?$$

$$0 > 6 \quad \text{False}$$

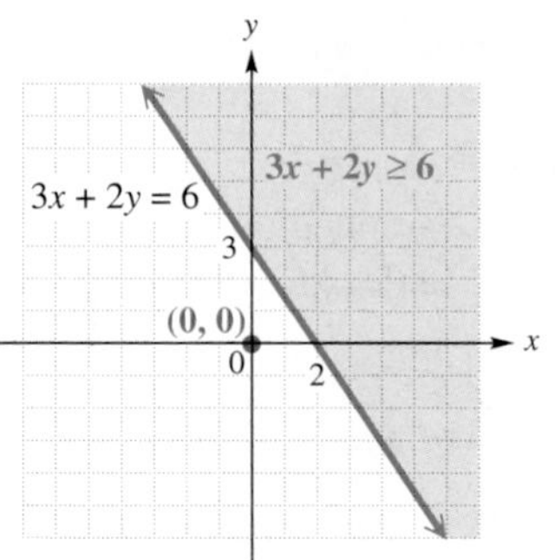

Figure 24

Since the result is false (0, 0) does not satisfy the inequality, and so the solution set includes all points on the other side of the line. This region is shaded in Figure 24.

If the inequality is written in the form $y > mx + b$ or $y < mx + b$, the inequality symbol indicates which half-plane to shade.

If $y > mx + b$, shade ***above*** the boundary line;
if $y < mx + b$, shade ***below*** the boundary line.

EXAMPLE 2 Graphing a Linear Inequality

Graph $x - 3y > 4$.

First graph the boundary line, shown in Figure 25. The points of the boundary line do not belong to the inequality $x - 3y > 4$ (since the inequality symbol is $>$, not $\geq$). For this reason, the line is dashed. Now solve the inequality for y.

$$x - 3y > 4$$
$$-3y > -x + 4$$
$$y < \frac{x}{3} - \frac{4}{3} \qquad \text{Multiply by } -\tfrac{1}{3}\text{; change } > \text{ to } <.$$

Because of the *less than* symbol, we should shade *below* the line. As a check, we can choose a test point not on the line, say (1, 2), and substitute for x and y in the original inequality.

$$1 - 3(2) > 4 \qquad ?$$
$$-5 > 4 \qquad \text{False}$$

This result agrees with our decision to shade below the line. The solution set, graphed in Figure 25, includes only those points in the shaded half-plane (not those on the line).

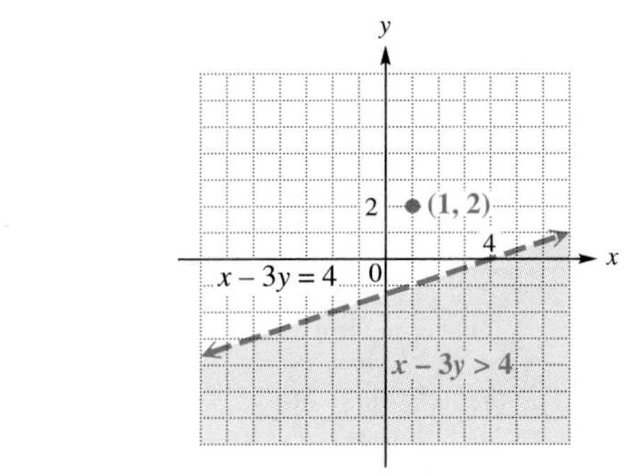

Figure 25

OBJECTIVE 2 Graph the intersection of two linear inequalities. In Section 2.6 we discussed how the words "and" and "or" are used with compound inequalities. In that section, the inequalities had one variable. Those ideas can be extended to include inequalities in two variables. A pair of inequalities joined with the word "and" is interpreted as the intersection of the solutions of the inequalities. The graph of the intersection of two or more inequalities is the region of the plane where all points satisfy all of the inequalities at the same time.

EXAMPLE 3 Graphing the Intersection of Two Inequalities

Graph $2x + 4y \geq 5$ and $x \geq 1$.

To begin, we graph each of the two inequalities $2x + 4y \geq 5$ and $x \geq 1$ separately. The graph of $2x + 4y \geq 5$ is shown in Figure 26(a), and the graph of $x \geq 1$ is shown in Figure 26(b).

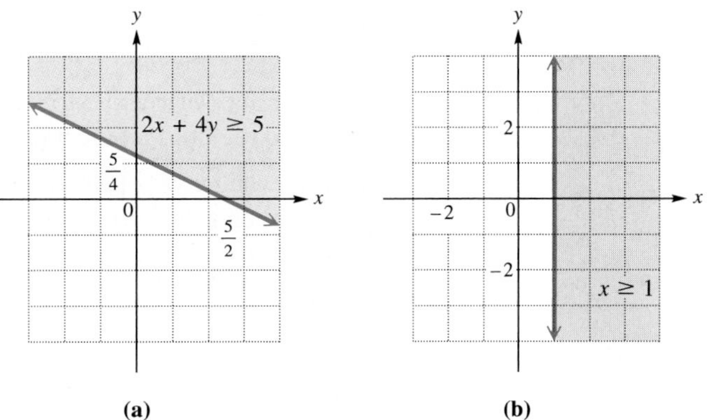

(a) (b)

Figure 26

In practice, the two graphs in Figure 26 are graphed on the same axes. Then we use heavy shading to identify the intersection of the graphs, as shown in Figure 27.

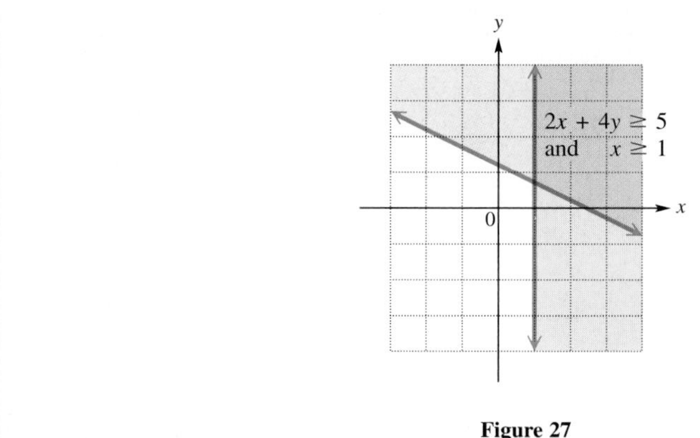

Figure 27

CONNECTIONS

Many realistic problems involve inequalities. For example, suppose a factory can have *no more than* 200 workers on a shift, but must have *at least* 100 and must manufacture *at least* 3000 units at minimum cost. The managers need to know how many workers should be on a shift in order to produce the required units at minimal cost. *Linear programming* is a method for finding the optimal (best possible) solution that meets all the conditions for such problems. The first step in solving linear programming problems with two variables is to express the conditions (constraints) as inequalities, graph the system of inequalities, and identify the region that satisfies all the inequalities at once.

FOR DISCUSSION OR WRITING

Let x represent the number of workers and y represent the number of units manufactured.

1. Write three inequalities expressing the conditions in the problem given above.
2. Graph the inequalities from Item 1 and shade the intersection.
3. The cost per worker is \$50 per day and the cost to manufacture 1 unit is \$100. Write an expression representing the total daily cost, C.
4. Find values of x and y for several points in or on the boundary of the shaded region. Include any "corner points."
5. Of the values of x and y that you chose in Item 4, which gives the least cost when substituted in the cost equation from Item 3? What does your answer mean in terms of the given problem? Is your answer reasonable? Explain why.

OBJECTIVE 3 Use a graphing calculator to solve linear inequalities. In Section 3.3 we saw that the x-intercept of the graph of the line $y = mx + b$ indicates the solution of the equation $mx + b = 0$. We can extend this observation to find solutions of the associated inequalities $mx + b > 0$ and $mx + b < 0$. The solution set of $mx + b > 0$ is the set of all x-values for which the graph of $y = mx + b$ is *above* the x-axis. (We consider points above because the symbol is $>$.) On the other hand, the solution set of $mx + b < 0$ is the set of all x-values for which the graph of $y = mx + b$ is *below* the x-axis. (We consider points below because the symbol is $<$.) Therefore, once we know the solution set of the equation and have the graph of the line, we can determine the solution sets of the corresponding inequalities.

In Figure 28, the x-intercept of $y = 3x - 9$ is $(3, 0)$. Therefore, as shown in Section 3.3,

$$\text{the solution set of } 3x - 9 = 0 \text{ is } \{3\}.$$

Because the graph of y lies above the x-axis for x-values greater than 3,

$$\text{the solution set of } 3x - 9 > 0 \text{ is } (3, \infty).$$

Because the graph lies below the x-axis for x-values less than 3,

$$\text{the solution set of } 3x - 9 < 0 \text{ is } (-\infty, 3).$$

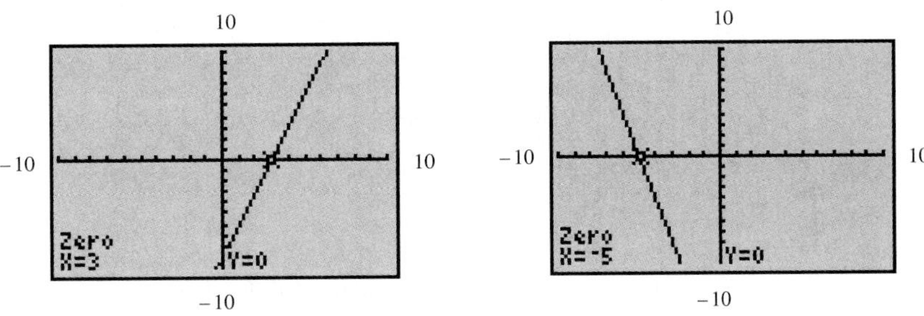

Figure 28

Figure 29

Suppose that we wish to solve the equation $-2(3x + 1) = -2x + 18$, and the associated inequalities $-2(3x + 1) > -2x + 18$ and $-2(3x + 1) < -2x + 18$. We begin by rewriting the equation so that the right side is equal to 0:

$$-2(3x + 1) + 2x - 18 = 0.$$

Graphing

$$y = -2(3x + 1) + 2x - 18$$

yields the x-intercept $(-5, 0)$ as shown in Figure 29. The first inequality listed above is equivalent to $y > 0$. Because the line lies *above* the x-axis for x-values less than -5,

$$\text{the solution set of } -2(3x + 1) > -2x + 18 \text{ is } (-\infty, -5).$$

Because the line lies *below* the x-axis for x-values greater than -5,

$$\text{the solution set of } -2(3x + 1) < -2x + 18 \text{ is } (-5, \infty).$$

3.4 EXERCISES

In Exercises 1–4, fill in the first blank with either solid *or* dashed. *Fill in the second blank with* above *or* below.

1. The boundary of the graph of $y \le -x + 2$ will be a ______ line, and the shading will be ______ the line.

2. The boundary of the graph of $y < -x + 2$ will be a ______ line, and the shading will be ______ the line.

3. The boundary of the graph of $y > -x + 2$ will be a ______ line, and the shading will be ______ the line.

4. The boundary of the graph of $y \geq -x + 2$ will be a ______ line, and the shading will be ______ the line.
5. In your own words, describe the steps used to graph a linear inequality. Use examples.
6. Compare and contrast the steps for graphing a linear inequality in two variables with those used to graph a linear inequality in one variable.

Graph each linear inequality in two variables. See Examples 1 and 2.

7. $x + y \leq 2$
8. $x + y \leq -3$
9. $4x - y < 4$
10. $3x - y < 3$
11. $x + 3y \geq -2$
12. $x + 4y \geq -3$
13. $x + y > 0$
14. $x + 2y > 0$
15. $x - 3y \leq 0$
16. $x - 5y \leq 0$
17. $y < x$
18. $y \leq 4x$

Graph each compound inequality. See Example 3.

19. $x + y \leq 1$ and $x \geq 1$
20. $x - y \geq 2$ and $x \geq 3$
21. $2x - y \geq 2$ and $y < 4$
22. $3x - y \geq 3$ and $y < 3$
23. $x + y > -5$ and $y < -2$
24. $6x - 4y < 10$ and $y > 2$

Use the method described in Section 2.7 to write the absolute value inequality as an "and" statement. Then solve each compound inequality and graph its solution set in the rectangular coordinate plane.

25. $|x| < 3$
26. $|y| < 5$
27. $|x + 1| < 2$
28. $|y - 3| < 2$

When a compound inequality involves the word or, *the graph is found by graphing each individual inequality and then taking the* union *of the two. For example, the graph of*

$$2x + 4y \geq 5 \quad \text{or} \quad x \geq 1$$

is shown here. Use this idea to graph each compound inequality.

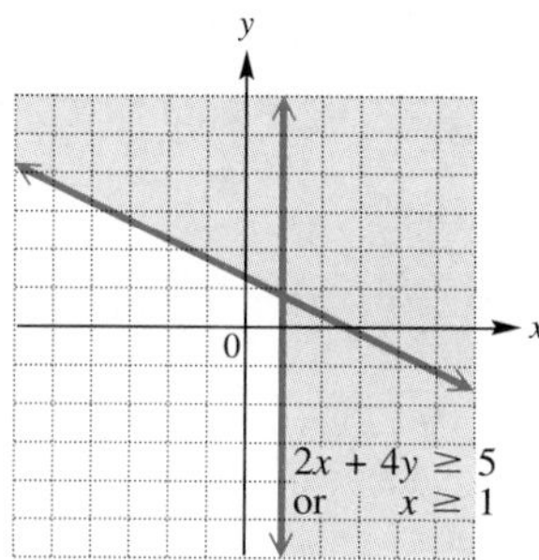

29. $x - y \geq 1$ or $y \geq 2$
30. $x + y \leq 2$ or $y \geq 3$
31. $x - 2 > y$ or $x < 1$
32. $x + 3 < y$ or $x > 3$
33. $3x + 2y < 6$ or $x - 2y > 2$
34. $x - y \geq 1$ or $x + y \leq 4$

TECHNOLOGY INSIGHTS (EXERCISES 35–42)

The graph of a linear equation $y = mx + b$ is shown on a graphing calculator screen, along with the x-value of the x-intercept of the line.

Use the screen to solve **(a)** $y = 0$, **(b)** $y < 0$, *and* **(c)** $y > 0$. *See Objective 3.*

35.

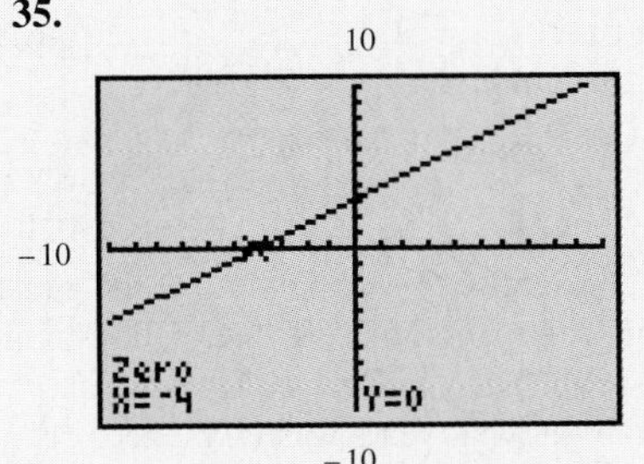

36.

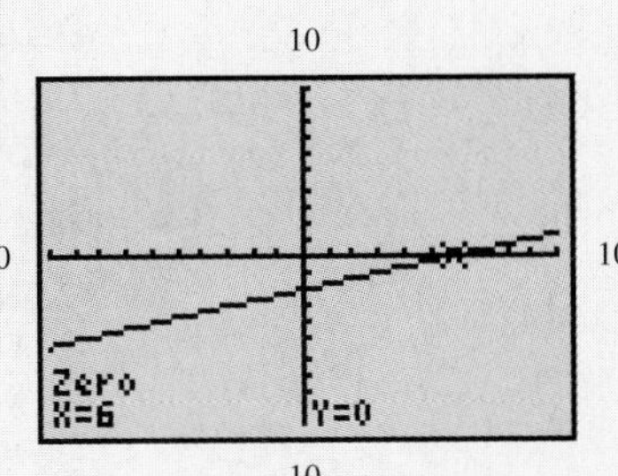

37.

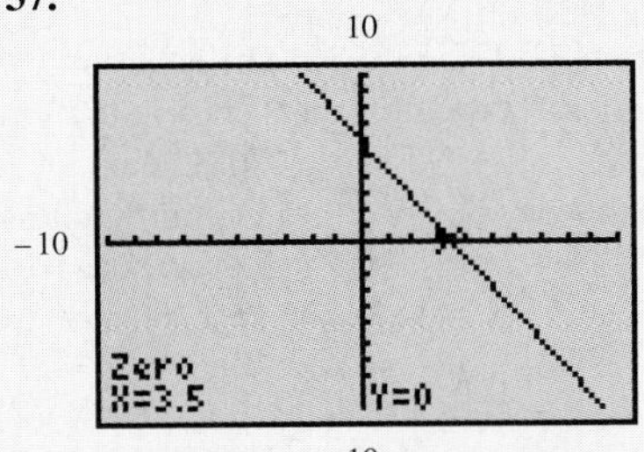

38.

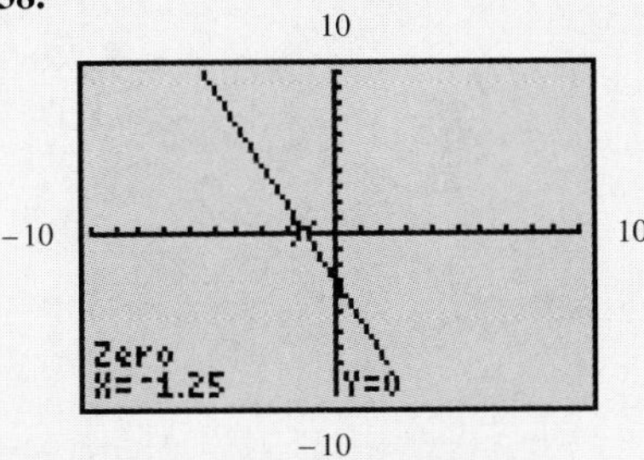

Match each inequality with its calculator-generated graph. (Hint: Use the slope, y-intercept, and inequality symbol in making your choice.)

39. $y \leq 3x - 6$

40. $y \geq 3x - 6$

41. $y \leq -3x - 6$

42. $y \geq -3x - 6$

A.

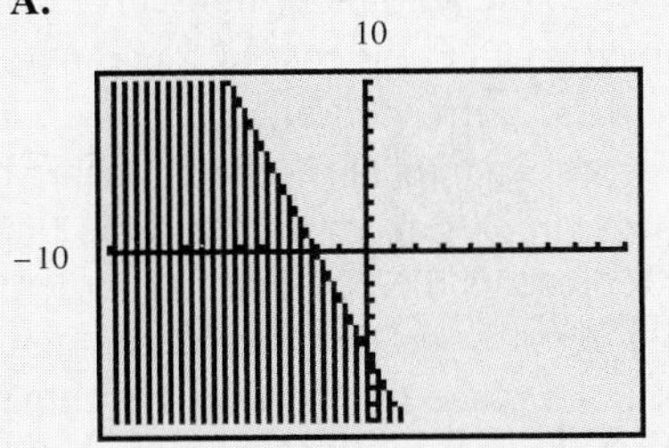

B.

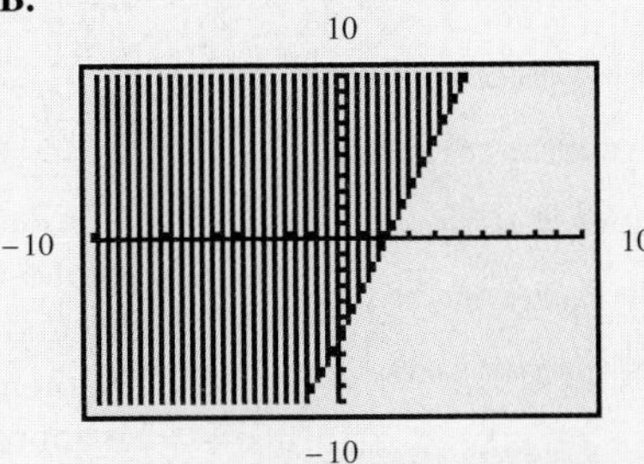

C.

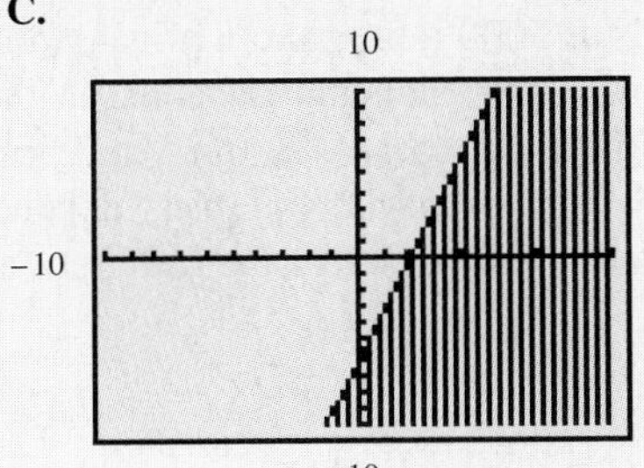

D.

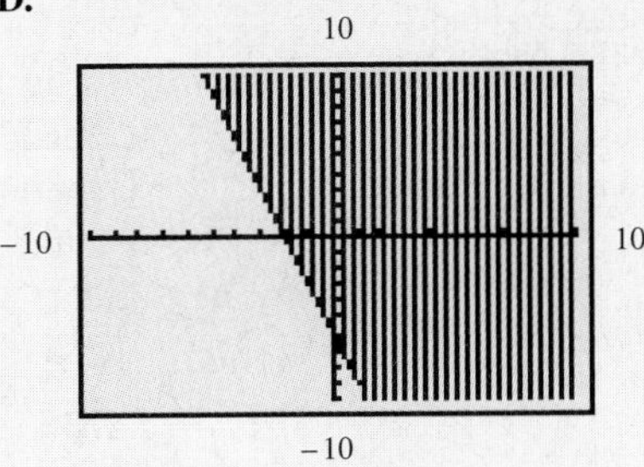

Solve the equation in part (a) and the associated inequalities in parts (b) and (c) using the methods of Chapter 2. Then graph the left side as y in the standard viewing window of a graphing calculator, and explain how the graph supports your answers in parts (a), (b), and (c).

43. **(a)** $5x + 3 = 0$
(b) $5x + 3 > 0$
(c) $5x + 3 < 0$

44. **(a)** $6x + 3 = 0$
(b) $6x + 3 > 0$
(c) $6x + 3 < 0$

45. **(a)** $-8x - (2x + 12) = 0$
(b) $-8x - (2x + 12) \geq 0$
(c) $-8x - (2x + 12) \leq 0$

46. **(a)** $-4x - (2x + 18) = 0$
(b) $-4x - (2x + 18) \geq 0$
(c) $-4x - (2x + 18) \leq 0$

Work each problem.

47. In Section 3.3 Exercise 69, we found that the equation $y = 1.86x + 78.24$ was a reasonable approximation of U.S. energy consumption y (in quadrillion BTUs) for the years 1991–1996. The years are coded so that $x = 1$ represents 1991 and so on. According to the equation, in what years was U.S. energy consumption greater than or equal to 85 quadrillion BTUs? (*Hint:* y represents U.S. energy consumption. Substitute for y and solve for x.)

48. Refer to Section 3.3 Exercise 70. The equation $y = .335x + 5.515$ approximates nuclear power production (in quadrillion BTUs) in the years 1993–1995. The years are coded so that $x = 3$ represents 1993 and so on. Use an inequality to find the years between 1992 and 1996 when nuclear power production was less than or equal to 7 quadrillion BTUs.

3.5 Introduction to Functions

OBJECTIVES

1. Define relation and function.
2. Find domain and range.
3. Identify functions.
4. Use function notation.
5. Identify linear functions.

FOR EXTRA HELP

SSG Sec. 3.5
SSM Sec. 3.5
Pass the Test Software
InterAct Math Tutorial Software
Video 5

It is often useful to describe one quantity in terms of another; for example, the growth of a plant is related to the amount of light it receives, the demand for a product is related to the price of the product, the cost of a trip is related to the distance traveled, and so on. To represent these corresponding quantities, it is helpful to use ordered pairs.

For example, we can indicate the relationship between the demand for a product and its price by writing ordered pairs in which the first number represents the price and the second number represents the demand. The ordered pair (5, 1000) then could indicate a demand for 1000 items when the price of the item is \$5. Since the demand depends on the price charged, we place the price first and the demand second. The ordered pair is an abbreviation for the sentence "If the price is 5 (dollars), then the demand is for 1000 (items)." Similarly, the ordered pairs (3, 5000) and (10, 250) show that a price of \$3 produces a demand for 5000 items, and a price of \$10 produces a demand for 250 items.

In this example, the demand depends on the price of the item. For this reason, demand is called the *dependent variable,* and price is called the *independent variable.* Generalizing, if the value of the variable y depends on the value of the variable x, then y is the **dependent variable** and x the **independent variable.**

Dependent variable

$$(x, y)$$

Independent variable

OBJECTIVE 1 Define relation and function. Since related quantities can be written using ordered pairs, the concept of *relation* can be defined as follows.

Relation

A **relation** is a set of ordered pairs.

For example, the sets

$$F = \{(1, 2), (-2, 5), (3, -1)\} \quad \text{and} \quad G = \{(-4, 1), (-2, 1), (-2, 0)\}$$

are both relations. A special kind of relation, called a *function,* is very important in mathematics and its applications.

Function

A **function** is a relation in which, for each value of the first component of the ordered pairs, there is *exactly one value* of the second component.

Of the two examples of a relation just given, only set F is a function, because for each x-value, there is exactly one y-value. In set $G,$ the last two ordered pairs have the same x-value paired with two different y-values, so G is a relation, but not a function.

In a function, there is *exactly one* value of the dependent variable, the second component, for each value of the independent variable, the first component. This is what makes functions so important in applications. It would not be as useful, for example, to know a price/demand relationship that gave more than one demand for a given price.

Another way to think of a functional relationship is to think of the independent variable as an input and the dependent variable as an output. A calculator is an input-output machine, for example. To find 8^2, we input 8, touch the squaring key, and see that the output is 64. Inputs and outputs can also be determined from a graph or a table.

A third way to describe a function is to give a rule that tells how to determine the dependent variable for a specific value of the independent variable. The rule may be given in words: the dependent variable is twice the independent variable. Usually the rule is an equation:

$$\underset{\substack{\uparrow \\ \text{Dependent} \\ \text{variable}}}{y} = 2\underset{\substack{\uparrow \\ \text{Independent} \\ \text{variable}}}{x}.$$

This is the most efficient way to define a function.

EXAMPLE 1 Determining Independent and Dependent Variables of a Function

Determine the independent and dependent variables for each of the following functions. Give an example of an ordered pair belonging to the function.

(a) The 1996 Summer Olympics medal winners in men's basketball are {(gold, United States), (silver, Yugoslavia), (bronze, Lithuania)} (*Source:* United States Olympic Committee).

The independent variable (the first component in each ordered pair) is the type of medal; the dependent variable (the second component) is the recipient. Any of the ordered pairs could be given as an example.

(b) An input-output machine that produces square roots

The independent variables (the inputs) are nonnegative real numbers, since the square root of a negative number is not a real number. The dependent variables are their nonnegative square roots. For example, (81, 9) belongs to this function.

(c) The graph in Figure 30, which shows the relationship between the number of gallons of water in a small swimming pool and time in hours

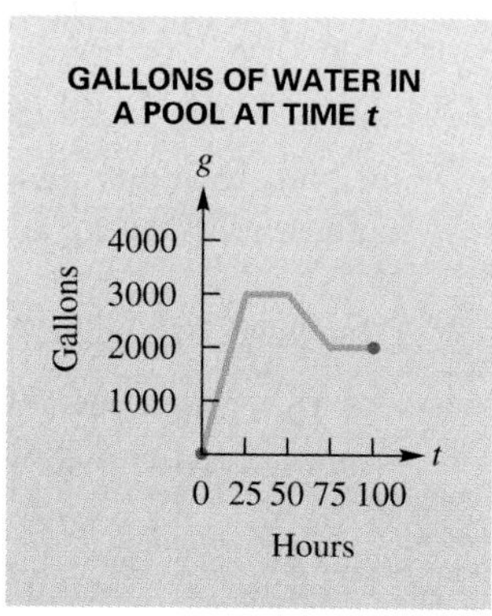

Figure 30

The independent variables are hours, and the dependent variables are the gallons of water in the pool. One ordered pair is (25, 3000).

(d) Petroleum imports in millions of barrels per day for selected years

U.S. Petroleum Imports

Year	*Imports*
1992	7.89
1993	8.62
1994	9.00
1995	8.83
1996	9.40

Source: U.S. Energy Department.

The independent variables are the years; the dependent variables are the imports. An example of an ordered pair is (1994, 9.00).

(e) $y = 3x + 4$

The independent variable is x, and the dependent variable is y. One ordered pair is $(\frac{1}{3}, 5)$.

OBJECTIVE 2 Find domain and range.

Domain and Range

In a relation, the set of all values of the independent variable (x) is the **domain;** the set of all values of the dependent variable (y) is the **range.**

EXAMPLE 2 Determining the Domain and Range of a Relation

Give the domain and range of each function in Example 1.

(a) The domain is the type of medal, {gold, silver, bronze}, and the range is the set of winning countries, {United States, Yugoslavia, Lithuania}.

(b) Here, the domain is restricted to nonnegative numbers: $[0, \infty)$. The range is also $[0, \infty)$.

(c) The domain is all possible values of t, the time in hours, which is the interval $[0, 100]$. The range is the number of gallons at time t, the interval $[0, 3000]$.

(d) The domain is the set of years, {1992, 1993, 1994, 1995, 1996}; the range is the set of imports (in millions of barrels per day), {7.89, 8.62, 9.00, 8.83, 9.40}.

(e) In the defining equation (or rule), $y = 3x + 4$, x can be any real number, so the domain is $\{x \mid x \text{ is a real number}\}$ or $(-\infty, \infty)$. (If the equation were $y = \frac{1}{x - 1}$, however, we would need to restrict the domain to all real numbers except 1, which would cause a 0 denominator. We will discuss this in more detail in a later chapter.) Since every real number y can be produced by some value of x, the range is also the set $\{y \mid y \text{ is a real number}\}$ or $(-\infty, \infty)$.

The **graph of a relation** is the graph of its ordered pairs. The graph gives a picture of the relation, which can be used to determine its domain and range.

EXAMPLE 3 Finding Domains and Ranges from Graphs

Give the domain and range of each relation.

(a)

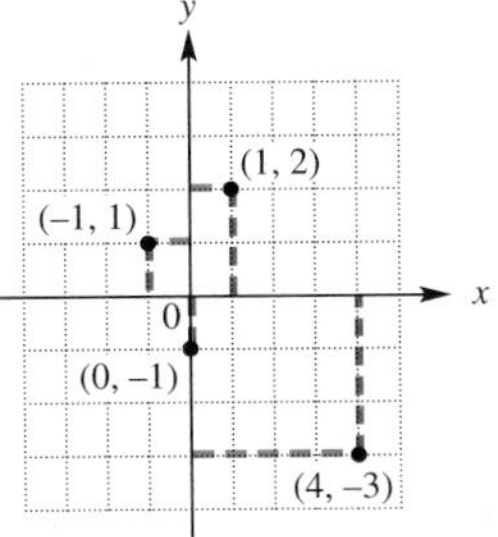

The domain is the set of x-values, $\{-1, 0, 1, 4\}$. The range is the set of y-values, $\{-3, -1, 1, 2\}$.

(b)

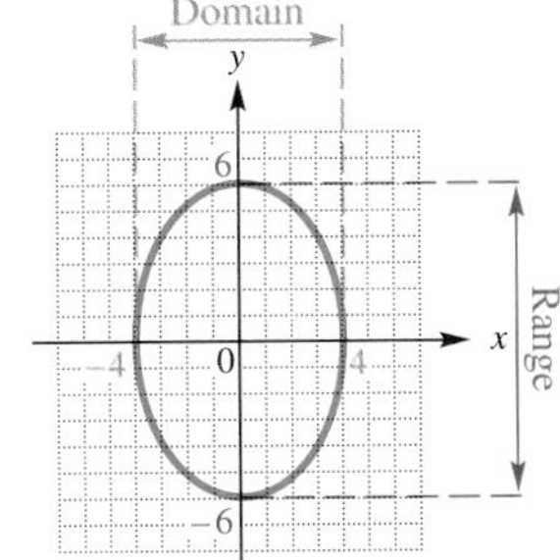

The x-values of the points on the graph include all numbers between -4 and 4, inclusive. The y-values include all numbers between -6 and 6, inclusive. Using interval notation,

the domain is $[-4, 4]$;
the range is $[-6, 6]$.

(c)

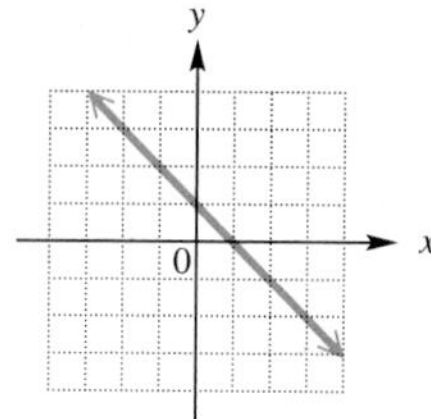

The arrowheads indicate that the line extends indefinitely left and right, as well as up and down. Therefore, both the domain and the range are the set of all real numbers, written $(-\infty, \infty)$.

(d)

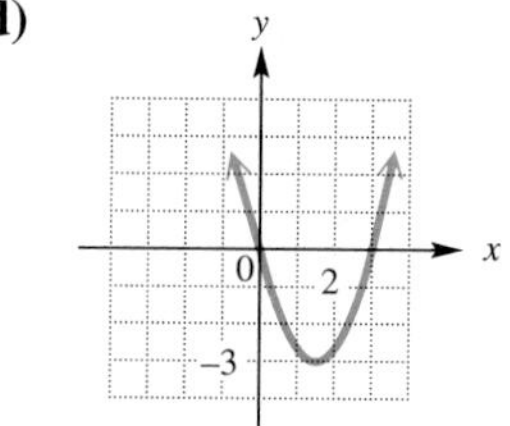

The arrowheads indicate that the graph extends indefinitely left and right, as well as upward. The domain is $(-\infty, \infty)$. Because there is a least y-value, -3, the range includes all numbers greater than or equal to -3, written $[-3, \infty)$.

Relations are often defined by equations, such as $y = 2x + 3$ and $y^2 = x$. It is sometimes necessary to determine the domain of a relation from its equation. In this book, the following agreement on the domain of a relation is assumed.

Agreement on Domain

The domain of a relation is assumed to be all real numbers that produce real numbers when substituted for the independent variable.

To illustrate this agreement, since any real number can be used as a replacement for x in $y = 2x + 3$, the domain of this function is the set of real numbers. As another example, the function defined by $y = \frac{1}{x}$ has all real numbers except 0 as domain, since y is undefined if $x = 0$. In general, the domain of a function defined by an algebraic expression is all real numbers, except those numbers that lead to division by 0 or an even root of a negative number.

OBJECTIVE 3 **Identify functions.** Most of the relations we have seen in the examples are functions—that is, each x-value corresponds to exactly one y-value. Now we look at ways to determine whether a given relation, defined algebraically, is a function.

In a function each value of x leads to only one value of y, so any vertical line drawn through the graph of a function must intersect the graph in at most one point. This is the **vertical line test for a function.**

Vertical Line Test

If a vertical line intersects the graph of a relation in more than one point, then the relation does not represent a function.

For example, the graph shown in Figure 31(a) is not the graph of a function, since a vertical line can intersect the graph in more than one point, while the graph in Figure 31(b) does represent a function.

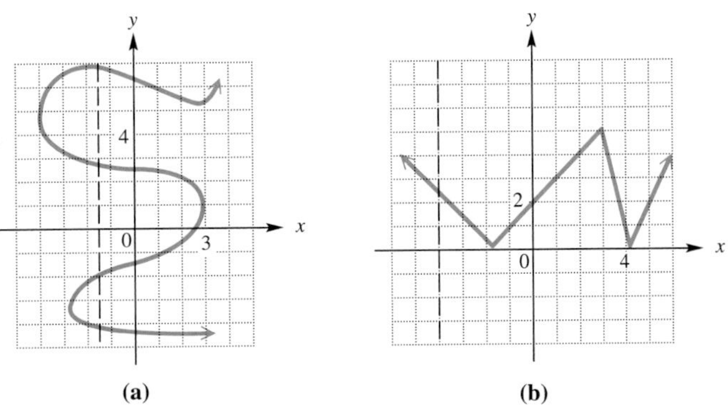

Figure 31

The vertical line test is a simple method for identifying a function defined by a graph. It is more difficult to decide whether a relation defined by an equation is a function. The next example gives some hints that may help.

EXAMPLE 4 Identifying Functions

Decide whether each of the following defines a function and give the domain.

(a) $y = \sqrt{2x - 1}$

Here, for any choice of x in the domain, there is exactly one corresponding value for y (the radical is a nonnegative number), so this equation defines a function. Since the radicand cannot be negative,

$$\begin{aligned} 2x - 1 &\geq 0 \\ 2x &\geq 1 \\ x &\geq \frac{1}{2}. \end{aligned}$$

The domain is $\left[\frac{1}{2}, \infty\right)$.

(b) $y^2 = x$

The ordered pairs (16, 4) and (16, −4) both satisfy this equation. Since one value of x, 16, corresponds to two values of y, 4 and −4, this equation does not define a function. Solving $y^2 = x$ for y gives $y = \sqrt{x}$ or $y = -\sqrt{x}$, which shows that two values of y correspond to each positive value of x. Because x is equal to the square of y, the values of x must always be nonnegative. The domain of the relation is $[0, \infty)$.

(c) $y \leq x - 1$

By definition, y is a function of x if a value of x leads to exactly one value of y. In this example, a particular value of x, say 1, corresponds to many values of y. The ordered pairs (1, 0), (1, −1), (1, −2), (1, −3), and so on, all satisfy the inequality. For this reason, an inequality does not define a function. Any number can be used for x, so the domain is the set of real numbers $(-\infty, \infty)$.

(d) $y = \dfrac{5}{x - 1}$

Given any value of x in the domain, we find y by subtracting 1, then dividing the result into 5. This process produces exactly one value of y for each value in the domain, so this equation defines a function. The domain includes all real numbers except those that make the denominator 0. We find these numbers by setting the denominator equal to 0 and solving for x.

$$x - 1 = 0$$
$$x = 1$$

Thus, the domain includes all real numbers except 1. In interval notation this is written as

$$(-\infty, 1) \cup (1, \infty).$$

In summary, three variations of the definition of function are given here.

Variations of the Definition of Function

1. A **function** is a relation in which, for each value of the first component of the ordered pairs, there is exactly one value of the second component.
2. A **function** is a set of ordered pairs in which no first component is repeated.
3. A **function** is a rule or correspondence that assigns exactly one range value to each domain value.

OBJECTIVE 4 Use function notation. When a function f is defined with a rule or an equation using x and y for the independent and dependent variables, we say "y is a function of x" to emphasize that y *depends on* x. We use the notation

$$y = f(x)$$

to express this. (In this special notation the parentheses do not indicate multiplication.) The letter f stands for *function.* For example, if $y = 2x - 7$, we write

$$f(x) = 2x - 7.$$

When you see the notation $f(x)$, remember that it is just another name for the dependent variable y. This **function notation** is useful for simplifying certain statements. For example, if $y = 9x - 5$, then replacing x with 2 gives

$$\begin{aligned} y &= 9 \cdot 2 - 5 \\ &= 18 - 5 \\ &= 13. \end{aligned}$$

The statement "if $x = 2$, then $y = 13$" is abbreviated with function notation as

$$f(2) = 13.$$

Read $f(2)$ as "f of 2" or "f at 2." Also, $f(0) = 9 \cdot 0 - 5 = -5$, and $f(-3) = 9(-3) - 5 = -32$.

These ideas and the symbols used to represent them can be explained as follows.

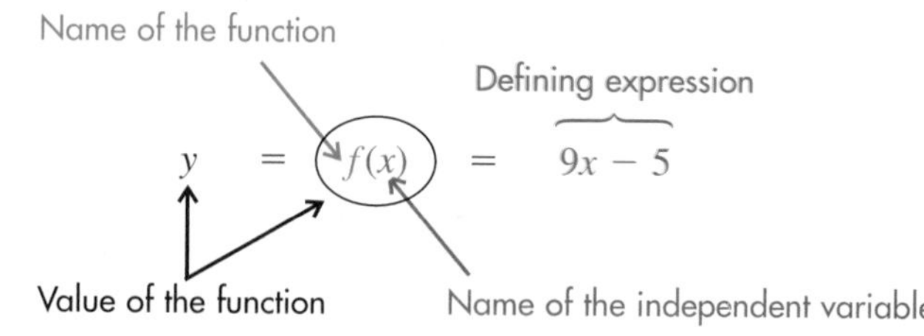

CAUTION The symbol $f(x)$ *does not* indicate "f times x," but represents the y-value for the indicated x-value. As shown above, $f(2)$ is the y-value that corresponds to the x-value 2.

EXAMPLE 5 Using Function Notation

Let $f(x) = -x^2 + 5x - 3$. Find the following.

(a) $f(2)$

Replace x with 2.

$$f(2) = -2^2 + 5 \cdot 2 - 3 = -4 + 10 - 3 = 3$$

(b) $f(-1)$

$$f(-1) = -(-1)^2 + 5(-1) - 3 = -1 - 5 - 3 = -9$$

(c) $f(q)$

Replace x with q.

$$f(q) = -q^2 + 5q - 3$$

The replacement of one variable with another is important in later courses.

Sometimes letters other than f, such as g, h, or capital letters F, G, and H are used to name functions.

EXAMPLE 6 Using Function Notation

Let $g(x) = 2x + 3$. Find and simplify the following.

(a) $g(a + 1)$

Replace x with $a + 1$.

$$g(a + 1) = 2(a + 1) + 3 = 2a + 2 + 3 = 2a + 5$$

(b) $G\left(\frac{1}{b + 4}\right)$

$$G\left(\frac{1}{b + 4}\right) = 2\left(\frac{1}{b + 4}\right) + 3 = \left(\frac{2}{b + 4}\right) + 3$$

Replacing the variable x with an algebraic expression, like $a + 1$ in Example 6(a), is called **composition of functions.** The result is called a **composite function.**

EXAMPLE 7 Forming a Composite Function

(a) Write $f(x) = 4x - 1$ and $g(x) = x^2$ as the composite function $g[f(x)]$.

First replace $f(x)$ with $4x - 1$. Then use the fact that $g(x) = x^2$.

$$g[f(x)] = g(4x - 1) = (4x - 1)^2$$

(b) Find $f[g(x)]$ for the functions in part (a).

Work from the inside out. Replace $g(x)$ with x^2. Then use function f.

$$f[g(x)] = f(x^2) = 4x^2 - 1$$

Notice that $g[f(x)] \neq f[g(x)]$. This is generally true.

If a function is defined by an equation with x and y, not with function notation, use the following steps to find $f(x)$.

Finding an Expression for $f(x)$

If an equation that defines a function is given with x and y, to find $f(x)$:

1. solve the equation for y;
2. replace y with $f(x)$.

EXAMPLE 8 Writing Equations Using Function Notation

Rewrite each equation using function notation; then find $f(-2)$ and $f(a)$.

(a) $y = x^2 + 1$

This equation is already solved for y. Since $y = f(x)$,

$$f(x) = x^2 + 1.$$

To find $f(-2)$, we let $x = -2$:

$$\begin{aligned} f(-2) &= (-2)^2 + 1 \\ &= 4 + 1 \\ &= 5. \end{aligned}$$

We find $f(a)$ by letting $x = a$: $f(a) = a^2 + 1$.

(b) $x - 4y = 5$

First solve $x - 4y = 5$ for y. Then replace y with $f(x)$.

$$\begin{aligned} x - 4y &= 5 \\ x - 5 &= 4y \\ y &= \frac{x-5}{4} \qquad \text{so} \qquad f(x) = \frac{1}{4}x - \frac{5}{4} \end{aligned}$$

Now find $f(-2)$ and $f(a)$

$$f(-2) = \frac{1}{4}(-2) - \frac{5}{4} = -\frac{7}{4}$$

and

$$f(a) = \frac{1}{4}a - \frac{5}{4}.$$

OBJECTIVE 5 Identify linear functions. Our first two-dimensional graphing was of straight lines. Linear equations (except for $x = c$) define *linear functions.*

Linear Function

A function that can be written in the form

$$f(x) = mx + b$$

for real numbers m and b is a **linear function.**

Recall from Section 3.3 that m is the slope of the line and $(0, b)$ is the y-intercept. A linear function of the form $f(x) = d$ is sometimes called a **constant function.** The domain of any linear function is $(-\infty, \infty)$. The range of a nonconstant linear function is $(-\infty, \infty)$, while the range of the constant function $f(x) = d$ is $\{d\}$.

In later chapters of this book, we will learn about several other types of functions.

3.5 EXERCISES

To work the following exercises, refer to Examples 1 and 2.

1. In your own words, define a function and give an example.
2. In your own words, define the domain of a function and give an example.
3. In an ordered pair of a relation, is the first element the independent or the dependent variable?

For each of the following relations, decide whether or not it is a function, and give the domain and range. See Examples 1–4. Use the vertical line test in Exercises 8 and 11–14.

4. $\{(1, 1), (1, -1), (2, 4), (2, -4), (3, 9), (3, -9)\}$
5. $\{(2, 5), (3, 7), (4, 9), (5, 11)\}$
6. The set containing the top five producers of hydroelectric power and the amount each produces, in billion kilowatt-hours, is {(Canada, 319.1), (United States, 268.2), (Brazil, 225.0), (Russia, 173.4), (China, 143.1)}. (*Source:* U.S. Department of Energy, *International Energy Annual,* 1993.)
7. An input-output machine accepts positive real numbers as input, and outputs both their positive and negative roots.

8.

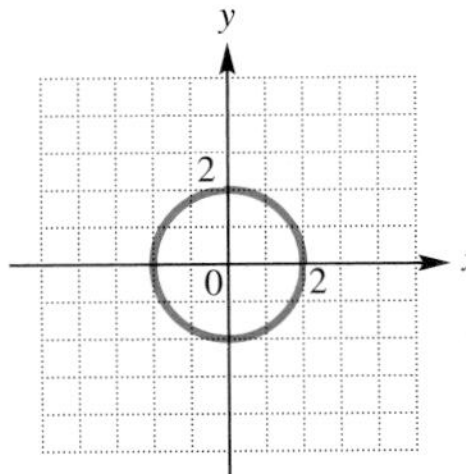

9. **Crude Oil Prices and Cost per Gallon of Gas at the Pump (1996)**

Unleaded regular	\$1.22
Unleaded premium	\$1.44
Crude oil	\$ .21

Source: U.S. Energy Department.

10. $x = |y|$

11.

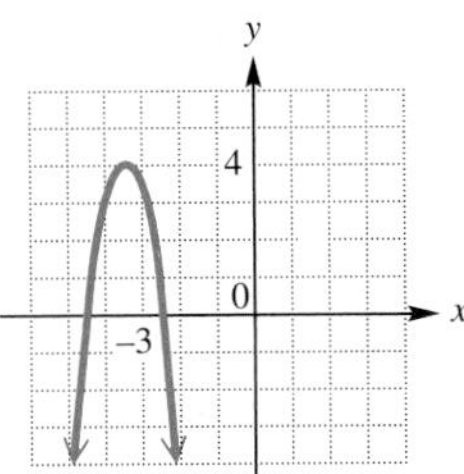

12.

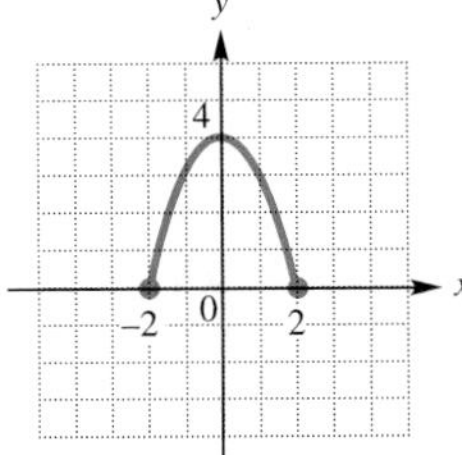

13.

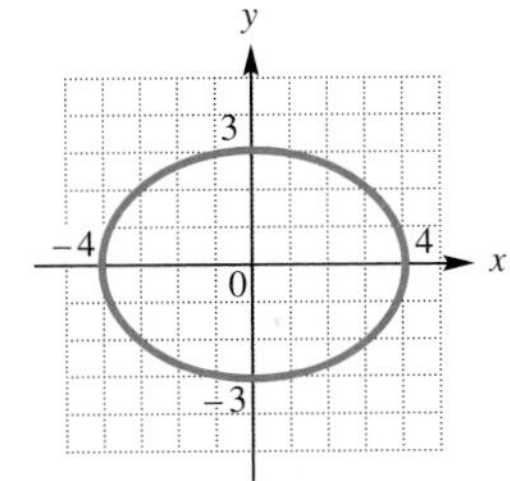

14.

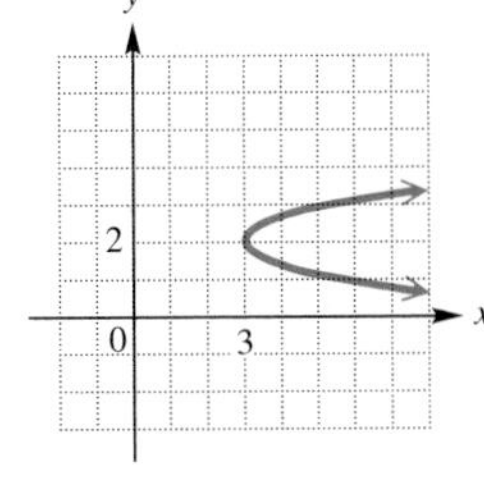

Decide whether the given relation defines y as a function of x. Give the domain. See Example 4.

15. $y = x^2$ **16.** $y = x^3$ **17.** $x = y^2$ **18.** $x = y^4$

19. $x + y < 4$ **20.** $x - y < 3$ **21.** $y = \sqrt{x}$ **22.** $y = -\sqrt{x}$

23. $xy = 1$ **24.** $xy = -3$ **25.** $y = \sqrt{4x + 2}$ **26.** $y = \sqrt{9 - 2x}$

27. $y = \dfrac{2}{x - 9}$ **28.** $y = \dfrac{-7}{x - 16}$

29. Refer to Example 1, Figure 30, to answer the questions.
(a) What numbers are possible values of the dependent variable?
(b) For how long is the water level increasing? Decreasing?
(c) How many gallons are in the pool after 90 hours?
(d) Call this function f. What is $f(0)$? What does it mean in this example?

30. The graph shows the daily megawatts of electricity used on a record-breaking summer day in Sacramento, California.

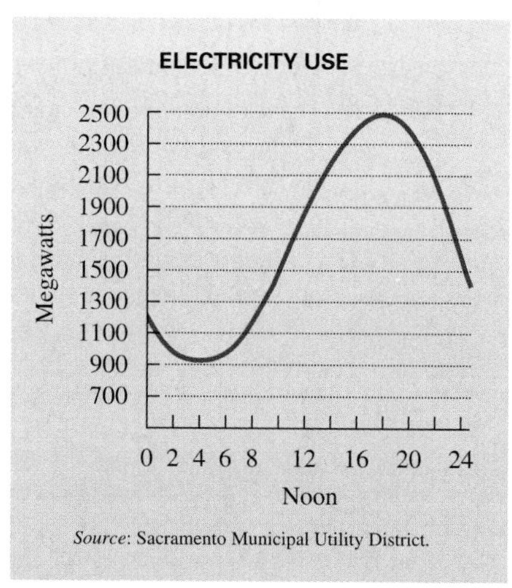

(a) Is this the graph of a function?
(b) What is the domain?
(c) Estimate the number of megawatts at 8 A.M.
(d) At what time was the most electricity used? The least electricity?

31. Give an example of a function from everyday life. (*Hint:* Fill in the blanks: ______ depends on ______, so ______ is a function of ______.)

32. Choose the correct response. The notation $f(3)$ means
(a) the variable f times 3 or $3f$.
(b) the value of the dependent variable when the independent variable is 3.
(c) the value of the independent variable when the dependent variable is 3.
(d) f equals 3.

Let $f(x) = -3x + 4$ *and* $g(x) = -x^2 + 4x + 1$. *Find the following. See Examples 5–7.*

33. $f(0)$ **34.** $f(-3)$ **35.** $f(-x)$ **36.** $f(x + 2)$

37. $g(10)$ **38.** $g(-1.5)$ **39.** $g\left(\frac{1}{2}\right)$ **40.** $g(k)$

41. $g(2p)$ **42.** $f[g(1)]$ **43.** $g[f(1)]$ **44.** $f[g(x)]$

45. Compare the answers to Exercises 42 and 43. Do you think that $f[g(x)]$ is, in general, equal to $g[f(x)]$?

46. Make up two linear functions f and g such that $f[g(2)] = 4$. (There are many ways to do this.)

An equation that defines y as a function of x is given. (a) Solve for y in terms of x, and replace y with the function notation f(x). (b) Find f(3). See Example 8.

47. $x + 3y = 12$ **48.** $x - 4y = 8$ **49.** $y + 2x^2 = 3$

50. $y - 3x^2 = 2$ **51.** $4x - 3y = 8$ **52.** $-2x + 5y = 9$

53. Fill in the blanks with the correct responses.

The equation $2x + y = 4$ has a straight ______ as its graph. One point that lies on the line is (3, ______). If we solve the equation for y and use function notation, we have a linear function $f(x) =$ ______. For this function, $f(3) =$ ______, meaning that the point (______, ______) lies on the graph of the function.

54. Which one of the following defines a linear function?

(a) $y = \frac{x - 5}{4}$ **(b)** $y = \frac{1}{x}$ **(c)** $y = x^2$ **(d)** $y = \sqrt{x}$

Sketch the graph of each linear function. Give the domain.

55. $f(x) = -2x + 5$ **56.** $g(x) = 4x - 1$ **57.** $h(x) = \frac{1}{2}x + 2$

58. $F(x) = -\frac{1}{4}x + 1$ **59.** $G(x) = 2x$ **60.** $H(x) = -3x$

61. $g(x) = -4$ **62.** $f(x) = 5$

63. **(a)** Suppose that a taxicab driver charges $1.50 per mile. Fill in the chart with the correct response for the price $f(x)$ she charges for a trip of x miles.

x	$f(x)$
0	
1	
2	
3	

(b) The linear function that gives a rule for the amount charged is $f(x) =$ ______.

(c) Graph this function for the domain $\{0, 1, 2, 3\}$.

64. Suppose that a package weighing x pounds costs $f(x)$ dollars to mail to a given location, where

$$f(x) = 2.75x.$$

(a) What is the value of $f(3)$?

(b) In your own words, describe what 3 and the value $f(3)$ mean in part (a), using the terminology *independent variable* and *dependent variable*.

Forensic scientists use the lengths of certain bones to calculate the height of a person. Two bones often used are the tibia (t), the bone from the ankle to the knee, and the femur (r), the bone from the knee to the hip socket. A person's height (h) is determined from the lengths of these bones using functions defined by the following formulas. All measurements are in centimeters.

For men: $h(r) = 69.09 + 2.24r$ or $h(t) = 81.69 + 2.39t$

For women: $h(r) = 61.41 + 2.32r$ or $h(t) = 72.57 + 2.53t$

65. Find the height of a man with a femur measuring 56 centimeters.

66. Find the height of a man with a tibia measuring 40 centimeters.

67. Find the height of a woman with a femur measuring 50 centimeters.

68. Find the height of a woman with a tibia measuring 36 centimeters.

Federal regulations set standards for the size of the quarters of marine mammals. A pool to house sea otters must have a volume of "the square of the sea otter's average adult length (in meters) multiplied by 3.14 *and by* .91 *meter." If x represents the sea otter's average adult length and $f(x)$ represents the volume of the corresponding pool size, this formula can be written as* $f(x) = (.91)(3.14)x^2$. *Find the volume of the pool for each of the following adult lengths (in meters). Round answers to the nearest hundredth.*

69. .8 **70.** 1.0 **71.** 1.2 **72.** 1.5

73. The linear function $f(x) = -183x + 40{,}034$ is a model for the number of U.S. post offices from 1990–1995, where $x = 0$ corresponds to 1990, $x = 1$ corresponds to 1991, and so on. Use this model to give the approximate number of post offices during the following years. (*Source:* U.S. Postal Service, *Annual Report of the Postmaster General and Comprehensive Statement on Postal Operations.*)

(a) 1991

(b) 1993

(c) 1995

(d) The graphing calculator screen shows a portion of the graph of $y = f(x)$ with the coordinates of a point on the graph displayed at the bottom of the screen. Interpret the meaning of the display in the context of this application.

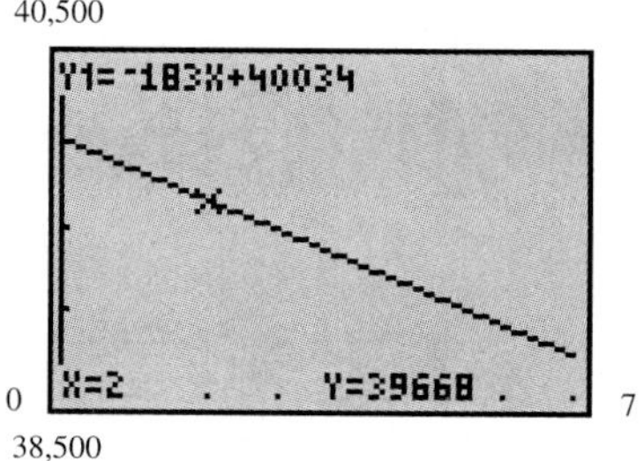

74. The linear function $f(x) = -6324x + 305{,}294$ is a model for U.S. defense budgets in millions of dollars from 1992–1996, where $x = 0$ corresponds to 1990, $x = 2$ corresponds to 1992, and so on. Use this model to approximate the defense budget for the following years. (*Source:* U.S. Office of Management and Budget.)

(a) 1993

(b) 1995

(c) 1996

(d) A portion of the graph of $y = f(x)$ is shown in the graphing calculator screen, with the coordinates of a point displayed at the bottom of the screen. Interpret the meaning of the display in the context of this application.

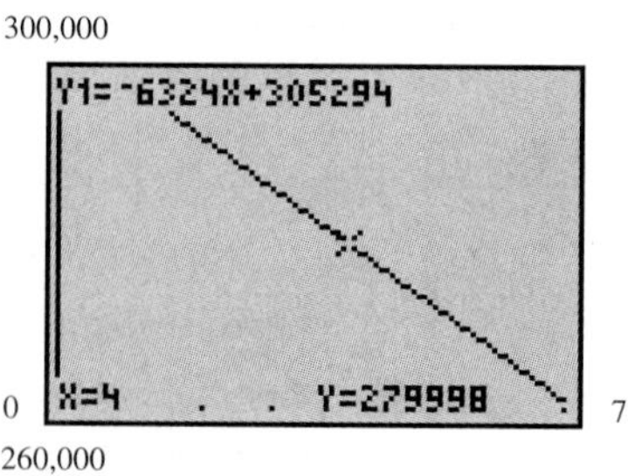

TECHNOLOGY INSIGHTS (EXERCISES 75–78)

75. The graphing calculator screen shows the graph of a linear function $y = f(x)$, along with the display of coordinates of a point on the graph. Use function notation to write what the display indicates.

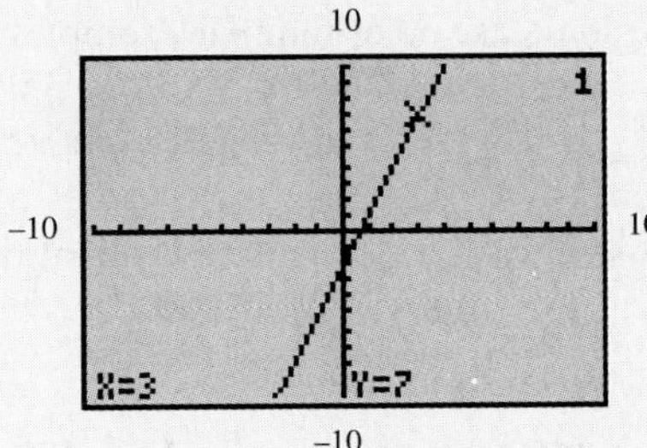

76. The table was generated by a graphing calculator for a linear function $Y_1 = f(x)$. Use the table to answer.

(a) What is $f(2)$?

(b) If $f(x) = -3.7$, what is the value of x?

(c) What is the slope of the line?

(d) What is the y-intercept of the line?

(e) Find the expression for $f(x)$.

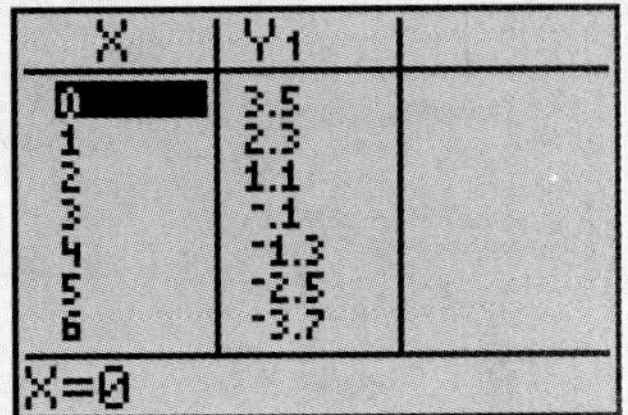

X	Y1
0	3.5
1	2.3
2	1.1
3	-.1
4	-1.3
5	-2.5
6	-3.7

X=0

77. The two screens show the graph of the same linear function $y = f(x)$. Find the expression for $f(x)$.

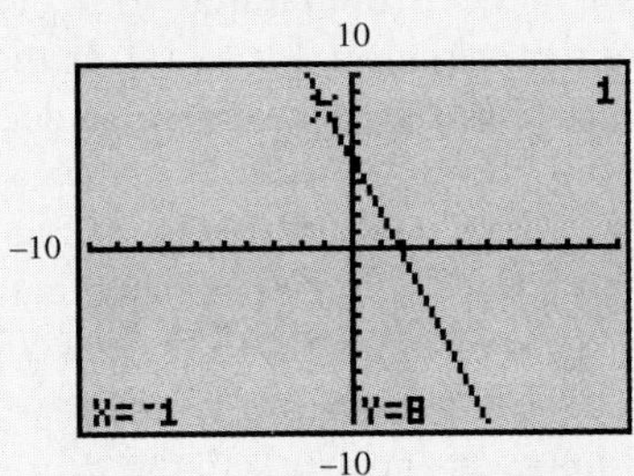

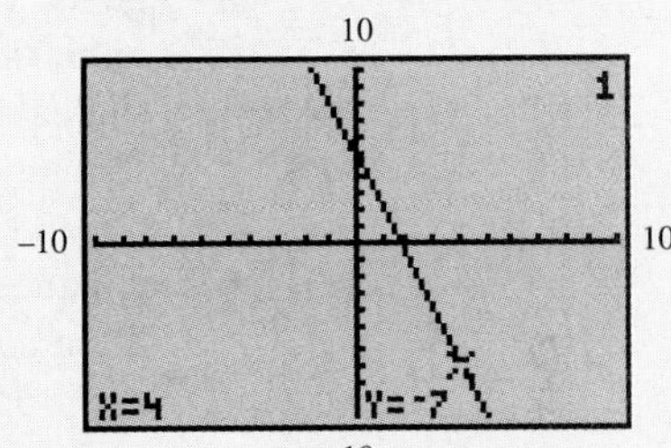

78. The formula for converting Celsius to Fahrenheit is $F = 1.8C + 32$. If we graph this formula as $y = f(x) = 1.8x + 32$ with a graphing calculator, we obtain the accompanying screen. The point $(-40, -40)$ lies on the graph, as indicated by the display. Interpret the meaning of this in the context of this exercise.

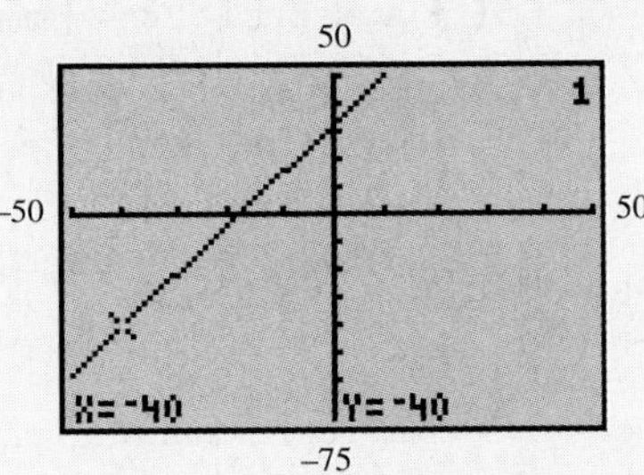

3.6 Variation

OBJECTIVES

1. Write an equation expressing direct variation.
2. Find the constant of variation and solve direct variation problems.
3. Solve inverse variation problems.
4. Solve joint variation problems.
5. Solve combined variation problems.

FOR EXTRA HELP

SSG Sec. 3.6
SSM Sec. 3.6
Pass the Test Software
InterAct Math Tutorial Software
Video 5

Certain types of functions are very common, especially in the physical sciences. These are functions where y depends on a multiple of x, or y depends on a number divided by x. In such situations, y is said to *vary directly* as x (in the first case) or *vary inversely* as x (in the second case). For example, the period of a pendulum varies directly as the square root of the length of the pendulum and inversely as the square root of the acceleration due to gravity. The distance formula, the simple interest formula, and formulas for area and volume are other familiar examples.

OBJECTIVE 1 Write an equation expressing direct variation. The circumference of a circle is given by the formula $C = 2\pi r$, where r is the radius of the circle. As the formula shows, the circumference is always a constant multiple of the radius (C is always found by multiplying r by the constant 2π). Because of this, the circumference is said to *vary directly* as the radius.

Direct Variation

y* varies directly as *x if there exists some constant k such that

$$y = kx.$$

Also, y is said to be **proportional** to x. The number k is called the **constant of variation.** In direct variation, for $k > 0$, as the value of x increases, the value of y also increases. Similarly, as x decreases, y also decreases.

OBJECTIVE 2 Find the constant of variation and solve direct variation problems. The direct variation equation defines a linear function. In applications, functions are often defined by variation equations. For example, if Tom earns \$8 per hour, his wages vary directly as, or are proportional to, the number of hours he works. If y represents his total wages and x the number of hours he has worked, then

$$y = 8x.$$

Here k, the constant of variation, is 8.

EXAMPLE 1 Finding the Constant of Variation

Miguel Hidalgo is paid an hourly wage. One week he worked 43 hours and was paid \$795.50. How much does he earn per hour?

Let h represent the number of hours he works and T represent his corresponding pay. Then, T varies directly as h, and

$$T = kh.$$

Here k represents Miguel's hourly wage. Since $T = 795.50$ when $h = 43$,

$$795.50 = 43k$$
$$\mathbf{k = 18.50} \quad \text{Use a calculator.}$$

His hourly wage is \$18.50. Thus, T and h are related by

$$\mathbf{T = 18.50h}.$$

In summary, follow these steps to solve a variation problem.

Solving a Variation Problem

Step 1 Write the variation equation.

Step 2 Substitute the initial values and solve for k.

Step 3 Rewrite the variation equation with the value of k from Step 2.

Step 4 Substitute the remaining values, solve for the unknown, and find the required answer.

EXAMPLE 2 Solving a Direct Variation Problem

Hooke's law for an elastic spring states that the distance a spring stretches is proportional to the force applied. If a force of 150 newtons* stretches a certain spring 8 centimeters, how much will a force of 400 newtons stretch the spring?

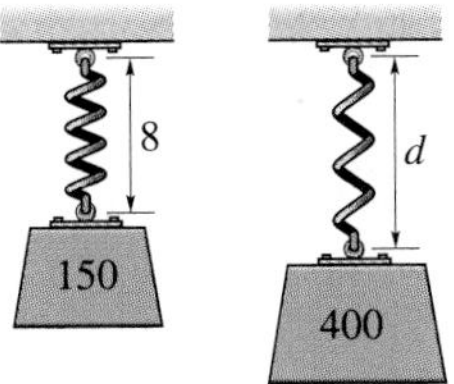

Figure 32

If d is the distance the spring stretches and f is the force applied, then $d = kf$ for some constant k. Since a force of 150 newtons stretches the spring 8 centimeters, we can use these values to find k.

$$d = kf \qquad \text{Formula}$$

$$8 = k \cdot 150 \qquad d = 8, f = 150$$

$$k = \frac{8}{150} = \frac{4}{75}, \qquad \text{Find } k.$$

so

$$d = \frac{4}{75}f.$$

For a force of 400 newtons,

$$d = \frac{4}{75}(400) \qquad \text{Let } f = 400.$$

$$= \frac{64}{3}.$$

The spring will stretch $\frac{64}{3}$ centimeters if a force of 400 newtons is applied.

The direct variation equation $y = kx$ is a linear equation. However, other kinds of variation involve other types of equations. For example, one variable can be directly proportional to a power of another variable.

*A newton is a unit of measure of force used in physics.

Direct Variation as a Power

y* varies directly as the *n*th power of *x if there exists a real number k such that

$$y = kx^n.$$

An example of direct variation as a power involves the area of a circle. The formula for the area of a circle is

$$A = \pi r^2.$$

Here, π is the constant of variation, and the area varies directly as the square of the radius.

EXAMPLE 3 Solving a Problem Involving Direct Variation as a Power

The distance a body falls from rest varies directly as the square of the time it falls (here we disregard air resistance). If an object falls 64 feet in 2 seconds, how far will it fall in 8 seconds?

If d represents the distance the object falls and t the time it takes to fall,

$$d = kt^2$$

for some constant k. To find the value of k, use the fact that the object falls 64 feet in 2 seconds.

$$d = kt^2 \quad \text{Formula}$$

$$64 = k(2)^2 \quad \text{Let } d = 64 \text{ and } t = 2.$$

$$k = 16 \quad \text{Find } k.$$

With this result, the variation equation becomes

$$d = 16t^2.$$

Now let $t = 8$ to find the number of feet the object will fall in 8 seconds.

$$d = 16(8)^2 \quad \text{Let } t = 8.$$

$$= 1024$$

The object will fall 1024 feet in 8 seconds. ■

OBJECTIVE 3 Solve inverse variation problems. In direct variation where $k > 0$, as x increases, y increases, and similarly as x decreases, y decreases. Another type of variation is *inverse variation.*

Inverse Variation

y* varies inversely as *x if there exists a real number k such that

$$y = \frac{k}{x}.$$

Also, ***y* varies inversely as the *n*th power of *x*** if there exists a real number k such that

$$y = \frac{k}{x^n}.$$

With inverse variation, for $k > 0$, as x increases, y decreases, and as x decreases, y increases. Notice that the inverse variation equation also defines a function. Since x is in the denominator, these functions are called *rational functions.*

An example of inverse variation can be found by looking at the formula for the area of a parallelogram. In its usual form, the formula is

$$A = bh.$$

Dividing both sides by b gives

$$h = \frac{A}{b}.$$

Here, h (height) varies inversely as b (base), with A (the area) serving as the constant of variation. For example, if a parallelogram has an area of 72 square inches, the values of b and h might be any of the following:

$$b = 2, h = 36 \qquad b = 12, h = 6$$

$$b = 3, h = 24 \qquad b = 9, \ h = 8$$

$$b = 4, h = 18 \qquad b = 8, \ h = 9.$$

Notice that in the first group, as b increases, h decreases. In the second group, as b decreases, h increases.

EXAMPLE 4 Solving an Inverse Variation Problem

The weight of an object above the earth varies inversely as the square of its distance from the center of the earth. A space vehicle in an elliptical orbit has a maximum distance from the center of the earth (apogee) of 6700 miles. Its minimum distance from the center of the earth (perigee) is 4090 miles. See Figure 33. If an astronaut in the vehicle weighs 57 pounds at its apogee, what does the astronaut weigh at its perigee?

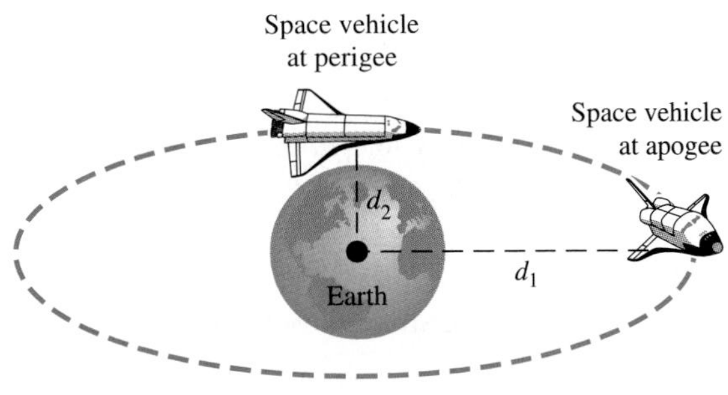

Figure 33

If w is the weight and d is the distance from the center of the earth, then

$$w = \frac{k}{d^2}$$

for some constant k.

At the apogee the astronaut weighs 57 pounds and the distance from the center of the earth is 6700 miles. Use these values to find k.

$$57 = \frac{k}{(6700)^2} \qquad \text{Let } w = 57 \text{ and } d = 6700.$$

$$k = 57(6700)^2$$

Then the weight at the perigee with $d = 4090$ miles is

$$w = \frac{57(6700)^2}{(4090)^2} \approx 153 \text{ pounds.}$$

OBJECTIVE 4 **Solve joint variation problems.** It is common for one variable to depend on several others. For example, if one variable varies as the product of several other variables (perhaps raised to powers), the first variable is said to **vary jointly** as the others.

EXAMPLE 5 Solving a Joint Variation Problem

The strength of a rectangular beam varies jointly as its width and the square of its depth. If the strength of a beam 2 inches wide by 10 inches deep is 1000 pounds per square inch, what is the strength of a beam 4 inches wide and 8 inches deep?

If S represents the strength, w the width, and d the depth, then

$$S = kwd^2$$

for some constant, k. Since $S = 1000$ if $w = 2$ and $d = 10$,

$$1000 = k(2)(10)^2. \qquad \text{Let } S = 1000, w = 2, \text{ and } d = 10.$$

Solving this equation for k gives

$$1000 = k \cdot 2 \cdot 100$$
$$1000 = 200k$$
$$k = 5,$$

so

$$S = 5wd^2.$$

Find S when $w = 4$ and $d = 8$ by substitution in $S = 5wd^2$.

$$S = 5(4)(8)^2 \qquad \text{Let } w = 4 \text{ and } d = 8.$$
$$= 1280$$

The strength of the beam is 1280 pounds per square inch.

OBJECTIVE 5 **Solve combined variation problems.** There are many combinations of direct and inverse variation. The final example shows a typical **combined variation** problem.

EXAMPLE 6 Solving a Combined Variation Problem

The body-mass index, or BMI, is used by physicians to assess a person's level of fatness.* The BMI varies directly as an individual's weight in pounds and inversely as the

*Source: *Reader's Digest,* October 1993.

square of the individual's height in inches. A person who weighs 118 pounds and is 64 inches tall has BMI 20. (The BMI is rounded to the nearest whole number.) Find the BMI of a person who weighs 165 pounds with height 70 inches.

Let B represent BMI, w weight, and h height. Then

$$B = \frac{kw}{h^2}. \quad \begin{array}{l}\leftarrow \text{BMI varies directly as the weight.} \\ \leftarrow \text{BMI varies inversely as the square of the height.}\end{array}$$

To find k, let $B = 20$, $w = 118$, and $h = 64$.

$$20 = \frac{k(118)}{64^2}$$

$$k = \frac{20(64^2)}{118} \quad \text{Multiply by } 64^2\text{; divide by } 118.$$

$$\approx 694 \quad \text{Use a calculator.}$$

Now find B when $k = 694$, $w = 165$, and $h = 70$.

$$B = \frac{694(165)}{70^2} \approx 23 \quad \text{(rounded)}$$

The required BMI is 23. A BMI from 20–26 is considered desirable.

3.6 EXERCISES

Determine whether each equation represents direct, inverse, joint, or combined variation.

1. $y = \frac{3}{x}$ **2.** $y = \frac{8}{x}$ **3.** $y = 10x^2$ **4.** $y = 2x^3$

5. $y = 3xz^4$ **6.** $y = 6x^3z^2$ **7.** $y = \frac{4x}{wz}$ **8.** $y = \frac{6x}{st}$

Solve each problem.

9. If x varies directly as y, and $x = 9$ when $y = 3$, find x when $y = 12$.

10. If x varies directly as y, and $x = 10$ when $y = 7$, find y when $x = 50$.

11. If z varies inversely as w, and $z = 10$ when $w = .5$, find z when $w = 8$.

12. If t varies inversely as s, and $t = 3$ when $s = 5$, find s when $t = 5$.

13. p varies jointly as q and r^2, and $p = 200$ when $q = 2$ and $r = 3$. Find p when $q = 5$ and $r = 2$.

14. f varies jointly as g^2 and h, and $f = 50$ when $g = 4$ and $h = 2$. Find f when $g = 3$ and $h = 6$.

15. For $k > 0$, if y varies directly as x, when x increases, y ________________, and when x decreases, y ________________.

16. For $k > 0$, if y varies inversely as x, when x increases, y ________________, and when x decreases, y ________________.

17. Explain the difference between inverse variation and direct variation.

18. What is meant by the constant of variation in a direct variation problem? If you were to graph the linear equation $y = kx$ for some nonnegative constant k, what role would the value of k play in the graph?

Solve each variation problem. Use a calculator as necessary. See Examples 1–6.

19. Todd bought 8 gallons of gasoline and paid \$8.79. To the nearest tenth of a cent, what is the price of gasoline per gallon?

20. Melissa gives horseback rides at Shadow Mountain Ranch. A 2.5-hour ride costs \$50.00. What is the price per hour?

21. The weight of an object on earth is directly proportional to the weight of that same object on the moon. A 200-pound astronaut would weigh 32 pounds on the moon. How much would a 50-pound dog weigh on the moon?

22. In the study of electricity, the resistance of a conductor of uniform cross-sectional area is directly proportional to its length. Suppose that the resistance of a certain type of copper wire is .640 ohm per 1000 feet. What is the resistance of 2500 feet of the wire?

23. The amount of water emptied by a pipe varies directly as the square of the diameter of the pipe. For a certain constant water flow, a pipe emptying into a canal will allow 200 gallons of water to escape in an hour. The diameter of the pipe is 6 inches. How much water would a 12-inch pipe empty into the canal in an hour, assuming the same water flow?

24. The pressure exerted by a certain liquid at a given point varies directly as the depth of the point beneath the surface of the liquid. The pressure at 30 meters is 80 newtons per square centimeter. What pressure is exerted at 50 meters?

25. For a body falling freely from rest (disregarding air resistance), the distance the body falls varies directly as the square of the time. If an object is dropped from the top of a tower 576 feet high and hits the ground in 6 seconds, how far did it fall in the first 4 seconds?

576 feet

26. The force required to compress a spring is proportional to the change in length of the spring. If a force of 20 newtons is required to compress a certain spring 2 centimeters, how much force is required to compress the spring from 20 centimeters to 8 centimeters?

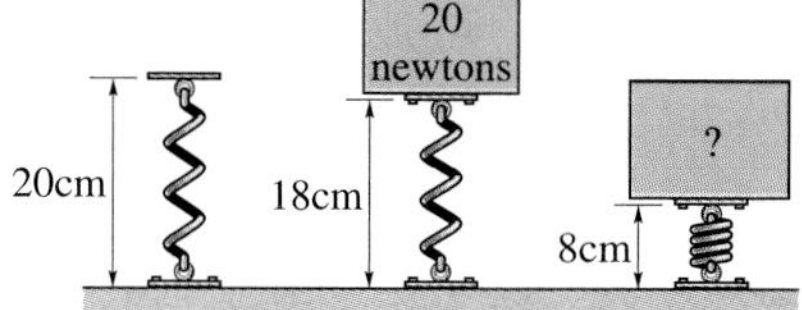

27. The illumination produced by a light source varies inversely as the square of the distance from the source. If the illumination produced 1 meter from a certain light source is 768 footcandles, find the illumination produced 6 meters from the same source.

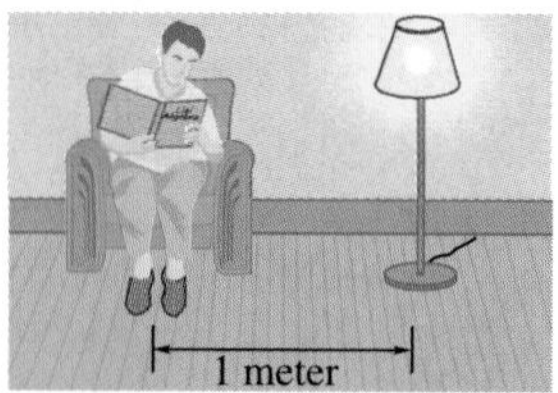

28. A meteorite approaching the earth has a velocity inversely proportional to the square root of its distance from the center of the earth. If the velocity is 5 kilometers per second when the distance is 8100 kilometers from the center of the earth, find the velocity at a distance of 6400 kilometers.

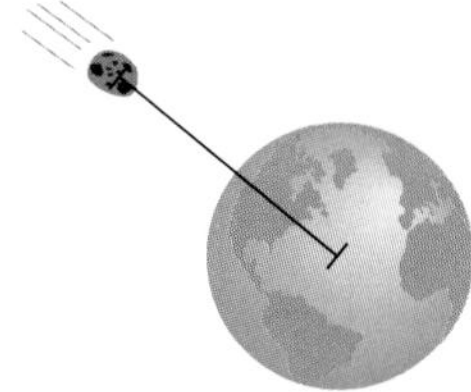

29. The frequency of a vibrating string varies inversely as its length. That is, a longer string vibrates fewer times in a second than a shorter string. Suppose a piano string 2 feet long vibrates 250 cycles per second. What frequency would a string 5 feet long have?

30. The force with which the earth attracts an object above the earth's surface varies inversely with the square of the distance of the object from the center of the earth. If an object 4000 miles from the center of the earth is attracted with a force of 160 pounds, find the force of attraction if the object were 6000 miles from the center of the earth.

31. The distance that a person can see to the horizon from a point above the surface of the earth varies directly as the square root of the height of that point (disregarding mountains, smog, and haze). If a person 144 meters above the surface of the earth can see for 18 kilometers to the horizon, how far can a person see to the horizon from a point 1600 meters high?

32. Natural gas provides 35.8% of U.S. energy. (*Source:* U.S. Energy Department.) The volume of a gas varies inversely as the pressure and directly as the temperature. (Temperature must be measured in *degrees Kelvin* (K), a unit of measurement used in physics.) If a certain gas occupies a volume of 1.3 liters at 300 K and a pressure of 18 newtons per square centimeter, find the volume at 340 K and a pressure of 24 newtons per square centimeter.

33. One source of renewable energy is wind, although as of 1995, it provided less than 5 trillion BTUs in the United States. (*Source:* U.S. Energy Information Administration, *Annual Energy Review.*) The force of the wind blowing on a vertical surface varies jointly as the area of the surface and the square of the velocity. If a wind of 40 miles per hour exerts a force of 50 pounds on a surface of $\frac{1}{2}$ square foot, how much force will a wind of 80 miles per hour place on a surface of 2 square feet?

34. It is shown in engineering that the maximum load a cylindrical column of circular cross-section can hold varies directly as the fourth power of the diameter and inversely as the square of the height. If a column 9 feet high and 3 feet in diameter will support a load of 8 tons, how great a load will be supported by a column 12 feet high and 2 feet in diameter?

35. The period of a pendulum varies directly as the square root of the length of the pendulum and inversely as the square root of the acceleration due to gravity. Find the period when the length is 4 feet and the acceleration due to gravity is 32 feet per second squared, if the period is 1.06π seconds when the length is 9 feet and the acceleration due to gravity is 32 feet per second squared.

36. The force needed to keep a car from skidding on a curve varies inversely as the radius of the curve and jointly as the weight of the car and the square of the speed. If 242 pounds of force keep a 2000-pound car from skidding on a curve of radius 500 feet at 30 miles per hour, what force would keep the same car from skidding on a curve of radius 750 feet at 50 miles per hour?

37. The maximum load of a horizontal beam that is supported at both ends varies directly as the width and the square of the height and inversely as the length between the supports. A beam 6 meters long, .1 meter wide, and .06 meter high supports a load of 360 kilograms. What is the maximum load supported by a beam 16 meters long, .2 meter wide, and .08 meter high?

38. The number of long-distance phone calls between two cities in a certain time period varies directly as the populations p_1 and p_2 of the cities, and inversely as the distance between them. If 80,000 calls are made between two cities 400 miles apart, with populations of 70,000 and 100,000, how many calls are made between cities with populations of 50,000 and 75,000 that are 250 miles apart?

39. NBA basketball player Chris Webber weighs 260 pounds and is 6 feet, 10 inches tall. Use the information in Example 6 to find his body-mass index. (*Source:* Internet: "The Unofficial Chris Webber Page.")

40. A body-mass index from 27 through 29 carries a slight risk of weight-related health problems, while one of 30 or more indicates a great increase in risk. Use your own height and weight and the information in Example 6 to determine whether you are at risk.

TECHNOLOGY INSIGHTS (EXERCISES 41 AND 42)

41. The graphing calculator screen shows a portion of the graph of a function $y = f(x)$ that satisfies the conditions for direct variation. What is $f(36)$?

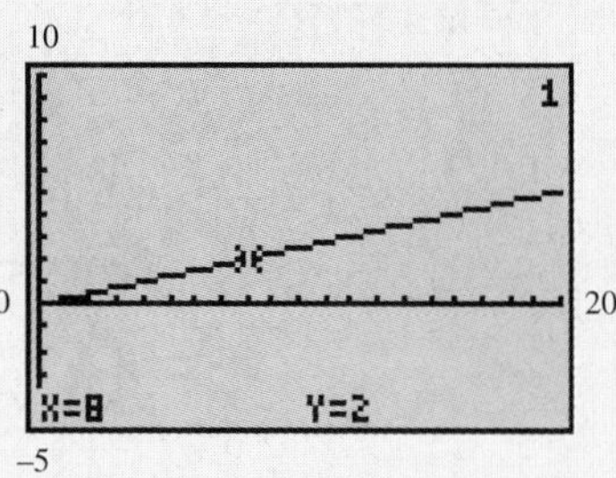

42. The accompanying table of points was generated by a graphing calculator. The points lie on the graph of a function $Y_1 = f(x)$ that satisfies the conditions for direct variation. What is $f(36)$?

X	Y1
0	0
1	1.5
2	3
3	4.5
4	6
5	7.5
6	9

X=0

RELATING CONCEPTS (EXERCISES 43–50)

A routine activity such as pumping gasoline can be related to many of the concepts studied in this chapter. Suppose that premium unleaded costs $1.25 per gallon.

Work Exercises 43–50 in order.

43. Zero gallons of gasoline cost $0.00, while 1 gallon costs $1.25. Represent these two pieces of information as ordered pairs of the form (gallons, price).

44. Use the information from Exercise 43 to find the slope of the line on which the two points lie.

45. Write the slope-intercept form of the equation of the line on which the two points lie.

46. Using function notation, if $f(x) = ax + b$ represents the line from Exercise 45, what are the values of a and b?

47. How does the value of a from Exercise 46 relate to gasoline in this situation? With relationship to the line, what do we call this number?

RELATING CONCEPTS (EXERCISES 43–50) (CONTINUED)

48. Why does the equation from Exercise 46 satisfy the conditions for direct variation? In the context of variation, what do we call the value of a?

49. The graph of the equation from Exercise 46 is shown in the accompanying graphing calculator screen, along with a display at the bottom of the screen. How is this display interpreted in the context of these exercises?

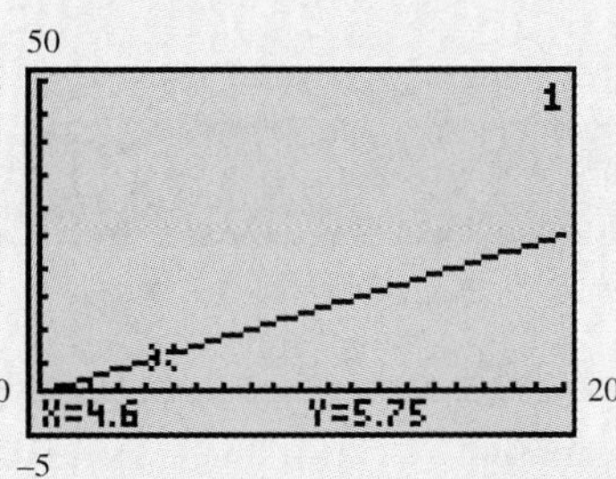

50. The accompanying table was generated by a graphing calculator, with Y_1 entered as the equation from Exercise 46. Interpret the entry for $x = 12$ in the context of these exercises.

X	Y_1
7	8.75
8	10
9	11.25
10	12.5
11	13.75
12	15
13	16.25

X=12

Did you make the connections between ordered pairs, slope, the equation of a line, a linear function, and direct variation?

CHAPTER 3 GROUP ACTIVITY

Choosing an Energy Source (or How to Get Your Water Hot)

Objective: Write and graph linear functions that model given data.

There are many different ways to heat water. In this activity you will look at three different energy sources that may be used to provide heat for a 40-gallon home water tank.

Have each student in your group choose one of the three types of water heaters listed in the table below.

Type of Hot Water Heater	*Size in Gallons*	*Price*	*Operating Cost per Month (manufacturer's estimate)*	*Hot Water Temperature*
Kenmore Economizer 6 —Electric	40	\$139.99	\$35.00	120°–130°
Kenmore Economizer 6 —Natural Gas	40	\$139.99	\$13.25	120°–130°
Sunbather Water Heater —Solar	40	\$950.00	\$0.00	*

Source: Jade Mountain 1999.

*No temperature listed but the ad says "Best for warm climates, preheating water, summer only in cold places, or when hot water needed only in afternoons and evenings."

(continued)

A. Using data for the water heater you selected, write a linear equation that represents total cost y of heating water with respect to time. Let x represent number of months.

B. Graph your equation using domain [0, 60] and range [0, 1000].

C. As a group, compare the graphs of your equations.
1. What are the y-intercepts?
2. How do the slopes compare? Which is the steepest? Which has a zero slope?

D. Discuss other factors to consider when choosing each type of water heater. Which of these water heaters would you choose to heat your home?

CHAPTER 3 SUMMARY

KEY TERMS

3.1 **ordered pair**
origin
x-axis
y-axis
rectangular (Cartesian) coordinate system
plot
coordinate
quadrant
graph of an equation
first-degree equation
linear equation in two variables
standard form
x-intercept
y-intercept

3.2 **rise**
run
slope

3.3 **point-slope form**
slope-intercept form

3.4 **linear inequality in two variables**
boundary line

3.5 **dependent variable**
independent variable
relation
function
domain
range
graph of a relation
vertical line test for a function
function notation
composition of functions
composite function
linear function
constant function

3.6 **vary directly**
proportional
constant of variation
vary inversely
vary jointly
combined variation

NEW SYMBOLS

(a, b) ordered pair

x_1 a specific value of the variable x (read "x-sub-one")

m slope

$f(x)$ function of x (read "f of x")

$f[g(x)]$ composite function

TEST YOUR WORD POWER

See how well you have learned the vocabulary in this chapter. Answers, with examples, are given at the bottom of the next page.

1. An **ordered pair** is a pair of numbers written
(a) in numerical order between brackets
(b) between parentheses or brackets
(c) between parentheses in which order is important
(d) between parentheses in which order does not matter.

2. The **coordinates** of a point are
(a) the numbers in the corresponding ordered pair
(b) the solution of an equation
(c) the values of the x- and y-intercepts
(d) the graph of the point.

3. A **linear equation in two variables** is an equation that can be written in the form
(a) $Ax + By < C$
(b) $ax = b$
(c) $y = x^2$
(d) $Ax + By = C$.

4. An **intercept** is
(a) the point where the x-axis and y-axis intersect
(b) a pair of numbers written between parentheses in which order matters
(c) one of the four regions determined by a rectangular coordinate system
(d) the point where a graph intersects the x-axis or the y-axis.

TEST YOUR WORD POWER

5. The **slope** of a line is
(a) the measure of the run over the rise of the line
(b) the distance between two points on the line
(c) the ratio of the change in y to the change in x along the line
(d) the horizontal change compared to the vertical change of two points on the line.

6. In a relationship between two variables x and y, the **independent variable** is
(a) x, if x depends on y
(b) x, if y depends on x
(c) either x or y
(d) the larger of x and y.

7. In a relationship between two variables x and y, the **dependent variable** is
(a) y, if y depends on x
(b) y, if x depends on y
(c) either x or y
(d) the smaller of x and y.

8. A **relation** is
(a) a set of ordered pairs
(b) the ratio of the change in y to the change in x along a line
(c) the set of all possible values of the independent variable
(d) all the second elements of a set of ordered pairs.

9. A **function** is
(a) the numbers in an ordered pair
(b) a set of ordered pairs in which each x-value corresponds to exactly one y-value
(c) a pair of numbers written between parentheses in which order matters
(d) the set of all ordered pairs that satisfy an equation.

10. The **domain** of a function is
(a) the set of all possible values of the dependent variable y
(b) a set of ordered pairs
(c) the difference between the x-values
(d) the set of all possible values of the independent variable x.

11. The **range** of a function is
(a) the set of all possible values of the dependent variable y
(b) a set of ordered pairs
(c) the difference between the y-values
(d) the set of all possible values of the independent variable x.

QUICK REVIEW

CONCEPTS	EXAMPLES
3.1 THE RECTANGULAR COORDINATE SYSTEM	
Finding Intercepts	The graph of $2x + 3y = 12$ has
To find the x-intercept, let $y = 0$.	x-intercept (6, 0)
To find the y-intercept, let $x = 0$.	and y-intercept (0, 4).
3.2 THE SLOPE OF A LINE	
If $x_2 \neq x_1$, then $m = \frac{\text{rise}}{\text{run}} = \frac{\text{change in } y}{\text{change in } x} = \frac{y_2 - y_1}{x_2 - x_1}$.	For $2x + 3y = 12$, $m = \frac{4 - 0}{0 - 6} = -\frac{2}{3}$.
A vertical line has undefined slope.	$x = 3$ has undefined slope.
A horizontal line has 0 slope.	$y = -5$ has $m = 0$.
Parallel lines have equal slopes.	$y = 2x + 3$, $m = 2$; $4x - 2y = 6$, $-2y = -4x + 6$, $y = 2x - 3$, $m = 2$. These lines are **parallel**.

(continued)

Answers to Test Your Word Power

1. (c) *Examples:* (0, 3), (3, 8), (4, 0) **2.** (a) *Example:* The point associated with the ordered pair (1, 2) has x-coordinate 1 and y-coordinate 2. **3.** (d) *Examples:* $3x + 2y = 6$, $x = y - 7$, $4x = y$ **4.** (d) *Example:* In Figure 3(b) of Section 3.1, the x-intercept is (3, 0) and the y-intercept is (0, 2). **5.** (c) *Example:* The line through (3, 6) and (5, 4) has slope $\frac{4-6}{5-3} = \frac{-2}{2} = -1$. **6.** (b) *Example:* See Item 7, which follows. **7.** (a) *Example:* When borrowing money, the amount you borrow (independent variable) determines the size of your payments (dependent variable). **8.** (a) *Example:* The set $\{(2, 0), (4, 3), (6, 6), (8, 9)\}$ defines a relation. **9.** (b) The relation given in Item 8 is a function since the x-value of each ordered pair corresponds to exactly one y-value. **10.** (d) *Example:* In the function in Item 8 above, the domain is the set of x-values, $\{2, 4, 6, 8\}$. **11.** (a) *Example:* In the function in Item 8 above, the range is the set of y-values, $\{0, 3, 6, 9\}$.

CONCEPTS	EXAMPLES
The slopes of perpendicular lines are negative reciprocals with a product of -1.	$y = 3x - 1$ $x + 3y = 4$ $m = 3$ $3y = -x + 4$ $y = -\frac{1}{3}x + \frac{4}{3}$ $m = -\frac{1}{3}$ These lines are **perpendicular.**

3.3 LINEAR EQUATIONS IN TWO VARIABLES

CONCEPTS	EXAMPLES
Point-Slope Form $y - y_1 = m(x - x_1)$	$y - 3 = 4(x - 5)$ $(5, 3)$ is on the line, $m = 4$.
Slope-Intercept Form $y = mx + b$	$y = 2x + 3$ $m = 2$, y-intercept is $(0, 3)$.
Vertical Line $x = c$	$x = -1$
Horizontal Line $y = d$	$y = 4$
Standard Form $Ax + By = C$	$2x - 5y = 8$

3.4 LINEAR INEQUALITIES IN TWO VARIABLES

CONCEPTS	EXAMPLES
Graphing a Linear Inequality *Step 1* Draw the graph of the line that is the boundary. Make the line solid if the inequality involves $\le$ or $\ge$; make the line dashed if the inequality involves $<$ or $>$.	Graph $2x - 3y \le 6$. Draw the graph of $2x - 3y = 6$. Use a solid line because of $\le$.
Step 2 Choose any point not on the line as a test point.	Choose $(1, 2)$. $2(1) - 3(2) = 2 - 6 \le 6$ True
Step 3 Shade the region that includes the test point if the test point satisfies the original inequality; otherwise, shade the region on the other side of the boundary line.	Shade the side of the line that includes $(1, 2)$. 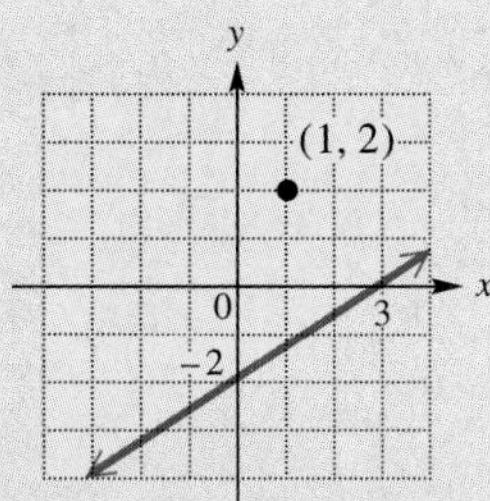

3.5 INTRODUCTION TO FUNCTIONS

CONCEPTS	EXAMPLES
A function is a set of ordered pairs such that for each first component there is one and only one second component. The set of first components is called the domain, and the set of second components is called the range.	$y = f(x) = x^2$ defines a function f, with domain $(-\infty, \infty)$ and range $[0, \infty)$.

CONCEPTS	EXAMPLES
To evaluate a function using function notation (that is, $f(x)$ notation) for a given value of x, substitute the value wherever x appears.	If $f(x) = x^2 - 7x + 12$, then $f(1) = 1^2 - 7(1) + 12 = 6$.
To write the equation that defines a function in function notation, solve the equation for y. Then replace y with $f(x)$.	Given: $2x + 3y = 12$. $3y = -2x + 12$ $y = -\frac{2}{3}x + 4$ $f(x) = -\frac{2}{3}x + 4$
3.6 VARIATION	
If there is some constant k such that: $y = kx^n$, then y varies directly as x^n, $y = \frac{k}{x^n}$, then y varies inversely as x^n.	The area of a circle **varies directly as** the square of the radius. $A = kr^2$ Pressure **varies inversely as** volume. $p = \frac{k}{V}$

CHAPTER 3 REVIEW EXERCISES

[3.1] *Complete the table of ordered pairs for each equation. Then graph the equation.*

1. $3x + 2y = 10$

x	y
0	
	0
2	
	−2

2. $x - y = 8$

x	y
2	
	−3
3	
	−2

Find the x- and y-intercepts and then graph each equation.

3. $4x - 3y = 12$

4. $5x + 7y = 28$

5. $2x + 5y = 20$

6. $x - 4y = 8$

7. Explain how the signs of the x- and y-coordinates of a point determine the quadrant in which the point lies.

[3.2] *Find the slope of each line.*

8. Through $(-1, 2)$ and $(4, -5)$

9. Through $(0, 3)$ and $(-2, 4)$

10. $y = 2x + 3$

11. $3x - 4y = 5$

12. $x = 5$

13. Parallel to $3y = 2x + 5$

14. Perpendicular to $3x - y = 4$

15. Through $(-1, 5)$ and $(-1, -4)$

16. The line containing the points shown in this table

X	Y_1
0	2
2	-.6667
4	-3.333
6	-6
8	-8.667
10	-11.33
12	-14

X=0

17. The line pictured in the two screens

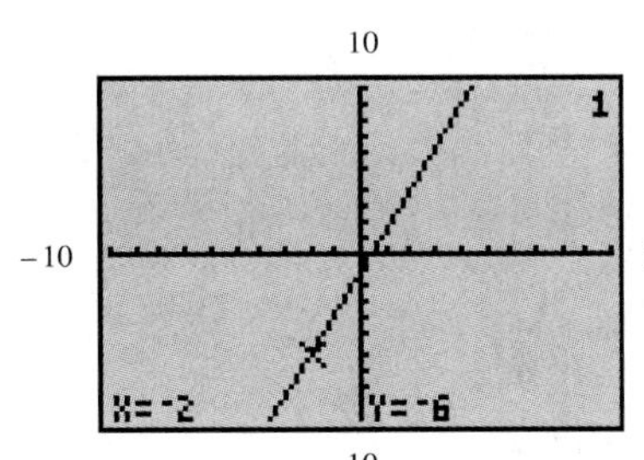

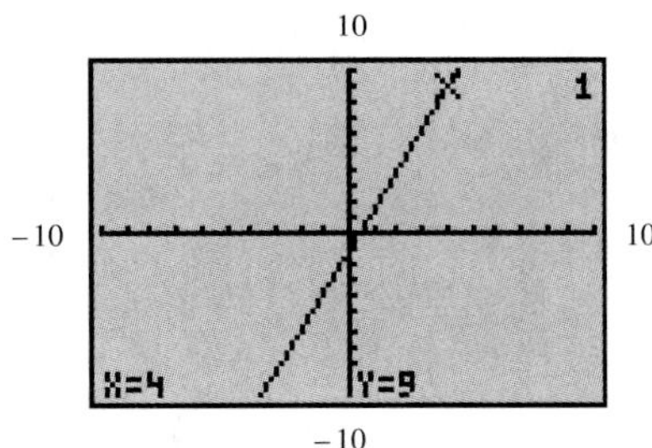

Tell whether each line has positive, negative, zero, or undefined slope.

18.

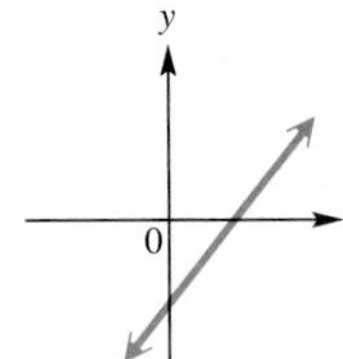

19.

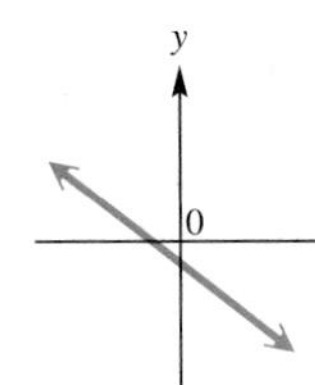

20.

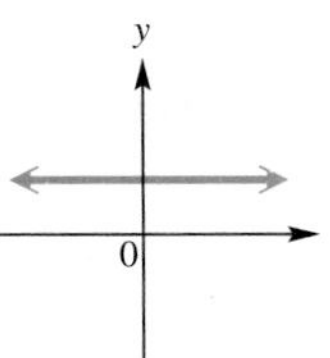

21.

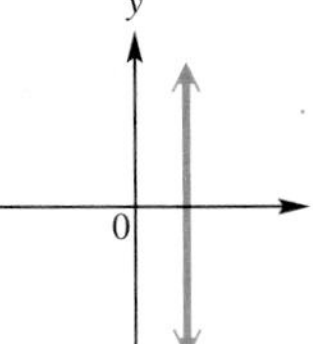

22. If a walkway rises 2 feet for every 10 feet on the horizontal, which of the following express its slope (or grade)? (There are several correct choices.)

(a) .2 **(b)** $\frac{2}{10}$ **(c)** $\frac{1}{5}$ **(d)** 20%

(e) 5 **(f)** $\frac{20}{100}$ **(g)** 500% **(h)** $\frac{10}{2}$

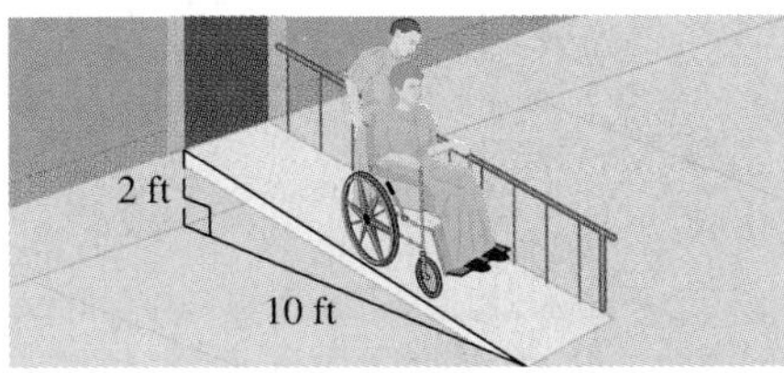

23. If the pitch of a roof is $\frac{1}{4}$, how many feet in the horizontal direction correspond to a rise of 3 feet?

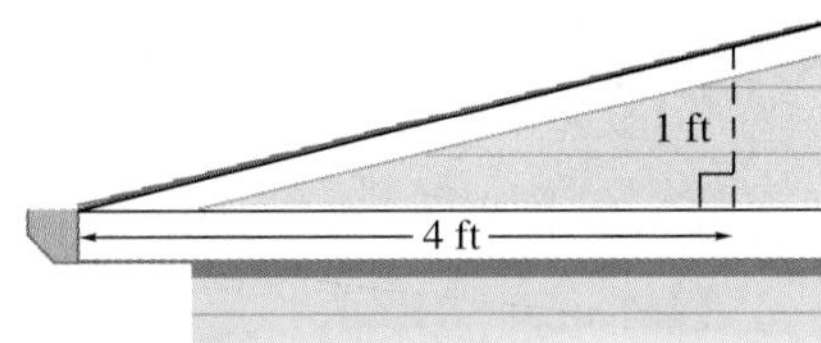

24. Family income in the United States has steadily increased for many years (primarily due to inflation). In 1970 the median family income was about \$10,000 a year. In 1995 it was about \$41,000 a year. Find the average rate of change of median family income over that period. (*Source:* Bureau of the Census.)

[3.3] *Write an equation in slope-intercept form for the line, if possible.*

25. Slope $-\frac{1}{3}$, y-intercept $(0, -1)$

26. Slope 0, y-intercept $(0, -2)$

27. Slope $-\frac{4}{3}$, through $(2, 7)$

28. Slope 3, through $(-1, 4)$

29. Vertical, through $(2, 5)$

30. Through $(2, -5)$ and $(1, 4)$

31. Through $(-3, -1)$ and $(2, 6)$

32. Parallel to $4x - y = 3$ and through $(7, -1)$

33. Perpendicular to $2x - 5y = 7$ and through $(4, 3)$

34. The line containing the points in the table accompanying Exercise 16

35. The line pictured in Exercise 17

The equation $y = 2.1x + 230$ *is a model that was recently used to predict the U.S. population, where* $x = 0$ *represents the year* 1980 *and* y *is the population in millions.*

36. According to this equation, what was the population in 1996?

37. According to this equation, in what year would the population have reached 247 million?

[3.4] *Graph the solution set of each inequality or compound inequality.*

38. $3x - 2y \leq 12$

39. $5x - y > 6$

40. $x \geq 2$

41. $2x + y \leq 1$ and $x \geq 2y$

[3.5] *In Exercises 42–44, give the domain and range of each relation. Identify any functions.*

42. $\{(-4, 2), (-4, -2), (1, 5), (1, -5)\}$

43.

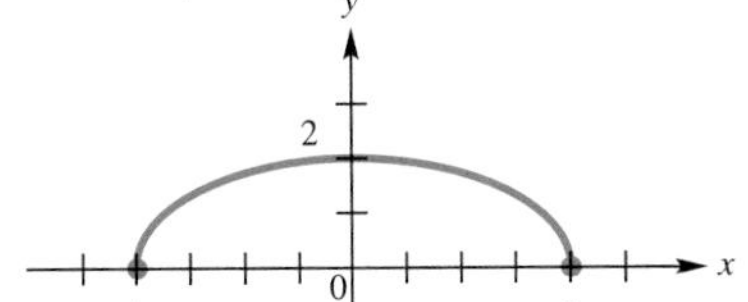

44. The number of small offices/home offices in 1996 for the top five states were {(California, 71,266), (New York, 50,101), (Texas, 48,010), (Pennsylvania, 42,142), (Washington, 38,240)}. (*Source:* Dun & Bradstreet's Cottage Industry File.)

Work the following problems.

45. The Top 6 countries in commercial nuclear power generation in 1995 are shown in the table.

Country	U.S.	France	Japan	Germany	Canada	Russia
Billion kilowatt-hours	706	377	286	154	100	99

Source: U.S. Bureau of the Census.

(a) Suppose, in a later year, the number of billion kilowatt-hours generated in Canada and Russia was the same, 101, while the other countries continued to generate the amounts shown in the table. For the relationship between the country and the amount of power generated to be a function, which must be the independent variable: the country or the number of kilowatt-hours? Explain why.

(b) Give the domain and range of the function of part (a).

46. Refer to the line graph in the chapter introduction. Is it the graph of a function? What type of function? Give two ordered pairs that are on the line. What are the domain and range?

Determine whether each equation or inequality defines y as a function of x. Give the domain in each case. Identify any linear functions.

47. $y = 3x - 3$ **48.** $y < x + 2$ **49.** $y = |x - 4|$

50. $y = \sqrt{4x + 7}$ **51.** $x = y^2$ **52.** $y = \dfrac{7}{x - 6}$

53. Explain the test that allows us to determine whether a graph is that of a function.

Given $f(x) = -2x^2 + 3x - 6$, *find each function value or expression.*

54. $f(0)$ **55.** $f(2.1)$ **56.** $f\left(-\dfrac{1}{2}\right)$

57. $f(k)$ **58.** $f[f(0)]$ **59.** $f(2p)$

60. The equation $2x^2 - y = 0$ defines y as a function of x. Rewrite it using $f(x)$ notation, and find $f(3)$.

61. Suppose that $2x - 5y = 7$ defines a function. If $y = f(x)$, which one of the following defines the same function?

(a) $f(x) = \dfrac{7 - 2x}{5}$ **(b)** $f(x) = \dfrac{-7 - 2x}{5}$

(c) $f(x) = \dfrac{-7 + 2x}{5}$ **(d)** $f(x) = \dfrac{7 + 2x}{5}$

62. Can the graph of a linear function have undefined slope? Explain.

RELATING CONCEPTS (EXERCISES 63–74)

Refer to the straight-line graph shown and ***work Exercises 63–74 in order.***

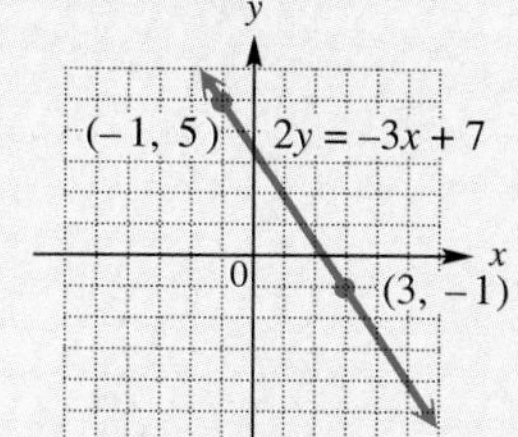

63. By just looking at the graph, how can you tell whether the slope is positive, negative, zero, or undefined?

64. Use the slope formula to find the slope of the line.

65. Find the x-intercept of the graph.

66. Find the y-intercept of the graph.

67. Use function notation to write the equation of the line. Use f to designate the function.

68. Find $f(8)$.

69. If $f(x) = -8$, what is the value of x?

70. Graph the solution set of $f(x) \geq 0$.

71. What is the solution set of $f(x) = 0$?

72. What is the solution set of $f(x) < 0$? (Use the graph and the result of Exercise 71.)

73. What is the solution set of $f(x) > 0$? (Use the graph and the result of Exercise 71.)

74. What is the slope of any line perpendicular to the line shown?

Did you make the connection between the two-variable linear functions in this chapter and the one-variable linear equations and inequalities in Chapter 2?

[3.6]

75. In which one of the following does y vary inversely as x?

(a) $y = 2x$ **(b)** $y = \dfrac{x}{3}$ **(c)** $y = \dfrac{3}{x}$ **(d)** $y = x^2$

Solve each problem.

76. The resistance in ohms of a platinum wire temperature sensor varies directly as the temperature in *degrees Kelvin* (K). If the resistance is .646 ohm at a temperature of 190 K, find the resistance at a temperature of 250 K.

77. For the subject in a photograph to appear in the same perspective in the photograph as in real life, the viewing distance must be properly related to the amount of enlargement. For a particular camera, the viewing distance varies directly as the amount of enlargement. A picture taken with this camera that is enlarged 5 times should be viewed from a distance of 250 millimeters. Suppose a print 8.6 times the size of the negative is made. From what distance should it be viewed?

78. The frequency of vibration, f, of a guitar string varies directly as the square root of the tension, t, and inversely as the length, L, of the string. If the frequency is 20 when the tension is 9 (in appropriate units) and the length is 30 inches, find f when the tension is doubled and the length remains the same.

CHAPTER 3 TEST

1. Complete the table of ordered pairs for the equation $2x - 3y = 12$.

x	y
1	
3	
	-4

2. Find the slope of the line through the points $(6, 4)$ and $(-4, -1)$.

Find the x- and y-intercepts, and graph each equation.

3. $3x - 2y = 20$

4. $y = 5$

5. $x = 2$

6. Describe how the graph of a line with undefined slope is situated in a rectangular coordinate system.

Determine whether each pair of lines is parallel, perpendicular, or neither.

7. $5x - y = 8$ and $5y = -x + 3$

8. $2y = 3x + 12$ and $3y = 2x - 5$

Find the equation of each line, and write it in slope-intercept form.

9. Through $(4, -1)$; $m = -5$

10. Through $(-3, 14)$; horizontal

11. Through $(-7, 2)$ and parallel to $3x + 5y = 6$

12. Through $(-7, 2)$ and perpendicular to $y = 2x$

13. The line shown in the figures (Look at the displays at the bottom.)

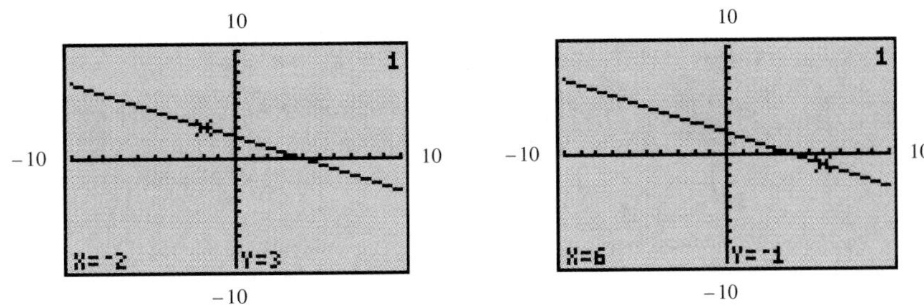

14. Which one of the following has positive slope and negative y-coordinate for its y-intercept?

(a) **(b)** **(c)** **(d)**

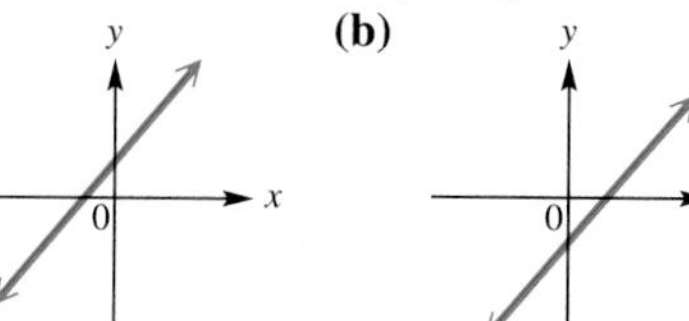
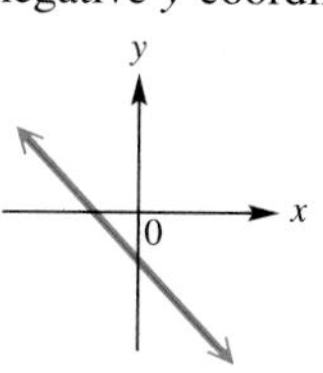

15. The linear equation $y = 1410x + 12{,}520$ provides a model for the number of cases served by the Child Support Enforcement program from 1990 to 1995, where $x = 0$ corresponds to 1990, $x = 1$ corresponds to 1991, and so on. Use this model to approximate the number of cases served during 1994. (*Source:* Office of Child Support Enforcement.)

16. What does the number 1410 in the equation in Exercise 15 refer to
(a) with respect to the graph? **(b)** in the context of the problem?

Graph each inequality or compound inequality.

17. $3x - 2y > 6$

18. $y < 2x - 1$ and $x - y < 3$

19. Which one of the following is the graph of a function?

(a) **(b)** **(c)** **(d)**

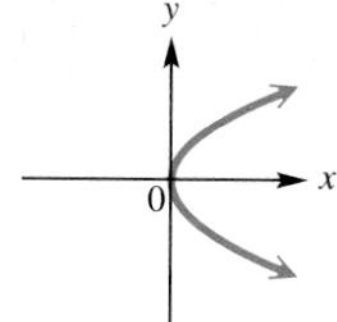
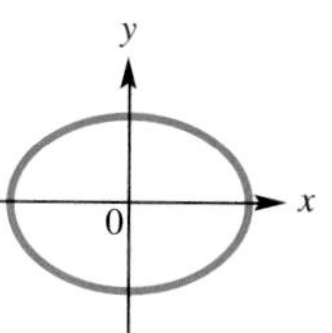
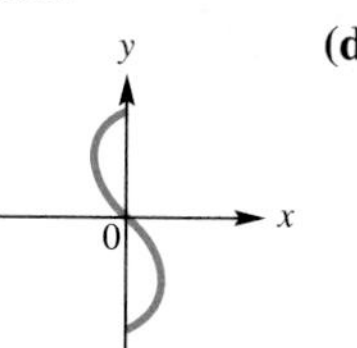
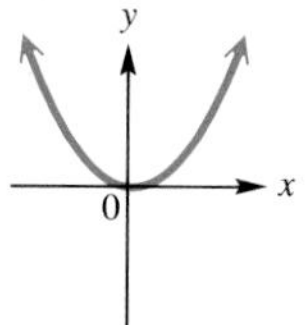

20. Which of the following does not define a function?
(a) $\{(0, 1), (-2, 3), (4, 8)\}$ **(b)** $y = 2x - 6$ **(c)** $y = \sqrt{x + 2}$
(d)

Input	Output
A	1
A	2
B	2
C	3

21. If $f(x) = -x^2 + 2x - 1$, find $f(1)$.

22. Graph the linear function $f(x) = \frac{2}{3}x - 1$. What is its domain? What is its range?

23. The deaths per 1000 population from 1990–1995 are shown in the table.

Death rate	8.6	8.6	8.5	8.8	8.8	8.8
Year	1990	1991	1992	1993	1994	1995

Source: U.S. National Center for Health Statistics.

Which of the following sets of ordered pairs from the table define(s) a function? Explain.
(a) {(Death rate, Year)} **(b)** {(Year, Death rate)}

24. The current in a simple electrical circuit is inversely proportional to the resistance. If the current is 80 amps when the resistance is 30 ohms, find the current when the resistance is 12 ohms.

25. The collision impact of an automobile varies jointly as its mass and the square of its speed. Suppose a 2000-pound car traveling at 55 miles per hour has a collision impact of 6.1. What is the collision impact of the same car at 65 miles per hour? Round to the nearest tenth.

CUMULATIVE REVIEW EXERCISES CHAPTERS 1–3

Decide whether each statement is always true, sometimes true, or never true. If the statement is sometimes true, give examples where it is true and where it is false.

1. The absolute value of a negative number equals the additive inverse of the number.

2. The quotient of two integers with nonzero denominator is a rational number.

3. The sum of two negative numbers is positive.

4. The sum of a positive number and a negative number is 0.

Find the value of each expression.

5. $-|-4|$

6. $|-12| - |-3|$

7. $\dfrac{3(\sqrt{16}) - (-1)(7)}{4 + (-6)}$

8. Simplify $2p - (4 + 3p) - 1 - p$.

Solve.

9. $3x + 2(x - 4) = -(2x + 5)$

10. $\dfrac{x - 1}{6} - \dfrac{x}{4} = \dfrac{2x + 4}{12}$

11. $V = \dfrac{1}{3}\pi r^2 h$ for h

Solve. Write each solution set in interval notation and graph it.

12. $-4 < 3 - 2k < 9$

13. $-.3x + 2.1(x - 4) \leq -6.6$

14. $x > 6$ and $x < 8$

15. $-5x + 1 \geq 11$ or $3x + 5 > 26$

Solve.

16. $|2k - 7| + 4 = 11$

17. $|3m + 6| \geq 0$

18. How are the solution sets of a linear equation and the two associated inequalities related?

Work each problem.

19. The lowest temperature ever recorded in Allagash, Maine, was $-55°$C, on January 14, 1999. What was the corresponding Fahrenheit temperature? (*Source:* National Climatic Data Center.)

20. For a 10-day period in early June 1993, projected sales of U.S.–built cars and trucks were 11.7 million, up from 10.9 million a year earlier. What percent increase does this represent?

21. If each side of a square were increased by 4 inches, the perimeter would be 8 inches less than twice the perimeter of the original square. Find the length of a side of the original square.

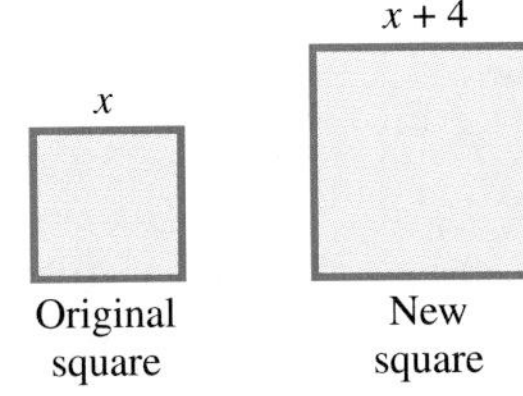

22. Find the x- and y-intercepts of the line with equation $3x + 5y = 12$ and graph the line.

23. Consider the points $A(-2, 1)$ and $B(3, -5)$.

(a) Find the slope of the line AB.

(b) Find the slope of a line perpendicular to line AB.

24. Graph the inequality $-2x + y < -6$.

25. For the function $f(x) = -3x + 6$,
(a) what is the domain? **(b)** what is $f(-6)$?

26. The table shows the percent of possible sunshine for the ten sunniest cities.

Which set is a function: {(Percent, City)} or {(City, Percent)}? Describe the elements of the domain of the function.

Sunniest U.S. Cities

Percent	*City*
90	Yuma, AZ
88	Redding, CA
86	Phoenix, AZ
85	Tucson, AZ
85	Las Vegas, NV
84	El Paso, TX
79	Fresno, CA
79	Reno, NV
78	Flagstaff, AZ
78	Sacramento, CA

Source: National Oceanic and Atmospheric Administration.

27. Does $2x - 7y = 14$ define a function? If so, write it using the function notation $f(x)$.

28. For $f(x) = 2x + \sqrt{x}$, find $f(4)$.

29. Which of these equations defines a linear function?

(a) $2x + 3 = -4y$ **(b)** $x^2 + y = 5 + x$ **(c)** $y - |x| = 3$ **(d)** $2y - 3 = \frac{6}{x}$

30. The cost of a pizza varies directly as the square of its radius. If a pizza with a 7-inch radius costs \$6.00, how much should a pizza with a 9-inch radius cost?

Systems of Linear Equations

Sports

Following the players' strike of 1994, Major League Baseball seemed to be in trouble with its fans. But the magic of the 1998 season, led by the assault on the season home run record by Mark McGwire and Sammy Sosa, the voluntary end to Cal Ripken's consecutive game streak, rookie pitcher Kerry Wood's 20-strikeout performance, David Wells's perfect game, and the New York Yankees' sweep of the World Series propelled baseball into the forefront of spectator sports once again. The biggest individual achievement in 1998 was that of Mark McGwire, who shattered Roger Maris's single-season home run record of 61 in 1961 by hitting 70 home runs. The all-time career home run leader is Hank Aaron with 744, followed by the immortal Babe Ruth with 714. After 12 seasons, McGwire has 499, and if he continues at his current rate, he will surpass both of them. From the graphs below, we can see how the three players' home run totals compare based on their years played in the major leagues. (Why do you think Babe Ruth's totals for his first few years were so low? The answer is on the next page.)

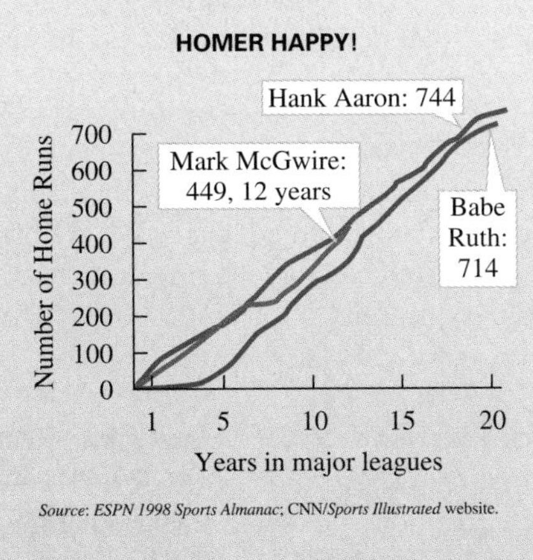

Presenting several graphs on the same set of axes allows us to compare different trends simultaneously. This is the idea behind a *system of equations.* This chapter presents several algebraic methods of solving linear equations. Many of the applications deal with sports, the theme of this chapter.

Visit our Web site at www.LialAlgebra.com

4.1 Linear Systems of Equations in Two Variables

OBJECTIVES

1. Decide whether an ordered pair is a solution of a linear system.
2. Solve linear systems by graphing.
3. Solve linear systems (with two equations and two unknowns) by elimination.
4. Solve linear systems (with two equations and two unknowns) by substitution.
5. Recognize how a graphing calculator is used to solve a linear system.

FOR EXTRA HELP

SSG Sec. 4.1
SSM Sec. 4.1
Pass the Test Software
InterAct Math Tutorial Software
Video 6

The graphs shown in Figure 1 appeared in the Sunday, July 10, 1994, sports section of the *Sacramento Bee.* During the middle of the baseball season, the San Francisco Giants had a .500 record on May 29, but began to slump, while the Oakland Athletics, playing at .265, started to improve. The graphs appear to intersect at the point representing (July 3, .425). As you might expect, the two teams had the same "winning percentage," .425, on July 3.

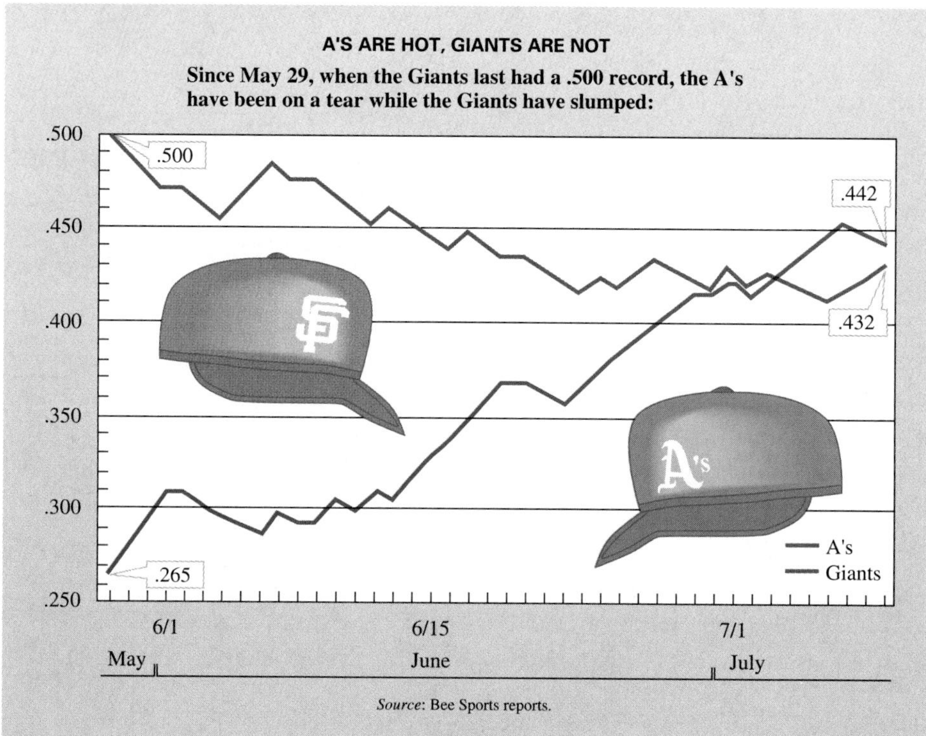

Figure 1

In the figure we can tell how each team was playing by reading the graphs. In algebra, information can also be described by an equation. We are often interested in finding the numbers that make two or more equations true at the same time. Such a set of equations is called a **system of equations.**

OBJECTIVE 1 Decide whether an ordered pair is a solution of a linear system. Recall from the previous chapter that the graph of a first-degree equation of the form $Ax + By = C$ is a straight line. For this reason, such an equation is called a linear equation. Two or more linear equations form a **linear system.** The **solution set of a linear system** of equations contains all ordered pairs that satisfy all the equations of the system at the same time.

Answer to the question in the chapter introduction: Early in his career, Babe Ruth was a pitcher and did not get as many at-bats as he would have had he played another position.

EXAMPLE 1 Deciding Whether an Ordered Pair Is a Solution

Decide whether the given ordered pair is a solution of the system.

(a) $x + y = 6$; $(4, 2)$
$4x - y = 14$

Replace x with 4 and y with 2 in each equation of the system.

$x + y = 6$		$4x - y = 14$	
$4 + 2 = 6$?		$4(4) - 2 = 14$?	
$6 = 6$	True	$14 = 14$	True

Since $(4, 2)$ makes both equations true, $(4, 2)$ is a solution of the system.

(b) $3x + 2y = 11$; $(-1, 7)$
$x + 5y = 36$

$3x + 2y = 11$		$x + 5y = 36$	
$3(-1) + 2(7) = 11$?		$-1 + 5(7) = 36$?	
$-3 + 14 = 11$	True	$-1 + 35 = 36$	False

The ordered pair $(-1, 7)$ is not a solution of the system, since it does not make *both* equations true.

OBJECTIVE 2 Solve linear systems by graphing. Sometimes we can estimate the solution set of a linear system of equations by graphing the equations of the system on the same axes and then estimating the coordinates of any point of intersection.

EXAMPLE 2 Solving a System by Graphing

Solve the following system by graphing.

$$x + y = 5$$
$$2x - y = 4$$

The graphs of these linear equations are shown in Figure 2. The graph suggests that the point of intersection is the ordered pair $(3, 2)$. Check this by substituting these values for x and y in both of the equations. As the check shows, the solution set of the system is $\{(3, 2)\}$.

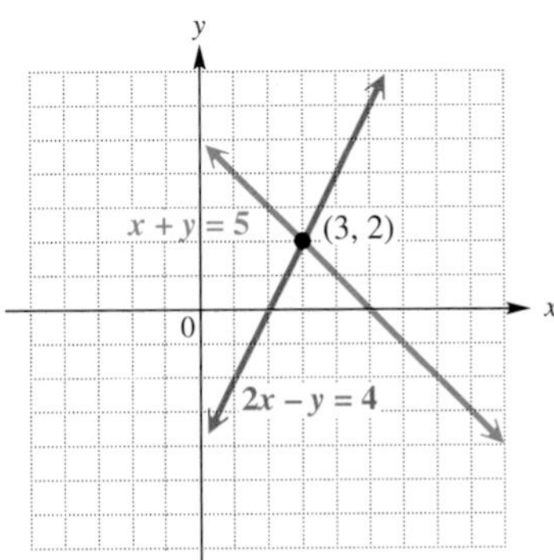

Figure 2

NOTE While we cannot always rely on graphing by hand to find accurate solutions, a graphing calculator can be used to find these solutions. This will be illustrated in Example 8 for the system given in Example 1.

Since the graph of a linear equation is a straight line, there are three possibilities for the solution set of a system of two linear equations, as shown in Figure 3.

Graphs of a Linear System

1. The two graphs intersect in a single point. The coordinates of this point give the only solution of the system. This is the most common case. See Figure 3(a).
2. The graphs are parallel lines. In this case the system is **inconsistent;** that is, there is no solution common to both equations of the system, and the solution set is $\emptyset$. See Figure 3(b).
3. The graphs are the same line. In this case the equations are **dependent,** since any solution of one equation of the system is also a solution of the other. The solution set is an infinite set of ordered pairs representing the points on the line. See Figure 3(c).

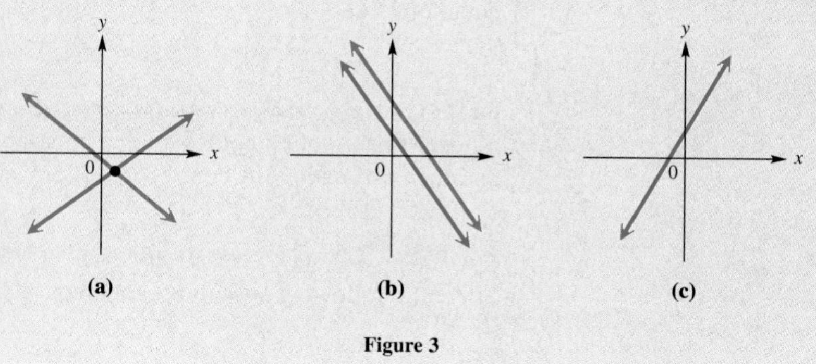

Figure 3

OBJECTIVE 3 Solve linear systems (with two equations and two unknowns) by elimination. In most cases we cannot rely on graphing to solve systems. There are algebraic methods to do this, and one such method, called the **elimination method,** is explained in the following examples. The elimination method involves combining the two equations of the system so that one variable is eliminated. This is done using the following fact.

$$\text{If } a = b \text{ and } c = d, \text{ then } a + c = b + d.$$

The general method of solving a system by the elimination method is summarized as follows.

Solving Linear Systems by Elimination

Step 1 **Put in standard form.** Write both equations in the form

$$Ax + By = C.$$

Step 2 **Make one pair of coefficients opposites.** Multiply one or both equations by appropriate numbers so that the sum of the coefficients of either x or y is zero.

Step 3 **Add.** Add the new equations to eliminate a variable. The sum should be an equation with just one variable.

Solving Linear Systems by Elimination (continued)

Step 4 **Solve.** Solve the equation from Step 3.

Step 5 **Find the other value.** Substitute the result of Step 4 into either of the given equations and solve for the other variable.

Step 6 **Find the solution set.** Check the solution in both of the given equations. Then write the solution set.

EXAMPLE 3 Solving a System by Elimination

Solve the system.

$$5x - 2y = 4 \quad \textbf{(1)}$$

$$2x + 3y = 13 \quad \textbf{(2)}$$

Step 1 Both equations are already in standard form.

Step 2 Our goal is to add the two equations so that one of the variables is eliminated. Suppose we wish to eliminate the variable x. Since the coefficients of x are *not* opposites, we must first transform one or both equations so that the coefficients *are* opposites. Then, when we combine the equations, the term with x will have a coefficient of 0, and we will be able to solve for y. We begin by multiplying equation (1) by 2 and equation (2) by -5.

$$10x - 4y = 8 \qquad \text{2 times each side of equation (1)}$$

$$-10x - 15y = -65 \qquad \text{−5 times each side of equation (2)}$$

Step 3 Now add the two equations to eliminate the x-terms.

$$\begin{array}{r} 10x - 4y = 8 \\ -10x - 15y = -65 \\ \hline -19y = -57 \end{array} \qquad \text{Add.}$$

Step 4 Solve the equation from Step 3 to get $y = 3$.

Step 5 To find x, we substitute 3 for y in either equation (1) or equation (2). Substituting in equation (2) gives

$$\begin{aligned} 2x + 3y &= 13 \\ 2x + 3(3) &= 13 && \text{Let } y = 3. \\ 2x + 9 &= 13 \\ 2x &= 4 && \text{Subtract 9.} \\ x &= 2. && \text{Divide by 2.} \end{aligned}$$

Step 6 The solution appears to be (2, 3). To check, substitute 2 for x and 3 for y in both equations (1) and (2).

$$\begin{aligned} 5x - 2y &= 4 && \text{(1)} \\ 5(2) - 2(3) &= 4 && ? \\ 10 - 6 &= 4 && ? \\ 4 &= 4 && \text{True} \end{aligned} \qquad \begin{aligned} 2x + 3y &= 13 && \text{(2)} \\ 2(2) + 3(3) &= 13 && ? \\ 4 + 9 &= 13 && ? \\ 13 &= 13 && \text{True} \end{aligned}$$

The solution set is $\{(2, 3)\}$.

EXAMPLE 4 Solving a System of Dependent Equations

Solve the system.

$$2x - y = 3 \quad \textbf{(3)}$$
$$6x - 3y = 9 \quad \textbf{(4)}$$

Multiply both sides of equation (3) by -3, and then add the result to equation (4).

$$\begin{aligned} -6x + 3y &= -9 && -3 \text{ times each side of equation (3)} \\ 6x - 3y &= 9 && (4) \\ \hline 0 &= 0 && \text{True} \end{aligned}$$

Adding these equations gives the true statement $0 = 0$, which indicates that the equations are dependent. Notice in the original system that equation (4) could be obtained from equation (3) by multiplying both sides of equation (3) by 3. Because of this, equations (3) and (4) are equivalent and have the same line as their graph. The solution set, the infinite set of ordered pairs on the line with equation $2x - y = 3$, is written as

$$\{(x, y) \mid 2x - y = 3\}.$$

EXAMPLE 5 Solving an Inconsistent System

Solve the system.

$$x + 3y = 4 \quad \textbf{(5)}$$
$$-2x - 6y = 3 \quad \textbf{(6)}$$

Multiply both sides of equation (5) by 2, and then add the result to equation (6).

$$\begin{aligned} 2x + 6y &= 8 && 2 \text{ times each side of equation (5)} \\ -2x - 6y &= 3 && (6) \\ \hline 0 &= 11 && \text{False} \end{aligned}$$

The result of the addition step here is a false statement, which shows that the system is inconsistent. The graphs of the equations of the system are parallel lines similar to the graphs in Figure 3(b). There are no ordered pairs that satisfy both equations, so the solution set of the system is $\emptyset$.

The results of Examples 4 and 5 are generalized as follows.

Special Cases of Linear Systems

If both variables are eliminated when a system of linear equations is solved,

1. there is no solution if the resulting statement is *false;*

2. there are infinitely many solutions if the resulting statement is *true.*

Slopes and y-intercepts can be used to decide if the graphs of a system of equations are parallel lines or if they coincide. In Example 4, writing each equation in slope-intercept form shows that both lines have slope 2 and y-intercept $(0, -3)$, so the graphs are the same line and the system has an infinite solution set.

In Example 5, both equations have slope $-\frac{1}{3}$, but y-intercepts $\left(0, \frac{4}{3}\right)$ and $\left(0, -\frac{1}{2}\right)$, showing that the graphs are two distinct parallel lines. Thus, the system has no solution.

OBJECTIVE 4 **Solve linear systems (with two equations and two unknowns) by substitution.** Linear systems can also be solved by the **substitution method.** This method is most useful for solving linear systems in which one variable has coefficient 1. As shown in a later chapter, the substitution method is the best choice for solving many *nonlinear* systems.

The method of solving a system by substitution is summarized as follows.

Solving Linear Systems by Substitution

Step 1 **Solve for one variable in terms of the other.** Solve one of the equations for either variable. (If one of the variables has coefficient 1 or -1, choose it, since the substitution method is usually easier this way.)

Step 2 **Substitute.** Substitute for that variable in the other equation. The result should be an equation with just one variable.

Step 3 **Solve.** Solve the equation from Step 2.

Step 4 **Find the other value.** Substitute the result from Step 3 into the equation from Step 1 to find the value of the other variable.

Step 5 **Find the solution set.** Check the solution in both of the given equations. Then write the solution set.

The next two examples illustrate this method.

EXAMPLE 6 **Solving a System by Substitution**

Solve the system.

$$3x + 2y = 13 \qquad \textbf{(7)}$$
$$4x - y = -1 \qquad \textbf{(8)}$$

Step 1 To use the substitution method, first solve one of the equations for either x or y. Since the coefficient of y in equation (8) is -1, it is easiest to solve for y in equation (8).

$$-y = -1 - 4x$$
$$y = 1 + 4x$$

Step 2 Substitute $1 + 4x$ for y in equation (7) to get an equation in x.

$$3x + 2y = 13$$
$$3x + 2(1 + 4x) = 13 \qquad \text{Let } y = 1 + 4x. \qquad \textbf{(9)}$$

Step 3 Solve for x in equation (9).

$$3x + 2 + 8x = 13 \qquad \text{Distributive property}$$
$$11x = 11 \qquad \text{Combine terms; subtract 2.}$$
$$x = 1 \qquad \text{Divide by 11.}$$

Step 4 Now solve for y. Since $y = 1 + 4x$, $y = 1 + 4(1) = 5$.

Step 5 Check the solution (1, 5) in both equations (7) and (8).

$$\begin{aligned} 3x + 2y &= 13 && (7) \\ 3(1) + 2(5) &= 13 \quad ? \\ 3 + 10 &= 13 \quad ? \\ 13 &= 13 && \text{True} \end{aligned} \qquad \begin{aligned} 4x - y &= -1 && (8) \\ 4(1) - 5 &= -1 \quad ? \\ 4 - 5 &= -1 \quad ? \\ -1 &= -1 && \text{True} \end{aligned}$$

The solution set is $\{(1, 5)\}$.

EXAMPLE 7 Solving a System by Substitution

Solve the system.

$$\frac{2}{3}x - \frac{1}{2}y = \frac{7}{6} \qquad \textbf{(10)}$$

$$3x - 2y = 6 \qquad \textbf{(11)}$$

This system will be easier to solve if we clear the fractions in equation (10). Multiply by the LCD 6.

$$6 \cdot \frac{2}{3}x - 6 \cdot \frac{1}{2}y = 6 \cdot \frac{7}{6}$$

$$4x - 3y = 7 \qquad \textbf{(12)}$$

Now the system consists of equations (11) and (12). To use the substitution method, one equation must be solved for one of the two variables. Solve equation (11) for x.

$$3x = 2y + 6$$

$$x = \frac{2y + 6}{3}$$

Substitute $\frac{2y + 6}{3}$ for x in equation (12).

$$4x - 3y = 7 \qquad (12)$$

$$4\left(\frac{2y + 6}{3}\right) - 3y = 7 \qquad \text{Let } x = \frac{2y + 6}{3}.$$

Multiply both sides of the equation by the common denominator 3 to eliminate the fraction.

$$\begin{aligned} 4(2y + 6) - 9y &= 21 && \text{Multiply by 3.} \\ 8y + 24 - 9y &= 21 && \text{Distributive property} \\ 24 - y &= 21 && \text{Combine terms.} \\ -y &= -3 && \text{Add } -24. \\ y &= 3 && \text{Divide by } -1. \end{aligned}$$

Since $x = \frac{2y + 6}{3}$ and $y = 3$,

$$x = \frac{2(3) + 6}{3} = \frac{6 + 6}{3} = 4.$$

A check verifies that the solution set is $\{(4, 3)\}$.

While the substitution method is not usually the best choice for solving a system like the one in Example 7, it is sometimes necessary to use it to solve a system of *nonlinear* equations.

OBJECTIVE 5 Recognize how a graphing calculator is used to solve a linear system. In Example 2 we showed how the system

$$x + y = 5$$
$$2x - y = 4$$

can be solved by graphing the two lines and finding their point of intersection. This can be done with a graphing calculator.

EXAMPLE 8 Finding the Solution Set of a System from a Graphing Calculator Screen

The solution set of the system

$$x + y = 5$$
$$2x - y = 4$$

can be found from the graphing calculator screen in Figure 4. The two lines were graphed by solving the first equation to get $y = 5 - x$ and the second to get $y = 2x - 4$. The coordinates of their point of intersection are displayed at the bottom of the screen, indicating that the solution set is $\{(3, 2)\}$. (Compare this graph to the one found in Figure 2.)

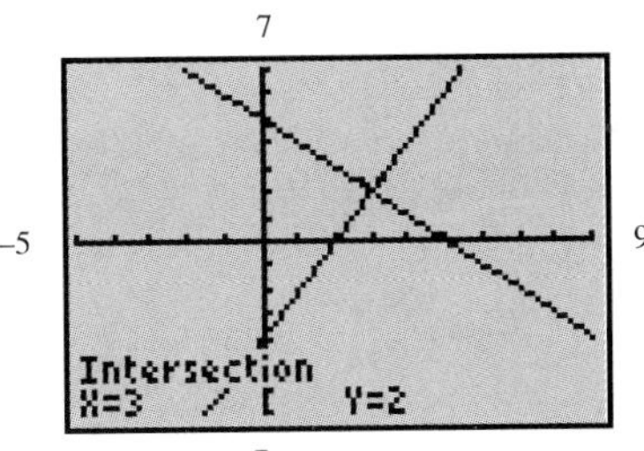

Figure 4

4.1 EXERCISES

1. Refer to Figure 1 and answer the following.
   - **(a)** If the graph for the San Francisco Giants were approximated by a straight line, would the line have positive slope or negative slope?
   - **(b)** Explain how your answer to part (a) relates to the general performance of the Giants during the months depicted on the graph.
   - **(c)** If the graph for the Oakland Athletics were approximated by a straight line, would the line have positive slope or negative slope?
   - **(d)** Explain how your answer to part (c) relates to the general performance of the Athletics during the months depicted on the graph.

2. Match each system with the correct graphical choice.

(a) $x + y = 6$
$x - y = 0$

(b) $x + y = -6$
$x - y = 0$

(c) $x + y = 0$
$x - y = -6$

(d) $x + y = 0$
$x - y = 6$

A.

B.

C.

D.

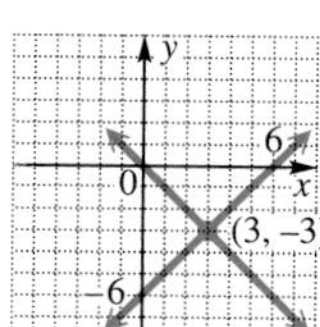

Decide whether each ordered pair is a solution of the given system. See Example 1.

3. $x + y = 6$ (5, 1)
$x - y = 4$

4. $x - y = 17$ (8, −9)
$x + y = -1$

5. $2x - y = 8$ (5, 2)
$3x + 2y = 20$

6. $3x - 5y = -12$ (−1, 2)
$x - y = 1$

Solve each system by graphing. See Example 2.

7. $x + y = 4$
$2x - y = 2$

8. $x + y = -5$
$-2x + y = 1$

Solve each system by elimination. If the system is inconsistent or has dependent equations, say so. See Examples 3–5.

9. $2x - 5y = 11$
$3x + y = 8$

10. $-2x + 3y = 1$
$-4x + y = -3$

11. $3x + 4y = -6$
$5x + 3y = 1$

12. $4x + 3y = 1$
$3x + 2y = 2$

13. $3x + 3y = 0$
$4x + 2y = 3$

14. $8x + 4y = 0$
$4x - 2y = 2$

15. $7x + 2y = 6$
$-14x - 4y = -12$

16. $x - 4y = 2$
$4x - 16y = 8$

17. $\frac{x}{2} + \frac{y}{3} = -\frac{1}{3}$
$\frac{x}{2} + 2y = -7$

18. $\frac{x}{5} + y = \frac{6}{5}$
$\frac{x}{10} + \frac{y}{3} = \frac{5}{6}$

19. $5x - 5y = 3$
$x - y = 12$

20. $2x - 3y = 7$
$-4x + 6y = 14$

21. Suppose that two linear equations are graphed on the same set of coordinate axes. Sketch what the graph might look like if the system has the given description.
(a) The system has a single solution. **(b)** The system has no solution.
(c) The system has infinitely many solutions.

22. Explain how to solve a system using the elimination method. Make up a system of your own to illustrate the steps.

Write the two equations of each system in slope-intercept form and then tell how many solutions the system has. Do not actually solve the system.

23. $3x + 7y = 4$
$6x + 14y = 3$

24. $-x + 2y = 8$
$4x - 8y = 1$

25. $2x = -3y + 1$
$6x = -9y + 3$

26. $5x = -2y + 1$
$10x = -4y + 2$

Solve each system by substitution. See Examples 6 and 7.

27. $4x + y = 6$
$y = 2x$

28. $2x - y = 6$
$y = 5x$

29. $3x - 4y = -22$
$-3x + y = 0$

30. $-3x + y = -5$
$x + 2y = 0$

31. $-x - 4y = -14$
$2x = y + 1$

32. $-3x - 5y = -17$
$4x = y - 8$

33. $5x - 4y = 9$
$3 - 2y = -x$

34. $6x - y = -9$
$4 + 7x = -y$

35. $x = 3y + 5$
$x = \frac{3}{2}y$

36. $x = 6y - 2$
$x = \frac{3}{4}y$

37. $\frac{1}{2}x + \frac{1}{3}y = 3$
$y = 3x$

38. $\frac{1}{4}x - \frac{1}{5}y = 9$
$y = 5x$

39. Explain how to solve a system using the substitution method. Make up a system of your own to illustrate the steps.

40. Refer to Example 3. What numbers might equations (1) and (2) be multiplied by so that adding the two equations eliminates y?

41. What will happen when you solve a system of two linear equations (in two variables) that has no solution using either elimination or substitution?

42. What will happen when you solve a system of two linear equations (in two variables) that has an infinite number of solutions using elimination or substitution?

In the system let $p = \frac{1}{x}$ and $q = \frac{1}{y}$. Substitute, solve for p and q, and then find x and y. (Hint: $\frac{3}{x} = 3 \cdot \frac{1}{x} = 3p$.)

43. $\frac{3}{x} + \frac{4}{y} = \frac{5}{2}$
$\frac{5}{x} - \frac{3}{y} = \frac{7}{4}$

44. $\frac{4}{x} - \frac{9}{y} = -1$
$-\frac{7}{x} + \frac{6}{y} = -\frac{3}{2}$

45. $\frac{2}{x} - \frac{5}{y} = \frac{3}{2}$
$\frac{4}{x} + \frac{1}{y} = \frac{4}{5}$

46. $\frac{2}{x} + \frac{3}{y} = \frac{11}{2}$
$-\frac{1}{x} + \frac{2}{y} = -1$

Solve by any method. Assume that a and b represent nonzero constants.

47. $ax + by = 2$
$-ax + 2by = 1$

48. $2ax - y = 3$
$y = 5ax$

49. $3ax + 2y = 1$
$-ax + y = 2$

50. $ax + by = c$
$ax - 2by = c$

RELATING CONCEPTS (EXERCISES 51–54)

Work Exercises 51–54 in order.

51. Solve the system.

$$3x + y = 6$$
$$-2x + 3y = 7$$

Use elimination or substitution.

52. For the first equation in the system of Exercise 51, solve for y and rename it $f(x)$. What special kind of function is f? (continued)

RELATING CONCEPTS (EXERCISES 51–54) (CONTINUED)

53. For the second equation in the system of Exercise 51, solve for y and rename it $g(x)$. What special kind of function is g?

54. Use the result of Exercise 51 to fill in the blanks with the appropriate responses:

Because the graphs of f and g are straight lines that are neither parallel nor coincide, they intersect in exactly _____ point. The coordinates of the point are (_____, _____). Using function notation, this is given by $f($_____$) =$ _____ and $g($_____$) =$ _____.

Did you make the connection between systems of linear equations and the graphs of linear functions?

TECHNOLOGY INSIGHTS (EXERCISES 55–58)

55. The table shown was generated by a graphing calculator. The functions defined by Y_1 and Y_2 are linear. Based on the table, what are the coordinates of the point of intersection of the graphs?

X	Y1	Y2
0	-7	-1
1	-6	-2
2	-5	-3
3	-4	-4
4	-3	-5
5	-2	-6
6	-1	-7

X=0

56. The table shown was generated by a graphing calculator. The functions defined by Y_1 and Y_2 are linear.

(a) Use the methods of Chapter 3 to find the equation for Y_1.

(b) Use the methods of Chapter 3 to find the equation for Y_2.

(c) Solve the system of equations formed by Y_1 and Y_2.

X	Y1	Y2
0	4	7
1	8	5
2	12	3
3	16	1
4	20	-1
5	24	-3
6	28	-5

X=0

57. The solution set of the system

$$y_1 = 3x - 5$$
$$y_2 = -4x + 2$$

is $\{(1, -2)\}$. Using slopes and y-intercepts, determine which one of the two calculator-generated graphs is the appropriate one for this system.

(a)

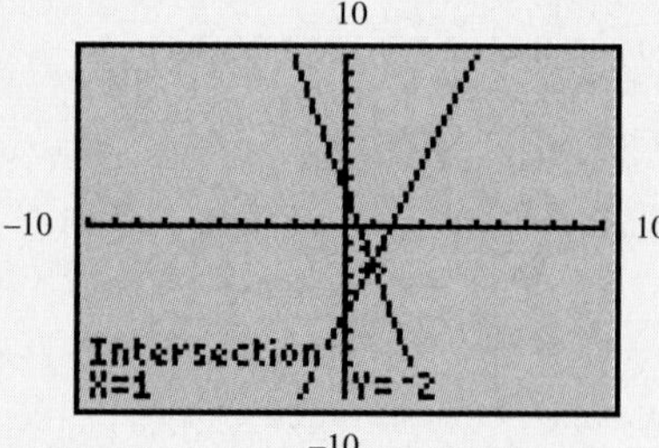

(b)

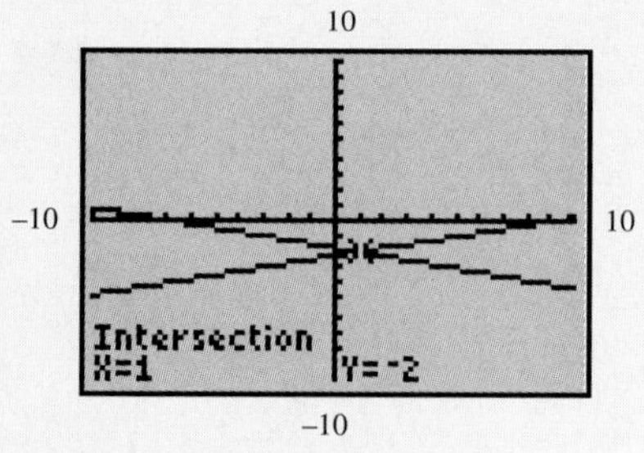

TECHNOLOGY INSIGHTS (EXERCISES 55–58) (CONTINUED)

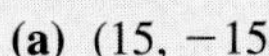

58. Which one of the ordered pairs listed could be the only possible solution of the system whose graphs are shown in the standard viewing window of a graphing calculator?

(a) $(15, -15)$
(b) $(15, 15)$
(c) $(-15, 15)$
(d) $(-15, -15)$

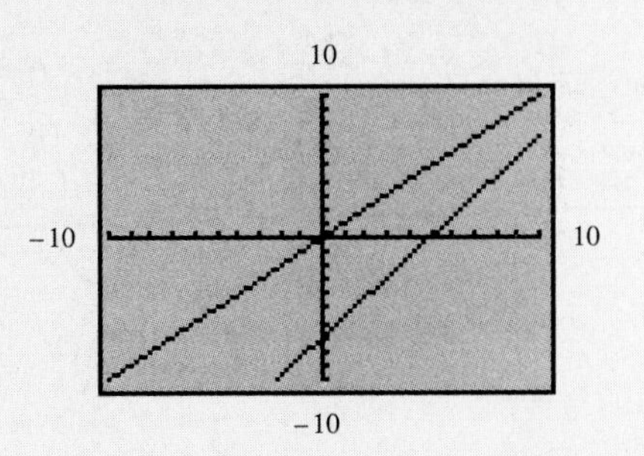

*For each system **(a)** solve by elimination or substitution and **(b)** use a graphing calculator to support your result. In part (b), be sure to solve each equation for y first. See Example 8.*

59. $x + y = 10$
$2x - y = 5$

60. $6x + y = 5$
$-x + y = -9$

61. $3x - 2y = 4$
$3x + y = -2$

62. $2x - 3y = 3$
$2x + 2y = 8$

Answer the questions in Exercises 63–65 by observing the graphs provided.

63. During the early 1990s, U.S. economic growth increased while that of Japan and Germany decreased. Use the accompanying graph to answer the questions.

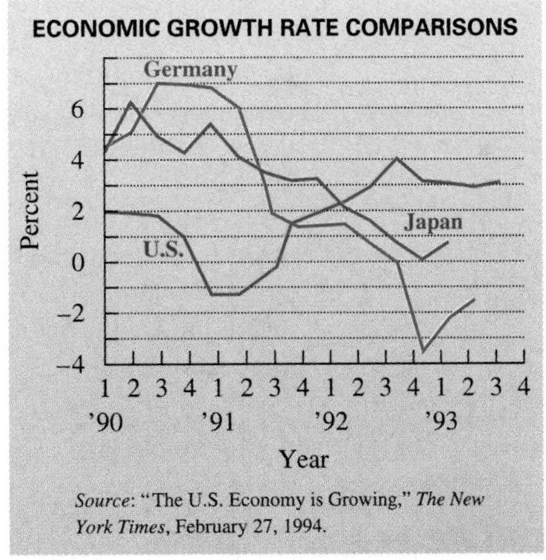

Source: "The U.S. Economy is Growing," *The New York Times*, February 27, 1994.

(a) In which quarter of which year did U.S. economic growth match that of Germany? What was the growth rate at that time?

(b) In which quarter of which year did U.S. economic growth match that of Japan? What was the growth rate at that time?

(c) How many times did the growth rate of Germany match that of Japan in the period shown? Between those times, which economy had a larger growth rate?

64. The accompanying graph shows the trends during the years 1966–1990 relating to bachelor's degrees awarded in the United States.

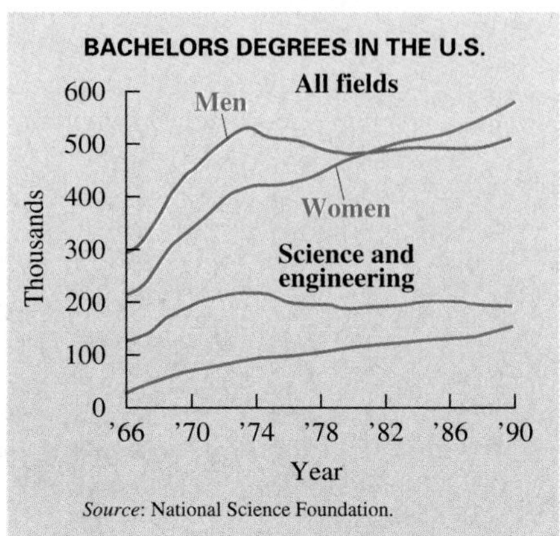

(a) Between what years shown on the horizontal axis did the number of degrees in all fields for men and women reach equal numbers?
(b) When the number of degrees for men and women reached equal numbers, what was that number (approximately)?

65. The accompanying graph shows how the production of vinyl LPs, audiocassettes, and compact discs (CDs) changed over the years from 1983 to 1993.
(a) In what year did cassette production and CD production reach equal levels? What was that level?
(b) Express as an ordered pair of the form (year, production level) the point of intersection of the graphs of LP production and CD production.

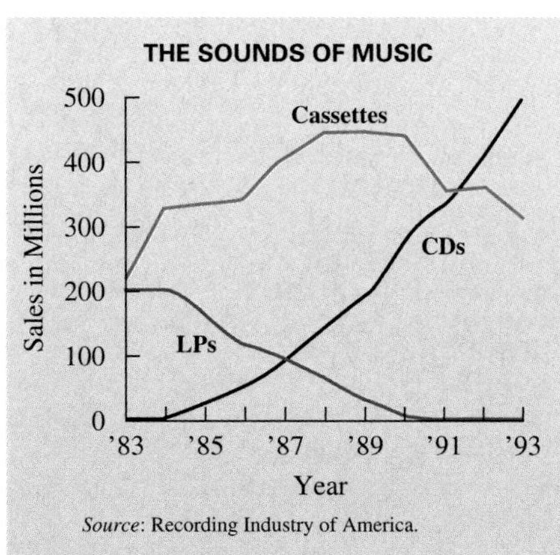

4.2 Linear Systems of Equations in Three Variables

OBJECTIVES

1. Understand the geometry of systems of three equations in three unknowns.
2. Solve linear systems (with three equations and three unknowns) by elimination.

A solution of an equation in three variables, such as $2x + 3y - z = 4$, is called an **ordered triple** and is written (x, y, z). For example, the ordered triples $(1, 1, 1)$ and $(10, -3, 7)$ are each solutions of $2x + 3y - z = 4$, since the numbers in these ordered triples satisfy the equation when used as replacements for x, y, and z, respectively.

In the rest of this chapter, the term *linear equation* is extended to first-degree equations of the form $Ax + By + Cz + \cdots + Dw = K$. For example, $2x + 3y - 5z = 7$ and $x - 2y - z + 3w - 2v = 8$ are linear equations, the first having three variables, and the second having five variables.

3 Solve linear systems (with three equations and three unknowns) where some of the equations have missing terms.

4 Solve special systems (with three equations and three unknowns).

FOR EXTRA HELP

SSG Sec. 4.2
SSM Sec. 4.2

Pass the Test Software

InterAct Math Tutorial Software

Video 6

OBJECTIVE 1 **Understand the geometry of systems of three equations in three unknowns.** In this section we discuss the solution of a system of linear equations in three variables such as

$$\begin{aligned} 4x + 8y + z &= 2 \\ x + 7y - 3z &= -14 \\ 2x - 3y + 2z &= 3. \end{aligned}$$

Theoretically, a system of this type can be solved by graphing. However, the graph of a linear equation with three variables is a *plane* and not a line. Since the graph of each equation of the system is a plane, which requires three-dimensional graphing, this method is not practical. However, it does illustrate the number of solutions possible for such systems, as Figure 5 shows.

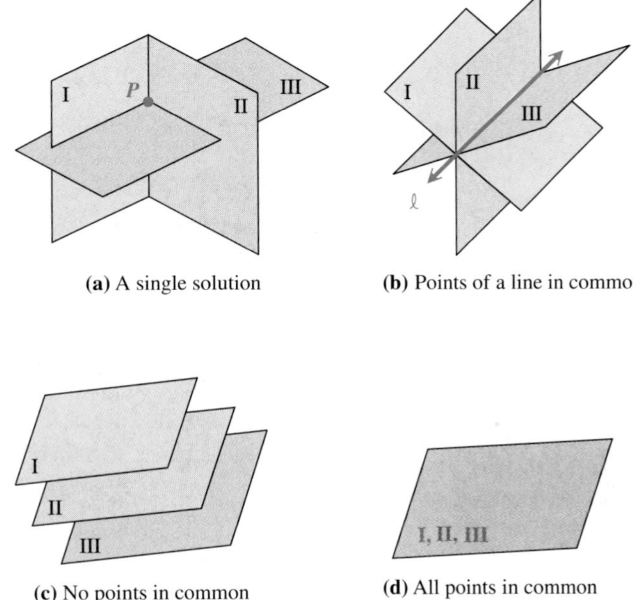

(a) A single solution

(b) Points of a line in common

(c) No points in common

(d) All points in common

Figure 5

Graphs of Linear Systems in Three Variables

1. The three planes may meet at a single, common point that is the solution of the system. See Figure 5(a).
2. The three planes may have the points of a line in common so that the set of points that satisfy the equation of the line is the solution of the system. See Figure 5(b).
3. The planes may have no points common to all three so that there is no solution for the system. See Figure 5(c).
4. The three planes may coincide so that the solution of the system is the set of all points on a plane. See Figure 5(d).

OBJECTIVE 2 Solve linear systems (with three equations and three unknowns) by elimination. Since graphing to find the solution set of a system of three equations in three variables is impractical, these systems are solved with an extension of the elimination method, summarized as follows.

Solving Linear Systems in Three Variables by Elimination

Step 1 **Eliminate a variable.** Use the elimination method to eliminate any variable from any two of the given equations. The result is an equation in two variables.

Step 2 **Eliminate the same variable again.** Eliminate the *same* variable from any *other* two equations. The result is an equation in the same two variables as in Step 1.

Step 3 **Eliminate a different variable and solve.** Use the elimination method to eliminate a second variable from the two equations in two variables that result from Steps 1 and 2. The result is an equation in one variable that gives the value of that variable.

Step 4 **Find a second value.** Substitute the value of the variable found in Step 3 into either of the equations in two variables to find the value of the second variable.

Step 5 **Find a third value.** Use the values of the two variables from Steps 3 and 4 to find the value of the third variable by substituting into any of the original equations.

Step 6 **Find the solution set.** Check the solution in all of the original equations. Then write the solution set.

EXAMPLE 1 Solving a System in Three Variables

Solve the system.

$$4x + 8y + z = 2 \quad \textbf{(1)}$$
$$x + 7y - 3z = -14 \quad \textbf{(2)}$$
$$2x - 3y + 2z = 3 \quad \textbf{(3)}$$

Step 1 As before, the elimination method involves eliminating a variable from the sum of two equations. The choice of which variable to eliminate is arbitrary. Suppose we decide to begin by eliminating z. To do this, multiply both sides of equation (1) by 3 and then add the result to equation (2).

$$\begin{array}{rll} 12x + 24y + 3z = & 6 & \text{Multiply both sides of (1) by 3.} \\ x + 7y - 3z = & -14 & \text{(2)} \\ \hline 13x + 31y \quad = & -8 & \text{Add.} \quad \textbf{(4)} \end{array}$$

Step 2 Equation (4) has only two variables. To get another equation without z, multiply both sides of equation (1) by -2 and add the result to equation (3). It is essential at this point to *eliminate the same variable, z.*

$$\begin{array}{rll} -8x - 16y - 2z = & -4 & \text{Multiply both sides of (1) by } -2. \\ 2x - 3y + 2z = & 3 & \text{(3)} \\ \hline -6x - 19y \quad = & -1 & \text{Add.} \quad \textbf{(5)} \end{array}$$

Step 3 Now solve the system of equations (4) and (5) for x and y. This step is possible only if the *same* variable is eliminated in the first two steps.

$$\begin{array}{rll} 78x + 186y = -48 & & \text{Multiply both sides of (4) by 6.} \\ -78x - 247y = -13 & & \text{Multiply both sides of (5) by 13.} \\ \hline -61y = -61 & & \text{Add.} \\ y = 1 & & \end{array}$$

Step 4 Substitute 1 for y in either equation (4) or (5). Choosing (5) gives

$$\begin{aligned} -6x - 19y &= -1 && (5) \\ -6x - 19(\mathbf{1}) &= -1 && \text{Let } y = 1. \\ -6x - 19 &= -1 \\ -6x &= 18 \\ x &= -3. \end{aligned}$$

Step 5 Substitute -3 for x and 1 for y in any one of the three given equations to find z. Choosing (1) gives

$$\begin{aligned} 4x + 8y + z &= 2 && (1) \\ 4(\mathbf{-3}) + 8(\mathbf{1}) + z &= 2 && \text{Let } x = -3 \text{ and } y = 1. \\ z &= 6. \end{aligned}$$

Step 6 It appears that the ordered triple $(-3, 1, 6)$ is the only solution of the system. Check that the solution satisfies all three equations of the system. We show the check here only for equation (1). The checks for equations (2) and (3) are requested in Exercise 2.

$$\begin{aligned} 4x + 8y + z &= 2 && (1) \\ 4(\mathbf{-3}) + 8(\mathbf{1}) + \mathbf{6} &= 2 \quad ? \\ -12 + 8 + 6 &= 2 \quad ? \\ 2 &= 2 && \text{True} \end{aligned}$$

Because $(-3, 1, 6)$ also satisfies equations (2) and (3), the solution set is $\{(-3, 1, 6)\}$. ■

OBJECTIVE 3 Solve linear systems (with three equations and three unknowns) where some of the equations have missing terms. When this happens, one elimination step can be omitted.

EXAMPLE 2 Solving a System of Equations with Missing Terms

Solve the system.

$$\begin{aligned} 6x - 12y &= -5 && \textbf{(6)} \\ 8y + z &= 0 && \textbf{(7)} \\ 9x - z &= 12 && \textbf{(8)} \end{aligned}$$

Since equation (8) is missing the variable y, one way to begin the solution is to eliminate y again with equations (6) and (7).

$$\begin{array}{rll} 12x - 24y = -10 & & \text{Multiply both sides of (6) by 2.} \\ 24y + 3z = 0 & & \text{Multiply both sides of (7) by 3.} \\ \hline 12x + 3z = -10 & & \text{Add.} \quad \textbf{(9)} \end{array}$$

Use this result, together with equation (8), to eliminate z. Multiply both sides of equation (8) by 3. This gives

$$\begin{aligned} 27x - 3z &= 36 && \text{Multiply both sides of (8) by 3.} \\ 12x + 3z &= -10 && \text{(9)} \\ \hline 39x \quad &= 26 && \text{Add.} \end{aligned}$$

$$x = \frac{26}{39} = \frac{2}{3}.$$

Substitution into equation (8) gives

$$\begin{aligned} 9x - z &= 12 && \text{(8)} \\ 9\left(\frac{2}{3}\right) - z &= 12 && \text{Let } x = \tfrac{2}{3}. \\ 6 - z &= 12 \\ z &= -6. \end{aligned}$$

Substitution of -6 for z in equation (7) gives

$$\begin{aligned} 8y + z &= 0 && \text{(7)} \\ 8y - 6 &= 0 && \text{Let } z = -6. \\ 8y &= 6 \\ y &= \frac{3}{4}. \end{aligned}$$

Check in each of the original equations of the system to verify that the solution set of the system is $\{(\frac{2}{3}, \frac{3}{4}, -6)\}$.

OBJECTIVE 4 **Solve special systems (with three equations and three unknowns).** Linear systems with three variables may be inconsistent or may include dependent equations. The next examples illustrate these cases.

EXAMPLE 3 **Solving an Inconsistent System with Three Variables**

Solve the system.

$$\begin{aligned} 2x - 4y + 6z &= 5 && \textbf{(10)} \\ -x + 3y - 2z &= -1 && \textbf{(11)} \\ x - 2y + 3z &= 1 && \textbf{(12)} \end{aligned}$$

Eliminate x by adding equations (11) and (12) to get the equation

$$y + z = 0.$$

Now, *eliminate x again,* using equations (10) and (12).

$$\begin{aligned} -2x + 4y - 6z &= -2 && \text{Multiply both sides of (12) by } -2. \\ 2x - 4y + 6z &= 5 && \text{(10)} \\ \hline 0 &= 3 && \text{False} \end{aligned}$$

The resulting false statement indicates that equations (10) and (12) have no common solution. Thus, the system is inconsistent and the solution set is $\emptyset$. The graph of this system would show at least two of the planes parallel to one another. (See Figure 5.)

NOTE If you get a false statement when adding as in Example 3, you do not need to go any further with the solution. Since two of the three planes are parallel, it is not possible for the three planes to have any common points.

EXAMPLE 4 **Solving a System of Dependent Equations with Three Variables**

Solve the system.

$$2x - 3y + 4z = 8 \qquad (13)$$

$$-x + \frac{3}{2}y - 2z = -4 \qquad (14)$$

$$6x - 9y + 12z = 24 \qquad (15)$$

Multiplying both sides of equation (13) by 3 gives equation (15). Multiplying both sides of equation (14) by -6 also gives equation (15). Because of this, the equations are dependent. All three equations have the same graph, as illustrated in Figure 5(d). The solution set is written $\{(x, y, z) \mid 2x - 3y + 4z = 8\}$. Although any one of the three equations could be used to write the solution set, we prefer to use the equation with coefficients that are integers with no common factor (except 1). This is similar to our choice of a standard form for a linear equation earlier.

The method discussed in this section can be extended to solve larger systems. For example, to solve a system of four equations in four variables, eliminate a variable from three pairs of equations to get a system of three equations in three unknowns. Then proceed as shown above.

4.2 EXERCISES

1. The two equations

$$x + y + z = 6$$
$$2x - y + z = 3$$

have a common solution of (1, 2, 3). Which one of the following equations would complete a system of three linear equations in three variables having solution set $\{(1, 2, 3)\}$?

(a) $3x + 2y - z = 1$ **(b)** $3x + 2y - z = 4$
(c) $3x + 2y - z = 5$ **(d)** $3x + 2y - z = 6$

2. Check that $(-3, 1, 6)$ is a solution for equations (2) and (3) in Example 1.

Solve each system of equations. See Example 1.

3. $3x + 2y + z = 8$
$2x - 3y + 2z = -16$
$x + 4y - z = 20$

4. $-3x + y - z = -10$
$-4x + 2y + 3z = -1$
$2x + 3y - 2z = -5$

5. $2x + 5y + 2z = 0$
$4x - 7y - 3z = 1$
$3x - 8y - 2z = -6$

6. $5x - 2y + 3z = -9$
$4x + 3y + 5z = 4$
$2x + 4y - 2z = 14$

7. $x + y - z = -2$
$2x - y + z = -5$
$-x + 2y - 3z = -4$

8. $x + 2y + 3z = 1$
$-x - y + 3z = 2$
$-6x + y + z = -2$

Solve each system of equations. See Example 2.

9. $2x - 3y + 2z = -1$
$x + 2y + z = 17$
$2y - z = 7$

10. $2x - y + 3z = 6$
$x + 2y - z = 8$
$2y + z = 1$

11. $4x + 2y - 3z = 6$
$x - 4y + z = -4$
$-x + 2z = 2$

12. $2x + 3y - 4z = 4$
$x - 6y + z = -16$
$-x + 3z = 8$

13. $2x + y = 6$
$3y - 2z = -4$
$3x - 5z = -7$

14. $4x - 8y = -7$
$4y + z = 7$
$-8x + z = -4$

15. Using your immediate surroundings, give an example of three planes that
(a) intersect in a single point **(b)** do not intersect
(c) intersect in infinitely many points.

16. Explain how you can determine algebraically that a system of three linear equations in three variables has no solution. Then do the same for infinitely many solutions.

Solve each system of equations. See Examples 1, 3, and 4.

17. $2x + 2y - 6z = 5$
$-3x + y - z = -2$
$-x - y + 3z = 4$

18. $-2x + 5y + z = -3$
$5x + 14y - z = -11$
$7x + 9y - 2z = -5$

19. $-5x + 5y - 20z = -40$
$x - y + 4z = 8$
$3x - 3y + 12z = 24$

20. $x + 4y - z = 3$
$-2x - 8y + 2z = -6$
$3x + 12y - 3z = 9$

21. $2x + y - z = 6$
$4x + 2y - 2z = 12$
$-x - \frac{1}{2}y + \frac{1}{2}z = -3$

22. $2x - 8y + 2z = -10$
$-x + 4y - z = 5$
$\frac{1}{8}x - \frac{1}{2}y + \frac{1}{8}z = -\frac{5}{8}$

23. $x + y - 2z = 0$
$3x - y + z = 0$
$4x + 2y - z = 0$

24. $2x + 3y - z = 0$
$x - 4y + 2z = 0$
$3x - 5y - z = 0$

Extend the method of this section to solve each system.

25. $x + y + z - w = 5$
$2x + y - z + w = 3$
$x - 2y + 3z + w = 18$
$-x - y + z + 2w = 8$

26. $3x + y - z + 2w = 9$
$x + y + 2z - w = 10$
$x - y - z + 3w = -2$
$-x + y - z + w = -6$

RELATING CONCEPTS (EXERCISES 27–36)

Suppose that on a distant planet a function of the form

$$f(x) = ax^2 + bx + c \quad (a \neq 0)$$

describes the height in feet of a projectile x seconds after it has been projected upward.

Work Exercises 27–36 in order *to see how this can be related to a system of three equations in three variables a, b, and c.*

27. After 1 second, the height of a certain projectile is 128 feet. Thus, $f(1) = 128$. Use this information to find one equation in the variables a, b, and c. (*Hint:* Substitute 1 for x and 128 for $f(x)$.)

28. After 1.5 seconds, the height is 140 feet. Find a second equation in a, b, and c.

29. After 3 seconds, the height is 80 feet. Find a third equation in a, b, and c.

RELATING CONCEPTS (EXERCISES 27–36) (CONTINUED)

30. Write a system of three equations in *a, b,* and *c,* based on your answers in Exercises 27–29. Solve the system.

31. What is the function *f* for this particular projectile?

32. What was the initial height of the projectile? (*Hint:* Find $f(0)$.)

33. The projectile reaches its maximum height in 1.625 seconds. Find its maximum height.

34. Verify that $f(3.25)$ also equals 0. What does this tell us about the projectile?

35. In Chapter 8 we discuss graphs of functions of the form $f(x) = ax^2 + bx + c$ $(a \neq 0)$. Use a system of equations to find the values of *a, b,* and *c* for the function of this form that satisfies $f(1) = 2, f(-1) = 0,$ and $f(-2) = 8$. Then write the expression for $f(x)$.

36. The accompanying table was generated by a graphing calculator for a function $Y_1 = ax^2 + bx + c$. Use any three points shown to find the values of *a, b,* and *c*. Then write the expression for Y_1.

X	Y1
1	8
2	15
3	24
4	35
5	48
6	63
7	80

X=1

Did you make the connection that a function can model the height of a propelled object? The height is a *function* of the time.

37. Discuss why it is necessary to eliminate the same variable in the first two steps of the elimination method with three equations and three variables.

38. In Step 3 of the elimination method for solving systems in three variables, does it matter which variable is eliminated? Explain.

4.3 Applications of Linear Systems of Equations

OBJECTIVES

1. Solve geometry problems using two variables.
2. Solve money problems using two variables.
3. Solve mixture problems using two variables.
4. Solve distance-rate-time problems using two variables.
5. Solve problems with three unknowns using a system of three equations.

Many applied problems involve more than one unknown quantity. Although most problems with two unknowns can be solved using just one variable, it often is easier to use two variables. To solve a problem with two unknowns, we must write two equations that relate the unknown quantities. The system formed by the pair of equations then can be solved using the methods of Section 4.1.

The following steps, based on the six-step problem-solving method first introduced in Chapter 2, give a strategy for solving problems using more than one variable.

Solving an Applied Problem by Writing a System of Equations

Step 1 **Determine what you are to find.** Assign a variable for each unknown and *write down* what it represents.

Step 2 **Write down other information.** If appropriate, draw a figure or a diagram and label it using the variables from Step 1. Make a chart if necessary to summarize the information.

(continued)

Solving an Applied Problem by Writing a System of Equations (continued)

Step 3 **Write a system of equations.** Write as many equations as there are unknowns.

Step 4 **Solve the system.** Use elimination or substitution to solve the system.

Step 5 **Answer the question(s).** Be sure you have answered all questions posed.

Step 6 **Check.** Check your solution(s) in the original problem. Be sure your answer makes sense.

OBJECTIVE 1 Solve geometry problems using two variables. Problems about the perimeter of a geometric figure often involve two unknowns and can be solved using a system of equations.

EXAMPLE 1 Finding Dimensions of a Soccer Field

Unlike football, where the dimensions of a playing field cannot vary, a rectangular soccer field may have a width between 50 and 100 yards and a length between 50 and 100 yards. Suppose that one particular field has a perimeter of 320 yards. Its length measures 40 yards more than its width. What are the dimensions of this field? (*Source:* Microsoft Encarta, *Soccer.*)

Step 1 We are asked to find the dimensions of the field. Let $L =$ the length and let $W =$ the width.

Step 2 Figure 6 shows a soccer field with the length labeled L and the width labeled W.

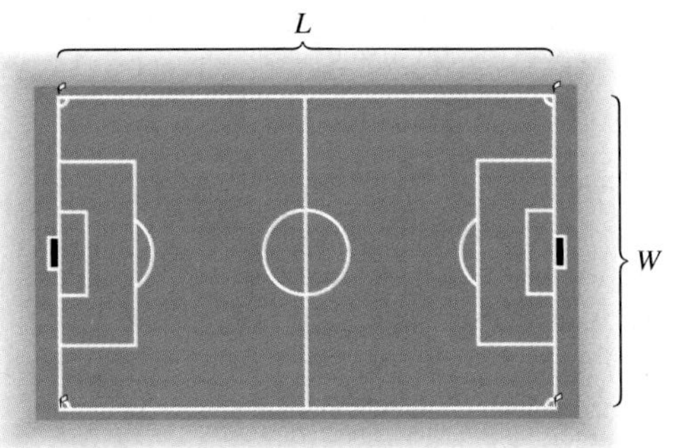

Figure 6

Step 3 Because the perimeter is 320 yards, one equation is found by using the perimeter formula:

$$2L + 2W = 320.$$

Because the length is 40 yards more than the width, we have

$$L = W + 40.$$

The system is, therefore,

$$2L + 2W = 320$$
$$L = W + 40.$$

Step 4 The second equation in the system is solved for L, so we can use substitution. Substitute $W + 40$ for L in the first equation, and solve for W.

$$
\begin{aligned}
2(W + 40) + 2W &= 320 && L = W + 40 \\
2W + 80 + 2W &= 320 && \text{Distributive property} \\
4W + 80 &= 320 && \text{Add.} \\
4W &= 240 && \text{Subtract 80.} \\
W &= \mathbf{60} && \text{Divide by 4.}
\end{aligned}
$$

Let $W = \mathbf{60}$ in the equation $L = W + 40$.

$$L = \mathbf{60} + 40 = \mathbf{100}.$$

Step 5 The length is **100** yards and the width is **60** yards.

Step 6 The perimeter of this soccer field is $2(\mathbf{100}) + 2(\mathbf{60}) = 320$ yards, and the length, 100 yards, is indeed 40 yards more than the width, since $\mathbf{100} - 40 = \mathbf{60}$. The solution is correct. ■

OBJECTIVE 2 Solve money problems using two variables. Professional sport ticket prices increase annually. Average per-ticket prices in three of the four major sports (football, basketball, and hockey) now exceed $30.00.

EXAMPLE 2 Solving a Problem about Ticket Prices

During the 1996–1997 National Hockey League and National Basketball Association seasons, 2 hockey tickets and 1 basketball ticket purchased at their average prices would have cost $110.40. One hockey ticket and 2 basketball tickets would have cost $106.32. What were the average ticket prices for the two sports? (*Source:* Team Marketing Report, Chicago.)

Let h represent the average price for a hockey ticket, and let b represent the average price for a basketball ticket. Because 2 hockey tickets and 1 basketball ticket cost a total of $110.40, one equation for the system is

$$2h + b = 110.40.$$

By similar reasoning, the second equation is

$$h + 2b = 106.32.$$

We must solve the following system:

$$
\begin{aligned}
2h + b &= 110.40 \\
h + 2b &= 106.32.
\end{aligned}
$$

To eliminate h, multiply the second equation by -2 and add.

$$
\begin{aligned}
2h + b &= 110.40 \\
-2h - 4b &= -212.64 \\
\hline
-3b &= -102.24 && \text{Add.} \\
b &= 34.08 && \text{Divide by } -3.
\end{aligned}
$$

To find the value of h, we can let $b = 34.08$ in the second equation.

$$h + 2(34.08) = 106.32 \quad \text{Let } b = 34.08.$$
$$h + 68.16 = 106.32 \quad \text{Multiply.}$$
$$h = 38.16 \quad \text{Subtract } 68.16.$$

Thus, one basketball ticket costs \$34.08 and one hockey ticket costs \$38.16. A check indicates that these values satisfy both equations of the system.

CONNECTIONS

Problems that can be solved by writing a system of equations have been of interest historically. The following problem appeared in a Hindu work that dates back to about 850 A.D.

> The mixed price of 9 citrons and 7 fragrant wood apples is 107; again, the mixed price of 7 citrons and 9 fragrant wood apples is 101. O you arithmetician, tell me quickly the price of a citron and the price of a wood apple here, having distinctly separated those prices well.
>
> (*Answer:* 8 for a citron and 5 for a wood apple.)

FOR DISCUSSION OR WRITING

What do you think is meant by "the mixed price" in the problem quoted above? Use the method discussed in this section to write a system of equations for this problem. Solve the system and compare your answer with the one given above.

OBJECTIVE 3 Solve mixture problems using two variables. We solved mixture problems earlier using one variable. For many mixture problems it seems more natural to use more than one variable and a system of equations.

EXAMPLE 3 Solving a Mixture Problem

How many ounces of 5% hydrochloric acid and of 20% hydrochloric acid must be combined to get 10 ounces of solution that is 12.5% hydrochloric acid?

Let x represent the number of ounces of 5% solution and y represent the number of ounces of 20% solution. A chart summarizes the given information.

Kind of Solution	Ounces of Solution	Ounces of Pure Acid
5%	x	$.05x$
20%	y	$.20y$
12.5%	10	$(.125)10$

Figure 7 also illustrates what is happening in the problem.

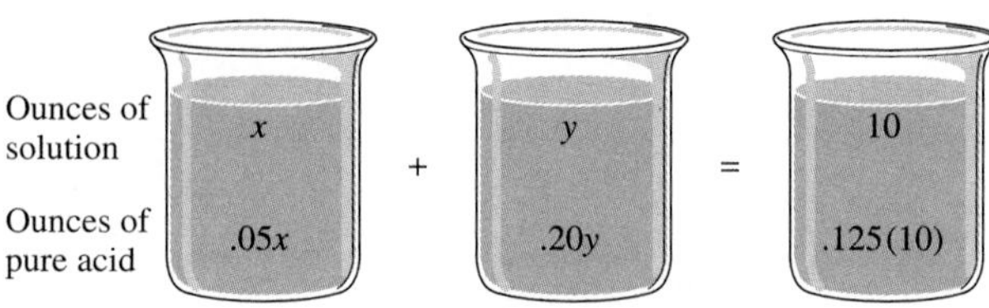

Figure 7

When x ounces of 5% solution and y ounces of 20% solution are combined, the total number of ounces is 10, so

$$x + y = 10. \qquad \textbf{(1)}$$

The ounces of acid in the 5% solution, $.05x$, plus the ounces of acid in the 20% solution, $.20y$, should equal the total number of ounces of acid in the mixture, which is $(.125)10$, or 1.25. That is,

$$.05x + .20y = 1.25. \qquad \textbf{(2)}$$

Eliminate x by first multiplying both sides of equation (2) by 100 to clear it of decimals, and then multiplying both sides of equation (1) by -5. Then add the results.

$$\begin{aligned} 5x + 20y &= 125 && \text{Multiply both sides of (2) by 100.} \\ -5x - 5y &= -50 && \text{Multiply both sides of (1) by } -5. \\ \hline 15y &= 75 && \text{Add.} \\ y &= 5 \end{aligned}$$

Since $y = 5$ and $x + y = 10$, x is also 5. Therefore, 5 ounces each of the 5% and the 20% solutions are required.

OBJECTIVE 4 Solve distance-rate-time problems using two variables. Motion applications require the distance formula, $d = rt$, where d is distance, r is rate (or speed), and t is time. These applications often lead to a system of equations, as in the next example.

EXAMPLE 4 Solving a Motion Problem

A car travels 250 kilometers in the same time that a truck travels 225 kilometers. If the speed of the car is 8 kilometers per hour faster than the speed of the truck, find both speeds.

A chart can be used to organize the information in problems about distance, rate, and time. Fill in the given information for each vehicle (in this case, distance) and use variables for the unknown speeds (rates) as follows.

	d	r	t
Car	250	x	
Truck	225	y	

The problem states that the car travels 8 kilometers per hour faster than the truck. Since the two speeds are x and y,

$$x = y + 8.$$

The chart shows nothing about time. To get an expression for time, solve the distance formula, $d = rt$, for t to get

$$\frac{d}{r} = t.$$

The two times can be written as $\frac{250}{x}$ and $\frac{225}{y}$. Since both vehicles travel for the same time,

$$\frac{250}{x} = \frac{225}{y}.$$

This is not a linear equation. However, multiplying both sides by xy gives

$$250y = 225x,$$

which is linear. Now solve the system.

$$x = y + 8 \qquad (3)$$
$$250y = 225x \qquad (4)$$

The substitution method can be used. Replace x with $y + 8$ in equation (4).

$$250y = 225(y + 8) \qquad \text{Let } x = y + 8.$$
$$250y = 225y + 1800 \qquad \text{Distributive property}$$
$$25y = 1800$$
$$y = 72$$

Since $x = y + 8$, the value of x is $72 + 8 = 80$. Check the solution in the original problem since one of the equations had variable denominators. Checking verifies that the speeds are 80 kilometers per hour for the car and 72 kilometers per hour for the truck.

OBJECTIVE 5 Solve problems with three unknowns using a system of three equations.

PROBLEM SOLVING

To solve applied problems with three or more unknowns, we extend the method given earlier for two unknowns. When three variables are used, three equations are necessary to find a solution.

EXAMPLE 5 Solving a Mixture Problem

A plant food is to be made from three chemicals. The mix must include 60% of the first two chemicals. The other two chemicals must be in a ratio of 4 to 3 by weight. How much of each chemical is needed to make 750 kilograms of the plant food?

First, choose variables to represent the three unknowns.

Let x = the number of kilograms of the first chemical;
y = the number of kilograms of the second chemical;
z = the number of kilograms of the third chemical.

Next, use the information in the problem to write three equations. To make 750 kilograms of the mix will require 60% of 750 kilograms of the first two chemicals, so

$$x + y = .60(750) = 450.$$

Since the ratio of the second and third chemicals is to be 4 to 3,

$$\frac{y}{z} = \frac{4}{3}.$$

Finally, the total amount of mix is to be 750 kilograms, so

$$x + y + z = 750.$$

Now, we must solve the system

$$x + y = 450$$
$$\frac{y}{z} = \frac{4}{3}$$
$$x + y + z = 750.$$

Use the method shown earlier to find the solution (50, 400, 300). The plant food should contain 50 kilograms of the first chemical, 400 kilograms of the second chemical, and 300 kilograms of the third chemical.

EXAMPLE 6 Solving a Business Production Problem

A company produces three color television sets, models X, Y, and Z. Each model X set requires 2 hours of electronics time, 2 hours of assembly time, and 1 hour of finishing time. Each model Y requires 1, 3, and 1 hours of electronics, assembly, and finishing time, respectively. Each model Z requires 3, 2, and 2 hours of the same time, respectively. There are 100 hours available for electronics, 100 hours available for assembly, and 65 hours available for finishing per week. How many of each model should be produced each week if all available time must be used?

Let $x =$ the number of model X produced per week;

$y =$ the number of model Y produced per week;

$z =$ the number of model Z produced per week.

Organize the information in a chart.

	Each Model X	*Each Model Y*	*Each Model Z*	*Totals*
Hours of electronics time	2	1	3	100
Hours of assembly time	2	3	2	100
Hours of finishing time	1	1	2	65

The x model X sets require $2x$ hours of electronics, the y model Y sets require $1y$ (or y) hours of electronics, and the z model Z sets require $3z$ hours of electronics. Since 100 hours are available for electronics,

$$2x + y + 3z = 100.$$

Similarly, from the fact that 100 hours are available for assembly,

$$2x + 3y + 2z = 100,$$

and the fact that 65 hours are available for finishing leads to the equation

$$x + y + 2z = 65.$$

Solve the system

$$\begin{aligned} 2x + y + 3z &= 100 \\ 2x + 3y + 2z &= 100 \\ x + y + 2z &= 65 \end{aligned}$$

to find $x = 15$, $y = 10$, and $z = 20$. The company should produce 15 model X, 10 model Y, and 20 model Z sets per week.

Notice the advantage of setting up the chart as in Example 6. By reading across, we can easily determine the coefficients and the constants in the system.

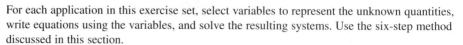

4.3 EXERCISES

For each application in this exercise set, select variables to represent the unknown quantities, write equations using the variables, and solve the resulting systems. Use the six-step method discussed in this section.

Use the techniques of this section to solve these sports-related problems.

1. During the 1996–1997 National Basketball Association regular season, the Chicago Bulls played 82 games. They won 56 more games than they lost. What was their win-loss record that year?

2. During the 1996–1997 National Basketball Association season, the Boston Celtics played 82 games. They lost 52 more games than they won. What was their win-loss record that year?

1996 NBA FINAL STANDINGS
EASTERN CONFERENCE

ATLANTIC DIVISION

Team	W	L
Miami	61	21
New York	57	25
Orlando	45	37
Washington	44	38
New Jersey	26	56
Philadelphia	22	60
Boston	—	—

CENTRAL DIVISION

Team	W	L
Chicago	—	—
Atlanta	56	26
Detroit	54	28
Charlotte	54	28
Cleveland	42	40
Indiana	39	43
Milwaukee	33	49
Toronto	30	52

Source: *Sports Illustrated 1998 Sports Almanac.*

3. During the 1996–1997 National Hockey League regular season, the Dallas Stars played 82 games. Together, their wins and losses totaled 74. They tied 18 fewer games than they lost. How many wins, losses, and ties did they have that year?

4. During the 1996–1997 National Hockey League season, the Boston Bruins played 82 games. Their losses and ties totaled 56, and they had 21 fewer wins than losses. How many wins, losses, and ties did they have that year?

1996-97 NHL FINAL STANDINGS
WESTERN CONFERENCE
CENTRAL DIVISION

	GP	W	L	T	GF	GA	Pts
Dallas	82	—	—	—	252	198	104
Detroit	82	38	26	18	253	197	94
Phoenix	82	38	37	7	240	243	83
St. Louis	82	36	35	11	236	239	83
Chicago	82	34	35	13	223	210	81
Toronto	82	30	44	8	230	273	68

EASTERN CONFERENCE
NORTHEAST DIVISION

	GP	W	L	T	GF	GA	Pts
Buffalo	82	40	30	12	237	208	92
Pittsburgh	82	38	36	8	285	280	84
Ottawa	82	31	36	15	226	234	77
Montreal	82	31	36	15	249	276	77
Hartford	82	32	39	11	226	256	75
Boston	82	—	—	—	234	300	61

Source: *Sports Illustrated 1998 Sports Almanac.*

The applications in Exercises 5–40 require solving systems with two variables, while those that follow require solving systems with three variables.

Solve each problem. See Example 1.

5. Pete and Venus measured the perimeter of a tennis court and found that it was 42 feet longer than it was wide, and had a perimeter of 228 feet. What were the length and the width of the tennis court?

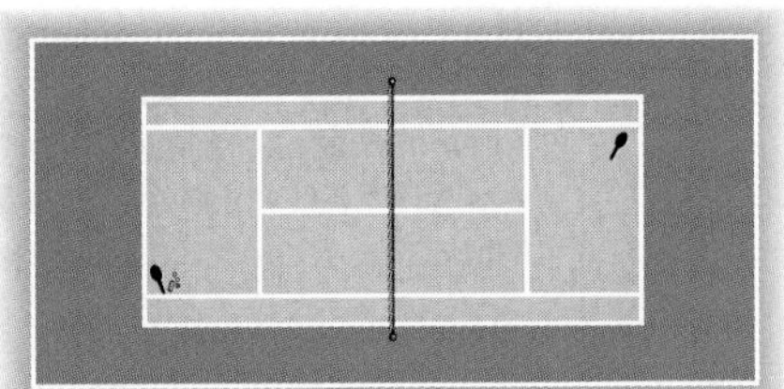

6. Scottie and Jamal found that the width of their basketball court was 44 feet less than the length. If the perimeter was 288 feet, what were the length and the width of their court?

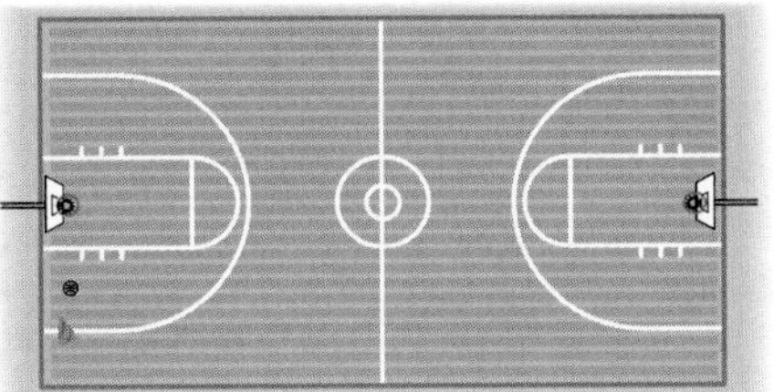

7. The length of a rectangle is 7 feet more than the width. If the length were decreased by 3 feet and the width were increased by 2 feet, the perimeter would be 32 feet. Find the length and width of the original rectangle.

8. The side of a square is 4 centimeters longer than the side of an equilateral triangle. The perimeter of the square is 24 centimeters more than the perimeter of the triangle. Find the lengths of a side of the square and a side of the triangle.

9. Find the measures of the angles marked x and y.

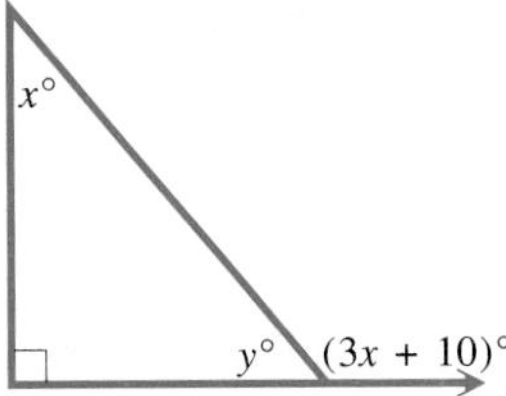

10. Find the measures of the angles marked x and y.

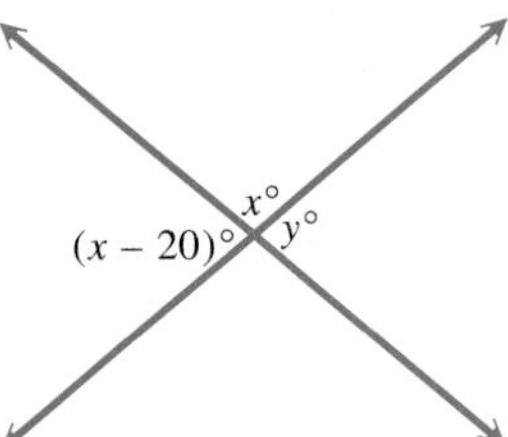

11. In a recent year, the number of daily newspapers in Texas was 52 more than the number of daily newspapers in Florida. Together the two states had a total of 134 dailies. How many did each state have? (*Source: Editor & Publisher International Yearbook.*)

12. In the United States, the number of nuclear power plants in the South exceeds that in the Midwest by 11. Together these two areas of the country have a total of 73 nuclear power plants. How many plants does each region have? (*Source:* U.S. Energy Information Administration.)

The Fan Cost Index (FCI) represents the cost of four average-price tickets, four small soft drinks, two small beers, four hot dogs, parking for one car, two game programs, and two souvenir caps to a sporting event. For example, in 1997, the FCI for Major League Baseball was \$105.63. This was by far the least for the four major professional sports. (*Source:* Team Marketing Report, Chicago.)

Solve each problem. See Example 2. Use the concept of FCI in Exercises 13 and 14.

13. For the 1996–1997 season, the FCI prices for the National Hockey League and the National Basketball Association totaled \$423.12. The hockey FCI was \$16.36 more than that of basketball. What were the FCIs for these sports?

14. For the 1996 season, the FCI prices for Major League Baseball and the National Football League totaled \$311.03. The football FCI was \$105.87 more than that of baseball. What were the FCIs for these sports?

15. Houston Community College has decided to supply its mathematics labs with color monitors. A trip to the local electronics outlet leads to the following information: 4 CGA monitors and 6 VGA monitors can be purchased for \$4600, while 6 CGA monitors and 4 VGA monitors will cost \$4400. What are the prices of a single CGA monitor and a single VGA monitor?

16. For his art class, Theodis bought 2 kilograms of dark clay and 3 kilograms of light clay, paying \$22 for the clay. He later needed 1 kilogram of dark clay and 2 kilograms of light clay, costing \$13 altogether. How much did he pay for each type of clay?

17. A factory makes use of two basic machines, A and B, which turn out two different products, yarn and thread. Each unit of yarn requires 1 hour on machine A and 2 hours on machine B, while each unit of thread requires 1 hour on A and 1 hour on B. Machine A runs 8 hours per day, while machine B runs 14 hours per day. How many units each of yarn and thread should the factory make to keep its machines running at capacity?

18. A biologist wants to grow two types of algae, green and brown. She has 15 kilograms of nutrient X and 26 kilograms of nutrient Y. A vat of green algae needs 2 kilograms of nutrient X and 3 kilograms of nutrient Y, while a vat of brown algae needs 1 kilogram of nutrient X and 2 kilograms of nutrient Y. How many vats of each type of algae should the biologist grow in order to use all the nutrients?

The formulas $p = br$ (percentage = base × rate) and $I = prt$ (simple interest = principal × rate × time) are used in the applications found in Exercises 23–34. To prepare for the use of these formulas, answer the questions in Exercises 19 and 20.

19. If a container of liquid contains 60 ounces of solution, what is the number of ounces of pure acid if the given solution contains the following acid concentrations?
(a) 10% **(b)** 25% **(c)** 40% **(d)** 50%

20. If $5000 is invested in an account paying simple annual interest, how much interest will be earned during the first year at the following rates?
(a) 2% **(b)** 3% **(c)** 4% **(d)** 3.5%

21. If a pound of turkey costs $.58, how much will x pounds cost?

22. If a ticket to the Kevin Costner movie *For Love of the Game* costs $7.00, and y tickets are sold, how much is collected from the sale?

Solve each problem. See Example 3.

23. How many gallons each of 25% alcohol and 35% alcohol should be mixed to get 20 gallons of 32% alcohol?

Kind of Solution	Gallons of Solution	Amount of Pure Alcohol
.25	x	$.25x$
.35	y	$.35y$
.32	20	.32(20)

24. How many liters each of 15% acid and 33% acid should be mixed to get 40 liters of 21% acid?

Kind of Solution	Liters of Solution	Amount of Pure Acid
.15	x	
.33	y	
.21	40	

25. Pure acid is to be added to a 10% acid solution to obtain 27 liters of a 20% acid solution. What amounts of each should be used?

26. A truck radiator holds 18 liters of fluid. How much pure antifreeze must be added to a mixture that is 4% antifreeze in order to fill the radiator with a mixture that is 20% antifreeze?

27. Joycelyn Lowe plans to mix pecan clusters that sell for $3.60 per pound with chocolate truffles that sell for $7.20 per pound to get a mixture that she can sell in Valentine boxes for $4.95 per pound. How much of the $3.60 clusters and the $7.20 truffles should she use to create 80 pounds of the mix?

	Number of Pounds	Price per Pound	Value of Candy
Pecan Clusters	x	3.60	$3.60x$
Chocolate Truffles	y	7.20	$7.20y$
Valentine Mixture	80	4.95	4.95(80)

28. A popular fruit drink is made by mixing fruit juices. Such a mixture with 50% juice is to be mixed with another mixture that is 30% juice to get 200 liters of a mixture that is 45% juice. How much of each should be used?

Kind of Juice	Number of Liters	Amount of Pure Fruit
.50	x	$.50x$
.30	y	
.45		

29. Tickets to a production of *Othello* at Nicholls State University cost $2.50 for general admission or $2.00 with a student identification. If 184 people paid to see a performance and $406 was collected, how many of each type of admission were sold?

30. A grocer plans to mix candy that sells for $1.20 a pound with candy that sells for $2.40 a pound to get a mixture that he plans to sell for $1.65 a pound. How much of the $1.20 and $2.40 candy should he use if he wants 80 pounds of the mix?

31. Jane Ann Lindstedt has been saving dimes and quarters. She has 94 coins in all. If the total value is $19.30, how many dimes and how many quarters does she have?

32. A teller at the Hibernia National Bank received a checking account deposit in twenty-dollar bills and fifty-dollar bills. She received a total of 70 bills, and the amount of the deposit was $3200. How many of each denomination were deposited?

33. A total of $3000 is invested, part at 2% simple interest and part at 4%. If the total annual return from the two investments is $100, how much is invested at each rate?

Principal	Rate	Interest
x	.02	$.02x$
y	.04	$.04y$
3000		100

34. An investor must invest a total of $15,000 in two accounts, one paying 4% annual simple interest, and the other 3%. If he wants to earn $550 annual interest, how much should he invest at each rate?

Principal	Rate	Interest
x	.04	
y	.03	
15,000		

The formula $d = rt$ (distance = rate × time) is used in the applications found in Exercises 37–40. To prepare for the use of this formula, answer the questions in Exercises 35 and 36.

35. If the speed of a killer whale is 25 miles per hour, and the whale swims for y hours, how many miles does the whale travel?

36. If the speed of a boat in still water is 10 miles per hour, and the speed of the current of a river is x miles per hour, what is the speed of the boat
(a) going upstream (that is, against the current) and
(b) going downstream (that is, with the current)?

Solve each problem. See Example 4.

37. A freight train and an express train leave towns 390 kilometers apart, traveling toward one another. The freight train travels 30 kilometers per hour slower than the express train. They pass one another 3 hours later. What are their speeds?

38. A train travels 150 kilometers in the same time that a plane covers 400 kilometers. If the speed of the plane is 20 kilometers per hour less than 3 times the speed of the train, find both speeds.

39. Braving blizzard conditions on the planet Hoth, Luke Skywalker sets out at top speed in his snow speeder for a rebel base 3600 miles away. He travels into a steady headwind, and makes the trip in 2 hours. Returning, he finds that the trip back, still at top speed but now with a tailwind, takes only 1.5 hours. Find the top speed of Luke's snow speeder and the speed of the wind.

40. In his motorboat, Nguyen travels upstream at top speed to his favorite fishing spot, a distance of 36 miles, in two hours. Returning, he finds that the trip downstream, still at top speed, takes only 1.5 hours. Find the speed of Nguyen's boat and the speed of the current.

Solve each problem involving three unknowns. See Examples 5 and 6. (In Exercises 43–46, remember that the sum of the measures of the angles of a triangle is 180°.*)*

41. Gil Troutman has a collection of tropical fish. For each fish, he paid either \$20, \$40, or \$65. The number of \$40 fish is one less than twice the number of \$20 fish. If there are 29 fish in all worth \$1150, how many of each kind of fish are in the collection?

42. A motorcycle manufacturer produces three different models: the Avalon, the Durango, and the Roadripper. Production restrictions require it to make, on a monthly basis, 10 more Roadrippers than the total of the other models, and twice as many Durangos as Avalons. The shop must produce a total of 490 cycles per month. How many cycles of each type should be made per month?

43. In the figure shown, $z = x + 10$ and $x + y = 100$. Determine a third equation involving x, y, and z, and then find the measures of the three angles.

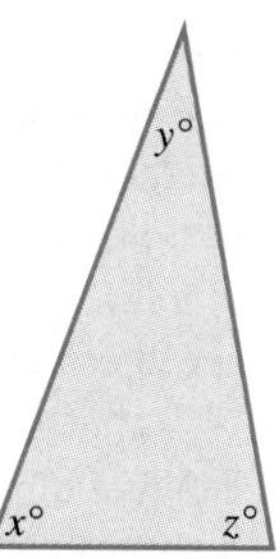

44. In the figure shown, x is 10 less than y and 20 less than z. Write a system of equations and find the measures of the three angles.

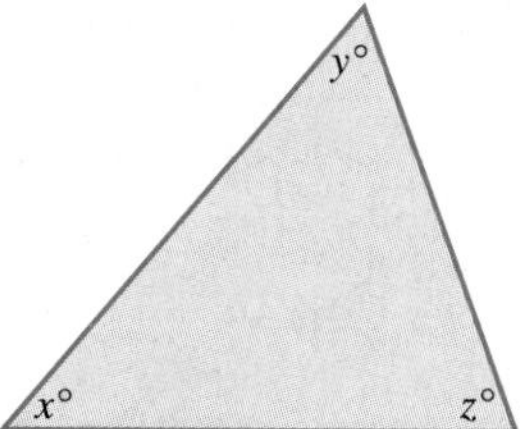

45. In a certain triangle, the measure of the second angle is 10° more than three times the first. The third angle measure is equal to the sum of the measures of the other two. Find the measures of the three angles.

46. The measure of the largest angle of a triangle is 12° less than the sum of the measures of the other two. The smallest angle measures 58° less than the largest. Find the measures of the angles.

47. The perimeter of a triangle is 70 centimeters. The longest side is 4 centimeters less than the sum of the other two sides. Twice the shortest side is 9 centimeters less than the longest side. Find the length of each side of the triangle.

48. The perimeter of a triangle is 56 inches. The longest side measures 4 inches less than the sum of the other two sides. Three times the shortest side is 4 inches more than the longest side. Find the lengths of the three sides.

49. A Mardi Gras trinket manufacturer supplies three wholesalers, A, B, and C. The output from a day's production is 320 cases of trinkets. She must send wholesaler A three times as many cases as she sends B, and she must send wholesaler C 160 cases less than she provides A and B together. How many cases should she send to each wholesaler to distribute the entire day's production to them?

50. A hardware supplier manufactures three kinds of clamps, types A, B, and C. Production restrictions require it to make 10 units more type C clamps than the total of the other types and twice as many type B clamps as type A. The shop must produce a total of 490 units of clamps per day. How many units of each type can be made per day?

51. The manager of a candy store wants to feature a special Easter candy mixture of jelly beans, small chocolate eggs, and marshmallow chicks. She plans to make 15 pounds of mix to sell at \$1 a pound. Jelly beans sell for \$.80 a pound, chocolate eggs for \$2 a pound, and marshmallow chicks for \$1 a pound. She will use twice as many pounds of jelly beans as eggs and chicks combined and fives times as many pounds of jelly beans as chocolate eggs. How many pounds of each candy should she use?

52. Three kinds of tickets are available for a Green Day concert: "up close," "in the middle," and "far out." "Up close" tickets cost \$10 more than "in the middle" tickets, while "in the middle" tickets cost \$10 more than "far out" tickets. Twice the cost of an "up close" ticket is \$20 more than 3 times the cost of a "far out" seat. Find the price of each kind of ticket.

RELATING CONCEPTS (EXERCISES 53–56)

In a later chapter we will see that an equation of the form

$$x^2 + y^2 + ax + by + c = 0$$

may have a circle as its graph. It is a fact from geometry that given three noncollinear points (that is, points that do not all lie on the same straight line), there will be a circle that contains them. For example, the points (4, 2), (−5, −2), and (0, 3) lie on the circle whose equation is

$$x^2 + y^2 - \frac{7}{5}x + \frac{27}{5}y - \frac{126}{5} = 0.$$

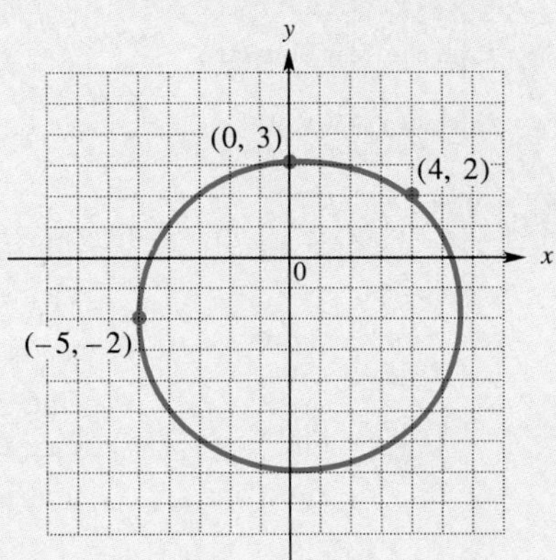

The circle is shown in the figure.

Work Exercises 53–56 in order, *so that the equation of the circle passing through the points* (2, 1), (−1, 0), *and* (3, 3) *can be found.*

53. Let $x = 2$ and $y = 1$ in the equation $x^2 + y^2 + ax + by + c = 0$ to find an equation in a, b, and c.

54. Let $x = -1$ and $y = 0$ to find a second equation in a, b, and c.

55. Let $x = 3$ and $y = 3$ to find a third equation in a, b, and c.

56. Solve the system of equations formed by your answers in Exercises 53–55 to find the values of a, b, and c. What is the equation of the circle?

Did you make the connection between solving a system and finding an equation of a circle?

57. Observe the graph of the circle in the preceding Relating Concepts exercises. Explain why the graph is not that of a function.

58. Make up a problem similar to the one in Exercise 41 or Exercise 42, and solve it. (*Hint:* Start with the answer, and write the problem to fit the answer.)

4.4 Solving Linear Systems of Equations by Matrix Methods

OBJECTIVES

1. Define a matrix.
2. Write the augmented matrix for a system.
3. Use row operations to solve a system with two equations.
4. Use row operations to solve a system with three equations.
5. Use row operations to solve special systems.

FOR EXTRA HELP

SSG Sec. 4.4
SSM Sec. 4.4
Pass the Test Software
InterAct Math Tutorial Software
Video 7

OBJECTIVE 1 Define a matrix. An ordered array of numbers such as

$$\begin{bmatrix} 2 & 3 & 5 \\ 7 & 1 & 2 \end{bmatrix}$$

is called a **matrix.** The numbers are called **elements** of the matrix. Matrices (the plural of *matrix*) are named according to the number of **rows** and **columns** they contain. The rows are read horizontally and the columns are read vertically. For example, the first row in the matrix above is 2 3 5 and the first column is $\begin{matrix} 2 \\ 7 \end{matrix}$. The matrix above is a 2×3 (read "two by three") matrix because it has 2 rows and 3 columns. The number of rows is given first, and then the number of columns.

The matrix

$$\begin{bmatrix} -1 & 0 \\ 1 & -2 \end{bmatrix}$$

is a 2×2 matrix, and the matrix

$$\begin{bmatrix} 8 & -1 & -3 \\ 2 & 1 & 6 \\ 0 & 5 & -3 \\ 5 & 9 & 7 \end{bmatrix}$$

is a 4×3 matrix. A **square matrix** is one that has the same number of rows as columns. The 2×2 matrix above is a square matrix.

Figure 8 shows how a graphing calculator displays the two matrices above. Work with matrices discussed in this section is made much easier by using technology when available.

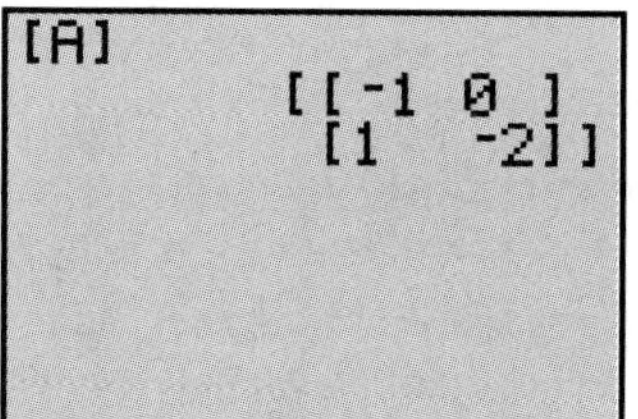

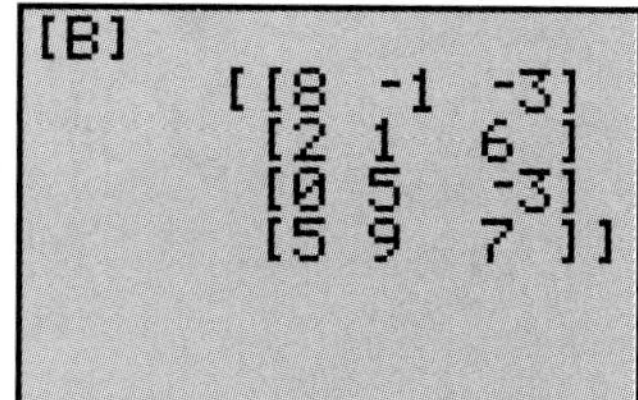

Figure 8

In this section we discuss a method of solving linear systems that uses matrices. This method is really just a very structured way of using the elimination method to solve a linear system. The advantage of this new method is that it can be done by a graphing calculator or a computer, allowing large systems of equations to be solved more easily.

OBJECTIVE 2 Write the augmented matrix for a system. To begin, we write an *augmented matrix* for the system. An **augmented matrix** has a vertical bar that separates the columns of the matrix into two groups. For example, to solve the system

$$\begin{aligned} x - 3y &= 1 \\ 2x + y &= -5, \end{aligned}$$

start with the augmented matrix

$$\left[\begin{array}{cc|c} 1 & -3 & 1 \\ 2 & 1 & -5 \end{array}\right].$$

Place the coefficients of the variables to the left of the bar, and the constants to the right. The bar separates the coefficients from the constants. The matrix is just a shorthand way of writing the system of equations, so the rows of the augmented matrix can be treated the same as the equations of a system of equations.

We know that exchanging the position of two equations in a system does not change the system. Also, multiplying any equation in a system by a nonzero number does not change the system. Comparable changes to the augmented matrix of a system of equations produce new matrices that correspond to systems with the same solutions as the original system.

The following **row operations** produce new matrices that lead to systems having the same solutions as the original system.

Matrix Row Operations

1. Any two rows of the matrix may be interchanged.
2. The numbers in any row may be multiplied by any nonzero real number.
3. Any row may be changed by adding to the numbers of the row the product of a real number and the corresponding numbers of another row.

Examples of these row operations follow.

Using row operation 1,

$$\begin{bmatrix} 2 & 3 & 9 \\ 4 & 8 & -3 \\ 1 & 0 & 7 \end{bmatrix} \quad \text{becomes} \quad \begin{bmatrix} 1 & 0 & 7 \\ 4 & 8 & -3 \\ 2 & 3 & 9 \end{bmatrix}$$

by interchanging row 1 and row 3.

Using row operation 2,

$$\begin{bmatrix} 2 & 3 & 9 \\ 4 & 8 & -3 \\ 1 & 0 & 7 \end{bmatrix} \quad \text{becomes} \quad \begin{bmatrix} 6 & 9 & 27 \\ 4 & 8 & -3 \\ 1 & 0 & 7 \end{bmatrix}$$

by multiplying the numbers in row 1 by 3.

Using row operation 3,

$$\begin{bmatrix} 2 & 3 & 9 \\ 4 & 8 & -3 \\ 1 & 0 & 7 \end{bmatrix} \quad \text{becomes} \quad \begin{bmatrix} 0 & 3 & -5 \\ 4 & 8 & -3 \\ 1 & 0 & 7 \end{bmatrix}$$

by multiplying the numbers in row 3 by -2 and adding them to the corresponding numbers in row 1.

The third row operation corresponds to the way we eliminated a variable from a pair of equations in the previous sections.

OBJECTIVE 3 Use row operations to solve a system with two equations. Row operations can be used to rewrite a matrix until it is the matrix of a system where the solution is easy to find. The goal is a matrix in the form

$$\begin{bmatrix} 1 & a & | & b \\ 0 & 1 & | & c \end{bmatrix} \quad \text{or} \quad \begin{bmatrix} 1 & a & b & | & c \\ 0 & 1 & d & | & e \\ 0 & 0 & 1 & | & f \end{bmatrix}$$

for systems with two or three equations, respectively. Notice that there are 1s down the diagonal from upper left to lower right and 0s below the 1s. A matrix written this way is said to be in **row echelon form.** When these matrices are rewritten as systems of equations, the value of one variable is known, and the rest can be found by substitution. The following examples illustrate the method.

EXAMPLE 1 Using Row Operations to Solve a System with Two Variables

Use row operations to solve the system.

$$x - 3y = 1$$
$$2x + y = -5$$

We start by writing the augmented matrix of the system.

$$\begin{bmatrix} 1 & -3 & | & 1 \\ 2 & 1 & | & -5 \end{bmatrix}$$

Now we use the various row operations to change this matrix into one that leads to a system that is easier to solve.

It is best to work by columns. We start with the first column and make sure that there is a 1 in the first row, first column position. There already is a 1 in this position. Next, we get 0s in every position below the first. To get a 0 in row two, column one, we use the third row operation and add to the numbers in row two the result of multiplying each number in row one by -2. (We abbreviate this as $-2R_1 + R_2$.) Row one remains unchanged.

$$\begin{bmatrix} 1 & -3 & | & 1 \\ 2 + \mathbf{1(-2)} & 1 + \mathbf{-3(-2)} & | & -5 + \mathbf{1(-2)} \end{bmatrix}$$

↑ Original number from row two ↑ -2 times number from row one

$$\begin{bmatrix} 1 & -3 & | & 1 \\ 0 & 7 & | & -7 \end{bmatrix} \quad -2R_1 + R_2$$

The matrix now has a 1 in the first position of column one, with 0s in every position below the first.

Now we go to column two. A 1 is needed in row two, column two. We get this 1 by using the second row operation, multiplying each number of row two by $\frac{1}{7}$.

$$\begin{bmatrix} 1 & -3 & | & 1 \\ 0 & 1 & | & -1 \end{bmatrix} \quad \tfrac{1}{7}R_2$$

This augmented matrix leads to the system of equations

$$\begin{aligned} \mathbf{1}x - 3y &= \mathbf{1} \\ \mathbf{0}x + \mathbf{1}y &= \mathbf{-1} \end{aligned} \quad \text{or} \quad \begin{aligned} x - 3y &= 1 \\ y &= -1. \end{aligned}$$

From the second equation, $y = -1$. We substitute -1 for y in the first equation to get

$$\begin{aligned} x - 3y &= 1 \\ x - 3(-1) &= 1 \\ x + 3 &= 1 \\ x &= -2. \end{aligned}$$

The solution set of the system is $\{(-2, -1)\}$. Check this solution by substitution in both equations.

If the augmented matrix of the system in Example 1 is entered as matrix A in a graphing calculator (Figure 9(a)) and the row echelon form of the matrix is found (Figure 9(b)), the system becomes

$$\begin{aligned} x + \frac{1}{2}y &= -\frac{5}{2} \\ y &= -1. \end{aligned}$$

While this system looks different from the one we obtained in Example 1, it is equivalent, since its solution set is also $\{(-2, -1)\}$.

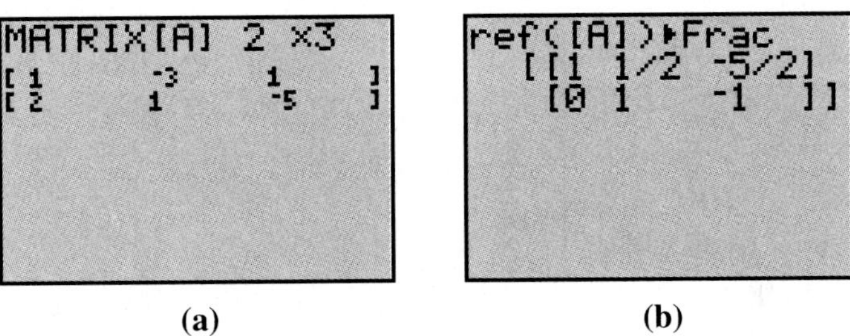

(a) (b)

Figure 9

CONNECTIONS

One of the beautiful aspects of mathematics is that there are often many different ways to solve the same problem. It is likely that, given a particular system of equations, different students will use a variety of methods to solve the system. Regardless of the method used, they should all get the same answer.

FOR DISCUSSION OR WRITING

1. Show that the system represented by the matrix in Figure 9(b) has the same solution set as the system in Example 1.
2. Make up another system, having $y = -1$ as one of the equations, that has the same solution set as the system in Example 1.

OBJECTIVE 4 Use row operations to solve a system with three equations. A linear system with three equations is solved in a similar way. We use row operations to get 1s down the diagonal from left to right and all 0s below each 1.

EXAMPLE 2 Using Row Operations to Solve a System with Three Variables

Use matrix methods to solve the system.

$$\begin{aligned} x - y + 5z &= -6 \\ 3x + 3y - z &= 10 \\ x + 3y + 2z &= 5 \end{aligned}$$

Start by writing the augmented matrix of the system.

$$\left[\begin{array}{rrr|r} 1 & -1 & 5 & -6 \\ 3 & 3 & -1 & 10 \\ 1 & 3 & 2 & 5 \end{array}\right]$$

This matrix already has 1 in row one, column one. Next get 0s in the rest of column one. First, add to row two the results of multiplying each number of row one by -3. This gives the matrix

$$\left[\begin{array}{rrr|r} 1 & -1 & 5 & -6 \\ 0 & 6 & -16 & 28 \\ 1 & 3 & 2 & 5 \end{array}\right]. \qquad -3R_1 + R_2$$

Now add to the numbers in row three the results of multiplying each number of row one by -1.

$$\left[\begin{array}{rrr|r} 1 & -1 & 5 & -6 \\ 0 & 6 & -16 & 28 \\ 0 & 4 & -3 & 11 \end{array}\right] \qquad -1R_1 + R_3$$

We get 1 in row two, column two by multiplying each number in row two by $\frac{1}{6}$.

$$\left[\begin{array}{rrr|r} 1 & -1 & 5 & -6 \\ 0 & 1 & -\frac{8}{3} & \frac{14}{3} \\ 0 & 4 & -3 & 11 \end{array}\right] \qquad \tfrac{1}{6}R_2$$

Get 0 in row three, column two by adding to row three the results of multiplying each number in row two by -4.

$$\left[\begin{array}{rrr|r} 1 & -1 & 5 & -6 \\ 0 & 1 & -\frac{8}{3} & \frac{14}{3} \\ 0 & 0 & \frac{23}{3} & -\frac{23}{3} \end{array}\right] \qquad -4R_2 + R_3$$

Finally, get 1 in row three, column three by multiplying each number in row three by $\frac{3}{23}$.

$$\left[\begin{array}{rrr|r} 1 & -1 & 5 & -6 \\ 0 & 1 & -\frac{8}{3} & \frac{14}{3} \\ 0 & 0 & 1 & -1 \end{array}\right] \qquad \tfrac{3}{23}R_3$$

This final matrix gives the system of equations

$$x - y + 5z = -6$$
$$y - \frac{8}{3}z = \frac{14}{3}$$
$$z = -1.$$

Substitute -1 for z in the second equation, to get

$$y - \frac{8}{3}z = \frac{14}{3}$$
$$y - \frac{8}{3}(-1) = \frac{14}{3}$$
$$y + \frac{8}{3} = \frac{14}{3}$$
$$y = 2.$$

Finally, substitute 2 for y and -1 for z in the first equation.

$$x - y + 5z = -6$$
$$x - 2 + 5(-1) = -6$$
$$x - 2 - 5 = -6$$
$$x = 1$$

The solution set of the original system is $\{(1, 2, -1)\}$. This solution should be checked by substitution in the system.

OBJECTIVE 5 **Use row operations to solve special systems.** In the final example we show how to recognize inconsistent systems or systems with dependent equations when solving these systems with row operations.

EXAMPLE 3 **Recognizing Inconsistent Systems or Dependent Equations**

Use row operations to solve each system.

(a) $2x - 3y = 8$
$-6x + 9y = 4$
Write the augmented matrix.

$$\left[\begin{array}{rr|r} 2 & -3 & 8 \\ -6 & 9 & 4 \end{array}\right]$$

Multiply the first row by $\frac{1}{2}$ to get 1 in row one, column one.

$$\left[\begin{array}{cc|c} 1 & -\dfrac{3}{2} & 4 \\ -6 & 9 & 4 \end{array}\right] \qquad \tfrac{1}{2}R_1$$

Multiply row one by 6 and add the results to row two.

$$\left[\begin{array}{cc|c} 1 & -\dfrac{3}{2} & 4 \\ 0 & 0 & 28 \end{array}\right] \qquad 6R_1 + R_2$$

The corresponding system of equations is

$$x - \frac{3}{2}y = 4$$

$$0 = 28,$$

which has no solution and so is inconsistent. The solution set is $\emptyset$.

(b) $-10x + 12y = 30$
$5x - 6y = -15$

The augmented matrix is

$$\left[\begin{array}{cc|c} -10 & 12 & 30 \\ 5 & -6 & -15 \end{array}\right].$$

Multiply the first row by $-\frac{1}{10}$.

$$\left[\begin{array}{cc|c} 1 & -\dfrac{6}{5} & -3 \\ 5 & -6 & -15 \end{array}\right] \qquad -\tfrac{1}{10}R_1$$

Multiply the first row by -5 and add the products to row two to get

$$\left[\begin{array}{cc|c} 1 & -\dfrac{6}{5} & -3 \\ 0 & 0 & 0 \end{array}\right]. \qquad -5R_1 + R_2$$

The corresponding system is

$$x - \frac{6}{5}y = -3$$

$$0 = 0,$$

which has dependent equations. Clearing fractions in the first equation, we write the solution set as $\{(x, y) \mid 5x - 6y = -15\}$.

CONNECTIONS

An extension of the matrix method described in this section involves transforming an augmented matrix into **reduced row echelon form.** This form has 1s down the main diagonal and 0s above and below this diagonal. For example, the matrix for the system in Example 2 could be transformed into the following:

$$\left[\begin{array}{ccc|c} 1 & 0 & 0 & 1 \\ 0 & 1 & 0 & 2 \\ 0 & 0 & 1 & -1 \end{array}\right].$$

This would indicate that an equivalent system is

$$\begin{aligned} x &= 1 \\ y &= 2 \\ z &= -1. \end{aligned}$$

The graphing calculator screens in Figures 10(a) and (b) indicate how easily this transformation can be obtained using technology.

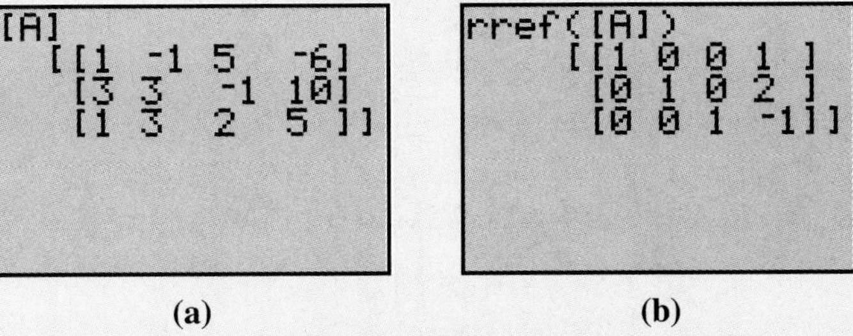

(a) (b)

Figure 10

FOR DISCUSSION OR WRITING

1. Write the reduced row echelon form for the matrix of the system in Example 1.

2. If transforming to reduced row echelon form leads to all 0s in the final row, what kind of system is represented?

4.4 EXERCISES

1. Consider the matrix $\begin{bmatrix} -2 & 3 & 1 \\ 0 & 5 & -3 \\ 1 & 4 & 8 \end{bmatrix}$ and answer the following.

(a) What are the elements of the second row?
(b) What are the elements of the third column?
(c) Is this a square matrix? Explain why or why not.
(d) Give the matrix obtained by interchanging the first and third rows.
(e) Give the matrix obtained by multiplying the first row by $-\frac{1}{2}$.
(f) Give the matrix obtained by multiplying the third row by 3 and adding to the first row.

2. Give the dimensions of each of the following matrices.

(a) $\begin{bmatrix} 3 & -7 \\ 4 & 5 \\ -1 & 0 \end{bmatrix}$ **(b)** $\begin{bmatrix} 4 & 9 & 0 \\ -1 & 2 & -4 \end{bmatrix}$

(c)

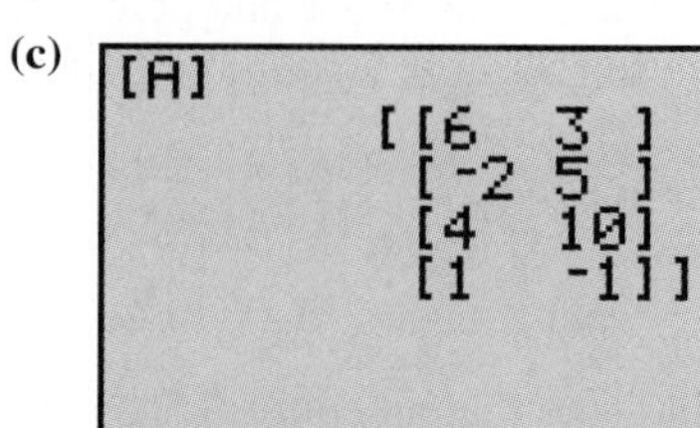

(d)

[B]
[[-1 7 1]
[0 2 1]
[-3 5 4]
[8 -9 1]]

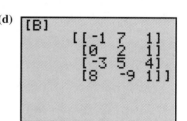

Complete the steps in the matrix solution of each system by filling in the boxes. Give the final system and the solution set. See Example 1.

3. $4x + 8y = 44$
$2x - y = -3$

$\left[\begin{array}{cc|c} 4 & 8 & 44 \\ 2 & -1 & -3 \end{array}\right]$

$\left[\begin{array}{cc|c} 1 & \square & \square \\ 2 & -1 & -3 \end{array}\right]$ $\frac{1}{4}R_1$

$\left[\begin{array}{cc|c} 1 & 2 & 11 \\ 0 & \square & \square \end{array}\right]$ $-2R_1 + R_2$

$\left[\begin{array}{cc|c} 1 & 2 & 11 \\ 0 & 1 & \square \end{array}\right]$ $-\frac{1}{5}R_2$

4. $2x - 5y = -1$
$3x + y = 7$

$\left[\begin{array}{cc|c} 2 & -5 & -1 \\ 3 & 1 & 7 \end{array}\right]$

$\left[\begin{array}{cc|c} 1 & -\frac{5}{2} & \square \\ 3 & 1 & 7 \end{array}\right]$ $\frac{1}{2}R_1$

$\left[\begin{array}{cc|c} 1 & -\frac{5}{2} & -\frac{1}{2} \\ 0 & \square & \square \end{array}\right]$ $-3R_1 + R_2$

$\left[\begin{array}{cc|c} 1 & -\frac{5}{2} & -\frac{1}{2} \\ 0 & 1 & \square \end{array}\right]$ $\frac{2}{17}R_2$

Use row operations to solve each system. See Examples 1 and 3.

5. $x + y = 5$
$x - y = 3$

6. $x + 2y = 7$
$x - y = -2$

7. $2x + 4y = 6$
$3x - y = 2$

8. $4x + 5y = -7$
$x - y = 5$

9. $3x + 4y = 13$
$2x - 3y = -14$

10. $5x + 2y = 8$
$3x - y = 7$

11. $-4x + 12y = 36$
$x - 3y = 9$

12. $2x - 4y = 8$
$-3x + 6y = 5$

13. Write a short explanation of each of the following. Include examples.
(a) matrix
(b) row of a matrix
(c) column of a matrix
(d) square matrix
(e) augmented matrix
(f) row operations on a matrix

14. Compare the use of the third row operation on a matrix and the elimination method of solving a system of linear equations. Give examples.

Complete the steps in the matrix solution of each system by filling in the boxes. Give the final system and the solution set. See Example 2.

15. $x + y - z = -3$
$2x + y + z = 4$
$5x - y + 2z = 23$

$$\left[\begin{array}{ccc|c} 1 & 1 & -1 & -3 \\ 2 & 1 & 1 & 4 \\ 5 & -1 & 2 & 23 \end{array}\right]$$

$$\left[\begin{array}{ccc|c} 1 & 1 & -1 & -3 \\ 0 & \square & \square & \square \\ 0 & \square & \square & \square \end{array}\right] \quad \begin{array}{l} \\ -2R_1 + R_2 \\ -5R_1 + R_3 \end{array}$$

$$\left[\begin{array}{ccc|c} 1 & 1 & -1 & -3 \\ 0 & 1 & \square & \square \\ 0 & -6 & 7 & 38 \end{array}\right] \quad -1R_2$$

$$\left[\begin{array}{ccc|c} 1 & 1 & -1 & -3 \\ 0 & 1 & -3 & -10 \\ 0 & 0 & \square & \square \end{array}\right] \quad 6R_2 + R_3$$

$$\left[\begin{array}{ccc|c} 1 & 1 & -1 & -3 \\ 0 & 1 & -3 & -10 \\ 0 & 0 & 1 & \square \end{array}\right] \quad -\tfrac{1}{11}R_3$$

16. $2x + y + 2z = 11$
$2x - y - z = -3$
$3x + 2y + z = 9$

$$\left[\begin{array}{ccc|c} 2 & 1 & 2 & 11 \\ 2 & -1 & -1 & -3 \\ 3 & 2 & 1 & 9 \end{array}\right]$$

$$\left[\begin{array}{ccc|c} 1 & \square & \square & \square \\ 2 & -1 & -1 & -3 \\ 3 & 2 & 1 & 9 \end{array}\right] \quad \tfrac{1}{2}R_1$$

$$\left[\begin{array}{ccc|c} 1 & \frac{1}{2} & 1 & \frac{11}{2} \\ 0 & \square & \square & \square \\ 0 & \square & \square & \square \end{array}\right] \quad \begin{array}{l} \\ -2R_1 + R_2 \\ -3R_1 + R_3 \end{array}$$

$$\left[\begin{array}{ccc|c} 1 & \frac{1}{2} & 1 & \frac{11}{2} \\ 0 & 1 & \square & \square \\ 0 & \frac{1}{2} & -2 & -\frac{15}{2} \end{array}\right] \quad -\tfrac{1}{2}R_2$$

$$\left[\begin{array}{ccc|c} 1 & \frac{1}{2} & 1 & \frac{11}{2} \\ 0 & 1 & \frac{3}{2} & 7 \\ 0 & 0 & \square & \square \end{array}\right] \quad -\tfrac{1}{2}R_2 + R_3$$

$$\left[\begin{array}{ccc|c} 1 & \frac{1}{2} & 1 & \frac{11}{2} \\ 0 & 1 & \frac{3}{2} & 7 \\ 0 & 0 & 1 & \square \end{array}\right] \quad -\tfrac{4}{11}R_3$$

Use row operations to solve each system. See Examples 2 and 3.

17. $x + y - 3z = 1$
$2x - y + z = 9$
$3x + y - 4z = 8$

18. $2x + 4y - 3z = -18$
$3x + y - z = -5$
$x - 2y + 4z = 14$

19. $x + y - z = 6$
$2x - y + z = -9$
$x - 2y + 3z = 1$

20. $x + 3y - 6z = 7$
$2x - y + 2z = 0$
$x + y + 2z = -1$

21. $x - y = 1$
$y - z = 6$
$x + z = -1$

22. $x + y = 1$
$2x - z = 0$
$y + 2z = -2$

23. $x - 2y + z = 4$
$3x - 6y + 3z = 12$
$-2x + 4y - 2z = -8$

24. $4x + 8y + 4z = 9$
$x + 3y + 4z = 10$
$5x + 10y + 5z = 12$

The augmented matrix for the system in Exercise 3 is shown in the graphing calculator screen on the top left as matrix A. The screen on the top right shows the row echelon form for A. Compare it to the matrix shown in the answer section for Exercise 3. The screen at the bottom shows the reduced row echelon form, and from this it can be determined by inspection that the solution set of the system is $\{(1, 5)\}$.

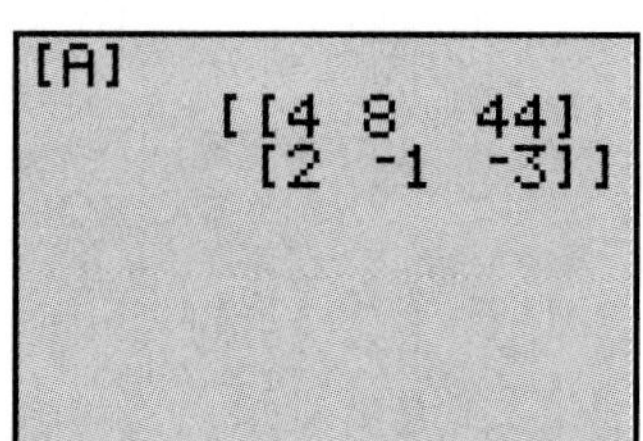

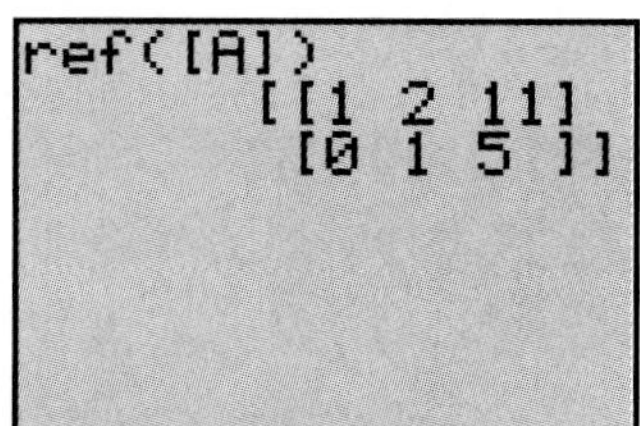

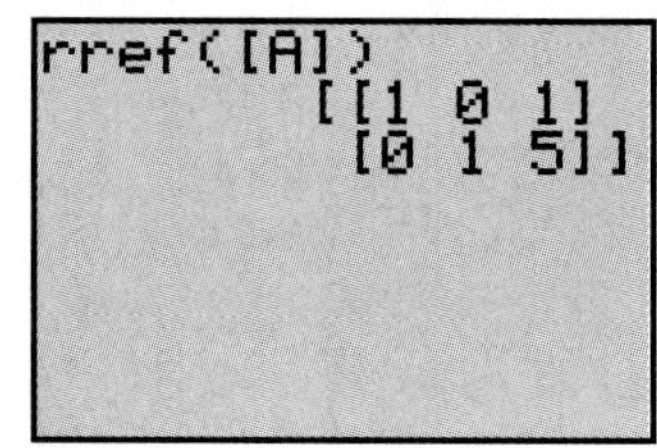

Use a graphing calculator and either one of the two matrix methods illustrated to solve each system.

25. $4x + y = 5$
$2x + y = 3$

26. $5x + 3y = 7$
$7x - 3y = -19$

27. $5x + y - 3z = -6$
$2x + 3y + z = 5$
$-3x - 2y + 4z = 3$

28. $x + y + z = 3$
$3x - 3y - 4z = -1$
$x + y + 3z = 11$

29. $x + z = -3$
$y + z = 3$
$x + y = 8$

30. $x - y = -1$
$-y + z = -2$
$x + z = -2$

4.5 Determinants and Cramer's Rule

OBJECTIVES

1. Evaluate 2×2 determinants.
2. Use expansion by minors to evaluate 3×3 determinants.
3. Use a graphing calculator to evaluate determinants.
4. Understand the derivation of Cramer's rule.
5. Apply Cramer's rule to solve linear systems.

Three methods for solving linear systems have now been presented: elimination, substitution, and a matrix method. A method of solving linear systems by using *determinants* is introduced later in this section.

Associated with every *square* matrix is a real number called the **determinant** of the matrix. A determinant is symbolized by the entries of the matrix placed between two vertical lines, such as

$$\begin{vmatrix} 2 & 3 \\ 7 & 1 \end{vmatrix} \quad \text{or} \quad \begin{vmatrix} 7 & 4 & 3 \\ 0 & 1 & 5 \\ 6 & 0 & 1 \end{vmatrix}.$$

Like matrices, determinants are named according to the number of rows and columns they contain. For example, the first determinant shown is a 2×2 (read "two by two") determinant. The second is a 3×3 determinant.

FOR EXTRA HELP
SSG Sec. 4.5
SSM Sec. 4.5
Pass the Test Software
InterAct Math Tutorial Software
Video 7

OBJECTIVE 1 Evaluate 2 × 2 determinants. The value of the 2×2 determinant

$$\begin{vmatrix} a & b \\ c & d \end{vmatrix}$$

is defined as follows.

Value of a 2 × 2 Determinant

$$\begin{vmatrix} a & b \\ c & d \end{vmatrix} = ad - bc$$

EXAMPLE 1 Evaluating a 2 × 2 Determinant

Evaluate the determinant.

$$\begin{vmatrix} -1 & -3 \\ 4 & -2 \end{vmatrix}$$

Here $a = -1$, $b = -3$, $c = 4$, and $d = -2$, and

$$\begin{vmatrix} -1 & -3 \\ 4 & -2 \end{vmatrix} = (-1)(-2) - (-3)(4) = 2 + 12 = 14.$$

A 3×3 determinant can be evaluated in a similar way.

Value of a 3 × 3 Determinant

$$\begin{vmatrix} a_1 & b_1 & c_1 \\ a_2 & b_2 & c_2 \\ a_3 & b_3 & c_3 \end{vmatrix} = (a_1b_2c_3 + b_1c_2a_3 + c_1a_2b_3) - (a_3b_2c_1 + b_3c_2a_1 + c_3a_2b_1)$$

This rule for evaluating a 3×3 determinant is hard to remember. A method for calculating a 3×3 determinant that is easier to use is based on the definition above. Rearranging terms and using the distributive property gives

$$\begin{vmatrix} a_1 & b_1 & c_1 \\ a_2 & b_2 & c_2 \\ a_3 & b_3 & c_3 \end{vmatrix} = a_1(b_2c_3 - b_3c_2) - a_2(b_1c_3 - b_3c_1) + a_3(b_1c_2 - b_2c_1). \quad \textbf{(1)}$$

Each of the quantities in parentheses represents a 2×2 determinant, which is that part of the 3×3 determinant remaining when the row and column of the multiplier are eliminated, as shown below.

$$a_1(b_2c_3 - b_3c_2) \quad \begin{vmatrix} a_1 & b_1 & c_1 \\ a_2 & b_2 & c_2 \\ a_3 & b_3 & c_3 \end{vmatrix}$$

$$a_2(b_1c_3 - b_3c_1) \quad \begin{vmatrix} a_1 & b_1 & c_1 \\ a_2 & b_2 & c_2 \\ a_3 & b_3 & c_3 \end{vmatrix}$$

$$a_3(b_1c_2 - b_2c_1) \quad \begin{vmatrix} a_1 & b_1 & c_1 \\ a_2 & b_2 & c_2 \\ a_3 & b_3 & c_3 \end{vmatrix}$$

These 2 × 2 determinants are called **minors** of the elements in the 3 × 3 determinant. In the determinant above, the minors of a_1, a_2, and a_3 are, respectively,

$$\begin{vmatrix} b_2 & c_2 \\ b_3 & c_3 \end{vmatrix}, \quad \begin{vmatrix} b_1 & c_1 \\ b_3 & c_3 \end{vmatrix}, \quad \begin{vmatrix} b_1 & c_1 \\ b_2 & c_2 \end{vmatrix}.$$

OBJECTIVE 2 Use expansion by minors to evaluate 3 × 3 determinants. A 3 × 3 determinant can be evaluated by multiplying each element in the first column by its minor and combining the products as shown in equation (1). This is called **expansion of the determinant by minors** about the first column.

EXAMPLE 2 Evaluating a 3 × 3 Determinant

Evaluate the determinant by expanding by minors about the first column.

$$\begin{vmatrix} 1 & 3 & -2 \\ -1 & -2 & -3 \\ 1 & 1 & 2 \end{vmatrix}$$

In this determinant, $a_1 = 1$, $a_2 = -1$, and $a_3 = 1$. Multiply each of these numbers by its minor and combine the three terms using the definition. Notice that the second term in the definition is *subtracted.*

$$\begin{aligned}\begin{vmatrix} 1 & 3 & -2 \\ -1 & -2 & -3 \\ 1 & 1 & 2 \end{vmatrix} &= 1\begin{vmatrix} -2 & -3 \\ 1 & 2 \end{vmatrix} - (-1)\begin{vmatrix} 3 & -2 \\ 1 & 2 \end{vmatrix} + 1\begin{vmatrix} 3 & -2 \\ -2 & -3 \end{vmatrix} \\ &= 1[(-2)(2) - (-3)(1)] + 1[(3)(2) - (-2)(1)] \\ &\quad + 1[(3)(-3) - (-2)(-2)] \\ &= 1(-1) + 1(8) + 1(-13) \\ &= -1 + 8 - 13 \\ &= -6\end{aligned}$$

To get equation (1) we could have rearranged terms in the definition of the determinant and factored out the three elements of the second or third columns or of any of the three rows. Therefore, expanding by minors about any row or any column results in the same value for a 3 × 3 determinant. To determine the correct signs for the terms of other expansions, the following **array of signs** is helpful.

Array of Signs for a 3 × 3 Determinant

$$\begin{matrix} + & - & + \\ - & + & - \\ + & - & + \end{matrix}$$

The signs alternate for each row and column beginning with + in the first row, first column position. For example, if the expansion is to be about the second column, the first term would have a minus sign associated with it, the second term a plus sign, and the third term a minus sign.

EXAMPLE 3 Evaluating a 3 × 3 Determinant by a Different Expansion

Evaluate the determinant of Example 2 by expansion by minors about the second column.

$$\begin{vmatrix} 1 & 3 & -2 \\ -1 & -2 & -3 \\ 1 & 1 & 2 \end{vmatrix} = -3\begin{vmatrix} -1 & -3 \\ 1 & 2 \end{vmatrix} + (-2)\begin{vmatrix} 1 & -2 \\ 1 & 2 \end{vmatrix} - 1\begin{vmatrix} 1 & -2 \\ -1 & -3 \end{vmatrix}$$

$$= -3(1) - 2(4) - 1(-5)$$

$$= -3 - 8 + 5$$

$$= -6$$

As expected, the result is the same as in Example 2.

OBJECTIVE 3 Use a graphing calculator to evaluate determinants. The graphing calculator function det(A) assigns to each square matrix A one and only one real number, the determinant of A.

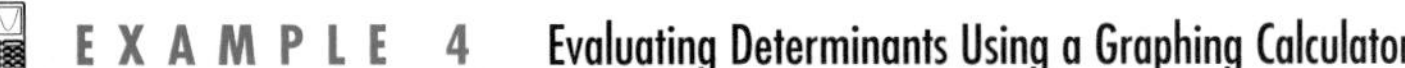

EXAMPLE 4 Evaluating Determinants Using a Graphing Calculator

Evaluate the determinants in Examples 1 and 2 using a graphing calculator.

Figure 11 shows how a graphing calculator displays the correct value for the determinant in Example 1. Similarly, Figure 12 supports the result of Example 2.

```
[A]
          [[-1  -3]
           [4   -2]]
det([A])
                 14
```

Figure 11

```
[B]
        [[1   3  -2]
         [-1 -2  -3]
         [1   1   2 ]]
det([B])
                  -6
```

Figure 12

CONNECTIONS

Determinants of larger dimensions (such as 4 × 4) can be evaluated by extending the concepts presented thus far. However, because of the tedious calculations and chance for error, they are usually evaluated by computer or graphing calculator. For example, the determinant

$$\begin{vmatrix} -1 & -2 & 3 & 2 \\ 0 & 1 & 4 & -2 \\ 3 & -1 & 4 & 0 \\ 2 & 1 & 0 & 3 \end{vmatrix}$$

is equal to -185, as shown in the graphing calculator screen in Figure 13.

(continued)

CONNECTIONS (CONTINUED)

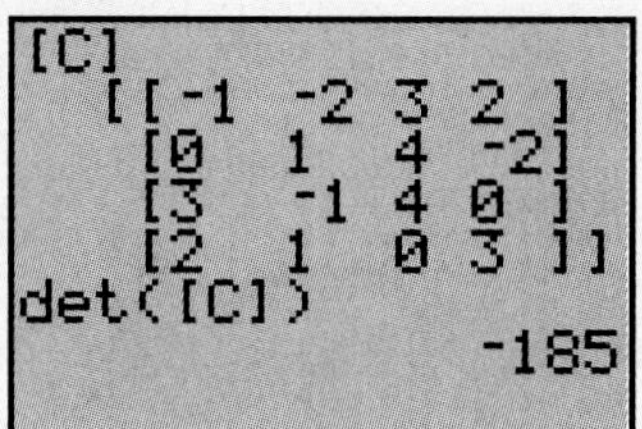

Figure 13

FOR DISCUSSION OR WRITING

Using the array of signs

$$\begin{matrix} + & - & + & - \\ - & + & - & + \\ + & - & + & - \\ - & + & - & + \end{matrix}$$

evaluate the determinant above by hand, expanding about the fourth row.

OBJECTIVE 4 Understand the derivation of Cramer's rule. Determinants can be used to solve a system of the form

$$a_1x + b_1y = c_1 \quad (1)$$
$$a_2x + b_2y = c_2. \quad (2)$$

The result will be a formula that can be used for any system of two equations with two unknowns. To get this general solution, we eliminate y and solve for x by first multiplying both sides of equation (1) by b_2 and both sides of equation (2) by $-b_1$. Then we add these results and solve for x.

$$\begin{aligned} a_1b_2x + b_1b_2y &= c_1b_2 && b_2 \text{ times both sides of equation (1)} \\ -a_2b_1x - b_1b_2y &= -c_2b_1 && -b_1 \text{ times both sides of equation (2)} \\ \hline (a_1b_2 - a_2b_1)x &= c_1b_2 - c_2b_1 \end{aligned}$$

$$x = \frac{c_1b_2 - c_2b_1}{a_1b_2 - a_2b_1} \quad (\text{if } a_1b_2 - a_2b_1 \neq 0)$$

To solve for y, we multiply both sides of equation (1) by $-a_2$ and both sides of equation (2) by a_1 and add.

$$\begin{aligned} -a_1a_2x - a_2b_1y &= -a_2c_1 && -a_2 \text{ times both sides of (1)} \\ a_1a_2x + a_1b_2y &= a_1c_2 && a_1 \text{ times both sides of (2)} \\ \hline (a_1b_2 - a_2b_1)y &= a_1c_2 - a_2c_1 \end{aligned}$$

$$y = \frac{a_1c_2 - a_2c_1}{a_1b_2 - a_2b_1}$$

Both numerators and the common denominator of these values for x and y can be written as determinants, since

$$a_1c_2 - a_2c_1 = \begin{vmatrix} a_1 & c_1 \\ a_2 & c_2 \end{vmatrix},$$

$$c_1b_2 - c_2b_1 = \begin{vmatrix} c_1 & b_1 \\ c_2 & b_2 \end{vmatrix},$$

and

$$a_1b_2 - a_2b_1 = \begin{vmatrix} a_1 & b_1 \\ a_2 & b_2 \end{vmatrix}.$$

Using these results, the solutions for x and y become

$$x = \frac{\begin{vmatrix} c_1 & b_1 \\ c_2 & b_2 \end{vmatrix}}{\begin{vmatrix} a_1 & b_1 \\ a_2 & b_2 \end{vmatrix}} \quad \text{and} \quad y = \frac{\begin{vmatrix} a_1 & c_1 \\ a_2 & c_2 \end{vmatrix}}{\begin{vmatrix} a_1 & b_1 \\ a_2 & b_2 \end{vmatrix}}, \quad \text{if} \begin{vmatrix} a_1 & b_1 \\ a_2 & b_2 \end{vmatrix} \neq 0.$$

For convenience, we denote the three determinants in the solution as

$$\begin{vmatrix} a_1 & b_1 \\ a_2 & b_2 \end{vmatrix} = D, \qquad \begin{vmatrix} c_1 & b_1 \\ c_2 & b_2 \end{vmatrix} = D_x, \qquad \begin{vmatrix} a_1 & c_1 \\ a_2 & c_2 \end{vmatrix} = D_y.$$

Note that the elements of D are the four coefficients of the variables in the given system; the elements of D_x are obtained by replacing the coefficients of x by the respective constants; the elements of D_y are obtained by replacing the coefficients of y by the respective constants.

These results are summarized as **Cramer's rule.**

Cramer's Rule for 2 × 2 Systems

Given the system

$$a_1x + b_1y = c_1$$
$$a_2x + b_2y = c_2,$$

with

$$a_1b_2 - a_2b_1 \neq 0,$$

then

$$x = \frac{\begin{vmatrix} c_1 & b_1 \\ c_2 & b_2 \end{vmatrix}}{\begin{vmatrix} a_1 & b_1 \\ a_2 & b_2 \end{vmatrix}} = \frac{D_x}{D} \quad \text{and} \quad y = \frac{\begin{vmatrix} a_1 & c_1 \\ a_2 & c_2 \end{vmatrix}}{\begin{vmatrix} a_1 & b_1 \\ a_2 & b_2 \end{vmatrix}} = \frac{D_y}{D}.$$

OBJECTIVE 5 Apply Cramer's rule to solve linear systems. To use Cramer's rule to solve a system of linear equations, find the three determinants, D, D_x, and D_y, and then write the necessary quotients for x and y.

CAUTION As indicated above, Cramer's rule does not apply if $D = a_1b_2 - a_2b_1$ is 0. When $D = 0$, the system is inconsistent or has dependent equations. For this reason, it is a good idea to evaluate D first.

EXAMPLE 5 Using Cramer's Rule for a 2 × 2 System

Use Cramer's rule to solve the system.

$$5x + 7y = -1$$
$$6x + 8y = 1$$

By Cramer's rule, $x = D_x/D$ and $y = D_y/D$. We will find D first, since if $D = 0$, Cramer's rule does not apply. If $D \neq 0$, then we will find D_x and D_y.

$$D = \begin{vmatrix} 5 & 7 \\ 6 & 8 \end{vmatrix} = 5(8) - 6(7) = \mathbf{-2}$$

$$D_x = \begin{vmatrix} -1 & 7 \\ 1 & 8 \end{vmatrix} = (-1)8 - 7(1) = \mathbf{-15}$$

$$D_y = \begin{vmatrix} 5 & -1 \\ 6 & 1 \end{vmatrix} = 5(1) - (-1)6 = \mathbf{11}$$

From Cramer's rule,

$$x = \frac{D_x}{D} = \frac{\mathbf{-15}}{\mathbf{-2}} = 7.5$$

and $$y = \frac{D_y}{D} = \frac{\mathbf{11}}{\mathbf{-2}} = -5.5.$$

The solution set is $\{(7.5, -5.5)\}$, as we can verify by checking in the given system.

Because graphing calculators can evaluate determinants, Cramer's rule can be applied using them. Figure 14 shows how the work of Example 5 is accomplished, with D the determinant of matrix A, D_x the determinant of matrix B, and D_y the determinant of matrix C.

```
[A]
        [[5 7]
         [6 8]]
[B]
       [[-1 7]
        [1  8]]
```

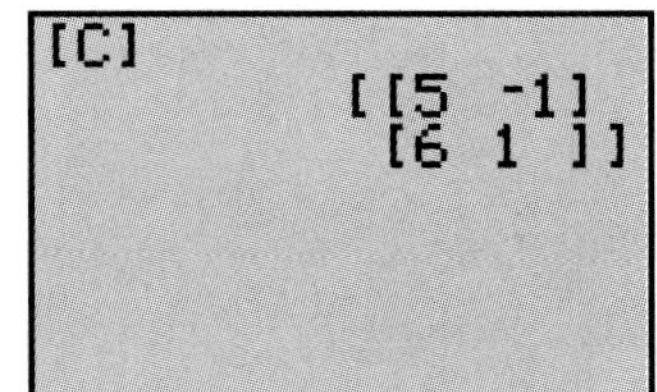

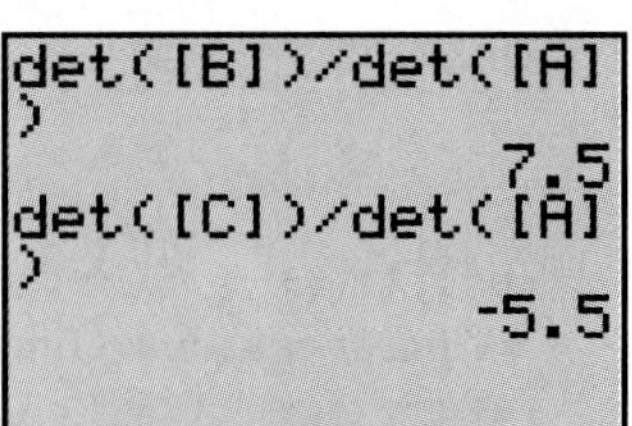

Figure 14

In a similar manner, Cramer's rule can be applied to systems of three linear equations with three variables.

Cramer's Rule for 3 × 3 Systems

Given the system

$$\begin{aligned} a_1x + b_1y + c_1z &= d_1 \\ a_2x + b_2y + c_2z &= d_2 \\ a_3x + b_3y + c_3z &= d_3, \end{aligned}$$

with
$$D = \begin{vmatrix} a_1 & b_1 & c_1 \\ a_2 & b_2 & c_2 \\ a_3 & b_3 & c_3 \end{vmatrix} \neq 0, \qquad D_x = \begin{vmatrix} d_1 & b_1 & c_1 \\ d_2 & b_2 & c_2 \\ d_3 & b_3 & c_3 \end{vmatrix},$$

$$D_y = \begin{vmatrix} a_1 & d_1 & c_1 \\ a_2 & d_2 & c_2 \\ a_3 & d_3 & c_3 \end{vmatrix}, \qquad D_z = \begin{vmatrix} a_1 & b_1 & d_1 \\ a_2 & b_2 & d_2 \\ a_3 & b_3 & d_3 \end{vmatrix},$$

then
$$x = \frac{D_x}{D}, \qquad y = \frac{D_y}{D}, \qquad z = \frac{D_z}{D}.$$

EXAMPLE 6 Using Cramer's Rule for a 3 × 3 System

Use Cramer's rule to solve the system.

$$\begin{aligned} x + y - z + 2 &= 0 \\ 2x - y + z + 5 &= 0 \\ x - 2y + 3z - 4 &= 0 \end{aligned}$$

To use Cramer's rule, we must rewrite the system in the form

$$\begin{aligned} x + y - z &= -2 \\ 2x - y + z &= -5 \\ x - 2y + 3z &= 4. \end{aligned}$$

We expand by minors about row 1 to find D.

$$D = \begin{vmatrix} 1 & 1 & -1 \\ 2 & -1 & 1 \\ 1 & -2 & 3 \end{vmatrix} = 1\begin{vmatrix} -1 & 1 \\ -2 & 3 \end{vmatrix} - 1\begin{vmatrix} 2 & 1 \\ 1 & 3 \end{vmatrix} + (-1)\begin{vmatrix} 2 & -1 \\ 1 & -2 \end{vmatrix}$$
$$= 1(-1) - 1(5) - 1(-3) = -3$$

Expanding D_x by minors about row 1 gives

$$D_x = \begin{vmatrix} -2 & 1 & -1 \\ -5 & -1 & 1 \\ 4 & -2 & 3 \end{vmatrix} = -2\begin{vmatrix} -1 & 1 \\ -2 & 3 \end{vmatrix} - 1\begin{vmatrix} -5 & 1 \\ 4 & 3 \end{vmatrix} + (-1)\begin{vmatrix} -5 & -1 \\ 4 & -2 \end{vmatrix}$$
$$= -2(-1) - 1(-19) - 1(14) = 7.$$

In the same way, $D_y = -22$ and $D_z = -21$, so that

$$x = \frac{D_x}{D} = \frac{7}{-3} = -\frac{7}{3}, \qquad y = \frac{D_y}{D} = \frac{-22}{-3} = \frac{22}{3}, \qquad z = \frac{D_z}{D} = \frac{-21}{-3} = 7.$$

The solution set is $\{(-\frac{7}{3}, \frac{22}{3}, 7)\}$.

As mentioned earlier, Cramer's rule does not apply when $D = 0$. The next example illustrates this case.

EXAMPLE 7 Determining When Cramer's Rule Does Not Apply

Show that Cramer's rule does not apply to the following system.

$$\begin{aligned} 2x - 3y + 4z &= 8 \\ 6x - 9y + 12z &= 24 \\ x + 2y - 3z &= 5 \end{aligned}$$

We need to show that $D = 0$. Here, expanding about column 1 gives

$$D = \begin{vmatrix} 2 & -3 & 4 \\ 6 & -9 & 12 \\ 1 & 2 & -3 \end{vmatrix} = 2\begin{vmatrix} -9 & 12 \\ 2 & -3 \end{vmatrix} - 6\begin{vmatrix} -3 & 4 \\ 2 & -3 \end{vmatrix} + 1\begin{vmatrix} -3 & 4 \\ -9 & 12 \end{vmatrix}$$

$$= 2(3) - 6(1) + 1(0)$$

$$= 0.$$

Since $D = 0$ here, Cramer's rule does not apply and we must use another method to solve the system.

Cramer's rule can be extended to 4×4 or larger systems. See a standard college algebra text for details.

4.5 EXERCISES

1. Which one of the following is the expression for the determinant $\begin{vmatrix} -2 & -3 \\ 4 & -6 \end{vmatrix}$?

   (a) $-2(-6) + (-3)(4)$ **(b)** $-2(-6) - 3(4)$

   (c) $-3(4) - (-2)(-6)$ **(d)** $-2(-6) - (-3)(4)$

2. Evaluate $\begin{vmatrix} 0 & 0 \\ 3 & -4 \end{vmatrix}$ and $\begin{vmatrix} 0 & 1 & 2 \\ 0 & -3 & 4 \\ 0 & 2 & 6 \end{vmatrix}$ and make a conjecture (educated guess) about the value of a determinant that has all 0s in a row or a column.

Evaluate each determinant. See Example 1.

3. $\begin{vmatrix} -2 & 5 \\ -1 & 4 \end{vmatrix}$

4. $\begin{vmatrix} 3 & -6 \\ 2 & -2 \end{vmatrix}$

5. $\begin{vmatrix} 1 & -2 \\ 7 & 0 \end{vmatrix}$

6. $\begin{vmatrix} -5 & -1 \\ 1 & 0 \end{vmatrix}$

7. $\begin{vmatrix} 0 & 4 \\ 0 & 4 \end{vmatrix}$

8. $\begin{vmatrix} 8 & -3 \\ 0 & 0 \end{vmatrix}$

Evaluate each determinant by expansion by minors about the first column. See Example 2.

9. $\begin{vmatrix} -1 & 2 & 4 \\ -3 & -2 & -3 \\ 2 & -1 & 5 \end{vmatrix}$

10. $\begin{vmatrix} 2 & -3 & -5 \\ 1 & 2 & 2 \\ 5 & 3 & -1 \end{vmatrix}$

11. $\begin{vmatrix} 1 & 0 & -2 \\ 0 & 2 & 3 \\ 1 & 0 & 5 \end{vmatrix}$

12. $\begin{vmatrix} 2 & -1 & 0 \\ 0 & -1 & 1 \\ 1 & 2 & 0 \end{vmatrix}$

13. Explain in your own words how to evaluate a 2×2 determinant. Illustrate with an example.

14. Explain in your own words the method of evaluating a 3 × 3 determinant. Illustrate with an example.

Evaluate each determinant by expansion by minors about any row or column. (Hint: The work is easier if you choose a row or a column with zeros.) See Example 3.

15. $\begin{vmatrix} 4 & 4 & 2 \\ 1 & -1 & -2 \\ 1 & 0 & 2 \end{vmatrix}$

16. $\begin{vmatrix} 3 & -1 & 2 \\ 1 & 5 & -2 \\ 0 & 2 & 0 \end{vmatrix}$

17. $\begin{vmatrix} 3 & 5 & -2 \\ 1 & -4 & 1 \\ 3 & 1 & -2 \end{vmatrix}$

18. $\begin{vmatrix} 0 & 0 & 3 \\ 4 & 0 & -2 \\ 2 & -1 & 3 \end{vmatrix}$

19. $\begin{vmatrix} 3 & 0 & -2 \\ 1 & -4 & 1 \\ 3 & 1 & -2 \end{vmatrix}$

20. $\begin{vmatrix} 1 & 1 & 2 \\ 5 & 5 & 7 \\ 3 & 3 & 1 \end{vmatrix}$

21. Explain why a determinant with a row or column of zeros has a value of zero.

RELATING CONCEPTS (EXERCISES 22–25)

Recall the formula for slope and the point-slope form of the equation of a line from Chapter 3.

Use these formulas to ***work Exercises 22–25 in order*** *and see how a determinant can be used in writing the equation of a line.*

22. Write the expression for the slope of a line passing through the points (x_1, y_1) and (x_2, y_2).

23. Using the expression from Exercise 22 as m, and the point (x_1, y_1), write the point-slope form of the equation of the line.

24. Using the equation obtained in Exercise 23, multiply both sides by $x_2 - x_1$, and write the equation so that 0 is on the right side.

25. Consider the *determinant equation*

$$\begin{vmatrix} x & y & 1 \\ x_1 & y_1 & 1 \\ x_2 & y_2 & 1 \end{vmatrix} = 0.$$

Expand by minors on the left and show that this determinant equation yields the same result that you obtained in Exercise 24.

Did you make the connection that a determinant with variables can represent one side of an equation?

TECHNOLOGY INSIGHTS (EXERCISES 26–29)

Predict the display the calculator will give for each determinant.

26.

```
[A]
        [[-6 8 ]
         [1  -3]]
det([A])
```

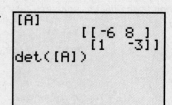

27.

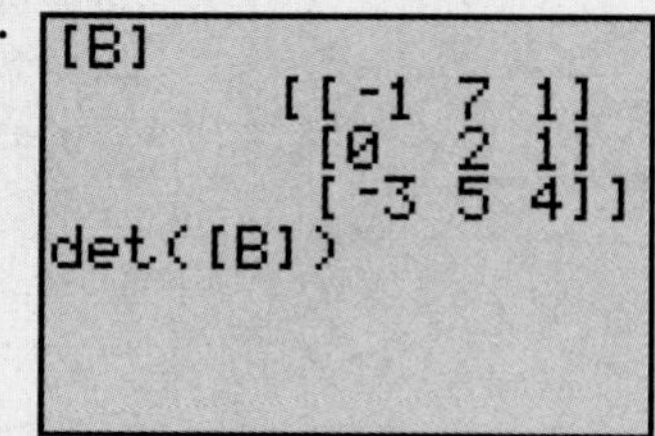

(continued)

TECHNOLOGY INSIGHTS (EXERCISES 26–29) (CONTINUED)

28.

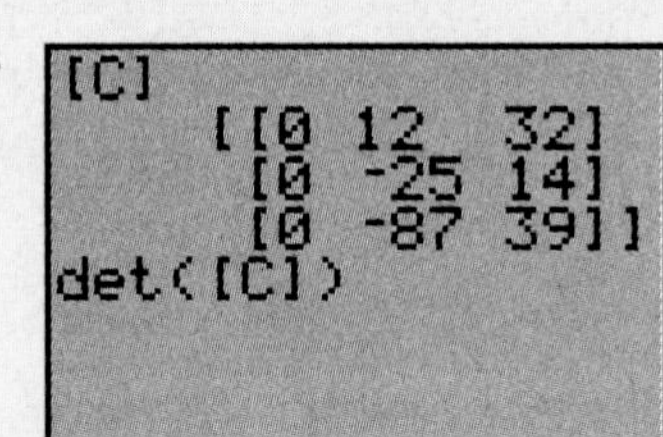

29.

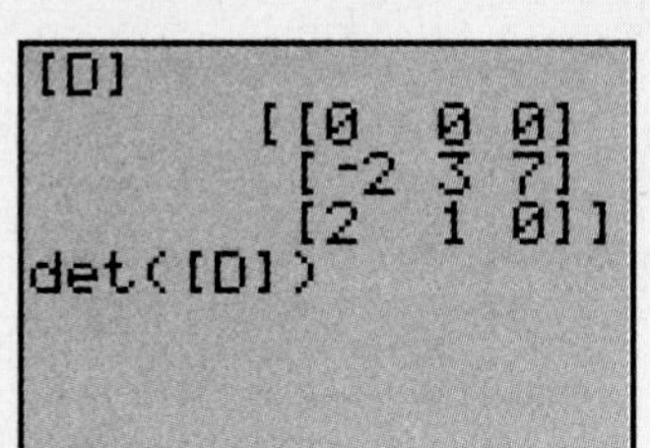

Use a graphing calculator with matrix capabilities to find each determinant.

30. $\begin{vmatrix} 1.5 & 2.6 & 9.3 \\ 5.2 & -1.4 & 8.6 \\ 0 & .7 & 1.2 \end{vmatrix}$

31. $\begin{vmatrix} \sqrt{5} & \sqrt{2} & -\sqrt{3} \\ \sqrt{7} & -\sqrt{6} & \sqrt{10} \\ -\sqrt{5} & -\sqrt{2} & \sqrt{17} \end{vmatrix}$ (To as many places as the calculator shows)

There is another method for evaluating a 3×3 determinant. Refer to Example 2, and carry over the first two columns to the right of the original determinant to get

$$\begin{vmatrix} 1 & 3 & -2 & 1 & 3 \\ -1 & -2 & -3 & -1 & -2 \\ 1 & 1 & 2 & 1 & 1 \end{vmatrix}.$$

Multiply along the diagonals as shown below, placing the product at the end of the arrow.

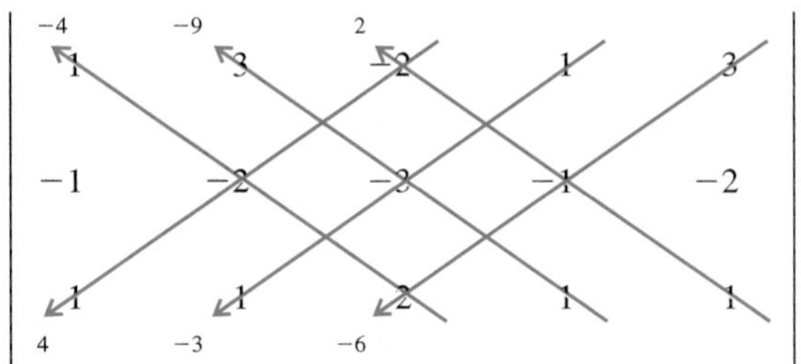

Add the top numbers: $-4 - 9 + 2 = -11$.

Add the bottom numbers: $4 - 3 - 6 = -5$.

Find the *difference* between these sums to obtain the final answer:

$$-11 - (-5) = -6.$$

Use this method to find each determinant in the indicated exercise.

32. Exercise 15 **33.** Exercise 16 **34.** Exercise 17

35. Exercise 18 **36.** Exercise 19 **37.** Exercise 20

Solve each equation by finding an expression for the determinant on the left, and then solving using the methods of Chapter 2.

38. $\begin{vmatrix} 4 & x \\ 2 & 3 \end{vmatrix} = 8$ **39.** $\begin{vmatrix} 5 & 3 \\ x & x \end{vmatrix} = 20$ **40.** $\begin{vmatrix} x & 4 \\ x & -3 \end{vmatrix} = 0$

41. Consider the system

$$\begin{aligned} 4x + 3y - 2z &= 1 \\ 7x - 4y + 3z &= 2 \\ -2x + y - 8z &= 0. \end{aligned}$$

Match each determinant in (a)–(d) with its correct representation from choices I, II, III, and IV.

(a) D

(b) D_x

(c) D_y

(d) D_z

I. $\begin{vmatrix} 1 & 3 & -2 \\ 2 & -4 & 3 \\ 0 & 1 & -8 \end{vmatrix}$ **II.** $\begin{vmatrix} 4 & 3 & 1 \\ 7 & -4 & 2 \\ -2 & 1 & 0 \end{vmatrix}$

III. $\begin{vmatrix} 4 & 1 & -2 \\ 7 & 2 & 3 \\ -2 & 0 & -8 \end{vmatrix}$ **IV.** $\begin{vmatrix} 4 & 3 & -2 \\ 7 & -4 & 3 \\ -2 & 1 & -8 \end{vmatrix}$

42. For the system

$$\begin{aligned} x + 3y - 6z &= 7 \\ 2x - y + z &= 1 \\ x + 2y + 2z &= -1, \end{aligned}$$

$D = -43$, $D_x = -43$, $D_y = 0$, and $D_z = 43$. What is the solution set of the system?

Use Cramer's rule to solve each linear system in two variables. See Example 5.

43. $3x + 5y = -5$
$-2x + 3y = 16$

44. $5x + 2y = -3$
$4x - 3y = -30$

45. $8x + 3y = 1$
$6x - 5y = 2$

46. $3x - y = 9$
$2x + 5y = 8$

47. $2x + 3y = 4$
$5x + 6y = 7$

48. $4x + 5y = 6$
$7x + 8y = 9$

49. Look at the coefficients and constants in the systems in Exercises 47 and 48. Notice that in both cases, the six numbers are consecutive integers. Make up a system having this same pattern for its coefficients and constants, and solve it using Cramer's rule. Compare the solutions in Exercises 47, 48, and here. What do you notice?

50. Use Cramer's rule to prove that the system

$$\begin{aligned} ax + (a + 1)y &= a + 2 \\ (a + 3)x + (a + 4)y &= a + 5, \quad \text{where } D \neq 0, \end{aligned}$$

has solution set $\{(-1, 2)\}$.

Use Cramer's rule where applicable to solve each linear system in three variables. See Examples 6 and 7.

51. $2x + 3y + 2z = 15$
$x - y + 2z = 5$
$x + 2y - 6z = -26$

52. $x - y + 6z = 19$
$3x + 3y - z = 1$
$x + 9y + 2z = -19$

53. $2x - 3y + 4z = 8$
$6x - 9y + 12z = 24$
$-4x + 6y - 8z = -16$

54. $7x + y - z = 4$
$2x - 3y + z = 2$
$-6x + 9y - 3z = -6$

55. $3x + 5z = 0$
$2x + 3y = 1$
$-y + 2z = -11$

56. $-x + 2y = 4$
$3x + y = -5$
$2x + z = -1$

57. $x - 3y = 13$
$2y + z = 5$
$-x + z = -7$

58. $-5x - y = -10$
$3x + 2y + z = -3$
$-y - 2z = -13$

RELATING CONCEPTS (EXERCISES 59–62)

In this section we have seen how determinants can be used to solve systems of equations. In Exercises 22–25, the earlier Relating Concepts group, we saw how a determinant can be used to write the equation of a line given two points on the line. Here, we show how a determinant can be used to find the area of a triangle if we know the coordinates of its vertices.

(continued)

RELATING CONCEPTS (EXERCISES 59–62) (CONTINUED)

Suppose that $A(x_1, y_1)$, $B(x_2, y_2)$, and $C(x_3, y_3)$ are the coordinates of the vertices of triangle ABC in the coordinate plane. Then it can be shown that the area of the triangle is given by the absolute value of

$$\frac{1}{2}\begin{vmatrix} x_1 & y_1 & 1 \\ x_2 & y_2 & 1 \\ x_3 & y_3 & 1 \end{vmatrix}.$$

Work Exercises 59–62 in order.

59. Sketch triangle ABC in the coordinate plane, given that the coordinates of A are $(0, 0)$, of B are $(-3, -4)$, and of C are $(2, -2)$.

60. Write the determinant expression described above that gives the area of triangle ABC described in Exercise 59.

61. Evaluate the absolute value of the determinant expression in Exercise 60 to find the area.

62. Use the determinant expression described above to find the area of the triangle with vertices at $(3, 8)$, $(-1, 4)$, and $(0, 1)$.

Did you make the connection that a geometric formula can be expressed using a determinant?

Use a graphing calculator and the approach described with Figure 14 to solve each of the following systems using Cramer's rule.

63.
$$\begin{aligned} x + 2y + z &= 10 \\ 2x - y - 3z &= -20 \\ -x + 4y + z &= 18 \end{aligned}$$

64.
$$\begin{aligned} 2x + y + 3z &= 1 \\ x - 2y + z &= -3 \\ -3x + y - 2z &= -4 \end{aligned}$$

65.
$$\begin{aligned} -8w + 4x - 2y + z &= -28 \\ -w + x - y + z &= -10 \\ w + x + y + z &= -4 \\ 27w + 9x + 3y + z &= 2 \end{aligned}$$

66.
$$\begin{aligned} 5w + 2x - 3y + z &= 4.7 \\ -2w + x + 2y - z &= -3.2 \\ w + 3x - y + 2z &= 2.1 \\ 2w + x - 5y + 3z &= 3.4 \end{aligned}$$

CHAPTER 4 GROUP ACTIVITY

Olympic Track and Field Results

Objective: Solve systems of equations to determine trends in Olympic track and field events.

In this activity you will compare trends for men and women in three Olympic track and field events over a span of thirty-six years. Use the data for winning distances (in meters) given in the following tables.

Gold Medal Results: Shot Put

Year	*1960*	*1964*	*1968*	*1972*	*1976*	*1980*	*1984*	*1988*	*1992*	*1996*
Men	19.68	20.33	20.54	21.18	21.05	21.35	21.26	22.47	21.70	21.62
Women	17.30	18.10	19.60	21.00	21.10	22.40	20.40	22.20	21.00	20.56

(continued)

Gold Medal Results: Javelin

Year	1960	1964	1968	1972	1976	1980	1984	1988	1992	1996
Men	84.64	82.66	90.10	90.48	94.58	91.20	86.76	84.28	89.66	88.16
Women	55.90	60.50	60.30	63.80	65.90	68.40	69.50	74.60	68.30	67.94

Gold Medal Results: Discus

Year	1960	1964	1968	1972	1976	1980	1984	1988	1992	1996
Men	59.18	61.00	63.78	64.40	67.50	66.64	66.60	68.82	65.12	69.40
Women	55.10	57.20	58.20	66.60	69.00	69.90	65.30	72.30	70.00	69.66

Source: Centre for Innovation in Mathematics Teaching; www.ex.ac.uk/cimt.

A. Have each person in the group select a different Olympic track and field event. Use the data for your selected event to do the following.

1. Plot the data from the table for your event on a graph, using the vertical axis for distances, and the horizontal axis for the years, where $x = 0$ represents 1960, and $4x$ equals the number of years after 1960. Use different colors to differentiate between the data for men and women.
2. Draw a straight line that would best fit the data.
3. Estimate the intersection point of the two lines; that is, the year when the distances for men and women will be the same. Record your estimate.

B. The equations below model the data for each table.

Shot put: $y = .25x + 19.68$ and $y = .46x + 17.3$
Javelin: $y = .39x + 84.64$ and $y = 1.55x + 55.9$
Discus: $y = .74x + 59.18$ and $y = 1.86x + 55.1$

Have each person solve the system of equations for the selected event.

C. As a group, write a paragraph that discusses your findings. Include answers to the following questions in your paragraph.

1. How closely did your estimates match the results from solving the systems of equations?
2. In which events might it be possible for women's distances to exceed men's distances?
3. Which method for solving the systems of equations—graphing, substitution, or elimination—was easiest? Explain.

CHAPTER 4 SUMMARY

KEY TERMS

4.1 system of equations
linear system
solution set of a linear system
inconsistent system
dependent equations
elimination method
substitution method

4.2 ordered triple

4.4 matrix
element of a matrix
row
column
square matrix
augmented matrix
row operations
row echelon form

4.5 determinant
minor
expansion by minors
array of signs
Cramer's rule

NEW SYMBOLS

Symbol	Meaning	Symbol	Meaning	Symbol	Meaning
(x, y, z)	ordered triple	$\begin{vmatrix} a & b \\ c & d \end{vmatrix}$	2×2 determinant	$\begin{vmatrix} a & b & c \\ d & e & f \\ g & h & i \end{vmatrix}$	3×3 determinant
$\begin{bmatrix} a & b & c \\ d & e & f \end{bmatrix}$	matrix with two rows, three columns				

TEST YOUR WORD POWER

See how well you have learned the vocabulary in this chapter. Answers, with examples, are given at the bottom of the page.

1. A **system of equations** consists of
(a) at least two equations with different variables
(b) two or more equations that have an infinite number of solutions
(c) two or more equations that are to be solved at the same time
(d) two or more inequalities that are to be solved.

2. The **solution set of a linear system** is
(a) all ordered pairs that satisfy one equation of the system
(b) all ordered pairs that satisfy all the equations of the system at the same time
(c) any ordered pair that satisfies one or more equations of the system
(d) the set of values that make all the equations of the system false.

3. A **matrix** is
(a) an ordered pair of numbers
(b) an array of numbers with the same number of rows and columns
(c) a pair of numbers written between brackets
(d) a rectangular array of numbers.

4. A matrix written in **row echelon form** has
(a) elements that are all 0
(b) elements that are all 1
(c) upper left to lower right diagonal elements of 1 with 0s below the 1s
(d) upper left to lower right diagonal elements of 0 with 1s below the 0s.

5. A **determinant** is
(a) a rectangular array of numbers
(b) a real number associated with a square matrix
(c) a matrix with the same number of rows and columns
(d) an ordered pair of numbers.

6. **Expansion by minors** is
(a) a method of evaluating a 3×3 or larger determinant
(b) a way to use row operations to produce new matrices
(c) a method of evaluating a 2×2 determinant
(d) a method of evaluating any determinant.

QUICK REVIEW

CONCEPTS	EXAMPLES
4.1 LINEAR SYSTEMS OF EQUATIONS IN TWO VARIABLES	
Solving Linear Systems by Elimination	Solve by elimination.
Step 1 Write both equations in the form $Ax + By = C$.	$5x + y = 2$ $2x - 3y = 11$
Step 2 Multiply one or both equations by appropriate numbers so that the sum of the coefficients of either x or y is zero.	To eliminate y, multiply the top equation by 3, and add.

Answers to Test Your Word Power

1. (c) *Example:* $3x - y = 3$, $2x + y = 7$ **2.** (b) *Example:* The ordered pair $(2, 3)$ satisfies both equations of the system given above, so $\{(2, 3)\}$ is the solution set of the system. **3.** (d) *Examples:* $\begin{bmatrix} 3 & -1 & 0 \\ 4 & 2 & 1 \end{bmatrix}$, $\begin{bmatrix} 1 & 2 \\ 4 & 3 \end{bmatrix}$, $[2 \quad 0 \quad -5 \quad 4]$ **4.** (c) *Example:* $\left[\begin{array}{ccc|c} 1 & -3 & \frac{1}{2} & 1 \\ 0 & 1 & 4 & -1 \\ 0 & 0 & 1 & 2 \end{array}\right]$ **5.** (b) *Examples:* $\begin{vmatrix} 1 & 2 \\ 4 & 3 \end{vmatrix}$, $\begin{vmatrix} 1 & 4 & 0 \\ 3 & -2 & -1 \\ 0 & 5 & 0 \end{vmatrix}$ **6.** (a) *Example:* See Section 4.5, Example 2.

CONCEPTS	EXAMPLES
Step 3 Add the new equations. The sum should be an equation with just one variable.	$15x + 3y = 6$ $2x - 3y = 11$ $17x = 17$
Step 4 Solve the equation from Step 3.	$x = 1$
Step 5 Substitute the result of Step 4 into either of the given equations and solve for the other variable.	Let $x = 1$ in the top equation, and solve for y. $5(1) + y = 2$ $y = -3$
Step 6 Check the solution in both of the given equations. Write the solution set.	Check to verify that $\{(1, -3)\}$ is the solution set.
Solving Linear Systems by Substitution	Solve by substitution. $4x - y = 7$ $3x + 2y = 30$
Step 1 Solve one of the equations for either variable.	Solve for y in the top equation. $y = 4x - 7$
Step 2 Substitute for that variable in the other equation. The result should be an equation with just one variable.	Substitute $4x - 7$ for y in the bottom equation, and solve for x. $3x + 2(4x - 7) = 30$ $3x + 8x - 14 = 30$
Step 3 Solve the equation from Step 2.	$11x = 44$ $x = 4$
Step 4 Substitute the result from Step 3 into the equation from Step 1 to find the value of the other variable.	Substitute 4 for x in the equation $y = 4x - 7$ to find that $y = 9$.
Step 5 Check the solution in both of the given equations. Write the solution set.	Check to see that $\{(4, 9)\}$ is the solution set.
4.2 LINEAR SYSTEMS OF EQUATIONS IN THREE VARIABLES	
Solving Linear Systems in Three Variables	Solve the system
Step 1 Use the elimination method to eliminate any variable from any two of the given equations. The result is an equation in two variables.	$x + 2y - z = 6$ $x + y + z = 6$ $2x + y - z = 7.$ Add the first and second equations; z is eliminated and the result is $2x + 3y = 12$.
Step 2 Eliminate the *same* variable from any *other* two equations. The result is an equation in the same two variables as in Step 1.	Eliminate z again by adding the second and third equations to get $3x + 2y = 13$. Now solve the system $2x + 3y = 12$ (*) $3x + 2y = 13.$

(continued)

CONCEPTS	EXAMPLES

Step 3 Use the elimination method to eliminate a second variable from the two equations in two variables that result from Steps 1 and 2. The result is an equation in one variable that gives the value of that variable.

To eliminate x, multiply the top equation by -3 and the bottom equation by 2.

$$\begin{aligned} -6x - 9y &= -36 \\ 6x + 4y &= 26 \\ \hline -5y &= -10 \\ y &= 2 \end{aligned}$$

Step 4 Substitute the value of the variable found in Step 3 into either of the equations in two variables to find the value of the second variable.

Let $y = 2$ in equation (*).

$$\begin{aligned} 2x + 3(2) &= 12 \\ 2x &= 6 \\ x &= 3 \end{aligned}$$

Step 5 Use the values of the two variables from Steps 3 and 4 to find the value of the third variable by substituting into any of the original equations.

Let $y = 2$ and $x = 3$ in any of the original equations to find $z = 1$.

Step 6 Check in all of the original equations. Write the solution set.

Verify that when $x = 3$, $y = 2$, and $z = 1$, all three equations are satisfied. The solution set is $\{(3, 2, 1)\}$.

4.3 APPLICATIONS OF LINEAR SYSTEMS OF EQUATIONS

Use the six-step problem solving-method.

The perimeter of a rectangle is 18 feet. The length is 3 feet more than twice the width. What are the dimensions of the rectangle?

Step 1 Choose a variable to represent each unknown value.

Let x represent the length and y represent the width.

Step 2 Make a figure, diagram, or chart if it will help.

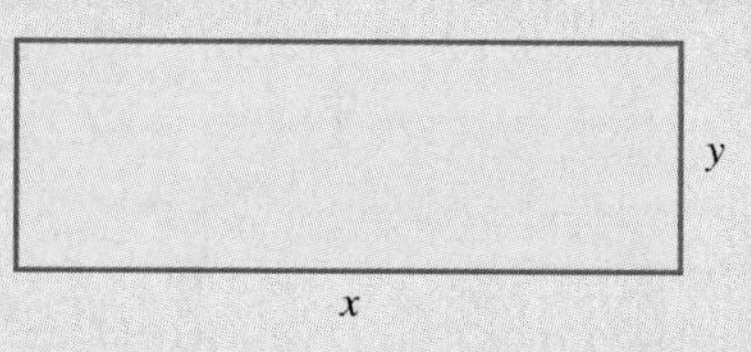

Step 3 Write two (three) equations that relate the unknowns.

From the perimeter formula, one equation is $2x + 2y = 18$. From the problem, another equation is $x = 3 + 2y$. Now solve the system

$$\begin{aligned} 2x + 2y &= 18 \\ x &= 3 + 2y. \end{aligned}$$

Step 4 Solve the system.

Substitute $3 + 2y$ for x in the top equation.

$$2(3 + 2y) + 2y = 18$$

Solve to get $y = 2$. If $y = 2$, then $x = 3 + 2(2) = 7$.

Step 5 Answer the question(s) asked.

The length is 7 feet and the width is 2 feet.

Step 6 Check your solution in the words of the problem.

The perimeter is $2(7) + 2(2) = 18$, and $3 + 2(2) = 7$, so the solution checks.

CONCEPTS	EXAMPLES

4.4 SOLVING LINEAR SYSTEMS OF EQUATIONS BY MATRIX METHODS

Matrix Row Operations

1. Any two rows of the matrix may be interchanged.

$$\begin{bmatrix} 1 & 5 & 7 \\ 3 & 9 & -2 \\ 0 & 6 & 4 \end{bmatrix} \text{ becomes } \begin{bmatrix} 3 & 9 & -2 \\ 1 & 5 & 7 \\ 0 & 6 & 4 \end{bmatrix} \quad \text{Interchange } R_1 \text{ and } R_2.$$

2. The numbers in any row may be multiplied by any nonzero real number.

$$\begin{bmatrix} 1 & 5 & 7 \\ 3 & 9 & -2 \\ 0 & 6 & 4 \end{bmatrix} \text{ becomes } \begin{bmatrix} 1 & 5 & 7 \\ 1 & 3 & -\frac{2}{3} \\ 0 & 6 & 4 \end{bmatrix} \quad \tfrac{1}{3}R_2$$

3. Any row may be changed by adding to the numbers of the row the product of a real number and the numbers of another row.

$$\begin{bmatrix} 1 & 5 & 7 \\ 3 & 9 & -2 \\ 0 & 6 & 4 \end{bmatrix} \text{ becomes } \begin{bmatrix} 1 & 5 & 7 \\ \mathbf{0} & \mathbf{-6} & \mathbf{-23} \\ 0 & 6 & 4 \end{bmatrix} \quad -3R_1 + R_2$$

A system can be solved by matrix methods. Write the augmented matrix, and use row operations to obtain a matrix in row echelon form.

Solve.

$$\begin{aligned} x + 3y &= 7 \\ 2x + y &= 4 \end{aligned}$$

$$\left[\begin{array}{cc|c} 1 & 3 & 7 \\ 2 & 1 & 4 \end{array}\right]$$

$$\left[\begin{array}{cc|c} 1 & 3 & 7 \\ 0 & -5 & -10 \end{array}\right] \quad -2R_1 + R_2$$

$$\left[\begin{array}{cc|c} 1 & 3 & 7 \\ 0 & 1 & 2 \end{array}\right] \quad -\tfrac{1}{5}R_2$$

$$\begin{aligned} x + 3y &= 7 \\ y &= 2 \end{aligned}$$

When $y = 2$, $x + 3(2) = 7$, so $x = 1$. The solution set is $\{(1, 2)\}$.

4.5 DETERMINANTS AND CRAMER'S RULE

Value of a 2 × 2 Determinant

$$\begin{vmatrix} a & b \\ c & d \end{vmatrix} = ad - bc.$$

Evaluate.

$$\begin{vmatrix} 3 & 4 \\ -2 & 6 \end{vmatrix} = (3)(6) - (4)(-2) = 26$$

Determinants larger than 2 × 2 are evaluated by expansion by minors about a column or row.

Array of Signs for a 3 × 3 Determinant

$$\begin{matrix} + & - & + \\ - & + & - \\ + & - & + \end{matrix}$$

Evaluate

$$\begin{vmatrix} 2 & -3 & -2 \\ -1 & -4 & -3 \\ -1 & 0 & 2 \end{vmatrix}$$

by expanding about the second column.

$$\begin{aligned} \begin{vmatrix} 2 & \mathbf{-3} & -2 \\ -1 & \mathbf{-4} & -3 \\ -1 & \mathbf{0} & 2 \end{vmatrix} &= -(-3)(-5) + (-4)(2) - (0)(-8) \\ &= -15 - 8 + 0 \\ &= -23 \end{aligned}$$

(continued)

CONCEPTS	EXAMPLES
Cramer's Rule for 2 × 2 Systems Given the system $$a_1x + b_1y = c_1$$ $$a_2x + b_2y = c_2$$ with $a_1b_2 - a_2b_1 = D \neq 0$, then $$x = \frac{\begin{vmatrix} c_1 & b_1 \\ c_2 & b_2 \end{vmatrix}}{\begin{vmatrix} a_1 & b_1 \\ a_2 & b_2 \end{vmatrix}} = \frac{D_x}{D}$$ and $$y = \frac{\begin{vmatrix} a_1 & c_1 \\ a_2 & c_2 \end{vmatrix}}{\begin{vmatrix} a_1 & b_1 \\ a_2 & b_2 \end{vmatrix}} = \frac{D_y}{D}.$$	Solve using Cramer's rule. $$x - 2y = -1$$ $$2x + 5y = 16$$ $$x = \frac{\begin{vmatrix} -1 & -2 \\ 16 & 5 \end{vmatrix}}{\begin{vmatrix} 1 & -2 \\ 2 & 5 \end{vmatrix}} = \frac{-5 + 32}{5 + 4} = \frac{27}{9} = 3$$ $$y = \frac{\begin{vmatrix} 1 & -1 \\ 2 & 16 \end{vmatrix}}{\begin{vmatrix} 1 & -2 \\ 2 & 5 \end{vmatrix}} = \frac{16 + 2}{5 + 4} = \frac{18}{9} = 2$$ The solution set is $\{(3, 2)\}$.
Cramer's Rule for 3 × 3 Systems Given the system $$a_1x + b_1y + c_1z = d_1$$ $$a_2x + b_2y + c_2z = d_2$$ $$a_3x + b_3y + c_3z = d_3$$ with $$D = \begin{vmatrix} a_1 & b_1 & c_1 \\ a_2 & b_2 & c_2 \\ a_3 & b_3 & c_3 \end{vmatrix} \neq 0, \quad D_x = \begin{vmatrix} d_1 & b_1 & c_1 \\ d_2 & b_2 & c_2 \\ d_3 & b_3 & c_3 \end{vmatrix},$$ $$D_y = \begin{vmatrix} a_1 & d_1 & c_1 \\ a_2 & d_2 & c_2 \\ a_3 & d_3 & c_3 \end{vmatrix}, \quad D_z = \begin{vmatrix} a_1 & b_1 & d_1 \\ a_2 & b_2 & d_2 \\ a_3 & b_3 & d_3 \end{vmatrix},$$ then $$x = \frac{D_x}{D}, \quad y = \frac{D_y}{D}, \quad z = \frac{D_z}{D}.$$	Solve using Cramer's rule. $$3x + 2y + z = -5$$ $$x - y + 3z = -5$$ $$2x + 3y + z = 0$$ Using the methods of expansion of minors, it can be shown that $D_x = 45$, $D_y = -30$, $D_z = 0$, and $D = -15$. Therefore, $$x = \frac{D_x}{D} = \frac{45}{-15} = -3,$$ $$y = \frac{D_y}{D} = \frac{-30}{-15} = 2,$$ $$z = \frac{D_z}{D} = \frac{0}{-15} = 0.$$ The solution set is $\{(-3, 2, 0)\}$.

CHAPTER 4 REVIEW EXERCISES

[4.1] *Various tools are used by business management to improve performance, but they go in and out of fashion, according to an article in the September 7, 1998, issue of* Fortune. *Three of these tools are strategic alliances, core competencies, and total quality management. Use the graph to answer the following.*

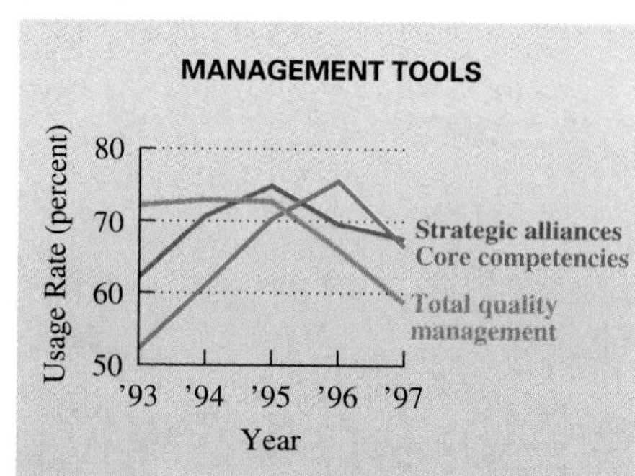

1. During what year was the usage rate of core competencies and total quality management the same? What was this rate?

2. On how many occasions did strategic alliances and core competencies reach the same usage level? Between these times, which one of these tools was more popular?

3. The figure shows graphs that represent the supply and demand for a certain brand of low-fat frozen yogurt at various prices per half-gallon.

The Fortunes of Frozen Yogurt

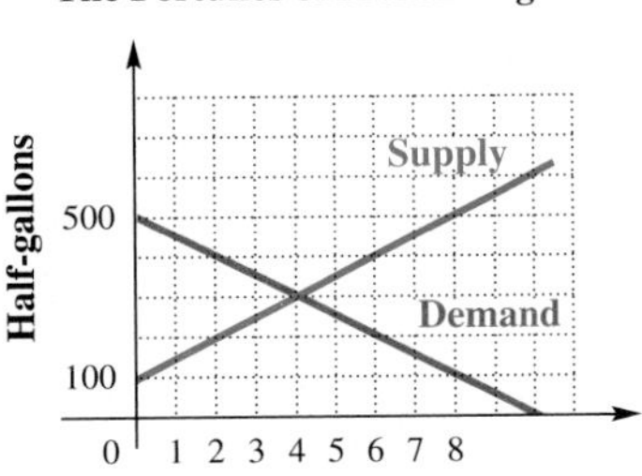

(a) At what price does supply equal demand?
(b) For how many half-gallons does supply equal demand?
(c) What are the supply and the demand at a price of \$2 per half-gallon?

Solve each system of equations by the elimination method. In Exercises 4 and 5, also graph the system.

4. $\begin{aligned} x + 3y &= 8 \\ 2x - y &= 2 \end{aligned}$

5. $\begin{aligned} x - 4y &= -4 \\ 3x + y &= 1 \end{aligned}$

6. $\begin{aligned} 6x + 5y &= 4 \\ -4x + 2y &= 8 \end{aligned}$

7. $\begin{aligned} \frac{x}{6} + \frac{y}{6} &= -\frac{1}{2} \\ x - y &= -9 \end{aligned}$

8. $\begin{aligned} 9x - y &= -4 \\ y &= x + 4 \end{aligned}$

9. $\begin{aligned} -3x + y &= 6 \\ y &= 6 + 3x \end{aligned}$

10. $\begin{aligned} 5x - 4y &= 2 \\ -10x + 8y &= 7 \end{aligned}$

Solve each system by the substitution method.

11. $\begin{aligned} 3x + y &= -4 \\ x &= \frac{2}{3}y \end{aligned}$

12. $\begin{aligned} -5x + 2y &= -2 \\ x + 6y &= 26 \end{aligned}$

13. State whether the graphed system is inconsistent or has dependent equations.

(a)

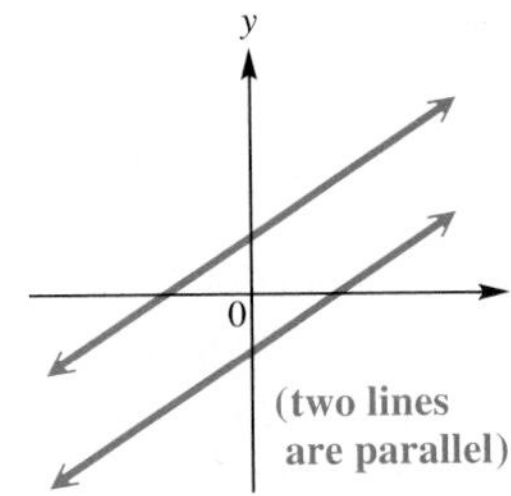

(b)

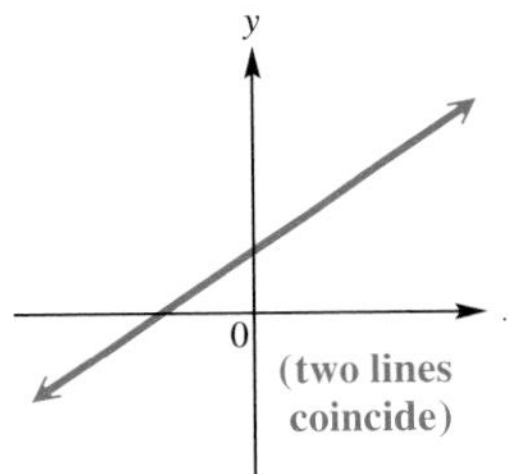

14. Explain why the system

$$y = 3x + 2$$
$$y = 3x - 4$$

has $\emptyset$ as its solution set without doing any algebraic work but answering based on your knowledge of the graphs of the two lines.

[4.2] *Solve each system.*

15. $2x + 3y - z = -16$
$x + 2y + 2z = -3$
$-3x + y + z = -5$

16. $4x - y = 2$
$3y + z = 9$
$x + 2z = 7$

17. $3x - y - z = -8$
$4x + 2y + 3z = 15$
$-6x + 2y + 2z = 10$

[4.3] *Solve each problem by writing a system of equations and then solving the system. Use the six-step problem-solving method.*

18. Professional basketball courts are uniform in size. They are rectangular, with a perimeter of 288 feet. Twice the width is 6 feet more than the length. What are the dimensions of such a court? (*Source:* Microsoft Encarta, *Basketball.*)

19. During the 1996 National Football League season, John Elway of the Denver Broncos threw 40 passes that resulted in either a touchdown or an interception. His number of touchdowns was 2 less than twice the number of interceptions. How many passes of each type did he throw? (*Source: The Wall Street Journal Almanac 1998.*)

20. A plane flies 560 miles in 1.75 hours traveling with the wind. The return trip later against the same wind takes the plane 2 hours. Find the speed of the plane and the speed of the wind. (*Hint:* Let x represent the speed of the plane and let y represent the speed of the wind. Use the chart.)

	r	t	d
With wind	$x + y$	1.75	
Against wind	$x - y$	2	

21. Sweet's Candy Store is offering a special mix for Valentine's Day. Ms. Sweet will mix some \$2-a-pound nuts with some \$1-a-pound chocolate candy to get 100 pounds of mix which she will sell at \$1.30 a pound. How many pounds of each should she use?

	Pounds	Price per Pound (in Dollars)	Value
Nuts	x	2	
Chocolate	y	1	
Total	100	1.30	

22. The sum of the measures of the angles of a triangle is 180°. The largest angle measures 10° less than the sum of the other two. The measure of the middle-sized angle is the average of the other two. Find the measures of the three angles.

23. Keshon Grant sells real estate. On three recent sales, he made 10% commission, 6% commission, and 5% commission. His total commissions on these sales were \$17,000, and he sold property worth \$280,000. If the 5% sale amounted to the sum of the other two, what were the three sales prices?

24. How many liters each of 8%, 10%, and 20% hydrogen peroxide should be mixed together to get 8 liters of 12.5% solution, if the amount of 8% solution used must be 2 liters more than the amount of 20% solution used?

25. In the great baseball year of 1961, Yankee teammates Mickey Mantle, Roger Maris, and John Blanchard combined for 136 home runs. Mantle hit 7 fewer than Maris. Maris hit 40 more than Blanchard. What were the home run totals for each player? (*Source:* Neft, David S. and Richard M. Cohen, *The Sports Encyclopedia: Baseball 1997.*)

[4.4] *Solve each system of equations by matrix methods.*

26. $2x + 5y = -4$
$4x - y = 14$

27. $6x + 3y = 9$
$-7x + 2y = 17$

28. $x + 2y - z = 1$
$3x + 4y + 2z = -2$
$-2x - y + z = -1$

29. $x + 3y = 7$
$3x + z = 2$
$y - 2z = 4$

[4.5]

30. Which one of the following determinants is equal to 0?

(a) $\begin{vmatrix} 3 & 2 \\ 2 & 3 \end{vmatrix}$ **(b)** $\begin{vmatrix} 4 & 2 \\ -3 & 2 \end{vmatrix}$ **(c)** $\begin{vmatrix} -1 & 1 \\ 8 & 8 \end{vmatrix}$ **(d)** $\begin{vmatrix} 1 & 2 \\ 6 & 12 \end{vmatrix}$

Evaluate each determinant.

31. $\begin{vmatrix} 2 & -9 \\ 8 & 4 \end{vmatrix}$

32. $\begin{vmatrix} 7 & 0 \\ 5 & -3 \end{vmatrix}$

33. $\begin{vmatrix} 2 & 10 & 4 \\ 0 & 1 & 3 \\ 0 & 6 & -1 \end{vmatrix}$

34. $\begin{vmatrix} -1 & 7 & 2 \\ 3 & 0 & 5 \\ -1 & 2 & 6 \end{vmatrix}$

35. Under what conditions can a system *not* be solved using Cramer's rule?

36. Why can't the system $\begin{aligned} 3x + 2y + z &= 0 \\ -x + y - 3z &= 1 \end{aligned}$ be solved using Cramer's rule?

Use Cramer's rule to solve each system of equations.

37. $\begin{aligned} 3x - 4y &= -32 \\ 2x + y &= -3 \end{aligned}$

38. $\begin{aligned} -4x + 3y &= -12 \\ 2x + 6y &= 15 \end{aligned}$

39. $\begin{aligned} 4x + y + z &= 11 \\ x - y - z &= 4 \\ y + 2z &= 0 \end{aligned}$

40. $\begin{aligned} -x + 3y - 4z &= 4 \\ 2x + 4y + z &= -14 \\ 3x - y + 2z &= -8 \end{aligned}$

RELATING CONCEPTS (EXERCISES 41–46)

Consider the system of equations

$$\begin{aligned} 2x + y + 3z &= 1 \\ x - 2y + z &= -3 \\ -3x + y - 2z &= -4. \end{aligned}$$

In Exercises 41–46 we examine several different ways of solving this system. ***Work the exercises in order.***

41. Eliminate x from the first and second equations, and eliminate x from the second and third equations. Write the resulting system of two equations in y and z.

42. Solve the system of two equations found in Exercise 41 using elimination.

43. Complete the solution of the original system using the results of Exercise 42.

44. Solve the original system using matrix row operations (Section 4.4). Verify that your solution is the same as the one found in Exercise 43.

45. Evaluate these determinants.

(a) $D = \begin{vmatrix} 2 & 1 & 3 \\ 1 & -2 & 1 \\ -3 & 1 & -2 \end{vmatrix}$ **(b)** $D_x = \begin{vmatrix} 1 & 1 & 3 \\ -3 & -2 & 1 \\ -4 & 1 & -2 \end{vmatrix}$

(c) $D_y = \begin{vmatrix} 2 & 1 & 3 \\ 1 & -3 & 1 \\ -3 & -4 & -2 \end{vmatrix}$ **(d)** $D_z = \begin{vmatrix} 2 & 1 & 1 \\ 1 & -2 & -3 \\ -3 & 1 & -4 \end{vmatrix}$

46. Use the results of Exercise 45 and Cramer's rule to solve the original system. Verify that your solution is the same as the one found in Exercises 43 and 44.

Did you make the connection that a system can be solved by various methods, all resulting in the same solution?

MIXED REVIEW EXERCISES

Solve by any method.

47. $\frac{2}{3}x + \frac{1}{6}y = \frac{19}{2}$
$\frac{1}{3}x - \frac{2}{9}y = 2$

48. $2x + 5y - z = 12$
$-x + y - 4z = -10$
$-8x - 20y + 4z = 31$

49. $x = 7y + 10$
$2x + 3y = 3$

50. $x + 4y = 17$
$-3x + 2y = -9$

51. $-7x + 3y = 12$
$5x + 2y = 8$

52. $2x - 5y = 8$
$3x + 4y = 10$

53. A local electronics store will sell 7 AC adaptors and 2 rechargeable flashlights for $86, or 3 AC adaptors and 4 rechargeable flashlights for $84. What is the price of a single AC adaptor and a single rechargeable flashlight?

54. During the Summer Olympics in Atlanta in 1996, Canada won a total of 22 medals, consisting of gold, silver, and bronze. There were 5 fewer gold medals than bronze, and 3 fewer bronze than silver. How many medals of each kind did Canada win? (*Source: The Universal Almanac 1997.*)

CHAPTER 4 TEST

Hank Aaron and Babe Ruth are the all-time home run leaders in the major leagues. If Mark McGwire continues at the pace he is going, he will hit more home runs than either of them. Use the graphs to answer the following.

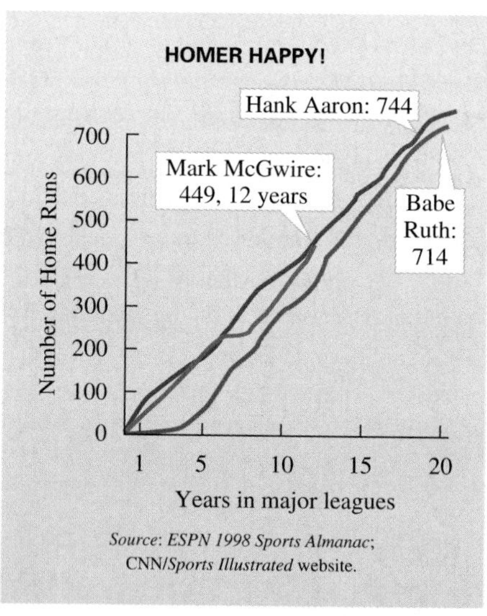

1. Was there any year in Babe Ruth's career that he had more home runs than Hank Aaron in the same year of Aaron's career? Explain.
2. After eight years, which player had the most home runs? Who had the fewest?
3. Use a graph to solve the system.

$$x + y = 7$$
$$x - y = 5$$

Solve each system by elimination.

4. $3x + y = 12$
$2x - y = 3$

5. $-5x + 2y = -4$
$6x + 3y = -6$

6. $3x + 4y = 8$
$8y = 7 - 6x$

7. $3x + 5y + 3z = 2$
$6x + 5y + z = 0$
$3x + 10y - 2z = 6$

Solve each system by substitution.

8. $2x - 3y = 24$
$y = -\frac{2}{3}x$

9. $12x - 5y = 8$
$3x = \frac{5}{4}y + 2$

Solve each problem by writing a system of equations.

10. Two cars start from points 420 miles apart and travel toward each other. They meet after 3.5 hours. Find the average speed of each car if one travels 30 miles per hour slower than the other.

11. A chemist needs 12 liters of a 40% alcohol solution. She must mix a 20% solution and a 50% solution. How many liters of each will be required to obtain what she needs?

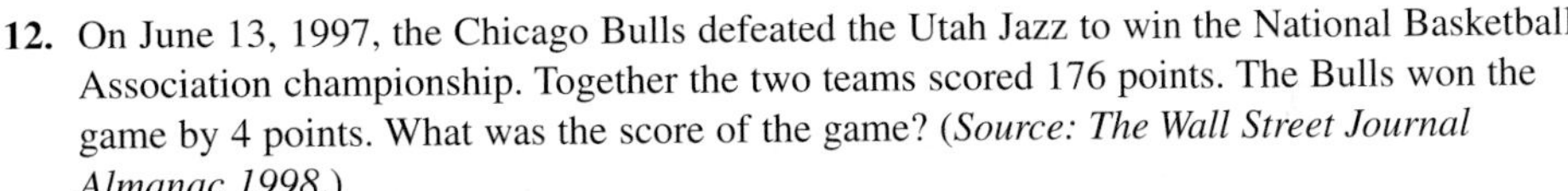

12. On June 13, 1997, the Chicago Bulls defeated the Utah Jazz to win the National Basketball Association championship. Together the two teams scored 176 points. The Bulls won the game by 4 points. What was the score of the game? (*Source: The Wall Street Journal Almanac 1998.*)

13. For an art project Kay bought 8 sheets of colored paper and 3 marker pens for \$6.50. She later needed 2 more sheets of colored paper and 2 different colored pens. These items cost \$3.00. Find the cost of 1 marker pen and 1 sheet of colored paper.

Solve each system by matrix methods.

14. $3x + 2y = 4$
$5x + 5y = 9$

15. $x + 3y + 2z = 11$
$3x + 7y + 4z = 23$
$5x + 3y - 5z = -14$

16. Use any method described in this chapter to solve the system.
$4x - 2y = -8$
$3y - 5z = 14$
$2x + z = -10$

Evaluate each determinant.

17. $\begin{vmatrix} 6 & -3 \\ 5 & -2 \end{vmatrix}$

18. $\begin{vmatrix} 4 & 1 & 0 \\ -2 & 7 & 3 \\ 0 & 5 & 2 \end{vmatrix}$

Solve each system by Cramer's rule.

19. $3x - y = -8$
$2x + 6y = 3$

20. $x + y + z = 6$
$2x - 2y + z = 5$
$-x + 3y + z = 0$

CUMULATIVE REVIEW EXERCISES CHAPTERS 1–4

Evaluate.

1. $(-3)^4$ **2.** -3^4 **3.** $-(-3)^4$ **4.** $\sqrt{.49}$

5. $-\sqrt{.49}$ **6.** $\sqrt{-.49}$ **7.** $\sqrt[3]{64}$ **8.** $\sqrt[3]{-64}$

Evaluate if $x = -4$, $y = 3$, *and* $z = 6$.

9. $|2x| + 3y - z^3$ **10.** $-5(x^3 - y^3)$

11. Which property of real numbers justifies the statement $5 + (3 \cdot 6) = 5 + (6 \cdot 3)$?

Solve each equation.

12. $7(2x + 3) - 4(2x + 1) = 2(x + 1)$ **13.** $|6x - 8| = 4$

14. $ax + by = cx + d$ for x **15.** $.04x + .06(x - 1) = 1.04$

Solve each inequality.

16. $\frac{2}{3}y + \frac{5}{12}y \leq 20$ **17.** $|3x + 2| \leq 4$ **18.** $|12t + 7| \geq 0$

Solve each problem.

19. On February 12, 1999, the U.S. Senate voted to acquit William Jefferson Clinton on both counts of impeachment (perjury and obstruction of justice). Of the 200 votes cast that day, there were 10 more "not guilty" votes than "guilty" votes. How many of each vote were there? (*Source:* MSNBC website, February 13, 1999.)

20. A triangle has an area of 42 square meters. The base is 14 meters long. Find the height of the triangle.

21. A jar contains only pennies, nickels, and dimes. The number of dimes is 1 more than the number of nickels, and the number of pennies is 6 more than the number of nickels. How many of each denomination can be found in the jar, if the total value is $4.80?

22. Two angles of a triangle have the same measure. The measure of the third angle is 4° less than twice the measure of each of the equal angles. Find the measures of the three angles.

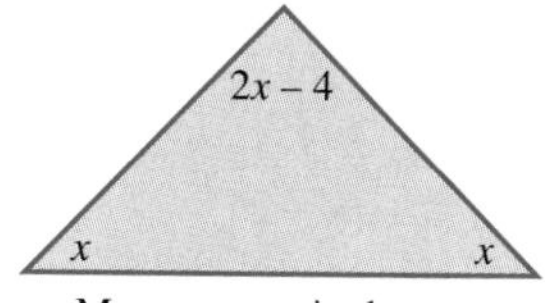

Measures are in degrees.

In Exercises 23–27, point A has coordinates $(-2, 6)$ *and point B has coordinates* $(4, -2)$.

23. What is the equation of the horizontal line through A?

24. What is the equation of the vertical line through B?

25. What is the slope of line AB?

26. What is the slope of a line perpendicular to line AB?

27. What is the standard form of the equation of line AB?

28. Graph the line having slope $\frac{2}{3}$ and passing through the point $(-1, -3)$.

29. Graph the inequality $-3x - 2y \leq 6$.

30. Given that $f(x) = x^2 + 3x - 6$, find each of the following.
(a) $f(-3)$ **(b)** $f(a)$

31. If y varies directly as x and $y = 5$ when $x = 12$, find y when $x = 42$.

Solve by any method.

32. $-2x + 3y = -15$
$4x - y = 15$

33. $x + y + z = 10$
$x - y - z = 0$
$-x + y - z = -4$

34. Evaluate: $\begin{vmatrix} 1 & 2 & 3 \\ 0 & 5 & 1 \\ -1 & 0 & 4 \end{vmatrix}$.

In Exercises 35–38, solve each problem using a system of equations. Use the method of your choice: elimination, substitution, matrix row operations, or Cramer's rule.

35. Mabel Johnston bought apples and oranges at DeVille's Grocery. She bought 6 pounds of fruit. Oranges cost \$.90 per pound, while apples cost \$.70 per pound. If she spent a total of \$5.20, how many pounds of each kind of fruit did she buy?

36. Two of the best-selling toys of 1996 were Tickle Me Elmo and Snacktime Kid. Based on their average retail prices, Elmo cost \$8.63 less than Kid, and together they cost \$63.89. What was the average retail price for each toy? (*Source:* NPD Group, Inc.)

37. Kenneth and Peggy are planning to move, and they need some cardboard boxes. They can buy 10 small and 20 large boxes for \$65, or 6 small and 10 large boxes for \$34. Find the cost of each size of box.

38. At the Chalmette Nut Shop, 6 pounds of peanuts and 12 pounds of cashews cost \$60, while 3 pounds of peanuts and 4 pounds of cashews cost \$22. Find the cost of each type of nut.

The graph shows a company's costs to produce computer parts and the revenue from the sale of computer parts.

39. At what production level does the cost equal the revenue? What is the revenue at that point?

40. Profit is revenue less cost. Estimate the profit on the sale of 1100 parts.

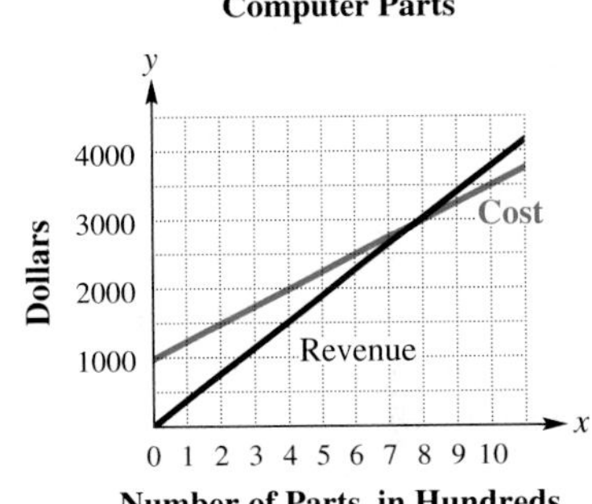

Exponents and Polynomials

Health Care

The health care industry, the theme of this chapter, is in a period of transition. Managed care has emerged as the leading weapon in the battle to contain ever-rising costs. This has led to a loss of both income and autonomy for medical professionals. One response has been an increase in unions for medical professionals. Another has been a number of legislative actions at both the national and state levels to curb managed care practices that have been particularly damaging in the eyes of patients and doctors alike.

The graph shows how average pay for registered nurses over the years from 1990 to 1996 has continued to increase, but at a progressively slower rate. For example, how much did average pay change from 1990 to 1991 and from 1995 to 1996? We will return to this example in Section 5.3.

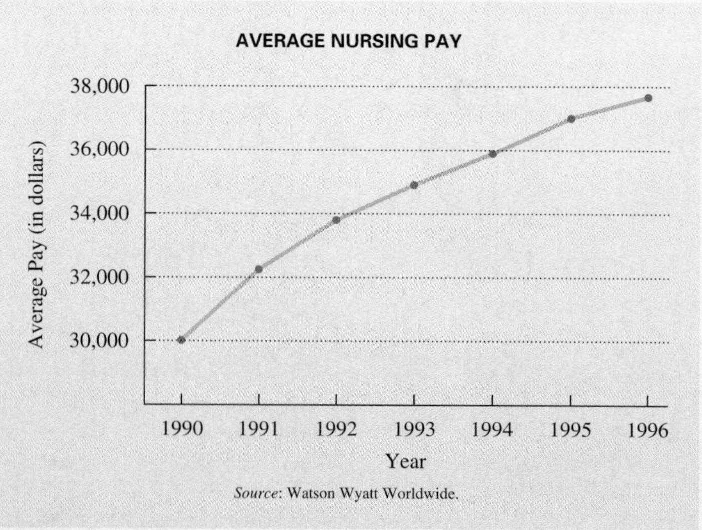

Visit our Web site at www.LialAlgebra.com

INTERNET

5.1 Integer Exponents and Scientific Notation

OBJECTIVES

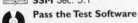

1. Use the product rule for exponents.
2. Define zero as an exponent and negative exponents.
3. Use the quotient rule for exponents.
4. Use the power rules for exponents.
5. Simplify exponential expressions.
6. Use the rules for exponents with scientific notation.

FOR EXTRA HELP

SSG Sec. 5.1
SSM Sec. 5.1

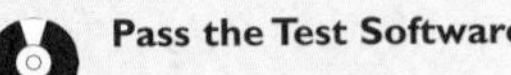
Pass the Test Software

InterAct Math Tutorial Software

Video 8

In Chapter 1, we used exponents to write products of repeated factors. Now, in this section, we give further definitions and rules of exponents.

OBJECTIVE 1 **Use the product rule for exponents.** There are several useful rules that simplify work with exponents. For example, the product $2^5 \cdot 2^3$ can be simplified as follows.

$$2^5 \cdot 2^3 = (2 \cdot 2 \cdot 2 \cdot 2 \cdot 2)(2 \cdot 2 \cdot 2) = 2^8 \qquad (5 + 3 = 8)$$

This result, that products of exponential expressions with the same base are found by adding exponents, is generalized as the **product rule for exponents.**

Product Rule for Exponents

If m and n are natural numbers and a is any real number, then

$$a^m \cdot a^n = a^{m+n}.$$

To see that the product rule is true, use the definition of an exponent as follows.

$$a^m = \underbrace{a \cdot a \cdot a \ldots a}_{a \text{ appears as a factor } m \text{ times.}} \qquad a^n = \underbrace{a \cdot a \cdot a \ldots a}_{a \text{ appears as a factor } n \text{ times.}}$$

From this,

$$\begin{aligned} a^m \cdot a^n &= \underbrace{a \cdot a \cdot a \ldots a}_{m \text{ factors}} \cdot \underbrace{a \cdot a \cdot a \ldots a}_{n \text{ factors}} \\ &= \underbrace{a \cdot a \cdot a \ldots a}_{(m+n) \text{ factors}} \\ a^m \cdot a^n &= a^{m+n}. \end{aligned}$$

EXAMPLE 1 **Using the Product Rule for Exponents**

Apply the product rule for exponents in each case.

(a) $3^4 \cdot 3^7 = 3^{4+7} = 3^{11}$

(b) $5^3 \cdot 5 = 5^3 \cdot 5^1 = 5^{3+1} = 5^4$

(c) $y^3 \cdot y^8 \cdot y^2 = y^{3+8+2} = y^{13}$

(d) $(5y^2)(-3y^4)$

Use the associative and commutative properties as necessary to multiply the numbers and multiply the variables.

$$\begin{aligned} (5y^2)(-3y^4) &= 5(-3)y^2y^4 \\ &= -15y^{2+4} \\ &= -15y^6 \end{aligned}$$

(e) $(7p^3q)(2p^5q^2) = 7(2)p^3p^5qq^2 = 14p^8q^3$

(f) $x^2 \cdot y^4$

Because the bases are not the same, the product rule does not apply.

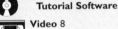

CAUTION Be careful in problems like Example 1(a) not to multiply the bases. Notice that $3^4 \cdot 3^7 \neq 9^{11}$. Remember to keep the *same* base and add the exponents.

OBJECTIVE 2 Define zero as an exponent and negative exponents. So far we have discussed only positive exponents. Let us consider how we might define a zero exponent. Suppose we multiply 4^2 by 4^0. By the product rule,

$$4^2 \cdot 4^0 = 4^{2+0} = 4^2.$$

For the product rule to hold true, 4^0 must equal 1, and so a^0 is defined as follows for any nonzero real number.

Zero Exponent

If a is any nonzero real number, then

$$a^0 = 1.$$

The symbol 0^0 is undefined.

EXAMPLE 2 Using Zero as an Exponent

Evaluate each expression.

(a) $12^0 = 1$

(b) $(-6)^0 = 1$

(c) $-6^0 = -(6^0) = -1$

(d) $5^0 + 12^0 = 1 + 1 = 2$

(e) $(8k)^0 = 1 \quad (k \neq 0)$

How should we define a negative exponent? Using the product rule again,

$$8^2 \cdot 8^{-2} = 8^{2+(-2)} = 8^0 = 1.$$

This indicates that 8^{-2} is the reciprocal of 8^2. But $\frac{1}{8^2}$ is the reciprocal of 8^2, and a number can have only one reciprocal. Thus, we must define 8^{-2} to equal $\frac{1}{8^2}$, so negative exponents are defined as follows.

Negative Exponent

For any natural number n and any nonzero real number a,

$$a^{-n} = \frac{1}{a^n}.$$

With this definition, the expression a^n is meaningful for any integer exponent n and any nonzero real number a.

A negative exponent does not imply a negative number; negative exponents lead to reciprocals, as shown below.

Expression	*Example*	
a^{-m}	$3^{-2} = \frac{1}{3^2} = \frac{1}{9}$	Not negative
$-a^{-m}$	$-3^{-2} = -\frac{1}{3^2} = -\frac{1}{9}$	Negative

EXAMPLE 3 Using Negative Exponents

In parts (a)–(f), write the expressions with only positive exponents. In parts (g) and (h), simplify the expression.

(a) $2^{-3} = \frac{1}{2^3} = \frac{1}{8}$

(b) $6^{-1} = \frac{1}{6^1} = \frac{1}{6}$

(c) $(5z)^{-3} = \frac{1}{(5z)^3}, \quad z \neq 0$

(d) $5z^{-3} = 5\left(\frac{1}{z^3}\right) = \frac{5}{z^3}, \quad z \neq 0$

(e) $-m^{-2} = -\frac{1}{m^2}, \quad m \neq 0$

(f) $(-m)^{-2} = \frac{1}{(-m)^2}, \quad m \neq 0$

(g) $3^{-1} + 4^{-1}$

Since $3^{-1} = \frac{1}{3}$ and $4^{-1} = \frac{1}{4}$,

$$3^{-1} + 4^{-1} = \frac{1}{3} + \frac{1}{4} = \frac{4}{12} + \frac{3}{12} = \frac{7}{12}.$$

(h) $5^{-1} - 2^{-1} = \frac{1}{5} - \frac{1}{2} = \frac{2}{10} - \frac{5}{10} = -\frac{3}{10}$

In Example 3(g), note that $3^{-1} + 4^{-1} \neq (3 + 4)^{-1}$. The expression on the left is equal to $\frac{7}{12}$, as shown in the solution, while the expression on the right is $7^{-1} = \frac{1}{7}$. Similar reasoning can be applied to part (h).

EXAMPLE 4 Using Negative Exponents

Evaluate each expression.

(a) $\frac{1}{2^{-3}} = \frac{1}{\frac{1}{2^3}} = 1 \div \frac{1}{2^3} = 1 \cdot \frac{2^3}{1} = 2^3 = 8$

(b) $\frac{2^{-3}}{3^{-2}} = \frac{\frac{1}{2^3}}{\frac{1}{3^2}} = \frac{1}{2^3} \cdot \frac{3^2}{1} = \frac{3^2}{2^3} = \frac{9}{8}$

Example 4 suggests the following generalizations.

Special Rules for Negative Exponents

If $a \neq 0$ and $b \neq 0$, $\quad \dfrac{1}{a^{-n}} = a^n \quad$ and $\quad \dfrac{a^{-n}}{b^{-m}} = \dfrac{b^m}{a^n}$.

OBJECTIVE 3 **Use the quotient rule for exponents.** A quotient, such as $\frac{a^8}{a^3}$, can be simplified in much the same way as a product. (In all quotients of this type, assume that the denominator is not zero.) Using the definition of an exponent,

$$\frac{a^8}{a^3} = \frac{a \cdot a \cdot a \cdot a \cdot a \cdot a \cdot a \cdot a}{a \cdot a \cdot a} = a \cdot a \cdot a \cdot a \cdot a = a^5.$$

Notice that $8 - 3 = 5$. In the same way,

$$\frac{a^3}{a^8} = \frac{a \cdot a \cdot a}{a \cdot a \cdot a \cdot a \cdot a \cdot a \cdot a \cdot a} = \frac{1}{a^5}.$$

Here again, $8 - 3 = 5$. These examples suggest the following quotient rule for exponents.

Quotient Rule for Exponents

If a is any nonzero real number and m and n are integers, then

$$\frac{a^m}{a^n} = a^{m-n}.$$

EXAMPLE 5 **Using the Quotient Rule for Exponents**

Apply the quotient rule for exponents, and write the result using only positive exponents.

Numerator exponent ↓ ↓ Denominator exponent

(a) $\dfrac{3^7}{3^2} = 3^{7-2} = 3^5$
↑ Minus sign

(b) $\dfrac{p^6}{p^2} = p^{6-2} = p^4 \quad (p \neq 0)$

(c) $\dfrac{k^7}{k^{12}} = k^{7-12} = k^{-5} = \dfrac{1}{k^5} \quad (k \neq 0)$

(d) $\dfrac{2^7}{2^{-3}} = 2^{7-(-3)}$

Since $7 - (-3) = 10$, $\dfrac{2^7}{2^{-3}} = 2^{10}$.

(e) $\dfrac{8^{-2}}{8^5} = 8^{-2-5} = 8^{-7} = \dfrac{1}{8^7}$

(f) $\dfrac{6}{6^{-1}} = \dfrac{6^1}{6^{-1}} = 6^{1-(-1)} = 6^2$

(g) $\dfrac{z^{-5}}{z^{-8}} = z^{-5-(-8)} = z^3 \quad (z \neq 0)$

(h) $\dfrac{a^3}{b^4}$

The quotient rule does not apply because the bases are different.

As seen in Example 5, be very careful when working with quotients that involve negative exponents in the denominator. Always be sure to write the numerator exponent, then a minus sign, and then the denominator exponent.

The product and quotient rules for exponents are summarized below.

Summary of Product and Quotient Rules

To multiply expressions such as a^m and a^n where the base is the same, keep the same base and add the exponents. To divide such expressions, keep the same base and subtract the exponent of the denominator from the exponent of the numerator.

OBJECTIVE 4 **Use the power rules for exponents.** The expression $(3^4)^2$ can be simplified as $(3^4)^2 = 3^4 \cdot 3^4 = 3^{4+4} = 3^8$, where $4 \cdot 2 = 8$. This example suggests the first of the **power rules for exponents;** the other two parts can be demonstrated with similar examples.

Power Rules for Exponents

If a and b are real numbers and m and n are integers, then

$$(a^m)^n = a^{mn}, \qquad (ab)^m = a^m b^m, \qquad \text{and} \qquad \left(\frac{a}{b}\right)^m = \frac{a^m}{b^m} \quad (b \neq 0).$$

The parts of this rule can be illustrated in the same way as the product rule. In the statements of rules for exponents, we always assume that zero never appears to a negative power.

EXAMPLE 6 Using the Power Rules for Exponents

Use a power rule in each case.

(a) $(p^8)^3 = p^{8 \cdot 3} = p^{24}$

(b) $\left(\frac{2}{3}\right)^4 = \frac{2^4}{3^4} = \frac{16}{81}$

(c) $(3y)^4 = 3^4 y^4 = 81y^4$

(d) $(6p^7)^2 = 6^2 p^{7 \cdot 2} = 6^2 p^{14} = 36p^{14}$

(e) $\left(\frac{-2m^5}{z}\right)^3 = \frac{(-2)^3 m^{5 \cdot 3}}{z^3} = \frac{(-2)^3 m^{15}}{z^3} = \frac{-8m^{15}}{z^3} \quad (z \neq 0)$

The reciprocal of a^n is $\frac{1}{a^n} = \left(\frac{1}{a}\right)^n$. Also, by definition, a^n and a^{-n} are reciprocals since

$$a^n \cdot a^{-n} = a^n \cdot \frac{1}{a^n} = 1.$$

Thus, since both are reciprocals of a^n,

$$a^{-n} = \left(\frac{1}{a}\right)^n.$$

Some examples of this result are

$$6^{-3} = \left(\frac{1}{6}\right)^3 \quad \text{and} \quad \left(\frac{1}{3}\right)^{-2} = 3^2.$$

The discussion above can be generalized as follows.

Special Rules for Negative Exponents

Any nonzero number raised to the negative nth power is equal to the reciprocal of that number raised to the nth power. That is, if $a \neq 0$ and $b \neq 0$,

$$a^{-n} = \left(\frac{1}{a}\right)^n \quad \text{and} \quad \left(\frac{a}{b}\right)^{-n} = \left(\frac{b}{a}\right)^n.$$

EXAMPLE 7 Using Negative Exponents with Fractions

Write the following expressions with only positive exponents and then evaluate.

(a) $\left(\frac{3}{7}\right)^{-2} = \left(\frac{7}{3}\right)^2 = \frac{49}{9}$

(b) $\left(\frac{4}{5}\right)^{-3} = \left(\frac{5}{4}\right)^3 = \frac{125}{64}$

The definitions and rules of this section are summarized below.

Definitions and Rules for Exponents

For all integers m and n and all real numbers a and b:

Product Rule $a^m \cdot a^n = a^{m+n}$

Quotient Rule $\frac{a^m}{a^n} = a^{m-n} \quad (a \neq 0)$

Zero Exponent $a^0 = 1 \quad (a \neq 0)$

Negative Exponent $a^{-n} = \frac{1}{a^n} \quad (a \neq 0)$

Power Rules $(a^m)^n = a^{mn}$

$(ab)^m = a^m b^m$

$\left(\frac{a}{b}\right)^m = \frac{a^m}{b^m} \quad (b \neq 0)$

Special Rules $a^{-n} = \left(\frac{1}{a}\right)^n \quad (a \neq 0)$

$\left(\frac{a}{b}\right)^{-n} = \left(\frac{b}{a}\right)^n \quad (a, b \neq 0)$

OBJECTIVE 5 Simplify exponential expressions. With the rules of exponents developed so far in this section, we can simplify expressions that involve one or more rules.

EXAMPLE 8 Using the Definitions and Rules for Exponents

Simplify each expression so that no negative exponents appear in the final result. Assume that all variables represent nonzero real numbers.

(a) $3^2 \cdot 3^{-5} = 3^{2+(-5)} = 3^{-3} = \dfrac{1}{3^3}$ or $\dfrac{1}{27}$

(b) $x^{-3} \cdot x^{-4} \cdot x^2 = x^{-3+(-4)+2} = x^{-5} = \dfrac{1}{x^5}$

(c) $(4^{-2})^{-5} = 4^{(-2)(-5)} = 4^{10}$

(d) $(x^{-4})^6 = x^{(-4)6} = x^{-24} = \dfrac{1}{x^{24}}$

(e) $\dfrac{x^{-4}y^2}{x^2y^{-5}} = \dfrac{x^{-4}}{x^2} \cdot \dfrac{y^2}{y^{-5}}$

$= x^{-4-2} \cdot y^{2-(-5)}$

$= x^{-6}y^7$

$= \dfrac{y^7}{x^6}$

(f) $(2^3x^{-2})^{-2} = (2^3)^{-2} \cdot (x^{-2})^{-2}$

$= 2^{-6}x^4$

$= \dfrac{x^4}{2^6}$ or $\dfrac{x^4}{64}$

There is often more than one way to simplify expressions like those in Example 8. For instance, we could simplify Example 8(e) as follows.

$$\frac{x^{-4}y^2}{x^2y^{-5}} = \frac{y^5y^2}{x^4x^2} \qquad \text{Use } \frac{a^{-n}}{b^{-m}} = \frac{b^m}{a^n}.$$

$$= \frac{y^7}{x^6} \qquad \text{Product rule}$$

OBJECTIVE 6 Use the rules for exponents with scientific notation. Scientists often need to use numbers that are very large or very small. For example, the number of one-celled organisms that will sustain a whale for a few hours is 400,000,000,000,000, and the shortest wavelength of visible light is approximately .0000004 meter. It is simpler to write these numbers using *scientific notation.*

In scientific notation, a number is written with the decimal point after the first nonzero digit and multiplied by a power of 10, as indicated in the following definition.

Scientific Notation

A number is written in **scientific notation** when it is expressed in the form

$$a \times 10^n$$

where $1 \le |a| < 10$, and n is an integer.

For example, in scientific notation,

$$8000 = 8 \times 1000 = 8 \times 10^3.$$

The following numbers are not in scientific notation.

$.230 \times 10^4$	46.5×10^{-3}
.230 is less than 1.	46.5 is greater than 10.

To write a number in scientific notation, use the steps given below. (If the number is negative, ignore the minus sign, go through these steps, and then attach a minus sign to the result.)

Converting to Scientific Notation

Step 1 **Position the decimal point.** Place a caret, ^, to the right of the first nonzero digit, where the decimal point will be placed.

Step 2 **Determine the numeral for the exponent.** Count the number of digits from the caret to the decimal point. This number gives the absolute value of the exponent on ten.

Step 3 **Determine the sign for the exponent.** Decide whether multiplying by 10^n should make the result of Step 1 larger or smaller. The exponent should be positive to make the result larger; it should be negative to make the result smaller.

It is helpful to remember that for $n \geq 1$, $10^{-n} < 1$ and $10^n \geq 10$.

EXAMPLE 9 **Writing a Number in Scientific Notation**

Write each number in scientific notation.

(a) 820,000

Place a caret to the right of the 8 (the first nonzero digit) to mark the new location of the decimal point.

$$8_{\wedge}20{,}000$$

Count from the caret to the decimal point, which is understood to be after the last 0.

8 20,000. ← Decimal point

↑

Count 5 places

Since the number 8.2 is to be made larger, the exponent on 10 is positive.

$$820{,}000 = 8.2 \times 10^5$$

(b) .0000072

Count from right to left.

.000007 2

6 places

Since the number 7.2 is to be made smaller, the exponent on 10 is negative.

$$.0000072 = 7.2 \times 10^{-6}$$

To convert a number written in scientific notation to standard notation, just work in reverse.

Converting from Scientific Notation

Multiplying a number by a positive power of 10 makes the number larger, so move the decimal point to the right if n is positive in 10^n.

Converting from Scientific Notation (continued)

Multiplying by a negative power of 10 makes a number smaller, so move the decimal point to the left if n is negative.

If n is zero, leave the decimal point where it is.

EXAMPLE 10 Converting from Scientific Notation to Standard Notation

Write the following numbers without scientific notation.

(a) 6.93×10^7

$$\underbrace{6.9300000}_{\text{7 places}}$$

The decimal point was moved 7 places to the right. (It was necessary to attach 5 zeros.)

$$6.93 \times 10^7 = 69{,}300{,}000$$

(b) 4.7×10^{-6}

$$\underbrace{000004.7}_{\text{6 places}}$$

The decimal point was moved 6 places to the left.

$$4.7 \times 10^{-6} = .0000047$$

(c) $1.083 \times 10^0 = 1.083$

When problems require operations with numbers that are very large and/or very small, it is often advantageous to write the numbers in scientific notation first, and then perform the calculations using the rules for exponents.

EXAMPLE 11 Using Scientific Notation in Computation

Find $\dfrac{1{,}920{,}000 \times .0015}{.000032 \times 45{,}000}$.

First, express all numbers in scientific notation.

$$\frac{1{,}920{,}000 \times .0015}{.000032 \times 45{,}000} = \frac{1.92 \times 10^6 \times 1.5 \times 10^{-3}}{3.2 \times 10^{-5} \times 4.5 \times 10^4}$$

Next, use the commutative and associative properties and the rules for exponents to simplify the expression.

$$\begin{aligned}\frac{1{,}920{,}000 \times .0015}{.000032 \times 45{,}000} &= \frac{1.92 \times 1.5 \times 10^6 \times 10^{-3}}{3.2 \times 4.5 \times 10^{-5} \times 10^4} \\ &= \frac{1.92 \times 1.5}{3.2 \times 4.5} \times 10^4 \\ &= .2 \times 10^4 \\ &= (2 \times 10^{-1}) \times 10^4 \\ &= 2 \times 10^3\end{aligned}$$

The expression is equal to 2×10^3, or 2000.

EXAMPLE 12 **Using Scientific Notation to Solve a Problem**

In 1985, the national health care expenditure was 428.2 billion dollars. By 1995, this figure had risen by a factor of 2.3; that is, it more than doubled in only ten years. (*Source:* U.S. Health Care Financing Administration.)

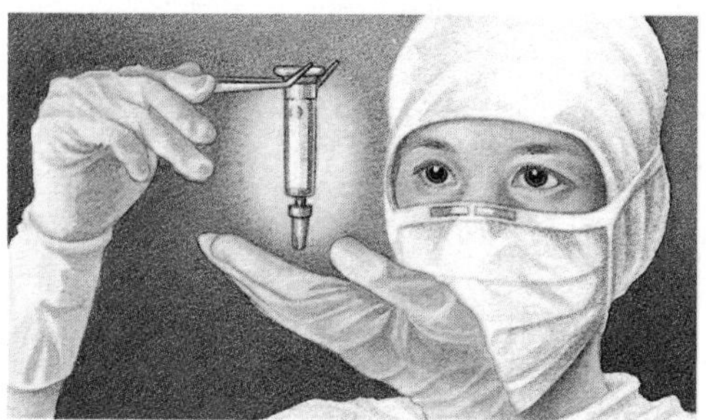

(a) Write the 1985 health care expenditure amount using scientific notation.

$$\begin{aligned}428.2 \text{ billion} = 428.2 \times 10^9 &= (4.282 \times 10^2) \times 10^9 \\ &= 4.282 \times 10^{11} \qquad \text{Add exponents.}\end{aligned}$$

In 1985, the expenditure was \$4.282 $\times 10^{11}$.

(b) What was the expenditure in 1995?

We must multiply the result in part (a) by 2.3.

$$\begin{aligned}(4.282 \times 10^{11}) \times 2.3 &= (2.3 \times 4.282) \times 10^{11} \qquad \text{Commutative and associative properties} \\ &= 9.849 \times 10^{11} \qquad \text{Round to three decimal places.}\end{aligned}$$

The 1995 expenditure was nearly \$984,900,000,000.

5.1 EXERCISES

Decide whether the expression has been simplified correctly. If not, correct it.

1. $(ab)^2 = ab^2$

2. $(5x)^3 = 5^3x^3$

3. $\left(\dfrac{4}{a}\right)^3 = \dfrac{4^3}{a} \quad (a \neq 0)$

4. $y^2 \cdot y^6 = y^{12}$

5. $x^3 \cdot x^4 = x^7$

6. $xy^0 = 0 \quad (y \neq 0)$

7. State the product and quotient rules for exponents in your own words. Give examples with your explanations.

8. State the three power rules for exponents in your own words. Give examples with your explanations.

Use the product and/or quotient rules as needed to simplify each expression. Write the answer with only positive exponents. Assume that all variables represent nonzero real numbers. See Examples 1–5.

9. $a^5 \cdot a^3$

10. $x^{15} \cdot x^4$

11. $y^5 \cdot y^4 \cdot y^{-3}$

12. $k^3 \cdot k^9 \cdot k^{-8}$

13. $(9x^2y^3)(-2x^3y^5)$

14. $(-3x^5y^4)(-5xy^2)$

15. $\dfrac{p^{19}}{p^5}$

16. $\dfrac{q^{13}}{q^7}$

17. $\dfrac{z^{-6}}{z^{-12}}$

18. $\dfrac{r^{-4}}{r^{-9}}$

19. $\dfrac{r^{13}r^{-4}r^{-3}}{r^{-2}r^{-5}r^0}$

20. $\dfrac{z^{-4}z^{-2}w^0}{z^3z^{-1}w^{-5}}$

21. $7k^2(-2k)(4k^{-5})^0$

22. $3a^2(-5a^{-6})(-2a)^0$

23. $-4(2x^3)(3x)$

24. $6(5z^3)(2zw^2)$

25. $\dfrac{(3pq)q^2}{6p^2q^4}$

26. $\dfrac{(-8xy)y^3}{4x^5y^4}$

27. $\dfrac{6x^{-5}y^{-2}}{(3x^{-3})(2x^{-2}y^{-2})}$

28. $\dfrac{-8(x^2y^{-4})}{(4xy^{-5})(-2xy)}$

29. Your friend evaluated $4^5 \cdot 4^2$ as 16^7. Explain to him why his answer is incorrect.

30. Consider the expressions $-a^n$ and $(-a)^n$. In some cases they are equal and in some cases they are not. Using $n = 2, 3, 4, 5$, and 6 and $a = 2$, draw a conclusion as to when they are equal and when they are opposites.

Evaluate. See Examples 1–5.

31. $\left(\dfrac{2}{3}\right)^2$

32. $\left(\dfrac{4}{3}\right)^3$

33. 4^{-3}

34. 5^{-2}

35. -4^{-3}

36. -5^{-2}

37. $(-4)^{-3}$

38. $(-5)^{-2}$

39. $\dfrac{1}{3^{-2}}$

40. $\dfrac{1}{6^{-1}}$

41. $\dfrac{-3^{-1}}{4^{-2}}$

42. $\dfrac{2^{-3}}{-3^{-1}}$

43. $\left(\dfrac{2}{3}\right)^{-3}$

44. $\left(\dfrac{5}{4}\right)^{-2}$

45. $3^{-1} + 2^{-1}$

46. $4^{-1} + 5^{-1}$

47. $6^{-1} - 4^{-1}$

48. $8^{-1} - 16^{-1}$

49. $(6 - 4)^{-1}$

50. $(8 - 16)^{-1}$

51. $\dfrac{3^{-5}}{3^{-2}}$

52. $\dfrac{2^{-4}}{2^{-3}}$

53. $\dfrac{9^{-1}}{-9}$

54. $\dfrac{8^{-1}}{-8}$

Evaluate each expression. Assume that all variables represent nonzero real numbers. See Example 2.

55. 25^0

56. 14^0

57. -7^0

58. -10^0

59. $(-7)^0$

60. $(-10)^0$

61. $-4^0 - m^0$

62. $-8^0 - k^0$

63. Your friend thinks that $(-3)^{-2}$ is a negative number. Why is she incorrect?

64. Which one of the following does not represent the reciprocal of x ($x \neq 0$)?

(a) x^{-1} **(b)** $\dfrac{1}{x}$ **(c)** $\left(\dfrac{1}{x^{-1}}\right)^{-1}$ **(d)** $-x$

Simplify the expression using only positive exponents. Assume that variables represent nonzero real numbers. See Examples 1–8.

65. $(2^{-3} \cdot 5^{-1})^3$

66. $(5^{-4} \cdot 6^{-2})^3$

67. $(5^{-4} \cdot 6^{-2})^{-3}$

68. $(2^{-5} \cdot 3^{-4})^{-1}$

69. $(k^2)^{-3}k^4$

70. $(x^3)^{-4}x^5$

71. $-4r^{-2}(r^4)^2$

72. $-2m^{-1}(m^3)^2$

73. $(5a^{-1})^4(a^2)^{-3}$

74. $(3p^{-4})^2(p^3)^{-1}$

75. $(z^{-4}x^3)^{-1}$

76. $(y^{-2}z^4)^{-3}$

77. $\dfrac{(p^{-2})^0}{5p^{-4}}$

78. $\dfrac{(m^4)^0}{9m^{-3}}$

79. $\dfrac{4a^5(a^{-1})^3}{(a^{-2})^{-2}}$

80. $\dfrac{12k^{-2}(k^{-3})^{-4}}{6k^5}$

81. $\dfrac{(-y^{-4})^2}{6(y^{-5})^{-1}}$

82. $\dfrac{2(-m^{-1})^{-4}}{9(m^{-3})^2}$

83. $\dfrac{(2k)^2m^{-5}}{(km)^{-3}}$

84. $\dfrac{(3rs)^{-2}}{3^2r^2s^{-4}}$

85. $\dfrac{(2k)^2k^3}{k^{-1}k^{-5}}(5k^{-2})^{-3}$

86. $\dfrac{(3r^2)^2r^{-5}}{r^{-2}r^3}(2r^{-6})^2$

87. $\left(\dfrac{3k^{-2}}{k^4}\right)^{-1} \cdot \dfrac{2}{k}$

88. $\left(\dfrac{7m^{-2}}{m^{-3}}\right)^{-2} \cdot \dfrac{m^3}{4}$

89. $\left(\dfrac{2p}{q^2}\right)^3\left(\dfrac{3p^4}{q^{-4}}\right)^{-1}$

90. $\left(\dfrac{5z^3}{2a^2}\right)^{-3}\left(\dfrac{8a^{-1}}{15z^{-2}}\right)^{-3}$

91. $\dfrac{2^2y^4(y^{-3})^{-1}}{2^5y^{-2}}$

92. $\dfrac{3^{-1}m^4(m^2)^{-1}}{3^2m^{-2}}$

93. $\dfrac{(2m^2p^3)^2(4m^2p)^{-2}}{(-3mp^4)^{-1}(2m^3p^4)^3}$

94. $\dfrac{(-5y^3z^4)^2(2yz^5)^{-2}}{10(y^4z)^3(3y^3z^2)^{-1}}$

95. $\dfrac{(-3y^3x^3)(-4y^4x^2)(x^2)^{-4}}{18x^3y^2(y^3)^3(x^3)^{-2}}$

96. $\dfrac{(2m^3x^2)^{-1}(3m^4x)^{-3}}{(m^2x^3)^3(m^2x)^{-5}}$

97. $\left(\dfrac{p^2q^{-1}}{2p^{-2}}\right)^2 \cdot \left(\dfrac{p^3 \cdot 4q^{-2}}{3q^{-5}}\right)^{-1} \cdot \left(\dfrac{pq^{-5}}{q^{-2}}\right)^3$

98. $\left(\dfrac{a^6b^{-2}}{2a^{-2}}\right)^{-1} \cdot \left(\dfrac{6a^{-2}}{5b^{-4}}\right)^2 \cdot \left(\dfrac{2b^{-1}a^2}{3b^{-2}}\right)^{-1}$

RELATING CONCEPTS (EXERCISES 99–102)

Consider the expression

$$\left(\frac{a^{-8}b^2}{a^{-5}b^{-4}}\right)^{-2} \quad (a, b \neq 0).$$

There are several ways to simplify the expression so that only positive exponents appear in the answer. However, all methods should yield the same answer.

Work Exercises 99–102 in order.

99. Simplify the expression by first writing the fraction in parentheses using only positive exponents. Then apply the exponent -2 using the power rule $\left(\frac{a}{b}\right)^m = \frac{a^m}{b^m}$. Finally, simplify this result.

100. Simplify the expression by first applying the exponent -2 to each factor within the parentheses. Then simplify this result.

101. How do your answers in Exercises 99 and 100 compare?

102. After seeing results from Exercises 99 and 100, how would you answer the question, "Which one of these methods is the correct way of simplifying the expression?"

Did you make the connection that some expressions can be simplified in different ways, all resulting in the correct answer?

Write each number in scientific notation. See Example 9.

103. 530 **104.** 1600 **105.** .830 **106.** .0072

107. .00000692 **108.** .875 **109.** $-38{,}500$ **110.** $-976{,}000{,}000$

Write each of the following in standard notation. See Example 10.

111. 7.2×10^4 **112.** 8.91×10^2 **113.** 2.54×10^{-3}

114. 5.42×10^{-4} **115.** -6×10^4 **116.** -9×10^3

117. 1.2×10^{-5} **118.** 2.7×10^{-6}

Use the rules for exponents to find each value. See Example 11.

119. $\dfrac{12 \times 10^4}{2 \times 10^6}$ **120.** $\dfrac{16 \times 10^5}{4 \times 10^8}$ **121.** $\dfrac{3 \times 10^{-2}}{12 \times 10^3}$

122. $\dfrac{5 \times 10^{-3}}{25 \times 10^2}$ **123.** $\dfrac{.05 \times 1600}{.0004}$ **124.** $\dfrac{.003 \times 40{,}000}{.00012}$

125. $\dfrac{20{,}000 \times .018}{300 \times .0004}$ **126.** $\dfrac{840{,}000 \times .03}{.00021 \times 600}$

 *Solve each problem. Use a calculator as necessary. See Example 12.*

127. In 1995, annual receipts/revenues for health service industries totaled \$382,553,000,000. Write this number in scientific notation. (*Source:* U.S. Bureau of the Census.)

128. In 1995, the number of persons in the United States with health insurance coverage was 223,733,000 and the number of persons without coverage was 40,582,000. Write these numbers in scientific notation. (*Source:* U.S. Bureau of the Census.)

129. On October 28, 1998, IBM announced that it was going to offer a computer that is capable of 3.9×10^8 operations per second. This was 15,000 times faster than the normal desktop computer at that time. What was the number of operations that the normal desktop could do? (*Source:* IBM.)

130. In the Powerball Lottery, a player must choose five numbers from 1 through 49 and one number from 1 through 42. It can be shown that there are about 8.009×10^7 different ways to do this. Suppose that a group of 2000 persons decides to purchase tickets for all these numbers and each ticket costs \$1.00. How much should each person expect to pay? (*Source:* www.powerball.com.)

131. A parsec, a unit of length used in astronomy, is 19×10^{12} miles. The mean distance of Uranus from the sun is 1.8×10^7 miles. How many parsecs is Uranus from the sun?

132. An inch is approximately 1.57828×10^{-5} mile. Find the reciprocal of this number to determine the number of inches in a mile.

133. The speed of light is approximately 3×10^{10} centimeters per second. How long will it take light to travel 9×10^{12} centimeters?

134. The average distance from Earth to the sun is 9.3×10^7 miles. How long would it take a rocket, traveling at 2.9×10^3 miles per hour, to reach the sun?

135. A *light-year* is the distance that light travels in one year. Find the number of miles in a light-year if light travels 1.86×10^5 miles per second.

136. Use the information given in the previous two exercises to find the number of minutes necessary for light from the sun to reach Earth.

137. **(a)** The planet Mercury has an average distance from the sun of 3.6×10^7 miles, while the mean distance of Venus to the sun is 6.7×10^7 miles. How long would it take a spacecraft traveling at 1.55×10^3 miles per hour to travel from Venus to Mercury? (Give your answer in hours, without scientific notation.)

(b) Use the information from part (a) to find the number of days it would take the spacecraft to travel from Venus to Mercury. Round your answer to the nearest whole number of days.

138. When the distance between the centers of the Moon and the Earth is 4.60×10^8 meters, an object on the line joining the centers of the Moon and the Earth exerts the same gravitational force on each when it is 4.14×10^8 meters from the center of the Earth. How far is the object from the center of the Moon at that point?

The graph shows the estimated annual number of Americans, by age and sex, experiencing heart attacks.

Use scientific notation to represent the numbers of people for the following categories.

139. Males between 29 and 44

140. Females between 29 and 44

141. Males between 45 and 64

142. Females between 45 and 64

143. Males 65 or older

144. Females 65 or older

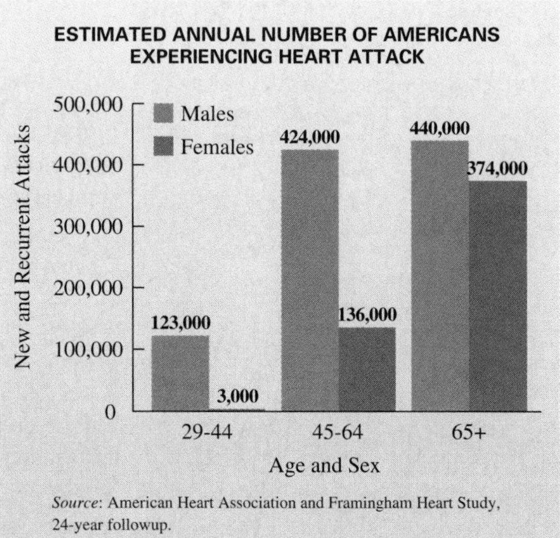

TECHNOLOGY INSIGHTS (EXERCISES 145–148)

The screen on the left shows how a graphing calculator displays 250,000 and .000000034 in scientific notation. When put in scientific mode, it will calculate and display results as shown in the screen on the right.

```
250000
                2.5E5
.000000034
               3.4E-8
```

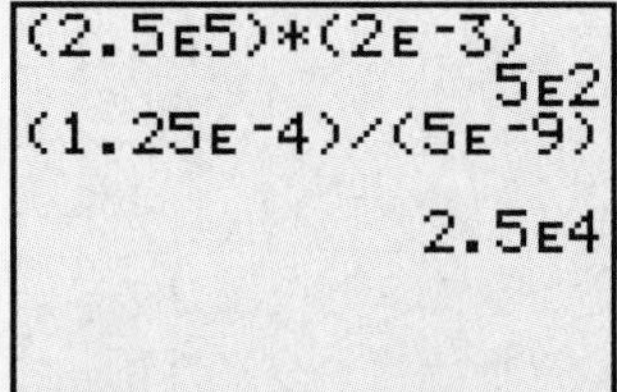

Predict the result the calculator will give for the following screens. (Use the usual scientific notation in writing your answer.)

145.

146.

TECHNOLOGY INSIGHTS (EXERCISES 145–148) (CONTINUED)

147.
```
(8.4E14)/(2.1E-3
)
```

148.
```
(2.5E10)/(2E-3)
```

5.2 Addition and Subtraction of Polynomials

OBJECTIVES

1 Know the basic definitions for polynomials.

2 Find the degree of a polynomial.

3 Add and subtract polynomials.

FOR EXTRA HELP

SSG Sec. 5.2
SSM Sec. 5.2

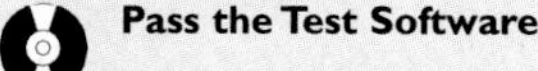
Pass the Test Software

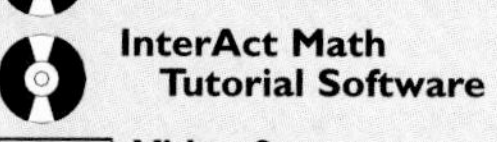
InterAct Math Tutorial Software

Video 8

OBJECTIVE 1 Know the basic definitions for polynomials. Just as whole numbers are the basis of arithmetic, *polynomials* are fundamental in algebra. To understand polynomials, we must review several words from Chapter 1. A *term* is a number, a variable, or the product or quotient of a number and one or more variables raised to nonnegative powers. Examples of terms include

$$4x, \quad \frac{1}{2}m^5 \left(\text{or } \frac{m^5}{2}\right), \quad -7z^9, \quad 6x^2z, \quad \text{and} \quad 9.$$

The number in the product is called the *numerical coefficient,* or just the *coefficient.* In the term $8x^3$, the coefficient is 8. In the term $-4p^5$, it is -4. The coefficient of the term k is understood to be 1. The coefficient of $-r$ is -1. More generally, any factor in a term is the coefficient of the product of the remaining factors. For example, $3x^2$ is the coefficient of y in the term $3x^2y$, and $3y$ is the coefficient of x^2 in $3x^2y$.

Any combination of variables or constants (numerical values) joined by the basic operations of addition, subtraction, multiplication, and division (except by 0), or taking roots is called an **algebraic expression.** The simplest kind of algebraic expression is a **polynomial.**

Polynomial

A **polynomial** is a term or a finite sum of terms in which all variables have whole number exponents and no variables appear in denominators.

Examples of polynomials include

$$3x - 5, \quad 4m^3 - 5m^2p + 8, \quad \text{and} \quad -5t^2s^3.$$

Even though the expression $3x - 5$ involves subtraction, it is called a sum of terms, since it could be written as $3x + (-5)$.

Some examples of expressions that are not polynomials are

$$x^{-1} + 3x^{-2}, \quad \sqrt{9 - x}, \quad \text{and} \quad \frac{1}{x}.$$

The first of these is not a polynomial because it has negative integer exponents, the second because it involves a variable under a radical, and the third because it contains a variable in the denominator.

Most of the polynomials used in this book contain only one variable. A polynomial containing only the variable x is called a **polynomial in x.** A polynomial in one variable is written in **descending powers** of the variable if the exponents on the terms of the polynomial decrease from left to right. For example,

$$x^5 - 6x^2 + 12x - 5$$

is a polynomial in descending powers of x. The term -5 in the polynomial above can be thought of as $-5x^0$, since $-5x^0 = -5(1) = -5$.

EXAMPLE 1 Writing Polynomials in Descending Powers

Write each of the following in descending powers of the variable.

(a) $y - 6y^3 + 8y^5 - 9y^4 + 12$
Write the polynomial as

$$8y^5 - 9y^4 - 6y^3 + y + 12.$$

(b) $-2 + m + 6m^2 - 4m^3$ would be written as $-4m^3 + 6m^2 + m - 2$.

Polynomials with a specific number of terms are so common that they are given special names. A polynomial of exactly three terms is a **trinomial,** and a polynomial with exactly two terms is a **binomial.** A single-term polynomial is a **monomial.** The table that follows gives examples.

Types of Polynomials	*Examples*
Monomials	$5x$, $7m^9$, -8, x^2y^2
Binomials	$3x^2 - 6$, $11y + 8$, $5a^2b + 3a$
Trinomials	$y^2 + 11y + 6$, $8p^3 - 7p + 2m$, $-3 + 2k^5 + 9z^4$
None of these	$p^3 - 5p^2 + 2p - 5$, $-9z^3 + 5c^3 + 2m^5 + 11r^2 - 7r$

OBJECTIVE 2 Find the degree of a polynomial. The **degree of a term** with one variable is the exponent on the variable. For example, the degree of $2x^3$ is 3, the degree of $-x^4$ is 4, and the degree of $17x$ is 1. The degree of a term in more than one variable is defined to be the sum of the exponents of the variables. For example, the degree of $5x^3y^7$ is 10, because $3 + 7 = 10$.

The greatest degree of any of the terms in a polynomial is called the **degree of the polynomial.** In most cases, we will be interested in finding the degree of a polynomial in one variable. For example, the degree of $4x^3 - 2x^2 - 3x + 7$ is 3, because the greatest degree of any term is 3 (the degree of $4x^3$).

EXAMPLE 2 Finding the Degree of a Polynomial

Find the degree of each polynomial.

Greatest exponent is 2.

(a) $9x^2 - 5x + 8$
The greatest exponent is 2, so the polynomial is of degree 2.

(b) $17m^9 + 8m^{14} - 9m^3$
This polynomial is of degree 14.

(c) $5x$
The degree is 1, since $5x = 5x^1$.

(d) -2
Since $-2 = -2x^0$, the degree is 0. Any constant term, other than zero, has degree 0.

(e) $5a^2b^5$
The degree is the sum of the exponents, $2 + 5 = 7$.

(f) $x^3y^9 + 12xy^4 + 7xy$
The degrees of the terms are 12, 5, and 2. Therefore, the degree of the polynomial is 12, which is the greatest degree of any term.

The number 0 has no degree, since 0 times a variable to any power is 0.

OBJECTIVE 3 Add and subtract polynomials. We use the distributive property to simplify polynomials by combining terms. For example, simplify $x^3 + 4x^2 + 5x^2 - 1$ as follows.

$$x^3 + 4x^2 + 5x^2 - 1 = x^3 + (4 + 5)x^2 - 1 \quad \text{Distributive property}$$
$$= x^3 + 9x^2 - 1$$

On the other hand, the terms in the polynomial $4x + 5x^2$ cannot be combined. As these examples suggest, only terms containing exactly the same variables to the same powers may be combined. As mentioned in Chapter 1, such terms are called *like terms.*

EXAMPLE 3 Combining Like Terms

Combine terms.

(a) $-5y^3 + 8y^3 - y^3$
Combine these like terms by the distributive property.

$$-5y^3 + 8y^3 - y^3 = (-5 + 8 - 1)y^3 = 2y^3$$

(b) $6x + 5y - 9x + 2y$
Use the associative and commutative properties to rewrite the expression with all the x-terms together and all the y-terms together.

$$6x + 5y - 9x + 2y = 6x - 9x + 5y + 2y$$

Now combine like terms.

$$= -3x + 7y$$

Since $-3x$ and $7y$ are unlike terms, no further simplification is possible.

(c) $5x^2y - 6xy^2 + 9x^2y + 13xy^2 = 5x^2y + 9x^2y - 6xy^2 + 13xy^2$
$$= 14x^2y + 7xy^2$$

CAUTION Remember that only *like terms* can be combined.

We use the following rule to add two polynomials.

Adding Polynomials

To add two polynomials, combine like terms.

EXAMPLE 4 Adding Polynomials

(a) Add: $(4k^2 - 5k + 2) + (-9k^2 + 3k - 7)$.

Use the commutative and associative properties to rearrange the polynomials so that like terms are together. Then use the distributive property to combine like terms.

$$\begin{aligned}(4k^2 - 5k + 2) + (-9k^2 + 3k - 7) &= 4k^2 - 9k^2 - 5k + 3k + 2 - 7\\ &= -5k^2 - 2k - 5\end{aligned}$$

(b) Add the two polynomials in part (a) vertically.

To add the polynomials vertically, line up like terms in columns. Then add by column.

$$\begin{array}{r} 4k^2 - 5k + 2 \\ -9k^2 + 3k - 7 \\ \hline -5k^2 - 2k - 5 \end{array}$$

EXAMPLE 5 Adding Polynomials

Add: $(3a^5 - 9a^3 + 4a^2) + (-8a^5 + 8a^3 + 2)$.

$$\begin{aligned}&(3a^5 - 9a^3 + 4a^2) + (-8a^5 + 8a^3 + 2)\\ &\quad = 3a^5 - 8a^5 - 9a^3 + 8a^3 + 4a^2 + 2\\ &\quad = -5a^5 - a^3 + 4a^2 + 2 \qquad \text{Combine like terms.}\end{aligned}$$

Add these same two polynomials vertically by placing like terms in columns.

$$\begin{array}{rrrr} 3a^5 & -\ 9a^3 & +\ 4a^2 & \\ -8a^5 & +\ 8a^3 & & +\ 2 \\ \hline -5a^5 & -\ \ a^3 & +\ 4a^2 & +\ 2 \end{array}$$

In Chapter 1, subtraction of real numbers was defined as

$$a - b = a + (-b).$$

That is, add the first number and the negative (or opposite) of the second. We can give a similar definition for subtraction of polynomials by defining the **negative of a polynomial** as that polynomial with every sign changed.

Subtracting Polynomials

To subtract two polynomials, add the first polynomial and the negative of the *second* polynomial.

EXAMPLE 6 **Subtracting Polynomials**

(a) Subtract: $(-6m^2 - 8m + 5) - (-5m^2 + 7m - 8)$.
Change every sign in the second polynomial and add.

$$(-6m^2 - 8m + 5) - (-5m^2 + 7m - 8)$$
$$= -6m^2 - 8m + 5 + 5m^2 - 7m + 8$$

Now add by combining like terms.

$$= -6m^2 + 5m^2 - 8m - 7m + 5 + 8$$
$$= -m^2 - 15m + 13$$

Check by adding the sum, $-m^2 - 15m + 13$, to the second polynomial. The result should be the first polynomial.

(b) Subtract the two polynomials in part (a) vertically.
Write the first polynomial above the second, lining up like terms in columns.

$$\begin{array}{r} -6m^2 - 8m + 5 \\ \underline{-5m^2 + 7m - 8} \end{array}$$

Change all the signs in the second polynomial, and add.

$$\begin{array}{rrrl} -6m^2 & - \;\; 8m & + \;\; 5 & \\ \underline{+5m^2} & \underline{- \;\; 7m} & \underline{+ \;\; 8} & \text{All signs changed} \\ -m^2 & - 15m & + 13 & \text{Add in columns.} \end{array}$$

5.2 EXERCISES

Write each polynomial in descending powers of the variable. See Example 1.

1. $2x^3 + x - 3x^2 + 4$

2. $3y^2 + y^4 - 2y^3 + y$

3. $4p^3 - 8p^5 + p^7$

4. $q^2 + 3q^4 - 2q + 1$

5. $-m^3 + 5m^2 + 3m^4 + 10$

6. $4 - x + 3x^2$

Identify each polynomial as a monomial, binomial, trinomial, or none of these. Also give the degree. See Example 2.

7. 25

8. 5

9. $7m - 22$

10. $-x^2 + 6x^5$

11. $2r^3 + 4r^2 + 5r$

12. $5z^2 - 6z + 7$

13. $-6p^4q - 3p^3q^2 + 2pq^3 - q^4$

14. $8s^3t - 4s^2t^2 + 2st^3 + 9$

15. Which one of the following is a trinomial in descending powers, having degree 6?
(a) $5x^6 - 4x^5 + 12$ **(b)** $6x^5 - x^6 + 4$
(c) $2x + 4x^2 - x^6$ **(d)** $4x^6 - 6x^4 + 9x^2 - 8$

16. Give an example of a polynomial of four terms in the variable x, having degree 5, written in descending powers, lacking a fourth degree term.

Give the numerical coefficient and the degree of each term. See Example 2.

17. $7z$

18. $3r$

19. $-15p^2$

20. $-27k^3$

21. x^4

22. y^6

23. $-mn^5$

24. $-a^5b$

Combine terms. See Examples 3–6.

25. $5z^4 + 3z^4$

26. $8r^5 - 2r^5$

27. $-m^3 + 2m^3 + 6m^3$

28. $3p^4 + 5p^4 - 2p^4$

29. $x + x + x + x + x$

30. $z - z - z + z$

31. $y^2 + 7y - 4y^2$

32. $2c^2 - 4 + 8 - c^2$

33. $2k + 3k^2 + 5k^2 - 7$

34. $4x^2 + 2x - 6x^2 - 6$

35. $n^4 - 2n^3 + n^2 - 3n^4 + n^3$

36. $2q^3 + 3q^2 - 4q - q^3 + 5q^2$

37. $[4 - (2 + 3m)] + (6m + 9)$

38. $[8a - (3a + 4)] - (5a - 3)$

39. $[(6 + 3p) - (2p + 1)] - (2p + 9)$

40. $[(4x - 8) - (-1 + x)] - (11x + 5)$

41. $(3p^2 + 2p - 5) + (7p^2 - 4p^3 + 3p)$

42. $(y^3 + 3y + 2) + (4y^3 - 3y^2 + 2y - 1)$

43. $(2x^5 - 2x^4 + x^3 - 1) + (x^4 - 3x^3 + 2)$

44. $(y^2 + 3y) + [2y - (5y^2 + 3y + 4)]$

45. $(9a - 5a) - [2a - (4a + 3)]$

46. Define *polynomial* in your own words. Give examples. Include the words *term, monomial, binomial,* and *trinomial* in your explanation.

47. Write a paragraph explaining how to add and subtract polynomials. Give examples.

48. Consider the exponent in the expression 3^6. Explain why the degree of 3^6 is not 6. What is its degree?

RELATING CONCEPTS (EXERCISES 49–54)

Our modern system of numeration, the Hindu-Arabic system, is a place value system that uses ten as its base. For example, the numeral 4353 means 4 thousands, 3 hundreds, 5 tens, and 3 ones. The two 3s in the numeral have different meanings because they appear in different place values.

A whole number in the Hindu-Arabic system can be written in expanded form using powers of ten. In the case of 4353, we have

$$(4 \times 10^3) + (3 \times 10^2) + (5 \times 10^1) + (3 \times 10^0).$$

If we let $x = 10$ in the polynomial $4x^3 + 3x^2 + 5x + 3$, we get 4353. Thus we see that our usual method of writing whole numbers is just a specific case of writing a polynomial in descending powers of the variable.

Work Exercises 49–54 in order.

49. Write 241 in expanded form.

50. If you were to add 4353 to 241 by hand, using a vertical format, why would the following *not* lead to the correct answer?

$$\begin{array}{r} 4353 \\ +\ 241 \\ \hline \end{array}$$

51. Explain the rule you learned years ago about how to set up an addition problem of whole numbers vertically.

52. If you were to add $4x^3 + 3x^2 + 5x + 3$ to $2x^2 + 4x + 1$ using a vertical format, why would the following *not* lead to the correct answer?

$$\begin{array}{r} 4x^3 + 3x^2 + 5x + 3 \\ +\ 2x^2 + 4x\ \ + 1 \\ \hline \end{array}$$

53. Explain the rule you learned in this section about how to add polynomials vertically.

54. While we recognize that $4x^2 + 7x + 3$ and $7x + 3 + 4x^2$ represent the same expression, we also recognize that 473 and 734 represent two different numbers. Comment on why this is so.

Did you make the connection that adding Hindu-Arabic numerals is a specific case of adding polynomials, with base ten replacing the variable?

Add or subtract as indicated. See Examples 4(b), 5(b), and 6(b).

55. Add.
$$\begin{array}{r} 21p - 8 \\ -9p + 4 \\ \hline \end{array}$$

56. Add.
$$\begin{array}{r} 15m - 9 \\ 4m + 12 \\ \hline \end{array}$$

57. Add.
$$\begin{array}{r} -12p^2 + 4p - 1 \\ 3p^2 + 7p - 8 \\ \hline \end{array}$$

58. Add.
$$\begin{array}{r} -6y^3 + 8y + 5 \\ 9y^3 + 4y - 6 \\ \hline \end{array}$$

59. Subtract.
$$\begin{array}{r} 6m^2 - 11m + 5 \\ -8m^2 + 2m - 1 \\ \hline \end{array}$$

60. Subtract.
$$\begin{array}{r} -4z^2 + 2z - 1 \\ 3z^2 - 5z + 2 \\ \hline \end{array}$$

61. Subtract.
$$\begin{array}{r} 5q^3 - 5q + 2 \\ -3q^3 + 2q - 9 \\ \hline \end{array}$$

62. Subtract.
$$\begin{array}{l} 6y^3 - 9y^2 \qquad + 8 \\ 4y^3 + 2y^2 + 5y \\ \hline \end{array}$$

Perform the operations. See Examples 3–6.

63. Subtract $4y^2 - 2y + 3$ from $7y^2 - 6y + 5$.

64. Subtract $-(-4x + 2z^2 + 3m)$ from $[(2z^2 - 3x + m) + (z^2 - 2m)]$.

65. $(-4m^2 + 3n^2 - 5n) - [(3m^2 - 5n^2 + 2n) + (-3m^2) + 4n^2]$

66. $[-(4m^2 - 8m + 4m^3) - (3m^2 + 2m + 5m^3)] + m^2$

67. $[-(y^4 - y^2 + 1) - (y^4 + 2y^2 + 1)] + (3y^4 - 3y^2 - 2)$

68. $(2p - [3p - 6]) - [(5p - (8 - 9p)) + 4p]$

69. $-(3z^2 + 5z - [2z^2 - 6z]) + [(8z^2 - [5z - z^2]) + 2z^2]$

70. $5k - (5k - [2k - (4k - 8k)]) + 11k - (9k - 12k)$

5.3 Polynomial Functions

OBJECTIVES

1. Recognize and evaluate polynomial functions.
2. Use a polynomial function to model data.
3. Add and subtract polynomial functions.
4. Graph basic polynomial functions.

FOR EXTRA HELP

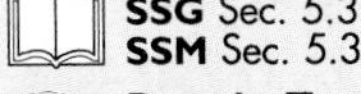
SSG Sec. 5.3
SSM Sec. 5.3

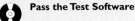

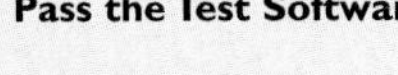
Pass the Test Software

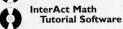

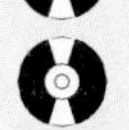
InterAct Math Tutorial Software

Video 8

OBJECTIVE 1 Recognize and evaluate polynomial functions. In Chapter 3 we studied linear (first-degree polynomial) functions, defined as $f(x) = mx + b$. Now we consider more general polynomial functions.

Polynomial Functions

A **polynomial function of degree *n*** is defined by

$$f(x) = a_n x^n + a_{n-1}x^{n-1} + \cdots + a_1 x + a_0,$$

for real numbers $a_n, a_{n-1}, \ldots, a_1$, and a_0, where $a_n \neq 0$ and n is a whole number.

Another way of describing a polynomial function is to say that it is a function defined by a polynomial in one variable, consisting of one or more terms. It is usually written in descending powers of the variable, and its degree is the degree of the polynomial that defines it.

As an example, suppose that we consider the polynomial $3x^2 - 5x + 7$, so that

$$f(x) = 3x^2 - 5x + 7.$$

If $x = -2$, then $f(x) = 3x^2 - 5x + 7$ takes on the value

$$\begin{aligned} f(-2) &= 3(-2)^2 - 5(-2) + 7 \qquad \text{Let } x = -2. \\ &= 3 \cdot 4 + 10 + 7 \\ &= 29. \end{aligned}$$

EXAMPLE 1 Evaluating a Polynomial Function

Let $f(x) = 4x^3 - x^2 + 5$. Find each of the following.

(a) $f(3)$

First, substitute 3 for x.

$$f(x) = 4x^3 - x^2 + 5$$
$$f(3) = 4 \cdot 3^3 - 3^2 + 5$$

Now use the order of operations from Chapter 1.

$$\begin{aligned} f(3) &= 4 \cdot 27 - 9 + 5 \\ &= 108 - 9 + 5 \\ &= 104 \end{aligned}$$

(b) $f(-4) = 4 \cdot (-4)^3 - (-4)^2 + 5$ Let $x = -4$.

$$= 4 \cdot (-64) - 16 + 5$$
$$= -267$$

While f is the most common letter used to represent functions, recall that other letters such as g and h are also frequently used. The capital letter P is often used for polynomial functions. Note that the function defined as $P(x) = 4x^3 - x^2 + 5$ yields the same ordered pairs as the function f in Example 1.

CONNECTIONS

The polynomial function defined by

$$f(x) = \frac{1}{24}x^4 - \frac{1}{4}x^3 + \frac{23}{24}x^2 - \frac{3}{4}x + 1$$

will give the number of interior regions formed in a circle if x points on the circumference are joined by all possible chords, for $x = 1, 2, 3, 4$, and 5. See Figure 1.

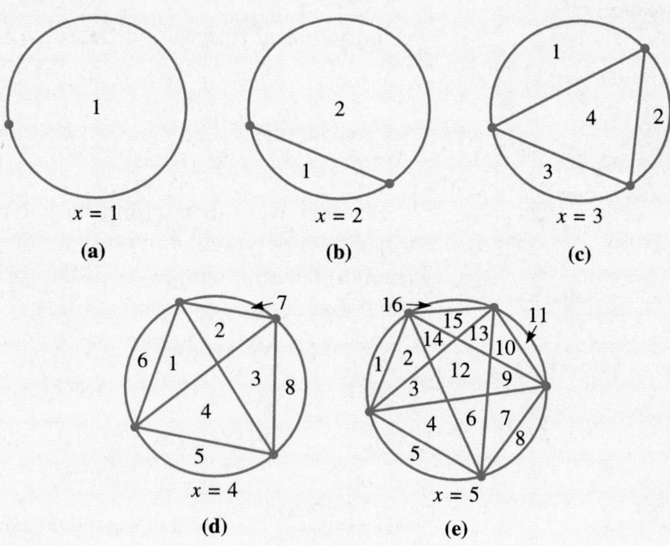

Figure 1

CONNECTIONS (CONTINUED)

For example, in (a) we have 1 point and since no chords can be drawn, we have only 1 interior region. In (b) there are 2 points and 2 interior regions are formed.

FOR DISCUSSION OR WRITING

1. Verify that $f(1) = 1$ and $f(2) = 2$.
2. Based on the appropriate figure alone, what should be the value of $f(3)$? $f(4)$? $f(5)$?
3. Verify your answers in Item 2 by evaluating $f(3)$, $f(4)$, and $f(5)$.
4. Observe a pattern in the results of the first three items. Use the pattern to predict $f(6)$. Does your prediction equal $f(6)$?

OBJECTIVE 2 **Use a polynomial function to model data.** Polynomial functions can be used to approximate data. They are usually valid for small intervals, and they allow us to predict (with caution) what might happen for values just outside the intervals. These intervals are often periods of years, as shown in Example 2.

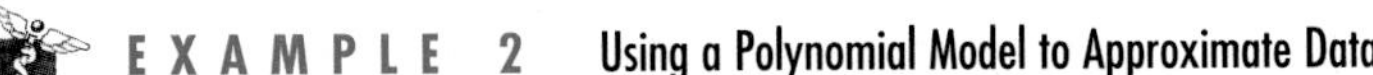

EXAMPLE 2 **Using a Polynomial Model to Approximate Data**

Average pay for registered nurses during the years 1990–1996 can be modeled by the polynomial function with

$$P(x) = -128.57x^2 + 2000x + 30{,}100,$$

where $x = 0$ corresponds to the year 1990, $x = 1$ corresponds to 1991, and so on, and $P(x)$ is in dollars. Use this function to approximate the pay for the year 1991. (*Source:* Watson Wyandotte Worldwide.)

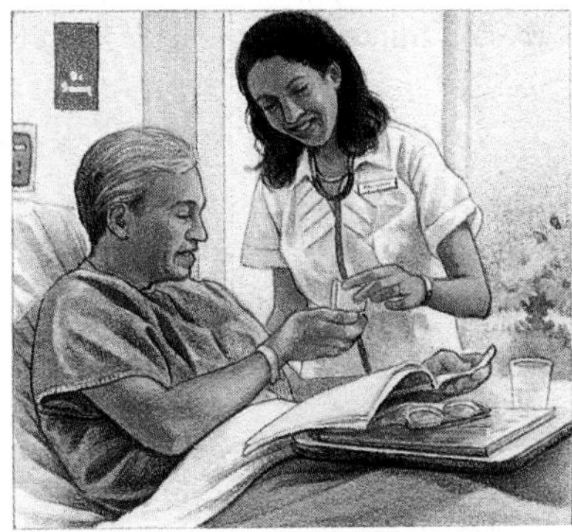

Since $x = 1$ corresponds to 1991, we must find $P(1)$:

$$\begin{aligned} P(1) &= -128.57(1)^2 + 2000(1) + 30{,}100 && \text{Let } x = 1. \\ &= 31{,}971.43. && \text{Evaluate.} \end{aligned}$$

Therefore, in 1991 a registered nurse earned on the average approximately \$32,000. ■

OBJECTIVE 3 **Add and subtract polynomial functions.** You are familiar with the operations of arithmetic for real numbers. The operations of addition, subtraction, multiplication, and division are also defined for functions. We now consider addition and subtraction of functions, using polynomial functions as examples.

Addition and Subtraction of Functions

If $f(x)$ and $g(x)$ define functions, their sum and difference are found as follows.

$$(f+g)(x) = f(x) + g(x)$$
$$(f-g)(x) = f(x) - g(x)$$

In both cases, the domain of the new function is the intersection of the domains of $f(x)$ and $g(x)$.

EXAMPLE 3 Adding and Subtracting Polynomial Functions

For $f(x) = x^2 - 3x + 7$ and $g(x) = -3x^2 - 7x + 7$, find **(a)** the sum and **(b)** the difference.

(a) $(f+g)(x) = f(x) + g(x)$

$= (x^2 - 3x + 7) + (-3x^2 - 7x + 7)$ Use the definition.

$= -2x^2 - 10x + 14$ Add the polynomials.

(b) $(f-g)(x) = f(x) - g(x)$

$= (x^2 - 3x + 7) - (-3x^2 - 7x + 7)$ Use the definition.

$= (x^2 - 3x + 7) + (3x^2 + 7x - 7)$ Change subtraction to addition.

$= 4x^2 + 4x$ Add.

OBJECTIVE 4 Graph basic polynomial functions. Functions were introduced in Section 3.5. Recall that each input (or x-value) of a function results in one output (or y-value). The simplest polynomial function is the **identity function,** defined by $f(x) = x$. The domain (set of x-values) of this function is all real numbers, $(-\infty, \infty)$, and it pairs each real number with itself. Therefore, the range (set of y-values) is also $(-\infty, \infty)$. Its graph is a straight line, as first seen in Chapter 3. (Notice that a *linear function* is a specific kind of polynomial function.) Figure 2 shows its graph. A table of selected ordered pairs is also given.

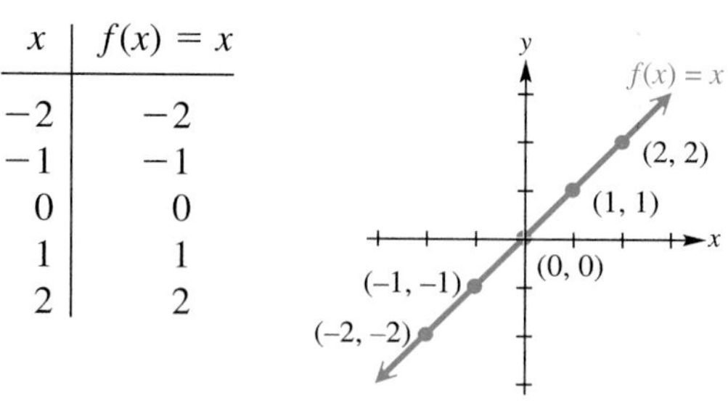

x	$f(x) = x$
-2	-2
-1	-1
0	0
1	1
2	2

Figure 2

Another polynomial function, defined by $f(x) = x^2$, is the **squaring function.** For this function, every real number is paired with its square. The input can be any real number, so the domain is $(-\infty, \infty)$. Since the square of any real number is nonnegative, the range is $[0, \infty)$. Its graph is a *parabola.* In later chapters we will investigate parabolas in more detail. Figure 3 shows the graph. Again, a table of selected ordered pairs is given.

x	$f(x) = x^2$
-2	4
-1	1
0	0
1	1
2	4

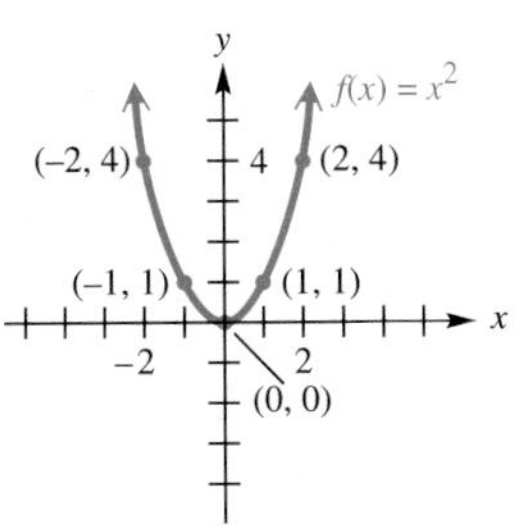

Figure 3

Another type of polynomial function, the **cubing function,** is defined by $f(x) = x^3$. Every real number is paired with its cube. The domain and the range are both $(-\infty, \infty)$. Its graph is neither a line nor a parabola. See Figure 4 and the table of ordered pairs. (Polynomial functions of degree 3 and greater are studied in detail in more advanced courses.)

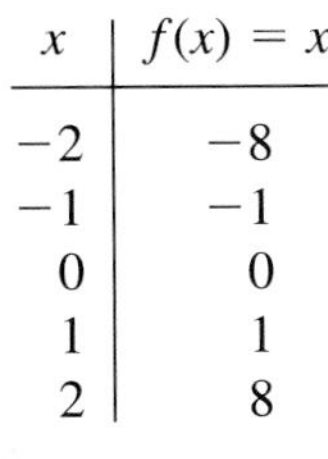

x	$f(x) = x^3$
-2	-8
-1	-1
0	0
1	1
2	8

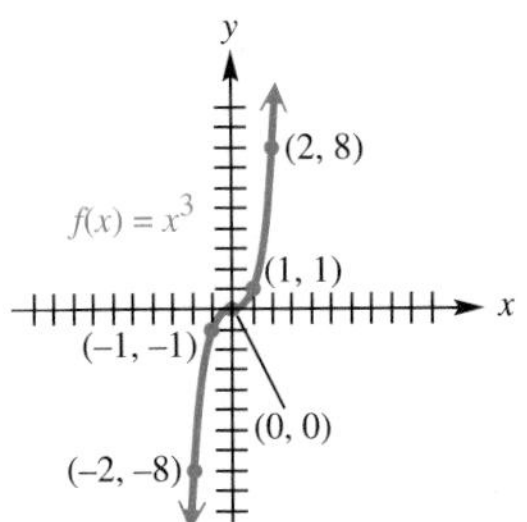

Figure 4

Example 4 shows how variations of these three polynomial functions can be graphed by plotting points.

EXAMPLE 4 Graphing Variations of the Identity, Squaring, and Cubing Functions

Graph each function by creating a table of ordered pairs. Give the domain and the range of each function by observing the graphs.

(a) $f(x) = 2x$

To find each range value, multiply the domain value by 2. Plot the points and join them with a straight line. See Figure 5. Both the domain and the range are $(-\infty, \infty)$.

x	$f(x) = 2x$
-2	-4
-1	-2
0	0
1	2
2	4

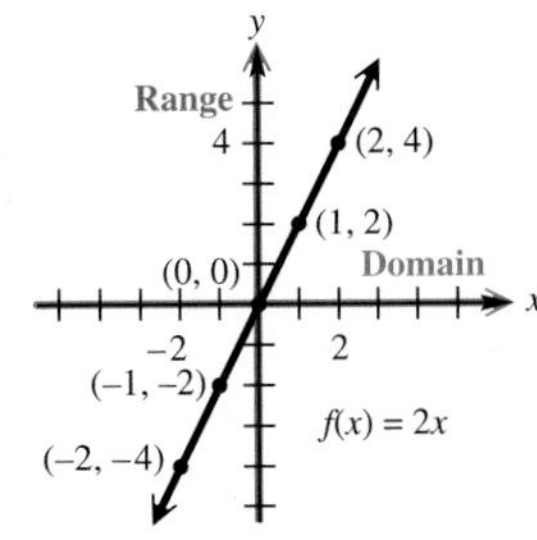

Figure 5

(b) $f(x) = -x^2$

For each input x, square it and then take its opposite. Plotting the points gives a parabola that opens downward. See the table and Figure 6. The domain is $(-\infty, \infty)$ and the range is $(-\infty, 0]$.

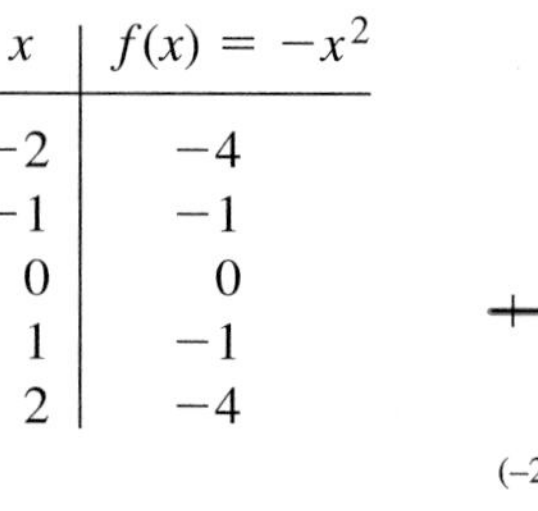

x	$f(x) = -x^2$
-2	-4
-1	-1
0	0
1	-1
2	-4

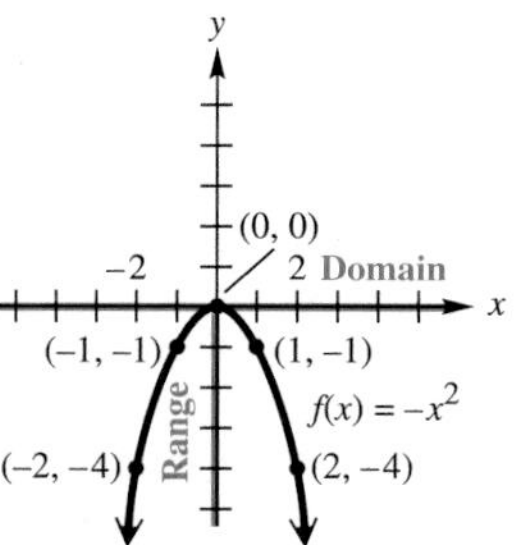

Figure 6

(c) $f(x) = x^3 - 2$

For this function, we cube the input and then subtract 2 from the result. The graph is that of the cubing function shifted 2 units downward. See the table and Figure 7. The domain and the range are both $(-\infty, \infty)$.

x	$f(x) = x^3 - 2$
-2	-10
-1	-3
0	-2
1	-1
2	6

Figure 7

Figure 8 shows how a graphing calculator plots the identity, squaring, and cubing functions in the standard window.

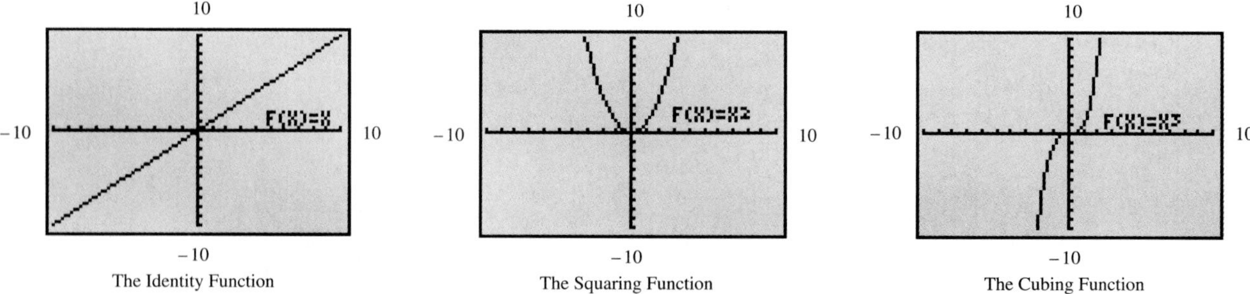

The Identity Function

The Squaring Function

The Cubing Function

Figure 8

5.3 EXERCISES

*For each polynomial function, find **(a)** $f(-1)$ and **(b)** $f(2)$. See Example 1.*

1. $f(x) = 6x - 4$

2. $f(x) = -2x + 5$

3. $f(x) = x^2 - 3x + 4$

4. $f(x) = 3x^2 + x - 5$

5. $f(x) = 5x^4 - 3x^2 + 6$

6. $f(x) = -4x^4 + 2x^2 - 1$

7. $f(x) = -x^2 + 2x^3 - 8$

8. $f(x) = -x^2 - x^3 + 11x$

TECHNOLOGY INSIGHTS (EXERCISES 9–12)

Graphing calculators are programmed to use function notation. For example, the left screen here shows $f(x) = -3x^2 + 2x - 1$ defined by Y_1, while the right screen shows that $f(2) = Y_1(2) = -9$.

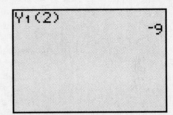

Four functions, defined by Y_1, Y_2, Y_3, and Y_4 are given in the following screen.

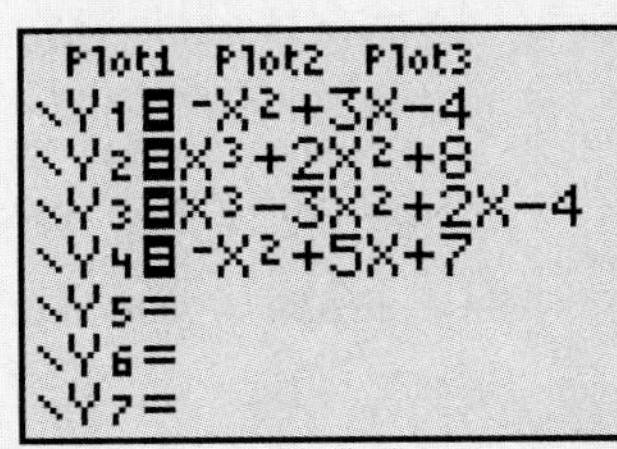

Use this screen to predict the calculator display for each of the following.

9. $Y_1(-3)$ **10.** $Y_2(4)$ **11.** $Y_3(6)$ **12.** $Y_4(-5)$

Solve each problem. See Example 2.

13. From 1980–1995, payments in millions of dollars for Medicaid can be modeled by the polynomial function with

$$f(x) = 411x^2 + 194x + 24{,}050,$$

where $x = 0$ corresponds to 1980. Approximate the amount of payments for each of the following years. (*Source:* U.S. Health Care Financing Administration.)
(a) 1980 **(b)** 1985 **(c)** 1990 **(d)** 1995

14. The number of recipients, in thousands, of Medicaid for the years 1980–1995 can be modeled by the polynomial function with

$$f(x) = 108x^2 - 673x + 21{,}823,$$

where $x = 0$ corresponds to 1980. Approximate the number of recipients for each of the following years. (*Source:* U.S. Health Care Financing Administration.)
(a) 1980 **(b)** 1985 **(c)** 1990 **(d)** 1995

15. The number of people, in thousands, enrolled in Health Maintenance Organizations (HMOs) during the period 1990–1995 can be modeled by the polynomial function with

$$f(x) = .39x^2 + .71x + 33.0,$$

where $x = 0$ corresponds to 1990, $x = 1$ corresponds to 1991, and so on. Use this model to approximate the number of people enrolled in each of the following years. (*Sources:* Interstudy and U.S. National Center for Health Statistics.)

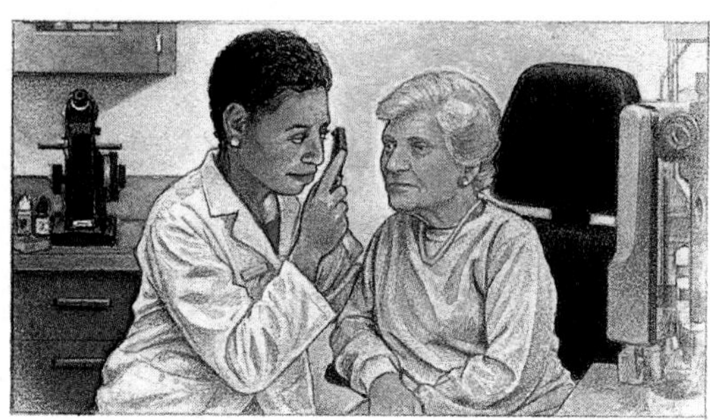

(a) 1990 **(b)** 1991 **(c)** 1992
(d) 1993 **(e)** 1994 **(f)** 1995

16. The number of medical doctors, in thousands, in the United States during the period 1990–1995 can be modeled by the polynomial function with

$$f(x) = 1.23x^2 + 13.9x + 616.7,$$

where $x = 0$ corresponds to 1990, $x = 1$ corresponds to 1991, and so on. Use this model to approximate the number of doctors in each of the following years. (*Source:* American Medical Association.)
(a) 1990 **(b)** 1991 **(c)** 1992
(d) 1993 **(e)** 1994 **(f)** 1995

*For each pair of functions, find **(a)** $(f + g)(x)$ and **(b)** $(f - g)(x)$. See Example 3.*

17. $f(x) = 5x - 10$, $g(x) = 3x + 7$

18. $f(x) = -4x + 1$, $g(x) = 6x + 2$

19. $f(x) = 4x^2 + 8x - 3$, $g(x) = -5x^2 + 4x - 9$

20. $f(x) = 3x^2 - 9x + 10$, $g(x) = -4x^2 + 2x + 12$

21. Construct two polynomial functions defined by $f(x)$, a polynomial of degree 3, and $g(x)$, a polynomial of degree 4. Find $(f - g)(x)$ and $(g - f)(x)$. Use your answers to decide whether subtraction of polynomial functions is a commutative operation.

22. Make up two polynomial functions defined by $f(x)$ and $g(x)$ such that $(f + g)(x) = 3x^3 - x + 3$.

Graph each function by creating a table of ordered pairs. Give the domain and the range. See Example 4.

23. $f(x) = -2x + 1$

24. $f(x) = 3x + 2$

25. $f(x) = -3x^2$

26. $f(x) = \frac{1}{2}x^2$

27. $f(x) = x^3 + 1$

28. $f(x) = -x^3 + 2$

5.4 Multiplication of Polynomials

OBJECTIVES

1. Multiply terms.
2. Multiply any two polynomials.
3. Multiply binomials.
4. Find the product of the sum and difference of two terms.
5. Find the square of a binomial.
6. Multiply polynomial functions.

FOR EXTRA HELP

SSG Sec. 5.4
SSM Sec. 5.4

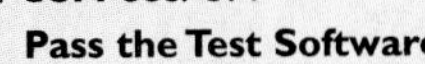
Pass the Test Software

InterAct Math Tutorial Software

Video 9

OBJECTIVE 1 **Multiply terms.** Recall that the product of the two terms $3x^4$ and $5x^3$ is found by using the commutative and associative properties, along with the rules for exponents.

$$\begin{aligned}(3x^4)(5x^3) &= 3 \cdot 5 \cdot x^4 \cdot x^3\\ &= 15x^{4+3}\\ &= 15x^7\end{aligned}$$

EXAMPLE 1 **Multiplying Monomials**

Find the following products.

(a) $(-4a^3)(3a^5) = (-4)(3)a^3 \cdot a^5 = -12a^8$

(b) $(2m^2z^4)(8m^3z^2) = (2)(8)m^2 \cdot m^3 \cdot z^4 \cdot z^2 = 16m^5z^6$

OBJECTIVE 2 **Multiply any two polynomials.** The distributive property can be used to extend this process to find the product of any two polynomials.

EXAMPLE 2 **Multiplying Polynomials**

Find the following products.

(a) $-2(8x^3 - 9x^2)$

Use the distributive property.

$$\begin{aligned}-2(8x^3 - 9x^2) &= -2(8x^3) - 2(-9x^2)\\ &= -16x^3 + 18x^2\end{aligned}$$

(b) $$\begin{aligned}5x^2(-4x^2 + 3x - 2) &= 5x^2(-4x^2) + 5x^2(3x) + 5x^2(-2)\\ &= -20x^4 + 15x^3 - 10x^2\end{aligned}$$

(c) $(3x - 4)(2x^2 + x)$

Use the distributive property to multiply each term of $2x^2 + x$ by $3x - 4$.

$$(3x - 4)(2x^2 + x) = (3x - 4)(2x^2) + (3x - 4)(x)$$

Here $3x - 4$ has been treated as a single expression so that the distributive property could be used. Now use the distributive property two more times.

$$\begin{aligned}&= 3x(2x^2) + (-4)(2x^2) + (3x)(x) + (-4)(x)\\ &= 6x^3 - 8x^2 + 3x^2 - 4x\\ &= 6x^3 - 5x^2 - 4x\end{aligned}$$

(d) $$\begin{aligned}2x^2(x + 1)(x - 3) &= 2x^2[(x + 1)(x) + (x + 1)(-3)]\\ &= 2x^2[x^2 + x - 3x - 3]\\ &= 2x^2(x^2 - 2x - 3)\\ &= 2x^4 - 4x^3 - 6x^2\end{aligned}$$

It is often easier to multiply polynomials by writing them vertically. To find the product from Example 2(c), $(3x - 4)(2x^2 + x)$, vertically, proceed as follows. (Notice how this process is similar to that of finding the product of two numbers, such as 24×78.)

1. Multiply x and $3x - 4$.

$$\begin{array}{rr} & 3x \quad - 4 \\ & 2x^2 + \quad x \\ \hline x(3x-4) \rightarrow & 3x^2 - 4x \end{array}$$

2. Multiply $2x^2$ and $3x - 4$. Line up like terms of the products in columns.

3. Combine like terms.

$$\begin{array}{rrrr} & & 3x & -4 \\ & & 2x^2 & + x \\ \hline & & 3x^2 & -4x \\ 2x^2(3x-4) \rightarrow & 6x^3 & -8x^2 & \\ \hline & 6x^3 & -5x^2 & -4x \end{array}$$

EXAMPLE 3 Multiplying Polynomials Vertically

Find the product vertically.

(a) $(5a - 2b)(3a + b)$

$$\begin{array}{rrrl} & 5a & -2b & \\ & 3a & +b & \\ \hline & 5ab & -2b^2 & \leftarrow b(5a-2b) \\ 15a^2 & -6ab & & \leftarrow 3a(5a-2b) \\ \hline 15a^2 & -ab & -2b^2 & \end{array}$$

(b) $(3m^3 - 2m^2 + 4)(3m - 5)$

$$\begin{array}{rrrrrl} & 3m^3 & -2m^2 & & +4 & \\ & & & 3m & -5 & \\ \hline & -15m^3 & +10m^2 & & -20 & -5 \text{ times } 3m^3 - 2m^2 + 4 \\ 9m^4 & -6m^3 & & +12m & & 3m \text{ times } 3m^3 - 2m^2 + 4 \\ \hline 9m^4 & -21m^3 & +10m^2 & +12m & -20 & \text{Combine like terms.} \end{array}$$

OBJECTIVE 3 Multiply binomials. When working with polynomials, the product of two binomials occurs repeatedly. There is a shortcut method for finding these products. Recall that a binomial is a polynomial with just two terms, such as $3x - 4$ or $2x + 3$. We can find the product of these binomials using the distributive property as follows.

$$\begin{aligned} (3x-4)(2x+3) &= 3x(2x+3) - 4(2x+3) \\ &= 3x(2x) + 3x(3) - 4(2x) - 4(3) \\ &= 6x^2 + 9x - 8x - 12 \end{aligned}$$

Before combining like terms to find the simplest form of the answer, let us check the origin of each of the four terms in the sum. First, $6x^2$ is the product of the two *first* terms.

$$(3x - 4)(2x + 3) \qquad 3x(2x) = 6x^2 \qquad \text{First terms}$$

To get $9x$, the *outside* terms are multiplied.

$$(3x - 4)(2x + 3) \qquad 3x(3) = 9x \qquad \text{Outside terms}$$

The term $-8x$ comes from the *inside* terms.

$$(3x - 4)(2x + 3) \qquad -4(2x) = -8x \qquad \text{Inside terms}$$

Finally, -12 comes from the *last* terms.

$$(3x - 4)(2x + 3) \qquad -4(3) = -12 \qquad \text{Last terms}$$

The product is found by combining these four results.

$$\begin{aligned}(3x - 4)(2x + 3) &= 6x^2 + 9x - 8x - 12\\ &= 6x^2 + x - 12\end{aligned}$$

To keep track of the order of multiplying these terms, we use the initials FOIL (First, Outside, Inside, Last). All the steps of the FOIL method can be done as follows. Try to do as many of these steps as possible in your head.

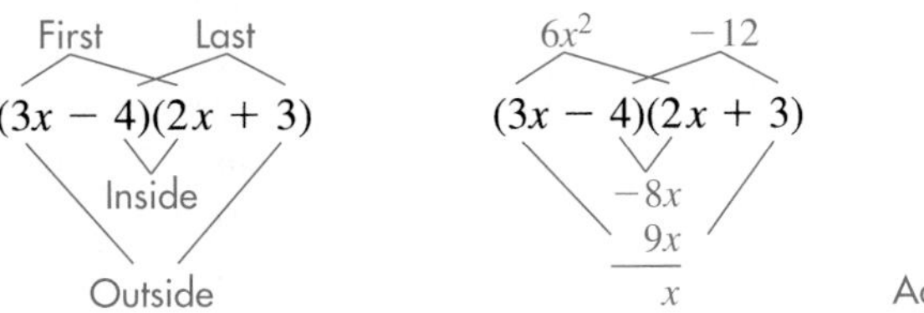

EXAMPLE 4 Using the FOIL Method

Use the FOIL method to find $(4m - 5)(3m + 1)$.

Find the product of the first terms.

$$(4m - 5)(3m + 1) \qquad 4m(3m) = \mathbf{12m^2}$$

Multiply the outside terms.

$$(4m - 5)(3m + 1) \qquad 4m(1) = \mathbf{4m}$$

Find the product of the inside terms.

$$(4m - 5)(3m + 1) \qquad -5(3m) = \mathbf{-15m}$$

Multiply the last terms.

$$(4m - 5)(3m + 1) \qquad -5(1) = \mathbf{-5}$$

Simplify by combining the four terms obtained above.

$$\begin{aligned}(4m - 5)(3m + 1) &= 12m^2 + 4m - 15m - 5\\ &= 12m^2 - 11m - 5\end{aligned}$$

The procedure can be written in compact form as follows.

$12m^2$ $\quad -5$

$(4m - 5)(3m + 1)$

$-15m$

$4m$

$-11m$ Add.

Combine these four results to get $12m^2 - 11m - 5$.

EXAMPLE 5 Using the FOIL Method

Use the FOIL method to multiply the binomials.

(a) $(6a - 5b)(3a + 4b) = 18a^2 + 24ab - 15ab - 20b^2$

$\uparrow$ First $\uparrow$ Outside $\uparrow$ Inside $\uparrow$ Last

$= 18a^2 + 9ab - 20b^2$

(b) $(2k + 3z)(5k - 3z) = 10k^2 - 6kz + 15kz - 9z^2$

$= 10k^2 + 9kz - 9z^2$

OBJECTIVE 4 **Find the product of the sum and difference of two terms.** Some types of binomial products occur frequently. By the FOIL method, the product of the sum and difference of the same two terms $(x + y)(x - y)$ is

$$(x + y)(x - y) = x^2 - xy + xy - y^2$$
$$= x^2 - y^2.$$

Product of the Sum and Difference of Two Terms

The **product of the sum and difference of two terms** is the difference of the squares of the terms, or

$$(x + y)(x - y) = x^2 - y^2.$$

EXAMPLE 6 Multiplying the Sum and Difference of Two Terms

Find the following products.

(a) $(p + 7)(p - 7) = p^2 - 7^2 = p^2 - 49$

(b) $(2r + 5)(2r - 5) = (2r)^2 - 5^2$

$= 2^2r^2 - 25$

$= 4r^2 - 25$

(c) $(6m + 5n)(6m - 5n) = (6m)^2 - (5n)^2$

$= 36m^2 - 25n^2$

(d) $2x^3(x + 3)(x - 3) = 2x^3(x^2 - 9) = 2x^5 - 18x^3$

The special product $(a + b)(a - b) = a^2 - b^2$ can be used to perform some multiplication problems. For example,

$$51 \times 49 = (50 + 1)(50 - 1)$$
$$= 50^2 - 1^2$$
$$= 2500 - 1$$
$$= 2499.$$

Once these patterns are recognized, multiplications like this can be done mentally.

OBJECTIVE 5 **Find the square of a binomial.** Another special binomial product is the *square of a binomial.* To find the square of $x + y$, or $(x + y)^2$, multiply $x + y$ and $x + y$.

$$(x + y)^2 = (x + y)(x + y) = x^2 + xy + xy + y^2$$
$$= x^2 + 2xy + y^2$$

A similar result is true for the square of a difference.

Square of a Binomial

The **square of a binomial** is the sum of the square of the first term, twice the product of the two terms, and the square of the last term.

$$(x + y)^2 = x^2 + 2xy + y^2$$
$$(x - y)^2 = x^2 - 2xy + y^2$$

EXAMPLE 7 **Squaring a Binomial**

Find the following products.

(a) $(m + 7)^2 = m^2 + 2 \cdot m \cdot 7 + 7^2$
$= m^2 + 14m + 49$

(b) $(p - 5)^2 = p^2 - 2 \cdot p \cdot 5 + 5^2$
$= p^2 - 10p + 25$

(c) $(2p + 3v)^2 = (2p)^2 + 2(2p)(3v) + (3v)^2$
$= 4p^2 + 12pv + 9v^2$

(d) $(3r - 5s)^2 = (3r)^2 - 2(3r)(5s) + (5s)^2$
$= 9r^2 - 30rs + 25s^2$

As the products in the definition of the square of a binomial show,

$$(x + y)^2 \neq x^2 + y^2.$$

More generally,

$$(x + y)^n \neq x^n + y^n.$$

CONNECTIONS

The special product

$$(a + b)^2 = a^2 + 2ab + b^2$$

can be illustrated geometrically using the diagram shown on the next page. The side of the large square has length $a + b$, so the area of the square is

$$(a + b)^2.$$

(continued)

CONNECTIONS (CONTINUED)

	a	b
a	Area: a^2	Area: ab
b	Area: ab	Area: b^2

The large square is made up of two smaller squares and two congruent rectangles. The sum of the areas of these figures is

$$a^2 + 2ab + b^2.$$

Since these expressions represent the same quantity, they must be equal, thus giving us the pattern for squaring a binomial.

FOR DISCUSSION OR WRITING

Draw a figure and give a similar proof for $(a - b)^2 = a^2 - 2ab + b^2$.

The special products of this section are now summarized.

Special Products

Product of the Sum and Difference of Two Terms	$(x + y)(x - y) = x^2 - y^2$
Square of a Binomial	$(x + y)^2 = x^2 + 2xy + y^2$ $(x - y)^2 = x^2 - 2xy + y^2$

We can use the patterns for the special products with more complicated products, as the following example shows.

EXAMPLE 8 Multiplying More Complicated Binomials

Use special products to multiply the following polynomials.

(a) $[(3p - 2) + 5q][(3p - 2) - 5q]$

$= (3p - 2)^2 - (5q)^2$ Product of sum and difference of terms

$= 9p^2 - 12p + 4 - 25q^2$ Square both quantities.

(b) $[(2z + r) + 1]^2 = (2z + r)^2 + 2(2z + r)(1) + 1^2$ Square of a binomial

$= 4z^2 + 4zr + r^2 + 4z + 2r + 1$ Square again; distributive property.

OBJECTIVE 6 Multiply polynomial functions. In the previous section we saw how functions can be added and subtracted. The product of two functions is defined as follows.

Multiplication of Functions

If $f(x)$ and $g(x)$ define functions, their product is found as follows.

$$(fg)(x) = f(x) \cdot g(x).$$

The domain of the product is the intersection of the domains of $f(x)$ and $g(x)$.

EXAMPLE 9 Multiplying Polynomial Functions

For $f(x) = 3x - 4$ and $g(x) = 2x^2 + x$, find the product.

From the explanation preceding Example 3, we see that the product of the two polynomials $f(x)$ and $g(x)$ is $6x^3 - 5x^2 - 4x$. Therefore,

$$(fg)(x) = 6x^3 - 5x^2 - 4x.$$

5.4 EXERCISES

Match each product on the left with the correct polynomial on the right.

1. $(2x - 5)(3x + 4)$ **A.** $6x^2 + 23x + 20$

2. $(2x + 5)(3x + 4)$ **B.** $6x^2 + 7x - 20$

3. $(2x - 5)(3x - 4)$ **C.** $6x^2 - 7x - 20$

4. $(2x + 5)(3x - 4)$ **D.** $6x^2 - 23x + 20$

Find each product. See Examples 1–5.

5. $(-2x^4)(4x^3)$ **6.** $(8y^4)(2y^8)$ **7.** $(14x^2y^3)(-2x^5y)$

8. $(-5m^3n^4)(4m^2n^5)$ **9.** $-5(3p^2 + 2p^3)$ **10.** $-6(5t^3 - 8t^4)$

11. $(3x + 7)(x - 4)$ **12.** $(4x - 7)(x + 2)$ **13.** $(2t + 3s)(3t - 2s)$

14. $(5r + 3u)(2r - 5u)$

15. $\begin{array}{r} 2y + 3 \\ \underline{3y - 4} \end{array}$

16. $\begin{array}{r} 5m - 3 \\ \underline{2m + 6} \end{array}$

17. $(5m - 3n)(5m + 3n)$ **18.** $(2k + 6q)(2k - 6q)$ **19.** $m(m + 5)(m - 8)$

20. $p(p - 6)(p + 4)$ **21.** $4z(2z + 1)(3z - 4)$ **22.** $2y(8y - 3)(2y + 1)$

23. $x^2(2x + 3)(2x - 3)$ **24.** $y^3(9y + 1)(9y - 1)$ **25.** $(2m + 3)(3m^2 - 4m - 1)$

26. $(4z - 2)(z^3 + 3z + 5)$

27. $\begin{array}{r} 6m^2 + 2m - 1 \\ \underline{2m + 3} \end{array}$

28. $\begin{array}{r} -y^2 + 2y + 1 \\ \underline{3y - 5} \end{array}$

29. $\begin{array}{r} 2z^3 - 5z^2 + 8z - 1 \\ \underline{4z + 3} \end{array}$

30. $\begin{array}{r} 3z^4 - 2z^3 + z - 5 \\ \underline{2z - 5} \end{array}$

31. $\begin{array}{r} -x^2 + 8x - 3 \\ \underline{2x^3 + 5x} \end{array}$

32. $\begin{array}{r} 2k^2 + 6k + 5 \\ \underline{-k^2 + 2k} \end{array}$

33. $\begin{array}{r} 2p^2 + 3p + 6 \\ \underline{3p^2 - 4p - 1} \end{array}$

34. $\begin{array}{r} 5y^2 - 2y + 4 \\ \underline{2y^2 + y + 3} \end{array}$

35. What type of polynomials can be multiplied by the FOIL method? Describe the method in your own words.

36. Make a list of special products. Give examples with solutions, and explain in your own words how to find these special products.

Find each product. See Example 6.

37. $(2p - 3)(2p + 3)$ **38.** $(3x - 8)(3x + 8)$ **39.** $(5m - 1)(5m + 1)$

40. $(6y + 3)(6y - 3)$ **41.** $(3a + 2c)(3a - 2c)$ **42.** $(5r - 4s)(5r + 4s)$

43. $\left(4x - \frac{2}{3}\right)\left(4x + \frac{2}{3}\right)$ **44.** $\left(3t + \frac{5}{4}\right)\left(3t - \frac{5}{4}\right)$ **45.** $(4m + 7n^2)(4m - 7n^2)$

46. $(2k^2 + 6h)(2k^2 - 6h)$ **47.** $(5y^3 + 2)(5y^3 - 2)$ **48.** $(3x^3 + 4)(3x^3 - 4)$

Find each square. See Example 7.

49. $(y - 5)^2$ **50.** $(a - 3)^2$ **51.** $(2p + 7)^2$

52. $(3z + 8)^2$ **53.** $(4n - 3m)^2$ **54.** $(5r - 7s)^2$

55. $\left(k - \frac{5}{7}p\right)^2$

56. $\left(q - \frac{3}{4}r\right)^2$

57. How do the expressions $(x + y)^2$ and $x^2 + y^2$ differ?

58. Find the product $101 \cdot 99$ using the special product $(a + b)(a - b) = a^2 - b^2$. (*Hint:* See the Note following Example 6.)

Find each product. See Example 8.

59. $[(5x + 1) + 6y]^2$

60. $[(3m - 2) + p]^2$

61. $[(2a + b) - 3]^2$

62. $[(4k + h) - 4]^2$

63. $[(2a + b) - 3][(2a + b) + 3]$

64. $[(m + p) + 5][(m + p) - 5]$

Find each product.

65. $(2a + b)(3a^2 + 2ab + b^2)$

66. $(m - 5p)(m^2 - 2mp + 3p^2)$

67. $(4z - x)(z^3 - 4z^2x + 2zx^2 - x^3)$

68. $(3r + 2s)(r^3 + 2r^2s - rs^2 + 2s^3)$

69. $(m^2 - 2mp + p^2)(m^2 + 2mp - p^2)$

70. $(3 + x + y)(-3 + x - y)$

71. $ab(a + b)(a + 2b)(a - 3b)$

72. $mp(m - p)(m - 2p)(2m + p)$

73. $(y + 2)^3$

74. $(z - 3)^3$

75. $(q - 2)^4$

76. $(r + 3)^4$

In Exercises 77–80, two expressions are given. Replace x with 3 and y with 4 to show that, in general, the two expressions do not equal each other.

77. $(x + y)^2$; $x^2 + y^2$

78. $(x + y)^3$; $x^3 + y^3$

79. $(x + y)^4$; $x^4 + y^4$

80. $(x + y)^5$; $x^5 + y^5$

81. A student claims that the two expressions in Exercise 77 *must* be equal to each other, because if we let $x = 0$ and $y = 1$, each expression simplifies to 1. How would you respond to the student's claim?

82. Which one of the following is equivalent to $(x + y)^{-2}$?

(a) $\frac{1}{(x + y)^{-2}}$ **(b)** $\frac{1}{x^{-2} + y^{-2}}$ **(c)** $x^{-2} + y^{-2}$ **(d)** $\frac{1}{x^2 + 2xy + y^2}$

Find the area of each figure. Express it as a polynomial in descending powers of the variable x.

83.

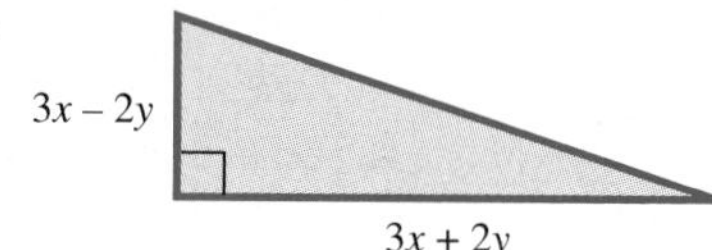

84.

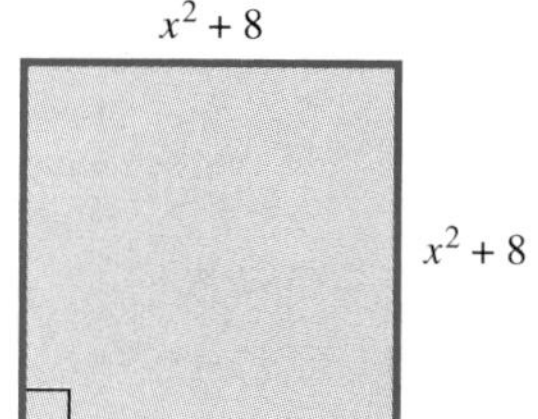

85.

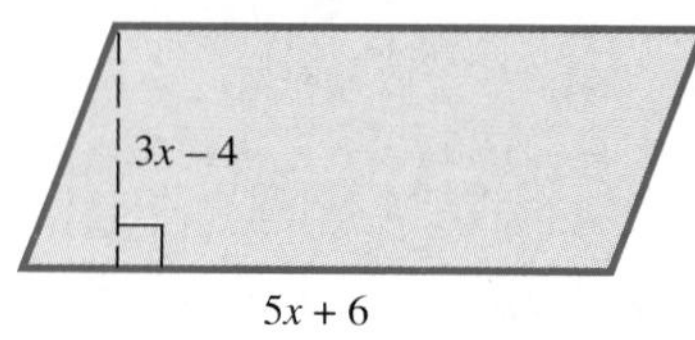

86.

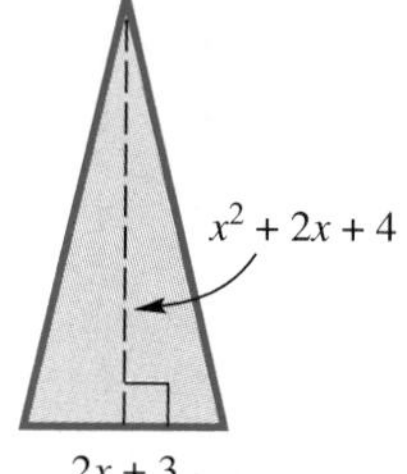

RELATING CONCEPTS (EXERCISES 87–94)

Consider the figure. **Work Exercises 87–94 in order.**

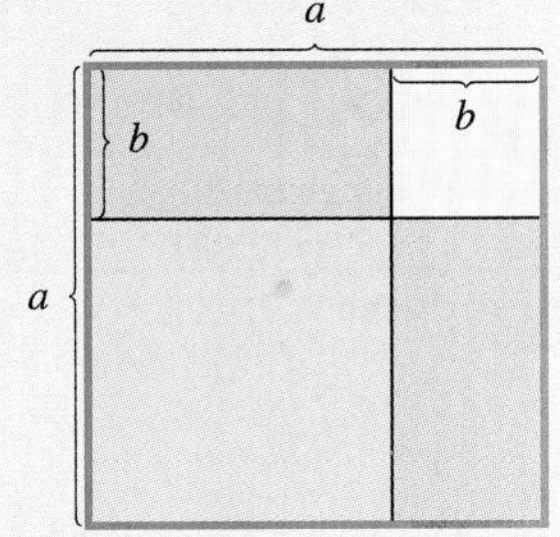

87. What is the length of each side of the blue square in terms of a and b?

88. What is the formula for the area of a square? Use the formula to write an expression, in the form of a product, for the area of the blue square.

89. Each green rectangle has an area of ______. Therefore, the total area in green is represented by the polynomial ______.

90. The yellow square has an area of ______.

91. The area of the entire colored region is represented by ______, because each side of the entire colored region has length ______.

92. The area of the blue square is equal to the area of the entire colored region minus the total area of the green squares minus the area of the yellow square. Write this as a simplified polynomial in a and b.

93. What must be true about the expressions for the area of the blue square you found in Exercises 88 and 92?

94. Write a statement of equality based on your answer in Exercise 93. How does this reinforce one of the main ideas of this section?

Did you make the connection between algebra and geometry that justifies the special product for the square of a binomial?

For each pair of functions, find the product $(fg)(x)$. See Example 9.

95. $f(x) = 2x, \quad g(x) = 5x - 1$

96. $f(x) = 3x, \quad g(x) = 6x - 8$

97. $f(x) = x + 1, \quad g(x) = 2x - 3$

98. $f(x) = x - 7, \quad g(x) = 4x + 5$

99. $f(x) = 2x - 3, \quad g(x) = 4x^2 + 6x + 9$

100. $f(x) = 3x + 4, \quad g(x) = 9x^2 - 12x + 16$

5.5 Greatest Common Factors; Factoring by Grouping

OBJECTIVES

1. Factor out the greatest common factor.
2. Factor by grouping.

Writing a polynomial as the product of two or more simpler polynomials is called **factoring** the polynomial. For example, the product of $3x$ and $5x - 2$ is $15x^2 - 6x$, and $15x^2 - 6x$ can be factored as the product $3x(5x - 2)$.

$$3x(5x - 2) = 15x^2 - 6x \qquad \text{Multiplication}$$

$$15x^2 - 6x = 3x(5x - 2) \qquad \text{Factoring}$$

Notice that both multiplication and factoring are examples of the distributive property, used in opposite directions. Factoring is the reverse of multiplying.

OBJECTIVE 1 Factor out the greatest common factor. The first step in factoring a polynomial is to find the *greatest common factor* for the terms of the polynomial. The **greatest common factor (GCF)** is the largest term that is a factor of all terms in the polynomial. For example, the greatest common factor for $8x + 12$ is 4, since 4 is the largest number that is a factor of both $8x$ and 12. Using the distributive property,

$$8x + 12 = 4(2x) + 4(3) = 4(2x + 3).$$

As a check, multiply 4 and $2x + 3$. The result should be $8x + 12$. This process is called **factoring out the greatest common factor.**

EXAMPLE 1 Factoring Out the Greatest Common Factor

Factor out the greatest common factor.

(a) $9z - 18$

Since 9 is the greatest common factor, factor 9 from each term.

$$9z - 18 = 9 \cdot z - 9 \cdot 2 = 9(z - 2)$$

(b) $56m + 35p = 7(8m + 5p)$

(c) $2y + 5$ There is no common factor other than 1.

(d) $12 + 24z = 12 \cdot 1 + 12 \cdot 2z$

$= 12(1 + 2z)$ 12 is the GCF.

(e) $r(3x + 2) - 8s(3x + 2)$

In this polynomial, the GCF is the binomial $(3x + 2)$.

$$r(3x + 2) - 8s(3x + 2) = (3x + 2)(r - 8s)$$

In Example 1(d), remember to write the factor 1. Always check answers by multiplying.

EXAMPLE 2 Factoring Out the Greatest Common Factor

Factor out the greatest common factor.

(a) $9x^2 + 12x^3$

The numerical part of the greatest common factor is 3, the largest number that divides into both 9 and 12. For the variable parts, x^2 and x^3, use the least exponent that appears on x; here the least exponent is 2. The GCF is $3x^2$.

$$9x^2 + 12x^3 = 3x^2(3) + 3x^2(4x)$$
$$= 3x^2(3 + 4x)$$

(b) $32p^4 - 24p^3 + 40p^5$

The greatest common numerical factor is 8. Since the least exponent on p is 3, the GCF is $8p^3$.

$$32p^4 - 24p^3 + 40p^5 = 8p^3(4p) + 8p^3(-3) + 8p^3(5p^2)$$
$$= 8p^3(4p - 3 + 5p^2)$$

(c) $3k^4 - 15k^7 + 24k^9 = 3k^4(1 - 5k^3 + 8k^5)$

(d) $24m^3n^2 - 18m^2n + 6m^4n^3$

The numerical part of the GCF is 6. Find the variable part by writing each variable with its least exponent. Here 2 is the least exponent that appears on m, while 1 is the least exponent on n. Finally, $6m^2n$ is the GCF.

$$\begin{aligned} 24m^3n^2 - 18m^2n + 6m^4n^3 &\\ &= \mathbf{6m^2n}(4mn) + \mathbf{6m^2n}(-3) + \mathbf{6m^2n}(m^2n^2) \\ &= \mathbf{6m^2n}(4mn - 3 + m^2n^2) \end{aligned}$$

(e) $25x^2y^3 + 30y^5 - 15x^4y^7 = 5y^3(5x^2 + 6y^2 - 3x^4y^4)$

When the coefficient of the term of greatest degree is negative, it is sometimes preferable to factor out the -1 that is understood along with the greatest common factor. The next example shows how this is done.

EXAMPLE 3 Factoring Out a Negative Greatest Common Factor

Factor

$$-a^3 + 3a^2 - 5a$$

in two ways.

First, a could be used as the common factor, giving

$$\begin{aligned} -a^3 + 3a^2 - 5a &= \mathbf{a}(-a^2) + \mathbf{a}(3a) + \mathbf{a}(-5) \\ &= a(-a^2 + 3a - 5). \end{aligned}$$

Alternatively, because of the leading negative sign, $-a$ could be used as the common factor.

$$\begin{aligned} -a^3 + 3a^2 - 5a &= \mathbf{-a}(a^2) + (\mathbf{-a})(-3a) + (\mathbf{-a})(5) \\ &= -a(a^2 - 3a + 5) \end{aligned}$$

Either answer is correct.

Example 3 showed two ways of factoring a polynomial. Sometimes in a particular problem there will be a reason to prefer one of these forms over the other, but both are correct. The answer section in this book will *usually* give the form where the common factor has a positive coefficient.

OBJECTIVE 2 **Factor by grouping.** Sometimes a polynomial has a greatest common factor of 1, but it still may be possible to factor the polynomial by using a process called **factoring by grouping.** We usually factor by grouping when a polynomial has more than three terms.

For example, to factor the polynomial

$$ax - ay + bx - by,$$

group the terms as follows.

Terms with common factors

$$\underset{\downarrow}{(ax - ay)} + \underset{\downarrow}{(bx - by)}$$

Then factor $ax - ay$ as $a(x - y)$ and factor $bx - by$ as $b(x - y)$ to get

$$ax - ay + bx - by = a(x - y) + b(x - y).$$

On the right, the common factor is $x - y$. The final factored form is

$$ax - ay + bx - by = (x - y)(a + b).$$

EXAMPLE 4 Factoring by Grouping

Factor $3x - 3y - ax + ay$.

Grouping terms as above gives

$$(3x - 3y) + (-ax + ay),$$

or

$$3(x - y) + a(-x + y).$$

There is no simple common factor here. However, if we factor out $-a$ instead of a in the second group of terms, we get

$$3(x - y) - a(x - y),$$

which equals

$$(x - y)(3 - a).$$

Check by multiplying.

In Example 4, different grouping would lead to the product

$$(a - 3)(y - x).$$

Verify by multiplying that this is also correct.

In some cases, the terms must be rearranged before the groupings are made.

EXAMPLE 5 Rearranging Terms before Factoring by Grouping

Factor $p^2q^2 - 10 - 2q^2 + 5p^2$.

The first two terms have no common factor except 1 (nor do the last two terms). Rearrange the terms as follows.

$$\begin{aligned} &(p^2q^2 - 2q^2) + (5p^2 - 10) && \text{Group terms.} \\ &= q^2(p^2 - 2) + 5(p^2 - 2) && \text{Factor out the common factors.} \\ &= (p^2 - 2)(q^2 + 5) && \text{Factor out } p^2 - 2. \end{aligned}$$

In Example 5, do not stop at the step

$$q^2(p^2 - 2) + 5(p^2 - 2).$$

This expression is *not in factored form* because it is a *sum* of two terms, $q^2(p^2 - 2)$ and $5(p^2 - 2)$, not a product.

EXAMPLE 6 Factoring by Grouping

Factor each polynomial by grouping.

(a) $6x^2 - 4x - 15x + 10$

Work as above. Note that we must factor -5 rather than 5 from the second group in order to get a common factor of $3x - 2$.

$$(6x^2 - 4x) + (-15x + 10) = 2x(3x - 2) - 5(3x - 2) \quad \text{Factor out } 2x \text{ and } -5.$$
$$= (3x - 2)(2x - 5) \quad \text{Factor out } (3x - 2).$$

(b) $6ax + 12bx + a + 2b$

Group the first two terms and the last two terms.

$$6ax + 12bx + a + 2b = (6ax + 12bx) + (a + 2b)$$

Now factor $6x$ from the first group, and use the identity property of multiplication to introduce the factor 1 in the second group.

$$(6ax + 12bx) + (a + 2b) = 6x(a + 2b) + 1(a + 2b)$$
$$= (a + 2b)(6x + 1) \quad \text{Factor out } (a + 2b).$$

Again, as in Example 1(d), remember to write the 1.

The steps used in factoring by grouping are listed below.

Factoring by Grouping

Step 1 **Group terms.** Collect the terms into groups so that each group has a common factor.

Step 2 **Factor within the groups.** Factor out the common factor in each group.

Step 3 **Factor the entire polynomial.** If each group now has a common factor, factor it out. If not, try a different grouping.

5.5 EXERCISES

1. Explain in your own words what it means to factor a polynomial. Discuss the two methods of factoring presented in this section, and include an example of each.
2. Write a paragraph explaining how multiplication of polynomials and factoring are related. Give an example.
3. What is the greatest common factor of the terms $7z^2(m + n)^4$ and $9z^3(m + n)^5$?
4. Why is $(2x + 9)(x - 4) + (2x + 9)(x + 3)$ not in factored form?

Factor out the greatest common factor. Simplify the factors, if possible. See Examples 1–3.

5. $12m + 60$
6. $15r - 27$
7. $8k^3 + 24k$
8. $9z^4 + 72z$
9. $xy - 5xy^2$
10. $5h^2j + hj$
11. $-4p^3q^4 - 2p^2q^5$
12. $-3z^5w^2 - 18z^3w^4$
13. $21x^5 + 35x^4 + 14x^3$
14. $6k^3 - 36k^4 + 48k^5$
15. $15a^2c^3 - 25ac^2 + 5ac$
16. $15y^3z^3 + 27y^2z^4 - 36yz^5$
17. $-27m^3p^5 + 36m^4p^3 - 72m^5p^4$
18. $-50r^4t^2 + 80r^3t^3 - 90r^2t^4$
19. $(m - 4)(m + 2) + (m - 4)(m + 3)$
20. $(z - 5)(z + 7) + (z - 5)(z + 9)$

21. $(2z - 1)(z + 6) - (2z - 1)(z - 5)$

22. $(3x + 2)(x - 4) - (3x + 2)(x + 8)$

23. $-y^5(r + w) - y^6(z + k)$

24. $-r^6(m + n) - r^7(p + q)$

25. $5(2 - x)^2 - (2 - x)^3 + 4(2 - x)$

26. $3(5 - x)^4 + 2(5 - x)^3 - (5 - x)^2$

27. $4(3 - x)^2 - (3 - x)^3 + 3(3 - x)$

28. $2(t - s) + 4(t - s)^2 - (t - s)^3$

29. $15(2z + 1)^3 + 10(2z + 1)^2 - 25(2z + 1)$

30. $6(a + 2b)^2 - 4(a + 2b)^3 + 12(a + 2b)^4$

31. $5(m + p)^3 - 10(m + p)^2 - 15(m + p)^4$

32. $-9a^2(p + q) - 3a^3(p + q)^2 + 6a(p + q)^3$

33. Which one of the following factored forms of

$$6x^3y^4 - 12x^5y^2 + 24x^4y^8$$

has the greatest common factor as one of the factors?

(a) $6x^3y^2(y^2 - 2x^2 + 4xy^6)$ **(b)** $6xy(x^2y^3 - 2x^4y + 4x^3y^7)$

(c) $2x^3y^2(3y^2 - 6x^2 + 12xy^6)$ **(d)** $6x^2y^2(xy^2 - 2x^3 + 4x^2y^6)$

34. When directed to factor the polynomial $4x^2y^5 - 8xy^3$ completely, a student wrote $2xy^3(2xy^2 - 4)$. When the teacher did not give him full credit, he complained because when his answer is multiplied out, the result is the original polynomial. Was the teacher justified in her grading? Why or why not?

Factor each polynomial twice. First, use a common factor with a positive coefficient, and then use a common factor with a negative coefficient. See Example 3.

35. $-2x^5 + 6x^3 + 4x^2$

36. $-5a^3 + 10a^4 - 15a^5$

37. $-32a^4m^5 - 16a^2m^3 - 64a^5m^6$

38. $-144z^{11}n^5 + 16z^3n^{11} - 32z^4n^7$

RELATING CONCEPTS (EXERCISES 39–46)

A natural number greater than 1 whose only natural number factors are 1 and itself is called a **prime number.** The first few prime numbers are 2, 3, 5, 7, 11, and 13. Any other natural number greater than 1 is called a **composite number.** A composite number is composed of prime number factors in one and only one way (if order of the factors is ignored). Some of the first few composite numbers and their prime factorizations are

$$4 = 2^2, \quad 6 = 2 \cdot 3, \quad 8 = 2^3, \quad 9 = 3^2, \quad 10 = 2 \cdot 5, \quad 12 = 2^2 \cdot 3.$$

One way to find the prime factorization of a composite number uses a *factor tree.* It does not matter how we start factoring—the prime factors will be the same. Here is the prime factorization of 360 by a factor tree:

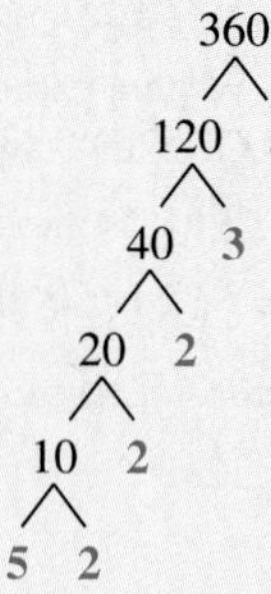

Therefore, the prime factorization of 360 is $2^3 \cdot 3^2 \cdot 5$.

Now use this idea to complete Exercises 39–46. ***Work them in order.***

39. Find the prime factorization of 60.

40. Find the prime factorization of 420.

RELATING CONCEPTS (EXERCISES 39–46) (CONTINUED)

41. Since $420 = 360 + 60$, the number 420 can be written as the sum of the prime factored forms of 360 (see the example) and 60 (see Exercise 39). Write 420 in this manner.

42. What is the GCF of 360 and 60? Give it in factored form.

43. Write your expression from Exercise 41 in factored form by factoring out the GCF you found in Exercise 42.

44. One of the factors obtained in Exercise 42 is a sum. Add the terms to find the prime number that it represents. (The sum will not always be a prime.)

45. Based on your answers in Exercises 43–44, what is the prime factored form of $360 + 60$?

46. Compare your answers in Exercises 40 and 45. What do you notice?

Did you make the connection that the greatest common factor of a composite number can be found by factoring two numbers that have it as a sum, and then factoring again?

Factor by grouping. See Examples 4–6.

47. $mx + 3qx + my + 3qy$

48. $2k + 2h + jk + jh$

49. $10m + 2n + 5mk + nk$

50. $3ma + 3mb + 2ab + 2b^2$

51. $m^2 - 3m - 15 + 5m$

52. $z^2 - 6z - 54 + 9z$

53. $p^2 - 4zq + pq - 4pz$

54. $r^2 - 9tw + 3rw - 3rt$

55. $3a^2 + 15a - 10 - 2a$

56. $7k + 2k^2 - 6k - 21$

57. $-15p^2 + 5pq - 6pq + 2q^2$

58. $-6r^2 + 9rs + 8rs - 12s^2$

59. $-3a^3 - 3ab^2 + 2a^2b + 2b^3$

60. $-16m^3 + 4m^2p^2 - 4mp + p^3$

61. $4 + xy - 2y - 2x$

62. $2ab^2 - 4 - 8b^2 + a$

63. $8 + 9y^4 - 6y^3 - 12y$

64. $x^3y^2 - 3 - 3y^2 + x^3$

65. $1 - a + ab - b$

66. $2ab^2 - 8b^2 + a - 4$

67. Refer to Exercise 65. The factored form as given in the answer section in the back of the text is $(1 - a)(1 - b)$. As mentioned in the text, sometimes other acceptable factored forms can be given. Which one of the following is *not* a factored form of $1 - a + ab - b$?
(a) $(a - 1)(b - 1)$
(b) $(-a + 1)(-b + 1)$
(c) $(-1 + a)(-1 + b)$
(d) $(1 - a)(b + 1)$

68. Refer to Exercise 66. One form of the answer is $(2b^2 + 1)(a - 4)$. Give two other acceptable factored forms of $2ab^2 - 8b^2 + a - 4$.

Factor out the given common factor. Assume all variables represent nonzero real numbers.

69. $3m^{-5} + m^{-3}$; common factor $= m^{-5}$

70. $k^{-2} + 2k^{-4}$; common factor $= k^{-4}$

71. $3p^{-3} + 2p^{-2}$; common factor $= p^{-3}$

72. $-5y^{-3} + 8y^{-2}$; common factor $= y^{-3}$

5.6 Factoring Trinomials

OBJECTIVES

1. Factor trinomials when the coefficient of the squared term is 1.
2. Factor trinomials when the coefficient of the squared term is not 1.
3. Use an alternative method for factoring trinomials.
4. Factor by substitution.

FOR EXTRA HELP

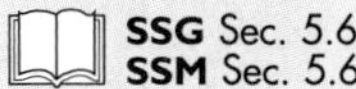
SSG Sec. 5.6
SSM Sec. 5.6

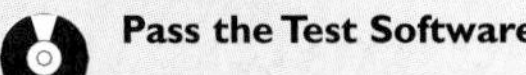
Pass the Test Software

InterAct Math Tutorial Software

Video 9

OBJECTIVE 1 Factor trinomials when the coefficient of the squared term is 1. We begin by finding the product of $x + 3$ and $x - 5$.

$$\begin{aligned}(x + 3)(x - 5) &= x^2 - 5x + 3x - 15 \\ &= x^2 - 2x - 15\end{aligned}$$

Also, by this result, the factored form of $x^2 - 2x - 15$ is $(x + 3)(x - 5)$.

$$\text{Factored form} \rightarrow (x + 3)(x - 5) \underset{\text{Factoring}}{\overset{\text{Multiplication}}{\rightleftarrows}} x^2 - 2x - 15 \leftarrow \text{Product}$$

Since multiplying and factoring are operations that "undo" each other, factoring trinomials involves using FOIL backwards. As shown below, the x^2-term came from multiplying x and x, and -15 came from multiplying 3 and -5.

Product of x and x is x^2.

$$(x + 3)(x - 5) = x^2 - 2x - 15$$

Product of 3 and -5 is -15.

We found the $-2x$ in $x^2 - 2x - 15$ by multiplying the outside terms, and then the inside terms, and adding.

Outside terms: $x(-5) = -5x$

$(x + 3)(x - 5)$ Add to get $-2x$.

Inside terms: $3 \cdot x = 3x$

Based on this example, follow these steps to factor a trinomial $x^2 + bx + c$, with 1 as the coefficient of the squared term.

Factoring $x^2 + bx + c$

Step 1 **Find pairs whose product is c.** Find all pairs of integers whose product is the third term of the trinomial (c).

Step 2 **Find pairs whose sum is b.** Choose the pair whose sum is the coefficient of the middle term (b).

If there are no such integers, the polynomial cannot be factored.

A polynomial that cannot be factored with integer coefficients is **prime.**

EXAMPLE 1 Factoring a Trinomial in $x^2 + bx + c$ Form

Factor each polynomial.

(a) $y^2 + 2y - 35$

Step 1 Find pairs of numbers whose product is -35.	*Step 2* Write sums of those numbers.	
$-35(1)$	$-35 + 1 = -34$	
$35(-1)$	$35 + (-1) = 34$	
$7(-5)$	$7 + (-5) = 2$	← Coefficient of the middle term
$5(-7)$	$5 + (-7) = -2$	

The required numbers are 7 and -5.

$$y^2 + 2y - 35 = (y + 7)(y - 5)$$

Check by finding the product of $y + 7$ and $y - 5$.

(b) $r^2 + 8r + 12$

Look for two numbers with a product of 12 and a sum of 8. Of all pairs of numbers having a product of 12, only the pair 6 and 2 has a sum of 8. Therefore,

$$r^2 + 8r + 12 = (r + 6)(r + 2).$$

Because of the commutative property, it would be equally correct to write $(r + 2)(r + 6)$. Check by multiplying. ■

EXAMPLE 2 Recognizing a Prime Polynomial

Factor $m^2 + 6m + 7$.

Look for two numbers whose product is 7 and whose sum is 6. Only two pairs of integers, 7 and 1 and -7 and -1, give a product of 7. Neither of these pairs has a sum of 6, so $m^2 + 6m + 7$ cannot be factored with integer coefficients, and is prime. ■

Factoring a trinomial that has more than one variable uses a similar process, as shown in the next example.

EXAMPLE 3 Factoring a Trinomial in Two Variables

Factor $p^2 + 6ap - 16a^2$.

Look for two expressions whose product is $-16a^2$ and whose sum is $6a$. The quantities $8a$ and $-2a$ have the necessary product and sum so

$$p^2 + 6ap - 16a^2 = (p + 8a)(p - 2a).$$

■

Sometimes a trinomial will have a common factor that should be factored out first, before using the procedure explained earlier.

EXAMPLE 4 Factoring a Trinomial with a Common Factor

Factor $16y^3 - 32y^2 - 48y$.

Start by factoring out the greatest common factor, $16y$.

$$16y^3 - 32y^2 - 48y = \mathbf{16y}(y^2 - 2y - 3)$$

To factor $y^2 - 2y - 3$, look for two integers whose product is -3 and whose sum is -2. The necessary integers are -3 and 1, with

$$16y^3 - 32y^2 - 48y = 16y(y - 3)(y + 1).$$

When factoring, always look for a common factor first. Remember to write the common factor as part of the answer.

OBJECTIVE 2 Factor trinomials when the coefficient of the squared term is not 1. We can use a generalization of the method shown in Objective 1 to factor a trinomial of the form $ax^2 + bx + c$, where $a \neq 1$. To factor $3x^2 + 7x + 2$, for example, first identify the values of a, b, and c.

$$\begin{array}{ccccc} ax^2 & + & bx & + & c \\ \downarrow & & \downarrow & & \downarrow \\ \mathbf{3x^2} & + & \mathbf{7x} & + & \mathbf{2} \end{array}$$

$$a = 3, \quad b = 7, \quad c = 2$$

The product ac is $3 \cdot 2 = 6$, so we must find integers having a product of 6 and a sum of 7 (since the middle term has coefficient 7). The necessary integers are 1 and 6, so we write $7x$ as $1x + 6x$, or $x + 6x$, giving

$$3x^2 + 7x + 2 = 3x^2 + \underbrace{\mathbf{x + 6x}}_{x + 6x = 7x} + 2.$$

Now we factor by grouping.

$$3x^2 + x + 6x + 2 = x(3x + 1) + 2(3x + 1)$$
$$3x^2 + 7x + 2 = (3x + 1)(x + 2)$$

EXAMPLE 5 Factoring a Trinomial in $ax^2 + bx + c$ Form

Factor $12r^2 - 5r - 2$.

Since $a = 12$, $b = -5$, and $c = -2$, the product ac is -24. The two integers whose product is -24 and whose sum is -5 are -8 and 3.

$$\begin{aligned} 12r^2 - \mathbf{5r} - 2 &= 12r^2 + \mathbf{3r} - \mathbf{8r} - 2 && \text{Write } -5r \text{ as } 3r - 8r. \\ &= 3r(4r + 1) - 2(4r + 1) && \text{Factor by grouping.} \\ 12r^2 - 5r - 2 &= (4r + 1)(3r - 2) && \text{Factor out the common factor.} \end{aligned}$$

OBJECTIVE 3 Use an alternative method for factoring trinomials. An alternative approach, the method of trying repeated combinations and using FOIL, is especially helpful when the product ac is large. This method is shown using the same polynomials as above.

EXAMPLE 6 Factoring a Trinomial in $ax^2 + bx + c$ Form

Factor each of the following.

(a) $3x^2 + 7x + 2$

To factor this polynomial, find the correct numbers to put in the blanks.

$$3x^2 + 7x + 2 = (\_\_\_\_x + \_\_\_\_)(\_\_\_\_x + \_\_\_\_)$$

Addition signs are used since all the signs in the polynomial indicate addition. The first two expressions have a product of $3x^2$, so they must be $3x$ and x.

$$3x^2 + 7x + 2 = (3x + \_\_\_\_)(x + \_\_\_\_)$$

The product of the two last terms must be 2, so the numbers must be 2 and 1. There is a choice. The 2 could be used with the $3x$ or with the x. Only one of these choices can give the correct middle term, $7x$. Use FOIL to try each one.

$(3x + 2)(x + 1)$	$(3x + 1)(x + 2)$
Outer: $3x$; Inner: $2x$	Outer: $6x$; Inner: x
$3x + 2x = 5x$	$6x + x = 7x$
Wrong middle term	Correct middle term

Therefore, $3x^2 + 7x + 2 = (3x + 1)(x + 2)$.

(b) $12r^2 - 5r - 2$

To reduce the number of trials, we note that the trinomial has no common factor. This means that neither of its factors can have a common factor. We should keep this in mind as we choose factors. Let us try 4 and 3 for the two first terms. If these do not work, we will make another choice.

$$12r^2 - 5r - 2 = (4r\_\_\_\_)(3r\_\_\_\_)$$

We do not know what signs to use yet. The factors of -2 are -2 and 1 or 2 and -1. Try both possibilities.

$(4r - 2)(3r + 1)$	$(4r - 1)(3r + 2)$
Wrong: $4r - 2$ has a common factor of 2.	Outer: $8r$; Inner: $-3r$
	$8r - 3r = 5r$
	Wrong middle term

The middle term on the right is $5r$, instead of the $-5r$ that is needed. We get $-5r$ by exchanging the signs in the factors.

$$(4r + 1)(3r - 2)$$

Outer: $-8r$; Inner: $3r$

$-8r + 3r = -5r$

Correct middle term

Thus, $12r^2 - 5r - 2 = (4r + 1)(3r - 2)$.

NOTE As shown in Example 6(b), if the terms of a polynomial have no common factor (except 1), then none of the terms of its factors can have a common factor. Remembering this will eliminate some potential factors.

This alternative method of factoring a trinomial $ax^2 + bx + c$, $a \neq 1$, is now summarized.

Factoring $ax^2 + bx + c$

Step 1 **Find pairs whose product is *a*.** Write all pairs of integer factors of the coefficient of the squared term (a).

Step 2 **Find pairs whose product is *c*.** Write all pairs of integer factors of the last term (c).

Step 3 **Choose inner and outer terms.** Use FOIL and various combinations of the factors from Steps 1 and 2 until the necessary middle term is found.

If no such combinations exist, the polynomial is prime.

It takes a great deal of practice using different methods to become proficient at factoring.

EXAMPLE 7 Factoring a Trinomial in Two Variables

Factor $18m^2 - 19mx - 12x^2$.

There is no common factor (except 1). Go through the steps to factor the trinomial. There are many possible factors of both 18 and -12. Let's try 6 and 3 for 18 and -3 and 4 for -12.

$(6m - 3x)(3m + 4x)$ Wrong: common factor

$(6m + 4x)(3m - 3x)$ Wrong: common factors

Since 6 and 3 do not work as factors of 18, try 9 and 2 instead, with -4 and 3 as factors of -12.

$(9m + 3x)(2m - 4x)$ Wrong: common factors

$(9m - 4x)(2m + 3x)$ — outer: $27mx$, inner: $-8mx$

$27mx + (-8mx) = 19mx$

The result on the right differs from the correct middle term only in sign, so exchange the signs in the factors. Check by multiplying.

$$18m^2 - 19mx - 12x^2 = (9m + 4x)(2m - 3x)$$

EXAMPLE 8 Factoring $ax^2 + bx + c$, $a < 0$

Factor $-3x^2 + 16x + 12$.

While it is possible to factor this polynomial directly, it is helpful to first factor out -1. Then proceed as in the earlier examples.

$$\begin{aligned} -3x^2 + 16x + 12 &= -1(3x^2 - 16x - 12) \\ &= -1(3x + 2)(x - 6) \end{aligned}$$

The factored form given in Example 8 can be written in other ways. Two of them are

$$(-3x - 2)(x - 6) \quad \text{and} \quad (3x + 2)(-x + 6).$$

Verify that these both give the original polynomial when multiplied.

OBJECTIVE 4 Factor by substitution. Sometimes we can factor a more complicated polynomial by making a substitution of one variable for an expression. This **method of substitution** is used when a particular polynomial appears to various powers in a more involved polynomial.

EXAMPLE 9 Factoring a Trinomial Using Substitution

Factor $2(x + 3)^2 + 5(x + 3) - 12$.

Since the binomial $x + 3$ appears to the powers of 2 and 1, let the substitution variable represent $x + 3$. We may choose any letter we wish except x. Let us choose y to equal $x + 3$.

$$\begin{aligned} 2(x + 3)^2 + 5(x + 3) - 12 &= 2y^2 + 5y - 12 && \text{Let } y = x + 3. \\ &= (2y - 3)(y + 4) && \text{Factor.} \end{aligned}$$

Now replace y with $x + 3$ to get

$$\begin{aligned} 2(x + 3)^2 + 5(x + 3) - 12 &= [2(x + 3) - 3][(x + 3) + 4] \\ &= (2x + 6 - 3)(x + 7) \\ &= (2x + 3)(x + 7). \end{aligned}$$

Remember to make the final substitution.

EXAMPLE 10 Factoring a Trinomial in $ax^4 + bx^2 + c$ Form

Factor $6y^4 + 7y^2 - 20$.

The variable y appears to powers in which the larger exponent is twice the smaller exponent. In a case such as this, let the substitution variable equal the smaller power. Here, let $m = y^2$. Since $y^4 = (y^2)^2 = m^2$, the given trinomial becomes

$$6m^2 + 7m - 20,$$

which is factored as

$$6m^2 + 7m - 20 = (3m - 4)(2m + 5).$$

Since $m = y^2$,

$$6y^4 + 7y^2 - 20 = (3y^2 - 4)(2y^2 + 5).$$

Some students feel comfortable enough about factoring to factor polynomials like the one in Example 10 directly, without using the substitution method.

5.6 EXERCISES

1. Match each polynomial in Column I with the correct factored form from Column II.

I	II
(a) $x^2 + x - 12$	**A.** $(2x + 7)(x - 5)$
(b) $x^2 - x - 12$	**B.** $(x - 4)(x + 3)$
(c) $2x^2 - 3x - 35$	**C.** $(3x + 8)(4x - 3)$
(d) $2x^2 + 3x - 35$	**D.** $(2x - 7)(x + 5)$
(e) $12x^2 + 23x - 24$	**E.** $(x + 4)(x - 3)$
(f) $12x^2 - 23x - 24$	**F.** $(3x - 8)(4x + 3)$

2. Explain in your own words how, when working Exercise 1, you can immediately narrow down your choices to two factored forms in the column on the right.

Factor each trinomial. (In Exercises 47–52, express your answers as shown in Example 8.) See Examples 1–8.

3. $a^2 - 2a - 15$

4. $m^2 - m - 56$

5. $p^2 + 11p + 24$

6. $k^2 + 11k + 30$

7. $r^2 - 15r + 36$

8. $t^2 - 13t + 30$

9. $a^2 - 2ab - 35b^2$

10. $z^2 + 8zw + 15w^2$

11. $y^2 - 8yq + 15q^2$

12. $k^2 - 11hk + 28h^2$

13. $x^2y^2 + 12xy + 18$

14. $p^2q^2 - 5pq - 18$

15. $6m^2 + 13m - 15$

16. $15y^2 - 17y - 18$

17. $10x^2 + 3x - 18$

18. $8k^2 + 34k + 35$

19. $20k^2 + 47k + 24$

20. $27z^2 + 42z - 5$

21. $15a^2 - 22ab + 8b^2$

22. $15p^2 + 24pq + 8q^2$

23. $40x^2 + xy + 6y^2$

24. $36m^2 - 60m + 25$

25. $25r^2 - 90r + 81$

26. $14c^2 - 17cd - 6d^2$

27. $6x^2z^2 + 5xz - 4$

28. $8m^2n^2 - 10mn + 3$

29. $24x^2 + 42x + 15$

30. $36x^2 + 18x - 4$

31. $15a^2 + 70a - 120$

32. $12a^2 + 10a - 42$

33. $4m^3 + 12m^2 - 40m$

34. $5z^3 + 45z^2 + 100z$

35. $11x^3 - 110x^2 + 264x$

36. $9k^3 + 36k^2 - 189k$

37. $2x^3y^3 - 48x^2y^4 + 288xy^5$

38. $6m^3n^2 - 24m^2n^3 - 30mn^4$

39. $18a^2 - 15a - 18$

40. $100r^2 - 90r + 20$

41. $6a^3 + 12a^2 - 90a$

42. $3m^4 + 6m^3 - 72m^2$

43. $13y^3 + 39y^2 - 52y$

44. $4p^3 + 24p^2 - 64p$

45. $12p^3 - 12p^2 + 3p$

46. $45t^3 + 60t^2 + 20t$

47. $-x^2 + 7x + 18$

48. $-p^2 - 7p - 12$

49. $-18a^2 + 17a + 15$

50. $-6x^2 + 23x + 4$

51. $-14r^3 + 19r^2 + 3r$

52. $-10h^3 + 29h^2 + 3h$

53. When a student was given the polynomial $4x^2 + 2x - 20$ to factor completely on a test, the student lost some credit when her answer was $(4x + 10)(x - 2)$. She complained to her teacher that when we multiply $(4x + 10)(x - 2)$, we get the original polynomial. Write a short explanation of why she lost some credit for her answer, even though the product is indeed $4x^2 + 2x - 20$.

54. Write an explanation as to why most people would find it more difficult to factor $36x^2 - 44x - 15$ than $37x^2 - 183x - 10$.

55. When factoring the polynomial $-4x^2 - 29x + 24$, Margo obtained $(-4x + 3)(x + 8)$, while Steve got $(4x - 3)(-x - 8)$. Who is correct? Explain your answer.

56. What are the only values of a for which $x^2 + ax - 3$ can be factored?

RELATING CONCEPTS (EXERCISES 57–64)

Refer to the note following Example 6 in this section. Then ***work Exercises 57–64 in order.***

57. Is 2 a factor of the composite number 45?

58. List all positive integer factors of 45. Is 2 a factor of any of these factors?

59. Is 3 a factor of 20?

60. List all positive integer factors of 20. Is 3 a factor of any of these factors?

61. Is 5 a factor of $10x^2 + 29x + 10$?

62. Factor $10x^2 + 29x + 10$. Is 5 a factor of either of its factors?

63. Suppose that k is an odd integer and you are asked to factor $2x^2 + kx + 8$. Why is $2x + 4$ not a possible choice in factoring this polynomial?

64. The polynomial $12y^2 - 11y - 15$ can be factored using the methods of this section. Explain why $3y + 15$ cannot be one of its factors.

Did you make the connection that if a prime p is not a factor of a composite number q, it cannot be a factor of any factor of q? (A similar statement can be made for polynominals.)

Factor the polynomial completely. See Examples 9 and 10.

65. $p^4 - 10p^2 + 16$

66. $k^4 + 10k^2 + 9$

67. $2x^4 - 9x^2 - 18$

68. $6z^4 + z^2 - 1$

69. $16x^4 + 16x^2 + 3$

70. $9r^4 + 9r^2 + 2$

71. $12p^6 - 32p^3r + 5r^2$

72. $2y^6 + 7xy^3 + 6x^2$

73. $10(k + 1)^2 - 7(k + 1) + 1$

74. $4(m - 5)^2 - 4(m - 5) - 15$

75. $3(m + p)^2 - 7(m + p) - 20$

76. $4(x - y)^2 - 23(x - y) - 6$

77. $a^2(a + b)^2 - ab(a + b)^2 - 6b^2(a + b)^2$

78. $m^2(m - p) + mp(m - p) - 2p^2(m - p)$

79. $p^2(p + q) + 4pq(p + q) + 3q^2(p + q)$

80. $2k^2(5 - y) - 7k(5 - y) + 5(5 - y)$

81. $z^2(z - x) - zx(x - z) - 2x^2(z - x)$

82. $r^2(r - s) - 5rs(s - r) - 6s^2(r - s)$

83. Describe the two kinds of trinomials whose factoring is discussed in this section. (*Hint:* See Objectives 1 and 2.) Give an example of each. Show an incorrect factored form and explain why it is incorrect. Then give the correct factored form.

84. Explain why the polynomial $x^2 + 4xy + 3$ does not have as a factored form $(x + 3y)(x + 1)$.

5.7 Special Factoring

OBJECTIVES

1. Factor a difference of two squares.
2. Factor a perfect square trinomial.

OBJECTIVE 1 **Factor a difference of two squares.** The two special products introduced in Section 5.4 are used in reverse when factoring. Recall that the product of the sum and difference of two terms leads to a **difference of two squares,** a pattern that occurs often when factoring.

3 Factor a difference of two cubes.

4 Factor a sum of two cubes.

FOR EXTRA HELP

SSG Sec. 5.7
SSM Sec. 5.7

Pass the Test Software

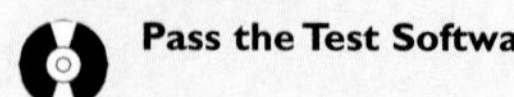
InterAct Math Tutorial Software

Video 10

Difference of Two Squares

$$x^2 - y^2 = (x + y)(x - y)$$

EXAMPLE 1 Factoring a Difference of Squares

Factor each difference of squares.

(a) $16m^2 - 49p^2 = (4m)^2 - (7p)^2$
$= (4m + 7p)(4m - 7p)$

(b) $81k^2 - 121a^2 = (9k)^2 - (11a)^2$
$= (9k + 11a)(9k - 11a)$

(c) $(m - 2p)^2 - 16 = (m - 2p)^2 - 4^2$
$= [(m - 2p) + 4][(m - 2p) - 4]$
$= (m - 2p + 4)(m - 2p - 4)$

Assuming no greatest common factor (except 1), it is not possible to factor (with real numbers) a *sum* of two squares such as $x^2 + 25$. In particular,

$$x^2 + y^2 \neq (x + y)^2.$$

OBJECTIVE 2 Factor a perfect square trinomial. Two other special products from Section 5.4 lead to the following rules for factoring.

Perfect Square Trinomial

$$x^2 + 2xy + y^2 = (x + y)^2$$
$$x^2 - 2xy + y^2 = (x - y)^2$$

The trinomial $x^2 + 2xy + y^2$ is the square of $x + y$. For this reason, the trinomial $x^2 + 2xy + y^2$ is called a **perfect square trinomial.** In this pattern, both the first and the last terms of the trinomial must be perfect squares. In the factored form, twice the product of the first and the last terms must give the middle term of the trinomial. It is important to understand these patterns in terms of words, since they occur with many different symbols (other than x and y).

$4m^2 + 20m + 25$	$p^2 - 8p + 64$
Square trinomial	Not a square trinomial; middle term should be $16p$.

EXAMPLE 2 Factoring Perfect Square Trinomials

Factor each of the following perfect square trinomials.

(a) $144p^2 - 120p + 25$

Here $144p^2 = (12p)^2$ and $25 = 5^2$. The sign on the middle term is $-$, so if $144p^2 - 120p + 25$ is a perfect square trinomial, it will have to be

$$(12p - 5)^2.$$

Take twice the product of the two terms to see if this is correct. We have

$$2(12p)(-5) = -120p,$$

which is the middle term of the given trinomial. Thus,

$$144p^2 - 120p + 25 = (12p - 5)^2.$$

(b) $4m^2 + 20mn + 49n^2$

If this is a square trinomial, it will equal $(2m + 7n)^2$. By the pattern in the box, if multiplied out, this squared binomial has a middle term of $2(2m)(7n) = 28mn$, which *does not equal* $20mn$. Verify that this trinomial cannot be factored by the methods of the previous section either. It is prime.

(c) $(r + 5)^2 + 6(r + 5) + 9 = [(r + 5) + 3]^2$

$$= (r + 8)^2,$$

since $2(r + 5)(3) = 6(r + 5)$, the middle term.

(d) $m^2 - 8m + 16 - p^2$

The first three terms here are the square of a binomial. Group them together, and factor as follows.

$$(m^2 - 8m + 16) - p^2 = (m - 4)^2 - p^2$$

The result is the difference of two squares. Factor again to get

$$(m - 4)^2 - p^2 = (m - 4 + p)(m - 4 - p).$$

Perfect square trinomials, of course, can be factored using the general methods shown earlier for other trinomials. The patterns given here provide a "shortcut."

OBJECTIVE 3 Factor a difference of two cubes. A **difference of two cubes,** $x^3 - y^3$, can be factored as follows.

Difference of Two Cubes

$$x^3 - y^3 = (x - y)(x^2 + xy + y^2)$$

We could check this pattern by finding the product of $x - y$ and $x^2 + xy + y^2$.

EXAMPLE 3 Factoring a Difference of Cubes

Factor each difference of cubes.

(a) $m^3 - 8 = m^3 - 2^3 = (m - 2)(m^2 + 2m + 2^2)$

Check: $(m - 2)(m^2 + 2m + 4)$ — products m^3, -8, $-2m$

Opposite of the product of the cube roots gives the middle term.

(b) $27x^3 - 8y^3 = (3x)^3 - (2y)^3$

$$= (3x - 2y)[(3x)^2 + (3x)(2y) + (2y)^2]$$

$$= (3x - 2y)(9x^2 + 6xy + 4y^2)$$

(c) $1000k^3 - 27n^3 = (10k)^3 - (3n)^3$

$$= (10k - 3n)[(10k)^2 + (10k)(3n) + (3n)^2]$$

$$= (10k - 3n)(100k^2 + 30kn + 9n^2)$$

OBJECTIVE 4 **Factor a sum of two cubes.** While an expression of the form $x^2 + y^2$ (a sum of two squares) cannot be factored with real numbers, a **sum of two cubes** is factored as follows.

Sum of Two Cubes

$$x^3 + y^3 = (x + y)(x^2 - xy + y^2)$$

To verify this result, find the product of $x + y$ and $x^2 - xy + y^2$. Compare this pattern with the pattern for a difference of two cubes.

The sign of the second term in the binomial factor of a sum or difference of cubes is *always the same* as the sign in the original polynomial. In the trinomial factor, the first and last terms are *always positive*; the sign of the middle term is *the opposite of* the sign of the second term in the binomial factor.

EXAMPLE 4 **Factoring a Sum of Cubes**

Factor each sum of cubes.

(a) $r^3 + 27 = r^3 + 3^3 = (r + 3)(r^2 - 3r + 3^2)$
$= (r + 3)(r^2 - 3r + 9)$

(b) $27z^3 + 125 = (3z)^3 + 5^3 = (3z + 5)[(3z)^2 - (3z)(5) + 5^2]$
$= (3z + 5)(9z^2 - 15z + 25)$

(c) $125t^3 + 216s^6 = (5t)^3 + (6s^2)^3$
$= (5t + 6s^2)[(5t)^2 - (5t)(6s^2) + (6s^2)^2]$
$= (5t + 6s^2)(25t^2 - 30ts^2 + 36s^4)$

A common error is to think that the xy-term has a coefficient of 2 when factoring the sum or difference of two cubes. Since there is no coefficient of 2, expressions of the form $x^2 + xy + y^2$ and $x^2 - xy + y^2$ cannot be factored further.

The special types of factoring in this section are summarized here. *These should be memorized.*

Special Types of Factoring

Difference of two squares	$x^2 - y^2 = (x + y)(x - y)$
Perfect square trinomial	$x^2 + 2xy + y^2 = (x + y)^2$ $x^2 - 2xy + y^2 = (x - y)^2$
Difference of two cubes	$x^3 - y^3 = (x - y)(x^2 + xy + y^2)$
Sum of two cubes	$x^3 + y^3 = (x + y)(x^2 - xy + y^2)$

5.7 EXERCISES

1. Match each binomial with its correct factored form.

(a) $x^2 - y^2$	**A.** $(x + y)(x^2 - xy + y^2)$
(b) $y^2 - x^2$	**B.** $(x - y)(x + y)$
(c) $x^3 - y^3$	**C.** $(x - y)(x^2 + xy + y^2)$
(d) $x^3 + y^3$	**D.** $(y - x)(y + x)$

2. Match each trinomial with its correct factored form.

(a) $x^2 + 2xy + y^2$	**A.** $(x^2 + y^2)^2$
(b) $x^2 - 2xy + y^2$	**B.** $(x - y)^2$
(c) $x^4 + 2x^2y^2 + y^4$	**C.** $(x + y)^2$
(d) $x^4 - 2x^2y^2 + y^4$	**D.** $(x + y)^2(x - y)^2$

3. Which of the following may be considered a sum or difference of cubes?
   (a) $64 + y^3$ **(b)** $125 - p^6$ **(c)** $9x^3 + 125$ **(d)** $(x + y)^3 - 1$

4. Which of the following may be considered a perfect square trinomial?
   (a) $a^2 - 8a - 16$ **(b)** $4t^2 + 20t + 25$ **(c)** $9r^4 + 30r^2 + 25$
   (d) $25z^2 - 45z + 81$

Factor each polynomial. See Examples 1–4.

5. $p^2 - 16$
6. $k^2 - 9$
7. $25x^2 - 4$
8. $36m^2 - 25$
9. $9a^2 - 49b^2$
10. $16c^2 - 49d^2$
11. $64m^4 - 4y^4$
12. $243x^4 - 3t^4$
13. $(y + z)^2 - 81$
14. $(h + k)^2 - 9$
15. $16 - (x + 3y)^2$
16. $64 - (r + 2t)^2$
17. $(p + q)^2 - (p - q)^2$
18. $(a + b)^2 - (a - b)^2$
19. $k^2 - 6k + 9$
20. $x^2 + 10x + 25$
21. $4z^2 + 4zw + w^2$
22. $9y^2 + 6yz + z^2$
23. $16m^2 - 8m + 1 - n^2$
24. $25c^2 - 20c + 4 - d^2$
25. $4r^2 - 12r + 9 - s^2$
26. $9a^2 - 24a + 16 - b^2$
27. $x^2 - y^2 + 2y - 1$
28. $-k^2 - h^2 + 2kh + 4$
29. $98m^2 + 84mn + 18n^2$
30. $80z^2 - 40zw + 5w^2$
31. $(p + q)^2 + 2(p + q) + 1$
32. $(x + y)^2 + 6(x + y) + 9$
33. $(a - b)^2 + 8(a - b) + 16$
34. $(m - n)^2 + 4(m - n) + 4$
35. $8x^3 - y^3$
36. $z^3 + 125p^3$
37. $64g^3 + 27h^3$
38. $27a^3 - 8b^3$
39. $24n^3 + 81p^3$
40. $250x^3 - 16y^3$
41. $(y + z)^3 - 64$
42. $(p - q)^3 + 125$
43. $64y^6 + 1$
44. $27r^6 + 1$
45. $1000x^9 - 27$
46. $729p^9 - 64$
47. $512t^6 - p^3$
48. $125y^6 + z^6$

RELATING CONCEPTS (EXERCISES 49–54)

The binomial $x^6 - y^6$ may be considered either as a difference of squares or a difference of cubes. ***Work Exercises 49–54 in order.***

49. Factor $x^6 - y^6$ by first factoring as a difference of squares. Then factor further by considering one of the factors as a sum of cubes and the other factor as a difference of cubes.

(continued)

RELATING CONCEPTS (EXERCISES 49–54) (CONTINUED)

50. Based on your answer in Exercise 49, fill in the blank with the correct factors so that $x^6 - y^6$ is factored completely:

$$x^6 - y^6 = (x - y)(x + y) \underline{\hspace{6cm}}.$$

51. Factor $x^6 - y^6$ by first factoring as a difference of cubes. Then factor further by considering one of the factors as a difference of squares.

52. Based on your answer in Exercise 51, fill in the blank with the correct factor so that $x^6 - y^6$ is factored:

$$x^6 - y^6 = (x - y)(x + y) \underline{\hspace{6cm}}.$$

53. Notice that the factor you wrote in the blank in Exercise 52 is a fourth-degree polynomial, while the two factors you wrote in the blank in Exercise 50 are both second-degree polynomials. What must be true about the product of the two factors you wrote in the blank in Exercise 50? Verify this.

54. If you have a choice of factoring as a difference of squares or a difference of cubes, how should you start to more easily obtain the factored form of the polynomial? Base the answer on your results in Exercises 49–53 and the methods of factoring explained in this section.

Did you make the connection that certain polynomials can be factored in different ways, but one way might be preferable to another?

Find a value for b or c so that each polynomial will be a perfect square.

55. $p^2 + 8p + c$

56. $y^2 - 16y + c$

57. $9z^2 + 30z + c$

58. $16r^2 - 24r + c$

59. $16x^2 + bx + 49$

60. $36y^2 + by + 1$

In some cases, the method of factoring by grouping can be combined with the methods of special factoring discussed in this section. For example, to factor $8x^3 + 4x^2 + 27y^3 - 9y^2$, we proceed as follows.

$$8x^3 + 4x^2 + 27y^3 - 9y^2 = (8x^3 + 27y^3) + (4x^2 - 9y^2)$$ Associative and commutative properties

$$= (2x + 3y)(4x^2 - 6xy + 9y^2) + (2x + 3y)(2x - 3y)$$ Factor within groups.

$$= (2x + 3y)[(4x^2 - 6xy + 9y^2) + (2x - 3y)]$$ Factor out the greatest common factor, $2x + 3y$.

$$= (2x + 3y)(4x^2 - 6xy + 9y^2 + 2x - 3y)$$ Combine terms.

In problems such as this, how we choose to group in the first step is essential to factoring correctly. If we reach a "dead end," then we should group differently and try again.

Use the method described above to factor each polynomial.

61. $27x^3 + 9x^2 + y^3 - y^2$

62. $125p^3 + 25p^2 + 8q^3 - 4q^2$

63. $1000k^3 + 20k - m^3 - 2m$

64. $27a^3 + 15a - 64b^3 - 20b$

65. $y^4 + y^3 + y + 1$

66. $8t^4 - 24t^3 + t - 3$

67. $10x^2 + 5x^3 - 10y^2 + 5y^3$

68. $64m^2 - 512m^3 - 81n^2 + 729n^3$

5.8 General Methods of Factoring

OBJECTIVES

1. Know the first step in trying to factor a polynomial.
2. Know the rules for factoring binomials.
3. Know the rules for factoring trinomials.
4. Know the rules for factoring polynomials of more than three terms.

FOR EXTRA HELP

SSG Sec. 5.8
SSM Sec. 5.8

Pass the Test Software

InterAct Math Tutorial Software

Video 10

At this point we have concluded the various kinds of factoring covered in this book. To become proficient at factoring, it helps to have a general plan for writing a given polynomial in factored form. The remainder of this section focuses on such a plan.

A polynomial is in **factored form** when the following conditions are satisfied.

1. The polynomial is written as a product of prime polynomials with integer coefficients.
2. All the polynomial factors are prime, except that a monomial factor need not be factored completely.

The order of the factors does not matter.

For example, $9x^2(x + 2)$ is the factored form of $9x^3 + 18x^2$. Because of the second rule above, it is not necessary to factor $9x^2$ as $3 \cdot 3 \cdot x \cdot x$.

OBJECTIVE 1 **Know the first step in trying to factor a polynomial.** The first step in factoring a polynomial is to factor out any common factor. This step is always the same, regardless of the number of terms in the polynomial.

EXAMPLE 1 **Factoring Out a Common Factor**

Factor each polynomial.

(a) $9p + 45 = 9(p + 5)$

(b) $5z^2 + 11z^3 + 9z^4 = z^2(5 + 11z + 9z^2)$

(c) $8m^2p^2 + 4mp = 4mp(2mp + 1)$

OBJECTIVE 2 **Know the rules for factoring binomials.** If the polynomial to be factored is a binomial, use one of the following rules.

For a **binomial** (two terms), check for the following.

Difference of two squares	$x^2 - y^2 = (x + y)(x - y)$
Difference of two cubes	$x^3 - y^3 = (x - y)(x^2 + xy + y^2)$
Sum of two cubes	$x^3 + y^3 = (x + y)(x^2 - xy + y^2)$

EXAMPLE 2 **Factoring Binomials**

Factor each polynomial, if possible.

(a) $64m^2 - 9n^2 = (8m)^2 - (3n)^2$ Difference of two squares

$= (8m + 3n)(8m - 3n)$

(b) $8p^3 - 27 = (2p)^3 - 3^3$ Difference of two cubes

$= (2p - 3)[(2p)^2 + (2p)(3) + 3^2]$

$= (2p - 3)(4p^2 + 6p + 9)$

(c) $100m^3 + 1 = (10m)^3 + 1^3$ Sum of two cubes

$= (10m + 1)[(10m)^2 - (10m)(1) + 1^2]$

$= (10m + 1)(100m^2 - 10m + 1)$

(d) $25m^2 + 121$ is prime. It is the sum of two squares.

OBJECTIVE 3 **Know the rules for factoring trinomials.** If the polynomial to be factored is a trinomial, proceed as follows.

For a **trinomial** (three terms), first see if it is a perfect square trinomial of the form

$$x^2 + 2xy + y^2 = (x + y)^2,$$

or

$$x^2 - 2xy + y^2 = (x - y)^2.$$

If it is not, use the methods of Section 5.6.

EXAMPLE 3 Factoring Trinomials

Factor each polynomial.

(a) $p^2 + 10p + 25 = (p + 5)^2$ Perfect square trinomial

(b) $49z^2 - 42z + 9 = (7z - 3)^2$ Perfect square trinomial

(c) $y^2 - 5y - 6 = (y - 6)(y + 1)$
The numbers -6 and 1 have a product of -6 and a sum of -5.

(d) $r^2 + 18r + 72 = (r + 6)(r + 12)$

(e) $2k^2 - k - 6 = (2k + 3)(k - 2)$
Use either method from Section 5.6.

(f) $28z^2 + 6z - 10 = 2(14z^2 + 3z - 5)$ Factor out the common factor.

$= 2(7z + 5)(2z - 1)$

OBJECTIVE 4 **Know the rules for factoring polynomials of more than three terms.** If the polynomial has more than three terms, try factoring by grouping.

EXAMPLE 4 Factoring Polynomials with More Than Three Terms

Factor each polynomial.

(a) $xy^2 - y^3 + x^3 - x^2y = y^2(x - y) + x^2(x - y)$

$= (x - y)(y^2 + x^2)$

(b) $20k^3 + 4k^2 - 45k - 9 = (20k^3 + 4k^2) - (45k + 9)$ Be careful with signs.

$= 4k^2(5k + 1) - 9(5k + 1)$

$= (5k + 1)(4k^2 - 9)$ $5k + 1$ is a common factor.

$= (5k + 1)(2k + 3)(2k - 3)$ Difference of two squares

(c) $4a^2 + 4a + 1 - b^2 = (4a^2 + 4a + 1) - b^2$ Associative property

$= (2a + 1)^2 - b^2$ Perfect square trinomial

$= (2a + 1 + b)(2a + 1 - b)$ Difference of two squares

(d) $8m^3 + 4m^2 - n^3 - n^2$

First, notice that the terms must be rearranged before grouping because

$$(8m^3 + 4m^2) - (n^3 + n^2) = 4m^2(2m + 1) - n^2(n + 1),$$

which cannot be factored further. Write the polynomial as follows.

$8m^3 + 4m^2 - n^3 - n^2$

$= (8m^3 - n^3) + (4m^2 - n^2)$ — Group the cubes and squares.

$= (2m - n)(4m^2 + 2mn + n^2) + (2m - n)(2m + n)$ — Factor each group.

$= (2m - n)(4m^2 + 2mn + n^2 + 2m + n)$ — Factor out the common factor.

The steps used in factoring a polynomial are summarized here.

Factoring a Polynomial

Step 1 **Factor out any common factor.**

Step 2 **If the polynomial is a binomial,** check to see if it is the difference of two squares, the difference of two cubes, or the sum of two cubes.

If the polynomial is a trinomial, check to see if it is a perfect square trinomial. If it is not, factor as in Section 5.6.

If the polynomial has more than three terms, try to factor by grouping.

Remember to check the factored form by multiplying.

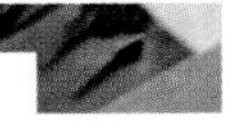

5.8 EXERCISES

Factor each polynomial. See Examples 1–4.

1. $100a^2 - 9b^2$

2. $10r^2 + 13r - 3$

3. $3p^4 - 3p^3 - 90p^2$

4. $k^4 - 16$

5. $3a^2pq + 3abpq - 90b^2pq$

6. $49z^2 - 16$

7. $225p^2 + 256$

8. $x^3 - 1000$

9. $6b^2 - 17b - 3$

10. $k^2 - 6k - 16$

11. $18m^3n + 3m^2n^2 - 6mn^3$

12. $6t^2 + 19tu - 77u^2$

13. $2p^2 + 11pq + 15q^2$

14. $40p - 32r$

15. $9m^2 - 45m + 18m^3$

16. $4k^2 + 28kr + 49r^2$

17. $54m^3 - 2000$

18. $mn - 2n + 5m - 10$

19. $2a^2 - 7a - 4$

20. $9m^2 - 30mn + 25n^2$

21. $kq - 9q + kr - 9r$

22. $56k^3 - 875$

23. $9r^2 + 100$

24. $16z^3x^2 - 32z^2x$

25. The polynomial in Exercise 23, $9r^2 + 100$, is an example of a sum of two squares. In general, a sum of two squares cannot be factored. Anne Kelly, a perceptive algebra student, commented that $9x^2 + 36y^2$ is a sum of two squares that *can* be factored. Factor this sum of two squares. Under what conditions can a sum of two squares be factored? Think of two other polynomials that qualify as a sum of two squares that can be factored, and then factor them.

26. A teacher once warned his students never to use ink when factoring. Do you think this was good advice? Why or why not?

Factor each polynomial. See Examples 1–4.

27. $x^4 - 625$

28. $2m^2 - mn - 15n^2$

29. $p^3 + 64$

30. $48y^2z^3 - 28y^3z^4$

31. $64m^2 - 625$

32. $14z^2 - 3zk - 2k^2$

33. $12z^3 - 6z^2 + 18z$

34. $225k^2 - 36r^2$

35. $256b^2 - 400c^2$

36. $z^2 - zp - 20p^2$

37. $1000z^3 + 512$

38. $64m^2 - 25n^2$

39. $10r^2 + 23rs - 5s^2$

40. $12k^2 - 17kq - 5q^2$

41. $24p^3q + 52p^2q^2 + 20pq^3$

42. $32x^2 + 16x^3 - 24x^5$

43. $48k^4 - 243$

44. $14x^2 - 25xq - 25q^2$

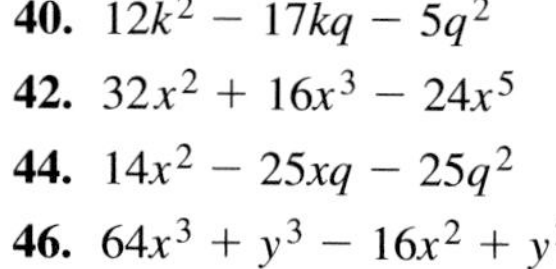

45. $m^3 + m^2 - n^3 - n^2$

46. $64x^3 + y^3 - 16x^2 + y^2$

47. $x^2 - 4m^2 - 4mn - n^2$

48. $4r^2 - s^2 - 2st - t^2$

49. The area of the rectangle shown is given by the polynomial $2W^2 + 9W$, where W represents the width. Express the length of the rectangle in terms of W.

50. The area of the square shown is given by the polynomial $4x^2 + 12xy + 9y^2$. Find the length of a side of the square in terms of x and y.

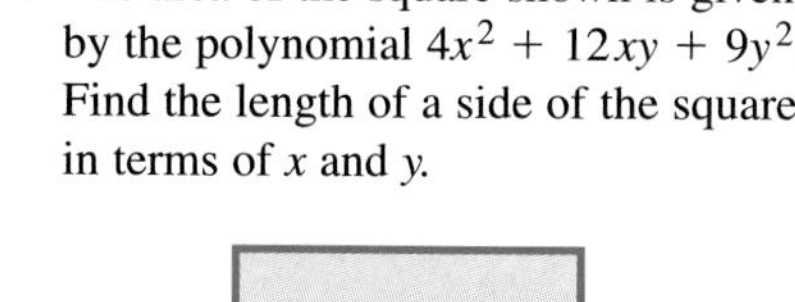

The area is $2W^2 + 9W$.

The area is $4x^2 + 12xy + 9y^2$.

51. $2x^2 - 2x - 40$

52. $27x^3 - 3y^3$

53. $(2m + n)^2 - (2m - n)^2$

54. $(3k + 5)^2 - 4(3k + 5) + 4$

55. $50p^2 - 162$

56. $y^2 + 3y - 10$

57. $12m^2rx + 4mnrx + 40n^2rx$

58. $18p^2 + 53pr - 35r^2$

59. $21a^2 - 5ab - 4b^2$

60. $x^2 - 2xy + y^2 - 4$

61. $x^2 - y^2 - 4$

62. $(5r + 2s)^2 - 6(5r + 2s) + 9$

63. $(p + 8q)^2 - 10(p + 8q) + 25$

64. $z^4 - 9z^2 + 20$

65. $21m^4 - 32m^2 - 5$

66. $(x - y)^3 - (27 - y)^3$

67. $(r + 2t)^3 + (r - 3t)^3$

68. $16x^3 + 32x^2 - 9x - 18$

69. $x^5 + 3x^4 - x - 3$

70. $x^{16} - 1$

71. $m^2 - 4m + 4 - n^2 + 6n - 9$

72. $x^2 + 4 + x^2y + 4y$

RELATING CONCEPTS (EXERCISES 73–76)

The polynomial in Exercise 72 factors as $(x^2 + 4)(1 + y)$. Students often forget to include the first term in the second binomial (that is, 1).

Work Exercises 73–76 in order, *to see a connection between this problem and a simpler one.*

73. Factor $a + ay$ by taking out the greatest common factor.

74. If you worked Exercise 73 correctly, you should have obtained $a(1 + y)$. Explain why the 1 must appear in the binomial.

75. Begin to factor $x^2 + 4 + x^2y + 4y$ by grouping the first two terms and the last two terms. Then complete the factorization.

RELATING CONCEPTS (EXERCISES 73–76) (CONTINUED)

76. If you worked Exercise 75 correctly, you should have obtained $(x^2 + 4)(1 + y)$. Remembering that $x^2 + 4$ is taking the place of a in Exercises 73 and 74, explain why 1 must appear in the second binomial.

Did you make the connection between the factor a in $a + ay$ and the factor $x^2 + 4$ in $(x^2 + 4) + (x^2 + 4)y$?

5.9 Solving Equations by Factoring

OBJECTIVES

1. Learn the zero-factor property.
2. Use the zero-factor property to solve equations.
3. Solve applied problems that require the zero-factor property.
4. Solve a polynomial equation using a graphing calculator.

FOR EXTRA HELP

SSG Sec. 5.9
SSM Sec. 5.9

Pass the Test Software

InterAct Math Tutorial Software

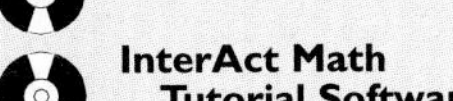
Video 10

The equations that we have solved so far in this book have been linear equations. Recall that in a linear equation, the greatest power of the variable is 1. In order to solve equations of degree greater than 1, other methods must be developed. One of these methods involves factoring.

OBJECTIVE 1 Learn the zero-factor property. Some equations can be solved by factoring. Solving equations by factoring depends on a special property of the number 0, called the **zero-factor property.**

Zero-Factor Property

If two numbers have a product of 0, then at least one of the numbers must be 0. That is, if $ab = 0$, then either $a = 0$ or $b = 0$.

To prove the zero-factor property, we first assume $a \neq 0$. (If a does equal 0, then the property is proved already.) If $a \neq 0$, then $\frac{1}{a}$ exists, and both sides of $ab = 0$ can be multiplied by $\frac{1}{a}$ to get

$$\frac{1}{a} \cdot ab = \frac{1}{a} \cdot 0$$

$$b = 0.$$

Thus, if $a \neq 0$, then $b = 0$, and the property is proved.

CAUTION If $ab = 0$, then $a = 0$ or $b = 0$. However, if $ab = 6$, for example, it is not necessarily true that $a = 6$ or $b = 6$; in fact, it is very likely that *neither* $a = 6$ *nor* $b = 6$. The zero-factor property works only for a product equal to *zero.*

OBJECTIVE 2 Use the zero-factor property to solve equations.

EXAMPLE 1 Using the Zero-Factor Property to Solve an Equation

Solve the equation $(x + 6)(2x - 3) = 0$.

Here the product of $x + 6$ and $2x - 3$ is 0. By the zero-factor property, this can be true only if $x + 6$ equals 0 or if $2x - 3$ equals 0. That is,

$$x + 6 = 0 \quad \text{or} \quad 2x - 3 = 0.$$

Solve these two equations.

$$x + 6 = 0 \quad \text{or} \quad 2x - 3 = 0$$
$$x = -6 \quad \text{or} \quad 2x = 3$$
$$x = \frac{3}{2}$$

To check these solutions, first replace x with -6 in the original equation. Then go back and replace x with $\frac{3}{2}$. This check shows that the solution set is $\{-6, \frac{3}{2}\}$.

Since the product $(x + 6)(2x - 3)$ equals $2x^2 + 9x - 18$, the equation of Example 1 has a squared term and is an example of a *quadratic equation.* A quadratic equation has degree 2.

Quadratic Equation

An equation that can be written in the form

$$ax^2 + bx + c = 0,$$

where $a \neq 0$, is a **quadratic equation.** The given form is called **standard form.**

Quadratic equations are discussed in more detail in a later chapter. Many quadratic equations can be solved by factoring.

EXAMPLE 2 Solving a Quadratic Equation by Factoring

Solve the equation $2x^2 + 3x = 2$.

The zero-factor property requires that a product of two factors equal 0. To get 0 on one side of the equals sign in this equation, subtract 2 from both sides.

$$2x^2 + 3x = 2$$
$$2x^2 + 3x - 2 = 0 \quad \text{Standard form}$$
$$(2x - 1)(x + 2) = 0 \quad \text{Factor on the left.}$$
$$2x - 1 = 0 \quad \text{or} \quad x + 2 = 0 \quad \text{Zero-factor property}$$
$$2x = 1 \quad \text{or} \quad x = -2 \quad \text{Solve each equation.}$$
$$x = \frac{1}{2}$$

The solution set is $\{\frac{1}{2}, -2\}$. Check by substituting in the original equation.

EXAMPLE 3 Solving a Quadratic Equation with a Missing Term

Solve $5z^2 - 25z = 0$.

This quadratic equation has a missing term. Comparing it with the standard form $ax^2 + bx + c = 0$ shows that $c = 0$. The zero-factor property still can be used, however, since $5z^2 - 25z = 5z(z - 5)$.

$$5z^2 - 25z = 0 \quad \text{Given equation}$$
$$5z(z - 5) = 0 \quad \text{Factor.}$$
$$5z = 0 \quad \text{or} \quad z - 5 = 0 \quad \text{Zero-factor property}$$
$$z = 0 \quad \text{or} \quad z = 5$$

The solutions are 0 and 5, as can be verified by substituting in the original equation. The solution set is $\{0, 5\}$.

EXAMPLE 4 **Solving an Equation Requiring Rewriting**

Solve $(2q + 1)(q + 1) = 2(1 - q) + 6$.

Put the equation in standard form $ax^2 + bx + c = 0$ by first multiplying on each side.

$$(2q + 1)(q + 1) = 2(1 - q) + 6$$
$$2q^2 + 3q + 1 = 2 - 2q + 6$$
$$2q^2 + 5q - 7 = 0 \qquad \text{Combine terms; get 0 on the right side.}$$
$$(2q + 7)(q - 1) = 0 \qquad \text{Factor.}$$
$$2q + 7 = 0 \quad \text{or} \quad q - 1 = 0 \qquad \text{Zero-factor property}$$
$$2q = -7 \quad \text{or} \quad q = 1$$
$$q = -\frac{7}{2}$$

Check that the solution set is $\left\{-\frac{7}{2}, 1\right\}$.

In summary, use the following steps to solve an equation by factoring.

Solving an Equation by Factoring

Step 1 **Write in standard form.** Rewrite the equation if necessary so that one side is 0.

Step 2 **Factor.** Factor the polynomial.

Step 3 **Use the zero-factor property.** Place each variable factor equal to zero, using the zero-factor property.

Step 4 **Find the solution(s).** Solve each equation formed in Step 3.

Step 5 **Check.** Check each solution in the original equation.

The equations we have solved so far in this section have all been quadratic equations. The zero-factor property can be extended to solve certain equations of degree 3 or higher, as shown in the next example.

EXAMPLE 5 **Solving an Equation of Degree 3**

Solve $-x^3 + x^2 = -6x$.

Start by adding $6x$ to both sides to get 0 on the right side.

$$-x^3 + x^2 + 6x = 0$$

To make the factoring step easier, multiply both sides by -1.

$$x^3 - x^2 - 6x = 0$$
$$x(x^2 - x - 6) = 0 \qquad \text{Factor.}$$
$$x(x - 3)(x + 2) = 0 \qquad \text{Factor the trinomial.}$$

Use the zero-factor property, extended to include the three variable factors.

$$x = 0 \quad \text{or} \quad x - 3 = 0 \quad \text{or} \quad x + 2 = 0$$
$$x = 3 \quad \text{or} \quad x = -2$$

The solution set is $\{0, 3, -2\}$.

OBJECTIVE 3 **Solve applied problems that require the zero-factor property.** The next example shows an application that leads to a quadratic equation. We continue to use the six-step problem-solving method introduced in Chapter 2.

EXAMPLE 6 **Using a Quadratic Equation in an Application**

Some surveyors are surveying a lot that is in the shape of a parallelogram. They find that the longer sides of the parallelogram are each 8 meters longer than the distance between them. The area of the lot is 48 square meters. Find the length of the longer sides and the distance between them.

Step 1 Let x represent the distance between the longer sides.

Step 2 Then $x + 8$ is the length of each longer side. See Figure 9.

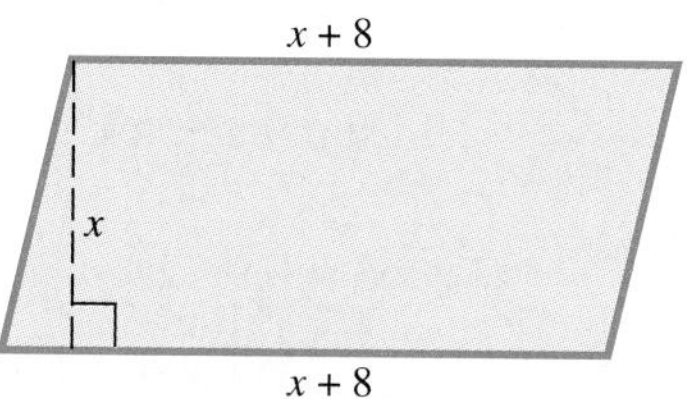

Figure 9

Step 3 The area of a parallelogram is given by $A = bh$, where b is the length of the longer side and h is the distance between the longer sides. Here $b = x + 8$ and $h = x$.

$$A = bh$$
$$48 = (x + 8)x \qquad \text{Let } A = 48,\ b = x + 8,\ h = x.$$

Step 4

$$48 = x^2 + 8x \qquad \text{Distributive property}$$
$$0 = x^2 + 8x - 48 \qquad \text{Subtract 48.}$$
$$0 = (x + 12)(x - 4) \qquad \text{Factor.}$$
$$x + 12 = 0 \quad \text{or} \quad x - 4 = 0 \qquad \text{Zero-factor property}$$
$$x = -12 \quad \text{or} \quad x = 4$$

Steps 5 and 6 A distance cannot be negative, so reject -12 as a solution. The only possible solution is 4, so the distance between the longer sides is 4 meters. The length of the longer sides is $4 + 8 = 12$ meters.

When applications lead to quadratic equations, a solution of the equation may not satisfy the physical requirements of the problem, as in Example 6. Reject such solutions.

EXAMPLE 7 Using a Quadratic Function in an Application

Quadratic functions are used to describe the distance a falling object or a propelled object travels in a specific time. For example, if a toy rocket is launched vertically upward from ground level with an initial velocity of 128 feet per second, then its height in feet after t seconds is a function defined by

$$h(t) = -16t^2 + 128t$$

if air resistance is neglected. After how many seconds will the rocket be 220 feet above the ground?

We must let $h(t) = 220$ and solve for t.

$220 = -16t^2 + 128t$	Let $h(t) = 220$.
$16t^2 - 128t + 220 = 0$	Standard form
$4t^2 - 32t + 55 = 0$	Divide by 4.
$(2t - 11)(2t - 5) = 0$	Factor.
$2t - 11 = 0$ or $2t - 5 = 0$	Zero-factor property
$t = 5.5$ or $t = 2.5$	

The rocket will reach a height of 220 feet twice: on its way up at 2.5 seconds and again on its way down at 5.5 seconds.

CONNECTIONS

In Section 5.3 we saw that the graph of $f(x) = x^2$ is a parabola. In general, the graph of $f(x) = ax^2 + bx + c$, $a \neq 0$, is a parabola, and the x-intercepts of its graph give the real number solutions of the equation $ax^2 + bx + c = 0$. In Figure 10, we show how a graphing calculator can locate these x-intercepts (called *zeros* of the function) for $Y_1 = f(x) = 2x^2 + 3x - 2$. Notice that this quadratic expression was found on the left side of the equation in Example 2 earlier in this section, where the equation was written in standard form.

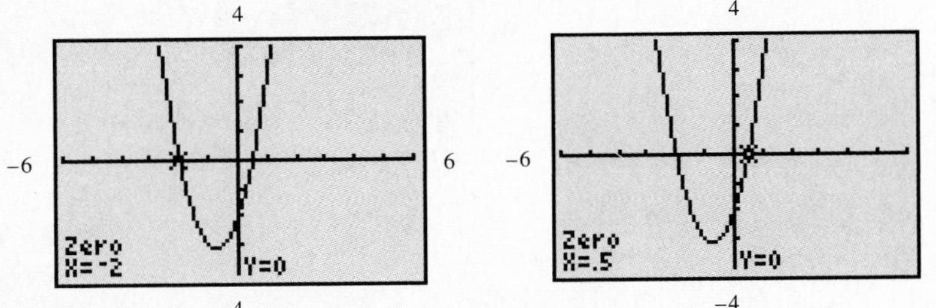

Figure 10

The x-intercepts (zeros) given with the graphs are the same as the solutions found in Example 2. This method of graphical solution can be used for any type of equation.

OBJECTIVE 4 Solve a polynomial equation using a graphing calculator. By graphing a polynomial function with a graphing calculator, we can find the x-intercepts of the graph. These values of x satisfy the equation $f(x) = 0$, where $f(x)$ is the polynomial that defines the function. Although linear equations can be solved algebraically as well as graphically, it is much more difficult to solve most polynomial equations algebraically. Since any polynomial equation can be written in the form $f(x) = 0$, graphing calculators allow us to approximate the solutions.

Recall that the squaring function, introduced in Section 5.3, is defined by a polynomial of degree 2, and the cubing function is defined by a polynomial of degree 3. It is important to know that the graph of a polynomial function of degree n will have no more than n x-intercepts. When solving a polynomial equation, to find all real solutions it is necessary to use a window that shows all intercepts. You may need to experiment with different windows to be sure that all x-intercepts are located. A graph that shows all x-intercepts and all "turning points" of the graph is called a **comprehensive graph.**

EXAMPLE 8 Solving a Polynomial Equation with a Graphing Calculator

Use a graphing calculator to find the solution set of $x^3 + 3x^2 - 4x - 12 = 0$.

The degree of the polynomial is 3, so there are no more than three x-intercepts. The window $[-5, 5]$ by $[-15, 15]$ provides a comprehensive graph, as seen in Figure 11. The three screens indicate that the x-intercepts of the graph of the function are $(-3, 0)$, $(-2, 0)$ and $(2, 0)$, and the solution set of the equation is $\{-3, -2, 2\}$.

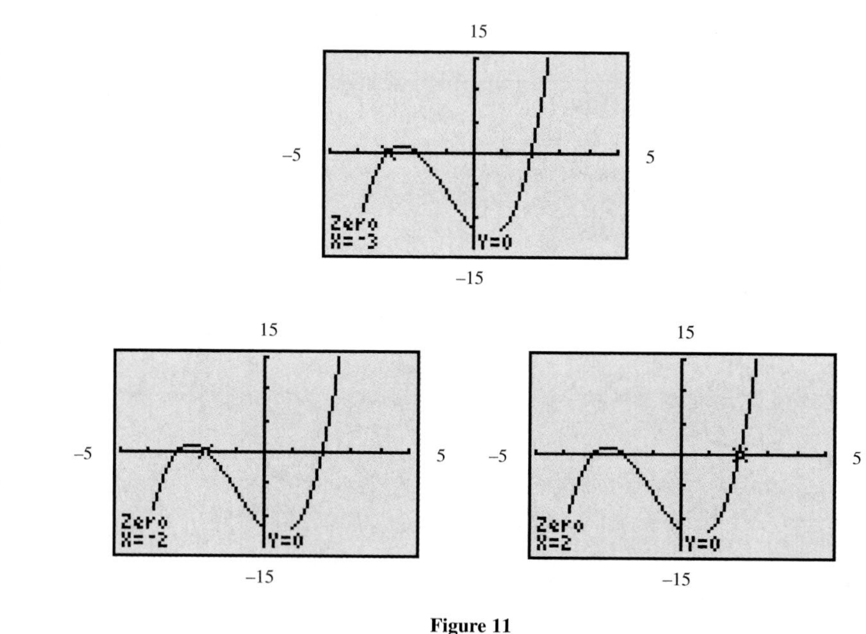

Figure 11

5.9 EXERCISES

1. Explain in your own words how the zero-factor property is used in solving a quadratic equation.
2. In solving the equation $4(x - 3)(x + 7) = 0$, a student writes $4 = 0$ or $x - 3 = 0$ or $x + 7 = 0$. Then the student becomes confused about how to handle the equation $4 = 0$. Explain what should be done, and then give the solution set.

Each of the following equations can be solved by the zero-factor property. However, there may be a step or steps required before setting each factor equal to 0. Explain what must be done, if anything, before setting each factor equal to 0.

3. $3x^2 - 8x = 0$
4. $x(x - 3) + 2(x - 3) = 0$
5. $x^2 - 4x = 12$
6. $3x^3 - 3x^2 = 216x$
7. $(x + 2)(2x - 9) = 0$
8. $(5 - x)(x + 3) = 0$

Solve each equation by using the zero-factor property. See Example 1.

9. $(x - 5)(x + 10) = 0$
10. $(y + 3)(y + 7) = 0$
11. $(2k - 5)(3k + 8) = 0$
12. $(3q - 4)(2q + 5) = 0$
13. $(m + 6)(4m - 3)(m - 1) = 0$
14. $(z - 2)(z - 7)(2z + 9) = 0$
15. $r(r - 4)(2r + 5) = 0$
16. $3x(3x - 5)(2x + 7) = 0$

Solve each equation by using the zero-factor property. See Examples 2 and 3.

17. $m^2 - 3m - 10 = 0$
18. $x^2 + x - 12 = 0$
19. $z^2 + 9z + 18 = 0$
20. $x^2 - 18x + 80 = 0$
21. $2x^2 = 7x + 4$
22. $2x^2 = 3 - x$
23. $15k^2 - 7k = 4$
24. $3c^2 + 3 = -10c$
25. $2y^2 - 12 - 4y = y^2 - 3y$
26. $3p^2 + 9p + 30 = 2p^2 - 2p$
27. $8m^2 - 72 = 0$
28. $6m^2 - 54 = 0$
29. $5k^2 + 3k = 0$
30. $9t^2 - 5t = 0$
31. $16x^2 + 24x + 9 = 0$
32. $9y^2 + 6y + 1 = 0$
33. $4x^2 = 9$
34. $16y^2 = 25$
35. $-3m^2 + 27 = 0$
36. $-2a^2 + 8 = 0$

Solve each equation by using the zero-factor property. See Example 4.

37. $(x - 3)(x + 5) = -7$
38. $(x + 8)(x - 2) = -21$
39. $(2x + 1)(x - 3) = 6x + 3$
40. $(3x + 2)(x - 3) = 7x - 1$
41. $(x + 3)(x - 6) = (2x + 2)(x - 6)$
42. $(2x + 1)(x + 5) = (x + 11)(x + 3)$
43. Explain why a quadratic equation of the form $ax^2 + bx = 0$ must have 0 in its solution set.
44. Explain why a third-degree equation of the form $ax^3 + bx^2 + cx = 0$ must have 0 in its solution set.

Solve each equation by using the zero-factor property. See Example 5.

45. $2x^3 - 9x^2 - 5x = 0$
46. $6x^3 - 13x^2 - 5x = 0$
47. $9t^3 = 16t$
48. $25y^3 = 64y$
49. $2r^3 + 5r^2 - 2r - 5 = 0$
50. $2p^3 + p^2 - 98p - 49 = 0$
51. A student tried to solve the equation in Exercise 47 by first dividing both sides by t, obtaining $9t^2 = 16$. She then solved the resulting equation by the zero-factor property to get the solution set $\left\{-\frac{4}{3}, \frac{4}{3}\right\}$. What was incorrect about her procedure?

52. Without actually solving each equation, determine which one of the following has 0 in its solution set.

(a) $4x^2 - 25 = 0$
(b) $x^2 + 2x - 3 = 0$
(c) $6x^2 + 9x + 1 = 0$
(d) $x^3 + 4x^2 = 3x$

TECHNOLOGY INSIGHTS (EXERCISES 53–56)

As shown in the Connections box following Example 7, the solutions of the quadratic equation $ax^2 + bx + c = 0$ $(a \neq 0)$ are represented on the graph of the quadratic function $f(x) = ax^2 + bx + c$ by the x-intercepts. For each equation, solve using the zero-factor property, and confirm that your solutions correspond to the x-intercepts (zeros) shown on the accompanying graphing calculator screens.

53. $2x^2 - 7x - 4 = 0$

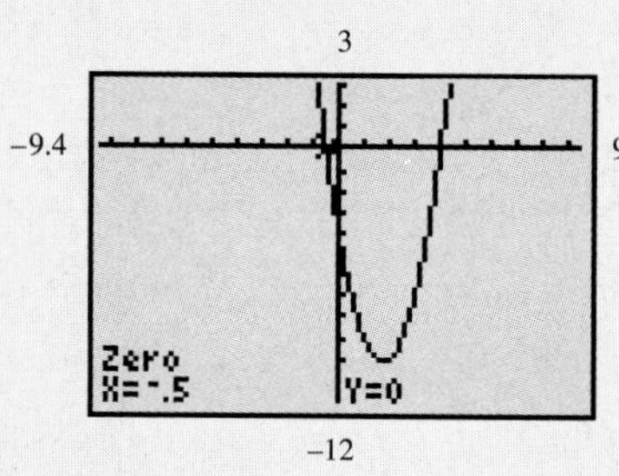

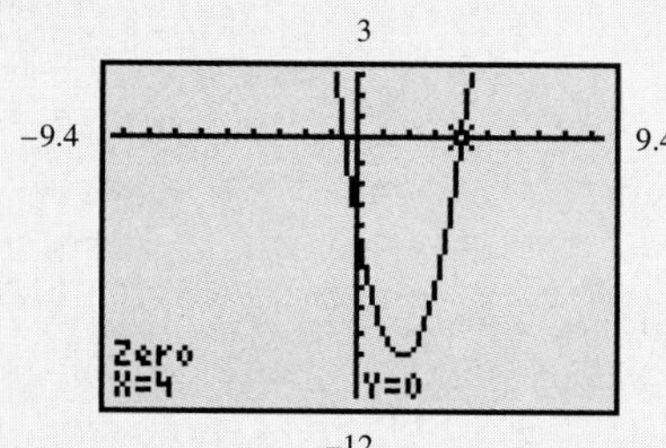

54. $2x^2 + 7x - 15 = 0$

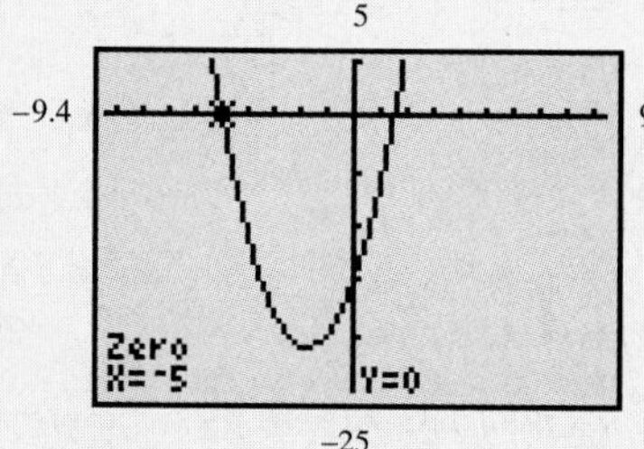

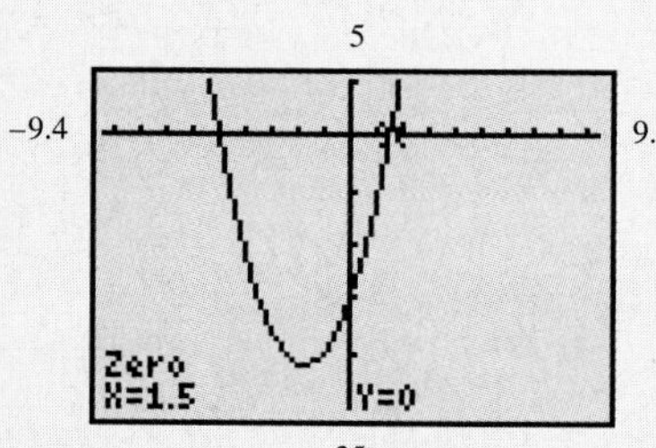

55. $-x^2 + 3x = -10$

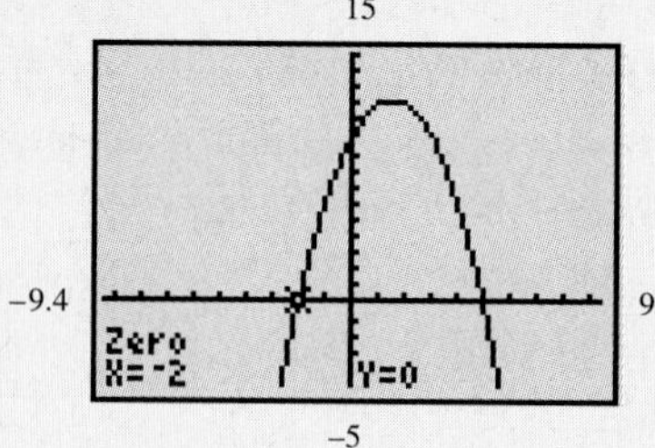

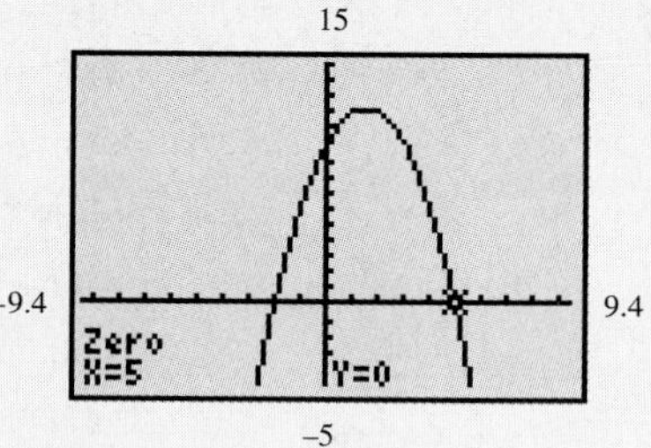

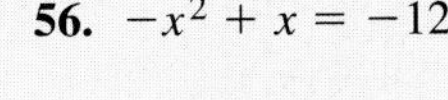

56. $-x^2 + x = -12$

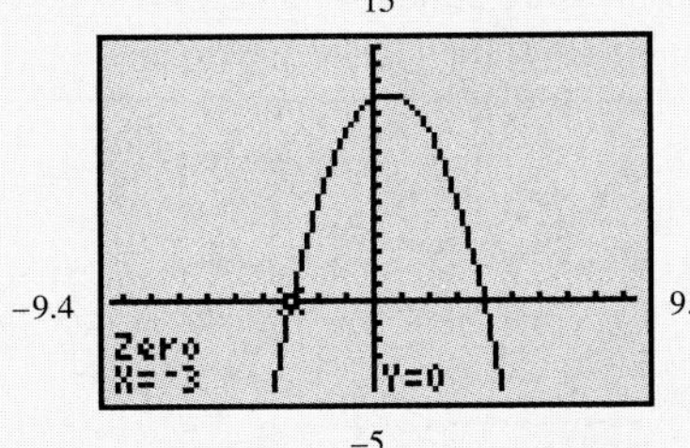

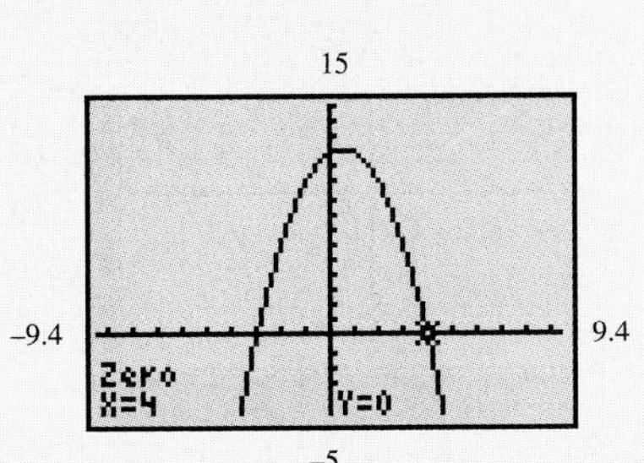

Solve each equation by using the zero-factor property.

57. $2(x - 1)^2 - 7(x - 1) - 15 = 0$

58. $4(2k + 3)^2 - (2k + 3) - 3 = 0$

59. $5(3a - 1)^2 + 3 = -16(3a - 1)$

60. $2(m + 3)^2 = 5(m + 3) - 2$

61. $(x - 1)^2 - (2x - 5)^2 = 0$

62. $(3y + 1)^2 - (y - 2)^2 = 0$

63. $(2k - 3)^2 = 16k^2$

64. $9p^2 = (5p + 2)^2$

Solve each problem by writing a quadratic equation and then solving it using the zero-factor property. See Examples 6 and 7.

65. A garden has an area of 320 square feet. Its length is 4 feet more than its width. What are the dimensions of the garden?

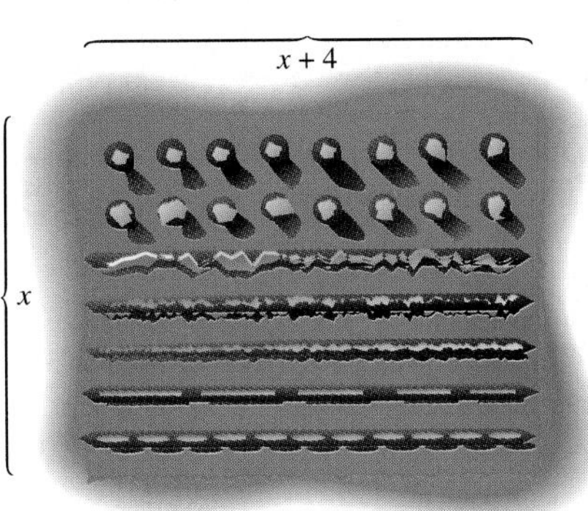

66. A square mirror has sides measuring 2 feet less than the sides of a square painting. If the difference between their areas is 32 square feet, find the lengths of the sides of the mirror and the painting.

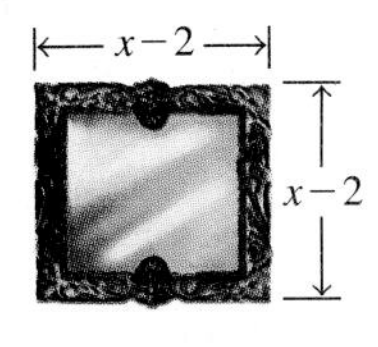

67. A sign has the shape of a triangle. The length of the base is 3 meters less than the height. What are the measures of the base and the height, if the area is 44 square meters?

68. The base of a parallelogram is 7 feet more than the height. If the area of the parallelogram is 60 square feet, what are the measures of the base and the height?

69. A farmer has 300 feet of fencing and wants to enclose a rectangular area of 5000 square feet. What dimensions should she use?

70. A rectangular landfill has an area of 30,000 square feet. Its length is 200 feet more than its width. What are the dimensions of the landfill?

71. Find two consecutive integers such that the sum of their squares is 61.

72. Find two consecutive integers such that their product is 72.

73. A box with no top is to be constructed from a piece of cardboard whose length measures 6 inches more than its width. The box is to be formed by cutting squares that measure 2 inches on each side from the four corners, and then folding up the sides. If the volume of the box will be 110 cubic inches, what are the dimensions of the piece of cardboard?

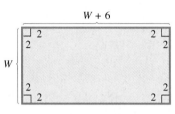

74. The surface area of the box with open top shown in the figure is 161 square inches. Find the dimensions of the base. (*Hint:* The surface area is a function defined by $S(x) = x^2 + 16x$.)

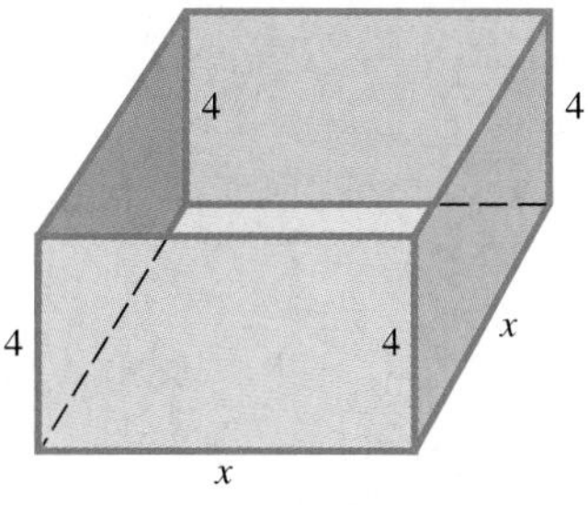

75. Refer to Example 7. After how many seconds will the rocket be 240 feet above the ground? 112 feet above the ground?

76. If an object is thrown upward with an initial velocity of 64 feet per second from a height of 80 feet, then its height in feet t seconds after it is thrown is a function defined by $f(t) = -16t^2 + 64t + 80$. How long after it is thrown will it hit the ground? (*Hint:* When it hits the ground, its height is 0 feet.)

77. If a baseball is dropped from a helicopter 625 feet above the ground, then its distance in feet from the ground t seconds later is a function defined by $f(t) = -16t^2 + 625$. How long after it is dropped will it hit the ground?

78. If a rock is dropped from a building 576 feet high, then its distance in feet from the ground t seconds later is a function defined by $f(t) = -16t^2 + 576$. How long after it is dropped will it hit the ground?

Graph the given function in the suggested window, and determine all real solutions of the equation $f(x) = 0$ (to the nearest hundredth if necessary to round off). See Example 8.

79. $f(x) = 4x^4 - 4x^2$, $[-3, 3]$ by $[-3, 6]$

80. $f(x) = -x^3 - 4x^2 - 3x$, $[-6, 2]$ by $[-6, 6]$

81. $f(x) = -2.47x^3 - 6.58x^2 - 3.33x + .14$, $[-5, 5]$ by $[-10, 10]$

82. $f(x) = .86x^3 - 5.24x^2 + 3.55x + 7.84$, $[-10, 10]$ by $[-10, 10]$

CHAPTER 5 GROUP ACTIVITY

Using a Function to Model AIDS Deaths

Objective: Create polynomial functions to model data from a graph; use the model to make predictions.

Throughout the 1980s and 1990s, AIDS became a major health issue in the United States and the rest of the world. In this activity you will create a simple polynomial function to model the death rate of male AIDS patients, ages 25–44. The following

(continued)

graph from the Centers for Disease Control and Prevention shows different causes of death over a 10-year period for men aged 25–44. The graph showing deaths from AIDS approximates a straight line and can be modeled by a linear function.

DEATH RATES IN MEN 25 TO 44 YEARS OF AGE

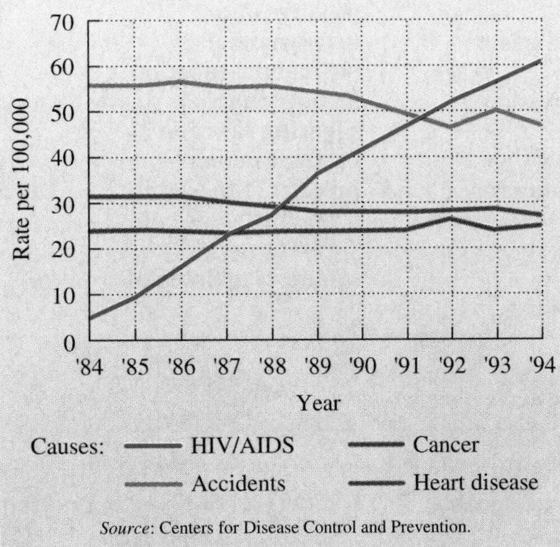

Source: Centers for Disease Control and Prevention.

A. Create polynomial models.

1. Look at the graph for HIV/AIDS deaths. Have each member of the group select two points on the graph. (*Note:* Pick points that you can estimate easily.) Write your points as ordered pairs, letting $x = 0$ represent 1984, $x = 1$ represent 1985, and so on.
2. Use the methods of Chapter 3 to write an equation of the line through the two points you selected.
3. Write your equation as a function.

B. Use your models to make predictions.

1. Compare functions within your group. Are they the same? Why or why not?
2. Select another point on the graph, one not used by anyone in your group to create their model. Find the value of each function at this point. Which function more accurately predicted the value shown on the graph?
3. Assume the death rate from AIDS continues to increase. Use your model to predict the death rate from AIDS for this age group of men in 1997.
4. In 1997 the total number of deaths from AIDS was 14,185.* The total population of the United States was 267,636,061. Use your model to find the rate of AIDS deaths per 100,000 for 1997. What does this imply about your prediction? What happened to the death rate from AIDS between 1994 and 1997?

**Source:* National Center for Health Statistics, U.S. Dept. of Health and Human Services.

CHAPTER 5 SUMMARY

KEY TERMS

5.1 **scientific notation**
5.2 **term**
numerical coefficient (coefficient)
algebraic expression
polynomial
polynomial in *x*
descending powers
trinomial
binomial
monomial
degree of a term
degree of a polynomial
negative of a polynomial
5.3 **polynomial function**
identity function
squaring function
cubing function
5.4 **product of the sum and difference of two terms**
square of a binomial
5.5 **factoring**
greatest common factor
factoring out the greatest common factor
factoring by grouping
5.6 **prime polynomial**
method of substitution
5.7 **difference of two squares**
perfect square trinomial
difference of two cubes
sum of two cubes
5.8 **factored form**
5.9 **zero-factor property**
standard form of a quadratic equation
comprehensive graph

TEST YOUR WORD POWER

See how well you have learned the vocabulary in this chapter. Answers, with examples, are given at the bottom of the page.

1. A **polynomial** is an algebraic expression made up of
(a) a term or a finite product of terms with positive coefficients and exponents
(b) the sum of two or more terms with whole number coefficients and exponents
(c) the product of two or more terms with positive exponents
(d) a term or a finite sum of terms with real coefficients and whole number exponents.

2. A **monomial** is a polynomial with
(a) only one term
(b) exactly two terms
(c) exactly three terms
(d) more than three terms.

3. A **binomial** is a polynomial with
(a) only one term
(b) exactly two terms
(c) exactly three terms
(d) more than three terms.

4. A **trinomial** is a polynomial with
(a) only one term
(b) exactly two terms
(c) exactly three terms
(d) more than three terms.

5. **FOIL** is a method for
(a) adding two binomials
(b) adding two trinomials
(c) multiplying two binomials
(d) multiplying two trinomials.

6. **Factoring** is
(a) a method of multiplying polynomials
(b) the process of writing a polynomial as a product
(c) the answer in a multiplication problem
(d) a way to add the terms of a polynomial.

7. The **difference of two squares** is a binomial
(a) that can be factored as the difference of two cubes
(b) that cannot be factored
(c) that is squared
(d) that can be factored as the product of the sum and difference of two terms.

8. A **perfect square trinomial** is a trinomial
(a) that can be factored as the square of a binomial
(b) that cannot be factored
(c) that is multiplied by a binomial
(d) where all terms are perfect squares.

9. A **quadratic equation** is a polynomial equation of
(a) degree one
(b) degree two
(c) degree three
(d) degree four.

Answers to Test Your Word Power

1. (d) *Example:* $5x^3 + 2x^2 - 7$ **2.** (a) *Examples:* $-4, 2x^3, 15a^2b$ **3.** (b) *Example:* $3t^3 + 5t$ **4.** (c) *Example:* $2a^2 - 3ab + b^2$
5. (c) *Example:* $(m + 4)(m - 3) = m(m) - 3m + 4m + 4(-3) = m^2 + m - 12$ (F O I L) **6.** (b) *Example:* $x^2 - 5x - 14 = (x - 7)(x + 2)$
7. (d) *Example:* $b^2 - 49$ is the difference of two squares, b^2 and 7^2. It can be factored as $(b + 7)(b - 7)$. **8.** (a) *Example:* $a^2 + 2a + 1$ is a perfect square trinomial; its factored form is $(a + 1)^2$. **9.** (b) *Examples:* $y^2 - 3y + 2 = 0, x^2 - 9 = 0, 2m^2 = 6m + 8$

QUICK REVIEW

CONCEPTS	EXAMPLES
5.1 INTEGER EXPONENTS AND SCIENTIFIC NOTATION	
Definitions and Rules for Exponents	Apply the rules of exponents.
Product Rule: $a^m \cdot a^n = a^{m+n}$	$3^4 \cdot 3^2 = 3^6$
Quotient Rule: $\dfrac{a^m}{a^n} = a^{m-n}$	$\dfrac{2^5}{2^3} = 2^2$
Negative Exponent: $a^{-n} = \dfrac{1}{a^n}$	$5^{-2} = \dfrac{1}{5^2}$
$\dfrac{a^{-n}}{b^{-m}} = \dfrac{b^m}{a^n}$	$\dfrac{5^{-3}}{4^{-6}} = \dfrac{4^6}{5^3}$
Zero Exponent: $a^0 = 1$	$27^0 = 1, \quad (-5)^0 = 1$
Power Rules:	
$(a^m)^n = a^{mn}$	$(6^3)^4 = 6^{12}$
$(ab)^m = a^m b^m$	$(5p)^4 = 5^4 p^4$
$\left(\dfrac{a}{b}\right)^n = \dfrac{a^n}{b^n}$	$\left(\dfrac{2}{3}\right)^5 = \dfrac{2^5}{3^5}$
$a^{-n} = \left(\dfrac{1}{a}\right)^n$	$4^{-3} = \left(\dfrac{1}{4}\right)^3$
$\left(\dfrac{a}{b}\right)^{-n} = \left(\dfrac{b}{a}\right)^n$	$\left(\dfrac{4}{7}\right)^{-2} = \left(\dfrac{7}{4}\right)^2$
Scientific Notation A number is in scientific notation when it is written as a product of a number between 1 and 10 (inclusive of 1) and an integer power of 10.	Write 23,500,000,000 in scientific notation. $23{,}500{,}000{,}000 = 2.35 \times 10^{10}$ Write 4.3×10^{-6} in standard notation. $4.3 \times 10^{-6} = .0000043$
5.2 ADDITION AND SUBTRACTION OF POLYNOMIALS	
Add or subtract polynomials by combining like terms.	Add: $(x^2 - 2x + 3) + (2x^2 - 8)$. $= 3x^2 - 2x - 5$ Subtract: $(5x^4 + 3x^2) - (7x^4 + x^2 - x)$. $= -2x^4 + 2x^2 + x$
5.3 POLYNOMIAL FUNCTIONS	
The graph of $f(x) = x$ is a line, and the graph of $f(x) = x^2$ is a parabola. The graph of $f(x) = x^3$ is neither of these. They define the identity, squaring, and cubing functions, respectively.	Graph the identity, squaring, and cubing functions.

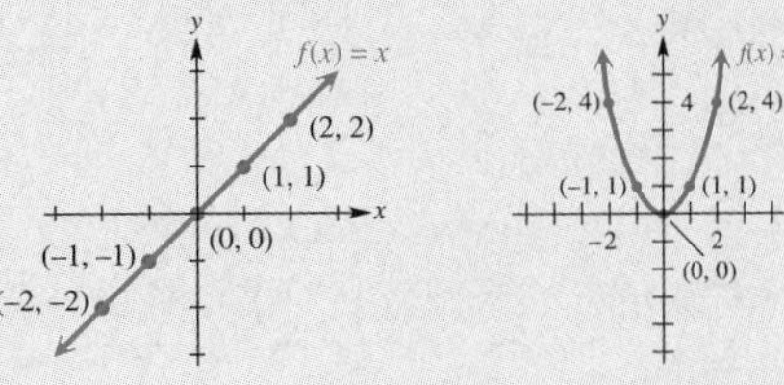

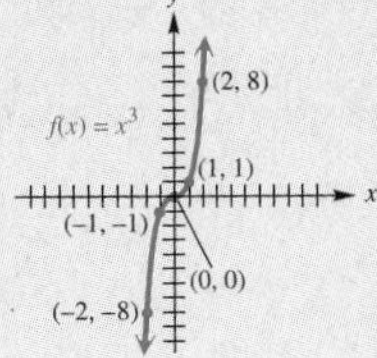

(continued)

CONCEPTS	EXAMPLES
5.4 MULTIPLICATION OF POLYNOMIALS	
To multiply two polynomials, multiply each term of one by each term of the other.	Multiply $(x^3 + 3x)(4x^2 - 5x + 2)$. $= 4x^5 + 12x^3 - 5x^4 - 15x^2 + 2x^3 + 6x$ $= 4x^5 - 5x^4 + 14x^3 - 15x^2 + 6x$
Special Products $(x + y)(x - y) = x^2 - y^2$ $(x + y)^2 = x^2 + 2xy + y^2$ $(x - y)^2 = x^2 - 2xy + y^2$	$(3m + 8)(3m - 8) = 9m^2 - 64$ $(5a + 3b)^2 = 25a^2 + 30ab + 9b^2$ $(2k - 1)^2 = 4k^2 - 4k + 1$
To multiply two binomials in general, use the FOIL method. Multiply the First terms, the Outside terms, the Inside terms, and the Last terms.	Multiply $(2x + 3)(x - 7)$. $= 2x(x) + 2x(-7) + 3x + 3(-7)$ $= 2x^2 - 14x + 3x - 21$ $= 2x^2 - 11x - 21$
5.5 GREATEST COMMON FACTORS; FACTORING BY GROUPING	
Factoring Out the Greatest Common Factor The product of the largest common numerical factor and the variable of lowest degree common to every term in a polynomial is the greatest common factor of the terms of the polynomial.	Factor $4x^2y - 50xy^2 = 2^2x^2y - 2 \cdot 5^2xy^2$. The greatest common factor is $2xy$. $4x^2y - 50xy^2 = 2xy(2x - 25y)$
Factoring by Grouping Group the terms so that each group has a common factor. Factor out the common factor in each group. If the groups now have a common factor, factor it out. If not, try a different grouping.	Factor by grouping. $5a - 5b - ax + bx = (5a - 5b) + (-ax + bx)$ $= 5(a - b) - x(a - b)$ $= (a - b)(5 - x)$
5.6 FACTORING TRINOMIALS	
To factor a trinomial, choose factors of the first term and factors of the last term. Then, place them in a pair of parentheses of this form: ( )( ). Try various combinations of the factors until the correct middle term of the trinomial is found.	Factor $15x^2 + 14x - 8$. The factors of $15x^2$ are $5x$ and $3x$ $15x$ and x. The factors of -8 are -4 and 2 4 and -2 -1 and 8 1 and -8. Various combinations of these factors lead to the correct factorization, $(5x - 2)(3x + 4)$. Check by multiplying, using the FOIL method.
5.7 SPECIAL FACTORING	
Difference of Two Squares $x^2 - y^2 = (x + y)(x - y)$	Factor. $4m^2 - 25n^2 = (2m)^2 - (5n)^2$ $= (2m + 5n)(2m - 5n)$

CONCEPTS	EXAMPLES
Perfect Square Trinomials	
$x^2 + 2xy + y^2 = (x + y)^2$ $x^2 - 2xy + y^2 = (x - y)^2$	$9y^2 + 6y + 1 = (3y + 1)^2$ $16p^2 - 56p + 49 = (4p - 7)^2$
Difference of Two Cubes	
$x^3 - y^3 = (x - y)(x^2 + xy + y^2)$	$8 - 27a^3 = (2 - 3a)(4 + 6a + 9a^2)$
Sum of Two Cubes	
$x^3 + y^3 = (x + y)(x^2 - xy + y^2)$	$64z^3 + 1 = (4z + 1)(16z^2 - 4z + 1)$
5.8 GENERAL METHODS OF FACTORING	
1. Factor out any common factors. 2. For a binomial, check for the difference of two squares, the difference of two cubes, or the sum of two cubes. For a trinomial, see if it is a perfect square. If not, factor as in Section 5.6. For more than three terms, try factoring by grouping.	Factor. $ak^3 + 2ak^2 - 9ak - 18a = a(k^3 + 2k^2 - 9k - 18)$ $= a[(k^3 + 2k^2) - (9k + 18)]$ $= a[k^2(k + 2) - 9(k + 2)]$ $= a[(k + 2)(k^2 - 9)]$ $= a(k + 2)(k - 3)(k + 3)$
5.9 SOLVING EQUATIONS BY FACTORING	
Step 1 Rewrite the equation if necessary so that one side is 0.	Solve. $2x^2 + 5x = 3$ $2x^2 + 5x - 3 = 0$ Standard form
Step 2 Factor the polynomial.	$(2x - 1)(x + 3) = 0$
Step 3 Set each factor equal to 0.	$2x - 1 = 0$ or $x + 3 = 0$
Step 4 Solve each equation.	$2x = 1$ or $x = -3$ $x = \frac{1}{2}$
Step 5 Check each solution.	A check verifies that the solution set is $\{\frac{1}{2}, -3\}$.

CHAPTER 5 REVIEW EXERCISES

[5.1] *Use the product rule and/or the quotient rule to simplify. Write the answer with only positive exponents. Assume that all variables represent nonzero real numbers.*

1. $(-3x^4y^3)(4x^{-2}y^5)$

2. $\dfrac{6m^{-4}n^3}{-3mn^2}$

3. $\dfrac{(5p^{-2}q)(4p^5q^{-3})}{2p^{-5}q^5}$

4. Explain the difference between the expressions $(-6)^0$ and -6^0.

Evaluate.

5. 4^3

6. $\left(\dfrac{1}{3}\right)^4$

7. $(-5)^3$

8. $\dfrac{2}{(-3)^{-2}}$

9. $\left(\dfrac{2}{3}\right)^{-4}$

10. $\left(\dfrac{5}{4}\right)^{-2}$

11. $5^{-1} + 6^{-1}$

12. $(5 + 6)^{-1}$

13. $-3^0 + 3^0$

14. Give an example to show that $(2a)^{-3}$ is not equal to $\dfrac{2}{a^3}$ in general by choosing a specific value for a.

Simplify. Write answers with only positive exponents. Assume that all variables represent positive real numbers.

15. $(3^{-4})^2$
16. $(x^{-4})^{-2}$
17. $(xy^{-3})^{-2}$
18. $(z^{-3})^3z^{-6}$
19. $(5m^{-3})^2(m^4)^{-3}$
20. $\dfrac{(3r)^2r^4}{r^{-2}r^{-3}}(9r^{-3})^{-2}$
21. $\left(\dfrac{5z^{-3}}{z^{-1}}\right)\dfrac{5}{z^2}$
22. $\left(\dfrac{6m^{-4}}{m^{-9}}\right)^{-1}\left(\dfrac{m^{-2}}{16}\right)$
23. $\left(\dfrac{3r^5}{5r^{-3}}\right)^{-2}\left(\dfrac{9r^{-1}}{2r^{-5}}\right)^3$
24. $\left(\dfrac{a^{-2}b^{-1}}{3a^2}\right)^{-2}\left(\dfrac{b^{-2}\cdot 3a^4}{2b^{-3}}\right)^{-2}\left(\dfrac{a^{-4}b^5}{a^3}\right)^{-2}$
25. Is $\left(\dfrac{a}{b}\right)^{-1} = \dfrac{a^{-1}}{b^{-1}}$ for all $a, b \neq 0$? If not, explain.
26. Is $(ab)^{-1} = ab^{-1}$ for all $a, b \neq 0$? If not, explain.
27. Give an example to show that $(x^2 + y^2)^2 \neq x^4 + y^4$ by choosing specific values for x and y.

Write in scientific notation.

28. 13,450
29. .0000000765
30. .138

Write without scientific notation.

31. 1.21×10^6
32. 5.8×10^{-3}

33. The median malpractice suit award in 1996 was \$568,000. Write this figure using scientific notation. (*Source:* Jury Verdict Research.)

Use scientific notation to compute. Give answers in both scientific notation and standard form.

34. $\dfrac{16 \times 10^4}{8 \times 10^8}$
35. $\dfrac{6 \times 10^{-2}}{4 \times 10^{-5}}$
36. $\dfrac{.0000000164}{.0004}$
37. $\dfrac{.0009 \times 12{,}000{,}000}{400{,}000}$

38. In a recent year, the estimated population of Luxembourg was 3.92×10^5. The population density was 400 people per square mile. Based on this information, what is the area of Luxembourg to the nearest square mile?

The population of Fresno, California, is approximately 3.45×10^5. *According to the 1994 World Almanac, the population density is* 5449 *per square mile.*

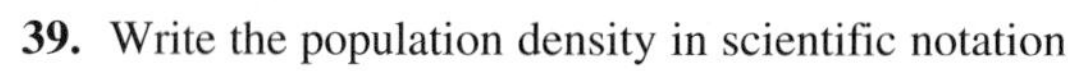

39. Write the population density in scientific notation.
40. To the nearest square mile, what is the area of Fresno?

[5.2] *Give the coefficient for each term.*

41. $14p^5$
42. $-z$
43. $504p^3r^5$

For the polynomial, **(a)** *write in descending powers,* **(b)** *identify as monomial, binomial, trinomial, or none of these, and* **(c)** *give the degree.*

44. $9k + 11k^3 - 3k^2$
45. $14m^6 + 9m^7$
46. $-5y^4 + 3y^3 + 7y^2 - 2y$
47. $-7q^5r^3$
48. Give an example of a polynomial in the variable x such that it has degree 5, is lacking a third-degree term, and is in descending powers of the variable.

Add or subtract as indicated.

49. Add.

$$\begin{array}{r} 3x^2 - 5x + 6 \\ \underline{-4x^2 + 2x - 5} \end{array}$$

50. Subtract.

$$\begin{array}{r} -5y^3 \qquad\quad + 8y - 3 \\ \underline{\qquad\qquad 4y^2 + 2y + 9} \end{array}$$

51. $(4a^3 - 9a + 15) - (-2a^3 + 4a^2 + 7a)$

52. $(3y^2 + 2y - 1) + (5y^2 - 11y + 6)$

53. Find the perimeter of the triangle.

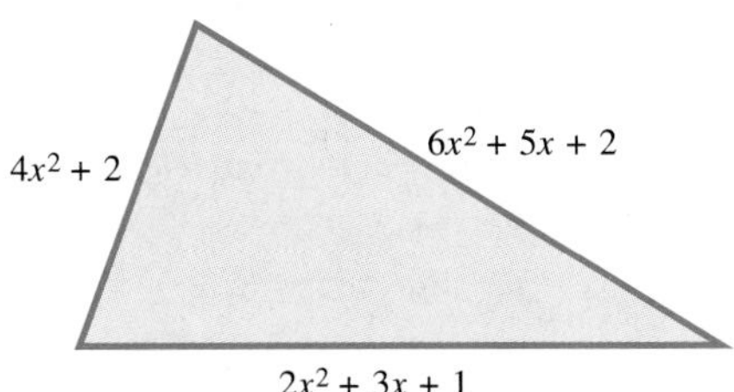

[5.3]

54. For the polynomial function defined by $f(x) = -2x^2 + 5x + 7$, find
(a) $f(-2)$ **(b)** $f(3)$.

55. For $f(x) = 2x + 3$ and $g(x) = 5x^2 - 3x + 2$, find each of the following.
(a) $(f + g)(x)$ **(b)** $(f - g)(x)$

56. The number of people employed in health service industries, in thousands, is approximated by the polynomial function with

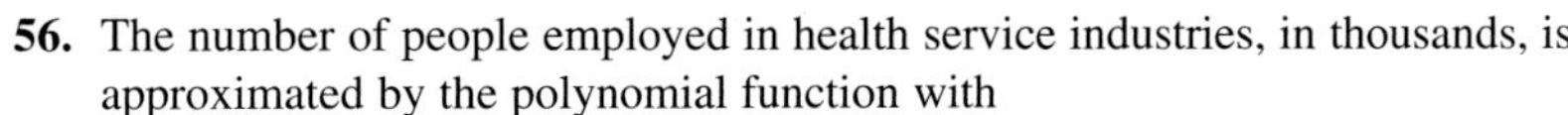

$$f(x) = 22x^2 + 243x + 8992,$$

where $x = 0$ represents 1994, $x = 1$ represents 1995, and $x = 2$ represents 1996. Find the number of employees for each of these years. (*Source:* U.S. Bureau of Labor Statistics.)

57. Graph each polynomial function defined as follows.
(a) $f(x) = -2x + 5$ **(b)** $f(x) = x^2 - 6$ **(c)** $f(x) = -x^3 + 1$

[5.4] *Find each product.*

58. $-6k(2k^2 + 7)$

59. $(3m - 2)(5m + 1)$

60. $(7y - 8)(2y + 3)$

61. $(3w - 2t)(2w - 3t)$

62. $(2p^2 + 6p)(5p^2 - 4)$

63. $(3q^2 + 2q - 4)(q - 5)$

64. $(3z^3 - 2z^2 + 4z - 1)(3z - 2)$

65. $(6r^2 - 1)(6r^2 + 1)$

66. $\left(z + \frac{3}{5}\right)\left(z - \frac{3}{5}\right)$

67. $(4m + 3)^2$

68. $(2n - 10)^2$

[5.5] *Factor out the greatest common factor.*

69. $12p^2 - 6p$

70. $21y^2 + 35y$

71. $12q^2b + 8qb^2 - 20q^3b^2$

72. $6r^3t - 30r^2t^2 + 18rt^3$

73. $(x + 3)(4x - 1) - (x + 3)(3x + 2)$

74. $(z + 1)(z - 4) + (z + 1)(2z + 3)$

Factor by grouping.

75. $4m + nq + mn + 4q$

76. $x^2 + 5y + 5x + xy$

77. $2m + 6 - am - 3a$

78. $2am - 2bm - ap + bp$

[5.6] *Factor completely.*

79. $3p^2 - p - 4$

80. $6k^2 + 11k - 10$

81. $12r^2 - 5r - 3$

82. $10m^2 + 37m + 30$

83. $10k^2 - 11kh + 3h^2$

84. $9x^2 + 4xy - 2y^2$

85. $24x - 2x^2 - 2x^3$

86. $6b^3 - 9b^2 - 15b$

87. $y^4 + 2y^2 - 8$

88. $2k^4 - 5k^2 - 3$

89. $p^2(p + 2)^2 + p(p + 2)^2 - 6(p + 2)^2$

90. $3(r + 5)^2 - 11(r + 5) - 4$

91. When asked to factor $x^2y^2 - 6x^2 + 5y^2 - 30$, a student gave the following incorrect answer: $x^2(y^2 - 6) + 5(y^2 - 6)$. Why is this answer incorrect? What is the correct answer?

92. If the area of this rectangle is represented by $4p^2 + 3p - 1$, what is the width in terms of p?

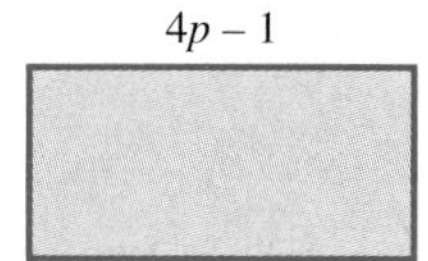

[5.7] *Factor completely.*

93. $16x^2 - 25$

94. $9t^2 - 49$

95. $x^2 + 14x + 49$

96. $9k^2 - 12k + 4$

97. $r^3 + 27$

98. $125x^3 - 1$

99. $m^6 - 1$

100. $x^8 - 1$

101. $x^2 + 6x + 9 - 25y^2$

102. $(a + b)^3 - (a - b)^3$

103. $x^5 - x^3 - 8x^2 + 8$

[5.9]

104. Which one of the following equations is not in a form that allows solving directly by the zero-factor property?

(a) $(2x + 9)(x - 3) + (4x + 7)(x - 3) = 0$

(b) $3x(x - 7) = 0$

(c) $(x - 5)(x + 3)(9x + 3) = 0$

(d) $(x - 4)^2 = 0$

105. For the equation in Exercise 104 that cannot be solved directly by the zero-factor property **(a)** put it in the form that does allow this procedure and **(b)** solve the equation.

Use the zero-factor property to solve each equation.

106. $(5x + 2)(x + 1) = 0$

107. $p^2 - 5p + 6 = 0$

108. $q^2 + 2q = 8$

109. $6z^2 = 5z + 50$

110. $6r^2 + 7r = 3$

111. $8k^2 + 14k + 3 = 0$

112. $-4m^2 + 36 = 0$

113. $6y^2 + 9y = 0$

114. $(2x + 1)(x - 2) = -3$

115. $(r + 2)(r - 2) = (r - 2)(r + 3) - 2$

116. $2x^3 - x^2 - 28x = 0$

117. $-t^3 - 3t^2 + 4t + 12 = 0$

118. $(r + 2)(5r^2 - 9r - 18) = 0$

Solve each problem.

119. A triangular wall brace has the shape of a right triangle. One of the perpendicular sides is 1 foot longer than twice the other. The area enclosed by the triangle is 10.5 square feet. Find the shorter of the perpendicular sides.

120. A rectangular parking lot has a length 20 feet more than its width. Its area is 2400 square feet. What are the dimensions of the lot?

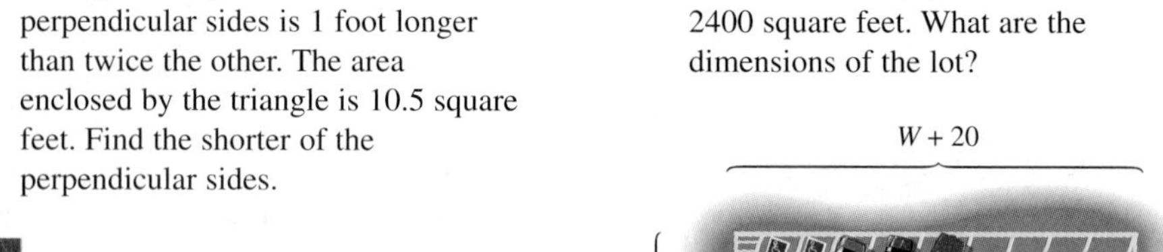

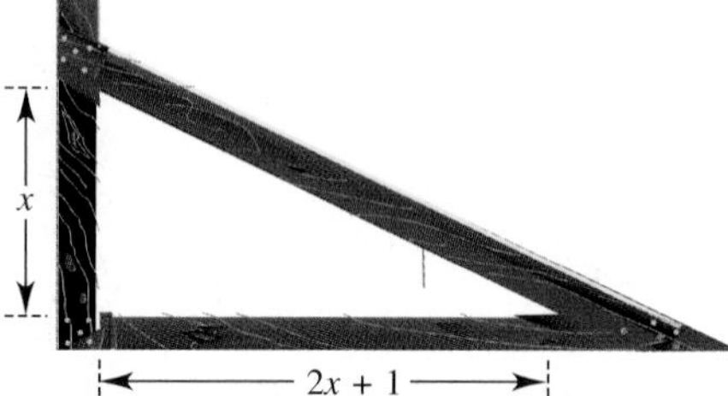

The area is 10.5 square feet.

The area is 2400 square feet.

A rock is thrown directly upward from ground level. After t seconds, its height is given by $f(t) = -16t^2 + 256t$ (if air resistance is neglected).

121. When will the rock return to the ground?

122. After how many seconds will it be 240 feet above the ground?

123. Why does the question in Exercise 122 have two answers?

MIXED REVIEW EXERCISES

Perform the indicated operations, then simplify. Write answers with only positive exponents. Assume all variables represent nonzero real numbers.

124. $(4x + 1)(2x - 3)$

125. $\dfrac{6^{-1}y^3(y^2)^{-2}}{6y^{-4}(y^{-1})}$

126. 5^{-3}

127. $(y^6)^{-5}(2y^{-3})^{-4}$

128. $(-5 + 11w) + (6 + 5w) + (-15 - 8w^2)$

129. $7p^5(3p^4 + p^3 + 2p^2)$

130. $\dfrac{(-z^{-2})^3}{5(z^{-3})^{-1}}$

131. $-(-3)^2$

132. $\dfrac{(5z^2x^3)^2(2zx^2)^{-1}}{(-10zx^{-3})^{-2}(3z^{-1}x^{-4})^2}$

133. $(2k - 1) - (3k^2 - 2k + 6)$

Factor completely.

134. $30a + am - am^2$

135. $11k + 12k^2$

136. $8 - a^3$

137. $9x^2 + 13xy - 3y^2$

138. $15y^3 + 20y^2$

139. $25z^2 - 30zm + 9m^2$

Solve.

140. $5x^2 - 17x - 12 = 0$

141. $x^3 - x = 0$

142. The length of a rectangular picture frame is 2 inches longer than its width. The area enclosed by the frame is 48 square inches. What is the width?

RELATING CONCEPTS (EXERCISES 143–148)

In this chapter we studied a variety of types of factoring. To show how several of these types may be used in factoring a polynomial, ***work Exercises 143–148 in order.*** *We will factor the polynomial*

$$x^{14} - x^2 - 4x^{13} + 4x + 4x^{12} - 4.$$

143. Start by grouping the first two terms, the next two terms, and the last two terms. (Be careful with signs when grouping the middle two terms.)

144. Factor out the greatest common factor within each group.

145. The three terms of the polynomial now have $x^{12} - 1$ as their GCF, so factor it out.

146. The polynomial is now factored, but not completely. Factor the trinomial as a perfect square trinomial, and factor the binomial as the difference of two squares.

147. One of the factors is now the sum of two cubes, and another is the difference of two squares. Factor these again.

148. In your result from Exercise 147, you should have obtained a sum of two cubes and a difference of two cubes. Factor again to obtain the complete factored form of the polynomial.

Did you make the connection that various types of factoring may be performed within the same problem, even though they are not obvious at first glance?

CHAPTER 5 TEST

1. Match the expression in Column I with its equivalent expression from Column II. Choices may be used once, more than once, or not at all.

I	II
(a) 7^{-2}	**A.** 1
(b) 7^0	**B.** $\frac{1}{9}$
(c) -7^0	**C.** $\frac{1}{49}$
(d) $(-7)^0$	**D.** -1
(e) -7^2	**E.** -49
(f) $7^{-1} + 2^{-1}$	**F.** $\frac{9}{14}$
(g) $(7 + 2)^{-1}$	**G.** $\frac{2}{7}$
(h) $\frac{7^{-1}}{2^{-1}}$	**H.** 0
(i) $(-7)^{-2}$	**I.** none of these

Simplify. Write answers with only positive exponents. Assume that all variables represent nonzero real numbers.

2. $(3x^{-2}y^3)^{-2}(4x^3y^{-4})$

3. $\frac{36r^{-4}(r^2)^{-3}}{6r^4}$

4. $\left(\frac{4p^2}{q^4}\right)^3\left(\frac{6p^8}{q^{-8}}\right)^{-2}$

5. $(-2x^4y^{-3})^0(-4x^{-3}y^{-8})^2$

6. (a) Write 9.1×10^{-7} without using scientific notation.

(b) Use scientific notation to simplify $\frac{2{,}500{,}000 \times .00003}{.05 \times 5{,}000{,}000}$. Write the answer in both standard form and scientific notation.

7. If $f(x) = -2x^2 + 5x - 6$ and $g(x) = 7x - 3$, find each of the following.
(a) $f(4)$ **(b)** $(f + g)(x)$ **(c)** $(f - g)(x)$

8. Graph the function defined by $f(x) = -2x^2 + 3$.

9. The number of first-year enrollments in osteopathy schools has risen in recent years. For the period 1990–1996, the number of students enrolled can be approximated by the polynomial function defined by $f(x) = .5x^2 + 68.9x + 1852$, where $x = 0$ corresponds to 1990, $x = 1$ corresponds to 1991, and so on. Based on this model, how many first-year students were there in 1996? (*Source:* U.S. National Center for Health Statistics.)

Perform the indicated operations.

10. $(4x^3 - 3x^2 + 2x - 5) - (3x^3 + 11x + 8) + (x^2 - x)$

11. $(5x - 3)(2x + 1)$

12. $(2m - 5)(3m^2 + 4m - 5)$

13. $(6x + y)(6x - y)$

14. $(3k + q)^2$

15. $[2y + (3z - x)][2y - (3z - x)]$

16. Explain why $(x^2 + 2y)p + 3(x^2 + 2y)$ is not in factored form. Then factor the polynomial.

Factor.

17. $11z^2 - 44z$

18. $3x + by + bx + 3y$

19. $4p^2 + 3pq - q^2$

20. $16a^2 + 40ab + 25b^2$

21. $y^3 - 216$

22. $9k^2 - 121j^2$

23. $6k^4 - k^2 - 35$

24. $27x^6 + 1$

25. Which one of the following is not a factored form of $-x^2 - x + 12$?
(a) $(3 - x)(x + 4)$ (b) $-(x - 3)(x + 4)$
(c) $(-x + 3)(x + 4)$ (d) $(x - 3)(-x + 4)$

Solve each equation by using the zero-factor property.

26. $3x^2 + 8x + 4 = 0$ 27. $10x^2 = 17x - 3$ 28. $5m(m - 1) = 2(1 - m)$

Solve each problem.

29. The area of the rectangle shown is 40 square inches. Find the length and the width of the rectangle.

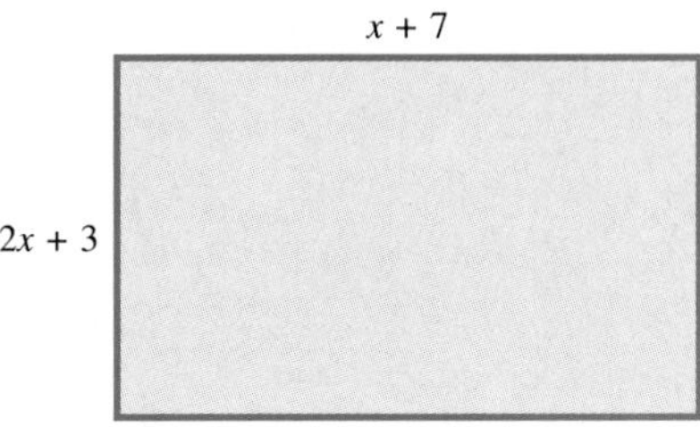

The area is 40 square inches.

30. A ball is thrown upward from ground level. After t seconds, its height in feet is a function defined by $f(t) = -16t^2 + 96t$. After how many seconds will it reach a height of 128 feet?

CUMULATIVE REVIEW EXERCISES CHAPTERS 1–5

Use the properties of real numbers to simplify.

1. $-2(m - 3)$ 2. $-(-4m + 3)$ 3. $3x^2 - 4x + 4 + 9x - x^2$

Evaluate if $p = -4$, $q = -2$, and $r = 5$.

4. $-3(2q - 3p)$
5. $8r^2 + q^2$
6. $\dfrac{\sqrt{r}}{-p + 2q}$
7. $\dfrac{5p + 6r^2}{p^2 + q - 1}$

Solve.

8. $2z - 5 + 3z = 4 - (z + 2)$
9. $\dfrac{3a - 1}{5} + \dfrac{a + 2}{2} = -\dfrac{3}{10}$
10. $-\dfrac{4}{3}d \geq -5$
11. $3 - 2(m + 3) < 4m$
12. $2k + 4 < 10$ and $3k - 1 > 5$
13. $2k + 4 > 10$ or $3k - 1 < 5$
14. $|5x + 3| - 10 = 3$
15. $|x + 2| < 9$
16. $|2y - 5| \geq 9$
17. $V = lwh$ for h

18. Two planes leave the Dallas-Fort Worth airport at the same time. One travels east at 550 miles per hour, and the other travels west at 500 miles per hour. Assuming no wind, how long will it take for the planes to be 2100 miles apart?

Plane	r	t	d
Eastbound	550	x	
Westbound	500	x	

19. Graph $4x + 2y = -8$.

20. Find the slope of the line passing through the points $(-4, 8)$ and $(-2, 6)$.

21. What is the slope of the line shown here?

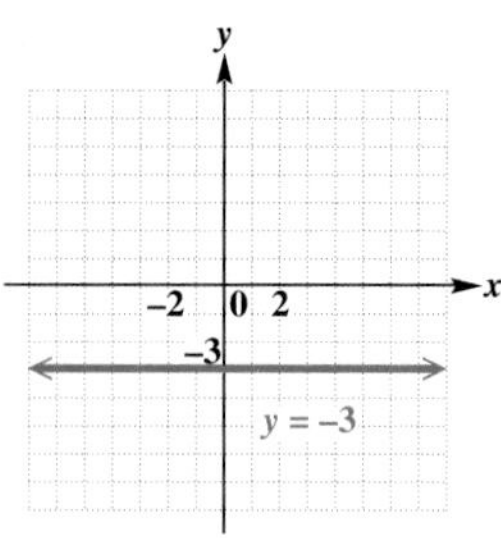

Use the function defined by $f(x) = 2x + 7$ to find each of the following.

22. $f(-4)$

23. The x-intercept of its graph

Solve each system.

24. $3x - 2y = -7$
$2x + 3y = 17$

25. $2x + 3y - 6z = 5$
$8x - y + 3z = 7$
$3x + 4y - 3z = 7$

26. Evaluate the determinant $\begin{vmatrix} -3 & 2 \\ -1 & 8 \end{vmatrix}$.

Perform the indicated operations. Assume that variables represent nonzero real numbers.

27. $(3x^2y^{-1})^{-2}(2x^{-3}y)^{-1}$

28. $\dfrac{5m^{-2}y^3}{3m^{-3}y^{-1}}$

Perform the indicated operations.

29. $(3x^3 + 4x^2 - 7) - (2x^3 - 8x^2 + 3x)$

30. $(7x + 3y)^2$

31. $(2p + 3)(5p^2 - 4p - 8)$

Factor.

32. $16w^2 + 50wz - 21z^2$

33. $4x^2 - 4x + 1 - y^2$

34. $4y^2 - 36y + 81$

35. $100x^4 - 81$

36. $8p^3 + 27$

Solve.

37. $(p - 1)(2p + 3)(p + 4) = 0$

38. $9q^2 = 6q - 1$

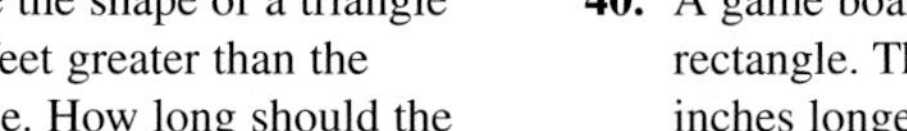

39. A sign is to have the shape of a triangle with a height 3 feet greater than the length of the base. How long should the base be if the area is to be 14 square feet?

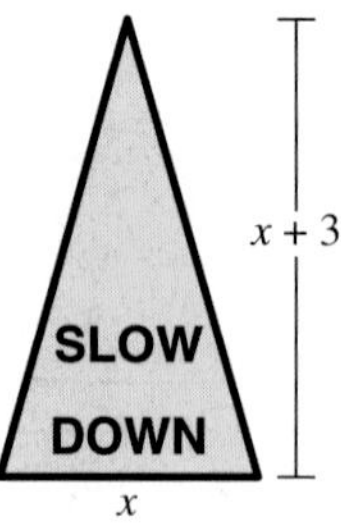

40. A game board has the shape of a rectangle. The longer sides are each 2 inches longer than the distance between them. The area of the board is 288 square inches. Find the length of the longer sides and the distance between them.

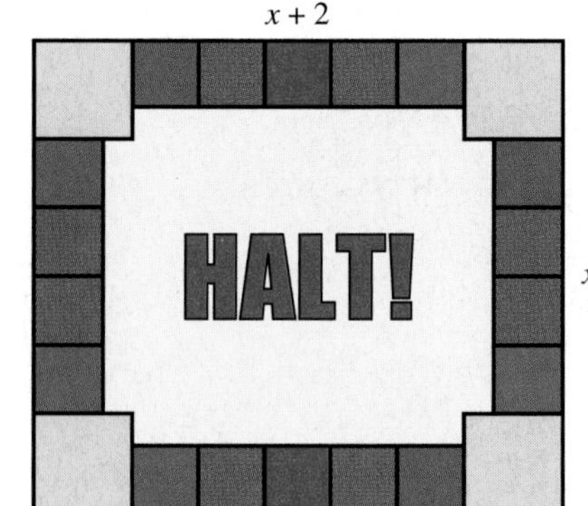

Rational Expressions

Rational expressions play the same role in algebra as rational numbers do in arithmetic. (Note the word "ratio" in the name.) The *rational function*

$$f(x)=\frac{8710x^2-69{,}400x+470{,}000}{1.08x^2-324x+82{,}200}$$

accurately models the braking distance for automobiles traveling at x miles per hour, where $20 \le x \le 70$.* A graph of this function is shown in the figure.

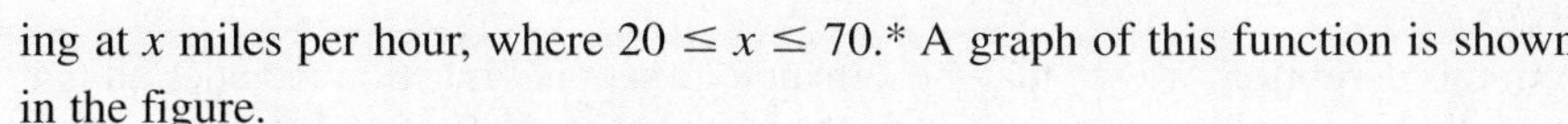

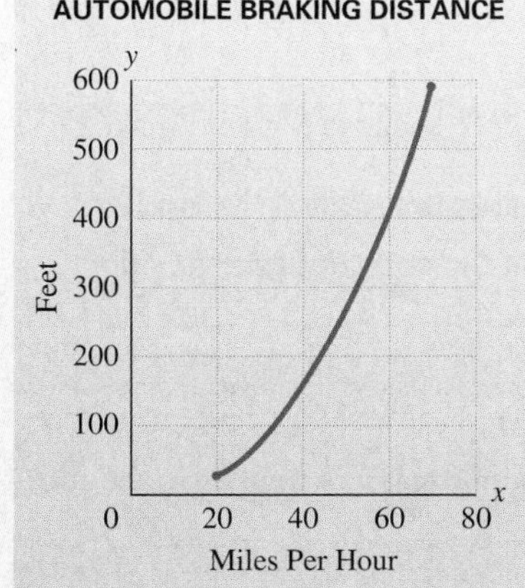

Notice that the graph, like that of a polynomial function, is a curve, not a straight line. From the graph, approximately what speed requires a braking distance of 300 feet?

Americans have had a love affair with their cars ever since the first automobiles hit the road early in the twentieth century. In recent years, the automotive industry has been changing rapidly. Imports have captured a bigger share of the U.S. automobile market. Cars have become more high tech. The theme of this chapter, transportation, will be discussed in other examples and exercises throughout the chapter.

Transportation

*Source: F. Mannering and W. Kilareski, *Principles of Highway Engineering and Traffic Control,* John Wiley & Sons, 1990.

Visit our Web site at www.LialAlgebra.com

6.1 Rational Expressions and Functions; Multiplication and Division

OBJECTIVES

1. Define rational expressions.
2. Define rational functions and describe their domains.
3. Write rational expressions in lowest terms.
4. Multiply rational expressions.
5. Find reciprocals for rational expressions.
6. Divide rational expressions.

FOR EXTRA HELP

SSG Sec. 6.1
SSM Sec. 6.1

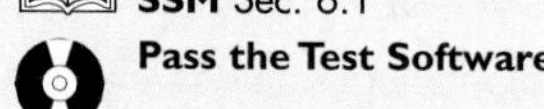
Pass the Test Software

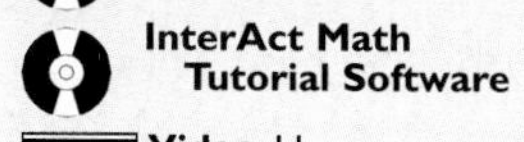
InterAct Math Tutorial Software

Video 11

OBJECTIVE 1 Define rational expressions. In arithmetic, a rational number is the quotient of two integers, with the denominator not 0. In algebra, a **rational expression** or *algebraic fraction* is the quotient of two polynomials, again with the denominator not 0. For example,

$$\frac{m+4}{m-2}, \quad \frac{8x^2-2x+5}{4x^2+5x}, \quad \text{and} \quad x^5 \left(\text{or } \frac{x^5}{1}\right)$$

are all rational expressions. In other words, rational expressions are the elements of the set

$$\left\{\frac{P}{Q} \,\middle|\, P, Q \text{ polynomials, with } Q \neq 0\right\}.$$

OBJECTIVE 2 Define rational functions and describe their domains. A function that is defined by a rational expression is called a **rational function** and has the form

$$f(x) = \frac{P(x)}{Q(x)},$$

where $Q(x) \neq 0$.

From this definition, we see that the domain of a rational function includes all real numbers except those that make $Q(x) = 0$. For example, the domain of

$$f(x) = \frac{2}{x-5}$$

includes all real numbers except 5, because 5 would make $Q(x) = x - 5$ equal to zero.

EXAMPLE 1 Finding Numbers That Are Not in the Domain

Find all numbers that are not in the domain of the function.

(a) $f(k) = \dfrac{3}{7k-14}$

The only values that cannot be used are those that make the denominator 0. To find these values, set the denominator equal to 0 and solve the resulting equation.

$$\begin{aligned} 7k - 14 &= 0 \\ 7k &= 14 \quad \text{Add 14.} \\ k &= 2 \quad \text{Divide by 7.} \end{aligned}$$

The number 2 cannot be used as a replacement for k; 2 is not in the domain of f.

(b) $g(p) = \dfrac{3+p}{p^2-4p+3}$

Set the denominator equal to 0 and solve the equation.

$$\begin{aligned} p^2 - 4p + 3 &= 0 \\ (p-3)(p-1) &= 0 \quad \text{Factor.} \end{aligned}$$

$$p - 3 = 0 \quad \text{or} \quad p - 1 = 0 \quad \text{Zero-factor property}$$

$$p = 3 \quad \text{or} \quad p = 1$$

The domain of g includes all real numbers except 3 and 1.

(c) $h(x) = \frac{8x + 2}{3}$

The denominator, 3, can never be 0, so the domain includes all real numbers.

OBJECTIVE 3 **Write rational expressions in lowest terms.** In arithmetic, we write the fraction $\frac{15}{20}$ in lowest terms by dividing the numerator and denominator by 5 to get $\frac{3}{4}$. We write rational expressions in lowest terms in a similar way, using the **fundamental principle of rational numbers.**

Fundamental Principle of Rational Numbers

If $\frac{a}{b}$ is a rational number and if c is any nonzero real number, then

$$\frac{a}{b} = \frac{ac}{bc}.$$

(The numerator and denominator of a rational number may either be multiplied or divided by the same nonzero number without changing the value of the rational number.)

Since $\frac{c}{c}$ is equivalent to 1, the fundamental principle is based on the identity property of multiplication.

A rational expression is a quotient of two polynomials, and since the value of a polynomial is a real number for all values of the variables for which it is defined, any statement that applies to rational numbers will also apply to rational expressions. We use the following steps to write rational expressions in lowest terms.

Writing in Lowest Terms

Step 1 **Factor.** Factor both numerator and denominator to find their greatest common factor (GCF).

Step 2 **Reduce.** Apply the fundamental principle.

EXAMPLE 2 **Writing Rational Expressions in Lowest Terms**

Write each rational expression in lowest terms.

(a) $\frac{8k}{16} = \frac{k \cdot 8}{2 \cdot 8} = \frac{k}{2}$

Here, the GCF of the numerator and denominator is 8. We then applied the fundamental principle.

(b) $\frac{8 + k}{16}$

This expression is in lowest terms. Because the numerator cannot be factored, the expression cannot be simplified further.

(c) $\frac{12x^3y^2}{6x^4y} = \frac{2y \cdot 6x^3y}{x \cdot 6x^3y} = \frac{2y}{x}$

Factor out the GCF, $6x^3y$. (Here 3 is the least exponent on x, and 1 the least exponent on y.)

(d) $\frac{a^2 - a - 6}{a^2 + 5a + 6}$

Start by factoring the numerator and denominator.

$$\frac{a^2 - a - 6}{a^2 + 5a + 6} = \frac{(a - 3)(a + 2)}{(a + 3)(a + 2)}$$

Divide the numerator and denominator by $a + 2$ to get

$$\frac{a^2 - a - 6}{a^2 + 5a + 6} = \frac{a - 3}{a + 3}.$$

(e) $\frac{y^2 - 4}{2y + 4} = \frac{(y + 2)(y - 2)}{2(y + 2)} = \frac{y - 2}{2}$

Be careful! When using the fundamental principle of rational numbers, only common *factors* may be divided. For example,

$$\frac{y - 2}{2} \neq y \quad \text{and} \quad \frac{y - 2}{2} \neq y - 1$$

because the 2 in $y - 2$ is not a *factor* of the numerator. Remember to *factor* before writing a fraction in lowest terms.

In the rational expression from Example 2(d),

$$\frac{a^2 - a - 6}{a^2 + 5a + 6}, \quad \text{or} \quad \frac{(a - 3)(a + 2)}{(a + 3)(a + 2)},$$

a can take any value except -3 or -2. In the rational expression

$$\frac{a - 3}{a + 3}$$

a cannot equal -3. Because of this,

$$\frac{a^2 - a - 6}{a^2 + 5a + 6} = \frac{a - 3}{a + 3}$$

for all values of a except -3 or -2. From now on such statements of equality will be made with the understanding that they apply only for those real numbers that make neither denominator equal 0, and we will no longer state these restrictions.

EXAMPLE 3 Writing Rational Expressions in Lowest Terms

Write each rational expression in lowest terms.

(a) $\frac{m - 3}{3 - m}$

In this rational expression, the numerator and denominator are opposites. The given expression can be written in lowest terms by writing the denominator as $3 - m = -1(m - 3)$, giving

$$\frac{m - 3}{3 - m} = \frac{m - 3}{-1(m - 3)} = \frac{1}{-1} = -1.$$

(b) $\dfrac{r^2 - 16}{4 - r} = \dfrac{(r + 4)(r - 4)}{4 - r}$

$= \dfrac{(r + 4)(r - 4)}{\mathbf{-1(r - 4)}}$ Write $4 - r$ as $-1(r - 4)$.

$= \dfrac{r + 4}{-1}$ Lowest terms

$= -(r + 4)$ or $-r - 4$

Working as in Examples 3(a) and (b), the quotient

$$\frac{a}{-a} \quad (a \neq 0)$$

can be simplified as

$$\frac{a}{-a} = \frac{a}{-1(a)} = \frac{1}{-1} = -1.$$

Based on this result,

$$\frac{q - 7}{7 - q} = -1, \quad \text{and} \quad \frac{-5a + 2b}{5a - 2b} = -1,$$

but

$$\frac{r - 2}{r + 2}$$

cannot be simplified further.

OBJECTIVE 4 Multiply rational expressions. To multiply rational expressions, follow these steps. (In practice, we usually simplify before multiplying.)

Multiplying Rational Expressions

Step 1 **Factor.** Factor all numerators and denominators as completely as possible.

Step 2 **Reduce.** Apply the fundamental principle.

Step 3 **Multiply.** Multiply remaining factors in the numerator and remaining factors in the denominator.

Step 4 **Check.** Check to be sure the product is in lowest terms.

EXAMPLE 4 Multiplying Rational Expressions

Multiply.

(a) $\dfrac{3x^2}{5} \cdot \dfrac{10}{x^3} = \dfrac{\mathbf{3x^2 \cdot 10}}{\mathbf{5 \cdot x^3}} = \dfrac{30x^2}{5x^3} = \dfrac{6 \cdot 5x^2}{x \cdot 5x^2} = \dfrac{6}{x}$

Apply the fundamental principle using $5x^2$ to write the product in lowest terms. Notice that common factors in the numerator and denominator can be divided out *before* multiplying the numerator factors and the denominator factors as follows.

$$\frac{3x^2}{5} \cdot \frac{10}{x^3} = \frac{3x^2}{5} \cdot \frac{2 \cdot 5}{x \cdot x^2} = \frac{6}{x}$$

This is the most efficient way to multiply fractions.

(b) $\dfrac{5p-5}{p} \cdot \dfrac{3p^2}{10p-10}$

Factor where possible.

$$\frac{5p-5}{p} \cdot \frac{3p^2}{10p-10} = \frac{5(p-1)}{p} \cdot \frac{3p \cdot p}{2 \cdot 5(p-1)} \quad \text{Factor.}$$

$$= \frac{1}{1} \cdot \frac{3p}{2} \quad \text{Lowest terms}$$

$$= \frac{3p}{2} \quad \text{Multiply.}$$

(c) $\dfrac{k^2+2k-15}{k^2-4k+3} \cdot \dfrac{k^2-k}{k^2+k-20} = \dfrac{(k+5)(k-3)}{(k-3)(k-1)} \cdot \dfrac{k(k-1)}{(k+5)(k-4)}$

$$= \frac{k}{k-4}$$

(d) $(p-4) \cdot \dfrac{3}{5p-20}$

$$(p-4) \cdot \frac{3}{5p-20} = \frac{p-4}{1} \cdot \frac{3}{5p-20} \quad \text{Write } p-4 \text{ as } \tfrac{p-4}{1}.$$

$$= \frac{p-4}{1} \cdot \frac{3}{5(p-4)} \quad \text{Factor.}$$

$$= \frac{3}{5}$$

(e) $\dfrac{a^2b^3c^4}{(ab^2)^2c} \cdot \dfrac{(a^2b)^3c^2}{(abc^3)^2}$

$$\frac{a^2b^3c^4}{(ab^2)^2c} \cdot \frac{(a^2b)^3c^2}{(abc^3)^2} = \frac{a^2b^3c^4}{a^2b^4c} \cdot \frac{a^6b^3c^2}{a^2b^2c^6} \quad \text{Power rules for exponents}$$

Use the definition of multiplication and then the product and quotient rules for exponents.

$$= \frac{a^8b^6c^6}{a^4b^6c^7} \quad \text{Product rule for exponents}$$

$$= a^{8-4}b^{6-6}c^{6-7} \quad \text{Quotient rule for exponents}$$

$$= a^4b^0c^{-1}$$

$$= \frac{a^4}{c} \quad b^0 = 1;\ c^{-1} = \tfrac{1}{c}$$

(f) $\dfrac{x^2+2x}{x+1} \cdot \dfrac{x^2-1}{x^3+x^2}$

$$\frac{x^2+2x}{x+1} \cdot \frac{x^2-1}{x^3+x^2} = \frac{x(x+2)}{x+1} \cdot \frac{(x+1)(x-1)}{x^2(x+1)} \quad \text{Factor where possible.}$$

$$= \frac{x(x+2)(x+1)(x-1)}{x^2(x+1)(x+1)} \quad \text{Multiply.}$$

$$\frac{x^2+2x}{x+1} \cdot \frac{x^2-1}{x^3+x^2} = \frac{(x+2)(x-1)}{x(x+1)} \quad \text{Lowest terms}$$

OBJECTIVE 5 **Find reciprocals for rational expressions.** Recall that rational numbers $\frac{a}{b}$ and $\frac{c}{d}$ are reciprocals of each other if they have a product of 1. The **reciprocal** of a rational expression can be defined in the same way: Two rational expressions are reciprocals of each other if they have a product of 1. Recall that 0 has no reciprocal. The chart shows several rational expressions and their reciprocals. In each case, check that the product of the rational expression and its reciprocal is 1.

Rational Expression	*Reciprocal*
$\frac{5}{k}$	$\frac{k}{5}$
$\frac{m^2 - 9m}{2}$	$\frac{2}{m^2 - 9m}$
$\frac{0}{4}$	undefined

The examples in the chart suggest the following procedure.

Reciprocal

To find the reciprocal of a nonzero rational expression, invert the rational expression.

OBJECTIVE 6 **Divide rational expressions.** Dividing rational expressions is like dividing rational numbers.

Dividing Rational Expressions

To divide two rational expressions, *multiply* the first by the reciprocal of the second.

EXAMPLE 5 **Dividing Rational Expressions**

Divide.

(a) $\frac{2z}{9} \div \frac{5z^2}{18} = \frac{2z}{9} \cdot \frac{18}{5z^2}$ Multiply by the reciprocal of the divisor.

$= \frac{4}{5z}$

(b) $\frac{8k - 16}{3k} \div \frac{3k - 6}{4k^2} = \frac{8k - 16}{3k} \cdot \frac{4k^2}{3k - 6}$ Multiply by the reciprocal.

$= \frac{8(k - 2)}{3k} \cdot \frac{4k^2}{3(k - 2)}$ Factor.

$= \frac{32k}{9}$

(c) $\dfrac{5m^2 + 17m - 12}{3m^2 + 7m - 20} \div \dfrac{5m^2 + 2m - 3}{15m^2 - 34m + 15}$

$$= \frac{(5m - 3)(m + 4)}{(m + 4)(3m - 5)} \div \frac{(5m - 3)(m + 1)}{(3m - 5)(5m - 3)}$$

$$= \frac{(5m - 3)(m + 4)}{(m + 4)(3m - 5)} \cdot \frac{(3m - 5)(5m - 3)}{(5m - 3)(m + 1)}$$

$$= \frac{5m - 3}{m + 1}$$

EXAMPLE 6 Dividing Rational Expressions

Divide $\dfrac{m^2pq^3}{mp^4}$ by $\dfrac{m^5p^2q}{mpq^2}$.

Use the definitions of division and multiplication and the properties of exponents.

$$\frac{m^2pq^3}{mp^4} \div \frac{m^5p^2q}{mpq^2} = \frac{m^2pq^3}{mp^4} \cdot \frac{mpq^2}{m^5p^2q}$$

$$= \frac{m^3p^2q^5}{m^6p^6q}$$

$$= \frac{q^4}{m^3p^4}$$

6.1 EXERCISES

Rational expressions often can be written in lowest terms in seemingly *different ways. For example,*

$$\frac{y - 3}{-5} \quad \textit{and} \quad \frac{-y + 3}{5}$$

look different, but we get the second quotient by multiplying the first by -1 *in both the numerator and denominator. As practice in recognizing equivalent rational expressions, match the expressions in Exercises 1–6 with their equivalents in choices A–F.*

1. $\dfrac{x - 3}{x + 4}$ **A.** $\dfrac{-x - 3}{4 - x}$

2. $\dfrac{x + 3}{x - 4}$ **B.** $\dfrac{-x - 3}{-x - 4}$

3. $\dfrac{x - 3}{x - 4}$ **C.** $\dfrac{3 - x}{-x - 4}$

4. $\dfrac{x + 3}{x + 4}$ **D.** $\dfrac{-x + 3}{-x + 4}$

5. $\dfrac{3 - x}{x + 4}$ **E.** $\dfrac{x - 3}{-x - 4}$

6. $\dfrac{x + 3}{4 - x}$ **F.** $\dfrac{-x - 3}{x - 4}$

7. In Example 1(a), we showed that the domain of the rational function $f(k) = \dfrac{3}{7k - 14}$ does not include 2. Explain in your own words why this is so. In general, how do we find the value or values excluded from the domain of a rational function?

8. The domain of the rational function $g(x) = \dfrac{x + 1}{x^2 + 3}$ includes all real numbers. Explain why this is so.

Find all numbers that are not in the domain of the function. See Example 1.

9. $f(z) = \dfrac{z}{z - 7}$ **10.** $f(r) = \dfrac{r}{r + 3}$ **11.** $f(p) = \dfrac{6p - 5}{7p + 1}$

12. $f(x) = \dfrac{8x - 3}{2x + 7}$ **13.** $f(x) = \dfrac{12x + 3}{x}$ **14.** $f(x) = \dfrac{9x + 8}{x}$

15. $f(x) = \dfrac{3x + 1}{2x^2 + x - 6}$ **16.** $f(x) = \dfrac{2x + 4}{3x^2 + 11x - 42}$ **17.** $f(x) = \dfrac{x + 2}{14}$

18. $f(x) = \dfrac{x - 9}{26}$ **19.** $f(x) = \dfrac{2x^2 - 3x + 4}{3x^2 + 8}$ **20.** $f(x) = \dfrac{9x^2 - 8x + 3}{4x^2 + 1}$

21. **(a)** Identify the two *terms* in the numerator and the two *terms* in the denominator of the rational expression $\dfrac{x^2 + 4x}{x + 4}$.

(b) Describe the steps you would use to rewrite this rational expression in lowest terms. (*Hint:* It simplifies to x.)

22. Only one of the following rational expressions can be simplified. Which one is it?

(a) $\dfrac{x^2 + 2}{x^2}$ **(b)** $\dfrac{x^2 + 2}{2}$ **(c)** $\dfrac{x^2 + y^2}{y^2}$ **(d)** $\dfrac{x^2 - 5x}{x}$

A ratio is a quotient of two quantities. Ratios provide a way to compare two quantities. Use the graphs to write the required ratios in Exercises 23 and 24.

23. Write the ratios of domestic sales to total U.S. sales in 1992 and in 1996 as fractions and then as decimals to the nearest hundredth, using a calculator. In which year were U.S. domestic sales a larger fraction of total sales?

24. Write the ratios of sales of imports to total U.S. sales in 1992 and 1996 as fractions and then as decimals to the nearest hundredth, using a calculator. In which year were import sales a larger fraction of U.S. total sales?

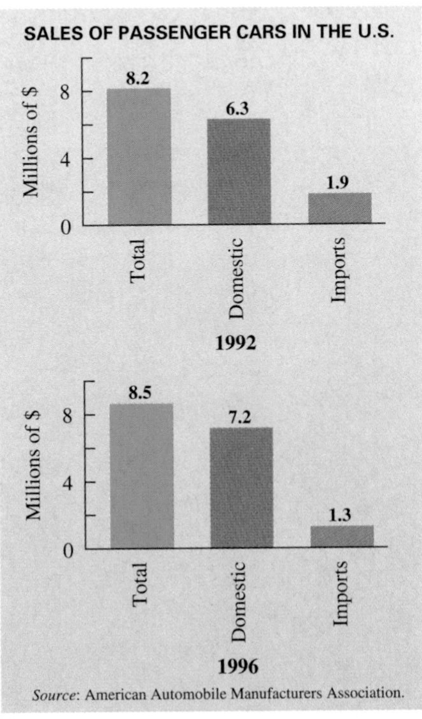

Source: American Automobile Manufacturers Association.

Write each rational expression in lowest terms. See Example 2.

25. $\dfrac{24x^2y^4}{18xy^5}$ **26.** $\dfrac{36m^4n^3}{24m^2n^5}$ **27.** $\dfrac{(x+4)(x-3)}{(x+5)(x+4)}$

28. $\dfrac{(2x+7)(x-1)}{(2x+3)(2x+7)}$ **29.** $\dfrac{4x(x+3)}{8x^2(x-3)}$ **30.** $\dfrac{5y^2(y+8)}{15y(y-8)}$

31. $\dfrac{3x+7}{3}$ **32.** $\dfrac{4x-9}{4}$ **33.** $\dfrac{6m+18}{7m+21}$

34. $\dfrac{5r-20}{3r-12}$ **35.** $\dfrac{3z^2+z}{18z+6}$ **36.** $\dfrac{2x^2-5x}{16x-40}$

37. $\dfrac{2t+6}{t^2-9}$ **38.** $\dfrac{5s-25}{s^2-25}$ **39.** $\dfrac{x^2+2x-15}{x^2+6x+5}$

40. $\dfrac{y^2-5y-14}{y^2+y-2}$ **41.** $\dfrac{8x^2-10x-3}{8x^2-6x-9}$ **42.** $\dfrac{12x^2-4x-5}{8x^2-6x-5}$

43. $\dfrac{a^3+b^3}{a+b}$ **44.** $\dfrac{r^3-s^3}{r-s}$ **45.** $\dfrac{2c^2+2cd-60d^2}{2c^2-12cd+10d^2}$

46. $\dfrac{3s^2-9st-54t^2}{3s^2-6st-72t^2}$ **47.** $\dfrac{ac-ad+bc-bd}{ac-ad-bc+bd}$ **48.** $\dfrac{2xy+2xw+y+w}{2xy+y-2xw-w}$

49. Only one of the following rational expressions is *not* equivalent to

$$\frac{x-3}{4-x}.$$

Which one is it?

(a) $\dfrac{3-x}{x-4}$ **(b)** $\dfrac{x+3}{4+x}$ **(c)** $-\dfrac{3-x}{4-x}$ **(d)** $-\dfrac{x-3}{x-4}$

50. Which of the following rational expressions equals -1?

(a) $\dfrac{2x+3}{2x-3}$ **(b)** $\dfrac{2x-3}{3-2x}$ **(c)** $\dfrac{2x+3}{3+2x}$ **(d)** $\dfrac{2x+3}{-2x-3}$

Write each rational expression in lowest terms. See Example 3.

51. $\dfrac{7-b}{b-7}$ **52.** $\dfrac{r-13}{13-r}$ **53.** $\dfrac{x^2-y^2}{y-x}$ **54.** $\dfrac{m^2-n^2}{n-m}$

55. $\dfrac{(a-3)(x+y)}{(3-a)(x-y)}$ **56.** $\dfrac{(8-p)(x+2)}{(p-8)(x-2)}$ **57.** $\dfrac{5k-10}{20-10k}$

58. $\dfrac{7x-21}{63-21x}$ **59.** $\dfrac{a^2-b^2}{a^2+b^2}$ **60.** $\dfrac{p^2+q^2}{p^2-q^2}$

61. Explain in a few words how to multiply rational expressions. Give an example.

62. Explain in a few words how to divide rational expressions. Give an example.

Multiply or divide as indicated. See Examples 4–6.

63. $\dfrac{x^3}{3y}\cdot\dfrac{9y^2}{x^5}$ **64.** $\dfrac{a^4}{5b^2}\cdot\dfrac{25b^4}{a^3}$ **65.** $\dfrac{5a^4b^2}{16a^2b}\div\dfrac{25a^2b}{60a^3b^2}$

66. $\dfrac{s^3t^2}{10s^2t^4}\div\dfrac{8s^4t^2}{5t^6}$ **67.** $\dfrac{(5pq^2)^2}{60p^3q^6}\div\dfrac{5p^2q^2}{16p^2q^3}$ **68.** $\dfrac{(6s^2t)^2}{(s^4t^5)^3}\div\dfrac{s^3t^2x^2}{3x^5t}$

69. $\dfrac{4x}{8x+4}\cdot\dfrac{14x+7}{6}$ **70.** $\dfrac{12x-20}{5x}\cdot\dfrac{6}{9x-15}$

71. $\dfrac{p^2-25}{4p}\cdot\dfrac{2}{5-p}$ **72.** $\dfrac{a^2-1}{4a}\cdot\dfrac{2}{1-a}$

73. $\dfrac{m^2 - 49}{m + 1} \div \dfrac{7 - m}{m}$

74. $\dfrac{k^2 - 4}{3k^2} \div \dfrac{2 - k}{11k}$

75. $\dfrac{12x - 10y}{3x + 2y} \cdot \dfrac{6x + 4y}{10y - 12x}$

76. $\dfrac{9s - 12t}{2s + 2t} \cdot \dfrac{3s + 3t}{4t - 3s}$

77. $\dfrac{x^2 - 25}{x^2 + x - 20} \cdot \dfrac{x^2 + 7x + 12}{x^2 - 2x - 15}$

78. $\dfrac{t^2 - 49}{t^2 + 4t - 21} \cdot \dfrac{t^2 + 8t + 15}{t^2 - 2t - 35}$

79. $\dfrac{6x^2 + 5xy - 6y^2}{12x^2 - 11xy + 2y^2} \div \dfrac{4x^2 - 12xy + 9y^2}{8x^2 - 14xy + 3y^2}$

80. $\dfrac{8a^2 - 6ab - 9b^2}{6a^2 - 5ab - 6b^2} \div \dfrac{4a^2 + 11ab + 6b^2}{9a^2 + 12ab + 4b^2}$

81. $\dfrac{3k^2 + 17kp + 10p^2}{6k^2 + 13kp - 5p^2} \div \dfrac{6k^2 + kp - 2p^2}{6k^2 - 5kp + p^2}$

82. $\dfrac{16c^2 + 24cd + 9d^2}{16c^2 - 16cd + 3d^2} \div \dfrac{16c^2 - 9d^2}{16c^2 - 24cd + 9d^2}$

83. $\left(\dfrac{6k^2 - 13k - 5}{k^2 + 7k} \div \dfrac{2k - 5}{k^3 + 6k^2 - 7k}\right) \cdot \dfrac{k^2 - 5k + 6}{3k^2 - 8k - 3}$

84. $\left(\dfrac{2x^3 + 3x^2 - 2x}{3x - 15} \div \dfrac{2x^3 - x^2}{x^2 - 3x - 10}\right) \cdot \dfrac{5x^2 - 10x}{3x^2 + 12x + 12}$

85. $\dfrac{a^2(2a + b) + 6a(2a + b) + 5(2a + b)}{3a^2(a + 2b) - 2a(a + 2b) - (a + 2b)} \div \dfrac{a + 1}{a - 1}$

86. $\dfrac{2x^2(x - 3z) - 5x(x - 3z) + 2(x - 3z)}{4x^2(3z - x) - 11x(3z - x) + 6(3z - x)} \div \dfrac{4x + 1}{4x - 3}$

TECHNOLOGY INSIGHTS (EXERCISES 87–90)

The calculator screen shows tables of values for $Y_1 = \dfrac{x^2 + x - 2}{x + 2} = \dfrac{(x + 2)(x - 1)}{x + 2}$ and $Y_2 = x - 1$, as shown at the bottom of the screen. Notice that the values in the table are the same for Y_1 and Y_2 except when $x = -2$, since -2 makes the denominator of Y_1 equal 0.

X	Y1	Y2
-5	-6	-6
-4	-5	-5
-3	-4	-4
-2	ERROR	-3
-1	-2	-2
0	-1	-1
1	0	0

Y2=X−1

In each of the following exercises, determine the expression that defines Y_1.

87.

X	Y1	Y2
-2	-5	-5
-1	-4	-4
0	-3	-3
1	-2	-2
2	ERROR	-1
3	0	0
4	1	1

Y2=X−3

88.

X	Y1	Y2
-4	0	0
-3	1	1
-2	2	2
-1	ERROR	3
0	4	4
1	5	5
2	6	6

Y2=X+4

(continued)

TECHNOLOGY INSIGHTS (EXERCISES 87-90) (CONTINUED)

89.

X	Y1	Y2
-1	4	4
0	5	5
1	6	6
2	ERROR	7
3	8	8
4	9	9
5	10	10

Y2=X+5

90.

X	Y1	Y2
-7	-10	-10
-6	-9	-9
-5	-8	-8
-4	ERROR	-7
-3	-6	-6
-2	-5	-5
-1	-4	-4

Y2=X-3

6.2 Addition and Subtraction of Rational Expressions

OBJECTIVES

1. Add and subtract rational expressions with the same denominator.
2. Find a least common denominator.
3. Add and subtract rational expressions with different denominators.

FOR EXTRA HELP

SSG Sec. 6.2
SSM Sec. 6.2

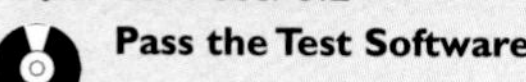
Pass the Test Software

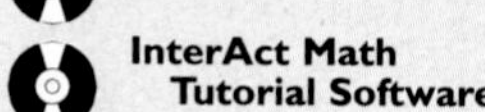
InterAct Math Tutorial Software

Video 11

OBJECTIVE 1 Add and subtract rational expressions with the same denominator. The following steps, used to add or subtract rational numbers, are also used to add or subtract rational expressions.

Adding or Subtracting Rational Expressions

Step 1 **If the denominators are the same,** add or subtract the numerators. Place the result over the common denominator.

If the denominators are different, first find the least common denominator. Write all rational expressions with this least common denominator, and then add or subtract the numerators. Place the result over the common denominator.

Step 2 **Simplify.** Write all answers in lowest terms.

EXAMPLE 1 Adding and Subtracting Rational Expressions with the Same Denominators

Add or subtract as indicated.

(a) $\dfrac{3y}{5} + \dfrac{x}{5} = \dfrac{3y + x}{5}$

The denominators of these rational expressions are the same, so just add the numerators, and place the sum over the common denominator.

(b) $\dfrac{7}{2r^2} - \dfrac{11}{2r^2} = \dfrac{7 - 11}{2r^2} = \dfrac{-4}{2r^2} = -\dfrac{2}{r^2}$ Lowest terms

Subtract the numerators since the denominators are the same, and keep the common denominator.

(c) $\dfrac{m}{m^2 - p^2} + \dfrac{p}{m^2 - p^2} = \dfrac{m + p}{m^2 - p^2}$ Add numerators; keep the common denominator.

$= \dfrac{m + p}{(m + p)(m - p)}$ Factor.

$= \dfrac{1}{m - p}$ Lowest terms

(d) $$\frac{4}{x^2+2x-8} + \frac{x}{x^2+2x-8} = \frac{4+x}{x^2+2x-8}$$
$$= \frac{4+x}{(x-2)(x+4)}$$
$$= \frac{1}{x-2}$$

OBJECTIVE 2 **Find a least common denominator.** We add or subtract rational expressions with different denominators by first writing them with a common denominator, usually the **least common denominator (LCD).**

Finding the Least Common Denominator

Step 1 **Factor.** Factor each denominator.

Step 2 **Find the least common denominator.** The LCD is the product of all different factors from each denominator, with each factor raised to the *greatest* power that occurs in any denominator.

EXAMPLE 2 Finding Least Common Denominators

Find the least common denominator for each pair of denominators.

(a) $5xy^2$, $2x^3y$

Each denominator is already factored.

$$5xy^2 = 5 \cdot x \cdot y^2$$
$$2x^3y = 2 \cdot x^3 \cdot y$$

Greatest exponent on x is 3.

$$\text{LCD} = 5 \cdot 2 \cdot x^3 \cdot y^2$$ ← Greatest exponent on y is 2.

$$= 10x^3y^2$$

(b) $k-3$, k

The LCD, an expression divisible by both $k-3$ and k, is

$$k(k-3).$$

It is often best to leave a least common denominator in factored form.

(c) y^2-2y-8, y^2+3y+2

Factor the denominators to get

$$y^2-2y-8 = (y-4)(y+2)$$
$$y^2+3y+2 = (y+2)(y+1).$$

The LCD, divisible by both polynomials, is

$$(y-4)(y+2)(y+1).$$

(d) $8z-24$, $5z^2-15z$

$$8z-24 = 8(z-3) \quad \text{and} \quad 5z^2-15z = 5z(z-3) \qquad \text{Factor.}$$

The LCD is $40z(z-3)$.

OBJECTIVE 3 Add and subtract rational expressions with different denominators. Before adding or subtracting two rational expressions, we write each expression with the least common denominator by multiplying its numerator and denominator by the factors needed to get the LCD. This procedure is valid because we are multiplying each rational expression by a form of 1, the identity element for multiplication.

EXAMPLE 3 Adding and Subtracting Rational Expressions with Different Denominators

Add or subtract as indicated.

(a) $\frac{5}{2p} + \frac{3}{8p}$

The LCD for $2p$ and $8p$ is $8p$. To write the first rational expression with a denominator of $8p$, multiply by $\frac{4}{4}$.

$$\frac{5}{2p} + \frac{3}{8p} = \frac{5 \cdot 4}{2p \cdot 4} + \frac{3}{8p} \qquad \text{Fundamental principle}$$

$$= \frac{20}{8p} + \frac{3}{8p}$$

$$= \frac{20 + 3}{8p} \qquad \text{Add numerators.}$$

$$= \frac{23}{8p}$$

(b) $\frac{6}{r} - \frac{5}{r - 3}$

The LCD is $r(r - 3)$. Rewrite each rational expression with this denominator.

$$\frac{6}{r} - \frac{5}{r - 3} = \frac{6(r - 3)}{r(r - 3)} - \frac{r \cdot 5}{r(r - 3)} \qquad \text{Fundamental principle}$$

$$= \frac{6r - 18}{r(r - 3)} - \frac{5r}{r(r - 3)} \qquad \text{Distributive property}$$

$$= \frac{6r - 18 - 5r}{r(r - 3)} \qquad \text{Subtract numerators.}$$

$$= \frac{r - 18}{r(r - 3)} \qquad \text{Combine terms in numerator.}$$

CAUTION One of the most common sign errors in algebra occurs when a rational expression with two or more terms in the numerator is being subtracted. Remember that in this situation, the subtraction sign must be distributed to *every* term in the numerator of the fraction that follows it. Read Example 4 carefully to see how this is done.

EXAMPLE 4 Subtracting Rational Expressions

Subtract.

(a) $\dfrac{3}{k+1} - \dfrac{k-2}{k+3}$

The LCD is $(k+1)(k+3)$.

$$\frac{3}{k+1} - \frac{k-2}{k+3} = \frac{3(k+3)}{(k+1)(k+3)} - \frac{(k-2)(k+1)}{(k+3)(k+1)} \quad \text{Fundamental principle}$$

$$= \frac{3k+9}{(k+1)(k+3)} - \frac{k^2-k-2}{(k+3)(k+1)} \quad \text{Multiply in numerator.}$$

$$= \frac{3k+9-(k^2-k-2)}{(k+1)(k+3)} \quad \text{Subtract.}$$

$$= \frac{3k+9-k^2+k+2}{(k+1)(k+3)} \quad \text{Distributive property}$$

$$= \frac{-k^2+4k+11}{(k+1)(k+3)} \quad \text{Combine terms.}$$

(b) $$\frac{1}{q-1} - \frac{1}{q+1} = \frac{1(q+1)}{(q-1)(q+1)} - \frac{1(q-1)}{(q+1)(q-1)} \quad \text{Get a common denominator.}$$

$$= \frac{(q+1)-(q-1)}{(q-1)(q+1)} \quad \text{Subtract.}$$

$$= \frac{q+1-q+1}{(q-1)(q+1)}$$

$$= \frac{2}{(q-1)(q+1)} \quad \text{Combine terms.}$$

(c) $\dfrac{m+4}{m^2-2m-3} - \dfrac{2m-3}{m^2-5m+6}$

$$= \frac{m+4}{(m-3)(m+1)} - \frac{2m-3}{(m-3)(m-2)} \quad \text{Factor each denominator.}$$

The LCD is $(m-3)(m+1)(m-2)$.

$$= \frac{(m+4)(m-2)}{(m-3)(m+1)(m-2)} - \frac{(2m-3)(m+1)}{(m-3)(m-2)(m+1)}$$

$$= \frac{m^2+2m-8}{(m-3)(m+1)(m-2)} - \frac{2m^2-m-3}{(m-3)(m-2)(m+1)} \quad \text{Multiply in numerator.}$$

$$= \frac{m^2+2m-8-(2m^2-m-3)}{(m-3)(m+1)(m-2)} \quad \text{Subtract.}$$

$$= \frac{m^2+2m-8-2m^2+m+3}{(m-3)(m+1)(m-2)} \quad \text{Distributive property}$$

$$= \frac{-m^2+3m-5}{(m-3)(m+1)(m-2)} \quad \text{Combine terms.}$$ ■

EXAMPLE 5 Adding Rational Expressions with Opposites in Denominators

Add.

$$\frac{a}{(a-1)^2} + \frac{2a}{1-a^2} = \frac{a}{(a-1)^2} + \frac{2a}{(1-a)(1+a)}$$ Factor denominators.

$$= \frac{a}{(a-1)^2} + \frac{-1 \cdot 2a}{-1(1-a)(1+a)}$$ $a - 1$ and $1 - a$ are opposites.

$$= \frac{a}{(a-1)^2} + \frac{-2a}{(a-1)(a+1)}$$

$$= \frac{a(a+1)}{(a-1)^2(a+1)} + \frac{-2a(a-1)}{(a-1)(a+1)(a-1)}$$

$$= \frac{a^2+a}{(a-1)^2(a+1)} + \frac{-2a^2+2a}{(a-1)(a+1)(a-1)}$$ Multiply in numerator.

$$= \frac{a^2+a-2a^2+2a}{(a-1)^2(a+1)}$$ Add numerators.

$$= \frac{-a^2+3a}{(a-1)^2(a+1)}$$ Combine terms.

6.2 EXERCISES

RELATING CONCEPTS (EXERCISES 1-6)

Work the following exercises in order.

1. Let $x = 4$ and $y = 2$. Evaluate $\frac{1}{x} + \frac{1}{y}$.

2. Let $x = 4$ and $y = 2$. Evaluate $\frac{1}{x + y}$.

3. Are the answers for Exercises 1 and 2 the same? What can you conclude?

4. Let $x = 3$ and $y = 5$. Evaluate $\frac{1}{x} - \frac{1}{y}$.

5. Let $x = 3$ and $y = 5$. Evaluate $\frac{1}{x - y}$.

6. Are the answers for Exercises 4 and 5 the same? What can you conclude?

Did you make the connection that you *cannot* find the sum or difference of two rational expressions with the same numerator by adding or subtracting their denominators and using the common numerator?

Add or subtract as indicated. Write the answer in lowest terms. See Example 1.

7. $\frac{7}{t} + \frac{2}{t}$

8. $\frac{5}{r} + \frac{9}{r}$

9. $\frac{11}{5x} - \frac{1}{5x}$

10. $\frac{7}{4y} - \frac{3}{4y}$

11. $\frac{5x + 4}{6x + 5} + \frac{x + 1}{6x + 5}$

12. $\frac{6y + 12}{4y + 3} + \frac{2y - 6}{4y + 3}$

13. $\frac{x^2}{x + 5} - \frac{25}{x + 5}$

14. $\frac{y^2}{y + 6} - \frac{36}{y + 6}$

15. $\frac{-3p + 7}{p^2 + 7p + 12} + \frac{8p + 13}{p^2 + 7p + 12}$

16. $\frac{5x + 6}{x^2 + x - 20} + \frac{4 - 3x}{x^2 + x - 20}$

17. $\frac{a^3}{a^2 + ab + b^2} - \frac{b^3}{a^2 + ab + b^2}$

18. $\frac{p^3}{p^2 - pq + q^2} + \frac{q^3}{p^2 - pq + q^2}$

19. Write a step-by-step method for adding or subtracting rational expressions that have a common denominator. Illustrate with an example.

20. Write a step-by-step method for adding or subtracting rational expressions that have different denominators. Give an example.

Find the least common denominator for each group of denominators. See Example 2.

21. $18x^2y^3$, $24x^4y^5$

22. $24a^3b^4$, $18a^5b^2$

23. $z - 2$, z

24. $k + 3$, k

25. $2y + 8$, $y + 4$

26. $3r - 21$, $r - 7$

27. $6x + 18$, $5x + 15$

28. $4c + 12$, $6c + 18$

29. $m + n$, $m - n$

30. $r + s$, $r - s$

31. $\frac{x + 8}{x^2 - 3x - 4}, \frac{-9}{x + x^2}$

32. $\frac{y + 1}{y^2 - 8y + 12}, \frac{-3}{y^2 - 6y}$

33. $\frac{t}{2t^2 + 7t - 15}, \frac{t}{t^2 + 3t - 10}$

34. $\frac{s}{s^2 - 3s - 4}, \frac{s}{3s^2 + s - 2}$

35. $\frac{y}{2y + 6}, \frac{3}{y^2 - 9}, \frac{6}{y}$

36. $\frac{x}{9x + 18}, \frac{-1}{x^2 - 4}, \frac{5}{x}$

37. $\frac{5}{6x}, \frac{3}{x^2}, \frac{7}{x + 1}$

38. $\frac{12}{11y}, \frac{5}{y^5}, \frac{9}{y - 1}$

39. One student added two rational expressions and obtained the answer $\frac{3}{5 - y}$. Another student obtained the answer $\frac{-3}{y - 5}$ for the same problem. Is it possible that both answers are correct? Explain.

40. What is *wrong* with the following work?

$$\frac{x}{x+2} - \frac{4x-1}{x+2} = \frac{x-4x-1}{x+2} = \frac{-3x-1}{x+2}$$

Add or subtract as indicated. Write the answer in lowest terms. See Examples 3–5.

41. $\frac{8}{t} + \frac{7}{3t}$

42. $\frac{5}{x} + \frac{9}{4x}$

43. $\frac{5}{12x^2y} - \frac{11}{6xy}$

44. $\frac{7}{18a^3b^2} - \frac{2}{9ab}$

45. $\frac{1}{x-1} - \frac{1}{x}$

46. $\frac{3}{x-3} - \frac{1}{x}$

47. $\frac{3a}{a+1} + \frac{2a}{a-3}$

48. $\frac{2x}{x+4} + \frac{3x}{x-7}$

49. $\frac{3x+2}{4-x} + \frac{5-3x}{x-4}$

50. $\frac{5t+3}{2-t} + \frac{1-5t}{t-2}$

51. $\frac{-3w+2z}{w-z} - \frac{4w-z}{z-w}$

52. $\frac{2b-5a}{a-b} - \frac{6a-b}{b-a}$

53. $\frac{4x}{x-1} - \frac{2}{x+1} - \frac{4}{x^2-1}$

54. $\frac{4}{x+3} - \frac{x}{x-3} - \frac{18}{x^2-9}$

55. $\frac{5}{x-2} + \frac{1}{x} + \frac{2}{x^2-2x}$

56. $\frac{5x}{x-3} + \frac{2}{x} + \frac{6}{x^2-3x}$

57. $\frac{3x}{x+1} + \frac{4}{x-1} - \frac{6}{x^2-1}$

58. $\frac{5x}{x+3} + \frac{x+2}{x} - \frac{6}{x^2+3x}$

59. $\frac{4}{x+1} + \frac{1}{x^2-x+1} - \frac{12}{x^3+1}$

60. $\frac{5}{x+2} + \frac{2}{x^2-2x+4} - \frac{60}{x^3+8}$

61. $\frac{2x+4}{x+3} + \frac{3}{x} - \frac{6}{x^2+3x}$

62. $\frac{4x+1}{x+5} - \frac{2}{x} + \frac{10}{x^2+5x}$

63. Add $\frac{2}{m+1} + \frac{5}{m}$ using each of the common denominators given below. Be sure all answers are given in lowest terms.
(a) $m^2(m+1)$ **(b)** $m(m+1)$ **(c)** $m(m+1)^2$ **(d)** $m^2(m+1)^2$

64. Is it necessary to find the *least* common denominator when adding or subtracting fractions, or would any common denominator work? What advantage is there in using the *least* common denominator? See Exercise 63.

Add or subtract as indicated. Write the answer in lowest terms.

65. $\frac{3}{(p-2)^2} - \frac{5}{p-2} + 4$

66. $\frac{8}{(3r-1)^2} + \frac{2}{3r-1} - 6$

67. $\frac{3}{x^2-5x+6} - \frac{2}{x^2-x-2}$

68. $\frac{2}{m^2-4m+4} + \frac{3}{m^2+m-6}$

69. $\frac{5x-y}{x^2+xy-2y^2} - \frac{3x+2y}{x^2+5xy-6y^2}$

70. $\frac{6x+5y}{6x^2+5xy-4y^2} - \frac{x+2y}{9x^2-16y^2}$

71. $\frac{r+s}{3r^2+2rs-s^2} - \frac{s-r}{6r^2-5rs+s^2}$

72. $\frac{3y}{y^2+yz-2z^2} + \frac{4y-1}{y^2-z^2}$

RELATING CONCEPTS (EXERCISES 73–76)

Students often confuse the steps used to multiply (or divide) rational expressions with the steps for adding (or subtracting) rational expressions. Consider the following problems.

Problem A *Addition*

$$\frac{-x}{4xy + 3y^2} + \frac{8x + 6y}{16x^2 - 9y^2}$$

Problem B *Multiplication*

$$\frac{-x}{4xy + 3y^2} \cdot \frac{8x + 6y}{16x^2 - 9y^2}$$

The following exercises compare the steps for performing these operations on the expressions. ***Work them in order.***

73. **(a)** Factor the denominators in Problem A.
(b) Factor all numerators and denominators in Problem B.

74. **(a)** In Problem A, find the least common denominator, and write each fraction with this LCD.
(b) In Problem B, write a single fraction with the product of the numerator factors in the numerator and the product of the denominator factors in the denominator.

75. **(a)** Add the numerators in Problem A and write the sum over the LCD. Write in lowest terms, if necessary.
(b) Use the fundamental principle to write Problem B in lowest terms.

76. Discuss the similarities and differences in these two processes.

Did you make the connection on how to differentiate between adding two rational expressions versus multiplying them?

Work each problem.

77. A *concours d'elegance* is a competition in which a maximum of 100 points is awarded to a car based on its general attractiveness. The function defined by the rational expression

$$c(x) = \frac{1010}{49(101 - x)} - \frac{10}{49}$$

approximates the cost, in thousands of dollars, of restoring a car so that it will win x points.

(a) Simplify the expression for $c(x)$ by performing the indicated subtraction.
(b) Use the simplified expression to determine how much it would cost to win 95 points.

78. A *cost-benefit model* expresses the cost of an undertaking in terms of the benefits received. One cost-benefit model gives the cost in thousands of dollars to remove x percent of a certain pollutant as $c(x) = \frac{6.7x}{100 - x}$. Another model produces the relationship $c(x) = \frac{6.5x}{102 - x}$.
(a) What is the cost found by averaging the two models? (*Hint:* The average of two quantities is half their sum.)
(b) Using the two given models and your answer to part (a), find the cost to the nearest dollar to remove 95% ($x = 95$) of the pollutant.
(c) Average the two costs in part (b) from the given models. What do you notice about this result compared to the cost using the average of the two models?

6.3 Complex Fractions

OBJECTIVES

1 Simplify complex fractions by simplifying numerator and denominator.

2 Simplify complex fractions by multiplying by a common denominator.

3 Simplify rational expressions with negative exponents.

FOR EXTRA HELP

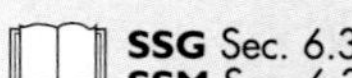
SSG Sec. 6.3
SSM Sec. 6.3

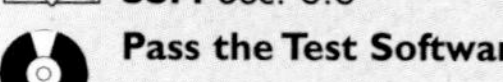
Pass the Test Software

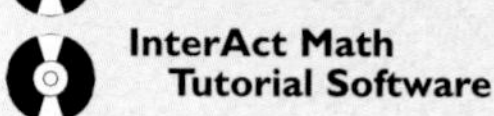
InterAct Math Tutorial Software

Video 11

A **complex fraction** is an expression having a fraction in the numerator, denominator, or both. Examples of complex fractions include

$$\frac{1+\frac{1}{x}}{2},\quad \frac{\frac{4}{y}}{6-\frac{3}{y}},\quad \text{and}\quad \frac{\frac{m^2-9}{m+1}}{\frac{m+3}{m^2-1}}.$$

OBJECTIVE 1 Simplify complex fractions by simplifying numerator and denominator. There are two different methods for simplifying complex fractions.

Simplifying Complex Fractions: *Method 1*

Step 1 Simplify the numerator and denominator separately, as much as possible.

Step 2 Divide by multiplying the numerator by the reciprocal of the denominator.

Step 3 Simplify the resulting fraction, if possible.

In Step 2, we are treating the complex fraction as a quotient of two rational expressions and dividing. Before performing this step, be sure that both the numerator and denominator are single fractions.

EXAMPLE 1 Simplifying Complex Fractions by Method 1

Use Method 1 to simplify each complex fraction.

(a) $\dfrac{\frac{x+1}{x}}{\frac{x-1}{2x}}$

Both the numerator and the denominator are already simplified, so multiply the numerator by the reciprocal of the denominator.

$$\frac{\frac{x+1}{x}}{\frac{x-1}{2x}} = \frac{x+1}{x} \div \frac{x-1}{2x} \qquad \text{Write as a division problem.}$$

$$= \frac{x+1}{x} \cdot \frac{2x}{x-1} \qquad \text{Reciprocal of } \frac{x-1}{2x}$$

$$= \frac{2(x+1)}{x-1}$$

(b) $$\frac{2+\frac{1}{y}}{3-\frac{2}{y}} = \frac{\frac{2y}{y}+\frac{1}{y}}{\frac{3y}{y}-\frac{2}{y}} = \frac{\frac{2y+1}{y}}{\frac{3y-2}{y}}$$ Simplify numerator and denominator.

$$= \frac{2y+1}{y} \cdot \frac{y}{3y-2}$$ Reciprocal of $\frac{3y-2}{y}$

$$= \frac{2y+1}{3y-2}$$

OBJECTIVE 2 Simplify complex fractions by multiplying by a common denominator. The second method for simplifying complex fractions uses the identity property for multiplication.

Simplifying Complex Fractions: *Method 2*

Step 1 Multiply the numerator and denominator of the complex fraction by the least common denominator of the fractions in the numerator and the fractions in the denominator of the complex fraction.

Step 2 Simplify the resulting fraction, if possible.

EXAMPLE 2 Simplifying Complex Fractions by Method 2

Use Method 2 to simplify each complex fraction.

(a) $$\frac{2+\frac{1}{y}}{3-\frac{2}{y}}$$

Multiply the numerator and denominator by the LCD of all the fractions in the numerator and the denominator of the complex fraction. (This is the same as multiplying by 1.) Here the LCD is y.

$$\frac{2+\frac{1}{y}}{3-\frac{2}{y}} = \frac{2+\frac{1}{y}}{3-\frac{2}{y}} \cdot 1 = \frac{\left(2+\frac{1}{y}\right)\cdot y}{\left(3-\frac{2}{y}\right)\cdot y}$$ Multiply numerator and denominator by y, since $\frac{y}{y} = 1$.

$$= \frac{2\cdot y+\frac{1}{y}\cdot y}{3\cdot y-\frac{2}{y}\cdot y}$$ Use the distributive property.

$$= \frac{2y+1}{3y-2}$$

Compare this method of solution with that used in Example 1(b) above.

(b) $\dfrac{2p + \dfrac{5}{p-1}}{3p - \dfrac{2}{p}}$

The LCD is $p(p-1)$.

$$\frac{2p + \dfrac{5}{p-1}}{3p - \dfrac{2}{p}} = \frac{2p[p(p-1)] + \dfrac{5}{p-1} \cdot p(p-1)}{3p[p(p-1)] - \dfrac{2}{p} \cdot p(p-1)}$$

$$= \frac{2p[p(p-1)] + 5p}{3p[p(p-1)] - 2(p-1)}$$

$$= \frac{2p^3 - 2p^2 + 5p}{3p^3 - 3p^2 - 2p + 2}$$

OBJECTIVE 3 **Simplify rational expressions with negative exponents.** Rational expressions and complex fractions often involve negative exponents. To simplify such expressions, we begin by rewriting the expressions with only positive exponents.

EXAMPLE 3 **Simplifying Expressions with Negative Exponents**

Simplify each of the following.

(a) $\dfrac{m^{-1} + p^{-2}}{2m^{-2} - p^{-1}}$

First write the expression with only positive exponents using the definition of a negative exponent.

$$\frac{m^{-1} + p^{-2}}{2m^{-2} - p^{-1}} = \frac{\dfrac{1}{m} + \dfrac{1}{p^2}}{\dfrac{2}{m^2} - \dfrac{1}{p}}$$

Note that the 2 in $2m^{-2}$ is not raised to the -2 power, so $2m^{-2} = \dfrac{2}{m^2}$. Simplify the complex fraction by multiplying numerator and denominator by the LCD, m^2p^2.

$$\frac{\dfrac{1}{m} + \dfrac{1}{p^2}}{\dfrac{2}{m^2} - \dfrac{1}{p}} = \frac{m^2p^2 \cdot \dfrac{1}{m} + m^2p^2 \cdot \dfrac{1}{p^2}}{m^2p^2 \cdot \dfrac{2}{m^2} - m^2p^2 \cdot \dfrac{1}{p}}$$

$$= \frac{mp^2 + m^2}{2p^2 - m^2p}$$

(b) $\dfrac{k^{-1}}{k^{-1}+1} = \dfrac{\frac{1}{k}}{\frac{1}{k}+1}$ Write with positive exponents.

$= \dfrac{k \cdot \frac{1}{k}}{k \cdot \frac{1}{k} + k \cdot 1}$ Use Method 2.

$= \dfrac{1}{1+k}$

CONNECTIONS

An infinite complex fraction, such as

$$1 + \cfrac{1}{1 + \cfrac{1}{1 + \cdots}}$$

is called a *continued fraction.* Its value can be approximated by adding more and more terms as follows.

$$1,\ 1 + \frac{1}{1},\ 1 + \cfrac{1}{1 + \cfrac{1}{1}},\ 1 + \cfrac{1}{1 + \cfrac{1}{1+1}},\ 1 + \cfrac{1}{1 + \cfrac{1}{1 + \cfrac{1}{1+1}}},\ \ldots$$

These approximations form the sequence

$$\frac{1}{1}, \frac{2}{1}, \frac{3}{2}, \frac{5}{3}, \frac{8}{5}, \frac{13}{8}, \frac{21}{13}, \ldots.$$

Both the numerators and the denominators form another sequence, called the *Fibonacci sequence,* which is found in nature, such as in the number of spirals in many pine cones and pineapples, as well as in the number of bees in succeeding generations. The sequence of fractions approaches a number called the *golden ratio.* This ratio has been widely used in art and architecture as producing the most pleasing ratio of length to width in a figure.

FOR DISCUSSION OR WRITING

1. Find the next number in the sequence of fractions given above. Use a calculator to write the fractions in the sequence, including the ones you find, in decimal form.
2. To the nearest hundredth, what decimal number do they seem to approach?

6.3 EXERCISES

1. Explain in your own words the two methods of simplifying complex fractions.
2. Method 2 of simplifying complex fractions says that we can multiply both the numerator and the denominator of the complex fraction by the same nonzero expression. What property of real numbers from Section 1.3 justifies this method?

Use either method to simplify each complex fraction. See Examples 1 and 2.

3. $\dfrac{\frac{12}{x-1}}{\frac{6}{x}}$
4. $\dfrac{\frac{24}{t+4}}{\frac{6}{t}}$
5. $\dfrac{\frac{k+1}{2k}}{\frac{3k-1}{4k}}$
6. $\dfrac{\frac{1-r}{4r}}{\frac{-1-r}{8r}}$
7. $\dfrac{\frac{4z^2x^4}{9}}{\frac{12x^2z^5}{15}}$
8. $\dfrac{\frac{3y^2x^3}{8}}{\frac{9y^3x^4}{16}}$
9. $\dfrac{\frac{1}{x}+1}{-\frac{1}{x}+1}$
10. $\dfrac{\frac{2}{k}-1}{\frac{2}{k}+1}$
11. $\dfrac{\frac{3}{x}+\frac{3}{y}}{\frac{3}{x}-\frac{3}{y}}$
12. $\dfrac{\frac{4}{t}-\frac{4}{s}}{\frac{4}{t}+\frac{4}{s}}$
13. $\dfrac{\frac{8x-24y}{10}}{\frac{x-3y}{5x}}$
14. $\dfrac{\frac{10x-5y}{12}}{\frac{2x-y}{6y}}$
15. $\dfrac{\frac{x^2-16y^2}{xy}}{\frac{1}{y}-\frac{4}{x}}$
16. $\dfrac{\frac{2}{s}-\frac{3}{t}}{\frac{4t^2-9s^2}{st}}$
17. $\dfrac{y-\frac{y-3}{3}}{\frac{4}{9}+\frac{2}{3y}}$
18. $\dfrac{p-\frac{p+2}{4}}{\frac{3}{4}-\frac{5}{2p}}$
19. $\dfrac{\frac{x+2}{x}+\frac{1}{x+2}}{\frac{5}{x}+\frac{x}{x+2}}$
20. $\dfrac{\frac{y+3}{y}-\frac{4}{y-1}}{\frac{y}{y-1}+\frac{1}{y}}$

Simplify each expression, using only positive exponents in your answer. See Example 3.

21. $\dfrac{1}{x^{-2}+y^{-2}}$
22. $\dfrac{1}{p^{-2}-q^{-2}}$
23. $\dfrac{x^{-2}+y^{-2}}{x^{-1}+y^{-1}}$
24. $\dfrac{x^{-1}-y^{-1}}{x^{-2}-y^{-2}}$
25. $(r^{-1}+s^{-1})^{-1}$
26. $((2k)^{-1}+(4s)^{-1})^{-1}$
27. **(a)** Start with the complex fraction $\dfrac{\frac{3}{mp}-\frac{4}{p}+\frac{8}{m}}{2m^{-1}-3p^{-1}}$ and write it so that there are no negative exponents in your expression.

 (b) Explain why $\dfrac{\frac{3}{mp}-\frac{4}{p}+\frac{8}{m}}{\frac{1}{2m}-\frac{1}{3p}}$ would *not* be a correct response in part (a).

 (c) Simplify the complex fraction in part (a).

28. Is $\dfrac{m^{-1} + n^{-1}}{m^{-2} + n^{-2}} = \dfrac{m^2 + n^2}{m + n}$ a true statement? Explain why or why not.

Simplify.

29. $1 - \dfrac{3}{3 - \dfrac{1}{2y}}$

30. $2 - \dfrac{1}{2 + \dfrac{1}{x}}$

31. $\dfrac{1}{p + \dfrac{1}{p + \dfrac{1}{1 + p}}}$

32. $1 - \dfrac{1}{1 - \dfrac{1}{1 - \dfrac{1}{x - 1}}}$

RELATING CONCEPTS (EXERCISES 33–38)

Simplifying a complex fraction by Method 1 is a good way to review the methods of adding, subtracting, multiplying, and dividing rational expressions. Method 2 gives a good review of the fundamental principle of rational expressions.

Refer to the complex fraction below and ***work the following exercises in order.***

$$\frac{\dfrac{4}{m} + \dfrac{m + 2}{m - 1}}{\dfrac{m + 2}{m} - \dfrac{2}{m - 1}}$$

33. Add the fractions in the numerator.

34. Subtract as indicated in the denominator.

35. Divide your answer from Exercise 33 by your answer from Exercise 34.

36. Go back to the original complex fraction and find the LCD of all denominators.

37. Multiply the numerator and denominator of the complex fraction by your answer from Exercise 36.

38. Your answers for Exercises 35 and 37 should be the same. Write a paragraph comparing the two methods. Which method do you prefer? Explain why.

Did you make the connection between the operations on rational expressions and complex fractions?

6.4 Division of Polynomials

OBJECTIVES

1. Divide a polynomial by a monomial.
2. Divide a polynomial by a polynomial of two or more terms.
3. Divide polynomial functions.

OBJECTIVE 1 Divide a polynomial by a monomial. In the previous chapter, we added, subtracted, and multiplied polynomials. We now discuss polynomial division, beginning with division by a monomial. (Recall that a monomial is a single term, such as $8x$, $-9m^4$, or $11y^2$.)

Dividing by a Monomial

To divide a polynomial by a monomial, divide each term in the polynomial by the monomial, and then write each quotient in lowest terms.

EXAMPLE 1 Dividing a Polynomial by a Monomial

Divide $15x^2 - 12x + 6$ by 3.

Divide each term of the polynomial by 3. Then write the result in lowest terms.

$$\frac{15x^2 - 12x + 6}{3} = \frac{15x^2}{3} - \frac{12x}{3} + \frac{6}{3}$$
$$= 5x^2 - 4x + 2$$

Check this answer by multiplying it by the divisor, 3. If you are correct, you should get $15x^2 - 12x + 6$ as the result.

$$\underset{\text{Divisor}}{3}\underset{\text{Quotient}}{(5x^2 - 4x + 2)} = \underset{\text{Original polynomial}}{15x^2 - 12x + 6}$$

EXAMPLE 2 Dividing a Polynomial by a Monomial

Find each quotient.

(a) $$\frac{5m^3 - 9m^2 + 10m}{5m^2} = \frac{5m^3}{5m^2} - \frac{9m^2}{5m^2} + \frac{10m}{5m^2}$$ Divide each term by $5m^2$.

$$= m - \frac{9}{5} + \frac{2}{m}$$ Write in lowest terms.

This result is not a polynomial; the quotient of two polynomials need not be a polynomial.

(b) $$\frac{8xy^2 - 9x^2y + 6x^2y^2}{x^2y^2} = \frac{8xy^2}{x^2y^2} - \frac{9x^2y}{x^2y^2} + \frac{6x^2y^2}{x^2y^2}$$
$$= \frac{8}{x} - \frac{9}{y} + 6$$

OBJECTIVE 2 Divide a polynomial by a polynomial of two or more terms. Earlier, we saw that the quotient of two polynomials can be found by factoring and then dividing out any common factors. For instance,

$$\frac{2x^2 + 5x - 3}{4x - 2} = \frac{(2x - 1)(x + 3)}{2(2x - 1)} = \frac{x + 3}{2}.$$

When the polynomials in a quotient of two polynomials have no common factors or cannot be factored, they can be divided by a process very similar to that for dividing one whole number by another. The following examples show how this is done.

EXAMPLE 3 Dividing a Polynomial by a Polynomial

Divide $2m^2 + m - 10$ by $m - 2$.

Write the problem, making sure that both polynomials are written in descending powers of the variables.

$$m - 2\overline{)2m^2 + m - 10}$$

Divide the first term of $2m^2 + m - 10$ by the first term of $m - 2$. Since $\frac{2m^2}{m} = 2m$, place this result above the division line.

$$\begin{array}{r} 2m \phantom{{}+m-10} \\ m-2\overline{)2m^2+\ m-10} \end{array} \quad \leftarrow \text{Result of } \tfrac{2m^2}{m}$$

Multiply $m - 2$ and $2m$, and write the result below $2m^2 + m - 10$.

$$\begin{array}{r} 2m \phantom{{}+m-10} \\ m-2\overline{)2m^2+\ m-10} \\ \underline{2m^2-4m}\phantom{{}-10} \end{array} \quad \leftarrow 2m(m-2) = 2m^2-4m$$

Now subtract $2m^2 - 4m$ from $2m^2 + m$. Do this by mentally changing the signs on $2m^2 - 4m$ and *adding*.

$$\begin{array}{r} 2m \phantom{{}+m-10} \\ m-2\overline{)2m^2+\ m-10} \\ \underline{2m^2-4m}\phantom{{}-10} \\ 5m\phantom{{}-10} \end{array} \quad \leftarrow \text{Subtract.}$$

Bring down -10 and continue by dividing $5m$ by m.

$$\begin{array}{r} 2m+\ 5 \\ m-2\overline{)2m^2+\ m-10} \\ \underline{2m^2-4m}\phantom{{}-10} \\ 5m-10 \\ \underline{5m-10} \\ 0 \end{array} \quad \begin{array}{l} \leftarrow \tfrac{5m}{m} = 5 \\ \\ \\ \leftarrow \text{Bring down } -10. \\ \leftarrow 5(m-2) = 5m-10 \\ \leftarrow \text{Subtract.} \end{array}$$

Finally, $\frac{2m^2 + m - 10}{m - 2} = 2m + 5$. Check by multiplying $m - 2$ and $2m + 5$. The result should be $2m^2 + m - 10$. Since there is no remainder, this quotient could have been found by factoring and writing in lowest terms.

EXAMPLE 4 Dividing a Polynomial with a Missing Term

Divide $3x^3 - 2x + 5$ by $x - 3$.

Make sure that $3x^3 - 2x + 5$ is in descending powers of the variable. Add a term with 0 coefficient as a placeholder for the missing x^2 term.

$$x-3\overline{)3x^3+0x^2-2x+5} \quad \leftarrow \text{Missing term } (0x^2)$$

Start with $\frac{3x^3}{x} = 3x^2$.

$$\begin{array}{r} 3x^2 \phantom{{}-2x+5} \\ x-3\overline{)3x^3+0x^2-2x+5} \\ \underline{3x^3-9x^2}\phantom{{}-2x+5} \end{array} \quad \begin{array}{l} \leftarrow \tfrac{3x^3}{x} = 3x^2 \\ \\ \leftarrow 3x^2(x-3) \end{array}$$

Subtract by changing the signs on $3x^3 - 9x^2$ and adding.

$$\begin{array}{r} 3x^2 \phantom{{}-2x+5} \\ x-3\overline{)3x^3+0x^2-2x+5} \\ \underline{3x^3-9x^2}\phantom{{}-2x+5} \\ 9x^2\phantom{{}-2x+5} \end{array} \quad \leftarrow \text{Subtract.}$$

Bring down the next term.

$$
\begin{array}{rl}
 & 3x^2 \\
x - 3 \,\big)\! & \overline{3x^3 + 0x^2 - 2x + 5} \\
 & \underline{3x^3 - 9x^2} \\
 & \qquad 9x^2 - 2x \qquad \leftarrow \text{Bring down } -2x.
\end{array}
$$

In the next step, $\dfrac{9x^2}{x} = 9x$.

$$
\begin{array}{rl}
 & 3x^2 + 9x \qquad \leftarrow \tfrac{9x^2}{x} = 9x \\
x - 3 \,\big)\! & \overline{3x^3 + 0x^2 - 2x + 5} \\
 & \underline{3x^3 - 9x^2} \\
 & \qquad 9x^2 - 2x \\
 & \qquad \underline{9x^2 - 27x} \qquad \leftarrow 9x(x - 3) \\
 & \qquad\qquad 25x + 5 \qquad \leftarrow \text{Subtract and bring down 5.}
\end{array}
$$

Finally, $\dfrac{25x}{x} = 25$.

$$
\begin{array}{rl}
 & 3x^2 + 9x + 25 \qquad \leftarrow \tfrac{25x}{x} = 25 \\
x - 3 \,\big)\! & \overline{3x^3 + 0x^2 - 2x + 5} \\
 & \underline{3x^3 - 9x^2} \\
 & \qquad 9x^2 - 2x \\
 & \qquad \underline{9x^2 - 27x} \\
 & \qquad\qquad 25x + 5 \\
 & \qquad\qquad \underline{25x - 75} \qquad \leftarrow 25(x - 3) \\
 & \qquad\qquad\qquad 80 \qquad \leftarrow \text{Subtract.}
\end{array}
$$

We write the remainder, 80, as the numerator of the fraction $\dfrac{80}{x - 3}$. In summary,

$$\frac{3x^3 - 2x + 5}{x - 3} = 3x^2 + 9x + 25 + \frac{80}{x - 3}.$$

Check by multiplying $x - 3$ and $3x^2 + 9x + 25$ and adding 80 to the result. You should get $3x^3 - 2x + 5$.

Remember the + sign when adding the remainder to a quotient.

EXAMPLE 5 Getting a Fractional Coefficient in the Quotient

Divide $2p^3 + 5p^2 + p - 2$ by $2p + 2$.

$$
\begin{array}{rl}
 & p^2 + \frac{3}{2}p - 1 \qquad \leftarrow \tfrac{3p^2}{2p} = \tfrac{3}{2}p \\
2p + 2 \,\big)\! & \overline{2p^3 + 5p^2 + p - 2} \\
 & \underline{2p^3 + 2p^2} \\
 & \qquad 3p^2 + p \\
 & \qquad \underline{3p^2 + 3p} \\
 & \qquad\qquad -2p - 2 \\
 & \qquad\qquad \underline{-2p - 2} \\
 & \qquad\qquad\qquad 0
\end{array}
$$

Since the remainder is 0, the quotient is $p^2 + \dfrac{3}{2}p - 1$.

EXAMPLE 6 Dividing by a Polynomial with a Missing Term

Divide $6r^4 + 9r^3 + 2r^2 - 8r + 7$ by $3r^2 - 2$.

The polynomial $3r^2 - 2$ has a missing term. Write it as $3r^2 + 0r - 2$ and divide as usual.

$$\begin{array}{r} 2r^2 + 3r + 2 \\ 3r^2 + 0r - 2 \overline{)6r^4 + 9r^3 + 2r^2 - 8r + 7} \\ \underline{6r^4 + 0r^3 - 4r^2} \quad\quad\quad \\ 9r^3 + 6r^2 - 8r \quad \\ \underline{9r^3 + 0r^2 - 6r} \quad \\ 6r^2 - 2r + 7 \\ \underline{6r^2 + 0r - 4} \\ -2r + 11 \end{array}$$

Since the degree of the remainder, $-2r + 11$, is less than the degree of $3r^2 - 2$, the division process is now finished. The result is written

$$2r^2 + 3r + 2 + \frac{-2r + 11}{3r^2 - 2}.$$

When dividing a polynomial by a monomial, do not confuse the two methods shown in this section. The long division method is not used to divide by a monomial.

Remember the following steps when dividing a polynomial by a polynomial of two or more terms.

1. Be sure the terms in both polynomials are in descending powers.
2. Write any missing terms with a 0 placeholder.

OBJECTIVE 3 Divide polynomial functions. In Chapter 5, we saw how polynomial functions are added, subtracted, and multiplied. At the beginning of this chapter, we defined a rational expression as the quotient of functions defined by polynomials $P(x)$ and $Q(x)$, with $Q(x) \neq 0$. This suggests the following definition for the quotient of two functions.

Division of Functions

If $f(x)$ and $g(x)$ define functions, their quotient is

$$\left(\frac{f}{g}\right)(x) = \frac{f(x)}{g(x)}.$$

The domain of the quotient function is the intersection of the domains of $f(x)$ and $g(x)$, excluding any values of x where $g(x) = 0$.

EXAMPLE 7 Dividing Polynomial Functions

For $f(x) = 2x^2 + x - 10$ and $g(x) = x - 2$, find the quotient $\left(\frac{f}{g}\right)(x)$.

In Example 3, we found the quotient $(2m^2 + m - 10) \div (m - 2) = 2m + 5$. Thus,

$$\left(\frac{f}{g}\right)(x) = \frac{f(x)}{g(x)} = \frac{2x^2 + x - 10}{x - 2} = 2x + 5.$$

Since 2 makes the divisor $g(x) = x - 2$ equal 0, x cannot equal 2.

6.4 EXERCISES

Complete each statement with the correct word or words.

1. We find the quotient of two monomials by using the ______ rule for ______.
2. To divide a polynomial by a monomial, divide ______ of the polynomial by the ______.
3. When dividing polynomials that are not monomials, first write them in ______.
4. If a polynomial in a division problem has a missing term, insert a term with ______ as a placeholder.

Divide. See Examples 1 and 2.

5. $\frac{9y^2 + 12y - 15}{3y}$
6. $\frac{80r^2 - 40r + 10}{10r}$
7. $\frac{15m^3 + 25m^2 + 30m}{5m^2}$
8. $\frac{64x^3 - 72x^2 + 12x}{8x^3}$
9. $\frac{14m^2n^2 - 21mn^3 + 28m^2n}{14m^2n}$
10. $\frac{24h^2k + 56hk^2 - 28hk}{16h^2k^2}$
11. $\frac{8wxy^2 + 3wx^2y + 12w^2xy}{4wx^2y}$
12. $\frac{12ab^2c + 10a^2bc + 18abc^2}{6a^2bc}$

Complete the division.

13. $3r - 1 \overline{) 3r^3 - 22r^2 + 25r - 6}$ with quotient r^2
$3r^3 - r^2$
$-21r^2$

14. $2b - 5 \overline{) 6b^3 - 7b^2 - 4b - 40}$ with quotient $3b^2$
$6b^3 - 15b^2$
$8b^2$

Divide. See Examples 3–6.

15. $\frac{y^2 + 3y - 18}{y + 6}$
16. $\frac{q^2 + 4q - 32}{q - 4}$
17. $\frac{3t^2 + 17t + 10}{3t + 2}$
18. $\frac{2k^2 - 3k - 20}{2k + 5}$
19. $(2z^3 - 5z^2 + 6z - 15) \div (2z - 5)$
20. $(3p^3 + p^2 + 18p + 6) \div (3p + 1)$
21. $(4x^3 + 9x^2 - 10x + 3) \div (4x + 1)$
22. $(10z^3 - 26z^2 + 17z - 13) \div (5z - 3)$
23. $\frac{6x^3 - 19x^2 + 14x - 15}{3x^2 - 2x + 4}$
24. $\frac{8m^3 - 18m^2 + 37m - 13}{2m^2 - 3m + 6}$
25. $\frac{4k^4 + 6k^3 + 3k - 1}{2k^2 + 1}$
26. $\frac{6y^4 + 4y^3 + 4y - 6}{3y^2 + 2y - 3}$
27. $(9z^4 - 13z^3 + 23z^2 - 10z + 8) \div (z^2 - z + 2)$
28. $(2q^4 + 5q^3 - 11q^2 + 11q - 20) \div (2q^2 - q + 2)$

29. $\dfrac{p^3 - 1}{p - 1}$

30. $\dfrac{8a^3 + 1}{2a + 1}$

31. If $P(x) = 2x^3 + 15x^2 + 28x$ and if $Q(x) = x^2 + 4x$, find $\left(\dfrac{P}{Q}\right)(x)$.

32. If $P(x) = 2x^3 + 15x^2 + 13x - 63$ and if $Q(x) = 2x + 9$, find $\left(\dfrac{P}{Q}\right)(x)$.

33. We can check a division problem by multiplying the quotient by the divisor. If the division is correct, the product is the dividend.

$$\begin{array}{r} c \leftarrow \text{Quotient} \\ \text{Divisor} \rightarrow a\overline{)b} \leftarrow \text{Dividend} \end{array}$$

Use this method to check the following division problem: $(4a^3 + 5a^2 + 4a + 5) \div (4a + 5) = a^2 + a + 1$. If the quotient is incorrect, find the correct quotient, and check it.

34. Look at Exercise 17. Factor the numerator and write the fraction in lowest terms. Compare your result with the quotient found in Exercise 17. What can you conclude?

Divide.

35. $\left(2x^2 - \dfrac{7}{3}x - 1\right) \div (3x + 1)$

36. $\left(m^2 + \dfrac{7}{2}m + 3\right) \div (2m + 3)$

37. $\left(3a^2 - \dfrac{23}{4}a - 5\right) \div (4a + 3)$

38. $\left(3q^2 + \dfrac{19}{5}q - 3\right) \div (5q - 2)$

Solve each problem.

39. The volume of a box is $2p^3 + 15p^2 + 28p$. The height is p and the length is $p + 4$; find the width.

40. Suppose a car goes $2m^3 + 15m^2 + 13m - 63$ kilometers in $2m + 9$ hours. Find its rate of speed.

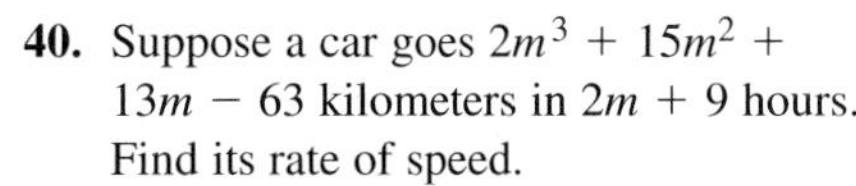

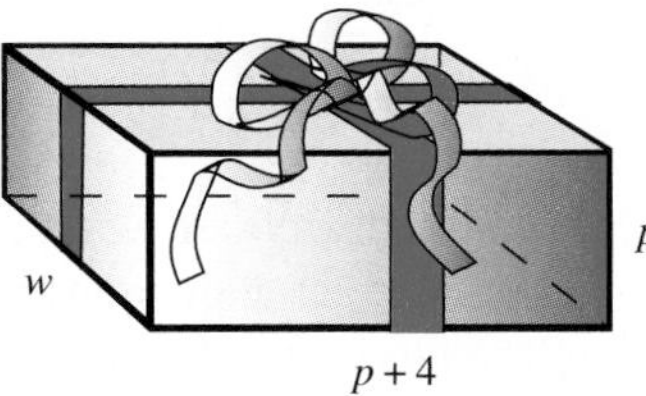

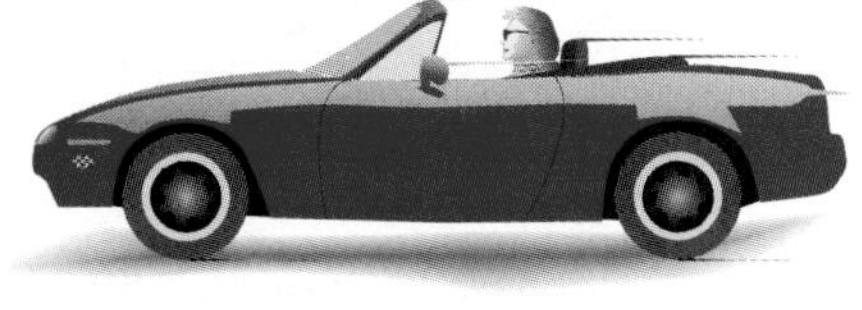

41. For $P(x) = x^3 - 4x^2 + 3x - 5$, find $P(-1)$. Then divide $P(x)$ by $D(x) = x + 1$. Compare the remainder to $P(-1)$. What do these results suggest?

42. Let $P(x) = 4x^3 - 8x^2 + 13x - 2$, and $D(x) = 2x - 1$. Use division to find polynomials $Q(x)$ and $R(x)$ so that $P(x) = Q(x) \cdot D(x) + R(x)$.

For each pair of functions, find the quotient $\left(\dfrac{f}{g}\right)(x)$ *and give any x-values that are not in the domain of the quotient function. See Example 7.*

43. $f(x) = 10x^2 - 2x$, $g(x) = 2x$

44. $f(x) = 18x^2 - 24x$, $g(x) = 3x$

45. $f(x) = 2x^2 - x - 3$, $g(x) = x + 1$

46. $f(x) = 4x^2 - 23x - 35$, $g(x) = x - 7$

47. $f(x) = 8x^3 - 27$, $g(x) = 2x - 3$

48. $f(x) = 27x^3 - 64$, $g(x) = 3x + 4$

6.5 Synthetic Division

OBJECTIVES

1. Use synthetic division to divide by a polynomial of the form $x - k$.
2. Use the remainder theorem to evaluate a polynomial.
3. Decide whether a given number is a solution of an equation.

FOR EXTRA HELP

SSG Sec. 6.5
SSM Sec. 6.5

Pass the Test Software

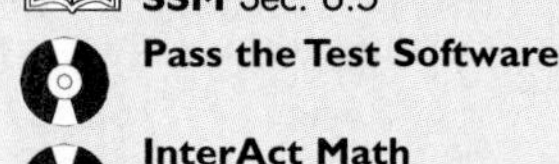
InterAct Math Tutorial Software

Video 12

OBJECTIVE 1 Use synthetic division to divide by a polynomial of the form $x - k$. Many times when one polynomial is divided by a second, the second polynomial is of the form $x - k$, where the coefficient of the x term is 1. There is a shortcut way for doing these divisions. To see how this shortcut works, look first below left, where the division of $3x^3 - 2x + 5$ by $x - 3$ is shown. Notice that 0 was inserted for the missing x^2 term.

$$
\begin{array}{rrrrr}
 & 3x^2 & +\ 9x & +\ 25 & \\
x - 3\,) & 3x^3 + 0x^2 & -\ 2x & +\ 5 & \\
 & 3x^3 - 9x^2 & & & \\
 & 9x^2 & -\ 2x & & \\
 & 9x^2 & -\ 27x & & \\
 & & 25x & +\ 5 & \\
 & & 25x & -\ 75 & \\
 & & & 80 &
\end{array}
\qquad
\begin{array}{rrrrr}
 & 3 & 9 & 25 & \\
1 - 3\,) & 3 & 0 & -2 & 5 \\
 & 3 & -9 & & \\
 & & 9 & -2 & \\
 & & 9 & -27 & \\
 & & & 25 & 5 \\
 & & & 25 & -75 \\
 & & & & 80
\end{array}
$$

On the right, exactly the same division is shown written without the variables. This is why it is *essential* to use 0 as a placeholder in synthetic division. All the numbers in color on the right are repetitions of the numbers directly above them, so they may be omitted, as shown on the left below.

$$
\begin{array}{rrrrr}
 & 3 & 9 & 25 & \\
1 - 3\,) & 3 & 0 & -2 & 5 \\
 & & -9 & & \\
 & & 9 & -2 & \\
 & & & -27 & \\
 & & & 25 & 5 \\
 & & & & -75 \\
 & & & & 80
\end{array}
\qquad
\begin{array}{rrrrr}
 & 3 & 9 & 25 & \\
1 - 3\,) & 3 & 0 & -2 & 5 \\
 & & -9 & & \\
 & & 9 & & \\
 & & & -27 & \\
 & & & 25 & \\
 & & & & -75 \\
 & & & & 80
\end{array}
$$

The numbers in color on the left are again repetitions of the numbers directly above them; they too may be omitted, as shown on the right above.

Now the problem can be condensed. If the 3 in the dividend is brought down to the beginning of the bottom row, the top row can be omitted, since it duplicates the bottom row.

$$
\begin{array}{rrrrr}
1 - 3\,) & 3 & 0 & -2 & 5 \\
 & & -9 & -27 & -75 \\
\hline
 & 3 & 9 & 25 & 80
\end{array}
$$

Finally, the 1 at the upper left can be omitted. Also, to simplify the arithmetic, subtraction in the second row is replaced by addition. We compensate for this by changing the -3 at the upper left to its additive inverse, 3. The result of doing all this is shown below.

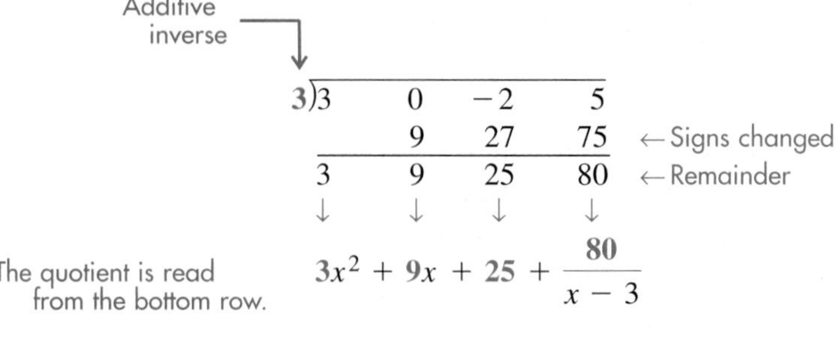

The first three numbers in the bottom row are the coefficients of the quotient polynomial with degree 1 less than the degree of the dividend. The last number gives the remainder.

Synthetic Division

This shortcut procedure is called **synthetic division.** It is used only when dividing a polynomial by a binomial of the form $x - k$.

EXAMPLE 1 Using Synthetic Division

Use synthetic division to divide $5x^2 + 16x + 15$ by $x + 2$.

As mentioned above, we use synthetic division only when dividing by a polynomial of the form $x - k$. Get $x + 2$ in this form by writing it as

$$x + 2 = x - (-2),$$

where $k = -2$. Now write the coefficients of $5x^2 + 16x + 15$, placing -2 to the left.

$x + 2$ leads to -2.

$$-2\overline{)5 \quad 16 \quad 15} \leftarrow \text{Coefficients}$$

Bring down the 5, and multiply: $-2 \cdot 5 = -10$.

$$\begin{array}{r|rrr} -2 & 5 & 16 & 15 \\ & & -10 & \\ \hline & 5 & & \end{array}$$

Add 16 and -10, getting 6. Multiply 6 and -2 to get -12.

$$\begin{array}{r|rrr} -2 & 5 & 16 & 15 \\ & & -10 & -12 \\ \hline & 5 & 6 & \end{array}$$

Add 15 and -12, getting 3.

$$\begin{array}{r|rrr} -2 & 5 & 16 & 15 \\ & & -10 & -12 \\ \hline & 5 & 6 & 3 \end{array} \leftarrow \text{Remainder}$$

The result is read from the bottom row.

$$\frac{5x^2 + 16x + 15}{x + 2} = 5x + 6 + \frac{3}{x + 2}$$

EXAMPLE 2 Using Synthetic Division with Missing Terms

Use synthetic division to find $(-4x^5 + x^4 + 6x^3 + 2x^2 + 50) \div (x - 2)$.

Use the steps given above, inserting a 0 for the missing x term.

$$\begin{array}{r|rrrrrr} 2 & -4 & 1 & 6 & 2 & 0 & 50 \\ & & -8 & -14 & -16 & -28 & -56 \\ \hline & -4 & -7 & -8 & -14 & -28 & -6 \end{array}$$

Read the result from the bottom row.

$$\frac{-4x^5 + x^4 + 6x^3 + 2x^2 + 50}{x - 2} = -4x^4 - 7x^3 - 8x^2 - 14x - 28 + \frac{-6}{x - 2}$$

OBJECTIVE 2 Use the remainder theorem to evaluate a polynomial. We can use synthetic division to evaluate polynomials. For example, in the synthetic division of Example 2, where the polynomial was divided by $x - 2$, the remainder was -6.

Replacing x in the polynomial with 2 gives

$$\begin{aligned} -4x^5 + x^4 + 6x^3 + 2x^2 + 50 &= -4 \cdot 2^5 + 2^4 + 6 \cdot 2^3 + 2 \cdot 2^2 + 50 \\ &= -4 \cdot 32 + 16 + 6 \cdot 8 + 2 \cdot 4 + 50 \\ &= -128 + 16 + 48 + 8 + 50 \\ &= -6, \end{aligned}$$

the same number as the remainder; that is, dividing by $x - 2$ produced a remainder equal to the result when x is replaced with 2. This always happens, as the following remainder theorem states.

Remainder Theorem

If the polynomial $P(x)$ is divided by $x - k$, then the remainder is equal to $P(k)$.

This result is proved in more advanced courses.

EXAMPLE 3 Using the Remainder Theorem

Let $P(x) = 2x^3 - 5x^2 - 3x + 11$. Find $P(-2)$.

Use the remainder theorem; divide $P(x)$ by $x - (-2)$.

$$\begin{array}{r|rrrr} -2 & 2 & -5 & -3 & 11 \\ & & -4 & 18 & -30 \\ \hline & 2 & -9 & 15 & \mathbf{-19} \end{array} \leftarrow \text{Remainder}$$

By this result, $P(-2) = -19$.

OBJECTIVE 3 Decide whether a given number is a solution of an equation. The remainder theorem also can be used to show that a given number is a solution of an equation.

EXAMPLE 4 Using the Remainder Theorem

Show that -5 is a solution of the equation

$$2x^4 + 12x^3 + 6x^2 - 5x + 75 = 0.$$

One way to show that -5 is a solution is by substituting -5 for x in the equation. However, an easier way is to use synthetic division and the remainder theorem given above.

$$\text{Proposed solution} \rightarrow \begin{array}{r|rrrrr} \mathbf{-5} & 2 & 12 & 6 & -5 & 75 \\ & & -10 & -10 & 20 & -75 \\ \hline & 2 & 2 & -4 & 15 & \mathbf{0} \end{array} \leftarrow \text{Remainder}$$

Since the remainder is 0, the polynomial has a value of 0 when $k = -5$, and so -5 is a solution of the given equation.

The synthetic division above also shows that $x - (-5)$ divides the polynomial with 0 remainder. Thus $x - (-5) = x + 5$ is a *factor* of the polynomial and

$$2x^4 + 12x^3 + 6x^2 - 5x + 75 = (x + 5)(2x^3 + 2x^2 - 4x + 15).$$

The second factor is the quotient polynomial found in the last row of the synthetic division.

CONNECTIONS

The procedure in Example 4 is exactly how we use a graphing calculator to find the solution of an equation by determining the x-intercepts of the graph. The screen shows the graph of $P(x) = 2x^4 + 12x^3 + 6x^2 - 5x + 75$ and shows that one value of x that makes $P(x) = 0$ is -5. This agrees with our result in Example 4.

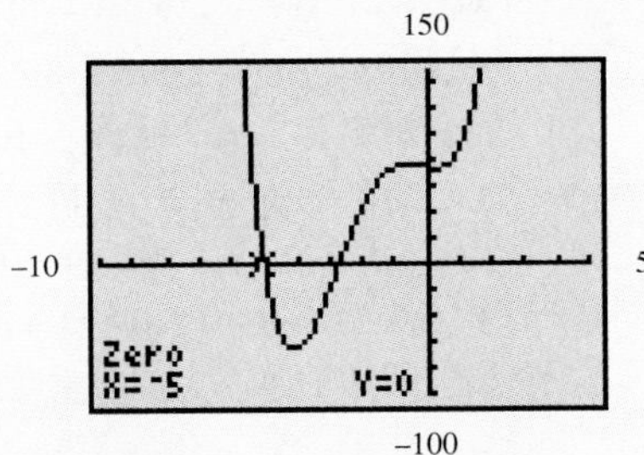

FOR DISCUSSION OR WRITING

Estimate the other x-intercept to the nearest tenth. Verify your answer using synthetic division.

6.5 EXERCISES

1. What is the purpose of synthetic division?
2. What type of polynomial divisors may be used with synthetic division?

Use synthetic division to perform the division. See Examples 1 and 2.

3. $\dfrac{x^2 - 6x + 5}{x - 1}$

4. $\dfrac{x^2 - 4x - 21}{x + 3}$

5. $\dfrac{4m^2 + 19m - 5}{m + 5}$

6. $\dfrac{3k^2 - 5k - 12}{k - 3}$

7. $\dfrac{2a^2 + 8a + 13}{a + 2}$

8. $\dfrac{4y^2 - 5y - 20}{y - 4}$

9. $(p^2 - 3p + 5) \div (p + 1)$

10. $(z^2 + 4z - 6) \div (z - 5)$

11. $\dfrac{4a^3 - 3a^2 + 2a - 3}{a - 1}$

12. $\dfrac{5p^3 - 6p^2 + 3p + 14}{p + 1}$

13. $(x^5 - 2x^3 + 3x^2 - 4x - 2) \div (x - 2)$
14. $(2y^5 - 5y^4 - 3y^2 - 6y - 23) \div (y - 3)$
15. $(-4r^6 - 3r^5 - 3r^4 + 5r^3 - 6r^2 + 3r + 3) \div (r - 1)$
16. $(2t^6 - 3t^5 + 2t^4 - 5t^3 + 6t^2 - 3t - 2) \div (t - 2)$
17. $(-3y^5 + 2y^4 - 5y^3 - 6y^2 - 1) \div (y + 2)$
18. $(m^6 + 2m^4 - 5m + 11) \div (m - 2)$

19. $\dfrac{y^3 + 1}{y - 1}$

20. $\dfrac{z^4 + 81}{z - 3}$

Use the remainder theorem to find P(k). See Example 3.

21. $P(x) = 2x^3 - 4x^2 + 5x - 3;\quad k = 2$

22. $P(y) = y^3 + 3y^2 - y + 5;\quad k = -1$

23. $P(r) = -r^3 - 5r^2 - 4r - 2;\quad k = -4$

24. $P(z) = -z^3 + 5z^2 - 3z + 4;\quad k = 3$

25. $P(y) = 2y^3 - 4y^2 + 5y - 33;\quad k = 3$

26. $P(x) = x^3 - 3x^2 + 4x - 4;\quad k = 2$

27. Explain why a zero remainder in synthetic division of $P(x)$ by $x - k$ indicates that k is a solution of the equation $P(x) = 0$.

28. Explain why it is important to insert zeros as placeholders for missing terms before performing synthetic division.

Use synthetic division to decide whether the given number is a solution of the equation. See Example 4.

29. $x^3 - 2x^2 - 3x + 10 = 0;\quad x = -2$

30. $x^3 - 3x^2 - x + 10 = 0;\quad x = -2$

31. $m^4 + 2m^3 - 3m^2 + 8m - 8 = 0;\quad m = -2$

32. $r^4 - r^3 - 6r^2 + 5r + 10 = 0;\quad r = -2$

33. $3a^3 + 2a^2 - 2a + 11 = 0;\quad a = -2$

34. $3z^3 + 10z^2 + 3z - 9 = 0;\quad z = -2$

35. $2x^3 - x^2 - 13x + 24 = 0;\quad x = -3$

36. $5p^3 + 22p^2 + p - 28 = 0;\quad p = -4$

RELATING CONCEPTS (EXERCISES 37–41)

In Section 6.4 we saw the close connection between polynomial division and writing a quotient of polynomials in lowest terms after factoring the numerator. Now we can show a connection between dividing one polynomial by another and factoring the first polynomial. Let $P(x) = 2x^2 + 5x - 12$. ***Work the following exercises in order.***

37. Factor $P(x)$.

38. Solve $P(x) = 0$.

39. Find $P(-4)$ and $P\left(\frac{3}{2}\right)$.

40. Complete the following sentence. If $P(a) = 0$, then $x -$ _____ is a factor of $P(x)$.

41. Use the conclusion in Exercise 40 to decide whether $x - 3$ is a factor of $Q(x) = 3x^3 - 4x^2 - 17x + 6$. Factor $Q(x)$ completely.

Did you make the connection between solutions of the equation $P(x) = 0$ and the factors of $P(x)$?

TECHNOLOGY INSIGHTS (EXERCISES 42–45)

Use the graph to determine a solution of each equation. Then use your results to completely factor the polynomial using the method in the text that follows Example 4.

42. $2x^3 + 12x^2 + 24x + 16 = 0$

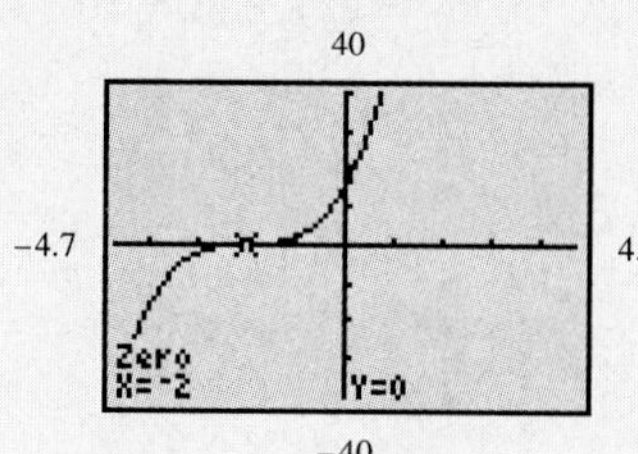

43. $x^3 - x^2 - 21x + 45 = 0$

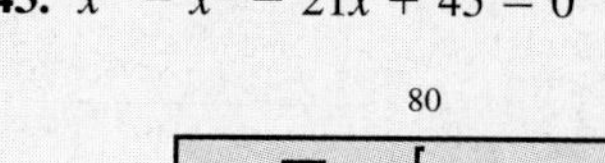

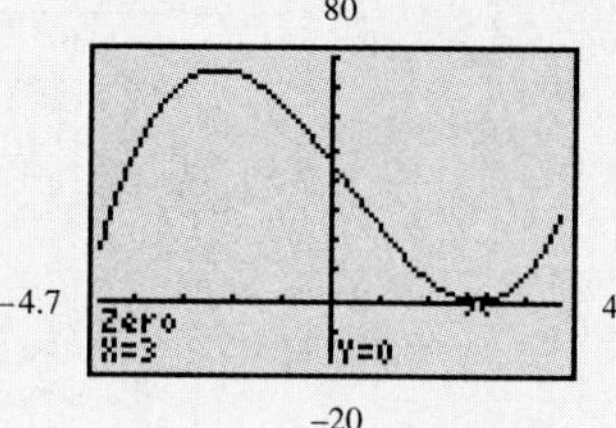

TECHNOLOGY INSIGHTS (EXERCISES 42–45) (CONTINUED)

44. $x^3 + 3x^2 - 10x - 24 = 0$

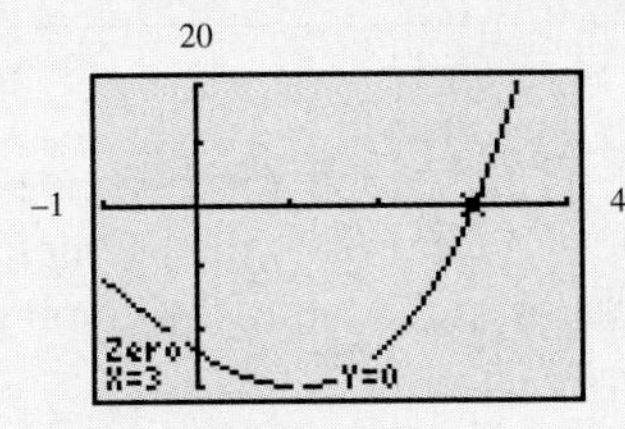

45. $x^3 + 3x^2 - 13x - 15 = 0$

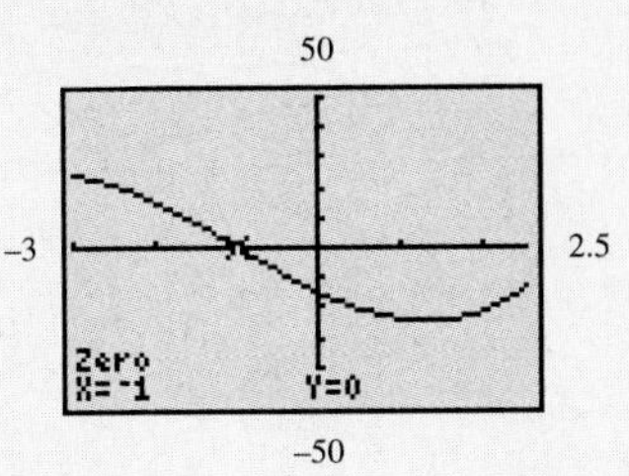

6.6 Graphs and Equations with Rational Expressions

OBJECTIVES

1. Recognize the graph of a rational function.
2. Determine the domain of a rational equation.
3. Solve rational equations.
4. Solve rational equations using a graphing calculator.

FOR EXTRA HELP

SSG Sec. 6.6
SSM Sec. 6.6

Pass the Test Software

InterAct Math Tutorial Software

Video 12

OBJECTIVE 1 Recognize the graph of a rational function. Because one or more values of x are excluded from the domain of most rational functions, their graphs are usually *discontinuous.* That is, there will be one or more breaks in the graph. For example, we use point plotting and observing the domain to graph the simple rational function

$$f(x) = \frac{1}{x}.$$

The domain of this function includes all real numbers except 0. Thus, there will be no point on the graph with $x = 0$. The vertical line with equation $x = 0$ is called an **asymptote** of the graph. We show some typical ordered pairs in the table for both negative and positive x-values.

x	-3	-2	-1	$-.5$	$-.25$	$-.1$	$.1$	$.25$	$.5$	1	2	3
y	$-\frac{1}{3}$	$-\frac{1}{2}$	-1	-2	-4	-10	10	4	2	1	$\frac{1}{2}$	$\frac{1}{3}$

Notice that the closer positive values of x are to 0, the larger y is. Similarly, the closer negative values of x are to 0, the smaller (more negative) y is. Using this observation, the domain excluding 0, and plotting the points found above produces the graph in Figure 1.

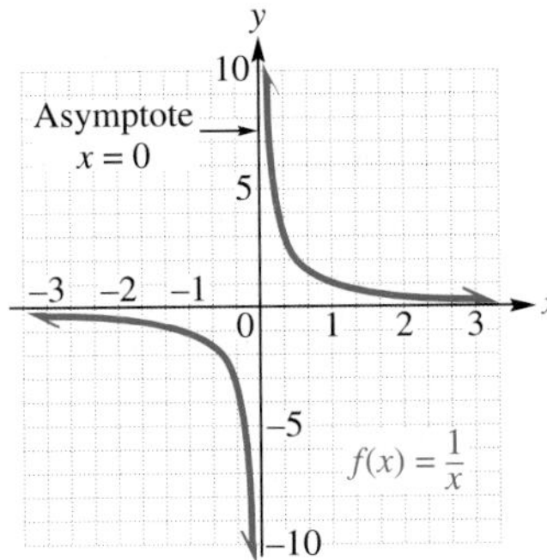

Figure 1

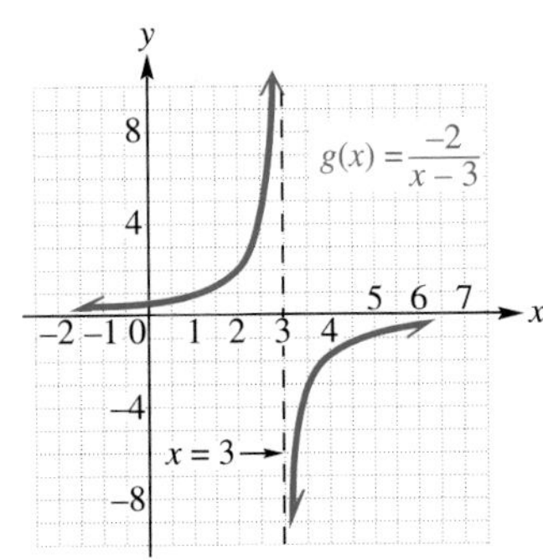

Figure 2

The graph of

$$g(x) = \frac{-2}{x - 3},$$

is shown in Figure 2. Some ordered pairs that belong to the function are shown in the table.

x	-2	-1	0	1	2	2.5	2.75	3.25	3.5	4	5	6	7
y	$\frac{2}{5}$	$\frac{1}{2}$	$\frac{2}{3}$	1	2	4	8	-8	-4	-2	-1	$-\frac{2}{3}$	$-\frac{1}{2}$

There is no point on the graph for $x = 3$, because 3 is excluded from the domain. The dashed line $x = 3$ represents the asymptote and is not part of the graph. As suggested by the points from the table, the graph gets closer to the asymptote $x = 3$ as the x-values get closer to 3.

The domain of a rational expression is the set of all possible values of the variable. Any value that makes the denominator 0 is excluded.

OBJECTIVE 2 **Determine the domain of a rational equation.** The **domain of a rational equation** is the intersection (overlap) of the domains of the rational expressions in the equation.

EXAMPLE 1 Determining the Domain of a Rational Equation

Find the domain of each equation.

(a) $\dfrac{2}{x} - \dfrac{3}{2} = \dfrac{7}{2x}$

The domains of the three rational terms of the equation are, in order, $x \neq 0$, $(-\infty, \infty)$, and $x \neq 0$. The graph in Figure 1 illustrates why 0 must be excluded from the domains of the first and third terms. The intersection of these three domains is all real numbers except 0, so that is the domain of the equation.

(b) $\dfrac{-2}{x-3} - \dfrac{3}{x+3} = \dfrac{12}{x^2-9}$

The domains of these three terms are, respectively, $x \neq 3$, $x \neq -3$, and $x \neq 3$ or -3. Here, the graph in Figure 2 shows why the first rational term has domain $x \neq 3$. The domain of the equation is the intersection of the three domains, all real numbers except 3 and -3.

OBJECTIVE 3 **Solve rational equations.** The easiest way to solve most equations involving rational expressions is to multiply all terms in the equation by the least common denominator. This step will clear the equation of all denominators, as the next examples show. *We can do this only with equations, not expressions.*

Because the first step in solving a rational equation is to multiply both sides of the equation by a common denominator, in many cases it is *necessary* to either check the solutions or verify that the solutions are in the domain.

CAUTION When both sides of an equation are multiplied by a *variable* expression, the resulting "solutions" may not satisfy the original equation. You *must* either determine and observe the domain or check all potential solutions in the original equation. It is wise to do both.

EXAMPLE 2 Solving an Equation with Rational Expressions

Solve $\frac{2}{x} - \frac{3}{2} = \frac{7}{2x}$.

The domain, which excludes 0, was found in Example 1(a). Multiply both sides of the equation by the LCD, $2x$.

$$2x\left(\frac{2}{x} - \frac{3}{2}\right) = 2x\left(\frac{7}{2x}\right)$$

$$2x\left(\frac{2}{x}\right) - 2x\left(\frac{3}{2}\right) = 2x\left(\frac{7}{2x}\right) \quad \text{Distributive property}$$

$$4 - 3x = 7 \quad \text{Multiply.}$$

$$-3x = 3 \quad \text{Subtract 4.}$$

$$x = -1 \quad \text{Divide by } -3.$$

To verify that -1 is a solution of the equation, replace x with -1 in the original equation. The solution set is $\{-1\}$.

EXAMPLE 3 Solving an Equation with No Solution

Solve $\frac{2}{x-3} - \frac{3}{x+3} = \frac{12}{x^2-9}$.

Using the concepts from Example 1, the domain excludes 3 and -3. Multiply both sides by the LCD, $(x + 3)(x - 3)$.

$$(x+3)(x-3) \cdot \frac{2}{x-3} - (x+3)(x-3) \cdot \frac{3}{x+3}$$

$$= (x+3)(x-3) \cdot \frac{12}{x^2-9}$$

$$2(x+3) - 3(x-3) = 12 \quad \text{Multiply.}$$

$$2x + 6 - 3x + 9 = 12 \quad \text{Distributive property}$$

$$-x + 15 = 12 \quad \text{Combine terms.}$$

$$-x = -3$$

$$x = 3$$

Since 3 is not in the domain, it cannot be a solution of the equation. Substituting 3 in the equation shows that

$$\frac{2}{3-3} - \frac{3}{3+3} \stackrel{?}{=} \frac{12}{3^2-9}$$

$$\frac{2}{0} - \frac{3}{6} \stackrel{?}{=} \frac{12}{0}.$$

Division by 0 is undefined, so the given equation has no solution, and, as we expected, the solution set is $\emptyset$.

EXAMPLE 4 Solving an Equation with Rational Expressions

Solve $\frac{3}{p^2+p-2} - \frac{1}{p^2-1} = \frac{7}{2(p^2+3p+2)}$.

Factor each denominator to find the least common denominator. The LCD is $2(p-1)(p+2)(p+1)$. The domain excludes 1, -2, and -1. Multiply both sides by the LCD.

$$2(p-1)(p+2)(p+1)\left[\frac{3}{(p+2)(p-1)}\right]$$
$$-\,2(p-1)(p+2)(p+1)\left[\frac{1}{(p+1)(p-1)}\right]$$
$$=2(p-1)(p+2)(p+1)\left[\frac{7}{2(p+2)(p+1)}\right]$$

Now simplify.

$$\begin{aligned} 2\cdot 3(p+1)-2(p+2) &= 7(p-1) && \text{Multiply.}\\ 6p+6-2p-4 &= 7p-7 && \text{Distributive property}\\ 4p+2 &= 7p-7 && \text{Combine terms.}\\ 9 &= 3p\\ 3 &= p \end{aligned}$$

Note that 3 is in the domain; substitute 3 for p in the original equation to check that the solution set is $\{3\}$.

EXAMPLE 5 Solving an Equation That Leads to a Quadratic Equation

Solve $\frac{2}{3x+1} = \frac{1}{x} - \frac{6x}{3x+1}$.

Since the denominator $3x+1$ cannot equal 0, $-\frac{1}{3}$ is excluded from the domain, as is 0. These are the only real numbers not in the domain. Multiply both sides of the given equation by $x(3x+1)$. The resulting equation is

$$2x = (3x+1) - 6x^2.$$

Since this equation is quadratic, to solve it, get 0 on the right side.

$$\begin{aligned} 6x^2-3x+2x-1 &= 0 && \text{Standard form}\\ 6x^2-x-1 &= 0 && \text{Combine terms.}\\ (3x+1)(2x-1) &= 0 && \text{Factor.} \end{aligned}$$

$$3x+1=0 \quad \text{or} \quad 2x-1=0 \qquad \text{Zero-factor property}$$

$$x=-\frac{1}{3} \quad \text{or} \quad x=\frac{1}{2}$$

Because $-\frac{1}{3}$ is not in the domain of the equation, it is not a solution. The solution set is $\left\{\frac{1}{2}\right\}$.

OBJECTIVE 4 Solve rational equations using a graphing calculator. Earlier, we saw how to use a graphing calculator to solve other types of equations. The procedure is similar with rational equations. If the equation has 0 on one side, the x-intercepts of the graph are the solutions of the equation. Because rational functions usually have values of x that are excluded from the domain, a calculator in connected mode may show a vertical line on the screen where an asymptote occurs. The dot mode of a graphing calculator will often give a more realistic picture.

The graph of

$$g(x) = \frac{x}{x+3}$$

is shown generated in connected mode in Figure 3(a) and dot mode in Figure 3(b). If dot mode is used, we must remember that, theoretically, the function is continuous (unbroken) on the intervals $(-\infty, -3)$ and $(-3, \infty)$. From the graph, we see that the only solution of the rational equation

$$\frac{x}{x+3} = 0$$

is the x-intercept, (0, 0) so the solution set is $\{0\}$.

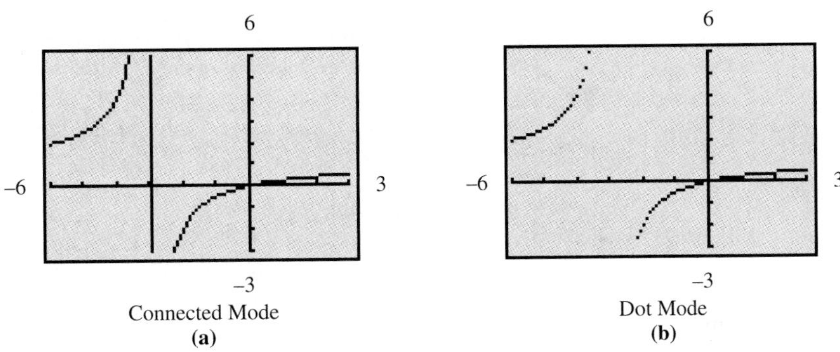

Figure 3

Recall from earlier chapters that if the equation does not have 0 on one side, we need to rewrite it so that it does.

EXAMPLE 6 Reading Solutions of a Rational Equation from a Graphing Calculator Screen

Two views of the graph of $f(x) = x^{-2} + x^{-1} - 1.5$ are shown in Figure 4. Use the graph to determine the solution set of $x^{-2} + x^{-1} - 1.5 = 0$.

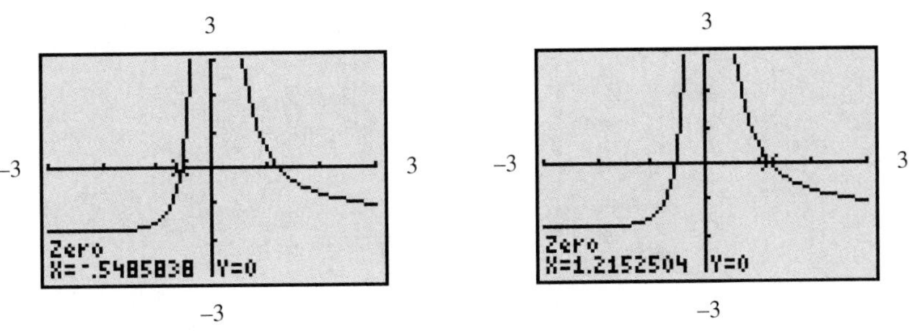

Figure 4

Look at the bottom of each calculator screen for the x-values for which $y = 0$, $-.5485838$ and 1.2152504. These are the solutions of the equation. The solution set is $\{-.5485838, 1.2152504\}$.

6.6 EXERCISES

To begin solving each equation in Exercises 1–4, what expression should be used to multiply on both sides?

1. $\frac{4}{x} + \frac{5}{x-1} = 10$

2. $\frac{10}{y+5} - \frac{5}{2y+3} = 20$

3. $\frac{m}{m+1} - \frac{2}{m+2} = \frac{m-1}{m+1}$

4. $\frac{3p}{p+5} + \frac{p-2}{p} = \frac{p+2}{p+5}$

5. Explain why -1 cannot be a solution of the equation in Exercise 3. What other number must be rejected as a potential solution?

6. Are there any potential solutions to the equation

$$\frac{x+7}{4} - \frac{x+3}{3} = \frac{x}{12}$$

that would have to be rejected? Explain why or why not.

From the explanation in this section, we know that any values that would cause a denominator to equal 0 must be excluded from the domain, and consequently as possible solutions of equations that have variable expressions in denominators. Without actually solving each equation, list all possible numbers that would have to be rejected if they appeared as potential solutions. See Example 1.

7. $\frac{1}{x+1} - \frac{1}{x-2} = 0$

8. $\frac{3}{x+4} - \frac{2}{x-9} = 0$

9. $\frac{5}{3x+5} - \frac{1}{x} = \frac{1}{2x+3}$

10. $\frac{6}{4x+7} - \frac{3}{x} = \frac{5}{6x-13}$

11. $\frac{3x+1}{x-4} = \frac{6x+5}{2x-7}$

12. $\frac{4x-1}{2x+3} = \frac{12x-25}{6x-2}$

Solve each equation. See Examples 2–5.

13. $\frac{4}{x} - \frac{6}{x} = \frac{2}{3}$

14. $\frac{10}{y} + \frac{5}{3y} = -\frac{7}{2}$

15. $\frac{x+8}{5} = \frac{6+x}{3}$

16. $\frac{r+1}{4} = \frac{1+2r}{5}$

17. $\frac{3x+1}{x-4} = \frac{6x+5}{2x-7}$

18. $\frac{4x-1}{2x+3} = \frac{12x-25}{6x-2}$

19. $\frac{-5}{2x} + \frac{3}{4x} = \frac{-7}{4}$

20. $\frac{6}{5x} - \frac{2}{3x} = \frac{-8}{45}$

21. $x - \frac{24}{x} = -2$

22. $p + \frac{15}{p} = -8$

23. $\frac{1}{y-1} + \frac{5}{12} = \frac{-4}{3y-3}$

24. $\frac{4}{m+2} - \frac{11}{9} = \frac{5}{3m+6}$

25. $\frac{3}{k+2} - \frac{2}{k^2-4} = \frac{1}{k-2}$

26. $\frac{3}{x-2} + \frac{21}{x^2-4} = \frac{14}{x+2}$

27. $\frac{1}{t+3}+\frac{4}{t+5}=\frac{2}{t^2+8t+15}$

28. $\frac{6}{w+3}+\frac{-7}{w-5}=\frac{-48}{w^2-2w-15}$

29. $\frac{2x}{x-3}+\frac{4}{x+3}=\frac{-24}{x^2-9}$

30. $\frac{2}{4x+7}+\frac{x}{3}=\frac{6}{12x+21}$

31. $\frac{7}{x-4}+\frac{3}{x}=\frac{-12}{x^2-4x}$

32. $\frac{5x+14}{x^2-9}=\frac{-2x^2-5x+2}{x^2-9}+\frac{2x+4}{x-3}$

33. $\frac{4x-7}{4x^2-9}=\frac{-2x^2+5x-4}{4x^2-9}+\frac{x+1}{2x+3}$

34. Make up an equation similar to the one in Exercise 13, and then solve it. (*Hint:* Start with the answer, and work backward.)

RELATING CONCEPTS (EXERCISES 35–40)

Consider the following *equation* and *expression.*

$$\frac{x}{2}+\frac{x}{3}=-5 \qquad \frac{x}{2}+\frac{x}{3}$$

↑ Equation ↑ Expression

A common student error is to confuse the two. Look for the equals sign to distinguish between them. Although they look very much alike, the steps we use to *solve* the equation or to *simplify* the expression are different. We begin the same way in each case by finding the least common denominator. The following exercises lead you through the steps for each problem.

Work these exercises in order.

35. Find the least common denominator for each problem.

36. **(a)** Multiply both sides of the equation by the LCD, and simplify both sides.
(b) Use the fundamental principle to rewrite both terms of the expression with the LCD.

37. **(a)** Combine terms on the left side of the equation. Solve for x.
(b) Combine terms in the expression by adding numerators and keeping the common denominator.

38. How do the answers differ in parts (a) and (b) of Exercise 37?

39. Explain the difference between *simplifying the expression*

$$\frac{2}{x+1}+\frac{3}{x-2}-\frac{6}{x^2-x-2}$$

and *solving the equation*

$$\frac{2}{x+1}+\frac{3}{x-2}=\frac{6}{x^2-x-2}.$$

40. What is wrong with the following problem? "Solve $\frac{2x+1}{3x-4}+\frac{1}{2x+3}$."

Did you make the connection between solving a rational equation and adding (or subtracting) rational expressions?

Solve each problem.

41. The amount of heating oil produced (in gallons per day) by an oil refinery is modeled by the rational function with

$$f(x) = \frac{125{,}000 - 25x}{125 + 2x},$$

where x is the amount of gasoline produced (in hundreds of gallons per day). Suppose the refinery must produce 300 gallons of heating oil per day to meet the needs of its customers.

(a) How much gasoline will be produced per day?

(b) What must be true of the amount of gasoline produced as the amount of heating oil produced increases?

42. The force required to keep a 2000-pound car, going 30 miles per hour, from skidding on a curve is given by

$$F(r) = \frac{225{,}000}{r},$$

where r is the radius of the curve in feet.

(a) What radius must a curve have if a force of 450 pounds is needed to keep the car from skidding?

(b) As the radius of the curve is lengthened, how is the force affected?

43. The average number of vehicles waiting in line to enter a sports arena parking area is modeled by the function with

$$w(x) = \frac{x^2}{2(1 - x)},$$

where x is a quantity between 0 and 1 known as the *traffic intensity.* To the nearest tenth, find the average number of vehicles waiting if the traffic intensity is

(a) .1 **(b)** .8 **(c)** .9.

(d) What happens to waiting time as traffic intensity increases?

(*Source:* F. Mannering and W. Kilareski, *Principles of Highway Engineering and Traffic Control,* John Wiley and Sons, 1990.)

44. The percent of deaths caused by smoking is modeled by the rational function with

$$p(x) = \frac{x - 1}{x},$$

where x is the number of times a smoker is more likely to die of lung cancer than a nonsmoker. This is called the *incidence rate.* For example, $x = 10$ means that a smoker is 10 times more likely than a nonsmoker to die of lung cancer.

(a) Find $p(x)$ if x is 10.

(b) For what values of x is $p(x) = 80\%$? (*Hint:* Change 80% to a decimal.)

(*Source:* A. Walker, *Observation and Inference: An Introduction to the Methods of Epidemiology,* Epidemiology Resources Inc., Newton Lower Falls, MA, 1991.)

TECHNOLOGY INSIGHTS (EXERCISES 45–48)

Two views of the graph of $f(x) = 7x^{-4} - 8x^{-2} + 1$ are shown.
Use the graphs to respond to Exercises 45 and 46. See Example 6.

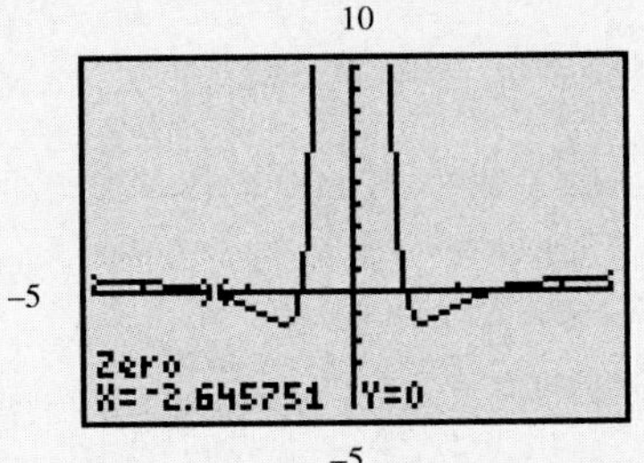

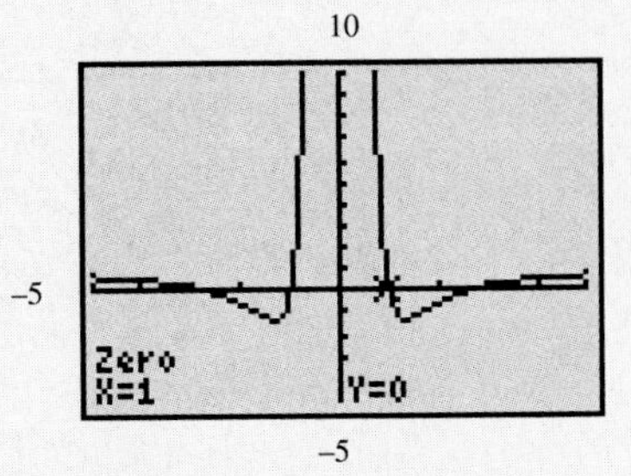

45. How many solutions does the equation $f(x) = 0$ have?

46. Give the solutions of $f(x) = 0$ (to the nearest hundredth if necessary).

In Exercises 47 and 48, use the graph to determine the solution set of the equation $f(x) = 0$. All solutions are integers.

47. $f(x) = \dfrac{x^3 - x^2 - 6x}{x^2 - 1}$

48. $f(x) = \dfrac{-x^3 - x^2 + 2x}{x^2}$

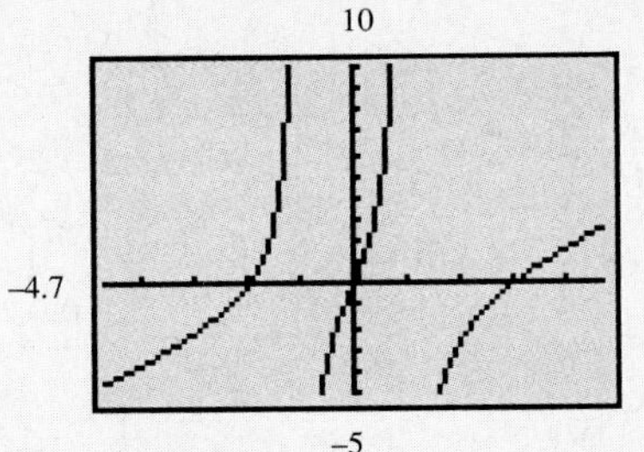

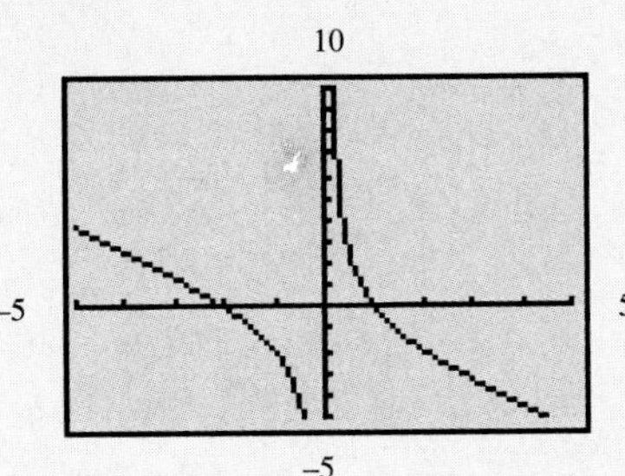

SUMMARY Exercises on Operations and Equations with Rational Expressions

The Relating Concepts in the exercises for Section 6.6 demonstrated the distinction between rational equations and rational expressions. Identify whether each exercise involves an equation *or an* expression. *Then perform the indicated operation or solve the given equation, as appropriate.*

1. $\dfrac{2}{x} - \dfrac{4}{3x} = 5$

2. $\dfrac{8x^4z}{12x^3z^2} \cdot \dfrac{7x}{3x^5}$

3. $\dfrac{4x - 20}{x^2 - 25} \cdot \dfrac{(x + 5)^2}{10}$

4. $\dfrac{6}{7x} - \dfrac{4}{x}$

5. $\dfrac{\dfrac{1}{x} + \dfrac{1}{y}}{\dfrac{1}{x} - \dfrac{1}{y}}$

6. $\dfrac{5}{7t} = \dfrac{52}{7} - \dfrac{3}{t}$

7. $\dfrac{x - 5}{3} + \dfrac{1}{3} = \dfrac{x - 2}{5}$

8. $\dfrac{7}{6x} + \dfrac{5}{8x}$

9. $\frac{4}{x} - \frac{8}{x+1} = 0$

10. $\dfrac{\frac{6}{x+1} - \frac{1}{x}}{\frac{2}{x} - \frac{4}{x+1}}$

11. $\frac{8}{r+2} - \frac{7}{4r+8}$

12. $\frac{x}{x+y} + \frac{2y}{x-y}$

13. $\frac{3p^2 - 6p}{p+5} \div \frac{p^2 - 4}{8p + 40}$

14. $\frac{x-2}{9} \cdot \frac{5}{8-4x}$

15. $\frac{a-4}{3} + \frac{11}{6} = \frac{a+1}{2}$

16. $\frac{b^2 + b - 6}{b^2 + 2b - 8} \cdot \frac{b^2 + 8b + 16}{3b + 12}$

17. $\frac{10z^2 - 5z}{3z^3 - 6z^2} \div \frac{2z^2 + 5z - 3}{z^2 + z - 6}$

18. $\frac{5}{x^2 - 2x} - \frac{3}{x^2 - 4}$

19. $\frac{6}{t+1} + \frac{4}{5t+5} = \frac{34}{15}$

20. $\frac{x^{-1} + y^{-1}}{y^{-1} - x^{-1}}$

21. $\dfrac{\frac{5}{x} - \frac{3}{y}}{\frac{9x^2 - 25y^2}{x^2y}}$

22. $\frac{-2}{a^2 + 2a - 3} - \frac{5}{3 - 3a} = \frac{4}{3a + 9}$

23. $\dfrac{2r^{-1} + 5s^{-1}}{\frac{4s^2 - 25r^2}{3rs}}$

24. $\frac{4y^2 - 13y + 3}{2y^2 - 9y + 9} \div \frac{4y^2 + 11y - 3}{6y^2 - 5y - 6}$

25. $\frac{8}{3k+9} - \frac{8}{15} = \frac{2}{5k+15}$

26. $\frac{3r}{r-2} = 1 + \frac{6}{r-2}$

27. $\frac{6z^2 - 5z - 6}{6z^2 + 5z - 6} \cdot \frac{12z^2 - 17z + 6}{12z^2 - z - 6}$

28. $\frac{-1}{3-x} - \frac{2}{x-3}$

29. $\dfrac{\frac{t}{4} - \frac{1}{t}}{1 + \frac{t+4}{t}}$

30. $\frac{2}{y+1} - \frac{3}{y^2 - y - 2} = \frac{3}{y-2}$

31. $\frac{7}{2x^2 - 8x} + \frac{3}{x^2 - 16}$

32. $\frac{3}{y-3} - \frac{3}{y^2 - 5y + 6} = \frac{2}{y-2}$

33. $\dfrac{2k + \frac{5}{k-1}}{3k - \frac{2}{k}}$

34. $1 + \frac{1}{x} = \frac{6}{x^2}$

6.7 Applications of Rational Expressions

We have seen models of applications that are rational expressions. In this section, we show additional examples of such models, including some formulas.

OBJECTIVE 1 Find the value of an unknown variable in a formula.

OBJECTIVES

1. Find the value of an unknown variable in a formula.
2. Solve a formula for a specified variable.
3. Solve applications using proportions.
4. Solve applications about distance, rate, and time.
5. Solve applications about work rates.

FOR EXTRA HELP

SSG Sec. 6.7
SSM Sec. 6.7

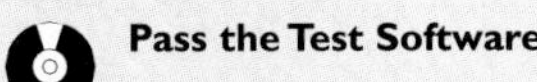
Pass the Test Software

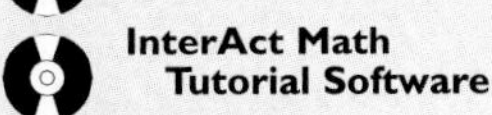
InterAct Math Tutorial Software

Video 12

EXAMPLE 1 Finding the Value of a Variable in a Formula

In physics, the focal length, f, of a lens is given by the formula

$$\frac{1}{f} = \frac{1}{p} + \frac{1}{q},$$

where p is the distance from the object to the lens and q is the distance from the lens to the image. See Figure 5. Find q if $p = 20$ centimeters and $f = 10$ centimeters.

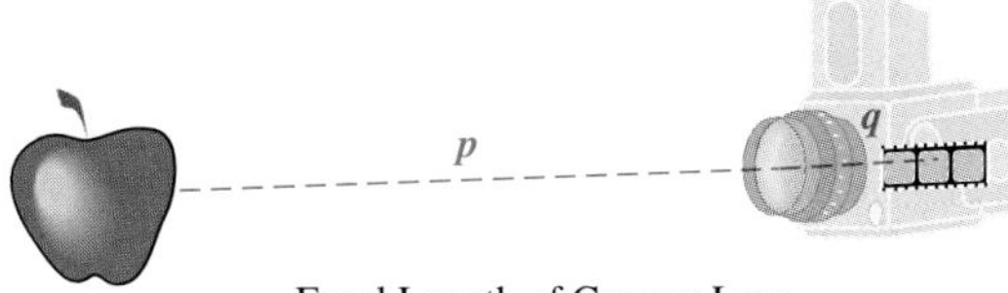

Focal Length of Camera Lens

Figure 5

Replace f with 10 and p with 20.

$$\frac{1}{f} = \frac{1}{p} + \frac{1}{q}$$

$$\frac{1}{10} = \frac{1}{20} + \frac{1}{q} \qquad \text{Let } f = 10, p = 20.$$

Multiply both sides by the least common denominator, $20q$.

$$20q \cdot \frac{1}{10} = 20q \cdot \frac{1}{20} + 20q \cdot \frac{1}{q}$$

$$2q = q + 20$$

$$q = 20$$

The distance from the lens to the image is 20 centimeters.

OBJECTIVE 2 Solve a formula for a specified variable. The goal, as we saw earlier, is to get the specified variable alone on one side of the equals sign.

EXAMPLE 2 Solving a Formula for a Specified Variable

Solve $\frac{1}{f} = \frac{1}{p} + \frac{1}{q}$ for p.

Begin by multiplying both sides by the common denominator fpq.

$$fpq \cdot \frac{1}{f} = fpq\left(\frac{1}{p} + \frac{1}{q}\right)$$

$$pq = fq + fp \qquad \text{Distributive property}$$

Get the terms with p (the specified variable) on the same side of the equation. To do this, subtract fp on both sides.

$$pq - fp = fq \quad \text{Subtract } fp.$$
$$p(q - f) = fq \quad \text{Distributive property; factor out } p.$$
$$p = \frac{fq}{q - f} \quad \text{Divide by } q - f.$$

EXAMPLE 3 Solving a Formula for a Specified Variable

Solve $I = \dfrac{nE}{R + nr}$ for n.

First, multiply both sides by $R + nr$.

$$(R + nr)I = (R + nr)\frac{nE}{R + nr}$$
$$RI + nrI = nE$$
$$RI = nE - nrI \quad \text{Subtract } nrI.$$
$$RI = n(E - rI) \quad \text{Distributive property; factor out } n.$$
$$\frac{RI}{E - rI} = n \quad \text{Divide by } E - rI.$$

Refer to the steps in Examples 2 and 3 that use the distributive property. This is a step that often gives students difficulty. Remember that the variable for which you are solving *must* be a factor on one side of the equation, so that in the last step both sides can be divided by the remaining factor.

PROBLEM SOLVING

We can now solve problems that translate into equations with rational expressions. To do so, we continue to use the six-step problem-solving method from Chapter 2.

OBJECTIVE 3 Solve applications using proportions. A **ratio** is a comparison of two quantities with the *same* units. The ratio of a to b may be written in any of the following ways.

$$a \text{ to } b, \qquad a : b, \qquad \text{or} \qquad \frac{a}{b}$$

Ratios are usually written as quotients in algebra. A **proportion** is a statement that two ratios are equal. Proportions are a useful and important type of rational equation.

EXAMPLE 4 Solving a Proportion

In 1990, 15 of every 100 Americans had no health insurance coverage. The population at that time was about 246 million. How many million had no health insurance?

Step 1 Let x = the number (in millions) who had no health insurance.

(Step 2 is not needed here.)

Step 3 To get an equation, we set up a proportion. The ratio x to 246 should equal the ratio 15 to 100. Write the proportion and solve the equation.

$$\frac{15}{100} = \frac{x}{246}$$

Step 4

$$24{,}600\left(\frac{15}{100}\right) = 24{,}600\left(\frac{x}{246}\right) \quad \text{Multiply by a common denominator.}$$

$$246(15) = 100x \quad \text{Simplify.}$$

$$3690 = 100x$$

$$x = 36.9$$

Step 5 There were 36.9 million Americans with no health insurance.

Step 6 Check that this number compared to 246 million is equivalent to $\frac{15}{100}$.

A comparison of two quantities with *different* units is called a **rate.** Two equal rates can be expressed as a proportion. It is important to be sure the two rates are expressed with the units in the same order.

EXAMPLE 5 Solving a Proportion Involving Rates

Marissa's car uses 10 gallons of gas to travel 210 miles. She has 5 gallons of gas in the car and she wants to know how much more gas she will need to drive 640 miles. If we assume the car continues to use gas at the same rate, how many more gallons will she need?

We can set up a proportion. Let $x =$ the additional amount of gas needed.

$$\frac{\text{gallons}}{\text{miles}} \quad \frac{10}{210} = \frac{5 + x}{640} \quad \frac{\text{gallons}}{\text{miles}}$$

The LCD is $10 \cdot 21 \cdot 64$.

$$10 \cdot 21 \cdot 64\left(\frac{10}{210}\right) = 10 \cdot 21 \cdot 64\left(\frac{5 + x}{640}\right)$$

$$64 \cdot 10 = 21(5 + x)$$

$$30.5 = 5 + x \quad \text{Divide 640 by 21; round to the nearest tenth.}$$

$$25.5 = x$$

Marissa will need 25.5 more gallons of gas. Check the answer in the words of the problem. The 25.5 gallons plus the 5 gallons equals 30.5 gallons.

$$\frac{30.5}{640} \approx .0476$$

$$\frac{10}{210} \approx .0476$$

Since the rates are equal, the solution is correct.

OBJECTIVE 4 Solve applications about distance, rate, and time. A familiar example of a rate is speed, which is the ratio of distance to time. The next examples use the distance formula $d = rt$ introduced in Chapter 2.

EXAMPLE 6 Solving a Problem about Distance, Rate, and Time

At the airport, Cheryl and Bill are walking to the gate (at the same speed) to catch their flight to Akron, Ohio. Since Bill wants a window seat, he steps onto the moving sidewalk and continues to walk while Cheryl uses the stationary sidewalk. If the sidewalk moves at 1 meter per second and Bill saves 50 seconds covering the 300-meter distance, what is their walking speed?

Step 1 Let x represent their walking speed in meters per second.

Step 2 Then Cheryl travels at x meters per second and Bill travels at $x + 1$ meters per second. Since Bill's time is 50 seconds less than Cheryl's time, express their times in terms of the known distances and the variable rates. Start with $d = rt$ and divide both sides by r to get

$$t = \frac{d}{r}.$$

For Cheryl, distance is 300 meters and the rate is x. Cheryl's time is

$$t = \frac{d}{r} = \frac{300}{x}.$$

Bill goes 300 meters at a rate of $x + 1$, so his time is

$$t = \frac{d}{r} = \frac{300}{x + 1}.$$

This information is summarized in the following chart.

	d	r	t
Cheryl	300	x	$\frac{300}{x}$
Bill	300	$x + 1$	$\frac{300}{x + 1}$

Step 3 Now use the information given in the problem about the times to write an equation.

Bill's time	is	Cheryl's time	less 50 seconds.
$\frac{300}{x + 1}$	$=$	$\frac{300}{x}$	$- 50$

Step 4 Multiply both sides by the LCD, $x(x + 1)$.

$$x(x+1)\left(\frac{300}{x+1}\right) = x(x+1)\left(\frac{300}{x} - 50\right)$$

$$300x = 300(x + 1) - 50x(x + 1)$$

$$300x = 300x + 300 - 50x^2 - 50x$$

$$0 = 50x^2 + 50x - 300 \quad \text{Subtract } 300x\text{; multiply by } -1.$$

$$0 = x^2 + x - 6 \quad \text{Divide both sides by 50.}$$

$$0 = (x + 3)(x - 2) \quad \text{Factor.}$$

$$x + 3 = 0 \quad \text{or} \quad x - 2 = 0$$

$$x = -3 \quad \text{or} \quad x = 2$$

Discard the negative answer.

Step 5 Their walking speed is 2 meters per second.

Step 6 Check the solution in the words of the original problem.

OBJECTIVE 5 Solve applications about work rates. Problems about work are closely related to the distance problems we discussed in Section 2.4.

PROBLEM SOLVING

People work at different rates. If the letters r, t, and A represent the rate at which the work is done, the time required, and the amount of work accomplished, respectively, then $A = rt$. Notice the similarity to the distance formula, $d = rt$. Amount of work is often measured in terms of jobs accomplished. Thus, if 1 job is completed, $A = 1$, and the formula gives

$$1 = rt$$

$$r = \frac{1}{t}$$

as the rate.

Rate of Work

If a job can be accomplished in t units of time, then the rate of work is

$$\frac{1}{t} \text{ job per unit of time.}$$

To solve a work problem, we begin by using this fact to express all rates of work.

EXAMPLE 7 Solving a Problem about Work

Letitia and Kareem are working on a neighborhood cleanup. Kareem can clean up all the trash in the area in 7 hours, while Letitia can do the same job in 5 hours. How long will it take them if they work together?

Let x = the number of hours it will take the two people working together. Just as we made a chart for the distance formula, $d = rt$, we can make a chart here for $A = rt$, with $A = 1$. Since $A = 1$, the rate for each person will be $\frac{1}{t}$ where t is the time it takes each person to complete the job alone. For example, since Kareem can clean up all the trash in 7 hours, his rate is $\frac{1}{7}$ of the job per hour. Similarly, Letitia's rate is $\frac{1}{5}$ of the job per hour. Fill in the chart as shown.

Worker	*Rate*	*Time Working Together*	*Fractional Part of the Job Done*
Kareem	$\frac{1}{7}$	x	$\frac{1}{7}x$
Letitia	$\frac{1}{5}$	x	$\frac{1}{5}x$

Since together they complete 1 job, the sum of the fractional parts accomplished by each of them should equal 1.

$$\begin{array}{ccccc} \text{Part done by Kareem} & + & \text{Part done by Letitia} & = & \text{1 whole job.} \\ \frac{1}{7}x & + & \frac{1}{5}x & = & 1 \end{array}$$

Solve this equation. The LCD is 35.

$$35\left(\frac{1}{7}x + \frac{1}{5}x\right) = 35 \cdot 1$$

$$5x + 7x = 35 \qquad (*)$$

$$12x = 35$$

$$x = \frac{35}{12}$$

Working together, Kareem and Letitia can do the entire job in $\frac{35}{12}$ hours, or 2 hours and 55 minutes. Check this result in the original problem. ■

There is another way to approach problems about work. For instance, in Example 7, x represents the number of hours it will take the two people working together to complete the entire job. In one hour, $\frac{1}{x}$ portion of the entire job will be completed. Kareem completes $\frac{1}{7}$ of the job in one hour, and Letitia completes $\frac{1}{5}$ of the job, so the sum of their rates should equal $\frac{1}{x}$. This gives the equation

$$\frac{1}{7} + \frac{1}{5} = \frac{1}{x}.$$

When both sides of this equation are multiplied by $35x$, the result is $5x + 7x = 35$. Notice that this is the same equation we got in Example 7 in the line marked (*). Thus the solution of the equation is the same with both approaches.

CONNECTIONS

It is very common for people to average two rates by adding and then dividing by two, as we find the average for two numbers. However, a rate is a different kind of average, and we cannot average rates as we do numbers. If a car travels from A to B at 40 miles per hour and returns at 60 miles per hour, what is its rate for the entire trip? The correct answer is not 50 miles per

CONNECTIONS (CONTINUED)

hour! Using the distance-rate-time relationship and letting x = the distance between A and B, we can simplify a complex fraction to find the correct answer.

$$\begin{aligned}\text{Rate for entire trip} &= \frac{\text{Total distance}}{\text{Total time}} \\ &= \frac{x + x}{\frac{x}{40} + \frac{x}{60}}\end{aligned}$$

FOR DISCUSSION OR WRITING

Simplify the complex fraction to find the rate for the entire trip. Notice that x (the distance) in the problem was eliminated. Why do you suppose the distance does not matter?

6.7 EXERCISES

Solve the following problems in your head. Use proportions in Exercises 1 and 2.

1. In a mathematics class, 3 of every 4 students are girls. If there are 20 students in the class, how many are girls? How many are boys?
2. In a certain southern state, sales tax on a purchase of \$1.50 is \$.12. What is the sales tax on a purchase of \$6.00?

3. If Marin can mow her yard in 2 hours, what is her rate (in job per hour)?
4. A van traveling from Atlanta to Detroit averages 50 miles per hour and takes 14 hours to make the trip. How far is it from Atlanta to Detroit?

In Exercises 5–8, a familiar formula is given. Give the letter of the choice that is an equivalent form of the given formula.

5. $p = br$ (percent)
   (a) $b = \frac{p}{r}$
   (b) $r = \frac{b}{p}$
   (c) $b = \frac{r}{p}$
   (d) $p = \frac{r}{b}$

6. $V = LWH$ (geometry)
   (a) $H = \frac{LW}{V}$
   (b) $L = \frac{V}{WH}$
   (c) $L = \frac{WH}{V}$
   (d) $W = \frac{H}{VL}$

7. $m = \frac{F}{a}$ (physics)
   (a) $a = mF$
   (b) $F = \frac{m}{a}$
   (c) $F = \frac{a}{m}$
   (d) $F = ma$

8. $I = \frac{E}{R}$ (electricity)
   (a) $R = \frac{I}{E}$
   (b) $R = IE$
   (c) $E = \frac{I}{R}$
   (d) $E = RI$

Solve each problem. See Example 1.

9. The gravitational force between two masses is given by

$$F = \frac{GMm}{d^2}.$$

Find M if $F = 10$, $G = 6.67 \times 10^{-11}$, $m = 1$, and $d = 3 \times 10^{-6}$.

10. A gas law in chemistry says that

$$\frac{PV}{T} = \frac{pv}{t}.$$

Suppose that $T = 300$, $t = 350$, $V = 9$, $P = 50$, and $v = 8$. Find p.

11. In work with electric circuits, the formula

$$\frac{1}{a} = \frac{1}{b} + \frac{1}{c}$$

occurs. Find b if $a = 8$ and $c = 12$.

12. A formula from anthropology says that

$$c = \frac{100b}{L}.$$

Find L if $c = 80$ and $b = 5$.

Solve each formula for the specified variable. See Examples 2 and 3.

13. $F = \dfrac{GMm}{d^2}$ for G (physics)

14. $F = \dfrac{GMm}{d^2}$ for M (physics)

15. $\dfrac{1}{a} = \dfrac{1}{b} + \dfrac{1}{c}$ for a (electricity)

16. $\dfrac{1}{a} = \dfrac{1}{b} + \dfrac{1}{c}$ for b (electricity)

17. $\dfrac{PV}{T} = \dfrac{pv}{t}$ for v (chemistry)

18. $\dfrac{PV}{T} = \dfrac{pv}{t}$ for T (chemistry)

19. $I = \dfrac{nE}{R + nr}$ for r (engineering)

20. $a = \dfrac{V - v}{t}$ for V (physics)

21. $A = \dfrac{1}{2}h(B + b)$ for b (mathematics)

22. $S = \dfrac{n}{2}(a + \ell)d$ for n (mathematics)

23. $\dfrac{E}{e} = \dfrac{R + r}{r}$ for r (engineering)

24. $y = \dfrac{x + z}{a - x}$ for x

25. To solve the equation $m = \dfrac{ab}{a - b}$ for a, what is the first step?

26. Suppose you get the equation

$$rp - rq = p + q$$

to be solved for r. What is the next step?

Use the figure to answer Exercises 27–30. Round numbers from the graph to the nearest .5 million.

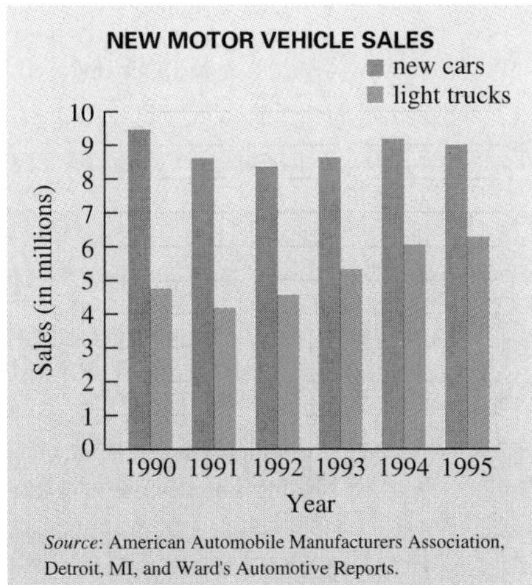

27. In 1991, what was the ratio of light truck sales to new car sales?

28. What was the ratio of light truck sales to new car sales in 1995?

29. What was the ratio of light truck sales in 1991 to light truck sales in 1995?

30. What new car sales in 1995 would have been in the same proportion to light truck sales as was the case in 1991?

Use a proportion to solve each problem. See Examples 4 and 5.

31. In 1997, 50 shares of common stock in Merck Company earned \$191.50. How much more would 75 shares of the stock have earned? (*Source:* Merck & Co., Inc., 1997 annual report.)

32. Seligman Communications and Information Fund, Inc. produced income of \$22,950 on an investment of \$100,000 in 1997. If the investment had been increased to \$260,000, how much more income would have been produced? (*Source:* Seligman Communications and Information Fund. Inc., 1997 annual report.)

33. Biologists tagged 500 fish in a lake on January 1. On February 1 they returned and collected a random sample of 400 fish, 8 of which had been previously tagged. Approximately how many fish does the lake have based on this experiment?

34. Suppose that in the experiment of Exercise 33, 10 of the previously tagged fish were collected on February 1. What would be the estimate of the fish population?

We list the six problem-solving steps in an abbreviated form here for quick reference.

Step 1 Determine what you are asked to find. Choose a variable and write down what it represents.

Step 2 Write down any other pertinent information. Express it using the variable, if appropriate. Use a chart or a diagram, if appropriate.

Step 3 Write an equation.

Step 4 Solve the equation.

Step 5 Answer the question(s) of the problem.

Step 6 Check your answer in the words of the original problem.

Go through the six steps to solve each problem in Exercises 35–40. See Example 6. Give the answers for each step in Exercise 35.

35. Driving from Tulsa to Detroit, Jeff averaged 50 miles per hour. He figured that if he had averaged 60 miles per hour, his driving time would have decreased 3 hours. How far is it from Tulsa to Detroit?

	d	r	t
Actual trip		50	t
Alternative trip		60	

36. A private plane traveled from San Francisco to a secret rendezvous. It averaged 200 miles per hour. On the return trip, the average speed was 300 miles per hour. If the total traveling time was 4 hours, how far from San Francisco was the secret rendezvous?

37. Johnny averages 30 miles per hour when he drives on the old highway to his favorite fishing hole, while he averages 50 miles per hour when most of his route is on the interstate. If both routes are the same length, and he saves 2 hours by traveling on the interstate, how far away is the fishing hole?

38. On the first part of a trip to Carmel traveling on the freeway, Marge averaged 60 miles per hour. On the rest of the trip, which was 10 miles longer than the first part, she averaged 50 miles per hour. Find the total distance to Carmel if the second part of the trip took 30 minutes more than the first part.

39. When enough typical waves in the Atlantic Ocean reach five meters in height (a condition called *head seas*), a container ship loses 6 knots of speed. (A knot is a nautical mile per hour. One nautical mile is about 6076 feet.) In head seas, a new, innovative FastShip loses only 2% of its calm-water speed. The speed of a FastShip in calm water is twice the speed of a container ship. A container ship and a FastShip leave New York at the same time and both encounter head seas. The FastShip arrives in London, a distance of about 2400 nautical miles, 72 hours later. The container ship takes 218 hours to reach London. To the nearest knot, find the calm-water speed of each type of ship. Let x and y represent the calm water speeds of the container ship and the FastShip, respectively. (*Source:* David Giles, "Faster Ships for the Future," *Scientific American,* October 1997, p. 129.)

	Distance	Rate	Time
Container ship		$x - 6$	
FastShip		$y - .02y$	72

40. Kellen's boat goes 12 miles per hour. Find the speed of the current of the river if she can go 6 miles upstream in the same amount of time she can go 10 miles downstream.

41. Explain the similarities between the methods of solving problems about distance, rate, and time and problems about work.

42. If one person takes 3 hours to do a job and another takes 4 hours to do the same job, why is "$3\frac{1}{2}$ hours" *not* a reasonable answer to the problem "How long will it take for them to do the job working together?"?

Solve each problem. See Example 7.

43. Lou can groom Jay Beckenstein's dogs in 8 hours, but it takes his business partner, Janet, only 5 hours to groom the same dogs. How long will it take them to groom Jay's dogs if they work together?

Worker	*Rate*	*Time Together*	*Part of the Job Done*
Lou	$\frac{1}{8}$	x	$\frac{1}{8}x$
Janet			

44. Bill and Julie want to pick up the mess that their grandson, J. W., has made in his playroom. Bill could do it in 15 minutes working alone. Julie, working alone, could clean it in 12 minutes. How long will it take them if they work together?

45. Bernard and Carolyn Goldstein are refinishing a table. Working alone, Bernard could do the job in 7 hours. If the two work together, the job takes 5 hours. How long will it take Carolyn to refinish the table if she works alone?

46. Mike can paint a room in 6 hours working alone. If Dee helps him, the job takes 4 hours. How long would it take Dee to do the job if she worked alone?

47. A winery has a barrel to hold chardonnay. An inlet pipe can fill the barrel in 9 hours, while an outlet pipe can empty it in 12 hours. How long will it take to fill the barrel if both the outlet and the inlet pipes are open?

48. If a vat of acid can be filled by an inlet pipe in 10 hours, and emptied by an outlet pipe in 20 hours, how long will it take to fill the vat if both pipes are open?

49. An inlet pipe can fill an artificial lily pond in 60 minutes, while an outlet pipe can empty it in 80 minutes. Through an error, both pipes are left open. How long will it take for the pond to fill?

50. Suppose that Hortense and Mort can clean their entire house in 7 hours, while their toddler, Mimi, just by being around, can completely mess it up in only 2 hours. If Hortense and Mort clean the house while Mimi is at her grandma's, and then start cleaning up after Mimi the minute she gets home, how long does it take from the time Mimi gets home until the whole place is a shambles?

In geometry, it is shown that two triangles with corresponding angles equal, called similar triangles, *have corresponding sides proportional. For example, in the figure, angle A = angle D, angle B = angle E, and angle C = angle F, so the triangles are similar. Then the following ratios of corresponding sides are equal.*

$$\frac{4}{6} = \frac{6}{9} = \frac{2x + 1}{2x + 5}$$

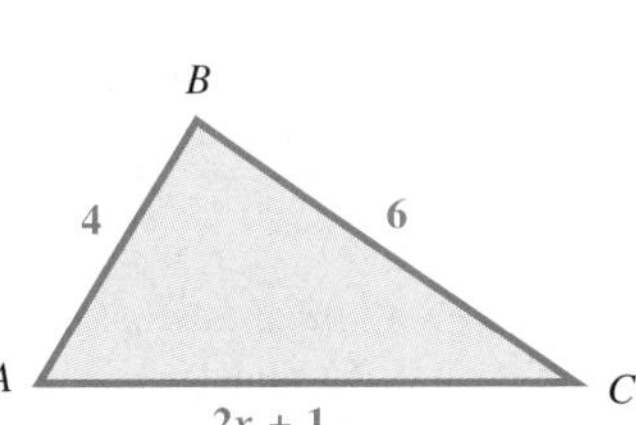

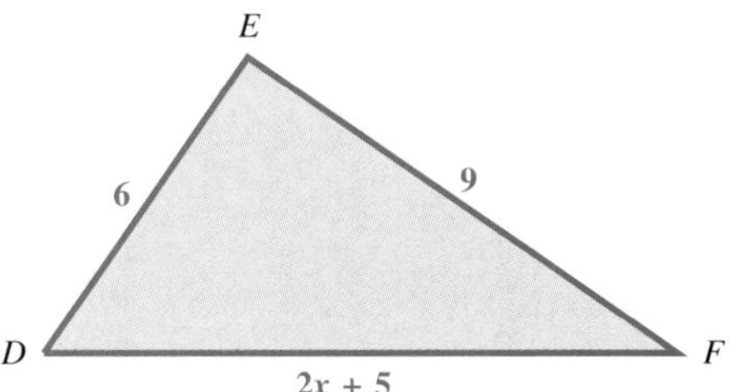

51. Solve for x using the proportion given above to find the lengths of the third sides of the triangles.

52. Suppose the triangles shown below are similar. Find y and the lengths of the two longest sides of each triangle.

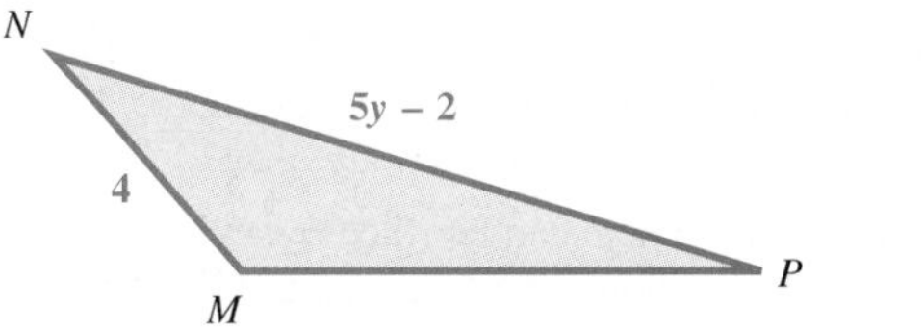

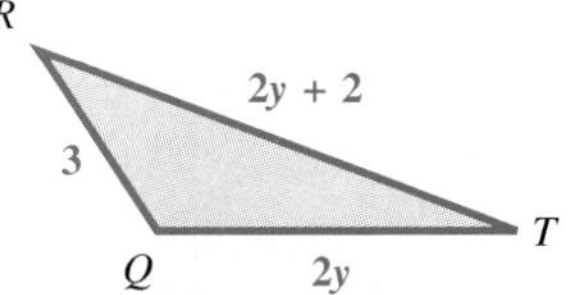

CHAPTER 6 GROUP ACTIVITY

Buying a Car

Objective: Use a complex fraction to calculate monthly car payments.

You are shopping for a sports car and have put aside a certain amount of money each month for a car payment. Your instructor will assign this amount to you. After looking through a variety of resources, you have narrowed your choices to the cars listed in the table.

Year/Make/Model	*Retail Price*	*Fuel Tank Size (in gallons)*	*Miles per Gallon (city)*	*Miles per Gallon (highway)*
1999 Mitsubishi 3000GT	$26,370	19.8	18	23
1999 Chevy Camaro	16,705	16.8	19	30
1999 Ford Mustang	16,470	15.7	20	27
1999 Pontiac Firebird	18,250	16.8	19	29
1999 BMW-Z3	36,200	13.5	19	26
1999 Toyota Celica	25,699	15.9	23	30

Source: www.excite.com/autos.

(continued)

As a group, work through the following steps to determine which car you can afford to buy.

A. Decide which cars you think are within your budget.

B. Select one of the cars you identified in part A. Have each member of the group calculate the monthly payment for this car using a different financing option. Use the formula given below, where P is principal, r is interest rate, and m is the number of monthly payments, along with the financing options table.

Financing Options

Time (in years)	*Interest Rate*
4	7.0%
5	8.5%
6	10.0%

$$\text{Monthly Payment} = \frac{\frac{Pr}{12}}{1 - \left(\frac{12}{12 + r}\right)^m}$$

C. Have each group member determine the amount of money paid in interest over the duration of the loan for their financing option.

D. Consider fuel expenses.

1. Assume you will travel an average of 75 miles in the city and 400 miles on the highway each week. How many gallons of gas will you need to buy each month?
2. Using typical prices for gas in your area at this time, how much money will you need to have available for buying gas?

E. Repeat parts B–D as necessary until your group can reach a consensus on the car you will buy and the financing option you will use. Write a paragraph to explain your choices.

CHAPTER 6 SUMMARY

KEY TERMS

6.1 **rational expression**
rational function
reciprocals

6.2 **least common denominator (LCD)**

6.3 **complex fraction**

6.5 **synthetic division**

6.6 **asymptote**
domain of a rational equation

6.7 **ratio**
proportion
rate

TEST YOUR WORD POWER

See how well you have learned the vocabulary in this chapter. Answers, with examples, are given at the bottom of the next page.

1. A **rational expression** is

(a) an algebraic expression made up of a term or the sum of a finite number of terms with real coefficients and whole number exponents

(b) a polynomial equation of degree two

(c) an expression with one or more fractions in the numerator, denominator, or both

(d) the quotient of two polynomials with denominator not zero.

(continued)

TEST YOUR WORD POWER

2. In a given set of fractions, the **least common denominator** is
(a) the smallest denominator of all the denominators
(b) the smallest expression that is divisible by all the denominators
(c) the largest integer that exactly divides the numerator and denominator of a fraction
(d) the largest denominator of all the denominators.

3. A **complex fraction** is
(a) an algebraic expression made up of a term or the sum of a finite number of terms with real coefficients and whole number exponents
(b) a polynomial equation of degree two
(c) an expression with one or more fractions in the numerator, denominator, or both
(d) the quotient of two polynomials with denominator not zero.

4. A **ratio**
(a) compares two quantities using a quotient
(b) says that two quotients are equal
(c) is a product of two quantities
(d) is a difference between two quantities.

5. A **proportion**
(a) compares two quantities using a quotient
(b) says that two quotients are equal
(c) is a product of two quantities
(d) is a difference between two quantities.

QUICK REVIEW

CONCEPTS	EXAMPLES
6.1 RATIONAL EXPRESSIONS AND FUNCTIONS; MULTIPLICATION AND DIVISION	
Writing Rational Expressions in Lowest Terms	Write in lowest terms.
Step 1 Factor the numerator and the denominator completely.	$\frac{2x+8}{x^2-16} = \frac{2(x+4)}{(x-4)(x+4)}$
Step 2 Apply the fundamental principle.	$= \frac{2}{x-4}$
Multiplying Rational Expressions Factor numerators and denominators. Apply the fundamental principle; replace all pairs of common factors in numerators and denominators by 1. Multiply remaining factors in the numerator and in the denominator.	Multiply. $\frac{x^2+2x+1}{x^2-1} \cdot \frac{5}{3x+3}$ $= \frac{(x+1)^2}{(x-1)(x+1)} \cdot \frac{5}{3(x+1)}$ $= \frac{5}{3(x-1)}$
Dividing Rational Expressions Multiply the first fraction by the reciprocal of the second.	Divide. $\frac{2x+5}{x-3} \div \frac{2x^2+3x-5}{x^2-9}$ $= \frac{2x+5}{x-3} \cdot \frac{(x+3)(x-3)}{(2x+5)(x-1)}$ $= \frac{x+3}{x-1}$

Answers to Test Your Word Power

1. (d) *Examples:* $-\frac{3}{4y^2}, \frac{5x^3}{x+2}, \frac{a+3}{a^2-4a-5}$ **2.** (b) *Example:* The LCD of $\frac{1}{x}, \frac{2}{3}$, and $\frac{5}{x+1}$ is $3x(x+1)$. **3.** (c) *Examples:* $\frac{\frac{2}{3}}{\frac{4}{7}}, \frac{x-\frac{1}{x}}{x+\frac{1}{y}}, \frac{\frac{2}{a+1}}{a^2-1}$ **4.** (a) *Example:* $\frac{7 \text{ inches}}{12 \text{ inches}}$ compares two quantities with the same units. **5.** (b) *Example:* The proportion $\frac{2}{3} = \frac{8}{12}$ states that the two ratios are equal.

CONCEPTS	EXAMPLES
6.2 ADDITION AND SUBTRACTION OF RATIONAL EXPRESSIONS	
Adding or Subtracting Rational Expressions If the denominators are the same, add or subtract the numerators. Place the result over the common denominator. If the denominators are different, write all rational expressions with the LCD. Then add or subtract the like fractions. Be sure answer is in lowest terms.	Subtract. $\frac{1}{x+6} - \frac{3}{x+2}$ $= \frac{x+2}{(x+6)(x+2)} - \frac{3(x+6)}{(x+6)(x+2)}$ $= \frac{x+2-3(x+6)}{(x+6)(x+2)}$ $= \frac{x+2-3x-18}{(x+6)(x+2)}$ $= \frac{-2x-16}{(x+6)(x+2)}$
6.3 COMPLEX FRACTIONS	
Simplifying Complex Fractions	Simplify the complex fraction.
Method 1 Simplify the numerator and denominator separately, as much as possible. Then multiply the numerator by the reciprocal of the denominator. Write the answer in lowest terms.	*Method 1* $\frac{\frac{1}{2}+\frac{1}{3}}{\frac{1}{4}-\frac{1}{2}} = \frac{\frac{3}{6}+\frac{2}{6}}{\frac{1}{4}-\frac{2}{4}}$ $= \frac{\frac{5}{6}}{\frac{-1}{4}} = \frac{5}{6} \cdot \frac{4}{-1}$ $= \frac{20}{-6} = -\frac{10}{3}$
Method 2 Multiply the numerator and denominator of the complex fraction by the least common denominator of all fractions appearing in the complex fraction. Then simplify the results.	*Method 2* $\frac{\frac{1}{2}+\frac{1}{3}}{\frac{1}{4}-\frac{1}{2}} = \frac{12\left(\frac{1}{2}\right)+12\left(\frac{1}{3}\right)}{12\left(\frac{1}{4}\right)-12\left(\frac{1}{2}\right)}$ $= \frac{6+4}{3-6} = \frac{10}{-3} = -\frac{10}{3}$
6.4 DIVISION OF POLYNOMIALS	
Dividing by a Monomial To divide a polynomial by a monomial, divide each term in the polynomial by the monomial, and then write each fraction in lowest terms.	Divide. $\frac{2x^3-4x^2+6x-8}{2x}$ $= \frac{2x^3}{2x} - \frac{4x^2}{2x} + \frac{6x}{2x} - \frac{8}{2x}$ $= x^2 - 2x + 3 - \frac{4}{x}$

(continued)

CONCEPTS	EXAMPLES
Dividing by a Polynomial Use the "long division" process.	Divide. $\dfrac{m^3 - m^2 + 2m + 5}{m + 1}$ $$\begin{array}{r} m^2 - 2m + 4 \\ m + 1\overline{)m^3 - m^2 + 2m + 5} \\ \underline{m^3 + m^2} \qquad\qquad \\ -2m^2 + 2m \qquad \\ \underline{-2m^2 - 2m} \qquad \\ 4m + 5 \\ \underline{4m + 4} \\ 1 \end{array} \leftarrow \text{Remainder}$$ The quotient is $$m^2 - 2m + 4 + \frac{1}{m + 1}.$$

6.6 GRAPHS AND EQUATIONS WITH RATIONAL EXPRESSIONS

The graph of a simple rational function may have one or more breaks. At such points, the graph will approach an asymptote.	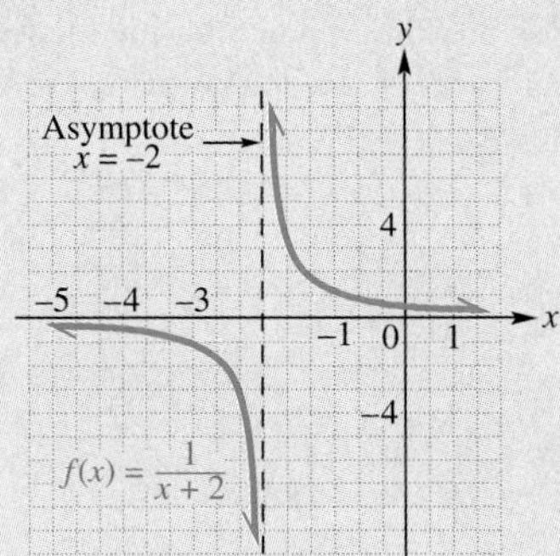
To solve an equation involving rational expressions, first determine the domain. Then multiply all the terms in the equation by the least common denominator. Solve the resulting equation. Each potential solution *must* be checked to see that it is in the domain of the equation.	Solve for x. $$\frac{1}{x} + x = \frac{26}{5}$$ Note that 0 is excluded from the domain. $5 + 5x^2 = 26x$ Multiply by $5x$. $5x^2 - 26x + 5 = 0$ $(5x - 1)(x - 5) = 0$ $x = \dfrac{1}{5}$ or $x = 5$ Both check. The solution set is $\left\{\dfrac{1}{5}, 5\right\}$.

6.7 APPLICATIONS OF RATIONAL EXPRESSIONS

To solve a formula for a particular variable, get that variable alone on one side by following the method described in Section 6.6.	Solve for L. $c = \dfrac{100b}{L}$ $cL = 100b$ Multiply by L. $L = \dfrac{100b}{c}$ Divide by c.

CONCEPTS	EXAMPLES
If an applied problem translates into an equation with rational expressions, solve the equation using the method described in Section 6.6.	If the 6.4-ounce tube of Crest toothpaste costs \$1.89, what should the 8.2-ounce tube cost? Let x represent the unknown cost. Use a proportion. $\frac{6.4}{1.89} = \frac{8.2}{x}$ $1.89x\left(\frac{6.4}{1.89}\right) = 1.89x\left(\frac{8.2}{x}\right)$ $6.4x = 15.498$ $x = 2.42$ To the nearest cent The 8.2-ounce tube should cost \$2.42. Check by comparing the two ratios.

CHAPTER 6 REVIEW EXERCISES

1. Write a few sentences defining the following terms and distinguishing among them: rational expression, fraction, complex fraction. Include examples.

[6.1] *Find all real numbers that are excluded from the domain.*

2. $g(t) = \frac{-7}{3t + 18}$

3. $f(r) = \frac{5r + 17}{r^2 - 7r + 10}$

4. List the steps you would use to write $\frac{x^2 + 3x}{5x + 15}$ in lowest terms.

Write in lowest terms.

5. $\frac{55m^4n^3}{10m^5n}$

6. $\frac{12x^2 + 6x}{24x + 12}$

7. $\frac{25m^2 - n^2}{25m^2 - 10mn + n^2}$

8. $\frac{r - 2}{4 - r^2}$

9. What is meant by the reciprocal of a rational expression?

Multiply or divide. Write the answer in lowest terms.

10. $\frac{25p^3q^2}{8p^4q} \div \frac{15pq^2}{16p^5}$

11. $\frac{w^2 - 16}{w} \cdot \frac{3}{4 - w}$

12. $\frac{z^2 - z - 6}{z - 6} \cdot \frac{z^2 - 6z}{z^2 + 2z - 15}$

13. $\frac{m^3 - n^3}{m^2 - n^2} \div \frac{m^2 + mn + n^2}{m + n}$

14. What is *wrong* with the following work?

$$\frac{x^2 + 5x}{x + 5} = \frac{x^2}{x} + \frac{5x}{5} = x + 5$$

What is the *correct* simplified form?

15. How would you explain finding the least common denominator to a classmate?

[6.2] *Find the least common denominator for each group of rational expressions.*

16. $\frac{5a}{32b^3}, \frac{31}{24b^5}$

17. $\frac{17}{9r^2}, \frac{5r - 3}{3r + 1}$

18. $\frac{4x - 9}{6x^2 + 13x - 5}, \frac{x + 15}{9x^2 + 9x - 4}$

Add or subtract as indicated.

19. $\dfrac{8}{z} - \dfrac{3}{2z^2}$

20. $\dfrac{5y + 13}{y + 1} - \dfrac{1 - 7y}{y + 1}$

21. $\dfrac{6}{5a + 10} + \dfrac{7}{6a + 12}$

22. $\dfrac{3r}{10r^2 - 3rs - s^2} + \dfrac{2r}{2r^2 + rs - s^2}$

[6.3] *Simplify each complex fraction.*

23. $\dfrac{\dfrac{3}{t} + 2}{\dfrac{4}{t} - 7}$

24. $\dfrac{\dfrac{4m^5n^6}{mn}}{\dfrac{8m^7n^3}{m^4n^2}}$

25. $\dfrac{\dfrac{3}{p} - \dfrac{2}{q}}{\dfrac{9q^2 - 4p^2}{qp}}$

[6.4] *Divide.*

26. $\dfrac{4y^3 - 12y^2 + 5y}{4y}$

27. $\dfrac{2p^3 + 9p^2 + 27}{2p - 3}$

28. $\dfrac{5p^4 + 15p^3 - 33p^2 - 9p + 18}{5p^2 - 3}$

[6.5] *Use synthetic division to perform each division.*

29. $\dfrac{3p^2 - p - 2}{p - 1}$

30. $(2k^3 - 5k^2 + 12) \div (k - 3)$

Use synthetic division to decide whether or not -5 is a solution of each equation.

31. $2w^3 + 8w^2 - 14w - 20 = 0$

32. $-3q^4 + 2q^3 + 5q^2 - 9q + 1 = 0$

Use synthetic division to evaluate $P(k)$ for the given value of k.

33. $P(x) = 3x^3 - 5x^2 + 4x - 1; \quad k = -1$

34. $P(z) = z^4 - 2z^3 - 9z - 5; \quad k = 3$

[6.6]

35. Which is the graph of a rational function?

(a)

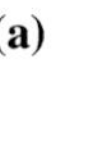

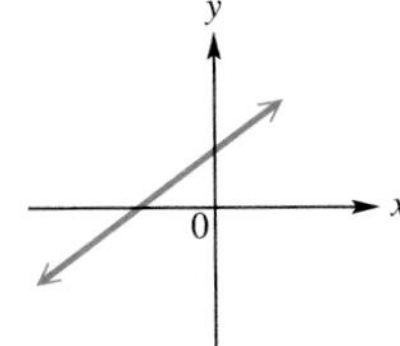

(b)

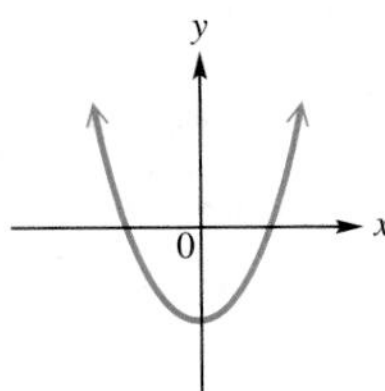

(c)

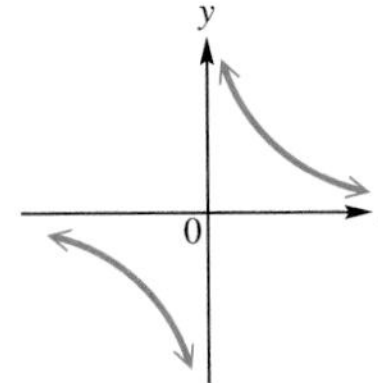

(d)

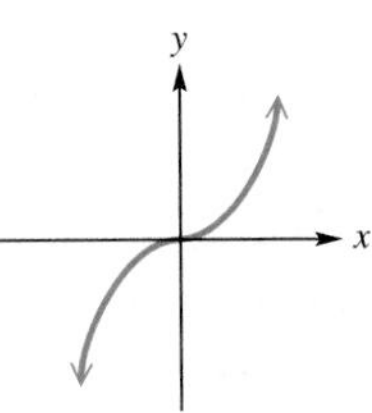

Solve each equation.

36. $\dfrac{1}{t + 4} + \dfrac{1}{2} = \dfrac{3}{2t + 8}$

37. $\dfrac{-5m}{m + 1} + \dfrac{m}{3m + 3} = \dfrac{56}{6m + 6}$

38. $\dfrac{2}{k - 1} - \dfrac{4k + 1}{k^2 - 1} = \dfrac{-1}{k + 1}$

39. $\dfrac{5}{x + 2} + \dfrac{3}{x + 3} = \dfrac{x}{x^2 + 5x + 6}$

40. After solving the equation

$$\frac{3}{x-3}-\frac{2}{x-2}=\frac{3}{x^2-5x+6},$$

a student got $x = 3$ as her final step. She could not understand why the answer in the back of the book was "$\emptyset$," because she checked her algebra several times and was sure that all her algebraic work was correct. Was she wrong or was the answer in the back of the book wrong? Explain.

41. Explain the difference between simplifying the expression $\frac{4}{x}+\frac{1}{2}-\frac{1}{3}$ and solving the equation $\frac{4}{x}+\frac{1}{2}=\frac{1}{3}$.

[6.7] *Work each problem.*

42. According to a law from physics, $\frac{1}{A}=\frac{1}{B}+\frac{1}{C}$. Find A if $B = 30$ and $C = 10$.

Solve each formula for the specified variable.

43. $F=\frac{GMm}{d^2}$ for m (physics)

44. $\mu=\frac{Mv}{M+m}$ for M (electronics)

A growing problem nationwide is the spread of wildlife into human habitats. The graphs show how the populations of deer in Georgia and black bears in Wisconsin have increased in recent years. Use these graphs for Exercises 45–49.

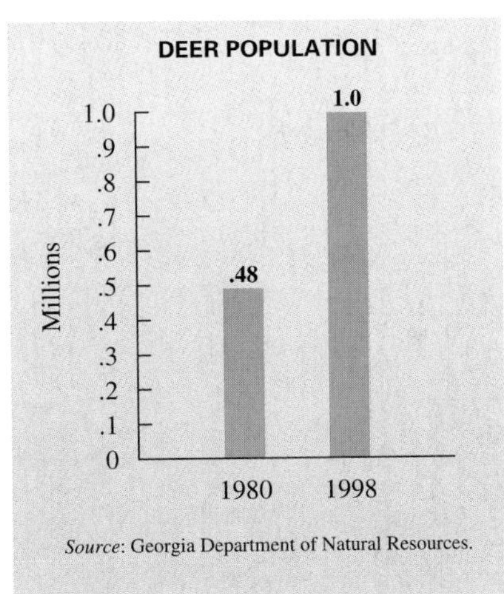

Source: Georgia Department of Natural Resources.

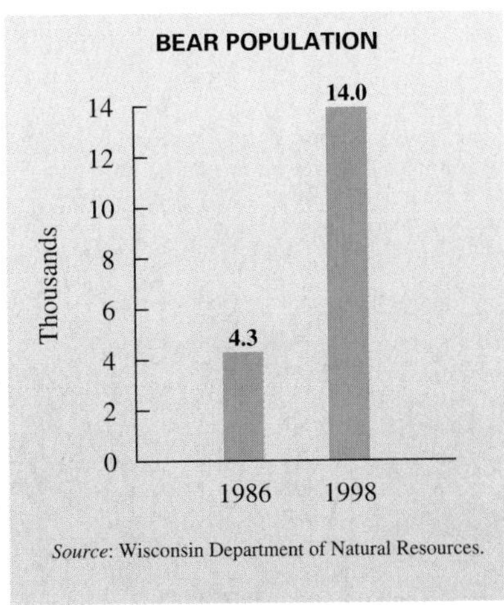

Source: Wisconsin Department of Natural Resources.

45. Find the ratio of the deer population in 1980 to the population in 1998.

46. What is the ratio of the bear population in 1998 to the population in 1986?

47. What was the average annual rate of change in the deer population from 1980 to 1998?

48. Find the average annual rate of change in the bear population from 1986 to 1998.

49. Which population has a higher growth rate?

50. An article in *Scientific American* predicts that, in the year 2050, 23,200 of the 58,000 passenger-kilometers per day in North America will be provided by high-speed trains. If the traffic volume in a typical region of North America is 15,000, how many passenger-kilometers per day will high-speed trains provide there? (*Source:* Andreas Schafer and David Victor, "The Past and Future of Global Mobility," *Scientific American,* October 1997.)

51. A river has a current of 4 kilometers per hour. Find the speed of Lynn McTernan's boat in still water if it goes 40 kilometers downstream in the same time that it takes to go 24 kilometers upstream.

52. A sink can be filled by a cold-water tap in 8 minutes, and filled by the hot-water tap in 12 minutes. How long would it take to fill the sink with both taps open?

53. Jane Estrella and Jason Jordan need to sort a pile of bottles at the recycling center. Working alone, Jane could do the entire job in 9 hours, while Jason could do the entire job in 6 hours. How long will it take them if they work together?

MIXED REVIEW EXERCISES

54. Find the least common denominator: $\frac{14}{6k+3}, \frac{7k^2+2k+1}{10k+5}, \frac{-11k}{18k+9}$.

Perform the indicated operations.

55. $\frac{2}{m} + \frac{5}{3m^2}$

56. $\frac{k^2-6k+9}{1-216k^3} \cdot \frac{6k^2+17k-3}{9-k^2}$

57. $\dfrac{\frac{-3}{x}+\frac{x}{2}}{1+\frac{x+1}{x}}$

58. $\frac{9x^2+46x+5}{3x^2-2x-1} \div \frac{x^2+11x+30}{x^3+5x^2-6x}$

59. $\frac{3x^{-1}-5}{6+x^{-1}}$

60. $\frac{9}{3-x} - \frac{2}{x-3}$

61. $\frac{4y+16}{30} \div \frac{2y+8}{5}$

62. $\frac{t^{-2}+s^{-2}}{t^{-1}-s^{-1}}$

63. $\frac{4a}{a^2-ab-2b^2} - \frac{6b-a}{a^2+4ab+3b^2}$

64. $(-a^4+19a^2+18a+15) \div (a+4)$

65. $\frac{12y^4+7y^2-2y+1}{3y^2+1}$

66. Use synthetic division to decide whether 3 is a solution of $7z^3-z^2+5z-3=0$.

Solve.

67. $\frac{x+3}{x^2-5x+4} - \frac{1}{x} = \frac{2}{x^2-4x}$

68. $A = \frac{Rr}{R+r}$ for r

69. $1 - \frac{5}{r} = \frac{-4}{r^2}$

70. $\frac{3x}{x-4} + \frac{2}{x} = \frac{48}{x^2 - 4x}$

71. The strength of a contact lens is given in units called diopters, and also in millimeters of arc. As the diopters increase, the millimeters of arc decrease. The rational function defined by

$$a = \frac{337}{d}$$

relates the arc measurement a to the diopter measurement d.

(a) What arc measurement will correspond to 40.5-diopter lenses?

(b) A lens with an arc measurement of 7.51 will provide what diopter strength? (*Source:* Bausch and Lomb.)

72. Taking traffic into account, an automobile can travel on the average 7 kilometers in the same time that an airplane can travel 100 kilometers. The average speed of an airplane is 558 kilometers per hour greater than that of an automobile. Find both speeds. (*Source:* Andreas Schafer and David Victor, "The Past and Future of Global Mobility," *Scientific American,* October 1997.)

	d	r	t
Automobile		x	
Airplane		$x + 558$	

73. The hot-water tap can fill a tub in 20 minutes. The cold-water tap takes 15 minutes to fill the tub. How long would it take to fill the tub with both taps open?

74. At a certain gasoline station, 3 gallons of unleaded gasoline cost $4.86. How much would 13 gallons of the same gasoline cost?

CHAPTER 6 TEST

1. Find all real numbers excluded from the domain of $f(k) = \frac{2k - 1}{3k^2 + 2k - 8}$.

2. Write $\frac{6x^2 - 13x - 5}{9x^3 - x}$ in lowest terms.

Multiply or divide.

3. $\frac{4x^2y^5}{7xy^8} \div \frac{8xy^6}{21xy}$

4. $\frac{y^2 - 16}{y^2 - 25} \cdot \frac{y^2 + 2y - 15}{y^2 - 7y + 12}$

5. $\frac{x^2 - 9}{x^3 + 3x^2} \div \frac{x^2 + x - 12}{x^3 + 9x^2 + 20x}$

6. Find the least common denominator for the following group of denominators: $t^2 + t - 6$, $t^2 + 3t$, t^2.

Add or subtract as indicated.

7. $\frac{7}{6t^2} - \frac{1}{3t}$

8. $\frac{9}{x-7} + \frac{4}{x+7}$

9. $\frac{6}{x+4} + \frac{1}{x+2} - \frac{3x}{x^2 + 6x + 8}$

Simplify each complex fraction.

10. $\dfrac{\dfrac{12}{r+4}}{\dfrac{11}{6r+24}}$

11. $\dfrac{\dfrac{1}{a}-\dfrac{1}{b}}{\dfrac{a}{b}-\dfrac{b}{a}}$

Divide.

12. $\dfrac{16p^3 - 32p^2 + 24p}{4p^2}$

13. $\dfrac{9q^4 - 18q^3 + 11q^2 + 10q - 10}{3q - 2}$

14. $\dfrac{6y^4 - 4y^3 + 5y^2 + 6y - 9}{2y^2 + 3}$

15. Use synthetic division to decide whether 4 is a solution of $x^4 - 8x^3 + 21x^2 - 14x - 24 = 0$.

16. Use synthetic division to divide $9x^5 + 40x^4 - 23x^3 + 8x^2 - 6x + 22$ by $x + 5$.

17. One of the following is an expression to be simplified by algebraic operations, and the other is an equation to be solved. Simplify the one that is an expression, and solve the one that is an equation.

 (a) $\dfrac{2x}{3} + \dfrac{x}{4} - \dfrac{11}{2}$ (b) $\dfrac{2x}{3} + \dfrac{x}{4} = \dfrac{11}{2}$

Solve each equation.

18. $\dfrac{1}{x} - \dfrac{4}{3x} = \dfrac{1}{x-2}$

19. $\dfrac{y}{y+2} - \dfrac{1}{y-2} = \dfrac{8}{y^2-4}$

20. Checking solutions of an equation in Chapters 1–5 verified that the algebraic steps were performed correctly. When an equation includes a term with a variable denominator, what additional reason *requires* that the solutions be checked?

21. Solve for the variable ℓ in this formula from the field of mathematics: $S = \dfrac{n}{2}(a + \ell)$.

Solve each problem.

22. Wayne can do a job in 9 hours, while Susan can do the same job in 5 hours. How long would it take them to do the job if they worked together?

23. The current of a river runs at 3 miles per hour. Nana's boat can go 36 miles downstream in the same time that it takes to go 24 miles upstream. Find the boat speed x in still water.

	d	r	t
Downstream	36	$x + 3$	
Upstream			

24. Biologists collected a sample of 600 fish from Lake Linda on May 1 and tagged each of them. When they returned on June 1, a new sample of 800 fish was collected, and 10 of these had been previously tagged. Use this experiment to determine the approximate fish population of Lake Linda.

25. In biology, the function defined by

$$g(x) = \frac{5x}{2 + x}$$

gives the growth rate of a population for x units of available food.

 (a) What amount of food (in appropriate units) would produce a growth rate of 3 units of growth per unit of food?

 (b) What is the growth rate if no food is available? (*Source:* J. Maynard Smith, *Models in Ecology*, Cambridge University Press, 1974.)

CUMULATIVE REVIEW EXERCISES CHAPTERS 1–6

Evaluate if $x = -4$, $y = 3$, and $z = 6$.

1. $|2x| + 3y - z^3$

2. $\dfrac{x(2x - 1)}{3y - z}$

Solve each equation.

3. $7(2x + 3) - 4(2x + 1) = 2(x + 1)$

4. $|6x - 8| - 4 = 0$

5. $ax + by = cx + d$ for x

Solve each inequality.

6. $\dfrac{2}{3}y + \dfrac{5}{12}y \leq 20$

7. $|3x + 2| \geq 4$

Solve each problem.

8. The popular vote (in millions) cast for president by major political parties in 1996 is shown in the graph. Democrats got 8.2 million more votes than Republicans. The total number of votes cast was 96.2 million. Find the number of votes received by the Democratic and Republican candidates.

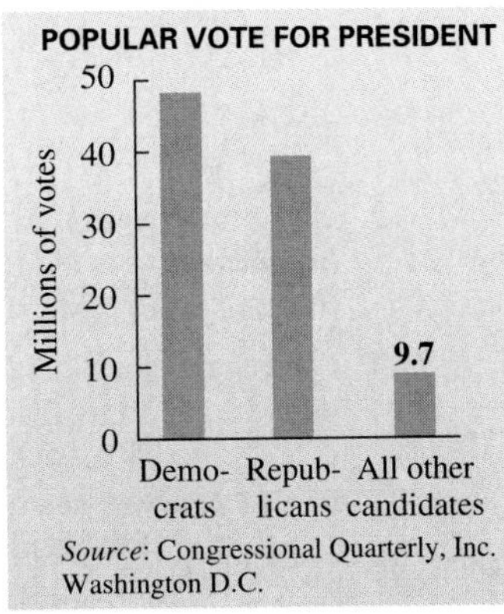

Source: Congressional Quarterly, Inc. Washington D.C.

9. A triangle has an area of 42 square meters. The base is 14 meters long. Find the height of the triangle.

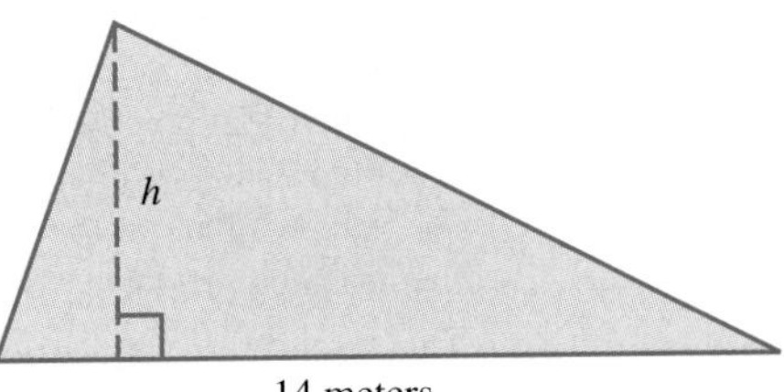

10. Graph $-4x + 2y = 8$ and give the intercepts.

Find the slope of the lines described in Exercises 11 and 12.

11. Through $(-5, 8)$ and $(-1, 2)$

12. Perpendicular to $4x - 3y = 12$

13. Write an equation of the line in Exercise 11 in the form $y = mx + b$.

Graph the solution set of each inequality.

14. $2x + 5y > 10$

15. $x - y \geq 3$ and $3x + 4y \leq 12$

Decide if each relation defined in Exercises 16–18 is a function, and give its domain and range.

16. **Selected U.S. Petroleum Imports in 1996**

Country	Thousand Barrels per Day
Venezuela	1657
Canada	1415
Saudi Arabia	1363
Mexico	1240

Source: U.S. Energy Department.

17.

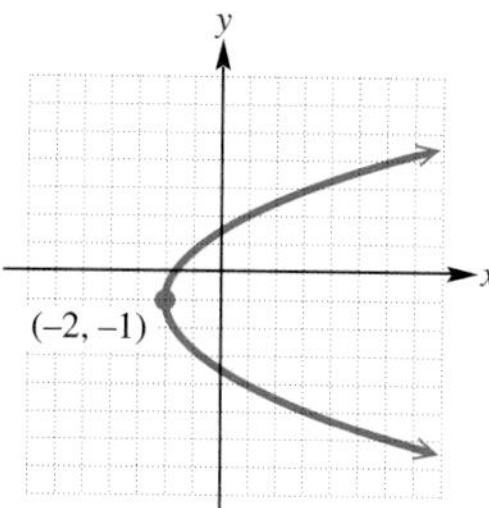

18. $y = -\sqrt{x + 2}$

19. Given the equation $5x - 3y = 8$,
(a) write it with the function notation $f(x)$.
(b) find $f(1)$.

20. Planets are approximately spherical in shape. The volume of a sphere is directly proportional to its diameter. Jupiter, with a diameter of about 1.4×10^5 kilometers, has a volume of 1.4×10^{15}. The diameter of Earth is about 1.3×10^4 kilometers. What is Earth's volume?

Solve each system.

21. **(a)** Use elimination.
(b) Use matrix row operations.

$$\begin{aligned} 4x - y &= -7 \\ 5x + 2y &= 1 \end{aligned}$$

22. Use any method.

$$\begin{aligned} x + y - 2z &= -1 \\ 2x - y + z &= -6 \\ 3x + 2y - 3z &= -3 \end{aligned}$$

23. In 1995, the total value of manufacturers' shipments of tobacco products was \$33 million. The value of shipments of cigarettes accounted for \$16.8 million more than twice the value of shipments of all other tobacco products. Use a system of equations to find the values (in millions of dollars to the nearest tenth) of shipments of cigarettes and shipments of all other tobacco products. (*Source:* U.S. Bureau of the Census.)

24. Evaluate the determinant using expansion by minors.

$$\begin{vmatrix} 2 & 0 & -3 \\ 5 & 1 & 0 \\ -2 & 0 & 8 \end{vmatrix}$$

Simplify. Write the answer with only positive exponents. Assume that all variables represent nonzero real numbers.

25. $\left(\dfrac{a^{-3}b^4}{a^2b^{-1}}\right)^{-2}$

26. $\left(\dfrac{m^{-4}n^2}{m^2n^{-3}}\right) \cdot \left(\dfrac{m^5n^{-1}}{m^{-2}n^5}\right)$

Perform the indicated operations.

27. $(3y^2 - 2y + 6) - (-y^2 + 5y + 12)$

28. $-6x^4(x^2 - 3x + 2)$

29. $(4f + 3)(3f - 1)$

30. $(7t^3 + 8)(7t^3 - 8)$

31. $\left(\frac{1}{4}x + 5\right)^2$

32. $(3x^3 + 13x^2 - 17x - 7) \div (3x + 1)$

33. Use synthetic division to divide $(2x^4 + 3x^3 - 8x^2 + x + 2)$ by $(x - 1)$.

Factor each polynomial completely.

34. $2x^2 - 13x - 45$

35. $100t^4 - 25$

36. $8p^3 + 125$

37. Solve the equation $3x^2 + 4x = 7$.

Write each rational expression in lowest terms.

38. $\frac{y^2 - 16}{y^2 - 8y + 16}$

39. $\frac{8x^2 - 18}{8x^2 + 4x - 12}$

Perform the indicated operations. Express the answer in lowest terms.

40. $\frac{2a^2}{a + b} \cdot \frac{a - b}{4a}$

41. $\frac{x + 4}{x - 2} + \frac{2x - 10}{x - 2}$

42. $\frac{2x}{2x - 1} + \frac{4}{2x + 1} + \frac{8}{4x^2 - 1}$

43. Solve the equation $\frac{-3x}{x + 1} + \frac{4x + 1}{x} = \frac{-3}{x^2 + x}$.

44. Solve the formula for q: $\frac{1}{f} = \frac{1}{p} + \frac{1}{q}$.

Solve each problem.

45. Lucinda can fly her plane 200 miles against the wind in the same time it takes her to fly 300 miles with the wind. The wind blows at 30 miles per hour. Find the speed of her plane in still air.

46. Machine A can complete a certain job in 2 hours. To speed up the work, Machine B, which could complete the job alone in 3 hours, is brought in to help. How long will it take the two machines to complete the job working together?

Roots and Radicals

Electronics

Early computers were based on mechanical switches. Computers based on electronic switches are much faster. The 1946 ENIAC, which had mechanical switches, occupied 2000 square feet, weighed 50 tons, and used 50,000 vacuum tubes. It could perform 10,000 multiplications per minute and had a memory of about 20 words. The best computer in 1997 was about 66 million times as fast as the 1951 UNIVAC, another vacuum tube computer with mechanical switches.*

As computers have become faster and gained more and more memory and software has become more sophisticated, personal computer use has grown worldwide. The figure shows a calculator graph of the radical equation

$$y = \frac{-.69 + \sqrt{.69^2 - 4(1.45)(24 - x)}}{2(1.45)}.$$

Here x represents worldwide shipments of personal computers in millions, and y represents the number of years after 1990 when x million shipments occurred. Thus, the ordered pair (50, 4), which satisfies the equation, means that 50 million shipments worldwide were made in 1994. What does (24, 0) mean?

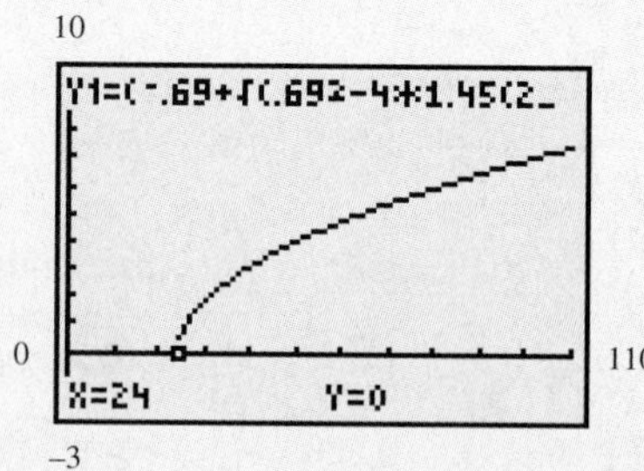

Other applications of electronics are introduced throughout this chapter.

*John W. Wright, General Editor, *The Universal Almanac, 1997,* Kansas City: Andrews and McMeel, p. 597.

Visit our Web site at www.LialAlgebra.com

7.1 Rational Exponents and Radicals

OBJECTIVES

1 Define $a^{1/n}$.

2 Use radical notation for nth roots.

3 Define $a^{m/n}$.

4 Convert between rational exponents and radicals.

5 Use the rules of exponents with rational exponents.

6 Use a calculator to find roots.

FOR EXTRA HELP

SSG Sec. 7.1
SSM Sec. 7.1

Pass the Test Software

InterAct Math Tutorial Software

Video 13

OBJECTIVE 1 **Define $a^{1/n}$.** In mathematics we often formulate definitions so that previously determined rules remain valid. For example, in Chapter 5 we defined 0 as an exponent in such a way that the rules for products, quotients, and powers would still be valid. Now we investigate rational numbers of the form $\frac{1}{n}$, where n is a natural number, as exponents. For the rules of exponents to remain valid, the product $(3^{1/2})^2 = 3^{1/2} \cdot 3^{1/2}$ should be found by adding exponents.

$$\begin{aligned}(3^{1/2})^2 &= 3^{1/2} \cdot 3^{1/2} \\ &= 3^{1/2+1/2} \\ &= 3^1 = 3\end{aligned}$$

From previous experience, the symbol $\sqrt{3}$ represents the positive real number whose square is 3. Since both $(3^{1/2})^2$ and $(\sqrt{3})^2$ are equal to 3, we should have

$$3^{1/2} = \sqrt{3}.$$

This suggests the following generalization.

nth Root, n Even

If n is an *even* positive integer and if a is positive, then $a^{1/n}$ is the positive real number whose nth power is a, and $a^{1/n}$ is the **principal nth root of a.**

EXAMPLE 1 **Evaluating Exponentials of the Form $a^{1/n}$, n Even**

Evaluate each exponential expression.

(a) $64^{1/2} = (8^2)^{1/2} = 8^1 = 8$

(b) $10{,}000^{1/4} = (10^4)^{1/4} = 10$

(c) $144^{1/2} = (12^2)^{1/2} = 12$

(d) $-144^{1/2} = -(144^{1/2}) = -12$

(e) $(-144)^{1/2}$ is not a real number; only nonnegative numbers have real square roots.

With *odd* values of n, the base may be negative.

nth Root, n Odd

If n is an *odd* positive integer, for any real number a, $a^{1/n}$ is the nth root of a.

Because only even powers have two roots, only even powers require a principal nth root, which is the positive root. Odd powers of a have the same sign as a itself.

EXAMPLE 2 **Evaluating Exponentials of the Form $a^{1/n}$, n Odd**

Evaluate each exponential expression.

(a) $64^{1/3} = (4^3)^{1/3} = 4^1 = 4$

(b) $(-64)^{1/3} = [(-4)^3]^{1/3} = (-4)^1 = -4$

(c) $(-32)^{1/5} = [(-2)^5]^{1/5} = -2$

(d) $\left(\frac{1}{8}\right)^{1/3} = \left[\left(\frac{1}{2}\right)^3\right]^{1/3} = \frac{1}{2}$

OBJECTIVE 2 Use radical notation for nth roots. The discussion of nth roots leads naturally to radical notation. It is common for nth roots to be written with a radical sign, first introduced in Chapter 1.

Radical Notation

If n is a positive integer greater than 1, and if $a^{1/n}$ is a real number, then

$$a^{1/n} = \sqrt[n]{a}.$$

The number a is the **radicand,** n is the **index** or **order,** and the expression $\sqrt[n]{a}$ is a **radical.** The second root of a number, $\sqrt[2]{a}$, is its principal *square root.* It is customary to omit the index 2 and write the square root of a as $\sqrt{a}$.

EXAMPLE 3 Simplifying Roots Using Radical Notation

Find each root that is a real number.

(a) $-\sqrt{25} = -5$

(b) $\sqrt{\frac{9}{16}} = \frac{3}{4}$

(c) $\sqrt[3]{125} = 5$, since $5^3 = 125$.

(d) $\sqrt[3]{-216} = -6$, since $(-6)^3 = -216$.

(e) $\sqrt[4]{-16}$ is not a real number.

As mentioned earlier, a positive number has *two* even nth roots, one positive and one negative. The symbol $\sqrt[n]{a}$ is used only for the *nonnegative* root, the principal nth root. The negative nth root is written $-\sqrt[n]{a}$. (Sometimes the two roots are written $\pm\sqrt[n]{a}$, read "positive or negative nth root of a.")

A square root of a^2 (where $a \neq 0$) is a number that can be squared to give a^2. This number is either a or $-a$. Since the symbol $\sqrt{a^2}$ represents the *nonnegative* square root, we must write $\sqrt{a^2}$ with absolute value bars, as $|a|$, because a may be a negative number.

$\sqrt{a^2}$

For any real number a, $\quad \sqrt{a^2} = |a|$.

This result can be generalized to any nth root.

$\sqrt[n]{a^n}$

If n is an **even** positive integer, $\quad \sqrt[n]{a^n} = |a|$,

and if n is an **odd** positive integer, $\quad \sqrt[n]{a^n} = a$.

EXAMPLE 4 Simplifying Roots with Absolute Value

Use absolute value as necessary to evaluate each radical expression.

(a) $\sqrt{5^2} = |5| = 5$ **(b)** $\sqrt{(-5)^2} = |-5| = 5$

(c) $\sqrt[6]{(-3)^6} = |-3| = 3$

(d) $\sqrt[5]{(-4)^5} = -4$ 5 is odd; it is not necessary to use absolute value.

(e) $-\sqrt[4]{(-9)^4} = -|-9| = -9$

OBJECTIVE 3 Define $a^{m/n}$. We can now define the more general expression $a^{m/n}$, where m/n is any rational number. Start with a rational number m/n where both m and n are positive integers, with m/n written in lowest terms. If the earlier rules of exponents are to remain valid for rational exponents, then we should have

$$a^{m/n} = a^{(1/n)m} = (a^{1/n})^m,$$

provided that $a^{1/n}$ is a real number. This result leads to the definition of $a^{m/n}$.

$a^{m/n}$

If m and n are positive integers with m/n in lowest terms, then

$$a^{m/n} = (a^{1/n})^m,$$

provided that $a^{1/n}$ is a real number. If $a^{1/n}$ is not a real number, then $a^{m/n}$ is not a real number.

The expression $a^{-m/n}$ is defined as follows.

$a^{-m/n}$

$$a^{-m/n} = \frac{1}{a^{m/n}} \quad (a \neq 0),$$

provided that $a^{m/n}$ is a real number.

EXAMPLE 5 Evaluating Exponentials of the Form $a^{m/n}$

Evaluate each exponential expression that is a real number.

(a) $36^{3/2} = (36^{1/2})^3 = 6^3 = 216$

(b) $125^{2/3} = (125^{1/3})^2 = 5^2 = 25$

(c) $-4^{5/2} = -(4^{5/2}) = -(4^{1/2})^5 = -(2)^5 = -32$

(d) $(-27)^{2/3} = [(-27)^{1/3}]^2 = (-3)^2 = 9$

(e) $(-100)^{3/2}$ is not a real number, since $(-100)^{1/2}$ is not a real number.

(f) $16^{-3/4} = \dfrac{1}{16^{3/4}} = \dfrac{1}{(16^{1/4})^3} = \dfrac{1}{2^3} = \dfrac{1}{8}$

Figure 1 shows two screens of a graphing calculator. These screens illustrate how a graphing calculator performs the evaluations seen in Example 5. (All results in the screens are rational numbers.)

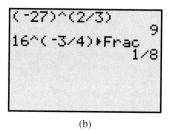

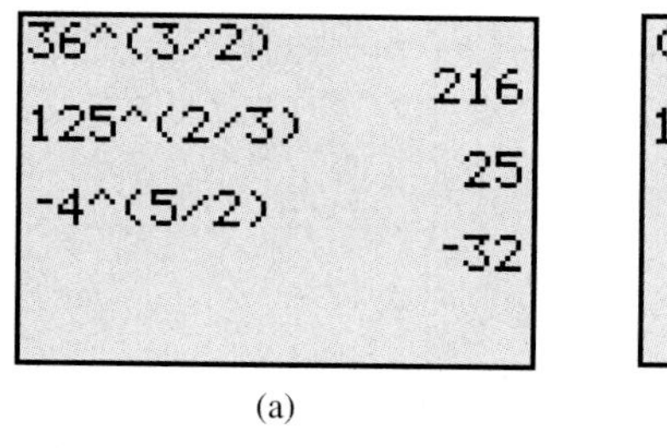

(a) See parts (a) - (c) of Example 5.

(b) See parts (d) and (f) of Example 5.

Figure 1

We get an alternative definition of $a^{m/n}$ by using the power rule for exponents a little differently than in the earlier definition. If all indicated roots are real numbers,

$$a^{m/n} = a^{m(1/n)} = (a^m)^{1/n},$$

so that

$$a^{m/n} = (a^m)^{1/n}.$$

$a^{m/n}$

If all indicated roots are real numbers, then $\quad a^{m/n} = (a^{1/n})^m = (a^m)^{1/n}.$

With this result, $a^{m/n}$ can be defined in either of two ways.

An expression such as $27^{2/3}$ can now be evaluated in two ways:

$$27^{2/3} = (27^{1/3})^2 = 3^2 = 9$$
$$27^{2/3} = (27^2)^{1/3} = 729^{1/3} = 9.$$

In most cases, it is easier to use $(a^{1/n})^m$.

OBJECTIVE 4 Convert between rational exponents and radicals. Until now, expressions of the form $a^{m/n}$ have been evaluated without introducing radical notation. However, we have seen that for appropriate values of a and n, $a^{1/n} = \sqrt[n]{a}$, and so it is not difficult to extend our discussion to using radical notation for $a^{m/n}$.

Radical Form of $a^{m/n}$

If all indicated roots are real numbers, $\quad a^{m/n} = (\sqrt[n]{a})^m \quad$ or $\quad a^{m/n} = \sqrt[n]{a^m}.$

EXAMPLE 6 Converting between Rational Exponents and Radicals

Write each of the following with radicals. Assume that all variables represent positive real numbers. Use the first definition in the box above.

(a) $13^{1/2} = \sqrt{13}$

(b) $6^{3/4} = (\sqrt[4]{6})^3$

(c) $9m^{5/8} = 9(\sqrt[8]{m})^5$

(d) $6x^{2/3} - (4x)^{3/5} = 6(\sqrt[3]{x})^2 - (\sqrt[5]{4x})^3$

(e) $r^{-2/3} = \dfrac{1}{r^{2/3}} = \dfrac{1}{(\sqrt[3]{r})^2}$

(f) $(a^2 + b^2)^{1/2} = \sqrt{a^2 + b^2}$ Note that $\sqrt{a^2 + b^2} \neq a + b$.

Replace all radicals with rational exponents. Simplify. Assume that all variables represent positive real numbers.

(g) $\sqrt{10} = 10^{1/2}$

(h) $\sqrt[4]{3^8} = 3^{8/4} = 3^2 = 9$

(i) $\sqrt[6]{z^6} = z$

In Example 6(i), it was not necessary to use absolute value bars, since the directions specifically stated that the variable represents a positive real number. Since the absolute value of the positive real number z is z itself, the answer is simply z. When working exercises with radicals, we will often assume that variables represent positive real numbers, which will eliminate the need for absolute value.

OBJECTIVE 5 **Use the rules of exponents with rational exponents.** The definition of rational exponents allows us to apply the familiar rules for exponents first introduced in Chapter 5.

Rules for Rational Exponents

Let r and s be rational numbers. For all real numbers a and b for which the indicated expressions exist:

$$a^r \cdot a^s = a^{r+s} \qquad a^{-r} = \frac{1}{a^r} \qquad \frac{a^r}{a^s} = a^{r-s} \qquad \left(\frac{a}{b}\right)^{-r} = \frac{b^r}{a^r}$$

$$(a^r)^s = a^{rs} \qquad (ab)^r = a^r b^r \qquad \left(\frac{a}{b}\right)^r = \frac{a^r}{b^r} \qquad a^{-r} = \left(\frac{1}{a}\right)^r.$$

EXAMPLE 7 **Applying Rules for Rational Exponents**

Write with only positive exponents. Assume that all variables represent positive real numbers.

(a) $2^{1/2} \cdot 2^{1/4} = 2^{1/2+1/4} = 2^{3/4}$

(b) $\dfrac{5^{2/3}}{5^{7/3}} = 5^{2/3-7/3} = 5^{-5/3} = \dfrac{1}{5^{5/3}}$

(c) $\dfrac{(x^{1/2}y^{2/3})^4}{y} = \dfrac{(x^{1/2})^4(y^{2/3})^4}{y}$ Power rule

$= \dfrac{x^2y^{8/3}}{y^1}$ Power rule

$= x^2y^{8/3-1}$ Quotient rule

$= x^2y^{5/3}$

(d) $m^{3/4}(m^{5/4} - m^{1/4}) = m^{3/4} \cdot m^{5/4} - m^{3/4} \cdot m^{1/4}$ Distributive property

$= m^{3/4+5/4} - m^{3/4+1/4}$ Product rule

$= m^{8/4} - m^{4/4}$

$= m^2 - m$

Do not make the common mistake of multiplying exponents in the first step.

Use the rules of exponents in problems like those in Example 7. Do not convert the expressions to radical form.

EXAMPLE 8 Applying Rules for Rational Exponents

Replace all radicals with rational exponents, and then apply the rules for rational exponents. Leave answers in exponential form. Assume that all variables represent positive real numbers.

(a) $\sqrt[3]{x^2} \cdot \sqrt[4]{x} = x^{2/3} \cdot x^{1/4}$ Convert to rational exponents.

$= x^{2/3+1/4}$ Use product rule for exponents.

$= x^{8/12+3/12}$ Get a common denominator.

$= x^{11/12}$

(b) $\dfrac{\sqrt{x^3}}{\sqrt[3]{x^2}} = \dfrac{x^{3/2}}{x^{2/3}} = x^{3/2-2/3} = x^{5/6}$

(c) $\sqrt{\sqrt[4]{z}} = \sqrt{z^{1/4}} = (z^{1/4})^{1/2} = z^{1/8}$

OBJECTIVE 6 Use a calculator to find roots. So far in this section we have discussed definitions, properties, and rules for rational exponents and radicals. While numbers such as $\sqrt{9}$ and $\sqrt[3]{-8}$ are rational, numbers involving radicals are often irrational numbers. To find approximations of roots such as $\sqrt{15}$, $\sqrt[3]{10}$, and $\sqrt[4]{2}$, we usually use basic scientific or graphing calculators. The methods for finding approximations differ among makes and models, and you should always consult your owner's manual for keystroke instructions. You should be aware that graphing calculators often differ from basic scientific calculators in the order in which keystrokes are made. Using a calculator, we can find

$$\sqrt{15} \approx 3.872983346, \quad \sqrt[3]{10} \approx 2.15443469, \quad \sqrt[4]{2} \approx 1.189207115$$

where the symbol $\approx$ means "is approximately equal to." In this book we will usually show approximations rounded to three decimal places. Thus, we would write

$$\sqrt{15} \approx 3.873, \quad \sqrt[3]{10} \approx 2.154, \quad \text{and} \quad \sqrt[4]{2} \approx 1.189.$$

Figure 2 shows how the above approximations are displayed on a TI-83 graphing calculator. In Figure 2(a), eight or nine decimal places are shown, while in Figure 2(b), the number of decimal places is fixed at three.

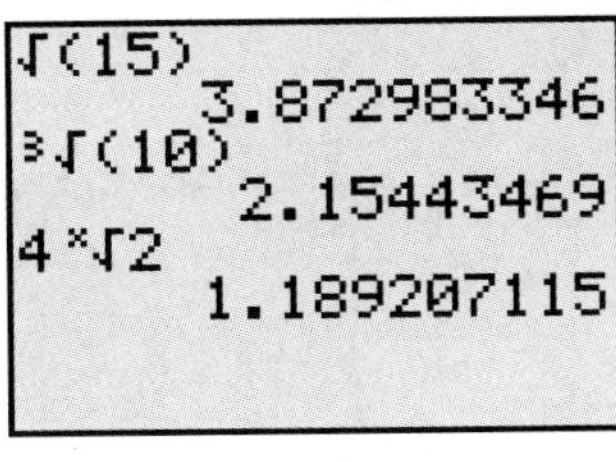

(a)

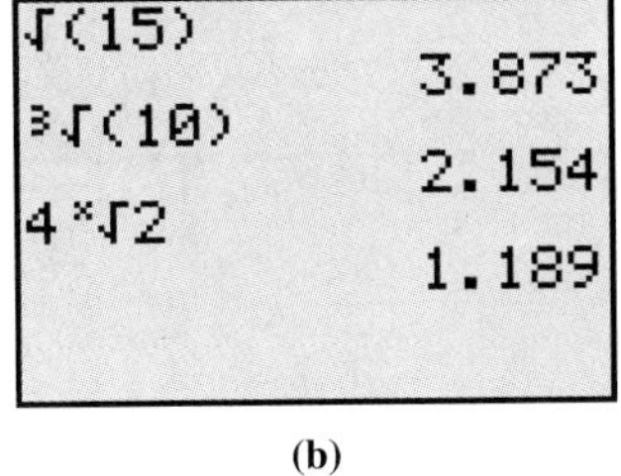

(b)

Figure 2

EXAMPLE 9 Finding Approximations for Roots

Use a calculator to verify that the following approximations are correct.

(a) $\sqrt{39} \approx 6.245$ **(b)** $-\sqrt{72} \approx -8.485$

(c) $\sqrt[3]{93} \approx 4.531$ **(d)** $\sqrt[4]{39} \approx 2.499$

If your calculator does not have a specific key for finding a cube root, fourth root, or higher root, use the fact that $\sqrt[n]{a} = a^{1/n}$ and then use the exponential key. For example, in part (c) we can enter $93^{1/3}$ and in part (d) we can use $39^{1/4}$ or $39^{.25}$. ■

Figure 3 shows several options for finding various roots and powers using a graphing calculator. The same results can be found using the root-finding and exponential keys of a scientific calculator.

```
5 ˣ√10847
                6.413
10847^(1/5)
                6.413
10847^.2
                6.413
```

$\sqrt[5]{10{,}847}$

(a)

```
1.97^(5/2)
                5.447
1.97^2.5
                5.447
(√(1.97))^5
                5.447
```

$1.97^{5/2}$

(b)

Figure 3

The final example of this section shows that formulas may involve roots.

EXAMPLE 10 Using Roots to Calculate Resonant Frequency

In electronics, the resonant frequency f of a circuit may be found by the formula

$$f = \frac{1}{2\pi\sqrt{LC}},$$

where f is in cycles per second, L is in henrys, and C is in farads.* Find the resonant frequency f if $L = 5 \times 10^{-4}$ henrys and $C = 3 \times 10^{-10}$ farads. Give your answer to the nearest thousand.

*Henrys and farads are units of measure in electronics.

Find the value of f when $L = 5 \times 10^{-4}$ and $C = 3 \times 10^{-10}$.

$$f = \frac{1}{2\pi\sqrt{LC}} \qquad \text{Given formula}$$

$$= \frac{1}{2\pi\sqrt{(5 \times 10^{-4})(3 \times 10^{-10})}} \qquad \text{Substitute.}$$

$$\approx 411{,}000 \qquad \text{Use a calculator and approximate.}$$

The resonant frequency f is approximately 411,000 cycles per second. ■

7.1 EXERCISES

Match each expression from Column I with the equivalent choice from Column II.

I	II
1. $2^{1/2}$	**A.** -4
2. $(-27)^{1/3}$	**B.** 8
3. $-\sqrt{16}$	**C.** $\sqrt{2}$
4. $\sqrt{-16}$	**D.** $-\sqrt{6}$
5. $(-32)^{1/5}$	**E.** -3
6. $(-32)^{2/5}$	**F.** $\sqrt{6}$
7. $4^{3/2}$	**G.** 4
8. $\sqrt[4]{6^2}$	**H.** -2
9. $-6^{1/2}$	**I.** 6
10. $36^{.5}$	**J.** not a real number

Simplify each expression involving rational exponents. See Examples 1, 2, and 5.

11. $169^{1/2}$ **12.** $121^{1/2}$ **13.** $729^{1/3}$ **14.** $512^{1/3}$

15. $16^{1/4}$ **16.** $625^{1/4}$ **17.** $\left(\frac{64}{81}\right)^{1/2}$ **18.** $\left(\frac{8}{27}\right)^{1/3}$

19. $(-27)^{1/3}$ **20.** $(-32)^{1/5}$ **21.** $100^{3/2}$ **22.** $64^{3/2}$

23. $-4^{5/2}$ **24.** $-32^{3/5}$ **25.** $(-144)^{1/2}$ **26.** $(-36)^{1/2}$

27. $64^{-3/2}$ **28.** $81^{-3/2}$ **29.** $\left(-\frac{8}{27}\right)^{-2/3}$ **30.** $\left(-\frac{64}{125}\right)^{-2/3}$

31. Explain why $(-64)^{1/2}$ is not a real number, while $-64^{1/2}$ is a real number.

32. Explain why $a^{1/n}$ is defined to be equal to $\sqrt[n]{a}$ when $\sqrt[n]{a}$ is real.

Find each root if it is a real number. Use a calculator as needed. See Examples 3 and 4.

33. $\sqrt{36}$ **34.** $\sqrt{100}$ **35.** $\sqrt{\frac{64}{81}}$ **36.** $\sqrt{\frac{100}{9}}$

37. $-\sqrt{-169}$ **38.** $-\sqrt{-400}$ **39.** $\sqrt[3]{216}$ **40.** $\sqrt[3]{343}$

41. $\sqrt[3]{-64}$ **42.** $\sqrt[3]{-125}$ **43.** $-\sqrt[3]{512}$ **44.** $-\sqrt[3]{1000}$

45. $-\sqrt[4]{81}$ **46.** $-\sqrt[4]{256}$ **47.** $\sqrt[4]{-16}$ **48.** $\sqrt[4]{-81}$

49. $\sqrt[6]{(-2)^6}$ **50.** $\sqrt[6]{(-4)^6}$ **51.** $\sqrt[5]{(-9)^5}$ **52.** $\sqrt[5]{(-8)^5}$

53. $\sqrt{x^2}$ **54.** $-\sqrt{x^2}$ **55.** $\sqrt[3]{x^3}$ **56.** $-\sqrt[3]{x^3}$

57. $\sqrt[3]{x^{15}}$ **58.** $\sqrt[4]{k^{20}}$

Write with radicals. Assume that all variables represent positive real numbers. See Example 6.

59. $12^{1/2}$ **60.** $3^{1/2}$ **61.** $8^{3/4}$

62. $7^{2/3}$ **63.** $(9q)^{5/8} - (2x)^{2/3}$ **64.** $(3p)^{3/4} + (4x)^{1/3}$

65. $(2m)^{-3/2}$ **66.** $(5y)^{-3/5}$ **67.** $(2y + x)^{2/3}$

68. $(r + 2z)^{3/2}$ **69.** $(3m^4 + 2k^2)^{-2/3}$ **70.** $(5x^2 + 3z^3)^{-5/6}$

71. Show that, in general, $\sqrt{a^2 + b^2} \neq a + b$ by replacing a with 3 and b with 4.

72. Suppose someone claims that $\sqrt[n]{a^n + b^n}$ must equal $a + b$, since when $a = 1$ and $b = 0$, a true statement results:

$$\sqrt[n]{a^n + b^n} = \sqrt[n]{1^n + 0^n} = \sqrt[n]{1^n} = 1 = 1 + 0$$
$$= a + b.$$

Write an explanation of why this is faulty reasoning.

Simplify by first converting to rational exponents. Assume that all variables represent positive real numbers. See Example 6.

73. $\sqrt{2^{12}}$ **74.** $\sqrt{5^{10}}$ **75.** $\sqrt[3]{4^9}$ **76.** $\sqrt[4]{6^8}$ **77.** $\sqrt{x^{20}}$

78. $\sqrt{r^{50}}$ **79.** $\sqrt[3]{x} \cdot \sqrt{x}$ **80.** $\sqrt[4]{y} \cdot \sqrt[5]{y^2}$ **81.** $\dfrac{\sqrt[3]{t^4}}{\sqrt[5]{t^4}}$ **82.** $\dfrac{\sqrt[4]{w^3}}{\sqrt[6]{w}}$

Use the rules of exponents to simplify each expression. Write all answers with positive exponents. Assume that all variables represent positive real numbers. See Example 7.

83. $3^{1/2} \cdot 3^{3/2}$ **84.** $6^{4/3} \cdot 6^{2/3}$ **85.** $\dfrac{64^{5/3}}{64^{4/3}}$

86. $\dfrac{125^{7/3}}{125^{5/3}}$ **87.** $y^{7/3} \cdot y^{-4/3}$ **88.** $r^{-8/9} \cdot r^{17/9}$

89. $\dfrac{k^{1/3}}{k^{2/3} \cdot k^{-1}}$ **90.** $\dfrac{z^{3/4}}{z^{5/4} \cdot z^{-2}}$ **91.** $(27x^{12}y^{15})^{2/3}$

92. $(64p^4q^6)^{3/2}$ **93.** $\dfrac{(x^{2/3})^2}{(x^2)^{7/3}}$ **94.** $\dfrac{(p^3)^{1/4}}{(p^{5/4})^2}$

95. $\dfrac{m^{3/4}n^{-1/4}}{(m^2n)^{1/2}}$ **96.** $\dfrac{(a^2b^5)^{-1/4}}{(a^{-3}b^2)^{1/6}}$ **97.** $\dfrac{p^{1/5}p^{7/10}p^{1/2}}{(p^3)^{-1/5}}$

98. $\dfrac{z^{1/3}z^{-2/3}z^{1/6}}{(z^{-1/6})^3}$ **99.** $\left(\dfrac{b^{-3/2}}{c^{-5/3}}\right)^2(b^{-1/4}c^{-1/3})^{-1}$ **100.** $\left(\dfrac{m^{-2/3}}{a^{-3/4}}\right)^4(m^{-3/8}a^{1/4})^{-2}$

101. $\left(\dfrac{p^{-1/4}q^{-3/2}}{3^{-1}p^{-2}q^{-2/3}}\right)^{-2}$ **102.** $\left(\dfrac{2^{-2}w^{-3/4}x^{-5/8}}{w^{3/4}x^{-1/2}}\right)^{-3}$ **103.** $p^{2/3}(p^{1/3} + 2p^{4/3})$

104. $z^{5/8}(3z^{5/8} + 5z^{11/8})$ **105.** $k^{1/4}(k^{3/2} - k^{1/2})$ **106.** $r^{3/5}(r^{1/2} + r^{3/4})$

107. $6a^{7/4}(a^{-7/4} + 3a^{-3/4})$ **108.** $4m^{5/3}(m^{-2/3} - 4m^{-5/3})$

Write with rational exponents, and then apply the properties of exponents. Assume that all radicands represent positive real numbers. Give answers in exponential form. See Example 8.

109. $\sqrt[5]{x^3} \cdot \sqrt[4]{x}$ **110.** $\sqrt[6]{y^5} \cdot \sqrt[3]{y^2}$ **111.** $\dfrac{\sqrt{x^5}}{x^4}$ **112.** $\dfrac{\sqrt[3]{k^5}}{\sqrt[3]{k^7}}$

113. $\sqrt{y} \cdot \sqrt[3]{yz}$ **114.** $\sqrt[3]{xz} \cdot \sqrt{z}$ **115.** $\sqrt[4]{\sqrt[3]{m}}$ **116.** $\sqrt[3]{\sqrt{k}}$

Suppose that a rectangle has length $\sqrt{98}$ and width $\sqrt{26}$.

117. Which one of the following is the best estimate of its area?

(a) 2500 **(b)** 250

(c) 50 **(d)** 100

118. Which one of the following is the best estimate of its perimeter?

(a) 15 **(b)** 250

(c) 100 **(d)** 30

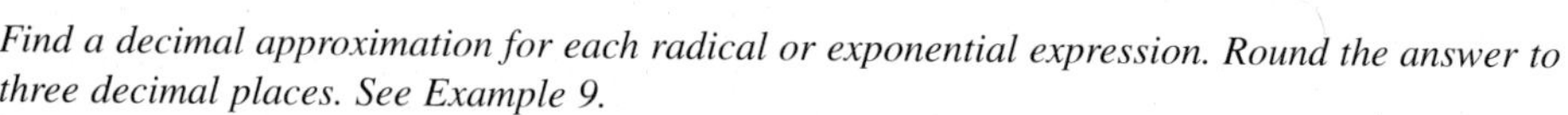

Find a decimal approximation for each radical or exponential expression. Round the answer to three decimal places. See Example 9.

119. $\sqrt{9483}$ **120.** $\sqrt{6825}$ **121.** $\sqrt{284.361}$ **122.** $\sqrt{846.104}$

123. $\sqrt{7}$ **124.** $\sqrt{11}$ **125.** $-\sqrt{82}$ **126.** $-\sqrt{91}$

127. $\sqrt[3]{423}$ **128.** $\sqrt[3]{555}$ **129.** $\sqrt[4]{100}$ **130.** $\sqrt[4]{250}$

131. $\sqrt[5]{23.8}$ **132.** $\sqrt[5]{98.4}$ **133.** $59^{2/3}$ **134.** $867^{7/8}$

135. $26^{-2/5}$ **136.** $104^{-5/4}$

Solve each problem. See Example 10.

137. Use the formula in Example 10 to calculate the resonant frequency of a circuit to the nearest thousand if $L = 7.237 \times 10^{-5}$ henrys and $C = 2.5 \times 10^{-10}$ farads.

138. The threshold weight T for a person is the weight above which the risk of death increases greatly. The threshold weight in pounds for men aged 40–49 is related to height in inches by the formula

$$h = 12.3\sqrt[3]{T}.$$

What height corresponds to a threshold weight of 216 pounds for a 43-year-old man? Round your answer to the nearest inch, and then to the nearest tenth of a foot.

139. According to an article in *The World Scanner Report* (August 1991), the distance D, in miles, to the horizon from an observer's point of view over water or "flat" earth is given by

$$D = \sqrt{2H}$$

where H is the height of the point of view, in feet. If a person whose eyes are 6 feet above ground level is standing at the top of a hill 44 feet above the "flat" earth, approximately how far to the horizon will she be able to see?

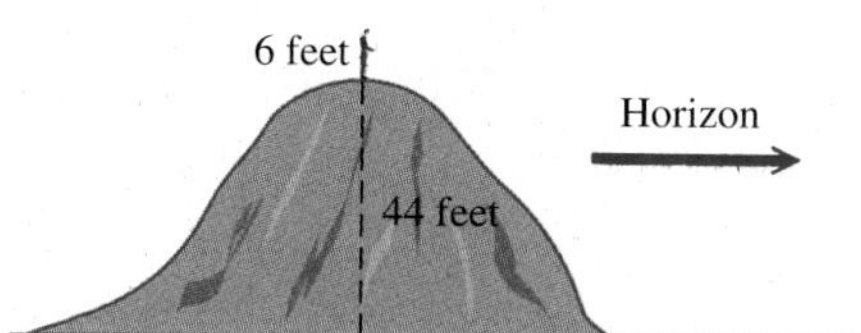

140. The time for one complete swing of a simple pendulum is

$$t = 2\pi\sqrt{\frac{L}{g}},$$

where t is time in seconds, L is the length of the pendulum in feet, and g, the force due to gravity, is about 32 feet per second squared. Find the time of a complete swing of a 2-foot pendulum to the nearest tenth of a second.

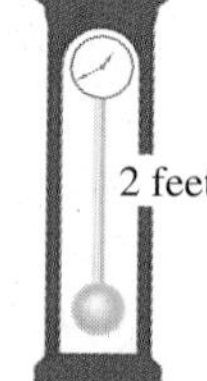

141. Meteorologists can determine the duration of a storm by using the formula

$$T = .07D^{3/2},$$

where D is the diameter of the storm in miles and T is the time in hours. Find the duration of a storm with a diameter of 16 miles. Round your answer to the nearest tenth of an hour.

142. *Heron's formula* gives a method of finding the area of a triangle if the lengths of its sides are known. Suppose that a, b, and c are the lengths of the sides. Let s denote one-half of the perimeter of the triangle (called the *semiperimeter*); that is,

$$s = \frac{1}{2}(a + b + c).$$

Then the area of the triangle is

$$A = \sqrt{s(s - a)(s - b)(s - c)}.$$

Find the area of the Bermuda Triangle, if the "sides" of this triangle measure approximately 850 miles, 925 miles, and 1300 miles. Give your answer to the nearest thousand square miles.

143. The Vietnam Veterans' Memorial in Washington, D.C., is in the shape of an unenclosed isosceles triangle with equal sides of length 246.75 feet. If the triangle were enclosed, the third side would have length 438.14 feet. Use Heron's formula from the previous exercise to find the area of this enclosure to the nearest hundred square feet. (*Source:* Information pamphlet obtained at the Vietnam Veterans' Memorial.)

246.75 feet

246.75 feet

438.14 feet

Not to scale

144. The formula

$$I = \sqrt{\frac{2P}{L}}$$

relates the coefficient of self-induction L (in henrys), the energy P stored in an electronic circuit (in joules) and the current I (in amps). Find I if $P = 120$ and $L = 80$.

7.2 Simplifying Radical Expressions

OBJECTIVES

1. Use the product and quotient rules for radicals.
2. Simplify radicals.
3. Simplify radicals by using different indexes.

OBJECTIVE 1 **Use the product and quotient rules for radicals.** We now develop rules for multiplying and dividing radicals that have the same index. It is natural to wonder whether the product of two nth root radicals is equal to the nth root of the product of the radicands. For example, is $\sqrt{36 \cdot 4} = \sqrt{36} \cdot \sqrt{4}$? To find out, we simply do the computations:

$$\sqrt{36 \cdot 4} = \sqrt{144} = 12$$

$$\sqrt{36} \cdot \sqrt{4} = 6 \cdot 2 = 12.$$

Notice that in both cases the result is the same. This is an example of a general rule, the **product rule for radicals.**

4 Use the Pythagorean formula to find the length of a side of a right triangle.

5 Use the distance formula.

FOR EXTRA HELP

SSG Sec. 7.2
SSM Sec. 7.2

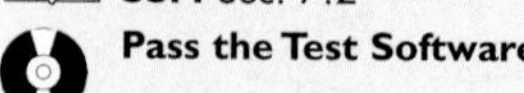

Pass the Test Software

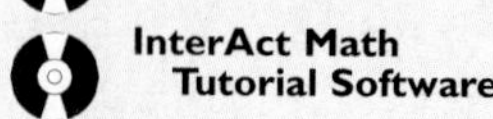

InterAct Math Tutorial Software

Video 13

Product Rule for Radicals

If $\sqrt[n]{a}$ and $\sqrt[n]{b}$ are real numbers and n is a natural number, then

$$\sqrt[n]{a} \cdot \sqrt[n]{b} = \sqrt[n]{ab}.$$

(The product of two radicals is the radical of the product.)

We justify the product rule using rational exponents. Since $\sqrt[n]{a} = a^{1/n}$ and $\sqrt[n]{b} = b^{1/n}$,

$$\sqrt[n]{a} \cdot \sqrt[n]{b} = a^{1/n} \cdot b^{1/n} = (ab)^{1/n} = \sqrt[n]{ab}.$$

CAUTION Use the product rule only when the radicals have the *same* indexes.

EXAMPLE 1 Using the Product Rule

Multiply. Assume that all variables represent positive real numbers.

(a) $\sqrt{5} \cdot \sqrt{7} = \sqrt{5 \cdot 7} = \sqrt{35}$

(b) $\sqrt{11} \cdot \sqrt{p} = \sqrt{11p}$

(c) $\sqrt{7} \cdot \sqrt{11xyz} = \sqrt{77xyz}$

(d) $\sqrt[3]{3} \cdot \sqrt[3]{12} = \sqrt[3]{3 \cdot 12} = \sqrt[3]{36}$

(e) $\sqrt[4]{8y} \cdot \sqrt[4]{3r^2} = \sqrt[4]{24yr^2}$

(f) $\sqrt[4]{2} \cdot \sqrt[5]{2}$ cannot be multiplied using the product rule, because the indexes (4 and 5) are different.

The quotient rule for radicals is similar to the product rule.

Quotient Rule for Radicals

If $\sqrt[n]{a}$ and $\sqrt[n]{b}$ are real numbers, $b \neq 0$, and n is a natural number, then

$$\sqrt[n]{\frac{a}{b}} = \frac{\sqrt[n]{a}}{\sqrt[n]{b}}.$$

(The radical of a quotient is the quotient of the radicals.)

The quotient rule can be justified, like the product rule, using the properties of exponents. It, too, is used only when the radicals have the same indexes.

EXAMPLE 2 Using the Quotient Rule

Simplify. Assume that all variables represent positive real numbers.

(a) $\sqrt{\frac{16}{25}} = \frac{\sqrt{16}}{\sqrt{25}} = \frac{4}{5}$

(b) $\sqrt{\frac{7}{36}} = \frac{\sqrt{7}}{\sqrt{36}} = \frac{\sqrt{7}}{6}$

(c) $\sqrt[3]{-\frac{8}{125}} = \sqrt[3]{\frac{-8}{125}} = \frac{\sqrt[3]{-8}}{\sqrt[3]{125}} = \frac{-2}{5} = -\frac{2}{5}$

(d) $\sqrt[3]{\frac{7}{216}} = \frac{\sqrt[3]{7}}{\sqrt[3]{216}} = \frac{\sqrt[3]{7}}{6}$

(e) $\sqrt[5]{\dfrac{x}{32}} = \dfrac{\sqrt[5]{x}}{\sqrt[5]{32}} = \dfrac{\sqrt[5]{x}}{2}$

(f) $\sqrt[3]{\dfrac{m^6}{125}} = \dfrac{\sqrt[3]{m^6}}{\sqrt[3]{125}} = \dfrac{m^2}{5}$

OBJECTIVE 2 **Simplify radicals.** The product and quotient rules are used to simplify radicals. A radical is **simplified** if the following four conditions are met.

Simplified Radical

1. The radicand has no factor raised to a power greater than or equal to the index.
2. Exponents in the radicand and the index of the radical have no common factor (except 1).
3. The radicand has no fractions.
4. No denominator contains a radical.

EXAMPLE 3 **Simplifying Roots of Numbers**

Simplify.

(a) $\sqrt{24}$

Check to see if 24 is divisible by a perfect square (square of a natural number), such as 4, 9, ... Choose the largest perfect square that divides into 24. The largest such number is 4. Write 24 as the product of 4 and 6, and then use the product rule.

$$\sqrt{24} = \sqrt{4 \cdot 6} = \sqrt{4} \cdot \sqrt{6} = 2\sqrt{6}$$

(b) $\sqrt{108}$

The number 108 is divisible by the perfect square 36. If this is not obvious, try factoring 108 into prime factors.

$$\begin{aligned} \sqrt{108} &= \sqrt{2^2 \cdot 3^3} \\ &= \sqrt{2^2 \cdot 3^2 \cdot 3} && \text{Factor.} \\ &= 2 \cdot 3 \cdot \sqrt{3} && \text{Product rule} \\ &= 6\sqrt{3} \end{aligned}$$

(c) $\sqrt{10}$

No perfect square (other than 1) divides into 10, so $\sqrt{10}$ cannot be simplified further.

(d) $\sqrt[3]{16}$

Look for the largest perfect *cube* that divides into 16. The number 8 satisfies this condition, so write 16 as $8 \cdot 2$.

$$\sqrt[3]{16} = \sqrt[3]{8 \cdot 2} = \sqrt[3]{8} \cdot \sqrt[3]{2} = 2\sqrt[3]{2}$$

(e)
$$\begin{aligned} \sqrt[4]{162} &= \sqrt[4]{81 \cdot 2} && \text{81 is a perfect 4th power.} \\ &= \sqrt[4]{81} \cdot \sqrt[4]{2} && \text{Product rule} \\ &= 3\sqrt[4]{2} \end{aligned}$$

EXAMPLE 4 Simplifying Roots of Variable Expressions

Simplify. Assume that all variables represent positive real numbers.

(a) $\sqrt{16m^3} = \sqrt{16m^2 \cdot m}$ $16m^2$ is the largest perfect square that divides $16m^3$.

$= \sqrt{16m^2} \cdot \sqrt{m}$

$= 4m\sqrt{m}$

No absolute value bars are needed around m because of the assumption that all variables represent *positive* real numbers.

(b) $\sqrt{200k^7q^8} = \sqrt{10^2 \cdot 2 \cdot (k^3)^2 \cdot k \cdot (q^4)^2}$ Factor.

$= \sqrt{10^2 \cdot (k^3)^2 \cdot (q^4)^2 \cdot 2 \cdot k}$

$= 10k^3q^4\sqrt{2k}$ Remove perfect square factors.

(c) $\sqrt[3]{8x^4y^5} = \sqrt[3]{(8x^3y^3)(xy^2)}$ $8x^3y^3$ is the largest perfect cube that divides $8x^4y^5$.

$= \sqrt[3]{8x^3y^3} \cdot \sqrt[3]{xy^2}$

$= 2xy\sqrt[3]{xy^2}$

(d) $\sqrt[4]{32y^9} = \sqrt[4]{(16y^8)(2y)}$ $16y^8$ is the largest 4th power that divides $32y^9$.

$= \sqrt[4]{16y^8} \cdot \sqrt[4]{2y}$

$= 2y^2\sqrt[4]{2y}$

From Example 4 we see that if a variable is raised to a power with an exponent divisible by 2, it is a perfect square. If it is raised to a power with an exponent divisible by 3, it is a perfect cube. In general, if it is raised to a power with an exponent divisible by n, it is a perfect nth power.

OBJECTIVE 3 Simplify radicals by using different indexes. The conditions for a simplified radical given earlier state that an exponent in the radicand and the index of the radical should have no common factor. The next example shows how to simplify radicals with such common factors.

EXAMPLE 5 Simplify Radicals by Using Smaller Indexes

Simplify. Assume that all variables represent positive real numbers.

(a) $\sqrt[9]{5^6}$

Write this radical using rational exponents and then write the exponent in lowest terms. Express the answer as a radical.

$$\sqrt[9]{5^6} = 5^{6/9} = 5^{2/3} = \sqrt[3]{5^2} \quad \text{or} \quad \sqrt[3]{25}$$

(b) $\sqrt[4]{p^2} = p^{2/4} = p^{1/2} = \sqrt{p}$

Recall the assumption $p > 0$.

These examples suggest the following rule.

If m is an integer, n and k are natural numbers, and all indicated roots exist,

$$\sqrt[kn]{a^{km}} = \sqrt[n]{a^m}.$$

Since the product and quotient rules apply only when radicals have the same indexes, we can multiply and divide radicals with different indexes by using rational exponents.

EXAMPLE 6 Multiplying Radicals with Different Indexes

Simplify $\sqrt{7} \cdot \sqrt[3]{2}$.

The indexes, 2 and 3, are different. Since they have a least common index of 6, use rational exponents to write each radical as a sixth root.

$$\sqrt{7} = 7^{1/2} = 7^{3/6} = \sqrt[6]{7^3} = \sqrt[6]{343}$$
$$\sqrt[3]{2} = 2^{1/3} = 2^{2/6} = \sqrt[6]{2^2} = \sqrt[6]{4}$$

Therefore,

$$\sqrt{7} \cdot \sqrt[3]{2} = \sqrt[6]{343} \cdot \sqrt[6]{4} = \sqrt[6]{1372}. \qquad \text{Product rule}$$

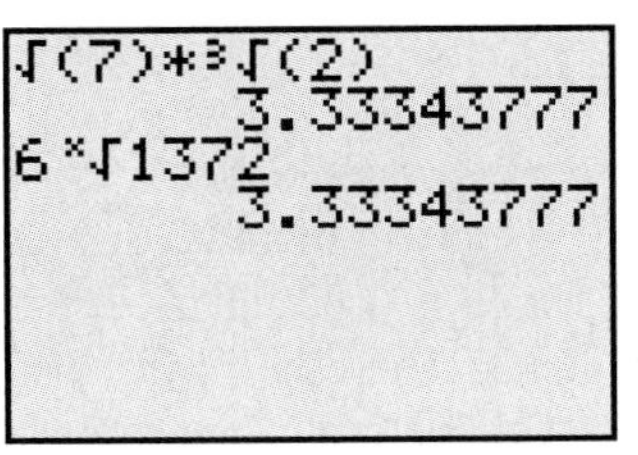

Figure 4

Results such as the one in Example 6 can be supported using a calculator, as shown in Figure 4. Notice that the calculator gives the same approximation for the initial product and the final radical that we obtained.

The computation in Figure 4 is not *proof* that the two expressions are equal. The algebra in Example 6, however, is valid proof of their equality.

OBJECTIVE 4 Use the Pythagorean formula to find the length of a side of a right triangle. The **Pythagorean formula** relates the lengths of the three sides of a right triangle.

Pythagorean Formula

If c is the length of the longest side of a right triangle and a and b are the lengths of the shorter sides, then

$$c^2 = a^2 + b^2.$$

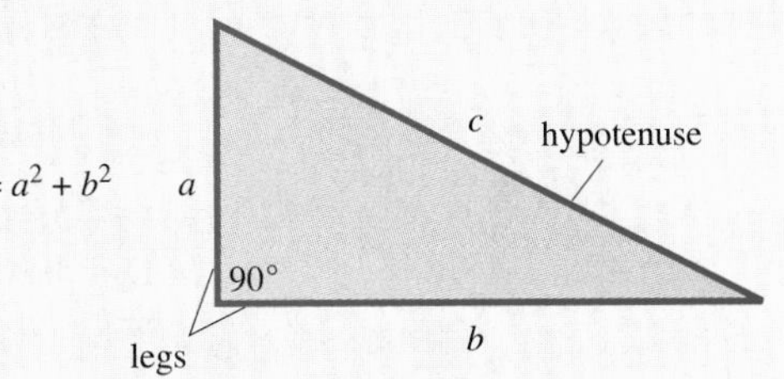

The longest side is the **hypotenuse** and the two shorter sides are the **legs** of the triangle. The hypotenuse is the side opposite the right angle.

EXAMPLE 7 Using the Pythagorean Formula

Use the Pythagorean formula to find the length of the hypotenuse in the triangle in Figure 5.

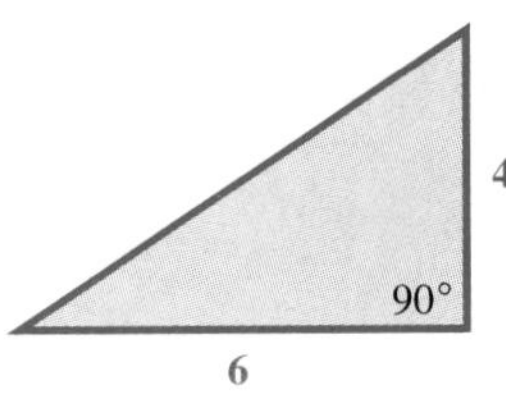

Figure 5

By the formula, the length of the hypotenuse is

$$\begin{aligned} c &= \sqrt{a^2 + b^2} \\ &= \sqrt{4^2 + 6^2} && \text{Let } a = 4 \text{ and } b = 6. \\ &= \sqrt{16 + 36} \\ &= \sqrt{52} = \sqrt{4 \cdot 13} && \text{Factor.} \\ &= \sqrt{4} \cdot \sqrt{13} && \text{Product rule} \\ &= 2\sqrt{13}. \end{aligned}$$

When using the equation $c^2 = a^2 + b^2$, be sure that the length of the hypotenuse is substituted for c, and that the lengths of the legs are substituted for a and b. Errors often occur because values are substituted incorrectly.

CONNECTIONS

The Pythagorean formula is undoubtedly one of the most widely used and oldest formulas we have. It is very important in trigonometry, which is used in surveying, drafting, engineering, navigation, and many other fields. There is evidence that the Babylonians knew the concept quite well. Although attributed to Pythagoras, it was known to every surveyor from Egypt to China for a thousand years before Pythagoras. In the 1939 movie *The Wizard of Oz,* the Scarecrow asks the Wizard for a brain. When the Wizard presents him with a diploma granting him a Th.D. (Doctor of Thinkology), the Scarecrow recites the following:

> The sum of the square roots of any two sides of an isosceles triangle is equal to the square root of the remaining side. . . .
> Oh joy! Rapture! I've got a brain.

FOR DISCUSSION OR WRITING

Did the Scarecrow recite the Pythagorean formula? (An *isosceles triangle* is a triangle with two equal sides.) Is his statement true? Explain.

EXAMPLE 8 Using a Formula from Electronics

The impedance Z of an alternating series circuit is given by the formula

$$Z = \sqrt{R^2 + X^2},$$

where R is the resistance and X is the reactance, both in ohms. Find the value of the impedance if $R = 40$ ohms and $X = 30$ ohms.

Substitute 40 for R and 30 for X in the formula.

$$\begin{aligned} Z &= \sqrt{R^2 + X^2} && \text{Given formula} \\ &= \sqrt{40^2 + 30^2} && \text{Let } R = 40 \text{ and } X = 30. \\ &= \sqrt{1600 + 900} \\ &= \sqrt{2500} \\ &= 50 \end{aligned}$$

The impedance is 50 ohms.

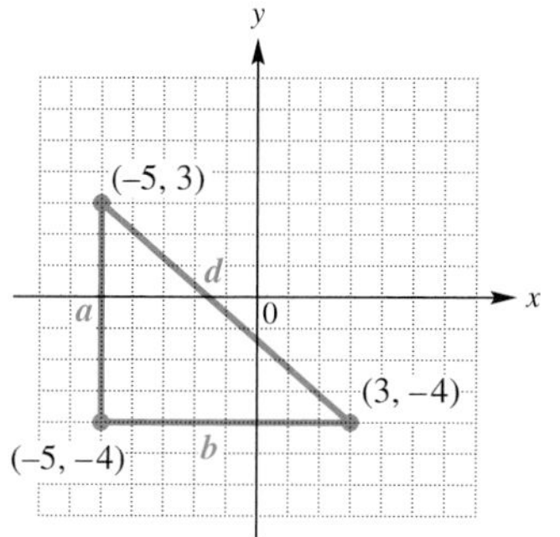

Figure 6

OBJECTIVE 5 Use the distance formula. An important result in algebra is derived by using the Pythagorean formula. The **distance formula** allows us to find the distance between two points in the coordinate plane, or the length of the line segment joining those two points. Figure 6 shows the points $(3, -4)$ and $(-5, 3)$. The vertical line through $(-5, 3)$ and the horizontal line through $(3, -4)$ intersect at the point $(-5, -4)$. Thus, the point $(-5, -4)$ becomes the vertex of the right angle in a right triangle. By the Pythagorean formula, the square of the length of the hypotenuse, d, of the right triangle in Figure 6 is equal to the sum of the squares of the lengths of the two legs a and b:

$$d^2 = a^2 + b^2.$$

The length a is the difference between the coordinates of the endpoints. Since the x-coordinate of both points is -5, the side is vertical, and we can find a by finding the difference between the y-coordinates. Subtract -4 from 3 to get a positive value for a.

$$a = 3 - (-4) = 7$$

Similarly, find b by subtracting -5 from 3.

$$b = 3 - (-5) = 8$$

Substituting these values into the formula we have

$$d^2 = a^2 + b^2$$
$$d^2 = 7^2 + 8^2 \qquad \text{Let } a = 7 \text{ and } b = 8.$$
$$d^2 = 49 + 64$$
$$d^2 = 113$$
$$d = \sqrt{113}. \qquad \text{Use the square root property.}$$

Therefore, the distance between $(-5, 3)$ and $(3, -4)$ is $\sqrt{113}$.

It is customary to leave the distance in radical form. Do not use a calculator to get an approximation unless you are specifically directed to do so.

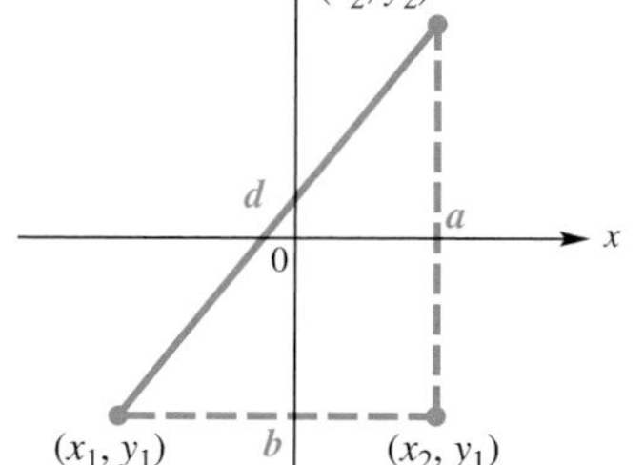

Figure 7

This result can be generalized. Figure 7 shows the two different points (x_1, y_1) and (x_2, y_2). To find a formula for the distance d between these two points, notice that the distance between (x_2, y_2) and (x_2, y_1) is given by $a = y_2 - y_1$, and the distance between (x_1, y_1) and (x_2, y_1) is given by $b = x_2 - x_1$. From the Pythagorean formula,

$$d^2 = (x_2 - x_1)^2 + (y_2 - y_1)^2,$$

and by using the square root property, we get the distance formula.

Distance Formula

The distance between the points (x_1, y_1) and (x_2, y_2) is

$$d = \sqrt{(x_2 - x_1)^2 + (y_2 - y_1)^2}.$$

This result is called the **distance formula.**

EXAMPLE 9 Using the Distance Formula

Find the distance between $(-3, 5)$ and $(6, 4)$.

When using the distance formula to find the distance between two points, designating the points as (x_1, y_1) and (x_2, y_2) is arbitrary. Let us choose $(x_1, y_1) = (-3, 5)$ and $(x_2, y_2) = (6, 4)$.

$$\begin{aligned} d &= \sqrt{(x_2 - x_1)^2 + (y_2 - y_1)^2} \\ &= \sqrt{(6 - (-3))^2 + (4 - 5)^2} \qquad x_2 = 6,\ y_2 = 4,\ x_1 = -3,\ y_1 = 5 \\ &= \sqrt{9^2 + (-1)^2} \\ &= \sqrt{82} \end{aligned}$$

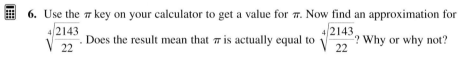

7.2 EXERCISES

Decide whether each statement is true or false by using the product rule explained in this section. Then support your answer by finding a calculator approximation for each expression.

1. $2\sqrt{12} = \sqrt{48}$ **2.** $\sqrt{72} = 2\sqrt{18}$

3. $3\sqrt{8} = 2\sqrt{18}$ **4.** $5\sqrt{72} = 6\sqrt{50}$

5. Which one of the following is *not* equal to $\sqrt{\frac{1}{2}}$? (Do not use calculator approximations.)

(a) $\sqrt{.5}$ **(b)** $\sqrt{\frac{2}{4}}$ **(c)** $\sqrt{\frac{3}{6}}$ **(d)** $\frac{\sqrt{4}}{\sqrt{16}}$

6. Use the π key on your calculator to get a value for π. Now find an approximation for $\sqrt[4]{\frac{2143}{22}}$. Does the result mean that π is actually equal to $\sqrt[4]{\frac{2143}{22}}$? Why or why not?

Multiply using the product rule. Assume that all variables represent positive numbers. See Example 1.

7. $\sqrt{5} \cdot \sqrt{6}$ **8.** $\sqrt{10} \cdot \sqrt{3}$

9. $\sqrt[3]{7x} \cdot \sqrt[3]{2y}$ **10.** $\sqrt[3]{9x} \cdot \sqrt[3]{4y}$

11. $\sqrt[4]{11} \cdot \sqrt[4]{3}$ **12.** $\sqrt[4]{6} \cdot \sqrt[4]{9}$

13. $\sqrt[3]{7} \cdot \sqrt[4]{3}$ **14.** $\sqrt[5]{8} \cdot \sqrt[6]{12}$

15. Explain the product rule for radicals in your own words. Give examples.

16. Explain the quotient rule for radicals in your own words. Give examples.

Simplify each radical. Assume that all variables represent positive real numbers. See Example 2.

17. $\sqrt{\frac{64}{121}}$ **18.** $\sqrt{\frac{16}{49}}$ **19.** $\sqrt{\frac{3}{25}}$ **20.** $\sqrt{\frac{13}{49}}$

21. $\sqrt{\frac{x}{25}}$ **22.** $\sqrt{\frac{k}{100}}$ **23.** $\sqrt{\frac{p^6}{81}}$ **24.** $\sqrt{\frac{w^{10}}{36}}$

25. $\sqrt[3]{\frac{27}{64}}$ **26.** $\sqrt[3]{\frac{216}{125}}$ **27.** $\sqrt[3]{-\frac{r^2}{8}}$ **28.** $\sqrt[3]{-\frac{t}{125}}$

29. $-\sqrt[4]{\frac{81}{x^4}}$ **30.** $-\sqrt[4]{\frac{625}{y^4}}$ **31.** $\sqrt[5]{\frac{1}{x^{15}}}$ **32.** $\sqrt[5]{\frac{32}{y^{20}}}$

Express each of the following in simplified form. See Example 3.

33. $\sqrt{28}$ **34.** $\sqrt{72}$ **35.** $-\sqrt{32}$ **36.** $-\sqrt{48}$

37. $\sqrt{300}$ **38.** $\sqrt{150}$ **39.** $\sqrt[3]{128}$ **40.** $\sqrt[3]{24}$

41. $\sqrt[3]{-16}$ **42.** $\sqrt[3]{-250}$ **43.** $\sqrt[3]{40}$ **44.** $\sqrt[3]{375}$

45. $-\sqrt[4]{512}$ **46.** $-\sqrt[4]{1250}$ **47.** $\sqrt[5]{64}$ **48.** $\sqrt[5]{128}$

49. A student claimed that $\sqrt[3]{14}$ is not in simplified form, since $14 = 8 + 6$, and 8 is a perfect cube. Was his reasoning correct? Why or why not?

50. Explain in your own words why $\sqrt[3]{k^4}$ is not a simplified radical. Then simplify it.

Express each of the following in simplified form. Assume that all variables represent positive real numbers. See Example 4.

51. $\sqrt{72k^2}$ **52.** $\sqrt{18m^2}$ **53.** $\sqrt[3]{\dfrac{81}{64}}$

54. $\sqrt[3]{\dfrac{32}{216}}$ **55.** $\sqrt{121x^6}$ **56.** $\sqrt{256z^{12}}$

57. $-\sqrt[3]{27t^{12}}$ **58.** $-\sqrt[3]{64y^{18}}$ **59.** $-\sqrt{100m^8z^4}$

60. $-\sqrt{25t^6s^{20}}$ **61.** $-\sqrt[3]{-125a^6b^9c^{12}}$ **62.** $-\sqrt[3]{-216y^{15}x^6z^3}$

63. $\sqrt[4]{\dfrac{1}{16}r^8t^{20}}$ **64.** $\sqrt[4]{\dfrac{81}{256}t^{12}u^8}$ **65.** $-\sqrt{13x^7y^8}$

66. $-\sqrt{23k^9p^{14}}$ **67.** $\sqrt[3]{8z^6w^9}$ **68.** $\sqrt[3]{64a^{15}b^{12}}$

69. $\sqrt[3]{-16z^5t^7}$ **70.** $\sqrt[3]{-81m^4n^{10}}$ **71.** $\sqrt[4]{81x^{12}y^{16}}$

72. $\sqrt[4]{81t^8u^{28}}$ **73.** $-\sqrt[4]{162r^{15}s^9}$ **74.** $-\sqrt[4]{32k^5m^9}$

75. $\sqrt{\dfrac{y^{11}}{36}}$ **76.** $\sqrt{\dfrac{v^{13}}{49}}$ **77.** $\sqrt[3]{\dfrac{x^{16}}{27}}$ **78.** $\sqrt[3]{\dfrac{y^{17}}{125}}$

Simplify each radical. Assume that all variables represent positive real numbers. See Example 5.

79. $\sqrt[4]{48^2}$ **80.** $\sqrt[4]{50^2}$ **81.** $\sqrt[4]{25}$

82. $\sqrt[6]{8}$ **83.** $\sqrt[10]{x^{25}}$ **84.** $\sqrt[12]{x^{44}}$

Simplify by first writing the radicals as radicals with the same index. Then multiply. Assume that all variables represent positive real numbers. See Example 6.

85. $\sqrt[3]{4} \cdot \sqrt{3}$ **86.** $\sqrt[3]{5} \cdot \sqrt{6}$ **87.** $\sqrt[4]{3} \cdot \sqrt[3]{4}$

88. $\sqrt[5]{7} \cdot \sqrt[7]{5}$ **89.** $\sqrt{x} \cdot \sqrt[3]{x}$ **90.** $\sqrt[3]{y} \cdot \sqrt[4]{y}$

TECHNOLOGY INSIGHTS (EXERCISES 91–94)

A graphing calculator can be used to test whether two quantities are equal. In the screen shown here, the first two lines of entries both represent true statements, and thus the calculator returns a 1 to indicate *true*. The third entry is *false*, and the calculator returns a 0. These can be verified algebraically using the rules for radicals found in this section.

```
√(8)=2√(2)
                1
³√(54)=3³√(2)
                1
√(18)=4√(2)
                0
```

(continued)

TECHNOLOGY INSIGHTS (EXERCISES 91–94) (CONTINUED)

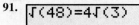

Determine whether the calculator should return a 1 or a 0 for each of the following screens.

91.

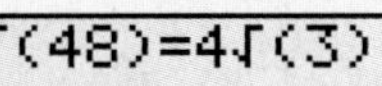

92.

93.

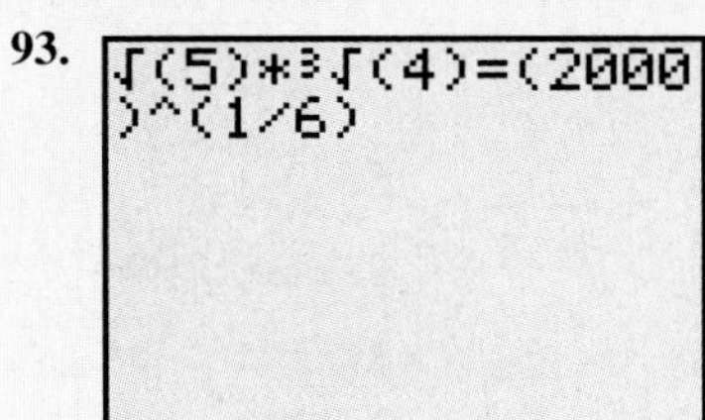

94.

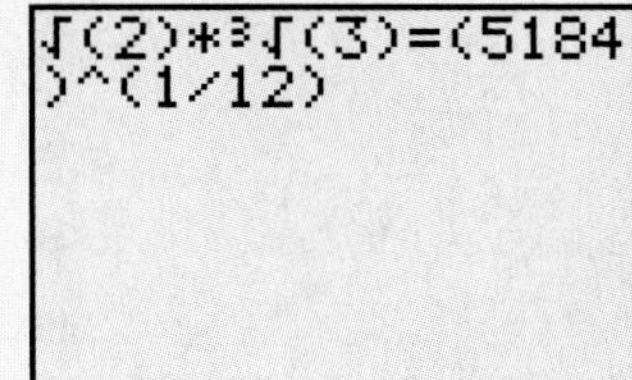

Find the missing length in each right triangle. Simplify the answer if necessary. (Hint: If one leg is unknown, write the formula as $a = \sqrt{c^2 - b^2}$ *or* $b = \sqrt{c^2 - a^2}$*.) See Example 7.*

95.

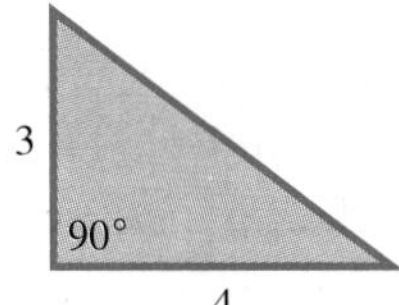

96.

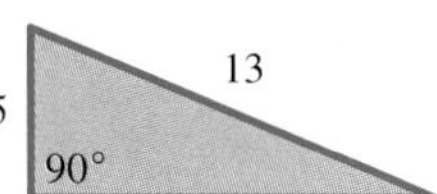

97.

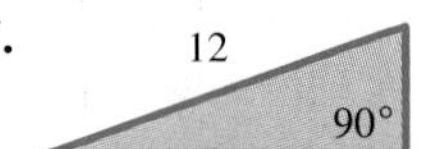

98.

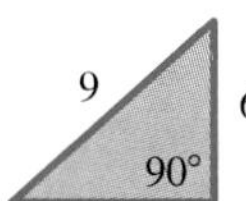

Solve each problem. See Example 8.

99. A Sanyo color television, model AVM-2755, has a rectangular screen with a 27-inch diameter. Its height is 16 inches. What is the width of the screen? (*Source:* Actual measurements of author's television.)

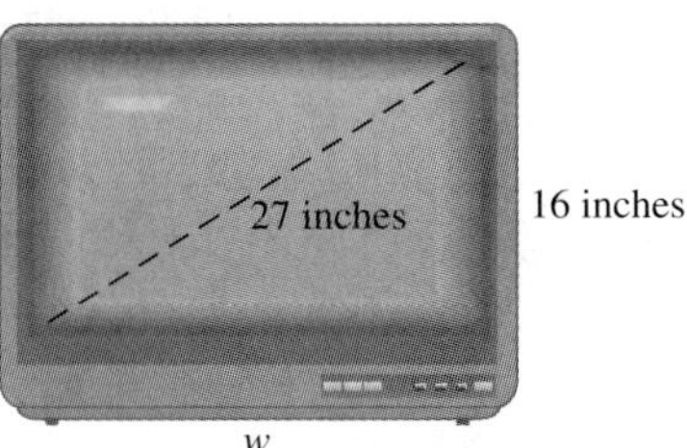

100. An RCA XL-100 Commercial Skip television has a rectangular screen that measures 8 inches by 11 inches. What is the diagonal measure of the screen? (*Source:* Actual measurements of author's television.)

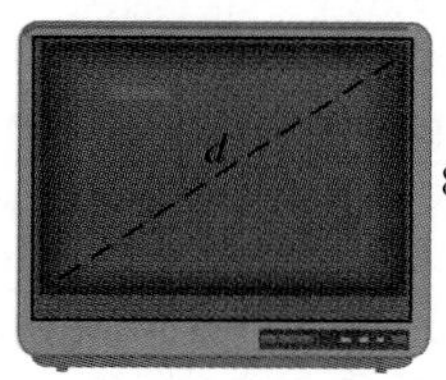

101. A formula from electronics dealing with impedance of parallel resonant circuits is

$$I = \frac{E}{\sqrt{R^2 + \omega^2 L^2}},$$

where the variables are in appropriate units. Find I if $E = 282$, $R = 100$, $L = 264$, and $\omega = 120\pi$. Give your answer to the nearest thousandth.

102. In the study of sound, one version of the law of tensions is

$$f_1 = f_2\sqrt{\frac{F_1}{F_2}}.$$

If $F_1 = 300$, $F_2 = 60$, and $f_2 = 260$, find f_1 to the nearest unit.

103. The illumination I, in footcandles, produced by a light source is related to the distance d, in feet, from the light source by the equation

$$d = \sqrt{\frac{k}{I}},$$

where k is a constant. If $k = 640$, how far from the light source will the illumination be 2 footcandles? Give the exact value, and then round to the nearest tenth of a foot.

104. The length of the diagonal of a box is given by

$$D = \sqrt{L^2 + W^2 + H^2},$$

where L, W, and H are the length, the width, and the height of the box. Find the length of the diagonal, D, of a box that is 4 feet long, 3 feet high, and 2 feet wide. Give the exact value, and then round to the nearest tenth of a foot.

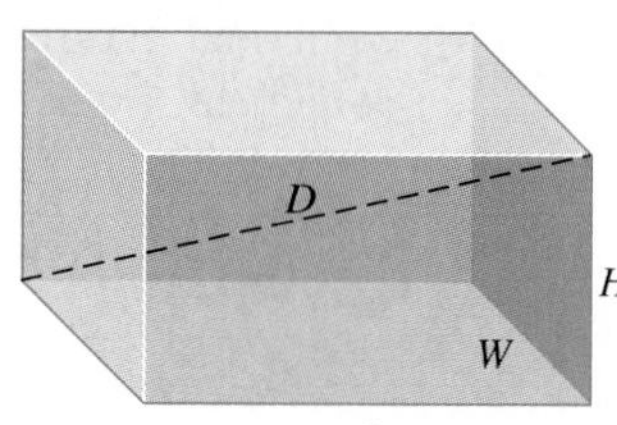

Find the distance between each pair of points. See Example 9.

105. $(-8, 2)$ and $(-4, 1)$

106. $(5, -3)$ and $(-1, -1)$

107. $(-1, 4)$ and $(5, 3)$

108. $(-6, 5)$ and $(3, -4)$

109. $(4.7, 2.3)$ and $(1.7, -1.7)$

110. $(-2.9, 18.2)$ and $(2.1, 6.2)$

111. $(x + y, y)$ and $(x - y, x)$

112. $(c, c - d)$ and $(d, c + d)$

113. As given in the text, the distance formula is expressed with a radical. Write the distance formula using rational exponents.

114. An alternative form of the distance formula is

$$d = \sqrt{(x_1 - x_2)^2 + (y_1 - y_2)^2}.$$

Compare this to the form given in this section, and explain why the two forms are equivalent.

Find the perimeter of each triangle. (Hint: *For Exercise 115,* $\sqrt{k} + \sqrt{k} = 2\sqrt{k}$.)

115.

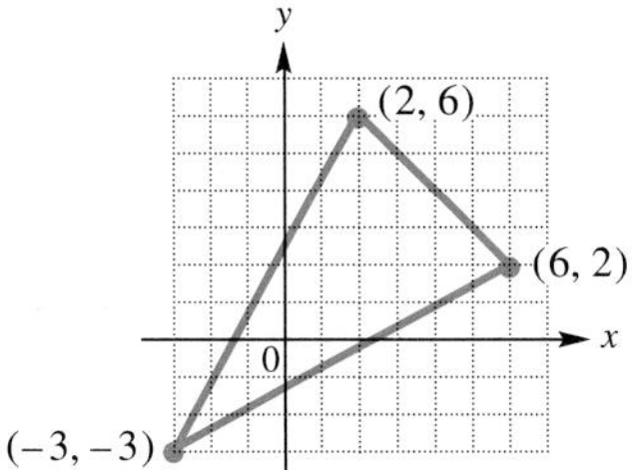

116.

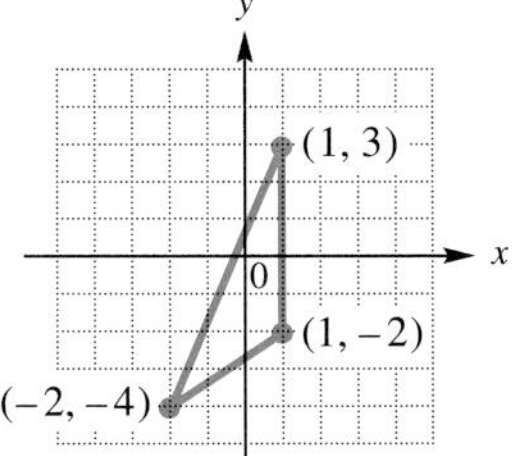

7.3 Addition and Subtraction of Radical Expressions

OBJECTIVES

1. Define a radical expression.
2. Simplify radical expressions involving addition and subtraction.

FOR EXTRA HELP

SSG Sec. 7.3
SSM Sec. 7.3
Pass the Test Software
InterAct Math Tutorial Software
Video 13

OBJECTIVE 1 Define a radical expression. A **radical expression** is an algebraic expression that contains radicals. For example,

$$\sqrt[4]{3} + \sqrt{6}, \qquad \sqrt{x + 2y} - 1, \qquad \text{and} \qquad \sqrt{8} - \sqrt{2r}$$

are radical expressions. The examples in the previous section discussed simplifying radical expressions that involve multiplication and division. Now we show how to simplify radical expressions that involve addition and subtraction.

OBJECTIVE 2 Simplify radical expressions involving addition and subtraction. An expression such as $4\sqrt{2} + 3\sqrt{2}$ can be simplified by using the distributive property.

$$4\sqrt{2} + 3\sqrt{2} = (4 + 3)\sqrt{2} = 7\sqrt{2}$$

As another example, $2\sqrt{3} - 5\sqrt{3} = (2 - 5)\sqrt{3} = -3\sqrt{3}$. This is very similar to simplifying $2x + 3x$ to $5x$ or $5y - 8y$ to $-3y$.

Only radical expressions with the same index and the same radicand may be combined. Expressions such as $5\sqrt{3} + 2\sqrt{2}$ and $3\sqrt{3} + 2\sqrt[3]{3}$ cannot be combined.

EXAMPLE 1 Adding and Subtracting Radical Expressions

Add or subtract the following radical expressions.

(a) $3\sqrt{24} + \sqrt{54}$

Begin by simplifying each radical. Then use the distributive property.

$$\begin{aligned} 3\sqrt{24} + \sqrt{54} &= 3\sqrt{4}\cdot\sqrt{6} + \sqrt{9}\cdot\sqrt{6} && \text{Product rule} \\ &= 3\cdot 2\sqrt{6} + 3\sqrt{6} \\ &= 6\sqrt{6} + 3\sqrt{6} \\ &= 9\sqrt{6} && \text{Combine terms.} \end{aligned}$$

(b) $$\begin{aligned} 2\sqrt{20x} - \sqrt{45x} &= 2\sqrt{4}\sqrt{5x} - \sqrt{9}\sqrt{5x} && \text{Product rule} \\ &= 2\cdot 2\sqrt{5x} - 3\sqrt{5x} \\ &= 4\sqrt{5x} - 3\sqrt{5x} \\ &= \sqrt{5x} \quad (\text{if } x \geq 0) \end{aligned}$$

(c) $2\sqrt{3} - 4\sqrt{5}$

Here the radicals differ, and are already simplified, so $2\sqrt{3} - 4\sqrt{5}$ cannot be simplified further.

Do not confuse the product rule with combining like terms. The root of a sum **does not equal** the sum of the roots. That is,

$$\sqrt{25} = \sqrt{9 + 16} \neq \sqrt{9} + \sqrt{16}.$$

$$\sqrt{25} = 5 \qquad \text{but} \qquad \sqrt{9} + \sqrt{16} = 3 + 4 = 7$$

EXAMPLE 2 Adding and Subtracting Radicals with Higher Indexes

Add or subtract the following radical expressions. Assume that all variables represent positive real numbers.

(a) $$\begin{aligned} 2\sqrt[3]{16} - 5\sqrt[3]{54} &= 2\sqrt[3]{8\cdot 2} - 5\sqrt[3]{27\cdot 2} \\ &= 2\sqrt[3]{8}\cdot\sqrt[3]{2} - 5\sqrt[3]{27}\cdot\sqrt[3]{2} \\ &= 2\cdot 2\cdot\sqrt[3]{2} - 5\cdot 3\cdot\sqrt[3]{2} \\ &= 4\sqrt[3]{2} - 15\sqrt[3]{2} \\ &= -11\sqrt[3]{2} \end{aligned}$$

(b) $$\begin{aligned} 2\sqrt[3]{x^2v} + \sqrt[3]{8x^5v^4} &= 2\sqrt[3]{x^2v} + \sqrt[3]{(8x^3v^3)x^2v} \\ &= 2\sqrt[3]{x^2y} + 2xy\sqrt[3]{x^2y} \\ &= (2 + 2xv)\sqrt[3]{x^2v} && \text{Distributive property} \end{aligned}$$

Remember to write the index when working with cube roots, fourth roots, and so on.

The next example shows how to simplify radical expressions involving fractions.

EXAMPLE 3 Adding and Subtracting Radical Expressions with Fractions

Perform the indicated operations.

(a)
$$2\sqrt{\frac{75}{16}} + 4\frac{\sqrt{8}}{\sqrt{32}} = 2\sqrt{\frac{25 \cdot 3}{16}} + 4\frac{\sqrt{4 \cdot 2}}{\sqrt{16 \cdot 2}}$$
$$= 2\left(\frac{5\sqrt{3}}{4}\right) + 4\left(\frac{2\sqrt{2}}{4\sqrt{2}}\right)$$
$$= \frac{5\sqrt{3}}{2} + 2 \qquad \text{Multiply.}$$
$$= \frac{5\sqrt{3}}{2} + \frac{4}{2} \qquad \text{Find the common denominator.}$$
$$= \frac{5\sqrt{3} + 4}{2}$$

(b)
$$10\sqrt[3]{\frac{5}{x^6}} - 3\sqrt[3]{\frac{4}{x^9}} = 10\frac{\sqrt[3]{5}}{\sqrt[3]{x^6}} - 3\frac{\sqrt[3]{4}}{\sqrt[3]{x^9}}$$
$$= \frac{10\sqrt[3]{5}}{x^2} - \frac{3\sqrt[3]{4}}{x^3}$$
$$= \frac{10x\sqrt[3]{5}}{x^3} - \frac{3\sqrt[3]{4}}{x^3} \qquad \text{Find the common denominator.}$$
$$= \frac{10x\sqrt[3]{5} - 3\sqrt[3]{4}}{x^3} \qquad (\text{if } x \neq 0)$$

A calculator can support some of the results obtained in the examples of this section. In Example 1(a), we simplified $3\sqrt{24} + \sqrt{54}$ to obtain $9\sqrt{6}$. The screen in Figure 8(a) shows that the approximations are the same, suggesting that our simplification was correct. Figure 8(b) shows support for the result of Example 2(a): $2\sqrt[3]{16} - 5\sqrt[3]{54} = -11\sqrt[3]{2}$. Figure 8(c) supports the result of Example 3(a).

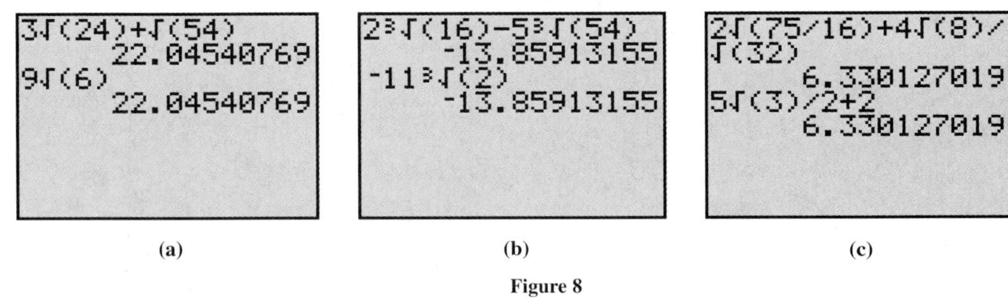

(a) (b) (c)

Figure 8

CONNECTIONS

A triangle that has whole number measures for the lengths of two sides may have an irrational number as the measure of the third side. For example, a right triangle with the two shorter sides measuring 1 and 2 units will have a longest side measuring $\sqrt{5}$ units. The ratio of the dimensions of the golden rectangle, considered to have the most pleasing dimensions of any rectangle, is irrational. To sketch a golden rectangle, begin with the square *ONRS*. Divide it into two equal parts by segment *MK*, as shown in the figure. Let *M* be the center of a circle with radius *MN*. Sketch the rectangle *PQRS*. This is a golden rectangle, with the property that

CONNECTIONS (CONTINUED)

if the original square is taken away, *PQNO* is still a golden rectangle. If the square with side *OP* is taken away, another golden rectangle results, and so on.

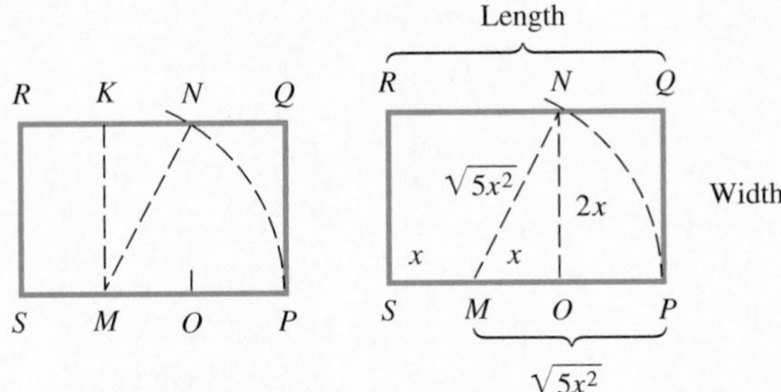

If the sides of the generating square have measure $2x$, then by the Pythagorean formula,

$$MN = \sqrt{x^2 + (2x)^2} = \sqrt{x^2 + 4x^2} = \sqrt{5x^2}.$$

Since *NP* is an arc of the circle with radius *MN*,

$$MP = MN = \sqrt{5x^2}.$$

The ratio of length to width is

$$\frac{\text{length}}{\text{width}} = \frac{x + \sqrt{5x^2}}{2x} = \frac{x + x\sqrt{5}}{2x} = \frac{x(1 + \sqrt{5})}{2x} = \frac{1 + \sqrt{5}}{2},$$

which is an irrational number.

FOR DISCUSSION OR WRITING

1. The golden rectangle has been widely used in art and architecture. See whether you can find some examples of its use. Use a calculator to approximate the ratio found above, called the *golden ratio.*

2. The sequence 1, 1, 2, 3, 5, 8, 13, 21, 34, 55, . . . is called the *Fibonacci sequence.* After the first two terms, both 1, every term is found by adding the two preceding terms. Form a sequence of ratios of the successive terms:

$\frac{1}{1}, \frac{2}{1}, \frac{3}{2}, \frac{5}{3}, \frac{8}{5}, \frac{13}{8}, \frac{21}{13}, \frac{34}{21}, \frac{55}{34}, \ldots$

Now use a calculator to find approximations of these ratios. What seems to be happening?

7.3 EXERCISES

1. Which one of the following sums could be simplified without first simplifying the individual radical expressions?
   (a) $\sqrt{50} + \sqrt{32}$ **(b)** $3\sqrt{6} + 9\sqrt{6}$ **(c)** $\sqrt[3]{32} - \sqrt[3]{108}$ **(d)** $\sqrt[5]{6} - \sqrt[5]{192}$

2. Let $a = 1$ and let $b = 64$.
   (a) Evaluate $\sqrt{a} + \sqrt{b}$. Then find $\sqrt{a + b}$. Are they equal?
   (b) Evaluate $\sqrt[3]{a} + \sqrt[3]{b}$. Then find $\sqrt[3]{a + b}$. Are they equal?
   (c) Complete the following: In general, $\sqrt[n]{a} + \sqrt[n]{b} \neq$ ________________, based on the observations in parts (a) and (b) of this exercise.

3. Even though the root indexes of the terms are not equal, the sum $\sqrt{64} + \sqrt[3]{125} + \sqrt[4]{16}$ can be simplified quite easily. What is this sum? Why can we add them so easily?

4. Explain why $28 - 4\sqrt{2}$ is not equal to $24\sqrt{2}$. (This error is a common one among algebra students.)

Simplify. Assume that all variables represent positive real numbers. See Examples 1–3.

5. $\sqrt{36} - \sqrt{100}$
6. $\sqrt{25} - \sqrt{81}$
7. $-2\sqrt{48} + 3\sqrt{75}$
8. $4\sqrt{32} - 2\sqrt{8}$
9. $\sqrt[3]{16} + 4\sqrt[3]{54}$
10. $3\sqrt[3]{24} - 2\sqrt[3]{192}$
11. $\sqrt[4]{32} + 3\sqrt[4]{2}$
12. $\sqrt[4]{405} - 2\sqrt[4]{5}$
13. $6\sqrt{18} - \sqrt{32} + 2\sqrt{50}$
14. $5\sqrt{8} + 3\sqrt{72} - 3\sqrt{50}$
15. $-2\sqrt{63} + 2\sqrt{28} + 2\sqrt{7}$
16. $-\sqrt{27} + 2\sqrt{48} - \sqrt{75}$
17. $2\sqrt{5} + 3\sqrt{20} + 4\sqrt{45}$
18. $5\sqrt{54} - 2\sqrt{24} - 2\sqrt{96}$
19. $8\sqrt{2x} - \sqrt{8x} + \sqrt{72x}$
20. $4\sqrt{18k} - \sqrt{72k} + \sqrt{50k}$
21. $3\sqrt{72m^2} - 5\sqrt{32m^2} - 3\sqrt{18m^2}$
22. $9\sqrt{27p^2} - 14\sqrt{108p^2} + 2\sqrt{48p^2}$
23. $-\sqrt[3]{54} + 2\sqrt[3]{16}$
24. $15\sqrt[3]{81} - 4\sqrt[3]{24}$
25. $2\sqrt[3]{27x} - 2\sqrt[3]{8x}$
26. $6\sqrt[3]{128m} + 3\sqrt[3]{16m}$
27. $5\sqrt[4]{32} + 3\sqrt[4]{162}$
28. $2\sqrt[4]{512} + 4\sqrt[4]{32}$
29. $3\sqrt[4]{x^5y} - 2x\sqrt[4]{xy}$
30. $2\sqrt[4]{m^9p^6} - 3m^2p\sqrt[4]{mp^2}$
31. $\sqrt[3]{64xy^2} + \sqrt[3]{27x^4y^5}$
32. $\sqrt[4]{625s^3t} - \sqrt[4]{81s^7t^5}$
33. $\sqrt{\frac{8}{9}} + \sqrt{\frac{18}{36}}$
34. $\sqrt{\frac{12}{16}} + \sqrt{\frac{48}{64}}$
35. $\frac{\sqrt{32}}{3} + \frac{2\sqrt{2}}{3} - \frac{\sqrt{2}}{\sqrt{9}}$
36. $\frac{\sqrt{27}}{2} - \frac{3\sqrt{3}}{2} + \frac{\sqrt{3}}{\sqrt{4}}$
37. $3\sqrt[3]{\frac{m^5}{27}} - 2m\sqrt[3]{\frac{m^2}{64}}$
38. $2a\sqrt[4]{\frac{a}{16}} - 5a\sqrt[4]{\frac{a}{81}}$

Follow the procedure described in the paragraph preceding Figure 8, and use a calculator to support the following statements.

39. $3\sqrt{32} - 2\sqrt{8} = 8\sqrt{2}$
40. $4\sqrt{12} - 7\sqrt{27} = -13\sqrt{3}$
41. $2\sqrt{40} + 6\sqrt{90} - 3\sqrt{160} = 10\sqrt{10}$
42. $5\sqrt{28} - 3\sqrt{63} + 2\sqrt{112} = 9\sqrt{7}$

Use estimation techniques to answer each of the following.

43. A rectangular electronic scoreboard for a sports arena has a length of $\sqrt{192}$ meters and a width of $\sqrt{48}$ meters. Choose the best estimate of its dimensions in meters.
(a) 14 by 7 (b) 5 by 7
(c) 14 by 8 (d) 15 by 8

44. If the base of a triangular liquid crystal display is $\sqrt{65}$ inches and its height is $\sqrt{26}$ inches, which one of the following is the best estimate for its area in square inches?

(a) 20 **(b)** 26 **(c)** 40 **(d)** 52

Work each problem. Give the answer as a simplified radical expression.

45. Find the perimeter of the triangle shown here.

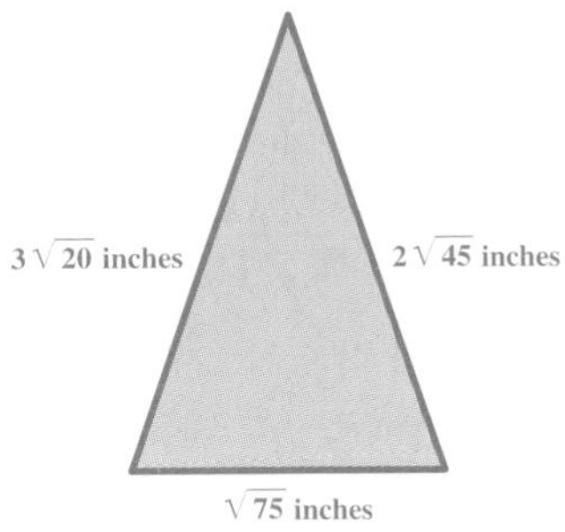

46. Find the perimeter of the rectangle shown here.

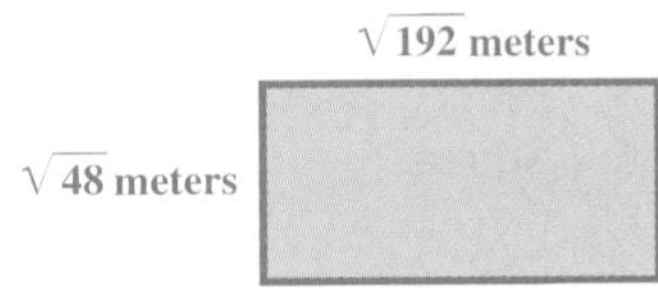

47. Find the perimeter of the computer graphic shown here.

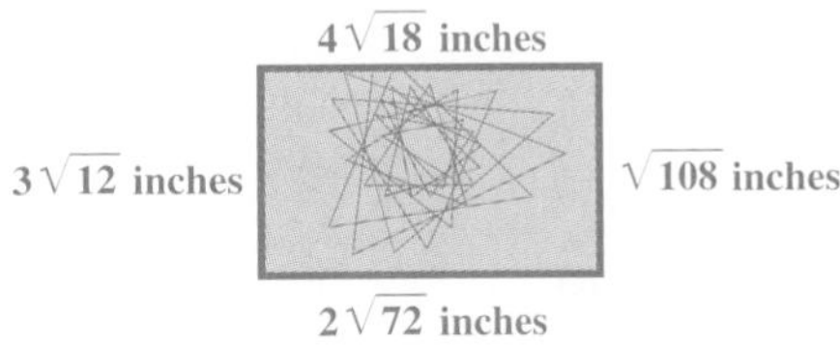

48. Find the area of the trapezoid shown here.

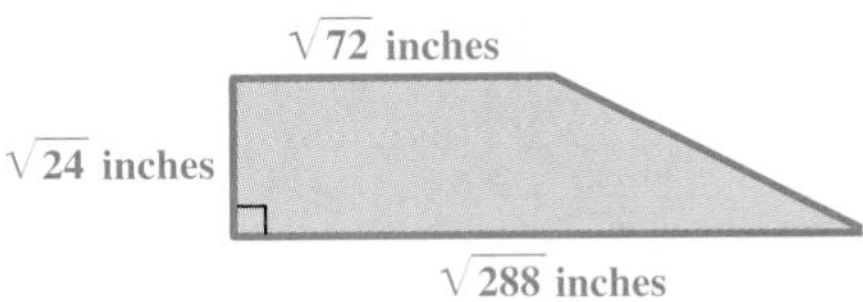

7.4 Multiplication and Division of Radical Expressions

OBJECTIVES

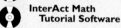

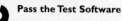

1. Multiply radical expressions.
2. Rationalize denominators with one radical term.
3. Rationalize denominators with binomials involving radicals.
4. Write radical quotients in lowest terms.

FOR EXTRA HELP

SSG Sec. 7.4
SSM Sec. 7.4
Pass the Test Software
InterAct Math Tutorial Software
Video 14

OBJECTIVE 1 Multiply radical expressions. We multiply binomial expressions involving radicals by using the FOIL (First, Outside, Inside, Last) method, which depends on the distributive property. For example, the product of the binomials $(\sqrt{5}+3)(\sqrt{6}+1)$ is found as follows.

$$\begin{aligned}(\sqrt{5}+3)(\sqrt{6}+1) &= \underset{\text{First}}{\sqrt{5}\cdot\sqrt{6}} + \underset{\text{Outside}}{\sqrt{5}\cdot 1} + \underset{\text{Inside}}{3\cdot\sqrt{6}} + \underset{\text{Last}}{3\cdot 1}\\ &= \sqrt{30}+\sqrt{5}+3\sqrt{6}+3\end{aligned}$$

This result cannot be simplified further.

EXAMPLE 1 Multiplying Binomials Involving Radical Expressions

Multiply.

(a) $$\begin{aligned}(7-\sqrt{3})(\sqrt{5}+\sqrt{2}) &= 7\sqrt{5}+7\sqrt{2}-\sqrt{3}\cdot\sqrt{5}-\sqrt{3}\cdot\sqrt{2} \quad \text{FOIL}\\ &= 7\sqrt{5}+7\sqrt{2}-\sqrt{15}-\sqrt{6}\end{aligned}$$

(b) $$\begin{aligned}(\sqrt{10}+\sqrt{3})(\sqrt{10}-\sqrt{3}) &= \sqrt{10}\sqrt{10}-\sqrt{10}\sqrt{3}+\sqrt{10}\sqrt{3}-\sqrt{3}\sqrt{3}\\ &= 10-3\\ &= 7\end{aligned}$$

Notice that this is an example of the kind of product that results in the difference of two squares:

$$(a+b)(a-b)=a^2-b^2.$$

Here, $a=\sqrt{10}$ and $b=\sqrt{3}$.

(c) $$\begin{aligned}(\sqrt{7}-3)^2 &= (\sqrt{7}-3)(\sqrt{7}-3)\\ &= \sqrt{7}\cdot\sqrt{7}-\sqrt{7}\cdot 3-3\sqrt{7}+3\cdot 3\\ &= 7-6\sqrt{7}+9\\ &= 16-6\sqrt{7}\end{aligned}$$

(d) $$\begin{aligned}(5-\sqrt[3]{3})(5+\sqrt[3]{3}) &= 5\cdot 5+5\sqrt[3]{3}-5\sqrt[3]{3}-\sqrt[3]{3}\cdot\sqrt[3]{3}\\ &= 25-\sqrt[3]{3^2}\\ &= 25-\sqrt[3]{9}\end{aligned}$$

(e) $$\begin{aligned}(\sqrt{k}+\sqrt{y})(\sqrt{k}-\sqrt{y}) &= (\sqrt{k})^2-(\sqrt{y})^2\\ &= k-y \quad (\text{if } k\ge 0 \text{ and } y\ge 0)\end{aligned}$$

OBJECTIVE 2 Rationalize denominators with one radical term. As defined earlier, a simplified radical expression will have no radical in the denominator. The origin of this agreement no doubt occurred before the days of high-speed calculation, when computation was a tedious process performed by hand. To see this, consider the radical expression $\frac{1}{\sqrt{2}}$. To find a decimal approximation by hand, it would be necessary to divide 1 by a decimal approximation for $\sqrt{2}$, such as 1.414. It would be much easier if the divisor were a whole number. This can be accomplished by multiplying $\frac{1}{\sqrt{2}}$ by 1 in the form $\frac{\sqrt{2}}{\sqrt{2}}$:

$$\frac{1}{\sqrt{2}} \cdot \frac{\sqrt{2}}{\sqrt{2}} = \frac{\sqrt{2}}{2}.$$

Now the computation would require dividing 1.414 by 2 to obtain .707, a much easier task.

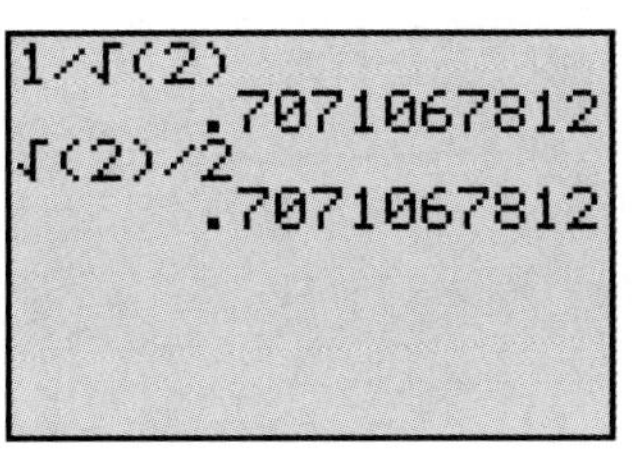

Figure 9

With today's technology, either form of this fraction can be approximated with the same number of keystrokes. See Figure 9, which shows how a calculator gives the same approximation for both forms of the expression.

It is still important to be able to find equivalent forms of radical expressions. A common way of "standardizing" the form of a radical expression is to have the denominator contain no radicals. The process of removing radicals from a denominator so that the denominator contains only rational numbers is called **rationalizing the denominator.**

EXAMPLE 2 Rationalizing Denominators with Square Roots

Rationalize each denominator.

(a) $\dfrac{3}{\sqrt{7}}$

Multiply numerator and denominator by $\sqrt{7}$. This is, in effect, multiplying by 1.

$$\frac{3}{\sqrt{7}} = \frac{3 \cdot \sqrt{7}}{\sqrt{7} \cdot \sqrt{7}}$$

Since $\sqrt{7} \cdot \sqrt{7} = \sqrt{7 \cdot 7} = \sqrt{49} = 7$,

$$\frac{3}{\sqrt{7}} = \frac{3\sqrt{7}}{7}.$$

The denominator is now a rational number.

(b) $\dfrac{5\sqrt{2}}{\sqrt{5}}$

Multiply the numerator and denominator by $\sqrt{5}$.

$$\frac{5\sqrt{2}}{\sqrt{5}} = \frac{5\sqrt{2} \cdot \sqrt{5}}{\sqrt{5} \cdot \sqrt{5}} = \frac{5\sqrt{10}}{5} = \sqrt{10}$$

(c) $\dfrac{6}{\sqrt{12}}$

Less work is involved if the radical in the denominator is simplified first.

$$\frac{6}{\sqrt{12}} = \frac{6}{\sqrt{4 \cdot 3}} = \frac{6}{2\sqrt{3}} = \frac{3}{\sqrt{3}}$$

Now rationalize the denominator by multiplying the numerator and the denominator by $\sqrt{3}$.

$$\frac{3 \cdot \sqrt{3}}{\sqrt{3} \cdot \sqrt{3}} = \frac{3\sqrt{3}}{3} = \sqrt{3}$$

EXAMPLE 3 Rationalizing Denominators in Roots of Fractions

Simplify each of the following.

(a) $\sqrt{\dfrac{18}{125}}$

$$\begin{aligned}\sqrt{\frac{18}{125}} &= \frac{\sqrt{18}}{\sqrt{125}} && \text{Quotient rule}\\ &= \frac{\sqrt{9 \cdot 2}}{\sqrt{25 \cdot 5}} && \text{Factor.}\\ &= \frac{3\sqrt{2}}{5\sqrt{5}} && \text{Product rule}\\ &= \frac{3\sqrt{2} \cdot \sqrt{5}}{5\sqrt{5} \cdot \sqrt{5}} && \text{Multiply by } \sqrt{5} \text{ in numerator and denominator.}\\ &= \frac{3\sqrt{10}}{5 \cdot 5} && \text{Product rule}\\ &= \frac{3\sqrt{10}}{25}\end{aligned}$$

(b) $\sqrt{\dfrac{50m^4}{p^5}}, \quad p > 0$

$$\begin{aligned}\sqrt{\frac{50m^4}{p^5}} &= \frac{\sqrt{50m^4}}{\sqrt{p^5}} && \text{Quotient rule}\\ &= \frac{\sqrt{25m^4 \cdot 2}}{\sqrt{p^4 \cdot p}}\\ &= \frac{5m^2\sqrt{2}}{p^2\sqrt{p}} && \text{Product rule}\\ &= \frac{5m^2\sqrt{2} \cdot \sqrt{p}}{p^2\sqrt{p} \cdot \sqrt{p}} && \text{Multiply by } \sqrt{p} \text{ in numerator and denominator.}\\ &= \frac{5m^2\sqrt{2p}}{p^2 \cdot p} && \text{Product rule}\\ &= \frac{5m^2\sqrt{2p}}{p^3}\end{aligned}$$

EXAMPLE 4 Rationalizing a Denominator with a Cube Root

Simplify $\sqrt[3]{\dfrac{27}{16}}$.

Use the quotient rule and simplify the numerator and denominator.

$$\sqrt[3]{\frac{27}{16}} = \frac{\sqrt[3]{27}}{\sqrt[3]{16}} = \frac{3}{\sqrt[3]{8} \cdot \sqrt[3]{2}} = \frac{3}{2\sqrt[3]{2}}$$

To get a rational denominator multiply the numerator and denominator by a number that will result in a perfect cube in the radicand in the denominator. Since $2 \cdot 4 = 8$, a perfect cube, multiply the numerator and denominator by $\sqrt[3]{4}$.

$$\frac{3}{2\sqrt[3]{2}} = \frac{3 \cdot \sqrt[3]{4}}{2 \cdot \sqrt[3]{2} \cdot \sqrt[3]{4}} = \frac{3\sqrt[3]{4}}{2\sqrt[3]{8}} = \frac{3\sqrt[3]{4}}{2 \cdot 2} = \frac{3\sqrt[3]{4}}{4}$$

It is easy to make mistakes in problems like the one in Example 4. A typical error is to multiply numerator and denominator by $\sqrt[3]{2}$, forgetting that

$$\sqrt[3]{2} \cdot \sqrt[3]{2} \neq 2.$$

You need *three* factors of 2 to get 2^3 under the radical. As shown in Example 4,

$$\sqrt[3]{2} \cdot \sqrt[3]{2} \cdot \sqrt[3]{2} = 2.$$

OBJECTIVE 3 **Rationalize denominators with binomials involving radicals.** Recall the special product

$$(a + b)(a - b) = a^2 - b^2.$$

In order to rationalize a denominator that contains a binomial expression (one that contains exactly two terms) involving radicals, such as

$$\frac{3}{1 + \sqrt{2}},$$

we must use conjugates. The **conjugate** of $1 + \sqrt{2}$ is $1 - \sqrt{2}$. In general, $a + b$ and $a - b$ are conjugates.

Rationalizing Binomial Denominators

Whenever a radical expression has a binomial with square root radicals in the denominator, rationalize by multiplying both the numerator and the denominator by the conjugate of the denominator.

For the expression $\frac{3}{1 + \sqrt{2}}$, rationalize the denominator by multiplying both the numerator and denominator by the conjugate of the denominator, $1 - \sqrt{2}$.

$$\frac{3}{1 + \sqrt{2}} = \frac{3(1 - \sqrt{2})}{(1 + \sqrt{2})(1 - \sqrt{2})}$$

Then $(1 + \sqrt{2})(1 - \sqrt{2}) = 1^2 - (\sqrt{2})^2 = 1 - 2 = -1$. Placing -1 in the denominator gives

$$\frac{3}{1 + \sqrt{2}} = \frac{3(1 - \sqrt{2})}{-1} = -3(1 - \sqrt{2}) \quad \text{or} \quad -3 + 3\sqrt{2}.$$

EXAMPLE 5 Rationalizing a Binomial Denominator

Rationalize the denominator in the following expressions. Assume that all variables represent positive real numbers.

(a) $\dfrac{5}{4-\sqrt{3}}$

Multiply numerator and denominator by the conjugate of the denominator, $4+\sqrt{3}$.

$$\frac{5}{4-\sqrt{3}} = \frac{5(4+\sqrt{3})}{(4-\sqrt{3})(4+\sqrt{3})}$$
$$= \frac{5(4+\sqrt{3})}{16-3}$$
$$= \frac{5(4+\sqrt{3})}{13}$$

Notice that the numerator is left in factored form. Doing this makes it easier to determine whether the expression can be reduced to lowest terms.

(b) $\dfrac{\sqrt{2}-\sqrt{3}}{\sqrt{5}+\sqrt{3}}$

Multiplication of both numerator and denominator by $\sqrt{5}-\sqrt{3}$ will rationalize the denominator.

$$\frac{\sqrt{2}-\sqrt{3}}{\sqrt{5}+\sqrt{3}} = \frac{(\sqrt{2}-\sqrt{3})(\sqrt{5}-\sqrt{3})}{(\sqrt{5}+\sqrt{3})(\sqrt{5}-\sqrt{3})}$$
$$= \frac{\sqrt{10}-\sqrt{6}-\sqrt{15}+3}{5-3}$$
$$= \frac{\sqrt{10}-\sqrt{6}-\sqrt{15}+3}{2}$$

(c) $$\frac{3}{\sqrt{5m}-\sqrt{p}} = \frac{3(\sqrt{5m}+\sqrt{p})}{(\sqrt{5m}-\sqrt{p})(\sqrt{5m}+\sqrt{p})}$$
$$= \frac{3(\sqrt{5m}+\sqrt{p})}{5m-p} \quad (5m \neq p)$$

CONNECTIONS

Sometimes it is desirable to rationalize the *numerator* in an expression. The procedure is similar to rationalizing the denominator. For example, to rationalize the numerator of

$$\frac{6-\sqrt{2}}{4},$$

we multiply by the conjugate of the numerator, $6+\sqrt{2}$.

$$\frac{6-\sqrt{2}}{4} = \frac{(6-\sqrt{2})(6+\sqrt{2})}{4(6+\sqrt{2})} = \frac{36-2}{4(6+\sqrt{2})} = \frac{34}{4(6+\sqrt{2})} = \frac{17}{2(6+\sqrt{2})}$$

In the final expression, the numerator is rationalized and the fraction is simplified.

CONNECTIONS (CONTINUED)

FOR DISCUSSION OR WRITING

Rationalize the numerators of the following expressions, assuming a and b are nonnegative real numbers.

1. $\dfrac{8\sqrt{5} - 1}{6}$ **2.** $\dfrac{3\sqrt{a} + \sqrt{b}}{b}$ **3.** $\dfrac{3\sqrt{a} + \sqrt{b}}{\sqrt{b} - \sqrt{a}}$

4. Rationalize the denominator of the expression in Exercise 3, and then describe the difference in the procedure you used from what you did in Exercise 3.

OBJECTIVE 4 Write radical quotients in lowest terms.

EXAMPLE 6 Writing a Radical Quotient in Lowest Terms

Write each expression in lowest terms.

(a) $\dfrac{6 + 2\sqrt{5}}{4}$

Factor the numerator, and then simplify.

$$\frac{6 + 2\sqrt{5}}{4} = \frac{2(3 + \sqrt{5})}{2 > 2} = \frac{3 + \sqrt{5}}{2}$$

Here is an alternative method for writing this expression in lowest terms.

$$\frac{6 + 2\sqrt{5}}{4} = \frac{6}{4} + \frac{2\sqrt{5}}{4} = \frac{3}{2} + \frac{\sqrt{5}}{2} = \frac{3 + \sqrt{5}}{2}$$

(b) $\dfrac{5y - \sqrt{8y^2}}{6y} = \dfrac{5y - 2y\sqrt{2}}{6y} = \dfrac{y(5 - 2\sqrt{2})}{6y} = \dfrac{5 - 2\sqrt{2}}{6}$ (if $y > 0$)

Notice that the final fraction cannot be reduced further because there is no common factor of 2 in the numerator.

CAUTION Refer to Example 6(a). Be careful to factor *before* writing a quotient in lowest terms.

7.4 EXERCISES

Match each part of a rule for a special product in Column I with the other part in Column II.

I	II
1. $(x + \sqrt{y})(x - \sqrt{y})$	**A.** $x - y$
2. $(\sqrt{x} + y)(\sqrt{x} - y)$	**B.** $x + 2y\sqrt{x} + y^2$
3. $(\sqrt{x} + \sqrt{y})(\sqrt{x} - \sqrt{y})$	**C.** $x - y^2$
4. $(\sqrt{x} + \sqrt{y})^2$	**D.** $x - 2\sqrt{xy} + y$
5. $(\sqrt{x} - \sqrt{y})^2$	**E.** $x^2 - y$
6. $(\sqrt{x} + y)^2$	**F.** $x + 2\sqrt{xy} + y$

7. Explain why $\sqrt[3]{x} \cdot \sqrt[3]{x}$ is not equal to x. What is it equal to?

8. Explain why $\sqrt[4]{x} \cdot \sqrt[4]{x}$ is not equal to x, but *is* equal to $\sqrt{x}$, for $x \geq 0$.

Multiply and then simplify each product. Assume that all variables represent positive real numbers. See Example 1.

9. $\sqrt{3}(\sqrt{12} - 4)$ **10.** $\sqrt{5}(\sqrt{125} - 6)$ **11.** $\sqrt{2}(\sqrt{18} - \sqrt{3})$

12. $\sqrt{5}(\sqrt{15} + \sqrt{5})$ **13.** $(\sqrt{6} + 2)(\sqrt{6} - 2)$ **14.** $(\sqrt{7} + 8)(\sqrt{7} - 8)$

15. $(\sqrt{12} - \sqrt{3})(\sqrt{12} + \sqrt{3})$ **16.** $(\sqrt{18} + \sqrt{8})(\sqrt{18} - \sqrt{8})$

17. $(\sqrt{3} + 2)(\sqrt{6} - 5)$ **18.** $(\sqrt{7} + 1)(\sqrt{2} - 4)$

19. $(\sqrt{3x} + 2)(\sqrt{3x} - 2)$ **20.** $(\sqrt{6y} - 4)(\sqrt{6y} + 4)$

21. $(2\sqrt{x} + \sqrt{y})(2\sqrt{x} - \sqrt{y})$ **22.** $(\sqrt{p} + 5\sqrt{s})(\sqrt{p} - 5\sqrt{s})$

23. $(4\sqrt{x} + 3)^2$ **24.** $(5\sqrt{p} - 6)^2$

25. $(9 - \sqrt[3]{2})(9 + \sqrt[3]{2})$ **26.** $(7 + \sqrt[3]{6})(7 - \sqrt[3]{6})$

27. $[(\sqrt{2} + \sqrt{3}) - \sqrt{6}][(\sqrt{2} + \sqrt{3}) + \sqrt{6}]$

28. $[(\sqrt{5} - \sqrt{2}) - \sqrt{3}][(\sqrt{5} - \sqrt{2}) + \sqrt{3}]$

29. $[(\sqrt{2} + \sqrt{3}) + \sqrt{5}]^2$

30. $[(\sqrt{6} - \sqrt{5}) + \sqrt{2}]^2$

31. The correct answer to Exercise 9 is $6 - 4\sqrt{3}$. Explain why this is not equal to $2\sqrt{3}$.

32. When we rationalize the denominator in the radical expression $\frac{1}{\sqrt{2}}$, we multiply both the numerator and the denominator by $\sqrt{2}$. What property of real numbers covered in Chapter 1 justifies this procedure?

Rationalize the denominator in each expression. Assume that all variables represent positive real numbers. See Examples 2 and 3.

33. $\frac{7}{\sqrt{7}}$ **34.** $\frac{11}{\sqrt{11}}$ **35.** $\frac{15}{\sqrt{3}}$ **36.** $\frac{12}{\sqrt{6}}$ **37.** $\frac{\sqrt{3}}{\sqrt{2}}$

38. $\frac{\sqrt{7}}{\sqrt{6}}$ **39.** $\frac{9\sqrt{3}}{\sqrt{5}}$ **40.** $\frac{3\sqrt{2}}{\sqrt{11}}$ **41.** $\frac{-6}{\sqrt{18}}$ **42.** $\frac{-5}{\sqrt{24}}$

43. $\sqrt{\frac{7}{2}}$ **44.** $\sqrt{\frac{10}{3}}$ **45.** $-\sqrt{\frac{7}{50}}$ **46.** $-\sqrt{\frac{13}{75}}$ **47.** $\sqrt{\frac{24}{x}}$

48. $\sqrt{\frac{52}{y}}$ **49.** $\frac{-8\sqrt{3}}{\sqrt{k}}$ **50.** $\frac{-4\sqrt{13}}{\sqrt{m}}$ **51.** $-\sqrt{\frac{150m^5}{n^3}}$ **52.** $-\sqrt{\frac{98r^3}{s^5}}$

53. $\sqrt{\frac{288x^7}{y^9}}$ **54.** $\sqrt{\frac{242t^9}{u^{11}}}$

55. Look again at the expression in Exercise 48. After writing it as a quotient of radicals, multiply both the numerator and the denominator by $\sqrt{y}$, and then obtain the final answer. Next, start over, multiplying both the numerator and the denominator by $\sqrt{y^3}$, to obtain the same answer. Which method do you prefer? Why?

56. Explain why $\frac{1}{\sqrt[3]{2}}$ would not be written with denominator rationalized if you multiply both the numerator and the denominator by $\sqrt[3]{2}$. By what should you multiply to achieve the desired result?

Simplify. Assume that all variables represent positive real numbers. See Example 4.

57. $\sqrt[3]{\frac{2}{3}}$ 58. $\sqrt[3]{\frac{4}{5}}$ 59. $\sqrt[3]{\frac{4}{9}}$ 60. $\sqrt[3]{\frac{5}{16}}$

61. $-\sqrt[3]{\frac{2p}{r^2}}$ 62. $-\sqrt[3]{\frac{6x}{y^2}}$ 63. $\sqrt[4]{\frac{16}{x}}$ 64. $\sqrt[4]{\frac{81}{y}}$

65. Explain the procedure you will use to rationalize the denominator of the expression in Exercise 66: $\frac{2}{4+\sqrt{3}}$. Would multiplying both the numerator and the denominator of this fraction by $4+\sqrt{3}$ lead to a rationalized denominator? Why or why not?

Rationalize the denominator in each expression. Assume that all variables represent positive real numbers and that no denominators are zero. See Example 5.

66. $\frac{2}{4+\sqrt{3}}$ 67. $\frac{6}{5+\sqrt{2}}$ 68. $\frac{6}{\sqrt{5}+\sqrt{3}}$

69. $\frac{12}{\sqrt{6}+\sqrt{3}}$ 70. $\frac{-4}{\sqrt{3}-\sqrt{7}}$ 71. $\frac{-3}{\sqrt{2}+\sqrt{5}}$

72. $\frac{1-\sqrt{2}}{\sqrt{7}+\sqrt{6}}$ 73. $\frac{-1-\sqrt{3}}{\sqrt{6}+\sqrt{5}}$ 74. $\frac{4\sqrt{x}}{\sqrt{x}-2\sqrt{y}}$

75. $\frac{5\sqrt{r}}{3\sqrt{r}+\sqrt{s}}$ 76. $\frac{\sqrt{x}-\sqrt{y}}{\sqrt{x}+\sqrt{y}}$ 77. $\frac{\sqrt{a}+\sqrt{b}}{\sqrt{a}-\sqrt{b}}$

78. If a and b are both positive numbers and $a^2 = b^2$, then $a = b$. Use this fact to show that $\frac{\sqrt{6}-\sqrt{2}}{4} = \frac{\sqrt{2-\sqrt{3}}}{2}$. Then use a calculator approximation to support your result.

Write each expression in lowest terms. Assume that all variables represent positive real numbers. See Example 6.

79. $\frac{25+10\sqrt{6}}{20}$ 80. $\frac{12-6\sqrt{2}}{24}$ 81. $\frac{16+4\sqrt{8}}{12}$

82. $\frac{12+9\sqrt{72}}{18}$ 83. $\frac{6x+\sqrt{24x^3}}{3x}$ 84. $\frac{11y+\sqrt{242y^5}}{22y}$

85. The following expression occurs in a certain standard problem in trigonometry:

$$\frac{1}{\sqrt{2}}\cdot\frac{\sqrt{3}}{2}-\frac{1}{\sqrt{2}}\cdot\frac{1}{2}.$$

Show that it simplifies to $\frac{\sqrt{6}-\sqrt{2}}{4}$. Then verify using a calculator approximation.

86. The following expression occurs in a certain standard problem in trigonometry:

$$\frac{\sqrt{3}+1}{1-\sqrt{3}}.$$

Show that it simplifies to $-2-\sqrt{3}$. Then verify using a calculator approximation.

RELATING CONCEPTS (EXERCISES 87–94)

In Chapter 5 we presented methods of factoring, where the terms in the factors were integers. For example, the binomial $x^2 - 9$ is a difference of two squares, and factors as $(x + 3)(x - 3)$. However, we can also use this pattern to factor any binomial if we allow square root radicals in the terms of the factors. For example, $t - 5$ can be factored as $(\sqrt{t} + \sqrt{5})(\sqrt{t} - \sqrt{5})$.

Similarly, we can factor any binomial as the sum or difference of cubes, using the patterns $a^3 + b^3 = (a + b)(a^2 - ab + b^2)$ and $a^3 - b^3 = (a - b)(a^2 + ab + b^2)$. For example, we can factor $y + 2$ and $y - 2$ as follows:

$$y + 2 = (\sqrt[3]{y} + \sqrt[3]{2})(\sqrt[3]{y^2} - \sqrt[3]{2y} + \sqrt[3]{4})$$
$$y - 2 = (\sqrt[3]{y} - \sqrt[3]{2})(\sqrt[3]{y^2} + \sqrt[3]{2y} + \sqrt[3]{4}).$$

Use these ideas to ***work Exercises 87–94 in order.***

87. Factor $x - 7$ as the difference of two squares.

88. Factor $x - 7$ as the difference of two cubes.

89. Factor $x + 7$ as the sum of two cubes.

90. Use the result of Exercise 87 to rationalize the denominator of $\dfrac{x + 3}{\sqrt{x} - \sqrt{7}}$.

91. Use the result of Exercise 88 to rationalize the denominator of $\dfrac{x + 3}{\sqrt[3]{x} - \sqrt[3]{7}}$.

92. Use the result of Exercise 89 to rationalize the denominator of $\dfrac{x + 3}{\sqrt[3]{x^2} - \sqrt[3]{7x} + \sqrt[3]{49}}$.

93. Factor the integer 2 as a difference of cubes by first writing it as $5 - 3$.

94. Use the result of Exercise 93 to rationalize the denominator of $\dfrac{2}{\sqrt[3]{5} - \sqrt[3]{3}}$.

Did you make the connection that factoring techniques studied earlier can be extended if we allow radicals to be used as terms within the factors?

7.5 Graphs and Equations with Radical Expressions

OBJECTIVES

1. Graph simple functions defined by radical expressions.
2. Solve radical equations using the power rule.
3. Solve radical equations that require additional steps.
4. Solve radical equations with indexes greater than 2.
5. Solve radical equations using a graphing calculator.
6. Use the power rule to solve a formula for a specified variable.

OBJECTIVE 1 Graph simple functions defined by radical expressions. We have seen in earlier chapters how to graph functions defined by polynomial and rational expressions. Now we examine the graphs of $f(x) = \sqrt{x}$ and $f(x) = \sqrt[3]{x}$, functions that are defined by radical expressions. Figure 10 shows the graph of the **square root function,** and Figure 11 shows the graph of the **cube root function.** A table of selected points is shown in each case.

x	$f(x) = \sqrt{x}$
0	0
1	1
4	2
9	3

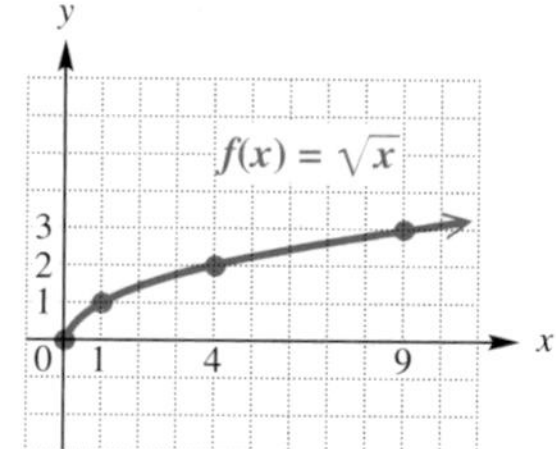

Figure 10

x	$f(x) = \sqrt[3]{x}$
-8	-2
-1	-1
0	0
1	1
8	2

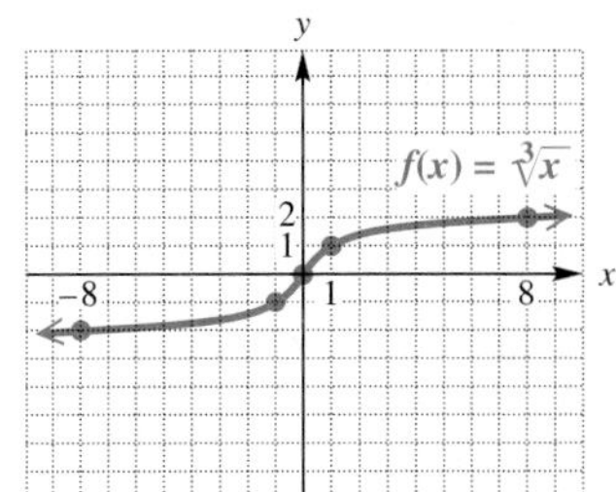

Figure 11

For the square root function, only nonnegative values can be used for x, so the domain is $[0, \infty)$. Because $\sqrt{x}$ takes on only nonnegative values, the range is also $[0, \infty)$. A different situation exists for the cube root function. Any real number (positive, negative, or zero) can be used for x, and $\sqrt[3]{x}$ takes on positive, negative, and zero values. Thus both the domain and the range of the cube root function are $(-\infty, \infty)$.

EXAMPLE 1 Graphing Functions Defined with Radicals

Graph each function by creating a table of values. Give the domain and the range.

(a) $f(x) = \sqrt{x - 3}$

A table of values is shown. The x-values were chosen in such a way that the function values are all integers. For the radicand to be nonnegative, we must have $x - 3 \geq 0$, or $x \geq 3$. Therefore the domain is $[3, \infty)$. Again, function values are positive or zero, so the range is $[0, \infty)$. The graph is shown in Figure 12.

x	$f(x) = \sqrt{x - 3}$
3	$\sqrt{3 - 3} = 0$
4	$\sqrt{4 - 3} = 1$
7	$\sqrt{7 - 3} = 2$

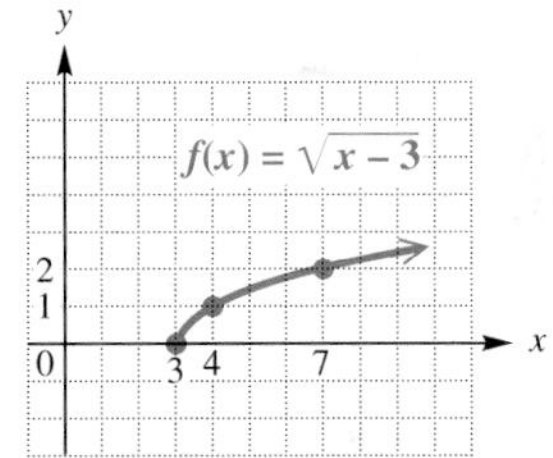

Figure 12

(b) $f(x) = \sqrt[3]{x} + 2$

See the table and Figure 13. Both the domain and the range are $(-\infty, \infty)$.

x	$f(x) = \sqrt[3]{x} + 2$
-8	$\sqrt[3]{-8} + 2 = 0$
-1	$\sqrt[3]{-1} + 2 = 1$
0	$\sqrt[3]{0} + 2 = 2$
1	$\sqrt[3]{1} + 2 = 3$
8	$\sqrt[3]{8} + 2 = 4$

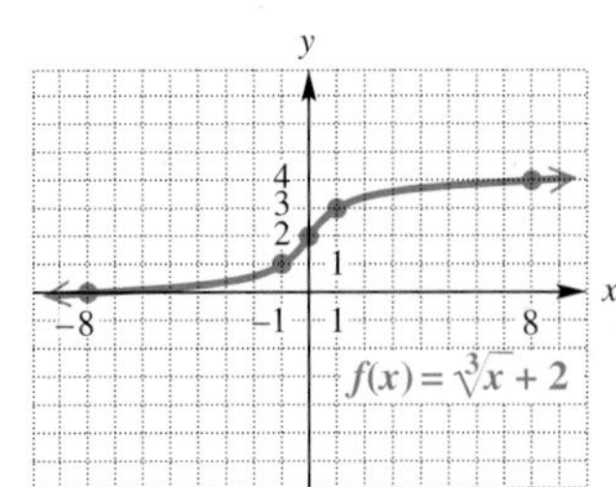

Figure 13

OBJECTIVE 2 **Solve radical equations using the power rule.** The equation $x = 1$ has only one solution. Its solution set is $\{1\}$. If we square both sides of this equation, another equation is obtained: $x^2 = 1$. This new equation has two solutions: -1 and 1. Notice that the solution of the original equation is also a solution of the squared equation. However, the squared equation has another solution, -1, that is *not* a solution of the original equation. When solving equations with radicals, we will use this idea of raising both sides to a power. It is an application of the *power rule.*

Power Rule for Solving Equations with Radicals

If both sides of an equation are raised to the same power, all solutions of the original equation are also solutions of the new equation.

Read the power rule carefully; it does *not* say that all solutions of the new equation are solutions of the original equation. They may or may not be. Solutions that do not satisfy the original equation are called **extraneous solutions**; they must be discarded.

When the power rule is used to solve an equation, **every solution of the new equation *must* be checked in the original equation.**

EXAMPLE 2 **Using the Power Rule**

Solve $\sqrt{3x + 4} = 8$.

Use the power rule and square both sides of the equation to get

$$\begin{aligned}(\sqrt{3x+4})^2 &= 8^2 \\ 3x + 4 &= 64 \\ 3x &= 60 \\ x &= 20.\end{aligned}$$

Check this proposed solution in the *original* equation.

$$\begin{aligned}\sqrt{3x+4} &= 8 && \text{Original equation} \\ \sqrt{3 \cdot 20 + 4} &= 8 \quad ? && \text{Let } x = 20. \\ \sqrt{64} &= 8 \quad ? && \\ 8 &= 8 && \text{True}\end{aligned}$$

Since 20 satisfies the *original* equation, the solution set is $\{20\}$.

The solution of the equation in Example 2 can be generalized to give a method for solving equations with radicals.

Solving an Equation with Radicals

Step 1 **Isolate the radical.** Make sure that one radical term is alone on one side of the equation.

Step 2 **Apply the power rule.** Raise each side of the equation to a power that is the same as the index of the radical.

Step 3 **Solve.** Solve the resulting equation; if it still contains a radical, repeat Steps 1 and 2.

Step 4 **Check.** It is essential that all potential solutions be checked in the original equation.

Remember Step 4 or you may get an incorrect solution set.

EXAMPLE 3 Using the Power Rule

Solve $\sqrt{5q-1}+3=0$.

Step 1 To get the radical alone on one side, subtract 3 from both sides.

$$\sqrt{5q-1}=-3$$

Step 2 Now square both sides.

$$(\sqrt{5q-1})^2=(-3)^2 \quad \text{Power rule}$$

Step 3

$$5q-1=9$$
$$5q=10$$
$$q=2$$

Step 4 Check the potential solution, 2, by substituting it in the original equation.

$$\sqrt{5q-1}+3=0 \quad \text{Original equation}$$
$$\sqrt{5\cdot 2-1}+3=0 \quad ? \quad \text{Let } q=2.$$
$$3+3=0 \quad \text{False}$$

This false result shows that 2 is not a solution of the original equation; it is extraneous. The solution set is $\emptyset$.

We could have determined after Step 1 that the equation in Example 3 had no solution. The equation $\sqrt{5q-1}=-3$ has no solution because the expression on the left cannot be negative.

OBJECTIVE 3 Solve radical equations that require additional steps. The next examples involve finding the square of a binomial. Recall that $(x+y)^2=x^2+2xy+y^2$.

EXAMPLE 4 Using the Power Rule; Squaring a Binomial

Solve $\sqrt{4 - x} = x + 2$.

Step 1 The radical is alone on the left side of the equation.

Step 2 Square both sides; the square of $x + 2$ is $(x + 2)^2 = x^2 + 4x + 4$. Thus, the new equation is quadratic, and we need to get 0 on one side.

$$(\sqrt{4 - x})^2 = (x + 2)^2 \qquad \text{Power rule}$$

Step 3

$$4 - x = x^2 + 4x + 4$$

↑ Twice the product of 2 and x

$$0 = x^2 + 5x \qquad \text{Subtract 4 and add } x.$$

$$0 = x(x + 5) \qquad \text{Factor.}$$

$$x = 0 \quad \text{or} \quad x + 5 = 0 \qquad \text{Zero-factor property}$$

$$x = -5$$

Step 4 Check each potential solution in the original equation.

If $x = 0$,

$$\sqrt{4 - x} = x + 2$$
$$\sqrt{4 - 0} = 0 + 2 \quad ?$$
$$\sqrt{4} = 2 \quad ?$$
$$2 = 2. \qquad \text{True}$$

If $x = -5$,

$$\sqrt{4 - x} = x + 2$$
$$\sqrt{4 - (-5)} = -5 + 2 \quad ?$$
$$\sqrt{9} = -3. \quad ?$$
$$3 = -3. \qquad \text{False}$$

The solution set is $\{0\}$. The other potential solution, -5, is extraneous. (In Objective 5 we will see a graphical reason why this extraneous value appeared.)

CAUTION When a radical equation requires squaring a binomial as in Example 4, remember to include the middle term.

$$(x + 2)^2 \neq x^2 + 4 \qquad (x + 2)^2 = x^2 + 4x + 4.$$

EXAMPLE 5 Using the Power Rule; Squaring a Binomial

Solve $\sqrt{m^2 - 4m + 9} = m - 1$.

Squaring both sides gives $(m - 1)^2 = m^2 - 2(m)(1) + 1^2$ on the right.

$$(\sqrt{m^2 - 4m + 9})^2 = (m - 1)^2$$

$$m^2 - 4m + 9 = m^2 - 2m + 1$$

↑ Twice the product of m and 1

Subtract m^2 and 1 from both sides. Then add $4m$ to both sides, to get

$$8 = 2m$$

$$4 = m.$$

Check this potential solution in the original equation.

$$\sqrt{m^2 - 4m + 9} = m - 1$$

$$\sqrt{4^2 - 4 \cdot 4 + 9} = 4 - 1 \quad ? \qquad \text{Let } m = 4.$$

$$3 = 3 \qquad \text{True}$$

The solution set of the original equation is $\{4\}$.

Sometimes we must isolate a radical and apply the power rule twice.

EXAMPLE 6 Using the Power Rule; Squaring Twice

Solve $\sqrt{5m + 6} + \sqrt{3m + 4} = 2$.

Start by getting one radical alone on one side of the equation. Do this by subtracting $\sqrt{3m + 4}$ from both sides.

$$\sqrt{5m + 6} = 2 - \sqrt{3m + 4}$$

Now square both sides.

$$(\sqrt{5m + 6})^2 = (2 - \sqrt{3m + 4})^2$$

$$5m + 6 = 4 - 4\sqrt{3m + 4} + (3m + 4)$$

↑ Twice the product of 2 and $\sqrt{3m + 4}$

This equation still contains a radical, so it will be necessary to square both sides again. Before doing this, isolate the radical term on the right.

$$5m + 6 = 8 + 3m - 4\sqrt{3m + 4}$$

$$2m - 2 = -4\sqrt{3m + 4} \qquad \text{Subtract 8 and } 3m.$$

$$m - 1 = -2\sqrt{3m + 4} \qquad \text{Divide by 2.}$$

Now square both sides again.

$$(m - 1)^2 = (-2\sqrt{3m + 4})^2$$

$$(m - 1)^2 = (-2)^2(\sqrt{3m + 4})^2 \qquad (ab)^2 = a^2b^2$$

$$m^2 - 2m + 1 = 4(3m + 4)$$

$$m^2 - 2m + 1 = 12m + 16 \qquad \text{Distributive property}$$

This equation is quadratic and may be solved with the zero-factor property. Start by getting 0 on one side of the equation; then factor.

$$m^2 - 14m - 15 = 0$$

$$(m - 15)(m + 1) = 0$$

By the zero-factor property,

$$m - 15 = 0 \quad \text{or} \quad m + 1 = 0$$

$$m = 15 \quad \text{or} \quad m = -1.$$

Check each of these potential solutions in the original equation. Only -1 checks, so the solution set, $\{-1\}$, has only one element.

OBJECTIVE 4 Solve radical equations with indexes greater than 2. The power rule also works for powers greater than 2.

EXAMPLE 7 Using the Power Rule for Powers Greater than 2

Solve $\sqrt[3]{z + 5} = \sqrt[3]{2z - 6}$.

Raise both sides to the third power.

$$(\sqrt[3]{z + 5})^3 = (\sqrt[3]{2z - 6})^3$$
$$z + 5 = 2z - 6$$
$$11 = z$$

Check this result in the original equation.

$$\sqrt[3]{z + 5} = \sqrt[3]{2z - 6}$$
$$\sqrt[3]{\mathbf{11} + 5} = \sqrt[3]{2 \cdot \mathbf{11} - 6} \quad ? \qquad \text{Let } z = 11.$$
$$\sqrt[3]{16} = \sqrt[3]{16} \qquad \text{True}$$

The solution set is $\{11\}$.

OBJECTIVE 5 Solve radical equations using a graphing calculator. In Example 5, with m as the variable, we solved the equation $\sqrt{x^2 - 4x + 9} = x - 1$ using algebraic methods. If we write this equation with one side equal to 0, we get

$$\sqrt{x^2 - 4x + 9} - x + 1 = 0.$$

Using a graphing calculator to graph the function with

$$f(x) = \sqrt{x^2 - 4x + 9} - x + 1,$$

we get the graph shown in Figure 14. Notice that its zero (x-value of the x-intercept) is 4, which is the solution we found in the example.

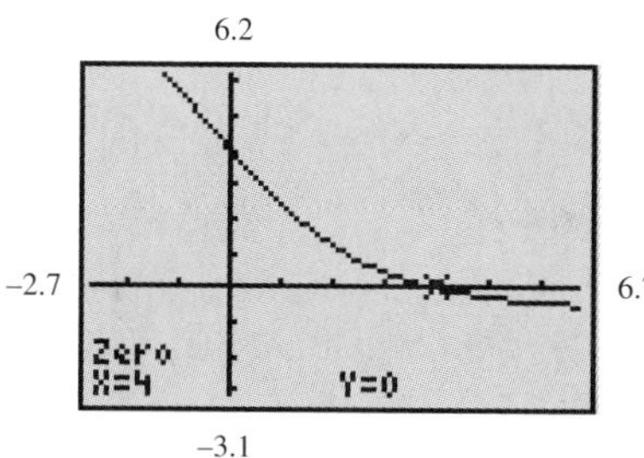

Figure 14

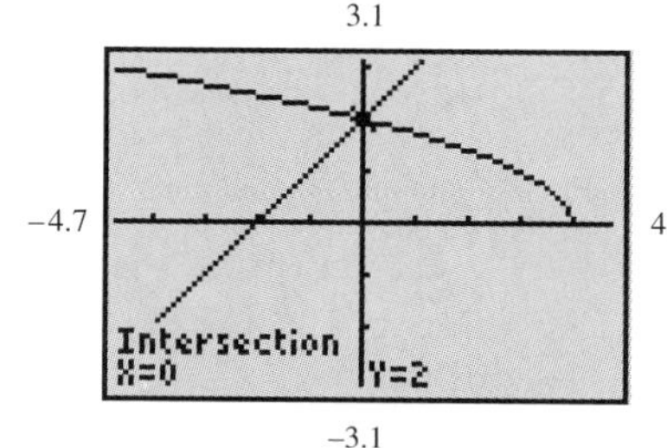

Figure 15

In Example 4, we found that the single solution of $\sqrt{4 - x} = x + 2$ is 0, with an extraneous value of -5. If we graph $f(x) = \sqrt{4 - x}$ and $g(x) = x + 2$ in the same window, we find that the x-coordinate of the point of intersection of the two graphs is 0, which is the solution of the equation. See Figure 15. We solved the equation in Example 4 by squaring both sides, obtaining the equation $4 - x = x^2 + 4x + 4$. In Figure 16, we show that the two functions defined by $f(x) = 4 - x$ and $g(x) = x^2 + 4x + 4$ have two points of intersection. The extraneous value -5 that we found in Example 4 shows up as an x-value of one of these points of intersection. However, our check showed that -5 was not a solution of the *original* equation (before the squaring step). Here we see a graphical interpretation of the extraneous value.

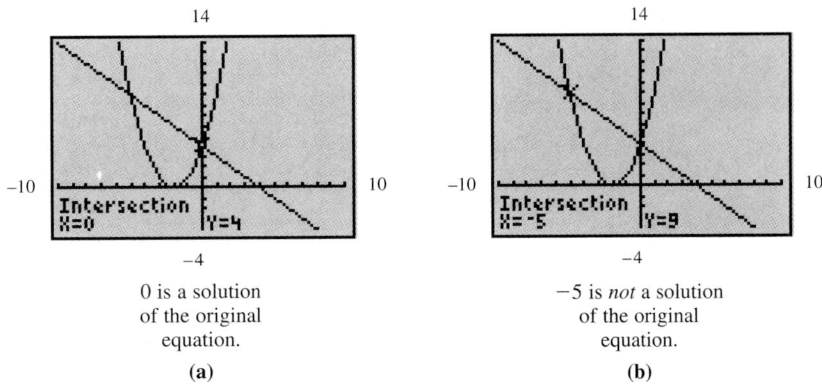

0 is a solution of the original equation.

(a)

-5 is *not* a solution of the original equation.

(b)

Figure 16

OBJECTIVE 6 Use the power rule to solve a formula for a specified variable. The final example uses a formula from electronics.

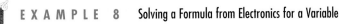

EXAMPLE 8 Solving a Formula from Electronics for a Variable

An important property of a radio frequency transmission line is its *characteristic impedance,* represented by Z and measured in ohms. If L and C are the inductance and capacitance, respectively, per unit of length of the line, then these quantities are related by the formula $Z = \sqrt{\frac{L}{C}}$. Solve this formula for C.

$$Z = \sqrt{\frac{L}{C}} \qquad \text{Given formula}$$

$$Z^2 = \frac{L}{C} \qquad \text{Use the power rule and square both sides.}$$

$$CZ^2 = L \qquad \text{Multiply by } C.$$

$$C = \frac{L}{Z^2} \qquad \text{Divide by } Z^2.$$

7.5 EXERCISES

Graph each function and give its domain and its range. See Example 1.

1. $f(x) = \sqrt{x + 3}$

2. $f(x) = \sqrt{x - 5}$

3. $f(x) = \sqrt{x} - 2$

4. $f(x) = \sqrt{x} + 4$

5. $f(x) = \sqrt[3]{x} - 3$

6. $f(x) = \sqrt[3]{x} + 1$

7. $f(x) = \sqrt[3]{x - 3}$

8. $f(x) = \sqrt[3]{x + 1}$

Check each equation to see if the given value for x is a solution.

9. $\sqrt{3x + 18} = x$
(a) 6 **(b)** -3

10. $\sqrt{3x - 3} = x - 1$
(a) 1 **(b)** 4

11. $\sqrt{x + 2} = \sqrt{9x - 2} - 2\sqrt{x - 1}$
(a) 2 **(b)** 7

12. $\sqrt{8x - 3} = 2x$
(a) $\frac{3}{2}$ **(b)** $\frac{1}{2}$

13. Is 9 a solution of the equation $\sqrt{x} = -3$? If not, what is the solution of this equation? Explain.

14. Before even attempting to solve $\sqrt{3x + 18} = x$, how can you be sure that the equation cannot have a negative solution?

Solve each equation. See Examples 2 and 3.

15. $\sqrt{x - 3} = 4$

16. $\sqrt{y + 2} = 5$

17. $\sqrt{3k - 2} - 8 = -2$

18. $\sqrt{4t + 7} - 14 = -5$

19. $\sqrt{x} + 9 = 0$

20. $\sqrt{w} + 4 = 0$

21. $\sqrt{3x - 6} - 3 = 0$

22. $\sqrt{7y + 11} - 5 = 0$

23. $\sqrt{6x + 2} - \sqrt{5x + 3} = 0$

24. $\sqrt{3 + 5x} - \sqrt{x + 11} = 0$

25. $3\sqrt{x} = \sqrt{8x + 9}$

26. $6\sqrt{p} = \sqrt{30p + 24}$

RELATING CONCEPTS (EXERCISES 27–32)

Solve the equations in Exercises 27–31 in order, *and then use a generalization to fill in the blanks in Exercise 32.*

27. $x = 3$

28. $x = -3$

29. $x^2 = 9$

30. $x^3 = 27$

31. $x^4 = 81$

32. Suppose both sides of $x = k$ are raised to the nth power.

(a) If n is even, the number of solutions of the new equation is ______________ the number of solutions of the original (more than/the same as/fewer than) equation.

(b) If n is odd, the number of solutions of the new equation is ______________ the number of solutions of the original (more than/the same as/fewer than) equation.

Did you make the connection between the number of solutions and the degree of the term in the equation?

33. Explain what is *wrong* with this first step in the solution process for $\sqrt{3x + 4} = 8 - x$. Then solve it correctly.

$$3x + 4 = 64 + x^2$$

34. Explain what is *wrong* with this first step in the solution process for the equation $\sqrt{5y + 6} - \sqrt{y + 3} = 3$. Then solve it correctly.

$$(5y + 6) + (y + 3) = 9$$

Solve each equation. See Examples 4 and 5.

35. $\sqrt{3x + 4} = 8 - x$

36. $\sqrt{5x + 1} = 2x - 2$

37. $\sqrt{13 + 4t} = t + 4$

38. $\sqrt{50 + 7k} = k + 8$

39. $\sqrt{r^2 - 15r + 15} + 5 = r$

40. $\sqrt{p^2 + 12p - 4} + 4 = p$

41. $\sqrt{3x + 7} - 3x = 5$

42. $\sqrt{4x + 13} - 2x = -1$

43. $\sqrt{4x + 2} - 4x = 0$

44. $\sqrt{4 - 2x} - 8 = 2x$

Solve each equation. See Example 6.

45. $\sqrt{r + 4} - \sqrt{r - 4} = 2$

46. $\sqrt{m + 1} - \sqrt{m - 2} = 1$

47. $\sqrt{11 + 2q} + 1 = \sqrt{5q + 1}$

48. $\sqrt{6 + 5y} - 3 = \sqrt{y + 3}$

49. $\sqrt{3 - 3p} - \sqrt{3p + 2} = 3$

50. $\sqrt{3x + 4} - \sqrt{2x - 4} = 2$

51. $\sqrt{5x - 6} = 2 + \sqrt{3x - 6}$

52. $\sqrt{x + 2} = 1 - \sqrt{x - 3}$

53. What is the smallest power to which you can raise both sides of the radical equation $\sqrt[3]{x + 3} = \sqrt[3]{5 + 4x}$ so that the radicals are eliminated?

54. What is the smallest power to which you can raise both sides of the radical equation $\sqrt{x + 3} = \sqrt[3]{10x + 14}$ so that the radicals are eliminated?

Solve each equation. See Example 7.

55. $\sqrt[3]{2x^2 + 3x - 7} = \sqrt[3]{2x^2 + 4x + 6}$

56. $\sqrt[3]{3y^2 - 4y + 6} = \sqrt[3]{3y^2 - 2y + 8}$

57. $\sqrt[3]{1 - 2k} - \sqrt[3]{-k - 13} = 0$

58. $\sqrt[3]{11 - 2t} - \sqrt[3]{-1 - 5t} = 0$

59. $\sqrt[4]{x - 1} + 2 = 0$

60. $\sqrt[4]{2k + 3} + 1 = 0$

61. $\sqrt[4]{x + 7} = \sqrt[4]{2x}$

62. $\sqrt[4]{y + 8} = \sqrt[4]{3y}$

For each equation, rewrite the expressions with rational exponents as radical expressions, and then solve using the procedures explained in this section.

63. $(2x - 9)^{1/2} = 2 + (x - 8)^{1/2}$

64. $(3w + 7)^{1/2} = 1 + (w + 2)^{1/2}$

65. $(2w - 1)^{2/3} - w^{1/3} = 0$

66. $(x^2 - 2x)^{1/3} - x^{1/3} = 0$

Solve each formula from electricity and radio for the indicated variable. See Example 8. (Source: Nelson M. Cooke, and Joseph B. Orleans, Mathematics Essential to Electricity and Radio, *McGraw Hill, 1943.)*

67. $V = \sqrt{\frac{2K}{m}}$ for K

68. $V = \sqrt{\frac{2K}{m}}$ for m

69. $f = \frac{1}{2\pi\sqrt{LC}}$ for L

70. $r = \sqrt{\frac{Mm}{F}}$ for F

Solve each problem.

71. A number of useful formulas involve radicals or radical expressions. Many occur in the mathematics needed for working with objects in space. The formula

$$N = \frac{1}{2\pi}\sqrt{\frac{a}{r}}$$

is used to find the rotational rate N of a space station. Here a is the acceleration and r represents the radius of the space station in meters. To find the value of r that will make N simulate the effect of gravity on Earth, the equation must be solved for r, using the required value of N. (*Source:* Bernice Kastner, *Space Mathematics,* NASA: 1972.)

(a) Solve the equation for r.

(b) Find the value of r that makes $N = .063$ rotation per second, if $a = 9.8$ meters per second squared.

(c) Find the value of r that makes $N = .04$ rotation per second, if $a = 9.8$ meters per second squared.

72. At an altitude of h feet above sea level or level ground, the distance d in miles that an observer can see an object is given by the function with

$$d(h) = \sqrt{\frac{3h}{2}}.$$

At what altitude must an observer be to see an object 20 miles away?

If x is the number of years since 1900, the equation $y = x^{.7}$ approximates the timber grown in the United States in billions of cubic feet. Let $x = 20$ represent 1920, $x = 52$ represent 1952, and so on.

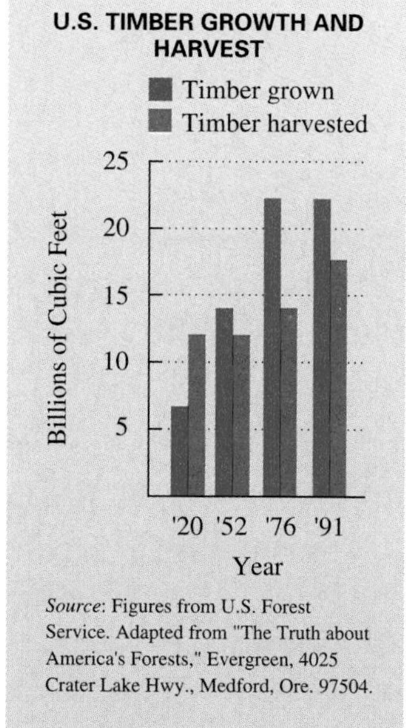

Source: Figures from U.S. Forest Service. Adapted from "The Truth about America's Forests," Evergreen, 4025 Crater Lake Hwy., Medford, Ore. 97504.

73. Replace x in the equation for each year shown in the figure and give the value of y. (Round answers to the nearest billion.)

74. Use the figure to estimate the amount of timber grown for each year shown.

75. Compare the values found from the equation with your estimates from the figure. Does the equation give a good approximation of the data from the figure? In which year is the approximation best?

76. From the figure, estimate the amount of timber harvested in each year shown.

77. Use the equation $y = x^{.62}$ to approximate the amount of timber harvested in each of the given years. (Round answers to the nearest billion.)

78. Compare your answers from Exercises 76 and 77. Does the equation give a good approximation? For which year is it poorest?

79. Some graphing calculators will not compute a value for expressions like $(-8)^{2/3}$, with a negative base raised to a power with an odd denominator and an even numerator. Check to see if your calculator is one of these. If it is, use the fact that $(-8)^{2/3} = [(-8)^{1/3}]^2$ to calculate it. What rule for exponents applies here?

80. Graph $y = x^{2/3}$ in the standard window of your calculator. Refer to the discussion in Exercise 79 if you get no graph for negative values of x.

Graphically support your answer to the indicated exercise.

81. Exercise 37

82. Exercise 38

83. Use a graphing calculator to solve $\sqrt{3 - 3x} = 3 + \sqrt{3x + 2}$. What is the domain of $y = \sqrt{3 - 3x} - 3 - \sqrt{3x + 2}$?

84. Use a graphing calculator with a window of $[-1, 4]$ by $[-1, 3]$ to solve $\sqrt{2\sqrt{7x + 2}} = \sqrt{3x + 2}$. What is the domain of $f(x) = \sqrt{2\sqrt{7x + 2}} - \sqrt{3x + 2}$?

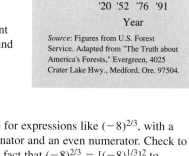

7.6 Complex Numbers

OBJECTIVES

1. Simplify numbers of the form $\sqrt{-b}$, where $b > 0$.
2. Recognize imaginary complex numbers.
3. Add and subtract complex numbers.
4. Multiply complex numbers.
5. Divide complex numbers.
6. Find powers of i.

FOR EXTRA HELP

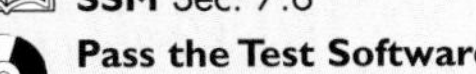

SSG Sec. 7.6
SSM Sec. 7.6

Pass the Test Software

InterAct Math Tutorial Software

Video 14

As discussed in Chapter 1, the set of real numbers includes many other number sets (the rational numbers, integers, and natural numbers, for example). In this section a new set of numbers is introduced that includes the set of real numbers, as well as numbers that are even roots of negative numbers, like $\sqrt{-2}$.

OBJECTIVE 1 Simplify numbers of the form $\sqrt{-b}$, where $b > 0$. The e quation $x^2 + 1 = 0$ has no real number solutions, since any solution must be a number whose square is -1. In the set of real numbers all squares are nonnegative numbers, because the product of either two positive numbers or two negative numbers is positive and $0^2 = 0$. To provide a solution for the equation $x^2 + 1 = 0$, a new number i is defined so that

$$i^2 = -1.$$

That is, i is a number whose square is -1, so $i = \sqrt{-1}$. This definition of i makes it possible to define any square root of a negative number as follows.

For any positive real number b, $\sqrt{-b} = i\sqrt{b}$.

EXAMPLE 1 Simplifying Square Roots of Negative Numbers

Write each number as a product of a real number and i.

(a) $\sqrt{-100} = i\sqrt{100} = 10i$ **(b)** $\sqrt{-2} = \sqrt{2}i = i\sqrt{2}$

CAUTION It is easy to mistake $\sqrt{2}i$ for $\sqrt{2i}$, with the i under the radical. For this reason, we often write $\sqrt{2}i$ as $i\sqrt{2}$.

When finding a product such as $\sqrt{-4} \cdot \sqrt{-9}$, we cannot use the product rule for radicals since that rule applies only when both radicals represent real numbers. For this reason, we *always* change $\sqrt{-b}$ $(b > 0)$ to the form $i\sqrt{b}$ before performing any multiplications or divisions. For example,

$$\sqrt{-4} \cdot \sqrt{-9} = i\sqrt{4} \cdot i\sqrt{9} = i \cdot 2 \cdot i \cdot 3 = 6i^2.$$

Since $i^2 = -1$,

$$6i^2 = 6(-1) = -6.$$

Using the product rule for radicals incorrectly gives a wrong answer.

$$\sqrt{-4} \cdot \sqrt{-9} = \sqrt{(-4)(-9)} = \sqrt{36} = 6 \qquad \text{INCORRECT}$$

EXAMPLE 2 Multiplying Square Roots of Negative Numbers

Multiply.

(a) $\sqrt{-3} \cdot \sqrt{-7} = i\sqrt{3} \cdot i\sqrt{7} = i^2\sqrt{3 \cdot 7} = (-1)\sqrt{21} = -\sqrt{21}$

(b) $\sqrt{-2} \cdot \sqrt{-8} = i\sqrt{2} \cdot i\sqrt{8} = i^2\sqrt{2 \cdot 8} = (-1)\sqrt{16} = (-1)4 = -4$

(c) $\sqrt{-5} \cdot \sqrt{6} = i\sqrt{5} \cdot \sqrt{6} = i\sqrt{30}$

The methods used to find products also apply to quotients.

EXAMPLE 3 Dividing Square Roots of Negative Numbers

Divide.

(a) $\dfrac{\sqrt{-75}}{\sqrt{-3}} = \dfrac{i\sqrt{75}}{i\sqrt{3}} = \sqrt{\dfrac{75}{3}} = \sqrt{25} = 5$

(b) $\dfrac{\sqrt{-32}}{\sqrt{8}} = \dfrac{i\sqrt{32}}{\sqrt{8}} = i\sqrt{\dfrac{32}{8}} = i\sqrt{4} = 2i$

OBJECTIVE 2 Recognize imaginary complex numbers. With the new number i and the real numbers, a new set of numbers can be formed that includes the real numbers as a subset. The *complex numbers* are defined as follows.

Complex Number

If a and b are real numbers, then any number of the form $a + bi$ is called a **complex number.**

In the complex number $a + bi$, the number a is called the **real part** and b is called the **imaginary part.** When $b = 0$, $a + bi$ is a real number, so the real numbers are a subset of the complex numbers. Complex numbers with $b \neq 0$ are called **imaginary numbers.*** In spite of their name, imaginary numbers are very useful in applications, particularly in work with electricity.

The relationships among the various sets of numbers discussed in this book are shown in Figure 17.

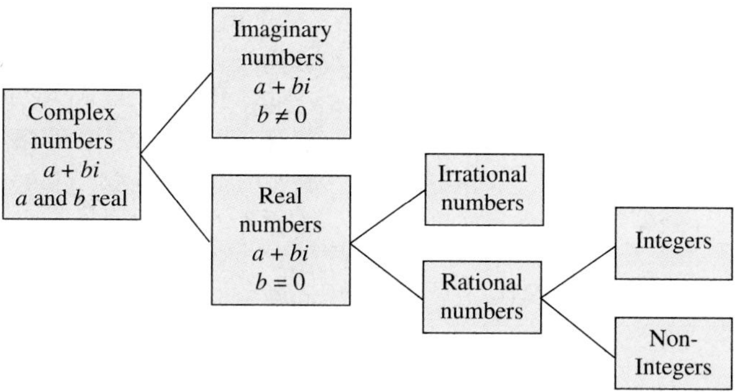

Figure 17

OBJECTIVE 3 Add and subtract complex numbers. The commutative, associative, and distributive properties for real numbers are also valid for complex numbers. To add complex numbers, add their real parts and add their imaginary parts.

*Some texts define bi as the imaginary part of the complex number $a + bi$. Also, imaginary numbers are sometimes defined as complex numbers with $a = 0$ and $b \neq 0$.

EXAMPLE 4 Adding Complex Numbers

Add.

(a) $(2 + 3i) + (6 + 4i) = (2 + 6) + (3 + 4)i$ Commutative, associative, and distributive properties

$= 8 + 7i$

(b) $5 + (9 - 3i) = (5 + 9) - 3i$

$= 14 - 3i$

To subtract complex numbers, subtract their real parts and subtract their imaginary parts.

EXAMPLE 5 Subtracting Complex Numbers

Subtract.

(a) $(6 + 5i) - (3 + 2i) = (6 - 3) + (5 - 2)i = 3 + 3i$

(b) $(7 - 3i) - (8 - 6i) = (7 - 8) + [-3 - (-6)]i = -1 + 3i$

(c) $(-9 + 4i) - (-9 + 8i) = [-9 - (-9)] + (4 - 8)i$

$= 0 - 4i$ or $-4i$

In Example 5(c), the answer was written as $0 - 4i$ and then as just $-4i$. A complex number written in the form $a + bi$, like $0 - 4i$, is in **standard form.** In this section, most answers will be given in standard form, but if a or b is 0, we consider answers such as a or bi to be in standard form.

OBJECTIVE 4 Multiply complex numbers. Complex numbers of the form $a + bi$ have the same form as a binomial, so we multiply two complex numbers by using the FOIL method for multiplying binomials. (Recall that FOIL stands for *First-Outside-Inside-Last.*)

EXAMPLE 6 Multiplying Complex Numbers

(a) Multiply $3 + 5i$ and $4 - 2i$.
Use the FOIL method.

$$(3 + 5i)(4 - 2i) = \underset{\text{First}}{3(4)} + \underset{\text{Outside}}{3(-2i)} + \underset{\text{Inside}}{5i(4)} + \underset{\text{Last}}{5i(-2i)}$$

Now simplify. (Remember that $i^2 = -1$.)

$$\begin{aligned}(3 + 5i)(4 - 2i) &= 12 - 6i + 20i - 10i^2 \\ &= 12 + 14i - 10(-1) \qquad \text{Let } i^2 = -1. \\ &= 12 + 14i + 10 \\ &= 22 + 14i\end{aligned}$$

(b) $(2 + 3i)(1 - 5i) = 2(1) + 2(-5i) + 3i(1) + 3i(-5i)$

$$= 2 - 10i + 3i - 15i^2$$
$$= 2 - 7i - 15(-1)$$
$$= 2 - 7i + 15$$
$$= 17 - 7i$$

(c) $4i(2 + 3i)$

Use the distributive property.

$$4i(2 + 3i) = 4i(2) + 4i(3i) = 8i + 12i^2 = 8i + 12(-1) = -12 + 8i$$

The two complex numbers $a + bi$ and $a - bi$ are called *conjugates* of each other. The product of a complex number and its conjugate is always a real number, as shown here.

$$(a + bi)(a - bi) = a \cdot a - abi + abi - b^2i^2$$
$$= a^2 - b^2(-1)$$
$$= a^2 + b^2$$

$$(a + bi)(a - bi) = a^2 + b^2$$

For example, $(3 + 7i)(3 - 7i) = 3^2 + 7^2 = 9 + 49 = 58$.

OBJECTIVE 5 **Divide complex numbers.** The quotient of two complex numbers should be a complex number. To write the quotient as a complex number, we need to eliminate i in the denominator. We use conjugates to do this.

EXAMPLE 7 Dividing Complex Numbers

Find the quotients.

(a) $\dfrac{8 + 9i}{5 + 2i}$

Multiply the numerator and the denominator by the conjugate of the denominator. The conjugate of $5 + 2i$ is $5 - 2i$.

$$\frac{8 + 9i}{5 + 2i} = \frac{(8 + 9i)(5 - 2i)}{(5 + 2i)(5 - 2i)}$$
$$= \frac{40 - 16i + 45i - 18i^2}{5^2 + 2^2}$$
$$= \frac{58 + 29i}{29}$$
$$= \frac{29(2 + i)}{29}$$
$$= 2 + i$$

Notice that this is just like rationalizing the denominator. The final result is given in standard form.

(b) $\dfrac{1+i}{i}$

The conjugate of i is $-i$. Multiply the numerator and the denominator by $-i$.

$$\begin{aligned}\frac{1+i}{i} &= \frac{(1+i)(-i)}{i(-i)}\\ &= \frac{-i-i^2}{-i^2}\\ &= \frac{-i-(-1)}{-(-1)}\\ &= \frac{-i+1}{1}\\ &= 1-i\end{aligned}$$

In Examples 4–7, we showed how complex numbers can be added, subtracted, multiplied, and divided using algebraic methods. Many current models of graphing calculators can perform these operations. Figure 18 shows how the computations in parts (a) of Examples 4–7 are carried out by a TI-83 calculator.

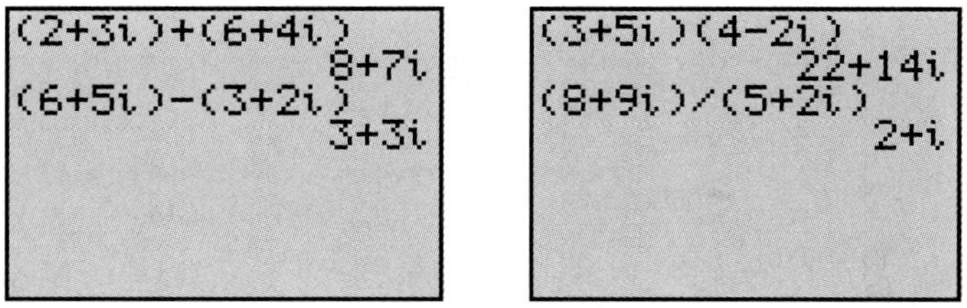

```
(2+3i)+(6+4i)
              8+7i
(6+5i)-(3+2i)
              3+3i
```

```
(3+5i)(4-2i)
            22+14i
(8+9i)/(5+2i)
               2+i
```

Figure 18

OBJECTIVE 6 Find powers of i. Because i^2 is defined to be -1, we can find higher powers of i, as shown in the following examples.

$i^3 = i \cdot i^2 = i(-1) = -i$ $\qquad$ $i^6 = i^2 \cdot i^4 = (-1) \cdot 1 = -1$

$i^4 = i^2 \cdot i^2 = (-1)(-1) = 1$ $\qquad$ $i^7 = i^3 \cdot i^4 = (-i) \cdot 1 = -i$

$i^5 = i \cdot i^4 = i \cdot 1 = i$ $\qquad$ $i^8 = i^4 \cdot i^4 = 1 \cdot 1 = 1$

A few powers of i are listed here.

Powers of i

$i^1 = i$	$i^5 = i$	$i^9 = i$	$i^{13} = i$
$i^2 = -1$	$i^6 = -1$	$i^{10} = -1$	$i^{14} = -1$
$i^3 = -i$	$i^7 = -i$	$i^{11} = -i$	$i^{15} = -i$
$i^4 = 1$	$i^8 = 1$	$i^{12} = 1$	$i^{16} = 1$

As these examples suggest, the powers of i rotate through the four numbers i, -1, $-i$, and 1. Larger powers of i can be simplified by using the fact that $i^4 = 1$. For example,

$$i^{75} = (i^4)^{18} \cdot i^3 = 1^{18} \cdot i^3 = 1 \cdot i^3 = i^3 = -i.$$

This example suggests a quick method for simplifying large powers of i.

EXAMPLE 8 Simplifying Powers of *i*

Find each power of i.

(a) $i^{12} = (i^4)^3 = 1^3 = 1$

(b) $i^{39} = i^{36} > i^3 = (i^4)^9 > i^3 = 1^9 > (-i) = -i$

(c) $i^{-2} = \dfrac{1}{i^2} = \dfrac{1}{-1} = -1$

(d) $i^{-1} = \dfrac{1}{i}$

To simplify this quotient, multiply the numerator and the denominator by the conjugate of i, which is $-i$.

$$i^{-1} = \frac{1}{i} = \frac{1(-i)}{i(-i)} = \frac{-i}{-i^2} = \frac{-i}{-(-1)} = \frac{-i}{1} = -i$$

7.6 EXERCISES

Decide whether each expression is equal to 1, -1, i, *or* $-i$.

1. $\sqrt{-1}$ **2.** $-\sqrt{-1}$ **3.** i^2 **4.** $-i^2$ **5.** $\dfrac{1}{i}$ **6.** $(-i)^2$

Write each number as a product of a real number and i. Simplify all radical expressions. See Example 1.

7. $\sqrt{-169}$ **8.** $\sqrt{-225}$ **9.** $-\sqrt{-144}$ **10.** $-\sqrt{-196}$

11. $\sqrt{-5}$ **12.** $\sqrt{-21}$ **13.** $\sqrt{-48}$ **14.** $\sqrt{-96}$

Multiply or divide as indicated. See Examples 2 and 3.

15. $\sqrt{-15} \cdot \sqrt{-15}$ **16.** $\sqrt{-19} \cdot \sqrt{-19}$ **17.** $\sqrt{-4} \cdot \sqrt{-25}$ **18.** $\sqrt{-9} \cdot \sqrt{-81}$

19. $\dfrac{\sqrt{-300}}{\sqrt{-100}}$ **20.** $\dfrac{\sqrt{-40}}{\sqrt{-10}}$ **21.** $\dfrac{\sqrt{-75}}{\sqrt{3}}$ **22.** $\dfrac{\sqrt{-160}}{\sqrt{10}}$

23. **(a)** Every real number is a complex number. Explain why this is so.
(b) Not every complex number is a real number. Give an example of this, and explain why this statement is true.

24. Explain how to perform addition, subtraction, multiplication, and division with complex numbers. Give examples.

Add or subtract as indicated. Write your answers in the form $a + bi$. *See Examples 4 and 5.*

25. $(6 + 2i) + (4 + 6i)$ **26.** $(1 + i) + (2 - 3i)$

27. $(3 + 2i) + (-4 + 5i)$ **28.** $(7 + 15i) + (-11 + 14i)$

29. $(5 - i) + (-5 + i)$

30. $(-2 + 6i) + (2 - 6i)$

31. $(4 + i) - (-3 - 2i)$

32. $(9 + i) - (3 + 2i)$

33. $(-3 - 4i) - (-1 - 4i)$

34. $(-2 - 3i) - (-5 - 3i)$

35. $(-4 + 11i) + (-2 - 4i) + (7 + 6i)$

36. $(-1 + i) + (2 + 5i) + (3 + 2i)$

37. $[(7 + 3i) - (4 - 2i)] + (3 + i)$

38. $[(7 + 2i) + (-4 - i)] - (2 + 5i)$

39. Fill in the blank with the correct response:

Because $(4 + 2i) - (3 + i) = 1 + i$, using the definition of subtraction, we can check this to find that $(1 + i) + (3 + i) =$ _______________.

40. Fill in the blank with the correct response:

Because $\frac{-5}{2 - i} = -2 - i$, using the definition of division, we can check this to find that $(-2 - i)(2 - i) =$ _______________.

Multiply. See Example 6.

41. $(3i)(27i)$

42. $(5i)(125i)$

43. $(-8i)(-2i)$

44. $(-32i)(-2i)$

45. $5i(-6 + 2i)$

46. $3i(4 + 9i)$

47. $(4 + 3i)(1 - 2i)$

48. $(7 - 2i)(3 + i)$

49. $(4 + 5i)^2$

50. $(3 + 2i)^2$

51. $2i(-4 - i)^2$

52. $3i(-3 - i)^2$

53. $(12 + 3i)(12 - 3i)$

54. $(6 + 7i)(6 - 7i)$

55. $(4 + 9i)(4 - 9i)$

56. $(7 + 2i)(7 - 2i)$

Write each expression in the form a + bi. See Example 7.

57. $\frac{2}{1 - i}$

58. $\frac{29}{5 + 2i}$

59. $\frac{-7 + 4i}{3 + 2i}$

60. $\frac{-38 - 8i}{7 + 3i}$

61. $\frac{8i}{2 + 2i}$

62. $\frac{-8i}{1 + i}$

63. $\frac{2 - 3i}{2 + 3i}$

64. $\frac{-1 + 5i}{3 + 2i}$

TECHNOLOGY INSIGHTS (EXERCISES 65–68)

Predict the answer that the calculator screen will provide for the given complex number operation entry.

65.
```
(8-5i)+2(11+6i)-
(1+i)²
```

66.
```
(4+3i)(2-6i)-2i(
5+7i)
```

67.
```
(26+32i)/(2+4i)-
(2i)(1+i)⁻¹
```

68.
```
(1+2i)³-(1+2i)²
```

RELATING CONCEPTS (EXERCISES 69–74)

Consider the following expressions:

Binomials	*Complex Numbers*
$x + 2, \quad 3x - 1$	$1 + 2i, \quad 3 - i$

When we add, subtract, or multiply complex numbers in standard form, the rules are the same as those for the corresponding operations on binomials. That is, we add or subtract like terms, and we use FOIL to multiply. Division, however, is comparable to division by the sum or difference of radicals, where we multiply by the conjugate to get a rational denominator. To express the quotient of two complex numbers in standard form, we also multiply by the conjugate of the denominator.

The following exercises illustrate these ideas. ***Work them in order.***

69. **(a)** Add the two binomials. **(b)** Add the two complex numbers.

70. **(a)** Subtract the second binomial from the first. **(b)** Subtract the second complex number from the first.

71. **(a)** Multiply the two binomials. **(b)** Multiply the two complex numbers.

72. **(a)** Rationalize the denominator: $\dfrac{\sqrt{3} - 1}{1 + \sqrt{2}}$.

(b) Write in standard form: $\dfrac{3 - i}{1 + 2i}$.

73. Explain why the answers for (a) and (b) in Exercise 71 do not correspond as the answers in Exercises 69–70 do.

74. Explain why the answers for (a) and (b) in Exercise 72 do not correspond as the answers in Exercises 69–70 do.

Did you make the connection that operations with binomials and operations with complex numbers have both similarities and differences?

75. Recall that if $a \neq 0$, $\dfrac{1}{a}$ is called the reciprocal of a. Use this definition to express the reciprocal of $5 - 4i$ in the form $a + bi$.

76. Recall that if $a \neq 0$, a^{-1} is defined to be $\dfrac{1}{a}$. Use this definition to express $(4 - 3i)^{-1}$ in the form $a + bi$.

Find each power of i. See Example 8.

77. i^{18} **78.** i^{26} **79.** i^{89} **80.** i^{48} **81.** i^{-5} **82.** i^{-17}

83. A student simplified i^{-18} as follows:

$$i^{-18} = i^{-18} \cdot i^{20} = i^{-18+20} = i^2 = -1.$$

Explain the mathematical justification for this correct work.

84. Explain why $(46 + 25i)(3 - 6i)$ and $(46 + 25i)(3 - 6i)(i^{12})$ must be equal. (Do not actually perform the computations.)

Ohm's law for the current I in a circuit with voltage E, resistance R, capacitance reactance X_c, and inductive reactance X_L is

$$I = \frac{E}{R + (X_L - X_c)i}.$$

Use this formula to find the unknown quantity in Exercises 85 and 86.

85. Find I if $E = 2 + 3i$, $R = 5$, $X_L = 4$, and $X_c = 3$.

86. Find E if $I = 1 - i$, $R = 2$, $X_L = 3$, and $X_c = 1$.

Complex numbers will appear again in this book in Chapter 8, when we study quadratic equations. The following exercises examine how a complex number can be a solution of a quadratic equation.

87. Show that $1 + 5i$ is a solution of $x^2 - 2x + 26 = 0$. Then show that its conjugate is also a solution.

88. Show that $3 + 2i$ is a solution of $x^2 - 6x + 13 = 0$. Then show that its conjugate is also a solution.

CHAPTER 7 GROUP ACTIVITY

Solar Electricity

Objective: Apply the Pythagorean formula.

In this activity you will determine the sizes of frames needed to support solar electric panels on a flat roof.

A. The following table gives three different solar modules by Solarex. Have each member of the group choose one of the solar panels.

Model	*Watts*	*Volts*	*Amps*	*Size in Inches*	*Cost in Dollars*
MSX-77	77	16.9	4.56	44 × 26	475
MSX-83	83	17.1	4.85	44 × 24	490
MSX-60	60	17.1	3.5	44 × 20	382

Source: Solarex table in *Jade Mountain* 1998 catalog.

B. In order to use your solar panel, you will need to make a wooden frame to support it. The sides of this frame will form a right triangle. The hypotenuse of the triangle will be the width of the solar panel you chose. Make a sketch and use the Pythagorean formula to find the dimensions of the legs for each frame given the following conditions. Round answers to the nearest tenth.

1. The legs have equal length.
2. One leg is twice the length of the other.
3. One leg is 3 times the length of the other.

C. Compare the different frame sizes for each panel. What factors might determine which of the above triangles you would use in your frame?

CHAPTER 7 SUMMARY

KEY TERMS

7.1 **principal *n*th root**
radicand
index (order)
radical
7.2 **simplified radical**
hypotenuse
legs (of a right triangle)
7.3 **radical expression**
7.4 **rationalizing the denominator**
conjugate
7.5 **square root function**
cube root function
extraneous solution
7.6 **complex number**
real part
imaginary part
imaginary numbers
standard form (of a complex number)

NEW SYMBOLS

$a^{1/n}$	a to the power $\frac{1}{n}$	$\pm$	positive or negative	$\approx$	is approximately equal to
$\sqrt[n]{a}$	principal nth root of a	$a^{m/n}$	a to the power $\frac{m}{n}$	i	a number whose square is -1

TEST YOUR WORD POWER

See how well you have learned the vocabulary in this chapter. Answers, with examples, are given at the bottom of the page.

1. A **radicand** is
(a) the index of a radical
(b) the number or expression under the radical sign
(c) the positive root of a number
(d) the radical sign.

2. The **Pythagorean formula** states that, in a right triangle,
(a) the sum of the measures of the angles is 180°
(b) the sum of the lengths of the two shorter sides equals the length of the longest side
(c) the longest side is opposite the right angle
(d) the square of the length of the longest side equals the sum of the s ıares of the lengths of the two shorter sides.

3. A **hypotenuse** is
(a) either of the two shorter sides of a triangle
(b) the shortest side of a triangle
(c) the side opposite the right angle in a triangle
(d) the longest side in any triangle.

4. **Rationalizing the denominator** is the process of
(a) eliminating fractions from a radical expression
(b) changing the denominator of a fraction from a radical to a rational number
(c) clearing a radical expression of radicals
(d) multiplying radical expressions.

5. An **extraneous solution** is a solution
(a) that does not satisfy the original equation
(b) that makes an equation true
(c) that makes an expression equal zero
(d) that checks in the original equation.

6. A **complex number** is
(a) a real number that includes a complex fraction
(b) a zero multiple of i
(c) a number of the form $a + bi$, where a and b are real numbers
(d) the square root of -1.

Answers to Test Your Word Power

1. (b) *Example:* In $\sqrt{3xy}$, $3xy$ is the radicand. **2.** (d) *Example:* In a right triangle where $a = 6$, $b = 8$, and $c = 10$, $6^2 + 8^2 = 10^2$. **3.** (c) *Example:* In a right triangle where the sides measure 9, 12, and 15 units, the hypotenuse is the side with measure 15 units. **4.** (b) *Example:* To rationalize the denominator of $\frac{5}{\sqrt{3}+1}$, multiply numerator and denominator by $\sqrt{3} - 1$ to get $\frac{5(\sqrt{3}-1)}{2}$. **5.** (a) *Example:* The potential solution 2 is extraneous in $\sqrt{5q - 1} + 3 = 0$. (See Example 3 of Section 7.5.) **6.** (c) *Examples:* -5 (or $-5 + 0i$), $7i$ (or $0 + 7i$), and $\sqrt{2} - 4i$.

QUICK REVIEW

CONCEPTS	EXAMPLES

7.1 RATIONAL EXPONENTS AND RADICALS

$a^{1/n} = b$ means $b^n = a$.
$a^{1/n}$ is the principal nth root of a.
$a^{1/n} = \sqrt[n]{a}$ whenever $\sqrt[n]{a}$ is a real number.
$\sqrt[n]{a^n} = |a|$ if n is even. $\sqrt[n]{a^n} = a$ if n is odd.

The two square roots of 64 are $\sqrt{64} = 8$ and $-\sqrt{64} = -8$. Of these, 8 is the principal square root of 64.

$$25^{1/2} = \sqrt{25} = 5$$
$$(-64)^{1/3} = \sqrt[3]{-64} = -4$$
$$\sqrt[3]{-27} = \sqrt[3]{(-3)^3} = -3 \qquad \sqrt[4]{(-2)^4} = |-2| = 2$$

If m and n are positive integers with m/n in lowest terms, then

$$a^{m/n} = (a^{1/n})^m$$

provided that $a^{1/n}$ is a real number.

$$8^{5/3} = (8^{1/3})^5 = 2^5 = 32$$

All the usual rules for exponents are valid for rational exponents.

$$5^{-1/2} \cdot 5^{1/4} = 5^{-1/2+1/4} = 5^{-1/4} = \frac{1}{5^{1/4}}$$
$$(y^{2/5})^{10} = y^4$$
$$\frac{x^{-1/3}}{x^{-1/2}} = x^{-1/3+1/2} = x^{1/6} \quad (x > 0)$$

7.2 SIMPLIFYING RADICAL EXPRESSIONS

Product and Quotient Rules for Radicals
If $\sqrt[n]{a}$ and $\sqrt[n]{b}$ are real numbers, and if n is a natural number,

$$\sqrt[n]{a} \cdot \sqrt[n]{b} = \sqrt[n]{ab}$$

and

$$\sqrt[n]{\frac{a}{b}} = \frac{\sqrt[n]{a}}{\sqrt[n]{b}} \quad (b \neq 0).$$

Apply the product and quotient rules.

$$\sqrt{3} \cdot \sqrt{7} = \sqrt{21}$$
$$\sqrt[5]{x^3y} \cdot \sqrt[5]{xy^2} = \sqrt[5]{x^4y^3}$$
$$\frac{\sqrt{x^5}}{\sqrt{x^4}} = \sqrt{\frac{x^5}{x^4}} = \sqrt{x} \quad (x > 0)$$

Simplified Radical
1. The radicand has no factor raised to a power greater than or equal to the index.
2. Exponents in the radicand and the index of the radical have no common factor (except 1). That is,
$$\sqrt[kn]{a^{km}} = \sqrt[n]{a^m}$$
if all roots exist.
3. The radicand has no fractions.
4. No denominator contains a radical.

Simplify each radical.

$$\sqrt{18} = \sqrt{9 \cdot 2} = 3\sqrt{2}$$
$$\sqrt[3]{54x^5y^3} = \sqrt[3]{27x^3y^3 \cdot 2x^2} = 3xy\sqrt[3]{2x^2}$$
$$\sqrt[9]{x^3} = x^{3/9} = x^{1/3} \quad \text{or} \quad \sqrt[3]{x}$$
$$\sqrt[8]{64} = \sqrt[8]{2^6} = \sqrt[4]{2^3} = \sqrt[4]{8}$$
$$\sqrt{\frac{7}{4}} = \frac{\sqrt{7}}{\sqrt{4}} = \frac{\sqrt{7}}{2}$$

Pythagorean Formula
In a right triangle with legs a and b and hypotenuse c,

$$a^2 + b^2 = c^2.$$

Find a.

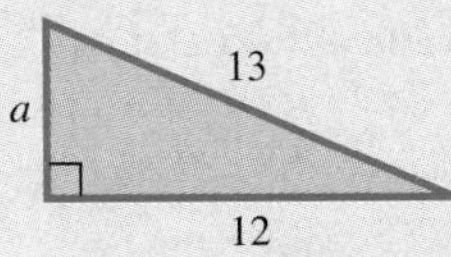

$$a^2 + 12^2 = 13^2$$
$$a^2 + 144 = 169$$
$$a^2 = 25$$
$$a = 5 \quad (a > 0)$$

Distance Formula
The distance d between the points (x_1, y_1) and (x_2, y_2) is given by

$$d = \sqrt{(x_2 - x_1)^2 + (y_2 - y_1)^2}.$$

Find the distance between the points $(2, -4)$ and $(5, 0)$.

$$d = \sqrt{(5-2)^2 + (0-(-4))^2}$$
$$= \sqrt{3^2 + 4^2}$$
$$= \sqrt{25}$$
$$= 5$$

(continued)

CONCEPTS	EXAMPLES

7.3 ADDITION AND SUBTRACTION OF RADICAL EXPRESSIONS

Only radical expressions with the same index and the same radicand may be combined.

Perform the indicated operations.

$$3\sqrt{17} + 2\sqrt{17} - 8\sqrt{17} = (3 + 2 - 8)\sqrt{17} = -3\sqrt{17}$$

$$\sqrt[3]{2} - \sqrt[3]{250} = \sqrt[3]{2} - 5\sqrt[3]{2} = -4\sqrt[3]{2}$$

7.4 MULTIPLICATION AND DIVISION OF RADICAL EXPRESSIONS

Radical expressions are multiplied by using the distributive property or the FOIL method. Special products from Chapter 5 may apply.

Multiply.

$$(\sqrt{2} + \sqrt{7})(\sqrt{3} - \sqrt{6}) = \sqrt{6} - \sqrt{12} + \sqrt{21} - \sqrt{42} = \sqrt{6} - 2\sqrt{3} + \sqrt{21} - \sqrt{42}$$

$$(\sqrt{5} - \sqrt{10})(\sqrt{5} + \sqrt{10}) = 5 - 10 = -5$$

$$(\sqrt{3} - \sqrt{2})^2 = 3 - 2\sqrt{3} \cdot \sqrt{2} + 2 = 5 - 2\sqrt{6}$$

Rationalize the denominator by multiplying both the numerator and denominator by the same expression.

Rationalize the denominator.

$$\frac{\sqrt{7}}{\sqrt{5}} = \frac{\sqrt{7}}{\sqrt{5}} \cdot \frac{\sqrt{5}}{\sqrt{5}} = \frac{\sqrt{35}}{5}$$

$$\frac{\sqrt[3]{2}}{\sqrt[3]{4}} = \frac{\sqrt[3]{2}}{\sqrt[3]{4}} \cdot \frac{\sqrt[3]{2}}{\sqrt[3]{2}} = \frac{\sqrt[3]{4}}{\sqrt[3]{8}} = \frac{\sqrt[3]{4}}{2}$$

$$\frac{4}{\sqrt{5} - \sqrt{2}} = \frac{4}{\sqrt{5} - \sqrt{2}} \cdot \frac{\sqrt{5} + \sqrt{2}}{\sqrt{5} + \sqrt{2}} = \frac{4(\sqrt{5} + \sqrt{2})}{5 - 2} = \frac{4(\sqrt{5} + \sqrt{2})}{3}$$

To write a quotient involving radicals, such as

$$\frac{5 + 15\sqrt{6}}{10}$$

in lowest terms, factor the numerator and denominator, and then divide both by the greatest common factor.

$$\frac{5 + 15\sqrt{6}}{10} = \frac{5(1 + 3\sqrt{6})}{5 \cdot 2} = \frac{1 + 3\sqrt{6}}{2}$$

7.5 GRAPHS AND EQUATIONS WITH RADICAL EXPRESSIONS

Functions Defined by Radical Expressions

The square root function with $f(x) = \sqrt{x}$ and the cube root function with $f(x) = \sqrt[3]{x}$ are two important functions defined by radical expressions.

The Square Root Function

The Cube Root Function

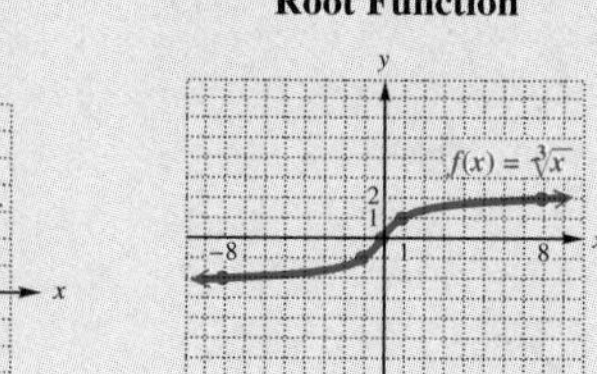

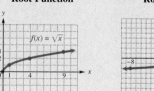

CONCEPTS	EXAMPLES
Solving Equations with Radicals *Step 1* Isolate one radical on one side of the equals sign. *Step 2* Raise each side of the equation to a power that is the same as the index of the radical. *Step 3* Solve the resulting equation; if it still contains a radical, repeat Steps 1 and 2. *Step 4* Check all potential solutions in the *original* equation. Potential solutions that do not check are *extraneous;* they are not part of the solution set.	Solve $\sqrt{2x+3} - x = 0$. $\sqrt{2x+3} = x$ $2x + 3 = x^2$ $x^2 - 2x - 3 = 0$ $(x-3)(x+1) = 0$ $x - 3 = 0$ or $x + 1 = 0$ $x = 3$ or $x = -1$ A check shows that 3 is a solution, but -1 is extraneous. The solution set is $\{3\}$.

7.6 COMPLEX NUMBERS

$i^2 = -1$ For any positive number b, $\sqrt{-b} = i\sqrt{b}$. To multiply $\sqrt{-3} \cdot \sqrt{-27}$, first change each factor to the form $i\sqrt{b}$, then multiply. The same procedure applies to quotients such as $\dfrac{\sqrt{-18}}{\sqrt{-2}}$.	Simplify. $\sqrt{-3} \cdot \sqrt{-27} = i\sqrt{3} \cdot i\sqrt{27}$ $= i^2\sqrt{81}$ $= -1 \cdot 9 = -9$ $\dfrac{\sqrt{-18}}{\sqrt{-2}} = \dfrac{i\sqrt{18}}{i\sqrt{2}} = \dfrac{\sqrt{18}}{\sqrt{2}} = \sqrt{9} = 3$
Adding and Subtracting Complex Numbers $(a+bi) + (c+di) = (a+c) + (b+d)i$ $(a+bi) - (c+di) = (a-c) + (b-d)i$	Perform the indicated operations. $(5+3i) + (8-7i) = 13 - 4i$ $(5+3i) - (8-7i) = (5-8) + (3+7)i$ $= -3 + 10i$
Multiplying and Dividing Complex Numbers Multiply complex numbers by using the FOIL method.	$(2+i)(5-3i) = 10 - 6i + 5i - 3i^2$ $= 10 - i - 3(-1)$ $= 10 - i + 3$ $= 13 - i$
Divide complex numbers by multiplying the numerator and the denominator by the conjugate of the denominator.	$\dfrac{20}{3+i} = \dfrac{20}{3+i} \cdot \dfrac{3-i}{3-i}$ $= \dfrac{20(3-i)}{9-i^2}$ $= \dfrac{20(3-i)}{10}$ $= 2(3-i)$ $= 6 - 2i$

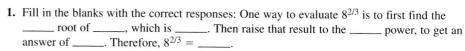

CHAPTER 7 REVIEW EXERCISES

[7.1]

1. Fill in the blanks with the correct responses: One way to evaluate $8^{2/3}$ is to first find the ______ root of ______, which is ______. Then raise that result to the ______ power, to get an answer of ______. Therefore, $8^{2/3} =$ ______.

2. Which one of the following is a positive number?
(a) $(-27)^{2/3}$ **(b)** $(-64)^{5/3}$ **(c)** $(-100)^{1/2}$ **(d)** $(-32)^{1/5}$

3. If a is a negative number and n is odd, what must be true about m for $a^{m/n}$ to be
(a) positive **(b)** negative?

4. If a is negative and n is even, what can be said about $a^{1/n}$?

Simplify.

5. $49^{1/2}$

6. $-121^{1/2}$

7. $16^{5/4}$

8. $-8^{2/3}$

9. $-\left(\frac{36}{25}\right)^{3/2}$

10. $\left(-\frac{1}{8}\right)^{-5/3}$

11. $\left(\frac{81}{10{,}000}\right)^{-3/4}$

12. Solve the Pythagorean formula $a^2 + b^2 = c^2$ for b, where $b > 0$.

13. Explain the relationship between the expressions $a^{m/n}$ and $\sqrt[n]{a^m}$. Give an example.

Find each root. Use a calculator as necessary.

14. $\sqrt{1764}$

15. $-\sqrt{289}$

16. $\sqrt[3]{216}$

17. $\sqrt[3]{-125}$

18. $-\sqrt[3]{27z^{12}}$

19. $\sqrt[5]{-32}$

20. Under what conditions is $\sqrt[n]{a}$ not a real number?

21. Simplify each radical so that no radicals appear. Assume that x represents any real number.
(a) $\sqrt{x^2}$ **(b)** $-\sqrt{x^2}$ **(c)** $\sqrt[3]{x^3}$

Write each expression as a radical.

22. $(m + 3n)^{1/2}$

23. $(3a + b)^{-5/3}$

Write each expression with a rational exponent.

24. $\sqrt{7^9}$

25. $\sqrt[5]{p^4}$

Use the rules for exponents to simplify each expression. Write the answer with only positive exponents. Assume that all variables represent positive real numbers.

26. $5^{1/4} \cdot 5^{7/4}$

27. $\frac{96^{2/3}}{96^{-1/3}}$

28. $\frac{(a^{1/3})^4}{a^{2/3}}$

29. $\frac{y^{-1/3} \cdot y^{5/6}}{y}$

30. $\left(\frac{z^{-1}x^{-3/5}}{2^{-2}z^{-1/2}x}\right)^{-1}$

31. $r^{-1/2}(r + r^{3/2})$

Simplify by first writing each radical in exponential form. Leave the answer in exponential form. Assume all variables represent positive real numbers.

32. $\sqrt[8]{s^4}$

33. $\sqrt[6]{r^9}$

34. $\frac{\sqrt{p^5}}{p^2}$

35. $\sqrt[4]{k^3} \cdot \sqrt{k^3}$

36. $\sqrt[3]{m^5} \cdot \sqrt[3]{m^8}$

37. $\sqrt[4]{\sqrt[3]{z}}$

38. $\sqrt{\sqrt{\sqrt{x}}}$

Use a calculator to find a decimal approximation for each number. Give the answer to the nearest thousandth.

39. $-\sqrt{47}$ **40.** $\sqrt[3]{-129}$ **41.** $\sqrt[4]{605}$

42. $500^{-3/4}$ **43.** $-500^{4/3}$ **44.** $-28^{-1/2}$

45. By the product rule for exponents, we know that $2^{1/4} \cdot 2^{1/5} = 2^{9/20}$. However, there is no exponent rule to simplify $3^{1/4} \cdot 2^{1/5}$. Why?

46. What is the best estimate of the area of the triangle shown here?
(a) 3600 **(b)** 30 **(c)** 60 **(d)** 360

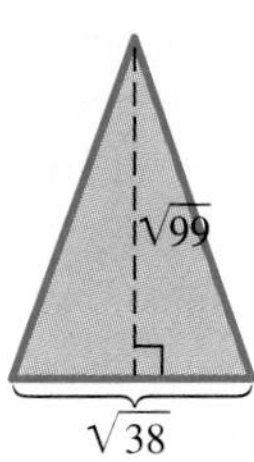

[7.2] *Simplify each radical. Assume that all variables represent positive real numbers.*

47. $\sqrt{6} \cdot \sqrt{11}$ **48.** $\sqrt{5} \cdot \sqrt{r}$ **49.** $\sqrt[3]{6} \cdot \sqrt[3]{5}$ **50.** $\sqrt[4]{7} \cdot \sqrt[4]{3}$

51. $\sqrt{20}$ **52.** $\sqrt{75}$ **53.** $-\sqrt{125}$ **54.** $\sqrt[3]{-108}$

55. $\sqrt{100y^7}$ **56.** $\sqrt[3]{64p^4q^6}$ **57.** $\sqrt[3]{108a^8b^5}$ **58.** $\sqrt[3]{632r^8t^4}$

59. $\sqrt{\frac{y^3}{144}}$ **60.** $\sqrt[3]{\frac{m^{15}}{27}}$ **61.** $\sqrt[3]{\frac{r^2}{8}}$ **62.** $\sqrt[4]{\frac{a^9}{81}}$

Simplify each radical expression.

63. $\sqrt[6]{15^3}$ **64.** $\sqrt[4]{p^6}$ **65.** $\sqrt[3]{2} \cdot \sqrt[4]{5}$ **66.** $\sqrt{x} \cdot \sqrt[5]{x}$

67. Find the missing length in the right triangle. Simplify the answer if applicable.

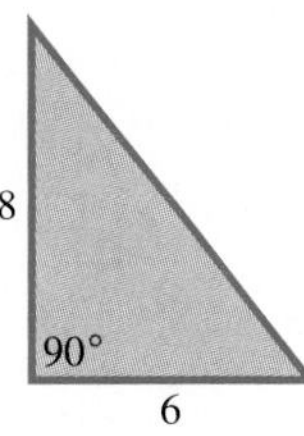

68. Find the distance between the points $(-4, 7)$ and $(10, 6)$.

[7.3] *Perform the indicated operations. Assume that all variables represent positive real numbers.*

69. $2\sqrt{8} - 3\sqrt{50}$ **70.** $8\sqrt{80} - 3\sqrt{45}$

71. $-\sqrt{27y} + 2\sqrt{75y}$ **72.** $2\sqrt{54m^3} + 5\sqrt{96m^3}$

73. $3\sqrt[3]{54} + 5\sqrt[3]{16}$ **74.** $-6\sqrt[4]{32} + \sqrt[4]{512}$

75. $\frac{3}{\sqrt{16}} - \frac{\sqrt{5}}{2}$ **76.** $\frac{4}{\sqrt{25}} + \frac{\sqrt{5}}{4}$

In Exercises 77 and 78, leave answers as simplified radicals.

77. Find the perimeter of a rectangular electronic billboard having sides of lengths shown in the figure.

$3\sqrt{48}$ feet

$4\sqrt{8}$ feet

$8\sqrt{2}$ feet

$6\sqrt{12}$ feet

78. Find the perimeter of a triangular electronic highway road sign having the dimensions shown in the figure.

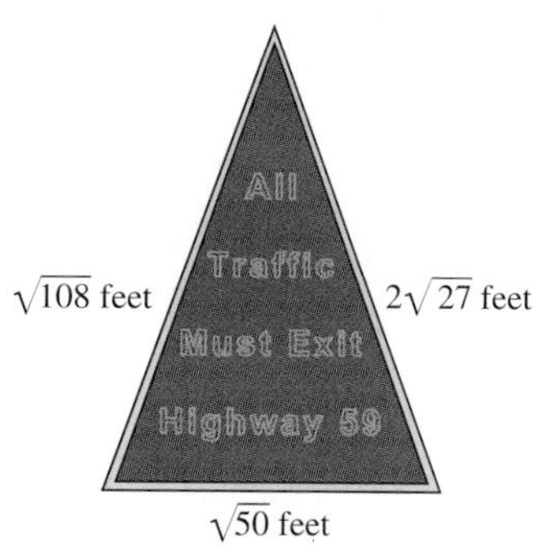

$\sqrt{108}$ feet

$2\sqrt{27}$ feet

$\sqrt{50}$ feet

[7.4] *Multiply.*

79. $(\sqrt{3} + 1)(\sqrt{3} - 2)$

80. $(\sqrt{7} + \sqrt{5})(\sqrt{7} - \sqrt{5})$

81. $(3\sqrt{2} + 1)(2\sqrt{2} - 3)$

82. $(\sqrt{13} - \sqrt{2})^2$

83. $(\sqrt[3]{2} + 3)(\sqrt[3]{4} - 3\sqrt[3]{2} + 9)$

84. $(\sqrt[3]{4y} - 1)(\sqrt[3]{4y} + 3)$

85. Use a calculator to show that the answer to Exercise 82, $15 - 2\sqrt{26}$, is not equal to $13\sqrt{26}$.

86. A friend wants to rationalize the denominator of the fraction $\frac{5}{\sqrt[3]{6}}$, and she decides to multiply the numerator and denominator by $\sqrt[3]{6}$. Why will her plan not work?

Rationalize each denominator. Assume that all variables represent positive real numbers.

87. $\frac{\sqrt{6}}{\sqrt{5}}$

88. $\frac{-6\sqrt{3}}{\sqrt{2}}$

89. $\frac{3\sqrt{7p}}{\sqrt{y}}$

90. $\sqrt{\frac{11}{8}}$

91. $-\sqrt[3]{\frac{9}{25}}$

92. $\sqrt[3]{\frac{108m^3}{n^5}}$

93. $\frac{1}{\sqrt{2} + \sqrt{7}}$

94. $\frac{-5}{\sqrt{6} - 3}$

Write in lowest terms.

95. $\frac{2 - 2\sqrt{5}}{8}$

96. $\frac{4 - 8\sqrt{8}}{12}$

[7.5] *Graph each function. Give the domain and the range.*

97. $f(x) = \sqrt{x - 1}$

98. $f(x) = \sqrt[3]{x} + 4$

Solve each equation.

99. $\sqrt{8y + 9} = 5$

100. $\sqrt{2z - 3} - 3 = 0$

101. $\sqrt{3m + 1} - 2 = -3$

102. $\sqrt{7z + 1} = z + 1$

103. $3\sqrt{m} = \sqrt{10m - 9}$

104. $\sqrt{p^2 + 3p + 7} = p + 2$

105. $\sqrt{a + 2} - \sqrt{a - 3} = 1$

106. $\sqrt[3]{5m - 1} = \sqrt[3]{3m - 2}$

107. In Exercise 101, when we add 2 and then square both sides of the equation and solve, the result is $m = 0$. Explain why the solution set is $\emptyset$.

108. Carpenters stabilize wall frames with a diagonal brace as shown in the figure. The length of the brace is given by

$L = \sqrt{H^2 + W^2}$.

(a) Solve this formula for H.

(b) If the bottom of the brace is attached 9 feet from the corner and the brace is 12 feet long, how far up the corner post should it be nailed? Give your answer to the nearest tenth of a foot.

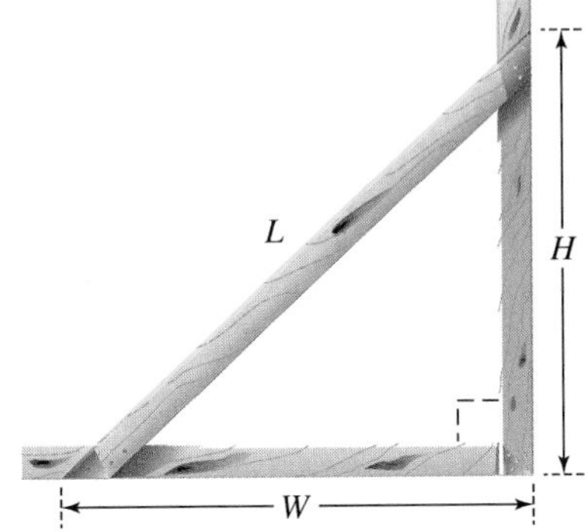

[7.6] *Write each expression as a product of a real number and i.*

109. $\sqrt{-25}$

110. $\sqrt{-200}$

111. If a is a positive real number, is $-\sqrt{-a}$ a real number?

Perform the indicated operations. Write each imaginary number answer in the form $a + bi$.

112. $(-2 + 5i) + (-8 - 7i)$

113. $(5 + 4i) - (-9 - 3i)$

114. $\sqrt{-5} \cdot \sqrt{-7}$

115. $\sqrt{-25} \cdot \sqrt{-81}$

116. $\dfrac{\sqrt{-72}}{\sqrt{-8}}$

117. $(2 + 3i)(1 - i)$

118. $(6 - 2i)^2$

119. $\dfrac{3 - i}{2 + i}$

120. $\dfrac{5 + 14i}{2 + 3i}$

Find each power of i.

121. i^{11}

122. i^{-10}

MIXED REVIEW EXERCISES

Simplify. Assume all variables represent positive real numbers.

123. $-\sqrt[4]{256}$

124. $1000^{-2/3}$

125. $\dfrac{z^{-1/5} \cdot z^{3/10}}{z^{7/10}}$

126. $\sqrt[4]{k^{24}}$

127. $\sqrt[3]{54z^9t^8}$

128. $-5\sqrt{18} + 12\sqrt{72}$

129. $8\sqrt[3]{x^3y^2} - 2x\sqrt[3]{y^2}$

130. $(\sqrt{5} - \sqrt{3})(\sqrt{7} + \sqrt{3})$

131. $\dfrac{-1}{\sqrt{12}}$

132. $\sqrt[3]{\dfrac{12}{25}}$

133. i^{-1000}

134. $\sqrt{-49}$

135. $(4 - 9i) + (-1 + 2i)$

136. $\dfrac{\sqrt{50}}{\sqrt{-2}}$

137. $\dfrac{3 + \sqrt{54}}{6}$

Solve each equation.

138. $\sqrt{x + 4} = x - 2$

139. $\sqrt[3]{2x - 9} = \sqrt[3]{5x + 3}$

140. $\sqrt{6 + 2y} - 1 = \sqrt{7 - 2y}$

CHAPTER 7 TEST

Simplify each expression. Assume that all variables represent positive numbers.

1. $\left(\frac{16}{25}\right)^{-3/2}$ **2.** $(-64)^{-4/3}$ **3.** $\frac{3^{2/5}x^{-1/4}y^{2/5}}{3^{-8/5}x^{7/4}y^{1/10}}$

4. $7^{3/4} \cdot 7^{-1/4}$ **5.** $\sqrt[3]{a^4} \cdot \sqrt[3]{a^7}$

6. Use a calculator to find an approximationto the nearest thousandth.
(a) $\sqrt{478}$ **(b)** $\sqrt[3]{-832}$ **(c)** $34^{1/4}$

7. Use the Pythagorean formula to find the exact length of side b in the figure.

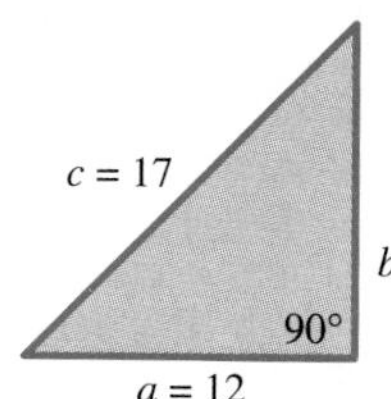

8. Find the distance between the points $(-4, 2)$ and $(2, 10)$.

Simplify each expression. Assume that all variables represent positive real numbers.

9. $\sqrt{54x^5y^6}$ **10.** $\sqrt[4]{32a^7b^{13}}$

11. $\sqrt{2} \cdot \sqrt[3]{5}$ **12.** $3\sqrt{20} - 5\sqrt{80} + 4\sqrt{500}$

13. $(7\sqrt{5} + 4)(2\sqrt{5} - 1)$ **14.** $(\sqrt{3} - 2\sqrt{5})^2$

15. $\frac{-5}{\sqrt{40}}$ **16.** $\frac{2}{\sqrt[3]{5}}$

17. $\frac{-4}{\sqrt{7} + \sqrt{5}}$

18. Write $\frac{6 + \sqrt{24}}{2}$ in lowest terms.

19. The following formula is used in physics, relating the velocity of sound V to the temperature T.

$$V = \frac{V_0}{\sqrt{1 - kT}}$$

(a) Find an approximation of V to the nearest tenth if $V_0 = 50$, $k = .01$, and $T = 30$. Use a calculator.
(b) Solve the formula for T.

20. Graph the function with $f(x) = \sqrt{x + 6}$ and give the domain and the range.

Solve each equation.

21. $\sqrt[3]{5x} = \sqrt[3]{2x - 3}$ **22.** $\sqrt{7 - x} + 5 = x$

23. Perform the indicated operations. Express the answers in the form $a + bi$.
(a) $(-2 + 5i) - (3 + 6i) - 7i$ **(b)** $(1 + 5i)(3 + i)$ **(c)** $\frac{7 + i}{1 - i}$

24. Simplify i^{37}.

25. Answer true or false to each of the following.
(a) $i^2 = -1$ **(b)** $i = \sqrt{-1}$ **(c)** $i = -1$ **(d)** $\sqrt{-3} = i\sqrt{3}$

CUMULATIVE REVIEW EXERCISES CHAPTERS 1–7

Evaluate each expression if $a = -3$, $b = 5$, and $c = -4$.

1. $|2a^2 - 3b + c|$

2. $\dfrac{(a + b)(a + c)}{3b - 6}$

Solve each equation.

3. $3(x + 2) - 4(2x + 3) = -3x + 2$

4. $\frac{1}{3}x + \frac{1}{4}(x + 8) = x + 7$

5. $.04y + .06(100 - y) = 5.88$

6. $|6x + 7| = 13$

7. $|-2x + 4| = |-2x - 3|$

8. Find the solution set of $-5 - 3(m - 2) < 11 - 2(m + 2)$. Give it in interval form.

9. Find the measures of the marked angles.

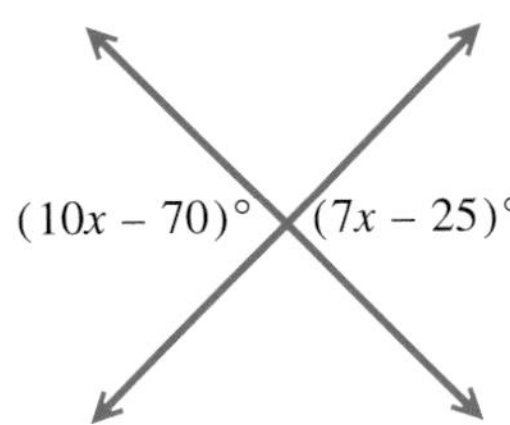

Solve each problem.

10. A piggy bank has 50 coins, all of which are nickels and quarters. The total value of the money is \$8.90. How many of each denomination are there in the bank?

11. How many liters of pure alcohol must be mixed with 40 liters of 18% alcohol to obtain a 22% alcohol solution?

12. Graph the equation $4x - 3y = 12$.

13. Find the slope of the line passing through the points $(-4, 6)$ and $(2, -3)$. Then find the equation of the line in the form $y = mx + b$.

14. If $f(x) = 3x - 7$, find $f(-10)$.

15. Solve the system by elimination or substitution:

$$3x - y = 23$$
$$2x + 3y = 8.$$

16. Solve the system by determinants and Cramer's rule:

$$x + y + z = 1$$
$$x - y - z = -3$$
$$x + y - z = -1.$$

Solve the problem by using a system of equations.

17. In 1997, if you had sent five 2-ounce letters and three 3-ounce letters using first-class mail, it would have cost you \$5.09. Sending three 2-ounce letters and five 3-ounce letters would have cost \$5.55. What was the 1997 postage rate for one 2-ounce letter and for one 3-ounce letter? (*Source:* U.S. Postal Service.)

Perform the indicated operations.

18. $(3k^3 - 5k^2 + 8k - 2) - (4k^3 + 11k + 7) + (2k^2 - 5k)$

19. $(8x - 7)(x + 3)$

20. $\dfrac{8z^3 - 16z^2 + 24z}{8z^2}$

21. $\dfrac{6y^4 - 3y^3 + 5y^2 + 6y - 9}{2y + 1}$

Factor each polynomial completely.

22. $2p^2 - 5pq + 3q^2$

23. $3k^4 + k^2 - 4$

24. $x^3 + 512$

Solve by factoring.

25. $2x^2 + 11x + 15 = 0$

26. $5t(t - 1) = 2(1 - t)$

27. What is the domain of $f(x) = \dfrac{2}{x^2 - 9}$?

Perform each operation and express the answer in lowest terms.

28. $\dfrac{y^2 + y - 12}{y^3 + 9y^2 + 20y} \div \dfrac{y^2 - 9}{y^3 + 3y^2}$

29. $\dfrac{1}{x + y} + \dfrac{3}{x - y}$

Simplify each complex fraction.

30. $\dfrac{\dfrac{-6}{x - 2}}{\dfrac{8}{3x - 6}}$

31. $\dfrac{\dfrac{1}{a} - \dfrac{1}{b}}{\dfrac{a}{b} - \dfrac{b}{a}}$

32. $\dfrac{x^{-1}}{y - x^{-1}}$

33. Natalie can ride her bike 4 miles per hour faster than her husband, Chuck. If Natalie can ride 48 miles in the same time that Chuck can ride 24 miles, what are their speeds?

34. Solve the equation $\dfrac{p + 1}{p - 3} = \dfrac{4}{p - 3} + 6$.

Write each expression in simplest form, using only positive exponents. Assume all variables represent positive real numbers.

35. $27^{-2/3}$

36. $\sqrt{200x^4}$

37. $\sqrt[3]{48x^5y^2}$

38. $\sqrt{50} + \sqrt{8}$

39. $\dfrac{1}{\sqrt{10} - \sqrt{8}}$

40. $(2\sqrt{x} + \sqrt{y})(-3\sqrt{x} - 4\sqrt{y})$

41. Solve the equation $\sqrt{3r - 8} = r - 2$.

42. The *fall speed,* in miles per hour, of a vehicle running off the road into a ditch is given by

$$S = \frac{2.74D}{\sqrt{h}},$$

where D is the horizontal distance traveled from the level surface to the bottom of the ditch and h is the height (or depth) of the ditch. What is the fall speed of a vehicle that traveled 32 feet horizontally into a 5-foot deep ditch?

Write in the form $a + bi$.

43. $(5 + 7i) - (3 - 2i)$

44. $\dfrac{6 - 2i}{1 - i}$

Quadratic Equations and Inequalities

8

In this chapter, we continue our study of quadratic equations and functions from Chapter 5. As mentioned there, quadratic equations are often good models for applied problems, particularly in astronomy, the theme of this chapter.

Astronomy

The German astronomer Johannes Kepler (1571–1630) discovered three laws of planetary motion. Kepler's third law, published in 1619, is expressed by the quadratic equation $P^2 = a^3$. Here, P is the *sidereal* or *orbital period* of a planet, the time from one orbital point back to that same point, in years. The variable a represents half the length of the major axis of the orbital ellipse in *astronomical units.* One astronomical unit (AU) is the average distance between Earth and the sun. The table gives a in astronomical units for selected planets.

Astronomical Units

Planet	Mercury	Venus	Mars	Jupiter	Saturn	Neptune
a (AU)	.387	.723	1.524	5.20	9.54	30.1

Source: James B. Kaler, *Astronomy! A Brief Edition,* Addison Wesley, 1997.

What is the value of a for Jupiter? To find the sidereal period for Jupiter, substitute 5.20 for a in the formula.

$$P^2 = a^3$$
$$P^2 = (5.20)^3$$
$$P^2 = 140.608$$

In Section 8.1 Exercises, the square root property is used to solve this quadratic equation.

Visit our Web site at www.LialAlgebra.com

8.1 Completing the Square

OBJECTIVES

1. Review the zero-factor property.
2. Learn the square root property.
3. Solve quadratic equations of the form $(ax - b)^2 = c$ by using the square root property.
4. Solve quadratic equations by completing the square.
5. Solve quadratic equations with complex solutions.

FOR EXTRA HELP

SSG Sec. 8.1
SSM Sec. 8.1

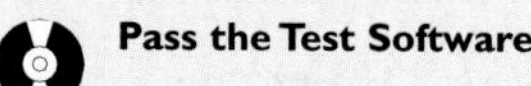

Pass the Test Software

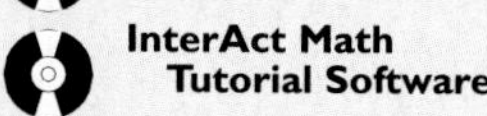

InterAct Math Tutorial Software

Video 15

We solved quadratic equations in Section 5.9 by using the zero-factor property. In this section we introduce additional methods for solving quadratic equations. Recall that a quadratic equation is a second-degree equation. For example,

$$m^2 + 4m - 5 = 0 \qquad \text{and} \qquad 3q^2 = 4q - 8$$

are quadratic equations, with the first equation in *standard form.*

OBJECTIVE 1 Review the zero-factor property.

Zero-Factor Property

If two numbers have a product of 0, then at least one of the numbers must be 0. That is, if $ab = 0$, then $a = 0$ or $b = 0$.

EXAMPLE 1 Using the Zero-Factor Property to Solve an Equation

Use the zero-factor property to solve $3x^2 - 5x = 28$.

$$3x^2 - 5x = 28 \qquad \text{Given equation}$$

$$3x^2 - 5x - 28 = 0 \qquad \text{Standard form}$$

$$(3x + 7)(x - 4) = 0 \qquad \text{Factor.}$$

$$3x + 7 = 0 \quad \text{or} \quad x - 4 = 0 \qquad \text{Zero-factor property}$$

$$3x = -7 \qquad\qquad x = 4 \qquad \text{Solve each linear equation.}$$

$$x = -\frac{7}{3}$$

The solution set is $\left\{-\frac{7}{3}, 4\right\}$.

OBJECTIVE 2 **Learn the square root property.** Although factoring is the simplest way to solve quadratic equations, not every quadratic equation can be solved easily by factoring. In this section and the next, we develop other methods of solving quadratic equations based on the following property.

Square Root Property

If $x^2 = b$, $b \geq 0$, then $x = \sqrt{b}$ or $x = -\sqrt{b}$.

The following steps justify the square root property.

$$x^2 = b$$

$$x^2 - b = 0 \qquad \text{Subtract } b \text{ on both sides.}$$

$$(x - \sqrt{b})(x + \sqrt{b}) = 0 \qquad \text{Factor; } b \geq 0.$$

$$x - \sqrt{b} = 0 \quad \text{or} \quad x + \sqrt{b} = 0 \qquad \text{Zero-factor property}$$

$$x = \sqrt{b} \quad \text{or} \quad x = -\sqrt{b} \qquad \text{Solve each equation.}$$

CAUTION Remember that if $b \neq 0$, using the square root property always produces *two* square roots, one positive and one negative.

EXAMPLE 2 Using the Square Root Property

Solve $r^2 = 5$.

From the square root property, since $(\sqrt{5})^2 = 5$,

$$r = \sqrt{5} \quad \text{or} \quad r = -\sqrt{5}$$

and the solution set is $\{\sqrt{5}, -\sqrt{5}\}$. ■

Recall from Chapter 7 that roots such as those in Example 2 are sometimes abbreviated with the symbol $\pm$ (read "positive or negative"); with this symbol the solutions in Example 2 would be written $\pm\sqrt{5}$.

OBJECTIVE 3 Solve quadratic equations of the form $(ax - b)^2 = c$ by using the square root property. For example, to solve the equation

$$(x - 5)^2 = 36,$$

substitute $(x - 5)^2$ for x^2 and 36 for b, using the square root property to get

$$x - 5 = 6 \quad \text{or} \quad x - 5 = -6$$

$$x = 11 \quad \text{or} \quad x = -1.$$

Check that both 11 and -1 satisfy the original equation, so the solution set is $\{11, -1\}$.

EXAMPLE 3 Using the Square Root Property

Solve $(2a - 3)^2 = 18$.

By the square root property,

$$2a - 3 = \sqrt{18} \quad \text{or} \quad 2a - 3 = -\sqrt{18},$$

from which

$$2a = 3 + \sqrt{18} \quad \text{or} \quad 2a = 3 - \sqrt{18}$$

$$a = \frac{3 + \sqrt{18}}{2} \quad \text{or} \quad a = \frac{3 - \sqrt{18}}{2}.$$

Since $\sqrt{18} = \sqrt{9 \cdot 2} = 3\sqrt{2}$, the solution set can be written

$$\left\{\frac{3 + 3\sqrt{2}}{2}, \frac{3 - 3\sqrt{2}}{2}\right\}.$$ ■

OBJECTIVE 4 Solve quadratic equations by completing the square. We use the square root property to solve any quadratic equation by writing it in the form $(x + k)^2 = n$. That is, the left side of the equation must be rewritten as a perfect square trinomial that can be factored as $(x + k)^2$ and the right side must be a constant. Rewriting the equation in this form is called **completing the square.**

For example,

$$m^2 + 8m + 10 = 0$$

is a quadratic equation that cannot be solved easily by factoring. To get a perfect square trinomial on the left side of the equation $m^2 + 8m + 10 = 0$, first subtract 10 from both sides:

$$m^2 + 8m = -10.$$

The left side should be a perfect square, say $(m + k)^2$. Since $(m + k)^2 = m^2 + 2mk + k^2$, comparing $m^2 + 8m$ with $m^2 + 2mk$ shows that

$$2mk = 8m$$
$$k = 4.$$

If $k = 4$, then $(m + k)^2$ becomes $(m + 4)^2$, or $m^2 + 8m + 16$. To get the necessary $+16$, add 16 on both sides.

$$m^2 + 8m = -10$$
$$m^2 + 8m + \mathbf{16} = -10 + \mathbf{16}$$

Now factor the left side, which should be a perfect square. Since $m^2 + 8m + 16$ factors as $(m + 4)^2$, the equation becomes

$$(m + 4)^2 = 6.$$

We can solve this equation with the square root property:

$$m + 4 = \sqrt{6} \qquad \text{or} \qquad m + 4 = -\sqrt{6},$$

leading to the solution set $\{-4 + \sqrt{6}, -4 - \sqrt{6}\}$.

Based on this example, to convert an equation of the form $x^2 + px = q$ into an equation of the form $(x + k)^2 = n$, add the square of half the coefficient of the first-degree term to both sides of the equation.

In summary, to find the solutions of a quadratic equation by completing the square, proceed as follows.

Completing the Square

To solve $ax^2 + bx + c = 0$ $(a \neq 0)$, use the following steps.

Step 1 **Divide by *a*.** If $a \neq 1$, divide both sides by a.

Step 2 **Rewrite the equation.** Rewrite the equation so that terms with variables are on one side of the equals sign, and the constant is on the other side.

Step 3 **Square half the coefficient of *x*.** Take half the coefficient of x (the first-degree term) and square it.

Step 4 **Add the square to both sides.**

Step 5 **Factor the perfect square trinomial.** One side should now be a perfect square trinomial. Factor it and write it as the square of a binomial. Simplify the other side.

Step 6 **Use the square root property.** Use the square root property to complete the solution.

NOTE Steps 1 and 2 can be done in either order. With some equations, it is more convenient to do Step 2 first.

EXAMPLE 4 Determining the Number to Be Used to Complete the Square

Determine the number that will complete the square to solve $2x^2 + 8x - 3 = 0$.

Use the first three steps in the box on the previous page.

$$2x^2 + 8x - 3 = 0 \quad \text{Given equation}$$

$$x^2 + 4x - \frac{3}{2} = 0 \quad \text{Step 1}$$

$$x^2 + 4x = \frac{3}{2} \quad \text{Step 2}$$

$$\left(\frac{1}{2} \cdot 4\right)^2 = 2^2 = 4 \quad \text{Step 3}$$

The number 4 should be added to both sides of the equation to continue the process of completing the square.

EXAMPLE 5 Solving a Quadratic Equation with $a = 1$ by Completing the Square

Solve $k^2 + 5k - 1 = 0$ by completing the square.

Follow the steps listed earlier. Since $a = 1$, Step 1 is not needed here. Begin by adding 1 to both sides.

$$k^2 + 5k = 1 \quad \text{Step 2}$$

Take half of the coefficient of the first-degree term and square the result.

$$\frac{1}{2} \cdot 5 = \frac{5}{2} \quad \text{and} \quad \left(\frac{5}{2}\right)^2 = \frac{25}{4} \quad \text{Step 3}$$

Add the square to each side of the equation to get

$$k^2 + 5k + \frac{25}{4} = 1 + \frac{25}{4}. \quad \text{Step 4}$$

Write the left side as a perfect square and add on the right. Then use the square root property.

$$\left(k + \frac{5}{2}\right)^2 = \frac{29}{4} \quad \text{Step 5}$$

$$k + \frac{5}{2} = \sqrt{\frac{29}{4}} \quad \text{or} \quad k + \frac{5}{2} = -\sqrt{\frac{29}{4}} \quad \text{Step 6}$$

$$k + \frac{5}{2} = \frac{\sqrt{29}}{2} \quad \text{or} \quad k + \frac{5}{2} = \frac{-\sqrt{29}}{2} \quad \text{Simplify.}$$

$$k = -\frac{5}{2} + \frac{\sqrt{29}}{2} \quad \text{or} \quad k = -\frac{5}{2} - \frac{\sqrt{29}}{2} \quad \text{Solve.}$$

$$k = \frac{-5 + \sqrt{29}}{2} \quad \text{or} \quad k = \frac{-5 - \sqrt{29}}{2} \quad \text{Combine terms.}$$

Check that the solution set is $\left\{\frac{-5 + \sqrt{29}}{2}, \frac{-5 - \sqrt{29}}{2}\right\}$.

EXAMPLE 6 Solving a Quadratic Equation with $a \neq 1$ by Completing the Square

Solve $2a^2 - 4a - 5 = 0$.

Follow the steps to complete the square. First divide both sides of the equation by 2 to make the coefficient of the second-degree term equal to 1.

$$a^2 - 2a - \frac{5}{2} = 0 \qquad \text{Step 1}$$

$$a^2 - 2a = \frac{5}{2} \qquad \text{Step 2}$$

$$\frac{1}{2}(-2) = -1 \text{ and } (-1)^2 = 1. \qquad \text{Step 3}$$

$$a^2 - 2a + 1 = \frac{5}{2} + 1 \qquad \text{Step 4}$$

$$(a - 1)^2 = \frac{7}{2} \qquad \text{Step 5}$$

$$a - 1 = \sqrt{\frac{7}{2}} \quad \text{or} \quad a - 1 = -\sqrt{\frac{7}{2}} \qquad \text{Step 6}$$

$$a = 1 + \sqrt{\frac{7}{2}} \quad \text{or} \quad a = 1 - \sqrt{\frac{7}{2}} \qquad \text{Solve.}$$

$$a = 1 + \frac{\sqrt{14}}{2} \quad \text{or} \quad a = 1 - \frac{\sqrt{14}}{2} \qquad \text{Rationalize denominators.}$$

Add the two terms in each solution as follows:

$$1 + \frac{\sqrt{14}}{2} = \frac{2}{2} + \frac{\sqrt{14}}{2} = \frac{2 + \sqrt{14}}{2}$$

$$1 - \frac{\sqrt{14}}{2} = \frac{2}{2} - \frac{\sqrt{14}}{2} = \frac{2 - \sqrt{14}}{2}.$$

The solution set is

$$\left\{\frac{2 + \sqrt{14}}{2}, \frac{2 - \sqrt{14}}{2}\right\}.$$

OBJECTIVE 5 Solve quadratic equations with complex solutions. In the equation $x^2 = b$, if $b < 0$, there will be two imaginary solutions. The square root property can be extended to complex numbers as follows.

If $x^2 = -b, b > 0$, then $x = i\sqrt{b}$ or $x = -i\sqrt{b}$.

EXAMPLE 7 Extending the Methods of Solving Quadratic Equations

Find all solutions of each equation.

(a) $x^2 = -15$

By the extension of the square root property, we have

$$x^2 = -15$$

$$x = \sqrt{-15} \quad \text{or} \quad x = -\sqrt{-15}$$

$$x = i\sqrt{15} \quad \text{or} \quad x = -i\sqrt{15}.$$

The solution set is $\{-i\sqrt{15}, i\sqrt{15}\}$.

(b) $(b + 2)^2 = -16$

$b + 2 = \sqrt{-16}$ or $b + 2 = -\sqrt{-16}$ Square root property

$b + 2 = 4i$ or $b + 2 = -4i$ $\sqrt{-16} = 4i$

$b = -2 + 4i$ or $b = -2 - 4i$

The solution set is $\{-2 + 4i, -2 - 4i\}$.

(c) $x^2 + 2x + 7 = 0$

Solve by completing the square.

$x^2 + 2x = -7$ Add -7.

$x^2 + 2x + 1 = -7 + 1$ $\left[\frac{1}{2} \cdot 2\right]^2 = 1^2 = 1$; add 1 to both sides.

$(x + 1)^2 = -6$ Factor on the left, combine terms on the right.

$x + 1 = \pm i\sqrt{6}$ Square root property

$x = -1 \pm i\sqrt{6}$ Subtract 1 from both sides.

The solution set is $\{-1 + i\sqrt{6}, -1 - i\sqrt{6}\}$.

8.1 EXERCISES

1. What would be your first step in solving $2x^2 + 8x = 9$ by completing the square?
2. Why would most students find the equation $x^2 + 4x = 20$ easier to solve by completing the square than the equation $5x^2 + 2x = 3$?
3. Of the two equations

$$(2x + 1)^2 = 5 \quad \text{and} \quad x^2 + 4x = 12,$$

one is more suitable for solving by the square root property, and the other is more suitable for solving by completing the square. Which method do you think most students would use for each equation?

4. Give a one-sentence description or explanation of each of the following.
   (a) quadratic equation in standard form
   (b) zero-factor property
   (c) square root property

What number must be added to each binomial to get a perfect square trinomial? See Example 4.

5. $x^2 + 8x$ **6.** $x^2 + 5x$ **7.** $x^2 - 9x$ **8.** $x^2 - 10x$

Use the square root property to solve each equation. (All solutions for these equations are real numbers.) See Examples 2 and 3.

9. $x^2 = 81$ **10.** $y^2 = 225$ **11.** $m^2 = 32$

12. $w^2 = 128$ **13.** $(x + 2)^2 = 25$ **14.** $(8 - y)^2 = 9$

15. $(1 - 3k)^2 = 7$ **16.** $(2x + 4)^2 = 10$ **17.** $(4p + 1)^2 = 24$

18. $(5k - 2)^2 = 12$

Find the imaginary number solutions of each equation. See Example 7.

19. $x^2 = -12$ **20.** $y^2 = -18$ **21.** $(r - 5)^2 = -3$

22. $(t + 6)^2 = -5$ **23.** $(6k - 1)^2 = -8$ **24.** $(4m - 7)^2 = -27$

Solve each equation by completing the square. (All solutions for these equations are real numbers.) See Examples 4–6.

25. $x^2 - 2x - 24 = 0$

26. $m^2 - 4m - 32 = 0$

27. $3y^2 + y = 24$

28. $4z^2 - z = 39$

29. $2k^2 + 5k - 2 = 0$

30. $3r^2 + 2r - 2 = 0$

31. $9x^2 - 24x = -13$

32. $25n^2 = 20n + 1$

Solve each equation by completing the square. (Some solutions for these equations are imaginary numbers.) See Examples 4, 5, 6, and 7(c).

33. $m^2 + 4m + 13 = 0$

34. $t^2 + 6t + 10 = 0$

35. $z^2 - \frac{4}{3}z = -\frac{1}{9}$

36. $p^2 - \frac{8}{3}p = -1$

37. $3r^2 + 4r + 4 = 0$

38. $4x^2 + 5x + 5 = 0$

39. $.1x^2 - .2x - .1 = 0$

40. $.1p^2 - .4p + .1 = 0$

41. $-m^2 - 6m - 12 = 0$

42. $-k^2 - 5k - 10 = 0$

43. What is wrong with the following "solution"?

$$(x - 2)(x + 1) = 5$$

$$x - 2 = 5 \quad \text{or} \quad x + 1 = 1 \qquad \text{Zero-factor property}$$

$$x = 7 \quad \text{or} \quad x = 0$$

44. A student was asked to solve the quadratic equation $x^2 = 16$ and did not get full credit for the solution set $\{4\}$. Why?

45. Why is the zero-factor property insufficient for solving *all* quadratic equations?

46. Can *any* quadratic equation be solved by completing the square?

Solve for x. Assume that a and b represent positive real numbers.

47. $x^2 - b = 0$

48. $x^2 = 4b$

49. $4x^2 = b^2 + 16$

50. $9x^2 - 25a = 0$

51. $(5x - 2b)^2 = 3a$

52. $x^2 - a^2 - 36 = 0$

Solve each problem.

53. In the chapter introduction, we saw that the sidereal period of Jupiter is *P*, where

$$P^2 = 140.608.$$

Find *P*.

54. Use the formula $P^2 = a^3$ and the value of *a* from the table in the chapter introduction to find the sidereal period of Venus.

55. The *orbital velocity* of a satellite is its speed along its orbit around Earth. This velocity v is given by the equation

$$v^2 = \frac{GM}{r},$$

where *GM*, which represents universal gravity times the mass of Earth, is 3.99×10^{11} cubic meters per second squared, and *r* is the radius of the circular orbit (the distance between the centers of the two bodies). *High Energy Astronomy Observatory (HEAO)* satellites have an average altitude above Earth of 4.3×10^5 meters, and the radius of Earth averages 6.37×10^6 meters. Because the numbers in this problem are very large, they have been written in scientific notation, which was discussed in Section 5.1. (*Source:* Bernice Kastner, *Space Mathematics,* NASA: pp. 136–137.)

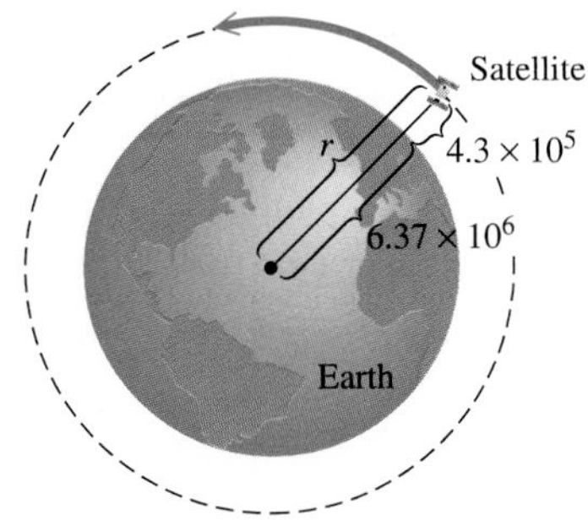

(a) Find the radius of orbit for an HEAO satellite.
(b) Use the formula to find the velocity of an HEAO satellite.

56. Refer to Exercise 55. The average distance between the centers of the moon and Earth is approximately 3.84×10^8 meters. Find the velocity of the moon around Earth.

RELATING CONCEPTS (EXERCISES 57–62)

The Greeks had a method of completing the square geometrically in which they literally changed a figure into a square. For example, to complete the square for $x^2 + 6x$, we begin with a square of side x, as in the figure. We add three rectangles of width 1 to the right side and the bottom to get a region with area $x^2 + 6x$. To fill in the corner (complete the square), we must add 9 1-by-1 squares as shown.

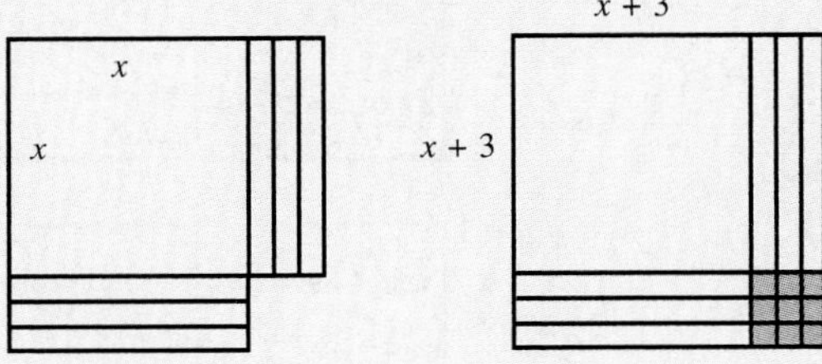

Work Exercises 57–62 in order.

57. What is the area of the original square?

58. What is the area of each strip?

59. What is the total area of the six strips?

60. What is the area of each small square in the corner of the second figure?

61. What is the total area of the small squares?

62. What is the area of the new, larger square?

Did you make the connection that the sum of the area of the first figure plus the area of the shaded region in the second figure equals the area of the large square?

TECHNOLOGY INSIGHTS (EXERCISES 63 AND 64)

In Section 5.3 we graphed the function defined by $y = x^2$ by plotting points. Two of the graphing calculator screens below show the intersection points of the graph of $y = x^2$ and the graph of the horizontal line $y = 5$. The other screen shows that $\sqrt{5} \approx 2.236068$, so the graphs intersect at $x = -\sqrt{5}$ and $x = \sqrt{5}$. This supports our solution $\pm\sqrt{5}$ for $x^2 = 5$ using the square root property in Example 1.

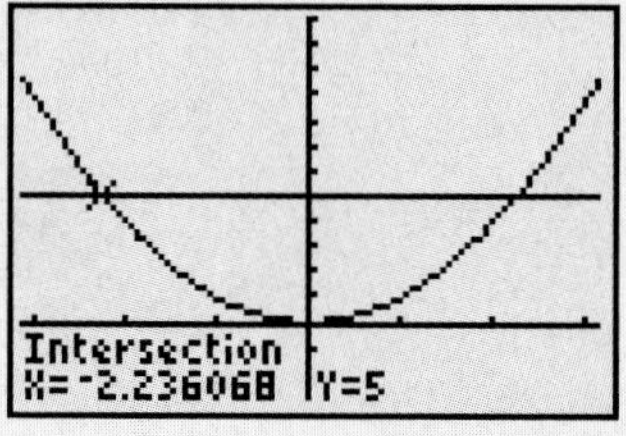

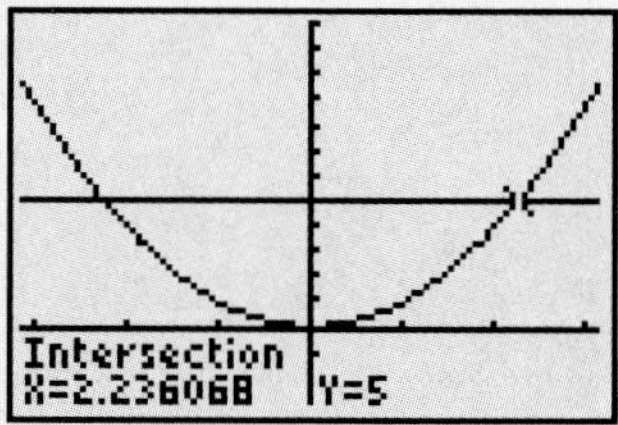

Use the screens in Exercises 63 and 64 to give the exact value of the solutions of $x^2 = b$.

63.

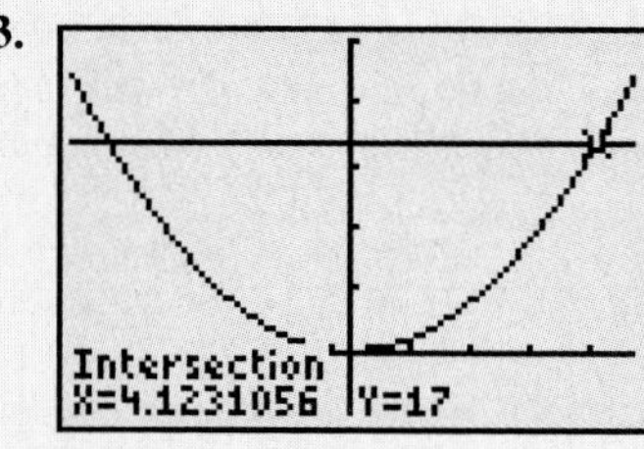

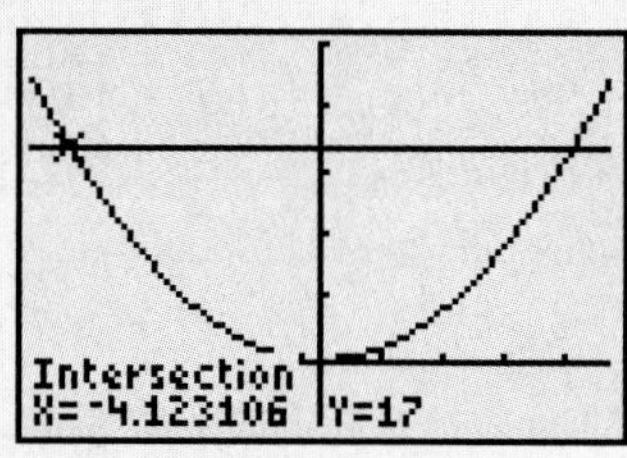

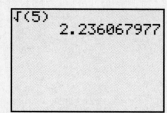

64.

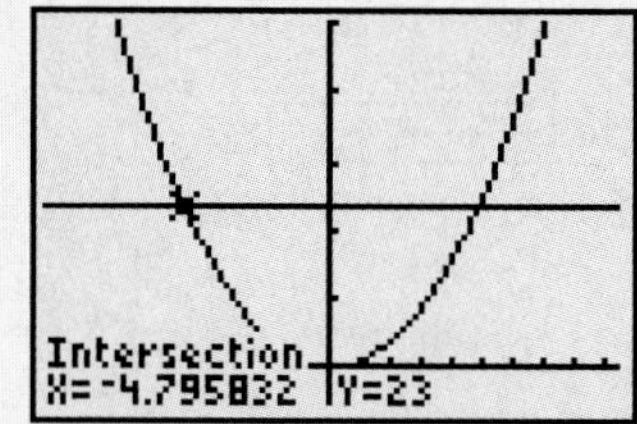

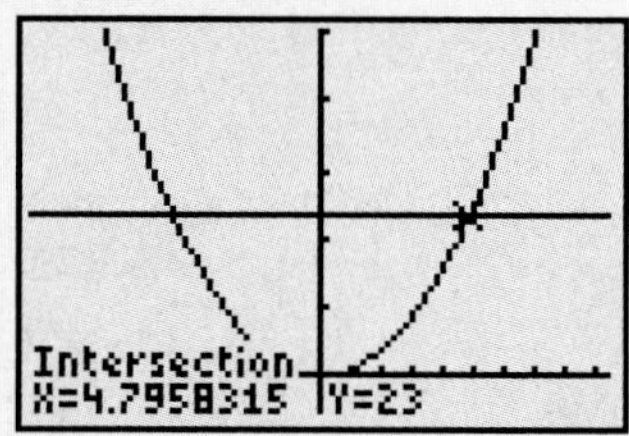

8.2 The Quadratic Formula

OBJECTIVES

1. Solve quadratic equations by using the quadratic formula.
2. Solve applications by using the quadratic formula.

The examples in the previous section showed that any quadratic equation can be solved by completing the square. However, completing the square can be tedious and time consuming. In this section we complete the square to solve the general quadratic equation $ax^2 + bx + c = 0$, where a, b, and c are complex numbers and $a \neq 0$. The solution of this general equation is a formula for finding the solution of any specific quadratic equation.

3 Use the discriminant to determine the number and type of solutions.

4 Use the discriminant to decide whether a quadratic trinomial can be factored.

FOR EXTRA HELP

SSG Sec. 8.2
SSM Sec. 8.2

Pass the Test Software

InterAct Math Tutorial Software

Video 15

To solve $ax^2 + bx + c = 0$ by completing the square (assuming $a > 0$ for now), we follow the steps given in Section 8.1.

$$ax^2 + bx + c = 0$$

$$x^2 + \frac{b}{a}x + \frac{c}{a} = 0 \qquad \text{Divide each side by } a. \text{ (Step 1)}$$

$$x^2 + \frac{b}{a}x = -\frac{c}{a} \qquad \text{Subtract } \tfrac{c}{a} \text{ from each side. (Step 2)}$$

$$\frac{1}{2}\left(\frac{b}{a}\right) = \frac{b}{2a}; \quad \left(\frac{b}{2a}\right)^2 = \frac{b^2}{4a^2} \qquad \text{(Step 3)}$$

$$x^2 + \frac{b}{a}x + \frac{b^2}{4a^2} = -\frac{c}{a} + \frac{b^2}{4a^2} \qquad \text{Add } \tfrac{b^2}{4a^2} \text{ to each side. (Step 4)}$$

$$x^2 + \frac{b}{a}x + \frac{b^2}{4a^2} = \left(x + \frac{b}{2a}\right)^2 \qquad \text{Write the left side as a square.}$$

$$\left(x + \frac{b}{2a}\right)^2 = \frac{b^2}{4a^2} + \frac{-c}{a} \qquad \text{Substitute on the left and rearrange the right side. (Step 5)}$$

$$= \frac{b^2}{4a^2} + \frac{-4ac}{4a^2} \qquad \text{Write with a common denominator.}$$

$$\left(x + \frac{b}{2a}\right)^2 = \frac{b^2 - 4ac}{4a^2} \qquad \text{Add fractions.}$$

$$x + \frac{b}{2a} = \sqrt{\frac{b^2 - 4ac}{4a^2}} \quad \text{or} \quad x + \frac{b}{2a} = -\sqrt{\frac{b^2 - 4ac}{4a^2}} \qquad \text{Square root property (Step 6)}$$

Since

$$\sqrt{\frac{b^2 - 4ac}{4a^2}} = \frac{\sqrt{b^2 - 4ac}}{\sqrt{4a^2}} = \frac{\sqrt{b^2 - 4ac}}{2a},$$

the result can be expressed as

$$x + \frac{b}{2a} = \frac{\sqrt{b^2 - 4ac}}{2a} \quad \text{or} \quad x + \frac{b}{2a} = \frac{-\sqrt{b^2 - 4ac}}{2a}$$

$$x = \frac{-b}{2a} + \frac{\sqrt{b^2 - 4ac}}{2a} \quad \text{or} \quad x = \frac{-b}{2a} - \frac{\sqrt{b^2 - 4ac}}{2a}$$

$$x = \frac{-b + \sqrt{b^2 - 4ac}}{2a} \quad \text{or} \quad x = \frac{-b - \sqrt{b^2 - 4ac}}{2a}.$$

If $a < 0$, the same two solutions are obtained. The result is the **quadratic formula,** often abbreviated as follows.

Quadratic Formula

The solutions of $ax^2 + bx + c = 0$ $(a \neq 0)$ are

$$x = \frac{-b \pm \sqrt{b^2 - 4ac}}{2a}.$$

CAUTION Notice in the quadratic formula that the square root is added to or subtracted from the value of $-b$ *before* dividing by $2a$.

OBJECTIVE 1 Solve quadratic equations by using the quadratic formula. First write the given equation in the form $ax^2 + bx + c = 0$; then identify the values of a, b, and c and substitute them into the quadratic formula, as shown in the next examples.

EXAMPLE 1 Using the Quadratic Formula to Find Rational Solutions

Solve $6x^2 - 5x - 4 = 0$.

This equation is in the required form to identify a, b, and c of the general quadratic equation. Here a, the coefficient of the second-degree term, is 6, while b, the coefficient of the first-degree term, is -5, and c, the constant, is -4. Substitute these values into the quadratic formula.

$$x = \frac{-b \pm \sqrt{b^2-4ac}}{2a}$$

$$x = \frac{-(-5) \pm \sqrt{(-5)^2-4(6)(-4)}}{2(6)} \qquad a = 6, b = -5, c = -4$$

$$= \frac{5 \pm \sqrt{25 + 96}}{12}$$

$$= \frac{5 \pm \sqrt{121}}{12} = \frac{5 \pm 11}{12}$$

This last statement leads to two solutions, one from + and one from −.

$$x = \frac{5+11}{12} = \frac{16}{12} = \frac{4}{3} \quad \text{or} \quad x = \frac{5-11}{12} = \frac{-6}{12} = -\frac{1}{2}$$

Check each of these solutions by substituting them in the original equation. The solution set is $\{\frac{4}{3}, -\frac{1}{2}\}$.

EXAMPLE 2 Using the Quadratic Formula to Find Irrational Solutions

Solve $4r^2 = 8r - 1$.

Rewrite the equation in standard form as $4r^2 - 8r + 1 = 0$ and identify $a = 4$, $b = -8$, and $c = 1$. Now use the quadratic formula.

$$r = \frac{-b \pm \sqrt{b^2 - 4ac}}{2a}$$

$$r = \frac{-(-8) \pm \sqrt{(-8)^2 - 4(4)(1)}}{2(4)} \qquad a = 4, b = -8, c = 1$$

$$= \frac{8 \pm \sqrt{64 - 16}}{8} = \frac{8 \pm \sqrt{48}}{8}$$

$$= \frac{8 \pm 4\sqrt{3}}{8} = \frac{4(2 \pm \sqrt{3})}{8} = \frac{2 \pm \sqrt{3}}{2}$$

The solution set is $\left\{\frac{2+\sqrt{3}}{2}, \frac{2-\sqrt{3}}{2}\right\}$.

The solutions of the equation in the next example are imaginary numbers.

EXAMPLE 3 Using the Quadratic Formula to Find Imaginary Solutions

Solve $(9q + 3)(q - 1) = -8$.

Every quadratic equation must be in standard form to begin its solution, whether we are factoring or using the quadratic formula. To put this equation in standard form, we first multiply and collect all nonzero terms on the left.

$$(9q + 3)(q - 1) = -8$$
$$9q^2 - 6q - 3 = -8$$
$$9q^2 - 6q + 5 = 0$$

Now, identify $a = 9$, $b = -6$, and $c = 5$, and use the quadratic formula.

$$q = \frac{-(-6) \pm \sqrt{(-6)^2 - 4(9)(5)}}{2(9)}$$
$$= \frac{6 \pm \sqrt{-144}}{18} = \frac{6 \pm 12i}{18} \qquad \sqrt{-144} = 12i$$
$$= \frac{6(1 \pm 2i)}{18} = \frac{1 \pm 2i}{3}$$

The solution set is $\left\{\frac{1}{3} + \frac{2}{3}i, \frac{1}{3} - \frac{2}{3}i\right\}$.

The solutions in Example 3 were written as $a + bi$, the standard form for complex numbers, as follows:

$$\frac{1 \pm 2i}{3} = \frac{1}{3} \pm \frac{2}{3}i.$$

OBJECTIVE 2 Solve applications by using the quadratic formula.

PROBLEM SOLVING

In Chapter 6 we solved problems about work rates. Recall, a person's work rate is $\frac{1}{t}$ part of the job per hour, where t is the time in hours required to do the complete job. Thus, the part of the job the person will do in x hours is $\frac{1}{t}x$. We continue to use the six-step problem-solving method from Chapter 2.

EXAMPLE 4 Solving a Work Problem

Two mechanics take 4 hours to repair a car. If each worked alone, one of them could do the job in 1 hour less time than the other. How long would it take the slower one to complete the job alone?

Steps 1 and 2 Let x represent the number of hours for the slower mechanic to complete the job alone. Then the faster mechanic could do the entire job in $x - 1$ hours. Together, they do the job in 4 hours. This information is shown in the following chart.

Worker	Rate	Time Working Together (in hours)	Fractional Part of the Job Done
Faster mechanic	$\frac{1}{x-1}$	4	$\frac{1}{x-1} \cdot 4$
Slower mechanic	$\frac{1}{x}$	4	$\frac{1}{x} \cdot 4$

Step 3 The sum of the fractional parts done by each should equal 1 (the whole job).

Part done by slower mechanic + Part done by faster mechanic = 1 whole job

$$\frac{4}{x} + \frac{4}{x-1} = 1$$

Step 4 Multiply both sides by the common denominator, $x(x - 1)$.

$$4(x-1) + 4x = x(x-1)$$

$$4x - 4 + 4x = x^2 - x \qquad \text{Distributive property}$$

$$0 = x^2 - 9x + 4 \qquad \text{Standard form}$$

$$x = \frac{9 \pm \sqrt{81 - 16}}{2} \qquad \text{Quadratic formula}$$

$$= \frac{9 \pm \sqrt{65}}{2}$$

In an applied problem like this, calculator approximations are more useful than exact values. To the nearest tenth,

$$x = \frac{9 + \sqrt{65}}{2} \approx 8.5 \quad \text{or} \quad x = \frac{9 - \sqrt{65}}{2} \approx .5.$$

Step 5 Only the solution 8.5 makes sense in the original problem. (Why?) Thus, the slower mechanic can do the job in about 8.5 hours and the faster in about $8.5 - 1 = 7.5$ hours.

Step 6 Check that these results satisfy the original problem.

OBJECTIVE 3 Use the discriminant to determine the number and type of solutions. The solutions of the quadratic equation $ax^2 + bx + c = 0$ are

$$x = \frac{-b \pm \sqrt{b^2 - 4ac}}{2a}.$$

If a, b, and c are integers, the type of solutions of a quadratic equation (that is, rational, irrational, or imaginary) is determined by the quantity under the square root sign, $b^2 - 4ac$. Because it distinguishes among the three types of solutions, the quantity $b^2 - 4ac$ is called the **discriminant.** By calculating the discriminant before solving a quadratic equation, we can predict whether the solutions will be rational numbers, irrational numbers, or imaginary numbers. This can be useful in an applied problem, for example, where irrational or imaginary number solutions are not acceptable. Also, if the discriminant is a perfect square (including 0), the equation can be solved by factoring. Otherwise, the quadratic formula should be used.

Discriminant

The discriminant of $ax^2 + bx + c = 0$ is given by $b^2 - 4ac$. If a, b, and c are integers, then the type of solution is determined as follows.

Discriminant	*Type of Solution*
Positive, and the square of an integer	Two different rational solutions
Positive, but not the square of an integer	Two different irrational solutions
Zero	One rational solution
Negative	Two different imaginary solutions

EXAMPLE 5 Using the Discriminant

Predict the number and type of solutions for the following equations.

(a) $6x^2 - x - 15 = 0$

We find the discriminant by evaluating $b^2 - 4ac$.

$$\begin{aligned} b^2 - 4ac &= (-1)^2 - 4(6)(-15) \qquad a = 6, b = -1, c = -15 \\ &= 1 + 360 = 361 \end{aligned}$$

A calculator shows that $361 = 19^2$, a perfect square. Since a, b, and c are integers, the solutions will be two different rational numbers, and the equation can be solved by factoring.

(b) $3m^2 - 4m = 5$

Rewrite the equation as $3m^2 - 4m - 5 = 0$ to find $a = 3$, $b = -4$, $c = -5$. The discriminant is

$$b^2 - 4ac = (-4)^2 - 4(3)(-5) = 16 + 60 = 76.$$

Because 76 is not the square of an integer, $\sqrt{76}$ is irrational. From this and from the fact that a, b, and c are integers, the equation will have two different irrational solutions, one using $\sqrt{76}$ and one using $-\sqrt{76}$.

(c) $4x^2 + x + 1 = 0$

Since $a = 4$, $b = 1$, and $c = 1$, the discriminant is

$$1^2 - 4(4)(1) = -15.$$

Since the discriminant is negative and a, b, and c are integers, this quadratic equation will have two imaginary number solutions. ■

EXAMPLE 6 Using the Discriminant

Find k so that $9x^2 + kx + 4 = 0$ will have only one rational number solution.

The equation will have only one rational number solution if the discriminant is 0. Here, since $a = 9$, $b = k$, and $c = 4$, the discriminant is

$$b^2 - 4ac = k^2 - 4(9)(4) = k^2 - 144.$$

Set this result equal to 0 and solve for k.

$$k^2 - 144 = 0$$
$$k^2 = 144$$
$$k = 12 \quad \text{or} \quad k = -12$$

The equation will have only one rational number solution if $k = 12$ or $k = -12$.

OBJECTIVE 4 **Use the discriminant to decide whether a quadratic trinomial can be factored.** A quadratic trinomial can be factored with rational coefficients only if the corresponding quadratic equation has rational solutions. Thus, the discriminant can be used to decide whether or not a given trinomial is factorable.

EXAMPLE 7 Deciding Whether a Trinomial Is Factorable

Decide whether or not the following trinomials can be factored.

(a) $24x^2 + 7x - 5$

To decide whether the solutions of $24x^2 + 7x - 5 = 0$ are rational numbers, we evaluate the discriminant.

$$b^2 - 4ac = 7^2 - 4(24)(-5) = 49 + 480 = 529 = 23^2$$

Since 529 is a perfect square, the solutions are rational numbers and the trinomial can be factored. Verify that

$$24x^2 + 7x - 5 = (3x - 1)(8x + 5).$$

(b) $11m^2 - 9m + 12$

The discriminant is

$$b^2 - 4ac = (-9)^2 - 4(11)(12) = -447.$$

This number is negative, so the corresponding quadratic equation has imaginary number solutions and therefore the trinomial cannot be factored.

CONNECTIONS

We develop here two interesting and useful properties of the solutions of a quadratic equation $ax^2 + bx + c = 0$, $a \neq 0$. If

$$x_1 = \frac{-b + \sqrt{b^2 - 4ac}}{2a} \quad \text{and} \quad x_2 = \frac{-b - \sqrt{b^2 - 4ac}}{2a},$$

then the sum of the solutions is

$$x_1 + x_2 = \frac{-b + \sqrt{b^2 - 4ac}}{2a} + \frac{-b - \sqrt{b^2 - 4ac}}{2a} = -\frac{b}{a}.$$

The product of the solutions is

$$x_1 x_2 = \left(\frac{-b + \sqrt{b^2 - 4ac}}{2a}\right)\left(\frac{-b - \sqrt{b^2 - 4ac}}{2a}\right).$$

CONNECTIONS (CONTINUED)

Using the rule $(x + y)(x - y) = x^2 - y^2$ with $x = -b$ and $y = \sqrt{b^2 - 4ac}$ gives the product of the numerators.

$$x_1x_2 = \frac{(-b)^2 - (\sqrt{b^2 - 4ac})^2}{(2a)^2} = \frac{c}{a}$$

FOR DISCUSSION OR WRITING

The sum and product of the solutions of a quadratic equation can be used to check the solutions you find.

1. Use the sum and product of the solutions to verify that the solutions of $5x^2 - 2x - 8 = 0$ are $\frac{1 + \sqrt{41}}{5}$ and $\frac{1 - \sqrt{41}}{5}$.
2. Verify that the solutions of $-2x^2 - 5ix = 12$ are $\frac{3}{2}i$ and $-4i$.
3. Fill in the missing steps above that show that the sum of the solutions is $-\frac{b}{a}$ and the product of the solutions is $\frac{c}{a}$.

8.2 EXERCISES

Decide whether each statement is true or false. If it is false, explain why.

1. The equations $x^2 + 3x - 4 = 0$ and $-x^2 - 3x + 4 = 0$ have the same solution set.
2. The quadratic formula cannot be used to solve $2y^2 - 5 = 0$.
3. The equation $4m^2 + 3m = 0$ can be solved by the quadratic formula.
4. The discriminant of $x^2 + x + 1 = 0$ is -3, so the equation has two imaginary solutions.
5. An equation of the form $x^2 + kx = 0$ has two real solutions, one of which is 0.
6. If the discriminant of a quadratic equation is 0, the equation has no solutions.

Use the quadratic formula to solve each equation. (All solutions for these equations are real numbers.) See Examples 1 and 2.

7. $m^2 - 8m + 15 = 0$
8. $x^2 + 3x - 28 = 0$
9. $2k^2 + 4k + 1 = 0$
10. $2y^2 + 3y - 1 = 0$
11. $2x^2 - 2x = 1$
12. $9t^2 + 6t = 1$
13. $x^2 + 18 = 10x$
14. $x^2 - 4 = 2x$
15. $-2t(t + 2) = -3$
16. $-3x(x + 2) = -4$
17. $(r - 3)(r + 5) = 2$
18. $(k + 1)(k - 7) = 1$
19. $p^2 + \frac{p}{3} = \frac{2}{3}$
20. $\frac{x^2}{4} - \frac{x}{2} = 1$
21. $4k(k + 1) = 1$
22. $(r - 1)(4r) = 19$
23. $(g + 2)(g - 3) = 1$
24. $(y - 5)(y + 2) = 6$
25. $3x^2 + 2x = 2$
26. $26r - 2 = 3r^2$
27. $y = \frac{5(5 - y)}{3(y + 1)}$
28. $k = \frac{k + 15}{3(k - 1)}$

29. What is wrong with the following "solution" of $5x^2 - 5x + 1 = 0$?

$$x = \frac{5 \pm \sqrt{25 - 4(5)(1)}}{2(5)} \qquad a = 5, b = -5, c = 1$$
$$= \frac{5 \pm \sqrt{5}}{10}$$
$$= \frac{1}{2} \pm \sqrt{5}$$

30. Is $\frac{-b \pm \sqrt{b^2 - 4ac}}{2a}$ equal to $-b \pm \frac{\sqrt{b^2 - 4ac}}{2a}$? Explain.

Use the quadratic formula to solve each equation. (All solutions for these equations are imaginary numbers.) See Example 3.

31. $k^2 + 47 = 0$
32. $x^2 + 19 = 0$
33. $r^2 - 6r + 14 = 0$
34. $t^2 + 4t + 11 = 0$
35. $4x^2 - 4x = -7$
36. $9x^2 - 6x = -7$
37. $x(3x + 4) = -2$
38. $y(2y + 3) = -2$
39. $\frac{x + 5}{2x - 1} = \frac{x - 4}{x - 6}$
40. $\frac{3x - 4}{2x - 5} = \frac{x + 5}{x + 2}$
41. $\frac{1}{x^2} + 1 = -\frac{1}{x}$
42. $\frac{4}{r^2} + 3 = \frac{1}{r}$

Solve each problem, using a calculator as necessary. Round your answer to the nearest tenth. (You may wish to review Objective 5 in Section 6.7 first.) See Example 4.

43. Working together, two people can cut a large lawn in 2 hours. One person can do the job alone in 1 hour less than the other. How long would it take the faster person to do the job? (*Hint:* x is the time of the faster person.)

Worker	Rate	Time Working Together	Fractional Part of the Job Done
Faster person	$\frac{1}{x}$	2	
Slower person		2	

44. A janitorial service provides two people to clean an office building. Working together, the two can clean the building in 5 hours. One person is new to the job and would take 2 hours longer than the other person to clean the building working alone. How long would it take the new worker to clean the building working alone?

Worker	Rate	Time Working Together	Fractional Part of the Job Done
Faster person			
Slower person			

45. Rusty and Nancy Brauner are planting flats of spring flowers. Working alone, Rusty would take 2 hours longer than Nancy to plant the flowers. Working together, they do the job in 12 hours. How long would it have taken each person working alone?

46. Jay Beckenstein can work through a stack of invoices in 1 hour less time than Colleen Manley Jones can. Working together they take $\frac{3}{2}$ hours. How long would it take each person working alone?

47. Two pipes can fill a tank in 4 hours when used together. Alone, one can fill the tank in .5 hour more than the other. How long will it take each pipe to fill the tank alone?

48. Mashari and Jamal are distributing brochures for a fund-raising campaign. Together they can complete the job in 3 hours. If Mashari could do the job alone in 1 hour more than Jamal, how long would it take Jamal working alone?

Work each problem, rounding answers to the nearest tenth. Objects are projected from Earth.

49. A ball is projected vertically upward from the ground. Its distance in feet from the ground in t seconds is $s = -16t^2 + 128t$. At what times will the ball be 213 feet from the ground?

50. A toy rocket is launched from ground level. Its distance in feet from the ground in t seconds is $s = -16t^2 + 208t$. At what times will the rocket be 550 feet from the ground?

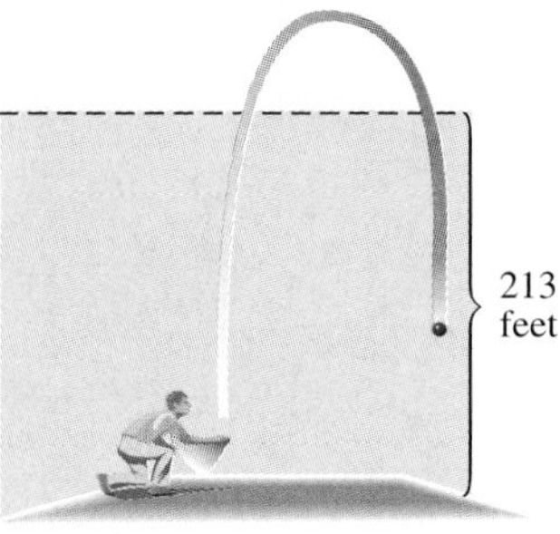

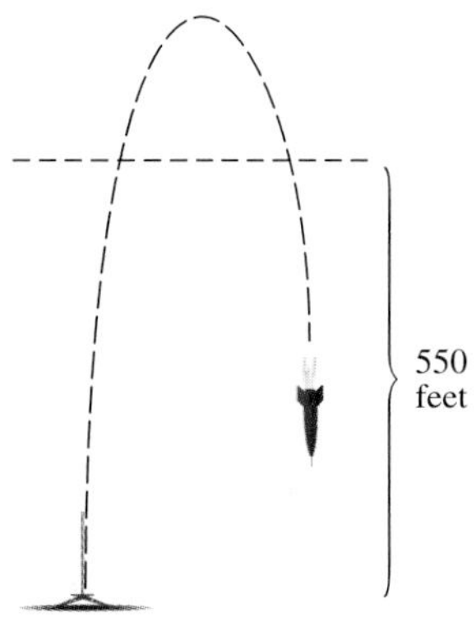

A rock is projected upward from ground level, and its distance from the ground in t seconds is $s = -16t^2 + 160t$. *Use algebra and a short explanation to answer Exercises 51 and 52.*

51. After how many seconds does it reach a height of 400 feet? How would you describe in words its position at this height?

52. After how many seconds does it reach a height of 425 feet? How would you interpret the mathematical result here?

Work each problem.

53. The graph shows sales (in thousands) of telephone answering machines in U.S. homes. The equation

$$f(x) = -88x^2 + 17{,}108x - 813{,}360$$

provides a close approximation of these sales in year x. Here, $x = 90$ represents 1990, $x = 91$ represents 1991, and so on. According to the equation, in what year did sales reach 16,000 thousand?

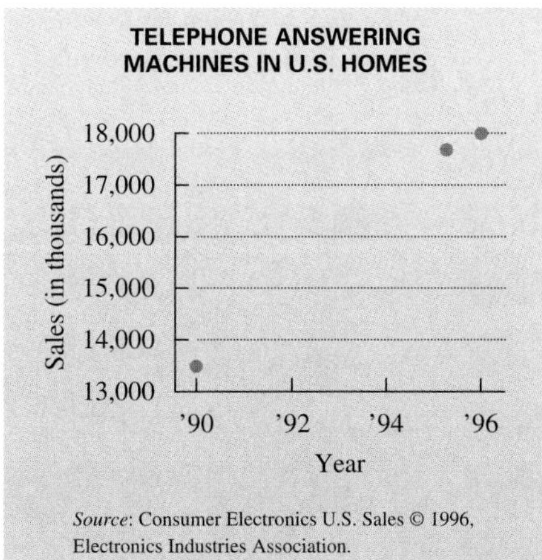

54. CD-ROM revenues for selected years are shown in the graph. The equation

$$y = 15.1x^2 - 2636x + 114{,}870$$

approximates these revenues (in millions of dollars) for year x. Again, $x = 91$ represents 1991, and so on. Assuming conditions remain the same, use the equation to estimate the year in which CD-ROM revenues reach $650 million.

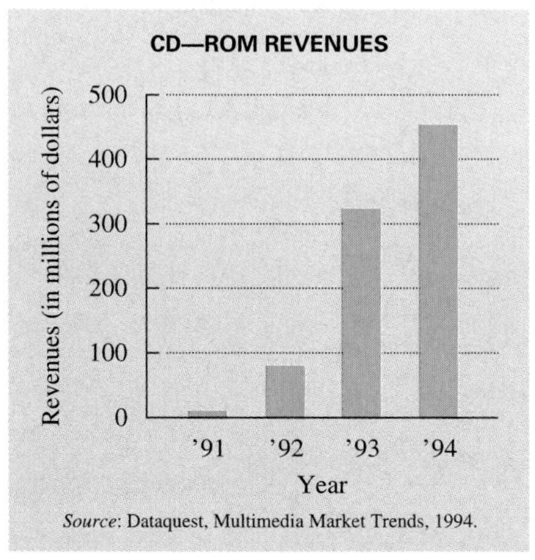

Source: Dataquest, Multimedia Market Trends, 1994.

Use the discriminant to determine whether each equation has solutions that are **(a)** *two different rational numbers,* **(b)** *exactly one rational number,* **(c)** *two different irrational numbers, or* **(d)** *two different imaginary numbers. See Example 5.*

55. $2x^2 - x + 1 = 0$ **56.** $4x^2 - 4x + 3 = 0$ **57.** $6m^2 + 7m - 3 = 0$

58. $7x^2 - 32x - 15 = 0$ **59.** $x^2 + 4x = -4$ **60.** $4y^2 + 36y = -81$

61. $9t^2 = 30t - 15$ **62.** $25k^2 = -20k + 2$

Find the value of a, b, or c so that each equation will have exactly one rational solution. See Example 6.

63. $p^2 + bp + 25 = 0$ **64.** $r^2 - br + 49 = 0$ **65.** $am^2 + 8m + 1 = 0$

66. $ay^2 + 24y + 16 = 0$ **67.** $9y^2 - 30y + c = 0$ **68.** $4m^2 + 12m + c = 0$

69. Is it possible for the solution of a quadratic equation with integer coefficients to include just one irrational number? Why?

70. Can the solution of a quadratic equation with integer coefficients include one real and one imaginary number? Why?

Use the discriminant to tell if each polynomial can be factored (using integer coefficients). If a polynomial can be factored, factor it. See Example 7.

71. $24x^2 - 34x - 45$ **72.** $36y^2 + 69y + 28$ **73.** $36x^2 + 21x - 24$

74. $18k^2 + 13k - 12$ **75.** $12x^2 - 83x - 7$ **76.** $16y^2 - 61y - 12$

RELATING CONCEPTS (EXERCISES 77–80)

We are usually interested in finding the solutions of a quadratic equation. Now we look at the opposite procedure: given two solutions, how do we find an equation that will yield them? There is a theorem which states that if a quadratic equation has real coefficients and $a + bi$ is a solution, then its conjugate, $a - bi$, must also be a solution.

RELATING CONCEPTS (EXERCISES 77–80) (CONTINUED)

Work Exercises 77–80 in order *to see how to find a quadratic equation with solutions* $1 + 5i$ *and* $1 - 5i$.

77. The equation $(x - r)(x - s) = 0$ has solutions r and s. Use $r = 1 + 5i$ and $s = 1 - 5i$, and write the equation making these substitutions. Do not multiply out the factors.

78. Distribute the negative sign in each factor in the equation from Exercise 77 so that the left-hand side is the product of the sum of two terms and the difference of two terms. One of these terms should be written as a binomial within parentheses.

79. Use the pattern $(m + n)(m - n) = m^2 - n^2$ and the fact that $i^2 = -1$ to write the equation from Exercise 78 in the form $ax^2 + bx + c = 0$.

80. Solve the equation from Exercise 79 using the quadratic formula to verify that the solutions are indeed $1 + 5i$ and $1 - 5i$.

Did you make the connection between solving an equation for its solutions and finding an equation given its solutions?

81. One solution of $4x^2 + bx - 3 = 0$ is $-\frac{5}{2}$. Find b and the other solution.

82. One solution of $3x^2 - 7x + c = 0$ is $\frac{1}{3}$. Find c and the other solution.

8.3 Equations Quadratic in Form

OBJECTIVES

1. Solve an equation with fractions by writing it in quadratic form.
2. Use quadratic equations to solve applied problems.
3. Solve an equation with radicals by writing it in quadratic form.
4. Solve an equation that is quadratic in form by substitution.

FOR EXTRA HELP

SSG Sec. 8.3
SSM Sec. 8.3
Pass the Test Software
InterAct Math Tutorial Software
Video 15

We have introduced four methods for solving quadratic equations written in standard form $ax^2 + bx + c = 0$. The following chart gives some advantages and disadvantages of each method.

Methods for Solving Quadratic Equations

Method	*Advantages*	*Disadvantages*
Factoring	Usually the fastest method	Not all polynomials are factorable; some factorable polynomials are hard to factor.
Square root property	Simplest method for solving equations of the form $(x + a)^2 = b$	Few equations are given in this form.
Completing the square	Can always be used, although most people prefer the quadratic formula (This procedure is useful in other areas of mathematics.)	It requires more steps than other methods.
Quadratic formula	Can always be used	It is more difficult than factoring because of the square root.

OBJECTIVE 1 Solve an equation with fractions by writing it in quadratic form. A variety of nonquadratic equations can be written in the form of a quadratic equation and solved by using one of the methods in the chart. As you solve the equations in this section try to decide which is the best method for each equation.

EXAMPLE 1 Writing an Equation with Fractions in Quadratic Form

Solve $\frac{1}{x} + \frac{1}{x-1} = \frac{7}{12}$.

Clear fractions by multiplying each term by the common denominator, $12x(x-1)$. (Note that the domain must be restricted to $x \neq 0$ and $x \neq 1$.)

$$12x(x-1)\frac{1}{x} + 12x(x-1)\frac{1}{x-1} = 12x(x-1)\frac{7}{12}$$

$$12(x-1) + 12x = 7x(x-1)$$

$$12x - 12 + 12x = 7x^2 - 7x \qquad \text{Distributive property}$$

$$24x - 12 = 7x^2 - 7x \qquad \text{Combine terms.}$$

A quadratic equation must be in standard form $ax^2 + bx + c = 0$ before it can be solved by factoring or the quadratic formula. Combine and rearrange terms so that one side is 0. Then solve the resulting equation by factoring.

$$0 = 7x^2 - 31x + 12 \qquad \text{Standard form}$$

$$0 = (7x-3)(x-4) \qquad \text{Factor.}$$

Using the zero-factor property gives the solutions $\frac{3}{7}$ and 4. Check by substituting these solutions in the original equation. The solution set is $\{\frac{3}{7}, 4\}$.

OBJECTIVE 2 Use quadratic equations to solve applied problems. Earlier we solved distance-rate-time (or motion) problems that led to linear equations or rational equations. Now we extend that work to motion problems that lead to quadratic equations.

EXAMPLE 2 Solving a Motion Problem

A riverboat for tourists averages 12 miles per hour in still water. It takes the boat 1 hour, 4 minutes to go 6 miles upstream and return. Find the speed of the current. See Figure 1.

Figure 1

For a problem about rate (or speed), we use the distance formula, $d = rt$.

Let
$$x = \text{the speed of the current;}$$
$$12 - x = \text{the rate upstream;}$$
$$12 + x = \text{the rate downstream.}$$

Rate upstream is the difference of the speed of the boat in still water and the speed of the current, or $12 - x$. Rate downstream is, in the same way, $12 + x$. To find time, rewrite $d = rt$ as

$$t = \frac{d}{r}.$$

This information was used to complete the following chart.

	d	r	t
Upstream	6	$12 - x$	$\frac{6}{12 - x}$
Downstream	6	$12 + x$	$\frac{6}{12 + x}$

Times in hours

Write total time, 1 hour and 4 minutes, as

$$1 + \frac{4}{60} = 1 + \frac{1}{15} = \frac{16}{15} \text{ hours.}$$

Since time upstream plus time downstream equals $\frac{16}{15}$ hours,

$$\frac{6}{12 - x} + \frac{6}{12 + x} = \frac{16}{15}.$$

Now multiply both sides of the equation by the common denominator $15(12 - x)(12 + x)$ and solve the resulting quadratic equation.

$$15(12 + x)6 + 15(12 - x)6 = 16(12 - x)(12 + x)$$
$$90(12 + x) + 90(12 - x) = 16(144 - x^2)$$
$$1080 + 90x + 1080 - 90x = 2304 - 16x^2 \quad \text{Distributive property}$$
$$2160 = 2304 - 16x^2 \quad \text{Combine terms.}$$
$$16x^2 = 144$$
$$x^2 = 9$$
$$x = 3 \quad \text{or} \quad x = -3 \quad \text{Square root property}$$

The speed of the current cannot be -3, so the solution is $x = 3$ miles per hour.

OBJECTIVE 3 Solve an equation with radicals by writing it in quadratic form.

EXAMPLE 3 Solving a Radical Equation That Leads to a Quadratic Equation

Solve $x + \sqrt{x} = 6$.

This equation is not quadratic. However, squaring both sides of the equation gives a quadratic equation that can be solved by factoring.

$$\sqrt{x} = 6 - x \quad \text{Isolate the radical on one side.}$$

$$x = 36 - 12x + x^2 \quad \text{Square both sides.}$$

$$0 = x^2 - 13x + 36 \quad \text{Get 0 on one side.}$$

$$0 = (x - 4)(x - 9) \quad \text{Factor.}$$

$$x - 4 = 0 \quad \text{or} \quad x - 9 = 0 \quad \text{Set each factor equal to 0.}$$

$$x = 4 \quad \text{or} \quad x = 9$$

Check both potential solutions in the *original* equation.

If $x = 4$,
$4 + \sqrt{4} = 6$?
$6 = 6$. True

If $x = 9$,
$9 + \sqrt{9} = 6$?
$12 = 6$. False

The solution set is $\{4\}$.

OBJECTIVE 4 Solve an equation that is quadratic in form by substitution. In Chapter 3, we discussed composition of functions of the form $g[f(x)]$, where g and f are functions. An equation that is written in composite form $a[f(x)]^2 + b[f(x)] + c = 0$, for $a \neq 0$, is called **quadratic in form.** For example, suppose

$$g(m) = 2(4m - 3)^2 + 7(4m - 3) + 5 = 0$$

and we let $f(m) = 4m - 3$. Then

$$g[f(m)] = 2[f(m)]^2 + 7[f(m)] + 5.$$

To simplify further, let $u = f(m)$; then

$$g(u) = 2u^2 + 7u + 5.$$

Solving this quadratic equation and substituting the solutions for u in $u = f(m) = 4m - 3$ will give the solutions for m.

EXAMPLE 4 Solving Equations That Are Quadratic in Form

Solve each equation.

(a) $2(4m - 3)^2 + 7(4m - 3) + 5 = 0$

Because of the repeated quantity $4m - 3$, this equation is quadratic in form with $f(m) = 4m - 3$. Let $f(m) = u$ and write

$$2(4m - 3)^2 + 7(4m - 3) + 5 = 0$$

as

$$2u^2 + 7u + 5 = 0. \quad \text{Let } 4m - 3 = u.$$

$$(2u + 5)(u + 1) = 0 \quad \text{Factor.}$$

By the zero-factor property, the solutions of $2u^2 + 7u + 5 = 0$ are

$$u = -\frac{5}{2} \quad \text{or} \quad u = -1.$$

To find m, substitute $4m - 3$ for u.

$$4m - 3 = -\frac{5}{2} \quad \text{or} \quad 4m - 3 = -1$$
$$4m = \frac{1}{2} \quad \text{or} \quad 4m = 2$$
$$m = \frac{1}{8} \quad \text{or} \quad m = \frac{1}{2}$$

The solution set of the original equation is $\{\frac{1}{8}, \frac{1}{2}\}$.

(b) $y^4 = 6y^2 - 3$

First write the equation as

$$y^4 - 6y^2 + 3 = 0 \quad \text{or} \quad (y^2)^2 - 6(y^2) + 3 = 0,$$

which is quadratic in form with $f(y) = y^2$. Substitute u for y^2 and u^2 for y^4 to get

$$u^2 - 6u + 3 = 0.$$

By the quadratic formula,

$$u = \frac{6 \pm \sqrt{36 - 12}}{2} \qquad a = 1, b = -6, c = 3$$
$$= \frac{6 \pm \sqrt{24}}{2}$$
$$= \frac{6 \pm 2\sqrt{6}}{2}$$
$$u = 3 \pm \sqrt{6}.$$

Find y by using the square root property as follows.

$$y^2 = 3 + \sqrt{6} \quad \text{or} \quad y^2 = 3 - \sqrt{6}$$
$$y = \pm\sqrt{3 + \sqrt{6}} \quad \text{or} \quad y = \pm\sqrt{3 - \sqrt{6}}$$

The solution set contains four numbers:

$$\{\sqrt{3 + \sqrt{6}}, -\sqrt{3 + \sqrt{6}}, \sqrt{3 - \sqrt{6}}, -\sqrt{3 - \sqrt{6}}\}.$$

EXAMPLE 5 Solving Equations That Are Quadratic in Form

Solve each equation.

(a) $4x^6 + 1 = 5x^3$

This equation is quadratic in form with $f(x) = x^3$. Let $x^3 = u$. Then $x^6 = u^2$. Substitute into the given equation.

$$4x^6 + 1 = 5x^3$$
$$4(x^3)^2 + 1 = 5x^3$$
$$4u^2 + 1 = 5u \qquad \text{Let } x^3 = u.$$
$$4u^2 - 5u + 1 = 0 \qquad \text{Get 0 on one side.}$$
$$(4u - 1)(u - 1) = 0 \qquad \text{Factor.}$$

$$u = \frac{1}{4} \quad \text{or} \quad u = 1$$

$$x^3 = \frac{1}{4} \quad \text{or} \quad x^3 = 1 \qquad u = x^3$$

From these equations,

$$x = \sqrt[3]{\frac{1}{4}} = \frac{\sqrt[3]{1}}{\sqrt[3]{4}} = \frac{1}{\sqrt[3]{4}} \cdot \frac{\sqrt[3]{2}}{\sqrt[3]{2}} = \frac{\sqrt[3]{2}}{2} \quad \text{or} \quad x = \sqrt[3]{1} = 1.$$

This method of substitution gives only real number solutions for equations with polynomials of degree $2n$ where n is odd. However, it gives all complex number solutions for equations with polynomials of degree $2n$ where n is even. The real number solution set of $4x^6 + 1 = 5x^3$ is $\left\{\frac{\sqrt[3]{2}}{2}, 1\right\}$.

(b) $2a^{2/3} - 11a^{1/3} + 12 = 0$

Let $a^{1/3} = u$; then $a^{2/3} = u^2$. Substitute into the given equation.

$$2(a^{1/3})^2 - 11a^{1/3} + 12 = 0$$

$$2u^2 - 11u + 12 = 0 \qquad \text{Let } a^{1/3} = u.$$

$$(2u - 3)(u - 4) = 0 \qquad \text{Factor.}$$

$$2u - 3 = 0 \quad \text{or} \quad u - 4 = 0$$

$$u = \frac{3}{2} \quad \text{or} \quad u = 4$$

$$a^{1/3} = \frac{3}{2} \quad \text{or} \quad a^{1/3} = 4 \qquad u = a^{1/3}$$

$$a = \left(\frac{3}{2}\right)^3 = \frac{27}{8} \quad \text{or} \quad a = 4^3 = 64 \qquad \text{Cube both sides.}$$

Check that the solution set is $\left\{\frac{27}{8}, 64\right\}$.

Some people prefer to solve equations like the one in Example 5(a) by directly factoring using the given equation. For example, the equation could be solved as follows.

$$4x^6 + 1 = 5x^3 \qquad \text{Given equation}$$

$$4x^6 - 5x^3 + 1 = 0 \qquad \text{Write in standard form.}$$

$$(4x^3 - 1)(x^3 - 1) = 0 \qquad \text{Factor.}$$

$$4x^3 - 1 = 0 \quad \text{or} \quad x^3 - 1 = 0 \qquad \text{Zero-factor property}$$

$$x^3 = \frac{1}{4} \quad \text{or} \quad x^3 = 1 \qquad \text{Solve each equation.}$$

Complete the solution as in Example 5(a).

8.3 EXERCISES

Based on the discussion and examples of this section, write a sentence describing the first step you would take in solving each of the following equations. Do not actually solve each equation.

1. $\frac{14}{x} = x - 5$

2. $\sqrt{1 + x} + x = 5$

3. $(r^2 + r)^2 - 8(r^2 + r) + 12 = 0$

4. $3t = \sqrt{16 - 10t}$

5. Of the four methods for solving quadratic equations, it is often said that the quadratic formula is the most efficient method. Do you agree or disagree with this statement? Explain.

6. What is the relationship between the method of completing the square in solving quadratic equations and the quadratic formula?

Solve by first clearing each equation of fractions. Check your answers. See Example 1.

7. $\frac{14}{x} = x - 5$

8. $\frac{-12}{x} = x + 8$

9. $1 - \frac{3}{x} - \frac{28}{x^2} = 0$

10. $4 - \frac{7}{r} - \frac{2}{r^2} = 0$

11. $\frac{1}{x} + \frac{2}{x + 2} = \frac{17}{35}$

12. $\frac{2}{m} + \frac{3}{m + 9} = \frac{11}{4}$

13. $\frac{2}{x + 1} + \frac{3}{x + 2} = \frac{7}{2}$

14. $\frac{4}{3 - y} + \frac{2}{5 - y} = \frac{26}{15}$

15. $\frac{3}{2x} - \frac{1}{2(x + 2)} = 1$

16. $\frac{4}{3x} - \frac{1}{2(x + 1)} = 1$

17. If it takes m hours to grade a set of papers, what is the grader's rate (in job per hour)?

18. If a boat goes 20 miles per hour in still water, and the rate of the current is t miles per hour, what is the rate of the boat when it travels upstream? Downstream?

Solve each problem by writing an equation with fractions and solving it. See Example 2.

19. On a windy day Yoshiaki found that he could go 16 miles downstream and then 4 miles back upstream at top speed in a total of 48 minutes. What was the top speed of Yoshiaki's boat if the speed of the current was 15 miles per hour?

20. Lekesha flew her plane for 6 hours at a constant speed. She traveled 810 miles with the wind, then turned around and traveled 720 miles against the wind. The wind speed was a constant 15 miles per hour. Find the speed of the plane.

21. In Canada, Medicine Hat and Cranbrook are 300 kilometers apart. Harry rides his Honda 20 kilometers per hour faster than Karen rides her Yamaha. Find Harry's average speed if he travels from Cranbrook to Medicine Hat in $1\frac{1}{4}$ hours less time than Karen.

22. The distance from Jackson to Lodi is about 40 miles, as is the distance from Lodi to Manteca. Rico drove from Jackson to Lodi during rush hour, stopped in Lodi for a root beer, and then drove on to Manteca at 10 miles per hour faster. Driving time for the entire trip was 88 minutes. Find his speed from Jackson to Lodi.

23. A washing machine can be filled in 6 minutes if both the hot and cold water taps are fully opened. To fill the washer with hot water alone takes 9 minutes longer than filling with cold water alone. How long does it take to fill the tank with cold water?

24. Two pipes together can fill a large tank in 2 hours. One of the pipes, used alone, takes 3 hours longer than the other to fill the tank. How long would each pipe take to fill the tank alone?

Find all solutions in Exercises 25–32 by first squaring. Check your answers. See Example 3.

25. $2x = \sqrt{11x + 3}$

26. $4x = \sqrt{6x + 1}$

27. $3y = (16 - 10y)^{1/2}$

28. $4t = (8t + 3)^{1/2}$

29. $p - 2\sqrt{p} = 8$

30. $k + \sqrt{k} = 12$

31. $m = \sqrt{\dfrac{6 - 13m}{5}}$

32. $r = \sqrt{\dfrac{20 - 19r}{6}}$

Solve each problem. Use 3.14 as an approximation for π.

33. Artificial gravity can be created in a space station by rotating all or part of the station. To achieve gravity equal to one-half the gravity of Earth with .063 rotation per second, the distance r in meters of a point in the station from the center of rotation is given by the equation $.126\pi r = \sqrt{4.7r}$. Find the value of r. (*Source:* Bernice Kastner, *Space Mathematics,* NASA: pp. 48–49.)

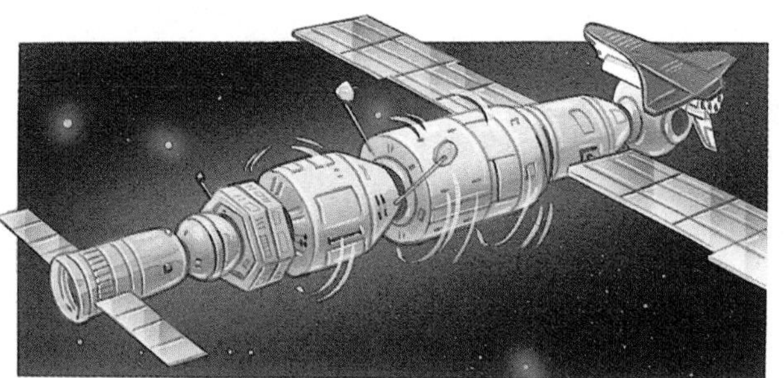

34. In Exercise 33, if Earth surface gravity is desired with a rotation rate of .04, the equation for r in meters is $.08\pi r = \sqrt{9.8r}$. Find r.

Find all solutions to each equation. Check your answers. See Examples 4 and 5.

35. $t^4 - 18t^2 + 81 = 0$

36. $y^4 - 8y^2 + 16 = 0$

37. $4k^4 - 13k^2 + 9 = 0$

38. $9x^4 - 25x^2 + 16 = 0$

39. $(x + 3)^2 + 5(x + 3) + 6 = 0$

40. $(k - 4)^2 + (k - 4) - 20 = 0$

41. $(t + 5)^2 + 6 = 7(t + 5)$

42. $3(m + 4)^2 - 8 = 2(m + 4)$

43. $2 + \dfrac{5}{3k - 1} = \dfrac{-2}{(3k - 1)^2}$

44. $3 - \dfrac{7}{2p + 2} = \dfrac{6}{(2p + 2)^2}$

45. $2 - 6(m - 1)^{-2} = (m - 1)^{-1}$

46. $3 - 2(x - 1)^{-1} = (x - 1)^{-2}$

Use substitution to solve each equation. Check your answers. See Example 5.

47. $4k^{4/3} - 13k^{2/3} + 9 = 0$

48. $9y^{2/5} = 16 - 10y^{1/5}$

49. $x^{2/3} + x^{1/3} - 2 = 0$

50. $3x^{2/3} - x^{1/3} - 24 = 0$

51. $2(1 + \sqrt{y})^2 = 13(1 + \sqrt{y}) - 6$

52. $(k^2 + k)^2 + 12 = 8(k^2 + k)$

53. $2x^4 + x^2 - 3 = 0$

54. $4k^4 + 5k^2 + 1 = 0$

55. What is wrong with the following "solution"?

$$2(m-1)^2 - 3(m-1) + 1 = 0$$
$$2u^2 - 3u + 1 = 0 \quad \text{Let } u = m - 1.$$
$$(2u-1)(u-1) = 0$$
$$2u - 1 = 0 \quad \text{or} \quad u - 1 = 0$$
$$u = \frac{1}{2} \qquad u = 1$$

Solution set: $\left\{\frac{1}{2}, 1\right\}$

56. Explain how to solve the equation

$$y + \sqrt{y} - 6 = 0$$

using both the method of Example 3 and the method of Example 4.

The equations in Exercises 57–64 are not grouped by type. Decide which method of solution applies, and then solve each equation. Give only real solutions. See Examples 1, 3, 4, and 5.

57. $x^4 - 16x^2 + 48 = 0$

58. $\left(x - \frac{1}{2}\right)^2 + 5\left(x - \frac{1}{2}\right) - 4 = 0$

59. $\sqrt{2x+3} = 2 + \sqrt{x-2}$

60. $\sqrt{m+1} = -1 + \sqrt{2m}$

61. $2m^6 + 11m^3 + 5 = 0$

62. $8x^6 + 513x^3 + 64 = 0$

63. $2 - (y-1)^{-1} = 6(y-1)^{-2}$

64. $3 = 2(p-1)^{-1} + (p-1)^{-2}$

RELATING CONCEPTS (EXERCISES 65–70)

Consider the following equation, which contains variable expressions in the denominators.

$$\frac{x^2}{(x-3)^2} + \frac{3x}{x-3} - 4 = 0.$$

Work Exercises 65–70 in order. *They all pertain to this equation.*

65. Why must 3 be excluded from the domain of this equation?

66. Multiply both sides of the equation by the LCD, $(x-3)^2$, and solve. There is only one solution—what is it?

67. Write the equation in a different manner so that it is quadratic in form, with the rational expression $\frac{x}{x-3}$ as the expression that you will substitute another variable for.

68. In your own words, explain why the expression $\frac{x}{x-3}$ cannot equal 1.

69. Solve the equation from Exercise 67 by making the substitution $t = \frac{x}{x-3}$. You should get two values for t. Why is one of them impossible for this equation?

70. Solve the equation $x^2(x-3)^{-2} + 3x(x-3)^{-1} - 4 = 0$ by letting $s = (x-3)^{-1}$. You should get two values for s. Why is this impossible for this equation?

Did you make the connection between the three solution methods and understand why one of the answers does not satisfy the equation?

8.4 Formulas and Applications

OBJECTIVES

1. Solve formulas for variables involving squares and square roots.
2. Solve applied problems about motion along a straight line.
3. Solve applied problems using the Pythagorean formula.
4. Solve applied problems using formulas for area.

FOR EXTRA HELP

SSG Sec. 8.4
SSM Sec. 8.4

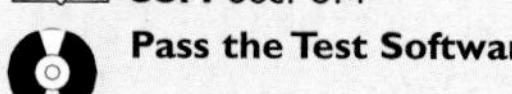
Pass the Test Software

InterAct Math Tutorial Software

Video 16

OBJECTIVE 1 **Solve formulas for variables involving squares and square roots.** The methods presented earlier in this chapter and the previous one can be used to solve such formulas.

EXAMPLE 1 **Solving for a Variable Involving a Square or a Square Root**

(a) Solve $w = \frac{kFr}{v^2}$ for v.

Multiply each side by v^2 to clear the equation of fractions. Then solve for v, using the square root property.

$$w = \frac{kFr}{v^2}$$

$$v^2 w = kFr \qquad \text{Multiply each side by } v^2.$$

$$v^2 = \frac{kFr}{w} \qquad \text{Divide each side by } w.$$

$$v = \pm\sqrt{\frac{kFr}{w}} \qquad \text{Square root property}$$

$$v = \frac{\pm\sqrt{kFr}}{\sqrt{w}} \cdot \frac{\sqrt{w}}{\sqrt{w}} = \frac{\pm\sqrt{kFrw}}{w} \qquad \text{Rationalize the denominator.}$$

(b) Solve $d = \sqrt{\frac{4A}{\pi}}$ for A.

Square both sides to eliminate the radical.

$$d = \sqrt{\frac{4A}{\pi}}$$

$$d^2 = \frac{4A}{\pi} \qquad \text{Square both sides.}$$

$$\pi d^2 = 4A \qquad \text{Multiply both sides by } \pi.$$

$$\frac{\pi d^2}{4} = A \qquad \text{Divide both sides by 4.}$$

Check the solution in the original equation. ■

EXAMPLE 2 **Solving for a Squared Variable**

Solve $s = 2t^2 + kt$ for t.

Since the equation has terms with t^2 and t, write it in standard form $ax^2 + bx + c = 0$, with t as the variable instead of x.

$$s = 2t^2 + kt$$

$$0 = 2t^2 + kt - s$$

Now use the quadratic formula with $a = 2$, $b = k$, and $c = -s$.

$$t = \frac{-k \pm \sqrt{k^2 - 4(2)(-s)}}{2(2)} \qquad a = 2, b = k, c = -s$$

$$= \frac{-k \pm \sqrt{k^2 + 8s}}{4}$$

The solutions are $t = \dfrac{-k + \sqrt{k^2 + 8s}}{4}$ and $t = \dfrac{-k - \sqrt{k^2 + 8s}}{4}$.

OBJECTIVE 2 **Solve applied problems about motion along a straight line.** In the next example we use a quadratic equation to describe the distance traveled by a propelled object.

EXAMPLE 3 Solving a Straight-Line Motion Problem

If a rock on Earth is thrown upward from a 144-foot building with an initial velocity of 112 feet per second, its position (in feet above the ground) is given by $s = -16t^2 + 112t + 144$, where t is time in seconds after it was thrown. When does it hit the ground?

When the rock hits the ground, its distance above the ground is 0. Find t when s is 0 by solving the equation

$$0 = -16t^2 + 112t + 144. \qquad \text{Let } s = 0.$$

$$0 = t^2 - 7t - 9 \qquad \text{Divide both sides by } -16.$$

$$t = \frac{7 \pm \sqrt{49 + 36}}{2} \qquad \text{Quadratic formula}$$

$$t = \frac{7 \pm \sqrt{85}}{2}$$

$$t \approx 8.1 \quad \text{or} \quad t \approx -1.1 \qquad \text{Use a calculator.}$$

Discard the negative solution. The rock will hit the ground about 8.1 seconds after it is thrown.

Remember to check all proposed solutions against the information of the original problem. Some solutions of the equation may not satisfy the physical conditions of the problem.

When an object moves under the influence of gravity (without air resistance) near the surface of a planet, its path has the shape of a parabola. Suppose an object is propelled upward at a 45° angle with an initial velocity of 30 miles per hour. Then the coordinates x and y of a point on the parabola in feet y at time x in seconds are given by

$$y = x - \frac{g}{1922}x^2,$$

where g is the force of gravity.

EXAMPLE 4 Solving a Problem about the Motion of a Propelled Object

Suppose a rocket is propelled from the surface of Mars at a 45° angle with an initial velocity of 30 miles per hour. See Figure 2. (*Source:* Bernice Kastner, *Space Mathematics,* NASA.)

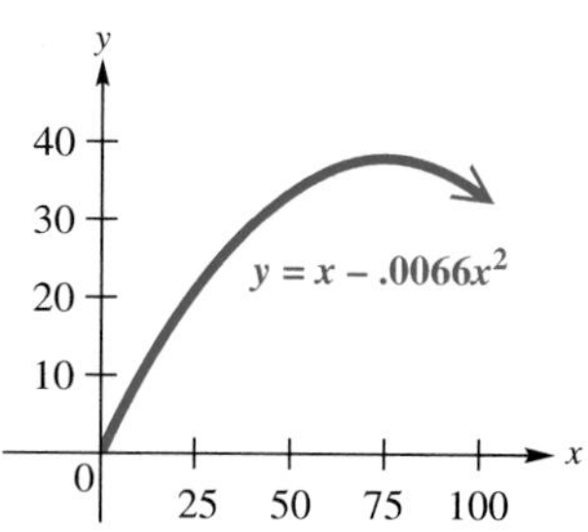

Figure 2

(a) For Mars, $g = 12.6$. Find the height of the rocket after 50 seconds.
Replacing g in the equation gives

$$y = x - \frac{12.6}{1922}x^2 \approx x - .0066x^2.$$

Now substitute 50 for x to get

$$y = 50 - .0066(50)^2 = 33.5 \text{ feet.}$$

The rocket reaches a height of 33.5 feet at 50 seconds.

(b) How long will it take for the rocket to reach 100 feet?
Let $y = 100$ and solve the equation

$$100 = x - .0066x^2.$$

Write the equation in standard form, and use the quadratic formula.

$$.0066x^2 - x + 100 = 0 \qquad \text{Get 0 on one side.}$$

$$x = \frac{1 \pm \sqrt{1 - 4(.0066)(100)}}{2(.0066)} \qquad \text{Let } a = .0066,\ b = -1,\ c = 100.$$

$$x = \frac{1 \pm \sqrt{1 - 2.64}}{.0132}$$

The quantity under the radical is negative. This indicates that the rocket never reaches a height of 100 feet.

OBJECTIVE 3 Solve applied problems using the Pythagorean formula. The Pythagorean formula, which was introduced in Chapter 7, is used in applications involving right triangles. Such problems often require solving a quadratic equation.

EXAMPLE 5 Using the Pythagorean Formula

Two cars left an intersection at the same time, one heading due north, and the other due west. Some time later, they were exactly 100 miles apart. The car headed north had gone 20 miles farther than the car headed west. How far had each car traveled?

Let x be the distance traveled by the car headed west. Then $x + 20$ is the distance traveled by the car headed north. These distances are shown in Figure 3.

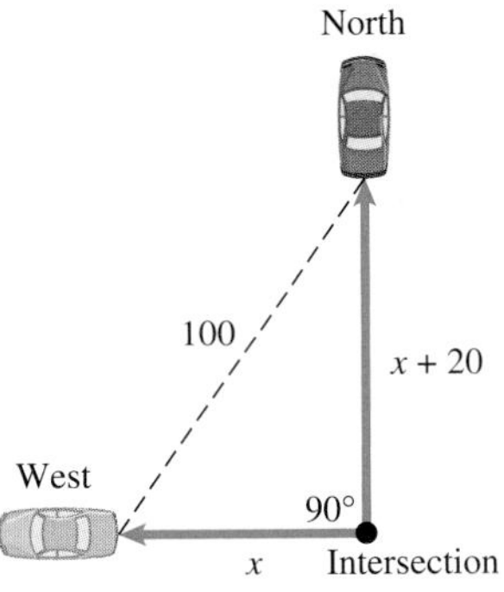

Figure 3

The cars are 100 miles apart, so the hypotenuse of the right triangle equals 100 and the two legs are equal to x and $x + 20$. Use the Pythagorean formula.

$$c^2 = a^2 + b^2$$
$$100^2 = x^2 + (x + 20)^2$$
$$10{,}000 = x^2 + x^2 + 40x + 400 \quad \text{Square the binomial.}$$
$$0 = 2x^2 + 40x - 9600 \quad \text{Get 0 on one side.}$$
$$0 = 2(x^2 + 20x - 4800) \quad \text{Factor out the common factor.}$$
$$0 = x^2 + 20x - 4800 \quad \text{Divide both sides by 2.}$$

Use the quadratic formula to find x.

$$x = \frac{-20 \pm \sqrt{400 - 4(1)(-4800)}}{2} \quad a = 1,\ b = 20,\ c = -4800$$
$$= \frac{-20 \pm \sqrt{19{,}600}}{2}$$
$$x = 60 \quad \text{or} \quad x = -80 \quad \text{Use a calculator.}$$

Discard the negative solution. The required distances are 60 miles and $60 + 20 = 80$ miles.

OBJECTIVE 4 Solve applied problems using formulas for area.

EXAMPLE 6 Solving an Area Problem

A rectangular reflecting pool in a park is 20 feet wide and 30 feet long. The park gardener wants to plant a strip of grass of uniform width around the edge of the pool. She has enough seed to cover 336 square feet. How wide will the strip be?

The pool is shown in Figure 4. If x represents the unknown width of the grass strip, the width of the large rectangle is given by $20 + 2x$ (the width of the pool plus two grass strips), and the length is given by $30 + 2x$.

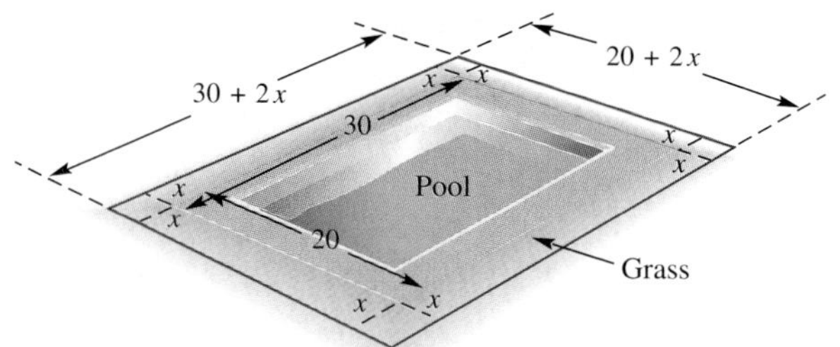

Figure 4

The area of the large rectangle is given by the product of its length and width, $(20 + 2x)(30 + 2x)$. The area of the pool is $20 \cdot 30 = 600$ square feet. The area of the large rectangle, minus the area of the pool, should equal the area of the grass strip. Since the area of the grass strip is to be 336 square feet, the equation is

$$\underset{\text{Area of rectangle}}{(20 + 2x)(30 + 2x)} - \underset{\text{Area of pool}}{600} = \underset{\text{Area of grass}}{336}$$

$$600 + 100x + 4x^2 - 600 = 336 \qquad \text{Multiply.}$$

$$4x^2 + 100x - 336 = 0 \qquad \text{Combine terms.}$$

$$x^2 + 25x - 84 = 0 \qquad \text{Divide by 4.}$$

$$(x + 28)(x - 3) = 0 \qquad \text{Factor.}$$

$$x = -28 \quad \text{or} \quad x = 3.$$

The width cannot be -28 feet, so the grass strip should be 3 feet wide.

8.4 EXERCISES

1. What is the first step in solving a formula that has the specified variable in the denominator?
2. What is the first step in solving a formula like $gw^2 = 2r$ for w?
3. What is the first step in solving a formula like $gw^2 = kw + 24$ for w?
4. Why is it particularly important to check all proposed solutions to an applied problem against the information in the original problem?

5. In a problem like the one in Example 3, is it possible to get two correct nonzero answers?
6. In Example 4(b), if the height is changed so that the quantity under the radical becomes .472, there are two correct nonzero answers. Why is this possible?

In Exercises 7 and 8, solve for m in terms of the other variables. Remember that m must be positive.

7.

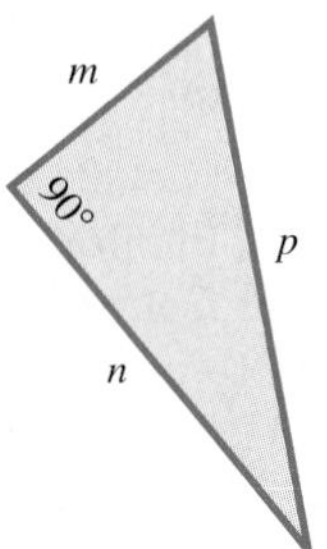

8.

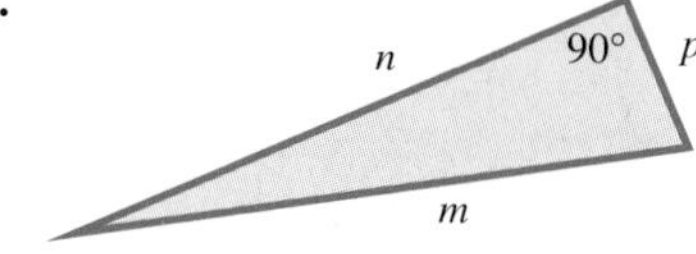

Solve each equation for the indicated variable. (While in practice we would often reject a negative value due to the physical nature of the quantity represented by the variable, leave ± in your answers here.) See Examples 1 and 2.

9. $d = kt^2$; for t

10. $s = kwd^2$; for d

11. $F = \dfrac{kA}{v^2}$; for v

12. $L = \dfrac{kd^4}{h^2}$; for h

13. $V = \dfrac{1}{3}\pi r^2 h$; for r

14. $V = \pi(r^2 + R^2)h$; for r

15. $At^2 + Bt = -C$; for t

16. $S = 2\pi rh + \pi r^2$; for r

17. $D = \sqrt{kh}$; for h

18. $F = \dfrac{k}{\sqrt{d}}$; for d

19. $p = \sqrt{\dfrac{k\ell}{g}}$; for ℓ

20. $p = \sqrt{\dfrac{k\ell}{g}}$; for g

21. If g is a positive number in the formula of Exercise 19, explain why k and ℓ must have the same sign in order for p to be a real number.

22. Refer to Example 2 of this section. Suppose that k and s both represent positive numbers.
(a) Which one of the two solutions given is positive?
(b) Which one is negative?
(c) How can you tell?

Solve each problem by using a quadratic equation. Use a calculator as necessary, and round the answer to the nearest tenth. See Examples 3 and 4.

23. The Mart Hotel in Dallas, Texas, is 400 feet high. Suppose that a ball is projected upward from the top of the Mart, and its position s in feet above the ground is given by the equation $s = -16t^2 + 45t + 400$, where t is the number of seconds elapsed. How long will it take for the ball to reach a height of 200 feet above the ground? (*Source: The World Almanac and Book of Facts.*)

24. The Toronto Dominion Center in Winnipeg, Manitoba, is 407 feet high. Suppose that a ball is projected upward from the top of the Center, and its position s in feet above the ground is given by the equation $s = -16t^2 + 75t + 407$, where t is the number of seconds elapsed. How long will it take for the ball to reach a height of 450 feet above the ground? (*Source: The World Almanac and Book of Facts.*)

25. Refer to the equations in Exercises 23 and 24. Suppose that the first sentence in each problem did not give the height of the building. How could you use the equation to determine the height of the building?

26. A search light moves horizontally back and forth along a wall with the distance of the light from a starting point at t minutes given by $s = 100t^2 - 300t$. How long will it take before the light returns to the starting point?

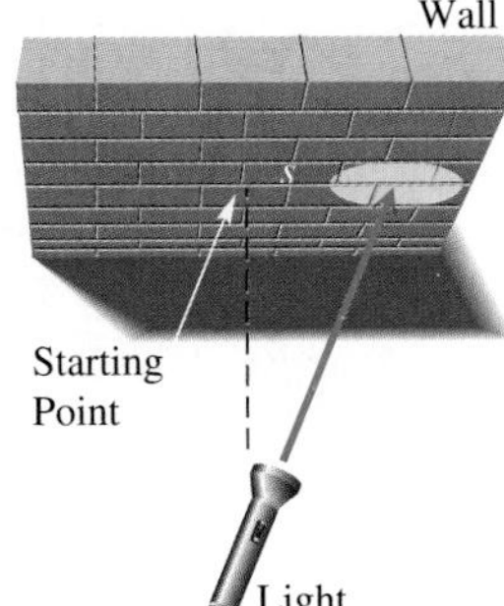

27. An object is projected directly upward from the ground. After t seconds its distance in feet above the ground is $s = 144t - 16t^2$.

(a) After how many seconds will the object be 128 feet above the ground? (*Hint:* Look for a common factor before solving the equation.)

(b) When does the object strike the ground?

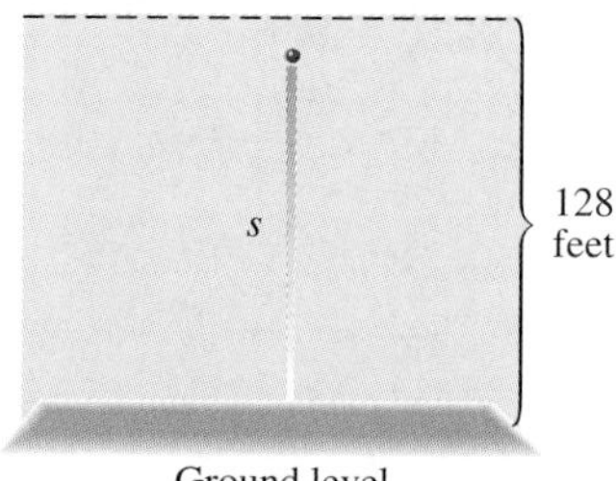

28. The formula $D = 100t - 13t^2$ gives the distance in feet a car going approximately 68 miles per hour will skid in t seconds. Find the time it would take for the car to skid 190 feet. (*Hint:* Your answer must be less than the time it takes the car to stop, which is 3.8 seconds.)

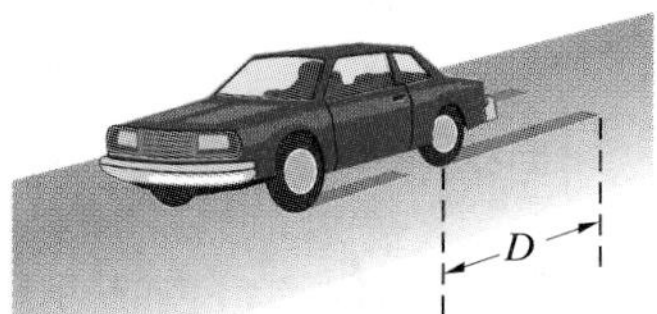

Refer to Example 4 for Exercises 29 and 30.

29. Suppose a space vehicle leaves the moon under the conditions in Example 4. The force of gravity on the moon is approximately 3320 feet per second squared. Answer the following using four significant digits.

(a) Write an equation for the height y of the space vehicle at x seconds.

(b) Find the height after .25 second and after .5 second.

(c) At what time(s) is the vehicle on the surface of the moon (that is, when is $y = 0$)?

(d) Explain why there are two answers for part (c).

30. Refer to Exercise 29. If a rock on the surface of the moon is propelled at an angle of 60° with an initial velocity of 60 miles an hour, the equation is $y = 1.727x - .0013x^2$ feet, where x is time in seconds.

(a) Find the height of the rock after 1 second and after 2 seconds.

(b) How long will it take the rock to return to the surface?

Use the Pythagorean formula to solve each problem. Use a calculator as necessary, and when appropriate, round answers to the nearest tenth. See Example 5.

31. Find the lengths of the sides of the triangle.

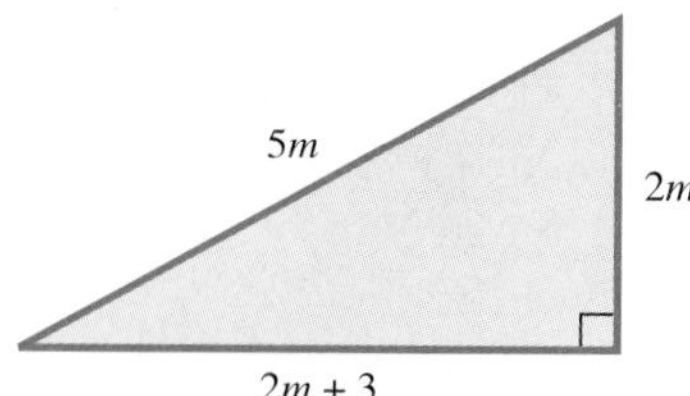

32. Find the lengths of the sides of the triangle.

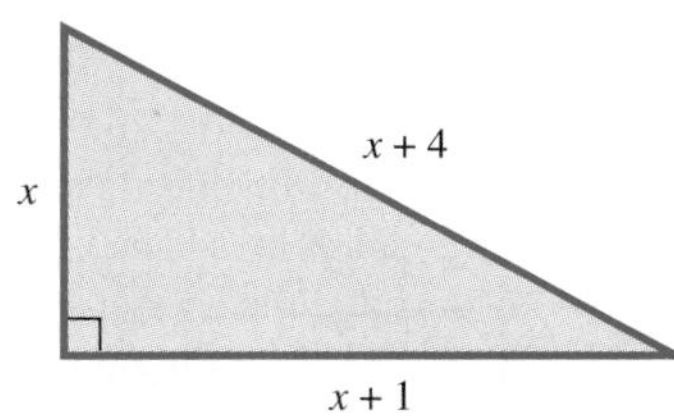

33. Refer to Exercise 23. Suppose that a wire is attached to the top of the Mart, and pulled tight. It is attached to the ground 100 feet from the base of the building, as shown in the figure. How long is the wire?

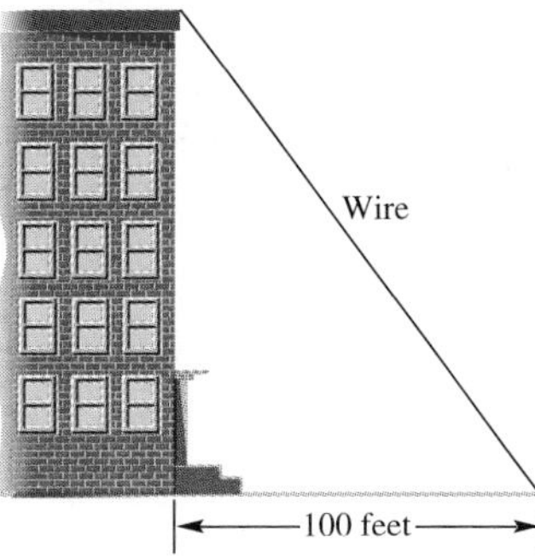

34. Refer to Exercise 24. Suppose that a wire is attached to the top of the Center, and pulled tight. The length of the wire is twice the distance between the base of the Center and the point on the ground where the wire is attached. How long is the wire?

35. Two ships leave port at the same time, one heading due south and the other heading due east. Several hours later, they are 170 miles apart. If the ship traveling south had traveled 70 miles farther than the other, how many miles had they each traveled?

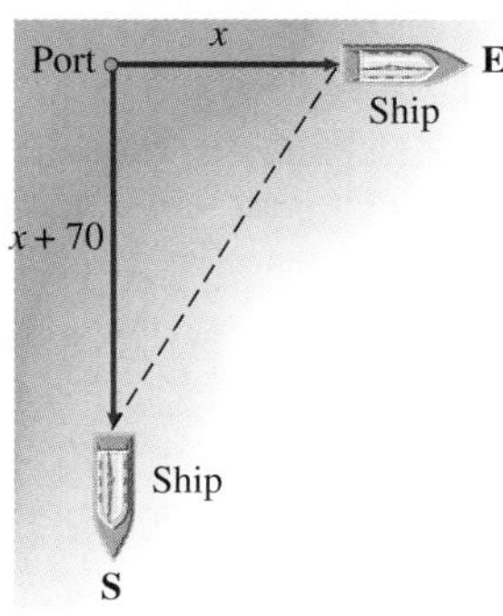

36. Kim Hobbs is flying a kite that is 30 feet farther above her hand than its horizontal distance from her. The string from her hand to the kite is 150 feet long. How high is the kite?

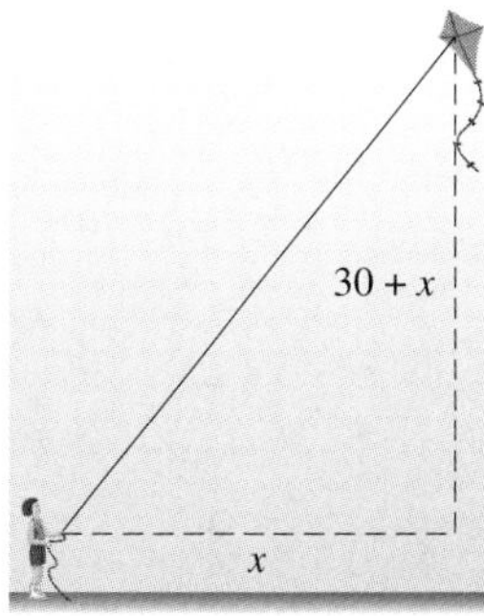

37. A toy manufacturer needs a piece of plastic in the shape of a right triangle with the longer leg 2 centimeters more than twice as long as the shorter leg, and the hypotenuse 1 centimeter more than the longer leg. How long should the three sides of the triangular piece be?

38. Michael Fuentes, a developer, owns a piece of land enclosed on three sides by streets, giving it the shape of a right triangle. The hypotenuse is 8 meters longer than the longer leg, and the shorter leg is 9 meters shorter than the hypotenuse. Find the lengths of the three sides of the property.

39. Two pieces of a large wooden puzzle fit together to form a rectangle with a length 1 centimeter less than twice the width. The diagonal, where the two pieces meet, is 2.5 centimeters in length. Find the length and width of the rectangle.

40. A 13-foot ladder is leaning against a house. The distance from the bottom of the ladder to the house is 7 feet less than the distance from the top of the ladder to the ground. How far is the bottom of the ladder from the house?

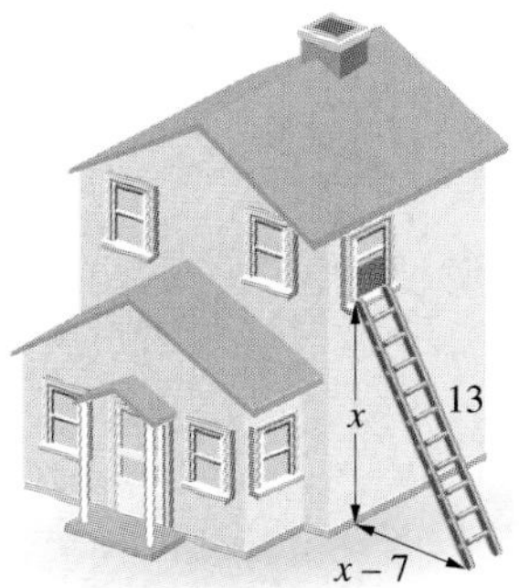

Use a quadratic equation to solve each problem. See Example 6.

41. Catarina and José want to buy a rug for a room that is 15 by 20 feet. They want to leave an even strip of flooring uncovered around the edges of the room. How wide a strip will they have if they buy a rug with an area of 234 square feet?

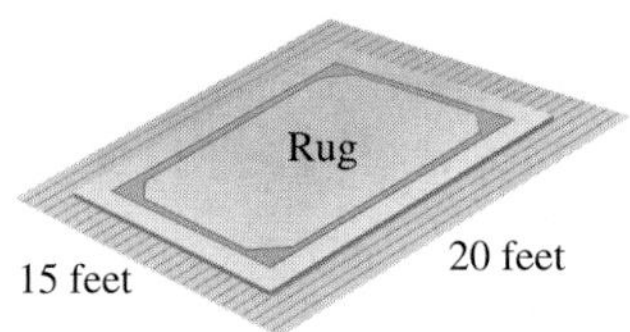

42. A club swimming pool is 30 feet wide and 40 feet long. The club members want an exposed aggregate border in a strip of uniform width around the pool. They have enough material for 296 square feet. How wide can the strip be?

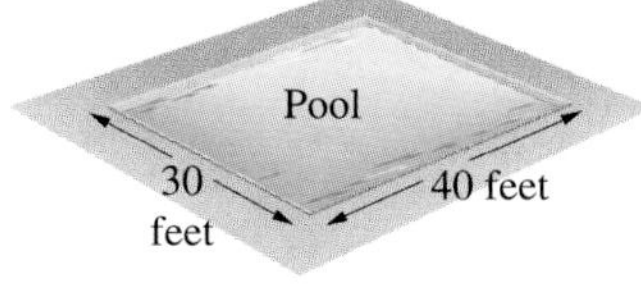

43. Arif's backyard is 20 by 30 meters. He wants to put a flower garden in the middle of the backyard, leaving a strip of grass of uniform width around the flower garden. To be happy, Arif must have 184 square meters of grass. Under these conditions what will the length and width of the garden be?

44. A rectangular piece of sheet metal has a length that is 4 inches less than twice the width. A square piece 2 inches on a side is cut from each corner. The sides are then turned up to form an uncovered box of volume 256 cubic inches. Find the length and width of the original piece of metal.

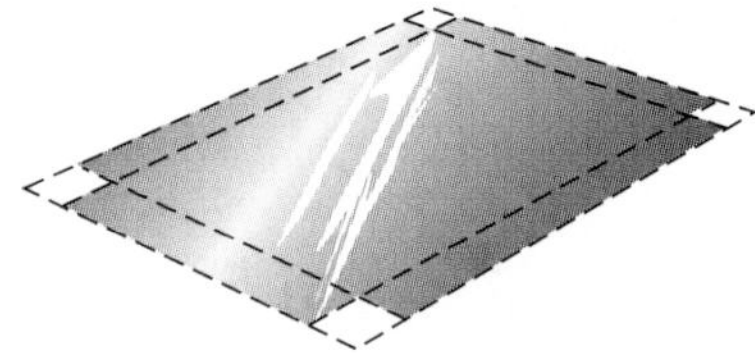

Solve each problem using a quadratic equation.

45. The formula $A = P(1 + r)^2$ gives the amount A in dollars that P dollars will grow to in 2 years at interest rate r (where r is given as a decimal), using compound interest. What interest rate will cause \$2000 to grow to \$2142.25 in 2 years?

46. If a square piece of cardboard has 3-inch squares cut from its corners and then has the flaps folded up to form an open-top box, the volume of the box is given by the formula $V = 3(x - 6)^2$, where x is the length of each side of the original piece of cardboard in inches. What original length would yield a box with a volume of 432 cubic inches?

47. A certain bakery has found that the daily demand for bran muffins is $\frac{3200}{p}$, where p is the price of a muffin in cents. The daily supply is $3p - 200$. Find the price at which supply and demand are equal.

48. In one area the demand for compact discs is $\frac{700}{P}$ per day, where P is the price in dollars per disc. The supply is $5P - 1$ per day. At what price does supply equal demand?

William Froude was a 19th century naval architect who used the expression $\frac{v^2}{g\ell}$ in shipbuilding. This expression, known as the Froude number, was also used by R. McNeill Alexander in his research on dinosaurs. (See "How Dinosaurs Ran," in Scientific American, *April 1991, pp. 130–136.) In Exercises 49 and 50, ℓ is given, as well as the value of the Froude number. Find the value of v (in meters per second). It is known that $g = 9.8$ meters per second squared.*

49. Rhinoceros: $\ell = 1.2$; Froude number $= 2.57$

50. Triceratops: $\ell = 2.8$; Froude number $= .16$

Recall that the corresponding sides of similar triangles are proportional. (Refer to Section 6.7 Exercises 51–52.) Use this fact to find the lengths of the indicated sides of each pair of similar triangles. Check all possible solutions in both triangles. Sides of a triangle cannot be negative.

51. Side AC

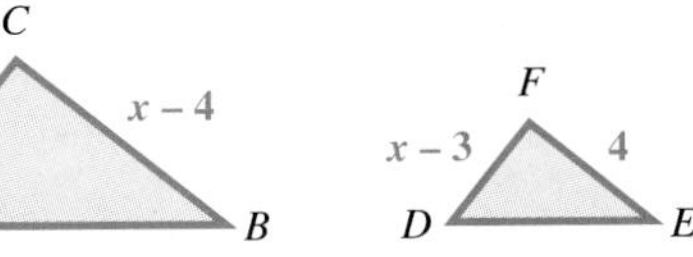

52. Side RQ

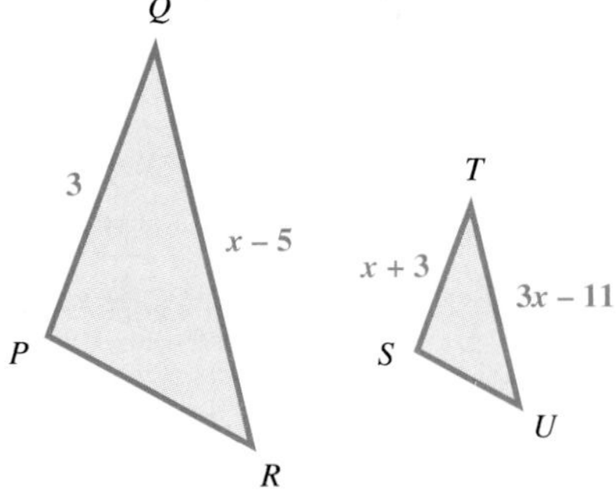

Solve each equation for the indicated variable.

53. $p = \frac{E^2R}{(r + R)^2}$; for R $(E > 0)$

54. $S(6S - t) = t^2$; for S

55. $10p^2c^2 + 7pcr = 12r^2$; for r

56. $S = vt + \frac{1}{2}gt^2$; for t

57. $LI^2 + RI + \frac{1}{c} = 0$; for I

58. $P = EI - RI^2$; for I

8.5 Graphs of Quadratic Functions

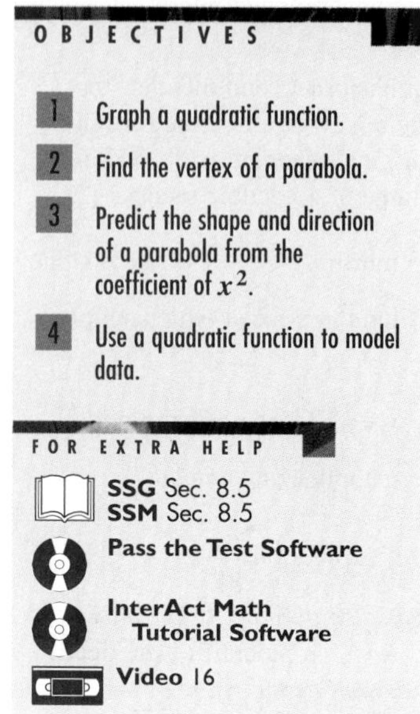

OBJECTIVES

1 Graph a quadratic function.

2 Find the vertex of a parabola.

3 Predict the shape and direction of a parabola from the coefficient of x^2.

4 Use a quadratic function to model data.

FOR EXTRA HELP

SSG Sec. 8.5
SSM Sec. 8.5

Pass the Test Software

InterAct Math Tutorial Software

Video 16

Polynomial functions were defined in Chapter 5, and we graphed a few simple second-degree polynomial functions there by point-plotting. In Figure 5, we repeat a table of ordered pairs for the simplest quadratic function, $y = x^2$, and the resulting graph.

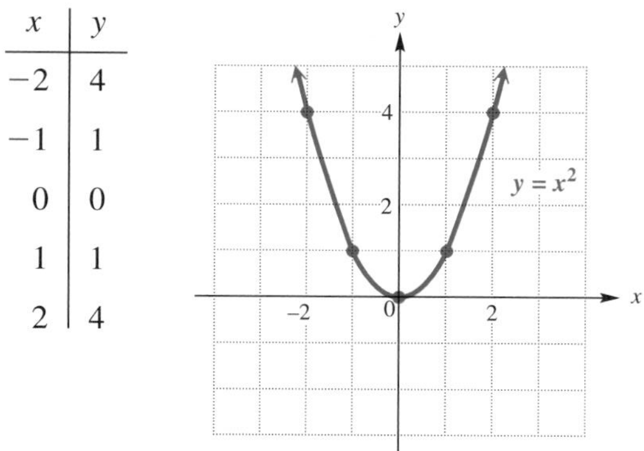

x	y
-2	4
-1	1
0	0
1	1
2	4

Figure 5

As mentioned in Chapter 5, this graph is called a **parabola.** The point (0, 0), the lowest point on the curve, is the **vertex** of this parabola. The vertical line through the vertex is the **axis** of this parabola. The parabola is **symmetric about its axis;** that is, if the graph were folded along the axis, the two portions of the curve would coincide. As Figure 5 suggests, the domain of the function defined by $y = x^2$ is $(-\infty, \infty)$, and, since y is always nonnegative, the range is $[0, \infty)$.

OBJECTIVE 1 Graph a quadratic function. We now consider the graphs of more general quadratic functions as defined here.

Quadratic Function

A function that can be written in the form

$$f(x) = ax^2 + bx + c$$

for real numbers a, b, and c, with $a \neq 0$, is a **quadratic function.**

The graph of any quadratic function is a parabola with a vertical axis. For the rest of this section and the next, we use the symbols y and $f(x)$ interchangeably when discussing parabolas.

OBJECTIVE 2 Find the vertex of a parabola. Parabolas need not have their vertices at the origin, as does $f(x) = x^2$. For example, to graph a parabola of the form $f(x) = x^2 + k$, start by selecting sample values of x that were used to graph $f(x) = x^2$. The corresponding values of $f(x)$ in $f(x) = x^2 + k$ differ by k from those of $f(x) = x^2$. For this reason, the graph of $f(x) = x^2 + k$ is *shifted,* or *translated,* k units vertically compared with that of $f(x) = x^2$.

EXAMPLE 1 Graphing a Parabola with a Vertical Shift

Graph $f(x) = x^2 - 2$.

This graph has the same shape as $f(x) = x^2$, but since k here is -2, the graph is shifted 2 units downward, with vertex at $(0, -2)$. Every function value is 2 less than the corresponding function value of $f(x) = x^2$. Plotting points gives the graph in Figure 6. Note that x can be any real number, so the domain is still $(-\infty, \infty)$; since the value of y (or $f(x)$) is always greater than or equal to -2, the range is $[-2, \infty)$. The graph of $f(x) = x^2$ is shown for comparison.

x	$f(x) = x^2$	$f(x) = x^2 - 2$
-2	4	2
-1	1	-1
0	0	-2
1	1	-1
2	4	2

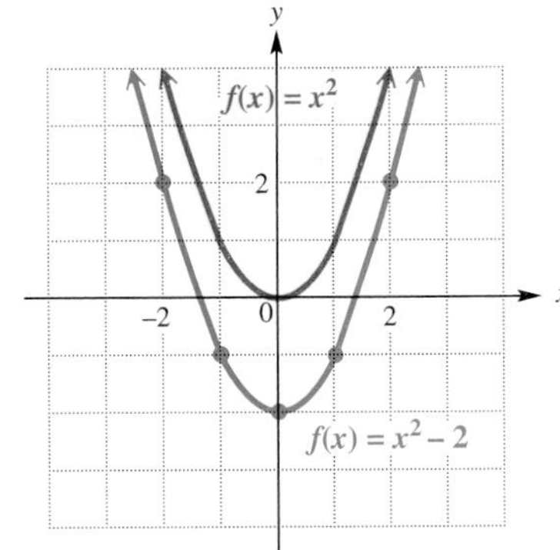

Figure 6

Vertical Shifts

The graph of $f(x) = x^2 + k$ is a parabola with the same shape as the graph of $f(x) = x^2$. The parabola is shifted k units upward if $k > 0$, and $|k|$ units downward if $k < 0$. The vertex is $(0, k)$.

The graph of $f(x) = (x - h)^2$ is also a parabola with the same shape as $f(x) = x^2$. The vertex of the parabola $f(x) = (x - h)^2$ is the lowest point on the parabola. The lowest point occurs here when $f(x)$ is 0. To get $f(x)$ equal to 0, we let $x = h$, so the vertex of $f(x) = (x - h)^2$ is at $(h, 0)$. Based on this, the graph of $f(x) = (x - h)^2$ is shifted, or translated, h units horizontally compared with that of $f(x) = x^2$.

EXAMPLE 2 Graphing a Parabola with a Horizontal Shift

Graph $f(x) = (x - 2)^2$.

When $x = 2$, then $f(x) = 0$, giving the vertex $(2, 0)$. The parabola $f(x) = (x - 2)^2$ has the same shape as $f(x) = x^2$ but is shifted 2 units to the right, as shown in Figure 7 on the next page. Again, the domain is $(-\infty, \infty)$; the range is $[0, \infty)$. As before, we show the graph of $f(x) = x^2$ for comparison.

x	$f(x) = (x - 2)^2$
0	4
1	1
2	0
3	1
4	4

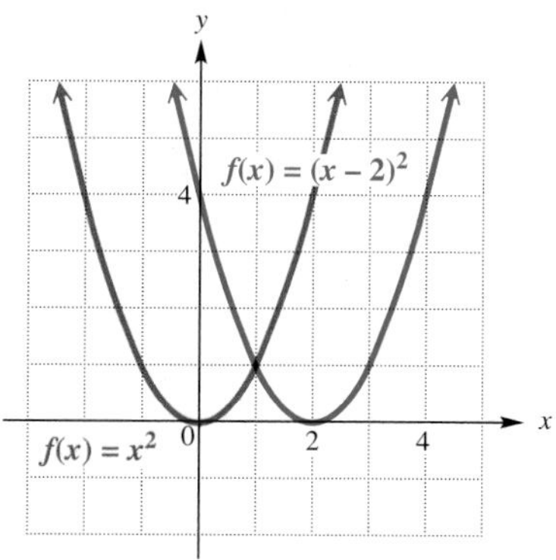

Figure 7

Horizontal Shifts

The graph of $f(x) = (x - h)^2$ is a parabola with the same shape as the graph of $f(x) = x^2$. The parabola is shifted h units horizontally: h units to the right if $h > 0$, and $|h|$ units to the left if $h < 0$. The vertex is $(h, 0)$.

CAUTION Errors frequently occur when horizontal shifts are involved. To determine the direction and magnitude of horizontal shifts, find the value that would cause the expression $x - h$ to equal 0. For example, the graph of $f(x) = (x - 5)^2$ would be shifted 5 units to the *right,* because $+5$ would cause $x - 5$ to equal 0. On the other hand, the graph of $f(x) = (x + 5)^2$ would be shifted 5 units to the *left,* because -5 would cause $x + 5$ to equal 0.

A parabola can have both a horizontal and a vertical shift, as in Example 3.

EXAMPLE 3 Graphing a Parabola with Horizontal and Vertical Shifts

Graph $f(x) = (x + 3)^2 - 2$.

This graph has the same shape as $f(x) = x^2$, but is shifted 3 units to the left (since $x + 3 = 0$ if $x = -3$), and 2 units downward (because of the -2). As shown in Figure 8, the vertex is at $(-3, -2)$. This function has domain $(-\infty, \infty)$ and range $[-2, \infty)$.

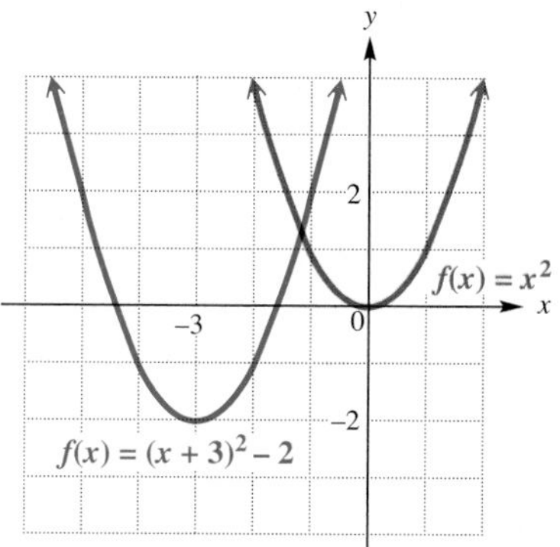

Figure 8

The characteristics of the graph of a parabola of the form $f(x) = (x - h)^2 + k$ are summarized as follows.

Vertex and Axis

The graph of $f(x) = (x - h)^2 + k$ is a parabola with the same shape as $f(x) = x^2$ and with vertex at (h, k). The axis is the vertical line $x = h$.

The vertical and horizontal shifts discussed in this section occur with other types of graphs as well. For example, the graph of $f(x) = \sqrt{x - h} + k$, with $h > 0$ and $k > 0$, would have the same shape as the graph of $f(x) = \sqrt{x}$ shifted h units to the right and k units up.

OBJECTIVE 3 **Predict the shape and direction of the graph of a parabola from the coefficient of x^2.** Not all parabolas open upward, and not all parabolas have the same shape as $f(x) = x^2$.

EXAMPLE 4 Graphing a Parabola That Opens Downward

Graph $f(x) = -\dfrac{1}{2}x^2$.

This parabola is shown in Figure 9. The coefficient $-\frac{1}{2}$ affects the shape of the graph; the $\frac{1}{2}$ makes the parabola wider (since the values of $\frac{1}{2}x^2$ increase more slowly than in $f(x) = x^2$), and the negative sign makes the parabola open downward. The graph is not shifted in any direction; the vertex is still (0, 0). Here, the vertex has the *largest* function value of any point on the graph. The domain is $(-\infty, \infty)$; the range is $(-\infty, 0]$.

x	y
-2	-2
-1	$-\frac{1}{2}$
0	0
1	$-\frac{1}{2}$
2	-2

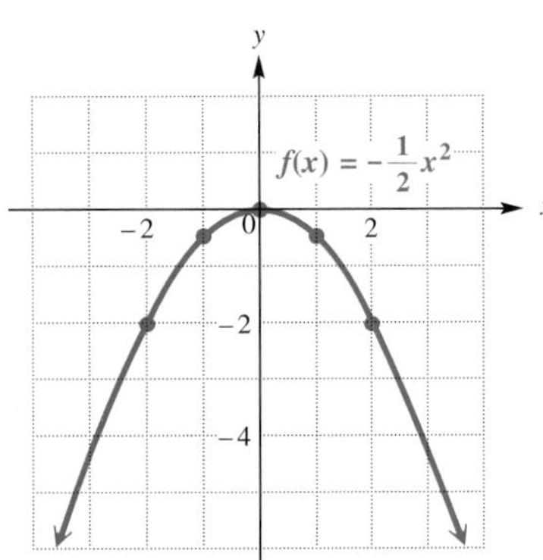

Figure 9

Some general principles concerning the graph of $f(x) = a(x - h)^2 + k$ are summarized as follows.

General Principles

1. The graph of the quadratic function

$$f(x) = a(x - h)^2 + k, \quad a \neq 0$$

is a parabola with vertex at (h, k) and the vertical line $x = h$ as axis.

2. The graph opens upward if a is positive and downward if a is negative.
3. The graph is wider than $f(x) = x^2$ if $0 < |a| < 1$. The graph is narrower than $f(x) = x^2$ if $|a| > 1$.

EXAMPLE 5 Using the General Principles to Graph a Parabola

Graph $f(x) = -2(x + 3)^2 + 4$.

The parabola opens downward (because $a < 0$), and is narrower than the graph of $f(x) = x^2$, since $|-2| = 2 > 1$, causing values of $f(x)$ to decrease more quickly than in $f(x) = -x^2$. This parabola has vertex $(-3, 4)$ as shown in Figure 10. To complete the graph, we plotted the ordered pairs $(-4, 2)$ and $(-2, 2)$. Notice that these two points are symmetric about the axis of the parabola. Symmetry can be used to find additional ordered pairs that satisfy the equation.

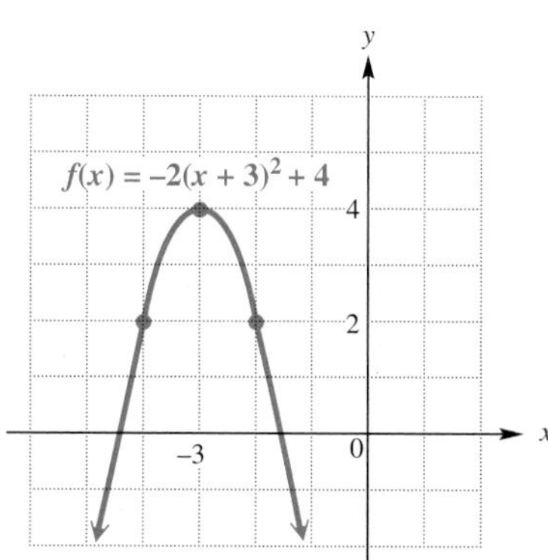

Figure 10

OBJECTIVE 4 Use a quadratic function to model data.

EXAMPLE 6 Using a Quadratic Function as a Model for the Rise in Multiple Births

The number of higher-order multiple births in the United States is rising. Let x represent the number of years since 1970 and y represent the rate of increase of higher-order multiples born per 100,000 births since 1971. The data are shown in the following table.

U.S. Higher-order Multiple Births

Year	x	y
1971	1	29.1
1976	6	35.0
1981	11	40.0
1986	16	47.0
1991	21	100.0
1996	26	152.6

Source: National Center for Health Statistics.

A graph of the ordered pairs (x, y) is shown in Figure 11(a). The graphs in this section suggest a parabola should approximate these points. Looking at the general shape suggested by the points in Figure 11(a), we see that the coefficient of x^2 should be positive, because the graph opens upward and should be less than 1, because the graph appears wider than the graph of $y = x^2$. Figure 11(b) shows the graph of $f(x) = .30x^2 - 3.52x + 37.3$, found by a method called *quadratic regression.* This quadratic function fits the set of data reasonably well.

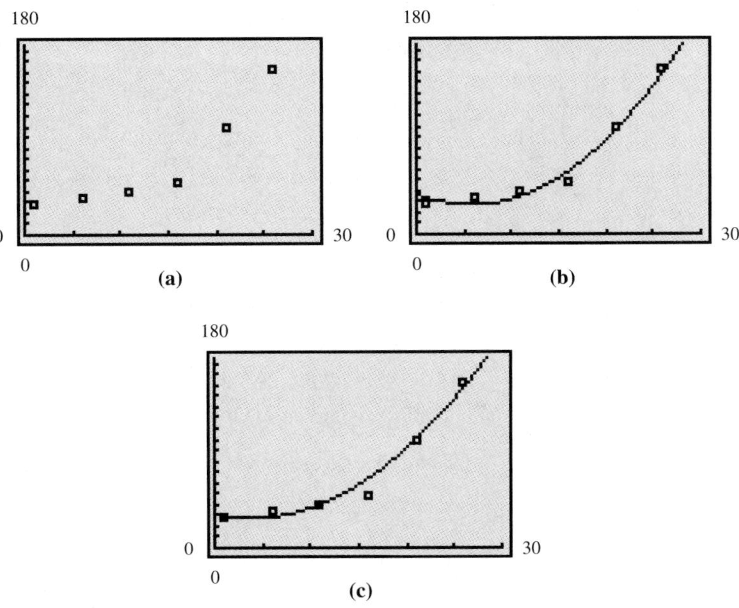

Figure 11

Another method used to find a quadratic function that fits these data is to choose three of the ordered pairs and use them to write a system of three equations. We want to find a, b, and c in the quadratic form $y = ax^2 + bx + c$. Choose any three points from the given data. We choose $(1, 29.1)$, $(11, 40)$, and $(21, 100)$. Substitute the values from these ordered pairs into the quadratic form $ax^2 + bx + c = y$ to get three equations.

$$\begin{aligned} a(1)^2 + b(1) + c &= 29.1 & \text{or} \quad a + b + c &= 29.1 \\ a(11)^2 + b(11) + c &= 40 & \text{or} \quad 121a + 11b + c &= 40 \\ a(21)^2 + b(21) + c &= 100 & \text{or} \quad 441a + 21b + c &= 100 \end{aligned}$$

Using one of the methods of Chapter 4 to solve this system of three equations in three variables, we find $a = .2455$, $b = -1.856$, and $c = 30.7105$, so the function we want is defined by $y = .2455x^2 - 1.856x + 30.7105$. This function is graphed with the data in Figure 11(c) and also appears to give a reasonably good approximation of the data.

8.5 EXERCISES

In Exercises 1–6, match each equation with the figure that most closely resembles its graph.

1. $g(x) = x^2 - 5$
2. $h(x) = -x^2 + 4$
3. $F(x) = (x - 1)^2$
4. $G(x) = (x + 1)^2$
5. $H(x) = (x - 1)^2 + 1$
6. $K(x) = (x + 1)^2 + 1$

A.
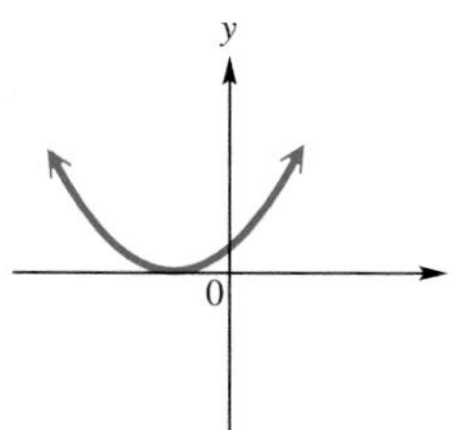

B.
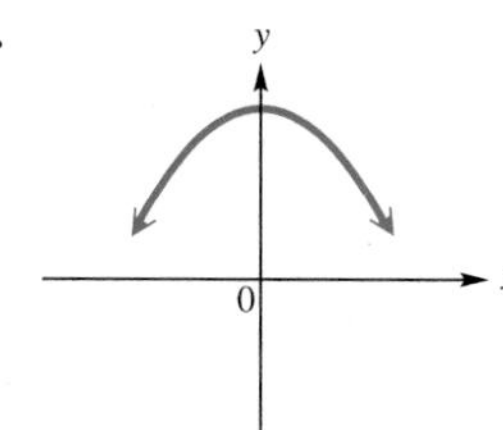

C.
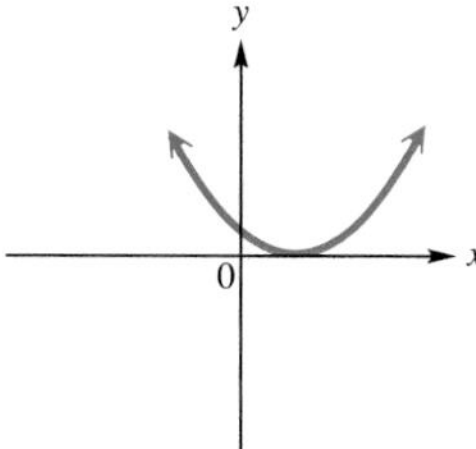

D.
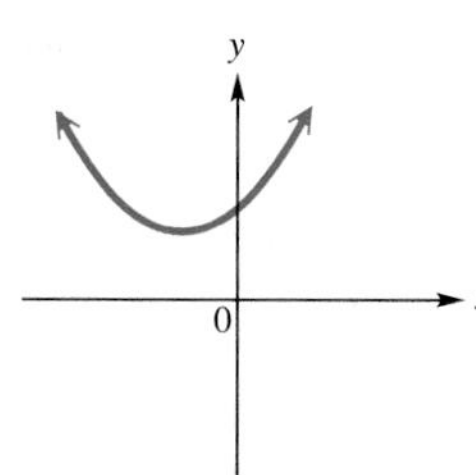

E.
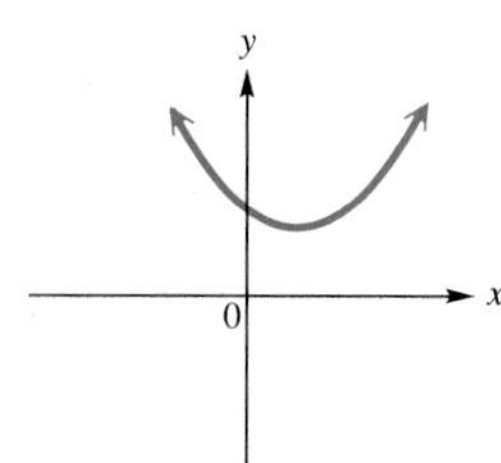

F.
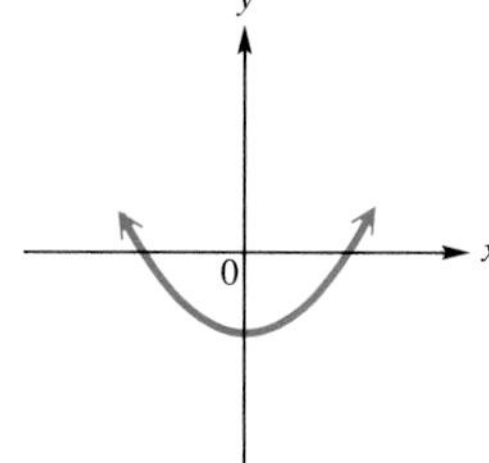

7. Explain in your own words the meaning of each term.
   (a) vertex of a parabola **(b)** axis of a parabola
8. Explain why the axis of the graph of a quadratic function cannot be a horizontal line.

Identify the vertex of the graph of each quadratic function. See Examples 1–3.

9. $f(x) = -3x^2$
10. $f(x) = -.5x^2$
11. $f(x) = x^2 + 4$
12. $f(x) = x^2 - 4$
13. $f(x) = (x - 1)^2$
14. $f(x) = (x + 3)^2$
15. $f(x) = (x + 3)^2 - 4$
16. $f(x) = (x - 5)^2 - 8$
17. Describe how the graph of each parabola in Exercises 15 and 16 is shifted compared to the graph of $y = x^2$.

For each quadratic function, tell whether the graph opens upward or downward, and tell whether the graph is wider, narrower, or the same as the graph of $f(x) = x^2$. See Example 4.

18. $f(x) = -2x^2$
19. $f(x) = -3x^2 + 1$
20. $f(x) = .5x^2$
21. $f(x) = \frac{2}{3}x^2 - 4$

22. What does the value of a in $f(x) = a(x - h)^2 + k$ tell you about the graph of the function compared to the graph of $y = x^2$?

23. For $f(x) = a(x - h)^2 + k$, in what quadrant is the vertex if:
(a) $h > 0, k > 0$; **(b)** $h > 0, k < 0$; **(c)** $h < 0, k > 0$; **(d)** $h < 0, k < 0$?

24. **(a)** What is the value of h if the graph of $f(x) = a(x - h)^2 + k$ has vertex on the y-axis?
(b) What is the value of k if the graph of $f(x) = a(x - h)^2 + k$ has vertex on the x-axis?

Sketch the graph of each parabola. Plot at least two points in addition to the vertex. Give the domain and range in Exercises 35–40. See Examples 1–5.

25. $f(x) = -2x^2$

26. $f(x) = \frac{1}{3}x^2$

27. $f(x) = x^2 - 1$

28. $f(x) = x^2 + 3$

29. $f(x) = -x^2 + 2$

30. $f(x) = 2x^2 - 2$

31. $f(x) = .5(x - 4)^2$

32. $f(x) = -2(x + 1)^2$

33. $f(x) = (x + 2)^2 - 1$

34. $f(x) = (x - 1)^2 + 2$

35. $f(x) = 2(x - 2)^2 - 4$

36. $f(x) = -2(x + 3)^2 + 4$

37. $f(x) = -.5(x + 1)^2 + 2$

38. $f(x) = -\frac{2}{3}(x + 2)^2 + 1$

39. $f(x) = 2(x - 2)^2 - 3$

40. $f(x) = \frac{4}{3}(x - 3)^2 - 2$

RELATING CONCEPTS (EXERCISES 41–46)

The procedures described in this section that allow the graph of $y = x^2$ to be shifted vertically and horizontally are applicable to other types of functions as well. In Section 3.5 we introduced linear functions (functions of the form $f(x) = ax + b$).

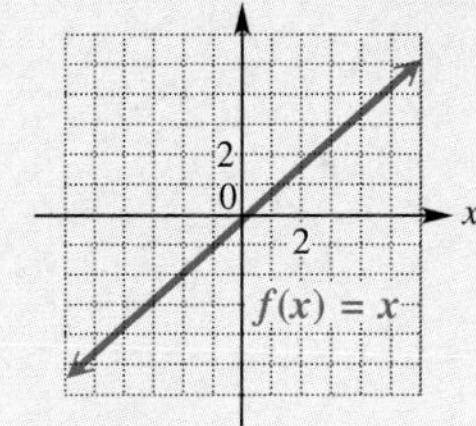

Consider the graph of the simplest linear function, $f(x) = x$, shown here, and then ***work through Exercises 41–46 in order.***

41. Based on the concepts of this section, how does the graph of $y = x^2 + 6$ compare to the graph of $y = x^2$ if a *vertical* shift is considered?

42. Graph the linear function $y = x + 6$.

43. Based on the concepts of Chapter 3, how does the graph of $y = x + 6$ compare to the graph of $y = x$ if a vertical shift is considered? (*Hint:* Look at the y-intercept.)

44. Based on the concepts of this section, how does the graph of $y = (x - 6)^2$ compare to the graph of $y = x^2$ if a *horizontal* shift is considered?

45. Graph the linear function $y = x - 6$.

46. Based on the concepts of Chapter 3, how does the graph of $y = x - 6$ compare to the graph of $y = x$ if a horizontal shift is considered? (*Hint:* Look at the x-intercept.)

Did you make the connection that horizontal and vertical shifts of the graphs of equations other than $y = x^2$ occur in a similar way? Horizontal shifts of $y = f(x)$ are indicated by $y = f(x - k)$ and vertical shifts are indicated by $y = f(x) + k$.

In the following exercise, the distance formula is used to develop the equation of a parabola.

47. A parabola can be defined as the set of all points in a plane equally distant from a given point and a given line not containing the point. (The point is called the *focus* and the line is called the *directrix*.) See the figure.

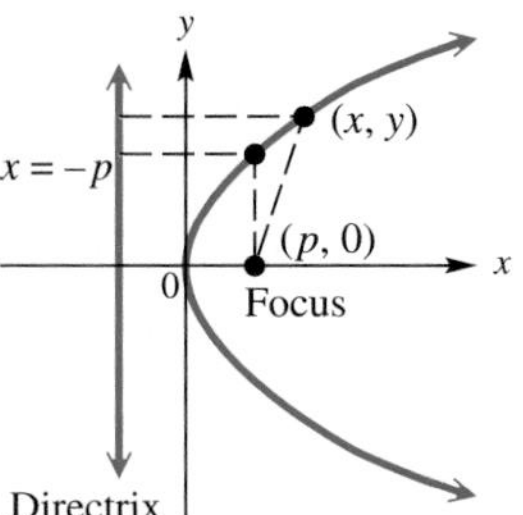

(a) Suppose (x, y) is to be on the parabola. Suppose the directrix has equation $x = -p$. Find the distance between (x, y) and the directrix. (The distance from a point to a line is the length of the perpendicular from the point to the line.)

(b) If $x = -p$ is the equation of the directrix, why should the focus have coordinates $(p, 0)$? (*Hint:* See the figure.)

(c) Find an expression for the distance from (x, y) to $(p, 0)$.

(d) Find an equation for the parabola of the figure. (*Hint:* Use the results of parts (a) and (c) and the fact that (x, y) is equally distant from the focus and the directrix.)

48. Use the equation derived in Exercise 47 to find an equation for a parabola with focus $(3, 0)$ and directrix with equation $x = -3$.

In Exercises 49 and 50, refer to Example 6.

49. The numbers of new AIDS patients who survived the first year for the years from 1991–1997 are shown in the table. In the year column, 1 represents 1991, 2 represents 1992, and so on.

AIDS Patients Who Survived the First Year

Year	Number of Patients
1	55
2	130
3	155
4	160
5	155
6	150
7	115

Source: HIV Health Services Planning Council.

(a) Plot the ordered pairs (year, number of patients).

(b) Use the graph to decide whether the coefficient a of x^2 in a quadratic model should be positive or negative.

(c) Determine a quadratic function that models these data by using a system of equations. Use the ordered pairs $(2, 130)$, $(3, 155)$, and $(7, 115)$.

(d) What would be an appropriate domain for $f(x)$? $\{1, 2, 3, 4, 5, 6, 7\}$? $(-\infty, \infty)$? Another choice?

50. In Section 8.2, Exercise 54, a quadratic equation is given that models CD-ROM revenues (in millions of dollars) for 1991–1994. The data used to get that equation are given here. The years are coded so that 1 represents 1991, and so on.

CD-ROM Revenues

Year	Millions of Dollars
1	11.9
2	81.5
3	324.5
4	454.5

Source: Dataquest, Multimedia Market Trends, 1994.

Repeat parts (a)–(d) of Exercise 49 for these data. In part (c) use the data points (1, 11.9), (2, 81.5), and (4, 454.5). In part (d) consider the set {1, 2, 3, 4} instead of {1, 2, 3, 4, 5, 6, 7}. Compare your answer with the equation given in Section 8.2, Exercise 54.

TECHNOLOGY INSIGHTS (EXERCISES 51–56)

In Chapter 3 we saw that the x-value of the x-intercept of the graph of the line $y = mx + b$ is the solution of the linear equation $mx + b = 0$. In the same way, the x-values of the x-intercepts of the graph of the parabola $y = ax^2 + bx + c$ are the real solutions of the quadratic equation $ax^2 + bx + c = 0$. In Exercises 51–54, the calculator-generated graphs show the x-values of the x-intercepts of the graph of the polynomial in the equation.

Use the graphs to solve each equation.

51. $x^2 - x - 20 = 0$

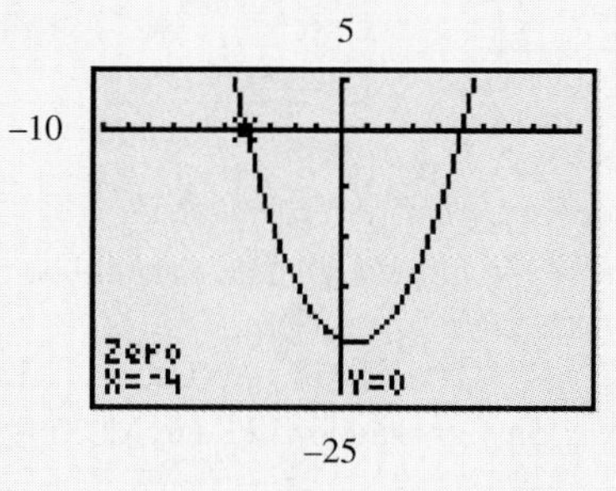

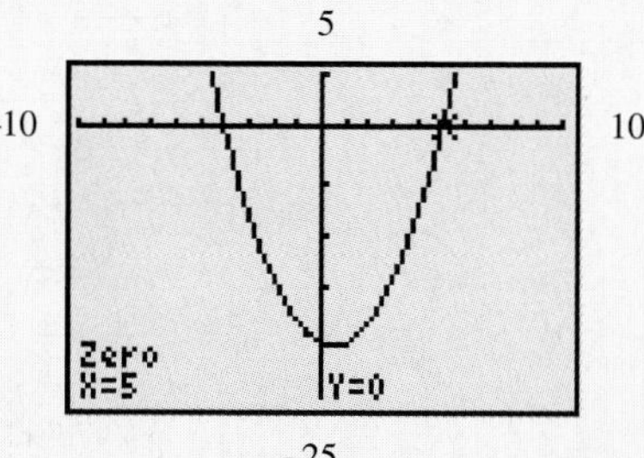

52. $x^2 + 9x + 14 = 0$

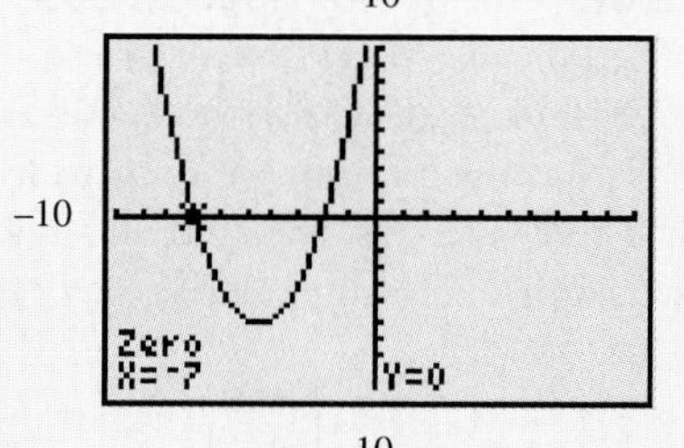

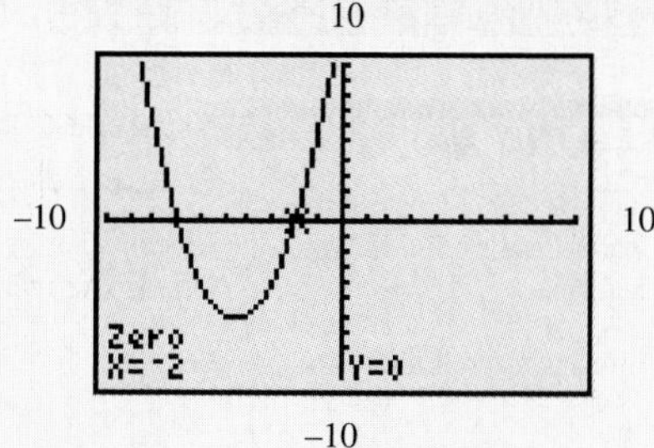

(continued)

TECHNOLOGY INSIGHTS (EXERCISES 51–56) (CONTINUED)

53. $-2x^2 + 5x + 3 = 0$

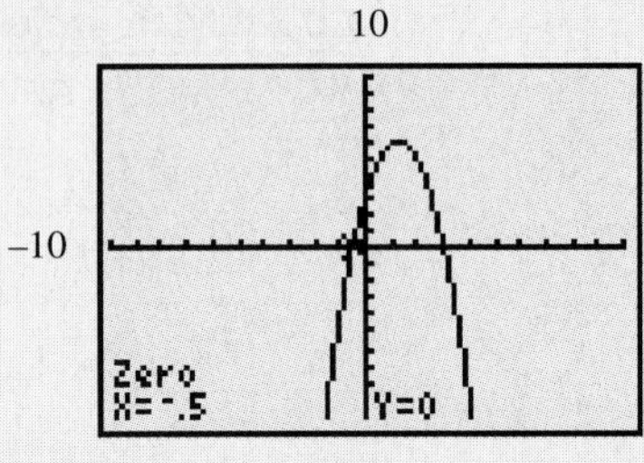

10
-10
10
Zero
X=3
Y=0
-10

54. $-8x^2 + 6x + 5 = 0$

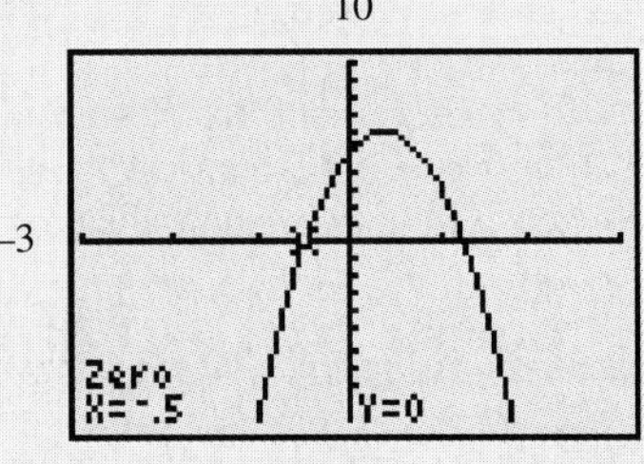

10
-3
3
Zero
X=1.25
Y=0
-10

The graph of a quadratic function with $y = f(x)$ is shown in the standard viewing window, without x-axis tick marks.

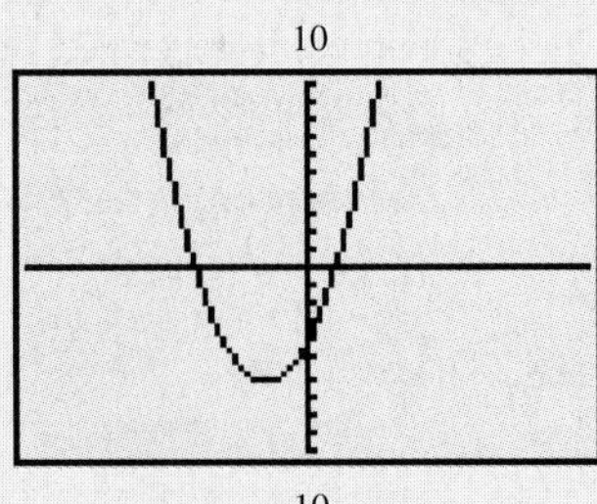

Refer to it to answer Exercises 55 and 56.

55. Which one of the following choices would be the only possible solution set for the equation $f(x) = 0$?

(a) $\{-4, 1\}$ **(b)** $\{1, 4\}$ **(c)** $\{-1, -4\}$ **(d)** $\{4, -1\}$

56. Explain why only one choice in Exercise 55 is possible.

8.6 More about Parabolas; Applications

OBJECTIVES

1. Find the vertex of a vertical parabola.
2. Graph a quadratic function.

OBJECTIVE 1 Find the vertex of a vertical parabola. When the equation of a parabola is given in the form $f(x) = ax^2 + bx + c$, we need to locate the vertex in order to sketch an accurate graph. There are two ways to do this: complete the square as shown in Examples 1 and 2, or use a formula derived by completing the square.

3 Use the discriminant to find the number of x-intercepts of a vertical parabola.

4 Use quadratic functions to solve problems involving maximum or minimum value.

5 Graph horizontal parabolas.

FOR EXTRA HELP

SSG Sec. 8.6
SSM Sec. 8.6

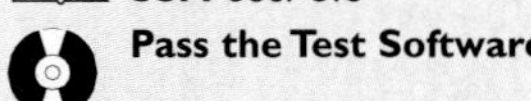

Pass the Test Software

InterAct Math Tutorial Software

Video 16

EXAMPLE 1 Completing the Square to Find the Vertex

Find the vertex of the graph of $f(x) = x^2 - 4x + 5$.

To find the vertex, express $x^2 - 4x + 5$ in the form $(x - h)^2 + k$ by completing the square. (This process was introduced in Section 8.1.) To simplify the notation, we replace $f(x)$ with y.

$$y = x^2 - 4x + 5$$

$$y - 5 = x^2 - 4x \qquad \text{Get the constant term on the left.}$$

$$y - 5 + 4 = x^2 - 4x + 4 \qquad \text{Half of } -4 \text{ is } -2;\ (-2)^2 = 4. \text{ Add 4 to both sides.}$$

$$y - 1 = (x - 2)^2 \qquad \text{Combine terms on the left and factor on the right.}$$

$$y = (x - 2)^2 + 1 \qquad \text{Add 1 to both sides.}$$

This form shows that the vertex of the parabola is $(2, 1)$.

EXAMPLE 2 Completing the Square to Find the Vertex When $a \neq 1$

Find the vertex of the graph of $y = -3x^2 + 6x - 1$.

Complete the square on $-3x^2 + 6x$. Because the x^2 term has a coefficient other than 1, divide both sides by this coefficient, and then proceed as in Example 1.

$$y = -3x^2 + 6x - 1$$

$$\frac{y}{-3} = x^2 - 2x + \frac{1}{3} \qquad \text{Divide both sides by } -3.$$

$$\frac{y}{-3} - \frac{1}{3} = x^2 - 2x \qquad \text{Get the constant term on the left.}$$

$$\frac{y}{-3} - \frac{1}{3} + 1 = x^2 - 2x + 1 \qquad \text{Half of } -2 \text{ is } -1;\ (-1)^2 = 1. \text{ Add 1 to both sides.}$$

$$\frac{y}{-3} + \frac{2}{3} = (x - 1)^2 \qquad \text{Combine terms on the left and factor on the right.}$$

$$\frac{y}{-3} = (x - 1)^2 - \frac{2}{3} \qquad \text{Subtract } \tfrac{2}{3} \text{ from both sides.}$$

$$y = -3(x - 1)^2 + 2 \qquad \text{Multiply by } -3 \text{ to get desired form.}$$

The vertex is $(1, 2)$.

To derive a formula for the vertex of the graph of the quadratic function $y = ax^2 + bx + c$, complete the square on the standard form of the equation. Going through the same steps as in Example 2 gives the equation

$$y = a\Bigg[x - \underbrace{\left(\frac{-b}{2a}\right)}_{h}\Bigg]^2 + \underbrace{\frac{4ac - b^2}{4a}}_{k}.$$

This equation shows that the vertex (h, k) can be expressed in terms of a, b, and c. However, it is not necessary to remember this expression for k, since it can be found by replacing x with $\frac{-b}{2a}$. Using function notation, if $y = f(x)$, the y-value of the vertex is $f\left(\frac{-b}{2a}\right)$.

Vertex Formula

The graph of the quadratic function with $f(x) = ax^2 + bx + c$ has vertex

$$\left(\frac{-b}{2a}, f\left(\frac{-b}{2a}\right)\right),$$

and the axis of the parabola is the line $x = \frac{-b}{2a}$.

EXAMPLE 3 Using the Formula to Find the Vertex

Use the vertex formula to find the vertex of the graph of the function

$$f(x) = x^2 - x - 6.$$

For this function, $a = 1$, $b = -1$, and $c = -6$. The x-coordinate of the vertex of the parabola is

$$\frac{-b}{2a} = \frac{-(-1)}{2(1)} = \frac{1}{2}.$$

The y-coordinate is $f\left(\frac{-b}{2a}\right) = f\left(\frac{1}{2}\right)$.

$$f\left(\frac{1}{2}\right) = \left(\frac{1}{2}\right)^2 - \frac{1}{2} - 6 = \frac{1}{4} - \frac{1}{2} - 6 = -\frac{25}{4}$$

Finally, the vertex is $\left(\frac{1}{2}, -\frac{25}{4}\right)$.

OBJECTIVE 2 Graph a quadratic function. We give a general approach for graphing any quadratic function here.

Graphing a Quadratic Function f

Step 1 **Find the y-intercept.** Find the y-intercept by evaluating $f(0)$.

Step 2 **Find the x-intercepts.** Find the x-intercepts, if any, by solving $f(x) = 0$.

Step 3 **Find the vertex.** Find the vertex either by using the formula or by completing the square.

Step 4 **Complete the graph.** Find and plot additional points as needed, using symmetry about the axis.

Verify that the graph opens upward (if $a > 0$) or opens downward (if $a < 0$).

EXAMPLE 4 Using the Steps for Graphing a Quadratic Function

Graph the quadratic function $f(x) = x^2 - x - 6$.

Step 1 Find the y-intercept.

$$f(x) = x^2 - x - 6$$
$$f(0) = 0^2 - 0 - 6 \quad \text{Find } f(0).$$
$$f(0) = -6$$

The y-intercept is $(0, -6)$.

Step 2 Find any x-intercepts.

$$f(x) = x^2 - x - 6$$

$$0 = x^2 - x - 6 \qquad \text{Let } f(x) = 0.$$

$$0 = (x - 3)(x + 2) \qquad \text{Factor.}$$

$$x - 3 = 0 \quad \text{or} \quad x + 2 = 0 \qquad \text{Set each factor equal to 0 and solve.}$$

$$x = 3 \quad \text{or} \quad x = -2$$

The x-intercepts are (3, 0) and (−2, 0).

Step 3 The vertex, found in Example 3, is $\left(\frac{1}{2}, -\frac{25}{4}\right)$.

Step 4 Plot the points found so far and additional points as needed using symmetry. The graph is shown in Figure 12.

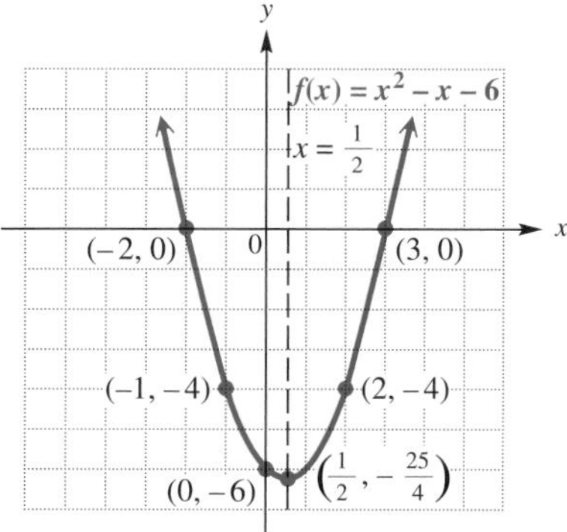

Figure 12

OBJECTIVE 3 **Use the discriminant to find the number of x-intercepts of a vertical parabola.** The graph of a quadratic function may have two x-intercepts, one x-intercept, or no x-intercepts, as shown in Figure 13. Recall from Section 8.2 that the value of $b^2 - 4ac$ is called the *discriminant* of the quadratic equation $ax^2 + bx + c = 0$. We can use it to determine the number of real solutions of a quadratic equation. In a similar way, we can use the discriminant of a quadratic *function* to determine the number of x-intercepts of its graph. If the discriminant is positive, the parabola will have two x-intercepts. If the discriminant is 0, there will be only one x-intercept, and it will be the vertex of the parabola. If the discriminant is negative, the graph will have no x-intercepts.

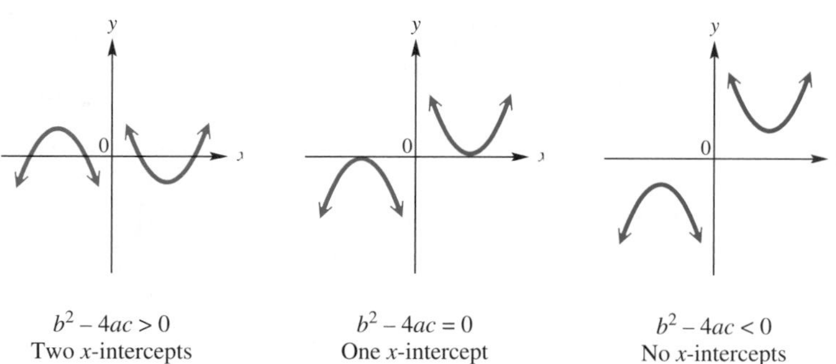

$b^2 - 4ac > 0$ Two x-intercepts | $b^2 - 4ac = 0$ One x-intercept | $b^2 - 4ac < 0$ No x-intercepts

Figure 13

EXAMPLE 5 Using the Discriminant to Determine the Number of x-intercepts

Determine the number of x-intercepts of the graph of each quadratic function. Use the discriminant.

(a) $f(x) = 2x^2 + 3x - 5$
The discriminant is $b^2 - 4ac$. Here $a = 2$, $b = 3$, and $c = -5$, so

$$b^2 - 4ac = 9 - 4(2)(-5) = 49.$$

Since the discriminant is positive, the parabola has two x-intercepts.

(b) $f(x) = -3x^2 - 1$
In this equation, $a = -3$, $b = 0$, and $c = -1$. The discriminant is

$$b^2 - 4ac = 0 - 4(-3)(-1) = -12.$$

The discriminant is negative and so the graph has no x-intercepts.

(c) $f(x) = 9x^2 + 6x + 1$
Here, $a = 9$, $b = 6$, and $c = 1$. The discriminant is

$$b^2 - 4ac = 36 - 4(9)(1) = 0.$$

The parabola has only one x-intercept (its vertex), since the value of the discriminant is 0.

OBJECTIVE 4 **Use quadratic functions to solve problems involving maximum or minimum value.** The vertex of a parabola is either the highest or the lowest point on the parabola. The y-value of the vertex gives the maximum or minimum value of y, while the x-value tells where that maximum or minimum occurs.

PROBLEM SOLVING

In many applied problems we must find the largest or smallest value of some quantity. When we can express that quantity as a quadratic function, the value of k in the vertex gives that optimum value.

EXAMPLE 6 Finding the Maximum Area of a Rectangular Region

A farmer has 120 feet of fencing. He wants to put a fence around a rectangular plot of land next to a river. Find the maximum area he can enclose.

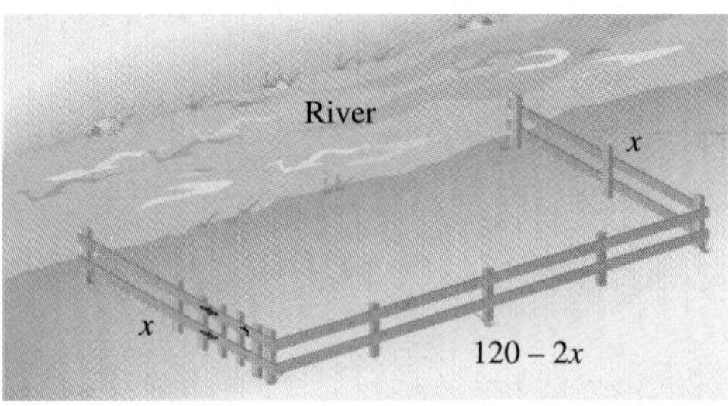

Figure 14

Figure 14 shows the plot. Let x represent the width of the plot. Since there are 120 feet of fencing,

$$x + x + \text{length} = 120 \quad \text{Sum of the sides is 120 feet.}$$
$$2x + \text{length} = 120 \quad \text{Combine terms.}$$
$$\text{length} = 120 - 2x. \quad \text{Subtract } 2x.$$

The area is given by the product of the width and length, so

$$A = x(120 - 2x) = 120x - 2x^2.$$

To determine the maximum area, find the vertex of the parabola $A = 120x - 2x^2$ using the vertex formula. Writing the equation in standard form as $A = -2x^2 + 120x$ shows that $a = -2$, $b = 120$, and $c = 0$, so

$$h = \frac{-b}{2a} = \frac{-120}{2(-2)} = \frac{-120}{-4} = 30$$
$$f(30) = -2(30)^2 + 120(30) = -2(900) + 3600 = 1800.$$

The graph is a parabola that opens downward, and its vertex is (30, 1800). Thus, the maximum area will be 1800 square feet. This area will occur if x, the width of the plot, is 30 feet.

Be careful when interpreting the meanings of the coordinates of the vertex. The first coordinate, x, gives the value for which the *function value* is a maximum or a minimum. Be sure to read the problem carefully to determine whether you are asked to find the value of the independent variable, the function value, or both.

Earlier in this chapter, we discussed the motion of a propelled object. Now we can find the maximum height of the object.

EXAMPLE 7 Finding the Maximum Height Attained by a Rocket

In Example 4, Section 8.4, we found the height of a rocket propelled from the surface of Mars to be given by $f(x) = x - .0066x^2$, where x is time in seconds for the rocket to reach y feet. Find the maximum height the rocket will reach. (*Source:* Bernice Kastner, *Space Mathematics,* NASA.)

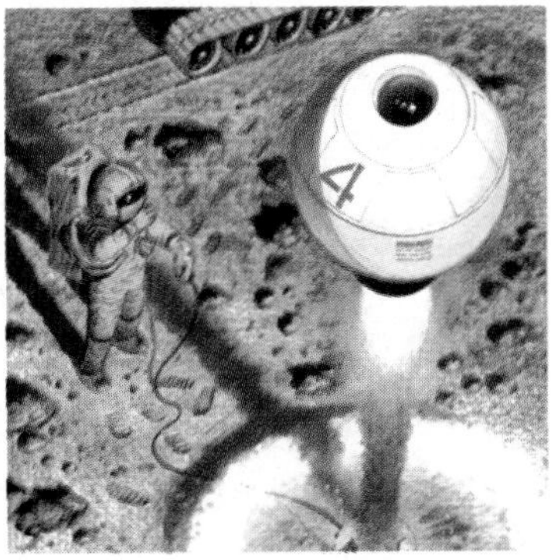

Use the formula for the vertex, with $a = -.0066$, $b = 1$, and $c = 0$.

$$h = \frac{-b}{2a} = \frac{-1}{2(-.0066)} \approx 75.8$$

This tells us that the maximum height is attained at 75.8 seconds. To find that height, calculate $f(75.8)$.

$$f(75.8) = 75.8 - .0066(75.8)^2$$
$$\approx 37.9 \qquad \text{Use a calculator.}$$

The rocket will attain a maximum height of approximately 37.9 feet.

OBJECTIVE 5 Graph horizontal parabolas. If x and y are exchanged in the equation $y = ax^2 + bx + c$, the equation becomes $x = ay^2 + by + c$. Because of the interchange of the roles of x and y, these parabolas are horizontal (with horizontal lines as axes), compared with the vertical ones graphed previously.

Graph of a Horizontal Parabola

The graph of $x = ay^2 + by + c$ or $x = a(y - k)^2 + h$ is a parabola with vertex at (h, k) and the horizontal line $y = k$ as axis. The graph opens to the right if a is positive and to the left if a is negative.

EXAMPLE 8 Graphing a Horizontal Parabola

Graph $x = (y - 2)^2 - 3$.

This graph has its vertex at $(-3, 2)$, since the roles of x and y are reversed. It opens to the right, the positive x-direction, and has the same shape as $y = x^2$. Plotting a few additional points gives the graph shown in Figure 15. Note that the graph is symmetric about its axis, $y = 2$.

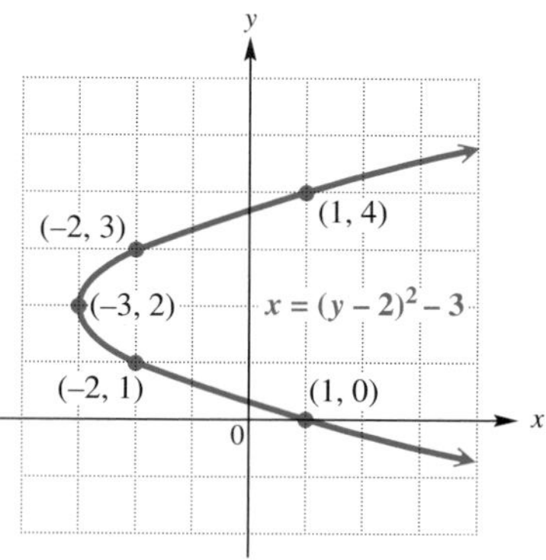

Figure 15

When a quadratic equation is given in the form $x = ay^2 + by + c$, completing the square on y changes the equation into a form in which the vertex can be more easily identified.

EXAMPLE 9 Completing the Square to Graph a Horizontal Parabola

Graph $x = -2y^2 + 4y - 3$.

Complete the square on the right to express the equation in the form $x = a(y - k)^2 + h$.

$$\frac{x}{-2} = y^2 - 2y + \frac{3}{2} \qquad \text{Divide by } -2.$$

$$\frac{x}{-2} - \frac{3}{2} = y^2 - 2y \qquad \text{Subtract } \tfrac{3}{2}.$$

$$\frac{x}{-2} - \frac{3}{2} + 1 = y^2 - 2y + 1 \qquad \text{Add 1.}$$

$$\frac{x}{-2} - \frac{1}{2} = (y - 1)^2 \qquad \text{Factor on the right; add on the left.}$$

$$\frac{x}{-2} = (y - 1)^2 + \frac{1}{2} \qquad \text{Add } \tfrac{1}{2}.$$

$$x = -2(y - 1)^2 - 1 \qquad \text{Multiply by } -2.$$

The coefficient -2 indicates the graph opens to the left (the negative x-direction) and is narrower than $y = x^2$. As shown in Figure 16, the vertex is $(-1, 1)$.

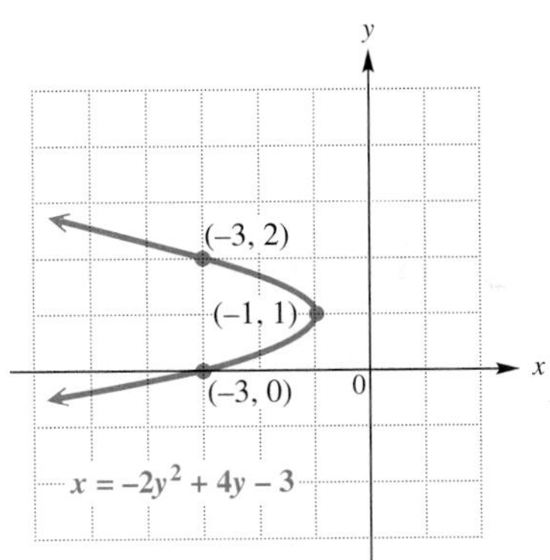

Figure 16

CAUTION Only quadratic equations solved for y (whose graphs are vertical parabolas) are examples of functions. The horizontal parabolas in Examples 8 and 9 are *not* graphs of functions, because they do not satisfy the vertical line test. Furthermore, the vertex formula does not apply to parabolas with horizontal axes.

In summary, the graphs of parabolas studied in this section and the previous one fall into the following categories.

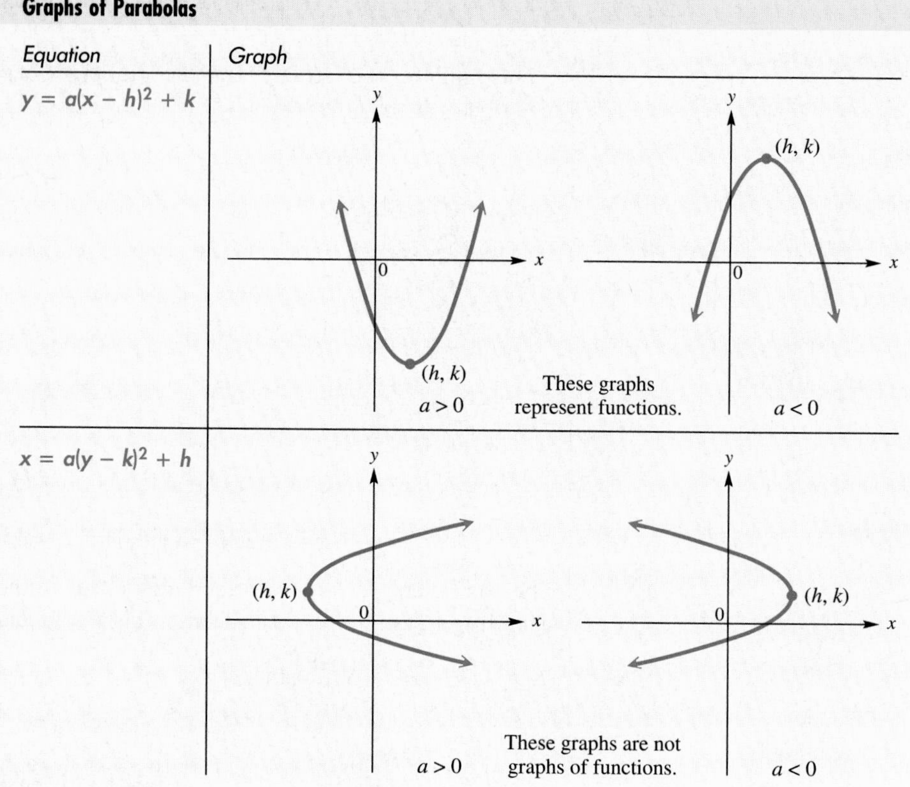

8.6 EXERCISES

1. How can you determine just by looking at the equation of a parabola whether it has a vertical or a horizontal axis?

2. Why can't the graph of a quadratic function be a parabola that opens to the left or to the right?

3. **(a)** How can you determine the number of x-intercepts of the graph of a quadratic function without graphing the function?
   (b) If the vertex of the graph of a quadratic function is $(1, -3)$, and the graph opens downward, how many x-intercepts does the graph have?

4. Explain how to graph a quadratic function given in the form $f(x) = ax^2 + bx + c$. Use $f(x) = x^2 - 3x - 4$ as an example.

Find the vertex of each parabola. Decide whether the graph opens upward, downward, to the left, or to the right, and state whether it is wider, narrower, or the same shape as the graph of $y = x^2$. If it is a vertical parabola, use the discriminant to determine the number of x-intercepts. See Examples 1–3, 5, 8, and 9.

5. $y = 2x^2 + 4x + 5$
6. $y = 3x^2 - 6x + 4$
7. $y = -x^2 + 5x + 3$
8. $x = -y^2 + 7y + 2$
9. $x = \frac{1}{3}y^2 + 6y + 24$
10. $x = .5y^2 + 10y - 5$

Graph the parabola using the techniques described in this section. If the graph represents a function, give its domain and range. See Examples 3, 4, 8, and 9.

11. $f(x) = x^2 + 8x + 10$
12. $f(x) = x^2 + 10x + 23$
13. $y = -2x^2 + 4x - 5$
14. $y = -3x^2 + 12x - 8$

15. $x = -\frac{1}{5}y^2 + 2y - 4$

16. $x = -.5y^2 - 4y - 6$

17. $x = 3y^2 + 12y + 5$

18. $x = 4y^2 + 16y + 11$

Use the concepts of this section to match the equation with its graph in Exercises 19–24.

A.

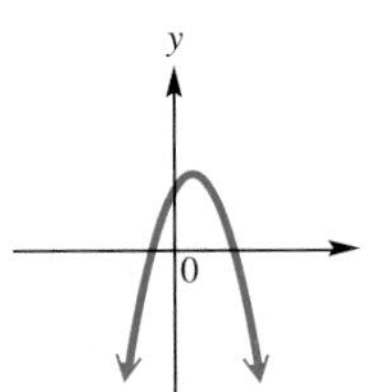

B.

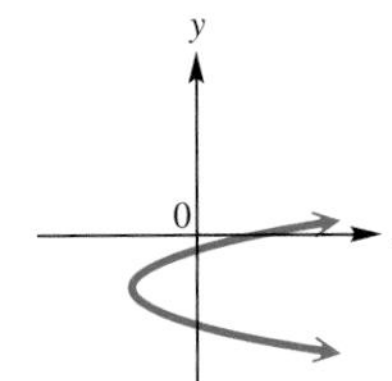

C.

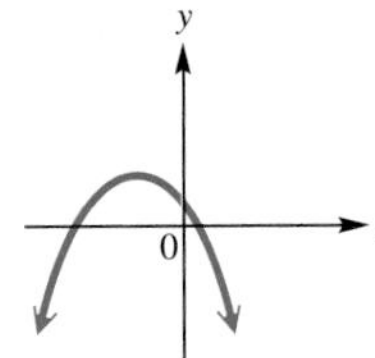

D.

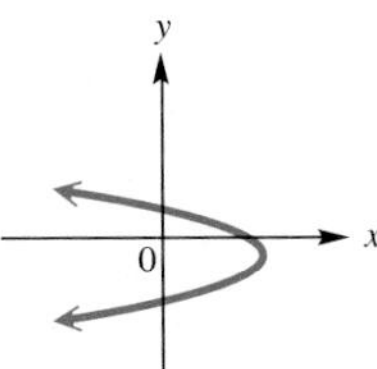

E.

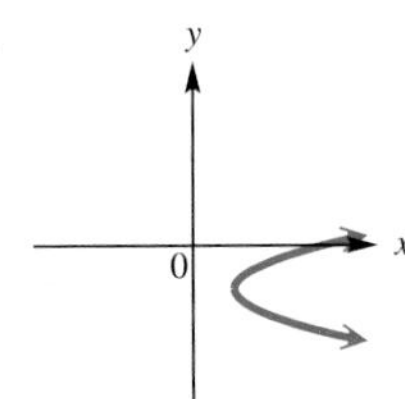

F.

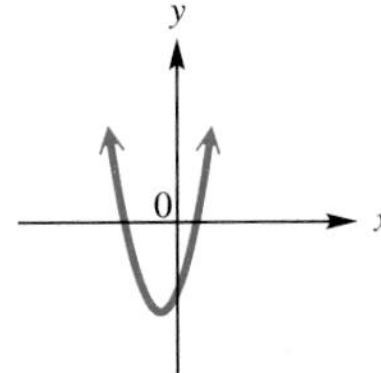

19. $y = 2x^2 + 4x - 3$

20. $y = -x^2 + 3x + 5$

21. $y = -\frac{1}{2}x^2 - x + 1$

22. $x = y^2 + 6y + 3$

23. $x = -y^2 - 2y + 4$

24. $x = 3y^2 + 6y + 5$

Solve each problem. See Examples 6 and 7.

25. Keisha Hughes has 100 meters of fencing material to enclose a rectangular exercise run for her dog. What width will give the enclosure the maximum area?

26. Morgan's Department Store wants to construct a rectangular parking lot on land bordered on one side by a highway. It has 280 feet of fencing that is to be used to fence off the other three sides. What should be the dimensions of the lot if the enclosed area is to be a maximum? What is the maximum area?

27. If an object on Earth is propelled upward with an initial velocity of 32 feet per second, then its height after t seconds is given by

$$h = 32t - 16t^2.$$

Find the maximum height attained by the object and the number of seconds it takes to hit the ground.

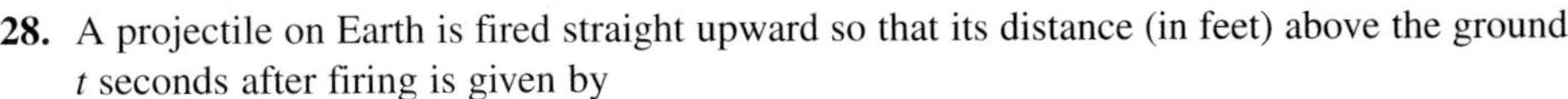

28. A projectile on Earth is fired straight upward so that its distance (in feet) above the ground t seconds after firing is given by

$$s(t) = -16t^2 + 400t.$$

Find the maximum height it reaches and the number of seconds it takes to reach that height.

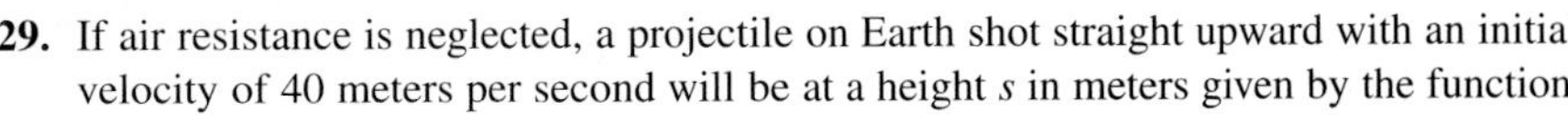

29. If air resistance is neglected, a projectile on Earth shot straight upward with an initial velocity of 40 meters per second will be at a height s in meters given by the function

$$s(t) = -4.9t^2 + 40t,$$

where t is the number of seconds elapsed after projection. After how many seconds will it reach its maximum height, and what is this maximum height? Round your answers to the nearest tenth.

30. A space robot is propelled from the moon with its distance in feet given by $f(x) = 1.727x - .0013x^2$ feet, where x is time in seconds. (See Section 8.4, Exercise 30.) Find the maximum height the robot can reach, and the time to get there. Give answers with four significant digits.

31. The bar graph shows how Social Security assets are expected to change as the number of retirees receiving benefits increases.

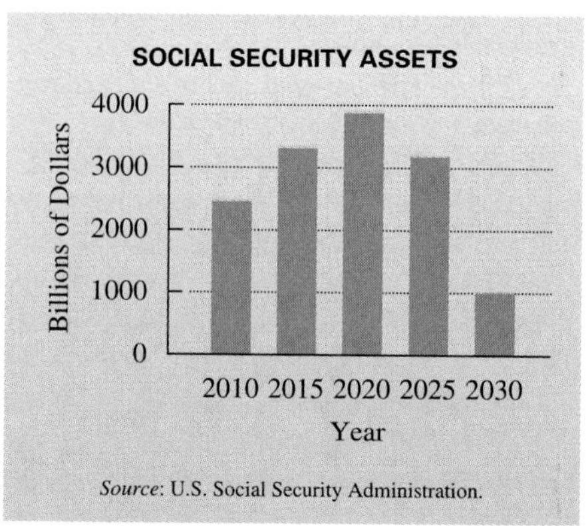

Source: U.S. Social Security Administration.

The graph suggests that a quadratic function would be a good fit to the data. Using a statistical technique, we determined that the data are approximated by the function with $f(x) = -20.57x^2 + 758.9x - 3140$. In the equation, $x = 10$ represents 2010, $x = 15$ represents 2015, and so on, and $f(x)$ is in billions of dollars.

(a) Explain why the coefficient of x^2 in the equation is negative, based on the graph.

(b) Algebraically determine the vertex of the graph, with coordinates to four significant digits.

(c) Interpret the answer to part (b) as it applies to the application.

32. The graph shows the performance of investment portfolios with different mixtures of U.S. and foreign investments for the period January 1, 1971, to December 31, 1996.

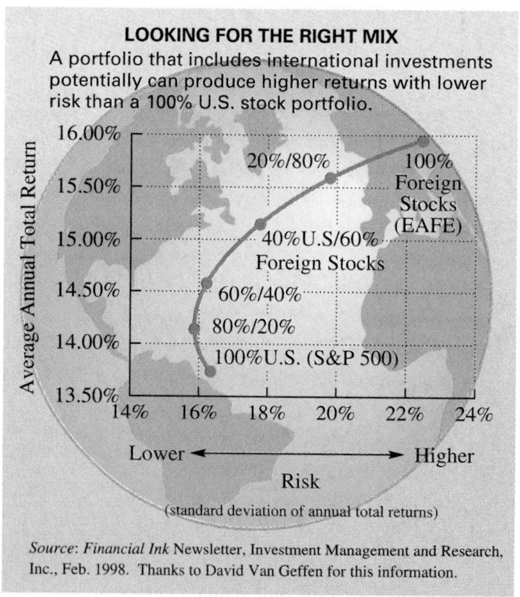

Source: *Financial Ink* Newsletter, Investment Management and Research, Inc., Feb. 1998. Thanks to David Van Geffen for this information.

Use the graph to answer the following questions.

(a) Is this the graph of a function? Explain.

(b) What investment mixture shown on the graph appears to represent the vertex? What relative amount of risk does this point represent? What return on investment does it provide?

(c) Which point on the graph represents the riskiest investment mixture? What return on investment does it provide?

33. A charter flight charges a fare of \$200 per person, plus \$4 per person for each unsold seat on the plane. If the plane holds 100 passengers and if x represents the number of unsold seats, find the following.

(a) An expression for the total revenue received for the flight (*Hint:* Multiply the number of people flying, $100 - x$, by the price per ticket.)

(b) The graph for the expression of part (a)

(c) The number of unsold seats that will produce the maximum revenue

(d) The maximum revenue

34. For a trip to a resort, a charter bus company charges a fare of \$48 per person, plus \$2 per person for each unsold seat on the bus. If the bus has 42 seats and x represents the number of unsold seats, find the following.

(a) An expression that defines the total revenue, R, from the trip (*Hint:* Multiply the total number riding, $42 - x$, by the price per ticket, $48 + 2x$)

(b) The graph of the function from part (a)

(c) The number of unsold seats that produces the maximum revenue

(d) The maximum revenue

TECHNOLOGY INSIGHTS (EXERCISES 35–38)

Graphing calculators are capable of determining the coordinates of "peaks" and "valleys" of graphs. In the case of quadratic functions, these peaks and valleys are the vertices, and are called maximum and minimum points. For example, the vertex of the graph of $f(x) = -x^2 - 6x - 13$ is $(-3, -4)$, as indicated in the display at the bottom of the screen. In this case, the vertex is a maximum point.

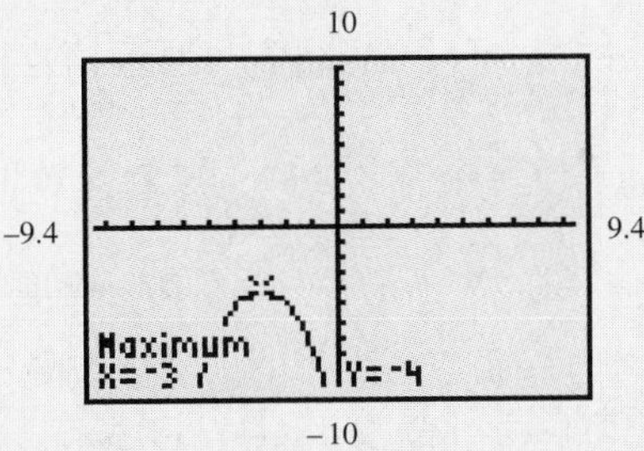

In Exercises 35–38, match the function with its calculator-generated graph by determining the vertex and using the display at the bottom of the screen.

A. **B.**

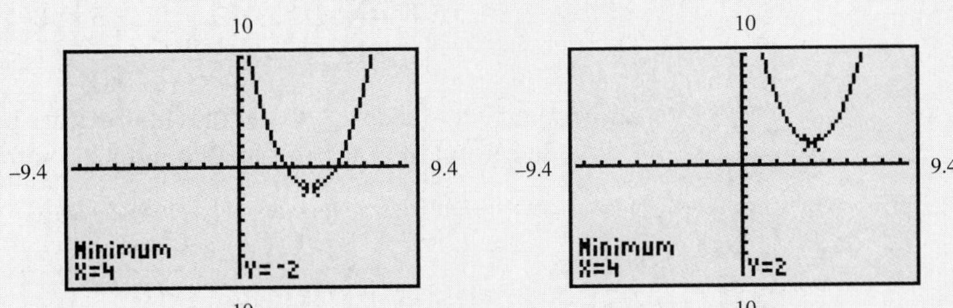

(continued)

TECHNOLOGY INSIGHTS (EXERCISES 35–38) (CONTINUED)

C.

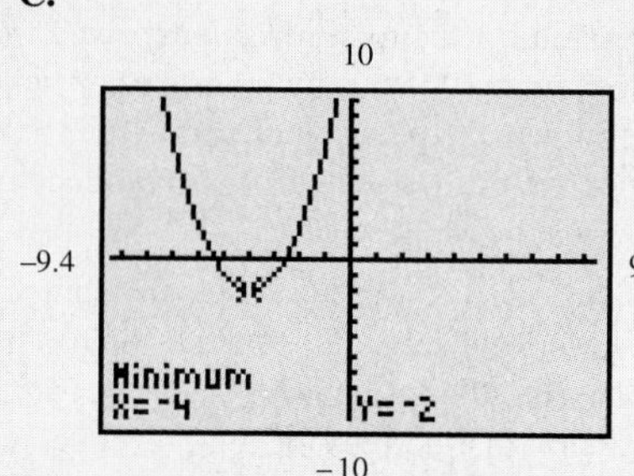

D.

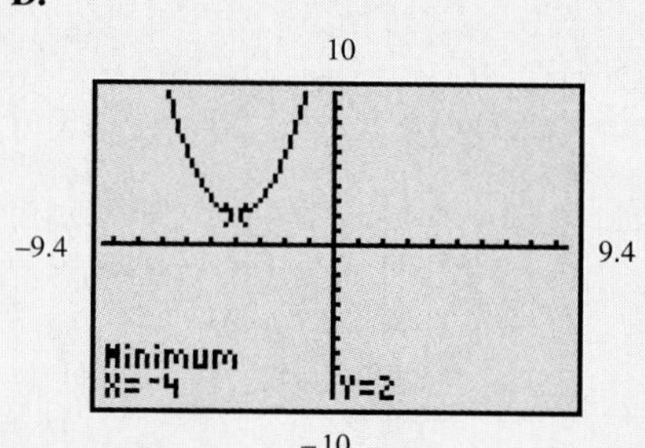

35. $f(x) = x^2 - 8x + 18$

36. $f(x) = x^2 + 8x + 18$

37. $f(x) = x^2 - 8x + 14$

38. $f(x) = x^2 + 8x + 14$

8.7 Nonlinear Inequalities

OBJECTIVES

1. Solve quadratic inequalities.
2. Solve polynomial inequalities of degree 3 or more.
3. Solve rational inequalities.

FOR EXTRA HELP

SSG Sec. 8.7
SSM Sec. 8.7

Pass the Test Software

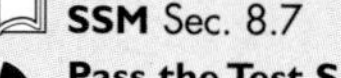

InterAct Math Tutorial Software

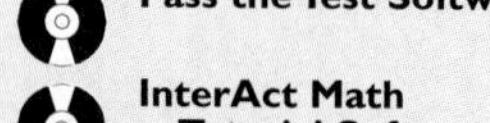

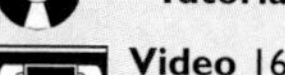

Video 16

We discussed methods of solving linear inequalities in Chapter 2 and methods of solving quadratic equations in this chapter. Now we combine these ideas to solve *quadratic inequalities.*

Quadratic Inequality

A **quadratic inequality** can be written in the form

$$ax^2 + bx + c < 0 \quad \text{or} \quad ax^2 + bx + c > 0,$$

where a, b, and c are real numbers, with $a \neq 0$.

As before, $<$ and $>$ may be replaced with $\leq$ and $\geq$ as necessary.

OBJECTIVE 1 Solve quadratic inequalities.

EXAMPLE 1 Solving a Quadratic Inequality

Solve $x^2 - x - 12 > 0$.

First solve the quadratic *equation*

$$x^2 - x - 12 = 0.$$

$$(x - 4)(x + 3) = 0 \qquad \text{Factor.}$$

$$x - 4 = 0 \quad \text{or} \quad x + 3 = 0 \qquad \text{Zero-factor property}$$

$$x = 4 \quad \text{or} \quad x = -3$$

The numbers 4 and -3 divide the number line into three regions, as shown in Figure 17. (Be careful to graph the smaller number on the left.)

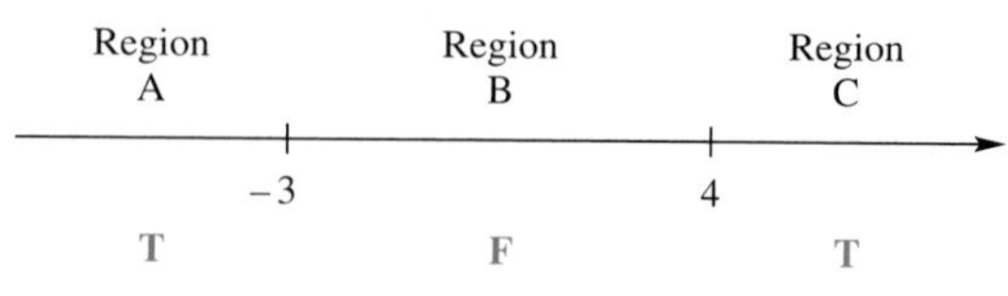

Figure 17

The numbers 4 and -3 are the only numbers that make the expression $x^2 - x - 12$ equal to 0. All other numbers make the expression either positive or negative, and the value of the expression can change from positive to negative or from negative to positive only on either side of a number that makes it 0. Therefore, if one number in a region satisfies the inequality, then all the numbers in that region will satisfy the inequality. Choose any number from Region A in Figure 17 (any number less than -3). Substitute this number for x in the original inequality. If the result is true, then all numbers in Region A satisfy the original inequality.

We choose -5 from Region A. Substitute -5 for x in the inequality.

$$x^2 - x - 12 > 0 \quad \text{Original inequality}$$
$$(-5)^2 - (-5) - 12 > 0 \quad ?$$
$$25 + 5 - 12 > 0 \quad ?$$
$$18 > 0 \quad \text{True}$$

Since -5 from Region A satisfies the inequality, all numbers from Region A are solutions.

We try 0 from Region B. If $x = 0$, then

$$0^2 - 0 - 12 > 0 \quad ?$$
$$-12 > 0. \quad \text{False}$$

The numbers in Region B are *not* solutions. Verify that the number 5 satisfies the inequality, so the numbers in Region C are also solutions to the inequality.

Based on these results (shown by the colored letters in Figure 17), the solution set includes the numbers in Regions A and C, as shown on the graph in Figure 18. The solution set is written

$$(-\infty, -3) \cup (4, \infty).$$

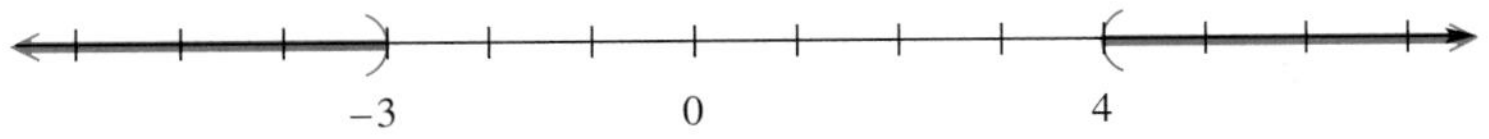

−3 0 4

Figure 18

In summary, we solve a quadratic inequality by following these steps.

Solving a Quadratic Inequality

Step 1 **Change the inequality to an equation and solve it.**

Step 2 **Use the solutions from Step 1 to determine regions.** Graph the numbers found in Step 1 on a number line. These numbers divide the number line into regions.

Step 3 **Find the intervals that satisfy the inequality.** Substitute a number from each region into the inequality to determine the regions that satisfy the inequality. All numbers in those regions are in the solution set. A graph of the solution set will usually look like one of these.

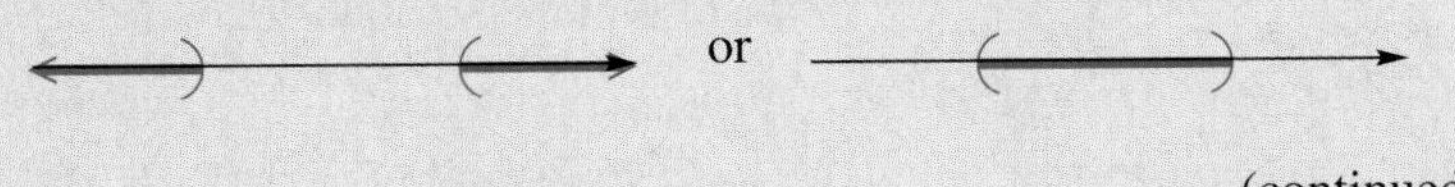

(continued)

Solving a Quadratic Inequality (continued)

Step 4 **Consider the endpoints separately.** The numbers from Step 1 are included in the solution set if the inequality is $\leq$ or $\geq$; they are not included if it is $<$ or $>$.

Special cases of quadratic inequalities may occur, as in the next example.

EXAMPLE 2 Solving Special Cases

Solve $(2y - 3)^2 > -1$.

Since $(2y - 3)^2$ is never negative, it is always greater than -1. Thus, the solution is the set of all real numbers $(-\infty, \infty)$. In the same way, there is no solution for $(2y - 3)^2 < -1$ and the solution set is $\emptyset$.

OBJECTIVE 2 Solve polynomial inequalities of degree 3 or more. Higher-degree polynomial inequalities that can be factored are solved in the same way as quadratic inequalities.

EXAMPLE 3 Solving a Third-Degree Polynomial Inequality

Solve $(x - 1)(x + 2)(x - 4) \leq 0$.

This is a *cubic* (third-degree) inequality rather than a quadratic inequality, but it can be solved by the method shown above and by extending the zero-factor property to more than two factors. Begin by setting the factored polynomial *equal* to 0 and solving the equation. (Step 1)

$$(x - 1)(x + 2)(x - 4) = 0$$

$$x - 1 = 0 \quad \text{or} \quad x + 2 = 0 \quad \text{or} \quad x - 4 = 0$$

$$x = 1 \quad \text{or} \quad x = -2 \quad \text{or} \quad x = 4$$

Locate the numbers -2, 1, and 4 on a number line as in Figure 19 to determine the regions A, B, C, and D. (Step 2)

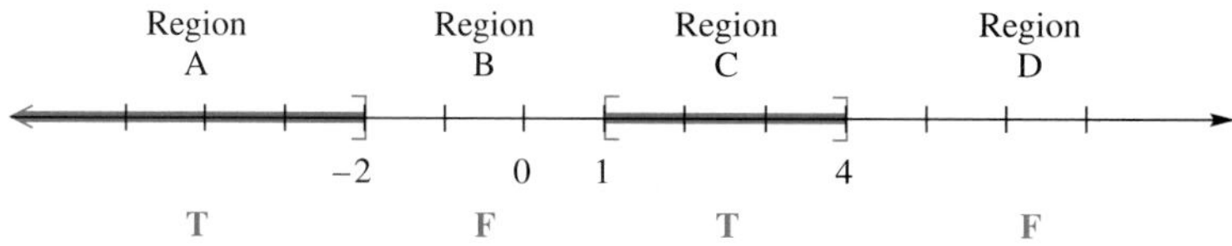

Figure 19

Substitute a number from each region into the original inequality to determine which regions satisfy the inequality. (Step 3) These results are shown below the number line in Figure 19. For example, in Region A, using $x = -3$ gives

$$(-3 - 1)(-3 + 2)(-3 - 4) \leq 0 \quad ?$$

$$(-4)(-1)(-7) \leq 0 \quad ?$$

$$-28 \leq 0. \quad \text{True}$$

The numbers in Region A are in the solution set. Verify that the numbers in Region C are also in the solution set, which is written

$$(-\infty, -2] \cup [1, 4].$$

Notice that the three endpoints are included in the solution set (Step 4), graphed in Figure 19.

OBJECTIVE 3 Solve rational inequalities. Inequalities that involve fractions, called **rational inequalities,** can change sign only at values where the corresponding equation is 0 or where the denominator is 0. Steps 1, 2, and 5 in the following list consider these two cases.

Solving Rational Inequalities

Step 1 **Change the inequality to an equation and solve it.**

Step 2 **Set the denominator equal to 0 and solve the equation.**

Step 3 **Divide the number line into regions.** Use the solutions from Steps 1 and 2 to divide the number line into regions.

Step 4 **Find the intervals that satisfy the inequality.** Test a number from each region by substituting it into the inequality to determine the intervals that satisfy the inequality.

Step 5 **Consider the endpoints.** Exclude any values that make the denominator 0.

Remember Steps 2 and 5. Any number that makes the denominator 0 *must* be excluded from the solution set, since there will be no point on the number line at that value.

EXAMPLE 4 Solving a Rational Inequality

Solve the inequality $\frac{-1}{p-3} \geq 1$.

Write the corresponding equation and solve it. (Step 1)

$$\frac{-1}{p-3} = 1$$

$$-1 = p - 3 \qquad \text{Multiply by the common denominator.}$$

$$2 = p$$

Find the number that makes the denominator 0. (Step 2)

$$p - 3 = 0$$

$$p = 3$$

These two numbers, 2 and 3, divide a number line into three regions. (Step 3) (See Figure 20.)

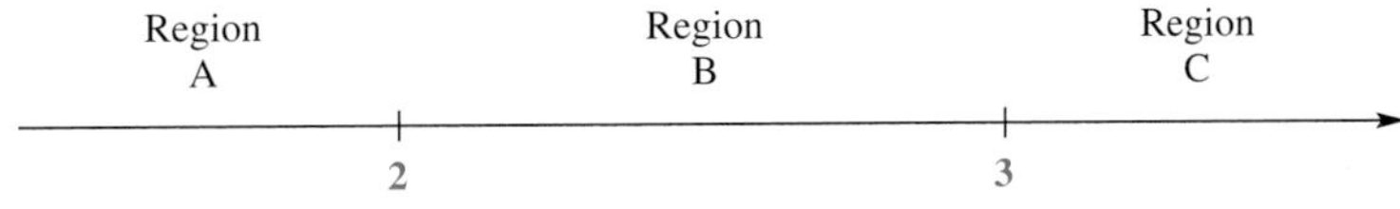

Figure 20

Testing one number from each region in the given inequality shows that the solution set is the interval [2, 3). (Step 4) Note that this interval does not include the number 3, because 3 makes the denominator of the original inequality equal to 0. (Step 5) A graph of the solution set is given in Figure 21.

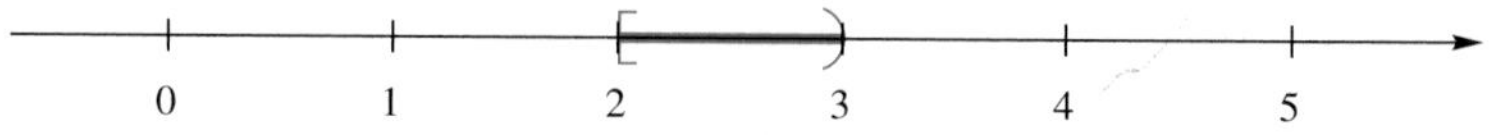

Figure 21

CONNECTIONS

As the examples in this section show, there is a close connection between equality and inequality. In Section 8.4, Example 4 discussed the motion of a rocket propelled from the surface of Mars. Under certain conditions, $y = x - .0066x^2$ gives the height of the rocket, in feet, x seconds after it leaves the surface. If we know the number of seconds it takes the rocket to reach 50 feet and return to the ground, we can determine the time intervals when it will be less than 50 feet high and more than 50 feet high. (*Source:* Bernice Kastner, *Space Mathematics,* NASA.)

FOR DISCUSSION OR WRITING

Use the quadratic formula and the first objective in this section to answer the following questions to the nearest tenth of a second.

1. At what times will the rocket be 20 feet above the surface? (*Hint:* Let $y = 20$ and solve the quadratic equation.)
2. In what time interval(s) will the rocket be more than 20 feet above the surface?
3. At what times will the rocket be at the surface? (*Hint:* Let $y = 0$ and solve the equation.)
4. In what time interval(s) will the rocket be less than 20 feet above the surface?

8.7 EXERCISES

Follow the steps in Exercises 1–4 to solve the inequality $2x^2 + x - 3 \leq 0$.

1. Change the inequality to an equation and solve it.
2. Indicate the regions determined by the solutions from Step 1 on a number line.
3. Determine the interval(s) that satisfy the inequality.
4. Determine whether the endpoints are included or not, and give the solution set in interval form.

Solve each inequality and graph the solution set. See Example 1. (Hint: In Exercises 17 and 18, use the quadratic formula.)

5. $(x + 1)(x - 5) > 0$
6. $(m + 6)(m - 2) > 0$
7. $(r + 4)(r - 6) < 0$
8. $(y + 4)(y - 8) < 0$
9. $x^2 - 4x + 3 \geq 0$
10. $m^2 - 3m - 10 \geq 0$
11. $10a^2 + 9a \geq 9$
12. $3r^2 + 10r \geq 8$
13. $9p^2 + 3p < 2$
14. $2y^2 + y < 15$
15. $6x^2 + x \geq 1$
16. $4y^2 + 7y \geq -3$
17. $y^2 - 6y + 6 \geq 0$
18. $3k^2 - 6k + 2 \leq 0$

19. Explain how you determine whether to include or exclude endpoints when solving a quadratic or higher-degree inequality.

20. The solution set of the inequality $x^2 + x - 12 < 0$ is the interval $(-4, 3)$. Without actually performing any work, give the solution set of the inequality $x^2 + x - 12 \geq 0$.

Solve each inequality. See Example 2.

21. $(4 - 3x)^2 \geq -2$ **22.** $(6y + 7)^2 \geq -1$ **23.** $(3x + 5)^2 \leq -4$ **24.** $(8t + 5)^2 \leq -5$

25. Explain how you would solve and graph the solution set of a quadratic inequality. Use an example.

Solve each inequality and graph the solution set. See Example 3.

26. $(p - 1)(p - 2)(p - 4) < 0$

27. $(2r + 1)(3r - 2)(4r + 7) < 0$

28. $(a - 4)(2a + 3)(3a - 1) \geq 0$

29. $(z + 2)(4z - 3)(2z + 7) \geq 0$

30. Without actually performing any work, give the solution set of the rational inequality $\frac{3}{x^2 + 1} > 0$. (*Hint:* Determine the sign of the numerator. Determine what the sign of the denominator *must* be. Then consider the inequality symbol.)

Solve each inequality and graph the solution set. See Example 4.

31. $\frac{x - 1}{x - 4} > 0$

32. $\frac{x + 1}{x - 5} > 0$

33. $\frac{2y + 3}{y - 5} \leq 0$

34. $\frac{3t + 7}{t - 3} \leq 0$

35. $\frac{8}{x - 2} \geq 2$

36. $\frac{20}{y - 1} \geq 1$

37. $\frac{3}{2t - 1} < 2$

38. $\frac{6}{m - 1} < 1$

39. $\frac{a}{a + 2} \geq 2$

40. $\frac{m}{m + 5} \geq 2$

41. $\frac{x}{x - 4} < 3$

42. $\frac{2y}{y + 1} > 4$

43. $\frac{4k}{2k - 1} < k$

44. $\frac{r}{r + 2} < 2r$

45. $\frac{2x - 3}{x^2 + 1} \geq 0$

46. $\frac{9x - 8}{4x^2 + 25} < 0$

47. $\frac{(3x - 5)^2}{x + 2} > 0$

48. $\frac{(5x - 3)^2}{2x + 1} \leq 0$

RELATING CONCEPTS (EXERCISES 49–52)

An alternative method for solving rational inequalities requires first writing the inequality with 0 on one side. The other side is then expressed as a single fraction.

Work Exercises 49–52 in order *to use the alternative method to solve the rational inequality in Example 4,*

$$\frac{-1}{p - 3} \geq 1.$$

49. Subtract 1 from both sides of the inequality to make one side equal to 0.

50. Rewrite the left side to get a common denominator for the two terms on the left.

51. To write the left side as a single fraction, perform the indicated subtraction and combine terms as needed.

52. Find the number that makes the numerator 0 and the number that makes the denominator 0. Complete the solution as in Example 4.

Did you make the connection between the two methods for solving rational inequalities? Which method do you prefer?

Work each problem. See the Connections box for this section.

53. The equation $y = x - 1.727x^2$, found in Section 8.4, Exercise 29, gives the height y of a space vehicle x seconds after leaving the moon. At what times will the height of the vehicle be greater than .140 foot?

54. In Section 8.4, Exercise 30, the height of a rock propelled from the surface of the moon is given as $y = 1.727x - .0013x^2$ feet, where x is time in seconds. The rock goes up and returns to the surface in approximately 1328 seconds. At what times will the height of the rock be less than 500 feet?

RELATING CONCEPTS (EXERCISES 55–58)

We know that the x-intercepts of the graph of a quadratic equation give the real solutions of the equation. We can also use these intercepts to solve the corresponding inequalities. The x-values of points on the graph that are *above* the x-axis give the solution set of the $>$ inequality. Similarly, the solution set of the $<$ inequality is found by locating the points on the graph that lie *below* the x-axis. For example, from the graph of $f(x) = x^2 - x - 12$ in the figure, the solution set of $x^2 - x - 12 > 0$ is $(-\infty, -3) \cup (4, \infty)$, and the solution set of $x^2 - x - 12 < 0$ is $(-3, 4)$.

$f(x) = x^2 - x - 12$

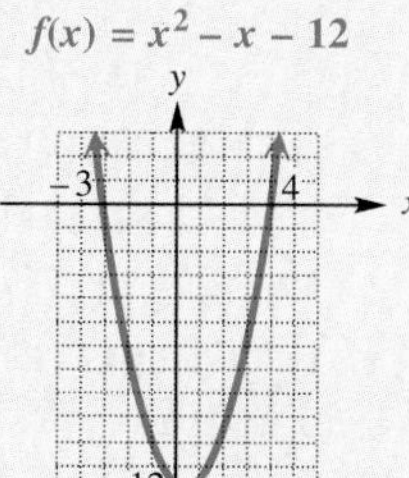

The graph is *above* the x-axis for $(-\infty, -3) \cup (4, \infty)$

In Exercises 55–58, the graph of a quadratic function f is given. Use only the graph to find the solution set of the equation or inequality. ***Work through parts (a)–(c) in order each time.***

55. $f(x) = x^2 - 4x + 3$

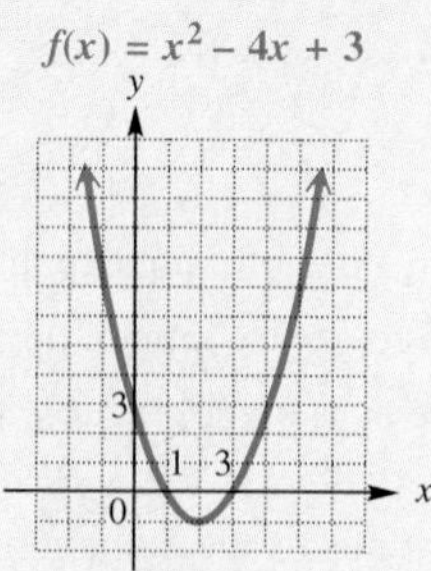

(a) $x^2 - 4x + 3 = 0$
(b) $x^2 - 4x + 3 > 0$
(c) $x^2 - 4x + 3 < 0$

56. $f(x) = 3x^2 + 10x - 8$

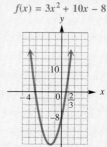

(a) $3x^2 + 10x - 8 = 0$
(b) $3x^2 + 10x - 8 \geq 0$
(c) $3x^2 + 10x - 8 < 0$

RELATING CONCEPTS (EXERCISES 55–58) (CONTINUED)

57. $f(x) = -x^2 + 3x + 10$

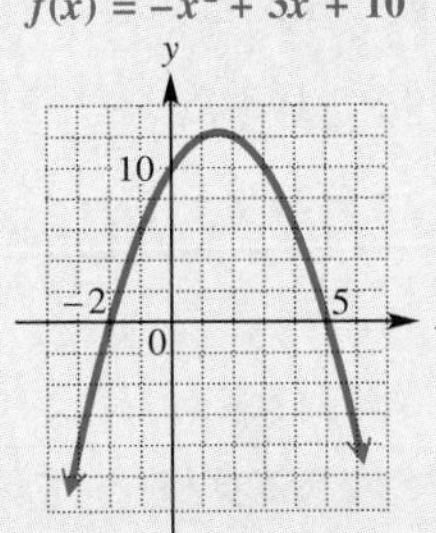

(a) $-x^2 + 3x + 10 = 0$
(b) $-x^2 + 3x + 10 \geq 0$
(c) $-x^2 + 3x + 10 \leq 0$

58. $f(x) = -2x^2 - x + 15$

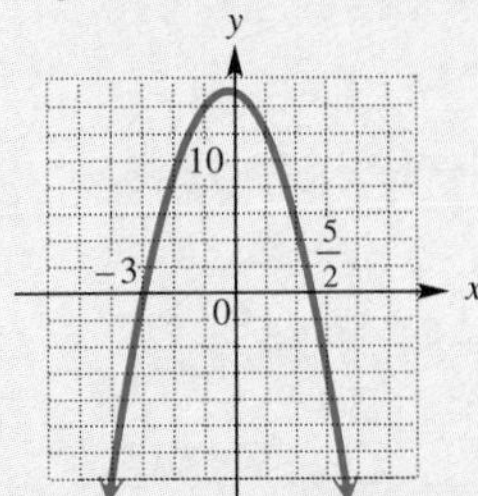

(a) $-2x^2 - x + 15 = 0$
(b) $-2x^2 - x + 15 \geq 0$
(c) $-2x^2 - x + 15 \leq 0$

Did you make the connection that the x-intercepts of the graph of a function where $y = f(x)$ give the endpoints of the solution sets of the inequalities $f(x) > 0$ and $f(x) < 0$?

Work each problem.

59. In Example 7 in Section 8.6, the height of a rocket, in feet, at time x, in seconds, is given as $f(x) = x - .0066x^2$. At what time(s) will the height of the rocket be less than 10 feet?

60. In Section 8.6, Exercise 30, the distance from the moon, in feet, of a space robot is given as $f(x) = 1.727x - .0013x^2$, where x is in seconds. At what time(s) is the robot more than 500 feet away?

CHAPTER 8 GROUP ACTIVITY

Finding the Path of a Comet

Objective: Find and graph an equation of a parabola with a given focus.

The orbit that a comet takes as it approaches the sun depends upon its velocity (as well as other factors). If the comet has enough velocity to escape the sun's pull, it takes either a parabolic orbit or a hyperbolic orbit. If it doesn't have enough velocity to escape the sun's pull, then it takes an elliptical orbit. Of course, a comet that has a parabolic or hyperbolic orbit will not come back around the sun again. Comets with elliptical orbits are the only ones we see again. For this activity we consider a comet with a parabolic orbit.

If the velocity of a comet equals escape velocity (that is, it is going just fast enough to get away from the sun), then its orbit will be parabolic. The sun is at the focus of the parabola. The vertex is the point where the comet is closest to the sun.

A. Make a sketch of the comet and the sun.

1. Place the vertex of the parabola at the origin with the focus on the y-axis and a horizontal directrix.
2. Assume that the comet is .75 astronomical unit from the sun at its closest point. (See the chapter introduction for a definition of an astronomical unit.)

(continued)

B. What are the coordinates of the focus?

C. What is the equation of the directrix?

D. Using the information from Section 8.5, Exercise 47, find an equation of the parabola.

E. Graph the parabola (as a vertical parabola). Include the focus and directrix on your graph.

CHAPTER 8 SUMMARY

KEY TERMS

8.2 quadratic formula	**8.5 parabola**	**axis**	**8.7 quadratic inequality**
discriminant	**vertex**	**quadratic function**	**rational inequality**
8.3 quadratic in form			

TEST YOUR WORD POWER

See how well you have learned the vocabulary in this chapter. Answers, with examples, are given at the bottom of the page.

1. The **quadratic formula** is
(a) a formula to find the number of solutions of a quadratic equation
(b) a formula to find the type of solutions of a quadratic equation
(c) the standard form of a quadratic equation
(d) a general formula for solving any quadratic equation.

2. The **discriminant** is
(a) the quantity under the radical in the quadratic formula
(b) the quantity in the denominator in the quadratic formula
(c) the solution set of a quadratic equation
(d) the result of using the quadratic formula.

3. A **quadratic function** is a function that can be written in the form
(a) $f(x) = mx + b$ for real numbers m and b
(b) $f(x) = \frac{P(x)}{Q(x)}$ where $Q(x) \neq 0$
(c) $f(x) = ax^2 + bx + c$ for real numbers a, b, c $(a \neq 0)$
(d) $f(x) = \sqrt{x}$ for $x \geq 0$.

4. A **parabola** is the graph of
(a) any equation in two variables
(b) a linear equation
(c) an equation of degree three
(d) a quadratic equation.

5. The **vertex** of a parabola is
(a) the point where the graph intersects the y-axis
(b) the points where the graph intersects the x-axis
(c) the lowest point on a parabola that opens up or the highest point on a parabola that opens down
(d) the origin.

6. The **axis** of a parabola is
(a) either the x-axis or the y-axis
(b) the vertical line (of a vertical parabola) or the horizontal line (of a horizontal parabola) through the vertex
(c) the lowest or highest point on the graph of a parabola
(d) a line through the origin.

7. A parabola is **symmetric about its axis** since
(a) its graph is near the axis
(b) its graph is identical when reflected across the axis
(c) its graph looks different on each side of the axis
(d) its graph intersects the axis.

Answers to Test Your Word Power

1. (d) *Example:* The solutions of $ax^2 + bx + c = 0$ $(a \neq 0)$ are $x = \frac{-b \pm \sqrt{b^2 - 4ac}}{2a}$. **2.** (a) *Example:* In the quadratic formula, the discriminant is $b^2 - 4ac$. **3.** (c) *Examples:* $f(x) = x^2 - 2$, $f(x) = (x + 4)^2 + 1$, $f(x) = x^2 - 4x + 5$ **4.** (d) *Examples:* See Figures 5–10 in Section 8.5. **5.** (c) *Example:* The graph of $y = (x + 3)^2$ has vertex $(-3, 0)$, which is the lowest point on the graph. **6.** (b) *Example:* The axis of $y = (x + 3)^2$ is the vertical line $x = -3$. **7.** (b) *Example:* Since the graph of $y = (x + 3)^2$ is symmetric about its axis $x = -3$, the points $(-2, 1)$ and $(-4, 1)$ are on the graph.

QUICK REVIEW

CONCEPTS	EXAMPLES

8.1 COMPLETING THE SQUARE

Square Root Property

(a) If $x^2 = b$, $b \geq 0$, then $x = \sqrt{b}$ or $x = -\sqrt{b}$.

Solve.

(a) $(x - 1)^2 = 8$

$$x - 1 = \pm\sqrt{8} = \pm 2\sqrt{2}$$
$$x = 1 \pm 2\sqrt{2}$$

Solution set: $\{1 \pm 2\sqrt{2}\}$

(b) If $x^2 = -b$, $b > 0$, then $x = i\sqrt{b}$ or $x = -i\sqrt{b}$.

(b) $x^2 = -5$

$$x = i\sqrt{5} \quad \text{or} \quad x = -i\sqrt{5}$$

Solution set: $\{\pm i\sqrt{5}\}$

Completing the Square

To solve $ax^2 + bx + c = 0$: If $a \neq 1$, divide both sides by a. Write the equation with variable terms on one side and the constant on the other. Find half the coefficient of x and square it. Add the square to both sides. One side should now be a perfect square. Write it as the square of a binomial. Use the square root property to complete the solution.

Solve.

$$2x^2 - 4x - 18 = 0$$
$$x^2 - 2x - 9 = 0 \quad \text{Divide by 2.}$$
$$x^2 - 2x = 9 \quad \text{Add 9.}$$
$$\left[\frac{1}{2}(-2)\right]^2 = (-1)^2 = 1$$
$$x^2 - 2x + 1 = 9 + 1 \quad \text{Add 1.}$$
$$(x - 1)^2 = 10 \quad \text{Factor.}$$
$$x - 1 = \pm\sqrt{10}$$
$$x = 1 \pm \sqrt{10}$$

Solution set: $\{1 \pm \sqrt{10}\}$

8.2 THE QUADRATIC FORMULA

Quadratic Formula

The solutions of $ax^2 + bx + c = 0$ $(a \neq 0)$ are

$$x = \frac{-b \pm \sqrt{b^2 - 4ac}}{2a}.$$

Solve $3x^2 + 5x + 2 = 0$.

$$x = \frac{-5 \pm \sqrt{5^2 - 4(3)(2)}}{2(3)}$$
$$x = -1 \quad \text{or} \quad x = -\frac{2}{3}$$

Solution set: $\left\{-1, -\frac{2}{3}\right\}$

The Discriminant

If a, b, and c are integers, then the discriminant, $b^2 - 4ac$, of $ax^2 + bx + c = 0$ determines the type of solutions as follows.

Discriminant	Solutions
Positive square of an integer	2 rational solutions
Positive, not square of an integer	2 irrational solutions
Zero	1 rational solution
Negative	2 imaginary solutions

For $x^2 + 3x - 10 = 0$, the discriminant is

$$3^2 - 4(1)(-10) = \mathbf{49}.$$

There are **2 rational** solutions.

For $2x^2 + 5x + 1 = 0$, the discriminant is

$$5^2 - 4(2)(1) = \mathbf{17}.$$

There are **2 irrational** solutions.

For $9x^2 - 6x + 1 = 0$, the discriminant is

$$(-6)^2 - 4(9)(1) = \mathbf{0}.$$

There is **1 rational** solution.

For $4x^2 + x + 1 = 0$, the discriminant is

$$1^2 - 4(4)(1) = \mathbf{-15}.$$

There are **2 imaginary** solutions.

(continued)

CONCEPTS	EXAMPLES
8.4 FORMULAS AND APPLICATIONS	
To solve a formula for a squared variable, proceed as follows.	
(a) The variable appears only to the second degree. Isolate the squared variable on one side of the equation, then use the square root property.	Solve $A = \frac{2mp}{r^2}$ for r. $r^2A = 2mp$ Multiply by r^2. $r^2 = \frac{2mp}{A}$ Divide by A. $r = \pm\sqrt{\frac{2mp}{A}}$ Take square roots. $r = \pm\frac{\sqrt{2mpA}}{A}$ Rationalize.
(b) The variable appears to the first and second degrees. Write the equation in standard quadratic form, then use the quadratic formula.	Solve $m^2 + rm = t$ for m. $m^2 + rm - t = 0$ Standard form $m = \frac{-r \pm \sqrt{r^2 - 4(1)(-t)}}{2(1)}$ $a = 1, b = r, c = -t$ $m = \frac{-r \pm \sqrt{r^2 + 4t}}{2}$
8.5 GRAPHS OF QUADRATIC FUNCTIONS	
1. The graph of the quadratic function with $f(x) = a(x - h)^2 + k$, $a \neq 0$, is a parabola with vertex at (h, k) and the vertical line $x = h$ as axis. 2. The graph opens upward if a is positive and downward if a is negative. 3. The graph is wider than $f(x) = x^2$ if $0 < \lvert a \rvert < 1$ and narrower if $\lvert a \rvert > 1$.	Graph $f(x) = -(x + 3)^2 + 1$. 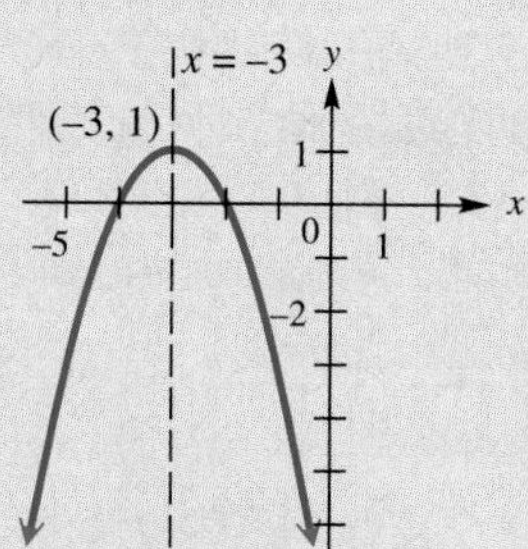
8.6 MORE ABOUT PARABOLAS; APPLICATIONS	
The vertex of the graph of $f(x) = ax^2 + bx + c$, $a \neq 0$, may be found by completing the square. The vertex has coordinates $\left(\frac{-b}{2a}, f\left(\frac{-b}{2a}\right)\right)$. To graph a quadratic function: Find the y-intercept by evaluating $f(0)$. Find any x-intercepts by solving $f(x) = 0$. Find the vertex either by using the formula or by completing the square. Find and plot any additional points as needed, using symmetry about the axis. Verify that the graph opens upward (if $a > 0$) or opens downward (if $a < 0$).	Graph $f(x) = x^2 + 4x + 3$. The vertex is $(-2, -1)$. Since $f(0) = 3$, the y-intercept is $(0, 3)$. The solutions of $x^2 + 4x + 3 = 0$ are -1 and -3, so the x-intercepts are $(-1, 0)$ and $(-3, 0)$. 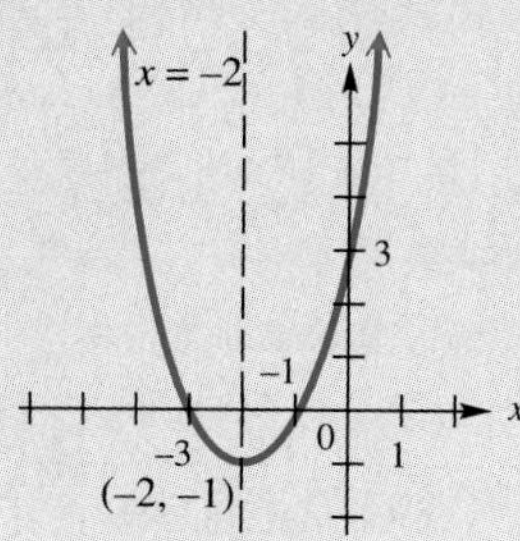

CONCEPTS	EXAMPLES

If the discriminant, $b^2 - 4ac$, is positive, the graph of $f(x) = ax^2 + bx + c$ has two x-intercepts; if zero, one x-intercept; if negative, no x-intercepts.

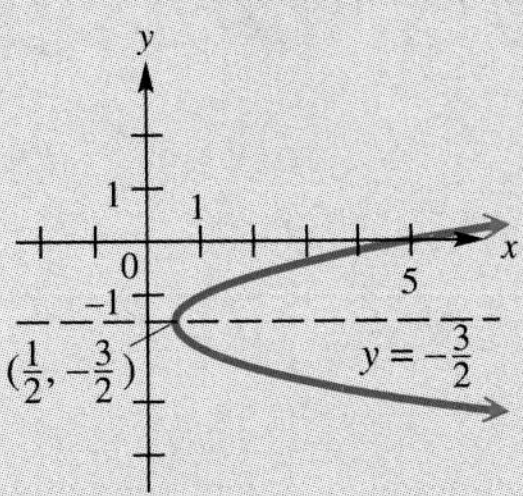

The graph of $x = ay^2 + by + c$ is a horizontal parabola, opening to the right if $a > 0$, or to the left if $a < 0$. Horizontal parabolas do not represent functions.

Graph $x = 2y^2 + 6y + 5$.
The graph is shown above.

8.7 NONLINEAR INEQUALITIES

Solving a Quadratic (or Higher Degree Polynomial) Inequality

Solve $2x^2 + 5x + 2 < 0$.

Step 1 Write the inequality as an equation and solve.

$$2x^2 + 5x + 2 = 0$$

$$x = -\frac{1}{2}, \qquad x = -2$$

Step 2 Graph the numbers found in Step 1 on a number line. These numbers divide the line into regions.

Step 3 Substitute a number from each region into the inequality to determine the intervals that belong in the solution set—those intervals containing numbers that make the inequality true.

$x = -3$ makes it false; $x = -1$ makes it true; $x = 0$ makes it false.

Step 4 Consider the endpoints separately.

Solution set: $\left(-2, -\frac{1}{2}\right)$

Solving a Rational Inequality

Solve $\frac{x}{x+2} \geq 4$.

Step 1 Write the inequality as an equation and solve the equation.

$\frac{x}{x+2} = 4$ leads to $x = -\frac{8}{3}$.

Step 2 Set the denominator equal to 0 and solve the equation.

$$x + 2 = 0$$
$$x = -2$$

Step 3 Use the solutions from Steps 1 and 2 to divide a number line into regions.

$-\frac{8}{3}$ -2

Step 4 Test a number from each region in the inequality to determine the regions that satisfy the inequality.

-4 makes it false; $-\frac{7}{3}$ makes it true; 0 makes it false.

Step 5 Exclude any values that make the denominator 0.

The solution set is $\left[-\frac{8}{3}, -2\right)$, since -2 makes the denominator 0.

CHAPTER 8 REVIEW EXERCISES

[8.1] *Solve each equation by either the square root property or the method of completing the square.*

1. $t^2 = 121$
2. $p^2 = 3$
3. $(2x + 5)^2 = 100$
*4. $(3k - 2)^2 = -25$
5. $x^2 + 4x = 15$
6. $2m^2 - 3m = -1$

[8.2] *Solve each equation by the quadratic formula.*

7. $2y^2 + y - 21 = 0$
8. $(t + 3)(t - 4) = -2$
9. $9p^2 = 42p - 49$
*10. $3p^2 = 2(2p - 1)$

11. A student wrote the following as the quadratic formula for solving $ax^2 + bx + c = 0$, $a \neq 0$:

$$x = -b \pm \frac{\sqrt{b^2 - 4ac}}{2a}.$$

Was this correct? If not, what is wrong with it?

Solve each problem. Use a calculator and round answers to the nearest tenth.

12. If a football is propelled upward on the moon at an angle of 60° with an initial velocity of 60 miles per hour, the equation $y = -.0013x^2 + 1.727x$ gives its height in feet at x seconds. At what time(s) will the football reach a height of 10 feet? (*Source:* Bernice Kastner, *Space Mathematics,* NASA.)

13. A paint-mixing machine has two inlet pipes. One takes 1 hour less than the other to fill the tank. Together they fill the tank in 3 hours. How long would it take each of them alone to fill the tank?

14. An old machine processes a batch of checks in one hour more time than a new one. How long would it take the old machine to process a batch of checks that the two machines together process in 2 hours?

Use the discriminant to predict whether the solutions to each equation are **(a)** *two distinct rational numbers;* **(b)** *exactly one rational number;* **(c)** *two distinct irrational numbers;* **(d)** *two distinct imaginary numbers.*

15. $a^2 + 5a + 2 = 0$
16. $4x^2 = 3 - 4x$
17. $4x^2 = 6x - 8$
18. $9z^2 + 30z + 25 = 0$

Use the discriminant to tell if each polynomial can be factored using integer coefficients. If a polynomial can be factored, factor it.

19. $24x^2 - 74x + 45$
20. $36x^2 + 69x - 34$

[8.3] *Solve each equation.*

21. $\frac{15}{x} = 2x - 1$
22. $\frac{1}{y} + \frac{2}{y + 1} = 2$
23. $8(3x + 5)^2 + 2(3x + 5) - 1 = 0$
24. $-2r = \sqrt{\frac{48 - 20r}{2}}$
25. $2x^{2/3} - x^{1/3} - 28 = 0$
*26. $5x^4 - 2x^2 - 3 = 0$

---

*Exercises identified with asterisks have imaginary number solutions.

Solve each problem.

27. Lisa Wunderle drove 8 miles to pick up her friend Laurie, and then drove 11 miles to a mall at a speed 15 miles per hour faster. If Lisa's total travel time was 24 minutes, what was her speed on the trip to pick up Laurie?

28. It takes Laketa De Mol 2 hours longer to complete a project for her boss than it takes Ed Moura. Working together, it would take them 3 hours. How long would it take each one to do the job alone? Round your answers to the nearest tenth of an hour.

29. Why can't the equation $x = \sqrt{2x + 4}$ have a negative solution?

[8.4] *Solve each formula for the indicated variable. (Give answers with ±.)*

30. $S = \dfrac{Id^2}{k}$; for d

31. $k = \dfrac{rF}{wv^2}$; for v

32. $2\pi R^2 + 2\pi RH - S = 0$; for R

Solve each problem.

33. The equation $s = 16t^2 + 15t$ gives the distance s in feet an object dropped off a building has fallen in t seconds. Find the time t when the object has fallen 25 feet. Round the answer to the nearest hundredth.

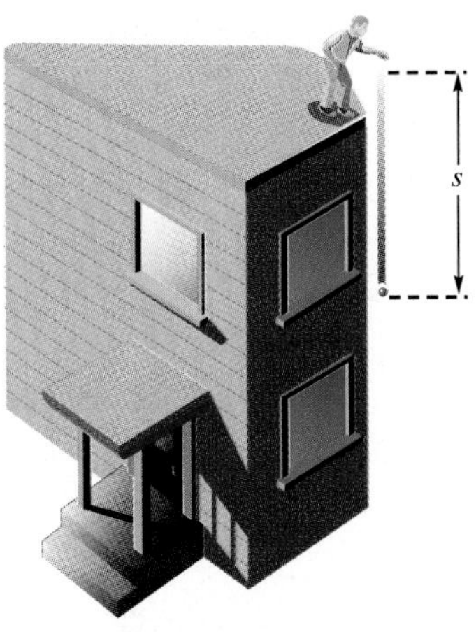

34. A large machine requires a part in the shape of a right triangle with a hypotenuse 9 feet less than twice the length of the longer leg. The shorter leg must be $\frac{3}{4}$ the length of the longer leg. Find the lengths of the three sides of the part.

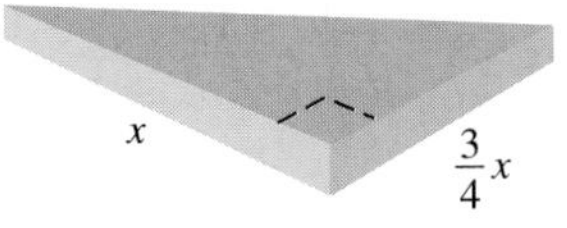

35. A rectangle has a length 2 meters more than its width. If one meter is cut from the length and one meter is added to the width, the resulting figure is a square with an area of 121 square meters. Find the dimensions of the original rectangle.

36. Nancy Mendoza wants to buy a mat for a photograph that measures 14 inches by 20 inches. She wants to have an even border around the picture when it is mounted on the mat. If the area of the mat she chooses is 352 square inches, how wide will the border be?

37. A lot is in the shape of a right triangle. The shortest side measures 50 meters. The longest side is 110 meters less than twice the middle side. How long is the middle side?

[8.5–8.6] *Identify the vertex of each parabola.*

38. $y = 3x^2 - 2$

39. $y = 6 - 2x^2$

40. $f(x) = -(x - 1)^2$

41. $f(x) = (x + 2)^2$

42. $y = (x - 3)^2 + 7$

43. $y = -3x^2 + 4x - 2$

Graph each parabola. Give the domain and range in Exercise 44.

44. $y = 4x^2 + 4x - 2$

45. $x = 2y^2 + 8y + 3$

46. If the discriminant of a quadratic function is negative, what do you know about the graph of the function?

47. Which one of the following would most closely resemble the graph of $f(x) = a(x - h)^2 + k$ if $a < 0$, $h > 0$, and $k < 0$?

(a)

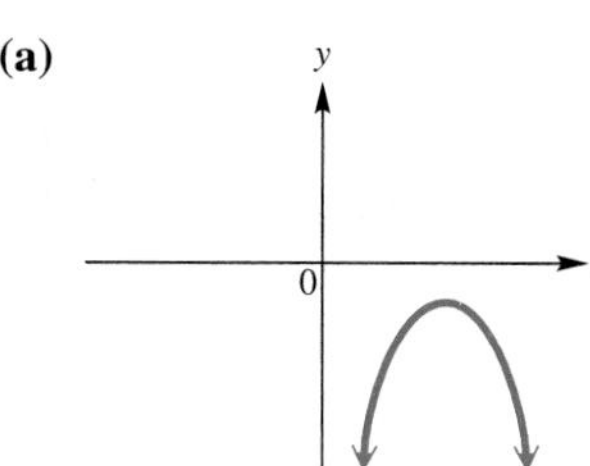

(b)

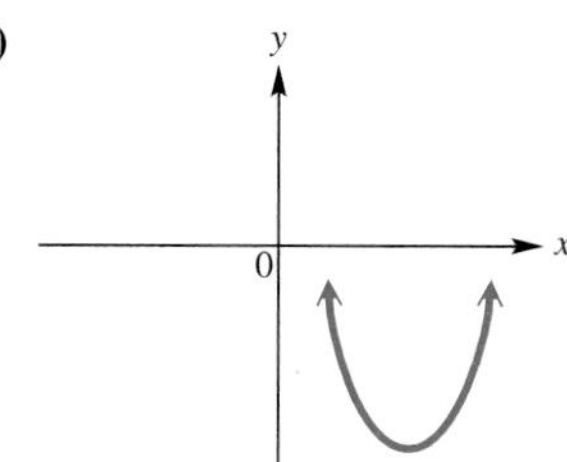

(c)

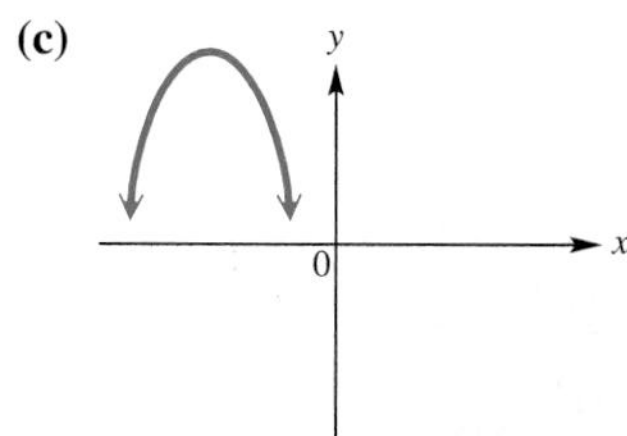

(d)

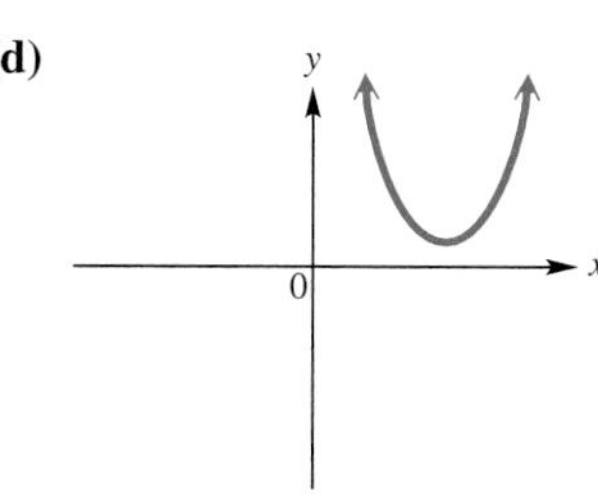

Work each problem.

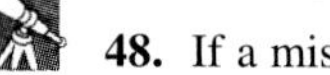

48. If a missile is propelled upward on Earth at an angle of 45° with an initial velocity of 30 miles per hour, the equation $f(x) = -.017x^2 + x$ gives the height of the missile in feet at x seconds. In how many seconds does the missile reach its maximum height? What is the maximum height?

49. Consumer spending for home video games in dollars per person per year is given in the table.

Consumer Spending for Home Video Games

Year	*Dollars*
1990	12.39
1992	13.08
1994	15.78
1996	19.43
1997	22.71

Source: Statistical Abstract of the United States, 1997.

Let $x = 0$ represent 1990, $x = 2$ represent 1992, and so on.

(a) Use the data for 1990, 1994, and 1997 in the quadratic form $ax^2 + bx + c = y$ to write a system of three equations.

(b) Solve the system from part (a) to get a quadratic function f that models these data.

50. Find the two numbers whose sum is 40 and whose product is a maximum.

[8.7]

51. The function with $f(x) = 2x^2 - 7x - 4$ was graphed in the standard viewing window of a graphing calculator and the two x-intercepts were located. See the figures.

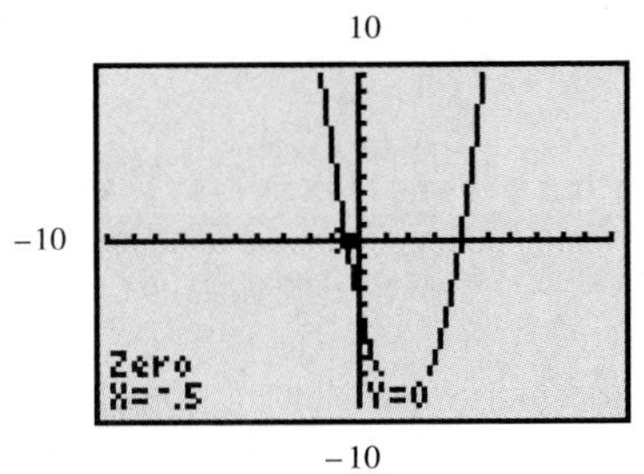

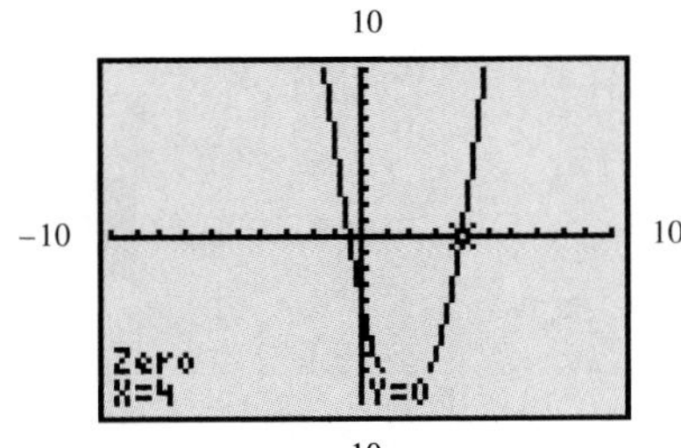

What is the solution set of

(a) $2x^2 - 7x - 4 = 0$? **(b)** $2x^2 - 7x - 4 > 0$? **(c)** $2x^2 - 7x - 4 \leq 0$?

Solve each inequality and graph the solution set.

52. $(x - 4)(2x + 3) > 0$

53. $x^2 + x \leq 12$

54. $2k^2 > 5k + 3$

55. $(4m + 3)^2 \leq -4$

56. $\dfrac{6}{2z - 1} < 2$

57. $\dfrac{3y + 4}{y - 2} \leq 1$

MIXED REVIEW EXERCISES

Solve.

58. $V = r^2 + R^2h$; for R

***59.** $3t^2 - 6t = -4$

***60.** $x^4 - 1 = 0$

61. $(b^2 - 2b)^2 = 11(b^2 - 2b) - 24$

62. $(r - 1)(2r + 3)(r + 6) < 0$

63. $\dfrac{2}{x - 4} + \dfrac{1}{x} = \dfrac{11}{5}$

64. $(3k + 11)^2 = 7$

65. $p = \sqrt{\dfrac{yz}{6}}$; for y

66. $(8k - 7)^2 \geq -1$

67. $-5x^2 = -8x + 3$

68. $6 + \dfrac{15}{s^2} = -\dfrac{19}{s}$

69. $\dfrac{-2}{x + 5} \leq -5$

Find the vertex and sketch the graph.

70. $y = \dfrac{4}{3}(x - 2)^2 + 1$

71. $x = 2(y + 3)^2 - 4$

72. $f(x) = -2x^2 + 8x - 5$

73. $x = -\dfrac{1}{2}y^2 + 6y - 14$

74. A student gave the following "solution" to the equation $b^2 = 12$.

$$b^2 = 12$$
$$b = \sqrt{12}$$
$$b = 2\sqrt{3}$$

What is wrong with this solution?

75. Explain how you would go about writing a quadratic equation whose solutions are -5 and 6. What is the standard form of one such equation?

76. The height (in feet) of a projectile t seconds after being fired from Earth into the air is given by $s(t) = -16t^2 + 160t$.
(a) Find the number of seconds required for the projectile to reach its maximum height.
(b) What is the maximum height?

77. Find the length and the width of a rectangle having a perimeter of 600 meters if the area is to be a maximum.

78. If you were to use the quadratic formula to solve $x^4 - 5x^2 + 6 = 0$, with $a = 1$, $b = -5$, and $c = 6$, what would you have to remember after you applied the formula?

RELATING CONCEPTS (EXERCISES 79–86)

Work Exercises 79–86 in order, *to see the connections between equations and inequalities.*

79. Use the methods of Chapter 2 to solve the equation or inequality, and graph the solution set.
(a) $3x - (4x + 2) = 0$ (b) $3x - (4x + 2) > 0$ (c) $3x - (4x + 2) < 0$

80. Use the methods of this chapter to solve the equation or inequality, and graph the solution set.
(a) $x^2 - 6x + 5 = 0$ (b) $x^2 - 6x + 5 > 0$ (c) $x^2 - 6x + 5 < 0$

81. Use the methods of Section 6.6 and Section 8.7 to solve the equation or inequality, and graph the solution set.
(a) $\dfrac{-5x + 20}{x - 2} = 0$ (b) $\dfrac{-5x + 20}{x - 2} > 0$ (c) $\dfrac{-5x + 20}{x - 2} < 0$

Review the definition of the union of two sets in Section 2.6, and then answer the questions in Exercises 82–86.

82. Form the union of the solution sets in Exercise 79. What is their union?

83. Repeat Exercise 82 for the solution sets in Exercise 80.

84. For the equation and inequalities in Exercise 81, what value of x cannot possibly be part of any of the solution sets? What is the union of the solution sets of the equation and inequalities in Exercise 81?

85. Fill in the blanks in the following statement: If we solve a linear, quadratic, or rational equation and the two inequalities associated with it, the union of the three solution sets will be ______; the only exception will be in the case of the rational equation and inequalities, where the number or numbers that cause the ________ to be zero will be deleted.

86. Suppose that the solution set of a quadratic equation is $\{-5, 3\}$ and the solution set of one of the associated inequalities is $(-\infty, -5) \cup (3, \infty)$. What is the solution set of the other associated inequality?

Did you make the connection that solutions of an inequality depend on the solutions of the corresponding equation?

CHAPTER 8 TEST

*Items marked * require knowledge of imaginary numbers.*

1. Solve $(7x + 3)^2 = 25$ by the square root property.

2. Solve $2x^2 + 4x = 8$ by completing the square.

Solve by the quadratic formula.

3. $2x^2 - 3x - 1 = 0$

***4.** $3t^2 - 4t = -5$

5. $3x = \sqrt{\dfrac{9x + 2}{2}}$

6. In Section 8.1, Exercise 55, we gave the formula $v^2 = \dfrac{GM}{r}$ for the orbital velocity of a satellite around Earth. As before, $GM = 3.99 \times 10^{11}$ cubic meters per second squared, and r is the radius of the circular orbit, the distance between the centers of the two bodies. The radius of Earth averages 6.37×10^6 meters. Most manned spacecraft are placed at altitudes of about 1.60×10^5 meters. (*Source:* Bernice Kastner, *Space Mathematics:* NASA, p. 137.)

(a) Find the radius of the circular orbit.

(b) Find the velocity needed for a body to stay in orbit around Earth at this altitude.

***7.** If k is a negative number, then which one of the following equations will have two imaginary solutions?

(a) $x^2 = 4k$ **(b)** $x^2 = -4k$ **(c)** $(x + 2)^2 = -k$ **(d)** $x^2 + k = 0$

8. Use the discriminant to predict the number and type of solutions of $2x^2 - 8x - 3 = 0$. Do not solve.

Solve by any method.

9. $3 - \dfrac{16}{x} - \dfrac{12}{x^2} = 0$

10. $9x^4 + 4 = 37x^2$

11. $12 = (2d + 1)^2 + (2d + 1)$

12. Solve for r: $S = 4\pi r^2$. (Leave $\pm$ in your answer.)

Solve each problem by writing a quadratic equation. Use any method to solve the equation.

13. Adam Bryer has a pool 24 feet long and 10 feet wide. He wants to construct a concrete walk around the pool. If he plans for the walk to be of uniform width and cover 152 square feet, what will be the width of the walk?

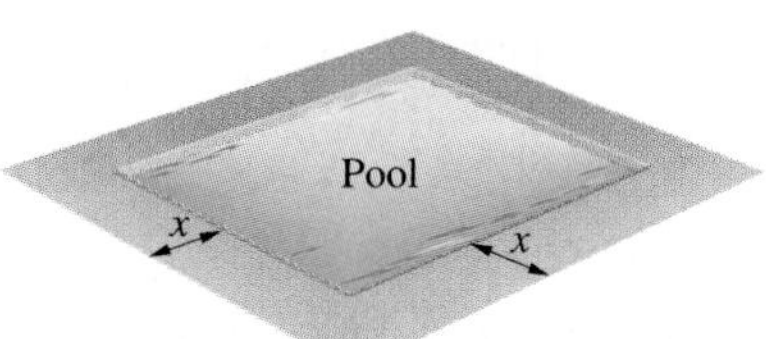

14. At a point 30 meters from the base of a tower, the distance to the top of the tower is 2 meters more than twice the height of the tower. Find the height of the tower.

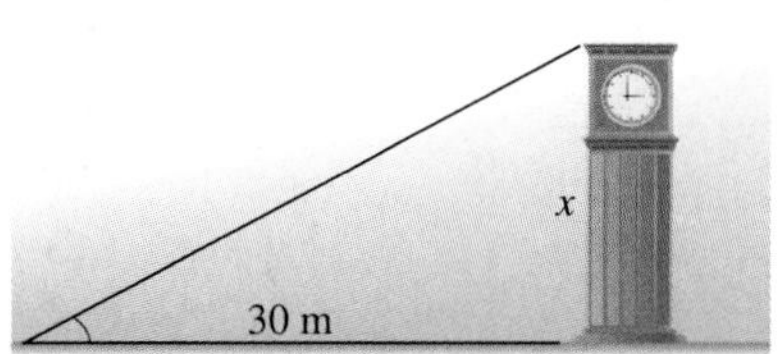

Graph each parabola. Identify the vertex. Identify the domain and range in Exercise 15.

15. $f(x) = \dfrac{1}{2}x^2 - 2$

16. $f(x) = -x^2 + 4x - 1$

17. $x = -(y - 2)^2 + 2$

18. The value (in millions) of the U.S. domestic salmon catch in the years 1990–1995 is closely approximated by the quadratic function with

$$f(x) = 22.56x^2 - 129.8x + 611.8.$$

Here, $x = 0$ represents 1990, $x = 1$ represents 1991, and so on. (*Source:* National Marine Fisheries Service.)

(a) Based on this model, how many salmon were caught in 1992?

(b) In what year did the catch reach a minimum? To the nearest million, from the model, how many salmon were caught that year? (Round your answer.)

Solve and graph each solution set.

19. $2x^2 + 7x > 15$

20. $\dfrac{5}{t - 4} \leq 1$

CUMULATIVE REVIEW EXERCISES CHAPTERS 1–8

1. Let $S = \left\{-\dfrac{7}{3}, -2, -\sqrt{3}, 0, .7, \sqrt{12}, \sqrt{-8}, 7, \dfrac{32}{3}\right\}$. List the elements of S that are elements of each set.

(a) integers **(b)** rational numbers

2. Simplify $2(-3)^2 + (-8)(-5) + (-17)$.

Solve each equation.

3. $7 - (4 + 3t) + 2t = -6(t - 2) - 5$

4. $|6x - 9| = |-4x + 2|$

5. $2x = \sqrt{\dfrac{5x + 2}{3}}$

6. $\dfrac{3}{x - 3} - \dfrac{2}{x - 2} = \dfrac{3}{x^2 - 5x + 6}$

7. $(r - 5)(2r + 3) = 1$

8. $b^4 - 5b^2 + 4 = 0$

Solve each inequality.

9. $-2x + 4 \leq -x + 3$

10. $|3y - 7| \leq 1$

11. $x^2 - 4x + 3 < 0$

12. $\dfrac{3}{y + 2} > 1$

Graph each relation. Tell whether or not each is a function, and if it is, give its domain and range.

13. $4x - 5y = 15$

14. $4x - 5y < 15$

15. $f(x) = -2(x - 1)^2 + 3$

16. Find the slope and intercepts of the line with equation $-2x + 7y = 16$.

Write an equation in slope-intercept form for the specified line in Exercises 17 and 18.

17. Through $(2, -3)$ and parallel to the line with equation $5x + 2y = 6$

18. Through $(-4, 1)$ and perpendicular to the line with equation $5x + 2y = 6$

19. The record track-qualifying speeds at North Carolina Motor Speedway since Richard Petty captured the first pole in 1965 are given in the table.

Qualifying Records

Year	*Speed (in mph)*
1965	116.26
1975	132.02
1985	141.85
1995	155.38
1998	156.36

Source: NASCAR.

Let $x = 0$ represent 1965, $x = 10$ represent 1975, and so on.

(a) Plot the ordered pairs (Year, Speed).

(b) Is this set of ordered pairs a function?

(c) Use the ordered pairs (0, 116.26) and (20, 141.85) to write a linear equation that models these data.

(d) Use your model to approximate the record speed for 1998 to the nearest hundredth. How does it compare to the actual value from the table?

20. Does the relation $x = 5$ define a function? Explain why or why not.

21. For the function with $f(x) = 2(x - 1)^2 - 5$, find
(a) $f(-2)$; **(b)** the domain and the range.

Solve each of the following.

22. $2x - 4y = 10$
$9x + 3y = 3$

23. $x + y + 2z = 3$
$-x + y + z = -5$
$2x + 3y - z = -8$

24. An excursion boat traveled 20 miles upriver (against the current) in 1 hour. The return trip downriver took .5 hour. Solve a system of equations to find the speed of the current and the speed of the boat in still water.

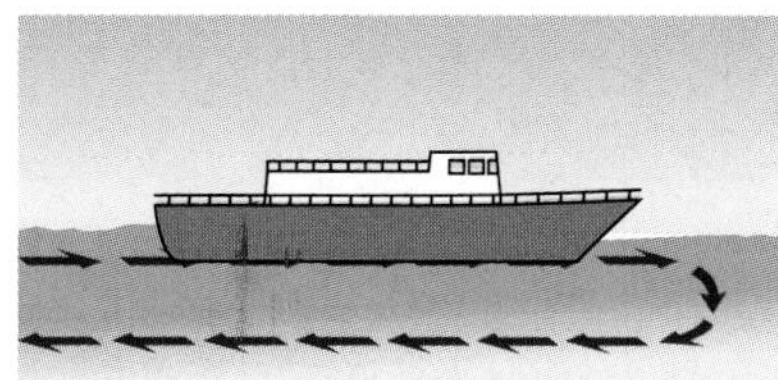

Perform the indicated operations.

25. $(7x + 4)(2x - 3)$

26. $\left(\frac{2}{3}t + 9\right)^2$

27. $(3t^3 + 5t^2 - 8t + 7) - (6t^3 + 4t - 8)$

28. Divide $4x^3 + 2x^2 - x + 26$ by $x + 2$.

Factor completely.

29. $16x - x^3$

30. $(3x + 2)^2 - 4(3x + 2) - 5$

31. $8x^3 + 27y^3$

32. $9x^2 - 30xy + 25y^2$

Perform the operations and express each answer in lowest terms. Assume denominators are nonzero.

33. $\dfrac{x^2 - 3x - 10}{x^2 + 3x + 2} \cdot \dfrac{x^2 - 2x - 3}{x^2 + 2x - 15}$

34. $\dfrac{3}{2 - k} - \dfrac{5}{k} + \dfrac{6}{k^2 - 2k}$

35. $\dfrac{\dfrac{r}{s} - \dfrac{s}{r}}{\dfrac{r}{s} + 1}$

Simplify each radical expression.

36. $\sqrt[3]{\dfrac{27}{16}}$

37. $\dfrac{2}{\sqrt{7} - \sqrt{5}}$

Solve each problem.

38. The perimeter of a rectangle is 20 inches and its area is 21 square inches. What are the dimensions of the rectangle?

39. Tri rode his bicycle for 12 miles and then walked an additional 8 miles. The total time for the trip was 5 hours. If his rate while walking was 10 miles per hour less than his rate while riding, what was each rate?

40. Two cars left an intersection at the same time, one heading due south and the other due east. Later they were exactly 95 miles apart. The car heading east had gone 38 miles less than twice as far as the car heading south. How far had each car traveled?

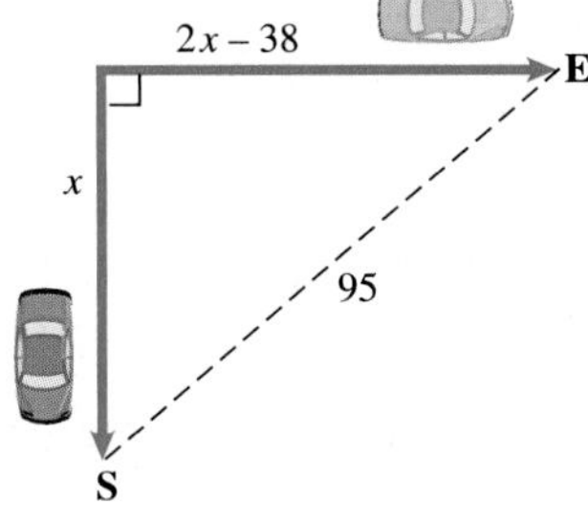

Inverse, Exponential, and Logarithmic Functions

9

Environment

One of the most important issues of our time is that of saving the environment. Industrial pollution of our air and water supplies, emission of "greenhouse gases," and climate changes as a result of global warming are concerns that must be faced now to save our planet for future generations.

Recycling is a way that we as individuals can help to protect the environment. One of the most common recyclables is plastic. The accompanying bar graph shows the increase in gross waste generated by plastics from 1960–1990. (The good news, however, is that the percentage of recovery also is increasing.) In which two 5-year periods did the waste more than double?

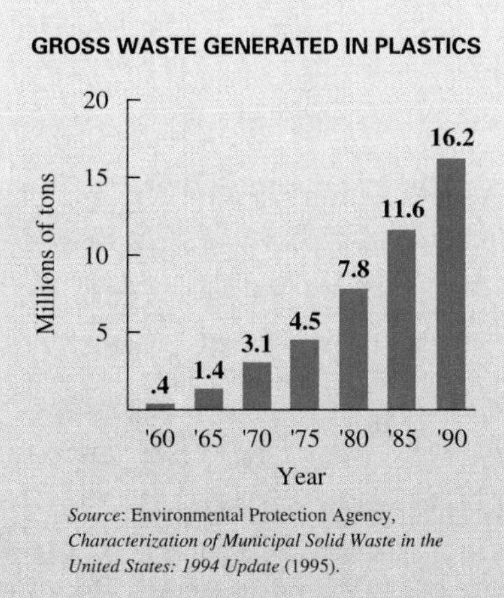

Source: Environmental Protection Agency, *Characterization of Municipal Solid Waste in the United States: 1994 Update* (1995).

A new type of function introduced in this chapter, the *exponential function,* can be used to model data like that shown in the graph. (See Exercise 15 in the Chapter 9 Review Exercises for an exponential function related to this graph.) Exponential functions are used throughout this chapter to analyze data involving climate and the environment, the theme of this chapter.

Visit our Web site at www.LialAlgebra.com

9.1 Inverse Functions

OBJECTIVES

1. Decide whether a function is one-to-one and, if it is, find its inverse.
2. Use the horizontal line test to determine whether a function is one-to-one.
3. Find the equation of the inverse of a function.
4. Graph the inverse f^{-1} from the graph of f.
5. Use a graphing calculator to graph inverse functions.

FOR EXTRA HELP

SSG Sec. 9.1
SSM Sec. 9.1

Pass the Test Software

InterAct Math Tutorial Software

Video 17

In this chapter we study two important types of functions, *exponential* and *logarithmic* functions. These functions are related in a special way. They are *inverses* of one another.

OBJECTIVE 1 Decide whether a function is one-to-one and, if it is, find its inverse. Suppose that G is the function $\{(-2, 2), (-1, 1), (0, 0), (1, 3), (2, 5)\}$. Another set of ordered pairs can be formed from G by interchanging the x- and y-values of each pair in G. Call this set F, with

$$F = \{(2, -2), (1, -1), (0, 0), (3, 1), (5, 2)\}.$$

To show that these two sets are related, F is called the *inverse* of G. For a function f to have an inverse, f must be *one-to-one*. In a **one-to-one function** each x-value corresponds to only one y-value and each y-value corresponds to just one x-value.

The function shown in Figure 1(a) is not one-to-one because the y-value 7 corresponds to *two* x-values, 2 and 3. That is, the ordered pairs (2, 7) and (3, 7) both appear in the function. The function in Figure 1(b) is one-to-one.

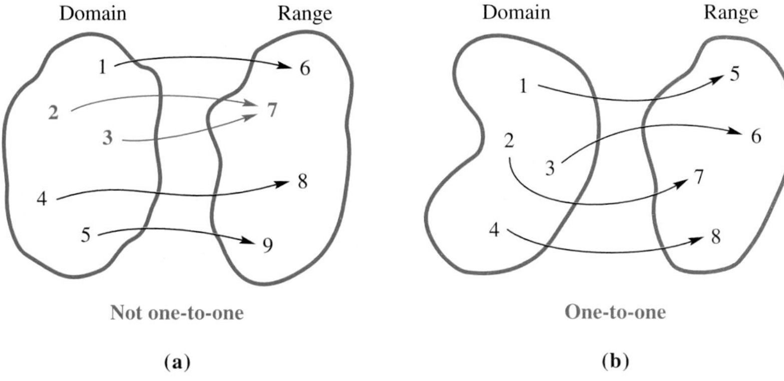

Figure 1

The *inverse* of any one-to-one function f is found by interchanging the components of the ordered pairs of f. The inverse of f is written f^{-1}. Read f^{-1} as "the inverse of f" or "f-inverse."

The symbol $f^{-1}(x)$ does not represent $\dfrac{1}{f(x)}$.

The definition of the inverse of a function follows.

Inverse of a Function

The **inverse** of a one-to-one function, f, written f^{-1}, is the set of all ordered pairs of the form (y, x) where (x, y) belongs to f. Since the inverse is formed by interchanging x and y, the domain of f becomes the range of f^{-1} and the range of f becomes the domain of f^{-1}.

For inverses f and f^{-1}, it follows that $f(f^{-1}(x)) = x$ and $f^{-1}(f(x)) = x$.

EXAMPLE 1 Deciding Whether a Function Is One-to-One

Decide whether each function is one-to-one. If it is, find the inverse function.

(a) $F = \{(-2, 1), (-1, 0), (0, 1), (1, 2), (2, 2)\}$

Each x-value in F corresponds to just one y-value. However, the y-value 2 corresponds to two x-values, 1 and 2. Also, the y-value 1 corresponds to both -2 and 0. Because some y-values correspond to more than one x-value, F is not one-to-one.

(b) $G = \{(3, 1), (0, 2), (2, 3), (4, 0)\}$

Every x-value in G corresponds to only one y-value, and every y-value corresponds to only one x-value, so G is a one-to-one function. The inverse function is found by interchanging the numbers in each ordered pair.

$$G^{-1} = \{(1, 3), (2, 0), (3, 2), (0, 4)\}$$

Notice how the domain and range of G become the range and domain, respectively, of G^{-1}.

(c) The Environmental Protection Agency has developed an indicator of air quality called the Pollutant Standard Index (PSI). If the PSI exceeds 100 on a particular day, that day is classified as unhealthy. The chart shows the number of unhealthy days in Chicago for the years 1988–1993.

Year	*Number of Unhealthy Days*
1988	21
1989	3
1990	3
1991	8
1992	6
1993	1

Source: Environmental Protection Agency.

Let f be the function defined in the table, with the years forming the domain and the numbers of unhealthy days forming the range. Then f is not one-to-one, because in two different years (1989 and 1990), the number of unhealthy days was the same, 3. ■

OBJECTIVE 2 **Use the horizontal line test to determine whether a function is one-to-one.** It may be difficult to decide whether a function is one-to-one just by looking at the equation that defines the function. However, by graphing the function and observing the graph, we can use the following *horizontal line test* to tell whether it is one-to-one.

Horizontal Line Test

A function is one-to-one if every horizontal line intersects the graph of the function at most once.

The horizontal line test follows from the definition of a one-to-one function. Any two points that lie on the same horizontal line have the same y-coordinate. No two ordered pairs that belong to a one-to-one function may have the same y-coordinate, and therefore no horizontal line will intersect the graph of a one-to-one function more than once.

EXAMPLE 2 Using the Horizontal Line Test

Use the horizontal line test to determine whether the graphs in Figures 2 and 3 are graphs of one-to-one functions.

(a)

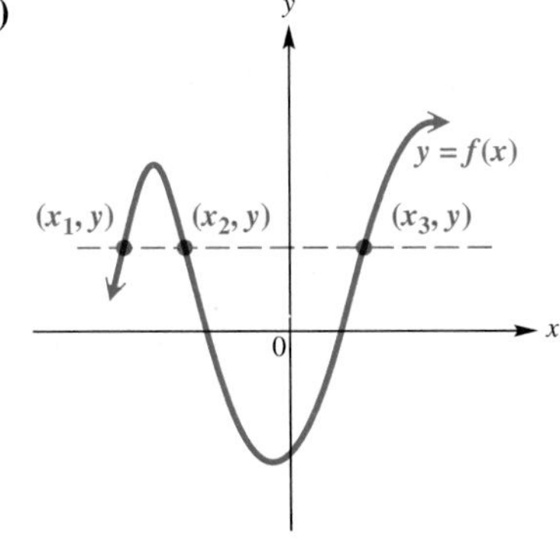

Figure 2

Because the horizontal line shown in Figure 2 intersects the graph in more than one point (actually three points in this case), the function is not one-to-one.

(b)

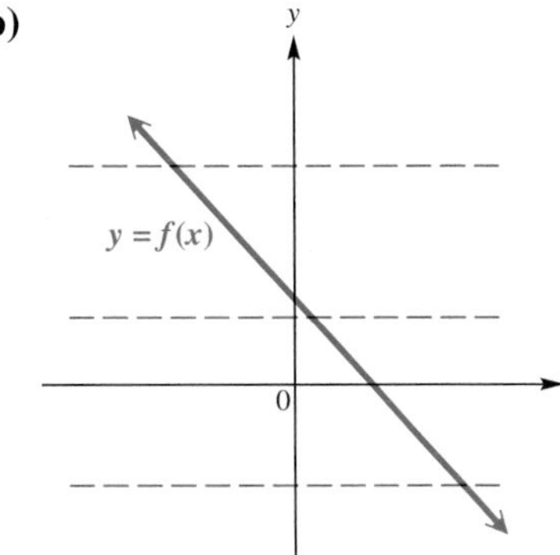

Figure 3

Every horizontal line will intersect the graph in Figure 3 in exactly one point. This function is one-to-one.

OBJECTIVE 3 Find the equation of the inverse of a function. By definition, the inverse of a function is found by interchanging the x- and y-values of each of its ordered pairs. The equation of the inverse of a function defined by $y = f(x)$ is found in the same way.

Finding the Equation of the Inverse of $y = f(x)$

For a one-to-one function f defined by an equation $y = f(x)$, find the defining equation of the inverse as follows.

Step 1 Interchange x and y.

Step 2 Solve for y.

Step 3 Replace y with $f^{-1}(x)$.

EXAMPLE 3 **Finding the Equation of the Inverse**

Decide whether each of the following defines a one-to-one function. If so, find the equation of the inverse.

(a) $f(x) = 2x + 5$

By definition, this is a one-to-one function. To find the inverse, let $y = f(x)$ so that

$$\begin{aligned} y &= 2x + 5 \\ x &= 2y + 5 && \text{Interchange } x \text{ and } y. \text{ (Step 1)} \\ 2y &= x - 5 && \text{Solve for } y. \text{ (Step 2)} \\ y &= \frac{x-5}{2}. \end{aligned}$$

From the last equation,

$$f^{-1}(x) = \frac{x-5}{2}, \qquad \text{(Step 3)}$$

which is a linear function. In the function $y = 2x + 5$, the value of y is found by starting with a value of x, multiplying by 2, and adding 5. The equation for the inverse has us *subtract* 5, and then *divide* by 2. This shows how an inverse is used to "undo" what a function does to the variable x.

(b) $y = x^2 + 2$

Both $x = 3$ and $x = -3$ correspond to $y = 11$. Because of the x^2-term, there are many pairs of x-values that each correspond to the same y-value. This means that the function defined by $y = x^2 + 2$ is not one-to-one and does not have an inverse.

If this were not noticed, following the steps given above for finding the equation of an inverse leads to

$$\begin{aligned} y &= x^2 + 2 \\ x &= y^2 + 2 && \text{Interchange } x \text{ and } y. \\ x - 2 &= y^2 \\ \pm\sqrt{x-2} &= y. && \text{Square root property} \end{aligned}$$

The last step shows that there are two y-values for each choice of $x > 2$, so the given function is not one-to-one and cannot have an inverse.

(c) $f(x) = (x - 2)^3$

Because of the cube, this is a one-to-one function. Find the inverse by replacing $f(x)$ with y and then interchanging x and y.

$$\begin{aligned} y &= (x-2)^3 \\ x &= (y-2)^3 \end{aligned}$$

Take the cube root on each side to solve for y.

$$\begin{aligned} \sqrt[3]{x} &= \sqrt[3]{(y-2)^3} \\ \sqrt[3]{x} &= y - 2 \\ \sqrt[3]{x} + 2 &= y \\ f^{-1}(x) &= \sqrt[3]{x} + 2 && \text{Replace } y \text{ with } f^{-1}(x). \end{aligned}$$

OBJECTIVE 4 Graph the inverse f^{-1} from the graph of f. Suppose the point (a, b) shown in Figure 4 belongs to a one-to-one function f. Then the point (b, a) would belong to f^{-1}. The line segment connecting (a, b) and (b, a) is perpendicular to and cut in half by the line $y = x$. The points (a, b) and (b, a) are "mirror images" of each other with respect to $y = x$. For this reason the graph of f^{-1} can be found from the graph of f by locating the mirror image of each point of f with respect to $y = x$.

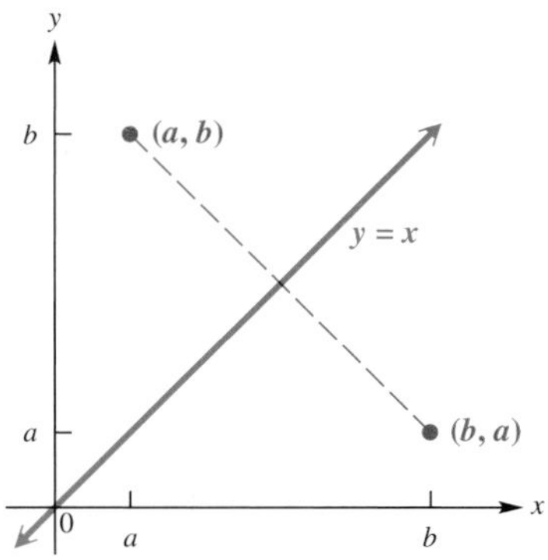

Figure 4

EXAMPLE 4 Graphing Inverses

Graph the inverses of the functions shown in Figure 5.

In Figure 5 the graphs of two functions are shown in blue and their inverses are shown in red. In each case, the graph of f^{-1} is symmetric to the graph of f with respect to the line $y = x$.

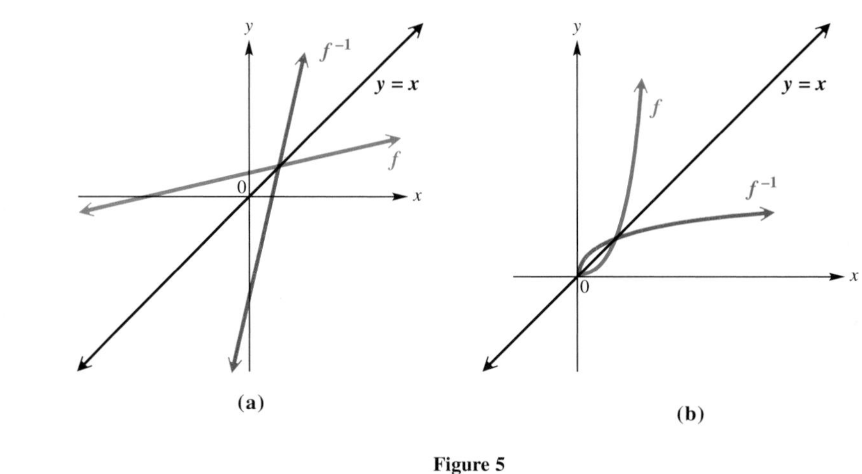

Figure 5

OBJECTIVE 5 Use a graphing calculator to graph inverse functions. We have described how inverses of one-to-one functions may be determined algebraically. We also explained how the graph of a one-to-one function f compares to the graph of its inverse f^{-1}: it is a reflection of the graph of f^{-1} across the line $y = x$. In Example 3 we showed that the inverse of the one-to-one function with $f(x) = 2x + 5$ is given by $f^{-1}(x) = \frac{x - 5}{2}$. If we use a square viewing window of a graphing calculator and graph $y_1 = f(x) = 2x + 5$,

$y_2 = f^{-1}(x) = \dfrac{x-5}{2}$, and $y_3 = x$, we can see how this reflection appears on the screen. See Figure 6.

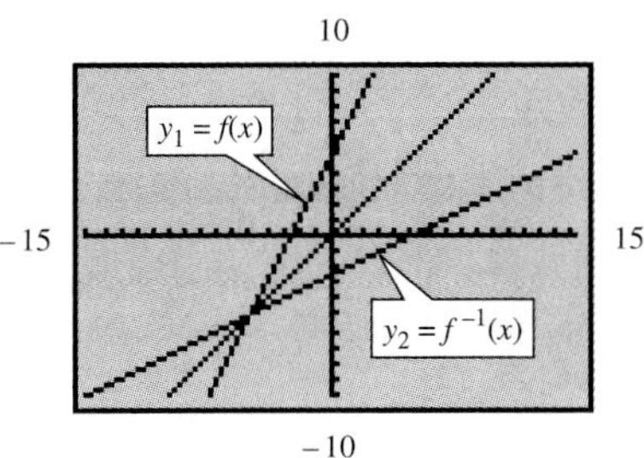

Figure 6

Some graphing calculators have the capability to "draw" the inverse of a function. Figure 7 shows the graphs of $f(x) = x^3 + 2$ and its inverse in a square viewing window.

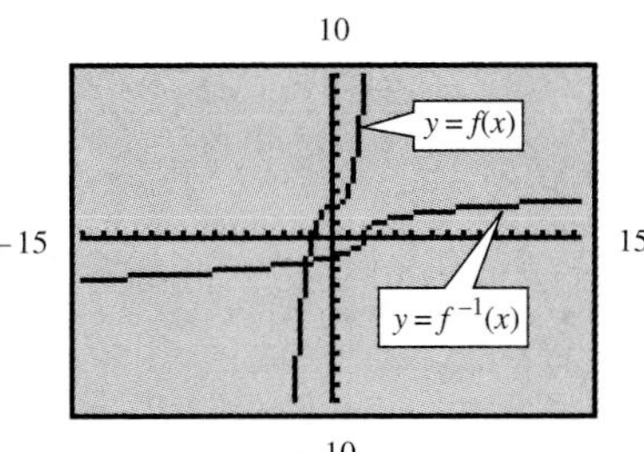

Figure 7

9.1 EXERCISES

1. The chart shows the number of uncontrolled hazardous waste sites that require further investigation to determine whether remedies are needed under the Superfund program. The ten states listed are the highest ranked in the United States.

   If this correspondence is considered to be a function that pairs each state with its number of uncontrolled waste sites, is it one-to-one? If not, explain why.

State	*Number*
New Jersey	108
Pennsylvania	101
California	94
New York	79
Michigan	75
Florida	53
Washington	50
Illinois	40
Wisconsin	40
Ohio	38

Source: U.S. Environmental Protection Agency.

2. The chart shows emissions of a major air pollutant, carbon monoxide, in the United States for the years 1990–1995.

Year	*Amount of Emissions in Thousand Short Tons*
1990	100,650
1991	97,376
1992	94,043
1993	94,133
1994	98,017
1995	92,099

Source: U.S. Environmental Protection Agency.

If this correspondence is considered to be a function that pairs each year with its emissions amount, is it one-to-one? If not, explain why.

3. The road mileage between Denver, Colorado, and several selected U.S. cities is shown in the table below.

City	*Distance to Denver in Miles*
Atlanta	1398
Dallas	781
Indianapolis	1058
Kansas City, MO	600
Los Angeles	1059
San Francisco	1235

If we consider this as a function that pairs each city with a distance, is it a one-to-one function? Why or why not? How could we change the answer to this question by adding 1 mile to one of the distances shown?

4. Suppose that you consider the set of ordered pairs (x, y) such that x represents a person in your mathematics class and y represents that person's mother. Explain how this function might not be a one-to-one function.

Choose the correct response from the given list.

5. If a function is made up of ordered pairs in such a way that the same y-value appears in a correspondence with two different x-values, then

(a) the function is one-to-one.
(b) the function is not one-to-one.
(c) its graph does not pass the vertical line test.
(d) it has an inverse function associated with it.

6. Which one of the following is a one-to-one function? Explain why the others are not, using specific examples.

(a) $f(x) = x$
(b) $f(x) = x^2$
(c) $f(x) = |x|$
(d) $f(x) = -x^2 + 2x - 1$

7. Only one of the graphs illustrates a one-to-one function. Which one is it?

(a)

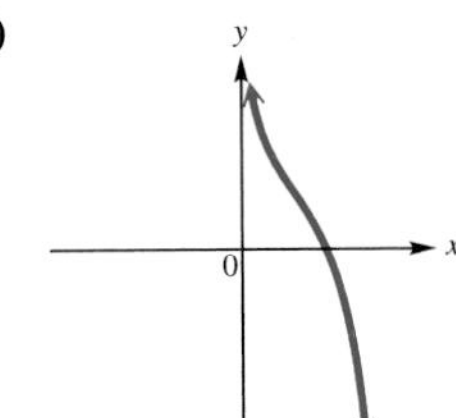

(b)

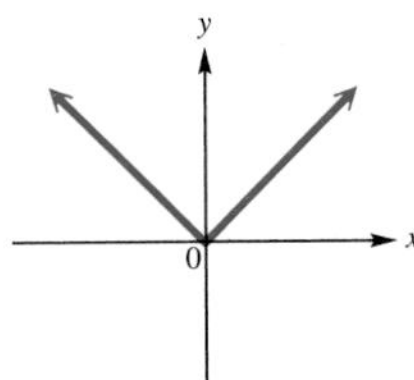

(c)

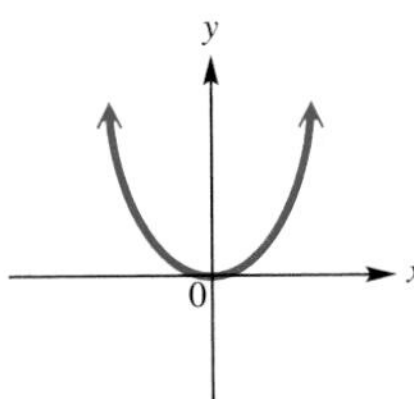

(d)

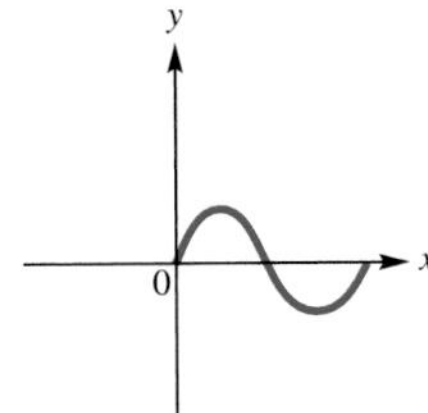

8. If a function f is one-to-one and the point (p, q) lies on the graph of f, then which one of the following *must* lie on the graph of f^{-1}?

(a) $(-p, q)$ **(b)** $(-q, -p)$ **(c)** $(p, -q)$ **(d)** (q, p)

If the function is one-to-one, find its inverse. See Examples 1–3.

9. $\{(3, 6), (2, 10), (5, 12)\}$

10. $\left\{(-1, 3), (0, 5), (5, 0), \left(7, -\frac{1}{2}\right)\right\}$

11. $\{(-1, 3), (2, 7), (4, 3), (5, 8)\}$

12. $\{(-8, 6), (-4, 3), (0, 6), (5, 10)\}$

13. $f(x) = 2x + 4$

14. $f(x) = 3x + 1$

15. $g(x) = \sqrt{x - 3}, \quad x \geq 3$

16. $g(x) = \sqrt{x + 2}, \quad x \geq -2$

17. $f(x) = 3x^2 + 2$

18. $f(x) = -4x^2 - 1$

19. $f(x) = x^3 - 4$

20. $f(x) = x^3 - 3$

Let $f(x) = 2^x$. We will see in the next section that this function is one-to-one. Find each of the following, always working part (a) before part (b).

21. **(a)** $f(3)$ **(b)** $f^{-1}(8)$

22. **(a)** $f(4)$ **(b)** $f^{-1}(16)$

23. **(a)** $f(0)$ **(b)** $f^{-1}(1)$

24. **(a)** $f(-2)$ **(b)** $f^{-1}\left(\frac{1}{4}\right)$

The graphs of some functions are given in Exercises 25–30. ***(a)*** *Use the horizontal line test to determine whether the function is one-to-one.* ***(b)*** *If the function is one-to-one, graph the inverse of the function. (Remember that if f is one-to-one and (a, b) is on the graph of f, then (b, a) is on the graph of f^{-1}.) See Example 4.*

25.

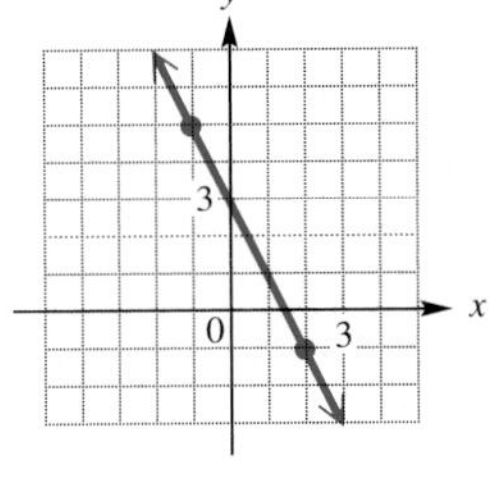

26.

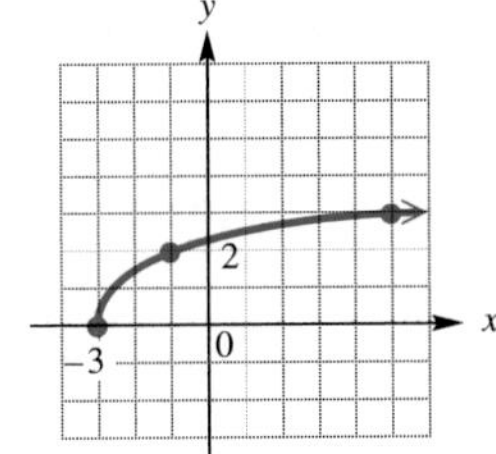

27.

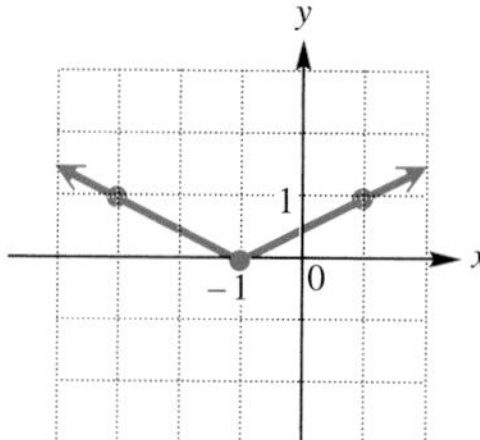

28.

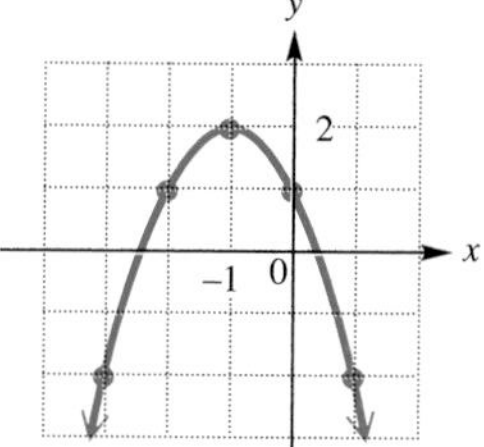

29.

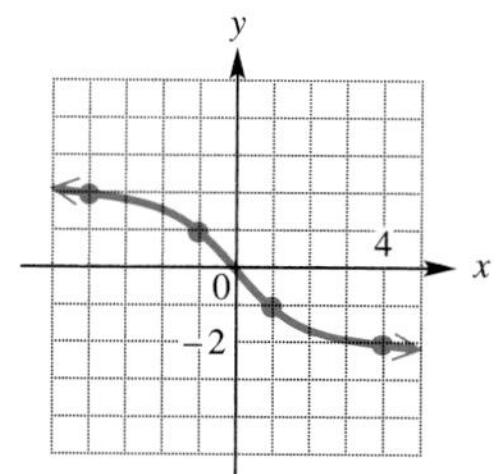

30.

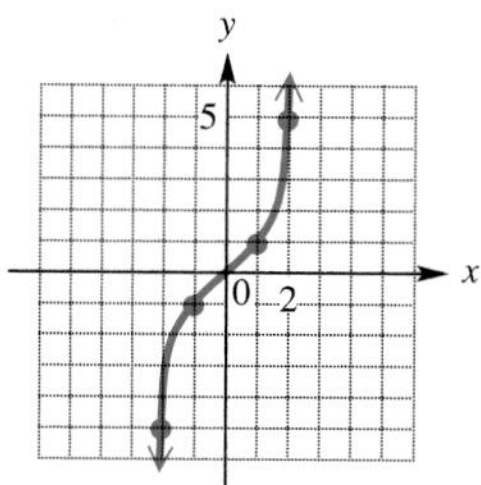

Each function defined in Exercises 31–38 is a one-to-one function. Graph the function as a solid line (or curve) and then graph its inverse on the same set of axes as a dashed line (or curve). In Exercises 35–38 you are given a table to complete so that graphing the function will be a bit easier. See Example 4.

31. $f(x) = 2x - 1$

32. $f(x) = 2x + 3$

33. $g(x) = -4x$

34. $g(x) = -2x$

35. $f(x) = \sqrt{x}, \quad x \geq 0$

x	$f(x)$
0	
1	
4	

36. $f(x) = -\sqrt{x}, \quad x \geq 0$

x	$f(x)$
0	
1	
4	

37. $f(x) = x^3 - 2$

x	$f(x)$
−1	
0	
1	
2	

38. $f(x) = x^3 + 3$

x	$f(x)$
−2	
−1	
0	
1	

RELATING CONCEPTS (EXERCISES 39–42)

Inverse functions are used by government agencies and other businesses to send and receive coded information. The functions they use are usually very complicated. A simple example might use the function defined by $f(x) = 2x + 5$. (Note that it is one-to-one.) Suppose that each letter of the alphabet is assigned a numerical value according to its position, as follows:

A	1	G	7	L	12	Q	17	V	22
B	2	H	8	M	13	R	18	W	23
C	3	I	9	N	14	S	19	X	24
D	4	J	10	O	15	T	20	Y	25
E	5	K	11	P	16	U	21	Z	26
F	6								

RELATING CONCEPTS (EXERCISES 39–42) (CONTINUED)

Using the function, the word ALGEBRA would be encoded as

7 29 19 15 9 41 7,

because $f(\text{A}) = f(1) = 2(1) + 5 = 7, f(\text{L}) = f(12) = 2(12) + 5 = 29$, and so on. The message would then be decoded by using the inverse of f, defined by $f^{-1}(x) = \frac{x - 5}{2}$.

For example, $f^{-1}(7) = \frac{7 - 5}{2} = 1 = \text{A}$, $f^{-1}(29) = \frac{29 - 5}{2} = 12 = \text{L}$, and so on.

Work Exercises 39–42 in order.

39. Suppose that you are an agent for a detective agency and you know that today's function for your code is defined by $f(x) = 4x - 5$. Find the rule for f^{-1} algebraically.

40. You receive the following coded message today.

47 95 23 67 −1 59 27 31 51 23 7 −1 43 7 79 43 −1 75 55 67
31 71 75 27 15 23 67 15 −1 75 15 71 75 75 27 31 51
23 71 31 51 7 15 71 43 31 7 15 11 3 67 15 −1 11

Use the letter/number assignment described earlier to decode the message.

41. Why is a one-to-one function essential in this encoding/decoding process?

42. Use $f(x) = x^3 + 4$ to encode your name, using the letter/number assignment described earlier.

Did you make the connection that inverse functions can be applied to the encoding/decoding process?

Each function defined below is one-to-one. Find the inverse algebraically, and then graph both the function and its inverse on the same square graphing calculator screen. See Objective 5.

43. $f(x) = 2x - 7$

44. $f(x) = -3x + 2$

45. $f(x) = x^3 + 5$

46. $f(x) = \sqrt[3]{x + 2}$

Some graphing calculators have the capability to draw the "inverse" of a function even if the function is not one-to-one; therefore, the inverse is not technically a function, but is a relation. For example, the graphs of $y = x^2$ and $x = y^2$ are shown in the accompanying square window.

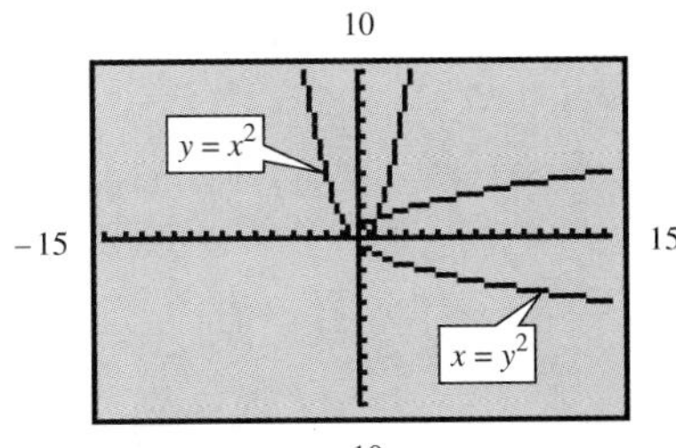

Read your instruction manual to see if your model has this capability. Draw both y_1 and its inverse in the same square window.

47. $y_1 = x^2 + 3x + 4$

48. $y_1 = x^3 - 9x$

49. Explain why the "inverse" of the function in Exercise 47 does not actually satisfy the definition of inverse as given in this section.

50. At what points do the graphs of $y = x^2$ and $x = y^2$ intersect? (See the graph above.) Verify this using algebraic methods.

9.2 Exponential Functions

OBJECTIVES

1. Identify exponential functions.
2. Graph exponential functions.
3. Solve exponential equations of the form $a^x = a^k$ for x.
4. Use exponential functions in applications.

FOR EXTRA HELP

SSG Sec. 9.2
SSM Sec. 9.2

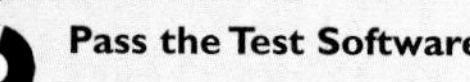
Pass the Test Software

InterAct Math Tutorial Software

Video 17

OBJECTIVE 1 Identify exponential functions. In Section 7.1, we showed how to evaluate 2^x for rational values of x. For example,

$$2^3 = 8, \qquad 2^{-1} = \frac{1}{2}, \qquad 2^{1/2} = \sqrt{2}, \qquad 2^{3/4} = \sqrt[4]{2^3} = \sqrt[4]{8}.$$

In more advanced courses it is shown that 2^x exists for all real number values of x, both rational and irrational. (Later in the chapter, methods are given for approximating the value of 2^x for irrational x.) The following definition of an exponential function assumes that a^x exists for all real numbers x.

Exponential Function

For $a > 0$ and $a \neq 1$, and all real numbers x,

$$F(x) = a^x$$

defines an **exponential function.**

NOTE The two restrictions on a in the definition of exponential function are important. The restriction that a must be positive is necessary so that the function can be defined for all real numbers x. For example, letting a be negative ($a = -2$, for instance) and letting $x = \frac{1}{2}$ would give the expression $(-2)^{1/2}$, which is not real. The other restriction, $a \neq 1$, is necessary because 1 raised to any power is equal to 1, and the function would then be the linear function with $F(x) = 1$.

OBJECTIVE 2 Graph exponential functions. We can graph exponential functions by finding several ordered pairs that belong to the function. Plotting these points and connecting them with a smooth curve gives the graph.

CAUTION Be sure to plot enough points to see how rapidly the graph rises.

EXAMPLE 1 Graphing an Exponential Function with $a > 1$

Graph the exponential function with $f(x) = 2^x$.

Choose some values of x and find the corresponding values of $f(x)$.

x	-3	-2	-1	0	1	2	3	4
$f(x) = 2^x$	$\frac{1}{8}$	$\frac{1}{4}$	$\frac{1}{2}$	1	2	4	8	16

Plotting these points and drawing a smooth curve through them gives the graph shown in Figure 8.

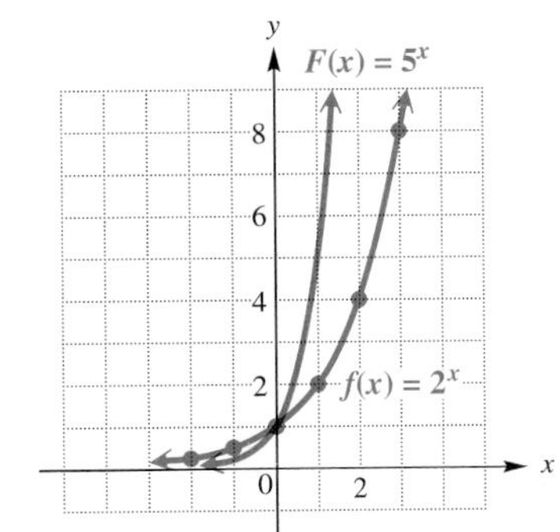

Figure 8

The graph in Figure 8 is typical of the graphs of exponential functions of the form $F(x) = a^x$, where $a > 1$. The larger the value of a, the faster the graph rises. To see this, compare the graph of $F(x) = 5^x$ with the graph of $f(x) = 2^x$ in Figure 8.

By the vertical line test, the graphs in Figure 8 represent functions. As these graphs suggest, the domain of an exponential function includes all real numbers. Since y is always positive, the range is $(0, \infty)$.

EXAMPLE 2 Graphing an Exponential Function with $a < 1$

Graph $g(x) = \left(\frac{1}{2}\right)^x$.

Again, find some points on the graph.

x	-3	-2	-1	0	1	2	3	4
$g(x) = \left(\frac{1}{2}\right)^x$	8	4	2	1	$\frac{1}{2}$	$\frac{1}{4}$	$\frac{1}{8}$	$\frac{1}{16}$

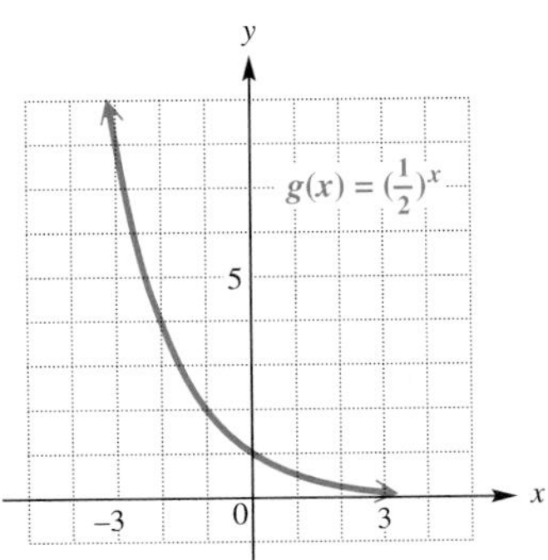

Figure 9

The graph, shown in Figure 9, is very similar to that of $f(x) = 2^x$, shown in Figure 8, except that here as x gets larger, *y decreases.* This graph is typical of the graph of a function of the form $F(x) = a^x$, where $0 < a < 1$.

The graph of $f(x) = 2^x$ and an accompanying table are shown in the graphing calculator screen in Figure 10(a). Compare this to the results of Example 1. Similarly, the graph of $g(x) = \left(\frac{1}{2}\right)^x$ and a table are shown in Figure 10(b); compare to Example 2.

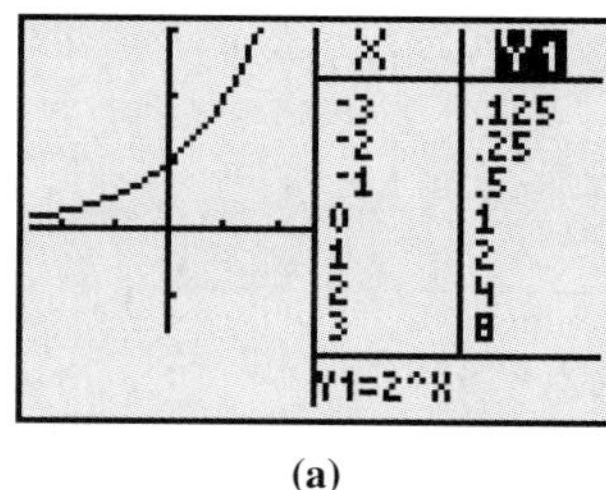

(a)

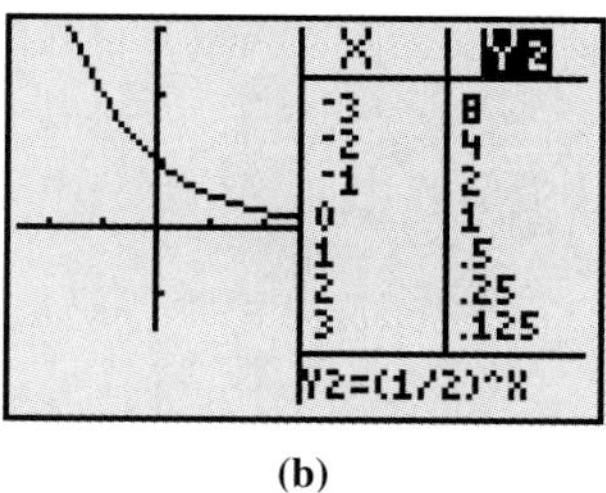

(b)

Figure 10

Based on Examples 1 and 2, we make the following generalizations about the graphs of exponential functions of the form $F(x) = a^x$.

Graph of $F(x) = a^x$

1. The graph will always contain the point (0, 1).
2. When $a > 1$, the graph will *rise* from left to right. When $0 < a < 1$, the graph will *fall* from left to right. In both cases, the graph goes from the second quadrant to the first.
3. The graph will approach the x-axis, but never touch it. (Recall from Chapter 6 that such a line is called an *asymptote.*)
4. The domain is $(-\infty, \infty)$ and the range is $(0, \infty)$.

EXAMPLE 3 **Graphing a More Complicated Exponential Function**

Graph $y = 3^{2x-4}$.

Find some ordered pairs. For example, if $x = 0$,

$$y = 3^{2(0)-4} = 3^{-4} = \frac{1}{81}.$$

Also, for $x = 2$,

$$y = 3^{2(2)-4} = 3^0 = 1.$$

These ordered pairs, $\left(0, \frac{1}{81}\right)$ and (2, 1), along with the ordered pairs $\left(1, \frac{1}{9}\right)$ and (3, 9), lead to the graph shown in Figure 11. The graph is similar to the graph of $f(x) = 2^x$ except that it is shifted to the right and rises more rapidly.

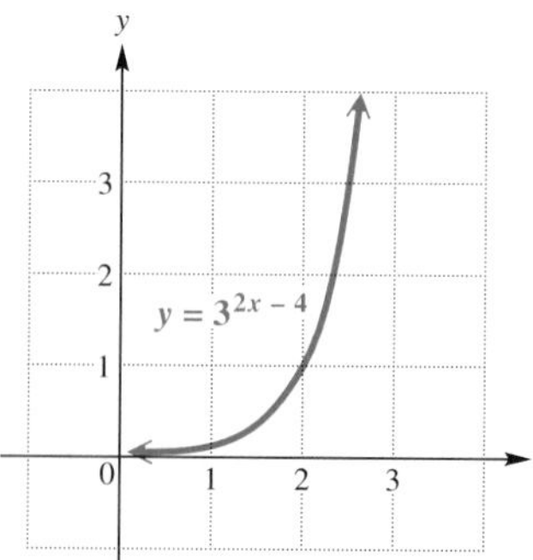

Figure 11

OBJECTIVE 3 Solve exponential equations of the form $a^x = a^k$ for x. Until now in this book, all equations that we have solved have had the variable as a base; all exponents have been constants. An **exponential equation** is an equation that has a variable in an exponent, such as

$$9^x = 27.$$

Because the exponential function defined by $F(x) = a^x$ is a one-to-one function, the following property can be used to solve many exponential equations.

Property for Solving Exponential Equations

For $a > 0$ and $a \neq 1$, if $a^x = a^y$ then $x = y$.

This property would not necessarily be true if $a = 1$.

To solve an exponential equation using this property, follow these steps.

Solving Exponential Equations

Step 1 **Each side must have the same base.** If the two sides of the equation do not have the same base, express each as a power of the same base.

Step 2 **Simplify exponents.** If necessary, use the rules of exponents to simplify the exponents.

Step 3 **Set exponents equal.** Use the above property to set the exponents equal to each other.

Step 4 **Solve.** Solve the equation obtained in Step 3.

The steps above cannot be applied to an exponential equation like

$$3^x = 12,$$

since Step 1 cannot easily be done. A method for solving such equations is given in Section 9.6.

EXAMPLE 4 Solving an Exponential Equation

Solve the equation $9^x = 27$.

We can use the property given above if both sides are changed to the same base. Since $9 = 3^2$ and $27 = 3^3$, the equation $9^x = 27$ is solved as follows.

$$(3^2)^x = 3^3 \quad \text{Substitute. (Step 1)}$$

$$3^{2x} = 3^3 \quad \text{Power rule for exponents (Step 2)}$$

$$2x = 3 \quad \text{If } a^x = a^y, \text{ then } x = y. \text{ (Step 3)}$$

$$x = \frac{3}{2} \quad \text{(Step 4)}$$

Check that the solution set is $\{\frac{3}{2}\}$ by substituting $\frac{3}{2}$ for x in the given equation.

OBJECTIVE 4 Use exponential functions in applications. Exponential functions frequently occur in applications describing growth or decay of some quantity.

EXAMPLE 5 Solving an Application of an Exponential Function

One result of the rapidly increasing world population is an increase of carbon dioxide in the air, which scientists believe may be contributing to global warming. Both population and carbon dioxide in the air are increasing exponentially. This means that the growth rate is continually increasing. The graph in Figure 12 shows the concentration of carbon dioxide (in parts per million) in the air.

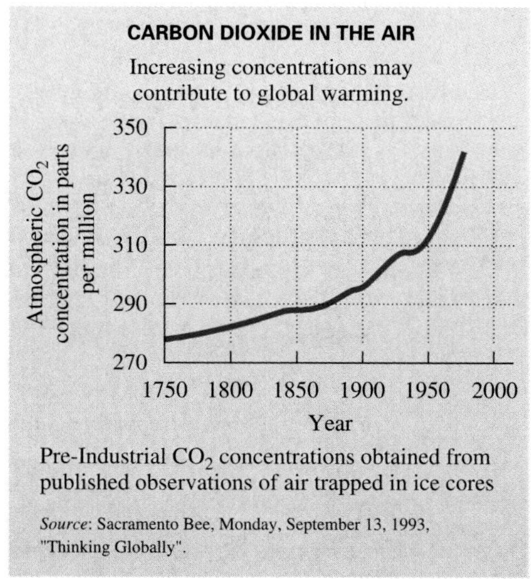

CARBON DIOXIDE IN THE AIR
Increasing concentrations may contribute to global warming.
Atmospheric CO_2 concentration in parts per million
350
330
310
290
270
1750 1800 1850 1900 1950 2000
Year
Pre-Industrial CO_2 concentrations obtained from published observations of air trapped in ice cores
Source: Sacramento Bee, Monday, September 13, 1993, "Thinking Globally".

Figure 12

The data are approximated by the function with

$$f(x) = 278(1.00084)^x,$$

where x is the number of years since 1750. Use this function and a calculator to approximate the concentration of carbon dioxide in parts per million for each year.

(a) 1900

Since x represents the number of years since 1750, in this case we have $x = 1900 - 1750 = 150$. Thus, evaluate $f(150)$.

$$\begin{aligned} f(150) &= 278(1.00084)^{150} && \text{Let } x = 150. \\ &\approx 315 \text{ parts per million} && \text{Use a calculator.} \end{aligned}$$

(b) 1950

Use $x = 1950 - 1750 = 200$: $f(200) \approx 329$ parts per million.

EXAMPLE 6 Applying an Exponential Function

The atmospheric pressure (in millibars) at a given altitude x, in meters, can be approximated by the function defined by

$$f(x) = 1038(1.000134)^{-x},$$

for values of x between 0 and 10,000. Because the base is greater than 1 and the coefficient of x in the exponent is negative, the function values decrease as x increases. This means that as the altitude increases, the atmospheric pressure decreases. (*Source:* A. Miller and J. Thompson, *Elements of Meteorology,* Charles E. Merrill Publishing Company, 1975.)

(a) According to this function, what is the pressure at ground level?
At ground level, $x = 0$, so

$$f(0) = 1038(1.000134)^{-0} = 1038(1) = 1038.$$

The pressure is 1038 millibars.

(b) What is the pressure at 5000 meters?
Use a calculator to find $f(5000)$.

$$f(5000) = 1038(1.000134)^{-5000} \approx 531$$

The pressure is approximately 531 millibars.

9.2 EXERCISES

Choose the correct response in Exercises 1–4.

1. Which one of the following points lies on the graph of $f(x) = 2^x$?

(a) $(1, 0)$ **(b)** $(2, 1)$ **(c)** $(0, 1)$ **(d)** $\left(\sqrt{2}, \frac{1}{2}\right)$

2. Which one of the following statements is true?
(a) The y-intercept of the graph of $f(x) = 10^x$ is $(0, 10)$.
(b) For any $a > 1$, the graph of $f(x) = a^x$ falls from left to right.
(c) The point $\left(\frac{1}{2}, \sqrt{5}\right)$ lies on the graph of $f(x) = 5^x$.
(d) The graph of $y = 4^x$ rises at a faster rate than the graph of $y = 10^x$.

3. The asymptote of the graph of $f(x) = a^x$
(a) is the x-axis. **(b)** is the y-axis.
(c) has equation $x = 1$. **(d)** has equation $y = 1$.

4. Which one of the following equations is that of the graph shown here?
(a) $y = 1000\left(\frac{1}{2}\right)^{.3x}$
(b) $y = 1000\left(\frac{1}{2}\right)^{x}$
(c) $y = 1000(2)^{.3x}$
(d) $y = 1000^x$

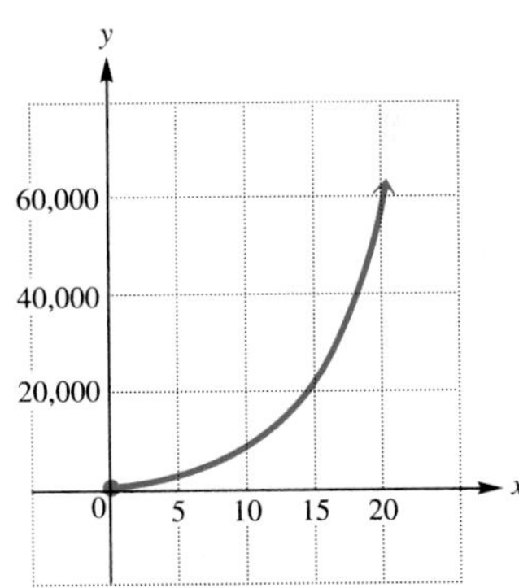

Graph each exponential function. See Examples 1–3.

5. $f(x) = 3^x$ **6.** $f(x) = 5^x$ **7.** $g(x) = \left(\frac{1}{3}\right)^x$ **8.** $g(x) = \left(\frac{1}{5}\right)^x$

9. $y = 4^{-x}$ **10.** $y = 6^{-x}$ **11.** $y = 2^{2x-2}$ **12.** $y = 2^{2x+1}$

13. **(a)** For an exponential function defined by $f(x) = a^x$, if $a > 1$, the graph ______ (rises/falls) from left to right. If $0 < a < 1$, the graph ______ (rises/falls) from left to right.

(b) Based on your answers in part (a), make a conjecture (an educated guess) concerning whether an exponential function defined by $f(x) = a^x$ is one-to-one. Then decide whether it has an inverse based on the concepts of Section 9.1.

14. In your own words, describe the characteristics of the graph of an exponential function. Use the exponential function defined by $f(x) = 3^x$ (Exercise 5) and the words asymptote, domain, and range in your explanation.

Solve each equation. See Example 4.

15. $100^x = 1000$ **16.** $8^x = 4$ **17.** $16^{2x+1} = 64^{x+3}$ **18.** $9^{2x-8} = 27^{x-4}$

19. $5^x = \frac{1}{125}$ **20.** $3^x = \frac{1}{81}$ **21.** $5^x = .2$ **22.** $10^x = .1$

23. $\left(\frac{3}{2}\right)^x = \frac{8}{27}$ **24.** $\left(\frac{4}{3}\right)^x = \frac{27}{64}$

Use the exponential key of a calculator to find an approximation to the nearest thousandth.

25. $12^{2.6}$ **26.** $13^{1.8}$ **27.** $.5^{3.921}$ **28.** $.6^{4.917}$ **29.** $2.718^{2.5}$ **30.** $2.718^{-3.1}$

31. Try to evaluate $(-2)^4$ on a scientific calculator. You may get an error message, since the exponential function key on many calculators does not allow negative bases. Discuss the concept introduced in this section that is closely related to this "peculiarity" of many scientific calculators.

32. Explain why the exponential equation $4^x = 6$ cannot be solved using the method explained in this section. Change 6 to another number that *will* allow the method of this section to be used, and then solve the equation.

The figure shown here accompanied the article "Is Our World Warming?" which appeared in the October 1990 issue of National Geographic. *It shows projected temperature increases using two graphs: one an exponential-type curve, and the other linear. From the figure, approximate the increase **(a)** for the exponential curve, and **(b)** for the linear graph for each of the following years.*

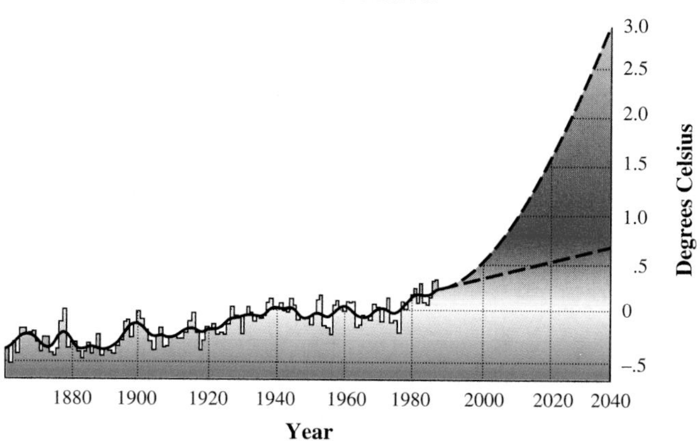

Graph, "Zero Equals Average Global Temperature for the Period 1950–1979." Dale D. Glasgow, © National Geographic Society. Reprinted by permission.

33. 2000 **34.** 2010 **35.** 2020 **36.** 2040

Solve each problem. See Examples 5 and 6.

37. Based on figures from 1950–1985, the number of worldwide carbon dioxide emissions in millions of short tons is approximated by the exponential function with

$$f(x) = 7147(1.0366)^x,$$

where $x = 0$ corresponds to 1950, $x = 5$ corresponds to 1955, and so on. (*Source: Carbon Dioxide Information Analysis Center,* 1994.)

(a) Use this model to approximate the number of emissions in 1950.
(b) Use this model to approximate the number of emissions in 1985.
(c) In 1990, the actual number of emissions was 25,010 million short tons. How does this compare to the number that the model provides?

38. Based on figures from 1960–1990, the gross waste generated by paper and paperboard products in millions of tons can be approximated by the exponential function with

$$f(x) = 31.28(1.028)^x,$$

where $x = 0$ corresponds to 1960, $x = 5$ corresponds to 1965, and so on. (*Source:* Environmental Protection Agency, *Characterization of Municipal Solid Waste in the United States: 1994 Update,* 1995.)

(a) Use this model to approximate the number of tons of this waste in 1960.
(b) Use this model to approximate the number of tons of this waste in 1985.
(c) In 1993, the actual number of millions of tons of this waste was 77.8. How does this compare to the number that the model provides?

39. The amount of radioactive material in an ore sample is given by the function defined by

$$A(t) = 100(3.2)^{-.5t},$$

where $A(t)$ is the amount present, in grams, of the sample t months after the initial measurement.
(a) How much was present at the initial measurement? (*Hint:* $t = 0$.)
(b) How much was present 2 months later?
(c) How much was present 10 months later?
(d) Graph the function.

40. A small business estimates that the value $V(t)$ of a copy machine is decreasing according to the function defined by

$$V(t) = 5000(2)^{-.15t},$$

where t is the number of years that have elapsed since the machine was purchased, and $V(t)$ is in dollars.
(a) What was the original value of the machine?
(b) What is the value of the machine 5 years after purchase? Give your answer to the nearest dollar.
(c) What is the value of the machine 10 years after purchase? Give your answer to the nearest dollar.
(d) Graph the function.

41. Refer to the function in Exercise 40. When will the value of the machine be \$2500? (*Hint:* Let $V(t) = 2500$, divide both sides by 5000, and use the method of Example 4.)

42. Refer to the function in Exercise 40. When will the value of the machine be \$1250?

The bar graph shows the average annual major league baseball player's salary for each year since free agency began. Using a technique from statistics, it was determined that the function with

$$S(x) = 74{,}741(1.17)^x$$

approximates the salary, where $x = 0$ *corresponds to 1976, and so on, up to* $x = 18$ *representing 1994. (Salary is in dollars.)*

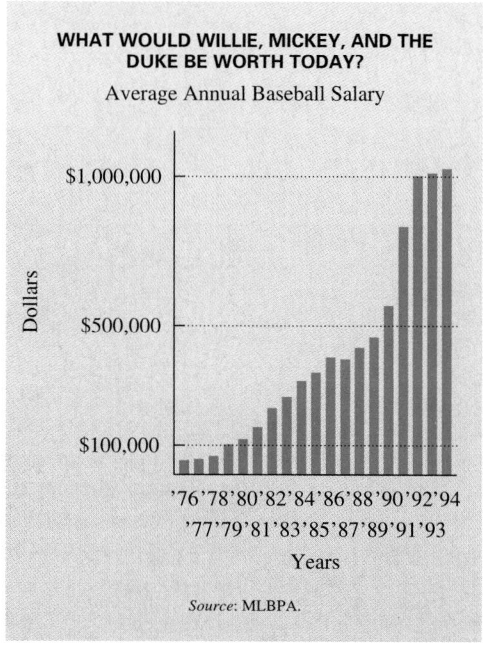

43. Based on this model, what was the average salary in 1986?

44. Based on the graph, in what year did the average salary first exceed $1,000,000?

RELATING CONCEPTS (EXERCISES 45–50)

In these exercises we examine several methods of simplifying the expression $16^{3/4}$.

Work Exercises 45–50 in order.

45. Write $16^{3/4}$ as a radical expression with the exponent outside the radical. Then simplify the expression.

46. Write $16^{3/4}$ as a radical expression with the exponent under the radical. Then simplify the expression.

47. Use a calculator to find the square root of 16^3. Now find the square root of that result.

48. Explain why the result in Exercise 47 is equal to $16^{3/4}$.

49. Predict the result a calculator will give when 16 is raised to the .75 power. Then check your answer by actually performing the operation on your calculator.

50. Write $\sqrt[100]{16^{75}}$ as an exponential expression. Then write the exponent in lowest terms and evaluate this radical expression.

Did you make the connection that the expression $16^{3/4}$ can be evaluated in several different ways?

TECHNOLOGY INSIGHTS (EXERCISES 51–52)

The number of tons, in millions, of solid waste generated in the United States during the period 1960–1990 is summarized in the following table.

Year	*Millions of Tons*
1960	87.8
1965	103.4
1970	121.9
1975	128.1
1980	151.5
1985	164.4
1990	195.7

Source: Environmental Protection Agency.

If $x = 0$ represents 1960, $x = 5$ represents 1965, and so on, the data can be plotted as points (0, 87.8), (5, 103.4), . . . , (30, 195.7) using a graphing calculator with statistical capability. See the graph on the left. Then, using techniques from statistics, an exponential curve can be determined to "fit" the points, as shown in the graph on the right.

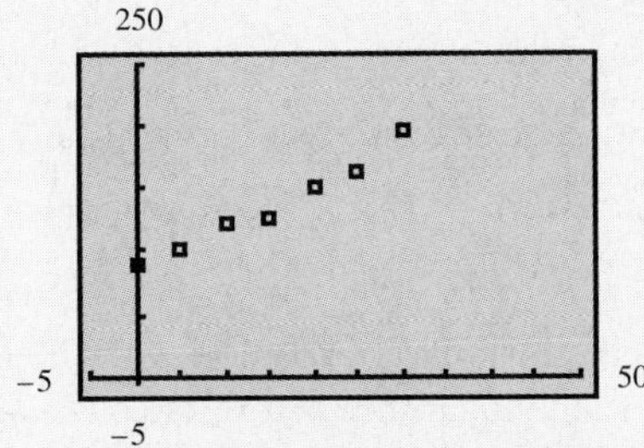

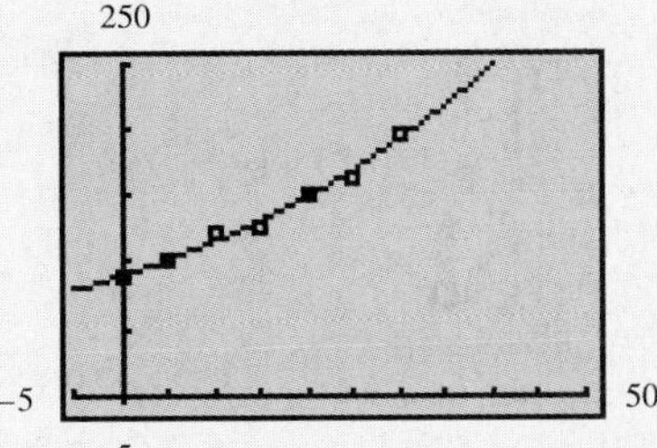

This curve is modeled by the equation

$$Y_1 = 90.11(1.0257)^x.$$

Refer to the screens in Exercises 51 and 52 and discuss the displays at the bottoms. How does the model differ from the actual figure in the table above?

51.

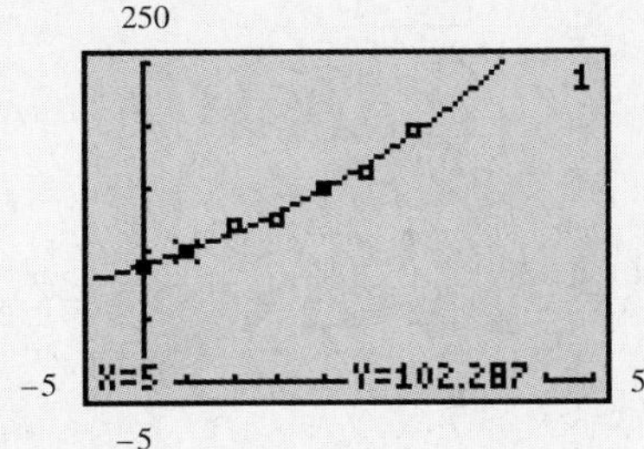

52.

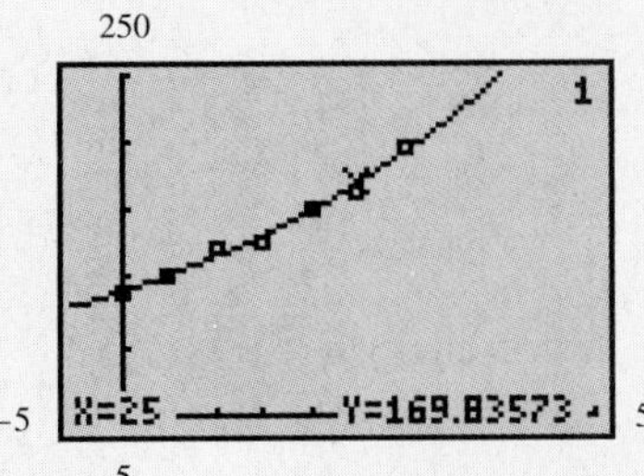

9.3 Logarithmic Functions

OBJECTIVES

1. Define a logarithm.
2. Convert between exponential and logarithmic forms.
3. Solve logarithmic equations of the form $\log_a b = k$ for a, b, or k.
4. Define and graph logarithmic functions.
5. Use logarithmic functions in applications.

FOR EXTRA HELP

SSG Sec. 9.3
SSM Sec. 9.3
Pass the Test Software
InterAct Math Tutorial Software
Video 17

The graph of $y = 2^x$ is the curve shown in **blue** in Figure 13. Since $y = 2^x$ is a one-to-one function, it has an inverse. Interchanging x and y gives $x = 2^y$, the inverse of $y = 2^x$. As we saw in Section 9.1, the graph of the inverse is found by reflecting the graph of $y = 2^x$ about the line $y = x$. The graph of $x = 2^y$ is shown as a **red** curve in Figure 13.

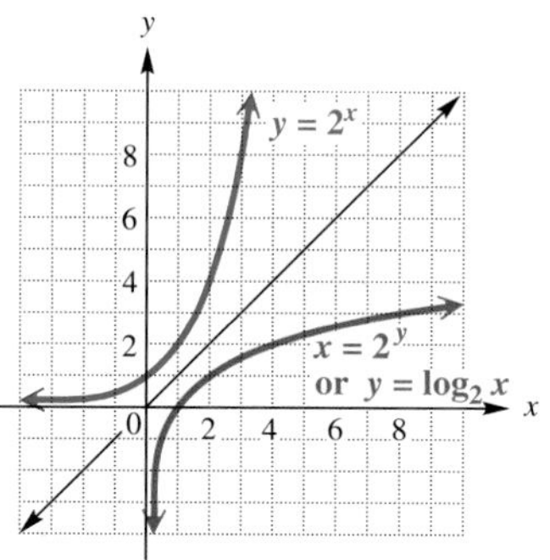

Figure 13

OBJECTIVE 1 **Define a logarithm.** We cannot solve the equation $x = 2^y$ for the dependent variable y with the methods presented up to now. The following definition is used to solve $x = 2^y$ for y.

Definition of Logarithm

For all positive numbers a, where $a \neq 1$, and all positive numbers x,

$$y = \log_a x \qquad \text{means the same as} \qquad x = a^y.$$

This key statement should be memorized. The abbreviation **log** is used for **logarithm.** Read $\log_a x$ as "the logarithm of x to the base a." To remember the location of the base and the exponent in each form, refer to the diagram that follows.

Exponent ↓
Logarithmic form: $y = \log_a x$
↑ Base

Exponent ↓
Exponential form: $x = a^y$
↑ Base

In working with logarithmic form and exponential form, remember the following.

Meaning of $\log_a x$

A **logarithm** is an exponent; $\log_a x$ is the exponent on the base a that yields the number x.

OBJECTIVE 2 **Convert between exponential and logarithmic forms.** We can use the definition of logarithm to write exponential statements in logarithmic form and logarithmic statements in exponential form. The chart below shows several pairs of equivalent statements.

Exponential Form	*Logarithmic Form*
$3^2 = 9$	$\log_3 9 = 2$
$\left(\frac{1}{5}\right)^{-2} = 25$	$\log_{1/5} 25 = -2$
$10^5 = 100{,}000$	$\log_{10} 100{,}000 = 5$
$4^{-3} = \frac{1}{64}$	$\log_4 \frac{1}{64} = -3$

OBJECTIVE 3 **Solve logarithmic equations of the form $\log_a b = k$ for a, b, or k.** A **logarithmic equation** is an equation with a logarithm in at least one term. We solve logarithmic equations of the form $\log_a b = k$ for any of the three variables by first writing the equation in exponential form.

EXAMPLE 1 **Solving Logarithmic Equations**

Solve the following equations.

(a) $\log_4 x = -2$

By the definition of logarithm, $\log_4 x = -2$ is equivalent to $4^{-2} = x$. Then

$$x = 4^{-2} = \frac{1}{4^2} = \frac{1}{16}.$$

The solution set is $\{\frac{1}{16}\}$.

(b) $\log_{1/2} 16 = y$

First write the statement in exponential form.

$$\log_{1/2} 16 = y$$

$$\left(\frac{1}{2}\right)^y = 16 \qquad \text{Convert to exponential form.}$$

$$(2^{-1})^y = 2^4 \qquad \text{Write with the same base.}$$

$$2^{-y} = 2^4 \qquad \text{Property of exponents}$$

$$-y = 4 \qquad \text{Set exponents equal.}$$

$$y = -4 \qquad \text{Multiply by } -1.$$

The solution set is $\{-4\}$. ■

For any positive real number b, we know that $b^1 = b$ and $b^0 = 1$. Writing these two statements in logarithmic form gives the following two properties of logarithms.

For any positive real number b, $b \neq 1$,

$$\log_b b = 1 \quad \text{and} \quad \log_b 1 = 0.$$

EXAMPLE 2 Using Properties of Logarithms

Use the two properties of logarithms above to simplify each logarithm.

(a) $\log_7 7 = 1$ **(b)** $\log_{\sqrt{2}} \sqrt{2} = 1$

(c) $\log_9 1 = 0$ **(d)** $\log_{.2} 1 = 0$

OBJECTIVE 4 Define and graph logarithmic functions. Now we define the logarithmic function with base a.

Logarithmic Function

If a and x are positive numbers, with $a \neq 1$, then

$$f(x) = \log_a x$$

defines the **logarithmic function with base *a*.**

To graph a logarithmic function, it is helpful to write it in exponential form first. Then plot selected ordered pairs to determine the graph.

EXAMPLE 3 Graphing a Logarithmic Function

Graph $y = \log_{1/2} x$.

By writing $y = \log_{1/2} x$ in its exponential form as $x = \left(\frac{1}{2}\right)^y$, we can identify ordered pairs that satisfy the equation. Here it is easier to choose values for y and find the corresponding values of x. See the table of ordered pairs.

x	y
$\frac{1}{4}$	2
$\frac{1}{2}$	1
1	0
2	−1
4	−2
8	−3

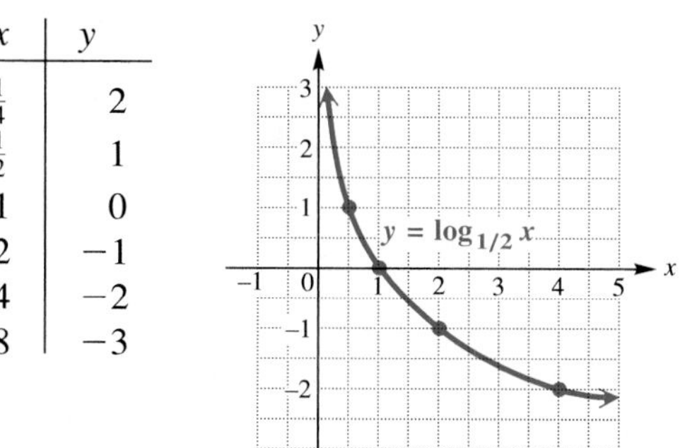

Figure 14

Plotting these points (be careful to get them in the right order) and connecting them with a smooth curve gives the graph in Figure 14. This graph is typical of logarithmic functions with $0 < a < 1$. The graph of $x = 2^y$ in Figure 13, which is equivalent to $y = \log_2 x$, is typical of graphs of logarithmic functions with base $a > 1$.

Based on the graphs of the functions $y = \log_2 x$ in Figure 13 and $y = \log_{1/2} x$ in Figure 14, we make the following generalizations about the graphs of logarithmic functions of the form $G(x) = \log_a x$.

Graph of $G(x) = \log_a x$

1. The graph will always contain the point (1, 0).
2. When $a > 1$, the graph will *rise* from left to right, from the fourth quadrant to the first. When $0 < a < 1$, the graph will *fall* from left to right, from the first quadrant to the fourth.
3. The graph will approach the y-axis, but never touch it. (It is an asymptote.)
4. The domain is $(0, \infty)$ and the range is $(-\infty, \infty)$.

Compare these generalizations to the similar ones for exponential functions in Section 9.2.

OBJECTIVE 5 **Use logarithmic functions in applications.** Logarithmic functions, like exponential functions, can be applied to real-world phenomena.

EXAMPLE 4 **Solving an Application of a Logarithmic Function**

The function defined by

$$f(x) = 27 + 1.105 \log_{10}(x + 1)$$

approximates the barometric pressure in inches of mercury at a distance of x miles from the eye of a typical hurricane. (*Source:* A. Miller and R. Anthes, *Meteorology,* 5th edition, Charles E. Merrill Publishing Company, 1985.)

(a) Approximate the pressure 9 miles from the eye of the hurricane.
We let $x = 9$, and find $f(9)$.

$$\begin{aligned} f(9) &= 27 + 1.105 \log_{10}(\mathbf{9} + 1) && \text{Let } x = 9. \\ &= 27 + 1.105 \log_{10} 10 && \text{Add in parentheses.} \\ &= 27 + 1.105(\mathbf{1}) && \log_{10} 10 = 1 \\ &= 28.105 && \text{Add.} \end{aligned}$$

The pressure 9 miles from the eye of the hurricane is 28.105 inches.

(b) Approximate the pressure 99 miles from the hurricane.

$$\begin{aligned} f(99) &= 27 + 1.105 \log_{10}(\mathbf{99} + 1) && \text{Let } x = 99. \\ &= 27 + 1.105 \log_{10} 100 && \text{Add in parentheses.} \\ &= 27 + 1.105(\mathbf{2}) && \log_{10} 100 = 2 \\ &= 29.21 \end{aligned}$$

The pressure 99 miles from the eye of the hurricane is 29.21 inches.

EXAMPLE 5 **Solving an Application of a Logarithmic Function**

Sales (in thousands of units) of a new product are approximated by the function with

$$S(t) = 100 + 30 \log_3(2t + 1),$$

where t is the number of years after the product is introduced.

(a) What were the sales after 1 year?
Find $S(1)$.

$$\begin{aligned} S(1) &= 100 + 30 \log_3(2 \cdot 1 + 1) && \text{Let } t = 1. \\ &= 100 + 30 \log_3 3 \\ &= 100 + 30(1) && \log_3 3 = 1 \\ &= 130 \end{aligned}$$

Sales were 130 thousand units after 1 year.

(b) Find the sales after 13 years.
Evaluate $S(13)$.

$$\begin{aligned} S(13) &= 100 + 30 \log_3(2 \cdot 13 + 1) && \text{Let } t = 13. \\ &= 100 + 30 \log_3 27 \\ &= 100 + 30(3) && \log_3 27 = 3 \\ &= 190 \end{aligned}$$

After 13 years, sales had increased to 190 thousand units.

(c) Graph $y = S(t)$.

Use the two ordered pairs (1, 130) and (13, 190) found above. Check that (0, 100) and (40, 220) also satisfy the equation. Use these ordered pairs and knowledge of the general shape of the graph of a logarithmic function to get the graph in Figure 15.

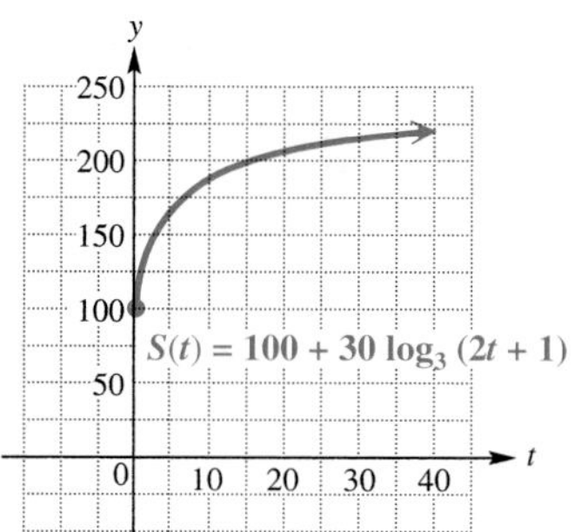

Figure 15

CONNECTIONS

In the United States, the intensity of an earthquake is rated using the *Richter scale.* The Richter scale rating of an earthquake of intensity x is given by

$$R = \log_{10} \frac{x}{x_0},$$

where x_0 is the intensity of an earthquake of a certain (small) size. Figure 16 shows Richter scale ratings for major Southern California earthquakes since 1920. As the figure indicates, earthquakes "come in bunches," and the 1990s have been an especially busy time.

CONNECTIONS (CONTINUED)

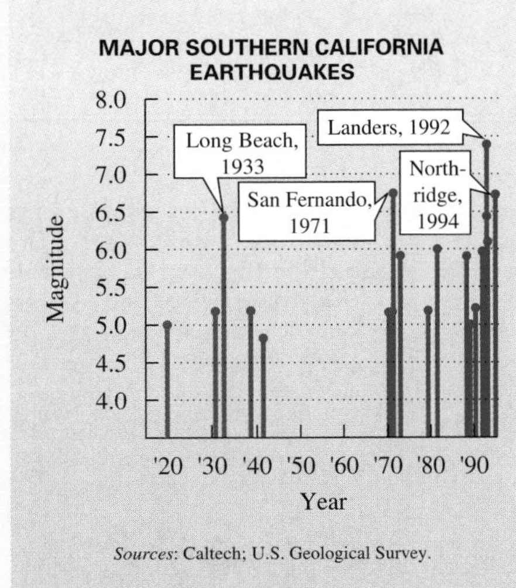

Figure 16

FOR DISCUSSION OR WRITING

Writing the logarithmic equation given above in exponential form, we get

$$10^R = \frac{x}{x_0} \quad \text{or} \quad x = 10^R x_0.$$

The 1994 Northridge earthquake had a Richter scale rating of 6.7; the Landers earthquake had a rating of 7.3. How much more powerful was the Landers earthquake than the Northridge earthquake? Compare the smallest rated earthquake in the figure (at 4.8) with the Landers quake. How much more powerful was the Landers quake?

9.3 EXERCISES

1. By definition, $\log_a x$ is the exponent to which the base a must be raised in order to obtain x. Use this definition to match the logarithm in Column I with its value from Column II. (*Example:* $\log_3 9$ is equal to 2, because 2 is the exponent to which 3 must be raised in order to obtain 9.)

I	II
(a) $\log_4 16$	**A.** -2
(b) $\log_3 81$	**B.** -1
(c) $\log_3\left(\frac{1}{3}\right)$	**C.** 2
(d) $\log_{10} .01$	**D.** 0
(e) $\log_5 \sqrt{5}$	**E.** $\frac{1}{2}$
(f) $\log_{13} 1$	**F.** 4

2. Match the logarithmic equation in Column I with the corresponding exponential equation from Column II.

I	II
(a) $\log_{1/3} 3 = -1$	**A.** $8^{1/3} = \sqrt[3]{8}$
(b) $\log_5 1 = 0$	**B.** $\left(\frac{1}{3}\right)^{-1} = 3$
(c) $\log_2 \sqrt{2} = \frac{1}{2}$	**C.** $4^1 = 4$
(d) $\log_{10} 1000 = 3$	**D.** $2^{1/2} = \sqrt{2}$
(e) $\log_8 \sqrt[3]{8} = \frac{1}{3}$	**E.** $5^0 = 1$
(f) $\log_4 4 = 1$	**F.** $10^3 = 1000$

Write in logarithmic form. See the table in Objective 2.

3. $4^5 = 1024$ **4.** $3^6 = 729$ **5.** $\left(\frac{1}{2}\right)^{-3} = 8$

6. $\left(\frac{1}{6}\right)^{-3} = 216$ **7.** $10^{-3} = .001$ **8.** $36^{1/2} = 6$

Write in exponential form. See the table in Objective 2.

9. $\log_4 64 = 3$ **10.** $\log_2 512 = 9$ **11.** $\log_{10} \frac{1}{10{,}000} = -4$

12. $\log_{100} 100 = 1$ **13.** $\log_6 1 = 0$ **14.** $\log_\pi 1 = 0$

15. When a student asked his teacher to explain how to evaluate $\log_9 3$ without showing any work, his teacher told him, "Think radically." Explain what the teacher meant by this hint.

16. A student told her teacher, "I know that $\log_2 1$ is the exponent to which 2 must be raised in order to obtain 1, but I can't think of any such number." How would you explain to the student that the value of $\log_2 1$ is 0?

Solve each equation for x. See Examples 1 and 2.

17. $x = \log_{27} 3$ **18.** $x = \log_{125} 5$ **19.** $\log_x 9 = \frac{1}{2}$ **20.** $\log_x 5 = \frac{1}{2}$

21. $\log_x 125 = -3$ **22.** $\log_x 64 = -6$ **23.** $\log_{12} x = 0$ **24.** $\log_4 x = 0$

25. $\log_x x = 1$ **26.** $\log_x 1 = 0$ **27.** $\log_x \frac{1}{25} = -2$ **28.** $\log_x \frac{1}{10} = -1$

29. $\log_8 32 = x$ **30.** $\log_{81} 27 = x$ **31.** $\log_\pi \pi^4 = x$ **32.** $\log_{\sqrt{2}} \sqrt{2}^9 = x$

33. $\log_6 \sqrt{216} = x$ **34.** $\log_4 \sqrt{64} = x$

If the point (p, q) is on the graph of $f(x) = a^x$ (for $a > 0$ and $a \neq 1$), then the point (q, p) is on the graph of $f^{-1}(x) = \log_a x$. Use this fact, and refer to the graphs required in Exercises 5–8 in Section 9.2 to graph each logarithmic function. See Example 3.

35. $y = \log_3 x$ **36.** $y = \log_5 x$ **37.** $y = \log_{1/3} x$ **38.** $y = \log_{1/5} x$

39. Explain why 1 is not allowed as a base for a logarithmic function.

40. Compare the summary of facts about the graph of $F(x) = a^x$ in Section 9.2 with the similar summary of facts about the graph of $G(x) = \log_a x$ in this section. Make a list of the facts that reinforce the concept that F and G are inverse functions.

41. The domain of $F(x) = a^x$ is $(-\infty, \infty)$, while the range is $(0, \infty)$. Therefore, since $G(x) = \log_a x$ defines the inverse of F, the domain of G is ____________, while the range of G is ____________.

42. The graphs of both $F(x) = 3^x$ and $G(x) = \log_3 x$ rise from left to right. Which one rises at a faster rate?

Use the graph at the right to predict the value of $f(t)$ for the given value of t.

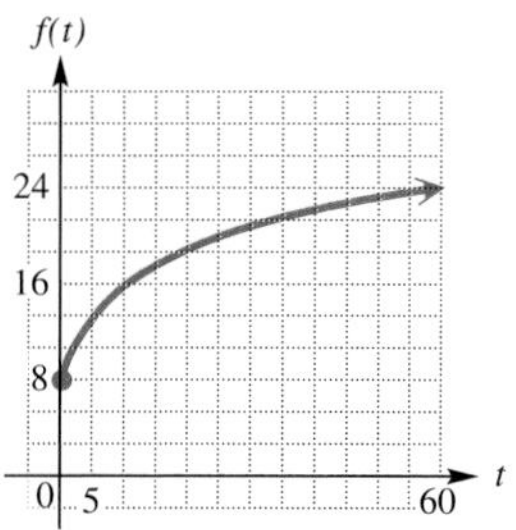

43. $t = 0$

44. $t = 10$

45. $t = 60$

46. Show that the points determined in Exercises 43–45 lie on the graph of $f(t) = 8 \log_5(2t + 5)$.

Solve each application of a logarithmic function. See Examples 4 and 5.

47. According to selected figures from 1981–1995, the number of Superfund hazardous waste sites in the United States can be approximated by the function with

$$f(x) = 11.34 + 317.01 \log_2 x,$$

where $x = 1$ corresponds to 1981, $x = 2$ to 1982, and so on. (*Source:* Environmental Protection Agency.)
(a) Use the function to approximate the number of sites in 1984.
(b) Use the function to approximate the number of sites in 1988.

48. According to selected figures from 1980–1993, the number of trillion cubic feet of dry natural gas consumed worldwide can be approximated by the function with

$$f(x) = 51.47 + 6.044 \log_2 x,$$

where $x = 1$ corresponds to 1980, $x = 2$ to 1981, and so on. (*Source:* Energy Information Administration.)
(a) Use the function to approximate consumption in 1980.
(b) Use the function to approximate consumption in 1987.

49. A study showed that the number of mice in an old abandoned house was approximated by the function with

$$M(t) = 6 \log_4(2t + 4),$$

where t is measured in months and $t = 0$ corresponds to January 1998. Find the number of mice in the house in
(a) January 1998
(b) July 1998
(c) July 2000.
(d) Graph the function.

50. A supply of hybrid striped bass were introduced into a lake in January 1990. Biologists researching the bass population over the next decade found that the number of bass in the lake was approximated by the function with

$$B(t) = 500 \log_3(2t + 3),$$

where $t = 0$ corresponds to January 1990, $t = 1$ to January 1991, $t = 2$ to January 1992, and so on. Use this function to find the bass population in
(a) January 1990
(b) January 1993
(c) January 2002.
(d) Graph the function for $0 \le t \le 12$.

As mentioned in Section 9.1, some graphing calculators have the capability of drawing the inverse of a function. For example, the two screens that follow show the graphs of $f(x) = 2^x$ and $g(x) = \log_2 x$. The graph of g was obtained by drawing the graph of f^{-1}, since $g(x) = f^{-1}(x)$. (Compare to Figure 13 in this section.)

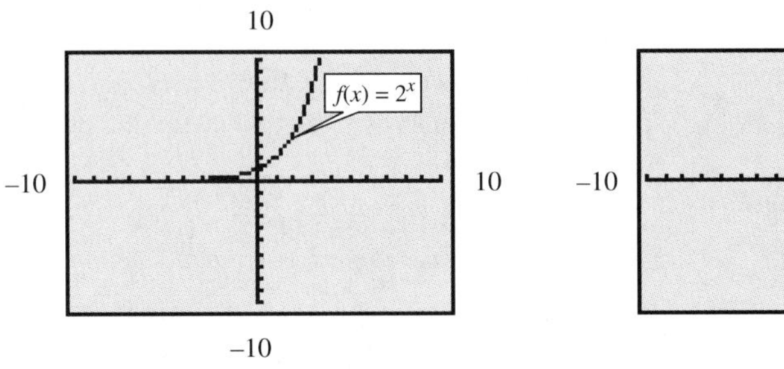

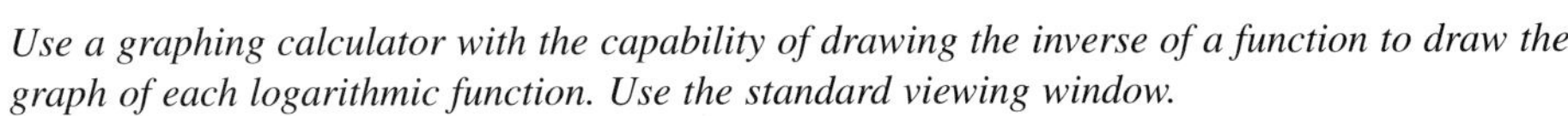

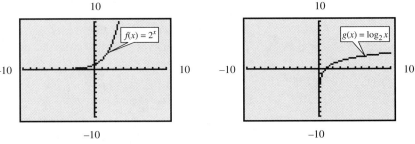

Use a graphing calculator with the capability of drawing the inverse of a function to draw the graph of each logarithmic function. Use the standard viewing window.

51. $g(x) = \log_3 x$ (Compare to Exercise 35.)

52. $g(x) = \log_5 x$ (Compare to Exercise 36.)

53. $g(x) = \log_{1/3} x$ (Compare to Exercise 37.)

54. $g(x) = \log_{1/5} x$ (Compare to Exercise 38.)

9.4 Properties of Logarithms

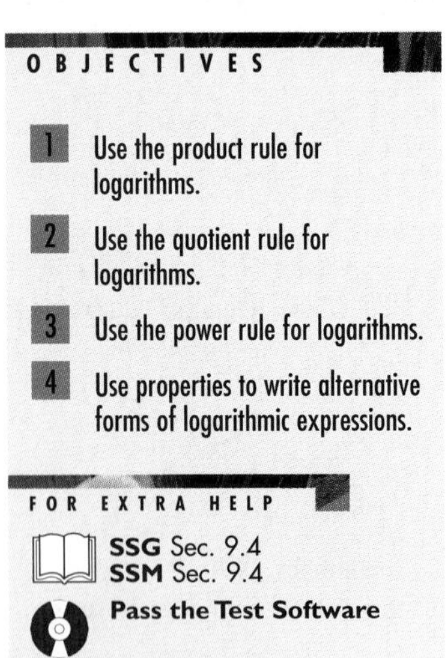

OBJECTIVES
1 Use the product rule for logarithms.
2 Use the quotient rule for logarithms.
3 Use the power rule for logarithms.
4 Use properties to write alternative forms of logarithmic expressions.
FOR EXTRA HELP
SSG Sec. 9.4
SSM Sec. 9.4
Pass the Test Software
InterAct Math Tutorial Software
Video 18

Logarithms have been used as an aid to numerical calculation for several hundred years. Today the widespread use of calculators has made the use of logarithms for calculation obsolete. However, logarithms are still very important in applications and in further work in mathematics.

OBJECTIVE 1 Use the product rule for logarithms. One way in which logarithms simplify problems is by changing a problem of multiplication into one of addition. This is done with the product rule for logarithms.

Product Rule for Logarithms

If x, y, and b are positive numbers, where $b \neq 1$, then

$$\log_b xy = \log_b x + \log_b y.$$

(The logarithm of a product is the sum of the logarithms of the factors.)

NOTE The word statement of the product rule can also be stated by replacing "logarithm" with "exponent," and the rule then becomes the familiar rule for multiplying exponential expressions: The *exponent* of a product is equal to the sum of the *exponents* of the factors.

To prove this rule, let $m = \log_b x$ and $n = \log_b y$, and recall that

$$\log_b x = m \quad \text{means} \quad b^m = x,$$
$$\log_b y = n \quad \text{means} \quad b^n = y.$$

Now consider the product xy.

$$
\begin{aligned}
xy &= b^m \cdot b^n && \text{Substitution} \\
xy &= b^{m+n} && \text{Product rule for exponents} \\
\log_b xy &= m + n && \text{Convert to logarithmic form.} \\
\log_b xy &= \log_b x + \log_b y && \text{Substitution}
\end{aligned}
$$

The last statement is the result we wished to prove.

EXAMPLE 1 Using the Product Rule

Use the product rule for logarithms to rewrite the following. Assume $x > 0$.

(a) $\log_5 6 \cdot 9$
By the product rule, $\log_5 6 \cdot 9 = \log_5 6 + \log_5 9$.

(b) $\log_7 8 + \log_7 12$

$$\log_7 8 + \log_7 12 = \log_7 8 \cdot 12 = \log_7 96$$

(c) $\log_3 3x$

$$
\begin{aligned}
\log_3 3x &= \log_3 3 + \log_3 x \\
\log_3 3x &= 1 + \log_3 x && \log_3 3 = 1
\end{aligned}
$$

(d) $\log_4 x^3$
Since $x^3 = x \cdot x \cdot x$,

$$
\begin{aligned}
\log_4 x^3 &= \log_4(x \cdot x \cdot x) \\
&= \log_4 x + \log_4 x + \log_4 x \\
&= 3 \log_4 x.
\end{aligned}
$$

OBJECTIVE 2 Use the quotient rule for logarithms. The rule for division is similar to the rule for multiplication.

Quotient Rule for Logarithms

If x, y, and b are positive numbers, where $b \neq 1$, then

$$\log_b \frac{x}{y} = \log_b x - \log_b y.$$

(The logarithm of a quotient is the difference between the logarithm of the numerator and the logarithm of the denominator.)

The proof of this rule is similar to the proof of the product rule.

EXAMPLE 2 Using the Quotient Rule

Use the quotient rule for logarithms to rewrite the following.

(a) $\log_4 \frac{7}{9} = \log_4 7 - \log_4 9$

(b) $\log_5 \frac{6}{x} = \log_5 6 - \log_5 x$, for $x > 0$.

(c) $\log_3 \frac{27}{5} = \log_3 27 - \log_3 5$

$= 3 - \log_3 5$ $\quad \log_3 27 = 3$

OBJECTIVE 3 **Use the power rule for logarithms.** There is also a rule for finding the logarithm of the power of a number.

Power Rule for Logarithms

If x and b are positive real numbers, where $b \neq 1$, and if r is any real number, then

$$\log_b x^r = r(\log_b x).$$

(The logarithm of a number to a power equals the exponent times the logarithm of the number.)

As examples of this rule,

$$\log_b m^5 = 5 \log_b m \quad \text{and} \quad \log_3 5^{3/4} = \frac{3}{4} \log_3 5.$$

To see an earlier illustration of this rule, refer to Example 1(d).

To prove the power rule, let

$$\begin{aligned}
\log_b x &= m \\
b^m &= x && \text{Convert to exponential form.} \\
(b^m)^r &= x^r && \text{Raise to the power } r. \\
b^{mr} &= x^r && \text{Power rule for exponents} \\
\log_b x^r &= mr && \text{Convert to logarithmic form.} \\
\log_b x^r &= rm \\
\log_b x^r &= r \log_b x. && m = \log_b x
\end{aligned}$$

This is the statement to be proved.

As a special case of the rule above, let $r = \frac{1}{p}$, so that

$$\log_b \sqrt[p]{x} = \log_b x^{1/p} = \frac{1}{p} \log_b x.$$

For example, using this result, with $x > 0$,

$$\log_b \sqrt[5]{x} = \log_b x^{1/5} = \frac{1}{5} \log_b x \quad \text{and} \quad \log_b \sqrt[3]{x^4} = \log_b x^{4/3} = \frac{4}{3} \log_b x.$$

Another special case is

$$\log_b \frac{1}{x} = -\log_b x$$

since $\frac{1}{x} = x^{-1}$.

EXAMPLE 3 Using the Power Rule

Use the power rule to rewrite each of the following. Assume $a > 0$, $b > 0$, $x > 0$, $a \neq 1$, and $b \neq 1$.

(a) $\log_3 5^2 = 2 \log_3 5$

(b) $\log_a x^4 = 4 \log_a x$

(c) $\log_b \sqrt{7}$

When using the power rule with logarithms of expressions involving radicals, begin by rewriting the radical expression with a rational exponent, as shown in Section 7.1.

$$\log_b \sqrt{7} = \log_b 7^{1/2} \qquad \sqrt{x} = x^{1/2}$$

$$= \frac{1}{2} \log_b 7 \qquad \text{Power rule}$$

(d) $\log_2 \sqrt[5]{x^2} = \log_2 x^{2/5} = \frac{2}{5} \log_2 x$

(e) $\log_3 \frac{1}{7} = -\log_3 7$

Two special properties involving both exponential and logarithmic expressions come directly from the fact that logarithmic and exponential functions are inverses of each other.

Special Properties

If $b > 0$ and $b \neq 1$, then

$$b^{\log_b x} = x \quad (x > 0) \qquad \text{and} \qquad \log_b b^x = x.$$

To prove the first statement, let

$$y = \log_b x.$$

$$b^y = x \qquad \text{Convert to exponential form.}$$

$$b^{\log_b x} = x \qquad \text{Replace } y \text{ with } \log_b x.$$

The proof of the second statement is similar.

EXAMPLE 4 Using the Special Properties

Find the value of the following logarithmic expressions.

(a) $\log_5 5^4$

Since $\log_b b^x = x$,

$$\log_5 5^4 = 4.$$

(b) $\log_3 9$

Since $9 = 3^2$,

$$\log_3 9 = \log_3 3^2 = 2.$$

The property $\log_b b^x = x$ was used in the last step.

(c) $4^{\log_4 10} = 10$

OBJECTIVE 4 Use properties to write alternative forms of logarithmic expressions. Doing so is important in solving equations with logarithms and in calculus.

EXAMPLE 5 Writing Logarithms in Alternative Forms

Use the properties of logarithms to rewrite each expression. Assume all variables represent positive real numbers.

(a) $\log_4 4x^3 = \log_4 4 + \log_4 x^3$ Product rule

$= 1 + 3\log_4 x$ $\log_4 4 = 1$; power rule

(b) $\log_7 \sqrt{\frac{p}{q}} = \log_7\left(\frac{p}{q}\right)^{1/2}$

$= \frac{1}{2}\log_7 \frac{p}{q}$ Power rule

$= \frac{1}{2}(\log_7 p - \log_7 q)$ Quotient rule

(c) $\log_5 \frac{a}{bc} = \log_5 a - \log_5 bc$ Quotient rule

$= \log_5 a - (\log_5 b + \log_5 c)$ Product rule

$= \log_5 a - \log_5 b - \log_5 c$

Notice the careful use of parentheses in the second step. Since we are subtracting the logarithm of a product, and it is being rewritten as a sum of two terms, parentheses *must* be placed around the sum.

(d) $3\log_b x + \frac{1}{2}\log_b y = \log_b x^3 + \log_b y^{1/2}$ Power rule

$= \log_b x^3\sqrt{y}$ Product rule; $y^{1/2} = \sqrt{y}$

(e) $\log_8(2p + 3r)$ cannot be rewritten by the properties of logarithms.

CAUTION

Remember that there is no property of logarithms to rewrite the logarithm of a *sum* or *difference*. For example, we *cannot* write $\log_b(x + y)$ in terms of $\log_b x$ and $\log_b y$. Also, $\log_b \frac{x}{y} \neq \frac{\log_b x}{\log_b y}$.

In the next example, we use numerical values for $\log_2 5$ and $\log_2 3$. While we use the equals sign to give these values, they are actually just approximations, since most logarithms of this type are irrational numbers. While it would be more correct to use the symbol $\approx$, we will simply use $=$ with the understanding that the values are correct to four decimal places.

EXAMPLE 6 Using the Properties of Logarithms with Numerical Values

Given that $\log_2 5 = 2.3219$ and $\log_2 3 = 1.5850$, evaluate the following.

(a) $\log_2 15$

$$\begin{aligned} \log_2 15 &= \log_2 3 \cdot 5 \\ &= \log_2 3 + \log_2 5 && \text{Product rule} \\ &= 1.5850 + 2.3219 \\ &= 3.9069 \end{aligned}$$

(b) $\log_2 .6$

$$\begin{aligned} \log_2 .6 &= \log_2 \frac{3}{5} && .6 = \tfrac{6}{10} = \tfrac{3}{5} \\ &= \log_2 3 - \log_2 5 && \text{Quotient rule} \\ &= 1.5850 - 2.3219 \\ &= -.7369 \end{aligned}$$

(c) $\log_2 27$

$$\begin{aligned} \log_2 27 &= \log_2 3^3 \\ &= 3 \log_2 3 && \text{Power rule} \\ &= 3(1.5850) \\ &= 4.7550 \end{aligned}$$

EXAMPLE 7 Deciding Whether Statements about Logarithms Are True

Decide whether each of the following statements is true or false.

(a) $\log_2 8 - \log_2 4 = \log_2 4$

Evaluate both sides.

$$\log_2 8 - \log_2 4 = \log_2 2^3 - \log_2 2^2 = 3 - 2 = 1$$
$$\log_2 4 = \log_2 2^2 = 2$$

The statement is false because $2 \neq 1$.

(b) $\log_3(\log_2 8) = \dfrac{\log_7 49}{\log_8 64}$

Evaluate both sides.

$$\log_3(\log_2 8) = \log_3(3) = 1$$
$$\frac{\log_7 49}{\log_8 64} = \frac{\log_7 7^2}{\log_8 8^2} = \frac{2}{2} = 1$$

The statement is true.

CONNECTIONS

Long before the days of calculators and computers, the search for making calculations easier was an ongoing process. Machines built by Charles Babbage and Blaise Pascal, a system of "rods" used by John Napier, and slide rules were the forerunners of today's electronic marvels. The invention of logarithms by John Napier in the sixteenth century was a great breakthrough in the search for easier methods of calculation.

Since logarithms are exponents, their properties allowed users of tables of common logarithms to multiply by adding, divide by subtracting, raise to powers by multiplying, and take roots by dividing. Although logarithms are no longer used for computations, they play an important part in higher mathematics.

FOR DISCUSSION OR WRITING

To multiply 458.3 by 294.6 using logarithms, we add $\log_{10} 458.3$ and $\log_{10} 294.6$, then find 10 to the sum. Perform this multiplication using the log* key and the 10^x key on your calculator. Check your answer by multiplying directly with your calculator. Try division, raising to a power, and taking a root by this method.

*In this text, the notation $\log x$ is used to mean $\log_{10} x$. This is also the meaning of the log key on calculators.

9.4 EXERCISES

Use the indicated rule of logarithms to complete each equation in Exercises 1–5.

1. $\log_{10}(3 \cdot 4) =$ ______________ (product rule)
2. $\log_{10}\left(\frac{3}{4}\right) =$ ______________ (quotient rule)
3. $\log_{10} 3^4 =$ ______________ (power rule)
4. $3^{\log_3 4} =$ ______________ (special property)
5. $\log_3 3^4 =$ ______________ (special property)
6. Evaluate $\log_2(8 + 8)$. Then evaluate $\log_2 8 + \log_2 8$. Are the results the same? How could you change the operation in the first expression to make the two expressions equal?

Use the properties of logarithms introduced in this section to express each logarithm as a sum or difference of logarithms, or as a single number if possible. Assume that all variables represent positive real numbers. See Examples 1–5.

7. $\log_7 \frac{4}{5}$
8. $\log_8 \frac{9}{11}$
9. $\log_2 8^{1/4}$
10. $\log_3 9^{3/4}$
11. $\log_4 \frac{3\sqrt{x}}{y}$
12. $\log_5 \frac{6\sqrt{z}}{w}$
13. $\log_3 \frac{\sqrt[3]{4}}{x^2 y}$
14. $\log_7 \frac{\sqrt[3]{13}}{pq^2}$
15. $\log_3 \sqrt{\frac{xy}{5}}$
16. $\log_6 \sqrt{\frac{pq}{7}}$
17. $\log_2 \frac{\sqrt[3]{x} \cdot \sqrt[5]{y}}{r^2}$
18. $\log_4 \frac{\sqrt[4]{z} \cdot \sqrt[5]{w}}{s^2}$
19. A student erroneously wrote $\log_a(x + y) = \log_a x + \log_a y$. When his teacher explained that this was indeed wrong, the student claimed that he had used the distributive property. Write a few sentences explaining why the distributive property does not apply in this case.

20. Write a few sentences explaining how the rules for multiplying and dividing powers of the same base are similar to the rules for finding logarithms of products and quotients.

Use the properties of logarithms introduced in this section to express each of the following as a single logarithm. Assume that all variables are defined in such a way that the variable expressions are positive, and bases are positive numbers not equal to 1. *See Examples 1–5.*

21. $\log_b x + \log_b y$

22. $\log_b 2 + \log_b z$

23. $3 \log_a m - \log_a n$

24. $5 \log_b x - \log_b y$

25. $(\log_a r - \log_a s) + 3 \log_a t$

26. $(\log_a p - \log_a q) + 2 \log_a r$

27. $3 \log_a 5 - 4 \log_a 3$

28. $3 \log_a 5 + \frac{1}{2} \log_a 9$

29. $\log_{10}(x + 3) + \log_{10}(x - 3)$

30. $\log_{10}(y + 4) + \log_{10}(y - 4)$

31. $3 \log_p x + \frac{1}{2} \log_p y - \frac{3}{2} \log_p z - 3 \log_p a$

32. $\frac{1}{3} \log_b x + \frac{2}{3} \log_b y - \frac{3}{4} \log_b s - \frac{2}{3} \log_b t$

To four decimal places, the values of $\log_{10} 2$ *and* $\log_{10} 9$ *are*

$$\log_{10} 2 = .3010 \qquad \log_{10} 9 = .9542.$$

Evaluate each logarithm by applying the appropriate rule or rules from this section. ***DO NOT USE A CALCULATOR FOR THESE EXERCISES.*** *See Example 6.*

33. $\log_{10} 18$

34. $\log_{10} \frac{9}{2}$

35. $\log_{10} \frac{2}{9}$

36. $\log_{10} 4$

37. $\log_{10} 36$

38. $\log_{10} 162$

39. $\log_{10} 3$

40. $\log_{10} \sqrt[5]{2}$

Decide whether each statement is true or false. See Example 7.

41. $\log_6 60 - \log_6 10 = 1$

42. $\log_3 7 + \log_3 \frac{1}{7} = 0$

43. $\frac{\log_{10} 7}{\log_{10} 14} = \frac{1}{2}$

44. $\frac{\log_{10} 10}{\log_{10} 100} = \frac{1}{10}$

45. Refer to the "NOTE" following the word statement of the product rule for logarithms in this section. Now, state the quotient rule in words, replacing "logarithm" with "exponent."

46. Explain why the statement for the power rule for logarithms requires that x be a positive real number.

47. Refer to Example 7(a). Change the left side of the equation using the quotient rule so that the statement becomes true, and simplify.

48. What is wrong with the following "proof" that $\log_2 16$ does not exist?

$$\begin{aligned} \log_2 16 &= \log_2(-4)(-4) \\ &= \log_2(-4) + \log_2(-4) \end{aligned}$$

Since the logarithm of a negative number is not defined, the final step cannot be evaluated, and so $\log_2 16$ does not exist.

RELATING CONCEPTS (EXERCISES 49–54)

Work Exercises 49–54 in order.

49. Evaluate $\log_3 81$.

50. Write the *meaning* of the expression $\log_3 81$.

(continued)

RELATING CONCEPTS (EXERCISES 49–54) (CONTINUED)

51. Evaluate $3^{\log_3 81}$.

52. Write the *meaning* of the expression $\log_2 19$.

53. Evaluate $2^{\log_2 19}$.

54. Keeping in mind that a logarithm is an exponent, and using the results from Exercises 49–53, what is the simplest form of the expression $k^{\log_k m}$?

Did you make the connection that a logarithm is an exponent?

9.5 Evaluating Logarithms

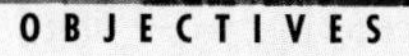

OBJECTIVES

1. Evaluate common logarithms by using a calculator.
2. Use common logarithms in an application.
3. Evaluate natural logarithms using a calculator.
4. Use exponential functions with base e and natural logarithms in applications.
5. Use the change-of-base rule.

FOR EXTRA HELP

SSG Sec. 9.5
SSM Sec. 9.5

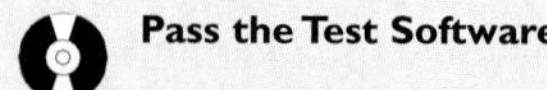

Pass the Test Software

InterAct Math Tutorial Software

Video 18

As mentioned earlier, logarithms are important in many applications of mathematics to everyday problems, particularly in biology, engineering, economics, and social science. In this section we find numerical approximations for logarithms. Traditionally, base 10 logarithms are used most often since our number system is base 10. Logarithms to base 10 are called **common logarithms** and $\log_{10} x$ is abbreviated as simply $\log x$, where the base is understood to be 10.

OBJECTIVE 1 Evaluate common logarithms by using a calculator. In the next example we give the results of evaluating some common logarithms using a calculator with a log key. (This may be a second function key on some calculators.) For simple scientific calculators, just enter the number, then press the log key. For graphing calculators, these steps are reversed. We will give all logarithms to four decimal places.

EXAMPLE 1 Evaluating Common Logarithms

Evaluate each logarithm using a calculator.

(a) log 327.1 **(b)** log 437,000 **(c)** log .0615

Figure 17 shows how a graphing calculator evaluates these logarithms. The calculator is set to give four decimal places.

Figure 17

Notice that log .0615 is found to be -1.2111, a negative result. The common logarithm of a number between 0 and 1 is always negative because the logarithm is the exponent on 10 that produces the number. For example,

$$10^{-1.2111} = .0615.$$

If the exponent (the logarithm) were positive, the result would be greater than 1, since $10^0 = 1$.

OBJECTIVE 2 Use common logarithms in an application. In chemistry, the **pH** of a solution is defined as follows.

Definition of pH

$$\text{pH} = -\log[H_3O^+],$$

where $[H_3O^+]$ is the hydronium ion concentration in moles per liter.

The pH is a measure of the acidity or alkalinity of a solution, with water, for example, having a pH of 7. In general, acids have pH numbers less than 7, and alkaline solutions have pH values greater than 7. It is customary to round pH values to the nearest tenth.

EXAMPLE 2 Using pH in an Application

Wetlands are classified as *bogs*, *fens*, *marshes*, and *swamps*. These classifications are based on pH values. A pH value between 6.0 and 7.5, such as that of Summerby Swamp in Michigan's Hiawatha National Forest, indicates that the wetland is a "rich fen." When the pH is between 4.0 and 6.0, it is a "poor fen," and if the pH falls to 3.0 or less, the wetland is a "bog." (*Source:* R. Mohlenbrock, "Summerby Swamp, Michigan," *Natural History*, March 1994.)

Suppose that the hydronium ion concentration of a sample of water from a wetland is 6.3×10^{-3}. How would this wetland be classified?

Use the definition of pH.

$$\begin{aligned} \text{pH} &= -\log(6.3 \times 10^{-3}) \\ &= -(\log 6.3 + \log 10^{-3}) && \text{Product rule} \\ &= -[.7993 - 3(1)] \\ &= -.7993 + 3 \approx 2.2 \end{aligned}$$

Since the pH is less than 3.0, the wetland is a bog.

EXAMPLE 3 Finding Hydronium Ion Concentration

Find the hydronium ion concentration of drinking water with a pH of 6.5.

$$\begin{aligned} \text{pH} = 6.5 &= -\log[H_3O^+] \\ \log[H_3O^+] &= -6.5 && \text{Multiply by } -1. \end{aligned}$$

Solve for $[H_3O^+]$ by writing the equation in exponential form, remembering that the base is 10.

$$[H_3O^+] = 10^{-6.5} = 3.2 \times 10^{-7}$$

OBJECTIVE 3 Evaluate natural logarithms using a calculator. The most important logarithms used in applications are **natural logarithms,** which have as base the number e. The number e is irrational, like π: $e \approx 2.7182818$. Logarithms to base e are called natural logarithms because they occur in biology and the social sciences in natural situations that involve growth or decay. The base e logarithm of x is written $\ln x$ (read "el en x"). A graph of $y = \ln x$, the natural logarithmic function, is given in Figure 18.

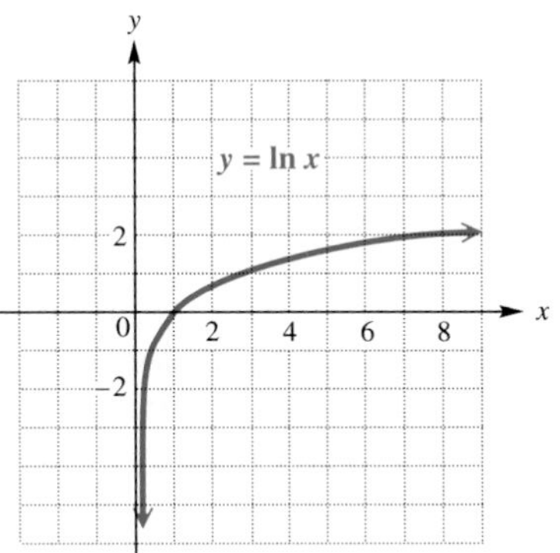

Figure 18

CONNECTIONS

The number e is a fundamental number in our universe. For this reason, e, like π, is called a *universal constant.* If there are intelligent beings elsewhere, they too will have to use e to do higher mathematics.

The letter e is used to honor Leonhard Euler, who published extensive results on the number in 1748. The first few digits of the decimal value of e are 2.7182818. Since it is an irrational number, its decimal expansion never terminates and never repeats.

The properties of e are used extensively in calculus and in higher mathematics. In Section 9.6 we see how it applies to growth and decay in the physical world.

FOR DISCUSSION OR WRITING

The value of e can be expressed as

$$e = 1 + \frac{1}{1} + \frac{1}{1 \cdot 2} + \frac{1}{1 \cdot 2 \cdot 3} + \frac{1}{1 \cdot 2 \cdot 3 \cdot 4} + \cdots.$$

Approximate e using 2 terms of this expression, then 3 terms, 4 terms, 5 terms, and 6 terms. How close is the approximation to the value of e given above with 6 terms? Does this infinite sum approach the value of e very quickly?

A calculator key labeled $\ln x$ is used to evaluate natural logarithms. If your calculator has an e^x key, but not a key labeled $\ln x$, find natural logarithms by entering the number, pressing the INV key, and then pressing the e^x key. This works because $y = e^x$ is the inverse function of $y = \ln x$ (or $y = \log_e x$).

EXAMPLE 4 Finding Natural Logarithms

Find each of the following logarithms to four decimal places.

(a) ln 192.7 **(b)** ln 10.84 **(c)** ln .5841

Figure 19 shows how a graphing calculator evaluates these natural logarithms. Like common logarithms, a number between 0 and 1 has a negative natural logarithm, as in the case of ln .5841.

```
ln(192.7)
               5.2611
ln(10.84)
               2.3832
ln(.5841)
               -.5377
```

Figure 19

OBJECTIVE 4 Use exponential functions with base *e* and natural logarithms in applications. One of the most common applications of exponential functions depends on the fact that in many situations involving growth or decay of a population, the amount or number of some quantity present at time t can be closely approximated by

$$y = y_0 e^{kt},$$

where y_0 is the amount or number present at time $t = 0$, k is a constant, and e is the base of natural logarithms.

EXAMPLE 5 Applying an Exponential Function

The *greenhouse effect* refers to the phenomenon whereby emissions of gases such as carbon dioxide, methane, and chlorofluorocarbons (CFCs) have the potential to alter the climate of the earth and destroy the ozone layer. Concentrations of CFC-12, used in refrigeration technology, in parts per billion (ppb) can be modeled by the exponential function with

$$f(x) = .48e^{.04x},$$

where $x = 0$ represents 1990. Use this function to approximate the concentration in 1998.

Since $x = 0$ represents 1990, $x = 8$ represents 1998. Evaluate $f(8)$ using a calculator.

$$f(8) = .48e^{.04(8)} = .48e^{.32} \approx .66$$

In 1998, the concentration of CFC-12 was about .66 ppb.

EXAMPLE 6 Applying Natural Logarithms

The number of years, $N(r)$, since two independently evolving languages split off from a common ancestral language is approximated by

$$N(r) = -5000 \ln r,$$

where r is the percent of words from the ancestral language common to both languages now. Find N if $r = 70\%$.

Write 70% as .7 and find $N(.7)$.

$$\begin{aligned} N(.7) &= -5000 \ln .7 \\ &\approx -5000(-.35667) \\ &\approx 1783 \end{aligned}$$

Approximately 1800 years have passed since the two languages separated.

OBJECTIVE 5 Use the change-of-base rule. A calculator can be used to approximate the values of common logarithms (base 10) or natural logarithms (base e). However, sometimes we need to use logarithms to other bases. The following rule is used to convert logarithms from one base to another.

Change-of-Base Rule

If $a > 0$, $a \neq 1$, $b > 0$, $b \neq 1$, and $x > 0$, then

$$\log_a x = \frac{\log_b x}{\log_b a}.$$

To help remember the change-of-base rule, notice that x is "above" a on both sides of the equation.

Any positive number other than 1 can be used for base b in the change-of-base rule, but usually the only practical bases are e and 10, since calculators give logarithms only for these two bases.

To prove the formula for change of base, let $\log_a x = m$.

$$\begin{aligned} \log_a x &= m \\ a^m &= x \qquad \text{Change to exponential form.} \end{aligned}$$

Since logarithmic functions are one-to-one, if all variables are positive and if $x = y$, then $\log_b x = \log_b y$.

$$\begin{aligned} \log_b(a^m) &= \log_b x && \text{Take logarithms on both sides.} \\ m \log_b a &= \log_b x && \text{Use the power rule.} \\ (\log_a x)(\log_b a) &= \log_b x && \text{Substitute for } m. \\ \log_a x &= \frac{\log_b x}{\log_b a} && \text{Divide both sides by } \log_b a. \end{aligned}$$

EXAMPLE 7 Using the Change-of-Base Rule

Find each logarithm using a calculator.

(a) $\log_5 12$

Use common logarithms and the rule for change of base.

$$\log_5 12 = \frac{\log 12}{\log 5} \approx 1.5440$$

(b) $\log_2 134$
Use natural logarithms and the change-of-base rule.

$$\log_2 134 = \frac{\ln 134}{\ln 2} \approx 7.0661$$

In Example 7, the final answers were obtained *without* rounding off the intermediate values. In general, it is best to wait until the final step to round off the answer; otherwise, a build-up of round-off error may cause the final answer to have an incorrect final decimal place digit.

9.5 EXERCISES

Choose the correct response in Exercises 1–4.

1. What is the base in the expression $\log x$?
(a) e **(b)** 1 **(c)** 10 **(d)** x

2. What is the base in the expression $\ln x$?
(a) e **(b)** 1 **(c)** 10 **(d)** x

3. Since $10^0 = 1$ and $10^1 = 10$, between what two consecutive integers is the value of $\log 5.6$?
(a) 5 and 6 **(b)** 10 and 11 **(c)** 0 and 1 **(d)** -1 and 0

4. Since $e^1 \approx 2.718$ and $e^2 \approx 7.389$, between what two consecutive integers is the value of $\ln 5.6$?
(a) 5 and 6 **(b)** 2 and 3 **(c)** 1 and 2 **(d)** 0 and 1

5. Without using a calculator, give the value of $\log 10^{19.2}$.

6. Without using a calculator, give the value of $\ln e^{\sqrt{2}}$.

You may need a calculator for the remaining exercises in this set.

Find each logarithm. Give an approximation to four decimal places. See Examples 1 and 4.

7. $\log 43$ **8.** $\log 98$ **9.** $\log 328.4$

10. $\log 457.2$ **11.** $\log .0326$ **12.** $\log .1741$

13. $\log(4.76 \times 10^9)$ **14.** $\log(2.13 \times 10^4)$ **15.** $\ln 7.84$

16. $\ln 8.32$ **17.** $\ln .0556$ **18.** $\ln .0217$

19. $\ln 388.1$ **20.** $\ln 942.6$ **21.** $\ln(8.59 \times e^2)$

22. $\ln(7.46 \times e^3)$ **23.** $\ln 10$ **24.** $\log e$

25. Use your calculator to find approximations of the following logarithms:
(a) $\log 356.8$ **(b)** $\log 35.68$ **(c)** $\log 3.568$.
(d) Observe your answers and make a conjecture concerning the decimal values of the common logarithms of numbers greater than 1 that have the same digits.

26. Let k represent the number of letters in your last name.
(a) Use your calculator to find $\log k$.
(b) Raise 10 to the power indicated by the number you found in part (a). What is your result?
(c) Use the concepts of Section 9.1 to explain why you obtained the answer you found in part (b). Would it matter what number you used for k to observe the same result?

27. Try to find $\log(-1)$ using a calculator. (If you have a graphing calculator, it should be in real number mode.) What happens? Explain why this happens.

Refer to Example 2. In Exercises 28–30, suppose that water from a wetland area is sampled and found to have the given hydronium ion concentration. Determine whether the wetland is a rich fen, *a* poor fen, *or a* bog.

28. 2.5×10^{-5} **29.** 2.5×10^{-2} **30.** 2.5×10^{-7}

Find the pH of the substance with the given hydronium ion concentration. See Example 2.

31. Ammonia, 2.5×10^{-12}

32. Sodium bicarbonate, 4.0×10^{-9}

33. Grapes, 5.0×10^{-5}

34. Tuna, 1.3×10^{-6}

Use the formula for pH to find the hydronium ion concentration of the substance with the given pH. See Example 3.

35. Human blood plasma, 7.4

36. Human gastric contents, 2.0

37. Spinach, 5.4

38. Bananas, 4.6

Solve each problem. See Examples 5 and 6.

39. The total expenditures in millions of current dollars for pollution abatement and control during the period from 1985 to 1993 can be approximated by the function with

$$P(x) = 70{,}967e^{.0526x},$$

where $x = 0$ corresponds to 1985, $x = 1$ to 1986, and so on. Approximate the expenditures for each of the following years. (*Source:* U.S. Bureau of Economic Analysis, *Survey of Current Business, May 1995.*)

(a) 1987 **(b)** 1990 **(c)** 1993

(d) What were the approximate expenditures for 1985?

40. The emission of the greenhouse gas nitrous oxide increased yearly during the first half of the 1990s. Based on figures during the period from 1990 to 1994, the emissions in thousands of metric tons can be modeled by the function with

$$N(x) = 446.5e^{.0118x},$$

where $x = 0$ corresponds to 1990, $x = 1$ to 1991, and so on. Approximate the emissions for each of the following years. (*Source:* U.S. Energy Information Administration, *Emission of Greenhouse Gases in the United States, annual.*)

(a) 1991 **(b)** 1992 **(c)** 1994

(d) What were the approximate emissions in 1990?

41. Based on selected figures obtained during the 1980s and 1990s, consumer expenditures on all types of books in the United States can be modeled by the function with

$$B(x) = 8768e^{.072x},$$

where $x = 0$ represents 1980, $x = 1$ represents 1981, and so on, and $B(x)$ is in millions of dollars. Approximate consumer expenditures for 1998. (*Source:* Book Industry Study Group.)

42. Based on selected figures obtained during the 1970s, 1980s, and 1990s, the total number of bachelor's degrees earned in the United States can be modeled by the function with

$$D(x) = 815{,}427e^{.0137x},$$

where $x = 1$ corresponds to 1971, $x = 10$ corresponds to 1980, and so on. Approximate the number of bachelor's degrees earned in 1994. (*Source:* U.S. National Center for Education Statistics.)

43. Suppose that the amount, in grams, of plutonium-241 present in a given sample is determined by the function with

$$A(t) = 2.00e^{-.053t},$$

where t is measured in years. Find the amount present in the sample after the given number of years.

(a) 4 **(b)** 10 **(c)** 20

(d) What was the initial amount present?

44. Suppose that the amount, in grams, of radium-226 present in a given sample is determined by the function with

$$A(t) = 3.25e^{-.00043t},$$

where t is measured in years. Find the amount present in the sample after the given number of years.

(a) 20 **(b)** 100 **(c)** 500

(d) What was the initial amount present?

For Exercises 45–48, refer to the function in Example 6.

45. Find $N(.9)$. **46.** Find $N(.3)$. **47.** Find $N(.5)$.

48. How many years have elapsed since the split if 80% of the words of the ancestral language are common to both languages today?

Use the change-of-base rule (with either common or natural logarithms) to find each logarithm to four decimal places. See Example 7.

49. $\log_6 12$ **50.** $\log_8 13$ **51.** $\log_{12} 6$

52. $\log_{13} 8$ **53.** $\log_{\sqrt{2}} \pi$ **54.** $\log_\pi \sqrt{2}$

55. Let m be the number of letters in your first name, and let n be the number of letters in your last name.

(a) In your own words, explain what $\log_m n$ means.

(b) Use your calculator to find $\log_m n$.

(c) Raise m to the power indicated by the number you found in part (b). What is your result?

56. The equation $5^x = 7$ cannot be solved using the methods described in Section 9.2. However, in solving this equation, we must find the exponent to which 5 must be raised in order to obtain 7: this is $\log_5 7$.

(a) Use the change-of-base rule and your calculator to find $\log_5 7$.

(b) Raise 5 to the number you found in part (a). What is your result?

(c) Using as many decimal places as your calculator gives, write the solution set of $5^x = 7$. (Equations of this type will be studied in more detail in Section 9.6.)

TECHNOLOGY INSIGHTS (EXERCISES 57–60)

57. The function defined by $P(x) = 70{,}967e^{.0526x}$, described in Exercise 39, is graphed in a graphing calculator-generated screen in the accompanying figure. Interpret the meanings of x and y in the display at the bottom of the screen in the context of Exercise 39.

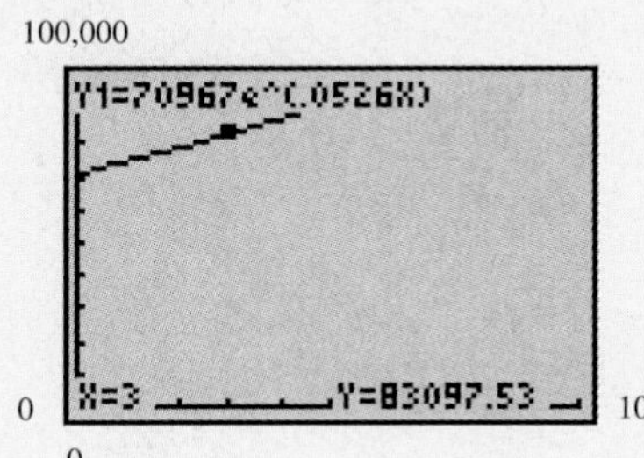

(continued)

TECHNOLOGY INSIGHTS (EXERCISES 57–60) (CONTINUED)

58. The function defined by $A(x) = 3.25e^{-.00043x}$, with $x = t$, described in Exercise 44, is graphed in a graphing calculator-generated screen in the accompanying figure. Interpret the meanings of x and y in the display at the bottom of the screen in the context of Exercise 44.

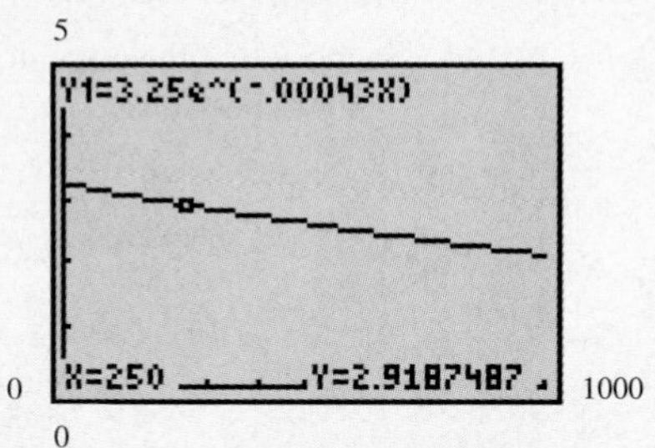

59. The screen shows a table of selected values for the function defined by

$$Y_1 = \left(1 + \frac{1}{x}\right)^x.$$

X	Y1
0	ERROR
1	2
10	2.5937
100	2.7048
1000	2.7169
10000	2.7181
100000	2.7183

Y1=(1+1/X)^X

(a) Why is there an error message for $x = 0$?

(b) What number does the function value seem to approach as x takes on larger and larger values?

(c) Use a calculator to evaluate this function for $x = 1{,}000{,}000$. What value do you get? Now evaluate $e = e^1$. How close are these two values?

(d) Make a conjecture: As the values of x approach infinity, the value of $\left(1 + \frac{1}{x}\right)^x$ approaches ______.

60. Here is another property of logarithms: For $b > 0$, $x > 0$, $b \neq 1$, $x \neq 1$,

$$\log_b x = \frac{1}{\log_x b}.$$

Now observe the accompanying calculator screen.

TECHNOLOGY INSIGHTS (EXERCISES 57–60) (CONTINUED)

(a) Without using a calculator, give a decimal representation for $\frac{1}{.4342944819}$.
Then support your answer using the reciprocal key of your calculator.

(b) Without using a calculator, give a decimal representation for $\frac{1}{2.302585093}$.
Then support your answer using the reciprocal key of your calculator.

Because graphing calculators are equipped with $\log x$ and $\ln x$ keys, it is possible to graph the functions defined by $f(x) = \log x$ and $g(x) = \ln x$ directly, as shown in the figures that follow.

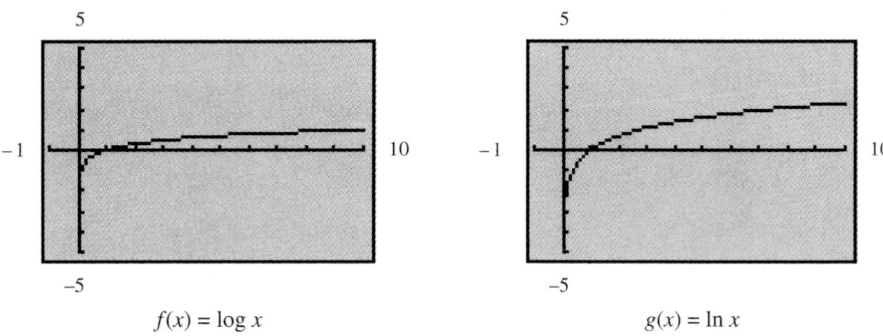

$f(x) = \log x$ $g(x) = \ln x$

In order to graph functions defined by logarithms to bases other than 10 or e, however, we must use the change-of-base rule. For example, to graph $y = \log_2 x$, we may enter y_1 as $\frac{\log x}{\log 2}$ or $\frac{\ln x}{\ln 2}$. This is shown in the figure that follows. (Compare it to the figure in the exercises of Section 9.3, where it was drawn using the fact that $y = \log_2 x$ is the inverse of $y = 2^x$.)

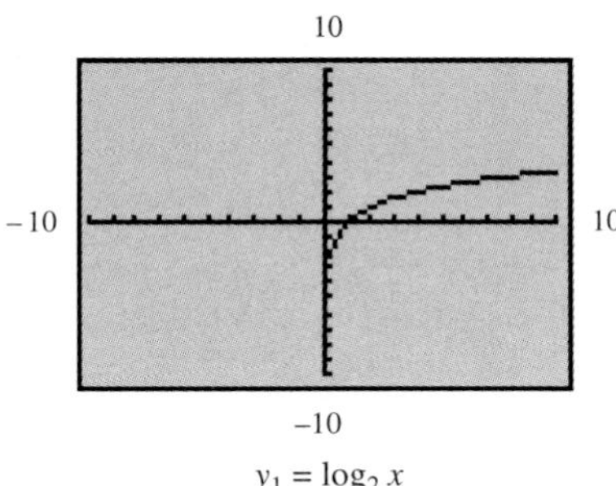

$y_1 = \log_2 x$

 *Use the change-of-base rule to graph each logarithmic function with a graphing calculator. Use a viewing window with* Xmin = −1, Xmax = 10, Ymin = −5, and Ymax = 5.

61. $g(x) = \log_3 x$

62. $g(x) = \log_5 x$

63. $g(x) = \log_{1/3} x$

64. $g(x) = \log_{1/5} x$

9.6 Exponential and Logarithmic Equations; Applications

OBJECTIVES

1. Solve equations involving variables in the exponents.
2. Solve equations involving logarithms.
3. Solve applications involving compound interest.
4. Solve applications involving exponential growth and decay.
5. Use a graphing calculator to solve exponential and logarithmic equations.

FOR EXTRA HELP

SSG Sec. 9.6
SSM Sec. 9.6
Pass the Test Software
InterAct Math Tutorial Software
Video 18

As mentioned earlier, exponential and logarithmic functions are important in many applications of mathematics. Using these functions in applications requires solving exponential and logarithmic equations. Some simple equations were solved in Sections 9.2 and 9.3. More general methods for solving these equations depend on the following properties.

Properties for Equation Solving

For all real numbers $b > 0$, $b \neq 1$, and any real numbers x and y:

1. If $x = y$, then $b^x = b^y$.
2. If $b^x = b^y$, then $x = y$.
3. If $x = y$, and $x > 0$, $y > 0$, then $\log_b x = \log_b y$.
4. If $x > 0$, $y > 0$, and $\log_b x = \log_b y$, then $x = y$.

We used property 2 to solve exponential equations in Section 9.2, and property 3 was used in the proof of the change-of-base rule in the previous section. We will refer to these properties by number throughout this section.

OBJECTIVE 1 **Solve equations involving variables in the exponents.** The first examples illustrate a general method for solving exponential equations using property 3.

EXAMPLE 1 **Solving an Exponential Equation**

Solve the equation $3^x = 12$. Give the answer in decimal form.

$$3^x = 12$$

$$\log 3^x = \log 12 \qquad \text{Property 3}$$

$$x \log 3 = \log 12 \qquad \text{Power rule for logarithms}$$

$$x = \frac{\log 12}{\log 3}$$

This quotient is the exact solution. To get a decimal approximation for the solution, we use a calculator. Correct to three decimal places, a calculator gives

$$x = 2.262,$$

and the solution set is $\{2.262\}$.

CAUTION Be careful: $\dfrac{\log 12}{\log 3}$ is *not* equal to log 4, since $\log 4 = .6021$, but $\dfrac{\log 12}{\log 3} = 2.262$.

When an exponential equation has e as the base, it is easiest to use base e logarithms.

EXAMPLE 2 Solving an Exponential Equation with Base e

Solve $e^{.003x} = 40$.

Take base e logarithms on both sides.

$$\ln e^{.003x} = \ln 40$$

$$.003x = \ln 40 \qquad \ln e^k = k$$

$$x = \frac{\ln 40}{.003} \qquad \text{Divide by .003.}$$

$$x \approx 1230 \qquad \text{Use a calculator.}$$

The solution set is $\{1230\}$. Check that $e^{.003(1230)} \approx 40$.

In summary, exponential equations can be solved by one of the following methods. (The method used depends on the form of the equation.)

Solving an Exponential Equation

Method 1 **Use property 3.** Take logarithms to the same base on each side and then use the power rule of logarithms or the special property $\log_b b^x = x$. (See Examples 1 and 2 above.)

Method 2 **Use property 2.** Write both sides as exponentials with the same base and then set the exponents equal. (See Section 9.2.)

OBJECTIVE 2 Solve equations involving logarithms. The properties of logarithms from Section 9.4 are useful here, as is using the definition of a logarithm to change it to exponential form.

EXAMPLE 3 Solving a Logarithmic Equation

Solve $\log_2(x + 5)^3 = 4$.

$$(x + 5)^3 = 2^4 \qquad \text{Convert to exponential form.}$$

$$(x + 5)^3 = 16$$

$$x + 5 = \sqrt[3]{16} \qquad \text{Take cube roots on both sides.}$$

$$x = -5 + \sqrt[3]{16}$$

$$x = -5 + 2\sqrt[3]{2} \qquad \text{Simplify the radical.}$$

Verify that the solution satisfies the equation, so the solution set is $\{-5 + 2\sqrt[3]{2}\}$.

Recall that the domain of $y = \log_b x$ is $(0, \infty)$. For this reason, it is always necessary to check that the solution of an equation with logarithms yields only logarithms of positive numbers in the original equation.

EXAMPLE 4 Solving a Logarithmic Equation

Solve $\log_2(x + 1) - \log_2 x = \log_2 8$.

$$\log_2(x + 1) - \log_2 x = \log_2 8$$

$$\log_2 \frac{x + 1}{x} = \log_2 8 \qquad \text{Quotient rule}$$

$$\frac{x + 1}{x} = 8 \qquad \text{Property 4}$$

$$8x = x + 1 \qquad \text{Multiply by } x.$$

$$x = \frac{1}{7} \qquad \text{Subtract } x \text{, divide by 7.}$$

Check this solution by substitution in the original equation. Here, both $x + 1$ and x must be positive. If $x = \frac{1}{7}$, this condition is satisfied, and the solution set is $\{\frac{1}{7}\}$.

EXAMPLE 5 Solving a Logarithmic Equation

Solve $\log x + \log(x - 21) = 2$.

For this equation, write the left side as a single logarithm. Then write in exponential form and solve the equation.

$$\log x + \log(x - 21) = 2$$

$$\log x(x - 21) = 2 \qquad \text{Product rule}$$

$$x(x - 21) = 10^2 \qquad \log x = \log_{10} x\text{; write in exponential form.}$$

$$x^2 - 21x = 100$$

$$x^2 - 21x - 100 = 0 \qquad \text{Standard form}$$

$$(x - 25)(x + 4) = 0 \qquad \text{Factor.}$$

$$x - 25 = 0 \quad \text{or} \quad x + 4 = 0 \qquad \text{Set each factor equal to 0.}$$

$$x = 25 \quad \text{or} \quad x = -4$$

The value -4 must be rejected as a solution, since it leads to the logarithm of a negative number in the original equation:

$$\log(-4) + \log(-4 - 21) = 2. \qquad \text{The left side is not defined.}$$

The only solution, therefore, is 25, and the solution set is $\{25\}$.

Do not reject a potential solution just because it is nonpositive. Reject any value that *leads to* the logarithm of a nonpositive number.

In summary, use the following steps to solve a logarithmic equation.

Solving a Logarithmic Equation

Step 1 **Get a single logarithm on one side.** Use the product rule or quotient rule of logarithms to do this.

Step 2 **(a) Use property 4.** If $\log_b x = \log_b y$, then $x = y$. (See Example 4.)

(b) Write the equation in exponential form. If $\log_b x = k$, then $x = b^k$. (See Examples 3 and 5.)

OBJECTIVE 3 **Solve applications involving compound interest.** So far in this book, problems involving applications of interest have been limited to the use of the simple interest formula, $I = prt$. In most cases, banks pay compound interest (interest paid on both principal and interest). The formula for compound interest is an important application of exponential functions.

Compound Interest Formula (for a Finite Number of Periods)

If P dollars is deposited in an account paying an annual rate of interest r compounded (paid) n times per year, the account will contain

$$A = P\left(1 + \frac{r}{n}\right)^{nt}$$

dollars after t years.

In the formula above, r is usually expressed as a decimal.

EXAMPLE 6 **Solving a Compound Interest Problem**

How much money will there be in an account at the end of 5 years if \$1000 is deposited at 6% compounded quarterly? (Assume no withdrawals are made.)

Since interest is compounded quarterly, $n = 4$. The other values given in the problem are $P = 1000$, $r = .06$ (since $6\% = .06$), and $t = 5$. Substitute into the compound interest formula to get the value of A.

$$A = 1000\left(1 + \frac{.06}{4}\right)^{4 \cdot 5} \quad \text{Substitute.}$$

$$A = 1000(1.015)^{20}$$

Now use the y^x key on a calculator, and round the answer to the nearest cent.

$$A = 1346.86$$

The account will contain \$1346.86. (The actual amount of interest earned is \$1346.86 − \$1000 = \$346.86. Why?)

Interest can be compounded annually, semiannually, quarterly, daily, and so on. The number of compounding periods can get larger and larger. If the value of n is allowed to approach infinity, we have an example of *continuous compounding.* However, the compound interest formula above cannot be used for continuous compounding, since there is no finite value for n. The formula for continuous compounding is an example of exponential growth involving the number e.

Continuous Compound Interest Formula

If a principal of P dollars is deposited at an annual rate of interest r compounded continuously for t years, the final amount on deposit is

$$A = Pe^{rt}.$$

EXAMPLE 7 Solving a Continuous Compound Interest Problem

In Example 6 we found that \$1000 invested for 5 years at 6% interest compounded quarterly would grow to \$1346.86.

(a) How much would this same investment grow to if interest were compounded continuously?

Use the formula for continuous compounding with $P = 1000$, $r = .06$, and $t = 5$.

$$\begin{aligned} A &= Pe^{rt} && \text{Formula} \\ &= 1000e^{.06(5)} && \text{Substitute.} \\ &= 1000e^{.30} \\ &= 1349.86 && \text{Use a calculator and round to the nearest cent.} \end{aligned}$$

Continuous compounding would cause the investment to grow to \$1349.86. Notice that this is \$3.00 more than the amount in Example 6, when interest was compounded quarterly.

(b) How long would it take for the initial investment to double its original amount? (This is called the **doubling time.**)

We must find the value of t that will cause A to be 2(\$1000) = \$2000.

$$\begin{aligned} A &= Pe^{rt} \\ 2000 &= 1000e^{.06t} && \text{Let } A = 2P = 2000. \\ 2 &= e^{.06t} && \text{Divide by 1000.} \\ \ln 2 &= .06t && \text{Take natural logarithms; } \ln e^k = k. \\ t &= \frac{\ln 2}{.06} && \text{Divide by .06.} \\ t &\approx 11.55 && \text{Use a calculator.} \end{aligned}$$

It would take about 11.55 years for the original investment to double.

CONNECTIONS

Work Example 7(b) using just P as the original principal and $2P$ as the final amount A.

FOR DISCUSSION OR WRITING

Comment on the statement: The original amount of the investment does not affect the doubling time.

OBJECTIVE 4 Solve applications involving exponential growth and decay. We saw some applications involving exponential growth and decay in Sections 9.2 and 9.5. In many cases, quantities grow or decay according to a function defined by an exponential expression with base e. (See Section 9.5, Example 5.) You have probably heard of the carbon-14 dating process used to determine the age of fossils. The method used is based on the exponential decay function.

EXAMPLE 8 Solving an Exponential Decay Problem

Carbon-14 is a radioactive form of carbon that is found in all living plants and animals. After a plant or animal dies, the radioactive carbon-14 disintegrates according to the function

$$y = y_0e^{-.000121t},$$

where t is time in years, and y is the amount of the sample at time t.

(a) If an initial sample contains $y_0 = 10$ grams of carbon-14, how many grams will be present after 3000 years?

Let $y_0 = 10$ and $t = 3000$ in the formula, and use a calculator.

$$y = 10e^{-.000121(3000)} \approx 6.96 \text{ grams}$$

(b) How long would it take for the initial sample to decay to half of its original amount? (This is called the **half-life.**)

Let $y = \frac{1}{2}(10) = 5$, and solve for t.

$$5 = 10e^{-.000121t} \qquad \text{Substitute.}$$

$$\frac{1}{2} = e^{-.000121t} \qquad \text{Divide by 10.}$$

$$\ln \frac{1}{2} = -.000121t \qquad \text{Take natural logarithms; } \ln e^k = k.$$

$$t = \frac{\ln \frac{1}{2}}{-.000121} \qquad \text{Divide by } -.000121.$$

$$t \approx 5728 \qquad \text{Use a calculator.}$$

The half-life is just over 5700 years. ■

OBJECTIVE 5 Use a graphing calculator to solve exponential and logarithmic equations. Earlier we saw that the x-intercepts of the graph of a function f correspond to the real solutions of the equation $f(x) = 0$. This idea was applied to linear and quadratic equations and can be extended to exponential and logarithmic equations as well. In Example 1, we solved the equation $3^x = 12$ algebraically using rules for logarithms and found the solution set to be $\{2.262\}$. This can be supported graphically by showing that the x-intercept of the graph of the function defined by $y = 3^x - 12$ corresponds to this solution. See Figure 20.

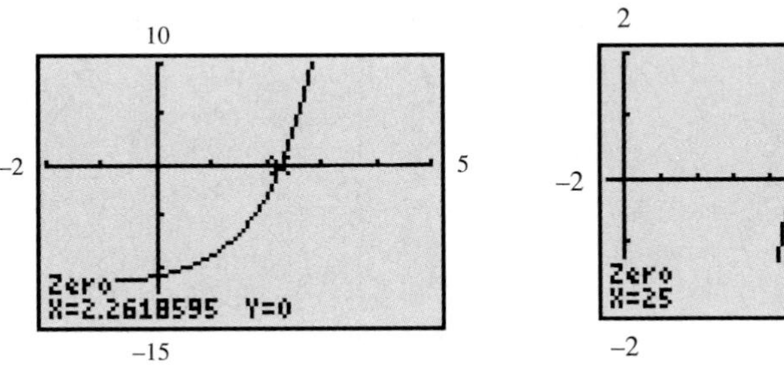

Figure 20

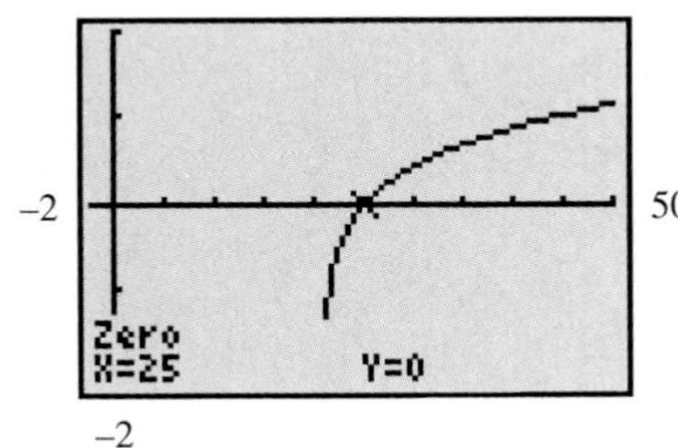

Figure 21

In Example 5, we solved $\log x + \log(x - 21) = 2$ and found the solution set to be $\{25\}$. Figure 21 shows that the x-intercept of the graph of the function defined by $y = \log x + \log(x - 21) - 2$ supports this result.

9.6 EXERCISES

RELATING CONCEPTS (EXERCISES 1-4)

In Section 9.2 we solved an equation such as

$$5^x = 125$$

by writing both sides as a power of the same base, setting exponents equal, and then solving the resulting equation. The equation above is solved as follows.

$5^x = 125$	Given equation
$5^x = 5^3$	$125 = 5^3$
$x = 3$	Set exponents equal.

Solution set: $\{3\}$

The method described in this section can also be used to solve this equation.

Work Exercises 1–4 in order, *to see how this is done.*

1. Take common logarithms of both sides, and write this equation.
2. Apply the power rule for logarithms on the left.
3. Get x alone on the left.
4. Use a calculator to find the decimal form of the solution. What is the solution set?

Did you make the connection that the method of solving exponential equations explained in this section can be used to solve the types studied in Section 9.2?

Solve each equation. Give solutions to three decimal places. See Example 1.

5. $7^x = 5$

6. $4^x = 3$

7. $9^{-x+2} = 13$

8. $6^{-t+1} = 22$

9. $2^{y+3} = 5^y$

10. $6^{m+3} = 4^m$

Use natural logarithms to solve each equation. Give solutions to three decimal places. See Example 2.

11. $e^{.006x} = 30$

12. $e^{.012x} = 23$

13. $e^{-.103x} = 7$

14. $e^{-.205x} = 9$

15. $100e^{.045x} = 300$

16. $500e^{-.003x} = 250$

17. Solve one of the equations in Exercises 11–14 using common logarithms rather than natural logarithms. (You should get the same solution.) Explain why using natural logarithms is a better choice.

18. If you were asked to solve $10^{.0025x} = 75$, would natural or common logarithms be a better choice? Explain your answer.

Solve each equation. Give exact solutions. See Example 3.

19. $\log_3(6x + 5) = 2$

20. $\log_5(12x - 8) = 3$

21. $\log_7(x + 1)^3 = 2$

22. $\log_4(y - 3)^3 = 4$

23. Suppose that in solving a logarithmic equation having the term $\log_4(x - 3)$, you obtain an apparent solution of 2. All algebraic work is correct. Explain why 2 must be rejected as a solution of the equation.

24. Suppose that in solving a logarithmic equation having the term $\log_7(3 - x)$, you obtain an apparent solution of -4. All algebraic work is correct. Should you reject -4 as a solution of the equation? Explain why or why not.

Solve each equation. Give exact solutions. See Examples 4 and 5.

25. $\log(6x + 1) = \log 3$
26. $\log(7 - x) = \log 12$
27. $\log_5(3t + 2) - \log_5 t = \log_5 4$
28. $\log_2(x + 5) - \log_2(x - 1) = \log_2 3$
29. $\log 4x - \log(x - 3) = \log 2$
30. $\log(-x) + \log 3 = \log(2x - 15)$
31. $\log_2 x + \log_2(x - 7) = 3$
32. $\log(2x - 1) + \log 10x = \log 10$
33. $\log 5x - \log(2x - 1) = \log 4$
34. $\log_3 x + \log_3(2x + 5) = 1$
35. $\log_2 x + \log_2(x - 6) = 4$
36. $\log_2 x + \log_2(x + 4) = 5$

Solve each problem. See Examples 6–8.

37. How much money will there be in an account at the end of 6 years if \$2000 is deposited at 4% compounded quarterly? (Assume no withdrawals are made.)

38. How much money will there be in an account at the end of 7 years if \$3000 is deposited at 3.5% compounded quarterly? (Assume no withdrawals are made.)

39. A sample of 400 grams of lead-210 decays to polonium-210 according to the function defined by
$$A(t) = 400e^{-.032t},$$
where t is time in years. How much lead will be left in the sample after 25 years?

40. How long will it take the initial sample of lead in Exercise 39 to decay to half of its original amount?

41. Find the amount of money in an account after 12 years if \$5000 is deposited at 7% annual interest compounded as follows.
(a) annually **(b)** semiannually **(c)** quarterly
(d) daily (Use $n = 365$.) **(e)** continuously

42. How much money will be in an account at the end of 8 years if \$4500 is deposited at 6% annual interest compounded as follows?
(a) annually **(b)** semiannually **(c)** quarterly
(d) daily (Use $n = 365$.) **(e)** continuously

43. How much money must be deposited today to become \$1850 in 40 years at 6.5% compounded continuously?

44. How much money must be deposited today to amount to \$1000 in 10 years at 5% compounded continuously?

45. Refer to Exercise 39 in Section 9.5. Assuming that the function continues to apply past 1993, in what year can we expect total expenditures to be 133,500 million dollars? (*Source:* U.S. Bureau of Economic Analysis, *Survey of Current Business, May 1995.*)

46. Refer to Exercise 40 in Section 9.5. Assuming that the function continues to apply past 1994, in what year can we expect nitrous oxide emissions to be 485 thousand metric tons? (*Source:* U.S. Energy Information Administration, *Emission of Greenhouse Gases in the United States, annual.*)

47. The concentration y of a drug in a person's system decreases according to the relationship
$$y = 2e^{-.125t},$$
where y is in appropriate units, and t is in hours. Find the amount of time that it will take for the concentration to be half of its original value.

48. The number y of ants in an anthill grows according to the function defined by

$$y = 300e^{.4t},$$

where t is time measured in days. Find the time it will take for the number of ants to double.

49. Radioactive strontium decays according to the function defined by

$$y = y_0e^{-.0239t},$$

where t is time in years.

(a) If an initial sample contains $y_0 = 5$ grams of radioactive strontium, how many grams will be present after 20 years?

(b) How many grams of the initial 5-gram sample will be present after 60 years?

(c) What is the half-life of radioactive strontium?

50. Plutonium-241 decays according to the function defined by

$$y = y_0e^{-.053t},$$

where t is time in years.

(a) If an initial sample contains $y_0 = 200$ grams of plutonium-241, how many grams will be present after 100 years?

(b) How many grams of the initial 200-gram sample will be present after 200 years?

(c) What is the half-life of plutonium-241?

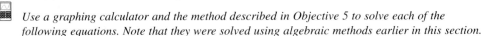

Use a graphing calculator and the method described in Objective 5 to solve each of the following equations. Note that they were solved using algebraic methods earlier in this section.

51. $7^x = 5$ (Exercise 5)

52. $4^x = 3$ (Exercise 6)

53. $\log(6x + 1) = \log 3$ (Exercise 25)

54. $\log(7 - x) = \log 12$ (Exercise 26)

CHAPTER 9 GROUP ACTIVITY

How Much Space Do We Need?

Objective: Use natural logarithms and exponential equations to calculate how long it will take to fully populate Earth with people.

Applications of exponential growth and decay were introduced in this chapter. The formula for exponential growth is:

$$P(t) = P_0e^{kt},$$

where $P(t)$ is population after t years, P_0 is initial population, k is annual growth rate, and t is number of years elapsed.

A. If Earth's population will double in 30 years at the current growth rate, what is this growth rate? (Express your answer as a percent.)

B. Earth has a total surface area of approximately 5.1×10^{14} square meters. Seventy percent of this surface area is rock, ice, sand, and open ocean. Another 8% of the total surface area, made up of tundra, lakes and streams, continental shelves, algae beds and reefs, and estuaries, is unfit for living space. The remaining area is suitable for growing food and for living space.

(continued)

1. Determine the surface area available for growing food.
2. Determine the surface area available for living space. Notice that the surface area available for living space is also considered available for growing food.

C. Suppose that each person needs 100 square meters of Earth's surface for living space. Earth's current population is approximately 5.5×10^9.

1. If none of the surface area available for living space is used for food, how long will it take for the livable surface of Earth to be covered with people? (Use the growth rate you found in part A and the surface area you found in Exercise B2.)
2. How much surface area would be left to grow food?

D. Measure a space that is one square meter in area. Discuss with your partner whether or not you would want to be packed this closely together on Earth. Take into account that many people live in high-rise apartment buildings and how that translates into surface area used per person.

E. Now suppose that for each person 100 square meters of Earth's surface is needed for living space and growing food.

1. Using the same population and growth rate as in part C, determine how long it will take to fill Earth with people. (Use the surface area from Exercise B2.)
2. Does 100 square meters per person for living space and growing food seem reasonable? Consider the following questions in your discussion.

   How much space do you think it takes to grow animals for food? To grow grains, nuts, fruits, and vegetables?

   Would food grow as well in desert areas, mountainous areas, or jungle areas?

   Would there be any space left for wild animals or natural plant life?

   Would there be any space left for shopping malls, movie theaters, concert halls, factories, office buildings, or parking lots?

3. Write a paragraph summarizing your results and your discussion.

CHAPTER 9 SUMMARY

KEY TERMS

9.1 **one-to-one function**
inverse of a function
9.2 **exponential function**
asymptote
exponential equation

9.3 **logarithm**
logarithmic equation
logarithmic function with base *a*

9.5 **common logarithm**
natural logarithm

9.6 **doubling time**
half-life

NEW SYMBOLS

$f^{-1}(x)$	the inverse of $f(x)$
$\log_a x$	the logarithm of x to the base a
$\log x$	common (base 10) logarithm of x
$\ln x$	natural (base e) logarithm of x
e	a constant, approximately 2.7182818

TEST YOUR WORD POWER

See how well you have learned the vocabulary in this chapter. Answers, with examples, are given at the bottom of the page.

1. In a **one-to-one function**
(a) each x-value corresponds to only one y-value
(b) each x-value corresponds to one or more y-values
(c) each x-value is the same as each y-value
(d) each x-value corresponds to only one y-value and each y-value corresponds to only one x-value.

2. If f is a one-to-one function, then the **inverse** of f is
(a) the set of all solutions of f
(b) the set of all ordered pairs formed by reversing the coordinates of the ordered pairs of f
(c) an equation involving an exponential expression
(d) the set of all ordered pairs that are the opposite (negative) of the coordinates of the ordered pairs of f.

3. An **exponential function** is a function defined by an expression of the form
(a) $f(x) = ax^2 + bx + c$ for real numbers a, b, c $(a \neq 0)$
(b) $f(x) = \log_a x$ for a and x positive numbers $(a \neq 1)$
(c) $f(x) = a^x$ for all real numbers x $(a > 0, a \neq 1)$
(d) $f(x) = \sqrt{x}$ for $x \geq 0$.

4. An **asymptote** is
(a) a line that a graph intersects just once
(b) a line that the graph of a function more and more closely approaches as the graph gets farther away from the origin
(c) the x-axis or y-axis
(d) a line about which a graph is symmetric.

5. A **logarithm** is
(a) an exponent
(b) a base
(c) an equation
(d) a term.

6. A **logarithmic function** is a function that is defined by an expression of the form
(a) $f(x) = ax^2 + bx + c$ for real numbers a, b, c $(a \neq 0)$
(b) $f(x) = \log_a x$ for a and x positive numbers $(a \neq 1)$
(c) $f(x) = a^x$ for all real numbers x $(a > 0, a \neq 1)$
(d) $f(x) = \sqrt{x}$ for $x \geq 0$.

QUICK REVIEW

CONCEPTS	EXAMPLES
9.1 INVERSE FUNCTIONS	
Horizontal Line Test If a horizontal line intersects the graph of a function in no more than one point, then the function is one-to-one.	Find f^{-1} if $f(x) = 2x - 3$. The graph of f is a straight line, so f is one-to-one by the horizontal line test.
Inverse Functions For a one-to-one function f defined by an equation $y = f(x)$, the defining equation of the inverse function f^{-1} is found by interchanging x and y, solving for y, and replacing y with $f^{-1}(x)$.	Interchange x and y in the equation $y = 2x - 3$. $x = 2y - 3$ Solve for y to get $y = \frac{1}{2}x + \frac{3}{2}$. Therefore, $f^{-1}(x) = \frac{1}{2}x + \frac{3}{2}$.

Answers to Test Your Word Power

1. (d) *Example:* The function $f = \{(0, 2), (1, -1), (3, 5), (-2, 3)\}$ is one-to-one. **2.** (b) *Example:* The inverse of the one-to-one function f defined above is $f^{-1} = \{(2, 0), (-1, 1), (5, 3), (3, -2)\}$. **3.** (c) *Examples:* $f(x) = 4^x$, $g(x) = \left(\frac{1}{2}\right)^x$, $h(x) = 2^{-x+3}$ **4.** (b) *Example:* The graph of $F(x) = 2^x$ has the x-axis $(y = 0)$ as an asymptote. **5.** (a) *Example:* $\log_a x$ is the exponent to which a must be raised to obtain x; $\log_3 9 = 2$ since $3^2 = 9$. **6.** (b) *Examples:* $y = \log_3 x$, $y = \log_{1/3} x$

CONCEPTS	EXAMPLES
The graph of f^{-1} is a mirror image of the graph of f with respect to the line $y = x$.	The graphs of a function f and its inverse f^{-1} are given here. 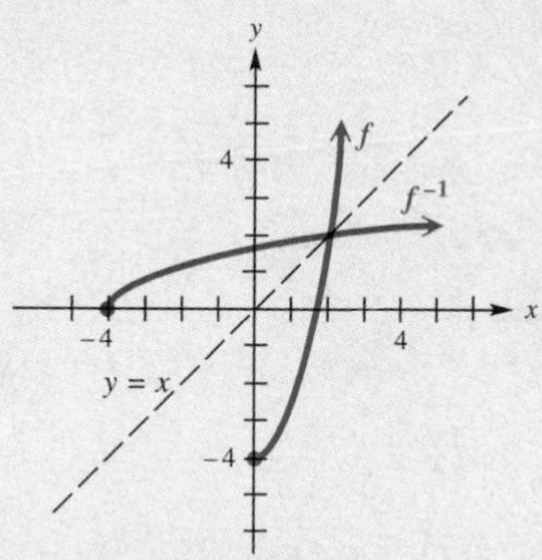

9.2 EXPONENTIAL FUNCTIONS

For $a > 0$, $a \neq 1$, $f(x) = a^x$ defines an exponential function with base a. **Graph of $F(x) = a^x$** The graph contains the point (0, 1). When $a > 1$, the graph rises from left to right. When $0 < a < 1$, the graph falls from left to right. The x-axis is an asymptote. The domain is $(-\infty, \infty)$; the range is $(0, \infty)$.	$f(x) = 3^x$ defines an exponential function with base 3. Its graph is shown here. 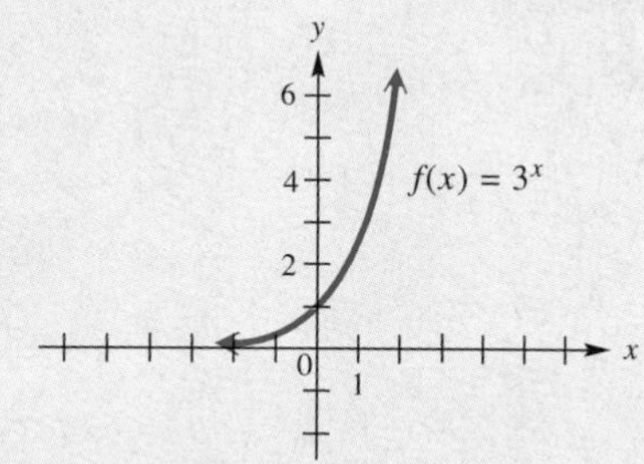

9.3 LOGARITHMIC FUNCTIONS

$y = \log_a x$ has the same meaning as $a^y = x$.	$y = \log_2 x$ means $x = 2^y$.
For $b > 0$, $b \neq 1$, $\log_b b = 1$ and $\log_b 1 = 0$.	$\log_3 3 = 1, \qquad \log_5 1 = 0$
For $a > 0$, $a \neq 1$, $x > 0$, $g(x) = \log_a x$ defines the logarithmic function with base a. **Graph of $G(x) = \log_a x$** The graph contains the point (1, 0). When $a > 1$, the graph rises from left to right. When $0 < a < 1$, the graph falls from left to right. The y-axis is an asymptote. The domain is $(0, \infty)$; the range is $(-\infty, \infty)$.	$g(x) = \log_3 x$ defines the logarithmic function with base 3. Its graph is shown here. 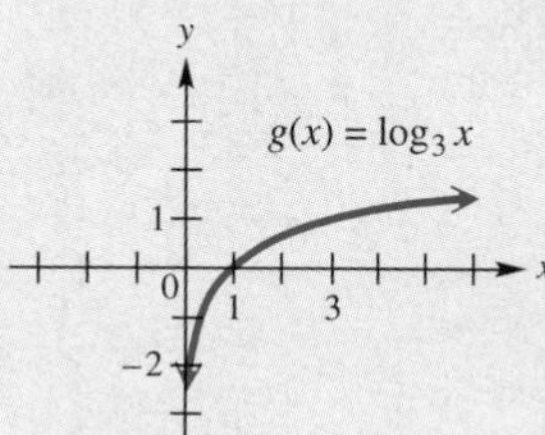

9.4 PROPERTIES OF LOGARITHMS

Product Rule $\log_a xy = \log_a x + \log_a y$	$\log_2 3m = \log_2 3 + \log_2 m \quad (m > 0)$
Quotient Rule $\log_a \frac{x}{y} = \log_a x - \log_a y$	$\log_5 \frac{9}{4} = \log_5 9 - \log_5 4$

(continued)

CONCEPTS	EXAMPLES
Power Rule	
$\log_a x^r = r \log_a x$	$\log_{10} 2^3 = 3 \log_{10} 2$
Special Properties	
$b^{\log_b x} = x$ and $\log_b b^x = x$	$6^{\log_6 10} = 10$ and $\log_3 3^4 = 4$

9.5 EVALUATING LOGARITHMS

CONCEPTS	EXAMPLES
Change-of-Base Rule If $a > 0$, $a \neq 1$, $b > 0$, $b \neq 1$, $x > 0$, then $\log_a x = \dfrac{\log_b x}{\log_b a}$.	Approximate $\log_3 17$ to four decimal places. $\log_3 17 = \dfrac{\ln 17}{\ln 3} = \dfrac{\log 17}{\log 3} \approx 2.5789$

9.6 EXPONENTIAL AND LOGARITHMIC EQUATIONS; APPLICATIONS

CONCEPTS	EXAMPLES
To solve exponential equations, use these properties ($b > 0$, $b \neq 1$). **1.** If $b^x = b^y$, then $x = y$.	Solve. $2^{3x} = 2^5$ $3x = 5$ $x = \dfrac{5}{3}$ The solution set is $\left\{\dfrac{5}{3}\right\}$.
2. If $x = y$ ($x > 0$, $y > 0$), then $\log_b x = \log_b y$.	Solve. $5^m = 8$ $\log 5^m = \log 8$ $m \log 5 = \log 8$ $m = \dfrac{\log 8}{\log 5} \approx 1.2920$ The solution set is $\{1.2920\}$.
To solve logarithmic equations, use these properties, where $b > 0$, $b \neq 1$, $x > 0$, $y > 0$. First use the properties of Section 9.4, if necessary, to get the equation in the proper form. **1.** If $\log_b x = \log_b y$, then $x = y$.	Solve. $\log_3 2x = \log_3(x + 1)$ $2x = x + 1$ $x = 1$ The solution set is $\{1\}$.
2. If $\log_b x = y$, then $b^y = x$.	Solve. $\log_2(3a - 1) = 4$ $3a - 1 = 2^4 = 16$ $3a = 17$ $a = \dfrac{17}{3}$ The solution set is $\left\{\dfrac{17}{3}\right\}$.

CHAPTER 9 REVIEW EXERCISES

[9.1] *Determine whether each graph is the graph of a one-to-one function.*

1.

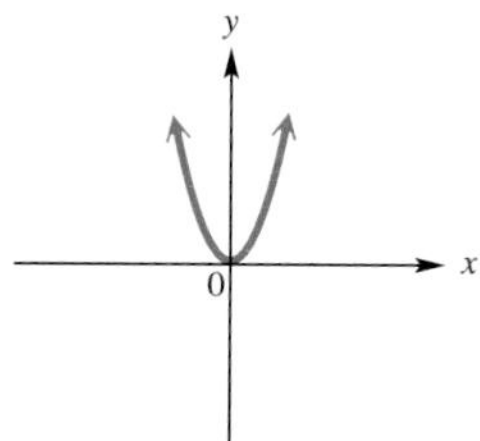

2.

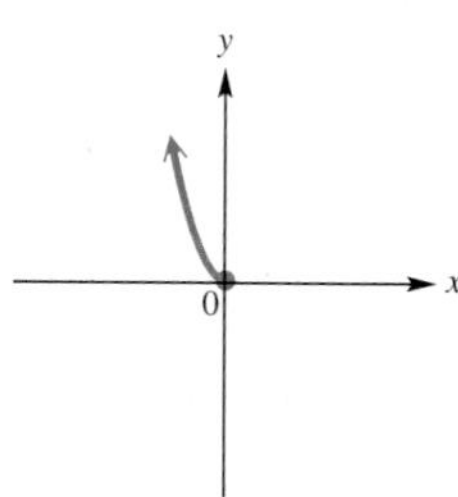

3. The chart lists the top five metropolitan areas with highest average levels of regulated particle pollution during the period from 1990–1994. (The measure is the concentration of particles 10 microns or smaller in diameter, down to 2.5 microns, measured in micrograms per cubic meter of air.) If the set of areas is the domain of the function and the set of particle pollution levels is the range, is this function one-to-one? Why or why not?

Metropolitan Area	*Average Particle Pollution*
1. Visalia-Tulare-Porterville, CA	60.4
2. Bakersfield, CA	54.8
3. Fresno, CA	51.7
4. Riverside-San Bernardino, CA	48.1
5. Stockton, CA	44.8

Sources: The New York Times, Environmental Protection Administration; Natural Resources Defense Council.

Determine whether each function is one-to-one. If it is, find its inverse.

4. $f(x) = -3x + 7$

5. $f(x) = \sqrt[3]{6x - 4}$

6. $f(x) = -x^2 + 3$

Each function graphed is one-to-one. Graph its inverse.

7.

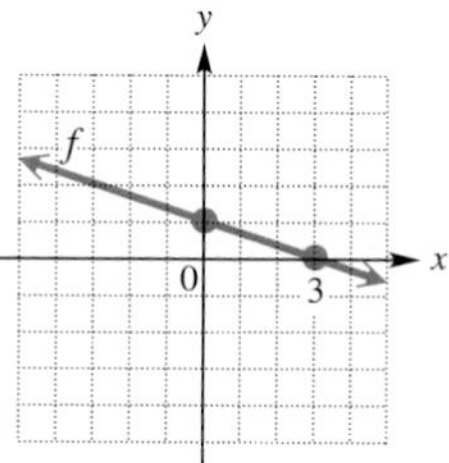

8.

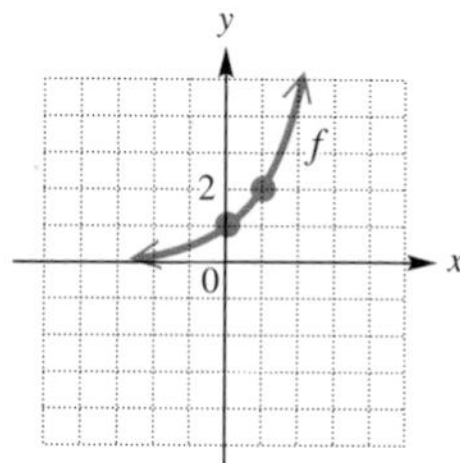

[9.2] *Graph each function.*

9. $f(x) = 3^x$

10. $f(x) = \left(\frac{1}{3}\right)^x$

11. $y = 3^{x+1}$

12. $y = 2^{2x+3}$

Solve each equation.

13. $4^{3x} = 8^{x+4}$

14. $\left(\frac{1}{27}\right)^{x-1} = 9^{2x}$

15. The gross wastes generated in plastics, in millions of tons, from 1960–1990 can be approximated by the exponential function with

$$W(x) = .67(1.123)^x,$$

where $x = 0$ corresponds to 1960, $x = 5$ to 1965, and so on. Use this function to approximate the plastic waste amounts for the following years. Compare your answers to the values from the bar graph in the chapter introduction. (*Source:* Environmental Protection Agency, *Characterization of Municipal Solid Waste in the United States: 1994 Update,* 1995.)
(a) 1965 **(b)** 1975 **(c)** 1990

[9.3] *Graph each function.*

16. $g(x) = \log_3 x$ (*Hint:* See Exercise 9.) **17.** $g(x) = \log_{1/3} x$ (*Hint:* See Exercise 10.)

Solve each equation.

18. $\log_8 64 = x$ **19.** $\log_2 \sqrt{8} = x$ **20.** $\log_7\left(\frac{1}{49}\right) = x$

21. $\log_4 x = \frac{3}{2}$ **22.** $\log_k 4 = 1$ **23.** $\log_b b^2 = 2$

24. In your own words, explain the meaning of $\log_b a$.

25. Based on the meaning of $\log_b a$, what is the simplest form of $b^{\log_b a}$?

26. A company has found that total sales, in thousands of dollars, are given by the function with

$$S(x) = 100 \log_2(x + 2),$$

where x is the number of weeks after a major advertising campaign was introduced. What were the total sales 6 weeks after the campaign was introduced? Graph the function.

[9.4] *Apply the properties of logarithms introduced in Section 9.4 to express each logarithm as a sum or difference of logarithms, or as a single number if possible. Assume that all variables represent positive real numbers.*

27. $\log_2 3xy^2$ **28.** $\log_4 \frac{\sqrt{x} \cdot w^2}{z}$

Use the properties of logarithms introduced in Section 9.4 to write each expression as a single logarithm. Assume that all variables represent positive real numbers, $b \neq 1$.

29. $\log_b 3 + \log_b x - 2 \log_b y$ **30.** $\log_3(x + 7) - \log_3(4x + 6)$

[9.5] *Evaluate each logarithm. Give approximations to four decimal places.*

31. log 28.9 **32.** log .257 **33.** ln 28.9 **34.** ln .257

Use the change-of-base rule (with either common or natural logarithms) to find each logarithm. Give approximations to four decimal places.

35. $\log_{16} 13$ **36.** $\log_4 12$

Use the formula $pH = -\log[H_3O^+]$ to find the pH of each substance with the given hydronium ion concentration.

37. Milk, 4.0×10^{-7} **38.** Crackers, 3.8×10^{-9}

39. If orange juice has a pH of 4.6, what is its hydronium ion concentration?

40. Suppose the quantity, measured in grams, of a radioactive substance present at time t is given by

$$Q(t) = 500e^{-.05t},$$

where t is measured in days. Find the quantity present at the following times.
(a) $t = 0$ **(b)** $t = 4$

[9.6] *Solve each equation. Give solutions to three decimal places.*

41. $3^x = 9.42$

42. $e^{.06x} = 3$

43. Which one of the following is *not* a solution of $7^x = 23$?

(a) $\dfrac{\log 23}{\log 7}$ (b) $\dfrac{\ln 23}{\ln 7}$ (c) $\log_7 23$ (d) $\log_{23} 7$

Solve each equation. Give exact solutions.

44. $\log_3(9x + 8) = 2$

45. $\log_3(p + 2) - \log_3 p = \log_3 2$

46. $\log(2x + 3) = \log 3x + 2$

47. $\log_4 x + \log_4(8 - x) = 2$

48. $\log_2 x + \log_2(x + 15) = 4$

49. Consider the logarithmic equation

$$\log(2x + 3) = \log x + 1.$$

(a) Solve the equation using properties of logarithms.

(b) If $Y_1 = \log(2x + 3)$ and $Y_2 = \log x + 1$, then the graph of $Y_1 - Y_2$ looks like this. Explain how the display at the bottom of the screen confirms the solution set found in part (a).

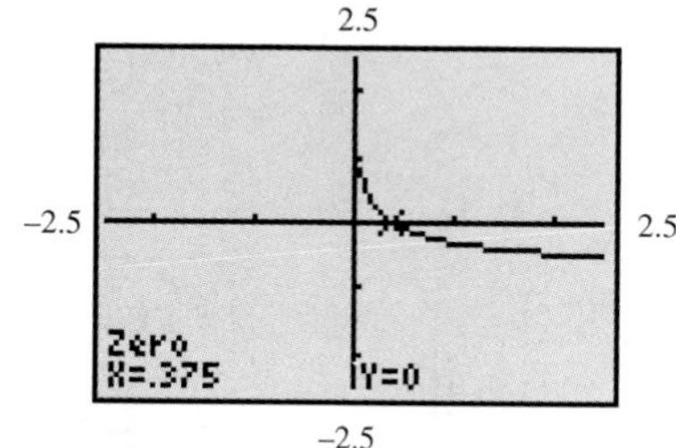

50. Explain the error in the following "solution" of the equation $\log x^2 = 2$.

$\log x^2 = 2$	Original equation
$2 \log x = 2$	Power rule for logarithms
$\log x = 1$	Divide both sides by 2.
$x = 10^1$	Write in exponential form.
$x = 10$	$10^1 = 10$

Solution set: {10}

Solve each problem. Use a calculator as necessary.

51. If \$20,000 is deposited at 7% annual interest compounded quarterly, how much will be in the account after 5 years, assuming no withdrawals are made?

52. How much will \$10,000 compounded continuously at 6% annual interest amount to in three years?

53. Which is a better plan?

Plan A: Invest \$1000 at 4% compounded quarterly for 3 years

Plan B: Invest \$1000 at 3.9% compounded monthly for 3 years

54. What is the half-life of the radioactive substance described in Exercise 40?

55. Based on selected figures from 1970–1995, the fractional part of the generation of municipal solid waste recovered can be approximated by the function with

$$R(x) = .0597e^{.0553x},$$

where $x = 0$ corresponds to 1970, $x = 10$ to 1980, and so on. Based on this model, what *percent* of municipal solid waste was recovered in 1990? (*Source:* Franklin Associates, Ltd., Prairie Village, KS, *Characterization of Municipal Solid Waste in the United States: 1995.*)

56. Recall from Example 6 in Section 9.5 that the number of years, $N(r)$, since two independently evolving languages split off from a common ancestral language is approximated by

$$N(r) = -5000 \ln r,$$

where r is the percent of words from the ancestral language common to both languages now. Find r if the split occurred 2000 years ago.

A machine purchased for business use depreciates, *or loses value, over a period of years. The value of the machine at the end of its useful life is called its scrap value. By one method of depreciation (where it is assumed a constant percentage of the value depreciates annually), the scrap value, S, is given by*

$$S = C(1 - r)^n,$$

where C is the original cost, n is the useful life in years, and r is the constant percent of depreciation.

57. Find the scrap value of a machine costing \$30,000, having a useful life of 12 years and a constant annual rate of depreciation of 15%.

58. A machine has a "half-life" of 6 years. Find the constant annual rate of depreciation.

RELATING CONCEPTS (EXERCISES 59–70)

Work Exercises 59–70 in order, *to see some of the relationships between exponential and logarithmic properties and functions.*

59. Complete the table, and graph the function $f(x) = 2^x$.

x	$f(x)$
-2	
-1	
0	
1	
2	
3	

60. Complete the table, and graph the function $g(x) = \log_2 x$.

x	$g(x)$
$\frac{1}{4}$	
$\frac{1}{2}$	
1	
2	
4	
8	

61. What do you notice about the ordered pairs found in Exercises 59 and 60? What do we call the functions f and g in relationship to each other?

62. Fill in the blanks with the word *vertical* or *horizontal:* The graph of f in Exercise 59 has a __________ asymptote, while the graph of g in Exercise 60 has a __________ asymptote.

63. Using properties of exponents, $2^2 \cdot 2^3 = 2^?$, because $\underline{\ ?\ } + \underline{\ ?\ } = \underline{\ ?\ }$.

64. It is a fact that $32 = 4 \cdot 8$. Therefore, using properties of logarithms, $\log_2 32 = \log_2$ _____ $+ \log_2$ _____.

65. Use the change-of-base rule to find an approximation for $\log_2 13$. Give as many digits as your calculator displays, and store this approximation in memory.

66. In your own words, explain what $\log_2 13$ means.

67. Simplify without using a calculator: $2^{\log_2 13}$.

68. Use the exponential key of your calculator to raise 2 to the power obtained in Exercise 65. What is the result? Why is this so?

69. Based on your result in Exercise 65, the point (13, _____) lies on the graph of $g(x) = \log_2 x$.

70. Use the method of Section 9.2 to solve the equation $2^{x+1} = 8^{2x+3}$.

Did you make the connections between logarithms and exponents?

MIXED REVIEW EXERCISES

Solve.

71. $\log_3(x + 9) = 4$

72. $\log_2 32 = x$

73. $\log_x \dfrac{1}{81} = 2$

74. $27^x = 81$

75. $2^{2x-3} = 8$

76. $\log_3(x + 1) - \log_3 x = 2$

77. $\log(3x - 1) = \log 10$

78. A small business estimates that the value of a copy machine is decreasing according to the function defined by

$$f(t) = 5000(2)^{-.15t},$$

where t is the number of years that have elapsed since the machine was purchased, and $f(t)$ is in dollars.

(a) What was the original value of the machine? (*Hint:* Find $f(0)$.)

(b) What is the value of the machine 5 years after purchase? Give your answer to the nearest dollar.

(c) What is the value of the machine 10 years after purchase? Give your answer to the nearest dollar.

CHAPTER 9 TEST

1. Decide whether each function is one-to-one.

(a) $f(x) = x^2 + 9$ **(b)**

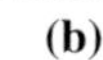

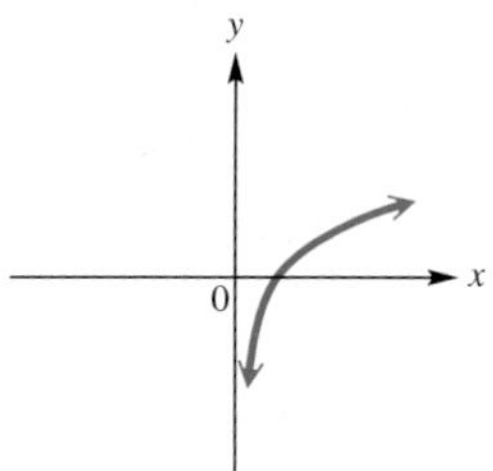

2. Find $f^{-1}(x)$ for the one-to-one function $f(x) = \sqrt[3]{x + 7}$.

3. Graph the inverse of f, given the graph of f at the right.

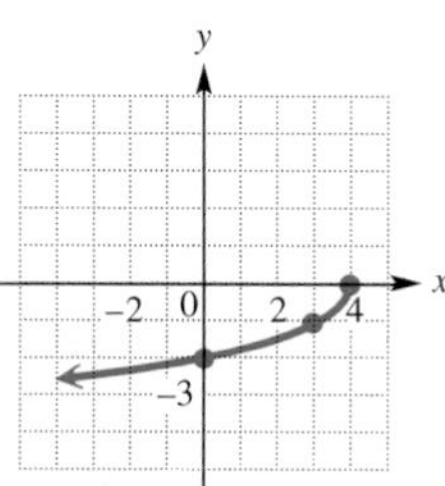

Graph each function.

4. $f(x) = 6^x$

5. $g(x) = \log_6 x$

6. Explain how the graph of the function in Exercise 5 can be obtained from the graph of the function in Exercise 4.

Solve each equation. Give the exact solution.

7. $5^x = \dfrac{1}{625}$

8. $2^{3x-7} = 8^{2x+2}$

9. A *toxic release* is an on-site discharge of a chemical toxic to the environment. Based on figures from 1990–1993, the toxic release inventory, in millions of pounds, can be approximated by the function with

$$R(x) = 2821(.9195)^x,$$

where $x = 0$ corresponds to 1990, $x = 1$ to 1991, and so on. (*Source:* U.S. Environmental Protection Agency, *1994 Toxics Release Inventory,* June 1996.)

(a) Use this model to approximate the toxic release inventory in 1990.

(b) Use this model to approximate the toxic release inventory in 1992.

(c) In 1993, the actual toxic release inventory was 2157.4 million pounds. How does this compare to the number that the model provides?

10. Write in logarithmic form: $4^{-2} = .0625$.

11. Write in exponential form: $\log_7 49 = 2$.

Solve each equation.

12. $\log_{1/2} x = -5$ **13.** $x = \log_9 3$ **14.** $\log_x 16 = 4$

15. Fill in the blanks with the correct responses: The value of $\log_2 32$ is ______. This means that if we raise ______ to the ______ power, the result is ______.

Use properties of logarithms to write each expression as a sum or difference of logarithms. Assume variables represent positive numbers.

16. $\log_3 x^2y$ **17.** $\log_5\left(\frac{\sqrt{x}}{yz}\right)$

Use properties of logarithms to write each expression as a single logarithm. Assume variables represent positive real numbers, $b \neq 1$.

18. $3 \log_b s - \log_b t$ **19.** $\frac{1}{4} \log_b r + 2 \log_b s - \frac{2}{3} \log_b t$

20. Use a calculator to approximate each logarithm to four decimal places.

(a) log 23.1 **(b)** ln .82

21. Use the change-of-base rule to express $\log_3 19$

(a) in terms of common logarithms.

(b) in terms of natural logarithms.

(c) correct to four decimal places.

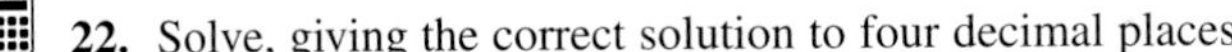

22. Solve, giving the correct solution to four decimal places.

$$3^x = 78$$

23. Solve $\log_8(x + 5) + \log_8(x - 2) = 1$.

24. Suppose that \$10,000 is invested at 4.5% annual interest, compounded quarterly. How much will be in the account in 5 years if no money is withdrawn?

25. Suppose that \$15,000 is invested at 5% annual interest, compounded continuously.

(a) How much will be in the account in 5 years if no money is withdrawn?

(b) How long will it take for the initial principal to double?

CUMULATIVE REVIEW EXERCISES CHAPTERS 1–9

Let $S = \left\{-\frac{9}{4}, -2, -\sqrt{2}, 0, .6, \sqrt{11}, \sqrt{-8}, 6, \frac{30}{3}\right\}$. *List the elements of S that are members of the set.*

1. Integers

2. Rational numbers

3. Irrational numbers

4. Real numbers

Simplify each expression.

5. $|-8| + 6 - |-2| - (-6 + 2)$

6. $-12 - |-3| - 7 - |-5|$

7. $2(-5) + (-8)(4) - (-3)$

Solve each equation or inequality.

8. $7 - (3 + 4a) + 2a = -5(a - 1) - 3$

9. $2m + 2 \leq 5m - 1$

10. $|2x - 5| = 9$

11. $|3p| - 4 = 12$

12. $|3k - 8| \leq 1$

13. $|4m + 2| > 10$

Graph.

14. $5x + 2y = 10$

15. $-4x + y \leq 5$

16. The graph indicates that the long-term debt of the Port of New Orleans has dropped from \$70,000,000 in 1986 to \$25,300,000 in 1994. What is the slope of the line graphed?

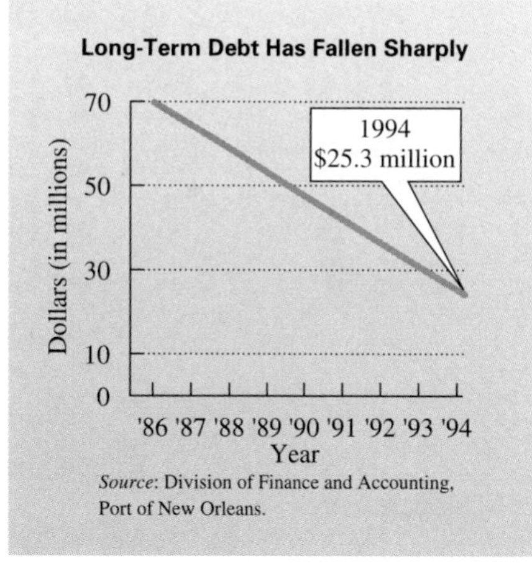

Source: Division of Finance and Accounting, Port of New Orleans.

17. Find the standard form of the equation of the line through $(5, -1)$ and parallel to the line with equation $3x - 4y = 12$.

Solve each system.

18. $5x - 3y = 14$
$2x + 5y = 18$

19. $x + 2y + 3z = 11$
$3x - y + z = 8$
$2x + 2y - 3z = -12$

20. Evaluate the determinant $\begin{vmatrix} -2 & -1 \\ 5 & 3 \end{vmatrix}$.

21. Candy worth \$1.00 per pound is to be mixed with 10 pounds of candy worth \$1.96 per pound to get a mixture that will be sold for \$1.60 per pound. How many pounds of the \$1.00 candy should be used?

Perform the indicated operations.

22. $(2p + 3)(3p - 1)$

23. $(4k - 3)^2$

24. $(3m^3 + 2m^2 - 5m) - (8m^3 + 2m - 4)$

25. Divide $6t^4 + 17t^3 - 4t^2 + 9t + 4$ by $3t + 1$.

Factor.

26. $8x + x^3$

27. $24y^2 - 7y - 6$

28. $5z^3 - 19z^2 - 4z$

29. $16a^2 - 25b^4$

30. $8c^3 + d^3$

31. $16r^2 + 56rq + 49q^2$

Perform the indicated operations.

32. $\dfrac{(5p^3)^4(-3p^7)}{2p^2(4p^4)}$

33. $\dfrac{x^2 - 9}{x^2 + 7x + 12} \div \dfrac{x - 3}{x + 5}$

34. $\dfrac{2}{k + 3} - \dfrac{5}{k - 2}$

35. $\dfrac{3}{p^2 - 4p} - \dfrac{4}{p^2 + 2p}$

Simplify.

36. $\sqrt{288}$

37. $2\sqrt{32} - 5\sqrt{98}$

38. Solve $\sqrt{2x + 1} - \sqrt{x} = 1$.

39. Multiply $(5 + 4i)(5 - 4i)$.

Solve each equation or inequality.

40. $3x^2 - x - 1 = 0$

41. $k^2 + 2k - 8 > 0$

42. $x^4 - 5x^2 + 4 = 0$

43. Find two numbers whose sum is 300 and whose product is a maximum.

44. Graph $f(x) = \dfrac{1}{3}(x - 1)^2 + 2$.

45. Graph $f(x) = 2^x$.

46. Solve $5^{x+3} = \left(\dfrac{1}{25}\right)^{3x+2}$.

47. Graph $f(x) = \log_3 x$.

48. Given that $\log_2 9 = 3.1699$, what is the value of $\log_2 81$?

49. Rewrite the following using the product, quotient, and power rules for logarithms:

$$\log \frac{x^3\sqrt{y}}{z}.$$

50. Let the number of bacteria present in a certain culture be given by

$$B(t) = 25{,}000e^{.2t},$$

where t is time measured in hours, and $t = 0$ corresponds to noon. Find, to the nearest hundred, the number of bacteria present at:

(a) noon; **(b)** 1 P.M.; **(c)** 2 P.M.; **(d)** 5 P.M.

Conic Sections

When a plane intersects an infinite cone at different angles, the figures formed by the intersections are called **conic sections.** See the figure. What is the difference in the angle of the plane that produces an ellipse from that of a circle? We investigated parabolas in Chapter 8. We will study the equations of the other conic sections in this chapter.

Aerospace

Johann Kepler (1571–1630) established the importance of the *ellipse* in 1609 when he discovered that orbits of the planets around the sun were elliptical, not circular. Comets and asteroids also have elliptical orbits around the sun. The orbits of the planets are nearly circular, while Halley's comet, for example, has an elliptical orbit that is long and narrow.

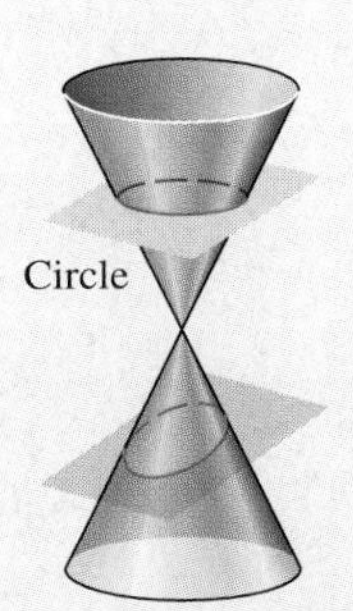

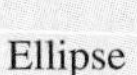

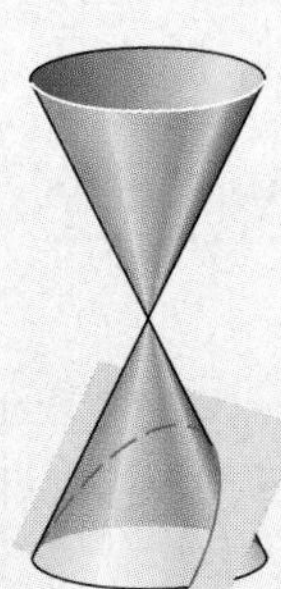

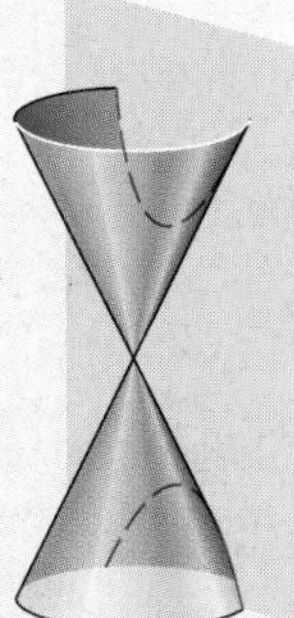

The theme of this chapter, aerospace, is closely related to the theme of Chapter 8, astronomy. Aerospace studies the design and navigation of space aircraft. In this chapter we will also present other applications of conic sections, such as elliptical gears, microwave antenna systems, and a location-finding system that uses the equation of a hyperbola.

Visit our Web site at www.LialAlgebra.com

10.1 The Circle and the Ellipse

OBJECTIVES

1 Find the equation of a circle given the center and radius.

2 Determine the center and radius of a circle given its equation.

3 Recognize the equation of an ellipse.

4 Graph ellipses.

5 Graph circles and ellipses using a graphing calculator.

FOR EXTRA HELP

SSG Sec. 10.1
SSM Sec. 10.1

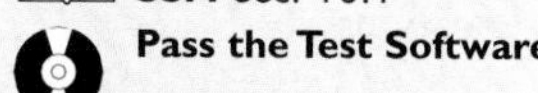

Pass the Test Software

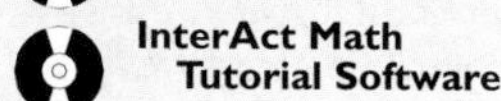

InterAct Math Tutorial Software

Video 19

OBJECTIVE 1 Find the equation of a circle given the center and radius. A **circle** is the set of all points in a plane that lie a fixed distance from a fixed point. The fixed point is called the **center,** and the fixed distance is called the **radius.** We use the distance formula derived earlier to find an equation of a circle.

EXAMPLE 1 Finding the Equation of a Circle and Graphing It

Find an equation of the circle with radius 3 and center at (0, 0), and graph it.

If the point (x, y) is on the circle, the distance from (x, y) to the center (0, 0) is 3. By the distance formula,

$$\sqrt{(x_2 - x_1)^2 + (y_2 - y_1)^2} = d$$
$$\sqrt{(x - 0)^2 + (y - 0)^2} = 3$$
$$x^2 + y^2 = 9. \quad \text{Square both sides.}$$

An equation of this circle is $x^2 + y^2 = 9$. The graph is shown in Figure 1.

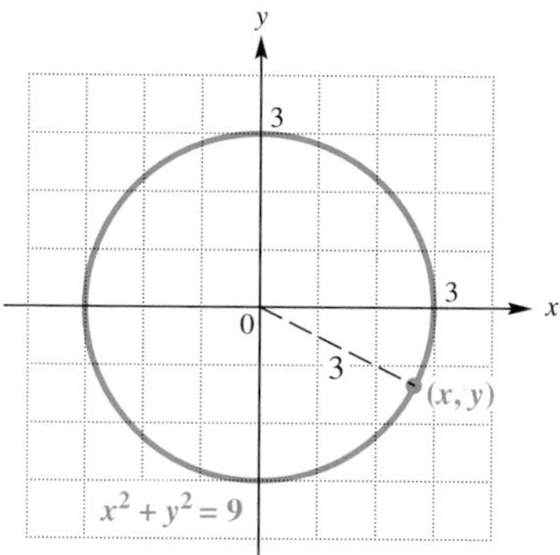

Figure 1

EXAMPLE 2 Finding the Equation of a Circle and Graphing It

Find an equation of the circle with center at $(4, -3)$ and radius 5, and graph it.

Use the distance formula again.

$$\sqrt{(x - 4)^2 + (y + 3)^2} = 5$$
$$(x - 4)^2 + (y + 3)^2 = 25$$

To graph the circle, plot the center $(4, -3)$, then move 5 units right, left, up, and down from the center. Draw a smooth curve through these four points, sketching one quarter of the circle at a time. The graph of this circle is shown in Figure 2.

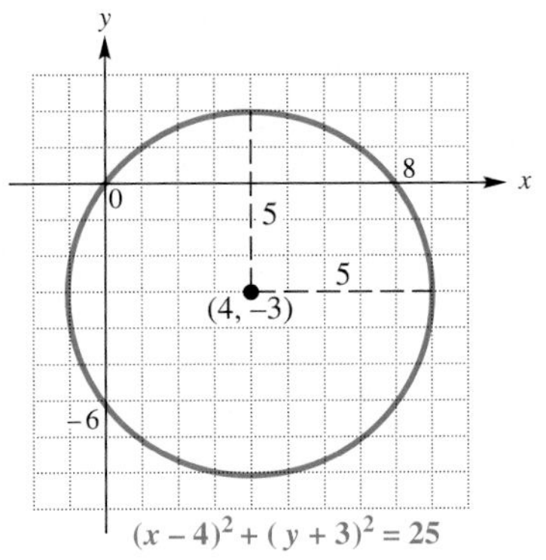

Figure 2

Examples 1 and 2 suggest the form of an equation of a circle with radius r and center at (h, k). If (x, y) is a point on the circle, the distance from the center (h, k) to the point (x, y) is r. Then by the distance formula,

$$\sqrt{(x-h)^2 + (y-k)^2} = r.$$

Squaring both sides gives us the following **center-radius form** of the equation of a circle.

Equation of a Circle (Center-Radius Form)

$$(x-h)^2 + (y-k)^2 = r^2$$

is an equation of a circle of radius r with center at (h, k).

EXAMPLE 3 Using the Center-Radius Form of the Equation of a Circle

Find an equation of the circle with center at $(-1, 2)$ and radius 4.

Use the center-radius form, with $h = -1$, $k = 2$, and $r = 4$.

$$(x-h)^2 + (y-k)^2 = r^2$$
$$[x-(-1)]^2 + (y-2)^2 = 4^2$$
$$(x+1)^2 + (y-2)^2 = 16$$

OBJECTIVE 2 Determine the center and radius of a circle given its equation. In the equation found in Example 2, multiplying out $(x-4)^2$ and $(y+3)^2$ and then combining like terms gives

$$(x-4)^2 + (y+3)^2 = 25$$
$$x^2 - 8x + 16 + y^2 + 6y + 9 = 25$$
$$x^2 + y^2 - 8x + 6y = 0.$$

This general form suggests that an equation with both x^2 and y^2 terms may represent a circle. The next example shows how to tell, by completing the square.

EXAMPLE 4 Completing the Square to Find the Center and Radius

Graph $x^2 + y^2 + 2x + 6y - 15 = 0$.

Since the equation has x^2 and y^2 terms with equal coefficients, its graph might be that of a circle. To find the center and radius, complete the square on x and y.

$$x^2 + y^2 + 2x + 6y = 15$$ Get the constant on the right.

$$(x^2 + 2x \quad) + (y^2 + 6y \quad) = 15$$ Rewrite in anticipation of completing the square.

$$\left[\frac{1}{2}(2)\right]^2 = 1 \qquad \left[\frac{1}{2}(6)\right]^2 = 9$$ Square half the coefficient of each middle term.

$$(x^2 + 2x + 1) + (y^2 + 6y + 9) = 15 + 1 + 9$$ Complete the square on both x and y.

$$(x + 1)^2 + (y + 3)^2 = 25$$ Factor on the left and add on the right.

The last equation shows that the graph is a circle with center at $(-1, -3)$ and radius 5. The graph is shown in Figure 3.

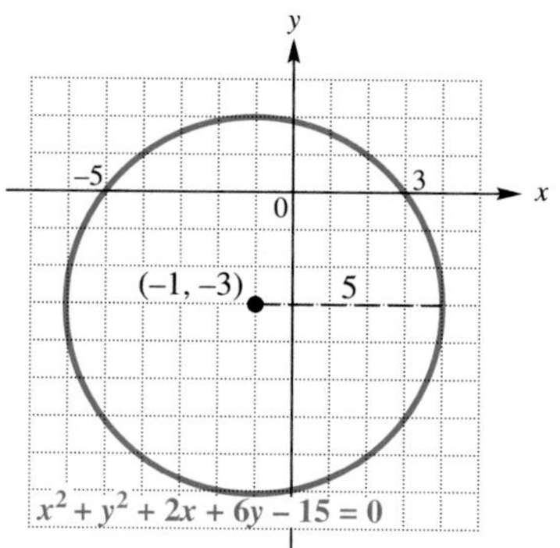

Figure 3

If the procedure of Example 4 leads to an equation of the form $(x - h)^2 + (y - k)^2 = 0$, the graph is the single point (h, k). If the constant on the right side is negative, the equation has no graph.

OBJECTIVE 3 Recognize the equation of an ellipse. An **ellipse** is the set of all points in a plane the sum of whose distances from two fixed points is constant. These fixed points are called **foci** (singular: *focus*). Figure 4 shows an ellipse whose foci are $(c, 0)$ and $(-c, 0)$, with x-intercepts $(a, 0)$ and $(-a, 0)$ and y-intercepts $(0, b)$ and $(0, -b)$. The origin is the **center** of the ellipse.

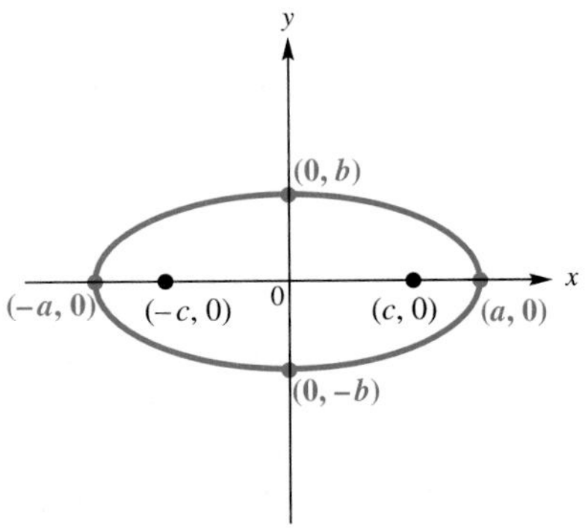

Figure 4

From the definition above, it can be shown by the distance formula that an ellipse has the following equation.

Equation of an Ellipse

The ellipse whose x-intercepts are $(a, 0)$ and $(-a, 0)$ and whose y-intercepts are $(0, b)$ and $(0, -b)$ has an equation of the form

$$\frac{x^2}{a^2} + \frac{y^2}{b^2} = 1.$$

The proof of this is outlined in the exercises. (See Exercise 51.) Note that a circle is a special case of an ellipse, where $a^2 = b^2$.

OBJECTIVE 4 **Graph ellipses.** To graph an ellipse, we plot the four intercepts and then sketch the ellipse through those points.

EXAMPLE 5 **Graphing an Ellipse**

Graph $\frac{x^2}{49} + \frac{y^2}{36} = 1$.

Here, $a^2 = 49$, so $a = \pm 7$, and the x-intercepts for this ellipse are $(7, 0)$ and $(-7, 0)$. Similarly, $b^2 = 36$, so $b = \pm 6$, and the y-intercepts are $(0, 6)$ and $(0, -6)$. Plotting the intercepts and sketching the ellipse through them gives the graph in Figure 5.

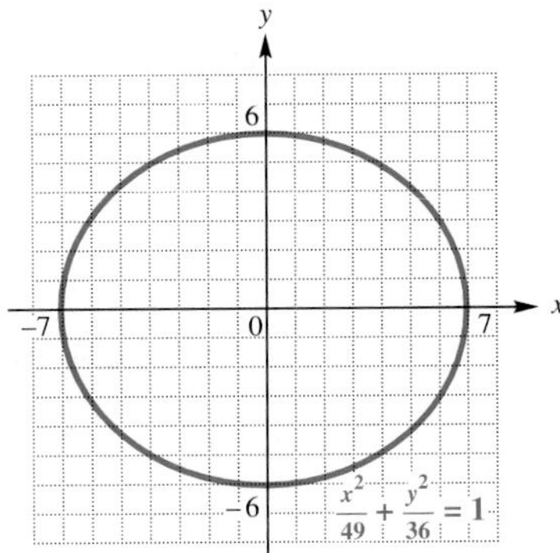

Figure 5

As with the graphs of parabolas and circles, the graph of an ellipse may be shifted horizontally and vertically, as in the next example.

EXAMPLE 6 Graphing an Ellipse Shifted Horizontally and Vertically

Graph $\dfrac{(x-2)^2}{25} + \dfrac{(y+3)^2}{49} = 1$.

Just as $(x-2)^2$ and $(y+3)^2$ would indicate that the center of a circle would be $(2, -3)$, so it is with this ellipse. Figure 6 shows that the graph goes through the four points $(2, 4)$, $(7, -3)$, $(2, -10)$, and $(-3, -3)$. The x-values of these points are found by adding $\pm a = \pm 5$ to 2, and the y-values come from adding $\pm b = \pm 7$ to -3.

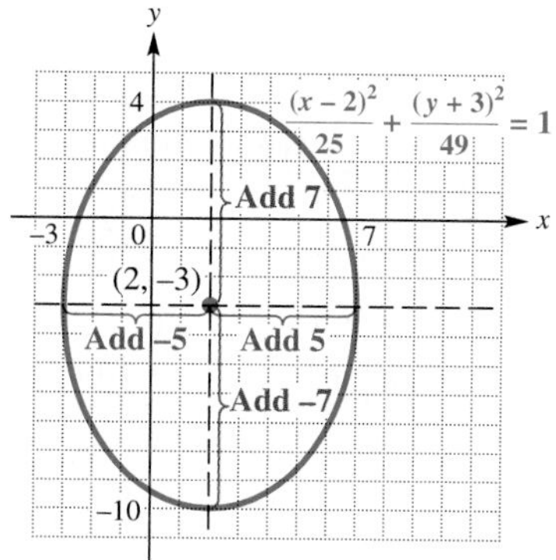

Figure 6

Notice that the graphs in this section are not graphs of functions. The only conic section whose graph is a function is the vertical parabola with equation $f(x) = ax^2 + bx + c$.

OBJECTIVE 5 Graph circles and ellipses using a graphing calculator. A graphing calculator in function mode cannot directly graph a circle or an ellipse. We must first solve the equation for y, getting two functions y_1 and y_2. The union of these two graphs is the graph of the entire figure. To get an undistorted screen, a *square window* must be used. (See your instruction manual for details.) For example, to graph $(x+3)^2 + (y+2)^2 = 25$, begin by solving for y.

$$
\begin{aligned}
(x+3)^2 + (y+2)^2 &= 25 \\
(y+2)^2 &= 25 - (x+3)^2 && \text{Subtract } (x+3)^2. \\
y + 2 &= \pm\sqrt{25 - (x+3)^2} && \text{Take square roots.} \\
y &= -2 \pm \sqrt{25 - (x+3)^2} && \text{Subtract 2.}
\end{aligned}
$$

The two functions to be graphed are

$$y_1 = -2 + \sqrt{25 - (x+3)^2} \quad \text{and} \quad y_2 = -2 - \sqrt{25 - (x+3)^2}.$$

See Figure 7. (*Note:* The two semicircles seem to be disconnected. This is because the graphs are nearly vertical at those points, and the calculator cannot show a true picture of the behavior there.)

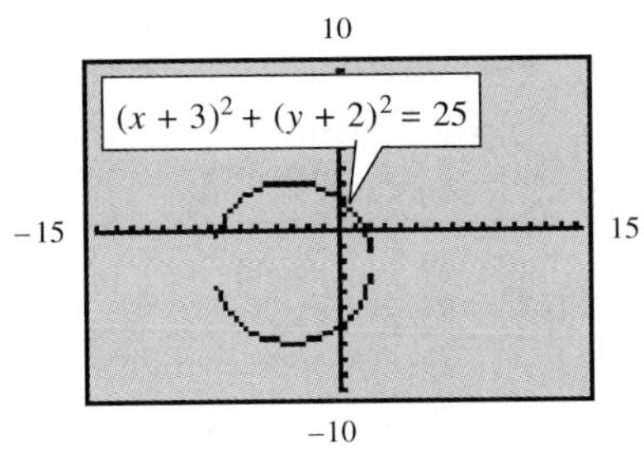

Square viewing window

Figure 7

CONNECTIONS

An interesting and perhaps surprising application of ellipses in our everyday life appears in gears. Elliptical bicycle gears are designed to respond to the legs' natural strengths and weaknesses. At the top and bottom of the powerstroke where the legs have the least leverage, the gear offers little resistance, but as the gear rotates, the resistance increases. This allows the legs to apply more power where it is most naturally available.

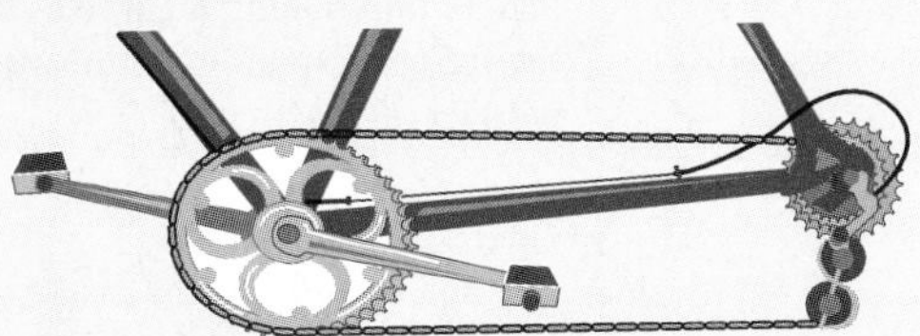

FOR DISCUSSION OR WRITING

A circle can be thought of as an ellipse in which $a = b$ in the equation. Explain how this fact distinguishes the operation of a circular gear from an elliptical gear.

10.1 EXERCISES

1. A circle has equation $x^2 + y^2 = 100$. What is the center of the circle? What is the radius?

2. The equation of an ellipse is $\frac{x^2}{16} + \frac{y^2}{7} = 1$. What is the center of the ellipse? What are the x-intercepts? The y-intercepts?

3. What is the center of the ellipse with equation $\frac{(y + 2)^2}{25} + \frac{(x + 3)^2}{49} = 1$?

4. What is the center of the circle with equation $(x - 4)^2 + (y + 1)^2 = 8$? What is the radius?

Write an equation of each circle with the given center and radius. See Examples 1–3.

5. $(0, 0)$; $r = 6$ **6.** $(0, 0)$; $r = 5$ **7.** $(-1, 3)$; $r = 4$

8. $(2, -2)$; $r = 3$ **9.** $(0, 4)$; $r = \sqrt{3}$ **10.** $(-2, 0)$; $r = \sqrt{5}$

11. Explain why a set of points that form an ellipse does not satisfy the definition of a function.

12. **(a)** How many points are there on the graph of $(x - 4)^2 + (y - 1)^2 = 0$? Explain your answer.

(b) How many points are there on the graph of $(x - 4)^2 + (y - 1)^2 = -1$? Explain your answer.

Find the center and the radius of each circle. See Example 4. (Hint: In Exercises 17 and 18 divide both sides by the greatest common factor.)

13. $x^2 + y^2 + 4x + 6y + 9 = 0$

14. $x^2 + y^2 - 8x - 12y + 3 = 0$

15. $x^2 + y^2 + 10x - 14y - 7 = 0$

16. $x^2 + y^2 - 2x + 4y - 4 = 0$

17. $3x^2 + 3y^2 - 12x - 24y + 12 = 0$

18. $2x^2 + 2y^2 + 20x + 16y + 10 = 0$

Graph each circle. Identify the center if it is not at the origin. See Examples 1, 2, and 4.

19. $x^2 + y^2 = 9$

20. $x^2 + y^2 = 4$

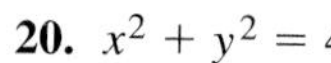

21. $2y^2 = 10 - 2x^2$

22. $3x^2 = 48 - 3y^2$

23. $(x + 3)^2 + (y - 2)^2 = 9$

24. $(x - 1)^2 + (y + 3)^2 = 16$

25. $x^2 + y^2 - 4x - 6y + 9 = 0$

26. $x^2 + y^2 + 8x + 2y - 8 = 0$

27. A circle can be drawn on a piece of posterboard by fastening one end of a string, pulling the string taut with a pencil, and tracing a curve as shown in the figure. Explain why this method works.

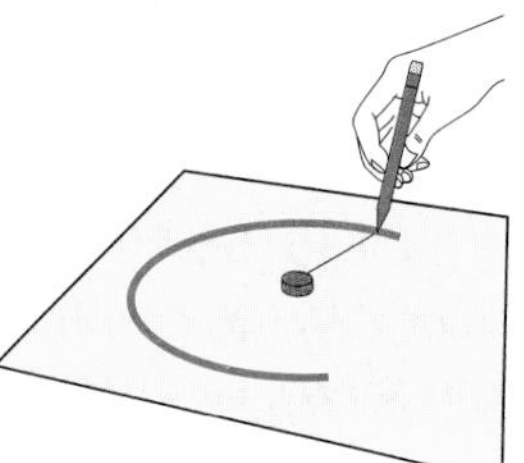

28. It is possible to sketch an ellipse on a piece of posterboard by fastening two ends of a length of string, pulling the string taut with a pencil, and tracing a curve, as shown in the drawing. Explain why this method works.

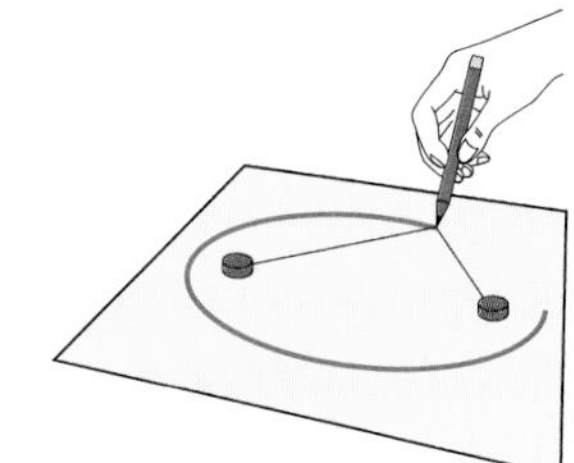

29. This figure shows the crawfish race held at the Crawfish Festival in Breaux Bridge, Louisiana. Explain why a circular "racetrack" is appropriate for such a race.

30. Discuss the similarities and differences between the equations of a circle and an ellipse.

Graph each ellipse. See Examples 5 and 6.

31. $\dfrac{x^2}{9} + \dfrac{y^2}{25} = 1$

32. $\dfrac{x^2}{9} + \dfrac{y^2}{16} = 1$

33. $\dfrac{x^2}{36} = 1 - \dfrac{y^2}{16}$

34. $\dfrac{x^2}{9} = 1 - \dfrac{y^2}{4}$

35. $\dfrac{y^2}{25} = 1 - \dfrac{x^2}{49}$

36. $\dfrac{y^2}{9} = 1 - \dfrac{x^2}{16}$

37. $\dfrac{(x+1)^2}{64} + \dfrac{(y-2)^2}{49} = 1$

38. $\dfrac{(x-4)^2}{9} + \dfrac{(y+2)^2}{4} = 1$

39. $\dfrac{(x-2)^2}{16} + \dfrac{(y-1)^2}{9} = 1$

40. $\dfrac{(x+3)^2}{25} + \dfrac{(y+2)^2}{36} = 1$

TECHNOLOGY INSIGHTS (EXERCISES 41–42)

41. The circle shown in the calculator-generated graph was created using function mode with a square viewing window. It is the graph of $(x+2)^2 + (y-4)^2 = 16$. What are the two functions y_1 and y_2 that were used to obtain this graph?

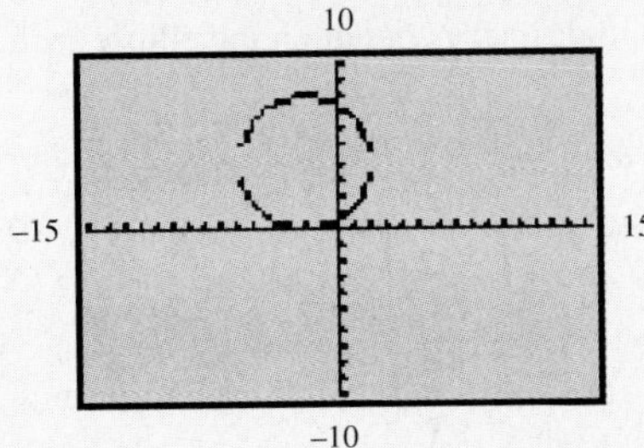

42. The ellipse shown in the calculator-generated graph was graphed using a graphing calculator in function mode, with a square viewing window. It is the graph of $\dfrac{x^2}{4} + \dfrac{y^2}{9} = 1$. What are the two functions y_1 and y_2 that were used to obtain this graph?

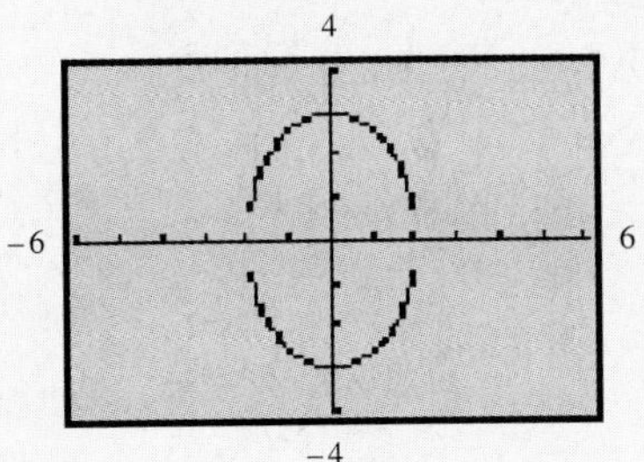

Use a graphing calculator in function mode to graph each circle or ellipse. Use a square viewing window. See Objective 5.

43. $x^2 + y^2 = 36$

44. $(x-2)^2 + y^2 = 49$

45. $\dfrac{x^2}{16} + \dfrac{y^2}{4} = 1$

46. $\dfrac{(x-3)^2}{25} + \dfrac{y^2}{9} = 1$

In Exercises 47–50, see Figure 4 and use the fact that $c^2 = a^2 - b^2$. (*Source for Exercises 47 and 48:* James B. Kaler, *Astronomy!*, Addison Wesley, 1997.)

47. The orbit of Mars is an ellipse with the sun at one focus. For x and y in millions of miles, the equation of the orbit is

$$\frac{x^2}{141.7^2} + \frac{y^2}{141.1^2} = 1.$$

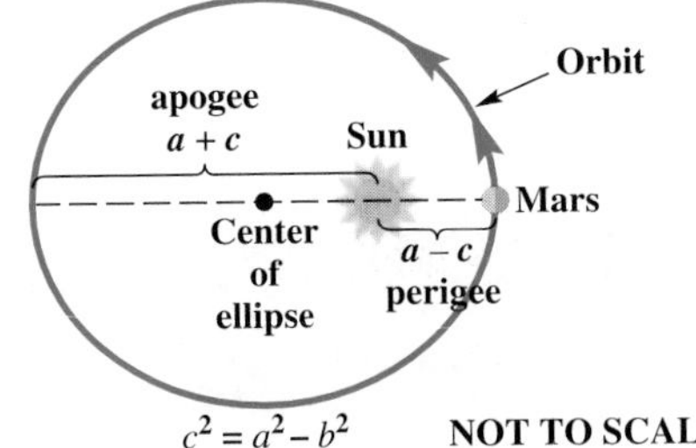

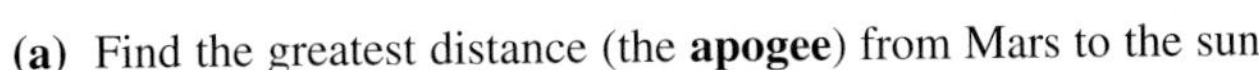

(a) Find the greatest distance (the **apogee**) from Mars to the sun.
(b) Find the smallest distance (the **perigee**) from Mars to the sun.

48. The orbit of Venus around the sun (one of the foci) is an ellipse with equation

$$\frac{x^2}{5013} + \frac{y^2}{4970} = 1,$$

where x and y are measured in millions of miles.

(a) Find the greatest distance between Venus and the sun.
(b) Find the smallest distance between Venus and the sun.

A *lithotripter* is a machine used to crush kidney stones using shock waves. The patient is placed in an elliptical tub with the kidney stone at one focus of the ellipse. A beam is projected from the other focus to the tub, so that it reflects to hit the kidney stone.

49. Suppose a lithotripter is based on the ellipse with equation

$$\frac{x^2}{36} + \frac{y^2}{9} = 1.$$

How far from the center of the ellipse must the kidney stone and the source of the beam be placed?

50. Rework Exercise 49 if the equation of the ellipse is $9x^2 + 4y^2 = 36$. (*Hint:* Write the equation in fraction form by dividing each term by 36.)

51. **(a)** Suppose that $(c, 0)$ and $(-c, 0)$ are the foci of an ellipse and that the sum of the distances from any point (x, y) on the ellipse to the two foci is $2a$. See the figure. Show that the equation of the resulting ellipse is

$$\frac{x^2}{a^2} + \frac{y^2}{a^2 - c^2} = 1.$$

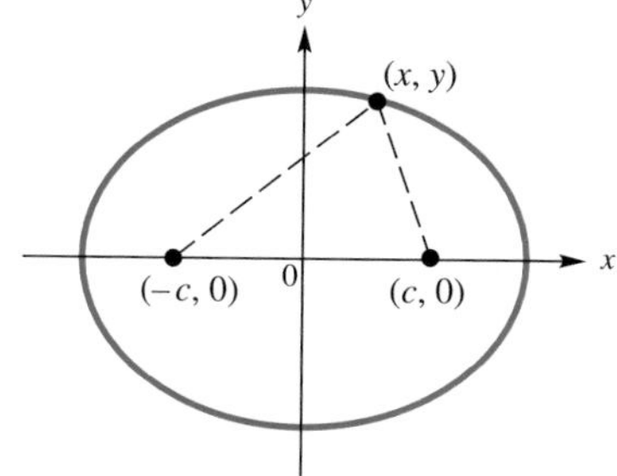

(b) Show that in the equation in part (a), the x-intercepts are $(a, 0)$ and $(-a, 0)$.

(c) Let $b^2 = a^2 - c^2$, and show that $(0, b)$ and $(0, -b)$ are the y-intercepts in the equation in part (a).

52. Use the result of Exercise 51(a) to find an equation of an ellipse with foci $(3, 0)$ and $(-3, 0)$, where the sum of the distances from any point of the ellipse to the two foci is 10.

10.2 The Hyperbola; More on Square Root Functions

OBJECTIVES

1. Recognize the equation of a hyperbola.
2. Graph hyperbolas by using asymptotes.

OBJECTIVE 1 **Recognize the equation of a hyperbola.** A **hyperbola** is the set of all points in a plane such that the absolute value of the *difference* of the distances from two fixed points (called *foci*) is constant. Figure 8 shows a hyperbola; using the distance formula and the definition above, we can show that this hyperbola has equation

$$\frac{x^2}{16} - \frac{y^2}{12} = 1.$$

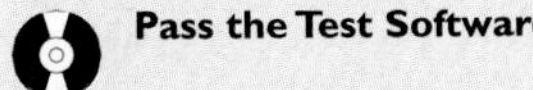

3 Identify conic sections by their equations.

4 Graph certain square root functions.

FOR EXTRA HELP

SSG Sec. 10.2
SSM Sec. 10.2

Pass the Test Software

InterAct Math Tutorial Software

Video 19

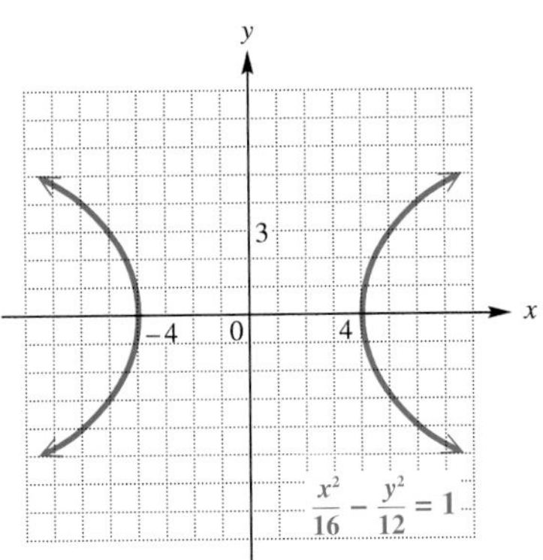

Figure 8

To graph this hyperbola, find the x-intercepts.

$$\frac{x^2}{16} - \frac{0^2}{12} = 1 \qquad \text{Let } y = 0.$$

$$\frac{x^2}{16} = 1$$

$$x^2 = 16 \qquad \text{Multiply by 16.}$$

$$x = \pm 4$$

The x-intercepts are (4, 0) and $(-4, 0)$. Now, find any y-intercepts.

$$\frac{0^2}{16} - \frac{y^2}{12} = 1 \qquad \text{Let } x = 0.$$

$$\frac{-y^2}{12} = 1$$

$$y^2 = -12 \qquad \text{Multiply by } -12.$$

Because there are no *real* solutions to $y^2 = -12$, the graph has no y-intercepts.

Figure 9 gives the graph of

$$\frac{y^2}{25} - \frac{x^2}{9} = 1.$$

Here the y-intercepts are (0, 5) and $(0, -5)$, and there are no x-intercepts.

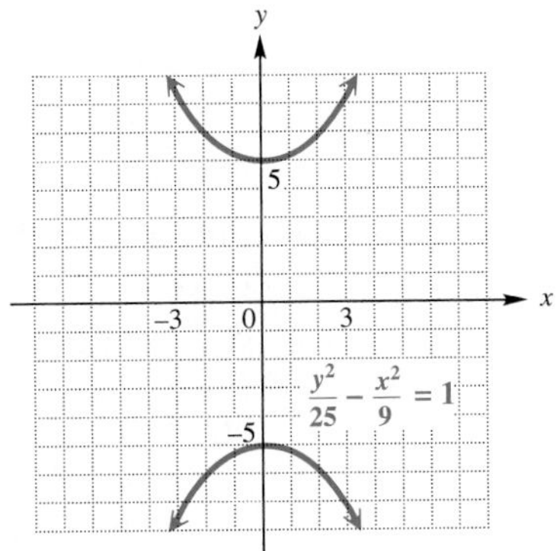

Figure 9

These examples suggest the following summary.

Equations of Hyperbolas

A hyperbola with x-intercepts $(a, 0)$ and $(-a, 0)$ has an equation of the form

$$\frac{x^2}{a^2} - \frac{y^2}{b^2} = 1,$$

and a hyperbola with y-intercepts $(0, b)$ and $(0, -b)$ has an equation of the form

$$\frac{y^2}{b^2} - \frac{x^2}{a^2} = 1.$$

OBJECTIVE 2 **Graph hyperbolas by using asymptotes.** The two branches of the graph of a hyperbola approach a pair of intersecting straight lines called **asymptotes.** See Figure 10. These lines are useful for sketching the graph of the hyperbola.

Asymptotes of Hyperbolas

The extended diagonals of the rectangle with corners at the points (a, b), $(-a, b)$, $(-a, -b)$, and $(a, -b)$ are the *asymptotes* of either of the hyperbolas

$$\frac{x^2}{a^2} - \frac{y^2}{b^2} = 1 \quad \text{or} \quad \frac{y^2}{b^2} - \frac{x^2}{a^2} = 1.$$

This rectangle is called the **fundamental rectangle.** Using the methods of Chapter 3 we could show that the equations of these asymptotes are

$$y = \frac{b}{a}x \quad \text{and} \quad y = -\frac{b}{a}x.$$

To graph either of the two forms of hyperbolas, $\frac{x^2}{a^2} - \frac{y^2}{b^2} = 1$ or $\frac{y^2}{b^2} - \frac{x^2}{a^2} = 1$, follow these steps.

Graphing a Hyperbola

Step 1 **Find the intercepts.** Locate the intercepts at $(a, 0)$ and $(-a, 0)$ if the x^2 term has a positive coefficient, or at $(0, b)$ and $(0, -b)$ if the y^2 term has a positive coefficient.

Step 2 **Find the fundamental rectangle.** Locate the corners of the fundamental rectangle at (a, b), $(-a, b)$, $(-a, -b)$, and $(a, -b)$.

Step 3 **Sketch the asymptotes.** The extended diagonals of the rectangle are the asymptotes of the hyperbola, and they have equations $y = \pm\frac{b}{a}x$.

Step 4 **Draw the graph.** Sketch each branch of the hyperbola through an intercept and approaching (but not touching) the asymptotes.

EXAMPLE 1 Graphing a Horizontal Hyperbola

Graph $\frac{x^2}{16} - \frac{y^2}{25} = 1$.

Step 1 Here $a = 4$ and $b = 5$. The x-intercepts are $(4, 0)$ and $(-4, 0)$.

Step 2 The four points $(4, 5)$, $(-4, 5)$, $(-4, -5)$, and $(4, -5)$ are the corners of the rectangle that determines the asymptotes, as shown in Figure 10.

Steps 3 and 4 The equations of the asymptotes are $y = \pm\frac{5}{4}x$, and the hyperbola approaches these lines as x and y get larger and larger in absolute value.

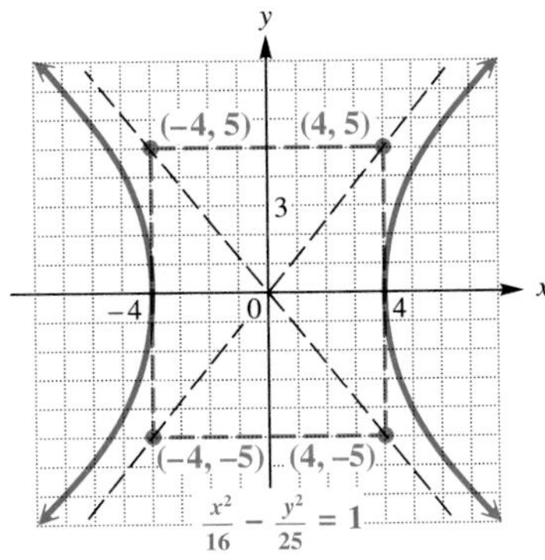

Figure 10

EXAMPLE 2 Graphing a Vertical Hyperbola

Graph $\frac{y^2}{49} - \frac{x^2}{16} = 1$.

This hyperbola has y-intercepts $(0, 7)$ and $(0, -7)$. The asymptotes are the extended diagonals of the rectangle with corners at $(4, 7)$, $(-4, 7)$, $(-4, -7)$, and $(4, -7)$. Their equations are $y = \pm\frac{7}{4}x$. See Figure 11.

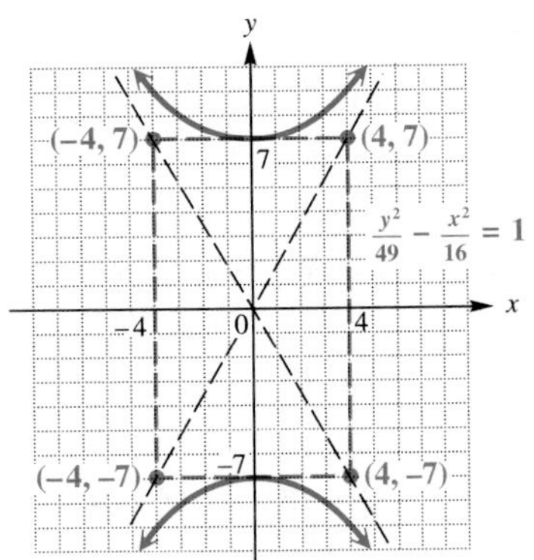

Figure 11

CAUTION When sketching the graph of a hyperbola, be sure that the branches do not touch the asymptotes.

Hyperbolas are graphed with a graphing calculator in much the same way as circles and ellipses, by first writing the equations of two root functions that combined are equivalent to the equation of the hyperbola. A square window gives a truer shape for hyperbolas, too.

CONNECTIONS

A hyperbola and a parabola are used together in one kind of microwave antenna system. The cross sections of the system consist of a parabola and a hyperbola with the focus of the parabola coinciding with one focus of the hyperbola.

The incoming microwaves that are parallel to the axis of the parabola are reflected from the parabola up toward the hyperbola and back to the other focus of the hyperbola, where the cone of the antenna is located to capture the signal.

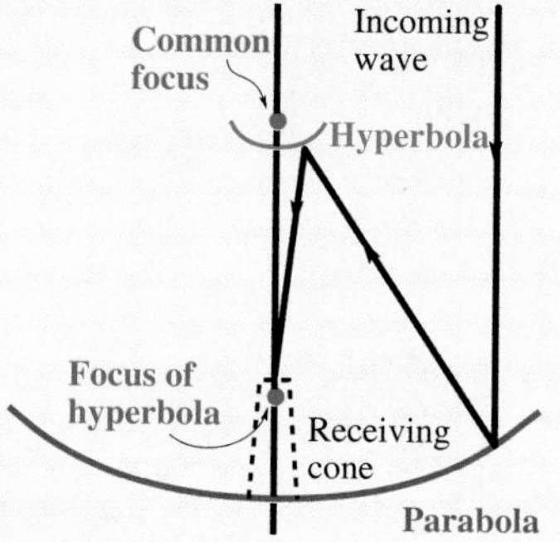

FOR DISCUSSION OR WRITING

The property of the parabola and the hyperbola that is used here is a "reflection property" of the foci. Explain why this name is appropriate.

OBJECTIVE 3 **Identify conic sections by their equations.** Rewriting a second-degree equation in one of the forms given for ellipses, hyperbolas, circles, or parabolas makes it possible to determine when the graph is one of these figures. A summary of the equations and graphs of the conic sections follows.

Equation	*Graph*	*Description*	*Identification*
$y = a(x - h)^2 + k$	$a > 0$, (h, k) Parabola	It opens upward if $a > 0$, downward if $a < 0$. The vertex is at (h, k).	x^2 term y is not squared.
$x = a(y - k)^2 + h$	$a > 0$, (h, k) Parabola	It opens to the right if $a > 0$, to the left if $a < 0$. The vertex is at (h, k).	y^2 term x is not squared.

Equation	Graph	Description	Identification
$(x - h)^2 + (y - k)^2 = r^2$	Circle	The center is at (h, k), and the radius is r.	x^2 and y^2 terms have the same positive coefficient.
$\frac{x^2}{a^2} + \frac{y^2}{b^2} = 1$	Ellipse	The x-intercepts are $(a, 0)$ and $(-a, 0)$. The y-intercepts are $(0, b)$ and $(0, -b)$.	x^2 and y^2 terms have different positive coefficients.
$\frac{x^2}{a^2} - \frac{y^2}{b^2} = 1$	Hyperbola	The x-intercepts are $(a, 0)$ and $(-a, 0)$. The asymptotes are found from (a, b), $(a, -b)$, $(-a, -b)$, and $(-a, b)$.	x^2 has a positive coefficient. y^2 has a negative coefficient.
$\frac{y^2}{b^2} - \frac{x^2}{a^2} = 1$	Hyperbola	The y-intercepts are $(0, b)$ and $(0, -b)$. The asymptotes are found from (a, b), $(a, -b)$, $(-a, -b)$, and $(-a, b)$.	y^2 has a positive coefficient. x^2 has a negative coefficient.

EXAMPLE 3 Identifying the Graph of a Given Equation

Identify the graph of each equation.

(a) $9x^2 = 108 + 12y^2$

Both variables are squared, so the graph is either an ellipse or a hyperbola. (This situation also occurs for a circle, which may be considered a special case of the ellipse.) To see which one it is, rewrite the equation so that the x and y terms are on one side of the equation and 1 is on the other.

$$9x^2 = 108 + 12y^2$$

$$9x^2 - 12y^2 = 108 \quad \text{Subtract } 12y^2.$$

$$\frac{x^2}{12} - \frac{y^2}{9} = 1 \quad \text{Divide by 108.}$$

Because of the minus sign, the graph of this equation is a hyperbola.

(b) $x^2 = y - 3$

Only one of the two variables, x, is squared, so this is the vertical parabola $y = x^2 + 3$.

(c) $x^2 = 9 - y^2$

Get the variable terms on the same side of the equation.

$$x^2 = 9 - y^2$$

$$x^2 + y^2 = 9 \qquad \text{Add } y^2.$$

This equation represents a circle with center at the origin and radius 3.

OBJECTIVE 4 Graph certain square root functions. Recall that no vertical line will intersect the graph of a function in more than one point. Thus, horizontal parabolas and all circles, ellipses, and the hyperbolas discussed in this chapter are examples of graphs that do not satisfy the conditions of a function. However, by considering only a part of the graph of each of these we have the graph of a function, as seen in Figure 12.

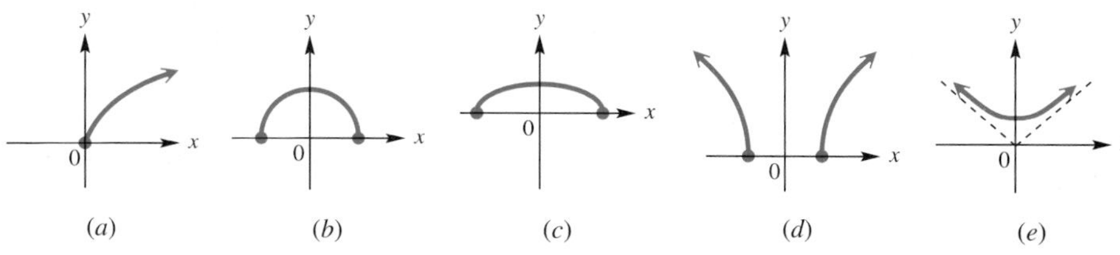

Figure 12

In parts (a), (b), (c), and (e) of Figure 12, the top portion of a conic section is shown (parabola, circle, ellipse, and hyperbola, respectively). In part (d), the top two portions of a hyperbola are shown. In each case, the graph is the graph of a function, since the graph satisfies the conditions of the vertical line test.

In Section 7.5, a square root function was defined as $f(x) = \sqrt{x}$. To get equations for the graphs shown in Figure 12, we extend that definition.

Square Root Function

A function of the form

$$f(x) = \sqrt{u}$$

for an algebraic expression u, with $u \geq 0$, is called a **square root function.**

EXAMPLE 4 Graphing a Square Root Function

Graph $f(x) = \sqrt{25 - x^2}$.

Replace $f(x)$ with y and square both sides to get the equation

$$y^2 = 25 - x^2, \qquad \text{or} \qquad x^2 + y^2 = 25.$$

This is the equation of a circle with center at (0, 0) and radius 5. Since $f(x)$, or y, represents a principal square root in the original equation, $f(x)$ must be nonnegative. This restricts the graph to the upper half of the circle, as shown in Figure 13. Use the graph and the vertical line test to verify that it is indeed a function.

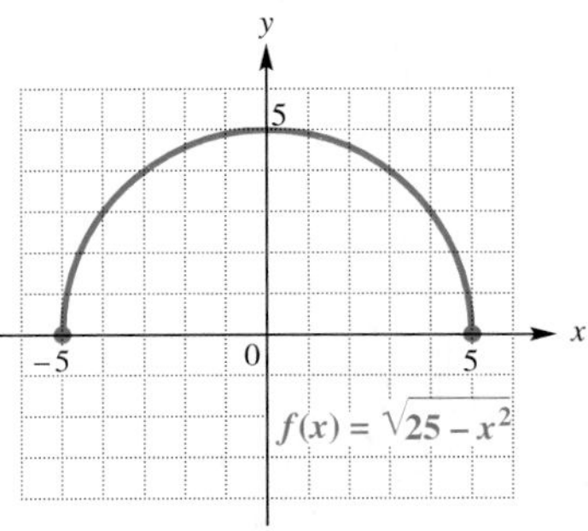

Figure 13

Refer to Figure 13. Had we wanted to graph the function

$$g(x) = -\sqrt{25 - x^2},$$

the graph would be the reflection of the graph of $f(x)$ across the x-axis. This would ensure nonpositive y-values, as indicated by the $-$ sign in the rule for $g(x)$.

Root functions, since they are functions, can be entered and graphed directly with a graphing calculator.

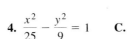

10.2 EXERCISES

Match each equation with the correct graph. See Examples 1 and 2, and Example 5 in Section 10.1.

1. $\dfrac{x^2}{25} + \dfrac{y^2}{9} = 1$

2. $\dfrac{x^2}{9} + \dfrac{y^2}{25} = 1$

3. $\dfrac{x^2}{9} - \dfrac{y^2}{25} = 1$

4. $\dfrac{x^2}{25} - \dfrac{y^2}{9} = 1$

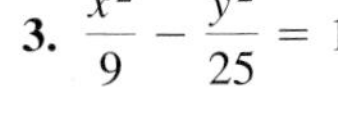

A.

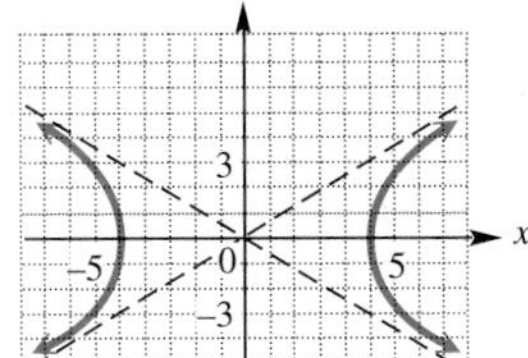

B.

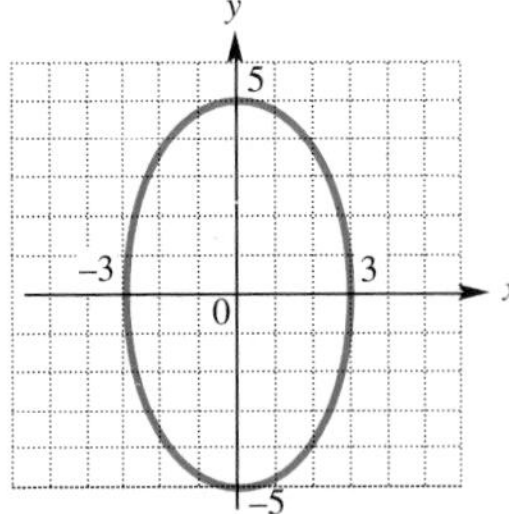

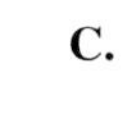

C.

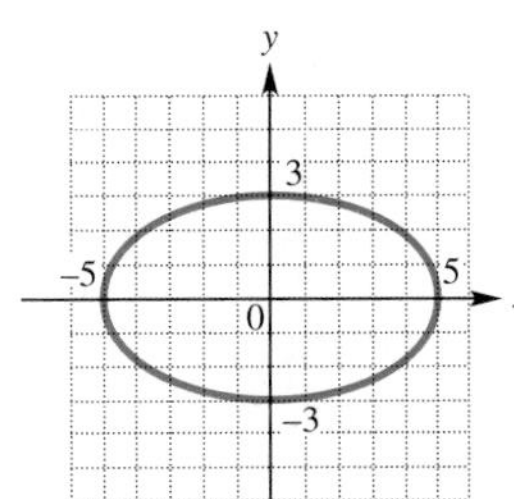

D.

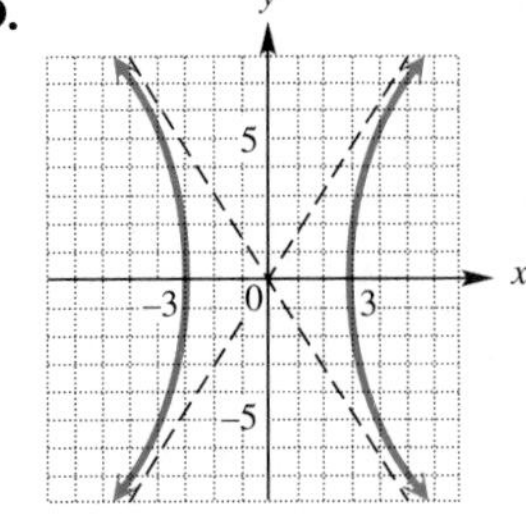

5. Write an explanation of how you can tell from the equation whether the branches of a hyperbola open up and down or open left and right.

6. Explain why the graph of a hyperbola of the type discussed in this section does not satisfy the conditions for the graph of a function.

Graph each hyperbola. See Examples 1 and 2.

7. $\dfrac{x^2}{16} - \dfrac{y^2}{9} = 1$

8. $\dfrac{y^2}{4} - \dfrac{x^2}{25} = 1$

9. $\dfrac{y^2}{9} - \dfrac{x^2}{9} = 1$

10. $\dfrac{x^2}{49} - \dfrac{y^2}{16} = 1$

11. $\dfrac{x^2}{25} - \dfrac{y^2}{36} = 1$

12. $\dfrac{y^2}{9} - \dfrac{x^2}{4} = 1$

Identify the graph of the equation as one of the four conic sections: parabolas, circles, ellipses, or hyperbolas. (It may be necessary to transform the equation into a more recognizable form.) Then sketch the graph of each equation. See Example 3.

13. $x^2 - y^2 = 16$

14. $x^2 + y^2 = 16$

15. $4x^2 + y^2 = 16$

16. $x^2 - 2y = 0$

17. $y^2 = 36 - x^2$

18. $9x^2 + 25y^2 = 225$

19. $9x^2 = 144 + 16y^2$

20. $x^2 + 9y^2 = 9$

21. $y^2 = 4 + x^2$

22. State in your own words the major difference between the definitions of *ellipse* and *hyperbola*.

Graph each square root function. See Example 4.

23. $f(x) = -\sqrt{36 - x^2}$

24. $f(x) = 3\sqrt{1 + \dfrac{x^2}{9}}$

25. $f(x) = \sqrt{\dfrac{x + 4}{2}}$

26. $f(x) = -2\sqrt{\dfrac{9 - x^2}{9}}$

Recall from Section 10.1 that the center of an ellipse may be shifted away from the origin. (See Example 6 in Section 10.1, for instance.) The same shifting process can be applied to hyperbolas. For example, the hyperbola

$$\frac{(x + 5)^2}{4} - \frac{(y - 2)^2}{9} = 1,$$

shown at the right, would have the same graph as $\dfrac{x^2}{4} - \dfrac{y^2}{9} = 1$, *but centered at* $(-5, 2)$. *Graph each hyperbola with center shifted away from the origin.*

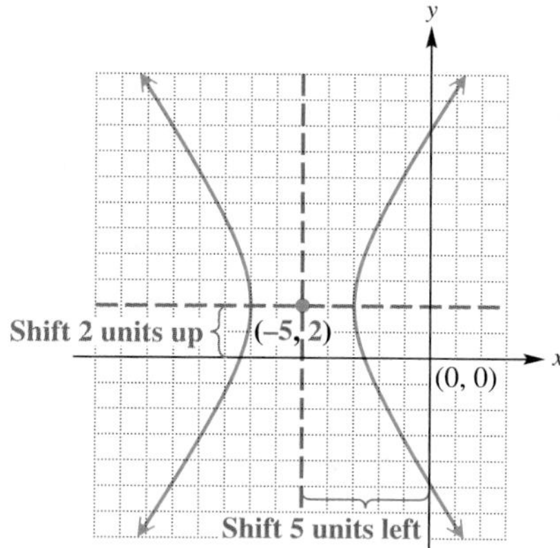

27. $\dfrac{(x - 2)^2}{4} - \dfrac{(y + 1)^2}{9} = 1$

28. $\dfrac{(x + 3)^2}{16} - \dfrac{(y - 2)^2}{25} = 1$

29. $\dfrac{y^2}{36} - \dfrac{(x - 2)^2}{49} = 1$

30. $\dfrac{(y - 5)^2}{9} - \dfrac{x^2}{25} = 1$

31. An arch has the shape of half an ellipse. The equation of the ellipse is $100x^2 + 324y^2 = 32{,}400$, where x and y are in meters.
(a) How high is the center of the arch?
(b) How wide is the arch across the bottom? (See the figure.)

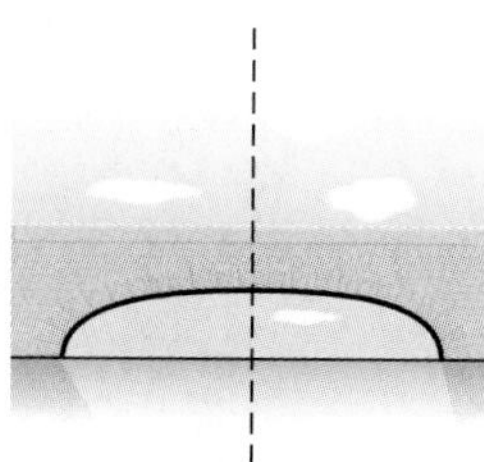

32. Two buildings in a sports complex are shaped and positioned like a portion of the branches of the hyperbola $400x^2 - 625y^2 = 250{,}000$, where x and y are in meters.
   (a) How far apart are the buildings at their closest point?
   (b) Find the distance d in the figure.

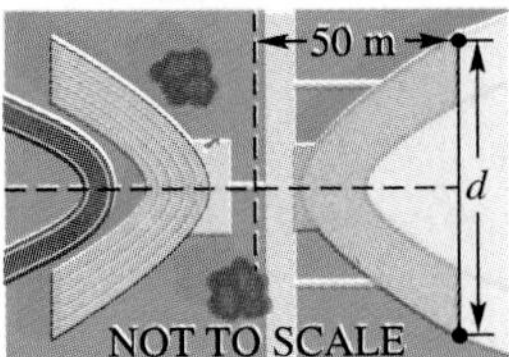

33. In rugby, after a *try* (similar to a touchdown in American football) the scoring team attempts a kick for extra points. The ball must be kicked from directly behind the point where the try was scored. The kicker can choose the distance but cannot move the ball sideways. It can be shown that the kicker's best choice is on the hyperbola with equation

$$\frac{x^2}{g^2} - \frac{y^2}{g^2} = 1,$$

where $2g$ is the distance between the goal posts. Since the hyperbola approaches its asymptotes, it is easier for the kicker to estimate points on the asymptotes instead of on the hyperbola. What are the asymptotes of this hyperbola? Why is it relatively easy to estimate them? (*Source:* Daniel C. Isaksen, "How to Kick a Field Goal," *The College Mathematics Journal,* vol. 27, no. 4, September 1996.)

34. When a satellite is launched into orbit, the shape of its trajectory is determined by its velocity. The trajectory will be hyperbolic if the velocity V, in meters per second, satisfies the inequality

$$V > \frac{2.82 \times 10^7}{\sqrt{D}},$$

where D is the distance, in meters, from the center of Earth. For what values of V will the trajectory be hyperbolic if $D = 42.5 \times 10^6$ meters? (*Source:* James B. Kaler, *Astronomy!,* Addison Wesley, 1997.)

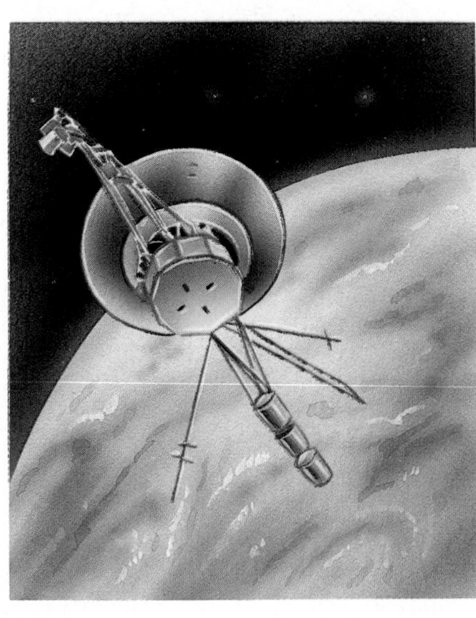

35. The percent of women in the work force has increased steadily for many years. The line graph shows the change for the period from 1975 to 1997, where $x = 75$ represents 1975, $x = 80$ represents 1980, and so on.

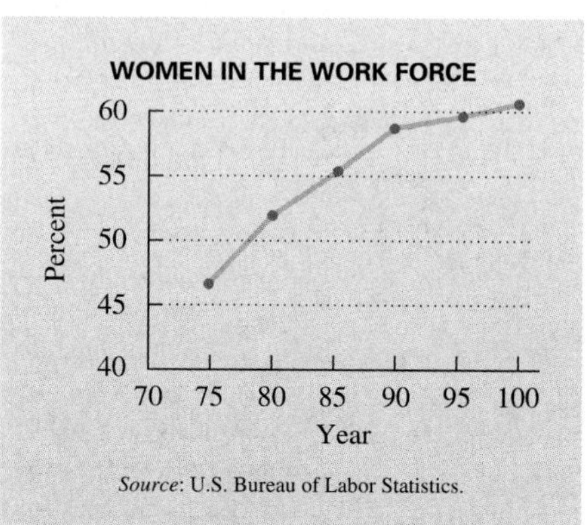

The graph resembles the upper branch of a horizontal hyperbola. Using statistical methods, we found the corresponding square root equation $y = .607\sqrt{383.9 + x^2}$, which closely approximates the line graph.

(a) According to the graph, what percent of women were in the work force in 1985?

(b) According to the equation, what percent of women worked in 1985? (Round to the nearest percent.)

36. Refer to Exercise 35. Use the equation to predict the year when the percent of women in the work force will be 62 percent. When will it be 65 percent?

TECHNOLOGY INSIGHTS (EXERCISES 37–38)

37. The hyperbola shown in the calculator-generated graph in the figure was graphed in function mode with a square viewing window. It is the graph of $\frac{x^2}{9} - y^2 = 1$. What are the two functions y_1 and y_2 that were used to obtain this graph?

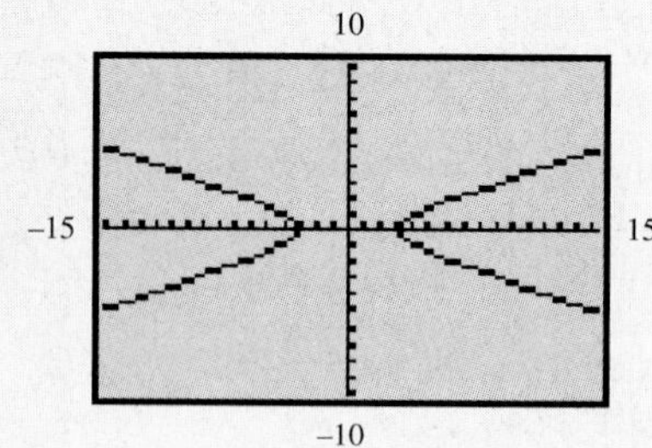

38. Repeat Exercise 37 for the graph of $\frac{y^2}{9} - x^2 = 1$, shown in the accompanying figure.

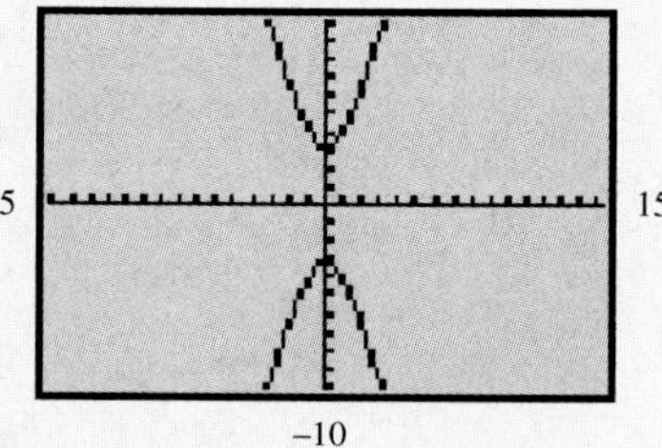

Use a graphing calculator in function mode to graph each hyperbola. Use a square viewing window.

39. $\frac{x^2}{25} - \frac{y^2}{49} = 1$

40. $\frac{x^2}{4} - \frac{y^2}{16} = 1$

41. $\frac{y^2}{9} - x^2 = 1$

42. $\frac{y^2}{36} - \frac{x^2}{4} = 1$

RELATING CONCEPTS (EXERCISES 43–48)

From the discussion in this section, we know that the graph of $\frac{x^2}{4} - y^2 = 1$ is a hyperbola. We know that the graph of this hyperbola approaches its asymptotes as x gets larger and larger.

Work Exercises 43–48 in order, *to see the relationship between the hyperbola and one of its asymptotes.*

43. Solve $\frac{x^2}{4} - y^2 = 1$ for y, and choose the positive square root.

44. Find the equation of the asymptote with positive slope.

45. Use a calculator to evaluate the y-coordinate of the point where $x = 50$ on the graph of the portion of the hyperbola represented by the equation obtained in Exercise 43. Round your answer to the nearest hundredth.

46. Find the y-coordinate of the point where $x = 50$ on the graph of the asymptote found in Exercise 44.

47. Compare your results in Exercises 45 and 46. How do they support the following statement? When $x = 50$, the graph of the function defined by the equation found in Exercise 43 lies *below* the graph of the asymptote found in Exercise 44.

48. What do you think will happen if we choose x-values larger than 50?

Did you make the connection between the graph of a hyperbola and the graphs of its asymptotes?

49. Suppose that a hyperbola has center at the origin, foci at $(-c, 0)$ and $(c, 0)$, and the absolute value of the difference between the distances from any point (x, y) of the hyperbola to the two foci is $2a$. See the figure. Let $b^2 = c^2 - a^2$, and show that an equation of the hyperbola is

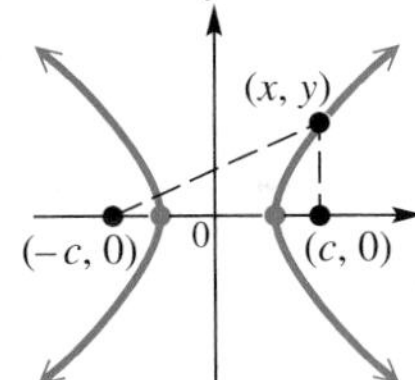

$$\frac{x^2}{a^2} - \frac{y^2}{b^2} = 1.$$

50. Use the result of Exercise 49 to find an equation of a hyperbola with center at the origin, foci at $(-2, 0)$ and $(2, 0)$, and the absolute value of the difference between the distances from any point of the hyperbola to the two foci equal to 2.

10.3 Nonlinear Systems of Equations

OBJECTIVES

1 Solve a nonlinear system by substitution.

2 Use the elimination method to solve a system with two second-degree equations.

An equation in which some terms have more than one variable or a variable of degree two or higher is called a **nonlinear equation.** A **nonlinear system of equations** includes at least one nonlinear equation.

When solving nonlinear systems, it helps to visualize the types of graphs of the equations of the system to determine the possible number of points of intersection. For

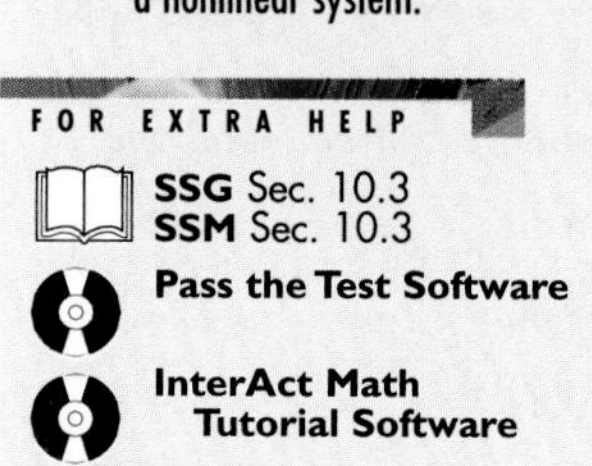

example, if a system includes two equations where the graph of one is a parabola and the graph of the other is a line, then there may be 0, 1, or 2 points of intersection. This is illustrated in Figure 14.

3 Solve a system that requires a combination of methods.
4 Use a graphing calculator to solve a nonlinear system.
FOR EXTRA HELP
SSG Sec. 10.3
SSM Sec. 10.3
Pass the Test Software
InterAct Math Tutorial Software
Video 19

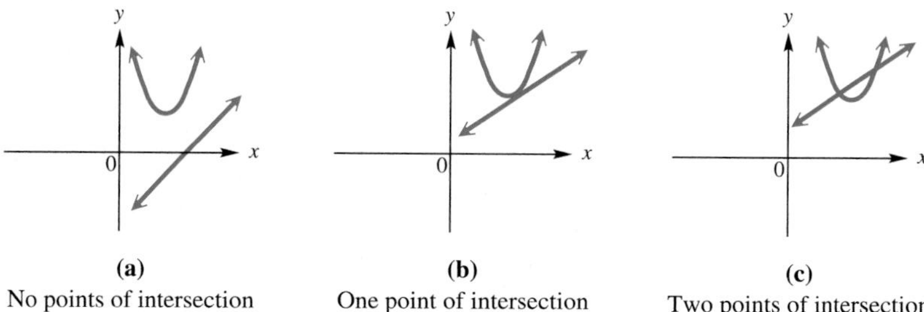

(a) No points of intersection

(b) One point of intersection

(c) Two points of intersection

Figure 14

OBJECTIVE 1 **Solve a nonlinear system by substitution.** We solve nonlinear systems by the elimination method, the substitution method, or a combination of the two. The substitution method is usually best when one of the equations is linear.

EXAMPLE 1 **Using Substitution When One Equation Is Linear**

Solve the system.

$$x^2 + y^2 = 9 \quad \textbf{(1)}$$
$$2x - y = 3 \quad \textbf{(2)}$$

The graph of (1) is a circle and the graph of (2) is a line. Visualizing the possible ways the graphs could intersect indicates that there may be 0, 1, or 2 points of intersection. It is best to solve the linear equation first for one of the two variables; then substitute the resulting expression into the nonlinear equation to obtain an equation in one variable.

$$2x - y = 3 \quad (2)$$
$$y = 2x - 3 \quad \textbf{(3)}$$

Substitute $2x - 3$ for y in equation (1).

$$x^2 + (2x - 3)^2 = 9$$
$$x^2 + 4x^2 - 12x + 9 = 9$$
$$5x^2 - 12x = 0$$
$$x(5x - 12) = 0 \quad \text{Common factor is } x.$$
$$x = 0 \quad \text{or} \quad x = \frac{12}{5} \quad \text{Zero-factor property}$$

Let $x = 0$ in equation (3) to get $y = -3$. If $x = \frac{12}{5}$, then $y = \frac{9}{5}$. The solution set of the system is $\{(0, -3), (\frac{12}{5}, \frac{9}{5})\}$. The graph of the system, shown in Figure 15, confirms the two points of intersection.

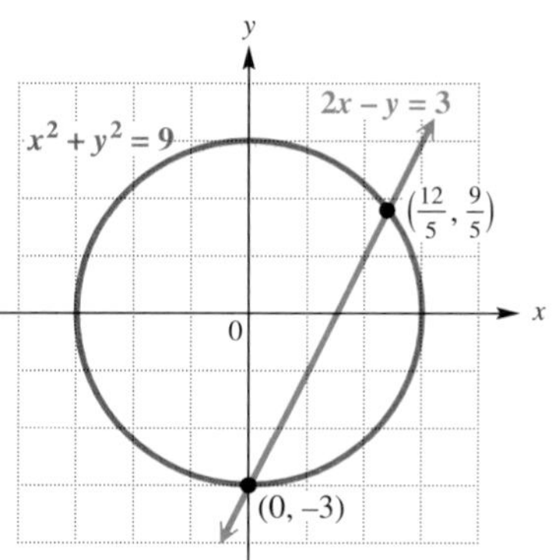

Figure 15

EXAMPLE 2 Using Substitution When One Equation Is Linear

Solve the system.

$$6x - y = 5 \quad (4)$$

$$xy = 4 \quad (5)$$

The graph of (4) is a line. We have not specifically mentioned equations like (5); however, it can be shown by plotting points that its graph is a hyperbola. Visualizing a line and a hyperbola indicates that there may be 0, 1, or 2 points of intersection. Since neither equation has a squared term here, solve either equation for one of the variables and then substitute the result into the other equation. Solving $xy = 4$ for x gives $x = \frac{4}{y}$. Substituting $\frac{4}{y}$ for x in equation (4) gives

$$6\left(\frac{4}{y}\right) - y = 5.$$

Clear fractions by multiplying both sides by y, noting the restriction that y cannot be 0. Then solve for y.

$$\frac{24}{y} - y = 5$$

$$24 - y^2 = 5y \qquad \text{Multiply by } y\ (y \neq 0).$$

$$0 = y^2 + 5y - 24$$

$$0 = (y - 3)(y + 8) \qquad \text{Factor.}$$

$$y = 3 \quad \text{or} \quad y = -8 \qquad \text{Zero-factor property}$$

Substitute these results into $x = \frac{4}{y}$ to obtain the corresponding values of x.

$$\text{If } y = 3, \text{ then } x = \frac{4}{3}. \quad \text{If } y = -8, \text{ then } x = -\frac{1}{2}.$$

The solution set has two ordered pairs: $\left\{\left(\frac{4}{3}, 3\right), \left(-\frac{1}{2}, -8\right)\right\}$. The graph in Figure 16 on the next page shows that there are two points of intersection.

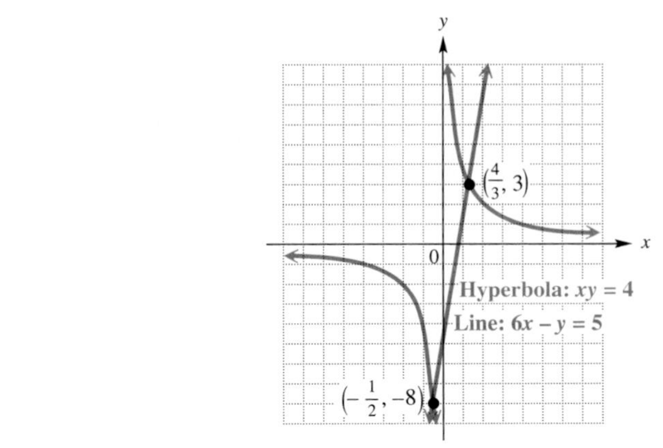

Figure 16

OBJECTIVE 2 **Use the elimination method to solve a system with two second-degree equations.** The elimination method is used when both equations are second degree.

EXAMPLE 3 Solving a Nonlinear System by Elimination

Solve the system.

$$x^2 + y^2 = 9 \qquad (6)$$
$$2x^2 - y^2 = -6 \qquad (7)$$

The graph of (6) is a circle, while the graph of (7) is a hyperbola. By analyzing the possibilities we conclude that there may be 0, 1, 2, 3, or 4 points of intersection. Adding the two equations will eliminate y, leaving an equation that can be solved for x.

$$\begin{aligned} x^2 + y^2 &= 9 \\ 2x^2 - y^2 &= -6 \\ \hline 3x^2 \quad &= 3 \end{aligned}$$

$$x^2 = 1$$

$$x = 1 \quad \text{or} \quad x = -1$$

Each value of x gives corresponding values for y when substituted into one of the original equations. Using equation (6) gives the following.

If $x = 1$,

$$(1)^2 + y^2 = 9$$
$$y^2 = 8$$
$$y = \sqrt{8} \quad \text{or} \quad y = -\sqrt{8}$$
$$y = 2\sqrt{2} \quad \text{or} \quad y = -2\sqrt{2}.$$

If $x = -1$,

$$(-1)^2 + y^2 = 9$$
$$y^2 = 8$$
$$y = 2\sqrt{2} \quad \text{or} \quad y = -2\sqrt{2}.$$

The solution set has four ordered pairs: $\{(1, 2\sqrt{2}), (1, -2\sqrt{2}), (-1, 2\sqrt{2}), (-1, -2\sqrt{2})\}$. Figure 17 shows the four points of intersection.

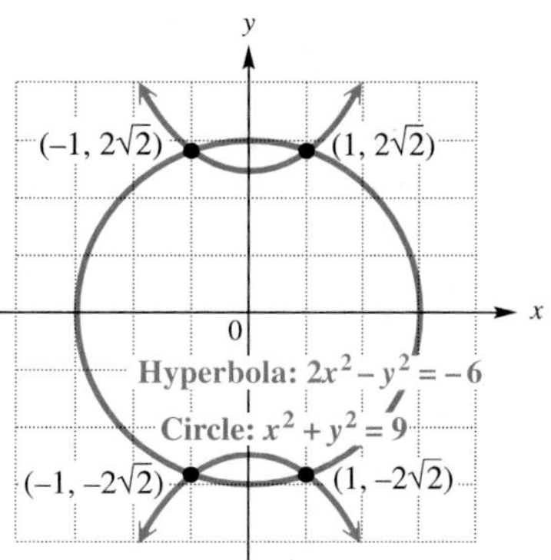

Figure 17

OBJECTIVE 3 **Solve a system that requires a combination of methods.** A system of second-degree equations may require a combination of methods to solve it.

EXAMPLE 4 **Solving a Nonlinear System by a Combination of Methods**

Solve the system.

$$x^2 + 2xy - y^2 = 7 \qquad \textbf{(8)}$$

$$x^2 - y^2 = 3 \qquad \textbf{(9)}$$

While we have not graphed equations like (8), its graph is a hyperbola. The graph of (9) is also a hyperbola. Two hyperbolas may have 0, 1, 2, 3, or 4 points of intersection. The elimination method can be used here in combination with the substitution method. Begin by eliminating the squared terms by multiplying both sides of equation (9) by -1 and then adding the result to equation (8).

$$\begin{array}{rcrcr} x^2 + 2xy & - & y^2 & = & 7 \\ -x^2 & + & y^2 & = & -3 \\ \hline 2xy & & & = & 4 \end{array}$$

Next, solve $2xy = 4$ for y. (Either variable would do.)

$$2xy = 4$$

$$y = \frac{2}{x} \qquad \textbf{(10)}$$

Now substitute $y = \frac{2}{x}$ into one of the original equations. It is easier to do this with equation (9).

$$x^2 - y^2 = 3 \qquad (9)$$

$$x^2 - \left(\frac{2}{x}\right)^2 = 3$$

$$x^2 - \frac{4}{x^2} = 3$$

$$x^4 - 4 = 3x^2 \qquad \text{Multiply by } x^2.$$

$$x^4 - 3x^2 - 4 = 0 \qquad \text{Subtract } 3x^2.$$

$$(x^2 - 4)(x^2 + 1) = 0 \qquad \text{Factor.}$$

$$x^2 - 4 = 0 \quad \text{or} \quad x^2 + 1 = 0$$
$$x^2 = 4 \quad \text{or} \quad x^2 = -1$$
$$x = 2 \quad \text{or} \quad x = -2 \qquad x = i \quad \text{or} \quad x = -i$$

Substituting the four values of x from above into equation (10) gives the corresponding values for y.

If $x = 2$, then $y = 1$. If $x = i$, then $y = -2i$.
If $x = -2$, then $y = -1$. If $x = -i$, then $y = 2i$.

Note that if you substitute the x-values found above into equations (8) or (9) instead of into equation (10), you get extraneous solutions. It is always wise to check all solutions in both of the given equations. There are four ordered pairs in the solution set, two with real values and two with imaginary values:

$$\{(2, 1), (-2, -1), (i, -2i), (-i, 2i)\}.$$

The graph of the system, shown in Figure 18, shows only the two real intersection points because the graph is in the real number plane. The two ordered pairs with imaginary components are solutions of the system, but do not appear on the graph.

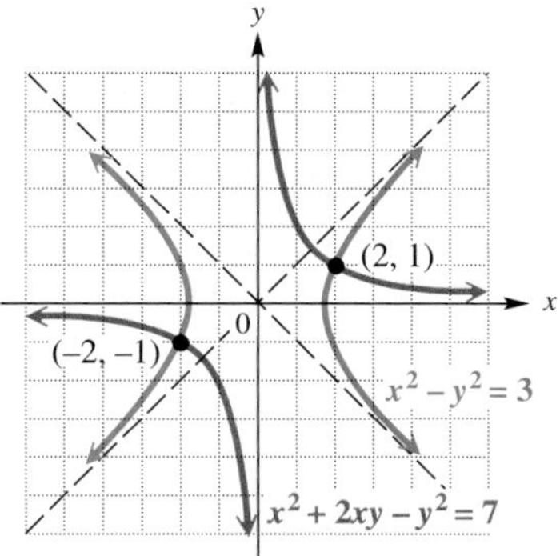

Figure 18

In the examples of this section, we analyzed the possible number of points of intersection of the graphs in each system. However, in Examples 2 and 4, we worked with equations whose graphs had not been studied. Keep in mind that it is not absolutely essential to visualize the number of points of intersection in order to solve the system. Visualizing the geometry of the graphs is only an aid to solving these systems.

OBJECTIVE 4 Use a graphing calculator to solve a nonlinear system. If the equations in a nonlinear system can be solved for y, then we can graph the equations of the system with a graphing calculator and use the capabilities of the calculator to identify all intersection points.

For instance, the two equations in Example 3 would require graphing the four separate functions

$$y_1 = \sqrt{9 - x^2}, \quad y_2 = -\sqrt{9 - x^2}, \quad y_3 = \sqrt{2x^2 + 6}, \quad \text{and} \quad y_4 = -\sqrt{2x^2 + 6}.$$

Figure 19 shows one of the points of intersection.

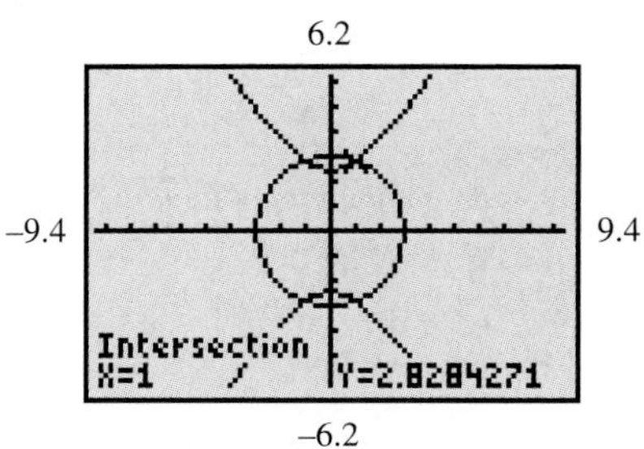

Figure 19

10.3 EXERCISES

1. Write an explanation of the steps you would use to solve the system

$$x^2 + y^2 = 25$$
$$y = x - 1$$

by the substitution method. Why would the elimination method not be appropriate for this system?

2. Write an explanation of the steps you would use to solve the system

$$x^2 + y^2 = 12$$
$$x^2 - y^2 = 13$$

by the elimination method.

Each sketch represents a system of equations. How many points are in its solution set?

3.

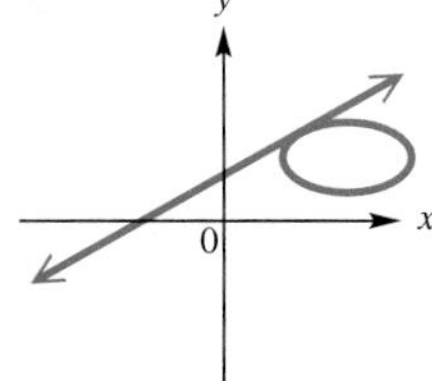

4.

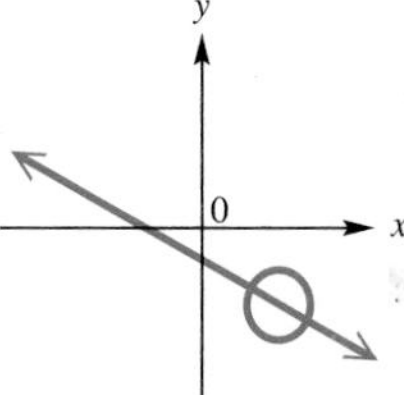

5.

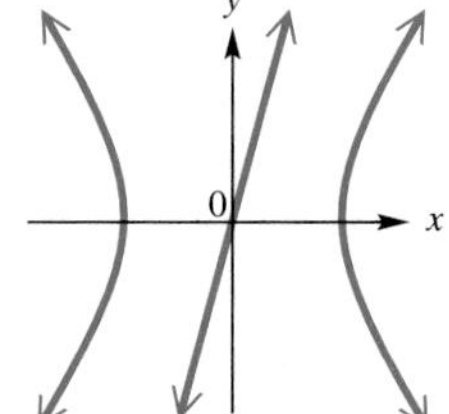

6.

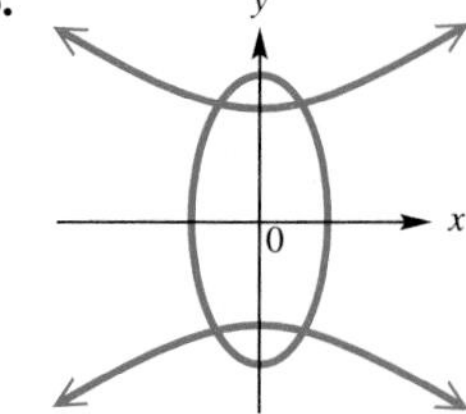

Suppose that a nonlinear system is composed of equations whose graphs are those described, and the number of points of intersection of the two graphs is as given. Make a sketch satisfying these conditions. (There may be more than one way to do this.)

7. a line and a circle; no points

8. a line and a circle; one point

9. a line and an ellipse; no points

10. a line and an ellipse; two points

11. a line and a hyperbola; no points

12. a line and a hyperbola; one point

13. a line and a hyperbola; two points

14. a circle and an ellipse; one point

15. a circle and an ellipse; four points

16. a parabola and an ellipse; one point

17. a parabola and an ellipse; four points

18. a parabola and a hyperbola; two points

Solve each system by the substitution method. See Examples 1 and 2.

19. $y = 4x^2 - x$
$y = x$

20. $y = x^2 + 6x$
$3y = 12x$

21. $y = x^2 + 6x + 9$
$x + y = 3$

22. $y = x^2 + 8x + 16$
$x - y = -4$

23. $x^2 + y^2 = 2$
$2x + y = 1$

24. $2x^2 + 4y^2 = 4$
$x = 4y$

25. $xy = 4$
$3x + 2y = -10$

26. $xy = -5$
$2x + y = 3$

27. $xy = -3$
$x + y = -2$

28. $xy = 12$
$x + y = 8$

29. $y = 3x^2 + 6x$
$y = x^2 - x - 6$

30. $y = 2x^2 + 1$
$y = 5x^2 + 2x - 7$

31. $2x^2 - y^2 = 6$
$y = x^2 - 3$

32. $x^2 + y^2 = 4$
$y = x^2 - 2$

Solve each system by the elimination method or a combination of the elimination and substitution methods. See Examples 3 and 4.

33. $3x^2 + 2y^2 = 12$
$x^2 + 2y^2 = 4$

34. $2x^2 + y^2 = 28$
$4x^2 - 5y^2 = 28$

35. $2x^2 + 3y^2 = 6$
$x^2 + 3y^2 = 3$

36. $6x^2 + y^2 = 9$
$3x^2 + 4y^2 = 36$

37. $2x^2 = 8 - 2y^2$
$3x^2 = 24 - 4y^2$

38. $5x^2 = 20 - 5y^2$
$2y^2 = 2 - x^2$

39. $x^2 + xy + y^2 = 15$
$x^2 + y^2 = 10$

40. $2x^2 + 3xy + 2y^2 = 21$
$x^2 + y^2 = 6$

41. $3x^2 + 2xy - 3y^2 = 5$
$-x^2 - 3xy + y^2 = 3$

42. $-2x^2 + 7xy - 3y^2 = 4$
$2x^2 - 3xy + 3y^2 = 4$

TECHNOLOGY INSIGHTS (EXERCISES 43–46)

In Exercises 43–46, nonlinear systems are given, along with a screen showing the coordinates of one of the points of intersection of the two graphs.

Solve each system using substitution or elimination to find the coordinates of the other *point of intersection.*

43. $y = x^2 + 1$
$x + y = 1$

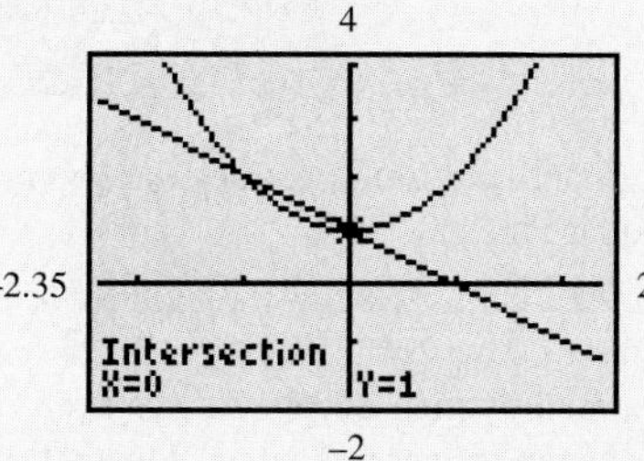

44. $y = -x^2$
$x + y = 0$

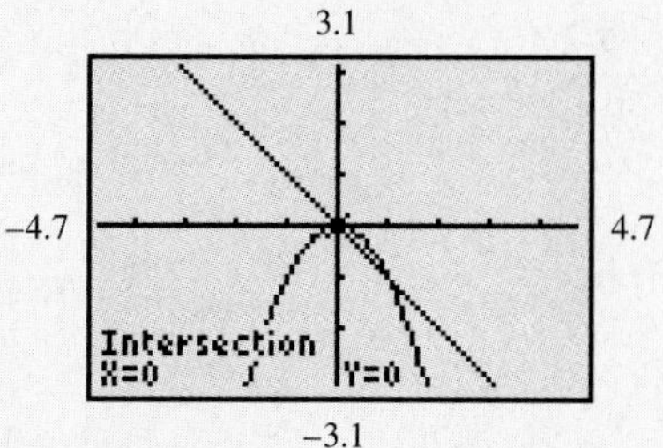

45. $y = \frac{1}{2}x^2$
$x + y = 4$

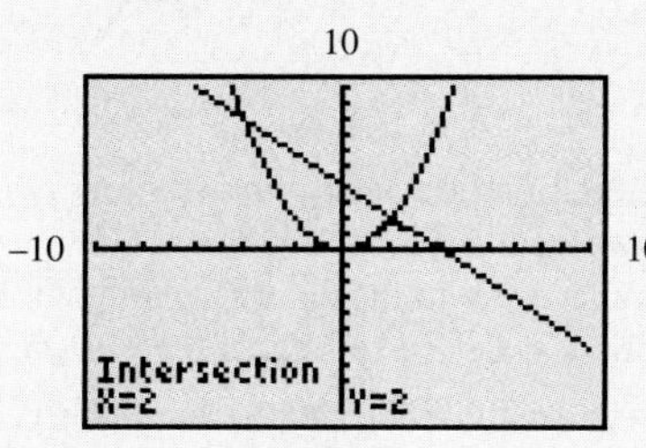

46. $y = -\frac{1}{3}x^2$
$2x - y = 9$

10
-10 10
Intersection
X=3 Y=-3
-40

Use a graphing calculator to solve each system. Then confirm your answer algebraically.

47. $xy = -6$
$x + y = -1$

48. $y = 2x^2 + 4x$
$y = -x^2 - 1$

RELATING CONCEPTS (EXERCISES 49–54)

To see how solving quadratic equations in one variable is related to solving a nonlinear system involving one quadratic equation and one linear equation, ***work Exercises 49–54 in order.***

49. Solve the quadratic equation $x^2 = 3x + 10$.

50. Graph the equations $y = x^2$ and $y = 3x + 10$ on the same set of coordinate axes.

51. Solve this nonlinear system using substitution.

$$y = x^2$$
$$y = 3x + 10$$

52. How do the x-coordinates of the solutions in Exercise 51 compare to the solutions of the equation in Exercise 49?

53. Graph the quadratic function $y = x^2 - 3x - 10$.

54. How do the x-values of the x-intercepts of the quadratic function in Exercise 53 compare to the x-coordinates of the points of intersection of the graphs in Exercise 50?

Did you make the connection that the solutions of a quadratic equation $ax^2 = bx + c$ are the x-values of the solutions of the system of equations $y = ax^2$ and $y = bx + c$?

Solve each problem by using a nonlinear system.

55. The area of a rectangular rug is 84 square feet and its perimeter is 38 feet. Find the length and width of the rug.

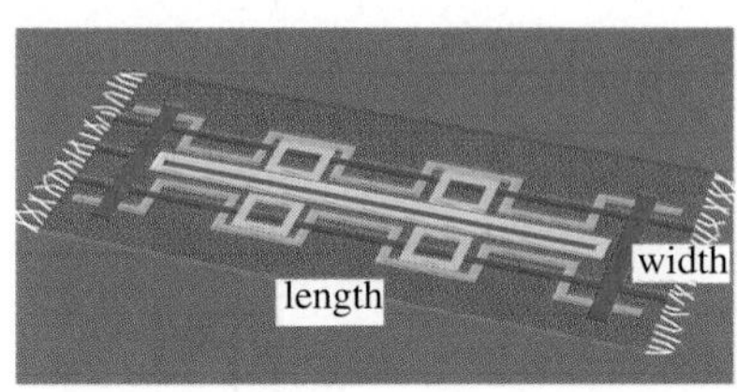

56. Find the length and width of a rectangular room whose perimeter is 50 meters and whose area is 100 square meters.

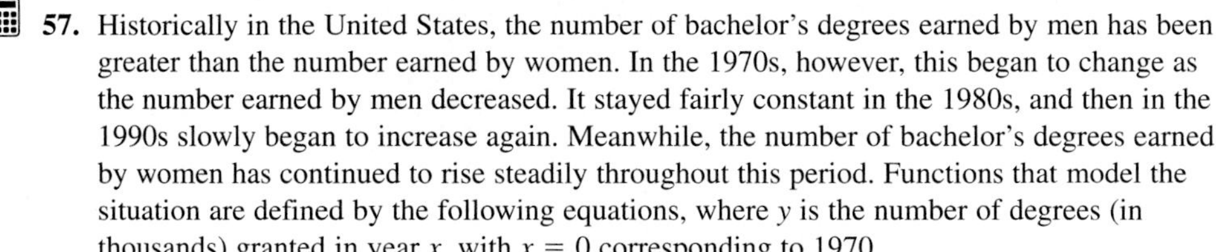

57. Historically in the United States, the number of bachelor's degrees earned by men has been greater than the number earned by women. In the 1970s, however, this began to change as the number earned by men decreased. It stayed fairly constant in the 1980s, and then in the 1990s slowly began to increase again. Meanwhile, the number of bachelor's degrees earned by women has continued to rise steadily throughout this period. Functions that model the situation are defined by the following equations, where y is the number of degrees (in thousands) granted in year x, with $x = 0$ corresponding to 1970.

Men: $y = .138x^2 + .064x + 451$

Women: $y = 12.1x + 334$

Solve this system of equations to find the year when the same number of bachelor's degrees was awarded to men and women. How many bachelor's degrees were awarded in that year? Give answers to the nearest whole number. (*Source:* U.S. National Center for Education Statistics, *Digest of Education Statistics,* annual.)

58. Andy Grove, chairman of chip maker Intel Corp., recently noted that decreasing prices for computers and stable prices for Internet access implied that the trend lines for these costs either have crossed or soon will. He predicted that the time is not far away when computers, like cell phones, may be given away to sell online time. To see this, assume a price of \$1000 for a computer, and let x represent the number of months it will be used. (*Source:* Elizabeth Corcoran, "Can Free Computers Be that Far Away?", *Washington Post,* from *The Sacramento Bee,* February 3, 1999.)

(a) Write an equation for the monthly cost y of the computer over this period.

(b) The average monthly online cost is about \$20. Assume this will remain constant and write an equation to express this cost.

(c) Solve the system of equations from parts (a) and (b). Interpret your answer in relation to the situation.

10.4 Second-Degree Inequalities; Systems of Inequalities

OBJECTIVES

1. Graph second-degree inequalities.
2. Graph the solution set of a system of inequalities.

OBJECTIVE 1 Graph second-degree inequalities. The linear inequality $3x + 2y \leq 5$ is graphed by first graphing the boundary line $3x + 2y = 5$. **Second-degree inequalities** such as $x^2 + y^2 \leq 36$ are graphed in the same way. The boundary of the inequality $x^2 + y^2 \leq 36$ is the graph of the equation $x^2 + y^2 = 36$, a circle with radius 6 and center at the origin, as shown in Figure 20. The inequality $x^2 + y^2 < 36$ will include either

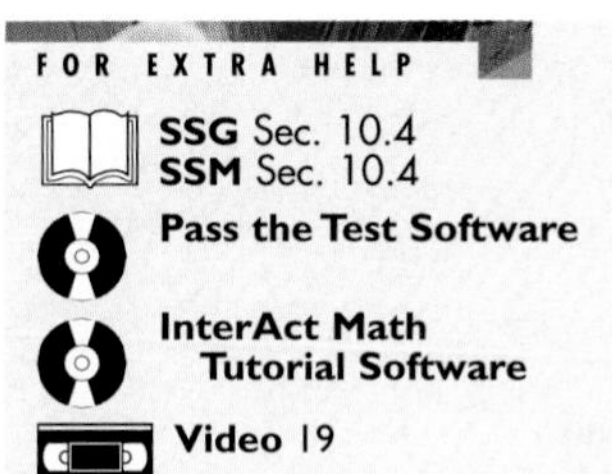

the points outside the circle or the points inside the circle. We decide which region to shade by substituting any point not on the circle, such as (0, 0), into the original inequality. Since $0^2 + 0^2 \le 36$ is a true statement, the original inequality includes the points inside the circle, the shaded region in Figure 20, and the boundary.

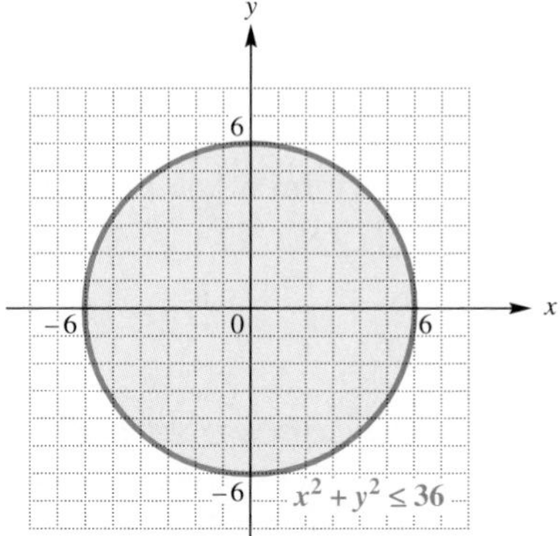

Figure 20

EXAMPLE 1 Graphing a Second-Degree Inequality

Graph $y < -2(x - 4)^2 - 3$.

The boundary, $y = -2(x - 4)^2 - 3$, is a parabola opening downward with vertex at (4, −3). Using the point (0, 0) as a test point gives

$$0 < -2(0 - 4)^2 - 3 \quad ?$$
$$0 < -32 - 3 \quad ?$$
$$0 < -35. \quad \text{False}$$

Because the final inequality is a false statement, the points in the region containing (0, 0) do not satisfy the inequality. Figure 21 shows the final graph; the parabola is drawn with a dashed line since the points of the parabola itself do not satisfy the inequality, and the region inside (or below) the parabola is shaded.

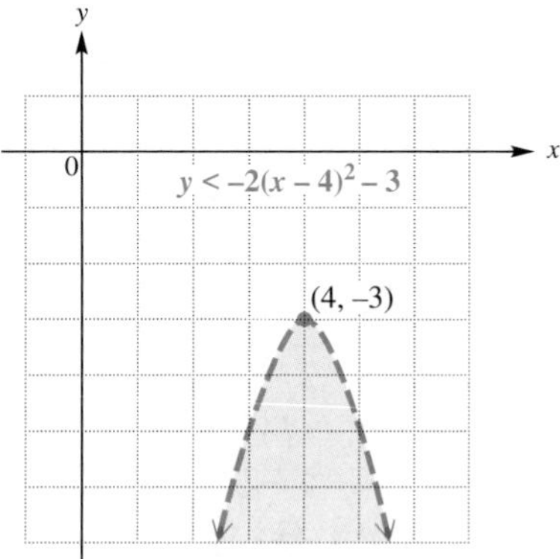

Figure 21

NOTE The origin is the test point of choice unless the graph actually passes through (0, 0).

EXAMPLE 2 Graphing a Second-Degree Inequality

Graph $16y^2 \leq 144 + 9x^2$.

First rewrite the inequality as follows.

$$16y^2 - 9x^2 \leq 144$$

$$\frac{y^2}{9} - \frac{x^2}{16} \leq 1 \qquad \text{Divide by 144.}$$

This form of the inequality shows that the boundary is the hyperbola

$$\frac{y^2}{9} - \frac{x^2}{16} = 1.$$

The desired region will be either the region between the branches of the hyperbola or the regions above the top branch and below the bottom branch. Test the region between the branches by choosing (0, 0) as a test point. Substitute into the original inequality.

$$16y^2 \leq 144 + 9x^2$$

$$16(0)^2 \leq 144 + 9(0)^2 \quad ?$$

$$0 \leq 144 + 0 \quad ?$$

$$0 \leq 144 \qquad \text{True}$$

Since the test point (0, 0) satisfies the inequality $16y^2 \leq 144 + 9x^2$, the region between the branches containing (0, 0) is shaded, as shown in Figure 22.

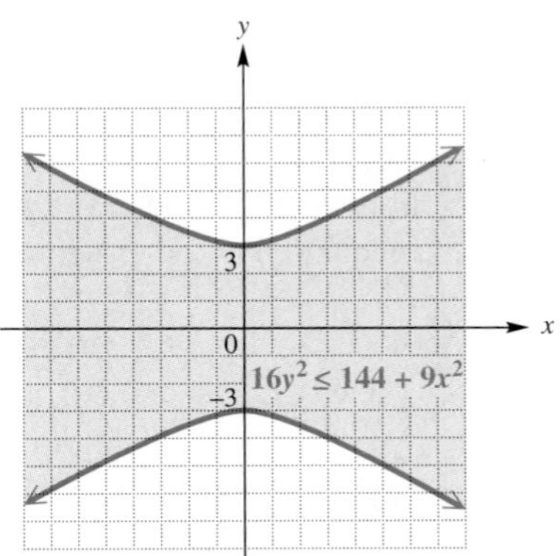

Figure 22

OBJECTIVE 2 Graph the solution set of a system of inequalities. If two or more inequalities are considered at the same time, at least one of which is nonlinear, we have a **nonlinear system of inequalities.** To find the solution set of a nonlinear system, we find the intersection of the graphs (solution sets) of the inequalities in the system. This is just an extension of the method used earlier to graph pairs of linear inequalities joined by the word *and.*

EXAMPLE 3 Graphing a Nonlinear System of Two Inequalities

Graph the solution set of

$$2x + 3y > 6$$
$$x^2 + y^2 < 16.$$

We begin by graphing the solution set of $2x + 3y > 6$. The boundary line is the graph of $2x + 3y = 6$ and is a dashed line because of the symbol $>$. The test point (0, 0) leads to a false statement in the inequality $2x + 3y > 6$, so we shade the region above the line, as shown in Figure 23.

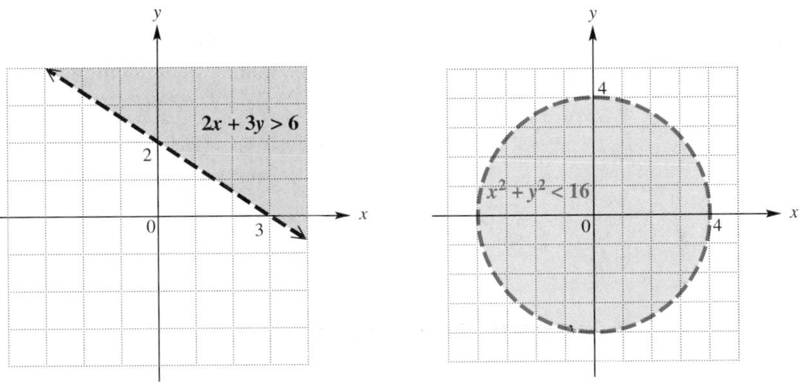

Figure 23

Figure 24

The graph of $x^2 + y^2 < 16$ is the interior of a dashed circle centered at the origin with radius 4. This is shown in Figure 24.

Finally, to get the graph of the solution set of the system, determine the intersection of the graphs of the two inequalities. The overlapping region in Figure 25 is the solution set.

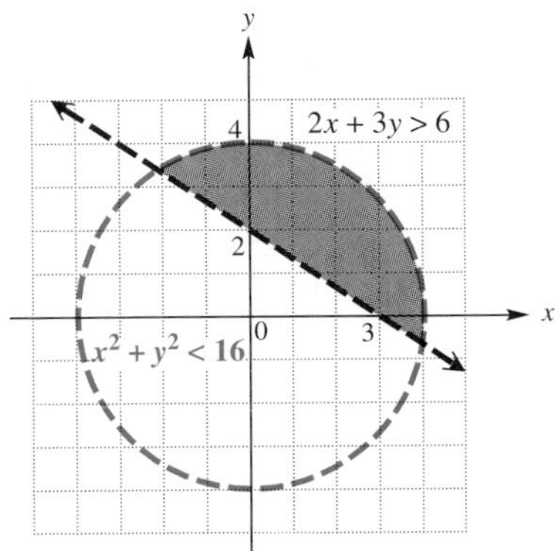

Figure 25

While the system in the following example does not contain a nonlinear inequality, it is different from those that we have solved previously.

EXAMPLE 4 Graphing a Linear System with Three Inequalities

Graph the solution set of the system

$$x + y < 1$$
$$y \le 2x + 3$$
$$y \ge -2.$$

Graph each inequality separately, on the same axes. The graph of $x + y < 1$ consists of all points below the dashed line $x + y = 1$. The graph of $y \le 2x + 3$ is the region below the solid line $y = 2x + 3$. Finally, the graph of $y \ge -2$ is the region above the solid horizontal line $y = -2$. The graph of the system, the intersection of these three graphs, is the triangular region enclosed by the three boundary lines in Figure 26, including two of its boundaries.

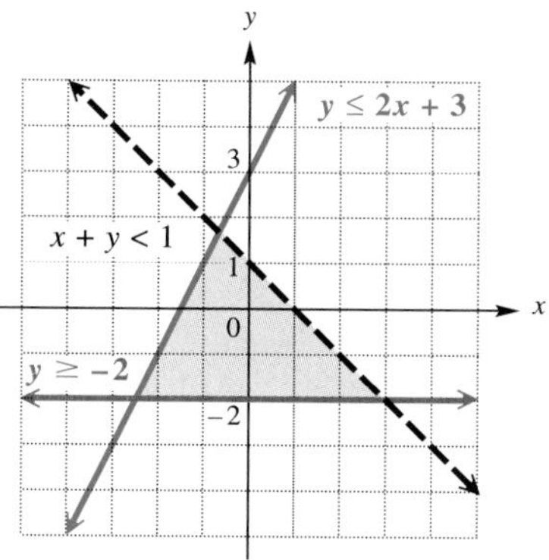

Figure 26

EXAMPLE 5 Graphing a Nonlinear System with Three Inequalities

Graph the solution set of the system

$$y \ge x^2 - 2x + 1$$
$$2x^2 + y^2 > 4$$
$$y < 4.$$

The graph of $y = x^2 - 2x + 1$ is a parabola with vertex at (1, 0). Those points above (or in the interior of) the parabola satisfy the condition $y > x^2 - 2x + 1$. Thus, points on the parabola or in the interior are in the solution set of $y \ge x^2 - 2x + 1$. The graph of the equation $2x^2 + y^2 = 4$ is an ellipse. We draw it with a dashed line. To satisfy the inequality $2x^2 + y^2 > 4$, a point must lie outside the ellipse. The graph of $y < 4$ includes all points below the dashed line $y = 4$. Finally, the graph of the system is the shaded region in Figure 27 that lies outside the ellipse, inside or on the boundary of the parabola, and below the line $y = 4$.

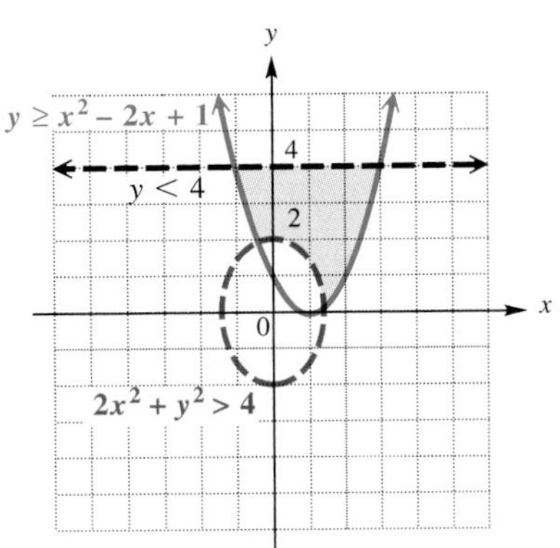

Figure 27

10.4 EXERCISES

1. Which one of the following is a description of the graph of the solution set of the system below?

$$x^2 + y^2 < 25$$
$$y > -2$$

(a) all points outside the circle $x^2 + y^2 = 25$ and above the line $y = -2$
(b) all points outside the circle $x^2 + y^2 = 25$ and below the line $y = -2$
(c) all points inside the circle $x^2 + y^2 = 25$ and above the line $y = -2$
(d) all points inside the circle $x^2 + y^2 = 25$ and below the line $y = -2$

2. Fill in each blank with the appropriate response. The graph of the system

$$y > x^2 + 1$$
$$\frac{x^2}{9} + \frac{y^2}{4} > 1$$
$$y < 5$$

consists of all points $\underset{\text{(above/below)}}{\underline{\quad ? \quad}}$ the parabola $y = x^2 + 1$, $\underset{\text{(inside/outside)}}{\underline{\quad ? \quad}}$ the ellipse $\frac{x^2}{9} + \frac{y^2}{4} = 1$, and $\underset{\text{(above/below)}}{\underline{\quad ? \quad}}$ the line $y = 5$.

3. Explain how to graph the solution set of a nonlinear inequality.

4. Explain how to graph the solution set of a system of nonlinear inequalities.

Match each nonlinear inequality with its graph.

5. $y \geq x^2 + 4$
6. $y \leq x^2 + 4$
7. $y < x^2 + 4$
8. $y > x^2 + 4$

A.
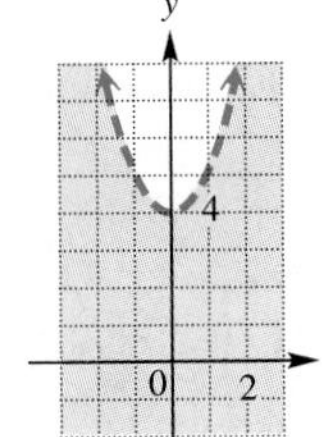

B.
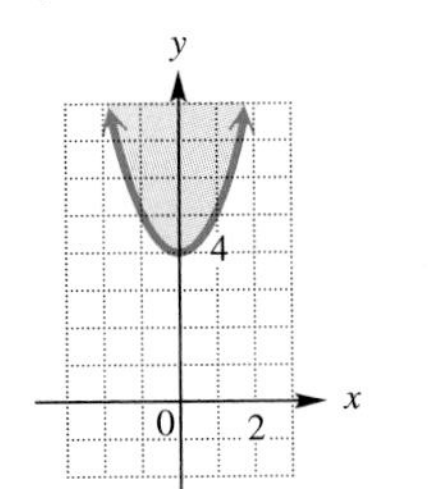

C.
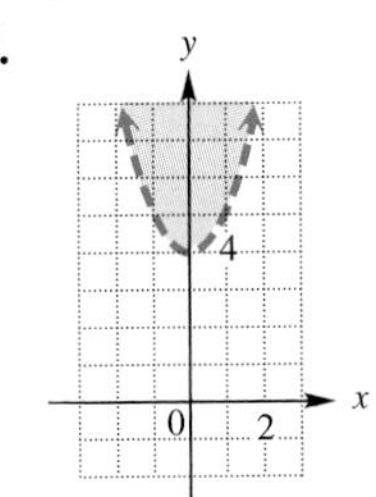

D.
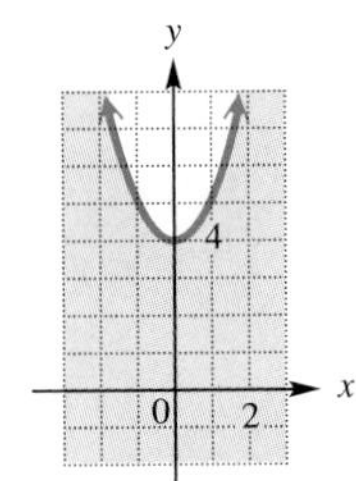

Graph each nonlinear inequality. See Examples 1 and 2.

9. $y^2 > 4 + x^2$

10. $y^2 \le 4 - 2x^2$

11. $y + 2 \ge x^2$

12. $x^2 \le 16 - y^2$

13. $2y^2 \ge 8 - x^2$

14. $x^2 \le 16 + 4y^2$

15. $y \le x^2 + 4x + 2$

16. $9x^2 < 16y^2 - 144$

17. $9x^2 > 16y^2 + 144$

18. $4y^2 \le 36 - 9x^2$

19. $x^2 - 4 \ge -4y^2$

20. $x \ge y^2 - 8y + 14$

21. $x \le -y^2 + 6y - 7$

22. $y^2 - 16x^2 \le 16$

Graph each system of inequalities. See Examples 3–5.

23. $2x + 5y < 10$
$x - 2y < 4$

24. $3x - y > -6$
$4x + 3y > 12$

25. $5x - 3y \le 15$
$4x + y \ge 4$

26. $4x - 3y \le 0$
$x + y \le 5$

27. $x \le 5$
$y \le 4$

28. $x \ge -2$
$y \le 4$

29. $y > x^2 - 4$
$y < -x^2 + 3$

30. $x^2 - y^2 \ge 9$
$\frac{x^2}{16} + \frac{y^2}{9} \le 1$

31. $y^2 - x^2 \ge 4$
$-5 \le y \le 5$

32. $x^2 + y^2 \ge 4$
$x + y \le 5$
$x \ge 0$
$y \ge 0$

33. $y \le -x^2$
$y \ge x - 3$
$y \le -1$
$x < 1$

34. $y < x^2$
$y > -2$
$x + y < 3$
$3x - 2y > -6$

For each nonlinear inequality in Exercises 35–42, a restriction is placed on one or both variables. For example, the graph of

$$x^2 + y^2 \le 4, \quad x \ge 0$$

would be as shown in the figure. Only the right half of the interior of the circle and its boundary is shaded, because of the restriction that x must be nonnegative. Graph each nonlinear inequality with restrictions.

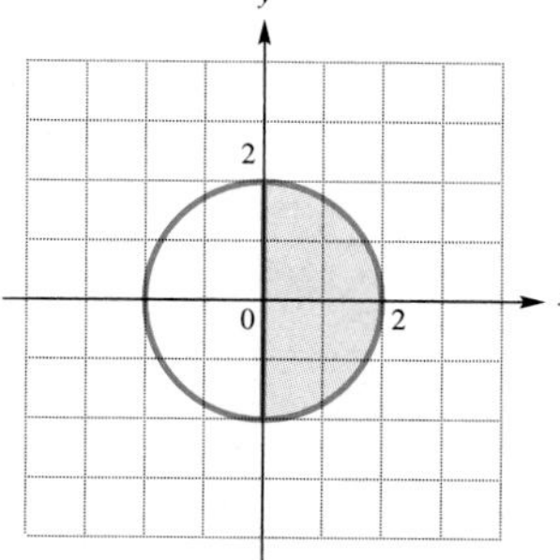

35. $x^2 + y^2 > 36, \quad x \ge 0$

36. $4x^2 + 25y^2 < 100, \quad y < 0$

37. $x < y^2 - 3, \quad x < 0$

38. $x^2 - y^2 < 4, \quad x < 0$

39. $4x^2 - y^2 > 16, \quad x < 0$

40. $x^2 + y^2 > 4, \quad y < 0$

41. $x^2 + 4y^2 \ge 1, \quad x \ge 0, y \ge 0$

42. $2x^2 - 32y^2 \le 8, \quad x \le 0, y \ge 0$

TECHNOLOGY INSIGHTS (EXERCISES 43–46)

Graphing calculators have the capability of shading above or below graphs, thus allowing them to illustrate nonlinear inequalities and systems of nonlinear inequalities. For example, the inequality discussed in Example 1 and graphed in Figure 21,

$$y < -2(x - 4)^2 - 3,$$

is shown in the accompanying screen.

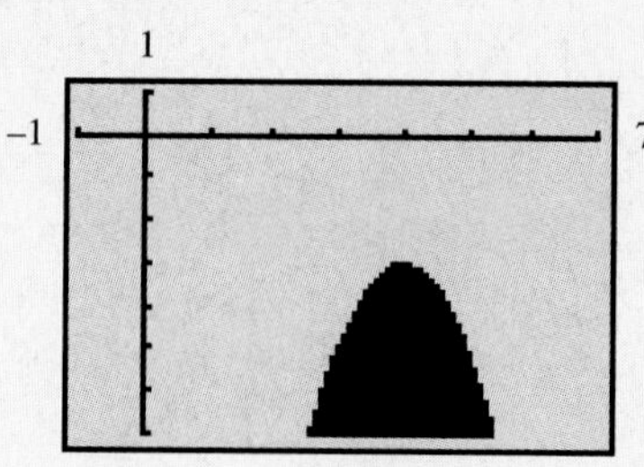

TECHNOLOGY INSIGHTS (EXERCISES 43–46) (CONTINUED)

Match each nonlinear inequality or system of nonlinear inequalities with its calculator-generated graph.

A.

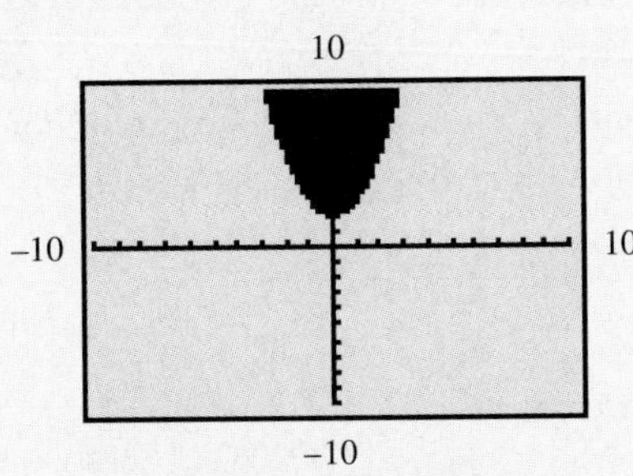

B.

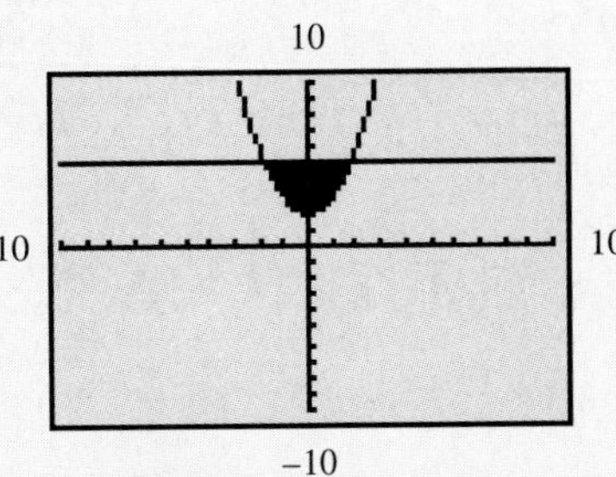

C.

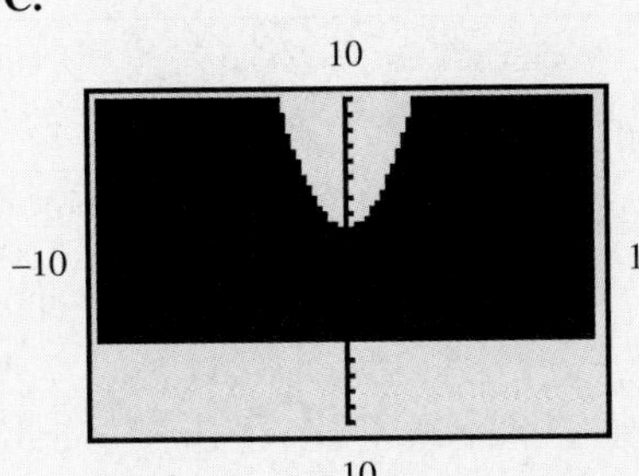

D.

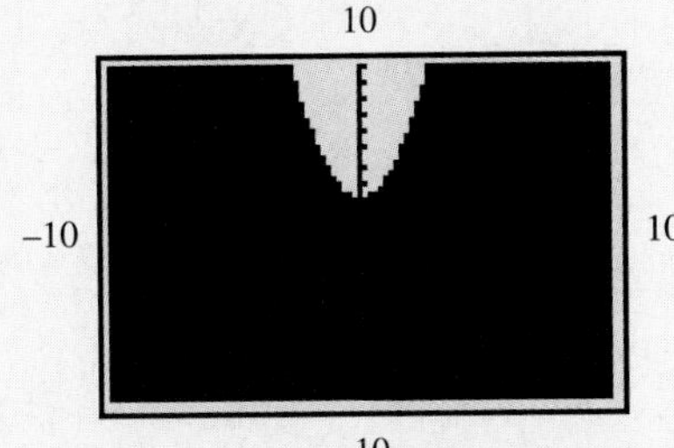

43. $y > x^2 + 2$

44. $y < x^2 + 2$

45. $y > x^2 + 2$
$y < 5$

46. $y < x^2 + 2$
$y > -5$

Use the shading feature of a graphing calculator to graph each system.

47. $y \geq x - 3$
$y \leq -x + 4$

48. $y \geq -x^2 + 5$
$y \leq x^2 - 3$

49. $y < x^2 + 4x + 4$
$y > -3$

50. $y > (x - 4)^2 - 3$
$y < 5$

CHAPTER 10 GROUP ACTIVITY

Finding the Paths of Natural Satellites

Objective: Write and graph equations of ellipses from given data.

(continued)

The moon, which orbits Earth, and Halley's comet, which orbits the sun, are both natural satellites. In Section 10.1, you solved problems where you were given equations of ellipses for the orbits of planets and were asked to find apogees (greatest distance from the sun) and perigees (smallest distance from the sun). This activity reverses the process; that is, given apogees and perigees you must find equations of ellipses.

A. Have each student choose a natural satellite from the chart. Predict the shape of the orbital ellipse for your satellite.

Moon or Comet	*Apogee*	*Perigee*
Moon	406.7 thousand kilometers from Earth	356.4 thousand kilometers from Earth
Halley's Comet	35 astronomical units* from the sun	.6 astronomical unit* from the sun

Source: World Book Encyclopedia 1998.
*One astronomical unit is the distance from Earth to the sun.

B. For your satellite, do the following.

1. Find values for a, b, and c. Note that apogee $= a + c$, perigee $= a - c$, and $c^2 = a^2 - b^2$.

2. Write the equation of an ellipse in the form $\frac{x^2}{a^2} + \frac{y^2}{b^2} = 1$.

3. Rewrite the equation so it can be graphed on a graphing calculator. (See Section 10.1, Objective 5.)

4. Graph your equation on a graphing calculator. Adjust the window setting in order to see the entire graph. Once the window is set correctly, get a square window to see the true shape of the ellipse.

C. Compare your graph with your partner's graph.

1. Do the graphs reflect the shapes you predicted in part A?

2. What window was used to graph each ellipse?

CHAPTER 10 SUMMARY

KEY TERMS

conic section

10.1 circle
center
radius
center-radius form
ellipse
foci

10.2 hyperbola
asymptotes of a hyperbola
fundamental rectangle
square root function

10.3 nonlinear equation
nonlinear system of equations
nonlinear system of inequalities

10.4 second-degree inequality
system of inequalities

TEST YOUR WORD POWER

See how well you have learned the vocabulary in this chapter. Answers, with examples, are given at the bottom of the page.

1. Conic sections are
(a) graphs of first-degree equations
(b) the result of two or more intersecting planes
(c) graphs of first-degree inequalities
(d) figures that result from the intersection of an infinite cone with a plane.

2. A **circle** is the set of all points in a plane
(a) the difference of whose distances from two fixed points is constant
(b) that lie a fixed distance from a fixed point
(c) the sum of whose distances from two fixed points is constant
(d) that make up the graph of any second-degree equation.

3. An **ellipse** is the set of all points in a plane
(a) the difference of whose distances from two fixed points is constant
(b) that lie a fixed distance from a fixed point
(c) the sum of whose distances from two fixed points is constant
(d) that make up the graph of any second-degree equation.

4. A **hyperbola** is the set of all points in a plane
(a) the difference of whose distances from two fixed points is constant
(b) that lie a fixed distance from a fixed point
(c) the sum of whose distances from two fixed points is constant
(d) that make up the graph of any second-degree equation.

5. A **nonlinear equation** is an equation
(a) that cannot be written in the form $Ax + By = C$, for real numbers A, B, C
(b) in which the terms have only one variable
(c) of degree one
(d) of a linear function.

6. A **nonlinear system of equations** is a system
(a) with at least one linear equation
(b) with two or more inequalities
(c) with at least one nonlinear equation
(d) with at least two linear equations.

7. A **second-degree inequality** is an inequality
(a) with at least one variable of degree two and no variable of degree greater than two
(b) with any variable of degree one or greater
(c) with at least two variables of degree two
(d) with a variable of degree two or greater.

8. A **system of inequalities** is a system
(a) with at least one linear inequality
(b) with two or more inequalities
(c) with at least one nonlinear equation
(d) with at least two linear equations.

QUICK REVIEW

CONCEPTS	EXAMPLES
10.1 THE CIRCLE AND THE ELLIPSE	
The circle with radius r and center at (h, k) has an equation of the form $$(x - h)^2 + (y - k)^2 = r^2.$$	The circle $(x + 2)^2 + (y - 3)^2 = 25$ has center $(-2, 3)$ and radius 5. 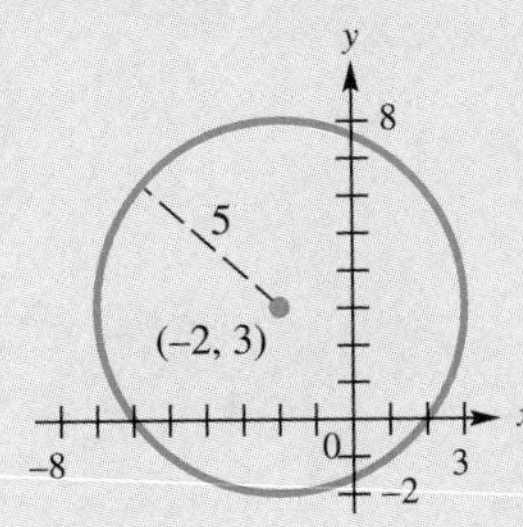

(continued)

Answers to Test Your Word Power

1. (d) *Example:* Parabolas, circles, ellipses, and hyperbolas are conic sections. **2.** (b) *Example:* See the graph of $x^2 + y^2 = 9$ in Figure 1 of Section 10.1. **3.** (c) *Example:* See the graph of $\frac{x^2}{49} + \frac{y^2}{36} = 1$ in Figure 5 of Section 10.1. **4.** (a) *Example:* See the graph of $\frac{x^2}{16} - \frac{y^2}{12} = 1$ in Figure 8 of Section 10.2. **5.** (a) *Examples:* $y = x^2 + 8x + 16$, $xy = 5$, $2x^2 - y^2 = 6$ **6.** (c) *Example:* $\begin{aligned} x^2 + y^2 &= 2 \\ 2x + y &= 1 \end{aligned}$

7. (a) *Examples:* $y > 3x^2 + 6x$, $x^2 - 4y^2 \le 16$ **8.** (b) *Example:* $\begin{aligned} x^2 + y^2 &\ge 4 \\ x + y &< 5 \end{aligned}$

CONCEPTS	EXAMPLES
The ellipse whose x-intercepts are $(a, 0)$ and $(-a, 0)$ and whose y-intercepts are $(0, b)$ and $(0, -b)$ has an equation of the form $$\frac{x^2}{a^2} + \frac{y^2}{b^2} = 1.$$	Graph $\frac{x^2}{9} + \frac{y^2}{4} = 1$. 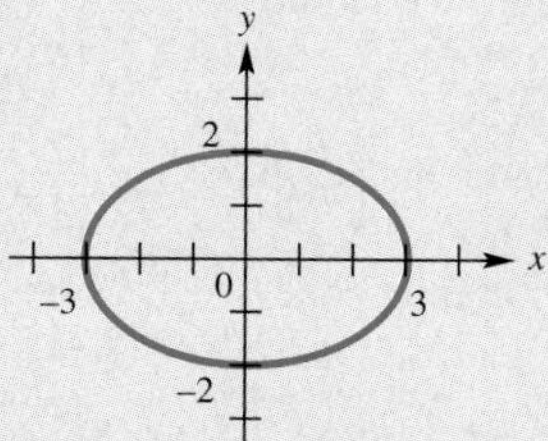
10.2 THE HYPERBOLA; MORE ON SQUARE ROOT FUNCTIONS	
A hyperbola with x-intercepts $(a, 0)$ and $(-a, 0)$ has an equation of the form $$\frac{x^2}{a^2} - \frac{y^2}{b^2} = 1,$$ and a hyperbola with y-intercepts $(0, b)$ and $(0, -b)$ has an equation of the form $$\frac{y^2}{b^2} - \frac{x^2}{a^2} = 1.$$	Graph $\frac{x^2}{4} - \frac{y^2}{4} = 1$. The graph has x-intercepts $(2, 0)$ and $(-2, 0)$. 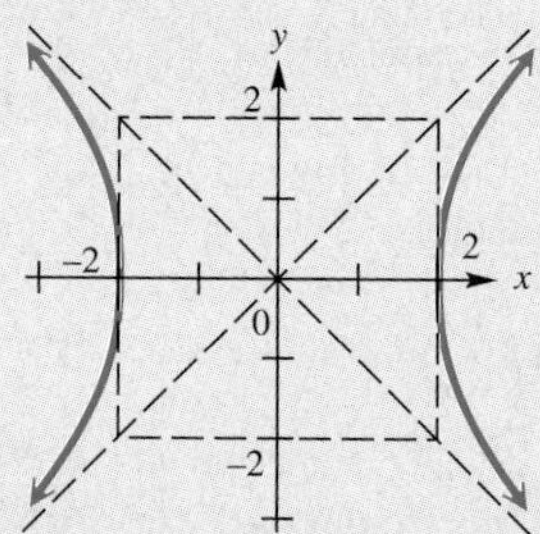
The extended diagonals of the fundamental rectangle with corners at the points (a, b), $(-a, b)$, $(-a, -b)$, and $(a, -b)$ are the asymptotes of these hyperbolas.	The fundamental rectangle has corners at $(2, 2)$, $(-2, 2)$, $(-2, -2)$, and $(2, -2)$.
To graph a square root function, square both sides so that the equation can be easily recognized. Then graph only the part indicated by the original equation.	Graph $y = -\sqrt{4 - x^2}$. Square both sides and rearrange terms to get $$x^2 + y^2 = 4.$$ This equation has a circle as its graph. However, graph only the lower half of the circle, since the original equation indicates that y cannot be positive. 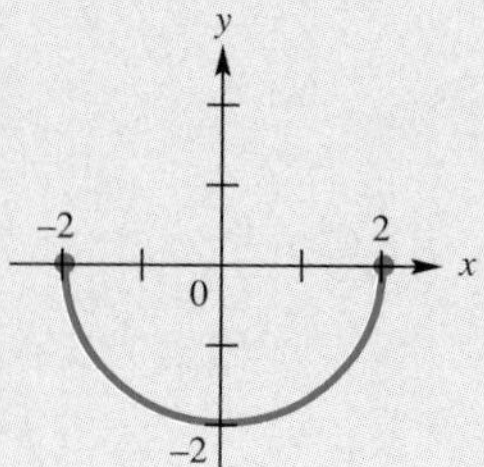

CONCEPTS	EXAMPLES

10.3 NONLINEAR SYSTEMS OF EQUATIONS

Nonlinear systems can be solved by the substitution method, the elimination method, or a combination of the two.

Solve the system

$$x^2 + 2xy - y^2 = 14$$
$$x^2 - y^2 = -16. \quad (*)$$

Multiply equation (*) by -1 and use elimination.

$$\begin{aligned} x^2 + 2xy - y^2 &= 14 \\ -x^2 \qquad + y^2 &= 16 \\ \hline 2xy \qquad &= 30 \\ xy &= 15 \end{aligned}$$

Solve for y to obtain $y = \frac{15}{x}$, and substitute into equation (*).

$$x^2 - \left(\frac{15}{x}\right)^2 = -16$$

$$x^2 - \frac{225}{x^2} = -16$$

$$x^4 + 16x^2 - 225 = 0 \qquad \text{Multiply by } x^2\text{; add } 16x^2.$$

$$(x^2 - 9)(x^2 + 25) = 0 \qquad \text{Factor.}$$

$$x = \pm 3 \quad \text{or} \quad x = \pm 5i \qquad \text{Zero-factor property}$$

Find corresponding y-values to get the solution set

$$\{(3, 5), (-3, -5), (5i, -3i), (-5i, 3i)\}.$$

10.4 SECOND-DEGREE INEQUALITIES; SYSTEMS OF INEQUALITIES

To graph a second-degree inequality, graph the corresponding equation as a boundary and use test points to determine which region(s) form the solution set. Shade the appropriate region(s).

Graph $y \geq x^2 - 2x + 3$.

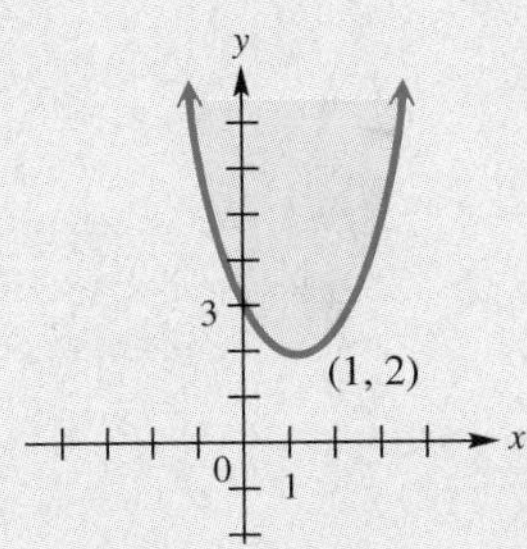

The solution set of a system of inequalities is the intersection of the solution sets of the individual inequalities.

Graph the solution set of the system

$$3x - 5y > -15$$
$$x^2 + y^2 \leq 25.$$

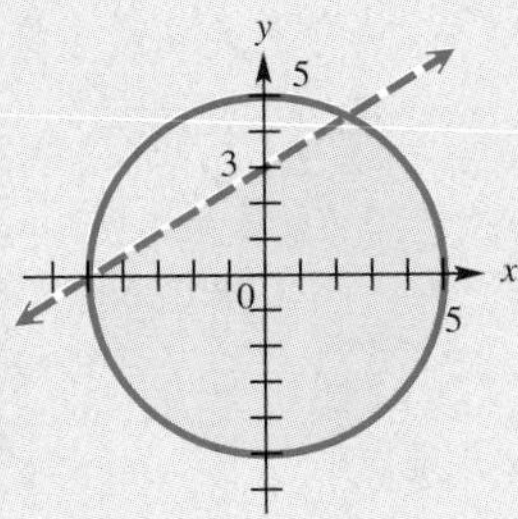

CHAPTER 10 REVIEW EXERCISES

[10.1] *Write an equation of each circle described or graphed.*

1. Center $(-2, 4)$, radius 3

2. 

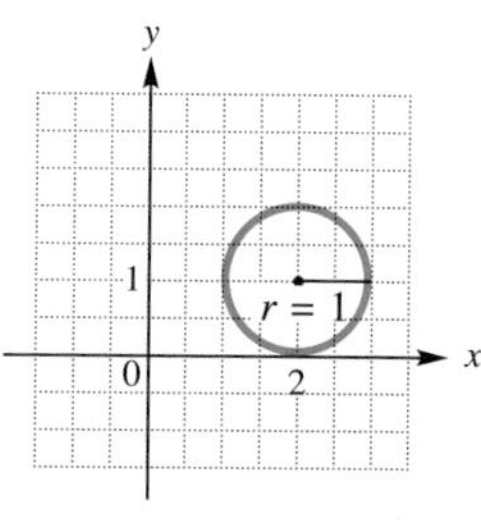

Find the center and radius of each circle.

3. $x^2 + y^2 + 6x - 4y - 3 = 0$

4. $x^2 + y^2 - 8x - 2y + 13 = 0$

5. $2x^2 + 2y^2 + 4x + 20y = -34$

6. $4x^2 + 4y^2 - 24x + 16y = 48$

[10.1–10.2] *Graph the following.*

7. $\dfrac{x^2}{16} + \dfrac{y^2}{9} = 1$

8. $\dfrac{x^2}{49} + \dfrac{y^2}{25} = 1$

9. $\dfrac{x^2}{16} - \dfrac{y^2}{25} = 1$

10. $\dfrac{y^2}{25} - \dfrac{x^2}{4} = 1$

11. $x^2 + 9y^2 = 9$

12. $f(x) = \sqrt{4 + x^2}$

13. $(x - 2)^2 + (y + 3)^2 = 16$

Identify the graph of each equation as a parabola, circle, ellipse, or hyperbola.

14. $y = 2x^2 - 3$

15. $y^2 = 2x^2 - 8$

16. $y^2 = 8 - 2x^2$

17. $x = y^2 + 4$

18. $x^2 + y^2 = 64$

19. A satellite is in an elliptical orbit around Earth with perigee altitude of 160 kilometers and apogee altitude of 16,000 kilometers. See the figure. (*Source:* Bernice Kastner, *Space Mathematics,* NASA.) Find the equation of the ellipse. (*Hint:* Refer to Section 10.1, Exercise 47.)

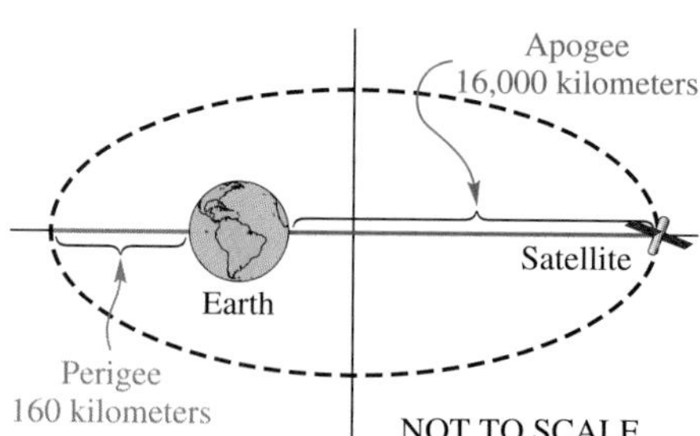

20. Ships and planes often use a location-finding system called LORAN. With this system, a radio transmitter at *M* (see the figure) sends out a series of pulses. When each pulse is received at transmitter *S,* it then sends out a pulse. A ship at *P* receives pulses from both *M* and *S.* A receiver on the ship measures the difference in the arrival times of the pulses. A special map gives hyperbolas that correspond to the differences in arrival times (which give the distances d_1 and d_2 in the figure). The ship can then be located as lying on a branch of a particular hyperbola. Suppose $d_1 = 80$ miles and $d_2 = 30$ miles, and the

distance between transmitters M and S is 100 miles. Use the definition to find an equation of the hyperbola the ship is located on.

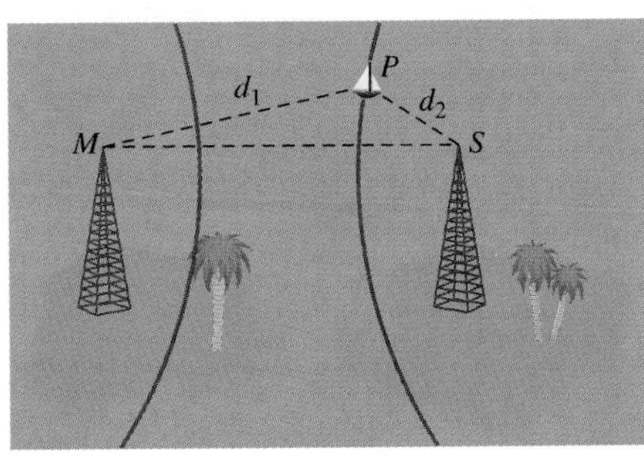

[10.3] *Solve each system.*

21. $2y = 3x - x^2$
$x + 2y = -12$

22. $y + 1 = x^2 + 2x$
$y + 2x = 4$

23. $x^2 + 3y^2 = 28$
$y - x = -2$

24. $xy = 8$
$x - 2y = 6$

25. $x^2 + y^2 = 6$
$x^2 - 2y^2 = -6$

26. $3x^2 - 2y^2 = 12$
$x^2 + 4y^2 = 18$

27. How many solutions are possible for a system of two equations whose graphs are a circle and a line?

28. How many solutions are possible for a system of two equations whose graphs are a parabola and a hyperbola?

29. Walker Lake is one of two remnants of a prehistoric lake that once covered northwestern Nevada. Since farmers began diverting the Walker River to their fields in the 1860s, the lake has shrunk by half. As it shrinks, the lake becomes more salty, and fish populations begin to die. The federal government and local residents are trying to find a way to guarantee the lake more water without disrupting farm economies. The graph shows the lowering lake level and the salt concentration over the years since 1880. Interpret the point of intersection of the two curves, addressing the year, altitude, and salinity (salt concentration) at that point.

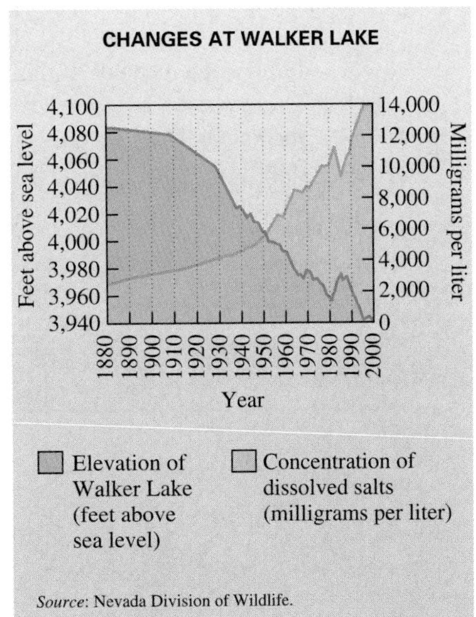

[10.4] *Graph each nonlinear inequality.*

30. $9x^2 \geq 16y^2 + 144$

31. $4x^2 + y^2 \geq 16$

32. $y < -(x + 2)^2 + 1$

Graph each system of inequalities.

33. $2x + 5y \leq 10$
$3x - y \leq 6$

34. $|x| \leq 2$
$|y| > 1$
$4x^2 + 9y^2 \leq 36$

35. $9x^2 \leq 4y^2 + 36$
$x^2 + y^2 \leq 16$

MIXED REVIEW EXERCISES

Graph.

36. $\dfrac{x^2}{64} + \dfrac{y^2}{25} = 1$

37. $\dfrac{y^2}{4} - 1 = \dfrac{x^2}{9}$

38. $x^2 + y^2 = 25$

39. $y = 2(x - 2)^2 - 3$

40. $f(x) = -\sqrt{16 - x^2}$

41. $f(x) = \sqrt{4 - x}$

42. $3x + 2y \geq 0$
$y \leq 4$
$x \leq 4$

43. $4y > 3x - 12$
$x^2 < 16 - y^2$

RELATING CONCEPTS (EXERCISES 44–49)

In Chapter 4 we presented several methods of solving systems of linear equations in three variables. Now these methods can be used to find the equation of a circle through three points in a plane that are not on the same line. The equation of a circle can be written in the form $x^2 + y^2 + ax + by + c = 0$ for some values of a, b, and c. We will find the equation of the circle through the points (2, 4), (5, 1), and (−1, 1).

Work Exercises 44–49 in order.

44. Determine one equation in a, b, and c by letting $x = 2$ and $y = 4$ in the general form given above. Write it with a, b, and c on the left and the constant on the right.

45. Repeat Exercise 44 for the point (5, 1).

46. Repeat Exercise 44 for the point (−1, 1).

47. Solve the system of equations formed by the equations found in Exercises 44–46, and give the equation of the circle that satisfies the conditions described above.

48. Use the methods of this chapter to find the center and the radius of the circle in Exercise 47.

49. Graph the circle found in Exercise 47.

Did you make the connection between solving a system of equations and writing the equation of a circle?

CHAPTER 10 TEST

1. Give an equation of the circle shown in the figure.

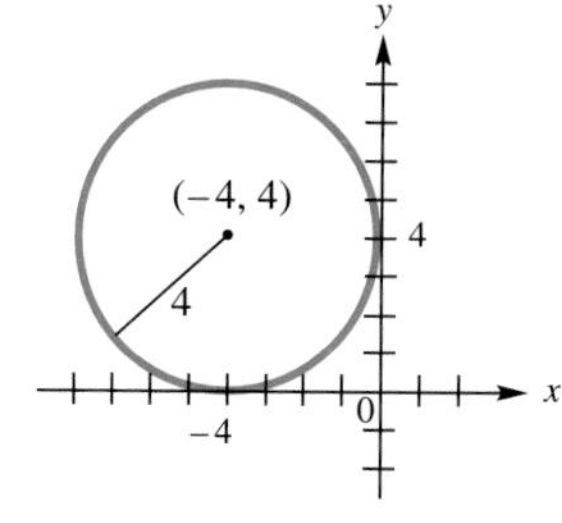

2. Find the coordinates of the center and the radius of the circle with equation $x^2 + y^2 + 8x - 2y = 8$.

Identify the graph of each equation as one of the following: parabola, hyperbola, ellipse, or circle.

3. $3x^2 + 3y^2 = 27$

4. $9x^2 + 4y^2 = 36$

5. $9x^2 = 36 + 4y^2$

6. $x = 36 - 4y^2$

Sketch the graph of each equation.

7. $x^2 + y^2 = 64$

8. $4x^2 + 9y^2 = 36$

9. $16y^2 - 4x^2 = 64$

10. $f(x) = \sqrt{16 - x^2}$

11. If a parabola and an ellipse are graphed in the same plane, what are the possible number of points of intersection of the graphs?

12. Sketch a parabola and an ellipse on the same set of axes so that there are two points of intersection of the graphs.

Solve each nonlinear system.

13. $2x - y = 9$
$xy = 5$

14. $x - 4 = 3y$
$x^2 + y^2 = 8$

15. $x^2 + y^2 = 25$
$x^2 - 2y^2 = 16$

Graph each inequality or system of inequalities.

16. $y \le x^2 - 2$

17. $2x - 5y \ge 12$
$3x + 4y \le 12$

18. $x^2 + 25y^2 \le 25$
$x^2 + y^2 \le 9$

The orbit of Mercury around the sun (a focus) is an ellipse with equation

$$\frac{x^2}{3352} + \frac{y^2}{3211} = 1,$$

where x and y are measured in million kilometers.

19. Find its apogee, its greatest distance from the sun. (*Hint:* Refer to Section 10.1, Exercise 47.)

20. Find its perigee, its smallest distance from the sun.

CUMULATIVE REVIEW EXERCISES CHAPTERS 1–10

1. Simplify $-10 + |-5| - |3| + 4$.

Solve.

2. $4 - (2x + 3) + x = 5x - 3$

3. $-4k + 7 \ge 6k + 1$

4. $|5m| - 6 = 14$

5. $|2p - 5| > 15$

6. Find the slope of the line through (2, 5) and (−4, 1).

7. Find the equation of the line through (−3, −2) and perpendicular to $2x - 3y = 7$.

Solve each system.

8. $3x - y = 12$
$2x + 3y = -3$

9. $x + y - 2z = 9$
$2x + y + z = 7$
$3x - y - z = 13$

10. $xy = -5$
$2x + y = 3$

Perform the indicated operations.

11. $(5y - 3)^2$

12. $(2r + 7)(6r - 1)$

13. $(8x^4 - 4x^3 + 2x^2 + 13x + 8) \div (2x + 1)$

Factor.

14. $12x^2 - 7x - 10$

15. $2y^4 + 5y^2 - 3$

16. $z^4 - 1$

17. $a^3 - 27b^3$

Simplify.

18. $\dfrac{5x - 15}{24} \cdot \dfrac{64}{3x - 9}$

19. $\dfrac{y^2 - 4}{y^2 - y - 6} \div \dfrac{y^2 - 2y}{y - 1}$

20. $\dfrac{5}{c + 5} - \dfrac{2}{c + 3}$

21. $\dfrac{p}{p^2 + p} + \dfrac{1}{p^2 + p}$

22. Kareem and Jamal want to clean an office they share. Kareem can do the job alone in 3 hours, while Jamal can do it alone in 2 hours. How long will it take them if they work together?

Simplify.

23. $\left(\dfrac{4}{3}\right)^{-1}$

24. $\dfrac{(2a)^{-2}a^4}{a^{-3}}$

25. $4\sqrt[3]{16} - 2\sqrt[3]{54}$

26. $\dfrac{3\sqrt{5x}}{\sqrt{2x}}, \quad x > 0$

27. $\dfrac{5 + 3i}{2 - i}$

Solve for real values of the variable.

28. $2\sqrt{k} = \sqrt{5k + 3}$

29. $10q^2 + 13q = 3$

30. $(4x - 1)^2 = 8$

31. $\log(2x) - \log(x - 1) = \log 3$

32. $2(x^2 - 3)^2 - 5(x^2 - 3) = 12$

33. $F = \dfrac{kwv^2}{r}$; for v

Graph.

34. $3x + y = 5$

35. $f(x) = x^2 - 4x + 5$

36. $f(x) = 3^{x - 1}$

37. $\dfrac{x^2}{4} - \dfrac{y^2}{16} = 1$

38. $\dfrac{x^2}{25} + \dfrac{y^2}{16} \leq 1$

Work each problem.

39. Under certain conditions, the stopping distance of a car traveling 25 miles per hour is 61.7 feet; for a car traveling 35 miles per hour, it is 106 feet. The stopping distance in feet can be described by the equation $y = ax^2 + bx$, where x is the speed in miles per hour. (*Source: National Traffic Safety Institute Student Workbook, 1993.*)

(a) Using the given pairs of values, write a system of equations with a and b as the variables.

(b) Solve the system from part (a). Give answers to the nearest thousandth.

(c) Use the result from part (b) to write a quadratic equation for the stopping distance.

(d) Use the equation from part (c) to find the stopping distance to the nearest tenth of a foot for a car traveling 55 miles per hour.

40. The bar graph shows historic and projected annual online retail sales (in billions of dollars) over the Internet.

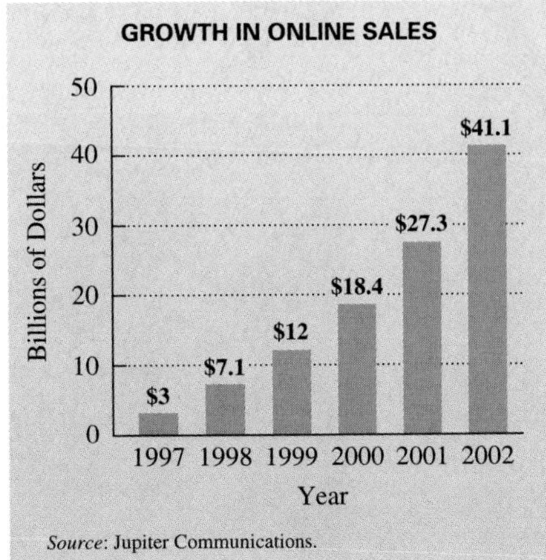

A reasonable model for sales y in billions of dollars is the exponential function defined by $y = 1.38(1.65)^x$. The years are coded so x is the number of years since 1995.

(a) Use the model to find sales in the year 2000. (*Hint:* Let $x = 5$.)

(b) Use the model to estimate sales in the year 2003.

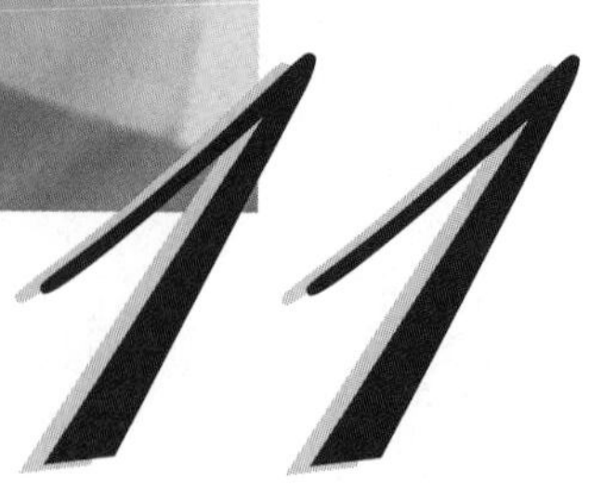

Sequences and Series

Banking/Finance

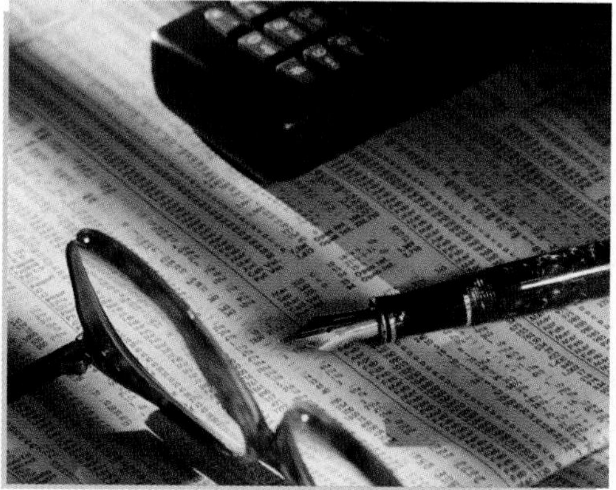

The popularity of mutual funds as a means of investing has shown continued growth over the past decade. The graph below indicates the number of mutual funds for each year during the period 1991–1995.

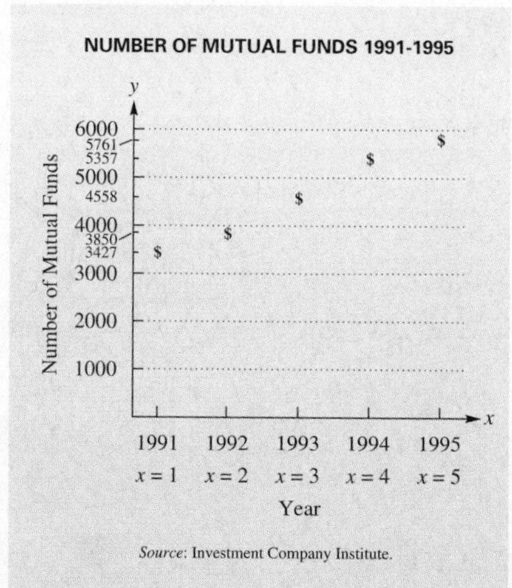

For the mutual fund graph, if we let $x = 1$ represent 1991, $x = 2$ represent 1992, and so on, the function

$$f(x) = -\frac{73}{6}x^4 + \frac{268}{3}x^3 - \frac{268}{3}x^2 + \frac{1489}{6}x + 3191$$

yields the exact function values (the number of funds). This function was determined using statistical methods and is based only on the positive integer inputs 1, 2, 3, 4, 5; thus, it can be used reliably only for these values. Evaluating this function for these five values yields the list, or *sequence*

$$3427, 3850, 4558, 5357, 5761.$$

Because these numbers are in a specific order, we know which year each one represents. For example, since 5357 appears in the fourth position, we know that it represents the fourth year, or 1994. What is $f(5)$? How would you interpret your result?

In this chapter we investigate sequences: functions whose domains consist of positive integers 1, 2, 3, ... The sum of the terms of a sequence is called a *series,* and we will investigate series as well. The example above, dealing with mutual funds, comes from the world of banking and finance, the theme of this chapter. We will return to this example in the exercises for Section 11.1.

11.1 Sequences and Series

OBJECTIVES

1. Find the terms of a sequence given the general term.
2. Find the general term of a sequence.
3. Use sequences to solve applied problems.
4. Use summation notation to evaluate a series.
5. Write a series using summation notation.
6. Find the arithmetic mean (average) of a group of numbers.

FOR EXTRA HELP

SSG Sec. 11.1
SSM Sec. 11.1

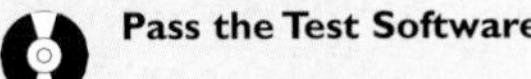
Pass the Test Software

InterAct Math Tutorial Software

Video 20

A **sequence** is a function whose domain is the set of natural numbers. Intuitively, a sequence is a list of numbers in which the order of their appearance is important. Sequences appear in many places in daily life. For instance, the interest portions of monthly loan payments made to pay off an automobile or home loan form a sequence.

In the Palace of the Alhambra, residence of the Moorish rulers of Granada, Spain, the Sultana's quarters feature an interesting architectural pattern. There are 2 matched marble slabs inlaid in the floor, 4 walls, an octagon (8-sided) ceiling, 16 windows, 32 arches, and so on. If this pattern is continued indefinitely, the set of numbers forms an *infinite sequence.*

Infinite Sequence

An **infinite sequence** is a function with the set of positive integers as the domain.

OBJECTIVE 1 **Find the terms of a sequence given the general term.** For any positive integer n, the function value (y-value) of a sequence is written as a_n (read "a sub-n") instead of $a(n)$ or $f(n)$. The function values $a_1, a_2, a_3, \ldots$, written in order, are the **terms** of the sequence, with a_1 the first term, a_2 the second term, and so on. The expression a_n, which defines the sequence, is called the **general term** of the sequence.

In the Palace of the Alhambra example given above, the first five terms of the sequence are

$$a_1 = 2, \quad a_2 = 4, \quad a_3 = 8, \quad a_4 = 16, \quad \text{and} \quad a_5 = 32.$$

The general term for this sequence is $a_n = 2^n$.

EXAMPLE 1 **Writing the Terms of a Sequence from the General Term**

Given an infinite sequence with $a_n = n + \frac{1}{n}$, find the following.

(a) The second term of the sequence

To get a_2, the second term, replace n with 2.

$$a_2 = 2 + \frac{1}{2} = \frac{5}{2}$$

(b) The fifth term

Replace n with 5.

$$a_5 = 5 + \frac{1}{5} = \frac{26}{5}$$

(c) $a_{10} = 10 + \frac{1}{10} = \frac{101}{10}$

(d) $a_{12} = 12 + \frac{1}{12} = \frac{145}{12}$

As mentioned earlier, a sequence is a special kind of function. Graphing calculators can be used to generate and graph sequences, as shown in Figure 1. The calculator must be in graphing dot mode, so the discrete points on the graph are not connected. Remember that the domain of a sequence consists only of natural numbers.

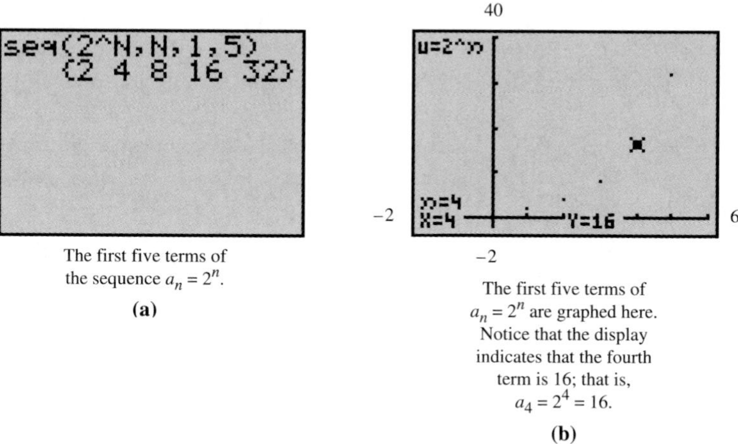

The first five terms of the sequence $a_n = 2^n$.

(a)

The first five terms of $a_n = 2^n$ are graphed here. Notice that the display indicates that the fourth term is 16; that is, $a_4 = 2^4 = 16$.

(b)

Figure 1

OBJECTIVE 2 Find the general term of a sequence. Sometimes we need to find a general term to fit the first few terms of a given sequence. There are no rules for finding the general term of a sequence from the first few terms. In fact, it is possible to give more than one general term that produce the same first three or four terms. However, in many examples, the terms may suggest a general term.

EXAMPLE 2 Finding the General Term of a Sequence

Find an expression for the general term a_n of the sequence

$$5, \quad 10, \quad 15, \quad 20, \quad 25, \ldots.$$

The first term is 5(1), the second is 5(2), and so on. By inspection, $a_n = 5n$ will produce the given first five terms. ■

One problem with using just a few terms to suggest a general term, as in Example 2, is that there may be more than one general term that gives the same first few terms.

OBJECTIVE 3 Use sequences to solve applied problems. Practical problems often involve *finite sequences.*

Finite Sequence

A **finite sequence** has a domain that includes only the first n positive integers.

For example, if n is 5, the domain is $\{1, 2, 3, 4, 5\}$, and the sequence has five terms.

EXAMPLE 3 Using a Sequence in an Application

Keshon borrows \$5000 and agrees to pay \$500 monthly, plus interest of 1% on the unpaid balance from the beginning of that month. Find the payments for the first four months and the remaining debt at the end of this period.

The payments and remaining balances are calculated as follows.

First month	Payment: $500 + .01(5000) = 550$ dollars
	Balance: $5000 - 500 = 4500$ dollars
Second month	Payment: $500 + .01(4500) = 545$ dollars
	Balance: $5000 - 2 \cdot 500 = 4000$ dollars
Third month	Payment: $500 + .01(4000) = 540$ dollars
	Balance: $5000 - 3 \cdot 500 = 3500$ dollars
Fourth month	Payment: $500 + .01(3500) = 535$ dollars
	Balance: $5000 - 4 \cdot 500 = 3000$ dollars

The payments for the first four months, in dollars, are

550, 545, 540, 535

and the remaining debt at the end of this period is 3000 dollars.

OBJECTIVE 4 Use summation notation to evaluate a series. By adding the terms of a sequence, we obtain a *series.*

Series

The indicated sum of the terms of a sequence is called a **series.**

For example, if we consider the sum of the payments listed in Example 3,

$$550 + 545 + 540 + 535,$$

we obtain a series that represents the total amount of payments for the first four months.

Since a sequence can be finite or infinite, there are finite or infinite series. One type of infinite series is discussed in Section 11.3, and the binomial theorem discussed in Section 11.4 defines an important finite series. In this section we discuss only finite series.

We use a compact notation, called **summation notation,** to write a series from the general term of the corresponding sequence. For example, the sum of the first six terms of the sequence with general term $a_n = 3n + 2$ is written with the Greek letter Σ (sigma) as

$$\sum_{i=1}^{6} (3i + 2).$$

We read this as "the sum from $i = 1$ to 6 of $3i + 2$." To find this sum, we replace the letter i in $3i + 2$ with 1, 2, 3, 4, 5, and 6, as follows.

$$\begin{aligned}\sum_{i=1}^{6} (3i + 2) &= (3 \cdot 1 + 2) + (3 \cdot 2 + 2) + (3 \cdot 3 + 2) \\ &\quad + (3 \cdot 4 + 2) + (3 \cdot 5 + 2) + (3 \cdot 6 + 2) \\ &= 5 + 8 + 11 + 14 + 17 + 20 \\ &= 75\end{aligned}$$

The letter i is called the **index of summation.**

CAUTION This use of i has no connection with the complex number i.

EXAMPLE 4 Evaluating a Series Written in Summation Notation

Write out the terms and evaluate each of the following.

(a) $\sum_{i=1}^{5}(i-4) = (1-4)+(2-4)+(3-4)+(4-4)+(5-4)$
$= -3-2-1+0+1$
$= -5$

(b) $\sum_{i=3}^{7} 3i^2 = 3(3)^2+3(4)^2+3(5)^2+3(6)^2+3(7)^2$
$= 27+48+75+108+147$
$= 405$

```
sum(seq(I-4,I,1,
5)
              -5
sum(seq(3I²,I,3,
7)
             405
```

Figure 2

Figure 2 shows how a graphing calculator can be used to obtain the results found in Example 4.

OBJECTIVE 5 Write a series using summation notation. In Example 4, we started with summation notation and wrote each series using + signs. It is possible to go in the other direction; that is, given a series, we can write it using summation notation. To do this, we observe a pattern in the terms and write the general term accordingly.

EXAMPLE 5 Writing a Series with Summation Notation

Write the following sums with summation notation.

(a) $2+5+8+11$

First, find a general term a_n that will give these four terms for a_1, a_2, a_3, and a_4. Inspection (and trial and error) shows that $3i-1$ will work for these four terms, since

$$3(1)-1=2$$
$$3(2)-1=5$$
$$3(3)-1=8$$
$$3(4)-1=11.$$

(Remember, there may be other expressions that also work. These four terms may be the first terms of more than one sequence.) Since i ranges from 1 to 4, write the sum as

$$2+5+8+11=\sum_{i=1}^{4}(3i-1).$$

(b) $8+27+64+125+216$

Since these numbers are the cubes of 2, 3, 4, 5, and 6,

$$8+27+64+125+216=\sum_{i=2}^{6} i^3.$$

OBJECTIVE 6 **Find the arithmetic mean (average) of a group of numbers.** The **arithmetic mean,** or **average,** of a group of numbers is defined as the sum of all the numbers divided by the number of numbers.

Arithmetic Mean or Average

The arithmetic mean, or average, of a group of numbers is symbolized $\bar{x}$ and is found by dividing the sum of the numbers by the number of numbers. That is,

$$\bar{x} = \frac{\sum_{i=1}^{n} x_i}{n}.$$

Here the values of x_i represent the individual numbers in the group, and n represents the number of numbers.

EXAMPLE 6 **Finding the Arithmetic Mean or Average**

The following table shows the number of companies listed on the New York Stock Exchange for each year during the period 1990–1996. What was the average number of listings for this seven-year period?

Year	*Number of Listings*
1990	1774
1991	1885
1992	2088
1993	2361
1994	2570
1995	2675
1996	2907

Source: New York Stock Exchange.

Let $x_1 = 1774$, $x_2 = 1885$, and so on. Since there are 7 numbers in the group, $n = 7$. Therefore,

$$\begin{aligned}\bar{x} &= \frac{\sum_{i=1}^{7} x_i}{7}\\ &= \frac{1774 + 1885 + 2088 + 2361 + 2570 + 2675 + 2907}{7}\\ &= 2323 \quad \text{(rounded to the nearest unit).}\end{aligned}$$

The average number of listings for this seven-year period was 2323.

CONNECTIONS

One of the most famous sequences in mathematics is the **Fibonacci sequence:**

$$1, 1, 2, 3, 5, 8, 13, 21, 34, 55, \ldots.$$

This sequence is named for the Italian mathematician Leonardo of Pisa (1170–1250), who was also known as Fibonacci. The Fibonacci sequence is found in numerous places in nature. For example, male honeybees hatch from eggs that have not been fertilized, so a male bee has only one parent, a female. On the other hand, female honeybees hatch from fertilized eggs, so a female has two parents, one male and one female. The number of ancestors in consecutive generations of bees follows the Fibonacci sequence. Successive terms in the sequence also appear in plants: in the daisy head, the pineapple, and the pine cone, for instance.

FOR DISCUSSION OR WRITING

1. See if you can discover the pattern in the Fibonacci sequence. (*Hint:* Can you explain how to find the next term, given the preceding two terms?)
2. Draw a tree showing the number of ancestors of a male bee in each generation following the description given above.

11.1 EXERCISES

1. Suppose that f is a function with domain all real numbers, where $f(x) = 2x + 4$. Suppose that an infinite sequence is defined by $a_n = 2n + 4$. Discuss the similarities and the differences between the function and the sequence. Give examples using each.
2. What is wrong with the following? For the sequence defined by $a_n = 2n + 4$, find $a_{1/2}$.

Write out the first five terms of each sequence. See Example 1.

3. $a_n = \dfrac{n+3}{n}$
4. $a_n = \dfrac{n+2}{n+1}$
5. $a_n = 3^n$
6. $a_n = 1^{n-1}$
7. $a_n = \dfrac{1}{n^2}$
8. $a_n = \dfrac{n^2}{n+1}$
9. $a_n = (-1)^n$
10. $a_n = (-1)^{2n-1}$

Find the indicated term for each sequence. See Example 1.

11. $a_n = -9n + 2;\ a_8$
12. $a_n = 3n - 7;\ a_{12}$
13. $a_n = \dfrac{3n+7}{2n-5};\ a_{14}$
14. $a_n = \dfrac{5n-9}{3n+8};\ a_{16}$
15. $a_n = (n+1)(2n+3);\ a_8$
16. $a_n = (5n-2)(3n+1);\ a_{10}$

Find a general term, a_n, for the given terms of each sequence. See Example 2.

17. 4, 8, 12, 16, . . .
18. −10, −20, −30, −40, . . .
19. $\dfrac{1}{3}, \dfrac{1}{9}, \dfrac{1}{27}, \dfrac{1}{81}, \ldots$
20. $\dfrac{1}{2}, \dfrac{2}{3}, \dfrac{3}{4}, \dfrac{4}{5}, \ldots$

Solve each applied problem by writing the first few terms of a sequence. See Example 3.

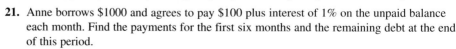

21. Anne borrows \$1000 and agrees to pay \$100 plus interest of 1% on the unpaid balance each month. Find the payments for the first six months and the remaining debt at the end of this period.

22. Larissa Perez is offered a new modeling job with a salary of $20{,}000 + 2500n$ dollars per year at the end of the nth year. Write a sequence showing her salary at the end of each of the first five years. If she continues in this way, what will her salary be at the end of the tenth year?

23. Suppose that an automobile loses $\frac{1}{5}$ of its value each year; that is, at the end of any given year, the value is $\frac{4}{5}$ of the value at the beginning of that year. If a car cost \$20,000 new, what is its value at the end of 5 years?

24. A certain car loses $\frac{1}{2}$ of its value each year. If this car cost \$40,000 new, what is its value at the end of 6 years?

Write out each series and evaluate it. See Example 4.

25. $\sum_{i=1}^{3} (i^2 + 2)$ **26.** $\sum_{i=1}^{4} i(i + 3)$ **27.** $\sum_{i=2}^{5} \frac{1}{i}$ **28.** $\sum_{i=0}^{4} \frac{i}{i + 1}$

29. $\sum_{i=1}^{6} (-1)^i$ **30.** $\sum_{i=1}^{5} (-1)^i \cdot i$ **31.** $\sum_{i=3}^{7} (i - 3)(i + 2)$ **32.** $\sum_{i=2}^{6} \frac{i^2 + 1}{2}$

Write out the terms of each series.

33. $\sum_{i=1}^{5} 2x \cdot i$ **34.** $\sum_{i=1}^{6} x^i$ **35.** $\sum_{i=1}^{5} i \cdot x^i$ **36.** $\sum_{i=2}^{6} \frac{x + i}{x - i}$

Write each series using summation notation. See Example 5.

37. $3 + 4 + 5 + 6 + 7$ **38.** $1 + 4 + 9 + 16$

39. $\frac{1}{2} + \frac{1}{3} + \frac{1}{4} + \frac{1}{5} + \frac{1}{6}$ **40.** $-1 + 2 - 3 + 4 - 5 + 6$

41. Explain the basic difference between a sequence and a series.

42. Evaluate $\sum_{i=1}^{3} 5i$ and $5 \sum_{i=1}^{3} i$. Notice that the sums are the same. Explain how the distributive property plays a role in assuring us that the two sums are equal.

Find the arithmetic mean for each collection of numbers. See Example 6.

43. 8, 11, 14, 9, 3, 6, 8 **44.** 10, 12, 8, 19, 23 **45.** 5, 9, 8, 2, 4, 7, 3, 2 **46.** 2, 1, 4, 8, 3, 7

Solve each problem. See Example 6.

47. As mentioned in the chapter introduction, the number of mutual funds available to investors for each year during the period 1991–1995 is given in the following table.

Year	Number of Funds Available
1991	3427
1992	3850
1993	4558
1994	5357
1995	5761

Source: Investment Company Institute.

To the nearest whole number, what was the average number of funds available during this period?

48. The total assets of mutual funds, in billions of dollars, for each year during the period 1992–1996 are shown in the following table.

Year	Assets (in billions of dollars)
1992	1646.3
1993	2075.4
1994	2161.5
1995	2820.4
1996	3539.2

Source: Investment Company Institute.

To the nearest tenth (in billions of dollars), what were the average assets during this period?

RELATING CONCEPTS (EXERCISES 49–56)

The following properties of series provide useful shortcuts for evaluating series.

If $a_1, a_2, a_3, \ldots, a_n$ and $b_1, b_2, b_3, \ldots, b_n$ are two sequences, and c is a constant, then for every positive integer n,

(a) $\sum_{i=1}^{n} c = nc$

(b) $\sum_{i=1}^{n} ca_i = c\sum_{i=1}^{n} a_i$

(c) $\sum_{i=1}^{n} (a_i + b_i) = \sum_{i=1}^{n} a_i + \sum_{i=1}^{n} b_i$

(d) $\sum_{i=1}^{n} (a_i - b_i) = \sum_{i=1}^{n} a_i - \sum_{i=1}^{n} b_i$.

Work these exercises in order.

49. Use property (c) to write $\sum_{i=1}^{6} (i^2 + 3i + 5)$ as the sum of three summations.

50. Use property (b) to rewrite the second summation from Exercise 49.

51. Use property (a) to rewrite the third summation from Exercise 49.

52. Rewrite $1 + 2 + 3 + 4 + \cdots + n = \frac{n(n + 1)}{2}$ in summation notation.

53. Rewrite $1^2 + 2^2 + 3^2 + 4^2 + \cdots + n^2 = \frac{n(n + 1)(2n + 1)}{6}$ in summation notation.

54. Use the summations you wrote in Exercises 52 and 53 and the properties given above to evaluate the three summations from Exercises 49–51. This gives the value of $\sum_{i=1}^{6} (i^2 + 3i + 5)$ without writing out all six terms.

55. Use the properties and summations given above to evaluate $\sum_{i=1}^{12} (i^2 - i)$.

56. Use the properties and summations given above to evaluate $\sum_{i=1}^{20} (2 + i - i^2)$.

Did you make the connection that evaluating series can be made easier using shortcuts that are true in general?

11.2 Arithmetic Sequences

OBJECTIVES

1. Find the common difference for an arithmetic sequence.
2. Find the general term of an arithmetic sequence.
3. Use an arithmetic sequence in an application.
4. Find any specified term or the number of terms of an arithmetic sequence.
5. Find the sum of a specified number of terms of an arithmetic sequence.

FOR EXTRA HELP

SSG Sec. 11.2
SSM Sec. 11.2

Pass the Test Software

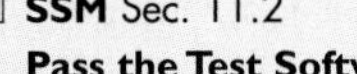

InterAct Math Tutorial Software

Video 20

OBJECTIVE 1 **Find the common difference for an arithmetic sequence.** In this section we introduce a special type of sequence that has many applications.

Arithmetic Sequence

A sequence in which each term after the first differs from the preceding term by a constant amount is called an **arithmetic sequence** or **arithmetic progression.**

For example, the sequence

$$6, 11, 16, 21, 26, \ldots$$

is an arithmetic sequence, since the difference between any two adjacent terms is always 5. The number 5 is called the **common difference** of the arithmetic sequence. The common difference, d, is found by subtracting any pair of terms a_n and a_{n+1}. That is,

$$d = a_{n+1} - a_n.$$

EXAMPLE 1 **Finding the Common Difference**

Find d for the arithmetic sequence

$$-11, -4, 3, 10, 17, 24, \ldots .$$

Since the sequence is arithmetic, d is the difference between any two adjacent terms. Choosing the terms 10 and 17 gives

$$\begin{aligned} d &= 17 - 10 \\ &= 7. \end{aligned}$$

The terms -11 and -4 would give $d = -4 - (-11) = 7$, the same result.

EXAMPLE 2 **Writing the Terms of a Sequence from the First Term and Common Difference**

Write the first five terms of the arithmetic sequence with first term 3 and common difference -2.

The second term is found by adding -2 to the first term 3, getting 1. For the next term, add -2 to 1, and so on. The first five terms are

$$3, 1, -1, -3, -5.$$

OBJECTIVE 2 **Find the general term of an arithmetic sequence.** Generalizing from Example 2, if we know the first term, a_1, and the common difference, d, of an arithmetic sequence, then the sequence is completely defined as

$$a_1, \quad a_2 = a_1 + d, \quad a_3 = a_1 + 2d, \quad a_4 = a_1 + 3d, \ldots .$$

Writing the terms of the sequence in this way suggests the following rule.

General Term of an Arithmetic Sequence

The general term of an arithmetic sequence with first term a_1 and common difference d is

$$a_n = a_1 + (n - 1)d.$$

Since $a_n = a_1 + (n - 1)d = dn + (a_1 - d)$ is a linear function in n, any linear expression of the form $kn + c$, where k and c are real numbers, defines an arithmetic sequence.

EXAMPLE 3 Finding the General Term of an Arithmetic Sequence

Find the general term for the arithmetic sequence

$$-9, -6, -3, 0, 3, 6, \ldots .$$

Then use the general term to find a_{20}.

Here the first term is $a_1 = -9$. To find d, subtract any two adjacent terms. For example,

$$d = -3 - (-6) = 3.$$

Now find a_n.

$a_n = a_1 + (n - 1)d$	Formula for a_n
$= -9 + (n - 1)(3)$	Let $a_1 = -9$, $d = 3$.
$= -9 + 3n - 3$	Distributive property
$a_n = 3n - 12$	Combine terms.

Thus, the general term is $a_n = 3n - 12$. To find a_{20}, let $n = 20$.

$$a_{20} = 3(20) - 12 = 60 - 12 = 48$$

OBJECTIVE 3 Use an arithmetic sequence in an application.

EXAMPLE 4 Applying an Arithmetic Sequence

Howie Sorkin's uncle decides to start a fund for Howie's education. He makes an initial contribution of \$3000 and each month deposits an additional \$500. Thus, after one month there will be \$3000 + \$500 = \$3500. How much will there be after 24 months? (Disregard any interest.)

The contributions can be described using an arithmetic sequence. After n months, the fund will contain

$$a_n = 3000 + 500n \text{ dollars.}$$

To find the amount in the fund after 24 months, find a_{24}.

$a_{24} = 3000 + 500(24)$	Let $n = 24$.
$= 3000 + 12{,}000$	Multiply.
$= 15{,}000$	Add.

The account will contain \$15,000 (disregarding interest) after 24 months.

OBJECTIVE 4 Find any specified term or the number of terms of an arithmetic sequence. The formula for the general term has four variables: a_n, a_1, n, and d. If we know any three of these, the formula can be used to find the value of the fourth variable. The next example shows how to find a particular term.

EXAMPLE 5 Finding a Specified Term

Find the indicated term for each of the following arithmetic sequences.

(a) $a_1 = -6, d = 12;\quad a_{15}$

Use the formula $a_n = a_1 + (n - 1)d$. Since we want $a_n = a_{15}$, $n = 15$.

$$\begin{aligned} a_{15} &= a_1 + (\mathbf{15} - 1)d && \text{Let } n = 15. \\ &= \mathbf{-6} + 14(\mathbf{12}) && \text{Let } a_1 = -6,\ d = 12. \\ &= 162 \end{aligned}$$

(b) $a_5 = 2$ and $a_{11} = -10;\quad a_{17}$

Any term can be found if a_1 and d are known. Use the formula for a_n with the two given terms.

$$\begin{aligned} a_5 &= a_1 + (5 - 1)d & a_{11} &= a_1 + (11 - 1)d \\ a_5 &= a_1 + 4d & a_{11} &= a_1 + 10d \\ 2 &= a_1 + 4d \quad a_5 = 2 & -10 &= a_1 + 10d \quad a_{11} = -10 \end{aligned}$$

This gives a system of two equations with two variables, a_1 and d. Find d by adding -1 times one equation to the other to eliminate a_1.

$$\begin{aligned} -10 &= a_1 + 10d \\ -2 &= -a_1 - 4d && \text{Multiply } 2 = a_1 + 4d \text{ by } -1. \\ \hline -12 &= 6d && \text{Add.} \\ -2 &= d && \text{Divide by 6.} \end{aligned}$$

Now find a_1 by substituting -2 for d into either equation.

$$\begin{aligned} -10 &= a_1 + 10(\mathbf{-2}) && \text{Let } d = -2. \\ -10 &= a_1 - 20 \\ \mathbf{10} &= a_1 \end{aligned}$$

Use the formula for a_n to find a_{17}.

$$\begin{aligned} a_{17} &= a_1 + (\mathbf{17} - 1)d && \text{Let } n = 17. \\ &= a_1 + 16d \\ &= \mathbf{10} + 16(\mathbf{-2}) && \text{Let } a_1 = 10,\ d = -2. \\ &= -22 \end{aligned}$$

Sometimes we need to find out how many terms are in a sequence as shown in the following example.

EXAMPLE 6 Finding the Number of Terms in a Sequence

Find the number of terms in the arithmetic sequence

$$-8, -2, 4, 10, \ldots, 52.$$

Let n represent the number of terms in the sequence. Since $a_n = 52$, $a_1 = -8$, and $d = -2 - (-8) = 6$, use the formula $a_n = a_1 + (n - 1)d$ to find n. Substituting the known values into the formula gives

$$\begin{aligned} a_n &= a_1 + (n-1)d && \\ \mathbf{52} &= \mathbf{-8} + (n-1)\mathbf{6} && \text{Let } a_n = 52,\ a_1 = -8,\ d = 6. \\ 52 &= -8 + 6n - 6 && \text{Distributive property} \\ 66 &= 6n && \text{Combine terms.} \\ n &= 11. && \text{Divide by 6.} \end{aligned}$$

The sequence has 11 terms.

OBJECTIVE 5 Find the sum of a specified number of terms of an arithmetic sequence. To find a formula for the sum, S_n, of the first n terms of an arithmetic sequence, we can write out the terms as

$$S_n = a_1 + (a_1 + d) + (a_1 + 2d) + \cdots + [a_1 + (n-1)d].$$

This same sum can be written in reverse as

$$S_n = a_n + (a_n - d) + (a_n - 2d) + \cdots + [a_n - (n-1)d].$$

Now add the corresponding terms of these two expressions for S_n to get

$$2S_n = (a_1 + a_n) + (a_1 + a_n) + (a_1 + a_n) + \cdots + (a_1 + a_n).$$

The right-hand side of this expression contains n terms, each equal to $a_1 + a_n$, so

$$2S_n = n(a_1 + a_n)$$

$$S_n = \frac{n}{2}(a_1 + a_n).$$

EXAMPLE 7 Finding the Sum of the First n Terms

Find the sum of the first five terms of the arithmetic sequence in which $a_n = 2n - 5$.

We can use the formula $S_n = \frac{n}{2}(a_1 + a_n)$ to find the sum of the first five terms. Here $n = 5$, $a_1 = 2(1) - 5 = -3$, and $a_5 = 2(5) - 5 = 5$. From the formula,

$$S_5 = \frac{5}{2}(-3 + 5) = \frac{5}{2}(2) = 5.$$

It is sometimes useful to express the sum of an arithmetic sequence, S_n, in terms of a_1 and d, the quantities that define the sequence. We can do this as follows. Since

$$S_n = \frac{n}{2}(a_1 + a_n) \qquad \text{and} \qquad a_n = a_1 + (n-1)d,$$

by substituting the expression for a_n into the expression for S_n, we get

$$S_n = \frac{n}{2}(a_1 + [a_1 + (n-1)d])$$

$$S_n = \frac{n}{2}[2a_1 + (n-1)d].$$

The following summary gives both of the alternative forms that may be used to find the sum of the first n terms of an arithmetic sequence.

Sum of the First n Terms of an Arithmetic Sequence

The sum of the first n terms of the arithmetic sequence with first term a_1, nth term a_n, and common difference d is

$$S_n = \frac{n}{2}(a_1 + a_n) \quad \text{or} \quad S_n = \frac{n}{2}[2a_1 + (n - 1)d].$$

EXAMPLE 8 Finding the Sum of the First n Terms

Find the sum of the first 8 terms of the arithmetic sequence having first term 3 and common difference -2.

Since the known values, $a_1 = 3$, $d = -2$, and $n = 8$, appear in the second formula for S_n, we use it.

$$S_n = \frac{n}{2}[2a_1 + (\text{n} - 1)d]$$

$$S_8 = \frac{8}{2}[2(3) + (8 - 1)(-2)] \qquad \text{Let } a_1 = 3,\ d = -2,\ n = 8.$$

$$= 4[6 - 14]$$

$$= -32$$

As mentioned above, linear expressions of the form $kn + c$, where k and c are real numbers, define an arithmetic sequence. For example, the sequences defined by $a_n = 2n + 5$ and $a_n = n - 3$ are arithmetic sequences. For this reason,

$$\sum_{i=1}^{n} (ki + c)$$

represents the sum of the first n terms of an arithmetic sequence having first term $a_1 = k(1) + c = k + c$ and general term $a_n = k(n) + c = kn + c$. We can find this sum with the first formula for S_n given above, as shown in the next example.

EXAMPLE 9 Using S_n to Evaluate a Summation

Find $\sum_{i=1}^{12} (2i - 1)$.

This is the sum of the first 12 terms of the arithmetic sequence having $a_n = 2n - 1$. This sum, S_{12}, is found with the formula for S_n,

$$S_n = \frac{n}{2}(a_1 + a_n).$$

Here $n = 12$, $a_1 = 2(1) - 1 = 1$, $a_{12} = 2(12) - 1 = 23$. Substitute these values into the formula to get

$$S_{12} = \frac{12}{2}(1 + 23) = 6(24) = 144.$$

Figure 3 shows how a graphing calculator supports the result of Example 9.

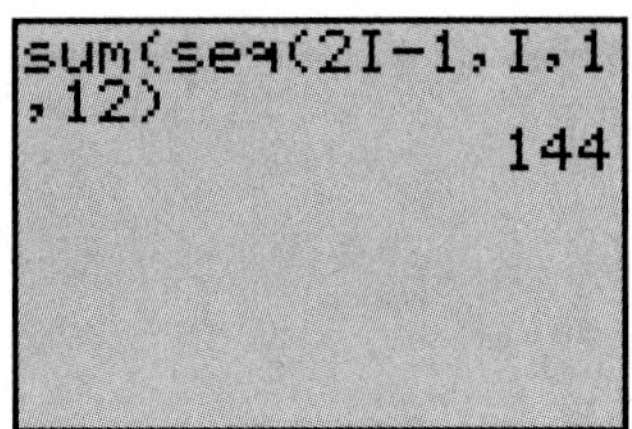

Figure 3

11.2 EXERCISES

1. Using several examples, explain the meaning of *arithmetic sequence.*
2. Can any two terms of an arithmetic sequence be used to find the common difference? Explain.

If the given sequence is arithmetic, find the common difference, d. If the sequence is not arithmetic, say so. See Example 1.

3. $1, 2, 3, 4, 5, \ldots$
4. $2, 5, 8, 11, \ldots$
5. $2, -4, 6, -8, 10, -12, \ldots$
6. $-6, -10, -14, -18, \ldots$
7. $-10, -5, 0, 5, 10, \ldots$
8. $1, 2, 4, 7, 11, 16, \ldots$
9. $3.42, 5.57, 7.72, 9.87, \ldots$
10. $1, \frac{3}{2}, 2, \frac{5}{2}, 3, \frac{7}{2}, \ldots$
11. $-\frac{5}{3}, -1, -\frac{1}{3}, \frac{1}{3}, \ldots$
12. $\frac{1}{2}, \frac{1}{3}, \frac{1}{4}, \frac{1}{5}, \frac{1}{6}, \ldots$

Use the formula for a_n to find the general term for each arithmetic sequence. See Example 3.

13. $a_1 = 2, d = 5$
14. $a_1 = 5, d = -3$
15. $3, \frac{15}{4}, \frac{9}{2}, \frac{21}{4}, \ldots$
16. $4, 14, 24, \ldots$
17. $-3, 0, 3, \ldots$
18. $-10, -5, 0, 5, 10, \ldots$

Find the indicated term for each arithmetic sequence. See Examples 2 and 5.

19. $a_1 = 4, d = 3;\ \ a_{25}$
20. $a_1 = 1, d = -\frac{1}{2};\ \ a_{12}$
21. $2, 4, 6, \ldots;\ \ a_{24}$
22. $1, 5, 9, \ldots;\ \ a_{50}$
23. $a_{12} = -45, a_{10} = -37;\ \ a_1$
24. $a_{10} = -2, a_{15} = -8;\ \ a_3$

Find the number of terms in each arithmetic sequence. See Example 6.

25. $3, 5, 7, \ldots, 33$
26. $2, \frac{3}{2}, 1, \frac{1}{2}, \ldots, -5$
27. $\frac{3}{4}, 3, \frac{21}{4}, \ldots, 12$
28. $4, 1, -2, \ldots, -32$
29. In the formulas for S_n, what does n represent?
30. Explain when you would use each of the two formulas for S_n.

RELATING CONCEPTS (EXERCISES 31–34)

The following exercises show how to find the sum $1 + 2 + 3 + \cdots + 99 + 100$ in an ingenious way.

Work these exercises in order.

31. Consider the following:

$$\begin{aligned} S &= 1 + 2 + 3 + \cdots + 99 + 100 \\ S &= 100 + 99 + 98 + \cdots + 2 + 1. \end{aligned}$$

Add the left sides of this equation. The result is ______. Add the columns on the right side. The sum ______ appears ______ times, so by multiplication, the sum of the right sides of the equations is ______.

32. Form an equation by setting the sum of the left sides equal to the sum of the right sides.

33. Solve the equation from Exercise 32 to find that the desired sum, S, is ______.

34. Find the sum $S = 1 + 2 + 3 + \cdots + 199 + 200$ using the procedure described in Exercises 31–33.

Did you make the connection that a sum of the form $1 + 2 + \cdots + n$ can be found using a pattern?

Find S_6 for each arithmetic sequence. See Examples 7 and 8.

35. $a_1 = 6, d = 3$ **36.** $a_1 = 5, d = 4$ **37.** $a_1 = 7, d = -3$

38. $a_1 = -5, d = -4$ **39.** $a_n = 4 + 3n$ **40.** $a_n = 9 + 5n$

Use a formula for S_n to evaluate each series. See Example 9.

41. $\sum_{i=1}^{10} (8i - 5)$ **42.** $\sum_{i=1}^{17} (i - 1)$ **43.** $\sum_{i=1}^{20} (2i - 5)$

44. $\sum_{i=1}^{10} \left(\frac{1}{2}i - 1\right)$ **45.** $\sum_{i=1}^{250} i$ **46.** $\sum_{i=1}^{2000} i$

Solve each applied problem. (Hint: *Determine whether you need to find a specific term of a sequence or the sum of the terms of a sequence immediately after reading the problem.) See Example 4.*

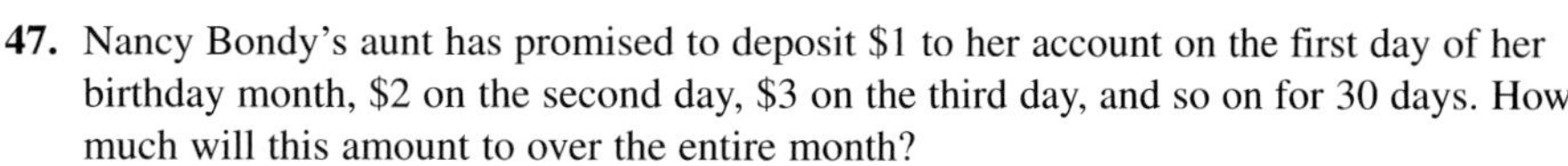

47. Nancy Bondy's aunt has promised to deposit \$1 to her account on the first day of her birthday month, \$2 on the second day, \$3 on the third day, and so on for 30 days. How much will this amount to over the entire month?

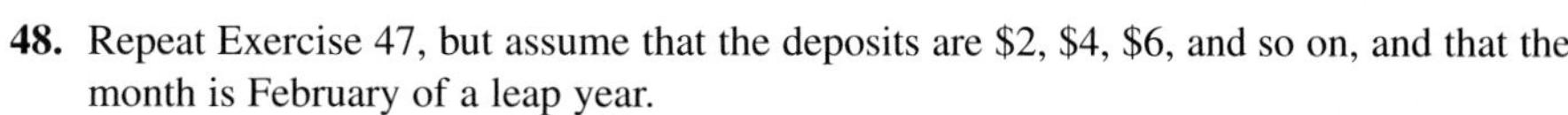

48. Repeat Exercise 47, but assume that the deposits are \$2, \$4, \$6, and so on, and that the month is February of a leap year.

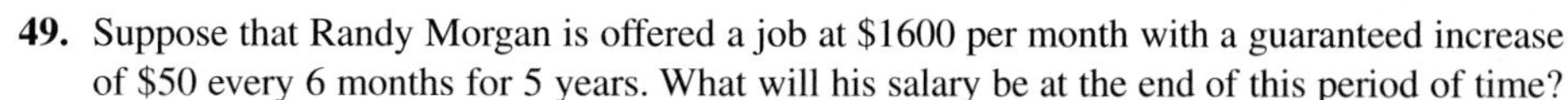

49. Suppose that Randy Morgan is offered a job at \$1600 per month with a guaranteed increase of \$50 every 6 months for 5 years. What will his salary be at the end of this period of time?

50. Repeat Exercise 49, but assume that the starting salary is \$2000 per month, and the guaranteed increase is \$100 every 4 months for 3 years.

51. A seating section in a theater-in-the-round has 20 seats in the first row, 22 in the second row, 24 in the third row, and so on for 25 rows. How many seats are there in the last row? How many seats are there in the section?

52. José Valdevielso has started on a fitness program. He plans to jog 10 minutes per day for the first week, and then add 10 minutes per day each week until he is jogging an hour each day. In which week will this occur? What is the total number of minutes he will run during the first four weeks?

53. A child builds with blocks, placing 35 blocks in the first row, 31 in the second row, 27 in the third row, and so on. Continuing this pattern, can she end with a row containing exactly 1 block? If not, how many blocks will the last row contain? How many rows can she build this way?

54. A stack of firewood has 28 pieces on the bottom, 24 on top of those, then 20, and so on. If there are 108 pieces of wood, how many rows are there? (*Hint:* $n \le 7$.)

RELATING CONCEPTS (EXERCISES 55–58)

Let $f(x) = mx + b$. ***Work these exercises in order.***

55. Find $f(1), f(2)$, and $f(3)$.

56. Consider the sequence $f(1), f(2), f(3), \ldots$. Is it an arithmetic sequence?

57. If the sequence is arithmetic, what is the common difference?

58. What is a_n for the sequence described in Exercise 56?

Did you make the connection between a linear function and an arithmetic sequence?

11.3 Geometric Sequences

OBJECTIVES

1. Find the common ratio of a geometric sequence.
2. Find the general term of a geometric sequence.
3. Find any specified term of a geometric sequence.
4. Find the sum of a specified number of terms of a geometric sequence.
5. Apply the formula for the future value of an ordinary annuity.
6. Find the sum of an infinite number of terms of certain geometric sequences.

FOR EXTRA HELP

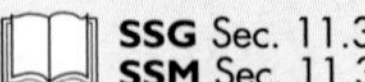

SSG Sec. 11.3
SSM Sec. 11.3

Pass the Test Software

InterAct Math Tutorial Software

Video 20

In an arithmetic sequence, each term after the first is found by *adding* a fixed number to the previous term. A *geometric sequence* is defined as follows.

Geometric Sequence

A **geometric sequence** or **geometric progression** is a sequence in which each term after the first is a constant multiple of the preceding term.

OBJECTIVE 1 **Find the common ratio of a geometric sequence.** We find the constant multiplier, called the **common ratio,** by dividing any term after the first by the preceding term. That is, the common ratio is

$$r = \frac{a_{n+1}}{a_n}.$$

For example,

$$2, 6, 18, 54, 162, \ldots$$

is a geometric sequence in which the first term, a_1, is 2 and the common ratio is

$$r = \frac{6}{2} = \frac{18}{6} = \frac{54}{18} = \frac{162}{54} = 3.$$

EXAMPLE 1 **Finding the Common Ratio**

Find r for the geometric sequence

$$15, \frac{15}{2}, \frac{15}{4}, \frac{15}{8}, \ldots.$$

To find r, choose any two adjacent terms and divide the second one by the first. Choosing the second and third terms of the sequence,

$$r = \frac{a_3}{a_2} = \frac{15}{4} \div \frac{15}{2} = \frac{1}{2}.$$

Any other two adjacent terms could have been used to find r. Additional terms of the sequence can be found by multiplying each successive term by $\frac{1}{2}$.

OBJECTIVE 2 **Find the general term of a geometric sequence.** The general term a_n of a geometric sequence $a_1, a_2, a_3, \ldots$ is expressed in terms of a_1 and r by writing the first few terms as

$$a_1, \quad a_2 = a_1 r, \quad a_3 = a_1 r^2, \quad a_4 = a_1 r^3, \ldots,$$

which suggests the following rule.

General Term of a Geometric Sequence

The general term of the geometric sequence with first term a_1 and common ratio r is

$$a_n = a_1 r^{n-1}.$$

Be careful to use the correct order of operations when finding $a_1 r^{n-1}$. The value of r^{n-1} must be found first. Then multiply the result by a_1.

EXAMPLE 2 **Finding the General Term**

Find the general term of the sequence in Example 1.

The first term is $a_1 = 15$ and the common ratio is $r = \frac{1}{2}$. Substituting into the formula for the general term gives

$$a_n = a_1 r^{n-1} = 15\left(\frac{1}{2}\right)^{n-1},$$

the required general term. Notice that it is not possible to simplify further, because the exponent must be applied before the multiplication can be done.

OBJECTIVE 3 **Find any specified term of a geometric sequence.** We can use the formula for the general term to find any particular term.

EXAMPLE 3 **Finding a Specified Term**

Find the indicated term for each geometric sequence.

(a) $a_1 = 4, r = -3; \quad a_6$

Let $n = 6$. From the general term $a_n = a_1 r^{n-1}$,

$a_6 = a_1 \cdot r^{6-1}$	Let $n = 6$.
$= 4 \cdot (-3)^5$	Let $a_1 = 4$, $r = -3$.
$= -972.$	Evaluate $(-3)^5$ first.

(b) $\frac{3}{4}, \frac{3}{8}, \frac{3}{16}, \ldots; \quad a_7$

Here, $r = \frac{1}{2}$, $a_1 = \frac{3}{4}$, and n is 7.

$$a_7 = \frac{3}{4} \cdot \left(\frac{1}{2}\right)^6 = \frac{3}{4} \cdot \frac{1}{64} = \frac{3}{256}$$

EXAMPLE 4 Writing the Terms of a Sequence

Write the first five terms of the geometric sequence whose first term is 5 and whose common ratio is $\frac{1}{2}$.

Using the formula $a_n = a_1 r^{n-1}$,

$$a_1 = 5,$$
$$a_2 = 5\left(\frac{1}{2}\right) = \frac{5}{2},$$
$$a_3 = 5\left(\frac{1}{2}\right)^2 = \frac{5}{4},$$
$$a_4 = 5\left(\frac{1}{2}\right)^3 = \frac{5}{8},$$
$$a_5 = 5\left(\frac{1}{2}\right)^4 = \frac{5}{16}.$$

OBJECTIVE 4 Find the sum of a specified number of terms of a geometric sequence. It is convenient to have a formula for the sum of the first n terms of a geometric sequence, S_n. We can develop a formula by first writing out S_n.

$$S_n = a_1 + a_1 r + a_1 r^2 + a_1 r^3 + \cdots + a_1 r^{n-1}$$

Next, we multiply both sides by r.

$$rS_n = a_1 r + a_1 r^2 + a_1 r^3 + a_1 r^4 + \cdots + a_1 r^n$$

We subtract the first result from the second.

$$\begin{aligned} rS_n - S_n = & (a_1 r - a_1) + (a_1 r^2 - a_1 r) + (a_1 r^3 - a_1 r^2) \\ & + (a_1 r^4 - a_1 r^3) + \cdots + (a_1 r^n - a_1 r^{n-1}) \end{aligned}$$

Using the commutative and associative properties to rearrange the terms on the right, we get

$$\begin{aligned} rS_n - S_n = & (a_1 r - a_1 r) + (a_1 r^2 - a_1 r^2) \\ & + (a_1 r^3 - a_1 r^3) + \cdots + (a_1 r^n - a_1) \end{aligned} \qquad \text{Distributive property}$$

so, if $r \neq 1$,

$$S_n = \frac{a_1 r^n - a_1}{r - 1} = \frac{a_1(r^n - 1)}{r - 1}. \qquad \text{Divide by } r - 1.$$

A summary of this discussion follows.

Sum of the First n Terms of a Geometric Sequence

The sum of the first n terms of the geometric sequence with first term a_1 and common ratio r is

$$S_n = \frac{a_1(r^n - 1)}{r - 1} \quad (r \neq 1).$$

If $r = 1$, $S_n = a_1 + a_1 + a_1 + \cdots + a_1 = na_1$.

Multiplying the formula for S_n by $\frac{-1}{-1}$ gives us an alternative form that is sometimes preferable.

$$S_n = \frac{a_1(r^n - 1)}{r - 1} \cdot \frac{-1}{-1} = \frac{a_1(1 - r^n)}{1 - r}$$

EXAMPLE 5 Finding the Sum of the First n Terms

Find the sum of the first six terms of the geometric sequence with first term -2 and common ratio 3.

Substitute $n = 6$, $a_1 = -2$, and $r = 3$ into the formula for S_n.

$$S_n = \frac{a_1(r^n - 1)}{r - 1}$$

$$S_6 = \frac{-2(3^6 - 1)}{3 - 1} \quad \text{Let } n = 6,\ a_1 = -2,\ r = 3.$$

$$= \frac{-2(729 - 1)}{2} \quad \text{Evaluate the exponential.}$$

$$= -728$$

A series of the form

$$\sum_{i=1}^{n} a \cdot b^i$$

represents the sum of the first n terms of a geometric sequence having first term $a_1 = a \cdot b^1 = ab$ and common ratio b. The next example illustrates this form.

EXAMPLE 6 Using the Formula for S_n to Find a Summation

Find $\sum_{i=1}^{4} 3 \cdot 2^i$.

Since the series is in the form

$$\sum_{i=1}^{n} a \cdot b^i,$$

it represents the sum of the first n terms of the geometric sequence with $a_1 = a \cdot b^1$ and $r = b$. The sum is found by using the formula

$$S_n = \frac{a_1(r^n - 1)}{r - 1}.$$

Here $n = 4$. Also, $a_1 = 6$ and $r = 2$. Now substitute into the formula for S_n.

$$S_4 = \frac{6(2^4 - 1)}{2 - 1} \qquad \text{Let } n = 4,\ a_1 = 6,\ r = 2.$$

$$= \frac{6(16 - 1)}{1} \qquad \text{Evaluate } 2^4.$$

$$= 90$$

Figure 4 shows how a graphing calculator can store the terms in a list, and then find the sum of these terms. This supports the result of Example 6.

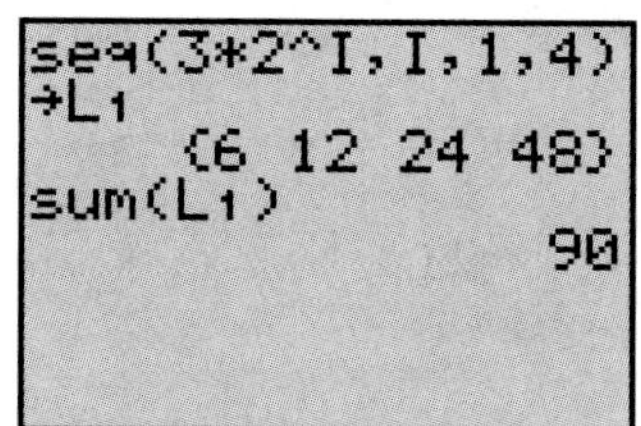

Figure 4

OBJECTIVE 5 Apply the formula for the future value of an ordinary annuity. A sequence of equal payments made at equal periods of time is called an **annuity.** If the payments are made at the end of the time period, and if the frequency of payments is the same as the frequency of compounding, the annuity is called an **ordinary annuity.** The time between payments is the **payment period,** and the time from the beginning of the first payment period to the end of the last period is called the **term of the annuity.** The **future value of the annuity,** the final sum on deposit, is defined as the sum of the compound amounts of all the payments, compounded to the end of the term.

For example, suppose \$1500 is deposited at the end of the year for the next 6 years in an account paying 8% per year compounded annually.

To find the future value of this annuity, look separately at each of the \$1500 payments. The first of these payments will produce a compound amount of

$$1500(1 + .08)^5 = 1500(1.08)^5.$$

Use 5 as the exponent instead of 6 since the money is deposited at the *end* of the first year and earns interest for only 5 years. The second payment of \$1500 will produce a compound amount of $1500(1.08)^4$. Continuing in this way and finding the sum of all the terms gives

$$1500(1.08)^5 + 1500(1.08)^4 + 1500(1.08)^3 + 1500(1.08)^2 + 1500(1.08)^1 + 1500.$$

(The last payment earns no interest at all.) Reading this in reverse order, we see that it is just the sum of the first six terms of a geometric sequence with $a_1 = 1500$, $r = 1.08$, and $n = 6$. Therefore, the sum is

$$\frac{a_1(r^n - 1)}{r - 1} = \frac{1500[(1.08)^6 - 1]}{1.08 - 1} = 11{,}003.89$$

or \$11,003.89.

We state the following formula without proof.

Future Value of an Ordinary Annuity

$$S = R\left[\frac{(1 + i)^n - 1}{i}\right]$$

where

S is future value,
R is the payment at the end of each period,
i is the interest rate per period,
n is the number of periods.

EXAMPLE 7 Applying the Formula for the Future Value of an Annuity

(a) Rocky Rhodes is an athlete who feels that his playing career will last 7 years. To prepare for his future, he deposits \$22,000 at the end of each year for 7 years in an account paying 6% compounded annually. How much will he have on deposit after 7 years?

His payments form an ordinary annuity with $R = 22{,}000$, $n = 7$, and $i = .06$. The future value of this annuity (by the formula above) is

$$S = 22{,}000\left[\frac{(1.06)^7 - 1}{.06}\right] = 184{,}664.43, \qquad \text{Use a calculator.}$$

or \$184,664.43.

(b) Experts say that the baby boom generation (born between 1946 and 1960) cannot count on a company pension or Social Security to provide a comfortable retirement, as their parents did. It is recommended that they start to save early and regularly. Judy Zahrndt, a baby boomer, has decided to deposit \$200 at the end of each month in an account that pays interest of 7.2% compounded monthly for retirement in 20 years. How much will be in the account at that time?

Because the interest is compounded monthly, $i = \frac{.072}{12}$. Also, $R = 200$ and $n = 12(20)$. The future value is

$$S = 200\left[\frac{\left(1 + \frac{.072}{12}\right)^{12(20)} - 1}{\frac{.072}{12}}\right] = 106{,}752.47,$$

or \$106,752.47. ■

OBJECTIVE 6 Find the sum of an infinite number of terms of certain geometric sequences. Now, consider an infinite geometric sequence such as

$$\frac{1}{3}, \frac{1}{6}, \frac{1}{12}, \frac{1}{24}, \frac{1}{48}, \ldots.$$

Can the sum of the terms of such a sequence be found somehow? The sum of the first two terms is

$$S_2 = \frac{1}{3} + \frac{1}{6} = \frac{1}{2} = .5.$$

In a similar manner,

$$S_3 = S_2 + \frac{1}{12} = \frac{1}{2} + \frac{1}{12} = \frac{7}{12} \approx .583,$$

$$S_4 = S_3 + \frac{1}{24} = \frac{7}{12} + \frac{1}{24} = \frac{15}{24} = .625,$$

$$S_5 = \frac{31}{48} \approx .64583,$$

$$S_6 = \frac{21}{32} = .65625,$$

$$S_7 = \frac{127}{192} \approx .6614583.$$

Each term of the geometric sequence is smaller than the preceding one, so each additional term is contributing less and less to the sum. In decimal form (to the nearest thousandth) the first seven terms and the tenth term are given below.

Term	a_1	a_2	a_3	a_4	a_5	a_6	a_7	a_{10}
Value	.333	.167	.083	.042	.021	.010	.005	.001

As the table suggests, the value of a term gets closer and closer to zero as the number of the term increases. To express this idea, we say that as n increases without bound (written $n \to \infty$), the limit of the term a_n is zero, written

$$\lim_{n \to \infty} a_n = 0.$$

A number that can be defined as the sum of an infinite number of terms of a geometric sequence can be found by starting with the expression for the sum of a finite number of terms:

$$S_n = \frac{a_1(r^n - 1)}{r - 1}.$$

If $|r| < 1$, then as n increases without bound the value of r^n gets closer and closer to zero. For example, in the infinite sequence discussed above, $r = \frac{1}{2} = .5$. The chart below shows how $r^n = (.5)^n$, given to the nearest thousandth, gets smaller as n increases.

n	1	2	3	4	5	6	7	10
r^n	.5	.25	.125	.063	.031	.016	.008	.001

As r^n approaches 0, $r^n - 1$ approaches $0 - 1 = -1$, and S_n approaches the quotient $\frac{-a_1}{r - 1}$. Thus,

$$\lim_{r^n \to 0} S_n = \lim_{r^n \to 0} \frac{a_1(r^n - 1)}{r - 1}$$

$$= \frac{a_1(0 - 1)}{r - 1}$$

$$= \frac{-a_1}{r - 1} = \frac{a_1}{1 - r}.$$

This limit is defined to be the sum of the infinite geometric sequence:

$$a_1 + a_1r + a_1r^2 + a_1r^3 + \cdots = \frac{a_1}{1-r}, \quad \text{if } |r| < 1.$$

What happens if $|r| > 1$? For example, suppose the sequence is

$$6, 12, 24, \ldots, 3(2)^n, \ldots.$$

In this kind of sequence, as n increases, the value of r^n also increases and so does the sum S_n. Since each new term adds a larger and larger amount to the sum, there is no limit to the value of S_n, and the sum S_n does not exist. A similar situation exists if $r = 1$.

In summary, the sum of the terms of an infinite geometric sequence is as follows.

Sum of the Terms of an Infinite Geometric Sequence

The sum S of the terms of an infinite geometric sequence with first term a_1 and common ratio r, where $|r| < 1$, is

$$S = \frac{a_1}{1-r}.$$

If $|r| \geq 1$, the sum does not exist.

EXAMPLE 8 Finding the Sum of the Terms of an Infinite Geometric Sequence

Find the sum of the terms of the infinite geometric sequence with $a_1 = 3$ and $r = -\frac{1}{3}$.

From the rule above, the sum is

$$S = \frac{a_1}{1-r} = \frac{3}{1-(-1/3)}$$
$$= \frac{3}{4/3}$$
$$= \frac{9}{4}.$$

In summation notation, the sum of an infinite geometric sequence is written as

$$\sum_{i=1}^{\infty} a_i.$$

For instance, the sum in Example 8 would be written

$$\sum_{i=1}^{\infty} 3\left(-\frac{1}{3}\right)^{i-1}.$$

EXAMPLE 9 Finding the Sum of the Terms of an Infinite Geometric Series

Find $\displaystyle\sum_{i=1}^{\infty} \left(\frac{1}{2}\right)^i$.

This is the infinite geometric series

$$\frac{1}{2} + \frac{1}{4} + \frac{1}{8} + \cdots,$$

with $a_1 = \frac{1}{2}$ and $r = \frac{1}{2}$. Since $|r| < 1$, we find the sum as follows.

$$S = \frac{a_1}{1-r} = \frac{\frac{1}{2}}{1-\frac{1}{2}} = \frac{\frac{1}{2}}{\frac{1}{2}} = 1$$

11.3 EXERCISES

1. Using several examples, explain the meaning of *geometric sequence.*
2. Explain why the sequence 5, 5, 5, 5, . . . can be considered either arithmetic or geometric.

If the given sequence is geometric, find the common ratio, r. If the sequence is not geometric, say so. See Example 1.

3. 4, 8, 16, 32, . . .
4. 5, 15, 45, 135, . . .
5. $\frac{1}{3}, \frac{2}{3}, \frac{3}{3}, \frac{4}{3}, \frac{5}{3}, \ldots$
6. $\frac{1}{3}, \frac{2}{3}, \frac{4}{3}, \frac{8}{3}, \ldots$
7. $1, -3, 9, -27, 81, \ldots$
8. $1, -3, 7, -11, \ldots$
9. $1, -\frac{1}{2}, \frac{1}{4}, -\frac{1}{8}, \frac{1}{16}, \ldots$
10. $\frac{2}{3}, \frac{2}{15}, \frac{2}{75}, \frac{2}{375}, \ldots$

Find a general term for each geometric sequence. See Example 2.

11. 5, 10, . . .
12. $-2, -6, \ldots$
13. $\frac{1}{9}, \frac{1}{3}, \ldots$
14. $-3, \frac{3}{2}, \ldots$
15. $10, -2, \ldots$
16. $-4, 8, \ldots$

Find the indicated term for each geometric sequence. See Example 3.

17. $2, 10, 50, \ldots; \quad a_{10}$
18. $-1, -3, -9, \ldots; \quad a_{15}$
19. $\frac{1}{2}, \frac{1}{6}, \frac{1}{18}, \ldots; \quad a_{12}$
20. $\frac{2}{3}, -\frac{1}{3}, \frac{1}{6}, \ldots; \quad a_{18}$
21. $a_3 = \frac{1}{2}, a_7 = \frac{1}{32}; \quad a_{25}$
22. $a_5 = 48, a_8 = -384; \quad a_{10}$

RELATING CONCEPTS (EXERCISES 23–26)

In Chapter 1 we learned that any repeating decimal is a rational number; that is, it can be expressed as a quotient of integers. Thus, the repeating decimal

$$.99999\ldots,$$

an endless string of 9s, must be a rational number.

Work Exercises 23–26 in order *to discover the surprising simplest form of this rational number.*

23. Use long division or your previous experience to write a repeating decimal representation for $\frac{1}{3}$.

RELATING CONCEPTS (EXERCISES 23–26) (CONTINUED)

24. Use long division or your previous experience to write a repeating decimal representation for $\frac{2}{3}$.

25. Because $\frac{1}{3} + \frac{2}{3} = 1$, the sum of the decimal representations in Exercises 23 and 24 must also equal 1. Line up the decimals in the usual vertical method for addition, and obtain the repeating decimal result. The value of this decimal is exactly 1.

26. The repeating decimal .99999 . . . can be written as the sum of the terms of a geometric sequence with $a_1 = .9$ and $r = .1$:

$$.99999\ldots = .9 + .9(.1) + .9(.1)^2 + .9(.1)^3 + .9(.1)^4 + .9(.1)^5 + \cdots.$$

Since $|.1| < 1$, this sum can be found using the formula $S = \frac{a_1}{1 - r}$. Use this formula to support the result you found another way in Exercises 23–25.

Did you make the connection that, although it may not seem to be true, the value of .99999 . . . is 1?

Use the formula for S_n to find the sum for each geometric sequence. See Examples 5 and 6. In Exercises 29–34, give the answer to the nearest thousandth.

27. $\frac{1}{3}, \frac{1}{9}, \frac{1}{27}, \frac{1}{81}, \frac{1}{243}$

28. $\frac{4}{3}, \frac{8}{3}, \frac{16}{3}, \frac{32}{3}, \frac{64}{3}, \frac{128}{3}$

29. $-\frac{4}{3}, -\frac{4}{9}, -\frac{4}{27}, -\frac{4}{81}, -\frac{4}{243}, -\frac{4}{729}$

30. $\frac{5}{16}, -\frac{5}{32}, \frac{5}{64}, -\frac{5}{128}, \frac{5}{256}$

31. $\sum_{i=1}^{7} 4\left(\frac{2}{5}\right)^i$

32. $\sum_{i=1}^{8} 5\left(\frac{2}{3}\right)^i$

33. $\sum_{i=1}^{10} (-2)\left(\frac{3}{5}\right)^i$

34. $\sum_{i=1}^{6} (-2)\left(-\frac{1}{2}\right)^i$

Solve each problem involving an ordinary annuity. See Example 7.

35. A father opened a savings account for his daughter on the day she was born, depositing \$1000. Each year on her birthday he deposits another \$1000, making the last deposit on her twenty-first birthday. If the account pays 9.5% interest compounded annually, how much is in the account at the end of the day on the daughter's twenty-first birthday?

36. A 45-year-old man puts \$1000 in a retirement account at the end of each quarter ($\frac{1}{4}$ of a year) until he reaches the age of 60. If the account pays 11% annual interest compounded quarterly, how much will be in the account at that time?

37. At the end of each quarter a 50-year-old woman puts \$1200 in a retirement account that pays 7% interest compounded quarterly. When she reaches age 60, she withdraws the entire amount and places it in a mutual fund that pays 9% interest compounded monthly. From then on she deposits \$300 in the mutual fund at the end of each month. How much is in the account when she reaches age 65?

38. John Bray deposits \$10,000 at the beginning of each year for 12 years in an account paying 5% compounded annually. He then puts the total amount on deposit in another account paying 6% compounded semiannually for another 9 years. Find the final amount on deposit after the entire 21-year period.

Find the sum, if it exists, of the terms of each infinite geometric sequence. See Examples 8 and 9.

39. $a_1 = 6, r = \frac{1}{3}$

40. $a_1 = 10, r = \frac{1}{5}$

41. $a_1 = 1000, r = -\frac{1}{10}$

42. $a_1 = 8500, r = \frac{3}{5}$

43. $\sum_{i=1}^{\infty} \frac{9}{8}\left(-\frac{2}{3}\right)^i$

44. $\sum_{i=1}^{\infty} \frac{3}{5}\left(\frac{5}{6}\right)^i$

45. $\sum_{i=1}^{\infty} \frac{12}{5}\left(\frac{5}{4}\right)^i$

46. $\sum_{i=1}^{\infty} \left(-\frac{16}{3}\right)\left(-\frac{9}{8}\right)^i$

*Solve each application. (*Hint: *Determine whether you need to find a specific term of a sequence or the sum of the terms of a sequence immediately after reading the problem.)*

47. A certain ball when dropped from a height rebounds $\frac{3}{5}$ of the original height. How high will the ball rebound after the fourth bounce if it was dropped from a height of 10 feet?

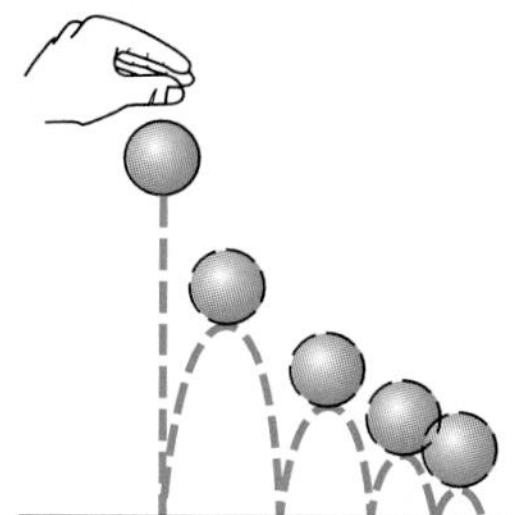

48. A fully wound yo-yo has a string 40 inches long. It is allowed to drop and on its first rebound, it returns to a height 15 inches lower than its original height. Assuming this "rebound ratio" remains constant until the yo-yo comes to rest, how far does it travel on its third trip up the string?

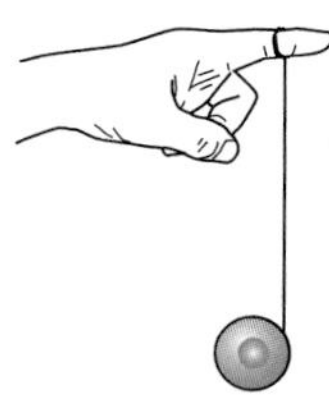

49. A particular substance decays in such a way that it loses half its weight each day. In how many days will 256 grams of the substance be reduced to 32 grams? How much of the substance is left after 10 days?

50. A tracer dye is injected into a system with an input and an excretion. After one hour $\frac{2}{3}$ of the dye is left. At the end of the second hour $\frac{2}{3}$ of the remaining dye is left, and so on. If one unit of the dye is injected, how much is left after 6 hours?

51. In a certain community the consumption of electricity has increased about 6% per year.
(a) If a community uses 1.1 billion units of electricity now, how much will it use five years from now?
(b) Find how many years it will take for the consumption to double.

52. Suppose the community in Exercise 51 reduces its increase in consumption to 2% per year.
(a) How much will it use five years from now?
(b) Find the number of years for the consumption to double.

53. A machine depreciates by $\frac{1}{4}$ of its value each year. If it cost $50,000 new, what is its value after 8 years?

54. Refer to Exercise 48. Theoretically, how far does the yo-yo travel before coming to rest?

RELATING CONCEPTS (EXERCISES 55–58)

Let $g(x) = ab^x$. ***Work Exercises 55–58 in order.***

55. Find $g(1)$, $g(2)$, and $g(3)$.

56. Consider the sequence $g(1), g(2), g(3), \ldots$. Is it a geometric sequence?

57. If the sequence is geometric, what is the common ratio?

58. What is a_n for the sequence described in Exercise 56?

Did you make the connection between an exponential function and a geometric sequence?

11.4 The Binomial Theorem

OBJECTIVES

1. Expand a binomial raised to a power.
2. Find any specified term of the expansion of a binomial.

FOR EXTRA HELP

SSG Sec. 11.4
SSM Sec. 11.4
Pass the Test Software
InterAct Math Tutorial Software
Video 20

OBJECTIVE 1 Expand a binomial raised to a power. Writing out the binomial expression $(x + y)^n$ for nonnegative integer values of n gives a family of expressions that is important in the study of mathematics and its applications. For example,

$$(x + y)^0 = 1,$$
$$(x + y)^1 = x + y,$$
$$(x + y)^2 = x^2 + 2xy + y^2,$$
$$(x + y)^3 = x^3 + 3x^2y + 3xy^2 + y^3,$$
$$(x + y)^4 = x^4 + 4x^3y + 6x^2y^2 + 4xy^3 + y^4,$$
$$(x + y)^5 = x^5 + 5x^4y + 10x^3y^2 + 10x^2y^3 + 5xy^4 + y^5.$$

Inspection shows that these expansions follow a pattern. By identifying the pattern we can write a general expression for $(x + y)^n$.

First, if n is a positive integer, each expansion after $(x + y)^0$ begins with x raised to the same power to which the binomial is raised. That is, the expansion of $(x + y)^1$ has a first term of x^1, the expansion of $(x + y)^2$ has a first term of x^2, the expansion of $(x + y)^3$ has a first term of x^3, and so on. Also, the last term in each expansion is y to this same power, so the expansion of $(x + y)^n$ should begin with the term x^n and end with the term y^n.

The exponents on x decrease by one in each term after the first, while the exponents on y, beginning with y in the second term, increase by one in each succeeding term. Thus the *variables* in the expansion of $(x + y)^n$ have the following pattern.

$$x^n, \quad x^{n-1}y, \quad x^{n-2}y^2, \quad x^{n-3}y^3, \ldots, xy^{n-1}, \quad y^n$$

This pattern suggests that the sum of the exponents on x and y in each term is n. For example, in the third term above, the variable part is $x^{n-2}y^2$ and the sum of the exponents, $n - 2$ and 2, is n.

Now examine the pattern for the *coefficients* in the terms of the expansions shown above. Writing the coefficients alone in a triangular pattern gives **Pascal's triangle,** in honor of the seventeenth-century mathematician Blaise Pascal, one of the first to use it extensively.

Pascal's Triangle

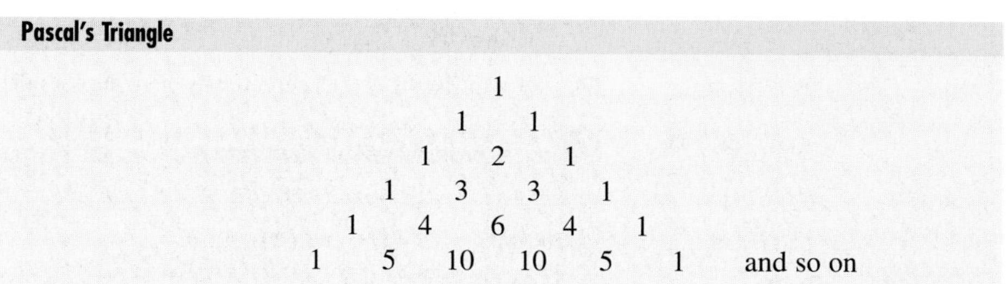

Arranging the coefficients in this way shows that each number in the triangle is the sum of the two numbers just above it (one to the right and one to the left). For example, starting with 1 as the first row, in the fifth row from the top, 1 is the sum of 1 (the only number above it), 4 is the sum of 1 and 3, while 6 is the sum of 3 and 3, and so on.

We get the coefficients for $(x + y)^6$ by attaching the seventh row to the table by adding pairs of numbers from the sixth row.

1 6 15 20 15 6 1

Use these coefficients to expand $(x + y)^6$ as

$$(x + y)^6 = x^6 + 6x^5y + 15x^4y^2 + 20x^3y^3 + 15x^2y^4 + 6xy^5 + y^6.$$

CONNECTIONS

Over the years, many interesting patterns have been discovered in Pascal's triangle. In the figure below, the triangular array is written in a different form. The indicated sums along the diagonals shown are the terms of the *Fibonacci sequence,* mentioned in Section 11.1. The presence of this sequence in the triangle apparently was not recognized by Pascal.

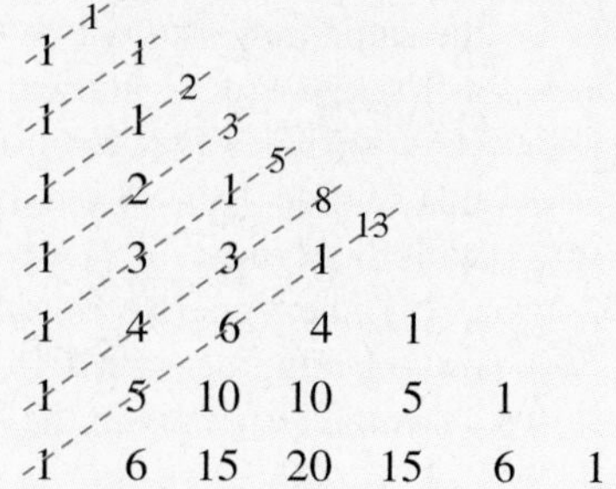

CONNECTIONS (CONTINUED)

Triangular numbers are found by counting the number of points in triangular arrangements of points. The first few triangular numbers are shown below.

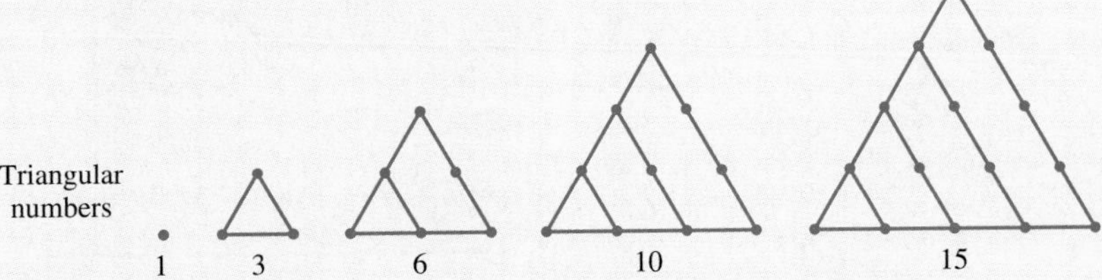

The number of points in these figures form the sequence 1, 3, 6, 10, . . . , a sequence that is found in Pascal's triangle, as shown in the next figure.

FOR DISCUSSION OR WRITING

1. Predict the next two numbers in the sequence of sums of the diagonals of Pascal's triangle.
2. Predict the next five numbers in the list of triangular numbers.
3. Describe other sequences that can be found in Pascal's triangle.

Although it is possible to use Pascal's triangle to find the coefficients of $(x + y)^n$ for any positive integer value of n, it is impractical for large values of n. A more efficient way to determine these coefficients is needed. It is helpful to use the following notational shorthand. The symbol $\boldsymbol{n!}$ (read "n factorial") is defined as follows.

n Factorial ($n!$)

For any positive integer n,

$$n(n-1)(n-2)(n-3) \cdots (2)(1) = n!.$$

For example,

$$3! = 3 \cdot 2 \cdot 1 = 6 \qquad \text{and} \qquad 5! = 5 \cdot 4 \cdot 3 \cdot 2 \cdot 1 = 120.$$

From the definition of n factorial, $n[(n-1)!] = n!$. If $n = 1$, then $1(0!) = 1! = 1$. Because of this, $0!$ is defined as

$$0! = 1.$$

Scientific and graphing calculators have the capability of computing factorials. The three factorial expressions above are shown in Figure 5(a). Figure 5(b) shows some larger factorials.

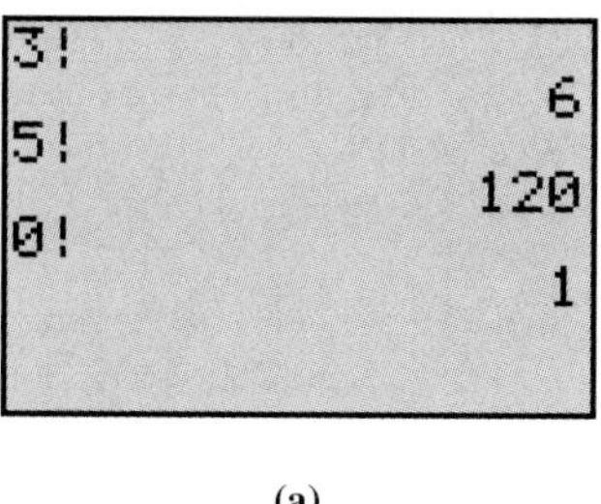

(a)

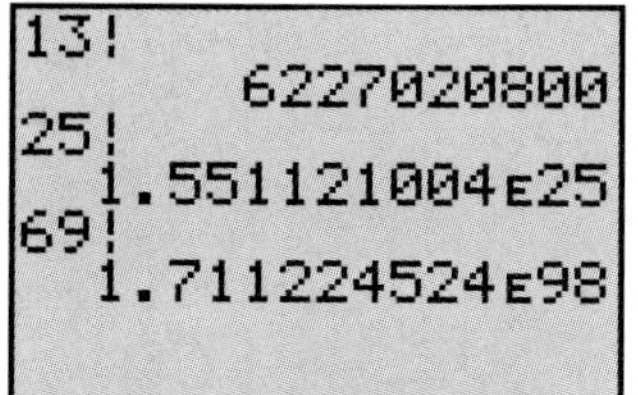

A graphing calculator with a 10-digit display will give the exact value of n! for $n \le 13$ and approximate values of n! for $14 \le n \le 69$.

(b)

Figure 5

EXAMPLE 1 Evaluating Expressions with n!

Find the value of each of the following.

(a) $\dfrac{5!}{4!\,1!} = \dfrac{5 \cdot 4 \cdot 3 \cdot 2 \cdot 1}{(4 \cdot 3 \cdot 2 \cdot 1)(1)} = 5$

(b) $\dfrac{5!}{3!\,2!} = \dfrac{5 \cdot 4 \cdot 3 \cdot 2 \cdot 1}{(3 \cdot 2 \cdot 1)(2 \cdot 1)} = \dfrac{5 \cdot 4}{2 \cdot 1} = 10$

(c) $\dfrac{6!}{3!\,3!} = \dfrac{6 \cdot 5 \cdot 4 \cdot 3 \cdot 2 \cdot 1}{(3 \cdot 2 \cdot 1)(3 \cdot 2 \cdot 1)} = \dfrac{6 \cdot 5 \cdot 4}{3 \cdot 2 \cdot 1} = 20$

(d) $\dfrac{4!}{4!\,0!} = \dfrac{4 \cdot 3 \cdot 2 \cdot 1}{(4 \cdot 3 \cdot 2 \cdot 1)(1)} = 1$

Now look again at the coefficients of the expansion

$$(x + y)^5 = x^5 + 5x^4y + 10x^3y^2 + 10x^2y^3 + 5xy^4 + y^5.$$

The coefficient of the second term is 5 and the exponents on the variables in that term are 4 and 1. From Example 1(a), $5!/(4!\,1!) = 5$. The coefficient of the third term is 10, and the exponents are 3 and 2. From Example 1(b), $5!/(3!\,2!) = 10$. Similar results hold true for the remaining terms. The first term can be written as $1x^5y^0$ and the last term can be written as $1x^0y^5$. Then the coefficient of the first term should be $5!/(5!\,0!) = 1$, and the coefficient of the last term would be $5!/(0!\,5!) = 1$. Generalizing, the coefficient for a term of $(x + y)^n$ in which the variable part is x^ry^{n-r} will be

$$\frac{n!}{r!(n-r)!}.$$

 The denominator factorials in the coefficient of a term are the same as the exponents on the variables in that term.

The expression $\dfrac{n!}{r!(n-r)!}$ is often represented by the symbol ${}_nC_r$. This comes from the fact that if we choose *combinations* of n things taken r at a time, the result is given

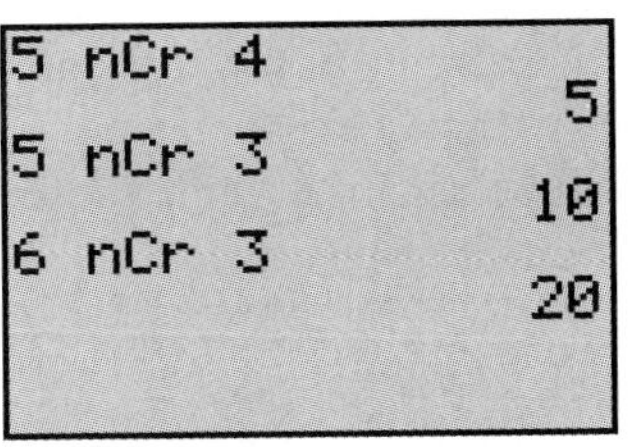

Figure 6

by that expression. A graphing calculator can evaluate this expression for particular values of n and r. Figure 6 shows how a calculator evaluates ${}_5C_4$, ${}_5C_3$, and ${}_6C_3$. Compare these results to parts (a), (b), and (c) of Example 1.

Summarizing this work gives the **binomial theorem,** or the **general binomial expansion.**

Binomial Theorem

For any positive integer n,

$$(x+y)^n = x^n + \frac{n!}{(n-1)!\,1!}x^{n-1}y + \frac{n!}{(n-2)!\,2!}x^{n-2}y^2 + \frac{n!}{(n-3)!\,3!}x^{n-3}y^3 + \cdots + \frac{n!}{1!(n-1)!}xy^{n-1} + y^n.$$

The binomial theorem can be written in summation notation as

$$(x+y)^n = \sum_{i=0}^{n} \frac{n!}{(n-i)!i!}x^{n-i}y^i.$$

NOTE The letter i is used here instead of r because we are using summation notation. It is not the imaginary number i.

EXAMPLE 2 Using the Binomial Theorem

Expand $(2m+3)^4$.

$$\begin{aligned}(2m+3)^4 &= (2m)^4 + \frac{4!}{3!\,1!}(2m)^3(3) + \frac{4!}{2!\,2!}(2m)^2(3)^2 + \frac{4!}{1!\,3!}(2m)(3)^3 + 3^4\\ &= 16m^4 + 4(8m^3)(3) + 6(4m^2)(9) + 4(2m)(27) + 81\\ &= 16m^4 + 96m^3 + 216m^2 + 216m + 81\end{aligned}$$

EXAMPLE 3 Using the Binomial Theorem

Expand $\left(a - \frac{b}{2}\right)^5$.

$$\begin{aligned}\left(a-\frac{b}{2}\right)^5 &= a^5 + \frac{5!}{4!\,1!}a^4\left(-\frac{b}{2}\right) + \frac{5!}{3!\,2!}a^3\left(-\frac{b}{2}\right)^2 + + \frac{5!}{2!\,3!}a^2\left(-\frac{b}{2}\right)^3\\ &\quad + \frac{5!}{1!\,4!}a\left(-\frac{b}{2}\right)^4 + \left(-\frac{b}{2}\right)^5\\ &= a^5 + 5a^4\left(-\frac{b}{2}\right) + 10a^3\left(\frac{b^2}{4}\right) + 10a^2\left(-\frac{b^3}{8}\right)\\ &\quad + 5a\left(\frac{b^4}{16}\right) + \left(-\frac{b^5}{32}\right)\\ &= a^5 - \frac{5}{2}a^4b + \frac{5}{2}a^3b^2 - \frac{5}{4}a^2b^3 + \frac{5}{16}ab^4 - \frac{1}{32}b^5\end{aligned}$$

CAUTION When the binomial is the *difference* of two terms as in Example 3, the signs of the terms in the expansion will alternate. Those terms with odd exponents on the second variable expression $\left(-\frac{b}{2}\right.$ in Example 3$\left.\right)$ will be negative, while those with even exponents on the second variable expression will be positive.

OBJECTIVE 2 Find any specified term of the expansion of a binomial. Any single term of a binomial expansion can be determined without writing out the whole expansion. For example, if $n \geq 10$, the tenth term of $(x + y)^n$ has y raised to the ninth power (since y has the power of 1 in the second term, the power of 2 in the third term, and so on). Since the exponents on x and y in any term must have a sum of n, the exponent on x in the tenth term is $n - 9$. These quantities, 9 and $n - 9$, determine the factorials in the denominator of the coefficient. Thus,

$$\frac{n!}{(n-9)!\,9!}x^{n-9}y^9$$

is the tenth term of $(x + y)^n$. A generalization of this idea follows.

rth Term of the Binomial Expansion

If $n \geq r - 1$, the rth term of the expansion of $(x + y)^n$ is

$$\frac{n!}{[n-(r-1)]!(r-1)!}x^{n-(r-1)}y^{r-1}.$$

This general expression is confusing. Remember to start with the exponent on y, which is 1 less than the term number r. Then subtract that exponent from n to get the exponent on $x : n - (r - 1)$. The two exponents are then used as the factorials in the denominator of the coefficient.

EXAMPLE 4 Finding a Single Term of a Binomial Expansion

Find the fourth term of $(a + 2b)^{10}$.

In the fourth term, $2b$ has an exponent of $4 - 1 = 3$ and a has an exponent of $10 - 3 = 7$. The fourth term is

$$\begin{aligned}\frac{10!}{7!\,3!}(a^7)(2b)^3 &= \frac{10 \cdot 9 \cdot 8}{3 \cdot 2 \cdot 1}(a^7)(8b^3)\\ &= 120a^7(8b^3)\\ &= 960a^7b^3.\end{aligned}$$

11.4 EXERCISES

Evaluate each of the following. See Example 1.

1. $2!$
2. $7!$
3. $\frac{6!}{4!\,2!}$
4. $\frac{7!}{3!\,4!}$
5. ${}_6C_2$
6. ${}_7C_4$
7. $\frac{4!}{0!\,4!}$
8. $\frac{5!}{5!\,0!}$

TECHNOLOGY INSIGHTS (EXERCISES 9–12)

Predict the answer that the calculator will display for the given entry.

9. 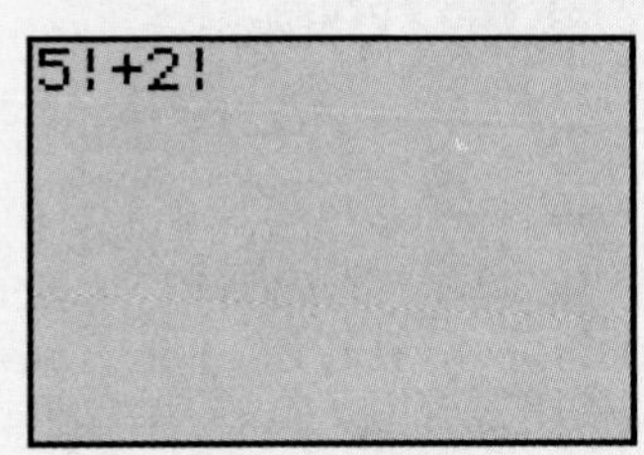

10. 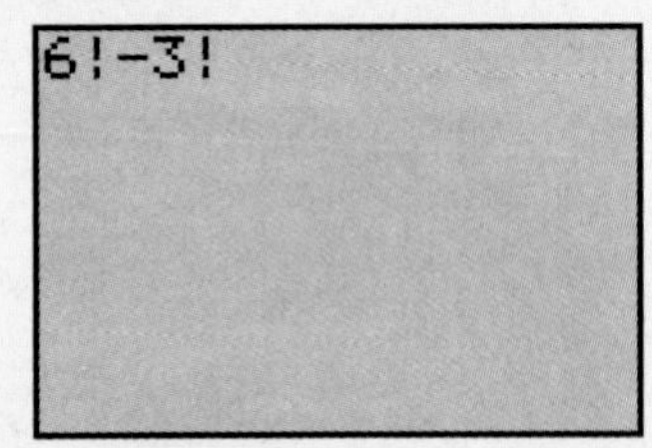

11. 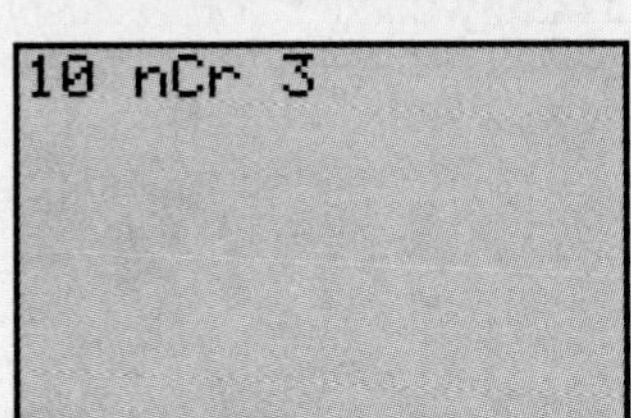

12. 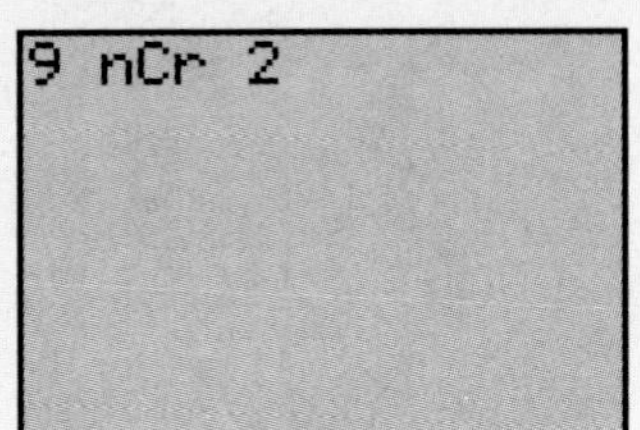

Use the binomial theorem to expand each of the following. See Examples 2 and 3.

13. $(m + n)^4$

14. $(x + r)^5$

15. $(a - b)^5$

16. $(p - q)^4$

17. $(2x + 3)^3$

18. $\left(\frac{x}{2} - y\right)^4$

19. $\left(\frac{x}{3} + 2y\right)^5$

20. $(x^2 + 1)^4$

21. $(mx - n^2)^3$

22. $(2p^2 - q^2)^3$

Write the first four terms of each binomial expansion.

23. $(r + 2s)^{12}$

24. $(m - n)^{20}$

25. $(3x - y)^{14}$

26. $(2p + 3q)^{11}$

27. $(t^2 + u^2)^{10}$

28. $(x^2 - y^2)^{15}$

Find the indicated term of each binomial expansion. See Example 4.

29. $(2m + n)^{10}$; fourth term

30. $(a - 3b)^{12}$; fifth term

31. $\left(x + \frac{y}{2}\right)^8$; seventh term

32. $(3p - 2q)^{15}$; eighth term

33. $(k - 1)^9$; third term

34. $(-4 - s)^{11}$; fourth term

35. The middle term of $(x^2 - 2y)^6$

36. The middle term of $(m^3 + 3)^8$

37. The term with x^9y^4 in $(3x^3 - 4y^2)^5$

38. The term with x^{10} in $\left(x^3 - \frac{2}{x}\right)^6$

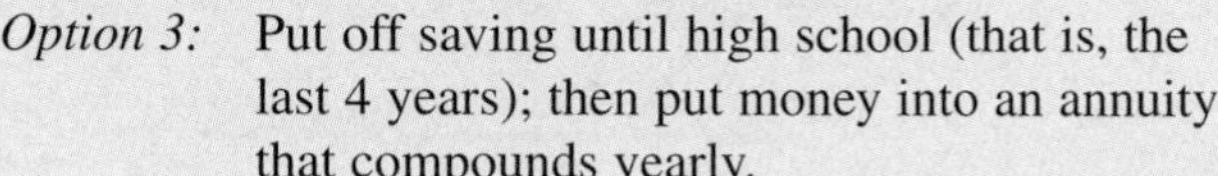

CHAPTER 11 GROUP ACTIVITY

Investing for the Future

Objective: Calculate compound interest; understand the effects of compounding monthly, quarterly and yearly.

In this chapter you have seen many different types of financing and investing options including loans, mutual funds, savings accounts, and annuities. In this activity you will analyze annuities with different periods of compound interest.

Consider a family that has a 10-year-old child, for whom they want to save for college. The family is considering three different savings options.

Option 1: An annuity that compounds quarterly.

Option 2: An annuity that compounds monthly.

Option 3: Put off saving until high school (that is, the last 4 years); then put money into an annuity that compounds yearly.

A. Have each member of your group calculate total savings for one of the three options. Use the following formula, where S is future value, R is payment amount made each *period,* i is annual interest rate divided by the number of periods per year, n is total number of compounding periods, along with the specific information given below for each option.

$$S = R\left[\frac{(1 + i)^n - 1}{i}\right]$$

Option 1: The family plans to save \$300 per quarter (3 months), the interest rate is 5%, and the annuity compounds quarterly. Savings will be for 8 years.

Option 2: The family plans to save \$100 per month, the interest rate is 5%, and the annuity compounds monthly. Again, savings will be for 8 years.

Option 3: The family plans to wait and save only the last 4 years. They will save \$2400 a year, the interest rate is 5%, and the annuity compounds yearly.

B. Compare your answers.

1. How much is being invested using each option?
2. Which option resulted in the largest amount of savings?
3. Explain why this option produced more savings.
4. What other considerations might be involved in deciding how to save?

CHAPTER 11 SUMMARY

KEY TERMS

11.1 **sequence**
infinite sequence
terms of a sequence
general term
finite sequence
series
summation notation
index of summation
arithmetic mean (average)

11.2 **arithmetic sequence (arithmetic progression)**
common difference

11.3 **geometric sequence (geometric progression)**
common ratio
annuity
ordinary annuity
payment period
future value of an annuity
term of an annuity

11.4 **Pascal's triangle**
binomial theorem (general binomial expansion)

NEW SYMBOLS

Symbol	Meaning
a_n	nth term of a sequence
$\sum_{i=1}^{n} a_i$	summation notation
S_n	sum of first n terms of a sequence
$\lim_{n \to \infty} a_n$	limit of a_n as n gets larger and larger
$\sum_{i=1}^{\infty} a_i\ a_i$	sum of an infinite number of terms
$n!$	n factorial
${}_nC_r$	binomial coefficient (combinations of n things taken r at a time)

TEST YOUR WORD POWER

See how well you have learned the vocabulary in this chapter. Answers, with examples, are given at the bottom of the page.

1. An **infinite sequence** is
(a) the values of a function
(b) a function whose domain is the set of natural numbers
(c) the sum of the terms of a function
(d) the average of a group of numbers.

2. A **series** is
(a) the sum of the terms of a sequence
(b) the product of the terms of a sequence
(c) the average of the terms of a sequence
(d) the function values of a sequence.

3. An **arithmetic sequence** is a sequence in which
(a) each term after the first is a constant multiple of the preceding term
(b) the numbers are written in a triangular array
(c) the terms are added
(d) each term after the first differs from the preceding term by a common amount.

4. A **geometric sequence** is a sequence in which
(a) each term after the first is a constant multiple of the preceding term
(b) the numbers are written in a triangular array
(c) the terms are multiplied
(d) each term after the first differs from the preceding term by a common amount.

5. The **common difference** is
(a) the average of the terms in a sequence
(b) the constant multiplier in a geometric sequence
(c) the difference between any two adjacent terms in an arithmetic sequence
(d) the sum of the terms of an arithmetic sequence.

6. The **common ratio** is
(a) the average of the terms in a sequence
(b) the constant multiplier in a geometric sequence
(c) the difference between any two adjacent terms in an arithmetic sequence
(d) the product of the terms of a geometric sequence.

Answers to Test Your Word Power

1. (b) *Example:* The ordered list of numbers 3, 6, 9, 12, 15, . . . is an infinite sequence. **2.** (a) *Example:* $3 + 6 + 9 + 12 + 15$, written in summation notation as $\sum_{i=1}^{5} 3i$, is a series. **3.** (d) *Example:* The sequence $-3, 2, 7, 12, 17, \ldots$ is arithmetic. **4.** (a) *Example:* The sequence 1, 4, 16, 64, 256, . . . is geometric. **5.** (c) *Example:* The common difference of the arithmetic sequence in Item 3 above is 5 since $2 - (-3) = 5$, $7 - 2 = 5$, $12 - 7 = 5$, and so on. **6.** (b) *Example:* The common ratio of the geometric sequence in Item 4 above is 4 since $\frac{4}{1} = \frac{16}{4} = \frac{64}{16} = \frac{256}{64} = 4$.

QUICK REVIEW

CONCEPTS	EXAMPLES
11.1 SEQUENCES AND SERIES	
Sequence	$1, \frac{1}{2}, \frac{1}{3}, \frac{1}{4}, \ldots, \frac{1}{n}$
General Term a_n	has general term $\frac{1}{n}$.
Series	The corresponding series is the *sum* $1 + \frac{1}{2} + \frac{1}{3} + \frac{1}{4} + \cdots + \frac{1}{n}$.
11.2 ARITHMETIC SEQUENCES	
Assume a_1 is the first term, a_n is the nth term, and d is the common difference.	The arithmetic sequence 2, 5, 8, 11, . . . has $a_1 = 2$.
Common Difference $d = a_{n+1} - a_n$	$d = 5 - 2 = 3$ (Any two successive terms could have been used.)
***n*th Term** $a_n = a_1 + (n - 1)d$	Suppose that $n = 10$. Then the 10th term is $a_{10} = 2 + (10 - 1)3 = 2 + 9 \cdot 3 = 29$.
Sum of the First *n* Terms $S_n = \frac{n}{2}(a_1 + a_n)$ or $S_n = \frac{n}{2}[2a_1 + (n - 1)d]$	The sum of the first 10 terms is $S_{10} = \frac{10}{2}(2 + a_{10}) = 5(2 + 29) = 5(31) = 155$ or $S_{10} = \frac{10}{2}[2(2) + (10 - 1)3] = 5(4 + 9 \cdot 3) = 5(4 + 27) = 5(31) = 155$.
11.3 GEOMETRIC SEQUENCES	
Assume a_1 is the first term, a_n is the nth term, and r is the common ratio.	The geometric sequence 1, 2, 4, 8, . . . has $a_1 = 1$.
Common Ratio $r = \frac{a_{n+1}}{a_n}$	$r = \frac{8}{4} = 2$ (Any two successive terms could have been used.)
***n*th Term** $a_n = a_1 r^{n-1}$	Suppose that $n = 6$. Then the 6th term is $a_6 = (1)(2)^{6-1} = 1(2)^5 = 32$.
Sum of the First *n* Terms $S_n = \frac{a_1(r^n - 1)}{r - 1} \quad (r \neq 1)$	The sum of the first 6 terms is $S_6 = \frac{1(2^6 - 1)}{2 - 1} = \frac{64 - 1}{1} = 63$.
Future Value of an Ordinary Annuity $S = R\left[\frac{(1 + i)^n - 1}{i}\right]$,	If \$5800 is deposited into an ordinary annuity at the end of each quarter for four years, and interest is earned at 6.4% compounded quarterly, then $R = \$5800,\ i = \frac{.064}{4} = .016,\ n = 4(4) = 16$,

CONCEPTS	EXAMPLES
where S is future value, R is the payment at the end of each period, i is the interest rate per period, and n is the number of periods.	and $S = 5800\left[\frac{(1+.016)^{16}-1}{.016}\right] = \$104{,}812.44.$
Sum of the Terms of an Infinite Geometric Sequence with $\lvert r \rvert < 1$ $S = \frac{a_1}{1-r}$	The sum S of the terms of an infinite geometric sequence with $a_1 = 1$ and $r = \frac{1}{2}$ is $S = \frac{1}{1-\frac{1}{2}} = \frac{1}{\frac{1}{2}} = 2.$

11.4 THE BINOMIAL THEOREM

For any positive integer n,

$$n(n-1)(n-2)\ldots(2)(1) = n!.$$

$$0! = 1$$

$$_nC_r = \frac{n!}{r!(n-r)!}$$

$$4! = 4 \cdot 3 \cdot 2 \cdot 1 = 24$$

$$_5C_3 = \frac{5!}{3!\,(5-3)!} = \frac{5!}{3!\,2!} = \frac{5 \cdot 4 \cdot 3 \cdot 2 \cdot 1}{3 \cdot 2 \cdot 1 \cdot 2 \cdot 1} = 10$$

General Binomial Expansion

For any positive integer n,

$$(x+y)^n = x^n + \frac{n!}{(n-1)!\,1!}x^{n-1}y + \frac{n!}{(n-2)!\,2!}x^{n-2}y^2 + \frac{n!}{(n-3)!\,3!}x^{n-3}y^3 + \cdots + \frac{n!}{1!(n-1)!}xy^{n-1} + y^n.$$

$$\begin{aligned}(2m+3)^4 &= (2m)^4 + \frac{4!}{3!\,1!}(2m)^3(3) + \frac{4!}{2!\,2!}(2m)^2(3)^2 + \frac{4!}{1!\,3!}(2m)(3)^3 + 3^4 \\ &= 2^4m^4 + 4(2)^3m^3(3) + 6(2)^2m^2(9) + 4(2m)(27) + 81 \\ &= 16m^4 + 12(8)m^3 + 54(4)m^2 + 216m + 81 \\ &= 16m^4 + 96m^3 + 216m^2 + 216m + 81\end{aligned}$$

***r*th Term of the Binomial Expansion of $(x + y)^n$**

$$\frac{n!}{[n-(r-1)]!\,(r-1)!}x^{n-(r-1)}y^{r-1}$$

The 8th term of $(a - 2b)^{10}$ is

$$\frac{10!}{3!\,7!}a^3(-2b)^7 = \frac{10 \cdot 9 \cdot 8}{3 \cdot 2 \cdot 1}a^3(-2)^7b^7 = 120(-128)a^3b^7 = -15{,}360a^3b^7.$$

CHAPTER 11 REVIEW EXERCISES

[11.1] *Write out the first four terms of each sequence.*

1. $a_n = 2n - 3$ **2.** $a_n = \frac{n-1}{n}$ **3.** $a_n = n^2$

4. $a_n = \left(\frac{1}{2}\right)^n$

5. $a_n = (n + 1)(n - 1)$

Write each series as a sum of terms.

6. $\sum_{i=1}^{5} i^2 x$

7. $\sum_{i=1}^{6} (i + 1)x^i$

Evaluate each series.

8. $\sum_{i=1}^{4} (i + 2)$

9. $\sum_{i=1}^{6} 2^i$

10. $\sum_{i=4}^{7} \frac{i}{i + 1}$

11. Find the arithmetic mean, or average, of the share volume of the five most active trading days on the New York Stock Exchange as of the end of 1997.

Date	*Volume (in thousands)*
Jan. 23, 1997	684,588
July 16, 1996	680,913
Dec. 20, 1996	654,110
June 20, 1997	652,945
July 16, 1997	652,848

Source: New York Stock Exchange.

[11.2–11.3] *Decide whether each sequence is arithmetic, geometric, or neither. If the sequence is arithmetic, find the common difference, d. If it is geometric, find the common ratio, r.*

12. 2, 5, 8, 11, . . .

13. −6, −2, 2, 6, 10, . . .

14. $\frac{2}{3}, -\frac{1}{3}, \frac{1}{6}, -\frac{1}{12}, \ldots$

15. −1, 1, −1, 1, −1, . . .

16. 64, 32, 8, $\frac{1}{2}$, . . .

17. 64, 32, 16, 8, . . .

18. 10, 8, 6, 4, . . .

[11.2] *Find the indicated term for each arithmetic sequence.*

19. $a_1 = -2, d = 5;\quad a_{16}$

20. $a_6 = 12, a_8 = 18;\quad a_{25}$

Find the general term for each arithmetic sequence.

21. $a_1 = -4, d = -5$

22. 6, 3, 0, −3, . . .

Find the number of terms in each arithmetic sequence.

23. 7, 10, 13, . . . , 49

24. 5, 1, −3, . . . , −79

Find S_8 for each arithmetic sequence.

25. $a_1 = -2, d = 6$

26. $a_n = -2 + 5n$

[11.3] *Find the general term for each geometric sequence.*

27. −1, −4, . . .

28. $\frac{2}{3}, \frac{2}{15}, \ldots$

Find the indicated term for each geometric sequence.

29. $2, -6, 18, \ldots; \quad a_{11}$

30. $a_3 = 20, a_5 = 80; \quad a_{10}$

Find each sum, if it exists.

31. $\sum_{i=1}^{5} \left(\frac{1}{4}\right)^i$

32. $\sum_{i=1}^{8} \frac{3}{4}(-1)^i$

33. $\sum_{i=1}^{\infty} 4\left(\frac{1}{5}\right)^i$

34. $\sum_{i=1}^{\infty} 2(3)^i$

[11.4] *Use the binomial theorem to expand each binomial.*

35. $(2p - q)^5$

36. $(x^2 + 3y)^4$

37. $(\sqrt{m} + \sqrt{n})^4$

38. Write the fourth term of the expansion of $(3a + 2b)^{19}$.

39. Write the twenty-third term of the expansion of $(-2k + 3)^{25}$.

MIXED REVIEW EXERCISES

Find the indicated term and S_{10} for each sequence.

40. a_{40}: arithmetic; $\quad 1, 7, 13, \ldots$

41. a_{10}: geometric; $\quad -3, 6, -12, \ldots$

42. a_9: geometric; $\quad a_1 = 1, r = -3$

43. a_{15}: arithmetic; $\quad a_1 = -4, d = 3$

Find the general term for each arithmetic or geometric sequence.

44. $2, 7, 12, \ldots$

45. $2, 8, 32, \ldots$

46. $27, 9, 3, \ldots$

47. $12, 9, 6, \ldots$

Solve each problem.

48. When Mary's sled goes down the hill near her home, she covers 3 feet in the first second, then for each second after that she goes 4 feet more than in the preceding second. If the distance she covers going down is 210 feet, how long does it take her to reach the bottom?

49. An ordinary annuity is set up so that \$672 is deposited at the end of each quarter for 7 years. The money earns 8% annual interest compounded quarterly. What is the future value of the annuity?

50. The school population in Pfleugerville has been dropping 3% per year. The current population is 50,000. If this trend continues, what will the population be in 6 years?

51. A pump removes $\frac{1}{2}$ of the liquid in a container with each stroke. What fraction of the liquid is left in the container after 7 strokes?

52. Consider the repeating decimal number .55555 . . . .
(a) Write it as the sum of the terms of an infinite geometric sequence.
(b) What is r for this sequence?
(c) Find this infinite sum, if it exists, and write it as a common fraction in lowest terms.

53. Can the sum of the terms of the infinite geometric sequence with $a_n = 5(2)^n$ be found? Explain.

54. Can any two terms of a geometric sequence be used to find the common ratio? Explain.

CHAPTER 11 TEST

Write the first five terms of each sequence described.

1. $a_n = (-1)^n + 1$

2. Arithmetic, with $a_1 = 4$ and $d = 2$

3. Geometric, with $a_4 = 6$ and $r = \frac{1}{2}$

Find a_4 for each sequence described.

4. Arithmetic, with $a_1 = 6$ and $d = -2$

5. Geometric, with $a_5 = 16$ and $a_7 = 9$

Find S_5 for each sequence described.

6. Arithmetic, with $a_2 = 12$ and $a_3 = 15$

7. Geometric, with $a_5 = 4$ and $a_7 = 1$

8. The share volume (in millions) of the five most active stocks on the American Stock Exchange in 1996 is shown in the following table.

Stock	*Share Volume (in millions)*
Viacom (Class B)	271.9
Trans World Airlines	199.8
Echo Bay Mines	198.5
IVAX	183.5
Ampex (Class A)	179.1

Source: American Stock Exchange.

To the nearest tenth of a million, what was the average share volume for these five stocks?

9. If \$4000 is deposited in an ordinary annuity at the end of each quarter for 7 years and earns 6% interest compounded quarterly, how much will be in the account at the end of this term?

10. Under what conditions does an infinite geometric series have a sum?

Find each sum that exists.

11. $\sum_{i=1}^{5} (2i + 8)$

12. $\sum_{i=1}^{6} (3i - 5)$

13. $\sum_{i=1}^{500} i$

14. $\sum_{i=1}^{3} \frac{1}{2}(4^i)$

15. $\sum_{i=1}^{\infty} \left(\frac{1}{4}\right)^i$

16. $\sum_{i=1}^{\infty} 6\left(\frac{3}{2}\right)^i$

17. Cheryl bought a new sewing machine for \$300. She agreed to pay \$20 per month for 15 months plus interest of 1% each month on the unpaid balance. Find the total cost of the machine.

18. During the summer months, the population of a certain insect colony triples each week. If there are 20 insects in the colony at the end of the first week in July, how many are present by the end of September? (Assume exactly four weeks in a month.)

19. Write the fifth term of $\left(2x - \frac{y}{3}\right)^{12}$.

20. Expand $(3k - 5)^4$.

CUMULATIVE REVIEW EXERCISES CHAPTERS 1–11

This set of exercises may be considered a final examination for the course.

Let $P = \left\{-\frac{8}{3}, 10, 0, \sqrt{13}, -\sqrt{3}, \frac{45}{15}, \sqrt{-7}, .82, -3\right\}$. *List the elements of P that are members of each set.*

1. Integers

2. Rational numbers

3. Irrational numbers

4. Real numbers

Simplify each expression.

5. $|-7| + 6 - |-10| - (-8 + 3)$

6. $-15 - |-4| - 10 - |-6|$

7. $4(-6) + (-8)(5) - (-9)$

Solve each equation or inequality.

8. $9 - (5 + 3a) + 5a = -4(a - 3) - 7$

9. $7m + 18 \leq 9m - 2$

10. $|4x - 3| = 21$

11. $\frac{x+3}{12} - \frac{x-3}{6} = 0$

12. $2x > 8$ or $-3x > 9$

13. $|2m - 5| \geq 11$

14. The president of InstaTune, a chain of franchised automobile tune-up shops, reports that people who buy a franchise and open a shop pay a weekly fee (in dollars) to company headquarters, according to the linear function $f(x) = .07x + 135$, where $f(x)$ is the fee and x is the total amount of money taken in during the week by the shop. Find the weekly fee if \$2000 is taken in for the week. (*Source: Business Week.*)

15. Find the slope of the line through $(4, -5)$ and $(-12, -17)$.

16. Find the standard form of the equation of the line through $(-2, 10)$ and parallel to the line with equation $3x + y = 7$.

Graph.

17. $x - 3y = 6$

18. $4x - y < 4$

Solve each system of equations.

19.
$$\begin{aligned} 2x + 5y &= -19 \\ -3x + 2y &= -19 \end{aligned}$$

20.
$$\begin{aligned} x + 2y + z &= 8 \\ 2x - y + 3z &= 15 \\ -x + 3y - 3z &= -11 \end{aligned}$$

Evaluate each determinant.

21. $\begin{vmatrix} -3 & -2 \\ 6 & 9 \end{vmatrix}$

22. $\begin{vmatrix} 2 & 4 & 1 \\ 1 & 3 & 6 \\ 2 & 3 & -1 \end{vmatrix}$

23. Nuts worth \$3 per pound are to be mixed with 8 pounds of nuts worth \$4.25 per pound to get a mixture that will be sold for \$4 per pound. How many pounds of the \$3 nuts should be used?

Perform the indicated operations.

24. $(4p + 2)(5p - 3)$

25. $(3k - 7)^2$

26. $(2m^3 - 3m^2 + 8m) - (7m^3 + 5m - 8)$

27. Divide $6t^4 + 5t^3 - 18t^2 + 14t - 1$ by $3t - 2$.

Factor.

28. $7x + x^3$

29. $14y^2 + 13y - 12$

30. $6z^3 + 5z^2 - 4z$

31. $49a^4 - 9b^2$

32. $c^3 + 27d^3$

33. $64r^2 + 48rq + 9q^2$

Solve each equation or inequality.

34. $2x^2 + x = 10$

35. $k^2 - k - 6 \leq 0$

Simplify.

36. $\left(\frac{2}{3}\right)^{-2}$

37. $\frac{(3p^2)^3(-2p^6)}{4p^3(5p^7)}$

38. What is the domain of the rational function $f(x) = \frac{2}{x^2 - 81}$?

Simplify.

39. $\frac{x^2 - 16}{x^2 + 2x - 8} \div \frac{x - 4}{x + 7}$

40. $\frac{5}{p^2 + 3p} - \frac{2}{p^2 - 4p}$

Solve.

41. $\frac{4}{x - 3} - \frac{6}{x + 3} = \frac{24}{x^2 - 9}$

42. $6x^2 + 5x = 8$

43. $\sqrt{3x - 2} = x$

44. Simplify $5\sqrt{72} - 4\sqrt{50}$.

45. Multiply $(8 + 3i)(8 - 3i)$.

46. Find $f^{-1}(x)$, if $f(x) = 9x + 5$.

47. Graph $g(x) = \left(\frac{1}{3}\right)^x$.

48. Solve $3^{2x-1} = 81$.

49. Graph $y = \log_{1/3} x$.

50. Solve $\log_8 x + \log_8(x + 2) = 1$.

Graph.

51. $f(x) = 2(x - 2)^2 - 3$

52. $\frac{x^2}{9} + \frac{y^2}{25} = 1$

53. $x^2 - y^2 = 9$

54. Find the equation of a circle with center at $(-5, 12)$ and radius 9.

55. Write the first five terms of the sequence defined by $a_n = 5n - 12$.

56. Find the sum of the first six terms of the arithmetic sequence with $a_1 = 8$ and $d = 2$.

57. Find the sum of the geometric series $15 - 6 + \frac{12}{5} - \frac{24}{25} + \cdots$.

58. Find the sum: $\sum_{i=1}^{4} 3i$.

59. Use the binomial theorem to expand $(2a - 1)^5$.

60. What is the fourth term in the expansion of $\left(3x^4 - \frac{1}{2}y^2\right)^5$?

Answers to Selected Exercises

TO THE STUDENT

If you need further help with algebra, you may want to obtain a copy of the *Student's Solution Manual* that goes with this book. It contains solutions to all of the odd-numbered exercises and all of the chapter test exercises. You also may want the *Student's Study Guide and Journal.* It has extra examples and exercises to complete, corresponding to each learning objective of the book. In addition, there is a practice test for each chapter. Your college bookstore either has these books or can order them for you.

In this section we provide the answers that we think most students will obtain when they work the exercises using the methods explained in the text. If your answer does not look exactly like the one given here, it is not necessarily wrong. In many cases there are equivalent forms of the answer that are correct. For example, if the answer section shows $\frac{3}{4}$ and your answer is .75, you have obtained the right answer but written it in a different (yet equivalent) form. Unless the directions specify otherwise, .75 is just as valid an answer as $\frac{3}{4}$.

In general, if your answer does not agree with the one given in the text, see whether it can be transformed into the other form. If it can, then it is the correct answer. If you still have doubts, talk with your instructor.

CHAPTER 1 REVIEW OF THE REAL NUMBER SYSTEM

SECTION 1.1 (PAGE 11)

EXERCISES **1.** true **3.** false; The additive inverse of a positive number is negative, and the additive inverse of 0 is 0. **5.** true **7.** $\{6, 7, 8, 9, \ldots\}$ **9.** $\{\ldots, -1, 0, 1, 2, 3, 4\}$ **11.** $\{10, 12, 14, 16, \ldots\}$ **13.** $\emptyset$ **15.** $\{-4, 4\}$ **17.** yes **19.** $\{x \mid x$ is a multiple of 4 greater than 0$\}$ **21.** $\{x \mid x$ is an even natural number less than or equal to 8$\}$ **23.** **(a)** $4, 5, 17, \frac{40}{2}$ (or 20) **(b)** $0, 4, 5, 17, \frac{40}{2}$ **(c)** $-8, 0, 4, 5, 17, \frac{40}{2}$ **(d)** $-8, -.6, 0, \frac{3}{4}, 4, 5, \frac{13}{2}, 17, \frac{40}{2}$ **(e)** $-\sqrt{5}, \sqrt{3}$ **(f)** All are real numbers except $\frac{1}{0}$. **(g)** $\frac{1}{0}$ **25.** 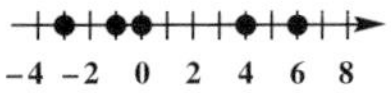

27. 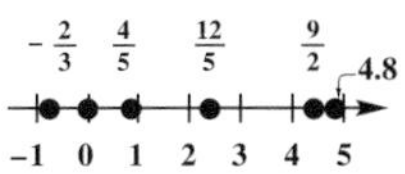

29. The graph of a number is a point on the number line. The coordinate of a point on the number line is the number that corresponds to the point. **31.** 8 **33.** -5 **35.** -2 **37.** -4.5 **39.** 5 **41.** 6 **43.** 22 **45.** 0 **47.** 1988 **49.** true **51.** false **53.** 1 **55.** 1 **57.** $2 > -3$; Both are true because -3 is less than 2 and 2 is greater than -3. **59.** $6 < 11$ **61.** $4 > x$ **63.** $3t - 4 \leq 10$ **65.** $5 \geq 5$ **67.** $-3 < t < 5$ **69.** $-3 \leq 3x < 4$ **71.** $5x + 3 \neq 0$ **73.** $3 \geq 2$ **75.** $-3 \leq -3$ **77.** $5 \nless 3$ **79.** Pacific Ocean, Indian Ocean, Caribbean Sea, South China Sea, Gulf of California **81.** true **83.** $(-2, \infty)$ 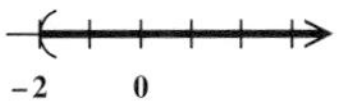

85. $(-\infty, 6]$ 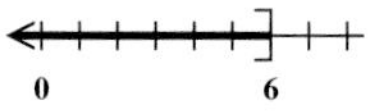

87. $(0, 3.5)$ 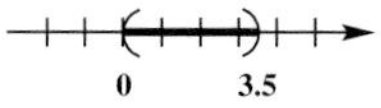

89. $[2, 7]$ 

91. $(-4, 3]$ 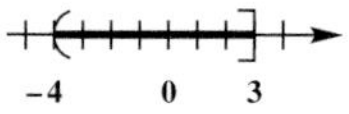

93. $(0, 3]$ 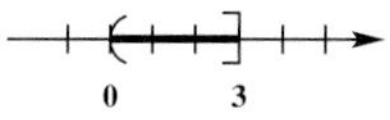

95. 1992, 1993, and 1994 **97.** $x > y$

SECTION 1.2 (PAGE 24)

CONNECTIONS **Page 21:** Answers will vary.

EXERCISES **1.** the numbers are additive inverses; $4 + (-4) = 0$ **3.** negative; $-7 + (-21) = -28$ **5.** the positive number has a larger absolute value; $15 + (-2) = 13$ **7.** the one with the smaller absolute value is subtracted from the one with the larger absolute value; $-15 - (-3) = -12$ **9.** negative; $(-5)(15) = -75$ **11.** 9 **13.** -19 **15.** $-\frac{19}{12}$ **17.** .187 **19.** 8 **21.** $-\frac{7}{4}$ **23.** 3.018 **25.** 6 **27.** $\frac{13}{2}$ or $6\frac{1}{2}$ **33.** 45 **35.** 180 **37.** 7.383 **39.** 4 **41.** $-\frac{9}{13}$ **43.** undefined **45.** true **47.** true **49.** false; $-4^6 = -(4^6) = -4096$, while $(-4)^6 = 4096$ **51.** true **53.** false; 3 is the base in -3^5. **55.** 11 **57.** .021952 **59.** $\frac{49}{100}$ **61.** -30 **63.** $\sqrt{16}$ is the positive square root of 16, namely 4; -4 is the negative square root of 16, written $-\sqrt{16}$.
The number of digits displayed will vary in Exercises 65 and 67.
65. 136.011029 **67.** 9.657121724 **69.** **(a)** not a real number **(b)** negative **71.** 29 **73.** -79 **75.** 39 **77.** -2 **79.** not a real number **81.** 13 **83.** $-\frac{32}{5}$ **85.** -1 **87.** 17 **89.** $-\frac{15}{238}$ **91.** -96 **93.** 112°F **95.** 11,331 feet **97.** $-11{,}478$ **99.** Answers will vary. **100.** It is less than 0. **101.** Answers will vary. **102.** It is less than 0; $<$

SECTION 1.3 (PAGE 33)

EXERCISES **1.** (b) **3.** (a) **5.** the commutative property says that the *order* in which terms are added or multiplied does not matter, while the associative property says that the *grouping* of the terms to be added or multiplied does not matter
7. $8k$ **9.** $-2r$ **11.** $-8z + 4w$ (cannot be simplified) **13.** $6a$ **15.** $2m + 2p$ **17.** $-12x + 12y$
19. $-10d - 5f$ **21.** $7x + 26$ **23.** $-6y + 3$ **25.** $-2k + 15$ **27.** -1 **29.** $2p + 7$ **31.** $-6z - 39$
33. 13 **35.** -16 **37.** $(5 + 8)x = 13x$ **39.** $(5 \cdot 9)r = 45r$ **41.** $9y + 5x$ **43.** 7 **45.** 0
47. $8(-4) + 8x = -32 + 8x$ **49.** 0 **51.** Answers will vary. One example is washing your face and brushing your teeth; one example is putting on your socks and putting on your shoes. **53.** $2 + 6 \cdot 5 = 2 + 30 = 32$, which does not equal $8 \cdot 5 = 40$. **55.** 1900 **57.** 75 **59.** 431 **61.** associative property **62.** associative property
63. commutative property **64.** associative property **65.** distributive property **66.** arithmetic facts
69. No. One example is $7 + (5 \cdot 3) = (7 + 5)(7 + 3)$, which is false.

CHAPTER 1 REVIEW EXERCISES (PAGE 40)

1. 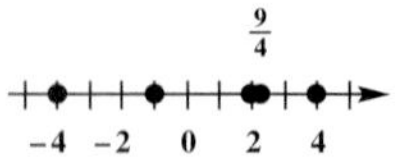

3. 16 **5.** -4 **7.** $-9, 0, 4$ **9.** All are real numbers. **11.** $\{0, 1, 2, 3\}$ **13.** false

15. $(-\infty, -5)$ 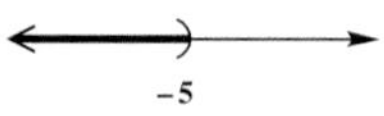

17. $\frac{41}{24}$ **19.** -3 **21.** -39 **23.** $\frac{23}{20}$ **25.** To subtract $a - b$, write it as the addition problem $a + (-b)$, and add. **27.** $\frac{2}{3}$ **29.** 3.21 **31.** true **33.** $\frac{5}{7 - 7}$ is undefined **35.** 2.89
37. not a real number **39.** -2 **41.** -30 **43.** $(4 + 6)^2 = 10^2 = 100$; $4^2 + 6^2 = 52$; $100 \neq 52$ **45.** $-4z$
47. $4p$ **49.** $6r + 18$ **51.** 0 **53.** $-18m$ **55.** -4 **57.** $13 + (-3) = 10$ **59.** $5x + 5z$ **61.** 1
63. $\frac{256}{625}$ **65.** -15 **67.** 31 **69.** $-3k + 4h$ **71.** -5 **73.** -11.408 **75.** 44 **77.** -116 **79.** -5

81. -29 **83.** $-6x + 6$ **85.** $-\frac{47}{3}$ **86.** It is greater than -16. **87. (a)** no **(b)** yes **88.** $\frac{47}{3}$ **89.** no
90. $\frac{2209}{9}$ **91.** $\frac{47}{3}$ **92.** $-\frac{3}{47}$ **93.** yes **94.** No, the new answer is $-\frac{29}{3}$.

CHAPTER 1 TEST (PAGE 43)

[1.1] **1.** 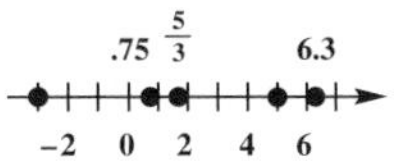
2. 0, 3, 12 **3.** $-1, 0, 3, 12$ **4.** $-1, -.5, 0, 3, 7.5, 12$ **5.** All are real numbers.
6. $(-\infty, -3)$ 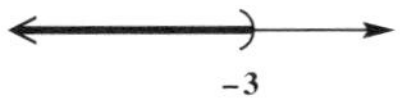
7. $(-4, 2]$ 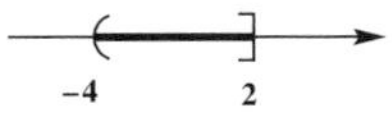
[1.2] **8.** 0 **9.** -26 **10.** 19 **11.** 1
12. $\frac{16}{7}$ **13.** $\frac{11}{23}$ [1.1] **14.** largest: 8.0; smallest: -11.4 **15.** largest: $|-11.4| = 11.4$; smallest: $|1.3| = 1.3$
[1.2] **16.** negative; $-6.8\% - 8.0\% = -14.8\%$ **17.** 14 **18.** -15 **19.** not a real number **20. (a)** a must be positive. **(b)** a must be negative. **(c)** a must be 0. **21.** $-\frac{6}{23}$ [1.3] **22.** $10k - 10$ **23.** It changes the sign of each term. The simplified form is $7r + 2$. **24.** B **25.** D **26.** A **27.** F **28.** C **29.** C **30.** E

CHAPTER 2 LINEAR EQUATIONS AND INEQUALITIES

SECTION 2.1 (PAGE 51)

CONNECTIONS **Page 51:** **1. (a)** 1940 **(b)** 1964 **2.** men: 37.22 seconds; women: 39.06 seconds

EXERCISES **1.** (a) and (c) **3.** Both sides are evaluated as 30, so 6 is a solution. **5.** solution set **7.** (b) **9.** $\{-1\}$ **11.** $\{3\}$ **13.** $\{-7\}$ **15.** $\{0\}$ **17.** $\left\{-\frac{5}{3}\right\}$ **19.** $\left\{-\frac{1}{2}\right\}$ **21.** $\{2\}$ **23.** $\{-2\}$ **25.** $\{7\}$ **27.** $\{2\}$ **29.** $\left\{\frac{3}{2}\right\}$ **31.** $\{-5\}$ **33.** $\{3\}$ **35.** Yes, you will get the same solution. The coefficients will be larger, but in the end, the solution will be the same. **37.** $\{-6\}$ **39.** $\{0\}$ **41.** $\{-5\}$ **43.** $\{100\}$ **45.** $\{7000\}$ **47.** $\{60\}$ **49.** (a) **51.** contradiction; $\emptyset$ **53.** conditional; $\{-8\}$ **55.** conditional; $\{0\}$ **57.** identity; {all real numbers} **59.** equivalent **61.** not equivalent; The solution sets are not the same. They are $\{-3\}$ and $\emptyset$, respectively. **63.** not equivalent; The solution sets $\{4\}$ and $\{-4, 4\}$ differ. **65. (a)** 9.75 million **(b)** 1991–1992 **67.** .8 million; 1992–1993

SECTION 2.2 (PAGE 58)

CONNECTIONS **Page 57:** .10 or 10%; .0676 or 6.76%; .1125 or 11.25%

EXERCISES **1. (a)** $3x = 5x + 8$ **(b)** $ct = bt + k$ **2. (a)** $3x - 5x = 8$ **(b)** $ct - bt = k$ **3. (a)** $-2x = 8$; distributive property **(b)** $(c - b)t = k$; distributive property **4. (a)** $x = -4$ **(b)** $t = \frac{k}{c - b}$ **5.** $b \neq c$; If $b = c$, the denominator becomes 0, and 0 is not allowed as the denominator of a fraction **6.** That equation had no variable in the denominator. **7.** (d) **9.** $t = \frac{d}{r}$ **11.** $b = \frac{A}{h}$ **13.** $a = P - b - c$ **15.** $h = \frac{2A}{b}$ **17.** $h = \frac{S - 2\pi r^2}{2\pi r}$ or $h = \frac{S}{2\pi r} - r$ **19.** $F = \frac{9}{5}C + 32$ **21.** $H = \frac{A - 2LW}{2W + 2L}$ **23.** $r = \frac{-2k - 3y}{a - 1}$ or $r = \frac{2k + 3y}{1 - a}$ **25.** $y = \frac{-x}{w - 3}$ or $y = \frac{x}{3 - w}$ **29.** about 3.7 hours **31.** $-40°F$ **33.** 230 meters **35.** 52 miles per hour **37.** 10 centimeters

39. The volume is 4 times as large. **41.** 8 feet **43.** perimeter **45.** 75% water, 25% alcohol **47.** 3% **49.** 1500 **51.** 12,250 **53.** \$10.51 **55.** \$45.66 **57.** 10% **59.** L = 10

SECTION 2.3 (PAGE 70)

CONNECTIONS **Page 69:** Our steps 1, 3, 4, and 6 correspond to Polya's steps.

EXERCISES **1.** expression **3.** equation **5.** expression **7.** yes **9.** $x - 18$ **11.** $(x - 9)(x + 6)$ **13.** $\frac{12}{x}, (x \neq 0)$ **17.** $2x + \frac{x}{6} = x - 8$ **19.** $12 - \frac{2}{3}x = 10$ **21.** *Step 1:* the number of shoppers at large chain bookstores *Step 2:* the number of shoppers at small chain/independent bookstores *Step 3:* x; $(x - 70)$ *Step 4:* 256 *Step 5:* 256; 186 *Step 6:* large chain shoppers: 70; small chain/independent shoppers: 442 **23.** Ingram Industries: \$11.5 billion; TLC Beatrice International: \$2.2 billion **25.** Ruth: 2873 hits; Hornsby: 2930 hits **27.** rentals: \$6.63 billion; sales: \$3.18 billion **31.** 1.5% **33.** \$4532 **35.** \$122.28 **37.** 4 liters **39.** $18\frac{2}{11}$ liters **41.** 5 liters **43.** \$4000 at 3%; \$8000 at 4% **45.** \$10,000 at 4.5%; \$19,000 at 3% **47.** \$58,000 **49.** 180 **51.** 180 **53.** 20°; 30°; 130° **55.** 65°; 115° **57. (a)** $800 - x$ **(b)** $800 - y$ **58. (a)** $.05x$; $.10(800 - x)$ **(b)** $.05y$; $.10(800 - y)$ **59. (a)** $.05x + .10(800 - x) = 800(.0875)$ **(b)** $.05y + .10(800 - y) = 800(.0875)$ **60. (a)** \$200 at 5%; \$600 at 10% **(b)** 200 liters of 5% acid; 600 liters of 10% acid **61.** The processes are the same. The amounts of money in Problem A correspond to the amounts of pure acid in Problem B.

SECTION 2.4 (PAGE 79)

EXERCISES **1.** No, the answers must be whole numbers because they represent the number of area codes in 1947. **3.** Most people will choose time, because that is the unknown. **5.** \$2.45 **7.** 52 miles per hour **9.** (c) **11.** 17 pennies, 17 dimes, 10 quarters **13.** 305 students, 105 nonstudents **15.** 54 Row 1 seats, 51 Row 2 seats **17.** $1\frac{3}{4}$ hours **19.** 10:00 A.M. **21.** 18 miles **23.** 8 hours **25.** $\frac{5}{6}$ hour **27.** width: 165 feet; length: 265 feet **29.** 850 miles, 925 miles, 1300 miles **31.** length: 60 meters; width: 30 meters **33.** 76 and 77 **35.** 19, 20, 21 **37.** \$425 **39.** length: 8 inches; width: 5 inches **41.** 20 heads

SECTION 2.5 (PAGE 91)

EXERCISES **1.** D **3.** B **5.** F **7.** Use a parenthesis when an endpoint is not included; use a bracket when it is included. **9.** $[5, \infty)$ 5 **11.** $(7, \infty)$ 7 **13.** $(-4, \infty)$ −4

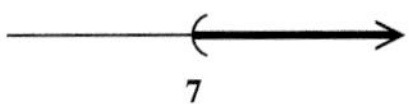

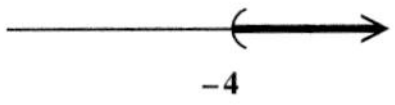

15. $(-\infty, -40]$ −40 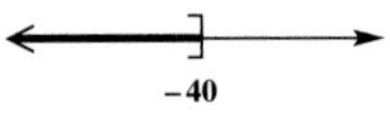 **17.** $(-\infty, 4]$  4 **19.** $\left(-\infty, -\frac{15}{2}\right)$ $-\frac{15}{2}$

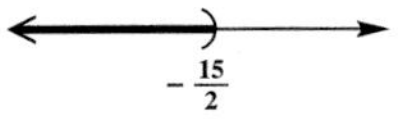

21. $\left[\frac{1}{2}, \infty\right)$ $\frac{1}{2}$ **23.** $(3, \infty)$ 3 **25.** $(-\infty, 4)$ 4

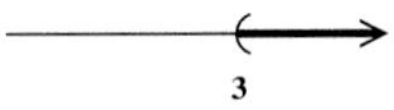

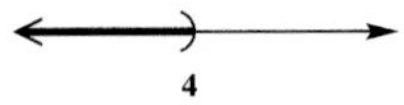

27. $\left(-\infty, \frac{23}{6}\right]$ $\frac{23}{6}$ 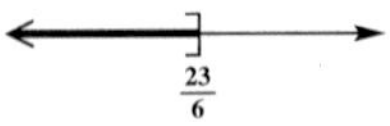 **29.** $\left(-\infty, \frac{76}{11}\right)$ $\frac{76}{11}$ **31.** $(-\infty, \infty)$ 0

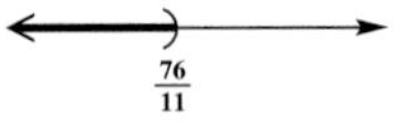

33. $\emptyset$ **35.** It is incorrect. The inequality symbol should be reversed only when multiplying or dividing by a negative number. Since 5 is positive, the inequality symbol should not be reversed. **37.** $(1, 11)$ 1 11

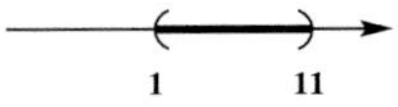

39. $[-14, 10]$ −14 10 **41.** $[-5, 6]$ −5 6 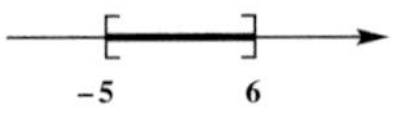 **43.** $\left[-\frac{14}{3}, 2\right]$ $-\frac{14}{3}$ 2

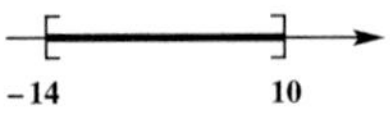

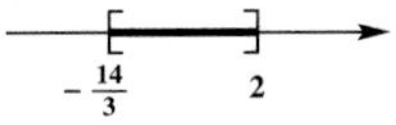

45. $\left[-\frac{1}{2}, \frac{35}{2}\right]$ $-\frac{1}{2}$ $\frac{35}{2}$ 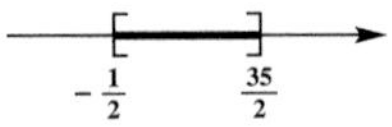 **47.** $\left(-\frac{1}{3}, \frac{1}{9}\right]$ $-\frac{1}{3}$ $\frac{1}{9}$ **49.** April, May, June, July

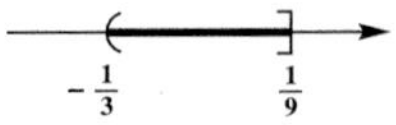

51. January, February, March, August, September, October, November, December **53.** 2 miles **55.** at least 80
57. 50 miles **59.** 26 tapes **61.** $\{-9\}$  **62.** $(-9, \infty)$

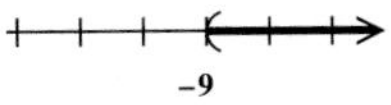

63. $(-\infty, -9)$ 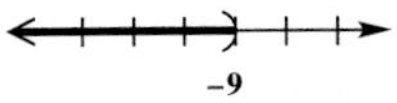 **64.** the set of all real numbers 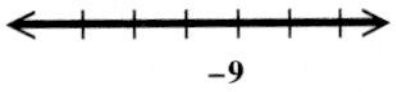 **65.** $(-\infty, -3)$

67. There is no such number y, since $4 \not< 1$. **69.** $\left[\frac{1}{2}, \infty\right)$ **71.** $(-6, -4)$

SECTION 2.6 (PAGE 99)

EXERCISES **1.** true **3.** false; The union is $(-\infty, 8) \cup (8, \infty)$. **5.** false; The intersection is $\emptyset$. **7.** $\{1, 3, 5\}$ or B **9.** $\{4\}$ or D **11.** $\emptyset$ **13.** $\{1, 2, 3, 4, 5, 6\}$ or A **15.** $\{1, 3, 5, 6\}$ **17.** $\{1, 4, 6\}$ **19.** Each is equal to $\{1\}$. This illustrates the associative property of set intersection. **21.** Many answers are possible. One example is the intersection of two streets is the region common to *both* streets. **23.** 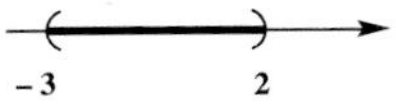 **25.**

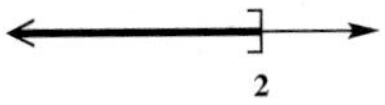

27. $[5, 9]$ 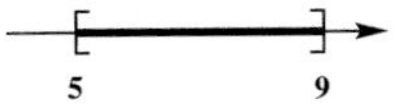 **29.** $(-3, -1)$ 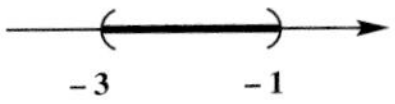 **31.** $(-\infty, 4]$

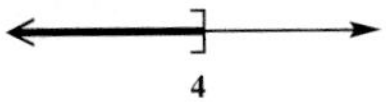

33. 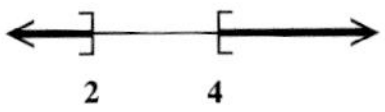 **35.** 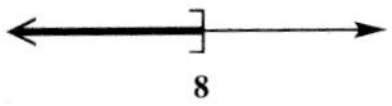 **37.** $(-\infty, -5) \cup (5, \infty)$

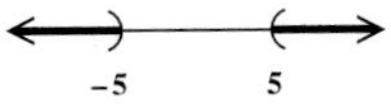

39. $(-\infty, -1) \cup (2, \infty)$ 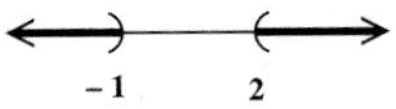 **41.** $[-4, -1]$ **43.** $[-9, -6]$ **45.** $(-\infty, 3)$ **47.** $[3, 9)$

49. intersection; $(-5, -1)$ 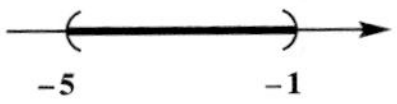 **51.** union; $(-\infty, 4)$

53. union; $(-\infty, 0] \cup [2, \infty)$ **55.** intersection; $[4, 12]$

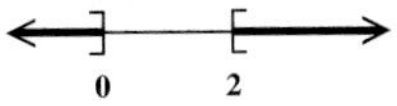

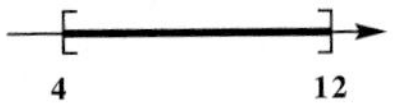

57. 1995 and 1996 **59.** Maria, Joe **60.** none of them **61.** none of them **62.** Luigi, Than **63.** Maria, Joe **64.** none of them

SECTION 2.7 (PAGE 107)

CONNECTIONS **Page 107:** The filled carton may contain between 30.4 and 33.6 ounces, inclusive.

EXERCISES **1. (a)** E **(b)** C **(c)** D **(d)** B **(e)** A **3. (a)** one **(b)** two **(c)** none **5.** $\{-12, 12\}$ **7.** $\{-5, 5\}$ **9.** $\{-6, 12\}$ **11.** $\{-4, 3\}$ **13.** $\left\{-3, \frac{11}{2}\right\}$ **15.** $\left\{-\frac{19}{2}, \frac{9}{2}\right\}$ **17.** $\{-10, -2\}$ **19.** $\left\{-8, \frac{32}{3}\right\}$ **21. (a)** use *or* **(b)** use *and* **(c)** use *or* **23.** $(-\infty, -3) \cup (3, \infty)$  **25.** $(-\infty, -4] \cup [4, \infty)$

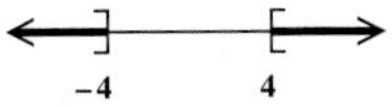

27. $(-\infty, -12) \cup (8, \infty)$ 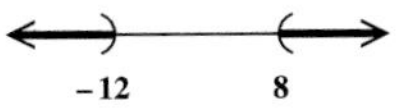 **29.** $(-\infty, -2) \cup (8, \infty)$

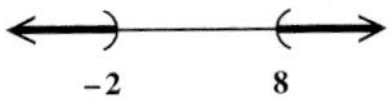

31. $[-3, 3]$  **33.** $(-4, 4)$  **35.** $[-12, 8]$

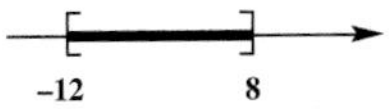

37. $[-2, 8]$ 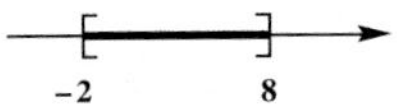 **39.** $(-\infty, -5) \cup (13, \infty)$

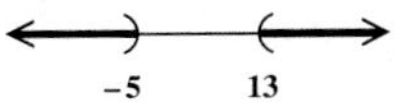

41. $(-\infty, -25) \cup (15, \infty)$ 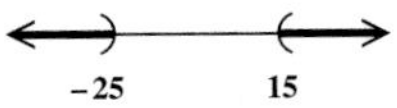 **43.** $\{-6, -1\}$ 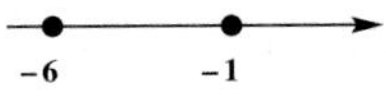 **45.** $\left[-\frac{10}{3}, 4\right]$

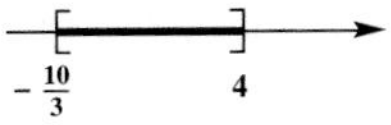

47. $\left[-\frac{7}{6}, -\frac{5}{6}\right]$ 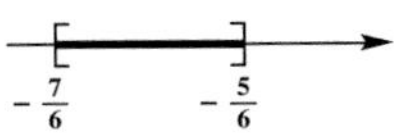 **49.** $\left(-\infty, -\frac{7}{3}\right] \cup [3, \infty)$ **51.** $|x - 4| = 9$ (or $|4 - x| = 9$) **53.** $\{-5, -3\}$ **55.** $(-\infty, -3) \cup (2, \infty)$ **57.** $[-10, 0]$ **59.** $\{-1, 3\}$ **61.** $\left\{-3, \frac{5}{3}\right\}$

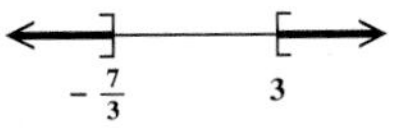

63. $\left\{-\frac{1}{3}, -\frac{1}{15}\right\}$ **65.** $\left\{-\frac{5}{4}\right\}$ **67.** $\emptyset$ **69.** $\left\{-\frac{1}{4}\right\}$ **71.** $\emptyset$ **73.** $(-\infty, \infty)$ **75.** $\left\{-\frac{3}{7}\right\}$ **77.** $(-\infty, \infty)$

79. $\left(-\infty, -\frac{7}{10}\right) \cup \left(-\frac{7}{10}, \infty\right)$ **81.** 460.2 feet **82.** Federal Office Building, City Hall, Kansas City Power and Light, Hyatt Regency **83.** Southwest Bell Telephone, City Center Square, Commerce Tower, Federal Office Building, City Hall, Kansas City Power and Light, Hyatt Regency **84. (a)** $|x - 460.2| \geq 75$ **(b)** $x \geq 535.2$ or $x \leq 385.2$ **(c)** Pershing Road Associates, AT&T Town Pavilion, One Kansas City Place **(d)** It makes sense because it includes all buildings *not* listed earlier. **85. (a)** $|x - 16| \leq .5$ **(b)** $15.5 \leq x \leq 16.5$ **(c)** $x \leq 15.5$ or $x \geq 16.5$

SUMMARY: EXERCISES ON SOLVING LINEAR AND ABSOLUTE VALUE EQUATIONS AND INEQUALITIES (PAGE 110)

1. $\{12\}$ **3.** $\{7\}$ **5.** $\emptyset$ **7.** $\left[-\frac{2}{3}, \infty\right)$ **9.** $\{-3\}$ **11.** $(-\infty, 5]$ **13.** $\{2\}$ **15.** $\emptyset$ **17.** $(-5.5, 5.5)$ **19.** $\left\{-\frac{96}{5}\right\}$ **21.** $(-\infty, -24)$ **23.** $\left\{\frac{7}{2}\right\}$ **25.** $(-\infty, \infty)$ **27.** $(-\infty, -4) \cup (7, \infty)$ **29.** $\left\{-\frac{1}{5}\right\}$ **31.** $\left[-\frac{1}{3}, 3\right]$ **33.** $\left\{-\frac{1}{6}, 2\right\}$ **35.** $(-\infty, -1] \cup \left[\frac{5}{3}, \infty\right)$ **37.** $\left\{-\frac{5}{2}\right\}$ **39.** $\left[-\frac{9}{2}, \frac{15}{2}\right]$ **41.** $(-\infty, \infty)$ **43.** $(-\infty, \infty)$ **45.** $\{-2\}$ **47.** $(-\infty, -1) \cup (2, \infty)$

CHAPTER 2 REVIEW EXERCISES (PAGE 116)

1. $\left\{-\frac{9}{5}\right\}$ **3.** $\left\{-\frac{7}{5}\right\}$ **5.** identity; $(-\infty, \infty)$ **7.** conditional; $\{0\}$ **9.** $b = \frac{2A - Bh}{h}$ or $b = \frac{2A}{h} - B$ **11.** $x = \frac{4}{3}(P + 12)$ or $x = \frac{4}{3}P + 16$ **13.** 3.5 years **15.** 104°F **17.** $9 - 2x$ **19.** 400 **21.** 9.8% **23.** 30 liters **25.** 80° **27.** 150°; 30° **29.** 15 pounds **31.** 46 miles per hour **33.** 6 inches; 12 inches; 16 inches **35.** $(-9, \infty)$ **37.** $\left(\frac{3}{2}, \infty\right)$ **39.** $[3, 5)$ **41.** any grade greater than or equal to 61% **43.** {a, c} **45.** {a, c, e, f, g} **47.** $(-\infty, 3)$ 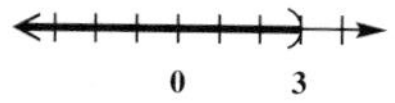

49. $(-\infty, -1] \cup (5, \infty)$ 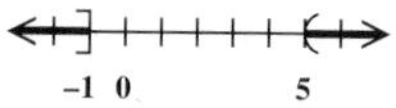

51. $(-3, 4)$ **53.** $(4, \infty)$ **55. (a)** managerial and professional specialty **(b)** managerial and professional specialty, mathematical and computer scientists **57.** $\left\{-\frac{1}{3}, 5\right\}$ **59.** $\left\{-\frac{3}{2}, \frac{1}{2}\right\}$ **61.** $\left\{-\frac{1}{2}\right\}$ **63.** $[-3, -2]$ **65.** $\left(-\infty, -\frac{8}{5}\right) \cup (2, \infty)$ **67.** $\left\{\frac{3}{11}\right\}$ **69.** $(-2, \infty)$ **71.** $[-2, 3)$ **73.** 10 feet **75.** 46, 47, 48 **77.** $\{300\}$ **79.** $\left(-\infty, \frac{14}{17}\right)$ **81.** $(-\infty, 1]$ **83.** \$22.50 **85.** $\left(-\infty, -\frac{13}{5}\right) \cup (3, \infty)$ **87.** 5 liters **89.** $\{30\}$ **91.** $\left\{1, \frac{11}{3}\right\}$ **93.** $\emptyset$ **95.** 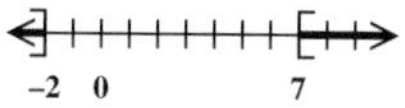

CHAPTER 2 TEST (PAGE 120)

[2.1] **1.** $\{-19\}$ **2.** $\{5\}$ **3.** $(-\infty, \infty)$ **4.** contradiction; $\emptyset$ [2.2] **5.** $v = \frac{S + 16t^2}{t}$ [2.3, 2.4] **6.** 3.2 hours **7.** 9696 residents **8.** Tanui: 2.14 hours; Pippig: 2.44 hours **9.** \$8000 at 3%; \$20,000 at 5% **10.** 40°; 40°; 100° **11. (a)** 22.803 million **(b)** 7.095 million [2.5] **12.** We must reverse the direction of the inequality symbol. **13.** $[1, \infty)$ 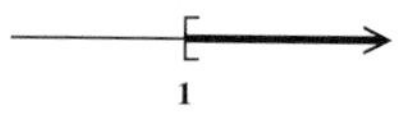

14. $(-\infty, 28)$ 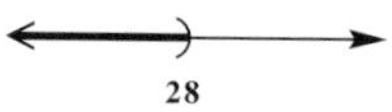

15. $[-3, 3]$ 

16. (c) **17.** 1992 and 1993; 1991 and 1992 [2.6] **18. (a)** $\{1, 5\}$ **(b)** $\{1, 2, 5, 7, 9, 12\}$ **19. (a)** $[2, 9)$ **(b)** $(-\infty, 3) \cup [6, \infty)$ [2.7] **20.** $\left\{-1, \frac{5}{2}\right\}$ **21.** $(-\infty, -1) \cup \left(\frac{5}{2}, \infty\right)$ **22.** $\left(-1, \frac{5}{2}\right)$ **23.** $\left\{-\frac{5}{7}, \frac{11}{3}\right\}$ **24.** $\left(\frac{1}{3}, \frac{7}{3}\right)$ **25. (a)** $\emptyset$ **(b)** $(-\infty, \infty)$ **(c)** $\emptyset$

CUMULATIVE REVIEW EXERCISES CHAPTERS 1–2 (PAGE 122)

[1.1] **1.** 9, 6 **2.** 0, 9, 6 **3.** $-8, 0, 9, 6$ **4.** $-8, -\frac{2}{3}, 0, \frac{4}{5}, 9, 6$ **5.** $-\sqrt{6}$ **6.** All are real numbers.

[1.2] **7.** $-\frac{22}{21}$ **8.** 8 **9.** 8 **10.** 0 **11.** -243 **12.** $\frac{216}{343}$ **13.** $-\frac{8}{27}$ **14.** -4096

15. $\sqrt{-36}$ is not a real number. **16.** $\frac{4+4}{4-4}$ is undefined. **17.** -16 **18.** -34 **19.** 184 **20.** $\frac{27}{16}$

[1.3] **21.** $-20r + 17$ **22.** $13k + 42$ **23.** commutative property **24.** distributive property **25.** inverse property

26. $-\frac{3}{2}$ [2.1] **27.** {5} **28.** {30} **29.** {15} [2.2] **30.** $b = P - a - c$

[2.5] **31.** $[-14, \infty)$ **32.** $\left[\frac{5}{3}, 3\right)$

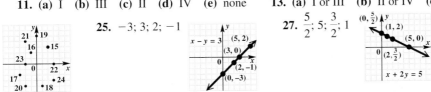

[2.6] **33.** $(-\infty, 0) \cup (2, \infty)$ [2.7] **34.** $\left(-\infty, -\frac{1}{7}\right] \cup [1, \infty)$

[2.3–2.4] **35.** \$5000 **36.** $6\frac{1}{3}$ grams **37.** 74 or greater **38.** $\frac{1}{8}$ hour **39.** 2 liters **40.** 9 pennies, 12 nickels, 8 quarters [2.2] **41.** 44 milligrams **42.** 20 drops [2.3] **43.** **(a)** 79 **(b)** 4.9% [2.2] **44.** 25.7

CHAPTER 3 GRAPHS, LINEAR EQUATIONS, AND FUNCTIONS

SECTION 3.1 (PAGE 132)

EXERCISES **1.** **(a)** 1991–1992 and 1993–1994 **(b)** 1992–1993 **(c)** 1993 **3.** a bar graph **5.** origin **7.** y; x **9.** two **11.** **(a)** I **(b)** III **(c)** II **(d)** IV **(e)** none **13.** **(a)** I or III **(b)** II or IV **(c)** II or IV **(d)** I or III

15–23.

25. -3; 3; 2; -1

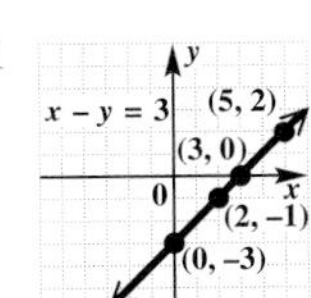

27. $\frac{5}{2}$; 5; $\frac{3}{2}$; 1

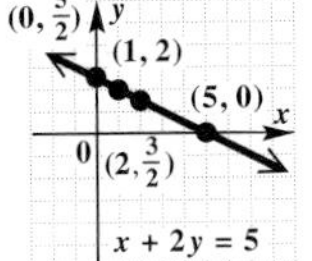

29. -4; 5; $-\frac{12}{5}$; $\frac{5}{4}$

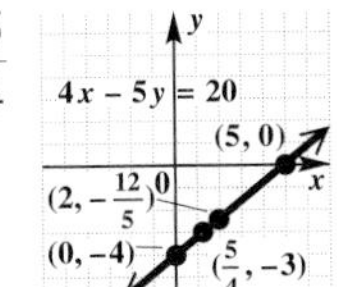

31. In quadrant III, both coordinates of the ordered pairs are negative. If $x + y = k$ and k is positive, then either x or y must be positive, because the sum of two negative numbers is negative.

33. (6, 0); (0, 4)

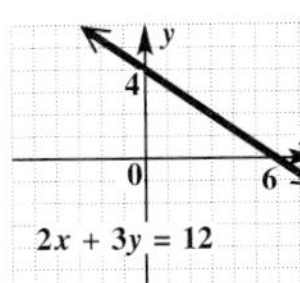

35. (6, 0); (0, -2)

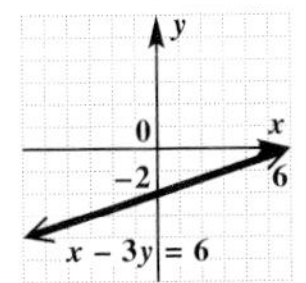

37. $\left(\frac{21}{2}, 0\right)$; $\left(0, -\frac{7}{3}\right)$

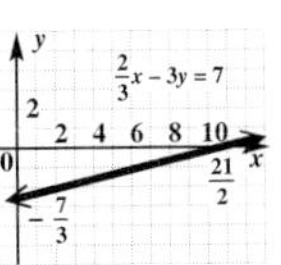

39. none; (0, 5)

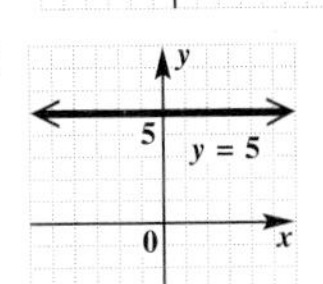

41. (2, 0); none

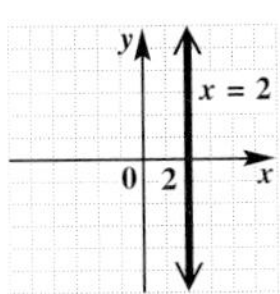

43. $(-4, 0)$; none

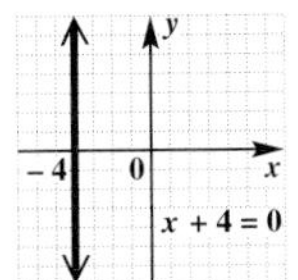

45. (0, 0); (0, 0)

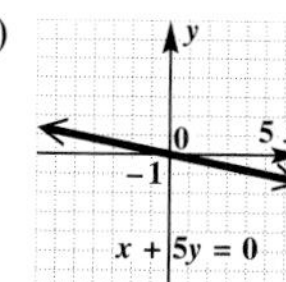

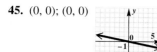

47. (0, 0); (0, 0)

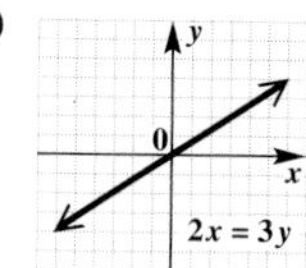

49. (0, 0); (0, 0)

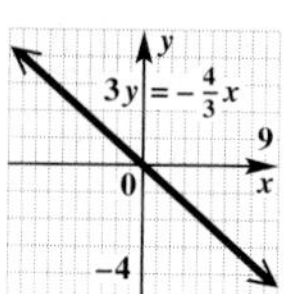

51. 154.6 miles per hour **53.** (c) **55.** **(a)** (1.5, 0) **(b)** (0, 3) **(c)** D **57.** The screen on the right is more useful because it shows the intercepts. **59.** (6, −2) **60.** (5, −2) **61.** (6, 0) **62.** (5, 0) **63.** 5; 0

64. The x-coordinate of M is the average of the x-coordinates of P and Q. The y-coordinate of M is the average of the y-coordinates of P and Q.

SECTION 3.2 (PAGE 145)

CONNECTIONS **PAGE 144:** 46 miles per hour; Make the time interval infinitesimally small. There are many reasons why the average speed would differ for different time intervals; for example, traffic conditions or stop lights.

EXERCISES **1.** (a), (b), and (d) **3.** 2 **5.** undefined **7.** 2 **9.** $\frac{5}{2}$ **11.** 0 **13.** (b) and (d) are correct. Choice (a) is wrong because the order of subtraction must be the same in the numerator and denominator. Choice (c) is wrong because slope is defined as the change in y divided by the change in x. **15.** 1 **17.** $\frac{9}{8}$ **19.** 0 **21.** 2 **23.** B

25. A **27.** $-\frac{1}{2}$

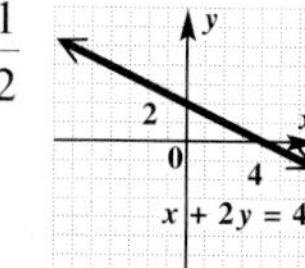

29. 1

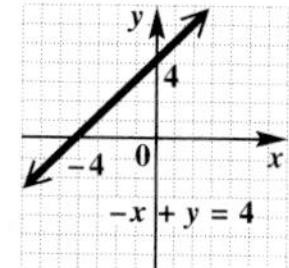

31. $-\frac{6}{5}$

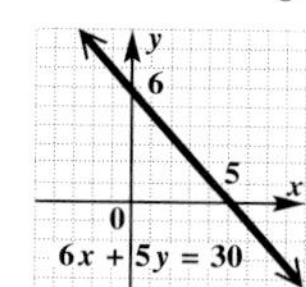

33. $\frac{5}{2}$

35. 4

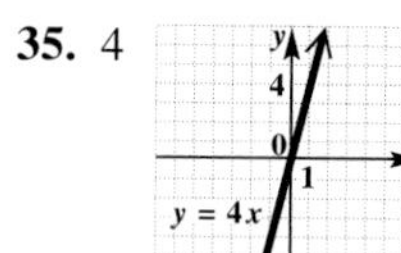

37. undefined

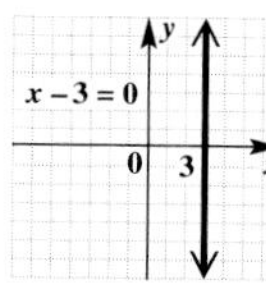

39. x; y **41.** Locate the point on a coordinate system. Write the slope as a fraction, if necessary. Move up or down the number of units in the numerator and left or right the number of units in the denominator to determine a second point. Draw the line through these two points. **43.**

45.

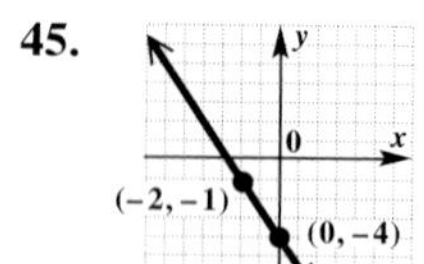

47.

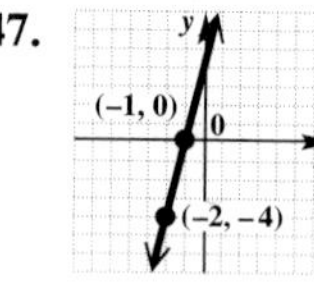

49.

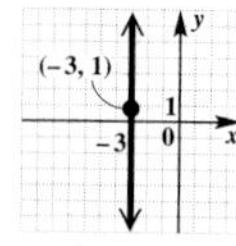

51.

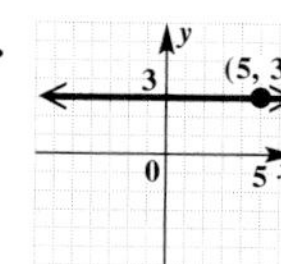

53. parallel **55.** neither

57. perpendicular **59.** neither **61.** parallel **63.** $\frac{7}{10}$ **65.** 4.2 billion minutes per year **67.** The average rate of change is the same, no matter which two ordered pairs are selected to calculate it. **69.** **(a)** −.66 million kilowatts **(b)** −1.1 million kilowatts **(c)** The graph is not a straight line, so the average rate of change varies for different pairs of years. **71.** 3 **73.** A is Y_1 and B is Y_2. **75.** Since the slopes of both pairs of opposite sides are equal, the figure is a parallelogram. **77.** $\frac{1}{3}$ **78.** $\frac{1}{3}$ **79.** $\frac{1}{3}$ **80.** $\frac{1}{3} = \frac{1}{3} = \frac{1}{3}$ is true. **81.** They are collinear. **82.** They are not collinear.

SECTION 3.3 (PAGE 159)

CONNECTIONS **PAGE 153:** approximately 12 million; 1992; Since $-.02$ is close to 0, the number of PCs doubles each year in the indicated years.

EXERCISES **1.** (a) **3.** (a) **5.** $3x + y = 10$ **7.** A **9.** C **11.** H **13.** B **15.** $y = -\frac{3}{4}x + \frac{5}{2}$ **17.** $y = -2x + 18$ **19.** $y = \frac{1}{2}x + \frac{13}{2}$ **21.** $y = 4x - 12$ **25.** $y = 5$ **27.** $x = 9$ **29.** $x = .5$ **31.** $y = 8$ **33.** $y = 2x - 2$ **35.** $y = -\frac{1}{2}x + 4$ **37.** $y = \frac{2}{13}x + \frac{6}{13}$ **39.** $y = 5$ **41.** $x = 7$ **43.** $y = -3$ **45.** $y = 5x + 15$ **47.** $y = -\frac{2}{3}x + \frac{4}{5}$ **49.** $y = \frac{2}{5}x + 5$ **51.** $y = \frac{2}{3}x + 1$ **53.** **(a)** $y = -x + 12$ **(b)** -1 **(c)** $(0, 12)$ **55.** **(a)** $y = -\frac{5}{2}x + 10$ **(b)** $-\frac{5}{2}$ **(c)** $(0, 10)$ **57.** **(a)** $y = \frac{2}{3}x - \frac{10}{3}$ **(b)** $\frac{2}{3}$ **(c)** $\left(0, -\frac{10}{3}\right)$ **59.** $y = 3x - 19$ **61.** $y = \frac{1}{2}x - 1$ **63.** $y = -\frac{1}{2}x + 9$ **65.** $y = 7$ **67.** $y = 3x + 15$; (1, 18), (5, 30), (10, 45) **69.** $y = 1.86x + 78.24$ **71.** **(a)** $y = 838.5x + 19{,}180.5$ **(b)** \$23,373; It is close to the actual value. **73.** **(a)** $y = -3x + 9$ **(b)** 3 **(c)** $\{3\}$ **75.** **(a)** $y = 4x + 2$ **(b)** $-.5$ **(c)** $\{-.5\}$ **77.** (d) **79.** 32; 212 **80.** (0, 32) and (100, 212) **81.** $\frac{9}{5}$ **82.** $F = \frac{9}{5}C + 32$ **83.** $C = \frac{5}{9}(F - 32)$ **84.** When the Celsius temperature is 50°, the Fahrenheit temperature is 122°.

SECTION 3.4 (PAGE 169)

CONNECTIONS **PAGE 168:** **1.** $x \le 200$, $x \ge 100$, $y \ge 3000$ **2.**

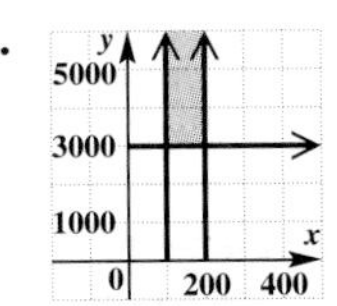

3. $C = 50x + 100y$

4. Some examples are (100, 5000), (150, 3000), and (150, 5000). The corner points are (100, 3000) and (200, 3000).
5. The least cost occurs when $x = 100$ and $y = 3000$. The company should use 100 workers and manufacture 3000 units to achieve the lowest possible cost.

EXERCISES **1.** solid; below **3.** dashed; above **7.**

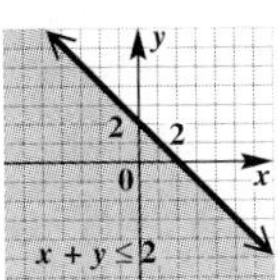

9.

11.

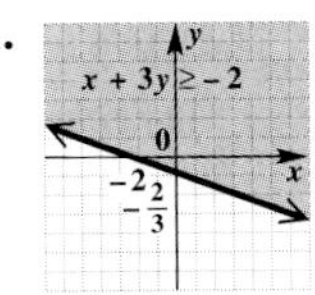

13.

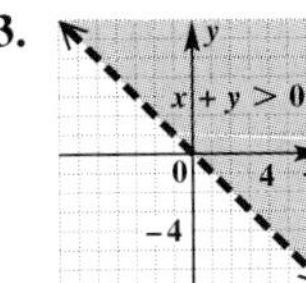

15.

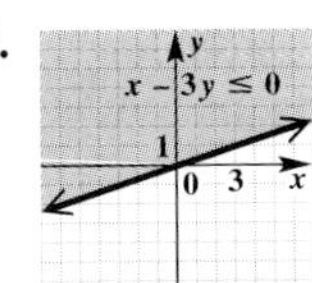

17.

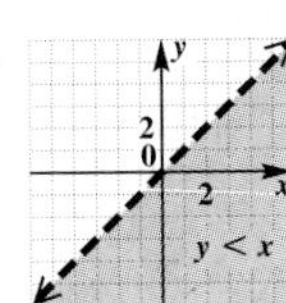

19.

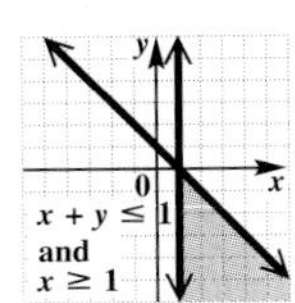

21.

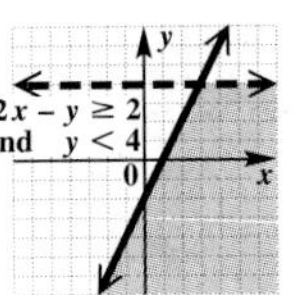

23.

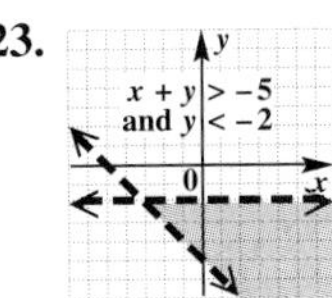

25. $-3 < x < 3$

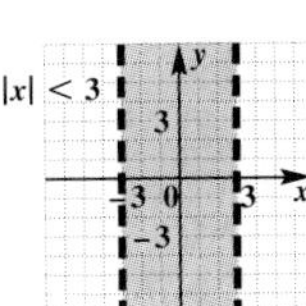

27. $-2 < x + 1 < 2$

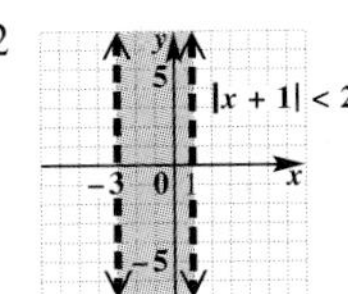

29. 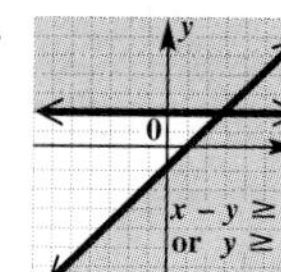

31. 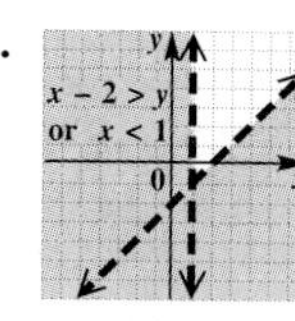

33. 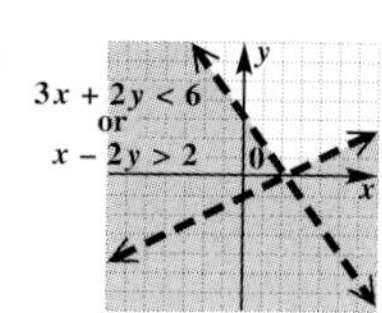

35. (a) $\{-4\}$ **(b)** $(-\infty, -4)$ **(c)** $(-4, \infty)$

37. (a) $\{3.5\}$ **(b)** $(3.5, \infty)$ **(c)** $(-\infty, 3.5)$ **39.** C **41.** A **43. (a)** $\{-.6\}$ **(b)** $(-.6, \infty)$ **(c)** $(-\infty, -.6)$ The graph of $y_1 = 5x + 3$ has x-intercept $(-.6, 0)$, supporting the result of part (a). The graph of y_1 lies *above* the x-axis for values of x *greater than* $-.6$, supporting the result of part (b). The graph of y_1 lies *below* the x-axis for values of x *less than* $-.6$, supporting the result of part (c).

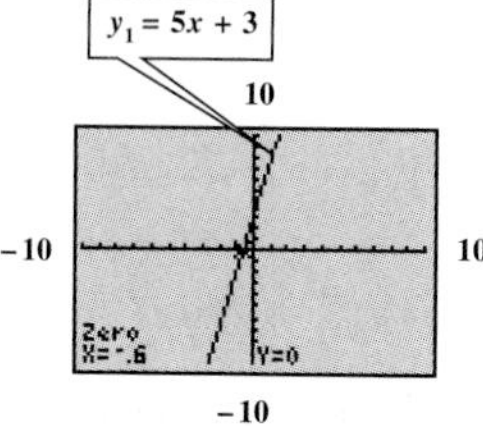

45. (a) $\{-1.2\}$ **(b)** $(-\infty, -1.2]$ **(c)** $[-1.2, \infty)$ The graph of $y_1 = -8x - (2x + 12)$ has x-intercept $(-1.2, 0)$, supporting the result of part (a). The graph of y_1 lies *above or on* the x-axis for values of x *less than or equal to* -1.2, supporting the result of part (b). The graph of y_1 lies *below or on* the x-axis for values of x *greater than or equal to* -1.2, supporting the result of part (c).

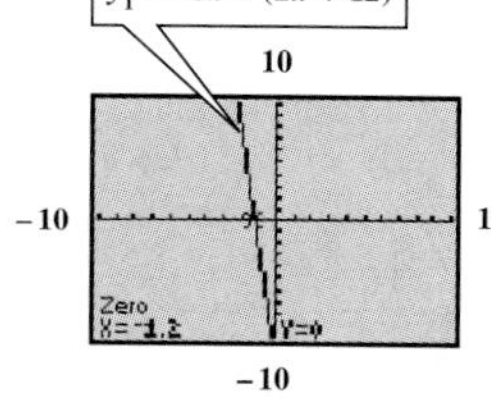

47. 1994–1996

SECTION 3.5 (PAGE 181)

EXERCISES **3.** the independent variable **5.** function; domain: $\{2, 3, 4, 5\}$; range: $\{5, 7, 9, 11\}$ **7.** not a function; domain: $(0, \infty)$; range: $(-\infty, 0) \cup (0, \infty)$ **9.** function; domain: {unleaded regular, unleaded premium, crude oil}; range: $\{1.22, 1.44, .21\}$ **11.** function; domain: $(-\infty, \infty)$; range: $(-\infty, 4]$ **13.** not a function; domain: $[-4, 4]$; range: $[-3, 3]$ **15.** function; domain: $(-\infty, \infty)$ **17.** not a function; domain: $[0, \infty)$ **19.** not a function; domain: $(-\infty, \infty)$ **21.** function; domain: $[0, \infty)$ **23.** function; domain: $(-\infty, 0) \cup (0, \infty)$ **25.** function; domain: $\left[-\frac{1}{2}, \infty\right)$ **27.** function; domain: $(-\infty, 9) \cup (9, \infty)$ **29. (a)** $[0, 3000]$ **(b)** 25 hours; 25 hours **(c)** 2000 gallons **(d)** $f(0) = 0$; The pool is empty at time zero. **31.** Here is one example. The cost of gasoline; number of gallons used; cost; number of gallons **33.** 4 **35.** $3x + 4$ **37.** -59 **39.** $\frac{11}{4}$ **41.** $-4p^2 + 8p + 1$ **43.** 4 **45.** No—in general, $f[g(x)] \neq g[f(x)]$. **47. (a)** $f(x) = \frac{12 - x}{3}$ **(b)** 3 **49. (a)** $f(x) = 3 - 2x^2$ **(b)** -15 **51. (a)** $f(x) = \frac{8 - 4x}{-3}$ **(b)** $\frac{4}{3}$ **53.** line; -2; $-2x + 4$; -2; 3; -2 **55.** domain: $(-\infty, \infty)$

57. domain: $(-\infty, \infty)$

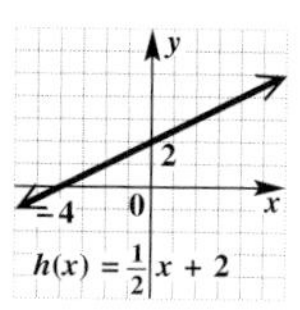

59. domain: $(-\infty, \infty)$

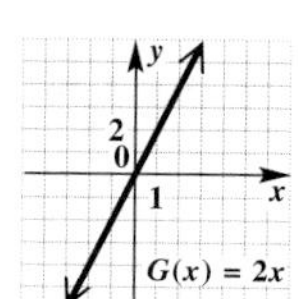

61. domain: $(-\infty, \infty)$

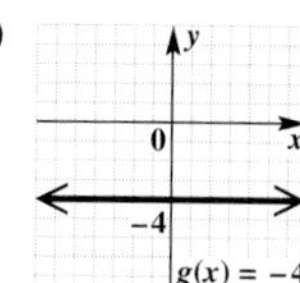

63. **(a)** \$0; \$1.50; \$3.00; \$4.50 **(b)** $1.50x$ **(c)**

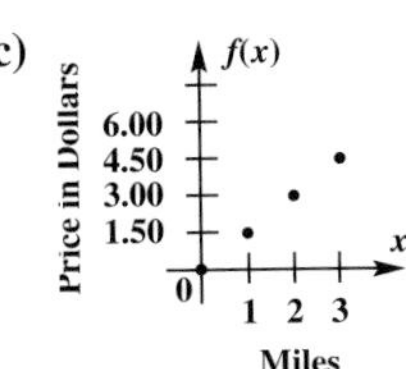

65. 194.53 centimeters **67.** 177.41 centimeters **69.** 1.83 cubic meters **71.** 4.11 cubic meters **73.** **(a)** 39,851 **(b)** 39,485 **(c)** 39,119 **(d)** In 1992, there were 39,668 post offices in the U.S. **75.** $f(3) = 7$ **77.** $f(x) = -3x + 5$

SECTION 3.6 (PAGE 191)

EXERCISES **1.** inverse **3.** direct **5.** joint **7.** combined **9.** 36 **11.** .625 **13.** $222\frac{2}{9}$ **15.** increases; decreases **17.** If y varies inversely as x, x is in the denominator; however, if y varies directly as x, x is in the numerator. Also, with inverse variation, as x increases, y decreases. With direct variation, y increases as x increases. **19.** $\$1.09\frac{9}{10}$ **21.** 8 pounds **23.** 800 gallons **25.** 256 feet **27.** $21\frac{1}{3}$ footcandles **29.** 100 cycles per second **31.** 60 kilometers **33.** 800 pounds **35.** $.71\pi$ seconds **37.** 480 kilograms **39.** 27 **41.** 9 **43.** (0, 0), (1, 1.25) **44.** 1.25 **45.** $y = 1.25x + 0$ or $y = 1.25x$ **46.** $a = 1.25$, $b = 0$ **47.** It is the price per gallon, and it is the slope of the line. **48.** It can be written in the form $y = kx$ (where $k = a$). The value of a is called the constant of variation. **49.** It means that 4.6 gallons cost \$5.75. **50.** It means that 12 gallons cost \$15.00.

CHAPTER 3 REVIEW EXERCISES (PAGE 199)

1.

x	y
0	5
$\frac{10}{3}$	0
2	2
$\frac{14}{3}$	−2

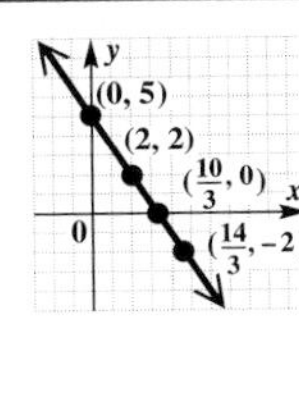

3. (3, 0); (0, −4)

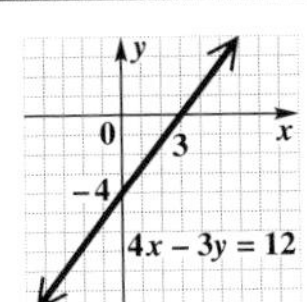

5. (10, 0); (0, 4)

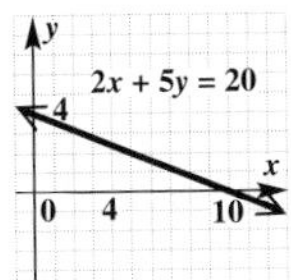

7. If both coordinates are positive, the point lies in quadrant I. If the first coordinate is negative and the second is positive, the point lies in quadrant II. To lie in quadrant III, the point must have both coordinates negative. To lie in quadrant IV, the first coordinate must be positive and the second must be negative. **9.** $-\frac{1}{2}$ **11.** $\frac{3}{4}$ **13.** $\frac{2}{3}$ **15.** undefined **17.** $\frac{5}{2}$ **19.** negative **21.** undefined **23.** 12 feet **25.** $y = -\frac{1}{3}x - 1$ **27.** $y = -\frac{4}{3}x + \frac{29}{3}$ **29.** $x = 2$ (Slope-intercept form is not possible.) **31.** $y = \frac{7}{5}x + \frac{16}{5}$ **33.** $y = -\frac{5}{2}x + 13$ **35.** $y = \frac{5}{2}x - 1$ **37.** 1988 **39.**

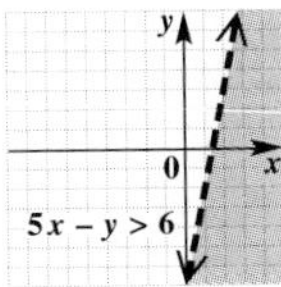

41.

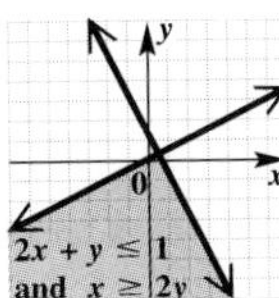

43. domain: $[-4, 4]$; range: $[0, 2]$; function **45.** **(a)** The independent variable must be the country, because, for each country, there is exactly one amount of power. On the other hand, for a specific amount of power, there would be more than one country generating that amount. **(b)** domain: {United States, France, Japan, Germany, Canada, Russia}; range: {101, 154, 286, 377, 706} **47.** function; domain: $(-\infty, \infty)$; linear function **49.** function; domain: $(-\infty, \infty)$ **51.** not a function; domain: $[0, \infty)$ **53.** If no vertical line intersects the graph in more than one point, then it is the graph of a function. **55.** −8.52 **57.** $-2k^2 + 3k - 6$ **59.** $-8p^2 + 6p - 6$ **61.** (c) **63.** Because it falls from left to right, the slope is negative. **64.** $-\frac{3}{2}$ **65.** $\left(\frac{7}{3}, 0\right)$ **66.** $\left(0, \frac{7}{2}\right)$ **67.** $f(x) = -\frac{3}{2}x + \frac{7}{2}$ **68.** $f(8) = -\frac{17}{2}$ **69.** $x = \frac{23}{3}$

70. 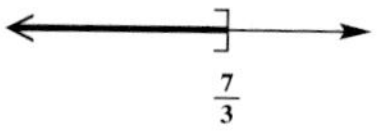

71. $\left\{\frac{7}{3}\right\}$ **72.** $\left(\frac{7}{3}, \infty\right)$ **73.** $\left(-\infty, \frac{7}{3}\right)$ **74.** $\frac{2}{3}$ **75.** (c) **77.** 430 millimeters

CHAPTER 3 TEST (PAGE 203)

[3.1] **1.**

x	y
1	$-\frac{10}{3}$
3	-2
0	-4

[3.2] **2.** $\frac{1}{2}$ [3.1, 3.3] **3.** $\left(\frac{20}{3}, 0\right)$; $(0, -10)$

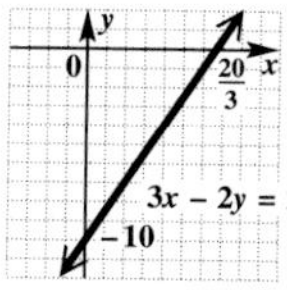

4. none; (0, 5)

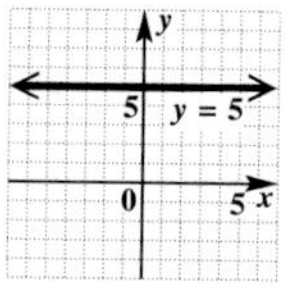

5. (2, 0); none

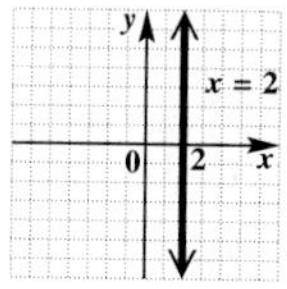

[3.2] **6.** It is a vertical line. **7.** perpendicular **8.** neither [3.3] **9.** $y = -5x + 19$

10. $y = 14$ **11.** $y = -\frac{3}{5}x - \frac{11}{5}$ **12.** $y = -\frac{1}{2}x - \frac{3}{2}$ **13.** $y = -\frac{1}{2}x + 2$ **14.** (b) **15.** 18,160

16. (a) It is the slope of the line. **(b)** It is the annual increase in the number of cases served. [3.4] **17.**

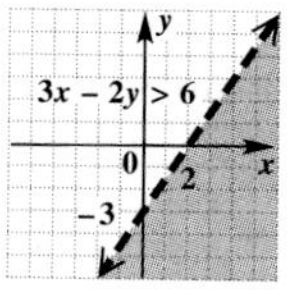

18.

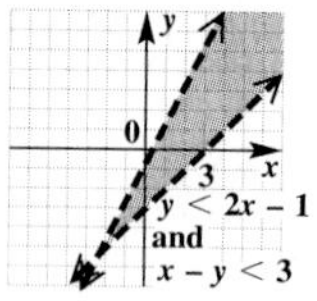

[3.5] **19.** (d) **20.** (d) **21.** 0 **22.** domain: $(-\infty, \infty)$; range: $(-\infty, \infty)$

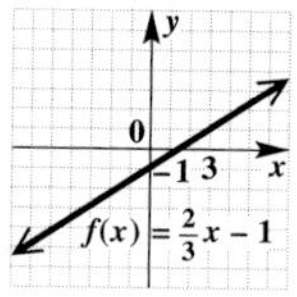

23. (b) The set in (a) includes, for example, the two ordered pairs (8.6, 1990) and (8.6, 1991). In a function, the independent variable (with value 8.6 here) cannot correspond to more than one dependent variable. [3.6] **24.** 200 amps **25.** 8.5

CUMULATIVE REVIEW EXERCISES CHAPTERS 1–3 (PAGE 205)

[1.1] **1.** always true **2.** always true [1.2] **3.** never true **4.** sometimes true; for example, $3 + (-3) = 0$, but $3 + (-1) = 2 \neq 0$ [1.1] **5.** -4 **6.** 9 [1.2] **7.** $-\frac{19}{2}$ [1.3] **8.** $-2p - 5$ [2.1] **9.** $\left\{\frac{3}{7}\right\}$ **10.** $\{-2\}$

[2.2] **11.** $h = \frac{3V}{\pi r^2}$ [2.5] **12.** $\left(-3, \frac{7}{2}\right)$

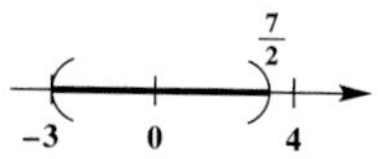

13. $(-\infty, 1]$

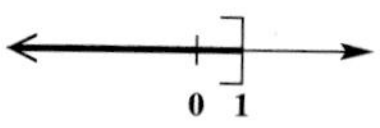

[2.6] **14.** (6, 8)

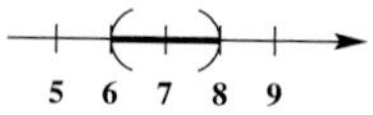

15. $(-\infty, -2] \cup (7, \infty)$

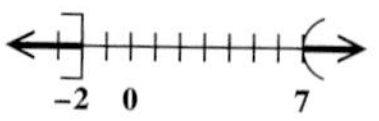

[2.7] **16.** $\{0, 7\}$ **17.** $(-\infty, \infty)$

[2.5] **18.** The union of the three solution sets is $(-\infty, \infty)$. [2.2] **19.** $-67°$F [2.3] **20.** 7.3% [2.4] **21.** 6 inches

[3.1] **22.** x-intercept: (4, 0); y-intercept: $\left(0, \frac{12}{5}\right)$

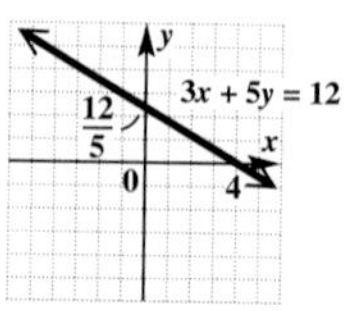

[3.2] **23. (a)** $-\frac{6}{5}$ **(b)** $\frac{5}{6}$ [3.4] **24.**

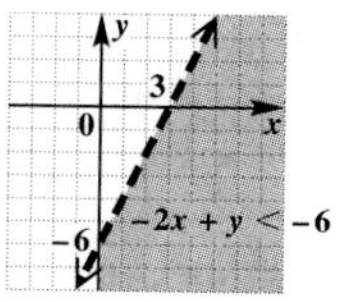

[3.5] **25. (a)** $(-\infty, \infty)$ **(b)** 24 **26.** {(City, Percent)}; The elements of the domain are the names of the cities.

27. yes; $f(x) = \frac{2}{7}x - 2$ **28.** 10 **29.** (a) [3.6] **30.** \$9.92

CHAPTER 4 SYSTEMS OF LINEAR EQUATIONS

SECTION 4.1 (PAGE 215)

EXERCISES **1. (a)** negative **(b)** The Giants' performance declined during those months, as indicated by the negative slope. As time passed, their winning percentage decreased. **(c)** positive **(d)** The Athletics' performance improved during those months, as indicated by the positive slope. As time passed, their winning percentage increased. **3.** yes **5.** no

7. $\{(2, 2)\}$ **9.** $\{(3, -1)\}$ **11.** $\{(2, -3)\}$ **13.** $\left\{\left(\frac{3}{2}, -\frac{3}{2}\right)\right\}$ **15.** $\{(x, y) \mid 7x + 2y = 6\}$; dependent equations **17.** $\{(2, -4)\}$ **19.** $\emptyset$; inconsistent system **21.** Answers will vary. **(a)**

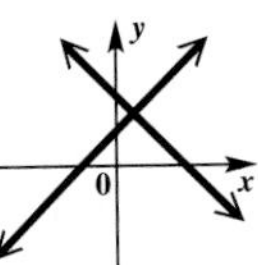

(b) **(c)**

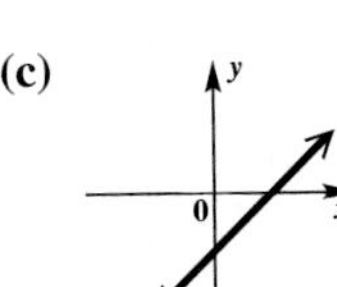

23. $y = -\frac{3}{7}x + \frac{4}{7}$; $y = -\frac{3}{7}x + \frac{3}{14}$; 0 **25.** both are $y = -\frac{2}{3}x + \frac{1}{3}$; infinitely many **27.** $\{(1, 2)\}$ **29.** $\left\{\left(\frac{22}{9}, \frac{22}{3}\right)\right\}$ **31.** $\{(2, 3)\}$ **33.** $\{(5, 4)\}$ **35.** $\left\{\left(-5, -\frac{10}{3}\right)\right\}$ **37.** $\{(2, 6)\}$ **41.** A false statement such as $0 = 1$ will occur. **43.** $\{(2, 4)\}$ **45.** $\{(4, -5)\}$ **47.** $\left\{\left(\frac{1}{a}, \frac{1}{b}\right)\right\}$ **49.** $\left\{\left(-\frac{3}{5a}, \frac{7}{5}\right)\right\}$ **51.** $\{(1, 3)\}$ **52.** $f(x) = -3x + 6$; linear **53.** $g(x) = \frac{2}{3}x + \frac{7}{3}$; linear **54.** one; 1; 3; 1; 3; 1; 3 **55.** $(3, -4)$ **57.** (a) **59. (a)** $\{(5, 5)\}$ **(b)**

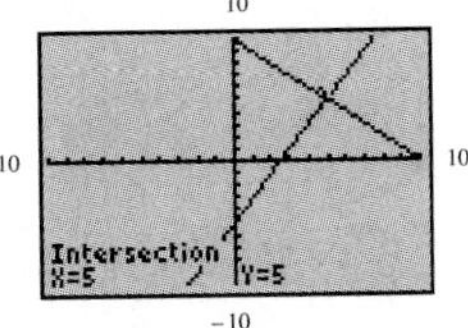

61. (a) $\{(0, -2)\}$ **(b)**

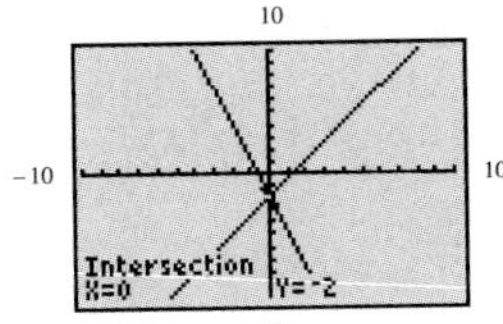

(c) two times; Germany **63. (a)** fourth quarter of 1991; about 1% **(b)** second quarter of 1992; about 2.2% **65. (a)** In 1991 they both reached the level of about 350 million. **(b)** (1987, 100 million)

SECTION 4.2 (PAGE 225)

EXERCISES **1.** (b) **3.** $\{(1, 4, -3)\}$ **5.** $\{(0, 2, -5)\}$ **7.** $\left\{\left(-\frac{7}{3}, \frac{22}{3}, 7\right)\right\}$ **9.** $\{(4, 5, 3)\}$ **11.** $\{(2, 2, 2)\}$ **13.** $\left\{\left(\frac{8}{3}, \frac{2}{3}, 3\right)\right\}$ **17.** $\emptyset$

The solution sets in Exercises 19–21 may be given in other equivalent forms.

19. $\{(x, y, z) \mid x - y + 4z = 8\}$ **21.** $\{(x, y, z) \mid 2x + y - z = 6\}$ **23.** $\{(0, 0, 0)\}$ **25.** $\{(2, 1, 5, 3)\}$

27. $128 = a + b + c$ **28.** $140 = 2.25a + 1.5b + c$ **29.** $80 = 9a + 3b + c$ **30.** $\begin{aligned} a + \quad b + c &= 128 \\ 2.25a + 1.5b + c &= 140 \\ 9a + \quad 3b + c &= 80 \end{aligned}$ $\{(-32, 104, 56)\}$

31. $f(x) = -32x^2 + 104x + 56$ **32.** 56 feet **33.** 140.5 feet **34.** It tells us the projectile hits the ground 3.25 seconds after it is projected. **35.** $a = 3, b = 1, c = -2; f(x) = 3x^2 + x - 2$ **36.** $a = 1, b = 4, c = 3; Y_1 = x^2 + 4x + 3$ **37.** If one were to eliminate *different* variables in the first two steps, the result would be two equations in three variables, and it would not be possible to solve for a single variable in the next step.

SECTION 4.3 (PAGE 234)

CONNECTIONS **PAGE 230:** "Mixed price" refers to the price of a mixture of the two products. The system is $9x + 7y = 107$, $7x + 9y = 101$, where x represents the price of a citron and y represents the price of a wood apple.

EXERCISES **1.** wins: 69; losses: 13 **3.** wins: 48; losses: 26; ties: 8 **5.** length: 78 feet; width: 36 feet **7.** length: 12 feet; width: 5 feet **9.** $x = 40$, $y = 50$, so the angles measure 40° and 50°. **11.** Texas: 93; Florida: 41 **13.** National Hockey League: \$219.74; National Basketball Association: \$203.38 **15.** CGA monitor: \$400; VGA monitor: \$500 **17.** 6 units of yarn; 2 units of thread **19. (a)** 6 ounces **(b)** 15 ounces **(c)** 24 ounces **(d)** 30 ounces **21.** $\$.58x$ **23.** 6 gallons of 25%; 14 gallons of 35% **25.** 3 liters of pure acid; 24 liters of 10% acid **27.** 50 pounds of \$3.60 clusters; 30 pounds of \$7.20 truffles **29.** 76 general admission; 108 with student identification **31.** 28 dimes; 66 quarters **33.** \$1000 at 2%; \$2000 at 4% **35.** $25y$ miles **37.** freight train: 50 kilometers per hour; express train: 80 kilometers per hour **39.** top speed: 2100 miles per hour; wind speed: 300 miles per hour **41.** 8 fish at \$20; 15 fish at \$40; 6 fish at \$65 **43.** $x + y + z = 180$; angle measures: 70°, 30°, 80° **45.** first: 20°; second: 70°; third: 90° **47.** shortest: 12 centimeters; middle: 25 centimeters; longest: 33 centimeters **49.** A: 180 cases; B: 60 cases; C: 80 cases **51.** 10 pounds of jelly beans; 2 pounds of chocolate eggs; 3 pounds of marshmallow chicks **53.** $2a + b + c = -5$ **54.** $-a + c = -1$ **55.** $3a + 3b + c = -18$ **56.** $a = 1, b = -7, c = 0; x^2 + y^2 + x - 7y = 0$ **57.** It fails the vertical line test.

SECTION 4.4 (PAGE 249)

CONNECTIONS **PAGE 245:** **1.** The solution set is also $\{(-2, -1)\}$. **2.** Answers will vary. One example is $\begin{aligned} 3x + y &= -7 \\ y &= -1. \end{aligned}$

PAGE 249: **1.** $\left[\begin{array}{cc|c} 1 & 0 & -2 \\ 0 & 1 & -1 \end{array}\right]$ **2.** a system of dependent equations

EXERCISES **1. (a)** 0, 5, −3 **(b)** 1, −3, 8 **(c)** yes; The number of rows is the same as the number of columns (three). **(d)** $\begin{bmatrix} 1 & 4 & 8 \\ 0 & 5 & -3 \\ -2 & 3 & 1 \end{bmatrix}$ **(e)** $\begin{bmatrix} 1 & -\frac{3}{2} & -\frac{1}{2} \\ 0 & 5 & -3 \\ 1 & 4 & 8 \end{bmatrix}$ **(f)** $\begin{bmatrix} 1 & 15 & 25 \\ 0 & 5 & -3 \\ 1 & 4 & 8 \end{bmatrix}$ **3.** $\left[\begin{array}{cc|c} 1 & 2 & 11 \\ 2 & -1 & -3 \end{array}\right]$; $\left[\begin{array}{cc|c} 1 & 2 & 11 \\ 0 & -5 & -25 \end{array}\right]$; $\left[\begin{array}{cc|c} 1 & 2 & 11 \\ 0 & 1 & 5 \end{array}\right]$; $x + 2y = 11, y = 5$; $\{(1, 5)\}$ **5.** $\{(4, 1)\}$ **7.** $\{(1, 1)\}$ **9.** $\{(-1, 4)\}$ **11.** $\emptyset$ **15.** $\left[\begin{array}{ccc|c} 1 & 1 & -1 & -3 \\ 0 & -1 & 3 & 10 \\ 0 & -6 & 7 & 38 \end{array}\right]$; $\left[\begin{array}{ccc|c} 1 & 1 & -1 & -3 \\ 0 & 1 & -3 & -10 \\ 0 & -6 & 7 & 38 \end{array}\right]$; $\left[\begin{array}{ccc|c} 1 & 1 & -1 & -3 \\ 0 & 1 & -3 & -10 \\ 0 & 0 & -11 & -22 \end{array}\right]$; $\left[\begin{array}{ccc|c} 1 & 1 & -1 & -3 \\ 0 & 1 & -3 & -10 \\ 0 & 0 & 1 & 2 \end{array}\right]$; $x + y - z = -3$, $y - 3z = -10$, $z = 2$; $\{(3, -4, 2)\}$ **17.** $\{(4, 0, 1)\}$ **19.** $\{(-1, 23, 16)\}$ **21.** $\{(3, 2, -4)\}$ **23.** $\{(x, y) \mid x - 2y + z = 4\}$ **25.** $\{(1, 1)\}$ **27.** $\{(-1, 2, 1)\}$ **29.** $\{(1, 7, -4)\}$

SECTION 4.5 (PAGE 260)

CONNECTIONS **PAGE 255:** −185

EXERCISES **1.** (d) **3.** −3 **5.** 14 **7.** 0 **9.** 59 **11.** 14 **15.** −22 **17.** 20 **19.** −5 **21.** By choosing that row or column to expand about, all terms will have a factor of 0, and so the sum of all these terms will be 0. **22.** $\dfrac{y_2 - y_1}{x_2 - x_1}$ **23.** $y - y_1 = \dfrac{y_2 - y_1}{x_2 - x_1}(x - x_1)$ **24.** $x_2y - x_1y - x_2y_1 - xy_2 + x_1y_2 + xy_1 = 0$ **25.** The result is the same as in Exercise 24. **27.** −18 **29.** 0 **31.** −22.04285452 **33.** 16 **35.** −12 **37.** 0 **39.** $\{10\}$ **41. (a)** IV **(b)** I **(c)** III **(d)** II **43.** $\{(-5, 2)\}$ **45.** $\left\{\left(\dfrac{11}{58}, -\dfrac{5}{29}\right)\right\}$ **47.** $\{(-1, 2)\}$

49. Answers will vary. One example is $\begin{aligned} 6x + 7y &= 8 \\ 9x + 10y &= 11. \end{aligned}$ Each system has the solution set $\{(-1, 2)\}$. **51.** $\{(-2, 3, 5)\}$

53. Cramer's rule does not apply. **55.** $\{(20, -13, -12)\}$ **57.** $\left\{\left(\frac{62}{5}, -\frac{1}{5}, \frac{27}{5}\right)\right\}$

59.

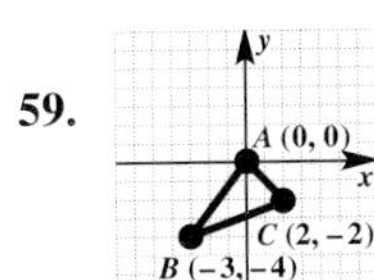

60. $\frac{1}{2}\begin{vmatrix} 0 & 0 & 1 \\ -3 & -4 & 1 \\ 2 & -2 & 1 \end{vmatrix}$ **61.** 7 **62.** 8 **63.** $\{(-1, 3, 5)\}$ **65.** $\{(1, -3, 2, -4)\}$

CHAPTER 4 REVIEW EXERCISES (PAGE 270)

1. In 1995, they both reached a rate of about 72%. **3. (a)** \$4 **(b)** 300 half-gallons **(c)** supply: 200 half-gallons; demand: 400 half-gallons **5.** $\{(0, 1)\}$

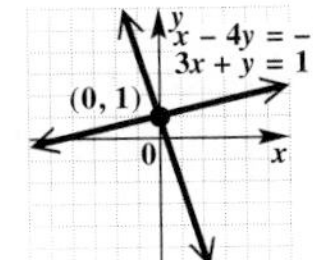

7. $\{(-6, 3)\}$ **9.** $\{(x, y) \mid -3x + y = 6\}$

11. $\left\{\left(-\frac{8}{9}, -\frac{4}{3}\right)\right\}$ **13. (a)** inconsistent **(b)** dependent equations **15.** $\{(1, -5, 3)\}$ **17.** $\emptyset$ **19.** touchdown passes: 26; interceptions: 14 **21.** 30 pounds of \$2-a-pound nuts; 70 pounds of \$1-a-pound candy **23.** \$40,000 at 10%; \$100,000 at 6%; \$140,000 at 5% **25.** Mantle: 54; Maris: 61; Blanchard: 21 **27.** $\{(-1, 5)\}$ **29.** $\{(1, 2, -1)\}$ **31.** 80 **33.** -38 **35.** Cramer's rule does not apply if $D = 0$. **37.** $\{(-4, 5)\}$ **39.** $\{(3, -2, 1)\}$ **41.** $\begin{aligned} 5y + z &= 7 \\ -5y + z &= -13 \end{aligned}$ **42.** $\{(2, -3)\}$ **43.** $\{(4, 2, -3)\}$ **44.** $\{(4, 2, -3)\}$ **45. (a)** -10 **(b)** -40 **(c)** -20 **(d)** 30 **46.** $\{(4, 2, -3)\}$ **47.** $\{(12, 9)\}$ **49.** $\{(3, -1)\}$ **51.** $\{(0, 4)\}$ **53.** AC adaptor: \$8; rechargeable flashlight: \$15

CHAPTER 4 TEST (PAGE 274)

[4.1] **1.** No; The graph for Babe Ruth lies completely below the graph for Aaron, indicating that Ruth's total was always lower than Aaron's. **2.** Aaron had the most and Ruth had the fewest. **3.** $\{(6, 1)\}$

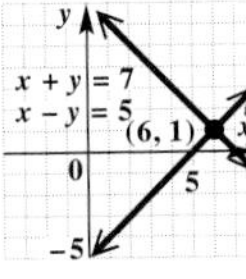

4. $\{(3, 3)\}$

5. $\{(0, -2)\}$ **6.** $\emptyset$ [4.2] **7.** $\left\{\left(-\frac{2}{3}, \frac{4}{5}, 0\right)\right\}$ **8.** $\{(6, -4)\}$ **9.** $\{(x, y) \mid 12x - 5y = 8\}$ [4.3] **10.** slower car: 45 miles per hour; faster car: 75 miles per hour **11.** 4 liters of 20%; 8 liters of 50% **12.** Bulls 90, Jazz 86 [4.1, 4.2] **13.** marker pen: \$1.10; colored paper sheet: \$.40 [4.4] **14.** $\left\{\left(\frac{2}{5}, \frac{7}{5}\right)\right\}$ **15.** $\{(-1, 2, 3)\}$

[4.1, 4.2, 4.4, 4.5] **16.** $\{(-3, -2, -4)\}$ [4.5] **17.** 3 **18.** 0 **19.** $\left\{\left(-\frac{9}{4}, \frac{5}{4}\right)\right\}$ **20.** $\{(5, 2, -1)\}$

CUMULATIVE REVIEW EXERCISES CHAPTERS 1–4 (PAGE 276)

[1.2] **1.** 81 **2.** -81 **3.** -81 **4.** .7 **5.** $-.7$ **6.** not a real number **7.** 4 **8.** -4 **9.** -199 **10.** 455 [1.3] **11.** commutative property [2.1] **12.** $\left\{-\frac{15}{4}\right\}$ [2.7] **13.** $\left\{\frac{2}{3}, 2\right\}$ [2.2] **14.** $x = \frac{d - by}{a - c}$ or $x = \frac{by - d}{c - a}$ [2.1] **15.** $\{11\}$ [2.5] **16.** $\left(-\infty, \frac{240}{13}\right]$ [2.7] **17.** $\left[-2, \frac{2}{3}\right]$ **18.** $(-\infty, \infty)$ [2.3, 2.4] **19.** not guilty: 105; guilty: 95 **20.** 6 meters **21.** pennies: 35; nickels: 29; dimes: 30 **22.** 46°, 46°, 88° [3.1] **23.** $y = 6$ **24.** $x = 4$ [3.2] **25.** $-\frac{4}{3}$ **26.** $\frac{3}{4}$ [3.3] **27.** $4x + 3y = 10$ [3.2] **28.**

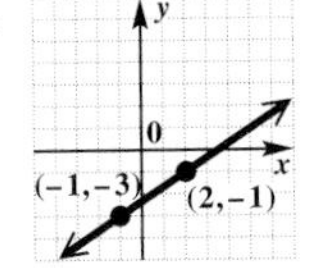

[3.4] **29.**

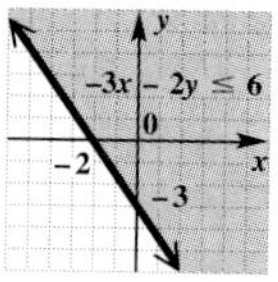

[3.5] **30.** **(a)** -6 **(b)** $a^2 + 3a - 6$ [3.6] **31.** 17.5 [4.1, 4.4, 4.5] **32.** $\{(3, -3)\}$

[4.2, 4.4, 4.5] **33.** $\{(5, 3, 2)\}$ [4.5] **34.** 33 [4.3] **35.** 5 pounds of oranges; 1 pound of apples **36.** Tickle Me Elmo: \$27.63; Snacktime Kid: \$36.26 **37.** small: \$1.50; large: \$2.50 **38.** peanuts: \$2 per pound; cashews: \$4 per pound [4.1] **39.** $x = 8$ or 800 items; \$3000 **40.** about \$500

CHAPTER 5 EXPONENTS AND POLYNOMIALS

SECTION 5.1 (PAGE 288)

EXERCISES **1.** incorrect; $(ab)^2 = a^2b^2$ **3.** incorrect; $\left(\frac{4}{a}\right)^3 = \frac{4^3}{a^3}$ **5.** correct **9.** a^8 **11.** y^6 **13.** $-18x^5y^8$ **15.** p^{14} **17.** z^6 **19.** r^{13} **21.** $-14k^3$ **23.** $-24x^4$ **25.** $\frac{1}{2pq}$ **27.** 1 **29.** $4^5 \cdot 4^2 = 4^7$, not 16^7. Do not multiply the bases together. **31.** $\frac{4}{9}$ **33.** $\frac{1}{64}$ **35.** $-\frac{1}{64}$ **37.** $-\frac{1}{64}$ **39.** 9 **41.** $-\frac{16}{3}$ **43.** $\frac{27}{8}$ **45.** $\frac{5}{6}$ **47.** $-\frac{1}{12}$ **49.** $\frac{1}{2}$ **51.** $\frac{1}{27}$ **53.** $-\frac{1}{81}$ **55.** 1 **57.** -1 **59.** 1 **61.** -2 **63.** $(-3)^{-2} = \frac{1}{(-3)^2} = \frac{1}{9}$, which is a positive number. **65.** $\frac{1}{2^9 \cdot 5^3}$ **67.** $5^{12} \cdot 6^6$ **69.** $\frac{1}{k^2}$ **71.** $-4r^6$ **73.** $\frac{625}{a^{10}}$ **75.** $\frac{z^4}{x^3}$ **77.** $\frac{p^4}{5}$ **79.** $\frac{4}{a^2}$ **81.** $\frac{1}{6y^{13}}$ **83.** $\frac{4k^5}{m^2}$ **85.** $\frac{4k^{17}}{125}$ **87.** $\frac{2k^5}{3}$ **89.** $\frac{8}{3pq^{10}}$ **91.** $\frac{y^9}{8}$ **93.** $-\frac{3}{32m^8p^4}$ **95.** $\frac{2}{3y^4}$ **97.** $\frac{3p^8}{16q^{14}}$ **99.** $\left(\frac{b^6}{a^3}\right)^{-2}; \frac{b^{-12}}{a^{-6}} = \frac{a^6}{b^{12}}$ **100.** $\frac{a^{16}b^{-4}}{a^{10}b^8}; \frac{a^{16-10}}{b^{8+4}} = \frac{a^6}{b^{12}}$ **101.** They are the same. **102.** "Both methods are correct." **103.** 5.3×10^2 **105.** 8.3×10^{-1} **107.** 6.92×10^{-6} **109.** -3.85×10^4 **111.** 72,000 **113.** .00254 **115.** $-60{,}000$ **117.** .000012 **119.** .06 **121.** .0000025 **123.** 200,000 **125.** 3000 **127.** 3.82553×10^{11} **129.** 26,000 or 2.6×10^4 **131.** approximately 9.474×10^{-7} parsec **133.** 300 seconds **135.** approximately 5.87×10^{12} miles **137.** **(a)** 20,000 hours **(b)** 833 days **139.** 1.23×10^5 **141.** 4.24×10^5 **143.** 4.4×10^5 **145.** 7.5×10^9 **147.** 4×10^{17}

SECTION 5.2 (PAGE 297)

EXERCISES **1.** $2x^3 - 3x^2 + x + 4$ **3.** $p^7 - 8p^5 + 4p^3$ **5.** $3m^4 - m^3 + 5m^2 + 10$ **7.** monomial; 0 **9.** binomial; 1 **11.** trinomial; 3 **13.** none of these; 5 **15.** (a) **17.** 7; 1 **19.** -15; 2 **21.** 1; 4 **23.** -1; 6 **25.** $8z^4$ **27.** $7m^3$ **29.** $5x$ **31.** $-3y^2 + 7y$ **33.** $8k^2 + 2k - 7$ **35.** $-2n^4 - n^3 + n^2$ **37.** $3m + 11$ **39.** $-p - 4$ **41.** $-4p^3 + 10p^2 + 5p - 5$ **43.** $2x^5 - x^4 - 2x^3 + 1$ **45.** $6a + 3$ **49.** $(2 \times 10^2) + (4 \times 10^1) + (1 \times 10^0)$ **50.** Corresponding place values are not aligned in columns. **51.** Line up corresponding place values in columns. **52.** Corresponding powers of the variable are not aligned in columns. **53.** Line up corresponding powers of the variable in columns. **54.** In the polynomials, the coefficients of the powers of x differ. In the numerals, the digits are in different places. **55.** $12p - 4$ **57.** $-9p^2 + 11p - 9$ **59.** $14m^2 - 13m + 6$ **61.** $8q^3 - 7q + 11$ **63.** $3y^2 - 4y + 2$ **65.** $-4m^2 + 4n^2 - 7n$ **67.** $y^4 - 4y^2 - 4$ **69.** $10z^2 - 16z$

SECTION 5.3 (PAGE 305)

CONNECTIONS **PAGE 300:** **1.** $f(1) = 1, f(2) = 2$ **2.** 4, 8, 16 **3.** $f(3) = 4, f(4) = 8, f(5) = 16$ **4.** Because the pattern 1, 2, 4, 8, 16 emerges, most people will predict 32, since the terms are doubling each time. However, $f(6) = 31$ (not 32). See the figure.

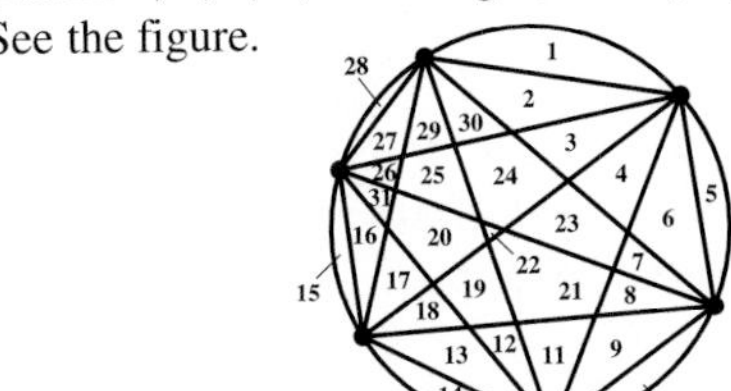

EXERCISES **1. (a)** -10 **(b)** 8 **3. (a)** 8 **(b)** 2 **5. (a)** 8 **(b)** 74 **7. (a)** -11 **(b)** 4 **9.** -22 **11.** 116 **13. (a)** $24,050 million **(b)** $35,295 million **(c)** $67,090 million **(d)** $119,435 million **15. (a)** 33.0 thousand **(b)** 34.1 thousand **(c)** 36.0 thousand **(d)** 38.6 thousand **(e)** 42.1 thousand **(f)** 46.3 thousand **17. (a)** $8x - 3$ **(b)** $2x - 17$ **19. (a)** $-x^2 + 12x - 12$ **(b)** $9x^2 + 4x + 6$ **23.**

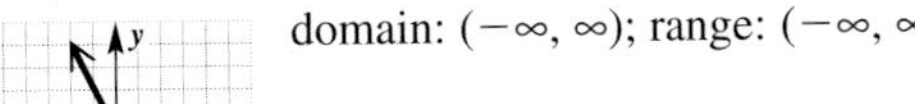

domain: $(-\infty, \infty)$; range: $(-\infty, \infty)$

25.

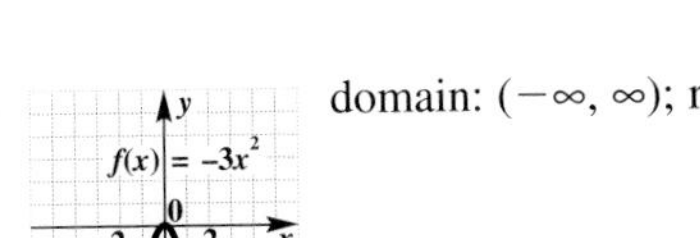

domain: $(-\infty, \infty)$; range: $(-\infty, 0]$

27.

domain: $(-\infty, \infty)$; range: $(-\infty, \infty)$

SECTION 5.4 (PAGE 313)

CONNECTIONS **PAGE 311:**

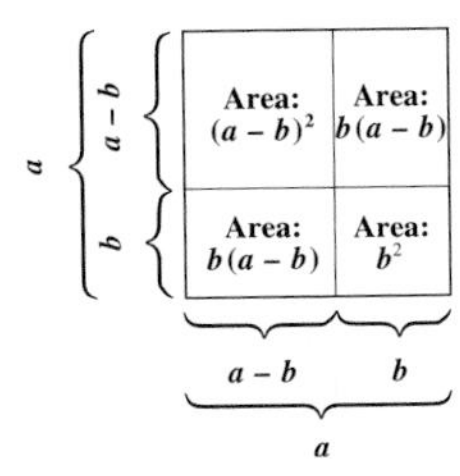

EXERCISES **1.** C **3.** D **5.** $-8x^7$ **7.** $-28x^7y^4$ **9.** $-15p^2 - 10p^3$ **11.** $3x^2 - 5x - 28$ **13.** $6t^2 + 5t - 6s^2$ **15.** $6y^2 + y - 12$ **17.** $25m^2 - 9n^2$ **19.** $m^3 - 3m^2 - 40m$ **21.** $24z^3 - 20z^2 - 16z$ **23.** $4x^4 - 9x^2$ **25.** $6m^3 + m^2 - 14m - 3$ **27.** $12m^3 + 22m^2 + 4m - 3$ **29.** $8z^4 - 14z^3 + 17z^2 + 20z - 3$ **31.** $-2x^5 + 16x^4 - 11x^3 + 40x^2 - 15x$ **33.** $6p^4 + p^3 + 4p^2 - 27p - 6$ **37.** $4p^2 - 9$ **39.** $25m^2 - 1$ **41.** $9a^2 - 4c^2$ **43.** $16x^2 - \frac{4}{9}$ **45.** $16m^2 - 49n^4$ **47.** $25y^6 - 4$ **49.** $y^2 - 10y + 25$ **51.** $4p^2 + 28p + 49$ **53.** $16n^2 - 24nm + 9m^2$ **55.** $k^2 - \frac{10}{7}kp + \frac{25}{49}p^2$ **57.** $(x + y)^2 = x^2 + 2xy + y^2$. The expression $x^2 + y^2$ is missing the $2xy$ term, so $(x + y)^2 \neq x^2 + y^2$ in general. **59.** $25x^2 + 10x + 1 + 60xy + 12y + 36y^2$ **61.** $4a^2 + 4ab + b^2 - 12a - 6b + 9$ **63.** $4a^2 + 4ab + b^2 - 9$ **65.** $6a^3 + 7a^2b + 4ab^2 + b^3$ **67.** $4z^4 - 17z^3x + 12z^2x^2 - 6zx^3 + x^4$ **69.** $m^4 - 4m^2p^2 + 4mp^3 - p^4$ **71.** $a^4b - 7a^2b^3 - 6ab^4$ **73.** $y^3 + 6y^2 + 12y + 8$ **75.** $q^4 - 8q^3 + 24q^2 - 32q + 16$ **77.** 49; 25; $49 \neq 25$ **79.** 2401; 337; $2401 \neq 337$ **81.** Although they are equal for this *particular* case, they are not equal *in general.* **83.** $\frac{9}{2}x^2 - 2y^2$ **85.** $15x^2 - 2x - 24$ **87.** $a - b$ **88.** $A = s^2$; $(a - b)^2$ **89.** $(a - b)b$ or $ab - b^2$; $2ab - 2b^2$ **90.** b^2 **91.** a^2; a **92.** $a^2 - (2ab - 2b^2) - b^2 = a^2 - 2ab + b^2$ **93.** They must be equal to each other. **94.** $(a - b)^2 = a^2 - 2ab + b^2$; This reinforces the special product for the square of a binomial difference. **95.** $10x^2 - 2x$ **97.** $2x^2 - x - 3$ **99.** $8x^3 - 27$

SECTION 5.5 (PAGE 319)

EXERCISES **3.** $z^2(m+n)^4$ **5.** $12(m+5)$ **7.** $8k(k^2+3)$ **9.** $xy(1-5y)$ **11.** $-2p^2q^4(2p+q)$ **13.** $7x^3(3x^2+5x+2)$ **15.** $5ac(3ac^2-5c+1)$ **17.** $-9m^3p^3(3p^2-4m+8m^2p)$ **19.** $(m-4)(2m+5)$ **21.** $11(2z-1)$ **23.** $-y^5(r+w+yz+yk)$ **25.** $(2-x)(10-x-x^2)$ **27.** $(3-x)(6+2x-x^2)$ **29.** $20z(2z+1)(3z+4)$ **31.** $5(m+p)^2(m+p-2-3m^2-6mp-3p^2)$ **33.** (a) **35.** $2x^2(-x^3+3x+2)$ or $-2x^2(x^3-3x-2)$ **37.** $16a^2m^3(-2a^2m^2-1-4a^3m^3)$ or $-16a^2m^3(2a^2m^2+1+4a^3m^3)$ **39.** $2^2 \cdot 3 \cdot 5$ **40.** $2^2 \cdot 3 \cdot 5 \cdot 7$ **41.** $2^3 \cdot 3^2 \cdot 5 + 2^2 \cdot 3 \cdot 5$ **42.** $2^2 \cdot 3 \cdot 5$ **43.** $2^2 \cdot 3 \cdot 5(2 \cdot 3 + 1)$ **44.** It represents 7. **45.** $2^2 \cdot 3 \cdot 5 \cdot 7$ **46.** The answers are the same. **47.** $(m+3q)(x+y)$ **49.** $(5m+n)(2+k)$ **51.** $(m-3)(m+5)$ **53.** $(p+q)(p-4z)$ **55.** $(a+5)(3a-2)$ **57.** $(-3p+q)(5p+2q)$ **59.** $(a^2+b^2)(-3a+2b)$ **61.** $(y-2)(x-2)$ **63.** $(3y-2)(3y^3-4)$ **65.** $(1-a)(1-b)$ **67.** (d) **69.** $m^{-5}(3+m^2)$ or $\frac{3+m^2}{m^5}$ **71.** $p^{-3}(3+2p)$ or $\frac{3+2p}{p^3}$

SECTION 5.6 (PAGE 328)

EXERCISES **1. (a)** E **(b)** B **(c)** A **(d)** D **(e)** C **(f)** F **3.** $(a-5)(a+3)$ **5.** $(p+8)(p+3)$ **7.** $(r-12)(r-3)$ **9.** $(a+5b)(a-7b)$ **11.** $(y-5q)(y-3q)$ **13.** prime **15.** $(6m-5)(m+3)$ **17.** $(5x-6)(2x+3)$ **19.** $(4k+3)(5k+8)$ **21.** $(3a-2b)(5a-4b)$ **23.** prime **25.** $(5r-9)^2$ **27.** $(2xz-1)(3xz+4)$ **29.** $3(4x+5)(2x+1)$ **31.** $5(a+6)(3a-4)$ **33.** $4m(m+5)(m-2)$ **35.** $11x(x-6)(x-4)$ **37.** $2xy^3(x-12y)^2$ **39.** $3(3a+2)(2a-3)$ **41.** $6a(a-3)(a+5)$ **43.** $13y(y+4)(y-1)$ **45.** $3p(2p-1)^2$ **47.** $-(x-9)(x+2)$ **49.** $-(9a+5)(2a-3)$ **51.** $-r(7r+1)(2r-3)$ **53.** There is a GCF of 2. She did not factor the polynomial *completely*. The factor $(4x+10)$ can be factored further as $2(2x+5)$, giving the final form as $2(2x+5)(x-2)$. **55.** They are both correct; in each case the product is $-4x^2-29x+24$. **57.** no **58.** 1, 3, 5, 9, 15, 45; no **59.** no **60.** 1, 2, 4, 5, 10, 20; no **61.** no **62.** $(5x+2)(2x+5)$; no **63.** Since k is odd, 2 is not a factor of $2x^2+kx+8$, and because 2 is a factor of $2x+4$, the binomial $2x+4$ cannot be a factor. **64.** 3 is a factor of $3y+15$, but it is not a factor of $12y^2-11y-15$. **65.** $(p^2-8)(p^2-2)$ **67.** $(2x^2+3)(x^2-6)$ **69.** $(4x^2+3)(4x^2+1)$ **71.** $(6p^3-r)(2p^3-5r)$ **73.** $(5k+4)(2k+1)$ **75.** $(3m+3p+5)(m+p-4)$ **77.** $(a+b)^2(a-3b)(a+2b)$ **79.** $(p+q)^2(p+3q)$ **81.** $(z-x)^2(z+2x)$

SECTION 5.7 (PAGE 333)

EXERCISES **1. (a)** B **(b)** D **(c)** C **(d)** A **3.** (a), (b), (d) **5.** $(p+4)(p-4)$ **7.** $(5x+2)(5x-2)$ **9.** $(3a+7b)(3a-7b)$ **11.** $4(4m^2+y^2)(2m+y)(2m-y)$ **13.** $(y+z+9)(y+z-9)$ **15.** $(4+x+3y) \cdot (4-x-3y)$ **17.** $4pq$ **19.** $(k-3)^2$ **21.** $(2z+w)^2$ **23.** $(4m-1+n)(4m-1-n)$ **25.** $(2r-3+s) \cdot (2r-3-s)$ **27.** $(x+y-1)(x-y+1)$ **29.** $2(7m+3n)^2$ **31.** $(p+q+1)^2$ **33.** $(a-b+4)^2$ **35.** $(2x-y)(4x^2+2xy+y^2)$ **37.** $(4g+3h)(16g^2-12gh+9h^2)$ **39.** $3(2n+3p)(4n^2-6np+9p^2)$ **41.** $(y+z-4)(y^2+2yz+z^2+4y+4z+16)$ **43.** $(4y^2+1)(16y^4-4y^2+1)$ **45.** $(10x^3-3)(100x^6+30x^3+9)$ **47.** $(8t^2-p)(64t^4+8t^2p+p^2)$ **49.** $(x^3-y^3)(x^3+y^3)$; $(x-y)(x^2+xy+y^2)(x+y)(x^2-xy+y^2)$ **50.** $(x^2+xy+y^2)(x^2-xy+y^2)$ **51.** $(x^2-y^2)(x^4+x^2y^2+y^4)$; $(x-y)(x+y)(x^4+x^2y^2+y^4)$ **52.** $x^4+x^2y^2+y^4$ **53.** The product must equal $x^4+x^2y^2+y^4$. Multiply $(x^2+xy+y^2)(x^2-xy+y^2)$ to verify this. **54.** Start by factoring as a difference of squares. **55.** 16 **57.** 25 **59.** -56 or 56 **61.** $(3x+y)(9x^2-3xy+y^2+3x-y)$ **63.** $(10k-m)(100k^2+10km+m^2+2)$ **65.** $(y+1)^2(y^2-y+1)$ **67.** $5(x+y)(2x-2y+x^2-xy+y^2)$

SECTION 5.8 (PAGE 337)

EXERCISES **1.** $(10a+3b)(10a-3b)$ **3.** $3p^2(p-6)(p+5)$ **5.** $3pq(a+6b)(a-5b)$ **7.** prime **9.** $(6b+1)(b-3)$ **11.** $3mn(3m+2n)(2m-n)$ **13.** $(2p+5q)(p+3q)$ **15.** $9m(m-5+2m^2)$ **17.** $2(3m-10)(9m^2+30m+100)$ **19.** $(2a+1)(a-4)$ **21.** $(k-9)(q+r)$ **23.** prime **27.** $(x-5)(x+5)(x^2+25)$ **29.** $(p+4)(p^2-4p+16)$ **31.** $(8m+25)(8m-25)$ **33.** $6z(2z^2-z+3)$ **35.** $16(4b+5c)(4b-5c)$ **37.** $8(5z+4)(25z^2-20z+16)$ **39.** $(5r-s)(2r+5s)$ **41.** $4pq(2p+q)(3p+5q)$ **43.** $3(4k^2+9)(2k+3)(2k-3)$ **45.** $(m-n)(m^2+mn+n^2+m+n)$ **47.** $(x-2m-n)(x+2m+n)$ **49.** $2w+9$ **51.** $2(x+4)(x-5)$ **53.** $8mn$ **55.** $2(5p+9)(5p-9)$ **57.** $4rx(3m^2+mn+10n^2)$

59. $(7a - 4b)(3a + b)$ **61.** prime **63.** $(p + 8q - 5)^2$ **65.** $(7m^2 + 1)(3m^2 - 5)$ **67.** $(2r - t)(r^2 - rt + 19t^2)$ **69.** $(x + 3)(x^2 + 1)(x + 1)(x - 1)$ **71.** $(m + n - 5)(m - n + 1)$ **73.** $a(1 + y)$ **74.** The 1 must appear so that when the distributive property is used to check, the term $a \cdot 1 = a$ will appear in the original binomial. **75.** $(x^2 + 4)(1 + y)$ **76.** If 1 did not appear, the terms x^2 and 4 would not appear in the product when the factoring was checked.

SECTION 5.9 (PAGE 345)

EXERCISES **3.** Factor $3x^2 - 8x$ as $x(3x - 8)$. **5.** Subtract 12 to get $x^2 - 4x - 12$ on the left, and then factor as $(x - 6)(x + 2)$. **7.** The equation is ready to solve as it is given. **9.** $\{5, -10\}$ **11.** $\left\{\frac{5}{2}, -\frac{8}{3}\right\}$ **13.** $\left\{-6, \frac{3}{4}, 1\right\}$ **15.** $\left\{0, 4, -\frac{5}{2}\right\}$ **17.** $\{-2, 5\}$ **19.** $\{-6, -3\}$ **21.** $\left\{-\frac{1}{2}, 4\right\}$ **23.** $\left\{-\frac{1}{3}, \frac{4}{5}\right\}$ **25.** $\{-3, 4\}$ **27.** $\{-3, 3\}$ **29.** $\left\{-\frac{3}{5}, 0\right\}$ **31.** $\left\{-\frac{3}{4}\right\}$ **33.** $\left\{-\frac{3}{2}, \frac{3}{2}\right\}$ **35.** $\{-3, 3\}$ **37.** $\{-4, 2\}$ **39.** $\left\{-\frac{1}{2}, 6\right\}$ **41.** $\{1, 6\}$ **43.** The left side factors as $x(ax + b)$. When the factor x is set equal to 0, the solution 0 is apparent. **45.** $\left\{-\frac{1}{2}, 0, 5\right\}$ **47.** $\left\{-\frac{4}{3}, 0, \frac{4}{3}\right\}$ **49.** $\left\{-\frac{5}{2}, -1, 1\right\}$ **51.** By dividing both sides by a variable expression, she "lost" the solution 0. **53.** $\{-.5, 4\}$ **55.** $\{-2, 5\}$ **57.** $\left\{-\frac{1}{2}, 6\right\}$ **59.** $\left\{-\frac{2}{3}, \frac{4}{15}\right\}$ **61.** $\{2, 4\}$ **63.** $\left\{-\frac{3}{2}, \frac{1}{2}\right\}$ **65.** width: 16 feet; length: 20 feet **67.** base: 8 meters; height: 11 meters **69.** 50 feet by 100 feet **71.** -6 and -5 or 5 and 6 **73.** length: 15 inches; width: 9 inches **75.** 3 seconds and 5 seconds; 1 second and 7 seconds **77.** $6\frac{1}{4}$ seconds **79.** $\{-1, 0, 1\}$ **81.** $\{-1.96, -.74, .04\}$

CHAPTER 5 REVIEW EXERCISES (PAGE 353)

1. $-12x^2y^8$ **3.** $\frac{10p^8}{q^7}$ **5.** 64 **7.** -125 **9.** $\frac{81}{16}$ **11.** $\frac{11}{30}$ **13.** 0 **15.** $\frac{1}{3^8}$ **17.** $\frac{y^6}{x^2}$ **19.** $\frac{25}{m^{18}}$ **21.** $\frac{25}{z^4}$ **23.** $\frac{2025}{8r^4}$ **25.** Yes, $\left(\frac{a}{b}\right)^{-1} = \frac{a^{-1}}{b^{-1}}$ for all $a, b \neq 0$. **27.** For example, let $x = 2$ and $y = 3$. Then $(x^2 + y^2)^2 = (2^2 + 3^2)^2 = 169$. $x^4 + y^4 = 2^4 + 3^4 = 97 \neq 169$. **29.** 7.65×10^{-8} **31.** 1,210,000 **33.** .63 **35.** 1.5×10^3; 1500 **37.** 2.7×10^{-2}; .027 **39.** 5.449×10^3 **41.** 14 **43.** 504 **45. (a)** $9m^7 + 14m^6$ **(b)** binomial **(c)** 7 **47. (a)** $-7q^5r^3$ **(b)** monomial **(c)** 8 **49.** $-x^2 - 3x + 1$ **51.** $6a^3 - 4a^2 - 16a + 15$ **53.** $12x^2 + 8x + 5$ **55. (a)** $5x^2 - x + 5$ **(b)** $-5x^2 + 5x + 1$ **57. (a)** $f(x) = -2x + 5$ **(b)** $f(x) = x^2 - 6$ **(c)** $f(x) = -x^3 + 1$ **59.** $15m^2 - 7m - 2$ **61.** $6w^2 - 13wt + 6t^2$ **63.** $3q^3 - 13q^2 - 14q + 20$ **65.** $36r^4 - 1$ **67.** $16m^2 + 24m + 9$ **69.** $6p(2p - 1)$ **71.** $4qb(3q + 2b - 5q^2b)$ **73.** $(x + 3)(x - 3)$ **75.** $(m + q)(4 + n)$ **77.** $(m + 3)(2 - a)$ **79.** $(3p - 4)(p + 1)$ **81.** $(3r + 1)(4r - 3)$ **83.** $(2k - h)(5k - 3h)$ **85.** $2x(4 + x)(3 - x)$ **87.** $(y^2 + 4)(y^2 - 2)$ **89.** $(p + 2)^2(p + 3)(p - 2)$ **91.** It is not factored because there are two terms: $x^2(y^2 - 6)$ and $5(y^2 - 6)$. The correct answer is $(y^2 - 6)(x^2 + 5)$. **93.** $(4x + 5)(4x - 5)$ **95.** $(x + 7)^2$ **97.** $(r + 3)(r^2 - 3r + 9)$ **99.** $(m + 1)(m^2 - m + 1)(m - 1)(m^2 + m + 1)$ **101.** $(x + 3 + 5y)(x + 3 - 5y)$ **103.** $(x + 1)(x - 1)(x - 2)(x^2 + 2x + 4)$ **105. (a)** $(x - 3)(6x + 16) = 0$ **(b)** $\left\{3, -\frac{8}{3}\right\}$ **107.** $\{2, 3\}$ **109.** $\left\{-\frac{5}{2}, \frac{10}{3}\right\}$ **111.** $\left\{-\frac{3}{2}, -\frac{1}{4}\right\}$ **113.** $\left\{-\frac{3}{2}, 0\right\}$ **115.** $\{4\}$ **117.** $\{-3, -2, 2\}$ **119.** 3 feet **121.** after 16 seconds **123.** The rock reaches a height of 240 feet once on its way up and once on its way down. **125.** $\frac{y^4}{36}$ **127.** $\frac{1}{16y^{18}}$ **129.** $21p^9 + 7p^8 + 14p^7$ **131.** -9 **133.** $-3k^2 + 4k - 7$ **135.** $k(11 + 12k)$ **137.** prime **139.** $(5z - 3m)^2$ **141.** $\{-1, 0, 1\}$ **143.** $(x^{14} - x^2) - (4x^{13} - 4x) + (4x^{12} - 4)$ **144.** $x^2(x^{12} - 1) - 4x(x^{12} - 1) + 4(x^{12} - 1)$ **145.** $(x^{12} - 1)(x^2 - 4x + 4)$ **146.** $(x^6 - 1)(x^6 + 1)(x - 2)^2$

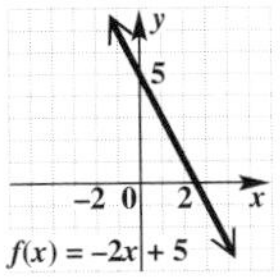

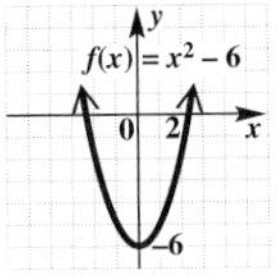

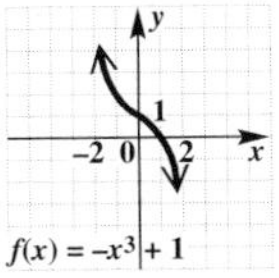

147. $(x^3 - 1)(x^3 + 1)(x^2 + 1)(x^4 - x^2 + 1)(x - 2)^2$ **148.** $(x - 1)(x^2 + x + 1)(x + 1)(x^2 - x + 1)(x^2 + 1) \cdot (x^4 - x^2 + 1)(x - 2)^2$

CHAPTER 5 TEST (PAGE 358)

[5.1] **1. (a)** C **(b)** A **(c)** D **(d)** A **(e)** E **(f)** F **(g)** B **(h)** G **(i)** C **2.** $\frac{4x^7}{9y^{10}}$ **3.** $\frac{6}{r^{14}}$ **4.** $\frac{16}{9p^{10}q^{28}}$ **5.** $\frac{16}{x^6y^{16}}$ **6. (a)** .00000091 **(b)** 3×10^{-4}; .0003 [5.3] **7. (a)** -18 **(b)** $-2x^2 + 12x - 9$ **(c)** $-2x^2 - 2x - 3$ **8.**

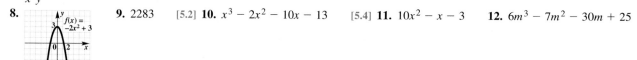

9. 2283 [5.2] **10.** $x^3 - 2x^2 - 10x - 13$ [5.4] **11.** $10x^2 - x - 3$ **12.** $6m^3 - 7m^2 - 30m + 25$ **13.** $36x^2 - y^2$ **14.** $9k^2 + 6kq + q^2$ **15.** $4y^2 - 9z^2 + 6zx - x^2$ [5.5] **16.** It is not in factored form because there are two terms: $(x^2 + 2y)p$ and $3(x^2 + 2y)$. The common factor is $x^2 + 2y$, and the factored form is $(x^2 + 2y)(p + 3)$. [5.5–5.8] **17.** $11z(z - 4)$ **18.** $(x + y)(3 + b)$ **19.** $(4p - q)(p + q)$ **20.** $(4a + 5b)^2$ **21.** $(y - 6)(y^2 + 6y + 36)$ **22.** $(3k + 11j)(3k - 11j)$ **23.** $(2k^2 - 5)(3k^2 + 7)$ **24.** $(3x^2 + 1)(9x^4 - 3x^2 + 1)$ **25.** (d) [5.9] **26.** $\left\{-2, -\frac{2}{3}\right\}$ **27.** $\left\{\frac{1}{5}, \frac{3}{2}\right\}$ **28.** $\left\{-\frac{2}{5}, 1\right\}$ **29.** length: 8 inches; width: 5 inches **30.** 2 seconds and 4 seconds

CUMULATIVE REVIEW EXERCISES CHAPTERS 1–5 (PAGE 359)

[1.2, 1.3] **1.** $-2m + 6$ **2.** $4m - 3$ **3.** $2x^2 + 5x + 4$ [1.2] **4.** -24 **5.** 204 **6.** undefined **7.** 10 [2.1] **8.** $\left\{\frac{7}{6}\right\}$ **9.** $\{-1\}$ [2.5] **10.** $\left(-\infty, \frac{15}{4}\right]$ **11.** $\left(-\frac{1}{2}, \infty\right)$ [2.6] **12.** $(2, 3)$ **13.** $(-\infty, 2) \cup (3, \infty)$ [2.7] **14.** $\left\{-\frac{16}{5}, 2\right\}$ **15.** $(-11, 7)$ **16.** $(-\infty, -2] \cup [7, \infty)$ [2.2] **17.** $h = \frac{V}{lw}$ [2.4] **18.** 2 hours [3.1] **19.**

[3.2] **20.** -1 **21.** 0 [3.5] **22.** -1 **23.** $\left(-\frac{7}{2}, 0\right)$ [4.1] **24.** $\{(1, 5)\}$ [4.2] **25.** $\{(1, 1, 0)\}$ [4.5] **26.** -22 [5.1] **27.** $\frac{y}{18x}$ **28.** $\frac{5my^4}{3}$ [5.2] **29.** $x^3 + 12x^2 - 3x - 7$ [5.4] **30.** $49x^2 + 42xy + 9y^2$ **31.** $10p^3 + 7p^2 - 28p - 24$ [5.5–5.8] **32.** $(2w + 7z)(8w - 3z)$ **33.** $(2x - 1 + y)(2x - 1 - y)$ **34.** $(2y - 9)^2$ **35.** $(10x^2 + 9)(10x^2 - 9)$ **36.** $(2p + 3)(4p^2 - 6p + 9)$ [5.9] **37.** $\left\{-4, -\frac{3}{2}, 1\right\}$ **38.** $\left\{\frac{1}{3}\right\}$ **39.** 4 feet **40.** longer sides: 18 inches; distance between: 16 inches

CHAPTER 6 RATIONAL EXPRESSIONS

SECTION 6.1 (PAGE 368)

EXERCISES **1.** C **3.** D **5.** E **9.** 7 **11.** $-\frac{1}{7}$ **13.** 0 **15.** $-2, \frac{3}{2}$ **17.** none **19.** none **23.** 1992: $\frac{6.3}{8.2} \approx .77$; 1996: $\frac{7.2}{8.5} \approx .85$; 1996 **25.** $\frac{4x}{3y}$ **27.** $\frac{x - 3}{x + 5}$ **29.** $\frac{x + 3}{2x(x - 3)}$ **31.** already in lowest terms **33.** $\frac{6}{7}$ **35.** $\frac{z}{6}$ **37.** $\frac{2}{t - 3}$ **39.** $\frac{x - 3}{x + 1}$ **41.** $\frac{4x + 1}{4x + 3}$ **43.** $a^2 - ab + b^2$ **45.** $\frac{c + 6d}{c - d}$ **47.** $\frac{a + b}{a - b}$

49. (b) **51.** -1

In Exercises 53–55, there are several other acceptable ways to express the answer.

53. $-(x + y)$ **55.** $-\frac{x+y}{x-y}$ **57.** $-\frac{1}{2}$ **59.** already in lowest terms **63.** $\frac{3y}{x^2}$ **65.** $\frac{3a^3b^2}{4}$ **67.** $\frac{4}{3pq}$ **69.** $\frac{7x}{6}$ **71.** $-\frac{p+5}{2p}$ (There are other ways.) **73.** $\frac{-m(m+7)}{m+1}$ (There are other ways.) **75.** -2 **77.** $\frac{x+4}{x-4}$ **79.** $\frac{2x+3y}{2x-3y}$ **81.** $\frac{k+5p}{2k+5p}$ **83.** $(k-1)(k-2)$ **85.** $\frac{(a+5)(2a+b)}{(3a+1)(a+2b)}$ **87.** $\frac{x^2-5x+6}{x-2}$ **89.** $\frac{x^2+3x-10}{x-2}$

SECTION 6.2 (PAGE 376)

EXERCISES **1.** $\frac{3}{4}$ **2.** $\frac{1}{6}$ **3.** no; We cannot find the sum $\frac{1}{x}+\frac{1}{y}$ by adding the denominators and keeping the common numerator. **4.** $\frac{2}{15}$ **5.** $-\frac{1}{2}$ **6.** no; We cannot find the difference $\frac{1}{x}-\frac{1}{y}$ by subtracting the denominators and keeping the common numerator. **7.** $\frac{9}{t}$ **9.** $\frac{2}{x}$ **11.** 1 **13.** $x-5$ **15.** $\frac{5}{p+3}$ **17.** $a-b$ **21.** $72x^4y^5$ **23.** $z(z-2)$ **25.** $2(y+4)$ **27.** $30(x+3)$ **29.** $(m+n)(m-n)$ **31.** $x(x-4)(x+1)$ **33.** $(t+5)(t-2)(2t-3)$ **35.** $2y(y+3)(y-3)$ **37.** $6x^2(x+1)$

39. Yes, they are both correct, because the expressions are equivalent. Multiplying $\frac{3}{5-y}$ by $\frac{-1}{-1}$ gives $\frac{-3}{y-5}$.

41. $\frac{31}{3t}$ **43.** $\frac{5-22x}{12x^2y}$ **45.** $\frac{1}{x(x-1)}$ **47.** $\frac{5a^2-7a}{(a+1)(a-3)}$ **49.** $\frac{-6x+3}{x-4}$ or $\frac{6x-3}{4-x}$ **51.** $\frac{w+z}{w-z}$ or $\frac{-w-z}{z-w}$ **53.** $\frac{2(2x-1)}{x-1}$ **55.** $\frac{6}{x-2}$ **57.** $\frac{3x-2}{x-1}$ **59.** $\frac{4x-7}{x^2-x+1}$ **61.** $\frac{2x+1}{x}$ **63.** (a)–(d) All equal $\frac{7m+5}{m(m+1)}$. **65.** $\frac{4p^2-21p+29}{(p-2)^2}$ **67.** $\frac{x+9}{(x-3)(x-2)(x+1)}$ **69.** $\frac{2x^2+21xy-10y^2}{(x+2y)(x-y)(x+6y)}$ **71.** $\frac{3r-2s}{(2r-s)(3r-s)}$ **73.** (a) $\frac{-x}{y(4x+3y)}+\frac{8x+6y}{(4x+3y)(4x-3y)}$ (b) $\frac{-x}{y(4x+3y)}\cdot\frac{2(4x+3y)}{(4x+3y)(4x-3y)}$ **74.** (a) $\frac{-x(4x-3y)}{y(4x+3y)(4x-3y)}+\frac{y(8x+6y)}{y(4x+3y)(4x-3y)}$ (b) $\frac{-2x(4x+3y)}{y(4x+3y)^2(4x-3y)}$ **75.** (a) $\frac{-4x^2+11xy+6y^2}{y(4x+3y)(4x-3y)}$ (b) $\frac{-2x}{y(4x+3y)(4x-3y)}$

76. In both cases, it is necessary to factor the denominators. In Problem A, the factored form is used to find the least common denominator; in Problem B, the numerator is also factored, and the factored forms are used to write the product in lowest terms.

77. (a) $c(x)=\frac{10x}{49(101-x)}$ (b) approximately 3.23 thousand dollars

SECTION 6.3 (PAGE 384)

CONNECTIONS **Page 383: 1.** $\frac{34}{21}$; 1, 2, 1.5, $1.\overline{6}$, 1.6, 1.625, 1.615384615, 1.619047619 **2.** They seem to approach a number close to 1.62.

EXERCISES **3.** $\frac{2x}{x-1}$ **5.** $\frac{2(k+1)}{3k-1}$ **7.** $\frac{5x^2}{9z^3}$ **9.** $\frac{1+x}{-1+x}$ **11.** $\frac{y+x}{y-x}$ **13.** $4x$ **15.** $x+4y$ **17.** $\frac{3y}{2}$ **19.** $\frac{x^2+5x+4}{x^2+5x+10}$ **21.** $\frac{x^2y^2}{y^2+x^2}$ **23.** $\frac{y^2+x^2}{xy^2+x^2y}$ or $\frac{y^2+x^2}{xy(y+x)}$ **25.** $\frac{rs}{s+r}$ **27.** (a) $\frac{\frac{3}{mp}-\frac{4}{p}+\frac{8}{m}}{\frac{2}{m}-\frac{3}{p}}$ (b) In the denominator, $2m^{-1}=\frac{2}{m}$, not $\frac{1}{2m}$, and $3p^{-1}=\frac{3}{p}$, not $\frac{1}{3p}$. (c) $\frac{3-4m+8p}{2p-3m}$ **29.** $\frac{-1}{6y-1}$ or $\frac{1}{1-6y}$ **31.** $\frac{p+p^2+1}{p^3+p^2+2p+1}$ **33.** $\frac{m^2+6m-4}{m(m-1)}$ **34.** $\frac{m^2-m-2}{m(m-1)}$ **35.** $\frac{m^2+6m-4}{m^2-m-2}$ **36.** $m(m-1)$

37. $\dfrac{m^2 + 6m - 4}{m^2 - m - 2}$ **38.** Method 1 involves simplifying the numerator and the denominator separately and then performing a division. Method 2 involves multiplying the fraction by a form of 1, the identity element for multiplication. (Preference will vary.)

SECTION 6.4 (PAGE 390)

EXERCISES **1.** quotient; exponents **3.** descending powers **5.** $3y + 4 - \dfrac{5}{y}$ **7.** $3m + 5 + \dfrac{6}{m}$ **9.** $n - \dfrac{3n^2}{2m} + 2$ **11.** $\dfrac{2y}{x} + \dfrac{3}{4} + \dfrac{3w}{x}$ **13.** $r^2 - 7r + 6$ **15.** $y - 3$ **17.** $t + 5$ **19.** $z^2 + 3$ **21.** $x^2 + 2x - 3 + \dfrac{6}{4x + 1}$ **23.** $2x - 5 + \dfrac{-4x + 5}{3x^2 - 2x + 4}$ **25.** $2k^2 + 3k - 1$ **27.** $9z^2 - 4z + 1 + \dfrac{-z + 6}{z^2 - z + 2}$ **29.** $p^2 + p + 1$ **31.** $2x + 7$ for $x = 0, -4$ **33.** The correct quotient is $a^2 + 1$. **35.** $\dfrac{2}{3}x - 1$ **37.** $\dfrac{3}{4}a - 2 + \dfrac{1}{4a + 3}$ **39.** $2p + 7$
41. -13; -13; They are the same, which suggests that when $P(x)$ is divided by $x - r$, the result is $P(r)$. Here, $r = -1$.
43. $5x - 1$; 0 **45.** $2x - 3$; -1 **47.** $4x^2 + 6x + 9$; $\dfrac{3}{2}$

SECTION 6.5 (PAGE 395)

CONNECTIONS **Page 395:** -2.7

EXERCISES **1.** Synthetic division provides a quick, easy way to divide a polynomial by a binomial of the form $x - k$.
3. $x - 5$ **5.** $4m - 1$ **7.** $2a + 4 + \dfrac{5}{a + 2}$ **9.** $p - 4 + \dfrac{9}{p + 1}$ **11.** $4a^2 + a + 3$
13. $x^4 + 2x^3 + 2x^2 + 7x + 10 + \dfrac{18}{x - 2}$ **15.** $-4r^5 - 7r^4 - 10r^3 - 5r^2 - 11r - 8 + \dfrac{-5}{r - 1}$
17. $-3y^4 + 8y^3 - 21y^2 + 36y - 72 + \dfrac{143}{y + 2}$ **19.** $y^2 + y + 1 + \dfrac{2}{y - 1}$ **21.** 7 **23.** -2 **25.** 0
27. By the remainder theorem, a zero remainder means that $P(k) = 0$; that is, k is a number that makes $P(x) = 0$.
29. yes **31.** no **33.** no **35.** yes **37.** $(2x - 3)(x + 4)$ **38.** $\left\{\dfrac{3}{2}, -4\right\}$ **39.** $P(-4) = 0$, $P\left(\dfrac{3}{2}\right) = 0$
40. a **41.** Yes, $x - 3$ is a factor. $Q(x) = (x - 3)(3x - 1)(x + 2)$ **43.** $(x - 3)(x^2 + 2x - 15) = (x - 3)^2(x + 5)$
45. $(x + 1)(x^2 + 2x - 15) = (x + 1)(x - 3)(x + 5)$

SECTION 6.6 (PAGE 402)

EXERCISES **1.** $x(x - 1)$ **3.** $(m + 1)(m + 2)$ **5.** Because -1 makes the denominators $m + 1$ equal 0, it is not in the domain of the equation. The other number excluded from the domain is -2. **7.** $-1, 2$ **9.** $-\dfrac{5}{3}, 0, -\dfrac{3}{2}$
11. $4, \dfrac{7}{2}$ **13.** $\{-3\}$ **15.** $\{-3\}$ **17.** $\emptyset$ **19.** $\{1\}$ **21.** $\{-6, 4\}$ **23.** $\left\{-\dfrac{23}{5}\right\}$ **25.** $\{5\}$ **27.** $\emptyset$
29. $\{-2\}$ **31.** $\emptyset$ **33.** $\left(-\infty, -\dfrac{3}{2}\right) \cup \left(-\dfrac{3}{2}, \dfrac{3}{2}\right) \cup \left(\dfrac{3}{2}, \infty\right)$ **35.** 6 **36. (a)** $3x + 2x = -30$ **(b)** $\dfrac{3x}{6} + \dfrac{2x}{6}$
37. (a) $\{-6\}$ **(b)** $\dfrac{5x}{6}$ **38.** One is a solution set; the other is an expression. **40.** The word "Solve" refers to finding the solution set of an equation. What appears here is not an equation, but an expression. "Solve" should be replaced by "Simplify" or "Add." **41. (a)** 14,000 gallons **(b)** It decreases. **43. (a)** 0 **(b)** 1.6 **(c)** 4.1 **(d)** The waiting time also increases.
45. four **47.** $\{-2, 0, 3\}$

SUMMARY: EXERCISES ON OPERATIONS AND EQUATIONS WITH RATIONAL EXPRESSIONS (PAGE 405)

1. equation; $\left\{\frac{2}{15}\right\}$ **3.** expression; $\frac{2(x+5)}{5}$ **5.** expression; $\frac{y+x}{y-x}$ **7.** equation; $\{7\}$ **9.** equation; $\{1\}$ **11.** expression; $\frac{25}{4(r+2)}$ **13.** expression; $\frac{24p}{p+2}$ **15.** equation; $\{0\}$ **17.** expression; $\frac{5}{3z}$ **19.** equation; $\{2\}$ **21.** expression; $\frac{-x}{3x+5y}$ **23.** expression; $\frac{3}{2s-5r}$ **25.** equation; $\left\{\frac{5}{4}\right\}$ **27.** expression; $\frac{2z-3}{2z+3}$ **29.** expression; $\frac{t-2}{8}$ **31.** expression; $\frac{13x+28}{2x(x+4)(x-4)}$ **33.** expression; $\frac{k(2k^2-2k+5)}{(k-1)(3k^2-2)}$

SECTION 6.7 (PAGE 413)

CONNECTIONS **Page 412:** 48 miles per hour

EXERCISES **1.** 15 girls, 5 boys **3.** $\frac{1}{2}$ job per hour **5.** (a) **7.** (d) **9.** 1.349 **11.** 24 **13.** $G = \frac{Fd^2}{Mm}$ **15.** $a = \frac{bc}{c+b}$ **17.** $v = \frac{PVt}{pT}$ **19.** $r = \frac{nE - IR}{In}$ **21.** $b = \frac{2A}{h} - B$ or $b = \frac{2A - Bh}{h}$ **23.** $r = \frac{eR}{E-e}$ **25.** Multiply both sides by $a - b$. **27.** $\frac{8}{17}$ **29.** $\frac{2}{3}$ **31.** $95.75 **33.** 25,000 **35.** *Step 1:* Find the distance from Tulsa to Detroit. Let x represent that distance.

Step 2:

d	r	t
x	50	$\frac{x}{50}$
x	60	$\frac{x}{60}$

Step 3: $\frac{x}{60} = \frac{x}{50} - 3$ *Step 4:* $x = 900$ *Step 5:* The distance is 900 miles.

Step 6: Check: 900 miles at 50 miles per hour takes 18 hours; 900 miles at 60 miles per hour takes 15 hours; $15 = 18 - 3$ as required. **37.** 150 miles **39.** container ship: 17 knots; FastShip: 34 knots **41.** To solve problems about distance, rate, and time, we use the formula $d = rt$. To solve problems about work, we use the similar formula $A = rt$, where A represents the part of the job completed. **43.** $\frac{40}{13}$ or $3\frac{1}{13}$ hours **45.** $17\frac{1}{2}$ hours **47.** 36 hours **49.** 240 minutes or 4 hours **51.** $x = \frac{7}{2}$; $AC = 8$; $DF = 12$

CHAPTER 6 REVIEW EXERCISES (PAGE 423)

3. 2, 5 **5.** $\frac{11n^2}{2m}$ **7.** $\frac{5m+n}{5m-n}$ **9.** The reciprocal of a rational expression is another rational expression, such that the two rational expressions have a product of 1. **11.** $\frac{-3(w+4)}{w}$ **13.** 1 **15.** To find the least common denominator of a group of rational expressions, factor each denominator. Include in the least common denominator the greatest power of each of these factors. **17.** $9r^2(3r+1)$ **19.** $\frac{16z-3}{2z^2}$ **21.** $\frac{71}{30(a+2)}$ **23.** $\frac{3+2t}{4-7t}$ **25.** $\frac{1}{3q+2p}$ **27.** $p^2 + 6p + 9 + \frac{54}{2p-3}$ **29.** $3p + 2$ **31.** yes **33.** -13 **35.** (c) **37.** $\{-2\}$ **39.** $\emptyset$ **41.** In simplifying the expression, we are combining terms to get a single fraction with a denominator of $6x$, while in solving the equation, we are finding a value for x that makes the equation true. **43.** $m = \frac{Fd^2}{GM}$ **45.** $\frac{12}{25}$ **47.** $\frac{.52}{18} \approx .03$ million per year **49.** the bear population **51.** 16 kilometers per hour

53. $\frac{18}{5}$ or $3\frac{3}{5}$ hours **55.** $\frac{6m+5}{3m^2}$ **57.** $\frac{x^2-6}{2(2x+1)}$ **59.** $\frac{3-5x}{6x+1}$ **61.** $\frac{1}{3}$ **63.** $\frac{5a^2+4ab+12b^2}{(a+3b)(a-2b)(a+b)}$

65. $4y^2+1+\frac{-2y}{3y^2+1}$ **67.** $\left\{\frac{1}{3}\right\}$ **69.** $\{1, 4\}$ **71. (a)** 8.32 **(b)** 44.9 **73.** $\frac{60}{7}$ or $8\frac{4}{7}$ minutes

CHAPTER 6 TEST (PAGE 427)

[6.1] **1.** $-2, \frac{4}{3}$ **2.** $\frac{2x-5}{x(3x-1)}$ **3.** $\frac{3x}{2y^8}$ **4.** $\frac{y+4}{y-5}$ **5.** $\frac{x+5}{x}$ [6.2] **6.** $t^2(t+3)(t-2)$ **7.** $\frac{7-2t}{6t^2}$ **8.** $\frac{13x+35}{(x-7)(x+7)}$ **9.** $\frac{4}{x+2}$ [6.3] **10.** $\frac{72}{11}$ **11.** $\frac{-1}{a+b}$ [6.4] **12.** $4p-8+\frac{6}{p}$ **13.** $3q^3-4q^2+q+4+\frac{-2}{3q-2}$ **14.** $3y^2-2y-2+\frac{12y-3}{2y^2+3}$ [6.5] **15.** yes **16.** $9x^4-5x^3+2x^2-2x+4+\frac{2}{x+5}$

[6.2, 6.6] **17. (a)** $\frac{11(x-6)}{12}$ **(b)** $\{6\}$ [6.6] **18.** $\left\{\frac{1}{2}\right\}$ **19.** $\{5\}$ **20.** A solution cannot make a denominator zero.

21. $\ell = \frac{2S}{n} - a$ or $\ell = \frac{2S-na}{n}$ [6.7] **22.** $\frac{45}{14}$ or $3\frac{3}{14}$ hours **23.** 15 miles per hour **24.** 48,000 **25. (a)** 3 units **(b)** 0

CUMULATIVE REVIEW EXERCISES CHAPTERS 1–6 (PAGE 429)

[1.2] **1.** -199 **2.** 12 [2.1] **3.** $\left\{-\frac{15}{4}\right\}$ [2.7] **4.** $\left\{\frac{2}{3}, 2\right\}$ [2.2] **5.** $x = \frac{d-by}{a-c}$ or $x = \frac{by-d}{c-a}$

[2.5] **6.** $\left(-\infty, \frac{240}{13}\right]$ [2.7] **7.** $(-\infty, -2) \cup \left[\frac{2}{3}, \infty\right)$ [2.3] **8.** Democrats: 47.3 million; Republicans: 39.2 million

[2.4] **9.** 6 meters [3.1] **10.** x-intercept: $(-2, 0)$; y-intercept: $(0, 4)$

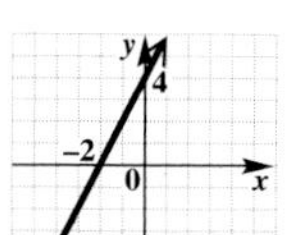

[3.2] **11.** $-\frac{3}{2}$

12. $-\frac{3}{4}$ [3.3] **13.** $y = -\frac{3}{2}x+\frac{1}{2}$ [3.4] **14.**

15.

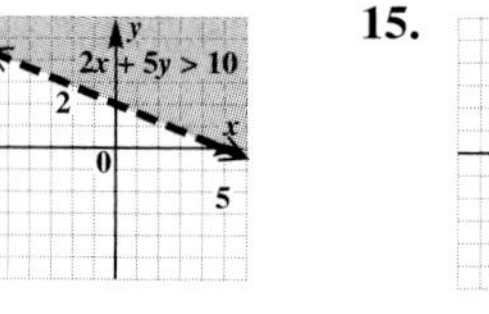

$x - y \geq 3$ and $3x + 4y \leq 12$

[3.5] **16.** function; domain: {Venezuela, Canada, Saudi Arabia, Mexico}; range: {1657, 1415, 1363, 1240}
17. not a function; domain: $[-2, \infty)$; range: $(-\infty, \infty)$ **18.** function; domain: $[-2, \infty)$; range: $(-\infty, 0]$
19. (a) $f(x) = \frac{5x-8}{3}$ or $f(x) = \frac{5}{3}x - \frac{8}{3}$ **(b)** -1 [3.6] **20.** 1.3×10^{14} [4.1, 4.4] **21. (a)** $\{(-1, 3)\}$ **(b)** $\{(-1, 3)\}$

[4.2] **22.** $\{(-2, 3, 1)\}$ [4.3] **23.** cigarette shipments: \$27.6 million; all other tobacco product shipments: \$5.4 million

[4.5] **24.** 10 [5.1] **25.** $\frac{a^{10}}{b^{10}}$ **26.** $\frac{m}{n}$ [5.2] **27.** $4y^2-7y-6$ [5.4] **28.** $-6x^6+18x^5-12x^4$

29. $12f^2+5f-3$ **30.** $49t^6-64$ **31.** $\frac{1}{16}x^2+\frac{5}{2}x+25$ [6.4] **32.** x^2+4x-7 [6.5] **33.** $2x^3+5x^2-3x-2$

[5.5] **34.** $(2x+5)(x-9)$ [5.6] **35.** $25(2t^2+1)(2t^2-1)$ **36.** $(2p+5)(4p^2-10p+25)$ [5.7] **37.** $\left\{-\frac{7}{3}, 1\right\}$

[6.1] **38.** $\frac{y+4}{y-4}$ **39.** $\frac{2x-3}{2(x-1)}$ **40.** $\frac{a(a-b)}{2(a+b)}$ [6.2] **41.** 3 **42.** $\frac{2(2x^2-3x+6)}{(2x+1)(2x-1)}$ [6.6] **43.** $\{-4\}$

[6.7] **44.** $q = \frac{fp}{p-f}$ or $q = \frac{-fp}{f-p}$ **45.** 150 miles per hour **46.** $\frac{6}{5}$ or $1\frac{1}{5}$ hours

CHAPTER 7 ROOTS AND RADICALS

SECTION 7.1 (PAGE 440)

EXERCISES **1.** C **3.** A **5.** H **7.** B **9.** D **11.** 13 **13.** 9 **15.** 2 **17.** $\frac{8}{9}$ **19.** -3 **21.** 1000 **23.** -32 **25.** not a real number **27.** $\frac{1}{512}$ **29.** $\frac{9}{4}$ **33.** 6 **35.** $\frac{8}{9}$ **37.** not a real number **39.** 6 **41.** -4 **43.** -8 **45.** -3 **47.** not a real number **49.** 2 **51.** -9 **53.** $|x|$ **55.** x **57.** x^5 **59.** $\sqrt{12}$ **61.** $(\sqrt[4]{8})^3$ **63.** $(\sqrt[8]{9q})^5 - (\sqrt[3]{2x})^2$ **65.** $\frac{1}{(\sqrt{2m})^3}$ **67.** $(\sqrt[3]{2y+x})^2$ **69.** $\frac{1}{(\sqrt[3]{3m^4+2k^2})^2}$ **71.** $\sqrt{a^2+b^2} = \sqrt{3^2+4^2} = 5$; $a + b = 3 + 4 = 7$; $5 \neq 7$ **73.** 64 **75.** 64 **77.** x^{10} **79.** $\sqrt[6]{x^5}$ **81.** $\sqrt[15]{t^8}$ **83.** 9 **85.** 4 **87.** y **89.** $k^{2/3}$ **91.** $9x^8y^{10}$ **93.** $\frac{1}{x^{10/3}}$ **95.** $\frac{1}{m^{1/4}n^{3/4}}$ **97.** p^2 **99.** $\frac{c^{11/3}}{b^{11/4}}$ **101.** $\frac{q^{5/3}}{9p^{7/2}}$ **103.** $p + 2p^2$ **105.** $k^{7/4} - k^{3/4}$ **107.** $6 + 18a$ **109.** $x^{17/20}$ **111.** $\frac{1}{x^{3/2}}$ **113.** $y^{5/6}z^{1/3}$ **115.** $m^{1/12}$ **117.** (c) **119.** 97.381 **121.** 16.863 **123.** 2.646 **125.** -9.055 **127.** 7.507 **129.** 3.162 **131.** 1.885 **133.** 15.155 **135.** .272 **137.** 1,183,000 cycles per second **139.** 10 miles **141.** 4.5 hours **143.** 24,900 square feet

SECTION 7.2 (PAGE 450)

CONNECTIONS **Page 448:** no; no; Answers will vary.

EXERCISES **1.** true: Both are equal to $4\sqrt{3}$ and approximately 6.92820323. **3.** true; Both are equal to $6\sqrt{2}$ and approximately 8.485281374. **5.** (d) **7.** $\sqrt{30}$ **9.** $\sqrt[3]{14xy}$ **11.** $\sqrt[4]{33}$ **13.** cannot be simplified using the product rule **17.** $\frac{8}{11}$ **19.** $\frac{\sqrt{3}}{5}$ **21.** $\frac{\sqrt{x}}{5}$ **23.** $\frac{p^3}{9}$ **25.** $\frac{3}{4}$ **27.** $-\frac{\sqrt[3]{r^2}}{2}$ **29.** $-\frac{3}{x}$ **31.** $\frac{1}{x^3}$ **33.** $2\sqrt{7}$ **35.** $-4\sqrt{2}$ **37.** $10\sqrt{3}$ **39.** $4\sqrt[3]{2}$ **41.** $-2\sqrt[3]{2}$ **43.** $2\sqrt[3]{5}$ **45.** $-4\sqrt[4]{2}$ **47.** $2\sqrt[5]{2}$ **49.** His reasoning was incorrect. Here 8 is a term and not a factor. **51.** $6k\sqrt{2}$ **53.** $\frac{3\sqrt[3]{3}}{4}$ **55.** $11x^3$ **57.** $-3t^4$ **59.** $-10m^4z^2$ **61.** $5a^2b^3c^4$ **63.** $\frac{1}{2}r^2t^5$ **65.** $-x^3y^4\sqrt{13x}$ **67.** $2z^2w^3$ **69.** $-2zt^2\sqrt[3]{2z^2t}$ **71.** $3x^3y^4$ **73.** $-3r^3s^2\sqrt[4]{2r^3s}$ **75.** $\frac{y^5\sqrt{y}}{6}$ **77.** $\frac{x^5\sqrt[3]{x}}{3}$ **79.** $4\sqrt{3}$ **81.** $\sqrt{5}$ **83.** $x^2\sqrt{x}$ **85.** $\sqrt[6]{432}$ **87.** $\sqrt[12]{6912}$ **89.** $\sqrt[6]{x^5}$ **91.** 1 **93.** 1 **95.** 5 **97.** $8\sqrt{2}$ **99.** 21.7 inches **101.** .003 **103.** $8\sqrt{5}$ feet; 17.9 feet **105.** $\sqrt{17}$ **107.** $\sqrt{37}$ **109.** 5 **111.** $\sqrt{5y^2 - 2xy + x^2}$ **113.** $d = [(x_2 - x_1)^2 + (y_2 - y_1)^2]^{1/2}$ **115.** $2\sqrt{106} + 4\sqrt{2}$

SECTION 7.3 (PAGE 457)

CONNECTIONS **Page 456:** **1.** 1.618033989 **2.** As one goes farther and farther into the sequence, the successive ratios appear to become closer and closer to the golden ratio. This is indeed the case.

EXERCISES **1.** (b) **3.** 15; Each radical expression simplifies to a whole number. **5.** -4 **7.** $7\sqrt{3}$ **9.** $14\sqrt[3]{2}$ **11.** $5\sqrt[4]{2}$ **13.** $24\sqrt{2}$ **15.** 0 **17.** $20\sqrt{5}$ **19.** $12\sqrt{2x}$ **21.** $-11m\sqrt{2}$ **23.** $\sqrt[3]{2}$ **25.** $2\sqrt[3]{x}$ **27.** $19\sqrt[4]{2}$ **29.** $x\sqrt[4]{xy}$ **31.** $(4 + 3xy)\sqrt[3]{xy^2}$ **33.** $\frac{7\sqrt{2}}{6}$ **35.** $\frac{5\sqrt{2}}{3}$ **37.** $\frac{m\sqrt[3]{m^2}}{2}$ **39.** Both are approximately 11.3137085. **41.** Both are approximately 31.6227766. **43.** (a) **45.** $12\sqrt{5} + 5\sqrt{3}$ inches **47.** $24\sqrt{2} + 12\sqrt{3}$ inches

SECTION 7.4 (PAGE 465)

CONNECTIONS **Page 464:** **1.** $\frac{319}{6(8\sqrt{5}+1)}$ **2.** $\frac{9a-b}{b(3\sqrt{a}-\sqrt{b})}$ **3.** $\frac{9a-b}{(\sqrt{b}-\sqrt{a})(3\sqrt{a}-\sqrt{b})}$

4. $\frac{(3\sqrt{a}+\sqrt{b})(\sqrt{b}+\sqrt{a})}{b-a}$; Instead of multiplying by the conjugate of the numerator, we use the conjugate of the denominator.

EXERCISES **1.** E **3.** A **5.** D **7.** $\sqrt[3]{x}\cdot\sqrt[3]{x}$ is not equal to x because the product rule leads to $\sqrt[3]{x^2}$, not $\sqrt[3]{x^3}$. **9.** $6-4\sqrt{3}$ **11.** $6-\sqrt{6}$ **13.** 2 **15.** 9 **17.** $3\sqrt{2}-5\sqrt{3}+2\sqrt{6}-10$ **19.** $3x-4$ **21.** $4x-y$ **23.** $16x+24\sqrt{x}+9$ **25.** $81-\sqrt[3]{4}$ **27.** $2\sqrt{6}-1$ **29.** $10+2\sqrt{6}+2\sqrt{10}+2\sqrt{15}$ **31.** The binomial $6-4\sqrt{3}$ is not equal to $2\sqrt{3}$ because the multiplication must be done before the subtraction. If the expression were $(6-4)\sqrt{3}$, then it would simplify to $2\sqrt{3}$. **33.** $\sqrt{7}$ **35.** $5\sqrt{3}$ **37.** $\frac{\sqrt{6}}{2}$ **39.** $\frac{9\sqrt{15}}{5}$ **41.** $-\sqrt{2}$ **43.** $\frac{\sqrt{14}}{2}$ **45.** $-\frac{\sqrt{14}}{10}$ **47.** $\frac{2\sqrt{6x}}{x}$ **49.** $\frac{-8\sqrt{3k}}{k}$ **51.** $\frac{-5m^2\sqrt{6mn}}{n^2}$ **53.** $\frac{12x^3\sqrt{2xy}}{y^5}$ **57.** $\frac{\sqrt[3]{18}}{3}$ **59.** $\frac{\sqrt[3]{12}}{3}$ **61.** $-\frac{\sqrt[3]{2pr}}{r}$ **63.** $\frac{2\sqrt[4]{x^3}}{x}$ **65.** Multiply both the numerator and the denominator by $4-\sqrt{3}$. No, it would not. The new denominator would be $(4+\sqrt{3})^2=19+8\sqrt{3}$, which is not rational. **67.** $\frac{6(5-\sqrt{2})}{23}$ **69.** $4(\sqrt{6}-\sqrt{3})$ **71.** $\sqrt{2}-\sqrt{5}$ **73.** $-\sqrt{6}+\sqrt{5}-3\sqrt{2}+\sqrt{15}$ **75.** $\frac{5\sqrt{r}(3\sqrt{r}-\sqrt{s})}{9r-s}$ **77.** $\frac{a+2\sqrt{ab}+b}{a-b}$ **79.** $\frac{5+2\sqrt{6}}{4}$ **81.** $\frac{4+2\sqrt{2}}{3}$ **83.** $\frac{6+2\sqrt{6x}}{3}$ **85.** Each expression is approximately equal to .2588190451. **87.** $(\sqrt{x}+\sqrt{7})(\sqrt{x}-\sqrt{7})$ **88.** $(\sqrt[3]{x}-\sqrt[3]{7})(\sqrt[3]{x^2}+\sqrt[3]{7x}+\sqrt[3]{49})$ **89.** $(\sqrt[3]{x}+\sqrt[3]{7})(\sqrt[3]{x^2}-\sqrt[3]{7x}+\sqrt[3]{49})$ **90.** $\frac{(x+3)(\sqrt{x}+\sqrt{7})}{x-7}$ **91.** $\frac{(x+3)(\sqrt[3]{x^2}+\sqrt[3]{7x}+\sqrt[3]{49})}{x-7}$ **92.** $\frac{(x+3)(\sqrt[3]{x}+\sqrt[3]{7})}{x+7}$ **93.** $(\sqrt[3]{5}-\sqrt[3]{3})(\sqrt[3]{25}+\sqrt[3]{15}+\sqrt[3]{9})$ **94.** $\sqrt[3]{25}+\sqrt[3]{15}+\sqrt[3]{9}$

SECTION 7.5 (PAGE 476)

EXERCISES **In Exercises 1–7, we give the domain and then the range.**

1. $[-3,\infty)$; $[0,\infty)$ **3.** $[0,\infty)$; $[-2,\infty)$ **5.** $(-\infty,\infty)$; $(-\infty,\infty)$

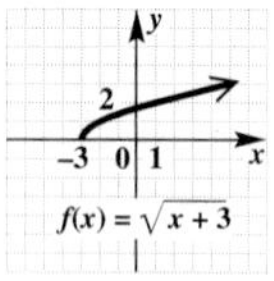

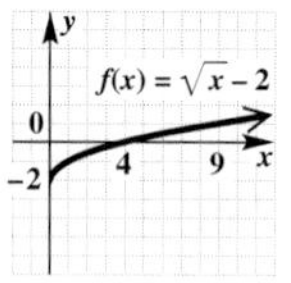

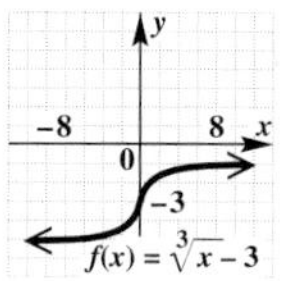

7. 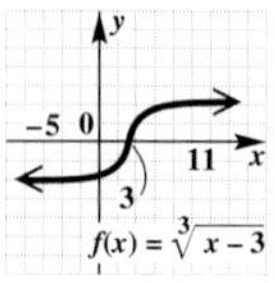 $(-\infty,\infty)$; $(-\infty,\infty)$ **9.** **(a)** yes **(b)** no **11.** **(a)** yes **(b)** no **13.** no; There is no solution. The radical expression, which is positive, cannot equal a negative number. **15.** $\{19\}$ **17.** $\left\{\frac{38}{3}\right\}$ **19.** $\emptyset$ **21.** $\{5\}$ **23.** $\{1\}$ **25.** $\{9\}$ **27.** $\{3\}$ **28.** $\{-3\}$ **29.** $\{\pm 3\}$ **30.** $\{3\}$ **31.** $\{\pm 3\}$ **32.** **(a)** more than **(b)** the same as **35.** $\{4\}$ **37.** $\{-3,-1\}$ **39.** $\emptyset$ **41.** $\{-1\}$ **43.** $\left\{\frac{1}{2}\right\}$ **45.** $\{5\}$ **47.** $\{7\}$ **49.** $\emptyset$ **51.** $\{2, 14\}$ **53.** 3 **55.** $\{-13\}$ **57.** $\{14\}$ **59.** $\emptyset$ **61.** $\{7\}$ **63.** $\{9, 17\}$ **65.** $\left\{\frac{1}{4}, 1\right\}$ **67.** $K=\frac{V^2m}{2}$

69. $L = \frac{1}{4\pi^2 f^2 C}$ **71.** **(a)** $r = \frac{a}{4\pi^2 N^2}$ **(b)** 62.5 meters **(c)** 155.1 meters **73.** 8 billion cubic feet; 16 billion cubic feet; 21 billion cubic feet; 24 billion cubic feet **75.** fairly good; 1976 **77.** 6 billion cubic feet; 12 billion cubic feet; 15 billion cubic feet; 16 billion cubic feet **79.** 4; the power rule **81.** $\{-3, -1\}$ **83.** $\emptyset$; $\left[-\frac{2}{3}, 1\right]$

SECTION 7.6 (PAGE 484)

EXERCISES **1.** i **3.** -1 **5.** $-i$ **7.** $13i$ **9.** $-12i$ **11.** $i\sqrt{5}$ **13.** $4i\sqrt{3}$ **15.** -15 **17.** -10 **19.** $\sqrt{3}$ **21.** $5i$ **23.** **(a)** Any real number a can be written as $a + 0i$, and this is a complex number with imaginary part 0. **(b)** A complex number such as $2 + 3i$, with nonzero imaginary part, is not real. **25.** $10 + 8i$ **27.** $-1 + 7i$ **29.** 0 **31.** $7 + 3i$ **33.** -2 **35.** $1 + 13i$ **37.** $6 + 6i$ **39.** $4 + 2i$ **41.** -81 **43.** -16 **45.** $-10 - 30i$ **47.** $10 - 5i$ **49.** $-9 + 40i$ **51.** $-16 + 30i$ **53.** 153 **55.** 97 **57.** $1 + i$ **59.** $-1 + 2i$ **61.** $2 + 2i$ **63.** $-\frac{5}{13} - \frac{12}{13}i$ **65.** $30 + 5i$ **67.** $8 - 3i$ **69.** **(a)** $4x + 1$ **(b)** $4 + i$ **70.** **(a)** $-2x + 3$ **(b)** $-2 + 3i$ **71.** **(a)** $3x^2 + 5x - 2$ **(b)** $5 + 5i$ **72.** **(a)** $-\sqrt{3} + \sqrt{6} + 1 - \sqrt{2}$ **(b)** $\frac{1}{5} - \frac{7}{5}i$ **73.** In parts (a) and (b) of Exercises 69 and 70, real and imaginary parts are added, just like coefficients of similar terms in the binomials, and the answers correspond. In Exercise 71, introducing $i^2 = -1$ when a product is found leads to answers that do not correspond. **74.** In parts (a) and (b) of Exercises 69 and 70, real and imaginary parts are added, just like coefficients of similar terms in binomials, and the answers correspond. In Exercise 72, introducing $i^2 = -1$ when performing the division leads to answers that do not correspond. **75.** $\frac{5}{41} + \frac{4}{41}i$ **77.** -1 **79.** i **81.** $-i$ **83.** Because $i^{20} = (i^4)^5 = 1^5 = 1$, multiplying by i^{20} is an application of the identity property for multiplication. **85.** $\frac{1}{2} + \frac{1}{2}i$ **87.** Substitute both $1 + 5i$ and $1 - 5i$ for x, and show that the result is $0 = 0$ in each case.

CHAPTER 7 REVIEW EXERCISES (PAGE 492)

1. cube (third); 8; 2; second; 4; 4 **3.** **(a)** m must be even. **(b)** m must be odd. **5.** 7 **7.** 32 **9.** $-\frac{216}{125}$ **11.** $\frac{1000}{27}$ **13.** The radical $\sqrt[n]{a^m}$ is equivalent to $a^{m/n}$. For example, $\sqrt[3]{8^2} = \sqrt[3]{64} = 4$, and $8^{2/3} = (8^{1/3})^2 = 2^2 = 4$. **15.** -17 **17.** -5 **19.** -2 **21.** **(a)** $|x|$ **(b)** $-|x|$ **(c)** x **23.** $\frac{1}{(\sqrt[3]{3a + b})^5}$ or $\frac{1}{\sqrt[3]{(3a + b)^5}}$ **25.** $p^{4/5}$ **27.** 96 **29.** $\frac{1}{y^{1/2}}$ **31.** $r^{1/2} + r$ **33.** $r^{3/2}$ **35.** $k^{9/4}$ **37.** $z^{1/12}$ **39.** -6.856 **41.** 4.960 **43.** -3968.503 **45.** The product rule for exponents applies only if the bases are the same. **47.** $\sqrt{66}$ **49.** $\sqrt[3]{30}$ **51.** $2\sqrt{5}$ **53.** $-5\sqrt{5}$ **55.** $10y^3\sqrt{y}$ **57.** $3a^2b\sqrt[3]{4a^2b^2}$ **59.** $\frac{y\sqrt{y}}{12}$ **61.** $\frac{\sqrt[3]{r^2}}{2}$ **63.** $\sqrt{15}$ **65.** $\sqrt[12]{2000}$ **67.** 10 **69.** $-11\sqrt{2}$ **71.** $7\sqrt{3y}$ **73.** $19\sqrt[3]{2}$ **75.** $\frac{3 - 2\sqrt{5}}{4}$ **77.** $16\sqrt{2} + 24\sqrt{3}$ feet **79.** $1 - \sqrt{3}$ **81.** $9 - 7\sqrt{2}$ **83.** 29 **85.** $4.801960973 \neq 66.28725368$ **87.** $\frac{\sqrt{30}}{5}$ **89.** $\frac{3\sqrt{7py}}{y}$ **91.** $-\frac{\sqrt[3]{45}}{5}$ **93.** $\frac{\sqrt{2} - \sqrt{7}}{-5}$ **95.** $\frac{1 - \sqrt{5}}{4}$ **97.**

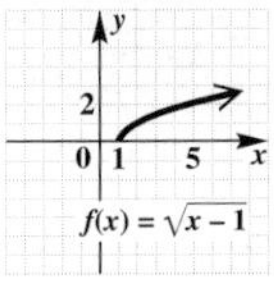

$f(x) = \sqrt{x - 1}$

domain: $[1, \infty)$; range: $[0, \infty)$ **99.** $\{2\}$ **101.** $\emptyset$ **103.** $\{9\}$ **105.** $\{7\}$ **107.** 0 is an extraneous solution introduced during the squaring step. **109.** $5i$ **111.** no **113.** $14 + 7i$

115. -45 **117.** $5 + i$ **119.** $1 - i$ **121.** $-i$ **123.** -4 **125.** $\frac{1}{z^{3/5}}$ **127.** $3z^3t^2\sqrt[3]{2t^2}$ **129.** $6x\sqrt[3]{y^2}$ **131.** $-\frac{\sqrt{3}}{6}$ **133.** 1 **135.** $3 - 7i$ **137.** $\frac{1 + \sqrt{6}}{2}$ **139.** $\{-4\}$

CHAPTER 7 TEST (PAGE 496)

[7.1] **1.** $\frac{125}{64}$ **2.** $\frac{1}{256}$ **3.** $\frac{9y^{3/10}}{x^2}$ **4.** $7^{1/2}$ or $\sqrt{7}$ **5.** $a^3\sqrt[3]{a^2}$ or $a^{11/3}$ **6.** **(a)** 21.863 **(b)** -9.405 **(c)** 2.415 [7.2] **7.** $b = \sqrt{145}$ **8.** 10 **9.** $3x^2y^3\sqrt{6x}$ **10.** $2ab^3\sqrt[4]{2a^3b}$ **11.** $\sqrt[6]{200}$ [7.3] **12.** $26\sqrt{5}$ [7.4] **13.** $66 + \sqrt{5}$ **14.** $23 - 4\sqrt{15}$ **15.** $-\frac{\sqrt{10}}{4}$ **16.** $\frac{2\sqrt[3]{25}}{5}$ **17.** $-2(\sqrt{7} - \sqrt{5})$ **18.** $3 + \sqrt{6}$ [7.5] **19.** **(a)** 59.8 **(b)** $T = \frac{V_0^2 - V^2}{-V^2k}$ or $T = \frac{V^2 - V_0^2}{V^2k}$ **20.** domain: $[-6, \infty)$; range: $[0, \infty)$

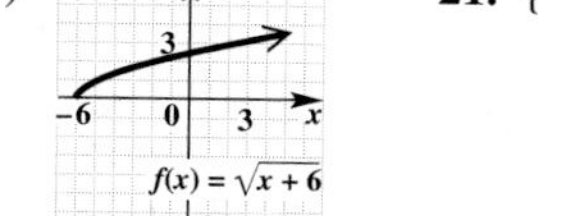

21. $\{-1\}$ **22.** $\{6\}$ [7.6] **23.** **(a)** $-5 - 8i$ **(b)** $-2 + 16i$ **(c)** $3 + 4i$ **24.** i **25.** **(a)** true **(b)** true **(c)** false **(d)** true

CUMULATIVE REVIEW EXERCISES CHAPTERS 1–7 (PAGE 497)

[1.2] **1.** 1 **2.** $-\frac{14}{9}$ [2.1] **3.** $\{-4\}$ **4.** $\{-12\}$ **5.** $\{6\}$ [2.7] **6.** $\left\{-\frac{10}{3}, 1\right\}$ **7.** $\left\{\frac{1}{4}\right\}$ [2.5] **8.** $(-6, \infty)$ [2.2] **9.** Both angles measure 80°. [2.3] **10.** 18 nickels; 32 quarters **11.** $2\frac{2}{39}$ liters [3.1] **12.**

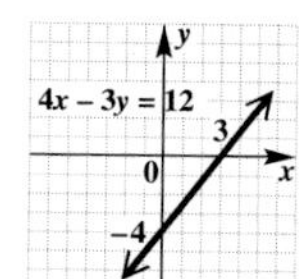

[3.2, 3.3] **13.** $-\frac{3}{2}$; $y = -\frac{3}{2}x$ [3.5] **14.** -37 [4.1] **15.** $\{(7, -2)\}$ [4.2] **16.** $\{(-1, 1, 1)\}$ [4.3] **17.** 2-ounce letter: \$.55; 3-ounce letter: \$.78 [5.2] **18.** $-k^3 - 3k^2 - 8k - 9$ [5.4] **19.** $8x^2 + 17x - 21$ [6.4] **20.** $z - 2 + \frac{3}{z}$ **21.** $3y^3 - 3y^2 + 4y + 1 + \frac{-10}{2y + 1}$ [5.6] **22.** $(2p - 3q)(p - q)$ [5.8] **23.** $(3k^2 + 4)(k - 1)(k + 1)$ [5.7] **24.** $(x + 8)(x^2 - 8x + 64)$ [5.9] **25.** $\left\{-3, -\frac{5}{2}\right\}$ **26.** $\left\{-\frac{2}{5}, 1\right\}$ [6.1] **27.** $\{x \mid x \neq \pm 3\}$ **28.** $\frac{y}{y + 5}$ [6.2] **29.** $\frac{4x + 2y}{(x + y)(x - y)}$ [6.3] **30.** $-\frac{9}{4}$ **31.** $\frac{-1}{a + b}$ **32.** $\frac{1}{xy - 1}$ [6.7] **33.** Natalie: 8 mph; Chuck: 4 mph [6.6] **34.** $\emptyset$ [7.1] **35.** $\frac{1}{9}$ [7.2] **36.** $10x^2\sqrt{2}$ **37.** $2x\sqrt[3]{6x^2y^2}$ [7.3] **38.** $7\sqrt{2}$ [7.4] **39.** $\frac{\sqrt{10} + 2\sqrt{2}}{2}$ **40.** $-6x - 11\sqrt{xy} - 4y$ [7.5] **41.** $\{3, 4\}$ **42.** 39.2 miles per hour [7.6] **43.** $2 + 9i$ **44.** $4 + 2i$

CHAPTER 8 QUADRATIC EQUATIONS AND INEQUALITIES

SECTION 8.1 (PAGE 505)

EXERCISES **1.** Divide both sides by 2. **3.** square root property for $(2x + 1)^2 = 5$; completing the square for $x^2 + 4x = 12$ **5.** 16 **7.** $\frac{81}{4}$ **9.** $\{-9, 9\}$ **11.** $\{-4\sqrt{2}, 4\sqrt{2}\}$ **13.** $\{-7, 3\}$ **15.** $\left\{\frac{1 + \sqrt{7}}{3}, \frac{1 - \sqrt{7}}{3}\right\}$

17. $\left\{\frac{-1+2\sqrt{6}}{4}, \frac{-1-2\sqrt{6}}{4}\right\}$ **19.** $\{-2i\sqrt{3}, 2i\sqrt{3}\}$ **21.** $\{5+i\sqrt{3}, 5-i\sqrt{3}\}$ **23.** $\left\{\frac{1+2i\sqrt{2}}{6}, \frac{1-2i\sqrt{2}}{6}\right\}$
25. $\{-4, 6\}$ **27.** $\left\{-3, \frac{8}{3}\right\}$ **29.** $\left\{\frac{-5+\sqrt{41}}{4}, \frac{-5-\sqrt{41}}{4}\right\}$ **31.** $\left\{\frac{4+\sqrt{3}}{3}, \frac{4-\sqrt{3}}{3}\right\}$
33. $\{-2+3i, -2-3i\}$ **35.** $\left\{\frac{2+\sqrt{3}}{3}, \frac{2-\sqrt{3}}{3}\right\}$ **37.** $\left\{\frac{-2+2i\sqrt{2}}{3}, \frac{-2-2i\sqrt{2}}{3}\right\}$ **39.** $\{1+\sqrt{2}, 1-\sqrt{2}\}$
41. $\{-3+i\sqrt{3}, -3-i\sqrt{3}\}$ **43.** The zero-factor property requires a product equal to 0. In this solution, the first step should have been to rewrite the equation with 0 on one side. **45.** Some quadratic polynomials cannot easily be factored.
47. $\{-\sqrt{b}, \sqrt{b}\}$ **49.** $\left\{\frac{-\sqrt{b^2+16}}{2}, \frac{\sqrt{b^2+16}}{2}\right\}$ **51.** $\left\{\frac{2b+\sqrt{3a}}{5}, \frac{2b-\sqrt{3a}}{5}\right\}$ **53.** 11.9 years
55. **(a)** 6.8×10^6 meters **(b)** approximately 242 meters per second **57.** x^2 **58.** x **59.** $6x$ **60.** 1 **61.** 9
62. $(x+3)^2$ or x^2+6x+9 **63.** $\pm\sqrt{17}$

SECTION 8.2 (PAGE 515)

CONNECTIONS **Page 514:** **1.** Since $\frac{1+\sqrt{41}}{5} + \frac{1-\sqrt{41}}{5} = \frac{2}{5} = -\frac{b}{a}$ and $\frac{1+\sqrt{41}}{5} \cdot \frac{1-\sqrt{41}}{5} = -\frac{8}{5} = \frac{c}{a}$, the solutions are correct. **2.** Since $\frac{3}{2}i + (-4i) = -\frac{5}{2}i = -\frac{b}{a}$ and $\frac{3}{2}i(-4i) = 6 = \frac{c}{a}$, the solutions are correct.

EXERCISES **1.** true **3.** true **5.** true **7.** $\{3, 5\}$ **9.** $\left\{\frac{-2+\sqrt{2}}{2}, \frac{-2-\sqrt{2}}{2}\right\}$ **11.** $\left\{\frac{1+\sqrt{3}}{2}, \frac{1-\sqrt{3}}{2}\right\}$
13. $\{5+\sqrt{7}, 5-\sqrt{7}\}$ **15.** $\left\{\frac{-2+\sqrt{10}}{2}, \frac{-2-\sqrt{10}}{2}\right\}$ **17.** $\{-1+3\sqrt{2}, -1-3\sqrt{2}\}$ **19.** $\left\{-1, \frac{2}{3}\right\}$
21. $\left\{\frac{-1+\sqrt{2}}{2}, \frac{-1-\sqrt{2}}{2}\right\}$ **23.** $\left\{\frac{1+\sqrt{29}}{2}, \frac{1-\sqrt{29}}{2}\right\}$ **25.** $\left\{\frac{-1+\sqrt{7}}{3}, \frac{-1-\sqrt{7}}{3}\right\}$
27. $\left\{\frac{-4+\sqrt{91}}{3}, \frac{-4-\sqrt{91}}{3}\right\}$ **29.** The last step is wrong. Because 5 is not a common factor in the numerator, the fraction cannot be reduced. The solutions are $\frac{5 \pm \sqrt{5}}{10}$. **31.** $\{-i\sqrt{47}, i\sqrt{47}\}$ **33.** $\{3+i\sqrt{5}, 3-i\sqrt{5}\}$
35. $\left\{\frac{1}{2}+\frac{\sqrt{6}}{2}i, \frac{1}{2}-\frac{\sqrt{6}}{2}i\right\}$ **37.** $\left\{-\frac{2}{3}+\frac{\sqrt{2}}{3}i, -\frac{2}{3}-\frac{\sqrt{2}}{3}i\right\}$ **39.** $\{4+3i\sqrt{2}, 4-3i\sqrt{2}\}$
41. $\left\{-\frac{1}{2}+\frac{\sqrt{3}}{2}i, -\frac{1}{2}-\frac{\sqrt{3}}{2}i\right\}$ **43.** 3.6 hours **45.** Rusty: 25.0 hours; Nancy: 23.0 hours **47.** 7.8 hours and 8.3 hours **49.** 2.4 seconds and 5.6 seconds **51.** It reaches its maximum height at 5 seconds, since this is the only time it reaches 400 feet. **53.** 1992 (Another answer, $x = 102$, leads to the year 2002, but that is too far from the given data to estimate.) **55.** (d) **57.** (a) **59.** (b) **61.** (c) **63.** -10 or 10 **65.** 16 **67.** 25 **69.** No, because an irrational solution occurs only if the discriminant is positive, but not the square of an integer. In that case, there will be two irrational solutions. **71.** The discriminant is 74^2, so it can be factored; $(6x+5)(4x-9)$ **73.** The discriminant, 3897, is not a perfect square, so it cannot be factored. **75.** The discriminant is 85^2, so it can be factored; $(12x+1)(x-7)$
77. $[x-(1+5i)][x-(1-5i)] = 0$ **78.** $[(x-1)-5i][(x-1)+5i] = 0$ **79.** $x^2-2x+26=0$
80. Use $a=1$, $b=-2$, and $c=26$. The solutions are $1+5i$ and $1-5i$. **81.** $b = \frac{44}{5}$; $x_2 = \frac{3}{10}$

SECTION 8.3 (PAGE 525)

EXERCISES **1.** Multiply by the LCD, x. **3.** Substitute for r^2+r. **7.** $\{-2, 7\}$ **9.** $\{-4, 7\}$ **11.** $\left\{-\frac{14}{17}, 5\right\}$
13. $\left\{-\frac{11}{7}, 0\right\}$ **15.** $\left\{\frac{-1+\sqrt{13}}{2}, \frac{-1-\sqrt{13}}{2}\right\}$ **17.** $\frac{1}{m}$ job per hour **19.** 25 miles per hour
21. 80 kilometers per hour **23.** 9 minutes **25.** $\{3\}$ **27.** $\left\{\frac{8}{9}\right\}$ **29.** $\{16\}$ **31.** $\left\{\frac{2}{5}\right\}$ **33.** 30 meters

35. $\{-3, 3\}$ **37.** $\left\{-\frac{3}{2}, -1, 1, \frac{3}{2}\right\}$ **39.** $\{-6, -5\}$ **41.** $\{-4, 1\}$ **43.** $\left\{-\frac{1}{3}, \frac{1}{6}\right\}$ **45.** $\left\{-\frac{1}{2}, 3\right\}$ **47.** $\left\{-\frac{27}{8}, -1, 1, \frac{27}{8}\right\}$ **49.** $\{-8, 1\}$ **51.** $\{25\}$ **53.** $\left\{-1, 1, -\frac{\sqrt{6}}{2}i, \frac{\sqrt{6}}{2}i\right\}$ **55.** The solutions given are for u. Each must be set equal to $m - 1$ and solved for m. The correct solutions are $\frac{3}{2}$ and 2. **57.** $\{-2\sqrt{3}, -2, 2, 2\sqrt{3}\}$ **59.** $\{3, 11\}$ **61.** $\left\{-\sqrt[3]{5}, -\frac{\sqrt[3]{4}}{2}\right\}$ **63.** $\left\{-\frac{1}{2}, 3\right\}$ **65.** It would cause both denominators to be 0, and division by 0 is undefined. **66.** The solution is $\frac{12}{5}$. **67.** $\left(\frac{x}{x-3}\right)^2 + 3\left(\frac{x}{x-3}\right) - 4 = 0$ **68.** The numerator can never equal the denominator, since the denominator is 3 less than the numerator. **69.** $\left\{\frac{12}{5}\right\}$; The values for t are -4 and 1. The value 1 is impossible because it leads to a contradiction (since $\frac{x}{x-3}$ is never equal to 1). **70.** $\left\{\frac{12}{5}\right\}$; The values for s are $\frac{1}{x}$ and $\frac{-4}{x}$. The value $\frac{1}{x}$ is impossible, since $\frac{1}{x} \neq \frac{1}{x-3}$ for all x.

SECTION 8.4 (PAGE 532)

EXERCISES **1.** Multiply both sides by the common denominator. **3.** Write it in standard form (with 0 on one side in decreasing powers of w). **5.** No. There is only one time after the rock starts moving when it hits the ground. **7.** $m = \sqrt{p^2 - n^2}$ **9.** $t = \frac{\pm\sqrt{dk}}{k}$ **11.** $v = \frac{\pm\sqrt{kAF}}{F}$ **13.** $r = \frac{\pm\sqrt{3\pi Vh}}{\pi h}$ **15.** $t = \frac{-B \pm \sqrt{B^2 - 4AC}}{2A}$ **17.** $h = \frac{D^2}{k}$ **19.** $\ell = \frac{p^2 g}{k}$ **21.** If g is positive, the only way to have a real value for p is to have $k\ell$ positive, since the quotient of two positive numbers is positive. If k and ℓ have different signs, their product is negative, leading to a negative radicand. **23.** 5.2 seconds **25.** Find s when $t = 0$. **27. (a)** 1 second and 8 seconds **(b)** 9 seconds after it is projected **29. (a)** $y = x - 1.727x^2$ **(b)** .006825 foot; .1421 foot **(c)** 0 seconds; .5790 second **(d)** One represents the time before the vehicle leaves the surface. The other represents the time it returns to the surface. **31.** 2.3, 5.3, 5.8 **33.** 412.3 feet **35.** eastbound ship: 80 miles; southbound ship: 150 miles **37.** 5 centimeters, 12 centimeters, 13 centimeters **39.** length: 2 centimeters; width: 1.5 centimeters **41.** 1 foot **43.** length: 26 meters; width: 16 meters **45.** .035 or 3.5% **47.** \$.80 **49.** 5.5 meters per second **51.** 5 or 14 **53.** $R = \frac{E^2 - 2pr \pm E\sqrt{E^2 - 4pr}}{2p}$ **55.** $r = \frac{5pc}{4}$ or $r = -\frac{2pc}{3}$ **57.** $I = \frac{-cR \pm \sqrt{c^2R^2 - 4cL}}{2cL}$

SECTION 8.5 (PAGE 544)

EXERCISES **1.** F **3.** C **5.** E **9.** (0, 0) **11.** (0, 4) **13.** (1, 0) **15.** (−3, −4) **17.** In Exercise 15, the parabola is shifted 3 units to the left and 4 units down. The parabola in Exercise 16 is shifted 5 units to the right and 8 units down. **19.** downward; narrower **21.** upward; wider **23. (a)** I **(b)** IV **(c)** II **(d)** III

25. **27.** **29.** **31.** **33.**

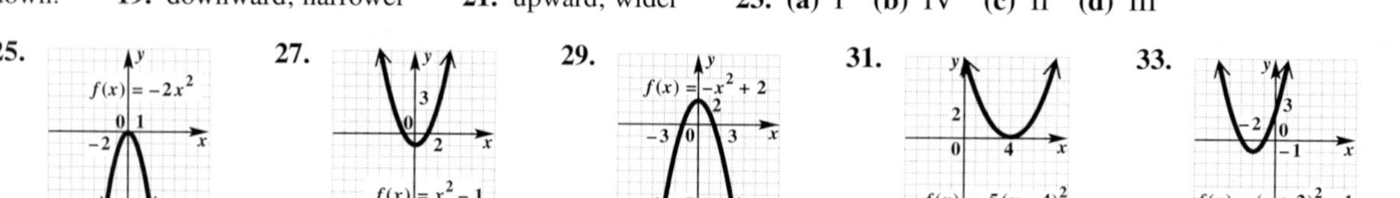

35. domain: $(-\infty, \infty)$; range: $[-4, \infty)$

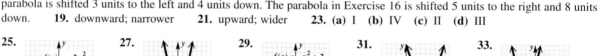

$f(x) = 2(x - 2)^2 - 4$

37. domain: $(-\infty, \infty)$; range: $(-\infty, 2]$

$f(x) = -.5(x + 1)^2 + 2$

39. domain: $(-\infty, \infty)$; range: $[-3, \infty)$

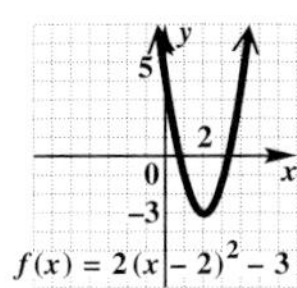

41. It is shifted 6 units upward. **42.**

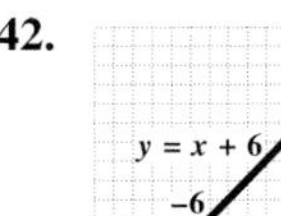

43. It is shifted 6 units upward. **44.** It is shifted 6 units to the right. **45.** $y = x - 6$

46. It is shifted 6 units to the right. **47. (a)** $|x + p|$ **(b)** The distance from the focus to the origin should equal the distance from the directrix to the origin. **(c)** $\sqrt{(x-p)^2 + y^2}$ **(d)** $y^2 = 4px$

49. (a) AIDS Patients Who Survived a Year **(b)** negative **(c)** $f(x) = -7x^2 + 60x + 38$ **(d)** Many answers are possible. Since the data give one y-value for each year, {1, 2, 3, 4, 5, 6, 7} may be most appropriate. **51.** $\{-4, 5\}$ **53.** $\{-.5, 3\}$ **55.** (a)

SECTION 8.6 (PAGE 556)

EXERCISES **1.** If x is squared, it has a vertical axis; if y is squared, it has a horizontal axis. **3. (a)** Use the discriminant of the function. If it is positive, there are two x-intercepts. If it is zero, there is one x-intercept (at the vertex), and if it is negative, there is no x-intercept. **(b)** none **5.** $(-1, 3)$; upward; narrower; no x-intercepts **7.** $\left(\frac{5}{2}, \frac{37}{4}\right)$; downward; same; two x-intercepts **9.** $(-3, -9)$; to the right; wider **11.** domain: $(-\infty, \infty)$; range: $[-6, \infty)$

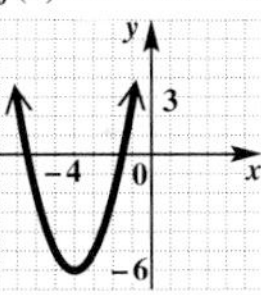

13. domain: $(-\infty, \infty)$; range: $(-\infty, -3]$

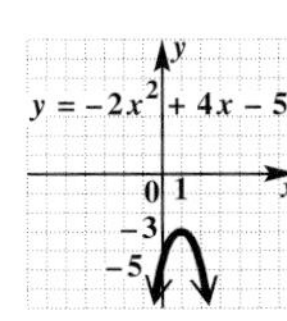

15. not a function

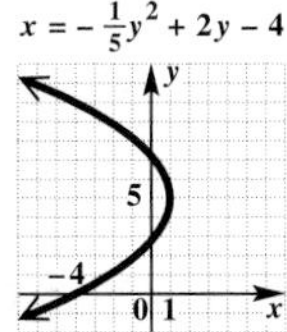

17. not a function

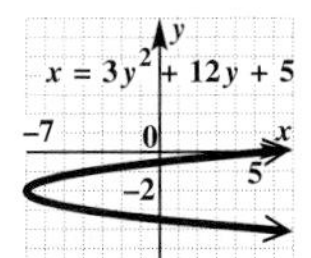

19. F **21.** C **23.** D **25.** 25 meters **27.** 16 feet; 2 seconds

29. 4.1 seconds; 81.6 meters **31. (a)** The coefficient of x^2 is negative, because the parabola opens downward. **(b)** (18.45, 3860) **(c)** In 2018 social security assets will reach their maximum value of \$3860 billion. **33. (a)** $R(x) = (100 - x)(200 + 4x) = 20{,}000 + 200x - 4x^2$ **(b)**

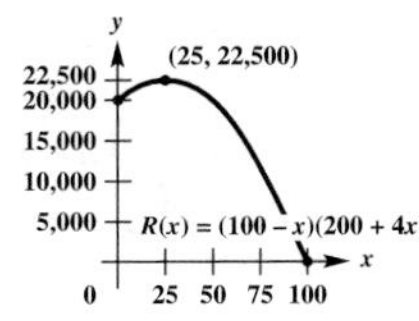

(c) 25 **(d)** \$22,500 **35.** B **37.** A

SECTION 8.7 (PAGE 564)

CONNECTIONS **Page 564:** **1.** 31.6 seconds and 119.9 seconds **2.** between 31.6 seconds and 119.9 seconds **3.** 0 seconds (when it is first propelled) and 151.5 seconds (when it hits the ground) **4.** between 0 and 31.6 seconds and also between 119.9 and 151.5 seconds

EXERCISES **1.** $\left\{-\frac{3}{2}, 1\right\}$ **3.** $\left(-\frac{3}{2}, 1\right)$ **5.** $(-\infty, -1) \cup (5, \infty)$ 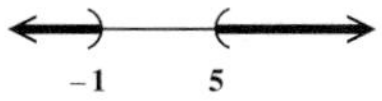 **7.** $(-4, 6)$ **9.** $(-\infty, 1] \cup [3, \infty)$ **11.** $\left(-\infty, -\frac{3}{2}\right] \cup \left[\frac{3}{5}, \infty\right)$ **13.** $\left(-\frac{2}{3}, \frac{1}{3}\right)$ **15.** $\left(-\infty, -\frac{1}{2}\right] \cup \left[\frac{1}{3}, \infty\right)$ **17.** $(-\infty, 3 - \sqrt{3}] \cup [3 + \sqrt{3}, \infty)$ **19.** Include the endpoints if the symbol is $\geq$ or $\leq$. Exclude the endpoints if the symbol is $>$ or $<$. **21.** $(-\infty, \infty)$ **23.** $\emptyset$ **27.** $\left(-\infty, -\frac{7}{4}\right) \cup \left(-\frac{1}{2}, \frac{2}{3}\right)$ **29.** $\left[-\frac{7}{2}, -2\right] \cup \left[\frac{3}{4}, \infty\right)$ **31.** $(-\infty, 1) \cup (4, \infty)$ **33.** $\left[-\frac{3}{2}, 5\right)$ **35.** $(2, 6]$ **37.** $\left(-\infty, \frac{1}{2}\right) \cup \left(\frac{5}{4}, \infty\right)$ **39.** $[-4, -2)$ **41.** $(-\infty, 4) \cup (6, \infty)$ **43.** $\left(0, \frac{1}{2}\right) \cup \left(\frac{5}{2}, \infty\right)$ **45.** $\left[\frac{3}{2}, \infty\right)$ **47.** $\left(-2, \frac{5}{3}\right) \cup \left(\frac{5}{3}, \infty\right)$ **49.** $\frac{-1}{p-3} - 1 \geq 0$ **50.** $\frac{-1}{p-3} - \frac{p-3}{p-3} \geq 0$ **51.** $\frac{-p+2}{p-3} \geq 0$ **52.** 2, 3; [2, 3) **53.** between approximately .237 and .342 second **55. (a)** $\{1, 3\}$ **(b)** $(-\infty, 1) \cup (3, \infty)$ **(c)** the open interval $(1, 3)$ **56. (a)** $\left\{-4, \frac{2}{3}\right\}$ **(b)** $(-\infty, -4] \cup \left[\frac{2}{3}, \infty\right)$ **(c)** the open interval $\left(-4, \frac{2}{3}\right)$ **57. (a)** $\{-2, 5\}$ **(b)** $[-2, 5]$ **(c)** $(-\infty, -2] \cup [5, \infty)$ **58. (a)** $\left\{-3, \frac{5}{2}\right\}$ **(b)** $\left[-3, \frac{5}{2}\right]$ **(c)** $(-\infty, -3] \cup \left[\frac{5}{2}, \infty\right)$ **59.** between 0 and 10.8 seconds and between 140.8 and 151.5 seconds

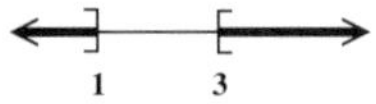

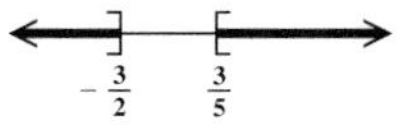

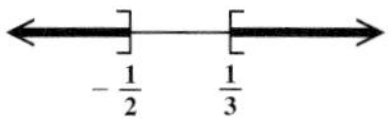

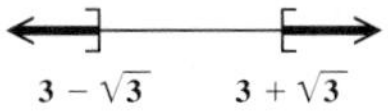

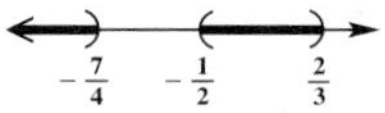

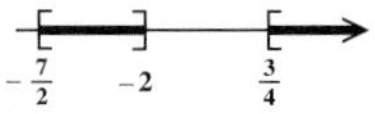

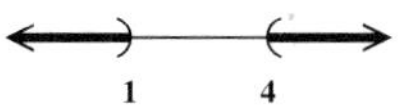

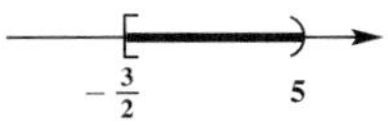

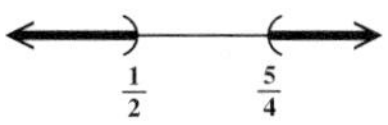

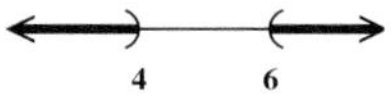

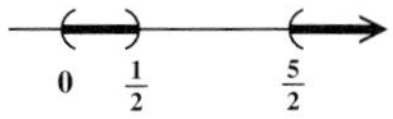

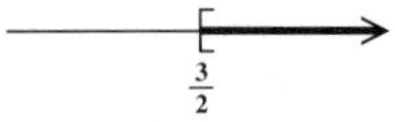

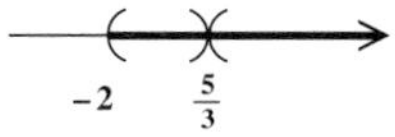

CHAPTER 8 REVIEW EXERCISES (PAGE 572)

1. $\{-11, 11\}$ **3.** $\left\{-\frac{15}{2}, \frac{5}{2}\right\}$ **5.** $\{-2 + \sqrt{19}, -2 - \sqrt{19}\}$ **7.** $\left\{-\frac{7}{2}, 3\right\}$ **9.** $\left\{\frac{7}{3}\right\}$ **11.** No. The term $-b$ and $\pm$ should be in the numerator. The correct formula is $x = \frac{-b \pm \sqrt{b^2 - 4ac}}{2a}$. **13.** 5.5 hours and 6.5 hours **15.** (c) **17.** (d) **19.** The discriminant is 34^2, so it can be factored; $(6x - 5)(4x - 9)$. **21.** $\left\{-\frac{5}{2}, 3\right\}$ **23.** $\left\{-\frac{11}{6}, -\frac{19}{12}\right\}$ **25.** $\left\{-\frac{343}{8}, 64\right\}$ **27.** 40 miles per hour **29.** Because x appears on the left side alone, and because it is equal to the nonnegative square root of $2x + 4$, it cannot be negative. **31.** $v = \frac{\pm\sqrt{rFkw}}{kw}$ **33.** .87 second **35.** 10 meters by

12 meters **37.** 120 meters **39.** (0, 6) **41.** (−2, 0) **43.** $\left(\frac{2}{3}, -\frac{2}{3}\right)$ **45.** **47.** (a)

49. **(a)** $c = 12.39$, $16a + 4b + c = 15.78$, $49a + 7b + c = 22.71$ **(b)** $f(x) = .2089x^2 + .0118x + 12.39$
51. **(a)** $\{-.5, 4\}$ **(b)** $(-\infty, -.5) \cup (4, \infty)$ **(c)** $[-.5, 4]$ **53.** $[-4, 3]$ **55.** $\emptyset$

57. $[-3, 2)$ **59.** $\left\{\frac{3 + i\sqrt{3}}{3}, \frac{3 - i\sqrt{3}}{3}\right\}$ **61.** $\{-2, -1, 3, 4\}$ **63.** $\left\{\frac{4}{11}, 5\right\}$ **65.** $y = \frac{6p^2}{z}$

67. $\left\{\frac{3}{5}, 1\right\}$ **69.** $\left(-5, -\frac{23}{5}\right]$ **71.** $(-4, -3)$

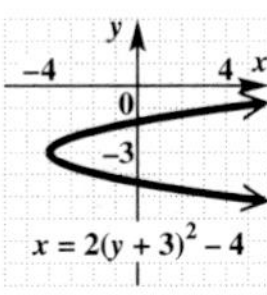

73. (4, 6)

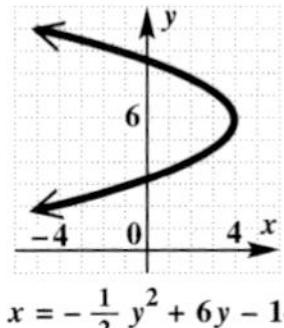

75. Work backwards as follows.

$x = -5$ or $x = 6$ Given
$x + 5 = 0$ or $x - 6 = 0$
$(x + 5)(x - 6) = 0$ Zero-factor property
$x^2 - x - 30 = 0$ Multiply the factors.

The equation $x^2 - x - 30 = 0$ has −5 and 6 as solutions.

77. length: 150 meters; width: 150 meters

79. **(a)** $\{-2\}$ **(b)** $(-\infty, -2)$ **(c)** $(-2, \infty)$

80. **(a)** $\{1, 5\}$ **(b)** $(-\infty, 1) \cup (5, \infty)$ **(c)** $(1, 5)$

81. **(a)** $\{4\}$ **(b)** $(2, 4)$ **(c)** $(-\infty, 2) \cup (4, \infty)$

82. $(-\infty, \infty)$ **83.** $(-\infty, \infty)$ **84.** 2; $(-\infty, 2) \cup (2, \infty)$ **85.** $(-\infty, \infty)$; denominator **86.** $(-5, 3)$

CHAPTER 8 TEST (PAGE 577)

[8.1] **1.** $\left\{-\frac{8}{7}, \frac{2}{7}\right\}$ **2.** $\{-1 + \sqrt{5}, -1 - \sqrt{5}\}$ [8.2] **3.** $\left\{\frac{3 + \sqrt{17}}{4}, \frac{3 - \sqrt{17}}{4}\right\}$ **4.** $\left\{\frac{2 + i\sqrt{11}}{3}, \frac{2 - i\sqrt{11}}{3}\right\}$

[8.3] **5.** $\left\{\frac{2}{3}\right\}$ **6.** **(a)** 6.53×10^6 meters **(b)** 247 meters per second [8.1] **7.** (a) [8.2] **8.** two irrational solutions

[8.1–8.3] **9.** $\left\{-\frac{2}{3}, 6\right\}$ **10.** $\left\{-2, -\frac{1}{3}, \frac{1}{3}, 2\right\}$ **11.** $\left\{-\frac{5}{2}, 1\right\}$ [8.4] **12.** $r = \frac{\pm\sqrt{\pi S}}{2\pi}$ **13.** 2 feet

14. 16 meters [8.5] **15.** vertex: (0, −2); domain: $(-\infty, \infty)$; range: $[-2, \infty)$

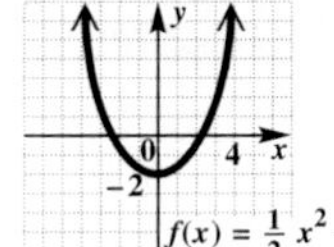

[8.6] **16.** vertex: (2, 3)

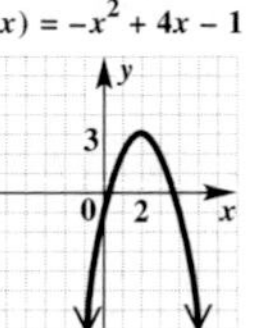

17. vertex: (2, 2)

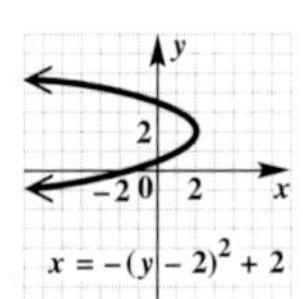

18. **(a)** 442 million **(b)** 1993; 425 million

[8.7] **19.** $(-\infty, -5) \cup \left(\frac{3}{2}, \infty\right)$

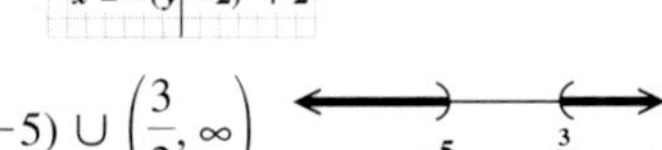

20. $(-\infty, 4) \cup [9, \infty)$

CUMULATIVE REVIEW EXERCISES CHAPTERS 1–8 (PAGE 578)

[1.1] **1. (a)** $-2, 0, 7$ **(b)** $-\frac{7}{3}, -2, 0, .7, 7, \frac{32}{3}$ [1.3] **2.** 41 [2.1] **3.** $\left\{\frac{4}{5}\right\}$ [2.7] **4.** $\left\{\frac{11}{10}, \frac{7}{2}\right\}$ [7.5] **5.** $\left\{\frac{2}{3}\right\}$

[6.6] **6.** $\emptyset$ [8.1–8.2] **7.** $\left\{\frac{7 + \sqrt{177}}{4}, \frac{7 - \sqrt{177}}{4}\right\}$ [8.3] **8.** $\{-2, -1, 1, 2\}$ [2.5] **9.** $[1, \infty)$ [2.7] **10.** $\left[2, \frac{8}{3}\right]$

[8.7] **11.** $(1, 3)$ **12.** $(-2, 1)$ [3.1] **13.** function; domain: $(-\infty, \infty)$; range: $(-\infty, \infty)$

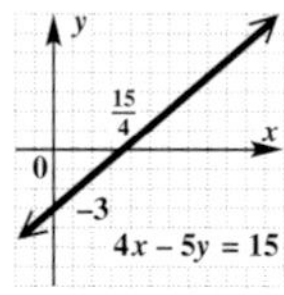

[3.4] **14.** not a function

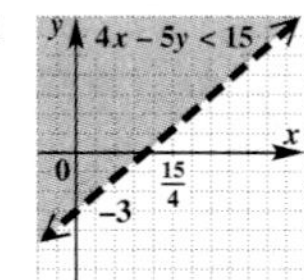

[8.5] **15.** function; domain: $(-\infty, \infty)$; range: $(-\infty, 3]$

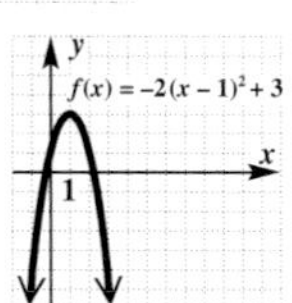

[3.2] **16.** $m = \frac{2}{7}$; x-intercept: $(-8, 0)$; y-intercept: $\left(0, \frac{16}{7}\right)$ [3.3] **17.** $y = -\frac{5}{2}x + 2$ **18.** $y = \frac{2}{5}x + \frac{13}{5}$

[3.5] **19. (a)** Qualifying Records **(b)** yes **(c)** $y = 1.279x + 116.26$ **(d)** 158.47; It is a little too high.

20. No, because the graph is a vertical line, which is not the graph of a function by the vertical line test. **21. (a)** 13 **(b)** domain: $(-\infty, \infty)$; range: $[-5, \infty)$ [4.1] **22.** $\{(1, -2)\}$ [4.2] **23.** $\{(3, -4, 2)\}$ [4.3] **24.** current: 10 miles per hour; boat: 30 miles per hour [5.4] **25.** $14x^2 - 13x - 12$ **26.** $\frac{4}{9}t^2 + 12t + 81$ [5.2] **27.** $-3t^3 + 5t^2 - 12t + 15$

[6.4] **28.** $4x^2 - 6x + 11 + \frac{4}{x + 2}$ [5.5–5.7] **29.** $x(4 + x)(4 - x)$ **30.** $9(x - 1)(x + 1)$

31. $(2x + 3y)(4x^2 - 6xy + 9y^2)$ **32.** $(3x - 5y)^2$ [6.1] **33.** $\frac{x - 5}{x + 5}$ [6.2] **34.** $-\frac{8}{k}$ [6.3] **35.** $\frac{r - s}{r}$

[7.1] **36.** $\frac{3\sqrt[3]{4}}{4}$ [7.4] **37.** $\sqrt{7} + \sqrt{5}$ [2.2] **38.** 7 inches by 3 inches [6.7] **39.** biking: 12 miles per hour; walking: 2 miles per hour [8.4] **40.** southbound car: 57 miles; eastbound car: 76 miles

CHAPTER 9 INVERSE, EXPONENTIAL, AND LOGARITHMIC FUNCTIONS

SECTION 9.1 (PAGE 587)

EXERCISES **1.** It is not one-to-one, because both Illinois and Wisconsin are paired with the same range element, 40. **5.** (b) **7.** (a) **9.** $\{(6, 3), (10, 2), (12, 5)\}$ **11.** not one-to-one **13.** $f^{-1}(x) = \frac{x - 4}{2}$ **15.** $g^{-1}(x) = x^2 + 3$, $x \geq 0$ **17.** not one-to-one **19.** $f^{-1}(x) = \sqrt[3]{x + 4}$ **21. (a)** 8 **(b)** 3 **23. (a)** 1 **(b)** 0

25. (a) one-to-one **(b)** **27. (a)** not one-to-one **29. (a)** one-to-one **(b)**

31.

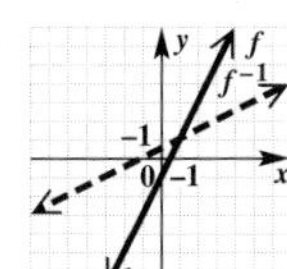

33.

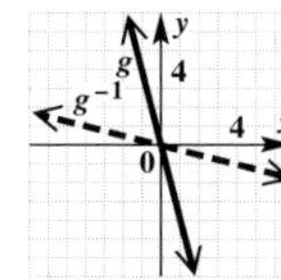

35.

x	$f(x)$
0	0
1	1
4	2

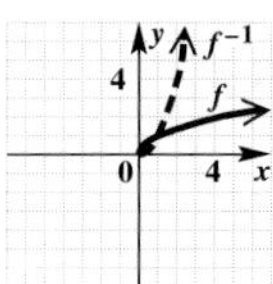

37.

x	$f(x)$
-1	-3
0	-2
1	-1
2	6

39. $f^{-1}(x) = \dfrac{x+5}{4}$ **40.** My graphing calculator is the greatest thing since sliced bread. **41.** If the function were not one-to-one, there would be ambiguity in some of the characters, as they could represent more than one letter.
42. Answers will vary. For example, Jane Doe is 1004 5 2748 129 68 3379 129.

43. $f^{-1}(x) = \dfrac{x+7}{2}$

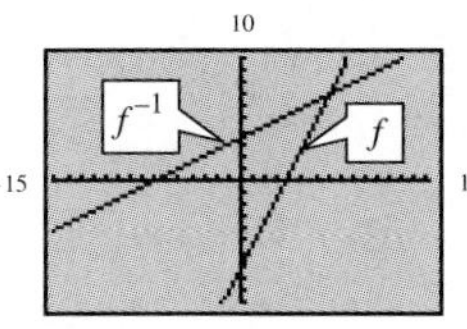

45. $f^{-1}(x) = \sqrt[3]{x-5}$

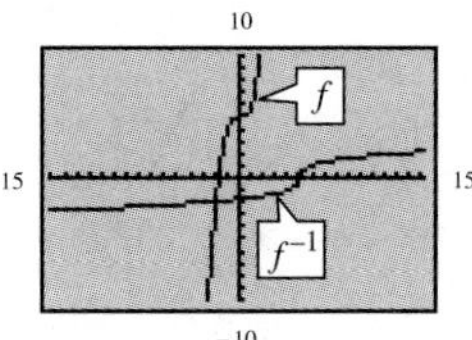

47.

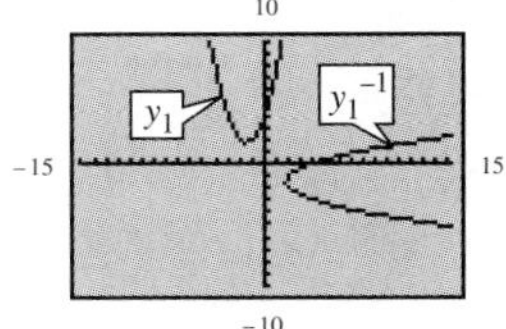

49. It is not a one-to-one function.

SECTION 9.2 (PAGE 597)

EXERCISES **1.** (c) **3.** (a) **5.**

7.

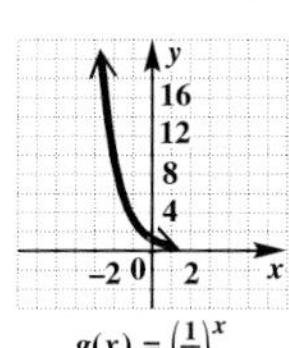

9.

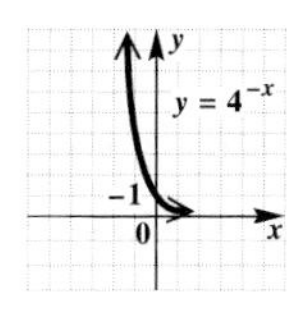

11.

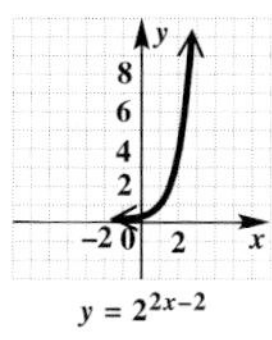

13. **(a)** rises; falls **(b)** It is one-to-one and thus has an inverse. **15.** $\left\{\frac{3}{2}\right\}$ **17.** $\{7\}$ **19.** $\{-3\}$ **21.** $\{-1\}$
23. $\{-3\}$ **25.** 639.545 **27.** .066 **29.** 12.179 **31.** In the definition of the exponential function defined by $F(x) = a^x$, a must be a positive number. This corresponds to the peculiarity that some scientific calculators do not allow negative bases. **33.** **(a)** .5°C **(b)** .35°C **35.** **(a)** 1.6°C **(b)** .5°C **37.** **(a)** 7147 million short tons **(b)** 25,149 million short tons **(c)** This is less than the 30,100 million short tons the model provides.
39. **(a)** 100 grams **(b)** 31.25 grams **(c)** .30 gram (to the nearest hundredth) **(d)**

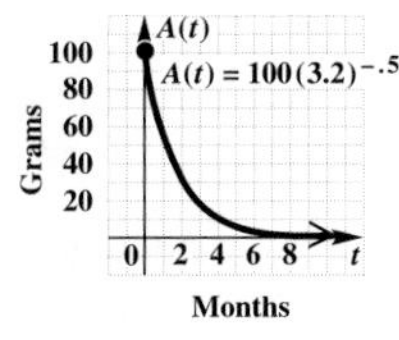

41. 6.67 years after it was purchased **43.** about \$360,000 **45.** $(\sqrt[4]{16})^3$; 8 **46.** $\sqrt[4]{16^3}$; 8 **47.** 64; 8
48. Because $\sqrt{\sqrt{x}} = (x^{1/2})^{1/2} = x^{1/4} = \sqrt[4]{x}$, the fourth root of 16^3 can be found by taking the square root twice.
49. 8 **50.** $16^{75/100}$; $16^{3/4} = (\sqrt[4]{16})^3 = 2^3 = 8$ **51.** The display indicates that for the year 1965, the model gives a value of 102.287 million tons, which is slightly less than the actual value of 103.4 million tons.

SECTION 9.3 (PAGE 607)

CONNECTIONS **Page 606:** almost 4 times as powerful; about 300 times as powerful

EXERCISES **1.** **(a)** C **(b)** F **(c)** B **(d)** A **(e)** E **(f)** D **3.** $\log_4 1024 = 5$ **5.** $\log_{1/2} 8 = -3$
7. $\log_{10} .001 = -3$ **9.** $4^3 = 64$ **11.** $10^{-4} = \dfrac{1}{10{,}000}$ **13.** $6^0 = 1$ **15.** By using the word "radically," the teacher

meant for him to consider roots. Because 3 is the square (2nd) root of 9, $\log_9 3 = \frac{1}{2}$. **17.** $\left\{\frac{1}{3}\right\}$ **19.** $\{81\}$ **21.** $\left\{\frac{1}{5}\right\}$

23. $\{1\}$ **25.** $\{x \mid x > 0, x \neq 1\}$ **27.** $\{5\}$ **29.** $\left\{\frac{5}{3}\right\}$ **31.** $\{4\}$ **33.** $\left\{\frac{3}{2}\right\}$ **35.** $y = \log_3 x$

37. $y = \log_{1/3} x$ **39.** Every power of 1 is equal to 1, and thus it cannot be used as a base. **41.** $(0, \infty)$; $(-\infty, \infty)$

43. 8 **45.** 24 **47.** **(a)** 645 sites **(b)** 962 sites **49.** **(a)** 6 **(b)** 12 **(c)** 18 **(d)**

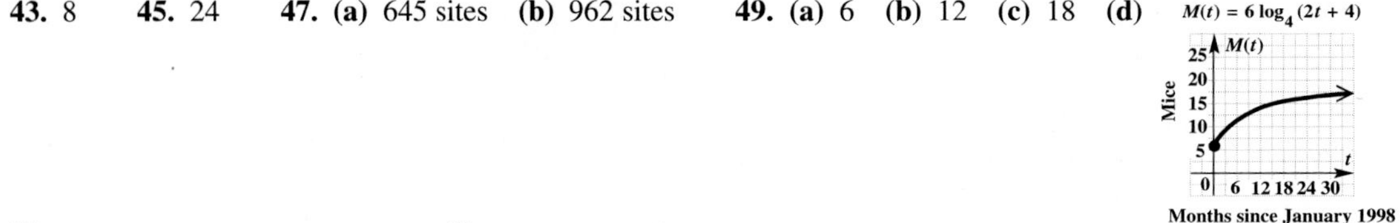

51. **53.**

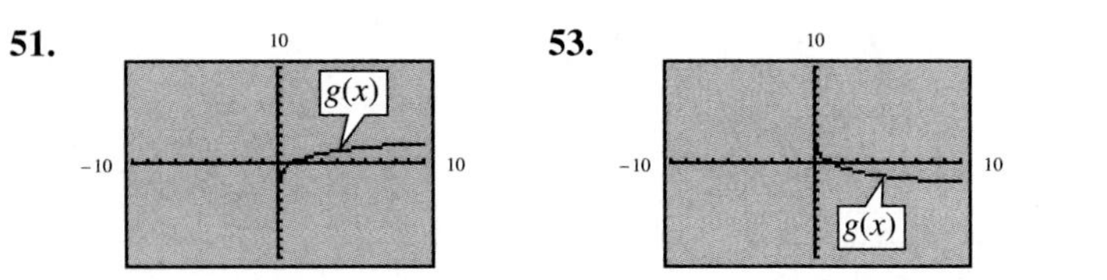

SECTION 9.4 (PAGE 616)

CONNECTIONS **Page 616:**

$$\begin{array}{r} \log_{10} 458.3 \approx 2.661149857 \\ + \log_{10} 294.6 \approx 2.469232743 \\ \hline \approx 5.130382600 \\ 10^{5.130382600} \approx 135{,}015.18 \end{array}$$

A calculator gives $(458.3)(294.6) = 135{,}015.18$.

EXERCISES **1.** $\log_{10} 3 + \log_{10} 4$ **3.** $4 \log_{10} 3$ **5.** 4 **7.** $\log_7 4 - \log_7 5$ **9.** $\frac{1}{4} \log_2 8$ or $\frac{3}{4}$ **11.** $\log_4 3 + \frac{1}{2} \log_4 x - \log_4 y$ **13.** $\frac{1}{3} \log_3 4 - 2 \log_3 x - \log_3 y$ **15.** $\frac{1}{2} \log_3 x + \frac{1}{2} \log_3 y - \frac{1}{2} \log_3 5$ **17.** $\frac{1}{3} \log_2 x + \frac{1}{5} \log_2 y - 2 \log_2 r$ **21.** $\log_b xy$ **23.** $\log_a \frac{m^3}{n}$ **25.** $\log_a \frac{rt^3}{s}$ **27.** $\log_a \frac{125}{81}$ **29.** $\log_{10}(x^2 - 9)$ **31.** $\log_p \frac{x^3 y^{1/2}}{z^{3/2} a^3}$ **33.** 1.2552 **35.** $-.6532$ **37.** 1.5562 **39.** .4771 **41.** true **43.** false **45.** The exponent of a quotient is the difference between the exponent of the numerator and the exponent of the denominator. **47.** $\log_2 8 - \log_2 4 = \log_2 \frac{8}{4} = \log_2 2 = 1$ **49.** 4 **50.** It is the exponent to which 3 must be raised in order to obtain 81. **51.** 81 **52.** It is the exponent to which 2 must be raised in order to obtain 19. **53.** 19 **54.** m

SECTION 9.5 (PAGE 623)

CONNECTIONS **Page 620:** 2; 2.5; $2.\overline{6}$; $2.708\overline{3}$; $2.71\overline{6}$; The difference is .0016151618. It approaches e fairly quickly.

EXERCISES **1.** (c) **3.** (c) **5.** 19.2 **7.** 1.6335 **9.** 2.5164 **11.** -1.4868 **13.** 9.6776 **15.** 2.0592 **17.** -2.8896 **19.** 5.9613 **21.** 4.1506 **23.** 2.3026 **25.** **(a)** 2.552424846 **(b)** 1.552424846 **(c)** .552424846 **(d)** The whole number parts will vary but the decimal parts are the same. **27.** An error message appears, because we cannot find the common logarithm of a negative number. **29.** bog **31.** 11.6 **33.** 4.3 **35.** 4.0×10^{-8} **37.** 4.0×10^{-6} **39.** **(a)** 78,840 million dollars **(b)** 92,316 million dollars **(c)** 108,095 million dollars **(d)** 70,967 million dollars **41.** 32,044 million dollars **43.** **(a)** 1.62 grams **(b)** 1.18 grams **(c)** .69 gram

(d) 2.00 grams **45.** 527 years **47.** 3466 years **49.** 1.3869 **51.** .7211 **53.** 3.3030 **55.** Answers will vary. Suppose the name is Jeffery Cole, with $m = 7$ and $n = 4$. **(a)** $\log_7 4$ is the exponent to which 7 must be raised in order to obtain 4. **(b)** .7124143742 **(c)** 4 **57.** It means that in 1988, expenditures were approximately 83,097 million dollars.

59. (a) The expression $\frac{1}{x}$ in the base cannot be evaluated, since division by 0 is not defined. **(b)** e **(c)** 2.718280469; 2.718281828; they differ in the sixth decimal place. **(d)** e **61.**

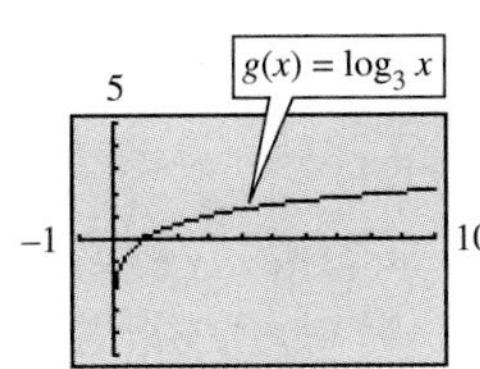

63.

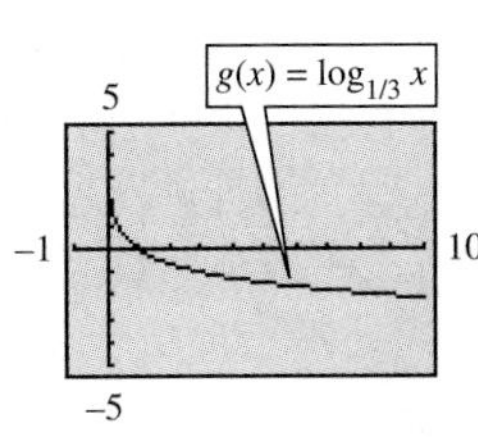

SECTION 9.6 (PAGE 634)

CONNECTIONS **Page 632:** When the equation is set up, the common factor P can be divided into both sides, and thus its particular value does not affect the solution of the equation.

EXERCISES **1.** $\log 5^x = \log 125$ **2.** $x \log 5 = \log 125$ **3.** $x = \frac{\log 125}{\log 5}$ **4.** $\frac{\log 125}{\log 5} = 3$; $\{3\}$ **5.** $\{.827\}$ **7.** $\{.833\}$ **9.** $\{2.269\}$ **11.** $\{566.866\}$ **13.** $\{-18.892\}$ **15.** $\{24.414\}$ **19.** $\left\{\frac{2}{3}\right\}$ **21.** $\{-1 + \sqrt[3]{49}\}$ **23.** The apparent solution, 2, causes the expression $\log_4(x - 3)$ to be $\log_4(-1)$, and we cannot take the logarithm of a negative number. **25.** $\left\{\frac{1}{3}\right\}$ **27.** $\{2\}$ **29.** $\emptyset$ **31.** $\{8\}$ **33.** $\left\{\frac{4}{3}\right\}$ **35.** $\{8\}$ **37.** \$2539.47 **39.** about 180 grams **41. (a)** \$11,260.96 **(b)** \$11,416.64 **(c)** \$11,497.99 **(d)** \$11,580.90 **(e)** \$11,581.83 **43.** \$137.41 **45.** 1997 **47.** 5.5 hours **49. (a)** 3.10 grams **(b)** 1.19 grams **(c)** 29 years

51. $\{.827\}$

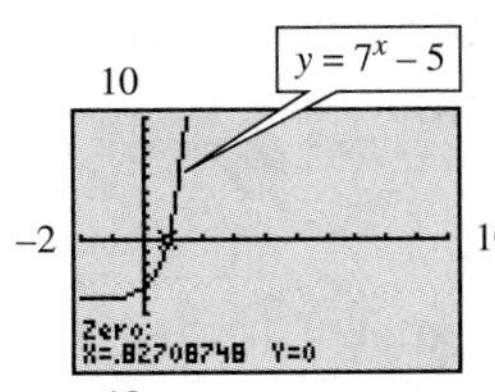

53. $\left\{\frac{1}{3}\right\}$ or $\{.333\}$

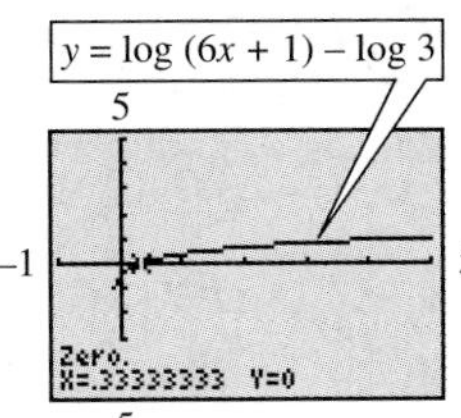

CHAPTER 9 REVIEW EXERCISES (PAGE 641)

1. not one-to-one **3.** one-to-one; Each element of the domain corresponds to only one element of the range, and vice versa.

5. $f^{-1}(x) = \frac{x^3 + 4}{6}$ **7.**

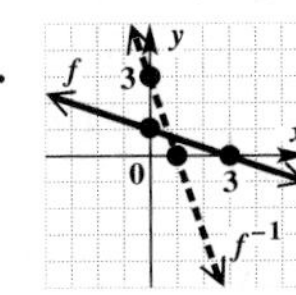

9.

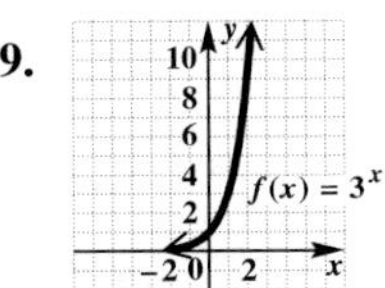

11.

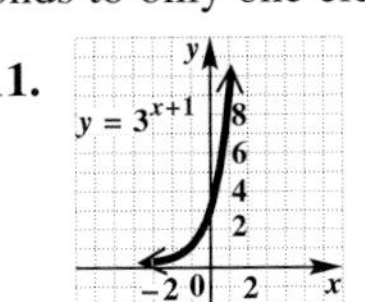

13. $\{4\}$

15. (a) 1.2 million tons (less than actual) **(b)** 3.8 million tons (less than actual) **(c)** 21.8 million tons (more than actual)

17. $g(x) = \log_{1/3} x$ **19.** $\left\{\frac{3}{2}\right\}$ **21.** $\{8\}$ **23.** $\{b \mid b > 0, b \neq 1\}$ **25.** a **27.** $\log_2 3 + \log_2 x + 2 \log_2 y$

29. $\log_b \frac{3x}{y^2}$ **31.** 1.4609 **33.** 3.3638 **35.** .9251 **37.** 6.4 **39.** 2.5×10^{-5} **41.** $\{2.042\}$ **43.** (d)

45. $\{2\}$ **47.** $\{4\}$ **49. (a)** $\left\{\frac{3}{8}\right\}$ **(b)** The x-value of the x-intercept is .375, the decimal equivalent of $\frac{3}{8}$.

51. \$28,295.56 **53.** Plan A is better, since it would pay \$2.92 more. **55.** about 18.04% **57.** \$4267

59. $\frac{1}{4}, \frac{1}{2}, 1, 2, 4, 8$

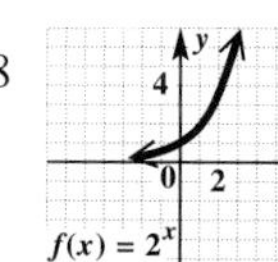

60. $-2, -1, 0, 1, 2, 3$

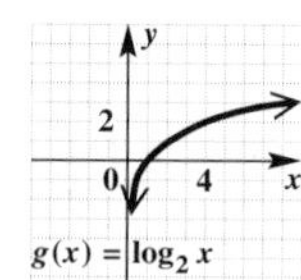

61. The roles of x and y are reversed. They are inverses. **62.** horizontal; vertical **63.** 5; 2; 3; 5 **64.** 4; 8 (or 8; 4) **65.** 3.700439718 (The number of displayed digits may vary.) **66.** $\log_2 13$ is the exponent to which 2 must be raised in order to obtain 13. **67.** 13 **68.** 13; The number in Exercise 65 is the exponent to which 2 must be raised in order to obtain 13. **69.** 3.700439718 **70.** $\left\{-\frac{8}{5}\right\}$ **71.** $\{72\}$ **73.** $\left\{\frac{1}{9}\right\}$ **75.** $\{3\}$ **77.** $\left\{\frac{11}{3}\right\}$

CHAPTER 9 TEST (PAGE 645)

[9.1] **1.** **(a)** not one-to-one **(b)** one-to-one **2.** $f^{-1}(x) = x^3 - 7$ **3.**

[9.2] **4.**

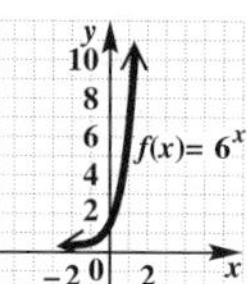

[9.3] **5.**

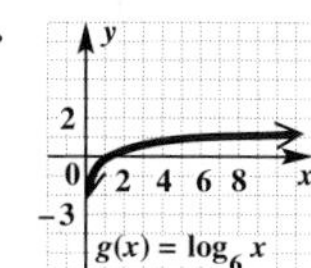

[9.1–9.3] **6.** Once the graph of $f(x) = 6^x$ is sketched, interchange the x- and y-values of its ordered pairs. The resulting points will be on the graph of $g(x) = \log_6 x$, since f and g are inverses. [9.2] **7.** $\{-4\}$ **8.** $\left\{-\frac{13}{3}\right\}$ **9.** **(a)** 2821 million pounds **(b)** 2385 million pounds **(c)** This is less than the 2193 million pounds that the model provides. [9.3] **10.** $\log_4 .0625 = -2$ **11.** $7^2 = 49$ **12.** $\{32\}$ **13.** $\left\{\frac{1}{2}\right\}$ **14.** $\{2\}$ **15.** 5; 2; 5th; 32 [9.4] **16.** $2 \log_3 x + \log_3 y$ **17.** $\frac{1}{2}\log_5 x - \log_5 y - \log_5 z$ **18.** $\log_b \frac{s^3}{t}$ **19.** $\log_b \frac{r^{1/4}s^2}{t^{2/3}}$ [9.5] **20.** **(a)** 1.3636 **(b)** $-.1985$ **21.** **(a)** $\frac{\log 19}{\log 3}$ **(b)** $\frac{\ln 19}{\ln 3}$ **(c)** 2.6801 [9.6] **22.** $\{3.9656\}$ **23.** $\{3\}$ **24.** \$12,507.51 **25.** **(a)** \$19,260.38 **(b)** approximately 13.9 years

CUMULATIVE REVIEW EXERCISES CHAPTERS 1–9 (PAGE 647)

[1.1] **1.** $-2, 0, 6, \frac{30}{3}$ (or 10) **2.** $-\frac{9}{4}, -2, 0, .6, 6, \frac{30}{3}$ (or 10) **3.** $-\sqrt{2}, \sqrt{11}$ **4.** $-\frac{9}{4}, -2, -\sqrt{2}, 0, .6, \sqrt{11}, 6, \frac{30}{3}$ (or 10) [1.2] **5.** 16 **6.** -27 **7.** -39 [2.1] **8.** $\left\{-\frac{2}{3}\right\}$ [2.5] **9.** $[1, \infty)$ [2.7] **10.** $\{-2, 7\}$ **11.** $\left\{\pm\frac{16}{3}\right\}$ **12.** $\left[\frac{7}{3}, 3\right]$ **13.** $(-\infty, -3) \cup (2, \infty)$ [3.1] **14.** [3.4] **15.**

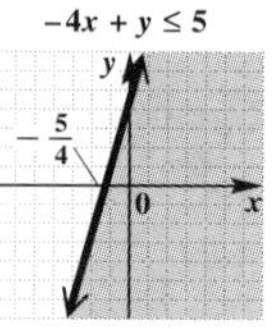

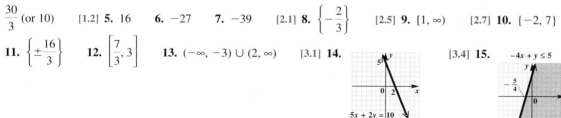

[3.2] **16.** $-5{,}587{,}500$ [3.3] **17.** $3x - 4y = 19$ [4.1] **18.** $\{(4, 2)\}$ [4.2] **19.** $\{(1, -1, 4)\}$ [4.5] **20.** -1 [4.3] **21.** 6 pounds [5.4] **22.** $6p^2 + 7p - 3$ **23.** $16k^2 - 24k + 9$ [5.2] **24.** $-5m^3 + 2m^2 - 7m + 4$ [6.4] **25.** $2t^3 + 5t^2 - 3t + 4$ [5.5] **26.** $x(8 + x^2)$ [5.6] **27.** $(3y - 2)(8y + 3)$ **28.** $z(5z + 1)(z - 4)$ [5.7] **29.** $(4a + 5b^2)(4a - 5b^2)$ **30.** $(2c + d)(4c^2 - 2cd + d^2)$ **31.** $(4r + 7q)^2$ [5.1] **32.** $-\frac{1875p^{13}}{8}$ [6.1] **33.** $\frac{x + 5}{x + 4}$ [6.2] **34.** $\frac{-3k - 19}{(k + 3)(k - 2)}$ **35.** $\frac{22 - p}{p(p - 4)(p + 2)}$ [7.2] **36.** $12\sqrt{2}$ [7.3] **37.** $-27\sqrt{2}$

[7.5] **38.** $\{0, 4\}$ [7.6] **39.** 41 [8.1, 8.2] **40.** $\left\{\frac{1 \pm \sqrt{13}}{6}\right\}$ [8.6] **41.** $(-\infty, -4) \cup (2, \infty)$

[8.3] **42.** $\{\pm 1, \pm 2\}$ [8.4] **43.** 150 and 150 [8.5] **44.**

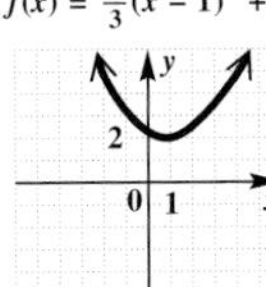

[9.2] **45.**

$f(x) = 2^x$

[9.6] **46.** $\{-1\}$ [9.3] **47.**

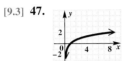

$f(x) = \log_3 x$

[9.4] **48.** 6.3398 **49.** $3 \log x + \frac{1}{2} \log y - \log z$

[9.6] **50.** **(a)** 25,000 **(b)** 30,500 **(c)** 37,300 **(d)** 68,000

CHAPTER 10 CONIC SECTIONS

SECTION 10.1 (PAGE 655)

CONNECTIONS **Page 655:** Answers will vary.

EXERCISES **1.** $(0, 0)$; 10 **3.** $(-3, -2)$ **5.** $x^2 + y^2 = 36$ **7.** $(x + 1)^2 + (y - 3)^2 = 16$ **9.** $x^2 + (y - 4)^2 = 3$ **11.** By the vertical line test the set is not a function, because a vertical line may intersect the graph of an ellipse in two points. **13.** center: $(-2, -3)$; radius: 2 **15.** center: $(-5, 7)$; radius: 9 **17.** center: $(2, 4)$; radius: 4 **19.**

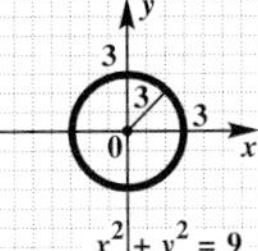

21.

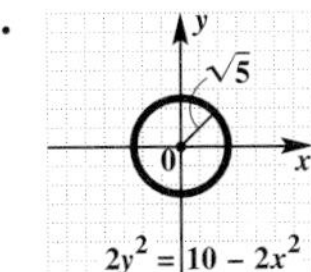

23. center: $(-3, 2)$

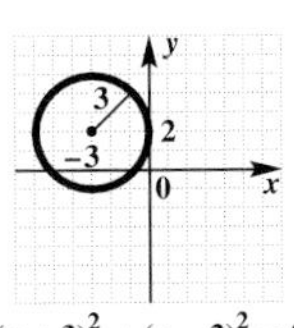

25. center: $(2, 3)$

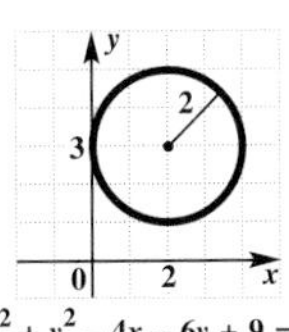

27. The taut string represents the constant radius, and the point where it is fastened serves as the center. **29.** A circular racetrack is most appropriate because the crawfish can move in any direction. Distance from the center determines the winner.

31.

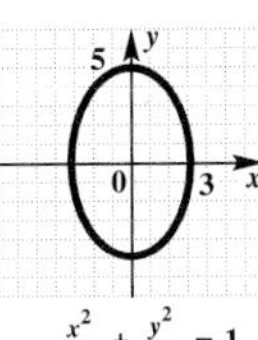

33.

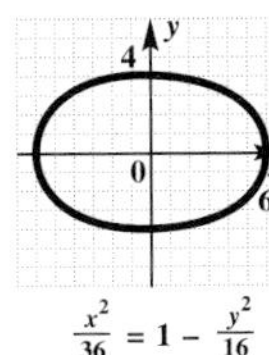

35.

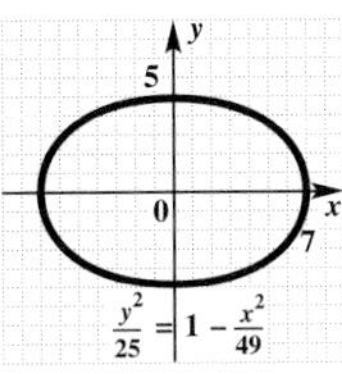

37.

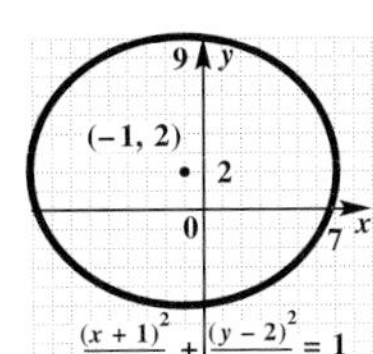

39.

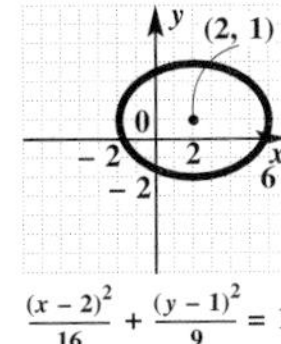

41. $y_1 = 4 + \sqrt{16 - (x + 2)^2}$, $y_2 = 4 - \sqrt{16 - (x + 2)^2}$ **43.**

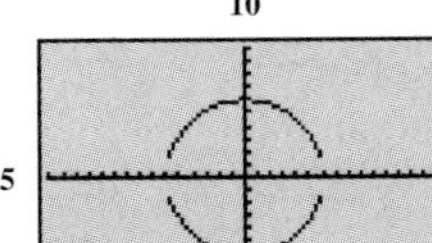
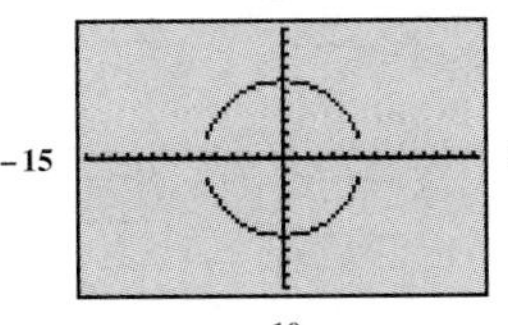

45.

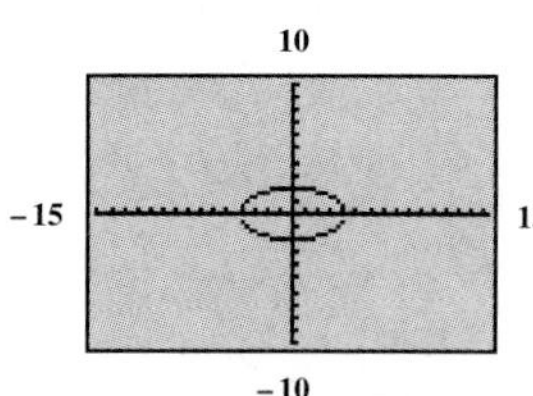

47. **(a)** 154.7 million miles **(b)** 128.7 million miles (Answers are rounded.)

49. $3\sqrt{3}$ units

SECTION 10.2 (PAGE 665)

CONNECTIONS **Page 662:** Answers will vary.

EXERCISES **1.** C **3.** D **5.** If the x^2 term is positive, the graph has only x-intercepts and opens left and right. If the y^2 term is positive, the graph has only y-intercepts and opens up and down. **7.**

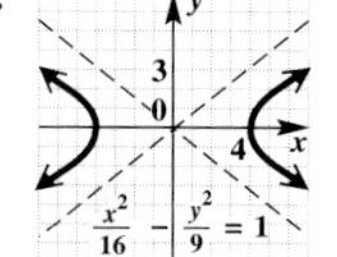

9.

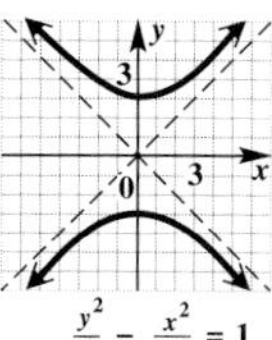

11.

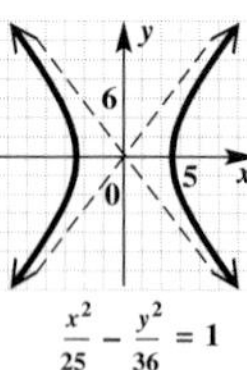

13. hyperbola

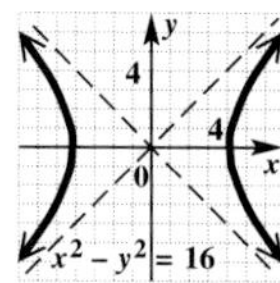

15. ellipse

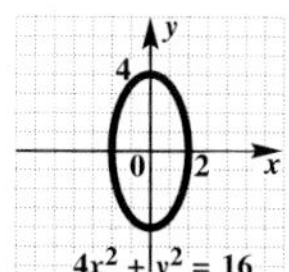

17. circle

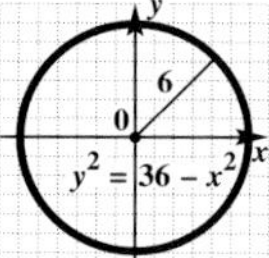

19. hyperbola

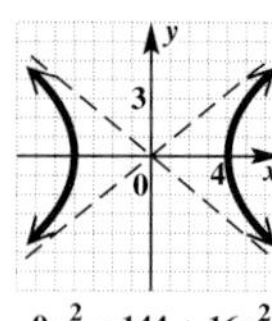

21. hyperbola

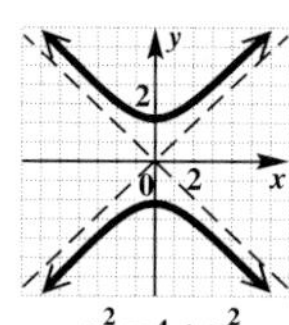

23.

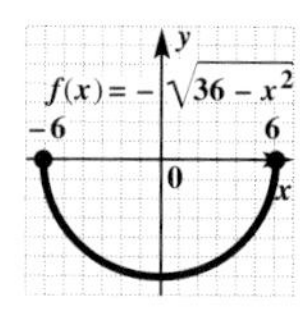

25.

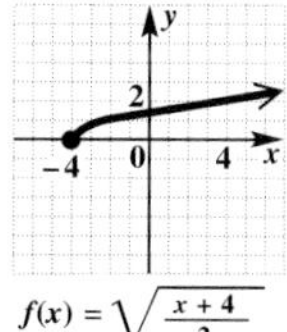

27. $\frac{(x-2)^2}{4} - \frac{(y+1)^2}{9} = 1$

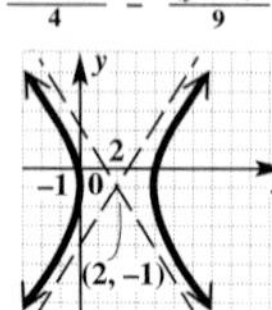

29. $\frac{y^2}{36} - \frac{(x-2)^2}{49} = 1$

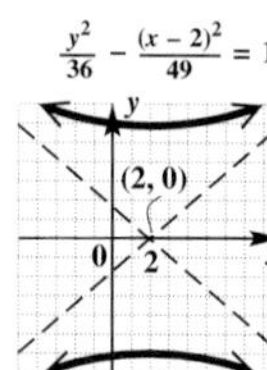

31. **(a)** 10 meters **(b)** 36 meters **33.** $y = \pm x$; The lines $y = \pm x$ form a 45° angle with the line through the goal posts. Most people can estimate a 45° angle fairly easily. **35.** **(a)** about 55 percent **(b)** 53 percent **37.** $y_1 = \sqrt{\frac{x^2}{9} - 1}$, $y_2 = -\sqrt{\frac{x^2}{9} - 1}$ **39.**

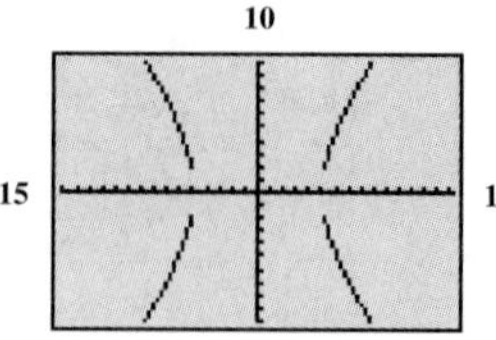

41.

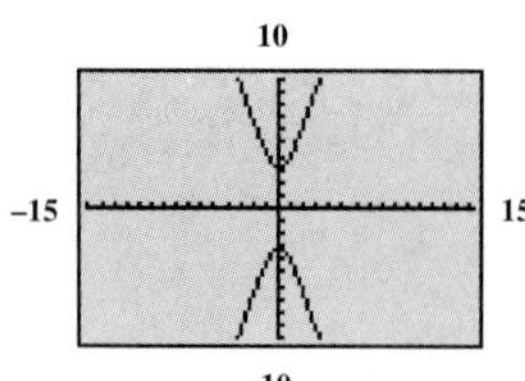

43. $y = \sqrt{\frac{x^2}{4} - 1}$ **44.** $y = \frac{1}{2}x$ **45.** $y \approx 24.98$ **46.** $y = 25$

47. Because $24.98 < 25$, the graph of $y = \sqrt{\frac{x^2}{4} - 1}$ lies below the graph of $y = \frac{1}{2}x$. **48.** The y-values on the hyperbola will approach the y-values on the line, but will always be less.

SECTION 10.3 (PAGE 675)

EXERCISES **3.** one **5.** none **7.** **9.** **11.** **13.**

15. **17.** **19.** $\left\{(0, 0), \left(\frac{1}{2}, \frac{1}{2}\right)\right\}$ **21.** $\{(-6, 9), (-1, 4)\}$ **23.** $\left\{\left(-\frac{1}{5}, \frac{7}{5}\right), (1, -1)\right\}$

25. $\left\{(-2, -2), \left(-\frac{4}{3}, -3\right)\right\}$ **27.** $\{(-3, 1), (1, -3)\}$ **29.** $\left\{\left(-\frac{3}{2}, -\frac{9}{4}\right), (-2, 0)\right\}$ **31.** $\{(-\sqrt{3}, 0), (\sqrt{3}, 0), (-\sqrt{5}, 2), (\sqrt{5}, 2)\}$ **33.** $\{(-2, 0), (2, 0)\}$ **35.** $\{(\sqrt{3}, 0), (-\sqrt{3}, 0)\}$ **37.** $\{(-2i\sqrt{2}, -2\sqrt{3}), (-2i\sqrt{2}, 2\sqrt{3}), (2i\sqrt{2}, -2\sqrt{3}), (2i\sqrt{2}, 2\sqrt{3})\}$ **39.** $\{(-\sqrt{5}, -\sqrt{5}), (\sqrt{5}, \sqrt{5})\}$ **41.** $\{(i, 2i), (-i, -2i), (2, -1), (-2, 1)\}$
43. $(-1, 2)$ **45.** $(-4, 8)$ **47.** $\{(2, -3), (-3, 2)\}$

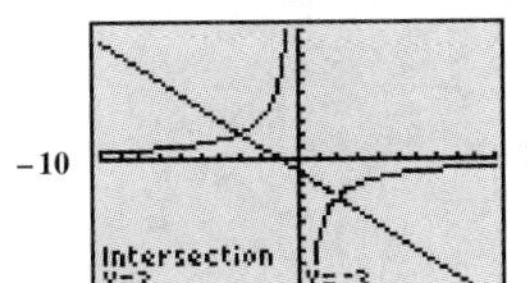

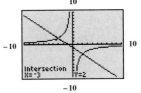

49. $\{-2, 5\}$ **50.**

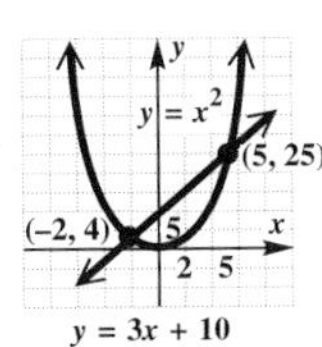

51. $\{(-2, 4), (5, 25)\}$ **52.** They are the same. **53.**

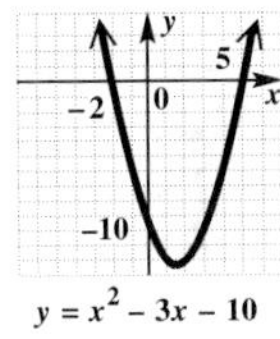

54. They are the same. **55.** length: 12 feet; width: 7 feet **57.** 1981; 467 thousand

SECTION 10.4 (PAGE 683)

EXERCISES **1.** (c) **5.** B **7.** A **9.** **11.**

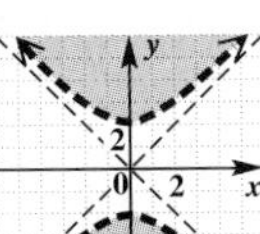

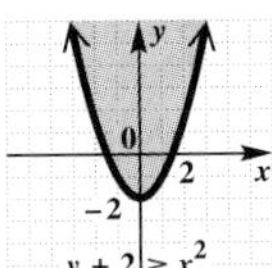

13.

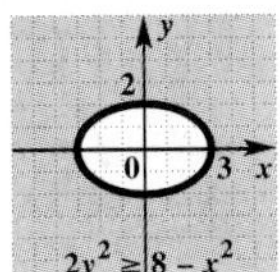

15.

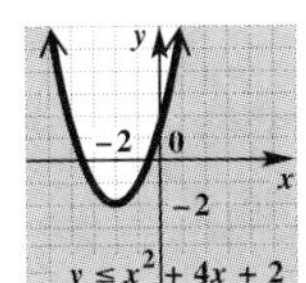

17.

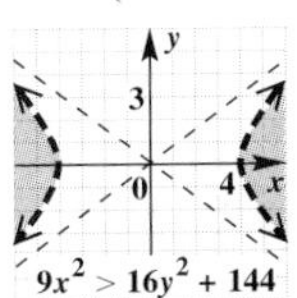

19.

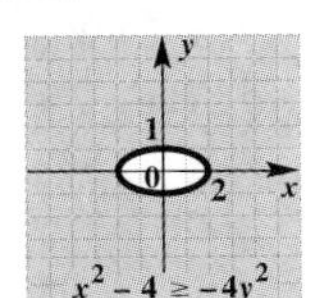

21.

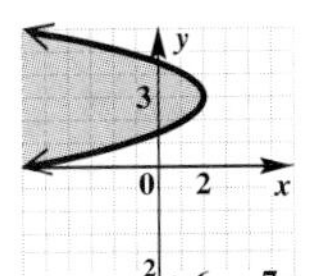

23.

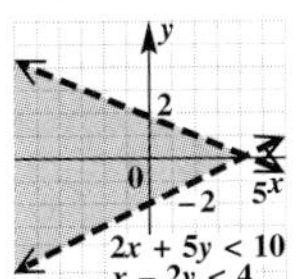

25.

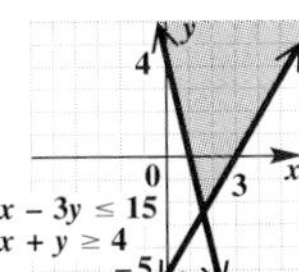

27.

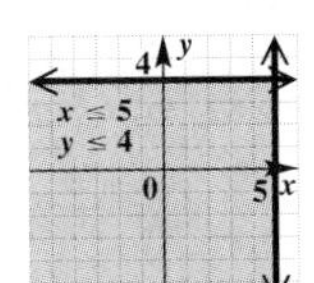

29.

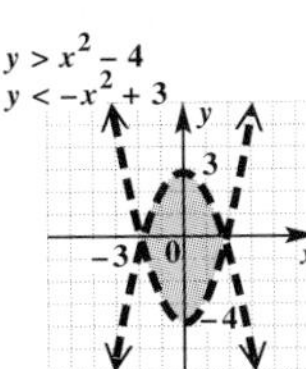

31.

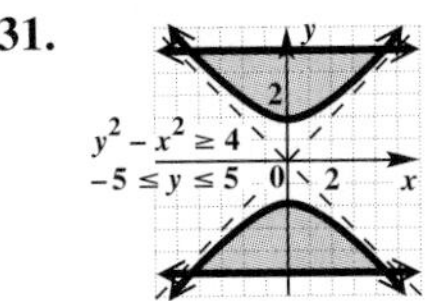

33.

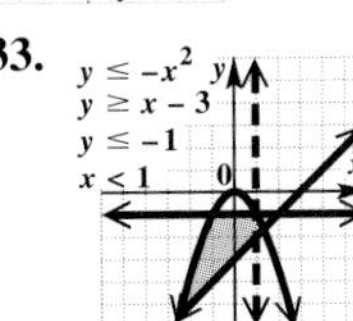

35. 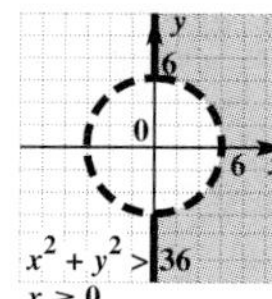

37. 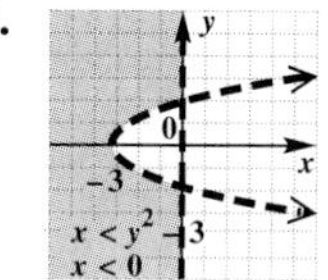

39. 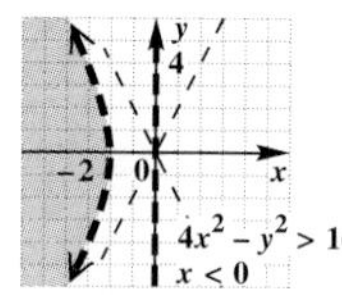

41. 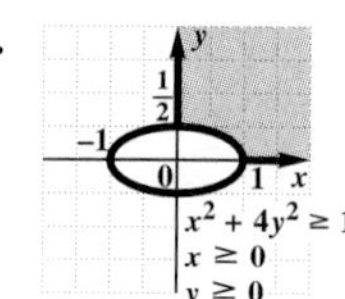

43. A 45. B

47. 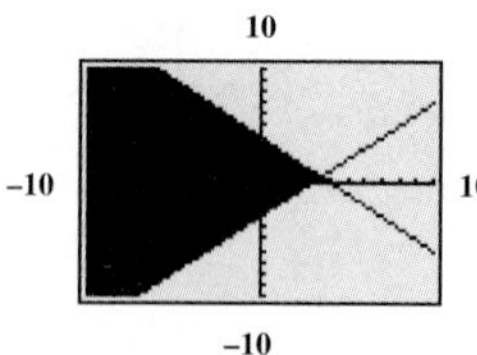

49. 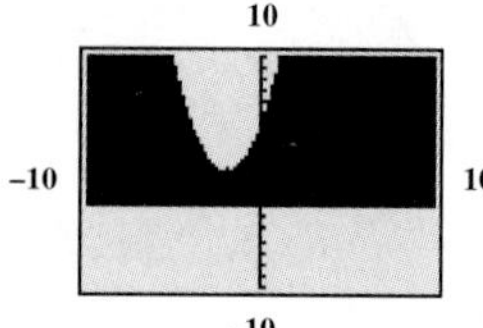

CHAPTER 10 REVIEW EXERCISES (PAGE 690)

1. $(x + 2)^2 + (y - 4)^2 = 9$ 3. center: $(-3, 2)$; radius: 4 5. center: $(-1, -5)$; radius: 3 7. 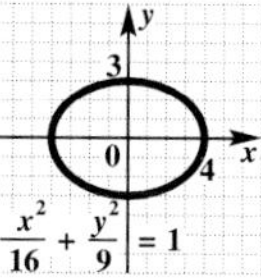

9. 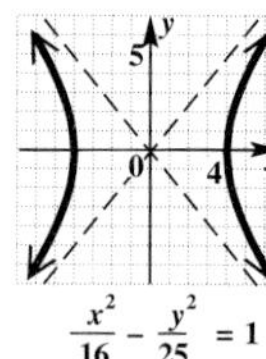

11. 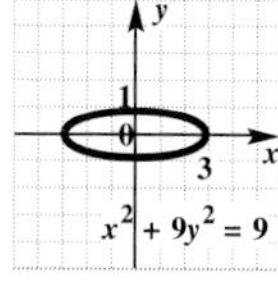

13. 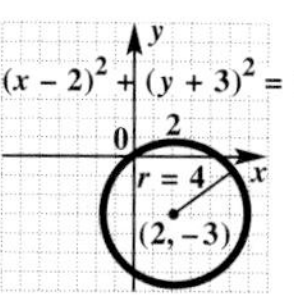

15. hyperbola 17. parabola

19. $\dfrac{x^2}{65{,}286{,}400} + \dfrac{y^2}{2{,}560{,}000} = 1$ 21. $\{(6, -9), (-2, -5)\}$ 23. $\{(4, 2), (-1, -3)\}$ 25. $\{(-\sqrt{2}, 2), (-\sqrt{2}, -2), (\sqrt{2}, -2), (\sqrt{2}, 2)\}$ 27. 0, 1, or 2 29. The graph shows that in about 1950 the altitude was reduced to approximately 4005 feet above sea level, causing the salinity to increase to about 6000 milligrams per liter. 31. 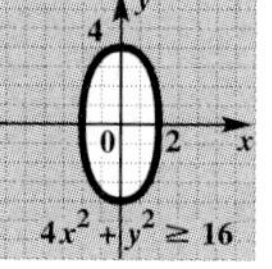

33. 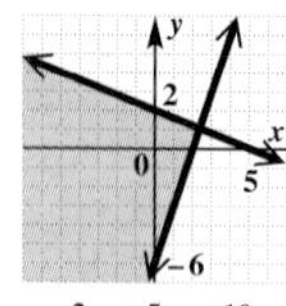

35. 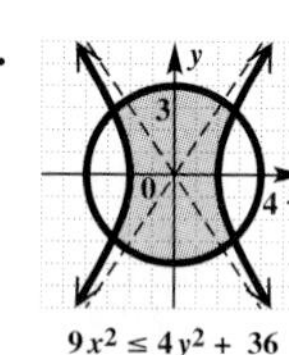

37. 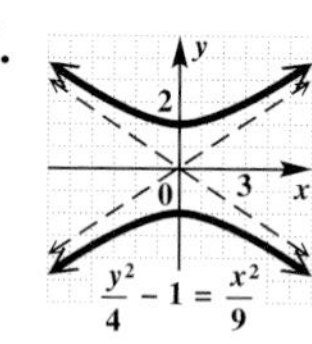

39. 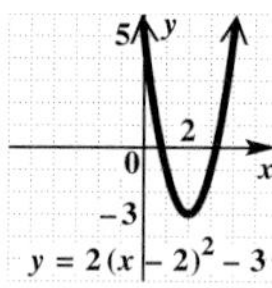

41. 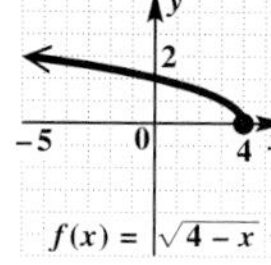

43. 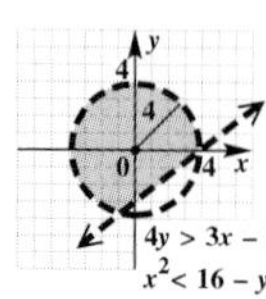

44. $2a + 4b + c = -20$ 45. $5a + b + c = -26$ 46. $-a + b + c = -2$

47. $\{(-4, -2, -4)\}$; $x^2 + y^2 - 4x - 2y - 4 = 0$ 48. center: $(2, 1)$; radius: 3 49. 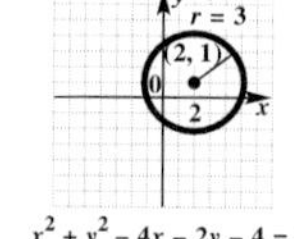

CHAPTER 10 TEST (PAGE 692)

[10.1] **1.** $(x + 4)^2 + (y - 4)^2 = 16$ **2.** center: $(-4, 1)$; radius: 5 [10.2] **3.** circle **4.** ellipse **5.** hyperbola **6.** parabola [10.1] **7.**

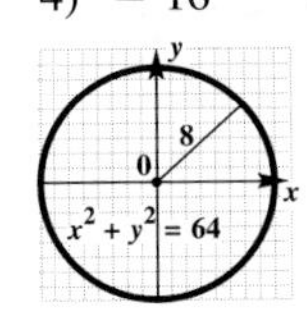

8.

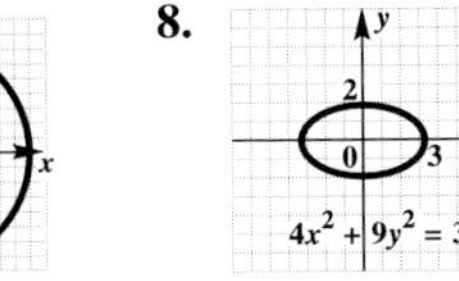

[10.2] **9.** $16y^2 - 4x^2 = 64$ **10.** $f(x) = \sqrt{16 - x^2}$

[10.3] **11.** 0, 1, 2, 3, or 4 **12.** **13.** $\left\{\left(-\frac{1}{2}, -10\right), (5, 1)\right\}$ **14.** $\left\{(-2, -2), \left(\frac{14}{5}, -\frac{2}{5}\right)\right\}$

15. $\{(-\sqrt{22}, -\sqrt{3}), (-\sqrt{22}, \sqrt{3}), (\sqrt{22}, -\sqrt{3}), (\sqrt{22}, \sqrt{3})\}$ [10.4] **16.** $y \le x^2 - 2$ **17.** $2x - 5y \ge 12$, $3x + 4y \le 12$

18. $x^2 + 25y^2 \le 25$, $x^2 + y^2 \le 9$ [10.1] **19.** 69.8 million kilometers **20.** 46.0 million kilometers

CUMULATIVE REVIEW EXERCISES CHAPTERS 1–10 (PAGE 693)

[1.1] **1.** -4 [2.1] **2.** $\left\{\frac{2}{3}\right\}$ [2.5] **3.** $\left(-\infty, \frac{3}{5}\right]$ [2.7] **4.** $\{-4, 4\}$ **5.** $(-\infty, -5) \cup (10, \infty)$ [3.2] **6.** $\frac{2}{3}$

[3.3] **7.** $3x + 2y = -13$ [4.1] **8.** $\{(3, -3)\}$ [4.2] **9.** $\{(4, 1, -2)\}$ [10.3] **10.** $\left\{(-1, 5), \left(\frac{5}{2}, -2\right)\right\}$

[5.4] **11.** $25y^2 - 30y + 9$ **12.** $12r^2 + 40r - 7$ [6.4] **13.** $4x^3 - 4x^2 + 3x + 5 + \frac{3}{2x + 1}$

[5.6] **14.** $(3x + 2)(4x - 5)$ **15.** $(2y^2 - 1)(y^2 + 3)$ [5.7] **16.** $(z^2 + 1)(z + 1)(z - 1)$ **17.** $(a - 3b)(a^2 + 3ab + 9b^2)$

[6.1] **18.** $\frac{40}{9}$ **19.** $\frac{y - 1}{y(y - 3)}$ [6.2] **20.** $\frac{3c + 5}{(c + 5)(c + 3)}$ **21.** $\frac{1}{p}$ [6.7] **22.** $\frac{6}{5}$ or $1\frac{1}{5}$ hours [5.1] **23.** $\frac{3}{4}$

24. $\frac{a^5}{4}$ [7.3] **25.** $2\sqrt[3]{2}$ [7.4] **26.** $\frac{3\sqrt{10}}{2}$ [7.6] **27.** $\frac{7}{5} + \frac{11}{5}i$ [7.5] **28.** $\emptyset$ [5.6] **29.** $\left\{\frac{1}{5}, -\frac{3}{2}\right\}$

[8.1, 8.2] **30.** $\left\{\frac{1 - 2\sqrt{2}}{4}, \frac{1 + 2\sqrt{2}}{4}\right\}$ [9.6] **31.** $\{3\}$ [8.3] **32.** $\left\{-\frac{\sqrt{6}}{2}, \frac{\sqrt{6}}{2}, -\sqrt{7}, \sqrt{7}\right\}$ [8.4] **33.** $v = \frac{\pm\sqrt{rFkw}}{kw}$

[3.1] **34.**

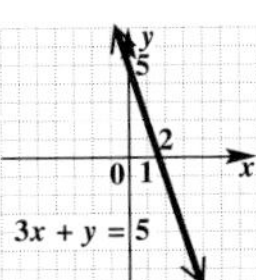

[8.5] **35.**

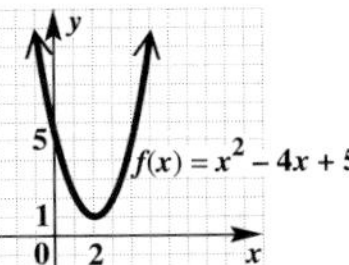

[9.2] **36.**

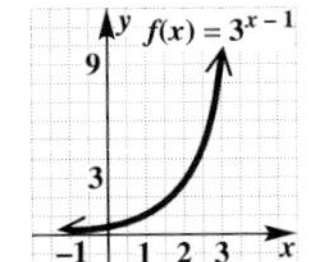

[10.2] **37.**

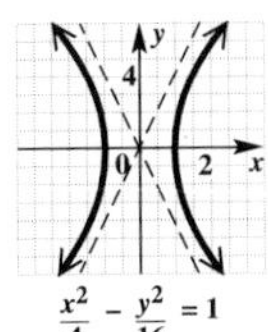

[10.4] **38.**

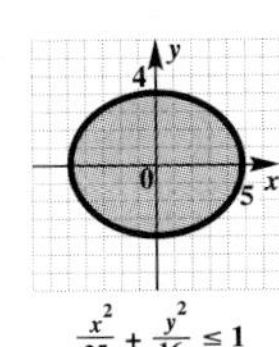

[4.3] **39.** **(a)** $61.7 = 625a + 25b$, $106 = 1225a + 35b$ **(b)** $a = .056$, $b = 1.067$

(c) $y = .056x^2 + 1.067x$ **(d)** 228.1 feet [9.6] **40.** **(a)** \$16.9 billion **(b)** \$75.8 billion

CHAPTER 11 SEQUENCES AND SERIES

SECTION 11.1 (PAGE 702)

CONNECTIONS **Page 702:** **1.** After the first term, each term is the sum of the two preceding terms. **2.** We show only four generations in this sketch.

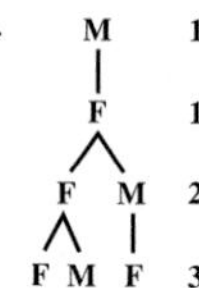

EXERCISES **3.** $4, \frac{5}{2}, 2, \frac{7}{4}, \frac{8}{5}$ **5.** 3, 9, 27, 81, 243 **7.** $1, \frac{1}{4}, \frac{1}{9}, \frac{1}{16}, \frac{1}{25}$ **9.** $-1, 1, -1, 1, -1$ **11.** -70 **13.** $\frac{49}{23}$

15. 171 **17.** $4n$ **19.** $\frac{1}{3^n}$ **21.** \$110, \$109, \$108, \$107, \$106, \$105; \$400 **23.** \$6554 **25.** $3 + 6 + 11 = 20$

27. $\frac{1}{2} + \frac{1}{3} + \frac{1}{4} + \frac{1}{5} = \frac{77}{60}$ **29.** $-1 + 1 - 1 + 1 - 1 + 1 = 0$ **31.** $0 + 6 + 14 + 24 + 36 = 80$

33. $2x + 4x + 6x + 8x + 10x$ **35.** $x + 2x^2 + 3x^3 + 4x^4 + 5x^5$

Answers may vary for Exercises 37 and 39.

37. $\sum_{i=1}^{5}(i + 2)$ **39.** $\sum_{i=1}^{5}\frac{1}{i + 1}$ **41.** A sequence is a list of terms in a specific order, while a series is the indicated sum of the terms of a sequence. **43.** $\frac{59}{7}$ **45.** 5 **47.** 4591 **49.** $\sum_{i=1}^{6} i^2 + \sum_{i=1}^{6} 3i + \sum_{i=1}^{6} 5$ **50.** $3\sum_{i=1}^{6} i$ **51.** $6 \cdot 5 = 30$

52. $\sum_{i=1}^{n} i = \frac{n(n + 1)}{2}$ **53.** $\sum_{i=1}^{n} i^2 = \frac{n(n + 1)(2n + 1)}{6}$ **54.** $91 + 63 + 30 = 184$ **55.** 572 **56.** -2620

SECTION 11.2 (PAGE 710)

EXERCISES **3.** $d = 1$ **5.** not arithmetic **7.** $d = 5$ **9.** $d = 2.15$ **11.** $d = \frac{2}{3}$ **13.** $a_n = 5n - 3$

15. $a_n = \frac{3}{4}n + \frac{9}{4}$ **17.** $a_n = 3n - 6$ **19.** 76 **21.** 48 **23.** -1 **25.** 16 **27.** 6 **29.** n represents the number of terms. **31.** $2S$; 101; 100; 10,100 **32.** $2S = 10{,}100$ **33.** 5050 **34.** 20,100 **35.** 81 **37.** -3 **39.** 87 **41.** 390 **43.** 320 **45.** 31,375 **47.** \$465 **49.** \$2100 per month **51.** 68; 1100 **53.** no; 3; 9 **55.** $m + b, 2m + b, 3m + b$ **56.** yes **57.** m **58.** $a_n = mn + b$

SECTION 11.3 (PAGE 720)

EXERCISES **3.** $r = 2$ **5.** not geometric **7.** $r = -3$ **9.** $r = -\frac{1}{2}$ **11.** $a_n = 5(2)^{n-1}$ **13.** $a_n = \frac{3^{n-1}}{9}$

15. $a_n = 10\left(-\frac{1}{5}\right)^{n-1}$ **17.** $2(5)^9$ **19.** $\frac{1}{2}\left(\frac{1}{3}\right)^{11}$ **21.** $2\left(\frac{1}{2}\right)^{24} = \frac{1}{2^{23}}$ **23.** .33333 . . . **24.** .66666 . . .

25. .99999 . . . **26.** $\frac{a_1}{1 - r} = \frac{.9}{1 - .1} = \frac{.9}{.9} = 1$; Therefore, .99999 . . . = 1. **27.** $\frac{121}{243}$ **29.** -1.997 **31.** 2.662

33. -2.982 **35.** \$66,988.91 **37.** \$130,159.72 **39.** 9 **41.** $\frac{10{,}000}{11}$ **43.** $-\frac{9}{20}$ **45.** does not exist

47. $10\left(\frac{3}{5}\right)^4 \approx 1.3$ feet **49.** 3 days; $\frac{1}{4}$ gram **51.** **(a)** $1.1(1.06)^5 \approx 1.5$ billion units **(b)** approximately 12 years

53. $\$50{,}000\left(\frac{3}{4}\right)^8 \approx \$5000$ **55.** ab, ab^2, ab^3 **56.** yes **57.** b **58.** $a_n = ab^n$

SECTION 11.4 (PAGE 728)

CONNECTIONS **Page 724:** **1.** 21 and 34 **2.** 15, 21, 28, 36, 45 **3.** Answers will vary.

EXERCISES **1.** 2 **3.** 15 **5.** 15 **7.** 1 **9.** 122 **11.** 120 **13.** $m^4 + 4m^3n + 6m^2n^2 + 4mn^3 + n^4$
15. $a^5 - 5a^4b + 10a^3b^2 - 10a^2b^3 + 5ab^4 - b^5$ **17.** $8x^3 + 36x^2 + 54x + 27$ **19.** $\frac{x^5}{243} + \frac{10x^4y}{81} + \frac{40x^3y^2}{27} + \frac{80x^2y^3}{9} + \frac{80xy^4}{3} + 32y^5$ **21.** $m^3x^3 - 3m^2n^2x^2 + 3mn^4x - n^6$ **23.** $r^{12} + 24r^{11}s + 264r^{10}s^2 + 1760r^9s^3$
25. $3^{14}x^{14} - 14(3^{13})x^{13}y + 91(3^{12})x^{12}y^2 - 364(3^{11})x^{11}y^3$ **27.** $t^{20} + 10t^{18}u^2 + 45t^{16}u^4 + 120t^{14}u^6$ **29.** $120(2^7)m^7n^3$
31. $\frac{7x^2y^6}{16}$ **33.** $36k^7$ **35.** $-160x^6y^3$ **37.** $4320x^9y^4$

CHAPTER 11 REVIEW EXERCISES (PAGE 733)

1. $-1, 1, 3, 5$ **3.** 1, 4, 9, 16 **5.** 0, 3, 8, 15 **7.** $2x + 3x^2 + 4x^3 + 5x^4 + 6x^5 + 7x^6$ **9.** 126
11. 665,081 (thousand) **13.** arithmetic; $d = 4$ **15.** geometric; $r = -1$ **17.** geometric; $r = \frac{1}{2}$
19. 73 **21.** $a_n = -5n + 1$ **23.** 15 **25.** 152 **27.** $a_n = -1(4)^{n-1}$ **29.** $2(-3)^{10} = 118{,}098$ **31.** $\frac{341}{1024}$
33. 1 **35.** $32p^5 - 80p^4q + 80p^3q^2 - 40p^2q^3 + 10pq^4 - q^5$ **37.** $m^2 + 4m\sqrt{mn} + 6mn + 4n\sqrt{mn} + n^2$
39. $-18{,}400(3)^{22}k^3$ **41.** $a_{10} = 1536$; $S_{10} = 1023$ **43.** $a_{15} = 38$; $S_{10} = 95$ **45.** $a_n = 2(4)^{n-1}$
47. $a_n = -3n + 15$ **49.** \$24,898.41 **51.** $\frac{1}{128}$ **53.** No, the sum cannot be found, because $r = 2$, and this value of r does not satisfy $|r| < 1$.

CHAPTER 11 TEST (PAGE 736)

[11.1] **1.** 0, 2, 0, 2, 0 [11.2] **2.** 4, 6, 8, 10, 12 [11.3] **3.** 48, 24, 12, 6, 3 [11.2] **4.** 0 [11.3] **5.** $\frac{64}{3}$ or $-\frac{64}{3}$
[11.2] **6.** 75 [11.3] **7.** 124 or 44 [11.1] **8.** 206.6 million [11.3] **9.** \$137,925.91 **10.** It has a sum if $|r| < 1$.
[11.2] **11.** 70 **12.** 33 **13.** 125,250 [11.3] **14.** 42 **15.** $\frac{1}{3}$ **16.** The sum does not exist. [11.1] **17.** \$324
[11.3] **18.** $20(3^{11}) = 3{,}542{,}940$ [11.4] **19.** $\frac{14{,}080x^8y^4}{9}$ **20.** $81k^4 - 540k^3 + 1350k^2 - 1500k + 625$

CUMULATIVE REVIEW EXERCISES CHAPTERS 1–11 (PAGE 737)

[1.1] **1.** $10, 0, \frac{45}{15}$ (or 3), -3 **2.** $-\frac{8}{3}, 10, 0, \frac{45}{15}$ (or 3), .82, -3 **3.** $\sqrt{13}, -\sqrt{3}$ **4.** all except $\sqrt{-7}$ [1.2] **5.** 8
6. -35 **7.** -55 [2.1] **8.** $\left\{\frac{1}{6}\right\}$ [2.5] **9.** $[10, \infty)$ [2.7] **10.** $\left\{-\frac{9}{2}, 6\right\}$ [2.1] **11.** $\{9\}$
[2.6] **12.** $(-\infty, -3) \cup (4, \infty)$ [2.7] **13.** $(-\infty, -3] \cup [8, \infty)$ [3.5] **14.** \$275 [3.2] **15.** $\frac{3}{4}$ [3.3] **16.** $3x + y = 4$
[3.1] **17.** [3.4] **18.** 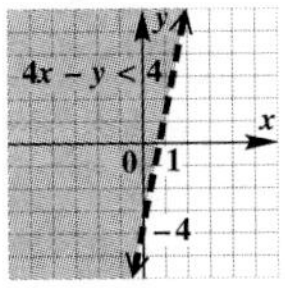 [4.1] **19.** $\{(3, -5)\}$ [4.2] **20.** $\{(2, 1, 4)\}$ [4.5] **21.** -15
22. 7 [4.3] **23.** 2 pounds [5.4] **24.** $20p^2 - 2p - 6$ **25.** $9k^2 - 42k + 49$ [5.2] **26.** $-5m^3 - 3m^2 + 3m + 8$
[6.4] **27.** $2t^3 + 3t^2 - 4t + 2 + \frac{3}{3t - 2}$ [5.5] **28.** $x(7 + x^2)$ [5.6] **29.** $(7y - 4)(2y + 3)$ **30.** $z(3z + 4)(2z - 1)$
[5.7] **31.** $(7a^2 + 3b)(7a^2 - 3b)$ **32.** $(c + 3d)(c^2 - 3cd + 9d^2)$ **33.** $(8r + 3q)^2$ [5.9] **34.** $\left\{-\frac{5}{2}, 2\right\}$

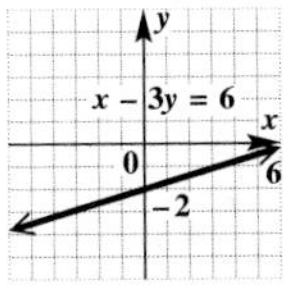

[8.7] **35.** $[-2, 3]$ [5.1] **36.** $\frac{9}{4}$ **37.** $-\frac{27p^2}{10}$ [6.1] **38.** $(-\infty, -9) \cup (-9, 9) \cup (9, \infty)$ **39.** $\frac{x+7}{x-2}$

[6.2] **40.** $\frac{3p-26}{p(p+3)(p-4)}$ [6.6] **41.** $\emptyset$ [8.2] **42.** $\left\{\frac{-5+\sqrt{217}}{12}, \frac{-5-\sqrt{217}}{12}\right\}$ [7.5] **43.** $\{1, 2\}$

[7.3] **44.** $10\sqrt{2}$ [7.6] **45.** 73 [9.1] **46.** $f^{-1}(x) = \frac{x-5}{9}$ [9.2] **47.**

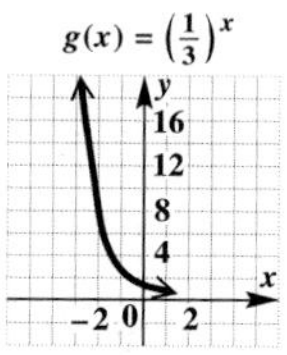

48. $\left\{\frac{5}{2}\right\}$

[9.3] **49.**

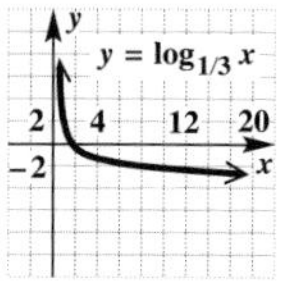

[9.6] **50.** $\{2\}$ [8.5] **51.**

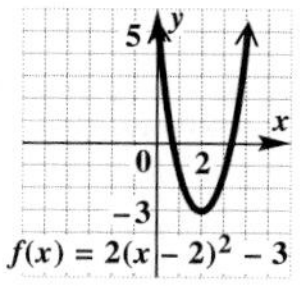

[10.1] **52.**

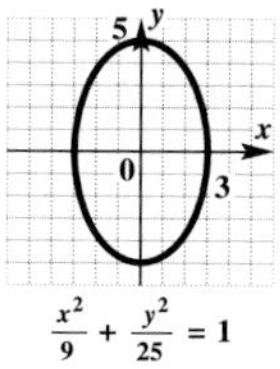

[10.2] **53.**

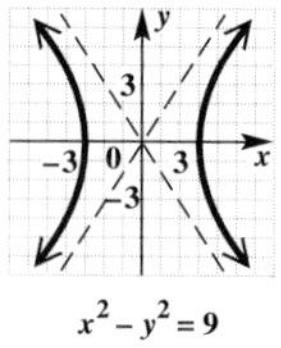

[10.1] **54.** $(x+5)^2 + (y-12)^2 = 81$ [11.1] **55.** $-7, -2, 3, 8, 13$ [11.2] **56.** 78

[11.3] **57.** $\frac{75}{7}$ [11.2] **58.** 30 [11.4] **59.** $32a^5 - 80a^4 + 80a^3 - 40a^2 + 10a - 1$ **60.** $-\frac{45x^8y^6}{4}$

Index

S

T

U

V

W

X

Y

Z

Index of Applications

Astronomy/Aerospace

Biology

Business

Chemistry

Construction

Consumer Applications

Economics

Education

Environment

Finance

Geometry

Government

Health/Life Sciences

Labor

Miscellaneous

Physics

Sports/Entertainment

Transportation

Formulas

Figure	*Formulas*	*Examples*
Square	Perimeter: $P = 4s$ Area: $A = s^2$	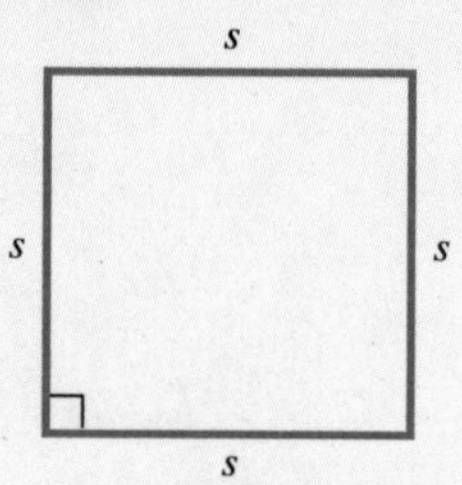
Rectangle	Perimeter: $P = 2L + 2W$ Area: $A = LW$	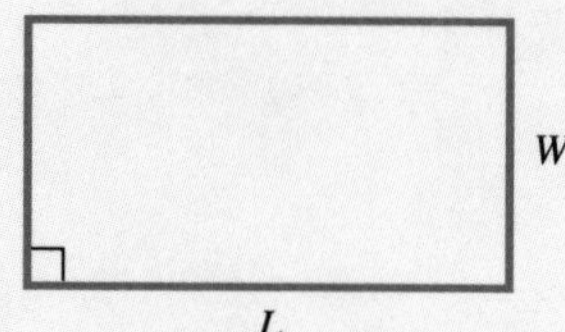
Triangle	Perimeter: $P = a + b + c$ Area: $A = \frac{1}{2}bh$	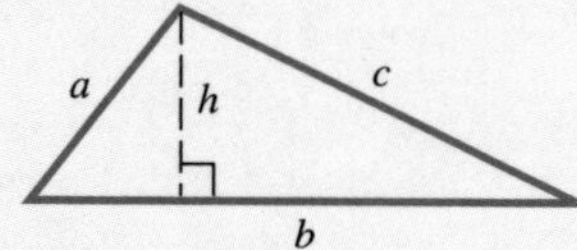
Pythagorean Formula (for right triangles)	In a right triangle with legs a and b and hypotenuse c, $c^2 = a^2 + b^2$.	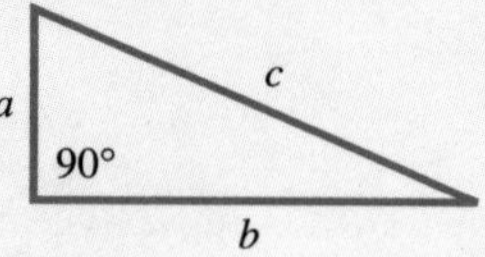
Sum of the Angles of a Triangle	$A + B + C = 180°$	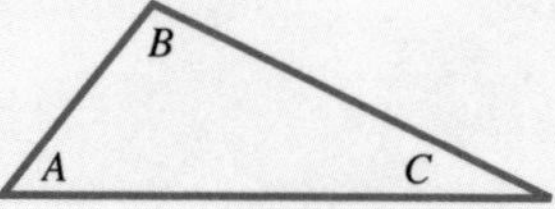
Circle	Diameter: $d = 2r$ Circumference: $C = 2\pi r$ $C = \pi d$ Area: $A = \pi r^2$	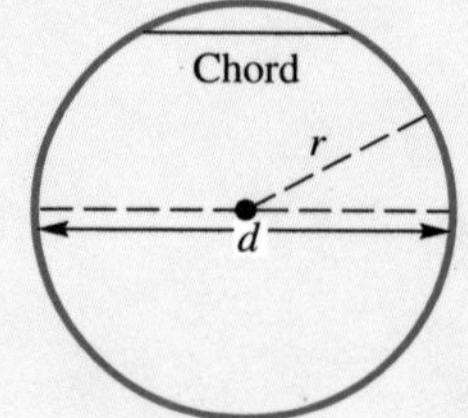
Parallelogram	Area: $A = bh$ Perimeter: $P = 2a + 2b$	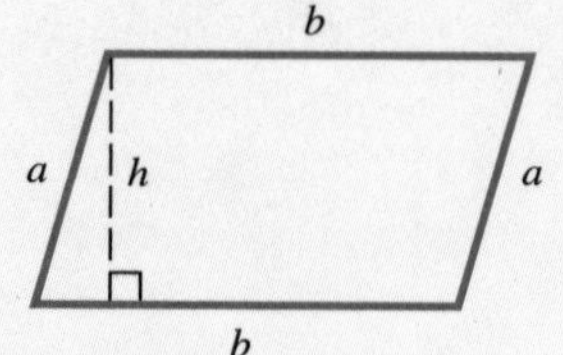
Trapezoid	Area: $A = \frac{1}{2}(B + b)h$ Perimeter: $P = a + b + c + B$	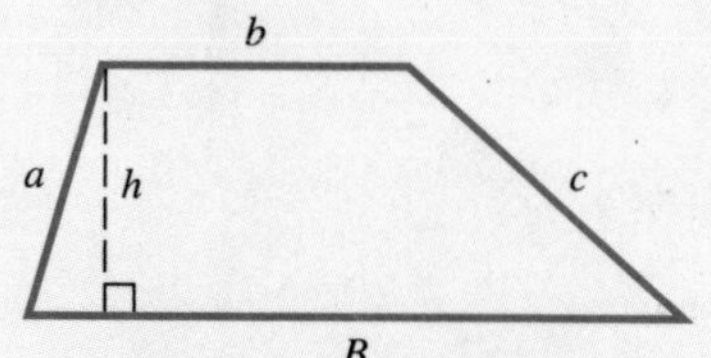